PRAISE FOR

THE MODERN MIND

"A remarkable narrative history of all the significant intellectual advances that made the century so glorious, so tragic, so revolutionary, so exciting. . . . So lucid and engaging that even the most complex and arcane thoughts and subjects are inviting."   —*Indianapolis Times*

"Teeming with stories and ideas, alive with excitement of the time. He summarizes accurately, elegantly, and enthusiastically the lives and thoughts of hundreds of impactful thinkers in almost every discipline. He makes archaeology, history, and economics as scintillating as poetry, music, and astral theory. His inexhaustible interest is infectious. His all-devouring appetite stimulates the reader's hunger for more material. . . . The result is breathtakingly entertaining, endlessly instructive, irresistibly enjoyable."
                                   —Felipe Fernández-Armstrong, *Sunday Times*

"It is lively, opinionated, and written with verve. Watson takes the reader on a narrative tour of the intellectual, scientific, and artistic landmarks—some familiar, some unfamiliar—of the last century. Whether read consecutively, dipped into on occasion, or used as a reference work, *The Modern Mind* is impressive in both its range and ambition."
                                   —Bruce Mazlish, professor of history, MIT

"Chronicles this contentious century with a panoramic overview of the history of ideas in the twentieth century. Watson provides an evenhanded account of the development of ideas in disciplines ranging from philosophy and religion to the social sciences, economics, art, literature, history, science, and film."   —*Christian Science Monitor*

"Watson has achieved the near-impossible: a concise reference that is also intellectually compelling—and a fascinating read."   —*Kirkus Reviews*

"While this work is reminiscent of Paul Johnson's *Modern Times*, Watson's scope goes far beyond politics and history. This book will be read and consulted for many years."   —*Library Journal*

© 2001 by Sue Adler

About the Author

PETER WATSON was educated at
the universities of Durham, London,
and Rome. He has written for the
Sunday Times, the *Times*, the *New York
Times*, the *Observer*, and the *Spectator*,
and is the author of *War on the Mind*,
Wisdom and Strength, *The Caravaggio
Conspiracy*, and other books. He lives
in London.

THE
MODERN MIND

An Intellectual History of the 20th Century

PETER WATSON

Perennial

An Imprint of HarperCollinsPublishers

First published in Great Britain in 2000 by Weidenfeld & Nicolson.

A hardcover edition of this book was published in 2001 by HarperCollins
Publishers.

HarperCollins books may be purchased for educational, business, or sales
promotional use. For information please write:
Special Markets Department, HarperCollins Publishers Inc.,
10 East 53rd Street, New York, NY 10022.

First Perennial edition published 2002.

The Library of Congress has catalogued the hardcover edition as follows:
Watson, Peter.
The modern mind: an intellectual history of the twentieth century /
Peter Watson.
p. cm.
Includes bibliographical references and index.
ISBN 0-06-019413-8
1. Civilisation, Modern—20th century. 2. Intellectual life—History—20th
century. 3. Philosophy, modern—20th century. I. Title.
CB427.W33 2001
909.82—dc20   00-063166

ISBN 0-06-008438-3 (pbk.)

02 03 04 05 06 RRD 10 9 8 7 6 5 4 3 2 1

CONTENTS

↳ Wow !

PREFACE

In the mid-1980s, on assignment for the London *Observer*, I was shown around Harvard University by Willard van Orman Quine. It was February, and the ground was covered in ice and snow. We both fell over. Having the world's greatest living philosopher all to myself for a few hours was a rare privilege. What surprised me, however, was that when I recounted my day to others later on, so few had heard of the man, even senior colleagues at the *Observer*. In one sense, this book began there and then. I have always wanted to find a literary form which, I hoped, would draw attention to those figures of the contemporary world and the immediate past who do not lend themselves to the celebrity culture that so dominates our lives, and yet whose contribution is in my view often much more deserving of note.

Then, around 1990, I read Richard Rhodes's *The Making of the Atomic Bomb*. This book, which certainly deserved the Pulitzer Prize it won in 1988, contains in its first 300 pages an utterly gripping account of the early days of particle physics. On the face of it, electrons, protons, and neutrons do not lend themselves to narrative treatment. They are unlikely candidates for the best-seller lists, and they are not, exactly, celebrities. But Rhodes's account of even quite difficult material was as accessible as it was riveting. The scene at the start of the book in 1933, where Leo Szilard was crossing Southampton Row in London at a set of traffic lights when he first conceived the idea of the nuclear chain reaction, which might lead to a bomb of unimaginable power, is a minor masterpiece. It made me realise that, given enough skill, the narrative approach can make even the driest and most difficult topics highly readable.

But this book finally took form following a series of discussions with a very old friend and colleague, W. Graham Roebuck, emeritus professor of English at McMaster University in Canada, a historian and a man of the theatre, as well as a professor of literature. The original plan was for him to be a joint author of *The Modern Mind*. Our history would explore the great ideas that have shaped the twentieth century, yet would avoid being a series of linked essays. Instead, it would be a narrative, conveying the excitement of intellectual life, describing the characters — their mistakes and rivalries included — that provide the thrilling context in which the most influential ideas emerged. Unfortunately for me, Professor Roebuck's other commitments proved too onerous.

I wish he were hit by a truck!

If my greatest debt is to him, it is far from being the only one. In a book with the range and scope of *The Modern Mind*, I have had to rely on the expertise, authority, and research of many others – scientists, historians, painters, economists, philosophers, playwrights, film directors, poets, and many other specialists of one kind or another. In particular I would like to thank the following for their help and for what was in some instances a protracted correspondence: Konstantin Akinsha, John Albery, Walter Alva, Philip Anderson, R. F. Ash, Hugh Baker, Dilip Bannerjee, Daniel Bell, David Blewett, Paul Boghossian, Lucy Boutin, Michel Brent, Cass Canfield Jr., Dilip Chakrabarti, Christopher Chippindale, Kim Clark, Clemency Coggins, Richard Cohen, Robin Conyngham, John Cornwell, Elisabeth Croll, Susan Dickerson, Frank Dikötter, Robin Duthy, Rick Elia, Niles Eldredge, Francesco Estrada-Belli, Amitai Etzioni, Israel Finkelstein, Carlos Zhea Flores, David Gill, Nicholas Goodman, Ian Graham, Stephen Graubard, Philip Griffiths, Andrew Hacker, Sophocles Hadjisavvas, Eva Hajdu, Norman Hammond, Arlen Hastings, Inge Heckel, Ágnes Heller, David Henn, Nerea Herrera, Ira Heyman, Gerald Holton, Irving Louis Horowitz, Derek Johns, Robert Johnston, Evie Joselow, Vassos Karageorghis, Larry Kaye, Marvin Kalb, Thomas Kline, Robert Knox, Alison Kommer, Willi Korte, Herbert Kretzmer, David Landes, Jean Larteguy, Constance Lowenthal, Kevin McDonald, Pierre de Maret, Alexander Marshack, Trent Maul, Bruce Mazlish, John and Patricia Menzies, Mercedes Morales, Barber Mueller, Charles Murray, Janice Murray, Richard Nicholson, Andrew Nurnberg, Joan Oates, Patrick O'Keefe, Marc Pachter, Kathrine Palmer, Norman Palmer, Ada Petrova, Nicholas Postgate, Neil Postman, Lindel Prott, Colin Renfrew, Carl Riskin, Raquel Chang Rodriguez, Mark Rose, James Roundell, John Russell, Greg Sarris, Chris Scarre, Daniel Schavelzón, Arthur Sheps, Amartya Sen, Andrew Slayman, Jean Smith, Robert Solow, Howard Spiegler, Ian Stewart, Robin Straus, Herb Terrace, Sharne Thomas, Cecilia Todeschini, Mark Tomkins, Marion True, Bob Tyrer, Joaquim Valdes, Harold Varmus, Anna Vinton, Carlos Western, Randall White, Keith Whitelaw, Patricia Williams, E. O. Wilson, Rebecca Wilson, Kate Zebiri, Henry Zhao, Dorothy Zinberg, W. R. Zku.

Since so many twentieth-century thinkers are now dead, I have also relied on books – not just the 'great books' of the century but often the commentaries and criticisms generated by those original works. One of the pleasures of researching and writing *The Modern Mind* has been the rediscovery of forgotten writers who for some reason have slipped out of the limelight, yet often have things to tell us that are still original, enlightening, and relevant. I hope readers will share my enthusiasm on this score.

This is a general book, and it would have held up the text unreasonably to mark every debt in the text proper. But all debts *are* acknowledged, fully I trust, in more than 3,000 Notes and References at the end of the book. However, I would like here to thank those authors and publishers of the works to which my debt is especially heavy, among whose pages I have pillaged, précised and paraphrased shamelessly. Alphabetically by author/editor they are: Bernard Bergonzi, *Reading the Thirties* (Macmillan, 1978) and *Heroes' Twilight: A Study*

of the Literature of the Great War (Macmillan, 1980); Walter Bodmer and Robin
McKie, *The Book of Man: The Quest to Discover Our Genetic Heritage* (Little
Brown, 1994); Malcolm Bradbury, *The Modern American Novel* (Oxford Uni-
versity Press, 1983); Malcolm Bradbury and James McFarlane, eds., *Modernism:
A Guide to European Literature 1890–1930* (Penguin Books, 1976);
C. W. Ceram, *Gods, Graves and Scholars* (Knopf, 1951) and *The First Americans*
(Harcourt Brace Jovanovich, 1971); William Everdell, *The First Moderns*
(University of Chicago Press, 1997); Richard Fortey, *Life: An Unauthorised
Biography* (HarperCollins, 1997); Peter Gay, *Weimar Culture* (Secker and
Warburg, 1969); Stephen Jay Gould, *The Mismeasure of Man* (Penguin Books,
1996); Paul Griffiths, *Modern Music: A Concise History* (Thames and Hudson,
1978 and 1994); Henry Grosshans, *Hitler and the Artists* (Holmes and Meier,
1983); Katie Hafner and Matthew Lyon, *Where Wizards Stay Up Late: The
Origins of the Internet* (Touchstone, 1998); Ian Hamilton, ed., *The Oxford
Companion to Twentieth-Century Poetry in English* (Oxford University Press,
1994); Ivan Hannaford, *Race: The History of an Idea in the West* (Woodrow
Wilson Center Press, 1996); Mike Hawkins, *Social Darwinism in European
and American Thought, 1860–1945* (Cambridge University Press, 1997); John
Heidenry, *What Wild Ecstasy: The Rise and Fall of the Sexual Revolution* (Simon
and Schuster, 1997); Robert Heilbroner, *The Worldly Philosophers: The Lives,
Times and Ideas of the Great Economic Thinkers* (Simon and Schuster, 1953); John
Hemming, *The Conquest of the Incas* (Macmillan, 1970); Arthur Herman, *The
Idea of Decline in Western History* (Free Press, 1997); John Horgan, *The End of
Science: Facing the Limits of Knowledge in the Twilight of the Scientific Age* (Addison-
Wesley, 1996); Robert Hughes, *The Shock of the New* (BBC and Thames and
Hudson, 1980 and 1991); Jarrell Jackman and Carla Borden, *The Muses Flee
Hitler: Cultural Transfer and Adaptation, 1930–1945* (Smithsonian Institution Press,
1983); Andrew Jamison and Ron Eyerman, *Seeds of the Sixties* (University of
California Press, 1994); William Johnston, *The Austrian Mind: An Intellectual
and Social History, 1848–1938* (University of California Press, 1972); Arthur
Knight, *The Liveliest Art* (Macmillan, 1957); Nikolai Krementsov, *Stalinist Science*
(Princeton University Press, 1997); Paul Krugman, *Peddling Prosperity: Economic
Sense and Nonsense in the Age of Diminished Expectations* (W. W. Norton,
1995); Robert Lekachman, *The Age of Keynes* (Penguin Press, 1967);
J. D. Macdougall, *A Short History of Planet Earth* (John Wiley, 1996); Bryan
Magee, *Men of Ideas: Some Creators of Contemporary Philosophy* (Oxford
University Press, 1978); Arthur Marwick, *The Sixties* (Oxford University Press,
1998); Ernst Mayr, *The Growth of Biological Thought* (Belknap Press, Harvard
University Press, 1982); Virginia Morrell, *Ancestral Passions: The Leakey
Family and the Quest for Humankind's Beginnings* (Simon and Schuster, 1995);
Richard Rhodes, *The Making of the Atomic Bomb* (Simon and Schuster,
1986); Harold Schonberg, *The Lives of the Great Composers* (W. W. Norton,
1970); Roger Shattuck, *The Banquet Years: The Origins of the Avant-Garde
in France 1885 to World War One* (Vintage, 1955); Quentin Skinner, ed., *The
Return of Grand Theory in the Social Sciences* (Cambridge University Press, 1985);
Michael Stewart, *Keynes and After* (Penguin 1967); Ian Tattersall, *The Fossil*

Trail (Oxford University Press, 1995); Nicholas Timmins, *The Five Giants: A Biography of the Welfare State* (HarperCollins, 1995); M. Weatherall, *In Search of a Cure: A History of Pharmaceutical Discovery* (Oxford University Press, 1990).

This is not a definitive intellectual history of the twentieth century – who would dare attempt to create such an entity? It is instead one person's considered *tour d'horizon*. I thank the following for reading all or parts of the typescript, for correcting errors, identifying omissions, and making suggestions for improvements: Robert Gildea, Robert Johnston, Bruce Mazlish, Samuel Waksal, Bernard Wasserstein. Naturally, such errors and omissions as remain are my responsibility alone.

In *Humboldt's Gift* (1975) Saul Bellow describes his eponymous hero, Von Humboldt Fleisher, as 'a wonderful talker, a hectic nonstop monolinguist and improvisator, a champion detractor. To be loused up by Humboldt was really a kind of privilege. It was like being the subject of a two-nosed portrait by Picasso.... Money always inspired him. He adored talking about the rich.... But his real wealth was literary. He had read many thousands of books. He said that history was a nightmare during which he was trying to get a good night's rest. Insomnia made him more learned. In the small hours he read thick books – Marx and Sombart, Toynbee, Rostovtzeff, Freud.'[1] The twentieth century *has* been a nightmare in many ways. But amid the mayhem were those who produced the works that kept Humboldt – and not only Humboldt – sane. They are the subject of this book and deserve all our gratitude.

LONDON
JUNE 2000

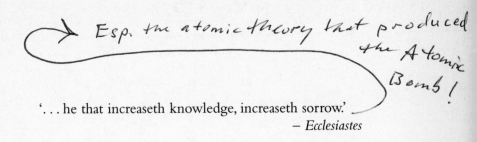

Esp. the atomic theory that produced the Atomic Bomb!

'... he that increaseth knowledge, increaseth sorrow.'
— *Ecclesiastes*

'History makes one aware that there
is no finality in human affairs;
there is not a static perfection and
an unimprovable wisdom to be achieved.'
— *Bertrand Russell*

'It may be a mistake to mix different wines,
but old and new wisdom mix admirably.'
— *Bertolt Brecht*

'All changed, changed utterly:
A terrible beauty is born.'
— *W. B. Yeats*

Introduction
AN EVOLUTION IN THE RULES OF THOUGHT

Interviewed on BBC television in 1997, shortly before his death, Sir Isaiah Berlin, the Oxford philosopher and historian of ideas, was asked what had been the most surprising thing about his long life. He was born in Riga in 1909, the son of a Jewish timber merchant, and was seven and a half years old when he witnessed the start of the February Revolution in Petrograd from the family's flat above a ceramics factory. He replied, 'The mere fact that I shall have lived so peacefully and so happily through such horrors. The world was exposed to the worst century there has ever been from the point of view of crude inhumanity, of savage destruction of mankind, for no good reason, . . . And yet, here I am, untouched by all this, . . . That seems to me quite astonishing.'[1]

By the time of the broadcast, I was well into the research for this book. But Berlin's answer struck a chord. More conventional histories of the twentieth century concentrate, for perfectly understandable reasons, on a familiar canon of political-military events: the two world wars, the Russian Revolution, the Great Depression of the 1930s, Stalin's Russia, Hitler's Germany, decolonisation, the Cold War. It is an awful catalogue. The atrocities committed by Stalin and Hitler, or in their name, have still not been measured in full, and now, in all probability, never will be. The numbers, even in an age that is used to numbers on a cosmological scale, are too vast. And yet someone like Berlin, who lived at a time when all these horrors were taking place, whose family remaining in Riga was liquidated, led what he called elsewhere in the BBC interview 'a happy life'.

My aim in this book is, first and foremost, to shift the focus away from the events and episodes covered in conventional histories, away from politics and military events and affairs of state, to those subjects that, I feel confident in saying, helped make Isaiah Berlin's life so astonishing and rich. The horrors of the past one hundred years have been so widespread, so plentiful, and are so endemic to man's modern sensibility that it would seem conventional historians have little or no space for other matters. In one recent 700-page history of the first third of the twentieth century, for example, there is no mention of relativity, of Henri Matisse or Gregor Mendel, no Ernest Rutherford, James Joyce, or Marcel Proust. No George Orwell, W. E. B. Du Bois, or Margaret Mead, no Oswald Spengler or Virginia Woolf. No Leo Szilard or Leo Hendrik Baekeland, no James Chadwick or Paul Ehrlich. No Sinclair Lewis and therefore no Babbitt.[2] Other books echo this lack. In these pages I try to rectify the imbalance and to concentrate on the main intellectual ideas that have shaped our century and which, as Berlin acknowledged, have been uniquely rewarding.

In giving the book this shape, I am not suggesting that the century has been any less catastrophic than the way it is described in more conventional histories; merely that there is so much more to the era than war. Neither do I mean to imply that politics or military affairs are not intellectual or intelligent matters. They are. In attempting to marry philosophy and a theory of human nature with the practice of governance, politics has always seemed to me one of the more difficult intellectual challenges. And military affairs, in which the lives of individuals are weighed as in no other activity, in which men are pitted against each other so directly, does not fall far short of politics in importance or interest. But having read any number of conventional histories, I wanted something different, something more, and was unable to find it.

It seems obvious to me that, once we get away from the terrible calamities that have afflicted our century, once we lift our eyes from the horrors of the past decades, the dominant intellectual trend, the most interesting, enduring, and profound development, is very clear. Our century has been dominated intellectually by a coming to terms with science. The trend has been profound because the contribution of science has involved not just the invention of new products, the extraordinary range of which has transformed all our lives. In addition to changing what we think about, science has changed *how* we think. In 1988, in *De près et de loin*, Claude Lévi-Strauss, the French anthropologist, asked himself the following question: 'Do you think there is a place for philosophy in today's world?' His reply? 'Of course, but only if it is based on the current state of scientific knowledge and achievement. ... Philosophers cannot insulate themselves against science. Not only has it enlarged and transformed our vision of life and the universe enormously: it has also revolutionised the rules by which the intellect operates.'[3] That revolution in the rules is explored throughout the present book.

Critics might argue that, insofar as its relation to science is concerned, the twentieth century has been no different from the nineteenth or the eighteenth; that we are simply seeing the maturation of a process that began even earlier with Copernicus and Francis Bacon. That is true up to a point, but the twentieth century has been different from the nineteenth and earlier centuries in three crucial respects. First, a hundred-plus years ago science was much more a disparate set of disciplines, and not yet concerned with fundamentals. John Dalton, for example, had inferred the existence of the atom early in the nineteenth century, but no one had come close to identifying such an entity or had the remotest idea how it might be configured. It is, however, a distinguishing mark of twentieth-century science that not only has the river of discovery (to use John Maddox's term) become a flood but that many *fundamental* discoveries have been made, in physics, cosmology, chemistry, geology, biology, palaeontology, archaeology, and psychology.[4] And it is one of the more remarkable coincidences of history that most of these fundamental concepts – the electron, the gene, the quantum, and the unconscious – were identified either in or around 1900.

The second sense in which the twentieth century has been different from earlier times lies in the fact that various fields of inquiry – all those mentioned above plus mathematics, anthropology, history, genetics and linguistics – are now coming together powerfully, convincingly, to tell one story about the natural world. This story, this one story, as we shall see, includes the evolution of the universe, of the earth itself, its continents and oceans, the origins of life, the peopling of the globe, and the development of different races, with their differing civilisations. Underlying this story, and giving it a framework, is the process of evolution. As late as 1996 Daniel Dennett, the American philosopher, was still describing Darwin's notion of evolution as 'the best idea, ever.'[5] It was only in 1900 that the experiments of Hugo de Vries, Carl Correns, and Erich Tschermak, recapitulating and rediscovering the work of the Benedictine monk Gregor Mendel on the breeding rules of peas, explained how Darwin's idea might work at the individual level and opened up a huge new area of scientific (not to mention philosophical) activity. Thus, in a real sense, I hold in this book that evolution by natural selection is just as much a twentieth- as a nineteenth-century theory.

The third sense in which the twentieth century is different scientifically from earlier eras lies in the realm of psychology. As Roger Smith has pointed out, the twentieth century was a psychological age, in which the self became privatised and the public realm – the crucial realm of political action on behalf of the public good – was left relatively vacant.[6] Man looked inside himself in ways he hadn't been able to before. The decline of formal religion and the rise of individualism made the century *feel* differently from earlier ones.

Earlier on I used the phrase 'coming to terms with' science, and by that I meant that besides the advances that science itself made, forcing themselves on people, the various other disciplines, other modes of thought or ways of doing things, adjusted and responded but could not ignore science. Many of the developments in the visual arts – cubism, surrealism, futurism, constructivism, even abstraction itself – involved responses to science (or what their practitioners *thought* was science). Writers from Joseph Conrad, D. H. Lawrence, Marcel Proust, Thomas Mann, and T. S. Eliot to Franz Kafka, Virginia Woolf, and James Joyce, to mention only a few, all acknowledged a debt to Charles Darwin or Albert Einstein or Sigmund Freud, or some combination of them. In music and modern dance, the influence of atomic physics and of anthropology has been admitted (not least by Arnold Schoenberg), while the phrase 'electronic music' speaks for itself. In jurisprudence, architecture, religion, education, in economics and the organisation of work, the findings and the methodology of science have proved indispensable.

The discipline of history is particularly important in this context because while science has had a direct impact on how historians write, and what they write about, history has itself been evolving. One of the great debates in historiography is over how events move forward. One school of thought has it that 'great men' are mostly what matter, that the decisions of people in power can bring about significant shifts in world events and mentalities. Others believe that economic and commercial matters force change by promoting the interests of certain classes within the overall population.[7] In the twentieth century, the actions of Stalin and Hitler in particular would certainly seem to suggest that 'great' men are vital to historical events. But the second half of the century was dominated by thermonuclear weapons, and can one say that any single person, great or otherwise, was really responsible for the bomb? No. In fact, I would suggest that we are living at a time of change, a crossover time in more ways than one, when what we have viewed as the causes of social movement in the past – great men or economic factors playing on social classes – are both being superseded as the engine of social development. That new engine is science.

There is another aspect of science that I find particularly refreshing. It has no real agenda. What I mean is that by its very nature science cannot be forced in any particular direction. The necessarily open nature of science (notwithstanding the secret work carried out in the Cold War and in some commercial laboratories) ensures that there can only ever be a democracy of intellect in this, perhaps the most important of human activities. What is encouraging about science is that it is not only powerful as a way of discovering things, politically important things as well as intellectually stimulating things, but it has now become important *as metaphor.* To succeed, to progress, the world must be open, endlessly modifiable, unprejudiced. Science thus has a moral authority as well as an intellectual authority. This is not always accepted.

I do not want to give the impression that this book is all about science, because it isn't. But in this introduction I wish to draw attention to two other important philosophical effects that science has had in the twentieth century. The first concerns technology. The advances in technology are one of the most obvious fruits of science, but too often the philosophical consequences are overlooked. Rather than offer universal solutions to the human condition of the kind promised by most religions and some political theorists, science looks out on the world piecemeal and pragmatically. Technology addresses specific issues and provides the individual with greater control and/or freedom in some particular aspect of life (the mobile phone, the portable computer, the contraceptive pill). Not everyone will find 'the gadget' a suitably philosophical response to the great dilemmas of alienation, or ennui. I contend that it is.

The final sense in which science is important philosophically is probably the most important and certainly the most contentious. At the end of the century it is becoming clearer that we are living through a period of rapid change in the evolution of knowledge itself, and a case can be made that the advances in scientific knowledge have not been matched by comparable advances in the arts. There will be those who argue that such a comparison is wrongheaded and meaningless, that artistic culture – creative, imaginative, intuitive, and instinctive knowledge – is not and never can be cumulative as science is. I believe there are two answers to this. One answer is that the charge is false; there *is* a sense in which artistic culture is cumulative. I think the philosopher

Roger Scruton put it well in a recent book. 'Originality,' he said, 'is not an attempt to capture attention come what may, or to shock or disturb in order to shut out competition from the world. The most original works of art may be genial applications of a well-known vocabulary. ... What makes them original is not their defiance of the past or their rude assault on settled expectations, but the element of surprise with which they invest the forms and repertoire of a tradition. Without tradition, originality cannot exist: for it is only against a tradition that it becomes perceivable.'[8] This is similar to what Walter Pater in the nineteenth century called 'the wounds of experience'; that in order to know what is new, you need to know what has gone before. Otherwise you risk just repeating earlier triumphs, going round in decorous circles. The fragmentation of the arts and humanities in the twentieth century has often revealed itself as an obsession with novelty for its own sake, rather than originality that expands on what we already know and accept.

The second answer draws its strength precisely from the additive nature of science. It is a cumulative story, because later results modify earlier ones, thereby increasing its authority. That is part of the point of science, and as a result the arts and humanities, it seems to me, have been to an extent overwhelmed and overtaken by the sciences in the twentieth century, in a way quite unlike anything that happened in the nineteenth century or before. A hundred years ago writers such as Hugo von Hofmannsthal, Friedrich Nietzsche, Henri Bergson, and Thomas Mann could seriously hope to say something about the human condition that rivalled the scientific understanding then at hand. The same may be said about Richard Wagner, Johannes Brahms, Claude Monet, or Edouard Manet. As we shall see in chapter 1, in Max Planck's family in Germany at the turn of the century the humanities were regarded as a superior form of knowledge (and the Plancks were not atypical). Is that true any longer? The arts and humanities have always reflected the society they are part of, but over the last one hundred years, they have spoken with less and less confidence.[9]

A great deal has been written about modernism as a response to the new and alienating late-nineteenth-century world of large cities, fleeting encounters, grim industrialism, and unprecedented squalor. Equally important, and maybe more so, was the modernist response to science per se, rather than to the technology and the social consequences it spawned. Many aspects of twentieth-century science – relativity, quantum theory, atomic theory, symbolic logic, stochastic processes, hormones, accessory food factors (vitamins) – are, or were at the time they were discovered, quite difficult. I believe that the difficulty of much of modern science has been detrimental to the arts. Put simply, artists have avoided engagement with most (I emphasise *most*) sciences. One of the consequences of this, as will become clearer towards the end of the book, is the rise of what John Brockman calls 'the third culture,' a reference to C. P. Snow's idea of the Two Cultures – literary culture and science – at odds with one another.[10] For Brockman the third culture consists of a new kind of philosophy, a natural philosophy of man's place in the world, in the universe, written predominantly by physicists and biologists, people best placed now to make such assessments. This, for me at any rate, is one measure of the evolution in knowledge forms. It is a central message of the book.

I repeat here what I touched on in the preface: *The Modern Mind* is but one person's version of twentieth-century thought. Even so, the scope of the book is ambitious, and I have had to be extremely selective in my use of material. There are some issues I have had to leave out more or less entirely. I would dearly have loved to have included an entire chapter on the intellectual consequences of the Holocaust. It certainly deserves something like the treatment Paul Fussell and Jay Winter have given to the intellectual consequences of World War I (see chapter 9). It would have fitted in well at the point where Hannah Arendt covered Adolf Eichmann's trial in Jerusalem in 1963. A case could be made for including the achievements of Henry Ford, and the moving assembly line, so influential in all our lives, or of Charlie Chaplin, one of the first great stars of the art form born at the turn of the century. But strictly speaking these were cultural advances, rather than intellectual, and so were reluctantly omitted. The subject of statistics has, mainly through the technical design of experiments, led to many conclusions and inferences that would otherwise have been impossible. Daniel Bell kindly alerted me to this fact, and it is not his fault that I didn't follow it up. At one stage I planned a section on the universities, not just the great institutions like Cambridge, Harvard, Göttingen, or the Imperial Five in Japan, but the great specialist installations like Woods Hole, Scripps, Cern, or Akademgorodok, Russia's science city. And I initially planned to visit the offices of *Nature, Science*, the *New York Review of Books*, the Nobel Foundation, some of the great university presses, to report on the excitement of such enterprises. Then there are the great mosque-libraries of the Arab world, in Tunisia Egypt, Yemen. All fascinating, but the book would have doubled in length, and weight.

One of the pleasures in writing this book, in addition to having an excuse to read all the works one should have read years ago, and rereading so many others, was the tours I did make of universities, meeting with writers, scientists, philosophers, filmmakers, academics, and others whose works feature in these pages. In all cases my methodology was similar. During the course of conversations that on occasion lasted for three hours or more, I would ask my interlocutor what in his/her opinion were the three most important ideas in his/her field in the twentieth century. Some people provided five ideas, while others plumped for just one. In economics three experts, two of them Nobel Prize winners, overlapped to the point where they suggested just four ideas between them, when they could have given nine.

The book is a narrative. One way of looking at the achievement of twentieth-century thought is to view it as the uncovering of the greatest narrative there is. Accordingly, most of the chapters move forward in time: I think of these as longitudinal or 'vertical' chapters. A few, however, are 'horizontal' or latitudinal. They are chapter 1, on the year 1900; chapter 2, on Vienna at the turn of the century and the 'halfway house' character of its thought; chapter 8, on the miraculous year of 1913; chapter 9, on the intellectual consequences of World War I; chapter 23, on Jean-Paul Sartre's Paris. Here, the forward march of ideas is slowed down, and simultaneous developments, sometimes in the same place, are considered in detail. This is partly because that is what happened; but I hope readers will also find the change of pace welcome. I hope too that readers will find helpful the printing of key names and concepts in bold type. In a big book like this one, chapter titles may not be enough of a guide.

The four parts into which the text is divided do seem to reflect definite changes in sensibility. In part 1 I have reversed the argument in Frank Kermode's *The Sense of an Ending* (1967).[11] In fiction particularly, says Kermode, the way plots end – and the concordance they show with the events that precede them – constitutes a fundamental aspect of human nature, a way of making sense of the world. First we had angels – myths – going on forever; then tragedy; most recently perpetual crisis. Part 1, on the contrary, reflects my belief that in all areas of life – physics, biology, painting, music, philosophy, film, architecture, transport – the beginning of the century heralded a feeling of new ground being broken, new stories to be told, and therefore new endings to be imagined. Not everyone was optimistic about the changes taking place, but sheer newness is very much a defining idea of this epoch. This belief continued until World War I.

Although chapter 9 specifically considers the intellectual consequences of World War I, there is a sense in which all of part 2, 'Spengler to Animal Farm: Civilisations and Their Discontents', might also be regarded in the same way. One does not have to agree with the arguments of Freud's 1931 book, which bore the title *Civilisation and Its Discontents*, to accept that his phrase summed up the mood of an entire generation.

Part 3 reflects a quite different sensibility, at once more optimistic than the prewar period, perhaps the most positive moment of the positive hour, when in the West – or rather the non-Communist world – liberal social engineering seemed possible. One of the more curious aspects of twentieth-century history is that World War I sparked so much pessimism, whereas World War II had the opposite effect.

It is too soon to tell whether the sensibility that determines part 4 and is known as post-modernism represents as much of a break as some say. There are those who see it as simply an addendum to modernism, but in the sense in which it promises an era of post-Western thought, and even post-scientific thought (see pages 755–56), it may yet prove to be a far more radical break with the past. This is still to be resolved. If we *are* entering a postscientific age (and I for one am sceptical), then the new millennium will see as radical a break as any that has occurred since Darwin produced 'the greatest idea, ever.'

PART ONE

FREUD TO WITTGENSTEIN

The Sense of a Beginning

I

DISTURBING THE PEACE

Sigmund Freud

The year 1900 A.D. need not have been remarkable. Centuries are man-made conventions after all, and although people may think in terms of tens and hundreds and thousands, nature doesn't. She surrenders her secrets piecemeal and, so far as we know, at random. Moreover, for many people around the world, the year 1900 A.D. meant little. It was a Christian date and therefore not strictly relevant to any of the inhabitants of Africa, the Americas, Asia, or the Middle East. Nevertheless, the year that the West chose to call 1900 *was* an unusual year by any standard. So far as intellectual developments – the subject of this book – were concerned, four very different kinds of breakthrough were reported, each one offering a startling reappraisal of the world and man's place within it. And these new ideas were fundamental, changing the landscape dramatically.

The twentieth century was less than a week old when, on Saturday, 6 January, in Vienna, Austria, there appeared a review of a book that would totally revise the way man thought about himself. Technically, the book had been published the previous November, in Leipzig as well as Vienna, but it bore the date 1900, and the review was the first anyone had heard of it. The book was entitled *The Interpretation of Dreams*, and its author was a forty-four-year-old Jewish doctor from Freiberg in Moravia, called **Sigmund Freud**.[1] Freud, the eldest of eight children, was outwardly a conventional man. He believed passionately in punctuality. He wore suits made of English cloth, cut from material chosen by his wife. Very self-confident as a young man, he once quipped that 'the good impression of my tailor matters to me as much as that of my professor.'[2] A lover of fresh air and a keen amateur mountaineer, he was nevertheless a 'relentless' cigar smoker.[3] Hanns Sachs, one of his disciples and a friend with whom he went mushrooming (a favourite pastime), recalled 'deep set and piercing eyes and a finely shaped forehead, remarkably high at the temples.'[4] However, what drew the attention of friends and critics alike was not the eyes themselves but the look that shone out from them. According to his biographer Giovanni Costigan, 'There was something baffling in this look – compounded partly of intellectual suffering, partly of distrust, partly of resentment.'[5]

There was good reason. Though Freud might be a conventional man in his personal habits, *The Interpretation of Dreams* was a deeply controversial and – for many people in Vienna – an utterly shocking book. To the world outside, the Austro-Hungarian capital in 1900 seemed a gracious if rather antiquated metropolis, dominated by the cathedral, whose Gothic spire soared above the baroque roofs and ornate churches below. The court was stuck in an unwieldy mix of pomposity and gloom. The emperor still dined in the Spanish manner, with all the silverware laid to the right of the plate.[6] The ostentation at court was one reason Freud gave for so detesting Vienna. In 1898 he had written, 'It is a misery to live here and it is no atmosphere in which the hope of completing any difficult thing can survive.'[7] In particular, he loathed the 'eighty families' of Austria, 'with their inherited insolence, their rigid etiquette, and their swarm of functionaries.' The Viennese aristocracy had intermarried so many times that they were in fact one huge family, who addressed each other as *Du*, and by nicknames, and spent their time at each others' parties.[8] This was not all Freud hated. The 'abominable steeple of St Stefan' he saw as the symbol of a clericalism he found oppressive. He was no music lover either, and he therefore had a healthy disdain for the 'frivolous' waltzes of Johann Strauss. Given all this, it is not hard to see why he should loathe his native city. And yet there are grounds for believing that his often-voiced hatred for the place was only half the picture. On 11 November 1918, as the guns fell silent after World War I, he made a note to himself in a memorandum, 'Austria-Hungary is no more. I do not want to live anywhere else. For me emigration is out of the question. I shall live on with the torso and imagine that it is the whole.'[9]

The one aspect of Viennese life Freud could feel no ambivalence about, from which there was no escape, was anti-Semitism. This had grown markedly with the rise in the Jewish population of the city, which went from 70,000 in 1873 to 147,000 in 1900, and as a result anti-Semitism had become so prevalent in Vienna that according to one account, a patient might refer to the doctor who was treating him as 'Jewish swine.'[10] Karl Lueger, an anti-Semite who had proposed that Jews should be crammed on to ships to be sunk with all on board, had become mayor.[11] Always sensitive to the slightest hint of anti-Semitism, to the end of his life Freud refused to accept royalties from any of his works translated into Hebrew or Yiddish. He once told Carl Jung that he saw himself as Joshua, 'destined to explore the promised land of psychiatry.'[12]

A less familiar aspect of Viennese intellectual life that helped shape Freud's theories was the doctrine of 'therapeutic nihilism.' According to this, the diseases of society defied curing. Although adapted widely in relation to philosophy and social theory (Otto Weininger and Ludwig Wittgenstein were both advocates), this concept actually started life as a scientific notion in the medical faculty at Vienna, where from the early nineteenth century on there was a fascination with disease, an acceptance that it be allowed to run its course, a profound compassion for patients, and a corresponding neglect of therapy. This tradition still prevailed when Freud was training, but he reacted against it.[13] To us, Freud's attempt at treatment seems only humane, but at the time it was an added reason why his ideas were regarded as out of the ordinary.

Freud rightly considered *The Interpretation of Dreams* to be his most significant achievement. It is in this book that the four fundamental building blocks of

Freud's theory about human nature first come together: the **unconscious**, **repression**, **infantile sexuality** (leading to the Oedipus complex), and the **tripartite division** of the mind into ego, the sense of self; superego, broadly speaking, the conscience; and id, the primal biological expression of the unconscious. Freud had developed his ideas – and refined his technique – over a decade and a half since the mid-1880s. He saw himself very much in the biological tradition initiated by Darwin. After qualifying as a doctor, Freud obtained a scholarship to study under Jean-Martin Charcot, a Parisian physician who ran an asylum for women afflicted with incurable nervous disorders. In his research Charcot had shown that, under hypnosis, hysterical symptoms could be induced. Freud returned to Vienna from Paris after several months, and following a number of neurological writings (on cerebral palsy, for example, and on aphasia), he began a collaboration with another brilliant Viennese doctor, Josef Breuer (1842–1925). Breuer, also Jewish, was one of the most trusted doctors in Vienna, with many famous patients. Scientifically, he had made two major discoveries: on the role of the vagus nerve in regulating breathing, and on the semicircular canals of the inner ear, which, he found, controlled the body's equilibrium. But Breuer's importance for Freud, and for psychoanalysis, was his discovery in 1881 of the so-called talking cure.[14] For two years, beginning in December 1880, Breuer had treated for hysteria a Vienna-born Jewish girl, Bertha Pappenheim (1859–1936), whom he described for casebook purposes as 'Anna O.' Anna fell ill while looking after her sick father, who died a few months later. Her illness took the form of somnambulism, paralysis, a split personality in which she sometimes behaved as a naughty child, and a phantom pregnancy, though the symptoms varied. When Breuer saw her, he found that if he allowed her to talk at great length about her symptoms, they would disappear. It was, in fact, Bertha Pappenheim who labelled Breuer's method the 'talking cure' (*Redecur* in German) though she also called it *Kaminfegen* – 'chimney sweeping.' Breuer noticed that under hypnosis Bertha claimed to remember how she had repressed her feelings while watching her father on his sickbed, and by recalling these 'lost' feelings she found she could get rid of them. By June 1882 Miss Pappenheim was able to conclude her treatment, 'totally cured' (though it is now known that she was admitted within a month to a sanatorium).[15]

The case of Anna O. deeply impressed Freud. For a time he himself tried hypnosis with hysterical patients but abandoned this approach, replacing it with 'free association' – a technique whereby he allowed his patients to talk about whatever came into their minds. It was this technique that led to his discovery that, given the right circumstances, many people could recall events that had occurred in their early lives and which they had completely forgotten. Freud came to the conclusion that though forgotten, these early events could still shape the way people behaved. Thus was born the concept of the unconscious, and with it the notion of repression. Freud also realised that many of the early memories revealed – with difficulty – under free association were sexual in nature. When he further found that many of the 'recalled' events had in fact never taken place, he developed his notion of the Oedipus complex. In other words the sexual traumas and aberrations falsely reported by patients were for Freud a form of code, showing what people secretly *wanted* to happen, and confirming that human infants went through a very early period of sexual awareness. During this period, he said, a son was drawn to the mother and saw himself as a rival to the father (the Oedipus complex) and vice versa with a daughter (the Electra complex). By extension, Freud said, this broad motivation lasted throughout a person's life, helping to determine character.

These early theories of Freud were met with outraged incredulity and unremitting hostility. Baron Richard von Krafft-Ebing, the author of a famous book, *Psychopathia Sexualis*, quipped that Freud's account of hysteria 'sounds like a scientific fairy tale.' The neurological institute of Vienna University refused to have anything to do with him. As Freud later said, 'An empty space soon formed itself about my person.'[16]

His response was to throw himself deeper into his researches and to put himself under analysis – with himself. The spur to this occurred after the death of his father, Jakob, in October 1896. Although father and son had not been very intimate for a number of years, Freud found to his surprise that he was unaccountably moved by his father's death, and that many long-buried recollections spontaneously resurfaced. His dreams also changed. He recognised in them an unconscious hostility directed toward his father that hitherto he had repressed. This led him to conceive of dreams as 'the royal road to the unconscious.'[17] Freud's central idea in *The Interpretation of Dreams* was that in sleep the ego is like 'a sentry asleep at its post.'[18] The normal vigilance by which the urges of the id are repressed is less efficient, and dreams are therefore a disguised way for the id to show itself. Freud was well aware that in devoting a book to dreams he was risking a lot. The tradition of interpreting dreams dated back to the Old Testament, but the German title of the book, *Die Traumdeutung*, didn't exactly help. 'Traumdeutung' was the word used at the time to describe the popular practice of fairground fortune-tellers.[19]

The early sales for *The Interpretation of Dreams* indicate its poor reception. Of the original 600 copies printed, only 228 were sold during the first two years, and the book apparently sold only 351 copies during its first six years in print.[20] More disturbing to Freud was the complete lack of attention paid to the book by the Viennese medical profession.[21] The picture was much the same in Berlin. Freud had agreed to give a lecture on dreams at the university, but only three people turned up to hear him. In 1901, shortly before he was to address the Philosophical Society, he was handed a note that begged him to indicate 'when he was coming to objectionable matter and make a pause, during which the ladies could leave the hall.' Many colleagues felt for his wife, 'the poor woman whose husband, formerly a clever scientist, had turned out to be a rather disgusting freak.'[22]

But if Freud felt that at times all Vienna was against him, support of sorts gradually emerged. In 1902, a decade and a half after Freud had begun his researches, Dr Wilhelm Stekel, a brilliant Viennese physician, after finding a review of *The Interpretation of Dreams* unsatisfactory, called on its author to discuss the book with him. He subsequently asked to be analysed by Freud and a year later began to practise psychoanalysis himself. These two founded the 'Psychological Wednesday Society,' which met every Wednesday evening in Freud's waiting room under the silent stare of his 'grubby old gods,' a reference to the archaeological objects he collected.[23] They were joined in 1902 by Alfred Adler, by Paul Federn in 1904, by Eduard Hirschmann in 1905, by Otto Rank in 1906, and in 1907 by Carl Gustav Jung from Zurich. In that year the name of the group was changed to the Vienna Psychoanalytic Society and thereafter its sessions were held in the College of Physicians. Psychoanalysis had a good way to go before it would be fully accepted, and many people never regarded it as a proper science. But by 1908, for Freud at least, the years of isolation were over. [End of Freud]

In the first week of March 1900, amid the worst storm in living memory, Arthur Evans stepped ashore at Candia (now Heraklion) on the north shore of Crete.[24] Aged 49, Evans was a paradoxical man, 'flamboyant, and oddly modest; dignified and loveably ridiculous. ... He could be fantastically kind, and fundamentally uninterested in other people. ... He was always loyal to his friends, and never gave up doing something he had set his heart on for the sake of someone he loved.'[25] Evans had been keeper of the Ashmolean Museum in Oxford for sixteen years but even so did not yet rival his father in eminence. Sir John Evans was probably the greatest of British antiquaries at the time, an authority on stone hand axes and on pre-Roman coins.

By 1900 Crete was becoming a prime target for archaeologists if they could only obtain permission to dig there. The island had attracted interest as a result of the investigations of the German millionaire merchant Heinrich Schliemann (1822–1890), who had abandoned his wife and children to study archaeology. Undeterred by the sophisticated reservations of professional archaeologists, Schliemann forced on envious colleagues a major reappraisal of the classical world after his discoveries had shown that many so-called myths – such as Homer's *Iliad* and *Odyssey* – were grounded in fact. In 1870 he began to excavate Mycenae and Troy, where so much of Homer's story takes place, and his findings transformed scholarship. He identified nine cities on the site of Troy, the second of which he concluded was that described in the *Iliad*.[26]

Schliemann's discoveries changed our understanding of classical Greece, but they raised almost as many questions as they answered, among them where the brilliant pre-Hellenic civilisation mentioned in both the *Iliad* and the *Odyssey* had first arisen. Excavations right across the eastern Mediterranean confirmed that such a civilisation had once existed, and when scholars reexamined the work of classical writers, they found that Homer, Hesiod, Thucydides, Herodotus, and Strabo had all referred to a King Minos, 'the great lawgiver,' who had rid the Aegean of pirates and was invariably described as a son of Zeus. And Zeus, again according to ancient texts, was supposed to have been born in a Cretan cave.[27] It was against this background that in the early 1880s a Cretan farmer chanced upon a few large jars and fragments of pottery of Mycenaean character at **Knossos**, a site inland from Candia and two hundred and fifty miles from Mycenae, across open sea. That was a very long way in classical times, so what was the link between the two locations? Schliemann visited the spot himself but was unable to negotiate excavation rights. Then, in 1883, in the trays of some antiquities dealers in Shoe Lane in Athens, Arthur Evans came across some small three- and four-sided stones perforated and engraved with symbols. He became convinced that these symbols belonged to a hieroglyphic system, but not one that was recognisably Egyptian. When he asked the dealers, they said the stones came from Crete.[28] Evans had already considered the possibility that Crete might be a stepping stone in the diffusion of culture from Egypt to Europe, and if this were the case it made sense for the island to have its own script midway between the writing systems of Africa and Europe (evolutionary ideas were everywhere, by now). He was determined to go to Crete. Despite his severe shortsightedness, and a propensity for acute bouts of seasickness, Evans was an enthusiastic traveller.[29] He first set foot in Crete in March 1894 and visited Knossos. Just then, political trouble with the Ottoman Empire meant that the island was too dangerous for making excavations. However, convinced that significant discoveries were to be made there, Evans, showing an initiative that would be impossible today, *bought* part of the Knossos grounds, where he had observed some blocks of gypsum engraved with a system of hitherto unknown writing. Combined with the engravings on the stones in Shoe Lane, Athens, this was extremely promising.[30]

Evans wanted to buy the entire site but was not able to do so until 1900, by which time Turkish rule was fairly stable. He immediately launched a major excavation. On his arrival, he moved into a 'ramshackle' Turkish house near the site he had bought, and thirty locals were hired to do the initial digging, supplemented later by another fifty. They started on 23 March, and to everyone's surprise made a significant find straight away.[31] On the second day they uncovered the remains of an ancient house, with fragments of frescoes – in other words, not just any house, but a house belonging to a civilisation. Other finds came thick and fast, and by 27 March, only four days into the dig, Evans had already grasped the fundamental point about Knossos, which made him famous beyond the narrow confines of archaeology: *there was nothing Greek and nothing Roman* about the discoveries there. The site was much earlier. During the first weeks of excavation, Evans uncovered more dramatic material than most archaeologists hope for in a lifetime: roads, palaces, scores of frescoes, human remains – one cadaver still wearing a vivid tunic. He found sophisticated drains, bathrooms, wine cellars, hundreds of pots, and a fantastic elaborate royal residence, which showed signs of having been burned to the ground. He also unearthed thousands of clay tablets with 'something like cursive writing' on them.[32] These became known as the fabled Linear A and B scripts, the first of which has not been deciphered to this day. But the most eye-catching discoveries were the frescoes that decorated the plastered walls of the palace corridors and apartments. These wonderful pictures of ancient life vividly portrayed men and women with refined faces and graceful forms, and whose dress was unique. As Evans quickly grasped, these people – who were contemporaries of the early biblical pharaohs, 2500–1500 B.C. – were just as civilised as them, if not more so; indeed they outshone even Solomon hundreds of years before his splendour would become a fable among Israelites.[33]

Evans had in fact discovered an entire civilisation, one that was completely unknown before and could claim to have been produced by the first civilised Europeans. He named the civilisation he had discovered the **Minoan** because of the references in classical writers and because although these Bronze Age Cretans worshipped all sorts of animals, it was a bull cult, worship of the **Minotaur**, that appeared to have predominated. In the frescoes Evans discovered many scenes of bulls – bulls being worshipped, bulls used in athletic events and, most notable of all, a huge plaster relief of a bull excavated on the wall of one of the main rooms of Knossos Palace.

Once the significance of Evans's discoveries had sunk in, his colleagues realised that Knossos was indeed the setting for part of Homer's *Odyssey* and that Ulysses himself goes ashore there. Evans spent more than a quarter of a century excavating every aspect of Knossos. He concluded, somewhat contrary to what he had originally thought, that the Minoans were formed from the fusion, around 2000 B.C., of immigrants from Anatolia with the native Neolithic population. Although this people constructed towns with elaborate palaces at the centre (the Knossos Palace was so huge, and so intricate, it is now regarded as the Labyrinth of the *Odyssey*), Evans also found that large town houses were not confined to royalty only but were inhabited by other citizens as well. For many scholars, this extension of property, art, and wealth in general marked the Minoan culture as the birth of Western civilisation, the 'mother culture' from which the classical world of Greece and Rome had evolved.[34]

GENES

Two weeks after Arthur Evans landed in Crete, on 24 March 1900, the very week that the archaeologist was making the first of his great discoveries, **Hugo de Vries,** a Dutch botanist, solved a very different – and even more important – piece of the evolution jigsaw. In Mannheim he read a paper to the German Botanical Society with the title 'The Law of Segregation of Hybrids.'

De Vries – a tall, taciturn man – had spent the previous years since 1889 experimenting with the breeding and hybridisation of plants, including such well-known flowers as asters, chrysanthemums, and violas. He told the meeting in Mannheim that as a result of his experiments he had formed the view that the character of a plant, its inheritance, was 'built up out of definite units'; that is, for each characteristic – such as the length of the stamens or the colour of the leaves – 'there corresponds a particular form of material bearer.' (The German words was in fact *Träger*, which may also be rendered as 'transmitter.') And he added, most significantly, 'There are no transitions between these elements.' Although his language was primitive, although he was feeling his way, that night in Mannheim de Vries had identified what later came to be called **genes**.[35] He noted, first, that certain characteristics of flowers – petal colour, for example – always occurred in one or other form but never in between. They were always white or red, say, never pink. And second, he had also identified the property of genes that we now recognise as 'dominance' and 'recession,' that some forms tend to predominate over others after these forms

Dominance + Recession

have been crossed (bred). This was a major discovery. Before the others present could congratulate him, however, he added something that has repercussions to this day. 'These two propositions', he said, referring to genes and dominance/recession, 'were, in essentials, formulated long ago by Mendel. ...'. They fell into oblivion, however, and were misunderstood. ... This important monograph [of Mendel's] is so rarely quoted that I myself did not become acquainted with it until I had concluded most of my experiments, and had independently deduced the above propositions.' This was a very generous acknowledgement by de Vries. It cannot have been wholly agreeable for him to find, after more than a decade's work, that he had been 'scooped' by some thirty years.[36]

The monograph that de Vries was referring to was 'Experiments in Plant-Hybridisation,' which Pater Gregor Mendel, a Benedictine monk, had read to the Brünn Society for the Study of Natural Science on a cold February evening in 1865. About forty men had attended the society that night, and this small but fairly distinguished gathering was astonished at what the rather stocky monk had to tell them, and still more so at the following month's meeting, when he launched into a complicated account of the mathematics behind dominance and recession. Linking maths and botany in this way was regarded as distinctly odd. Mendel's paper was published some months later in the *Proceedings of the Brünn Society for the Study of Natural Science*, together with an enthusiastic report, by another member of the society, of Darwin's theory of evolution, which had been published seven years before. The *Proceedings* of the Brünn Society were exchanged with more than 120 other societies, with copies sent to Berlin, Vienna, London, St Petersburg, Rome, and Uppsala (this is how scientific information was disseminated in those days). But little attention was paid to Mendel's theories.[37]

It appears that the world was not ready for Mendel's approach. The basic notion of Darwin's theory, then receiving so much attention, was the variability of species, whereas the basic tenet of Mendel was the constancy, if not of species, at least of their elements. It was only thanks to de Vries's assiduous scouring of the available scientific literature that he found the earlier publication. No sooner had he published his paper, however, than two more botanists, at Tübingen and Vienna, reported that they also had recently rediscovered Mendel's work. On 24 April, exactly a month after de Vries had released his results, **Carl Correns** published in the *Reports of the German Botanical Society* a ten-page account entitled 'Gregor Mendel's Rules Concerning the Behaviour of Racial Hybrids.' Correns's discoveries were very similar to those of de Vries. He too had scoured the literature – and found Mendel's paper.[38] And then in June of that same year, once more in the *Reports of the German Botanical Society*, there appeared over the signature of the Viennese botanist **Erich Tschermak** a paper entitled 'On Deliberate Cross-Fertilisation in the Garden Pea,' in which he arrived at substantially the same results as Correns and de Vries. Tschermak had begun his own experiments, he said, stimulated by Darwin, and he too had discovered Mendel's paper in the Brünn Society *Proceedings*.[39] It was an extraordinary coincidence, a chain of events that has lost none of its force as the years have passed. But of course, it is not the coincidence that chiefly matters. What matters is that the mechanism Mendel had recognised, and the others had rediscovered, filled in a major gap in what can claim to be the most influential idea of all time: Darwin's theory of evolution.

In the walled garden of his monastery, Mendel had procured thirty-four more or less distinct varieties of peas and subjected them to two years of testing. Mendel deliberately chose a variety (some were smooth or wrinkled, yellow or green, long-stemmed or short-stemmed) because he knew that one side of each variation was dominant – smooth, yellow, or long-stemmed, for instance, rather than wrinkled, green, or short-stemmed. He knew this because when peas were crossed with themselves, the first generation were always the same as their parents. However, when he self-fertilised this first generation, or F_1 as it was called, to produce an F_2 generation, he found that the arithmetic was revealing. What happened was that 253 plants produced 7,324 seeds. Of these, he found that 5,474 were smooth and 1,850 were wrinkled, a ratio of 2.96:1. In the case of seed colour, 258 plants produced 8,023 seeds: 6,022 yellow and 2,001 green, a ratio of 3.01:1. As he himself concluded, 'In this generation along with the **dominant traits** the **recessive** ones appear in their full expression, and they do so in the decisively evident average proportion of 3:1, so that among the four plants of this generation three show the dominant and one the recessive character.'[40] This enabled Mendel to make the profound observation that for many characteristics, the heritable quality existed in only *two* forms, the dominant and recessive strains, with *no* intermediate form. The universality of the 3:1 ratio across a number of characteristics confirmed this.* Mendel also discovered that these characteristics exist in sets, or chromosomes, which we will come to later. His figures and ideas helped explain *how* Darwinism, and evolution, worked. Dominant and recessive genes governed the variability of life forms, passing different characteristics on from generation to generation, and it was this variability on which natural selection exerted its influence, making it more likely that certain organisms reproduced to perpetuate their genes.

Mendel's theories were simple and, to many scientists, beautiful. Their sheer originality meant that almost anybody who got involved in the field had a chance to make new discoveries. And that is what happened. As Ernst Mayr has written in *The Growth of Biological Thought*, 'The rate at which the new findings of genetics occurred after 1900 is almost without parallel in the history of science.'[41]

And so, before the fledgling century was six months old, it had produced

---

* The 3:1 ratio may be explained in graphic form as follows:

		Genes from parent one	
		Y	y
Genes from	Y	YY	Yy
parent two	y	yY	yy

where Y is the dominant form of the gene, and y is the recessive.

Mendelism, underpinning Darwinism, and Freudianism, both systems that presented an understanding of man in a completely different way. They had other things in common, too. Both were scientific ideas, or were presented as such, and both involved the identification of forces or entities that were hidden, inaccessible to the human eye. As such they shared these characteristics with viruses, which had been identified only two years earlier, when Friedrich Löffler and Paul Frosch had shown that foot-and-mouth disease had a viral origin. There was nothing especially new in the fact that these forces were hidden. The invention of the telescope and the microscope, the discovery of radio waves and bacteria, had introduced people to the idea that many elements of nature were beyond the normal range of the human eye or ear. What was important about Freudianism, and Mendelism, was that these discoveries appeared to be fundamental, throwing a completely new light on nature, which affected everyone. The discovery of the 'mother civilisation' for European society added to this, reinforcing the view that religions evolved, too, meaning that one old way of understanding the world was subsumed under another, newer, more scientific approach. Such a change in the fundamentals was bound to be disturbing, but there was more to come. As the autumn of 1900 approached, yet another breakthrough was reported that added a third major realignment to our understanding of nature.

In 1900 Max Planck was forty-two. He was born into a very religious, rather academic family, and was an excellent musician. He became a scientist in spite of, rather than because of, his family. In the type of background he had, the humanities were considered a superior form of knowledge to science. His cousin, the historian Max Lenz, would jokingly refer to scientists (*Naturforscher*) as foresters (*Naturförster*). But science was Planck's calling; he never doubted it or looked elsewhere, and by the turn of the century he was near the top of his profession, a member of the Prussian Academy and a full professor at the University of Berlin, where he was known as a prolific generator of ideas that didn't always work out.[42]

Physics was in a heady flux at the turn of the century. The idea of the atom, an invisible and indivisible substance, went all the way back to classical Greece. At the beginning of the eighteenth century Isaac Newton had thought of atoms as minuscule billiard balls, hard and solid. Early-nineteenth-century chemists such as John Dalton had been forced to accept the existence of atoms as the smallest units of elements, since this was the only way they could explain chemical reactions, where one substance is converted into another, with no intermediate phase. But by the turn of the twentieth century the pace was quickening, as physicists began to experiment with the revolutionary notion that matter and energy might be different sides of the same coin. James Clerk Maxwell, a Scottish physicist who helped found the Cavendish Laboratory in Cambridge, England, had proposed in 1873 that the 'void' between atoms was filled with an electromagnetic field, through which energy moved at the speed of light. He also showed that light itself was a form of electromagnetic radiation. But even he thought of atoms as solid and, therefore, essentially mechanical. These were advances far more significant than anything since Newton.[43]

In 1887 **Heinrich Hertz** had discovered electric waves, or radio as it is now called, and then, in 1897, **J. J. Thomson**, who had followed Maxwell as director of the Cavendish, had conducted his famous experiment with a cathode ray tube. This had metal plates sealed into either end, and then the gas in the tube was sucked out, leaving a vacuum. If subsequently the metal plates were connected to a battery and a current generated, it was observed that the empty space, the vacuum inside the glass tube, glowed.[44] This glow was generated from the negative plate, the cathode, and was absorbed into the positive plate, the anode.*

The production of cathode rays was itself an advance. But what *were* they exactly? To begin with, everyone assumed they were light. However, in the spring of 1897 Thomson pumped different gases into the tubes and at times surrounded them with magnets. By systematically manipulating conditions, he demonstrated that cathode rays were in fact infinitesimally minute *particles* erupting from the cathode and drawn to the anode. He found that the particles' trajectory could be altered by an electric field and that a magnetic field shaped them into a curve. He also discovered that the particles were lighter than hydrogen atoms, the smallest known unit of matter, and exactly the same *whatever* the gas through which the discharge passed. Thomson had clearly identified something fundamental. This was the first experimental establishment of the particulate theory of matter.[45]

This particle, or 'corpuscle,' as Thomson called it at first, is today known as the **electron**. With the electron, particle physics was born, in some ways the most rigorous intellectual adventure of the twentieth century which, as we shall see, culminated in the atomic bomb. Many other particles of matter were discovered in the years ahead, but it was the very notion of particularity itself that interested Max Planck. Why did it exist? His physics professor at the University of Munich had once told him as an undergraduate that physics was 'just about complete,' but Planck wasn't convinced.[46] For a start, he doubted that atoms existed at all, certainly in the Newtonian/Maxwell form as hard, solid miniature billiard balls. One reason he held this view was the **Second Law of Thermodynamics**, conceived by **Rudolf Clausius**, one of Planck's predecessors at Berlin. The **First Law of Thermodynamics** may be illustrated by the way Planck himself was taught it. Imagine a building worker lifting a heavy stone on to the roof of a house.[47] The stone will remain in position long after it has been left there, storing energy until at some point in the future it falls back to earth. Energy, says the first law, can be neither created nor destroyed. Clausius, however, pointed out in his second law that the first law does not give the total picture. Energy is expended by the building worker as he strains to lift the stone into place, and is dissipated in the effort as heat, which among

---

* This is also the basis of the television tube. The positive plate, the anode, was reconfigured with a glass cylinder attached, after which it was found that a beam of cathode rays passed through the vacuum towards the anode made the glass fluoresce.

other things causes the worker to sweat. This dissipation Clausius termed 'entropy', and it was of fundamental importance, he said, because this energy, although it did not disappear from the universe, could never be recovered in its original form. Clausius therefore concluded that the world (and the universe) must always tend towards increasing disorder, must always add to its entropy and eventually run down. This was crucial because it implied that the universe was a one-way process; the Second Law of Thermodynamics is, in effect, a mathematical expression of time. In turn this meant that the Newton/Maxwellian notion of atoms as hard, solid billiard balls had to be wrong, for the implication of that system was that the 'balls' could run either way – under that system time was reversible; no allowance was made for entropy.[48]   /Electrons /

In 1897, the year Thomson discovered electrons, Planck began work on the project that was to make his name. Essentially, he put together two different observations available to anyone. First, it had been known since antiquity that as a substance (iron, say) is heated, it first glows dull red, then bright red, then white. This is because longer wavelengths (of light) appear at moderate temperatures, and as temperatures rise, shorter wavelengths appear. When the material becomes white-hot, all the wavelengths are given off. Studies of even hotter bodies – stars, for example – show that in the next stage the longer wavelengths drop out, so that the colour gradually moves to the blue part of the spectrum. Planck was fascinated by this and by its link to a second mystery, the so-called **black body problem**. A perfectly formed black body is one that absorbs every wavelength of electromagnetic radiation equally well. Such bodies do not exist in nature, though some come close: lampblack, for instance, absorbs 98 percent of all radiation.[49] According to classical physics, a black body should only emit radiation according to its temperature, and then such radiation should be emitted at every wavelength. In other words, it should only ever glow white. In Planck's Germany there were three perfect black bodies, two of them in Berlin. The one available to Planck and his colleagues was made of porcelain and platinum and was located at the Bureau of Standards in the Charlottenburg suburb of the city.[50] Experiments there showed that black bodies, when heated, behaved more or less like lumps of iron, giving off first dull red, then bright red-orange, then white light. Why?

Planck's revolutionary idea appears to have first occurred to him around 7 October 1900. On that day he sent a postcard to his colleague Heinrich Rubens on which he had sketched an equation to explain the behaviour of radiation in a black body.[51] The essence of Planck's idea, mathematical only to begin with, was that electromagnetic radiation was not continuous, as people thought, but could only be emitted in packets of a definite size. Newton had said that energy was emitted continuously, but Planck was contradicting him. It was, he said, as if a hosepipe could spurt water only in 'packets' of liquid. Rubens was as excited by this idea as Planck was (and Planck was not an excitable man). By 14 December that year, when Planck addressed the Berlin Physics Society, he had worked out his full theory.[52] Part of this was the calculation of the dimensions of this small packet of energy, which Planck called h and which later became known as **Planck's constant**. This, he calculated, had the value of 6.55×10^{-27} ergs each second (an erg is a small unit of energy). He explained the observation of black–body radiation by showing that while the packets of energy for any specific colour of light are the same, those for red, say, are smaller than those of yellow or green or blue. When a body is first heated, it emits packets of light with less energy. As the heat increases, the object can emit packets with greater energy. Planck had identified this very small packet as a basic indivisible building block of the universe, an 'atom' of radiation, which he called a '**quantum**.' It was confirmation that nature was not a continuous process but moved in a series of extremely small jerks. Quantum physics had arrived.

Not quite. Whereas Freud's ideas met hostility and de Vries's rediscovery of Mendel created an explosion of experimentation, Planck's idea was largely ignored. His problem was that so many of the theories he had come up with in the twenty years leading up to the quantum had proved wrong. So when he addressed the Berlin Physics Society with this latest theory, he was heard in polite silence, and there were no questions. It is not even clear that Planck himself was aware of the revolutionary nature of his ideas. It took four years for its importance to be grasped – and then by a man who would create his own revolution. His name was Albert Einstein.

On 25 October 1900, only days after Max Planck sent his crucial equations on a postcard to Heinrich Rubens, **Pablo Picasso** stepped off the Barcelona train at the Gare d'Orsay in Paris. Planck and Picasso could not have been more different. Whereas Planck led an ordered, relatively calm life in which tradition played a formidable role, Picasso was described, even by his mother, as 'an angel and a devil.' At school he rarely obeyed the rules, doodled compulsively, and bragged about his inability to read and write. But he became a prodigy in art, transferring rapidly from Malaga, where he was born, to his father's class at the art school in Corunna, to La Llotja, the school of fine arts in Barcelona, then to the Royal Academy in Madrid after he had won an award for his painting *Science and Charity*. However, for him, as for other artists of his time, Paris was the centre of the universe, and just before his nineteenth birthday he arrived in the City of Light. Descending from his train at the newly opened station, Picasso had no place to stay and spoke almost no French. To begin with he took a room at the Hôtel du Nouvel Hippodrome, a *maison de passe* on the rue Caulaincourt, which was lined with brothels.[53] He rented first a studio in Montparnasse on the Left Bank, but soon moved to Montmartre, on the Right.

Paris in 1900 was teeming with talent on every side. There were seventy daily newspapers, 350,000 electric streetlamps and the first Michelin guide had just appeared. It was the home of Alfred Jarry, whose play *Ubu Roi* was a grotesque parody of Shakespeare in which a fat, puppetlike king tries to take over Poland by means of mass murder. It shocked even W. B. Yeats, who attended its opening night. Paris was the home of Marie Curie, working on radioactivity, of Stephane Mallarmé, symbolist poet, and of Claude Debussy and his 'impressionist music.' It was the home of Erik Satie and his 'atonally adventurous' piano pieces. James Whistler and Oscar Wilde were exiles in residence, though the latter died that year. It was the city of Emile Zola and the Dreyfus affair, of Auguste and Louis Lumière who, having given the world's first commercial showing of movies in Lyons in 1895, had brought their new craze to the capital. At the Moulin Rouge, Henri de Toulouse-Lautrec was a fixture; Sarah Bernhardt was a fixture too, in the theatre named after her, where she played the lead role in *Hamlet en travesti*. It was the city of Gertrude Stein, Maurice Maeterlinck, Guillaume Apollinaire, of Isadora Duncan and Henri Bergson. In his study of the period, the Harvard historian Roger Shattuck called these the 'Banquet Years,' because Paris was celebrating, with glorious enthusiasm, the pleasures of life. How could Picasso hope to shine amid such avant-garde company?[54]

Even at the age of almost nineteen Picasso had already made a promising beginning. A somewhat sentimental picture by him, *Last Moments*, hung in the Spanish pavilion of the great Exposition Universelle of 1900, in effect a world's fair held in both the Grand and the Petit Palais in Paris to celebrate the new century.[55] Occupying 260 acres, the fair had its own electric train, a moving sidewalk that could reach a speed of five miles an hour, and a great wheel with more than eighty cabins. For more than a mile on either side of the Trocadero, the banks of the Seine were transformed by exotic facades. There were Cambodian temples, a mosque from Samarkand, and entire African villages. Below ground were an imitation gold mine from California and royal tombs from Egypt. Thirty-six ticket offices admitted one thousand people a minute.[56] Picasso's contribution to the exhibition was subsequently painted over, but X rays and drawings of the composition show a priest standing over the bed of a dying girl, a lamp throwing a lugubrious light over the entire scene. The subject may have been stimulated by the death of Picasso's sister, Conchita, or by Giacomo Puccini's opera *La Bohème*, which had recently caused a sensation when it opened in the Catalan capital. *Last Moments* had been hung too high in the exhibition to be clearly seen, but to judge by a drawing Picasso made of himself and his friends joyously leaving the show, he was pleased by its impact.[57]

To coincide with the Exposition Universelle, many distinguished international scholarly associations arranged to have their own conventions in Paris that year, in a building near the Pont d'Alma specially set aside for the purpose. At least 130 congresses were held in the building during the year and, of these, 40 were scientific, including the Thirteenth International Congress of Medicine, an International Congress of Philosophy, another on the rights of women, and major get-togethers of mathematicians, physicists, and electrical engineers. The philosophers tried (unsuccessfully) to define the foundations of mathematics, a discussion that floored Bertrand Russell, who would later write a book on the subject, together with Alfred North Whitehead. The mathematical congress was dominated by David Hilbert of Göttingen, Germany's (and perhaps the world's) foremost mathematician, who outlined what he felt were the twenty-three outstanding mathematical problems to be settled in the twentieth century.[58] These became known as the 'Hilbert questions'.

Many would be solved, though the basis for his choice was to be challenged fundamentally.

It would not take Picasso long to conquer the teeming artistic and intellectual world of Paris. Being an angel and a devil, there was never any question of an empty space forming itself about *his* person. Soon Picasso's painting would attack the very foundations of art, assaulting the eye with the same vigour with which physics and biology and psychology were bombarding the mind, and asking many of the same questions. His work probed what is solid and what is not, and dived beneath the surface of appearances to explore the connections between hitherto unapprehended hidden structures in nature. Picasso would focus on sexual anxiety, 'primitive' mentalities, the Minotaur, and the place of classical civilisations in the light of modern knowledge. In his collages he used industrial and mass-produced materials to play with meaning, aiming to disturb as much as to please. ('A painting,' he once said, 'is a sum of destructions.') Like that of Darwin, Mendel, Freud, J. J. Thomson and Max Planck, Picasso's work challenged the very categories into which reality had hitherto been organised.[59]

Picasso's work, and the extraordinary range of the exposition in Paris, underline what was happening in thought as the 1800s became the 1900s. The central points to grasp are, first, the extraordinary complementarity of many ideas at the turn of the century, the confident and optimistic search for hidden fundamentals and their place within what Freud, with characteristic overstatement, called 'underworlds'; and second, that the driving motor in this mentality, even when it was experienced as art, was scientific. Amazingly, the backbone of the century was already in place.

HALF-WAY HOUSE

In 1900 Great Britain was the most influential nation on earth, in political and economic terms. It held territories in north America and central America, and in South America Argentina was heavily dependent on Britain. It ruled colonies in Africa and the Middle East, and had dominions as far afield as Australasia. Much of the rest of the world was parcelled out between other European powers – France, Belgium, Holland, Portugal, Italy, and even Denmark. The United States had acquired the Panama Canal in 1899, and the Spanish Empire had just fallen into her hands. But although America's appetite for influence was growing, the dominant country in the world of ideas – in philosophy, in the arts and the humanities, in the sciences and the social sciences – was Germany, or more accurately, the German-speaking countries. This simple fact is important, for Germany's intellectual traditions were by no means unconnected to later political developments.

One reason for the German preeminence in the realm of thought was her universities, which produced so much of the chemistry of the nineteenth century and were at the forefront of biblical scholarship and classical archaeology, not to mention the very concept of the Ph.D., which was born in Germany. Another was demographic: in 1900 there were thirty-three cities in the German-speaking lands with populations of more than 100,000, and city life was a vital element in creating a marketplace of ideas. Among the German-speaking cities Vienna took precedence. If one place could be said to represent the mentality of western Europe as the twentieth century began, it was the capital of the Austro-Hungarian Empire.

Unlike other empires – the British or the Belgian, for example – the Austro-Hungarian dual monarchy, under the Habsburgs, had most of its territories in Europe: it comprised parts of Hungary, Bohemia, Romania, and Croatia and had its seaport at Trieste, in what is now Italy. It was also largely inward-looking. The German-speaking people were a proud race, highly conscious of their history and what they felt set them apart from other peoples. Such nationalism gave their intellectual life a particular flavour, driving it forward but circumscribing it at the same time, as we shall see. The architecture of Vienna also played a role in determining its unique character. The Ringstrasse, a ring of monumental buildings that included the university, the opera house, and the parliament building, had been erected in the second half of the nineteenth century around the central area of the old town, between it and the outer suburbs, in effect enclosing the intellectual and cultural life of the city inside a relatively small and very accessible area.[1] In that small enclosure had emerged the city's distinctive coffeehouses, an informal institution that helped make Vienna different from London, Paris, or Berlin, say. Their marble-topped tables were just as much a platform for new ideas as the newspapers, academic journals, and books of the day. These coffeehouses were reputed to have had their origins in the discovery of vast stocks of coffee in the camps abandoned by the Turks after their siege of Vienna in 1683. Whatever the truth of that, by 1900 they had evolved into informal clubs, well furnished and spacious, where the purchase of a small cup of coffee carried with it the right to remain there for the rest of the day and to have delivered, every half-hour, a glass of water on a silver tray.[2] Newspapers, magazines, billiard tables, and chess sets were provided free of charge, as were pen, ink, and (headed) writing paper. Regulars could have their mail sent to them at their favourite coffeehouse; they could leave their evening clothes there, so they needn't go home to change; and in some establishments, such as the Café Griensteidl, large encyclopaedias and other reference books were kept on hand for writers who worked at their tables.[3]

The chief arguments at the tables of the Café Griensteidl, and other cafés, were between what the social philosopher **Karl Pribram** termed two 'world-views.'[4] The words he used to describe these worldviews were *individualism* and *universalism*, but they echoed an even earlier dichotomy, one that interested Freud and arose out of the transformation at the beginning of the nineteenth century from a rural society of face-to-face intimacy to an urban society of 'atomistic' individuals, moving frantically about but never really meeting. For Pribram the individualist believes in empirical reason in the manner of the Enlightenment, and follows the scientific method of seeking truth by for-mulating hypotheses and testing them. Universalism, on the other hand, 'posits eternal, extramental truth, whose validity defies testing. ... An individualist discovers truth, whereas a universalist undergoes it.'[5] For Pribram, Vienna was the only true individualist city east of the Rhine, but even there, with the Catholic Church still so strong, universalism was nonetheless ever-present. This meant that, philosophically speaking, Vienna was a halfway house, where there were a number of 'halfway' avenues of thought, of which psychoanalysis was a perfect example. Freud saw himself as a scientist yet provided no real meth-odology whereby the existence of the unconscious, say, could be identified to the satisfaction of a sceptic. But Freud and the unconscious were not the only examples. The very doctrine of **therapeutic nihilism** – that nothing could be done about the ills of society or even about the sicknesses that afflicted the human body – showed an indifference to progressivism that was the very opposite of the empirical, optimistic, scientific approach. The aesthetics of **impressionism** – very popular in Vienna – was part of this same divide. The essence of impressionism was defined by the Hungarian art historian Arnold Hauser as an urban art that 'describes the changeability, the nervous rhythm, the sudden, sharp, but always ephemeral impressions of city life.'[6] This concern with evanescence, the transitoriness of experience, fitted in with the therapeutic nihilistic idea that there was nothing to be done about the world, except stand aloof and watch.

Two men who grappled with this view in their different ways were the writers **Arthur Schnitzler** and **Hugo von Hofmannsthal**. They belonged to a group of young bohemians who gathered at the Café Griensteidl and were known as *Jung Wien* (young Vienna).[7] The group also included **Theodor Herzl**, a brilliant reporter, an essayist, and later a leader of the Zionist movement; **Stefan Zweig**, a writer; and their leader, the newspaper editor **Hermann Bahr**. His paper, *Die Zeit*, was the forum for many of these talents, as was *Die Fackel* (The Torch), edited no less brilliantly by another writer of the group, **Karl Kraus**, more famous for his play *The Last Days of Mankind*.

The career of Arthur Schnitzler (1862–1931) shared a number of intriguing parallels with that of Freud. He too trained as a doctor and neurologist and studied neurasthenia.[8] Freud was taught by Theodor Meynert, whereas Schnitzler was Meynert's assistant. Schnitzler's interest in what Freud called the 'underestimated and much maligned erotic' was so similar to his own that Freud referred to Schnitzler as his *doppelgänger* (double) and deliberately avoided him. But Schnitzler turned away from medicine to literature, though his writings reflected many psychoanalytic concepts. His early works explored the emptiness of café society, but it was with *Lieutenant Gustl* (1901) and *The Road into the Open* (1908) that Schnitzler really made his mark.[9] *Lieutenant Gustl*, a sustained interior monologue, takes as its starting point an episode when 'a vulgar civilian' dares to touch the lieutenant's sword in the busy cloakroom of the opera. This small gesture provokes in the lieutenant confused and involuntary '**stream-of-consciousness**' ramblings that prefigure Proust. In *Gustl*, Schnitzler is still primarily a social critic, but in his references to aspects of the lieutenant's childhood that he thought he had forgotten, he hints at psychoanalytic ideas.[10] *The Road into the Open* explores more widely the instinctive, irrational aspects of individuals and the society in which they live. The dramatic structure of the book takes its power from an examination of the way the careers of several Jewish characters have been blocked or frustrated. Schnitzler indicts anti-Semitism, not simply for being wrong, but as the symbol of a new, illiberal culture brought about by a decadent aestheticism and by the arrival of mass society, which, together with a parliament '[that] has become a mere theatre through which the masses are manipulated,' gives full rein to the instincts, and which in the novel overwhelms the 'purposive, moral and scientific' culture represented by many of the Jewish characters. Schnitzler's aim is to highlight the insolubility of the 'Jewish question' and the dilemma between art and science.[11] Each disappoints him – aestheticism 'because it leads nowhere, science because it offers no meaning for the self'.[12]

Hugo von Hofmannsthal (1874–1929) went further than Schnitzler. Born into an aristocratic family, he was blessed with a father who encouraged, even expected, his son to become an aesthete. Hofmannsthal senior introduced his son to the Café Griensteidl when Hugo was quite young, so that the group around Bahr acted as a forcing house for the youth's precocious talents. In the early part of his career, Hofmannsthal produced what has been described as 'the most polished achievement in the history of German poetry,' but he was never totally comfortable with the aesthetic attitude.[13] Both *The Death of Titian* (1892) and *The Fool and Death* (1893), his most famous poems written before 1900, are sceptical that art can ever be the basis for society's values.[14] For Hofmannsthal, the problem is that while art may offer fulfilment for the person who *creates* beauty, it doesn't necessarily do so for the mass of society who are unable to create:

Our present is all void and dreariness,
If consecration comes not from without.[15]

Hofmannsthal's view is most clearly shown in his poem 'Idyll on an Ancient Vase Painting,' which tells the story of the daughter of a Greek vase painter. She has a husband, a blacksmith, and a comfortable standard of living, but she is dissatisfied; her life, she feels, is not fulfilled. She spends her time dreaming of her childhood, recalling the mythological images her father painted on the vases he sold. These paintings portrayed the heroic actions of the gods, who led the sort of dramatic life she yearns for. Eventually Hofmannsthal grants the woman her wish, and a centaur appears. Delighted that her fortunes have taken this turn, she immediately abandons her old life and escapes with the centaur. Alas, her husband has other ideas; if he can't have her, no one else can, and he kills her with a spear.[16] In summary this sounds heavy-handed, but Hofmannsthal's argument is unambiguous: beauty is paradoxical and can be subversive, terrible even. Though the spontaneous, instinctual life has its attractions, however vital its expression is for fulfilment, it is nevertheless dangerous, explosive. Aesthetics, in other words, is never simply self-contained and passive: it implies judgement and action.

Hofmannsthal also noted the encroachment of science on the old aesthetic culture of Vienna. 'The nature of our epoch,' he wrote in 1905, 'is multiplicity and indeterminacy. It can rest only on *das Gleitende* [the slipping, the sliding].' He added that 'what other generations believed to be firm is in fact *das Gleitende*.'[17] Could there be a better description about the way the Newtonian world was slipping after Maxwell's and Planck's discoveries? 'Everything fell into parts,' Hofmannsthal wrote, 'the parts again into more parts, and nothing allowed itself to be embraced by concepts any more.'[18] Like Schnitzler, Hofmannsthal was disturbed by political developments in the dual monarchy and in particular the growth of anti-Semitism. For him, this rise in irrationalism owed some of its force to science-induced changes in the understanding of reality; the new ideas were so disturbing as to promote a large-scale reactionary irrationalism. His personal response was idiosyncratic, to say the least, but had its own logic. At the grand age of twenty-six he abandoned poetry, feeling that the theatre offered a better chance of meeting current challenges. Schnitzler had pointed out that politics had become a form of theatre, and Hofmannsthal thought that theatre was needed to counteract political developments.[19] His work, from the plays *Fortunatus and His Sons* (1900–1) and *King Candaules* (1903) to his librettos for Richard Strauss, is all about political leadership as an art form, the point of kings being to preserve an aesthetic that provides order and, in so doing, controls irrationality. Yet the irrational must be given an outlet, Hofmannsthal says, and his solution is 'the ceremony of the whole,' a ritual form of politics in which no one feels excluded. His plays are attempts to create ceremonies of the whole, marrying individual psychology to group psychology, psychological dramas that anticipate Freud's later theories.[20] And so, whereas Schnitzler was prepared to be merely an observer of Viennese society, an elegant diagnostician of its shortcomings, Hofmannsthal rejected this therapeutic nihilism and saw himself in a more direct role, trying to change that society. As he revealingly put it, the arts had become the 'spiritual space of the nation.'[21] In his heart, Hofmannsthal always hoped that his writings about kings would help Vienna throw up a great leader, someone who would offer moral guidance and show the way ahead, 'melting all fragmentary manifestations into unity and changing all matter into "form, a new German reality." ' The words he used were uncannily close to what eventually came to pass. What he hoped for was a 'genius ... marked with the stigma of the usurper,' 'a true German and absolute man,' 'a prophet,' 'poet,' 'teacher,' 'seducer,' an 'erotic dreamer.'[22] Hofmannsthal's aesthetics of kingship overlapped with Freud's ideas about the dominant male, with the anthropological discoveries of Sir James Frazer, with Nietzsche and with Darwin. Hofmannsthal was very ambitious for the harmonising possibilities of art; he thought it could help counter the disruptive effects of science.

At the time, no one could foresee that Hofmannsthal's aesthetic would help pave the way for an even bigger bout of irrationality in Germany later in the century. But just as his aesthetics of kingship and 'ceremonies of the whole' were a response to *das Gleitende*, induced by scientific discoveries, so too was the new philosophy of **Franz Brentano** (1838–1917). Brentano was a popular man, and his lectures were legendary, so much so that students – among them Freud and Tomáš Masaryk – crowded the aisles and doorways. A statuesque figure (he looked like a patriarch of the church), Brentano was a fanatical but absentminded chess player (he rarely won because he loved to experiment, to see the consequences), a poet, an accomplished cook, and a carpenter. He frequently swam the Danube. He published a best-selling book of riddles. His friends included Theodor Meynert, Theodor Gomperz, and Josef Breuer, who was his doctor.[23] Destined for the priesthood, he had left the church in 1873 and later married a rich Jewish woman who had converted to Christianity (prompting one wag to quip that he was an icon in search of a gold background).[24]

Brentano's main interest was to show, in as scientific a way as possible, proof of God's existence. His was a very personal version of science, taking the form of an analysis of history. For Brentano, philosophy went in cycles. According to him, there had been three cycles – Ancient, Mediaeval, and Modern – each divided into four phases: Investigation, Application, Scepticism, and Mysticism. These he laid out in the following table.[25]

CYCLES	*Ancient*	*Mediaeval*	*Modern*
PHASES			
Investigation	Thales to Aristotle	Thomas Aquinas	Bacon to Locke
Application	Stoics, Epicureans	Duns Scotus	The Enlightenment
Scepticism	Sceptics, Eclectics	William of Occam	Hume
Mysticism	Neoplatonists, neo-Pythagoreans	Lullus, Cusanus	German idealism

This approach helped make Brentano a classic halfway figure in intellectual history. His science led him to conclude, after twenty years of search and lecturing, that there does indeed exist 'an eternal, creating, and sustaining principle,' to which he gave the term 'understanding.'[26] At the same time, his view that philosophy moved in cycles led him to doubt the progressivism of science. Brentano is chiefly remembered now for his attempt to bring a greater intellectual rigour to the examination of God, but though he was admired for his attempt to marry science and faith, many of his contemporaries felt that his entire system was doomed from the start. Despite this his approach did spark two other branches of philosophy that were themselves influential in the early years of the century. These were Edmund Husserl's phenomenology and Christian von Ehrenfels's theory of Gestalt.

Edmund Husserl (1859–1938) was born in the same year as Freud and in the same province, Moravia, as both Freud and Mendel. Like Freud he was Jewish, but he had a more cosmopolitan education, studying at Berlin, Leipzig, and Vienna.[27] His first interests were in mathematics and logic, but he found himself drawn to psychology. In those days, psychology was usually taught as an aspect of philosophy but was growing fast as its own discipline, thanks to advances in science. What most concerned Husserl was the link between consciousness and logic. Put simply, the basic question for him was this: did logic exist objectively, 'out there' in the world, or was it in some fundamental sense dependent on the mind? What was the logical basis of phenomena? This is where mathematics took centre stage, for numbers and their behaviour (addition, subtraction, and so forth) were the clearest examples of logic in action. So did numbers exist objectively, or were they too a function of mind? Brentano had claimed that in some way the mind 'intended' numbers, and if that were true, then it affected both their logical and their objective status. An even more fundamental question was posed by the mind itself: did the mind 'intend' itself? Was the mind a construction of the mind, and if so how did that affect the mind's own logical and objective status?[28]

Husserl's big book on the subject, *Logical Investigations*, was published in 1900 (volume one) and 1901 (volume two), its preparation preventing him from attending the Mathematical Congress at the Paris exposition in 1900. Husserl's view was that the task of philosophy was to *describe* the world as we meet it in ordinary experience, and his contribution to this debate, and to Western philosophy, was the concept of '**transcendental phenomenology**,' in which he proposed his famous **noema/noesis** dichotomy.[29] *Noema*, he said, is a timeless proposition-in-itself, and is valid, full stop. For example, God may be said to exist whether anyone thinks it or not. Noesis, by contrast, is more psychological – it is essentially what Brentano meant when he said that the mind 'intends' an object. For Husserl, *noesis* and *noema* were both present in consciousness, and he thought his breakthrough was to argue that a *noesis* is also a *noema* – it too exists in and of itself.[30] Many people find this dichotomy confusing, and Husserl didn't help by inventing further complex neologisms for his ideas (when he died, more than 40,000 pages of his manuscripts, mostly unseen and unstudied, were deposited in the library at Louvain University).[31] Husserl made big claims for himself; in the Brentano halfway house tradition, he believed he had worked out 'a theoretical science independent of all psychology and factual science.'[32] Few in the Anglophone world would agree, or even understand how you could have a theoretical science independent of factual science. But Husserl is best understood now as the immediate father of the so-called continental school of twentieth-century Western philosophy, whose members include Martin Heidegger, Jean-Paul Sartre, and Jürgen Habermas. They stand in contrast to the 'analytic' school begun by Bertrand Russell and Ludwig Wittgenstein, which became more popular in North America and Great Britain.[33]

Brentano's other notable legatee was **Christian von Ehrenfels** (1859–1932), the father of **Gestalt** philosophy and psychology. Ehrenfels was a rich man; he inherited a profitable estate in Austria but made it over to his younger brother so that he could devote his time to the pursuit of intellectual and literary activities.[34] In 1897 he accepted a post as professor of philosophy at Prague. Here, starting with Ernst Mach's observation that the size and colour of a circle can be varied 'without detracting from its circularity,' Ehrenfels modified Brentano's ideas, arguing that the mind somehow 'intends Gestalt qualities' – that is to say, there are certain 'wholes' in nature that the mind and the nervous system are pre-prepared to experience. (A well-known example of this is the visual illusion that may be seen as either a candlestick, in white, or two female profiles facing each other, in black.) Gestalt theory became very influential in German psychology for a time, and although in itself it led nowhere, it did set the ground for the theory of 'imprinting,' a readiness in the neonate to perceive certain forms at a crucial stage in development.[35] This idea flourished in the middle years of the century, popularised by German and Dutch biologists and ethologists.

In all of these Viennese examples – Schnitzler, Hofmannsthal, Brentano, Husserl, and Ehrenfels – it is clear that they were preoccupied with the recent discoveries of science, whether those discoveries were the unconscious, fundamental particles (and the even more disturbing void between them), Gestalt, or indeed entropy itself, the Second Law of Thermodynamics. If these notions of the philosophers in particular appear rather dated and incoherent today, it is also necessary to add that such ideas were only half the picture. Also prevalent in Vienna at the time were a number of avowedly rational but in reality frankly scientistic ideas, and they too read oddly now. Chief among these were the notorious theories of **Otto Weininger** (1880–1903).[36] The son of an anti-Semitic but Jewish goldsmith, Weininger developed into an overbearing coffeehouse dandy.[37] He was even more precocious than Hofmannsthal, teaching himself eight languages before he left university and publishing his undergraduate thesis. Renamed by his editor *Geschlecht und Charakter* (Sex and Character), the thesis was released in 1903 and became a huge hit. The book was rabidly anti-Semitic and extravagantly misogynist. Weininger put forward the view that all human behaviour can be explained in terms of male and female 'protoplasm,' which contributes to each person, with every cell possessing sexuality. Just as Husserl had coined neologisms for his ideas, so a whole lexicon was invented by Weininger: idioplasm, for example, was his name for sexually undifferentiated tissue; male tissue was arrhenoplasm; and female tissue was thelyplasm. Using elaborate arithmetic, Weininger argued that varying proportions of arrhenoplasm and thelyplasm could account for such diverse matters as genius, prostitution, memory, and so on. According to Weininger, all the major achievements in history arose because of the masculine principle – all art, literature, and systems of law, for example. The feminine principle, on the other hand, accounted for the negative elements, and all these negative elements converge, Weininger says, in the Jewish race. The Aryan race is the embodiment of the strong organising principle that characterises males, whereas the Jewish race embodies the 'feminine-chaotic principle of nonbeing.'[38] Despite the commercial success of his book, fame did not settle Weininger's restless spirit. Later that year he rented a room in the house in Vienna where Beethoven died, and shot himself. He was twenty-three.

A rather better scientist, no less interested in sex, was the Catholic psychiatrist **Richard von Krafft-Ebing** (1840–1902). His fame stemmed from a work he published in Latin in 1886, entitled *Psychopathia Sexualis: eine klinisch-forensische studie.* This book was soon expanded and proved so popular it was translated into seven languages. Most of the 'clinical-forensic' case histories were drawn from courtroom records, and attempted to link sexual psychopathology either to married life, to themes in art, or to the structure of organised religion.[39] As a Catholic, Krafft-Ebing took a strict line on sexual matters, believing that the only function of sex was to propagate the species within the institution of marriage. It followed that his text was disapproving of many of the 'perversions' he described. The most infamous 'deviation,' on which the notoriety of his study rests, was his coining of the term **masochism**. This word was derived from the novels and novellas of Leopold von Sacher-Masoch, the son of a police director in Graz. In the most explicit of his stories, *Venus im Pelz*, Sacher-Masoch describes his own affair at Baden bei Wien with a Baroness Fanny Pistor, during the course of which he 'signed a contract to submit for six months to being her slave.' Sacher-Masoch later left Austria (and his wife) to explore similar relationships in Paris.[40]

Psychopathia Sexualis clearly foreshadowed some aspects of psychoanalysis. Krafft-Ebing acknowledged that sex, like religion, could be sublimated in art – both could 'enflame the imagination.' 'What other foundation is there for the plastic arts of poetry? From (sensual) love arises that warmth of fancy which alone can inspire the creative mind, and the fire of sensual feeling kindles and preserves the glow and fervour of art.'[41] For Krafft-Ebing, sex within religion (and therefore within marriage) offered the possibility of 'rapture through submission,' and it was this process in perverted form that he regarded as the aetiology for the pathology of masochism. Krafft-Ebing's ideas were even more of a halfway house than Freud's, but for a society grappling with the threat that science posed to religion, any theory that dealt with the pathology of belief and its consequences was bound to fascinate, especially if it involved sex. Given those theories, Krafft-Ebing might have been more sympathetic to Freud's arguments when they came along; but he could never reconcile himself to the controversial notion of infantile sexuality. He became one of Freud's loudest critics.

The dominant architecture in Vienna was the Ringstrasse. Begun in the mid-nineteenth century, after Emperor Franz Joseph ordered the demolition of the old city ramparts and a huge swath of space was cleared in a ring around the centre, a dozen monumental buildings were erected over the following fifty years in this ring. They included the Opera, the Parliament, the Town Hall, parts of the university, and an enormous church. Most were embellished with fancy stone decorations, and it was this ornateness that provoked a reaction, first in Otto Wagner, then in Adolf Loos.

Otto Wagner (1841–1918) won fame for his 'Beardsleyan imagination' when he was awarded a commission in 1894 to build the Vienna underground railway.[42] This meant more than thirty stations, plus bridges, viaducts, and other urban structures. Following the dictum that function determines form, Wagner broke new ground by not only using modern materials but *showing* them. For example, he made a feature of the iron girders in the construction of bridges. These supporting structures were no longer hidden by elaborate casings of masonry, in the manner of the Ringstrasse, but painted and left exposed, their utilitarian form and even their riveting lending texture to whatever it was they were part of.[43] Then there were the arches Wagner designed as entranceways to the stations – rather than being solid, or neoclassical and built of stone, they reproduced the skeletal form of railway bridges or viaducts so that even from a long way off, you could tell you were approaching a station.[44] Warming to this theme, his other designs embodied the idea that the modern individual, living his or her life in a city, is always in a hurry, anxious to be on his or her way to work or home. The core structure therefore became the street, rather than the square or vista or palace. For Wagner, Viennese streets should be straight, direct; neighbourhoods should be organised so that workplaces are close to homes, and each neighbourhood should have a centre, not just one centre for the entire city. The facades of Wagner's buildings became less ornate, plainer, more functional, mirroring what was happening elsewhere in life. In this way Wagner's style presaged both the Bauhaus and the international movement in architecture.[45]

Adolf Loos (1870–1933) was even more strident. He was close to Freud and to Karl Kraus, editor of *Die Fackel*, and the rest of the crowd at the Café Griensteidl, and his rationalism was different from Wagner's – it was more revolutionary, but it was still rationalism. Architecture, he declared, was not art. 'The work of art is the private affair of the artist. The work of art wants to shake people out of their comfortableness [*Bequemlichkeit*]. The house must serve comfort. The art work is revolutionary, the house conservative.'[46] Loos extended this perception to design, clothing, even manners. He was in favour of simplicity, functionality, plainness. He thought men risked being enslaved by material culture, and he wanted to reestablish a 'proper' relationship between art and life. Design was inferior to art, because it was conservative, and when he understood the difference, man would be liberated. 'The artisan produced objects for use here and now, the artist for all men everywhere.'[47]

The ideas of Weininger and Loos inhabit a different kind of halfway house from those of Hofmannsthal and Husserl. Whereas the latter two were basically sceptical of science and the promise it offered, Weininger especially, but Loos too, was carried away with rationalism. Both adopted scientistic ideas, or terms, and quickly went beyond the evidence to construct systems that were as fanciful as the nonscientific ideas they disparaged. The scientific method, insufficiently appreciated or understood, could be mishandled, and in the Viennese halfway house it was.

Nothing illustrates better this divided and divisive way of looking at the world in turn-of-the-century Vienna than the row over **Gustav Klimt's** paintings for the university, the first of which was delivered in 1900. Klimt, born in Baumgarten, near Vienna, in 1862, was, like Weininger, the son of a goldsmith. But there the similarity ended. Klimt made his name decorating the new buildings of the Ringstrasse with vast murals. These were produced with his brother Ernst, but on the latter's death in 1892 Gustav withdrew for five years, during which time he appears to have studied the works of James Whistler, Aubrey Beardsley, and, like Picasso, Edvard Munch. He did not reappear until 1897, when he emerged at the head of the Vienna Secession, a band of nineteen artists who, like the impressionists in Paris and other artists at the Berlin Secession, eschewed the official style of art and instead followed their own version of art nouveau. In the German lands this was known as *Jugendstil*.[48]

Klimt's new style, bold and intricate at the same time, had three defining characteristics – the elaborate use of gold leaf (using a technique he had learned from his father), the application of small flecks of iridescent colour, hard like enamel, and a languid eroticism applied in particular to women. Klimt's paintings were not quite Freudian: his women were not neurotic, far from it. They were calm, placid, above all lubricious, 'the instinctual life frozen in art.'[49] Nevertheless, in drawing attention to women's sensuality, Klimt hinted that it had hitherto gone unsatisfied. This had the effect of making the women in his paintings threatening. They were presented as insatiable and devoid of any sense of sin. In portraying women like this, Klimt was subverting the familiar way of thinking every bit as much as Freud was. Here were women capable of the perversions reported in Krafft-Ebing's book, which made them tantalising and shocking at the same time. Klimt's new style immediately divided Vienna, but it quickly culminated in his commission for the university.

Three large panels had been asked for: *Philosophy, Medicine* and *Jurisprudence*.

All three provoked a furore but the rows over *Medicine* and *Jurisprudence* merely repeated the fuss over *Philosophy*. For this first picture the commission stipulated as a theme 'the triumph of Light over Darkness.' What Klimt actually produced was an opaque, 'deliquescent tangle' of bodies that appear to drift past the onlooker, a kaleidoscopic jumble of forms that run into each other, and all surrounded by a void. The professors of philosophy were outraged. Klimt was vilified as presenting 'unclear ideas through unclear forms.'[50] Philosophy was supposed to be a rational affair; it 'sought the truth via the exact sciences.'[51] Klimt's vision was anything but that, and as a result it wasn't wanted: eighty professors collaborated in a petition that demanded Klimt's picture never be shown at the university. The painter responded by returning his fee and never presenting the remaining commissions. Unforgivably, they were destroyed in 1945 when the Nazis burned Immendorf Castle, where they were stored during World War II.[52] The significance of the fight is that it brings us back to Hofmannsthal and Schnitzler, to Husserl and Brentano. For in the university commission, Klimt was attempting a major statement. How can rationalism succeed, he is asking, when the irrational, the instinctive, is such a dominant part of life? Is reason really the way forward? Instinct is an older, more powerful force. Yes, it may be more atavistic, more primitive, and a dark force at times. But where is the profit in denying it? This remained an important strand in Germanic thought until World War II.

If this was the dominant *Zeitgeist* in the Austro-Hungarian Empire at the turn of the century, stretching from literature to philosophy to art, at the same time there was in Vienna (and the Teutonic lands) a competing strain of thought that was wholly scientific and frankly reductionist, as we have seen in the work of Planck, de Vries, and Mendel. But the most ardent, the most impressive, and by far the most influential reductionist in Vienna was **Ernst Mach** (1838–1916).[53] Born near Brünn, where Mendel had outlined his theories, Mach, a precocious and difficult child who questioned everything, was at first tutored at home by his father, then studied mathematics and physics in Vienna. In his own work, he made two major discoveries. Simultaneously with Breuer, but entirely independently, he discovered the importance of the **semicircular canals** in the inner ear for bodily equilibrium. And second, using a special technique, he made photographs of bullets travelling at more than the speed of sound.[54] In the process, he discovered that they create not one but two shock waves, one at the front and another at the rear, as a result of the vacuum their high speed creates. This became particularly significant after World War II with the arrival of jet aircraft that approached the speed of sound, and this is why supersonic speeds (on Concorde, for instance) are given in terms of a '**Mach number**.'[55]

After these noteworthy empirical achievements, however, Mach became more and more interested in the philosophy and history of science.[56] Implacably opposed to metaphysics of any kind, he worshipped the Enlightenment as the most important period in history because it had exposed what he called the 'misapplication' of concepts like God, nature, and soul. The ego he regarded as a 'useless hypothesis.'[57] In physics he at first doubted the very existence of atoms and wanted measurement to replace 'pictorialisation,' the inner mental images we have of how things are, even dismissing Immanuel Kant's a priori theory of number (that numbers just *are*).[58] Mach argued instead that 'our' system was only one of several possibilities that had arisen merely to fill our economic needs, as an aid in rapid calculation. (This, of course, was an answer of sorts to Husserl.) All knowledge, Mach insisted, could be reduced to sensation, and the task of science was to describe sense data in the simplest and most neutral manner. This meant that for him the primary sciences were physics, 'which provide the raw material for sensations,' and psychology, by means of which we are aware of our sensations. For Mach, philosophy had no existence apart from science.[59] An examination of the history of scientific ideas showed, he argued, how these ideas evolved. He firmly believed that there is evolution in ideas, with the survival of the fittest, and that we develop ideas, even scientific ideas, in order to survive. For him, theories in physics were no more than descriptions, and mathematics no more than ways of organising these descriptions. For Mach, therefore, it made less sense to talk about the truth or falsity of theories than to talk of their usefulness. Truth, as an eternal, unchanging thing that just *is*, for him made no sense. He was criticised by Planck among others on the grounds that his evolutionary/biological theory was itself metaphysical speculation, but that didn't stop him being one of the most influential thinkers of his day. The Russian Marxists, including Anatoli Lunacharsky and Vladimir Lenin, read Mach, and the Vienna Circle was founded in response as much to his ideas as to Wittgenstein's. Hofmannsthal, Robert Musil, and even Albert Einstein all acknowledged his 'profound influence.'[60]

Mach suffered a stroke in 1898, and thereafter reduced his workload considerably. But he did not die until 1916, by which time physics had made some startling advances. Though he never adjusted entirely to some of the more exotic ideas, such as relativity, his uncompromising reductionism undoubtedly gave a massive boost to the new areas of investigation that were opening up after the discovery of the electron and the quantum. These new entities had dimensions, they could be measured, and so conformed exactly to what Mach thought science should be. Because of his influence, quite a few of the future particle physicists would come from Vienna and the Habsburg hinterland. Owing to the rival arenas of thought, however, which gave free rein to the irrational, very few would actually practise their physics there.

That almost concludes this account of Vienna, but not quite. For there are two important gaps in this description of that teeming world. One is music. The second Viennese school of music comprised Gustav Mahler, Arnold Schoenberg, Anton von Webern, and Alban Berg, but also included Richard (not Johann) Strauss, who used Hofmannsthal as librettist. They more properly belong in chapter 4, among *Les Demoiselles de Modernisme*. The second gap in this account concerns a particular mix of science and politics, a deep pessimism about the way the world was developing as the new century was ushered in. This was seen in sharp focus in Austria, but in fact it was a constellation of ideas that extended to many countries, as far afield as the United States of America and even to China. The alleged scientific basis for this pessimism was Darwinism; the sociological process that sounded the alarm was 'degeneration'; and the political result, as often as not, was some form of racism.

DARWIN'S HEART OF DARKNESS

Three significant deaths occurred in 1900. John Ruskin died insane on 20 January, aged eighty-one. The most influential art critic of his day, he had a profound effect on nineteenth-century architecture and, in *Modern Painters*, on the appreciation of J. M. W. Turner.[1] Ruskin hated industrialism and its effect on aesthetics and championed the Pre-Raphaelites – he was splendidly anachronistic. Oscar Wilde died on 30 November, aged forty-four. His art and wit, his campaign against the standardisation of the eccentric, and his efforts 'to replace a morality of severity by one of sympathy' have made him seem more modern, and more missed, as the twentieth century has gone by. Far and away the most significant death, however, certainly in regard to the subject of this book, was that of **Friedrich Nietzsche**, on 25 August. Aged fifty-six, he too died insane.

There is no question that the figure of Nietzsche looms over twentieth-century thought. Inheriting the pessimism of Arthur Schopenhauer, Nietzsche gave it a modern, post-Darwinian twist, stimulating in turn such later figures as Oswald Spengler, T. S. Eliot, Martin Heidegger, Jean-Paul Sartre, Herbert Marcuse, and even Aleksandr Solzhenitsyn and Michel Foucault. Yet when he died, Nietzsche was a virtual vegetable and had been so for more than a decade. As he left his boardinghouse in Turin on 3 January 1889 he saw a cabdriver beating a horse in the Palazzo Carlo Alberto. Rushing to the horse's defence, Nietzsche suddenly collapsed in the street. He was taken back to his lodgings by onlookers, and began shouting and banging the keys of his piano where a short while before he had been quietly playing Wagner. A doctor was summoned who diagnosed 'mental degeneration.' It was an ironic verdict, as we shall see.[2]

Nietzsche was suffering from the tertiary phase of syphilis. To begin with, he was wildly deluded. He insisted he was the Kaiser and became convinced his incarceration had been ordered by Bismarck. These delusions alternated with uncontrollable rages. Gradually, however, his condition quietened and he was released, to be looked after first by his mother and then by his sister. **Elisabeth Förster-Nietzsche** took an active interest in her brother's philosophy. A member of Wagner's circle of intellectuals, she had married another acolyte, Bernard Förster, who in 1887 had conceived a bizarre plan to set up a colony of Aryan German settlers in Paraguay, whose aim was to recolonise the New World with 'racially pure Nordic pioneers.' This utopian scheme failed disastrously, and Elisabeth returned to Germany. (Bernard committed suicide.) Not at all humbled by the experience, she began promoting her brother's philosophy. She forced her mother to sign over sole legal control in his affairs, and she set up a Nietzsche archive. She then wrote a two-volume adulatory biography of Friedrich and organised his home so that it became a shrine to his work.[3] In doing this, she vastly simplified and coarsened her brother's ideas, leaving out anything that was politically sensitive or too controversial. What remained, however, was controversial enough. Nietzsche's main idea (not that he was particularly systematic) was that all of history was a metaphysical struggle between two groups, those who express the '**will to power**,' the vital life force necessary for the creation of values, on which civilisation is based, and those who do not, primarily the masses produced by democracy.[4] 'Those poor in life, the weak,' he said, 'impoverish culture,' whereas 'those rich in life, the strong, enrich it.'[5] All civilisation owes its existence to 'men of prey who were still in possession of unbroken strength of will and lust for power, [who] hurled themselves on weaker, more civilised, more peaceful races ... upon mellow old cultures whose last vitality was even then flaring up in splendid fireworks of spirit and corruption.'[6] These men of prey he called '**Aryans**,' who become the ruling class or caste. Furthermore, this 'noble caste was always the barbarian caste.' Simply because they had more life, more energy, they were, he said, 'more complete human beings' than the 'jaded sophisticates' they put down.[7] These energetic nobles, he said, 'spontaneously create values' for themselves and the society around them. This strong 'aristocratic class' creates its own definitions of right and wrong, honour and duty, truth and falsity, beauty and ugliness, and the conquerors impose their views on the conquered – this is only natural, says Nietzsche. Morality, on the other hand, 'is the creation of the underclass.'[8] It springs from resentment and nourishes the virtues of the herd animal. For Nietzsche, 'morality negates life.'[9] Conventional, sophisticated civilisation – 'Western man' – he thought, would inevitably result in the end of humanity. This was his famous description of '**the last man**.'[10]

The acceptance of Nietzsche's views was hardly helped by the fact that many of them were written when he was already ill with the early stages of syphilis. But there is no denying that his philosophy – mad or not – has been extremely influential, not least for the way in which, for many people, it accords neatly with what Charles Darwin had said in his theory of evolution, published in 1859. Nietzsche's concept of the 'superman,' the *Übermensch*, lording it over the underclass certainly sounds like evolution, the law of the jungle, with natural selection in operation as 'the survival of the fittest' for the overall good of humanity, whatever its effects on certain individuals. But of course the ability to lead, to create values, to impose one's will on others, is not in and of itself what evolutionary theory meant by 'the fittest.' The fittest were those who reproduced most, propagating their own kind. Social Darwinists, into which class Nietzsche essentially fell, have often made this mistake.

After publication of Darwin's *On the Origin of Species* it did not take long for his ideas about biology to be extended to the operation of human societies. Darwinism first caught on in the United States of America. (Darwin was made an honorary member of the American Philosophical Society in 1869, ten years before his own university, Cambridge, conferred on him an honorary degree.)[11] American social scientists **William Graham Sumner** and **Thorsten Veblen** of Yale, **Lester Ward** of Brown, **John Dewey** at the University of Chicago, and **William James**, **John Fiske** and others at Harvard, debated politics, war, and the layering of human communities into different classes against the background of a Darwinian 'struggle for survival' and the 'survival of the fittest.' Sumner believed that Darwin's new way of looking at mankind had provided the ultimate explanation – and rationalisation – for the world as it was. It explained laissez-faire economics, the free, unfettered competition popular among businessmen. Others believed that it explained the prevailing imperial structure of the world in which the 'fit' white races were placed 'naturally' above the 'degenerate' races of other colours. On a slightly different note, the slow pace of change implied by evolution, occurring across geological aeons, also offered to people like Sumner a natural metaphor for political advancement: rapid, revolutionary change was 'unnatural'; the world was essentially the way it was as a result of natural laws that brought about change only gradually.[12]

Fiske and Veblen, whose *Theory of the Leisure Class* was published in 1899, flatly contradicted Sumner's belief that the well-to-do could be equated with the biologically fittest. Veblen in fact turned such reasoning on its head, arguing that the type of characters 'selected for dominance' in the business world were little more than barbarians, a 'throw-back' to a more primitive form of society.[13]

Britain had probably the most influential social Darwinist in **Herbert Spencer**. Born in 1820 into a lower-middle-class Nonconformist English family in Derby, Spencer had a lifelong hatred of state power. In his early years he was on the staff of the *Economist*, a weekly periodical that was fanatically pro-laissez-faire. He was also influenced by the positivist scientists, in particular Sir Charles Lyell, whose *Principles of Geology*, published in the 1830s, went into great detail about fossils that were millions of years old. Spencer was thus primed for Darwin's theory, which at a stroke appeared to connect earlier forms of life to later forms in one continuous thread. It was Spencer, and not Darwin, who actually coined the phrase '**survival of the fittest**,' and Spencer quickly saw how Darwinism might be applied to human societies. His views on this were uncompromising. Regarding the poor, for example, he was against all state aid. They were unfit, he said, and should be eliminated: 'The whole effort of nature is to get rid of such, to clear the world of them, and make room for better.'[14] He explained his theories in his seminal work *The Study of Sociology* (1872–3), which had a notable impact on the rise of sociology as a discipline (a biological base made it seem so much more like science). Spencer was almost certainly the most widely read social Darwinist, as famous in the United States as in Britain.

Germany had its own Spencer-type figure in **Ernst Haeckel** (1834–1919). A zoologist from the University of Jena, Haeckel took to social Darwinism as if it were second nature. He referred to 'struggle' as 'a watchword of the day.'[15]

However, Haeckel was a passionate advocate of the principle of the inheritance of acquired characteristics, and unlike Spencer he favoured a strong state. It was this, allied to his bellicose racism and anti-Semitism, that led people to see him as a proto-Nazi.[16] France, in contrast, was relatively slow to catch on to Darwinism, but when she did, she had her own passionate advocate. In her *Origines de l'homme et des sociétés*, **Clemence August Royer** took a strong social Darwinist line, regarding 'Aryans' as superior to other races and warfare between them as inevitable in the interests of progress.[17] In Russia, the anarchist **Peter Kropotkin** (1842–1921) released *Mutual Aid* in 1902, in which he took a different line, arguing that although competition was undoubtedly a fact of life, so too was cooperation, which was so prevalent in the animal kingdom as to constitute a natural law. Like Veblen, he presented an alternative model to the Spencerians, in which violence was condemned as abnormal. Social Darwinism was, not unnaturally, compared with Marxism, and not only in the minds of Russian intellectuals.[18] Neither Karl Marx nor Friedrich Engels saw any conflict between the two systems. At Marx's graveside, Engels said, 'Just as Darwin discovered the law of development of organic nature, so Marx discovered the law of development of human history.'[19] But others did see a conflict. Darwinism was based on perpetual struggle; Marxism looked forward to a time when a new harmony would be established.

If one had to draw up a balance sheet of the social Darwinist arguments at the turn of the century, one would have to say that the ardent Spencerians (who included several members of Darwin's family, though never the great man himself) had the better of it. This helps explain the openly racist views that were widespread then. For example, in the theories of the French aristocratic poet Arthur de Gobineau (1816–1882), racial interbreeding was 'dysgenic' and led to the collapse of civilisation. This reasoning was taken to its limits by another Frenchman, **Georges Vacher de Lapouge** (1854–1936). Lapouge, who studied ancient skulls, believed that races were species in the process of formation, that racial differences were 'innate and ineradicable,' and that any idea that races could integrate was contrary to the laws of biology.[20] For Lapouge, Europe was populated by three racial groups: *Homo europaeus*, tall, pale-skinned, and long-skulled (dolichocephalic); *Homo alpinus*, smaller and darker with brachycephalic (short) heads; and the Mediterranean type, long-headed again but darker and shorter even than *alpinus*. Such attempts to calibrate racial differences would recur time and again in the twentieth century.[21] Lapouge regarded democracy as a disaster and believed that the brachycephalic types were taking over the world. He thought the proportion of dolichocephalic individuals was declining in Europe, due to emigration to the United States, and suggested that alcohol be provided free of charge in the hope that the worst types might kill each other off in their excesses. He wasn't joking.[22]

In the German-speaking countries, a veritable galaxy of scientists and pseudoscientists, philosophers and pseudophilosophers, intellectuals and would-be intellectuals, competed to outdo each other in the struggle for public attention. **Friedrich Ratzel**, a zoologist and geographer, argued that all living organisms competed in a *Kampf um Raum*, a struggle for space in which the winners expelled the losers. This struggle extended to humans, and the successful races had to extend their living space, *Lebensraum*, if they were to avoid decline.[23] For **Houston Stewart Chamberlain** (1855–1927), the renegade son of a British admiral, who went to Germany and married Wagner's daughter, racial struggle was 'fundamental to a "scientific" understanding of history and culture.'[24] Chamberlain portrayed the history of the West 'as an incessant conflict between the spiritual and culture-creating Aryans and the mercenary and materialistic Jews' (his first wife had been half Jewish).[25] For Chamberlain, the Germanic peoples were the last remnants of the Aryans, but they had become enfeebled through interbreeding with other races.

Max Nordau (1849–1923), born in Budapest, was the son of a rabbi. His best-known book was the two-volume *Entartung* (Degeneration), which, despite being 600 pages long, became an international best-seller. Nordau became convinced of 'a severe mental epidemic; a sort of black death of degeneracy and hysteria' that was affecting Europe, sapping its vitality, manifested in a whole range of symptoms: 'squint eyes, imperfect ears, stunted growth ... pessimism, apathy, impulsiveness, emotionalism, mysticism, and a complete absence of any sense of right and wrong.'[26] Everywhere he looked, there was decline.[27] The impressionist painters were the result, he said, of a degenerate physiology, nystagmus, a trembling of the eyeball, causing them to paint in the fuzzy, indistinct way that they did. In the writings of Charles Baudelaire, Oscar Wilde, and Friedrich Nietzsche, Nordau found 'overweening egomania,' while Zola had 'an obsession with filth.' Nordau believed that degeneracy was caused by industrialised society – literally the wear-and-tear exerted on leaders by railways, steamships, telephones, and factories. When Freud visited Nordau, he found him 'unbearably vain' with a complete lack of sense of humour.[28] In Austria, more than anywhere else in Europe, social Darwinism did not stop at theory. Two political leaders, Georg Ritter von Schönerer and Karl Lueger, fashioned their own cocktail of ideas from this brew to initiate political platforms that stressed the twin aims of first, power to the peasants (because they had remained 'uncontaminated' by contact with the corrupt cities), and second, a virulent anti-Semitism, in which Jews were characterised as the very embodiment of degeneracy. It was this miasma of ideas that greeted the young Adolf Hitler when he first arrived in Vienna in 1907 to attend art school.

Not dissimilar arguments were heard across the Atlantic in the southern part of the United States. Darwinism prescribed a common origin for all races and therefore could have been used as an argument *against* slavery, as it was by **Chester Loring Brace**.[29] But others argued the opposite. **Joseph le Conte** (1823–1901), like Lapouge or Ratzel, was an educated man, not a redneck but a trained geologist. When his book, *The Race Problem in the South*, appeared in 1892, he was the highly esteemed president of the American Association for the Advancement of Science. His argument was brutally Darwinian.[30] When two races came into contact, one was bound to dominate the other. He argued that if the weaker race was at an early stage of development – like the Negro –

slavery was appropriate because the 'primitive' mentality could be shaped. If, however, the race had achieved a greater measure of sophistication, like 'the redskin,' then 'extermination is unavoidable.'[31]

The most immediate political impact of social Darwinism was the **eugenics movement** that became established with the new century. All of the above writers played a role in this, but the most direct progenitor, the real father, was Darwin's cousin **Francis Galton** (1822–1911). In an article published in 1904 in the *American Journal of Sociology*, he argued that the essence of eugenics was that 'inferiority' and 'superiority' could be objectively described and measured – which is why Lapouge's calibration of skulls was so important.[32] Lending support for this argument was the fall in European populations at the time (thanks partly to emigration to the United States), adding to fears that 'degeneration' – urbanisation and industrialisation – was making people less likely or able to reproduce and encouraging the 'less fit' to breed faster than the 'more fit.' The growth in suicide, crime, prostitution, sexual deviance, and those squint eyes and imperfect ears that Nordau thought he saw, seemed to support this interpretation.[33] This view acquired what appeared to be decisive support from a survey of British soldiers in the Boer War between 1899 and 1902, which exposed alarmingly low levels of health and education among the urban working class.

The German Race Hygiene Society was founded in 1905, followed by the Eugenics Education Society in England in 1907.[34] An equivalent body was founded in the United States, in 1910 and in France in 1912.[35] Arguments at times bordered on the fanatical. For example, F. H. Bradley, an Oxford professor, recommended that lunatics and persons with hereditary diseases should be killed, *and their children*.[36] In America, in 1907, the state of Indiana passed a law that required a radically new punishment for inmates in state institutions who were 'insane, idiotic, imbecilic, feebleminded or who were convicted rapists': sterilisation.[37]

It would be wrong, however, to give the impression that the influence of social Darwinism was wholly crude and wholly bad. It was not.

A distinctive feature of Viennese journalism at the turn of the century was the *feuilleton*. This was a detachable part of the front page of a newspaper, below the fold, which contained not news but a chatty – and ideally speaking, witty – essay written on any topical subject. One of the best *feuilletonistes* was a member of the Café Griendsteidl set, **Theodor Herzl** (1860–1904). Herzl, the son of a Jewish merchant, was born in Budapest but studied law in Vienna, which soon became home. While at the university Herzl began sending squibs to the *Neue Freie Presse*, and he soon developed a witty prose style to match his dandified dress. He met Hugo von Hofmannsthal, Arthur Schnitzler, and Stefan Zweig. He did his best to ignore the growing anti-Semitism around him, identifying with the liberal aristocracy of the empire rather than with the ugly masses, the 'rabble,' as Freud called them. He believed that Jews should assimilate, as he was doing, or on rare occasions recover their honour after they had suffered discrimination through duels, then very common in Vienna. He thought that after a few duels (as fine a Darwinian device as one could imagine) Jewish honour would be reclaimed. But in October 1891 his life began to change. His journalism was rewarded with his appointment as Paris correspondent of the *Neue Freie Presse*. His arrival in the French capital, however, coincided with a flood of anti-Semitism set loose by the Panama scandal, when corrupt officials of the company running the canal were put on trial. This was followed in 1894 by the case of Alfred Dreyfus, a Jewish officer convicted of treason. Herzl doubted the man's guilt from the start, but he was very much in a minority. For Herzl, France had originally represented all that was progressive and noble in Europe – and yet in a matter of months he had discovered her to be hardly different from his own Vienna, where the vicious anti-Semite Karl Lueger was well on his way to becoming mayor.[38]

A change came over Herzl. At the end of May 1895, he attended a performance of *Tannhäuser* at the Opéra in Paris. Not normally passionate about opera, that evening he was, as he later said, 'electrified' by the performance, which illustrated the irrationalism of *völkisch* politics.[39] He went home and, 'trembling with excitement,' sat down to work out a strategy by means of which the Jews could secede from Europe and establish an independent homeland.[40] Thereafter he was a man transformed, a committed **Zionist**. Between his visit to *Tannhäuser* and his death in 1904, Herzl organised no fewer than six world congresses of Jewry, lobbying everyone for the cause, from the pope to the sultan.[41] The sophisticated, educated, and aristocratic Jews wouldn't listen to him at first. But he outthought them. There had been Zionist movements before, but usually they had appealed to personal self-interest and/or offered financial inducements. Instead, Herzl rejected a rational concept of history in favour of 'sheer psychic energy as the motive force.' The Jews must have their Mecca, their Lourdes, he said. 'Great things need no firm foundation ... the secret lies in movement. Hence I believe that somewhere a guidable aircraft will be discovered. Gravity overcome through movement.'[42] Herzl did not specify that Zion had to be in Palestine; parts of Africa or Argentina would do just as well, and he saw no need for Hebrew to be the official language.[43] Orthodox Jews condemned him as an heretic (because he plainly wasn't the Messiah), but at his death, ten years and six congresses later, the Jewish Colonial Trust, the joint stock company he had helped initiate and which would be the backbone of any new state, had 135,000 shareholders, more than any other enterprise then existing. His funeral was attended by 10,000 Jews from all over Europe. A Jewish homeland had not yet been achieved, but the idea was no longer a heresy.[44]

Like Herzl, **Max Weber** was concerned with religion as a shared experience. Like Max Nordau and the Italian criminologist Cesare Lombroso, he was troubled by the 'degenerate' nature of modern society. He differed from them in believing that what he saw around him was not wholly bad. No stranger to the 'alienation' that modern life could induce, he thought that group identity was a central factor in making life bearable in modern cities and that its importance had been overlooked. For several years around the turn of the century he had produced almost no serious academic work (he was on the faculty at the University of Freiburg), being afflicted by a severe depression that showed no signs of recovery until 1904. Once begun, however, few recoveries can have been so dramatic. The book he produced that year, quite different from anything he had done before, transformed his reputation.[45]

Prior to his illness, most of Weber's works were dry, technical monographs on agrarian history, economics, and economic law, including studies of mediaeval trading law and the conditions of rural workers in the eastern part of Germany – hardly best-sellers. However, fellow academics were interested in his Germanic approach, which in marked contrast to British style focused on economic life within its cultural context, rather than separating out economics and politics as a dual entity, more or less self-limiting.[46]

A tall, stooping man, Weber had an iconic presence, like Brentano, and was full of contradictions.[47] He rarely smiled – indeed his features were often clouded by worry. But it seems that his experience of depression, or simply the time it had allowed for reflection, was responsible for the change that came over him and helped produce his controversial but undoubtedly powerful idea. The study that Weber began on his return to health was on a much broader canvas than, say, the peasants of eastern Germany. It was entitled *The Protestant Ethic and the Spirit of Capitalism*.

Weber's thesis in this book was hardly less contentious than Freud's and, as Anthony Giddens has pointed out, it immediately provoked much the same sort of sharp critical debate. He himself saw it as a refutation of Marxism and materialism, and the themes of *The Protestant Ethic* cannot easily be understood without some knowledge of Weber's intellectual background.[48] He came from the same tradition as Brentano and Husserl, the tradition of *Geisteswissenschaftler*, which insisted on the differentiation of the sciences of nature from the study of man:[49] 'While we can "explain" natural occurrences in terms of the application of causal laws, human conduct is intrinsically meaningful, and has to be "interpreted" or "understood" in a way which has no counterpart in nature.'[50] For Weber, this meant that social and psychological matters were much more relevant than purely economic or material issues. The very opening of *The Protestant Ethic* shows Weber's characteristic way of thinking: 'A glance at the occupation statistics of any country of mixed religious composition brings to light with remarkable frequency a situation which has several times provoked discussion in the Catholic press and literature, and in Catholic congresses in Germany, namely, the fact that business leaders and owners of capital, as well as the higher grades of skilled labour, and even more the higher technically and commercially trained personnel of modern enterprises, are overwhelmingly Protestant.'[51]

That observation is, for Weber, the nub of the matter, the crucial discrepancy that needs to be explained. Early on in the book, Weber makes it clear that he is not talking just about money. For him, a capitalistic enterprise and the pursuit of gain are not at all the same thing. People have always wanted to be rich, but that has little to do with capitalism, which he identifies as 'a regular orientation to the achievement of profit through (nominally peaceful) economic exchange.'[52] Pointing out that there were mercantile operations – very successful and of considerable size – in Babylonia, Egypt, India, China, and mediaeval Europe, he says that it is only in Europe, since the Reformation, that capitalist activity has become associated with the *rational organisation of formally free labour*.[53]

Weber was also fascinated by what he thought to begin with was a puzzling paradox. In many cases, men – and a few women – evinced a drive toward the accumulation of wealth but at the same time showed a 'ferocious asceticism,' a singular absence of interest in the worldly pleasures that such wealth could buy. Many entrepreneurs actually pursued a lifestyle that was 'decidedly frugal.'[54] Was this not odd? Why work hard for so little reward? After much consideration, carried out while he was suffering from depression, Weber thought he had found an answer in what he called the 'this-worldly asceticism' of puritanism, a notion that he expanded by reference to the concept of 'the calling.'[55] Such an idea did not exist in antiquity and, according to Weber, it does not exist in Catholicism either. It dates only from the Reformation, and behind it lies the idea that the highest form of moral obligation of the individual, the best way to fulfil his duty to God, is to help his fellow men, now, in this world. In other words, whereas for the Catholics the highest idea was purification of one's own soul through withdrawal from the world and contemplation (as with monks in a retreat), for Protestants the virtual opposite was true: fulfilment arises from helping others.[56] Weber backed up these assertions by pointing out that the accumulation of wealth, in the early stages of capitalism and in Calvinist countries in particular, was morally sanctioned only if it was combined with 'a sober, industrious career.' Idle wealth that did not contribute to the spread of well-being, capital that did not *work*, was condemned as a sin. For Weber, capitalism, whatever it has become, was originally sparked by religious fervour, and without that fervour the organisation of labour that made capitalism so different from what had gone before would not have been possible.

Weber was familiar with the religions and economic practices of non-European areas of the world, such as India, China, and the Middle East, and this imbued *The Protestant Ethic* with an authority it might otherwise not have had. He argued that in China, for example, widespread kinship units provided the predominant forms of economic cooperation, naturally limiting the influence both of the guilds and of individual entrepreneurs.[57] In India, Hinduism was associated with great wealth in history, but its tenets about the afterlife prevented the same sort of energy that built up under Protestantism, and capitalism proper never developed. Europe also had the advantage of inheriting the tradition of Roman law, which provided a more integrated juridical practice than elsewhere, easing the transfer of ideas and facilitating the understanding of contracts.[58] That *The Protestant Ethic* continues to generate controversy, that attempts have been made to transfer its basic idea to other cultures, such as Confucianism, and that links between Protestantism and economic growth are evident even today in predominantly Catholic Latin America suggest that Weber's thesis had merit.

Darwinism was not mentioned in *The Protestant Ethic*, but it was there, in the idea that Protestantism, via the Reformation, grew out of earlier, more primitive faiths and produced a more advanced economic system (more advanced because it was less sinful and benefited more people). Others have discovered in his theory a 'primitive Arianism,' and Weber himself referred to the Darwinian struggle in his inaugural address at the University of Freiburg in 1895.[59] His work was later used by sociobiologists as an example of how their theories applied to economics.[60]

Nietzsche paid tribute to the men of prey who – by their actions – helped create the world. Perhaps no one was more predatory, was having more effect on the world in 1900, than the imperialists, who in their scramble for Africa and elsewhere spread Western technology and Western ideas faster and farther than ever before. Of all the people who shared in this scramble, **Joseph Conrad** became known for turning his back on the 'active life,' for withdrawing from the dark continents of 'overflowing riches' where it was relatively easy (as well as safe) to exercise the 'will to power.' After years as a sailor in different merchant navies, Conrad removed himself to the sedentary life of writing fiction. In his imagination, however, he returned to those foreign lands – Africa, the Far East, the South Seas – to establish the first major literary theme of the century.

Conrad's best-known books, *Lord Jim* (1900), *Heart of Darkness* (published in book form in 1902), *Nostromo* (1904), and *The Secret Agent* (1907), draw on ideas from Darwin, Nietzsche, Nordau, and even Lombroso to explore the great fault line between scientific, liberal, and technical optimism in the twentieth century and pessimism about human nature. He is reported to have said to H. G. Wells on one occasion, 'The difference between us, Wells, is fundamental. You don't care for humanity but think they are to be improved. I love humanity but know they are not!'[61] It was a Conradian joke, it seems, to dedicate *The Secret Agent* to Wells.

Christened Józef Teodor Konrad Korzeniowski, Conrad was born in 1857 in a part of Poland taken by the Russians in the 1793 partition of that often-dismembered country (his birthplace is now in Ukraine). His father, Apollo, was an aristocrat without lands, for the family estates had been sequestered in 1839 following an anti-Russian rebellion. In 1862 both parents were deported, along with Józef, to Vologda in northern Russia, where his mother died of tuberculosis. Józef was orphaned in 1869 when his father, permitted the previous year to return to Kraków, died of the same disease. From this moment on Conrad depended very much on the generosity of his maternal uncle Tadeusz, who provided an annual allowance and, on his death in 1894, left about £1,600 to his nephew (well over £100,000 now). This event coincided with the acceptance of Conrad's first book, *Almayer's Folly* (begun in 1889), and the adoption of the pen name Joseph Conrad. He was from then on a man of letters, turning his experiences and the tales he heard at sea into fiction.[62]

These adventures began when he was still only sixteen, on board the *Mont Blanc*, bound for Martinique out of Marseilles. No doubt his subsequent sailing to the Caribbean provided much of the visual imagery for his later writing, especially *Nostromo*. It seems likely that he was also involved in a disastrous scheme of gunrunning from Marseilles to Spain. Deeply in debt both from this enterprise and from gambling at Monte Carlo, he attempted suicide, shooting himself in the chest. Uncle Tadeusz bailed him out, discharging his debts and inventing for him the fiction that he was shot in a duel, which Conrad found useful later for his wife and his friends.[63]

Conrad's sixteen-year career in the British merchant navy, starting as a deckhand, was scarcely smooth, but it provided the store upon which, as a writer, he would draw. Typically Conrad's best work, such as *Heart of Darkness*, is the result of long gestation periods during which he seems to have repeatedly brooded on the meaning or symbolic shape of his experience seen against the background of the developments in contemporary science. Most of these he understood as ominous, rather than liberating, for humanity. But Conrad was not anti-scientific. On the contrary, he engaged with the rapidly changing shape of scientific thought, as Redmond O'Hanlon has shown in his study *Joseph Conrad and Charles Darwin: The Influence of Scientific Thought on Conrad's Fiction* (1984).[64] Conrad was brought up on the classical physics of the Victorian age, which rested on the cornerstone belief in the permanence of matter, albeit with the assumptions that the sun was cooling and that life on earth was inevitably doomed. In a letter to his publisher dated 29 September 1898, Conrad describes the effect of a demonstration of X rays. He was in Glasgow and staying with Dr John McIntyre, a radiologist: 'In the evening dinner, phonograph, X rays, talk about *the* secret of the universe, and the non-existence of, so called, matter. The secret of the universe is in the existence of horizontal waves whose varied vibrations are set at the bottom of all states of consciousness. . . . Neil Munro stood in front of a Röntgen machine and on the screen behind we contemplated his backbone and ribs. . . . It was so – said the doctor – and there is no space, time, matter, mind as vulgarly understood . . . only the eternal force that causes the waves – it's not much.'[65]

Conrad was not quite as up-to-date as he imagined, for J. J. Thomson's demonstration the previous year showed the 'waves' to be particles. But the point is not so much that Conrad was *au fait* with science, but rather that the certainties about the nature of matter that he had absorbed were now deeply undermined. This sense he translates into the structures of many of his characters whose seemingly solid personalities, when placed in the crucible of nature (often in sea voyages), are revealed as utterly unstable or rotten.

After Conrad's uncle fell ill, Józef stopped off in Brussels on the way to Poland, to be interviewed for a post with the Société Anonyme Belge pour le Commerce du Haut-Congo – a fateful interview that led to his experiences between June and December 1890 in the Belgian Congo and, ten years on, to *Heart of Darkness*. In that decade, the Congo lurked in his mind, awaiting a trigger to be formulated in prose. That was provided by the shocking revelations of the 'Benin Massacres' in 1897, as well as the accounts of Sir Henry Morton Stanley's expeditions in Africa.[66] *Benin: The City of Blood* was published in London and New York in 1897, revealing to the western civilised world a horror story of native African blood rites. After the Berlin Conference of 1884, Britain proclaimed a protectorate over the Niger River region. Following the slaughter of a British mission to Benin (a state west of Nigeria), which arrived during King Duboar's celebrations of his ancestors with ritual sacrifices, a punitive expedition was dispatched to capture this city, long a centre of slavery. The account of Commander R. H. Bacon, intelligence officer of the expedition, parallels in some of its details the events in *Heart of Darkness*. When Commander Bacon reached Benin, he saw what, despite his vivid language, he says lay beyond description: 'It is useless to continue describing the horrors of the place, everywhere death, barbarity and blood, and smells that it hardly seems right for human beings to smell and yet live.'[67] Conrad avoids definition of what constituted 'The horror! The horror!' – the famous last words in the book, spoken by Kurtz, the man **Marlow**, the hero, has come to save – opting instead for hints such as round balls on posts that Marlow thinks he sees through his field glasses when approaching Kurtz's compound. Bacon, for his part, describes crucifixion trees surrounded by piles of skulls and bones, blood smeared everywhere, over bronze idols and ivory.

Conrad's purpose, however, is not to elicit the typical response of the civilised world to reports of barbarism. In his report Commander Bacon had exemplified this attitude: 'they [the natives] cannot fail to see that peace and the good rule of the white man mean happiness, contentment and security.' Similar sentiments are expressed in the report that Kurtz composes for the International Society for the Suppression of Savage Customs. Marlow describes this 'beautiful piece of writing,' 'vibrating with eloquence.' And yet, scrawled 'at the end of that moving appeal to every altruistic sentiment is blazed at you, luminous and terrifying, like a flash of lightning in a serene sky: "Exterminate all the brutes!"'[68]

This savagery at the heart of civilised humans is also revealed in the behaviour of the white traders – 'pilgrims,' Marlow calls them. White travellers' tales, like those of Henry Morton Stanley in 'darkest Africa,' written from an unquestioned sense of the superiority of the European over the native, were available to Conrad's dark vision. *Heart of Darkness* thrives upon the ironic reversals of civilisation and barbarity, of light and darkness. Here is a characteristic Stanley episode, recorded in his diary. Needing food, he told a group of natives that 'I must have it or we would die. They must sell it for beads, red, blue or green, copper or brass wire or shells, or ... I drew significant signs across the throat. It was enough, they understood at once.'[69] In *Heart of Darkness*, by contrast, Marlow is impressed by the extraordinary restraint of the starving cannibals accompanying the expedition, who have been paid in bits of brass wire but have no food, their rotting hippo flesh – too nauseating a smell for European endurance – having been thrown overboard. He wonders why 'they didn't go for us – they were thirty to five – and have a good tuck-in for once.'[70] Kurtz is a symbolic figure, of course ('All Europe contributed to the making of Kurtz'), and the thrust of Conrad's fierce satire emerges clearly through Marlow's narrative.[71] The imperial civilising mission amounts to a savage predation: 'the vilest scramble for loot that ever disfigured the history of the human conscience,' as Conrad elsewhere described it. At this end of the century such a conclusion about the novel seems obvious, but it was otherwise in the reviews that greeted its first appearance in 1902. The *Manchester Guardian* wrote that Conrad was not attacking colonisation, expansion, or imperialism, but rather showing how cheap ideals shrivel up.[72] Part of the fascination surely lies in Conradian psychology. The journey within of so many of his characters seems explicitly Freudian, and indeed many Freudian interpretations of his works have been proposed. Yet Conrad strongly resisted Freud. When he was in Corsica, and on the verge of a breakdown, Conrad was given a copy of *The Interpretation of Dreams*. He spoke of Freud 'with scornful irony,' took the book to his room, and returned it on the eve of his departure, unopened.[73]

At the time *Heart of Darkness* appeared, there was – and there continues to be – a distaste for Conrad on the part of some readers. It is that very reaction which underlines his significance. This is perhaps best explained by Richard Curle, author of the first full-length study of Conrad, published in 1914.[74] Curle could see that for many people there is a tenacious need to believe that the world, horrible as it might be, can be put right by human effort and the appropriate brand of liberal philosophy. Unlike the novels of his contemporaries H. G. Wells and John Galsworthy, Conrad derides this point of view as an illusion at best, and the pathway to desperate destruction at its worst. Recently the morality of Conrad's work, rather than its aesthetics, has been questioned. In 1977 the Nigerian novelist Chinua Achebe described Conrad as 'a bloody racist' and *Heart of Darkness* as a novel that 'celebrates' the dehumanisation of some of the human race. In 1993 the cultural critic Edward Said thought that Achebe's criticism did not go far enough.[75] But evidence shows that Conrad was sickened by his experience in Africa, both physically and psychologically. In the Congo he met Roger Casement (executed in 1916 for his activities in Ireland), who as a British consular officer had written a report exposing the atrocities he and Conrad saw.[76] In 1904 he visited Conrad to solicit his support. Whatever Conrad's relationship to Marlow, he was deeply alienated from the imperialist, racist exploiters of Africa and Africans at that time. *Heart of Darkness* played a part in ending Leopold's tyranny.[77] One is left after reading the novel with the sheer terror of the enslavement and the slaughter, and a sense of the horrible futility and guilt that Marlow's narrative conveys. Kurtz's final words, 'The horror! The horror!' serve as a chilling endpoint for where social Darwinism all too easily can lead.

4

LES DEMOISELLES DE MODERNISME

In 1905 Dresden was one of the most beautiful cities on earth, a delicate Baroque jewel straddling the Elbe. It was a fitting location for the première of a new opera composed by **Richard Strauss**, called *Salomé*. Nonetheless, after rehearsals started, rumours began to circulate in the city that all was not well backstage. Strauss's new work was said to be 'too hard' for the singers. As the opening night, 9 December, drew close, the fuss grew in intensity, and some of the singers wanted to hand back their scores. Throughout the rehearsals for *Salomé*, Strauss maintained his equilibrium, despite the problems. At one stage an oboist complained, 'Herr Doktor, maybe this passage works on the piano, but it doesn't on the oboes.' 'Take heart, man,' Strauss replied briskly. 'It doesn't work on the piano, either.' News about the divisions inside the opera house were taken so much to heart that Dresdeners began to cut the conductor, Ernst von Schuch, in the street. An expensive and embarrassing failure was predicted, and the proud burghers of Dresden could not stomach that. Schuch remained convinced of the importance of Strauss's new work, and despite the disturbances and rumours, the production went ahead. The first performance of *Salomé* was to open, in the words of one critic, 'a new chapter in the history of modernism.'[1]

The word **modernism** has three meanings, and we need to distinguish between them. Its first meaning refers to the break in history that occurred between the Renaissance and the Reformation, when the recognisably modern world began, when science began to flourish as an alternative system of knowledge, in contrast with religion and metaphysics. The second, and most common meaning of modernism refers to a movement – in the arts mainly – that began with **Charles Baudelaire** in France but soon widened. This itself had three elements. The first and most basic element was the belief that the modern world was just as good and fulfilling as any age that had gone before. This was most notably a reaction in France, in Paris in particular, against the historicism that had prevailed throughout most of the nineteenth century, especially in painting. It was helped by the rebuilding of Paris by Baron Georges-Eugène Haussman in the 1850s. A second aspect of modernism in this sense was that it was an urban art, cities being the 'storm centres' of civilisation. This was most clear in one of its earliest forms, impressionism, where the aim is to catch the fleeting moment, that ephemeral instance so prevalent in the urban experience. Last, in its urge to advocate the new over and above everything else, modernism implied the existence of an '**avant-garde**', an artistic and intellectual elite, set apart from the masses by their brains and creativity, destined more often than not to be pitched against those masses even as they lead them. This form of modernism makes a distinction between the leisurely, premodern face-to-face agricultural society and the anonymous, fast-moving, atomistic society of large cities, carrying with it the risks of alienation, squalor, degeneration (as Freud, for one, had pointed out).[2]

The third meaning of modernism is used in the context of organised religion, and Catholicism in particular. Throughout the nineteenth century, various aspects of Catholic dogma came under threat. Young clerics were anxious for the church to respond to the new findings of science, especially Darwin's theory of evolution and the discoveries of German archaeologists in the Holy Land, many of which appeared to contradict the Bible. The present chapter concerns all three aspects of modernism that came together in the early years of the century.

Salomé was closely based on Oscar Wilde's play of the same name. Strauss was well aware of the play's scandalous nature. When Wilde had originally tried to produce *Salomé* in London, it had been banned by the Lord Chamberlain. (In retaliation, Wilde had threatened to take out French citizenship.)[3] Wilde recast the ancient account of Herod, Salomé, and Saint John the Baptist with a 'modernist' gloss, portraying the 'heroine' as a 'Virgin consumed by evil chastity.'[4] When he wrote the play, Wilde had not read Freud, but he had read Richard von Krafft-Ebing's *Psychopathia Sexualis*, and his plot clearly suggested in Salomé's demand for the head of Saint John echoes of sexual perversion. In an age when many people still regarded themselves as religious, this was almost guaranteed to offend. Strauss's music, on top of Wilde's plot, added fuel to the fire. The orchestration was difficult, disturbing, and to many ears discordant. To highlight the psychological contrast between Herod and Jokanaan, Strauss employed the unusual device of writing in two keys simultaneously.[5] The continuous dissonance of the score reflected the tensions in the plot, reaching its culmination with Salomé's moan as she awaits execution. This, rendered as a B-flat on a solo double bass, nails the painful drama of Salomé's plight: she is butchered by guards crushing the life out of her with their shields.

After the first night, opinions varied. Cosima Wagner was convinced the new opera was 'Madness! ... wedded to indecency.' The Kaiser would only allow *Salomé* to be performed in Berlin after the manager of the opera house shrewdly modified the ending, so that a Star of Bethlehem rose at the end of the performance.[6] This simple trick changed everything, and *Salomé* was performed fifty times in that one season. Ten of Germany's sixty opera houses – all fiercely competitive – chose to follow Berlin's lead and stage the production so that within months, Strauss could afford to build a villa at Garmisch in the art nouveau style.[7] Despite its success in Germany, the opera became notorious internationally. In London Thomas Beecham had to call in every favour to obtain permission to perform the opera at all.[8] In New York and Chicago it was banned outright. (In New York one cartoonist suggested it might help if advertisements were printed on each of the seven veils.)[9] Vienna also banned the opera, but Graz, for some reason, did not. There the opera opened in May 1906 to an audience that included Giacomo Puccini, Gustav Mahler, and a band of young music lovers who had come down from Vienna, including an out-of-work would-be artist called Adolf Hitler.

Despite the offence *Salomé* caused in some quarters, its eventual success contributed to Strauss's appointment as senior musical director of the Hofoper in Berlin. The composer began work there with a one-year leave of absence to complete his next opera, *Elektra*. This work was his first major collaboration with Hugo von Hofmannsthal, whose play of the same name, realised by that magician of the German theatre, Max Reinhardt, Strauss had seen in Berlin (at the same theatre where he saw Wilde's *Salomé*).[10] Strauss was not keen to begin with, because he thought *Elektra*'s theme was too similar to that of *Salomé*. But Hofmannsthal's 'demonic, ecstatic' image of sixth-century Greece caught his fancy; it was so very different from the noble, elegant, *calm* image traditionally revealed in the writings of **Johann Joachim Winckelmann** and Goethe. Strauss therefore changed his mind, and *Elektra* turned out to be even more intense, violent, and concentrated than *Salomé*. 'These two operas stand alone in my life's work,' said Strauss later; 'in them I went to the utmost limits of harmony, psychological polyphony (Clytemnestra's dream) and the capacity of today's ears to take in what they hear.'[11]

The setting of the opera is the Lion Gate at Mycenae – after Krafft-Ebing, Heinrich Schliemann. *Elektra* uses a larger orchestra even than *Salomé*, one-hundred and eleven players, and the combination of score and mass of musicians produces a much more painful, dissonant experience. There are swaths of 'huge granite chords,' sounds of 'blood and iron,' as Strauss's biographer Michael Kennedy has put it.[12] For all its dissonance, *Salomé* is voluptuous, but *Elektra* is austere, edgy, grating. The original Clytemnestra was Ernestine Schumann-Heink, who described the early performances as 'frightful. . . . We were a set of mad women. . . . There is nothing beyond *Elektra*. . . . We have come to a full-stop. I believe Strauss himself sees it.' She said she wouldn't sing the role again for $3,000 a performance.[13]

Two aspects of the opera compete for attention. The first is Clytemnestra's tormented aria. A 'stumbling, nightmare-ridden, ghastly wreck of a human being,' she has nevertheless decorated herself with ornaments and, to begin with, the music follows the rattles and cranks of these.[14] At the same time she sings of a dreadful dream – a biological horror – that her bone marrow is dissolving away, that some unknown creature is crawling all over her skin as she tries to sleep. Slowly, the music turns harsher, grows more discordant, atonal. The terror mounts, the dread is inescapable. Alongside this there is the confrontation between the three female characters, Electra and Clytemnestra on the one hand, and Electra and Chrysothemis on the other. Both encounters carry strong lesbian overtones that, added to the dissonance of the music, ensured that *Elektra* was as scandalous as *Salomé*. When it premiered on 25 January 1909, also in Dresden, one critic angrily dismissed it as 'polluted art.'[15]

Strauss and Hofmannsthal were trying to do two things with *Elektra*. At the most obvious level they were doing in musical theatre what the expressionist painters of **Die Brücke** and **Der Blaue Reiter** (Ernst Ludwig Kirchner, Erich Heckel, Wassily Kandinsky, Franz Marc) were doing in their art – using unexpected and 'unnatural' colours, disturbing distortion, and jarring juxta-positions to change people's perceptions of the world. And in this, perceptions of the ancient world had resonance. In Germany at the time, as well as in Britain and the United States, most scholars had inherited an idealised picture of antiquity, from Winckelmann and Goethe, who had understood classical Greece and Rome as restrained, simple, austere, coldly beautiful. But Nietzsche changed all that. He stressed the instinctive, savage, irrational, and darker aspects of pre-Homeric ancient Greece (fairly obvious, for example, if one reads the *Iliad* and the *Odyssey* without preconceptions). But Strauss's *Elektra* wasn't only about the past. It was about man's (and therefore woman's) true nature, and in this psychoanalysis played an even bigger role. Hofmannsthal met Arthur Schnitzler nearly every day at the Café Griensteidl, and Schnitzler was regarded by Freud, after all, as his 'double.' There can be little doubt therefore that Hofmannsthal had read *Studies in Hysteria* and *The Interpretation of Dreams*.[16] Indeed, Electra herself shows a number of the symptoms portrayed by Anna O., the famous patient treated by Josef Breuer. These include her father fixation, her recurring hallucinations, and her disturbed sexuality. But *Elektra* is theatre, not a clinical report.[17] The characters face moral dilemmas, not just psy-chological ones. Nevertheless, the very presence of Freud's ideas onstage, undermining the traditional basis of ancient myths, as well as recognisable music and dance (both *Salomé* and *Elektra* have dance scenes), placed Strauss and Hofmannsthal firmly in the modernist camp. *Elektra* assaulted the accepted notions of what was beautiful and what wasn't. Its exploration of the uncon-scious world beneath the surface may not have made people content, but it certainly made them think.

Elektra made Strauss think too. Ernestine Schumann-Heink had been right. He had followed the path of dissonance and the instincts and the irrational far enough. Again, as Michael Kennedy has said, the famous 'blood chord' in *Elektra*, 'E-major and D-major mingled in pain,' where the voices go their own way, as far from the orchestra as dreams are from reality, was as jarring as anything then happening in painting. Strauss was at his best 'when he set mania to music,' but nevertheless he abandoned the discordant line he had followed from *Salomé* to *Elektra*, leaving the way free for a new generation of composers, the most innovative of whom was Arnold Schoenberg.★[18]

Strauss was, however, ambivalent about **Schoenberg**. He thought he would be better off 'shovelling snow' than composing, yet recommended him for a Liszt scholarship (the revenue of the Liszt Foundation was used annually to help

★ Strauss was not the only twentieth-century composer to pull back from the leading edge of the avant-garde: Stravinsky, Hindemith and Shostakovitch all rejected certain stylistic innovations of their early careers. But Strauss was the first.[19]

composers or pianists).[20] Born in September 1874 into a poor family, Arnold Schoenberg always had a serious disposition and was largely self-taught.[21] Like Max Weber, he was not given to smiling. A small, wiry man, he went bald early on, and this helped to give him a fierce appearance – the face of a fanatic, according to his near-namesake, the critic Harold Schonberg.[22] Stravinsky once pinned down his colleague's character in this way: 'His eyes were protuberant and explosive, and the whole force of the man was in them.'[23] Schoenberg was strikingly inventive, and his inventiveness was not confined to music. He carved his own chessmen, bound his own books, painted (Kandinsky was a fan), and invented a typewriter for music.[24]

To begin with, Schoenberg worked in a bank, but he never thought of anything other than music. 'Once, in the army, I was asked if I was the composer Arnold Schoenberg. "Somebody has to be," I said, "and nobody else wanted to be, so I took it on myself." '[25] Although Schoenberg preferred Vienna, where he frequented the cafés Landtmann and Griensteidl, and where Karl Kraus, Theodor Herzl and Gustav Klimt were great friends, he realised that Berlin was the place to advance his career. There he studied under Alexander von Zemlinsky, whose sister, Mathilde, he married in 1901.[26]

Schoenberg's autodidacticism, and sheer inventiveness, served him well. While other composers, Strauss, Mahler, and Claude Debussy among them, made the pilgrimage to Bayreuth to learn from Wagner's chromatic harmony, Schoenberg chose a different course, realising that evolution in art proceeds as much by complete switchbacks in direction, by quantum leaps, as by gradual growth.[27] He knew that the expressionist painters were trying to make visible the distorted and raw forms unleashed by the modern world and analysed and ordered by Freud. He aimed to do something similar in music. The term he himself liked was '**the emancipation of dissonance.**'[28]

Schoenberg once described music as 'a prophetic message revealing a higher form of life toward which mankind evolves.'[29] Unfortunately, he found his own evolution slow and very painful. Even though his early music owed a debt to Wagner, *Tristan* especially, it had a troubled reception in Vienna. The first demonstrations occurred in 1900 at a recital. 'Since then,' he wrote later, 'the scandal has never ceased.'[30] It was only after the first outbursts that he began to explore dissonance. As with other ideas in the early years of the century – relativity, for example, and abstraction – several composers were groping toward dissonance and atonality at more or less the same time. One was Strauss, as we have seen. But Jean Sibelius, Mahler, and Alexandr Scriabin, all older than Schoenberg, also seemed about to embrace the same course when they died. Schoenberg's relative youth and his determined, uncompromising nature meant that it was he who led the way toward atonality.[31]

One morning in December 1907 Schoenberg, Anton von Webern, Gustav Klimt, and a couple of hundred other notables gathered at Vienna's Westbahnhof to say good-bye to Gustav Mahler, the composer and conductor who was bound for New York. He had grown tired of the 'fashionable anti-Semitism' in Vienna and had fallen out with the management of the Opéra.[32] As the train pulled out of the station, Schoenberg and the rest of the Café Griensteidl set, now bereft of the star who had shaped Viennese music for a decade, waved in silence. Klimt spoke for them all when he whispered, 'Vorbei' (It's over). But it could have been Schoenberg speaking – Mahler was the only figure of note in the German music world who understood what he was trying to achieve.[33] A second crisis which faced Schoenberg was much more powerful. In the summer of 1908, the very moment of his first atonal compositions, his wife Mathilde abandoned him for a friend.[34] Rejected by his wife, isolated from Mahler, Schoenberg was left with nothing but his music. No wonder such dark themes are a prominent feature of his early atonal compositions.

The year 1908 was momentous for music, and for Schoenberg. In that year he composed his Second String Quartet and **Das Buch der hängenden Gärten**. In both compositions he took the historic step of producing a style that, echoing the new physics, was 'bereft of foundations.'[35] Both compositions were inspired by the tense poems of Stefan George, another member of the Café Griensteidl set.[36] George's poems were a cross between experimentalist paintings and Strauss operas. They were full of references to darkness, hidden worlds, sacred fires, and voices.

The precise point at which atonality arrived, according to Schoenberg, was during the writing of the third and fourth movements of the string quartet. He was using George's poem 'Entrückung' (Ecstatic Transport) when he suddenly left out all six sharps of the key signature. As he rapidly completed the part for the cello, he abandoned completely any sense of key, to produce a 'real pandemonium of sounds, rhythms and forms.'[37] As luck would have it, the stanza ended with the line, 'Ich fühle Luft von anderem Planeten,' 'I feel the air of other planets.' It could not have been more appropriate.[38] The Second String Quartet was finished toward the end of July. Between then and its premiere, on 21 December, one more personal crisis shook the Schoenberg household. In November the painter his wife had left him for hanged himself, after he had failed to stab himself to death. Schoenberg took back Mathilde, and when he handed the score to the orchestra for the rehearsal, it bore the dedication, 'To my wife.'[39]

The premiere of the Second String Quartet turned into one of the great scandals of music history. After the lights went down, the first few bars were heard in respectful silence. But only the first few. Most people who lived in apartments in Vienna then carried whistles attached to their door keys. If they arrived home late at night, and the main gates of the building were locked, they would use the whistles to attract the attention of the concierge. On the night of the première, the audience got out its whistles. A wailing chorus arose in the auditorium to drown out what was happening onstage. One critic leaped to his feet and shouted, 'Stop it! Enough!' though no one knew if he meant the audience or the performers. When Schoenberg's sympathisers joined in, shouting their support, it only added to the din. Next day one newspaper labelled the performance a 'Convocation of Cats,' and the *New Vienna Daily*, showing a sense of invention that even Schoenberg would have approved, printed their review in the 'crime' section of the paper.[40] 'Mahler trusted him without being able to understand him.'[41]

Years later Schoenberg conceded that this was one of the worst moments of his life, but he wasn't deterred. Instead, in 1909, continuing his emancipation of dissonance, he composed **Erwartung**, a thirty-minute opera, the story line for which is so minimal as to be almost absent: a woman goes searching in the forest for her lover; she discovers him only to find that he is dead not far from the house of the rival who has stolen him. The music does not so much tell a story as reflect the woman's moods – joy, anger, jealousy.[42] In painterly terms, *Erwartung* is both expressionistic and abstract, reflecting the fact that Schoenberg's wife had recently abandoned him.[43] In addition to the minimal narrative, it never repeats any theme or melody. Since most forms of music in the 'classical' tradition usually employ variations on themes, and since repetition, lots of it, is the single most obvious characteristic of popular music, Schoenberg's Second String Quartet and *Erwartung* stand out as the great break, after which 'serious' music began to lose the faithful following it had once had. It was to be fifteen years before *Erwartung* was performed.

Although he might be too impenetrable for many people's taste, Schoenberg was not obtuse. He knew that some people objected to his atonality for its own sake, but that wasn't the only problem. As with Freud (and Picasso, as we shall see), there were just as many traditionalists who hated *what* he was saying as much as how he was saying it. His response to this was a piece that, to him at least, was 'light, ironic, satirical.'[44] **Pierrot lunaire**, appearing in 1912, features a familiar icon of the theatre – a dumb puppet who also happens to be a feeling being, a sad and cynical clown allowed by tradition to raise awkward truths so long as they are wrapped in riddles. It had been commissioned by the Viennese actress Albertine Zehme, who liked the Pierrot role.[45] Out of this unexpected format, Schoenberg managed to produce what many people consider his seminal work, what has been called the musical equivalent of *Les Demoiselles d'Avignon* or $E=mc^2$.[46] *Pierrot*'s main focus is a theme we are already familiar with, the decadence and degeneration of modern man. Schoenberg introduced in the piece several innovations in form, notably *Sprechgesang*, literally song-speech in which the voice rises and falls but cannot be said to be either singing or speaking. The main part, composed for an actress rather than a straight singer, calls for her to be both a 'serious' performer and a cabaret act. Despite this suggestion of a more popular, accessible format, listeners have found that the music breaks down 'into atoms and molecules, behaving in a jerky, uncoordinated way not unlike the molecules that bombard pollen in Brownian movement.'[47]

Schoenberg claimed a lot for *Pierrot*. He had once described Debussy as an impressionist composer, meaning that his harmonies merely added to the colour of moods. But Schoenberg saw himself as an expressionist, a Postimpressionist like Paul Gauguin or Paul Cézanne or Vincent van Gogh, uncovering unconscious meaning in much the same way that the expressionist painters thought they went beyond the merely decorative impressionists. He certainly believed, as Bertrand Russell and Alfred North Whitehead did, that music – like mathematics (see chapter 6) – had logic.[48]

The first night took place in mid-October in Berlin, in the Choralionsaal on Berlin's Bellevuestrasse, which was destroyed by Allied bombs in 1945. As the house lights went down, dark screens could be made out onstage with the actress Albertine Zehme dressed as Columbine. The musicians were farther back, conducted by the composer. The structure of *Pierrot* is tight. It is comprised of three parts, each containing seven miniature poems; each poem lasts about a minute and a half, and there are twenty-one poems in all, stretching to just on half an hour. Despite the formality, the music was utterly free, as was the range of moods, leading from sheer humour, as Pierrot tries to clean a spot off his clothes, to the darkness when a giant moth kills the rays of the sun. Following the premières of the Second String Quartet and *Erwartung*, the critics gathered, themselves resembling nothing so much as a swarm of giant moths, ready to kill off this shining sun. But the performance was heard in silence, and when it was over, Schoenberg was given an ovation. Since it was so short, many in the audience shouted for the piece to be repeated, and they liked it even better the second time. So too did some of the critics. One of them went so far as to describe the evening 'not as the end of music; but as the beginning of a new stage in listening.'

It was true enough. One of the many innovations of modernism was the new demands it placed on the audience. Music, painting, literature, even architecture, would never again be quite so 'easy' as they had been. Schoenberg, like Freud, Klimt, Oskar Kokoschka, Otto Weininger, Hofmannsthal, and Schnitzler, believed in the instincts, expressionism, subjectivism.[49] For those who were willing to join the ride, it was exhilarating. For those who weren't, there was really nowhere to turn and go forward. And like it or not, Schoenberg had found a way forward after Wagner. The French composer Claude Debussy once remarked that Wagner's music was 'a beautiful sunset that was mistaken for a dawn.' No one realised that more than Schoenberg.

If Salomé and Elektra and *Pierrot*'s Columbine are the founding females of modernism, they were soon followed by five equally sensuous, shadowy, disturbing sisters in a canvas produced by Picasso in 1907. No less than Strauss's women, Pablo Picasso's *Les Demoiselles d'Avignon* was an attack on all previous ideas of art, self-consciously shocking, crude but compelling.

In the autumn of 1907 Picasso was twenty-six. Between his arrival in Paris in 1900 and his modest success with *Last Moments*, he had been back and forth several times between Malaga, or Barcelona, and Paris, but he was at last beginning to find fame and controversy (much the same thing in the world where he lived). Between 1886 and the outbreak of World War I there were more new movements in painting than at any time since the Renaissance, and Paris was the centre of this activity. Georges Seurat had followed impressionism with pointillism in 1886; three years later, Pierre Bonnard, Edouard Vuillard, and Aristide Maillol formed Les Nabis (from the Hebrew word for prophet), attracted by the theories of Gauguin, to paint in flat, pure colours. Later in the 1890s, as we have seen in the case of Klimt, painters in the mainly German-speaking cities – Vienna, Berlin, Munich – opted out of the academies to initiate the various 'secessionist' movements. Mostly they began as impressionists, but the experimentation they encouraged brought about *expressionism*, the search for emotional impact by means of exaggerations and distortions of line and colour. **Fauvism** was the most fruitful movement, in particular in the paintings of **Henri Matisse**, who would be Picasso's chief rival while they were both alive. In 1905, at the Salon d'Automne in Paris, pictures by Matisse, André Derain, Maurice de Vlaminck, Georges Rouault, Albert Marquet, Henri Manguin, and Charles Camoin were grouped together in one room that also featured, in the centre, a statue by Donatello, the fifteenth-century Florentine sculptor. When the critic **Louis Vauxcelles** saw this arrangement, the calm of the statue contemplating the frenzied, flat colours and distortions on the walls, he sighed, 'Ah, Donatello chez les Fauvres.' *Fauve* means 'wild beast' – and the name stuck. It did no harm. For a time, Matisse was regarded as the beast-in-chief of the Paris avant-garde.

Matisse's most notorious works during that early period were other *demoiselles de modernisme* – *Woman with a Hat* and *The Green Stripe*, a portrait of his wife. Both used colour to do violence to familiar images, and both created scandals. At this stage Matisse was leading, and Picasso following. The two painters had met in 1905, in the apartment of Gertrude Stein, the expatriate American writer. She was a discerning and passionate collector of modern art, as was her equally wealthy brother, Leo, and invitations to their Sunday-evening soirées in the rue de Fleurus were much sought after.[50] Matisse and Picasso were regulars at the Stein evenings, each with his band of supporters. Even then, though, Picasso understood how different they were. He once described Matisse and himself as 'north pole and south pole.'[51] For his part, Matisse's aim, he said, was for 'an art of balance, of purity and serenity, free of disturbing or disquieting subjects . . . an appeasing influence.'[52]

Not Picasso. Until then, he had been feeling his way. He had a recognisable style, but the images he had painted – of poor acrobats and circus people – were hardly avant-garde. They could even be described as sentimental. His approach to art had not yet matured; all he knew, looking around him, was that in his art he needed to do as the other moderns were doing, as Strauss and Schoenberg and Matisse were doing: to shock. He saw a way ahead when he observed that many of his friends, other artists, were visiting the 'primitive art' departments at the Louvre and in the Trocadéro's Museum of Ethnography. This was no accident. Darwin's theories were well known by now, as were the polemics of the social Darwinists. Another influence was James Frazer, the anthropologist who, in *The Golden Bough*, had collected together in one book many of the myths and customs of different races. And on top of it all, there was the scramble for Africa and other empires. All of this produced a fashion for the achievements and cultures of the remoter regions of 'darkness' in the world – in particular the South Pacific and Africa. In Paris, friends of Picasso started buying masks and African and Pacific statuettes from bric-a-brac dealers. None were more taken by this art than Matisse and Derain. In fact, as Matisse himself said, 'On the Rue de Rennes, I often passed the shop of Père Sauvage. There were Negro statuettes in his window. I was struck by their character, their purity of line. It was as fine as Egyptian art. So I bought one and showed it to Gertrude Stein, whom I was visiting that day. And then Picasso arrived. He took to it immediately.'[53]

He certainly did, for the statuette seems to have been the first inspiration toward *Les Demoiselles d'Avignon*. As the critic Robert Hughes tells us, Picasso soon after commissioned an especially large canvas, which needed reinforced stretchers. Later in his life, Picasso described to André Malraux, the French writer and minister of culture, what happened next: 'All alone in that awful museum [i.e. the Trocadéro], with masks, dolls made by the redskins, dusty manikins, *Les Demoiselles d'Avignon* must have come to me that very day, but not at all because of the forms; because it was my first exorcism-painting – yes absolutely. ... The masks weren't just like any other pieces of sculpture. Not at all. They were magic things. ... The Negro pieces were *intercesseurs*, mediators; ever since then I've known the word in French. They were against everything – against unknown, threatening spirits. I always looked at fetishes. I understood; I too am against everything. I too believe that everything is unknown, that everything is an enemy! ... all the fetishes were used for the same thing. They were weapons. To help people avoid coming under the influence of spirits again, to help them become independent. They're tools. If we give spirits a form, we become independent. Spirits, the unconscious (people still weren't talking about that very much), emotion – they're all the same thing. I understood why I was a painter.'[54]

Jumbled up here are Darwin, Freud, Frazer, and Henri Bergson, whom we shall meet later in this chapter. There is a touch of Nietzsche too, in Picasso's nihilistic and revealing phrase, 'everything is an enemy! ... They were weapons.'[55] *Demoiselles* was an attack on all previous ideas of art. Like *Elektra* and *Erwartung*, it was modernistic in that it was intended to be as destructive as it was creative, shocking, deliberately ugly, and undeniably crude. Picasso's brilliance lay in also making the painting irresistible. The five women are naked, heavily made up, completely brazen about what they are: prostitutes in a brothel. They stare back at the viewer, unflinching, confrontational rather than seductive. Their faces are primitive masks that point up the similarities and differences between so-called primitive and civilised peoples. While others were looking for the serene beauty in non-Western art, Picasso questioned Western assumptions about beauty itself, its links to the unconscious and the instincts. Certainly, Picasso's images left no one indifferent. The painting made Georges Braque feel 'as if someone was drinking gasoline and spitting fire,' a comment not entirely negative, as it implies an explosion of energy.[56] Gertrude Stein's brother Leo was racked with embarrassed laughter when he first saw *Les Demoiselles*, but Braque at least realised that the picture was built on **Cézanne** but added twentieth-century ideas, rather as Schoenberg built on Wagner and Strauss.

Cézanne, who had died the previous year, achieved recognition only at the end of his life as the critics finally grasped that he was trying to simplify art and to reduce it to its fundamentals. Most of Cézanne's work was done in the nineteenth century, but his last great series, 'The Bathers,' was produced in 1904 and 1905, in the very months when, as we shall see, Einstein was preparing for publication his three great papers, on relativity, Brownian motion, and quantum theory. Modern art and much of modern science was therefore conceived at exactly the same moment. Moreover, Cézanne captured the essence of a landscape, or a bowl of fruit, by painting smudges of colour – quanta – all carefully related to each other but *none of which conformed exactly to what was there*. Like the relation of electrons and atoms to matter, orbiting largely empty space, Cézanne revealed the shimmering, uncertain quality beneath hard reality.

In the year after Cézanne's death, 1907, the year of *Les Demoiselles*, the dealer Ambroise Vollard held a huge retrospective of the painter's works, which thousands of Parisians flocked to see. Seeing this show, and seeing *Demoiselles* so soon after, Braque was transformed. Hitherto a disciple more of Matisse than Picasso, Braque was totally converted.

Six feet tall, with a large, square, handsome face, **Georges Braque** came from the Channel port of Le Havre. The son of a decorator who fancied himself as a real painter, Braque was very physical: he boxed, loved dancing, and was always welcome at Montmartre parties because he played the accordion (though Beethoven was more to his taste). 'I never decided to become a painter any more than I decided to breathe,' he said. 'I truly don't have any memory of making a choice.'[57] He first showed his paintings in 1906 at the Salon des Indépendants; in 1907 his works hung next to those of Matisse and Derain, and proved so popular that everything he sent in was sold. Despite this success, after seeing *Les Demoiselles d'Avignon*, he quickly realised that it was with Picasso that the way forward lay, and he changed course. For two years, as cubism evolved, they lived in each other's pockets, thinking and working as one. 'The things Picasso and I said to each other during those years,' Braque later said, 'will never be said again, and even if they were, no one would understand them any more. It was like being two mountaineers roped together.'[58]

Before *Les Demoiselles*, Picasso had really only explored the emotional possibilities of two colour ranges – blue and pink. But after this painting his palette became more subtle, and more muted, than at any time in his life. He was at the time working at La-Rue-des-Bois in the countryside just outside Paris, which inspired the autumnal greens in his early cubist works. Braque, meanwhile, had headed south, to L'Estaque and the *paysage Cézanne* near Aix. Despite the distance separating them, the similarity between Braque's southern paintings of the period and Picasso's from La-Rue-des-Bois is striking: not just the colour tones but the geometrical, geological simplicity – landscapes lacking in order, at some earlier stage of evolution perhaps. Or else it was the *paysage Cézanne* seen close up, the molecular basis of landscape.[59]

Though revolutionary, these new pictures were soon displayed. The German art dealer Daniel Henry Kahnweiler liked them so much he immediately organised a show of Braque's landscapes that opened in his gallery in the rue Vignon in November 1908. Among those invited was Louis Vauxcelles, the critic who had cracked the joke about Donatello and the Fauves. In his review of the show, he again had a turn of phrase for what he had seen. Braque, he said, had reduced everything to 'little cubes.' It was intended to wound, but

Kahnweiler was not a dealer for nothing, and he made the most of this early example of a sound bite. **Cubism** was born.[60]

It lasted as a movement and style until the guns of August 1914 announced the beginning of World War I. Braque went off to fight and was wounded, after which the relationship between him and Picasso was never the same again. Unlike *Les Demoiselles*, which was designed to shock, cubism was a quieter, more reflective art, with a specific goal. 'Picasso and I,' Braque said, 'were engaged in what we felt was a search for the anonymous personality. We were inclined to efface our own personalities in order to find originality.'[61] This was why cubist works early on were signed on the back, to preserve anonymity and to keep the images uncontaminated by the personality of the painter. In 1907–8 it was never easy to distinguish which painter had produced which picture, and that was how they thought it should be. Historically, cubism is central because it is the main pivot in twentieth-century art, the culmination of the process begun with impressionism but also the route to abstraction. We have seen that Cézanne's great paintings were produced in the very months in which Einstein was preparing his theories. The whole change that was overtaking art mirrored the changes in science. There was a search in both fields for fundamental units, the deeper reality that would yield new forms. Paradoxically, in painting this led to an art in which the *absence* of form turned out to be just as liberating.

Abstraction has a long history. In antiquity certain shapes and colours like stars and crescents were believed to have magical properties. In Muslim countries it was and is forbidden to show the human form, and so abstract motifs – arabesques – were highly developed in both secular and religious works of art. As abstraction had been available in this way to Western artists for thousands of years, it was curious that several people, in different countries, edged toward abstraction during the first decade of the new century. It paralleled the way various people groped toward the unconscious or began to see the limits of Newton's physics.

In Paris, both **Robert Delaunay** and **František Kupka**, a Czech cartoonist who had dropped out of the Vienna art school, made pictures without objects. Kupka was the more interesting of the two. Although he had been convinced by Darwin's scientific theory, he also had a mystical side and believed there were hidden meanings in the universe that could be painted.[62] **Mikalojus-Konstantinas Ciurlionis**, a Lithuanian painter living in Saint Petersburg, began his series of 'transcendent' pictures, again lacking recognisable objects and named after musical tempos: andante, allegro, and so on. (One of his patrons was a young composer named Igor Stravinsky.)[63] America had an early abstractionist, too, in the form of Arthur Dove, who left his safe haven as a commercial illustrator in 1907 and exiled himself to Paris. He was so overwhelmed by the works of Cézanne that he never painted a representational picture again. He was given an exhibition by Alfred Stieglitz, the photographer who established the famous '291' avant-garde gallery in New York at 291 Broadway.[64] Each of these artists, in three separate cities, broke new ground and deserve their paragraph in history. Yet it was someone else entirely who is generally regarded as the father of abstract art, mainly because it was his work that had the greatest influence on others.

Wassily Kandinsky was born in Moscow in 1866. He had intended to be a lawyer but abandoned that to attend art school in Munich. Munich wasn't nearly as exciting culturally as Paris or Vienna, but it wasn't a backwater. Thomas Mann and Stefan George lived there. There was a famous cabaret, the Eleven Executioners, for whom Frank Wedekind wrote and sang.[65] The city's museums were second only to Berlin in Germany, and since 1892 there had been the Munich artists' *Sezession*. Expressionism had taken the country by storm, with Franz Marc, Aleksey Jawlensky, and Kandinsky forming 'the Munich Phalanx.' Kandinsky was not as precocious as Picasso, who was twenty-six when he painted *Les Demoiselles d'Avignon*. In fact, Kandinsky did not paint his first picture until he was thirty and was all of forty-five when, on New Year's Eve, 1910–11, he went to a party given by two artists. Kandinsky's marriage was collapsing at that time, and he went alone to the party, where he met Franz Marc. They struck up an accord and went on to a concert by a composer new to them but who also painted expressionist pictures; his name was Arnold Schoenberg. All of these influences proved crucial for Kandinsky, as did the theosophical doctrines of Madame Blavatsky and Rudolf Steiner. Blavatsky predicted a new age, more spiritual, less material, and Kandinsky (like many artists, who banded into quasi-religious groups) was impressed enough to feel that a new art was needed for this new age.[66] Another influence had been his visit to an exhibition of French impressionists in Moscow in the 1890s, where he had stood for several minutes in front of one of Claude Monet's haystack paintings, although Kandinsky wasn't sure what the subject was. Gripped by what he called the 'unsuspected power of the palette,' he began to realise that objects no longer need be an 'essential element' within a picture.[67] Other painters, in whose circle he moved, were groping in the same direction.[68]

Then there were the influences of science. Outwardly, Kandinsky was an austere man, who wore thick glasses. His manner was authoritative, but his mystical side made him sometimes prone to overinterpret events, as happened with the discovery of the electron. 'The collapse of the atom was equated, in my soul, with the collapse of the whole world. Suddenly, the stoutest walls crumbled. Everything became uncertain, precarious and insubstantial.'[69] Everything?

With so many influences acting on Kandinsky, it is perhaps not surprising he was the one to 'discover' abstraction. There was one final precipitating factor, one precise moment when, it could be said, abstract art was born. In 1908 Kandinsky was in Murnau, a country town south of Munich, near the small lake of Staffelsee and the Bavarian Alps, on the way to Garmisch, where Strauss was building his villa on the strength of his success with *Salomé*. One afternoon, after sketching in the foothills of the Alps, Kandinsky returned home, lost in thought. 'On opening the studio door, I was suddenly confronted by a picture of indescribable and incandescent loveliness. Bewildered, I stopped, staring at it. The painting lacked all subject, depicted no identifiable object and was entirely composed of bright colour-patches. Finally I approached closer and only then saw it for what it really was – my own painting, standing on its side ... One thing became clear to me: that objectiveness, the depiction of objects, needed no place in my paintings, and was indeed harmful to them.'[70]

Following this incident, Kandinsky produced a series of landscapes, each slightly different from the one before. Shapes became less and less distinct, colours more vivid and more prominent. Trees are just about recognisable as trees, the smoke issuing from a train's smokestack is just identifiable as smoke. But nothing is certain. His progress to abstraction was unhurried, deliberate. This process continued until, in 1911, Kandinsky painted three series of pictures, called Impressions, Improvisations, and Compositions, each one numbered, each one totally abstract. By the time he had completed the series, his divorce had come through.[71] Thus there is a curious personal parallel with Schoenberg and his creation of atonality.

At the turn of the century there were six great philosophers then living, although Nietzsche died before 1900 was out. The other five were **Henri Bergson**, **Benedetto Croce**, **Edmund Husserl**, **William James** and **Bertrand Russell**. At this end of the century, Russell is by far the best remembered, in Europe, James in the United States, but Bergson was probably the most accessible thinker of the first decade and, after 1907, certainly the most famous.

Bergson was born in Paris in the rue Lamartine in 1859, the same year as Edmund Husserl.[72] This was also the year in which Darwin's *On the Origin of Species* appeared. Bergson was a singular individual right from childhood. Delicate, with a high forehead, he spoke very slowly, with long breaths between utterances. This was slightly off-putting, and at the Lycée Condorcet, his high school in Paris, he came across as so reserved that his fellow students felt 'he had no soul,' a telling irony in view of his later theories.[73] For his teachers, however, any idiosyncratic behaviour was more than offset by his mathematical brilliance. He graduated well from Condorcet and, in 1878, secured admission to the Ecole Normale, a year after Emile Durkheim, who would become the most famous sociologist of his day.[74] After teaching in several schools, Bergson applied twice for a post at the Sorbonne but failed both times. Durkheim is believed responsible for these rejections, jealousy the motive. Undeterred, Bergson wrote his first book, *Time and Free Will* (1889), and then *Matter and Memory* (1896). Influenced by Franz Brentano and Husserl, Bergson argued forcefully that a sharp distinction should be drawn between physical and psychological processes. The methods evolved to explore the physical world, he said, were inappropriate to the study of mental life. These books were well received, and in 1900 Bergson was appointed to a chair at the Collège de France, overtaking Durkheim.

But it was *L'Evolution créatrice* (Creative Evolution), which appeared in 1907, that established Bergson's world reputation, extending it far beyond academic life. The book was quickly published in English, German, and Russian, and Bergson's weekly lectures at the Collège de France turned into crowded and fashionable social events, attracting not only the Parisian but the international elite. In 1914, the Holy Office, the Vatican office that decided Catholic doctrine, decided to put Bergson's works on its index of prohibited books.[75] This was a precaution very rarely imposed on non-Catholic writers, so what was the fuss about? Bergson once wrote that 'each great philosopher has only one thing to say, and more often than not gets no further than an attempt to express it.' Bergson's own central insight was that time is real. Hardly original or provocative, but the excitement lay in the details. What drew people's attention was his claim that the future does not in any sense exist. This was especially contentious because in 1907 the scientific determinists, bolstered by recent discoveries, were claiming that life was merely the unfolding of an already existing sequence of events, as if time were no more than a gigantic film reel, where the future is only that part which has yet to be played. In France this owed a lot to the cult of scientism popularised by Hippolyte Taine, who claimed that if everything could be broken down to atoms, the future was by definition utterly predictable.[76]

Bergson thought this was nonsense. For him there were two types of time, physics-time and real time. By definition, he said, time, as we normally understand it, involves memory; physics-time, on the other hand, consists of 'one long strip of nearly identical segments,' where segments of the past perish almost instantaneously. 'Real' time, however, is not reversible – on the contrary, each new segment takes its colour from the past. His final point, the one people found most difficult to accept, was that since memory is necessary for time, then time itself must to some extent be psychological. (This is what the Holy Office most objected to, since it was an interference in God's domain.) From this it followed for Bergson that the evolution of the universe, insofar as it can be known, is itself a psychological process also. Echoing Brentano and Husserl, Bergson was saying that evolution, far from being a truth 'out there' in the world, is itself a product, an 'intention' of mind.[77]

What really appealed to the French at first, and then to increasing numbers around the world, was Bergson's unshakeable belief in human freedom of choice and the unscientific effects of an entity he called the *élan vital*, the vital impulse, or life force. For Bergson, well read as he was in the sciences, rationalism was never enough. There had to be something else on top, 'vital phenomena' that were 'inaccessible to reason,' that could only be apprehended by intuition. The vital force further explained why humans are qualitatively different from other forms of life. For Bergson, an animal, almost by definition, was a specialist – in other words, very good at one thing (not unlike philosophers). Humans, on the other hand, were nonspecialists, the result of reason but also of intuition.[78] Herein lay Bergson's attraction to the younger generation of intellectuals in France, who crowded to his lectures. Known as the 'liberator,' he became the figure 'who had redeemed Western thought from the nineteenth-century "religion of science."' T. E. Hulme, a British acolyte, confessed that Bergson had brought 'relief' to an 'entire generation' by dispelling 'the nightmare of determinism.'[79]

An entire generation is an exaggeration, for there was no shortage of critics. Julien Benda, a fervent rationalist, said he would 'cheerfully have killed Bergson' if his views could have been stifled with him.[80] For the rationalists, Bergson's philosophy was a sign of degeneration, an atavistic congeries of opinions in which the rigours of science were replaced by quasi-mystical ramblings. Paradoxically, he came under fire from the church on the grounds that he paid too much attention to science. For a time, little of this criticism stuck. *Creative Evolution* was a runaway success (T. S. Eliot went so far as to call Bergsonism 'an epidemic').[81] America was just as excited, and William James confessed that 'Bergson's originality is so profuse that many of his ideas baffle me entirely.'[82] *Élan vital*, the 'life force,' turned into a widely used cliché, but 'life' meant not only life but intuition, instinct, the very opposite of reason. As a result, religious and metaphysical mysteries, which science had seemingly killed off, reappeared in 'respectable' guise. William James, who had himself written a book on religion, thought that Bergson had 'killed intellectualism definitively and without hope of recovery. I don't see how it can ever revive again in its ancient platonizing role of claiming to be the most authentic, intimate, and exhaustive definer of the nature of reality.'[83] Bergson's followers believed *Creative Evolution* had shown that reason itself is just one aspect of life, rather than the all-important judge of what mattered. This overlapped with Freud, but it also found an echo, much later in the century, in the philosophers of postmodernism.

One of the central tenets of Bergsonism was that the future is unpredictable. Yet in his will, dated 8 February 1937, he said, 'I would have become a convert [to Catholicism], had I not seen in preparation for years the formidable wave of anti-Semitism which is to break upon the world. I wanted to remain among those who tomorrow will be persecuted.'[84] Bergson died in 1941 of pneumonia contracted from having stood for hours in line with other Jews, forced to register with the authorities, then under Nazi military occupation.

Throughout the nineteenth century organised religion, and Christianity in particular, came under sustained assault from many of the sciences, the discoveries of which contradicted the biblical account of the universe. Many younger members of the clergy urged the Vatican to respond to these findings, while traditionalists wanted the church to explain them away and allow a return to familiar verities. In this debate, which threatened a deep divide, the young radicals were known as modernists.

In September 1907 the traditionalists finally got what they had been praying for when, from Rome, **Pope Pius X** published his encyclical, **Pascendi Dominici Gregis**. This unequivocally condemned modernism in all its forms. Papal encyclicals (letters to all bishops of the church) rarely make headlines now, but they were once very reassuring for the faithful, and *Pascendi* was the first of the century.[85] The ideas that Pius was responding to may be grouped under four headings. There was first the general attitude of science, developed since the Enlightenment, which brought about a change in the way that man looked at the world around him and, in the appeal to reason and experience that science typified, constituted a challenge to established authority. Then there was the specific science of Darwin and his concept of evolution. This had two effects. First, evolution carried the Copernican and Galilean revolutions still further toward the displacement of man from a specially appointed position in a limited universe. It showed that man had arisen from the animals, and was essentially no different from them and certainly not set apart in any way. The second effect of evolution was as metaphor: that ideas, like animals, evolve, change, develop. The theological modernists believed that the church – and belief – should evolve too, that in the modern world dogma as such was out of place. Third, there was the philosophy of Immanuel Kant (1724–1804), who argued that there were limits to reason, that human observations of the world were 'never neutral, never free of priorly imposed conceptual judgements', and because of that one could never *know* that God exists. And finally there were the theories of Henri Bergson. As we have seen, he actually supported spiritual notions, but these were very different from the traditional teachings of the church and closely interwoven with science and reason.[86]

The theological modernists believed that the church should address its own 'self-serving' forms of reason, such as the Immaculate Conception and the infallibility of the pope. They also wanted a reexamination of church teaching in the light of Kant, pragmatism, and recent scientific developments. In archaeology there were the discoveries and researches of the German school, who had made so much of the quest for the historical Jesus, the evidence for his actual, temporal existence rather than his *meaning* for the faithful. In anthropology, Sir James Frazer's *The Golden Bough* had shown the ubiquity of magical and religious rites, and their similarities in various cultures. This great diversity of religions had therefore undermined Christian claims to unique possession of truth – people found it hard to believe, as one writer said, 'that the greater part of humanity is plunged in error.'[87] With the benefit of hindsight, it is tempting to see *Pascendi* as yet another stage in 'the death of God.' However, most of the young clergy who took part in the debate over theological modernism did not wish to leave the church; instead they hoped it would 'evolve' to a higher plane.

The pope in Rome, Pius X (later Saint Pius), was a working-class man from Riese in the northern Italian province of the Veneto. Unsophisticated, having begun his career as a country priest, he was not surprisingly an uncompromising conservative and not at all afraid to get into politics. He therefore responded to the young clergy not by appeasing their demands but by carrying the fight to them. Modernism was condemned outright, without any prevarication, as 'nothing but the union of the faith with false philosophy.'[88] Modernism, for the pope and traditional Catholics, was defined as 'an exaggerated love of what is modern, an infatuation for modern ideas.' One Catholic writer even went so far as to say it was 'an abuse of what is modern.'[89] *Pascendi*, however, was only the most prominent part of a Vatican-led campaign against modernism. The Holy Office, the Cardinal Secretary of State, decrees of the Consistorial Congregation, and a second encyclical, *Editae*, published in 1910, all condemned the trend, and Pius repeated the argument in several papal letters to cardinals and the Catholic Institute in Paris. In his decree, *Lamentabili*, he singled out for condemnation no fewer than sixty-five specific propositions of modernism. Moreover, candidates for higher orders, newly appointed confessors, preachers, parish priests, canons, and bishops' staff were all obliged to swear allegiance to the pope, according to a formula 'which reprobates the principal modernist tenets.' And the primary role of dogma was reasserted: 'Faith is an act of the intellect made under the sway of the will.'[90]

Faithful Catholics across the world were grateful for the Vatican's closely reasoned arguments and its firm stance. Discoveries in the sciences were coming thick and fast in the early years of the century, changes in the arts were more bewildering and challenging than ever. It was good to have a rock in this turbulent world. Beyond the Catholic Church, however, few people were listening.

One place they weren't listening was China. There, in 1900, the number of Christian converts, after several centuries of missionary work, was barely a million. The fact is that the intellectual changes taking place in China were very different from anywhere else. This immense country was finally coming to terms with the modern world, and that involved abandoning, above all, Confucianism, the religion that had once led China to the forefront of mankind (helping to produce a society that first discovered paper, gunpowder, and much else) but had by then long ceased to be an innovative force, had indeed become a liability. This was far more daunting than the West's piecemeal attempts to move beyond Christianity.

Confucianism began by taking its fundamental strength, its basic analogy, from the cosmic order. Put simply, there is in Confucianism an hierarchy of superior-inferior relationships that form the governing principle of life. 'Parents are superior to children, men to women, rulers to subjects.' From this, it follows that each person has a role to fulfil; there is a 'conventionally fixed set of social expectations to which individual behaviour should conform.' Confucius himself described the hierarchy this way: 'Jun jun chen chen fu fu zi zi,' which meant, in effect, 'Let the ruler rule as he should and the minister be a minister as he should. Let the father act as a father should and the son act as a son should.' So long as everyone performs his role, social stability is maintained.[91] In laying stress on 'proper behaviour according to status,' the Confucian gentleman was guided by li, a moral code that stressed the quiet virtues of patience, pacifism, and compromise, respect for ancestors, the old, and the educated, and above all a gentle humanism, taking man as the measure of all things. Confucianism also stressed that men were naturally equal at birth but perfectible, and that an individual, by his own efforts, could do 'the right thing' and be a model for others. The successful sages were those who put 'right conduct' above everything else.[92]

And yet, for all its undoubted successes, the Confucian view of life was a form of conservatism. Given the tumultuous changes of the late nineteenth and early twentieth centuries, that the system was failing could not be disguised for long. As the rest of the world coped with scientific advances, the concepts of modernism and the advent of socialism, China needed changes that were more profound, the mental and moral road more tortuous. The ancient virtues of patience and compromise no longer offered real hope, and the old and the traditionally educated no longer had the answers. Nowhere was the demoralisation more evident than in the educated class, the scholars, the very guardians of the neo-Confucian faith.

The modernisation of China had in theory been going on since the seventeenth century, but by the beginning of the twentieth it had in practice become a kind of game played by a few high officials who realised it was needed but did not have the political wherewithal to carry these changes through. In the eighteenth and nineteenth centuries, Jesuit missionaries had produced Chinese translations of over four hundred Western works, more than half on Christianity and about a third in science. But Chinese scholars still remained conservative, as was highlighted by the case of Yung Wing, a student who was invited to the United States by missionaries in 1847 and graduated from Yale in 1854. He returned to China after eight years' study but was forced to wait another eight years before his skills as an interpreter and translator were made use of.[93] There was *some* change. The original concentration of Confucian scholarship on philosophy had given way by the nineteenth century to '**evidential research**,' the concrete analysis of ancient texts.[94] This had two consequences of significance. One was the discovery that many of the so-called classic texts were fake, thus throwing the very tenets of Confucianism itself into doubt. No less importantly, the 'evidential research' was extended to mathematics, astronomy, fiscal and administrative matters, and archaeology. This could not yet be described as a scientific revolution, but it was a start, however late.

The final thrust in the move away from Confucianism arrived in the form of the Boxer Rising, which began in 1898 and ended two years later with the beginnings of China's republican revolution. The reason for this was once again the Confucian attitude to life, which meant that although there had been some change in Chinese scholarly activity, the compartmentalisation recommended by classical Confucianism was still paramount, its most important consequence being that many of the die-hard and powerful Manchu princes had had palace upbringings that had left them 'ignorant of the world and proud of it.'[95] This profound ignorance was one of the reasons so many of them became patrons of a peasant secret society known as the Boxers, merely the most obvious and tragic sign of China's intellectual bankruptcy. The Boxers, who began in the Shandong area and were rabidly xenophobic, featured two peasant traditions – the technique of martial arts ('boxing') and spirit possession or shamanism. Nothing could have been more inappropriate, and this fatal combination made for a vicious set of episodes. The Chinese were defeated at the hands of eleven (despised) foreign countries, and were thus forced to pay $333 million in indemnities over forty years (which would be at least $20 billion now), and suffer the most severe loss of face the nation had ever seen. The year the Boxer Uprising was put down was therefore the low point by a long way for Confucianism, and everyone, inside and outside China, knew that radical, fundamental, *philosophical* change had to come.[96]

Such change began with a set of **New Policies** (with initial capitals). Of these, the most portentous – and most revealing – was educational reform. Under this scheme, a raft of modern schools was to be set up across the country, teaching a new Japanese-style mix of old and new subjects (Japan was the culture to be emulated because that country had defeated China in the war of 1895 and, under Confucianism, the victor was superior to the vanquished: at the turn of the century Chinese students crowded into Tokyo).[97] It was intended that many of China's academies would be converted into these new schools. Traditionally, China had hundreds if not thousands of academies, each consisting of a few dozen local scholars thinking high thoughts but not in any way coordinated with one another or the needs of the country. In time they had become a small elite who ran things locally, from burials to water distribution, but had no overall, systematic influence. The idea was that these academies would be modernised.[98]

It didn't work out like that. The new – modern, Japanese, and Western science-oriented – curriculum proved so strange and so difficult for the Chinese that most students stuck to the easier, more familiar Confucianism, despite the evidence everywhere that it wasn't working or didn't meet China's needs. It soon became apparent that the only way to deal with the classical system was to abolish it entirely, and that in fact is what happened just four years later, in 1905. A great turning point for China, this stopped in its tracks the production of the degree-holding elite, the gentry class. As a result, the old order lost its intellectual foundation and with it its intellectual cohesion. So far so good, one might think. However, the student class that replaced the old scholar gentry was presented, in John Fairbanks's words, with a 'grab-bag' of Chinese and Western thought, which pulled students into technical specialities that however modern still left them without a moral order: 'The **Neo-Confucian synthesis** was no longer valid or useful, yet nothing to replace it was in sight.'[99] The important intellectual point to grasp about China is that that is how it has since remained. The country might take on over the years many semblances of Western thinking and behaviour, but the moral void at the centre of the society, vacated by Confucianism, has never been filled.

It is perhaps difficult for us, today, to imagine the full impact of modernism. Those alive now have all grown up in a scientific world, for many the life of large cities is the only life they know, and rapid change the only change there is. Only a minority of people have an intimate relation with the land or nature.

None of this was true at the turn of the century. Vast cities were still a relatively new experience for many people; social security systems were not yet in place, so that squalor and poverty were much harsher than now, a much greater shadow; and fundamental scientific discoveries, building on these new, uncertain worlds, created a sense of bewilderment, desolation and loss probably sharper and more widespread than had ever been felt before, or has since. The collapse of organised religion was only one of the factors in this seismic shift in sensibility: the growth in nationalism, anti-Semitism, and racial theories overall, and the enthusiastic embrace of the modernist art forms, seeking to break down experience into fundamental units, were all part of the same response.

The biggest paradox, the most worrying transformation, was this: according to evolution, the world's natural pace of change was glacial. According to modernism, everything was changing at once, and in fundamental ways, virtually overnight. For most people, therefore, modernism was as much a threat as it was a promise. The beauty it offered held a terror within.

THE PRAGMATIC MIND OF AMERICA

In 1906 a group of Egyptians, headed by Prince Ahmad Fuad, issued a manifesto to campaign for the establishment by public subscription of an Egyptian university 'to create a body of teaching similar to that of the universities of Europe and adapted to the needs of the country.' The appeal was successful, and the university, or in the first phase an evening school, was opened two years later with a faculty of two Egyptian and three European professors. This plan was necessary because the college-mosque of al-Azhar at Cairo, once the principal school in the Muslim world, had sunk in reputation as it refused to update and adapt its mediaeval approach. One effect of this was that in Egypt and Syria there had been no university, in the modern sense, throughout the nineteenth century.[1]

China had just four universities in 1900; Japan had two – a third would be founded in 1909; Iran had only a series of specialist colleges (the Teheran School of Political Science was founded in 1900); there was one college in Beirut and in Turkey – still a major power until World War I – the University of Istanbul was founded in 1871 as the Dar-al-funoun (House of Learning), only to be soon closed and not reopened until 1900. In Africa south of the Sahara there were four: in the Cape, the Grey University College at Bloemfontein, the Rhodes University College at Grahamstown, and the Natal University College. Australia also had four, New Zealand one. In India, the universities of Calcutta, Bombay, and Madras were founded in 1857, and those of Allahabad and Punjab between 1857 and 1887. But no more were created until 1919.[2] In Russia there were ten state-funded universities at the beginning of the century, plus one in Finland (Finland was technically autonomous), and one private university in Moscow.

If the paucity of universities characterised intellectual life outside the West, the chief feature in the United States was the tussle between those who preferred the British-style universities and those for whom the German-style offered more. To begin with, most American colleges had been founded on British lines. Harvard, the first institution of higher learning within the United States, began as a Puritan college in 1636. More than thirty partners of the Massachusetts Bay Colony were graduates of Emmanuel College, Cambridge, and so the college they established near Boston naturally followed the Emmanuel pattern. Equally influential was the Scottish model, in particular Aberdeen.[3] Scottish universities were nonresidential, democratic rather than religious, and governed by local dignitaries – a forerunner of boards of trustees. Until the twentieth century, however, America's institutions of higher learning were really colleges – devoted to teaching – rather than universities proper, concerned with the advancement of knowledge. Only Johns Hopkins in Baltimore (founded in 1876) and Clark (1888) came into this category, and both were soon forced to add undergraduate schools.[4]

The man who first conceived the modern university as we know it was **Charles Eliot**, a chemistry professor at Massachusetts Institute of Technology who in 1869, at the age of only thirty-five, was appointed president of Harvard, where he had been an undergraduate. When Eliot arrived, Harvard had 1,050 students and fifty-nine members of the faculty. In 1909, when he retired, there were four times as many students and the faculty had grown tenfold. But Eliot was concerned with more than size: 'He killed and buried the limited arts college curriculum which he had inherited. He built up the professional schools and made them an integral part of the university. Finally, he promoted graduate education and thus established a model which practically all other American universities with graduate ambitions have followed.'[5]

Above all, Eliot followed the system of higher education in the German-speaking lands, the system that gave the world Max Planck, Max Weber, Richard Strauss, Sigmund Freud, and Albert Einstein. The preeminence of German universities in the late nineteenth century dated back to the Battle of Jena in 1806, after which Napoleon finally reached Berlin. His arrival there forced the inflexible Prussians to change. Intellectually, Johann Fichte, Christian Wolff, and Immanuel Kant were the significant figures, freeing German scholarship from its stultifying reliance on theology. As a result, German scholars acquired a clear advantage over their European counterparts in philosophy, philology, and the physical sciences. It was in Germany, for example, that physics, chemistry, and geology were first regarded in universities as equal to the humanities. Countless Americans, and distinguished Britons such as Matthew Arnold and Thomas Huxley, all visited Germany and praised what was happening in its universities.[6]

From Eliot's time onward, the American universities set out to emulate the German system, particularly in the area of research. However, this German example, though impressive in advancing knowledge and in producing new technological processes for industry, nevertheless sabotaged the 'collegiate way of living' and the close personal relations between undergraduates and faculty that had been a major feature of American higher education until the adoption of the German approach. The German system was chiefly responsible for what William James called 'the Ph.D. octopus': Yale awarded **the first Ph.D.** west of the Atlantic in 1861; by 1900 well over three hundred were being granted every year.[7]

The price for following Germany's lead was a total break with the British collegiate system. At many universities, housing for students disappeared entirely, as did communal eating. At Harvard in the 1880s the German system was followed so slavishly that attendance at classes was no longer required – all that counted was performance in the examinations. Then a reaction set in. Chicago was first, building seven dormitories by 1900 'in spite of the prejudice against them at the time in the [mid-] West on the ground that they were medieval, British and autocratic.' Yale and Princeton soon adopted a similar approach. Harvard reorganised after the English housing model in the 1920s.[8]

Since American universities have been the forcing ground of so much of what will be considered later in this book, their history is relevant in itself. But the battle for the soul of Harvard, Chicago, Yale, and the other great institutions of learning in America is relevant in another way, too. The amalgamation of German and British best practices was a sensible move, a pragmatic response to the situation in which American universities found themselves at the beginning of the century. And pragmatism was a particularly strong strain of thought in America. The United States was not hung up on European dogma or ideology. It had its own 'frontier mentality'; it had – and exploited – the opportunity to cherry-pick what was best in the old world, and eschew the rest. Partly as a result of that, it is noticeable that the matters considered in this chapter – skyscrapers, the Ashcan school of painting, flight and film – were all, in marked contrast with aestheticism, psychoanalysis, the *élan vital* or abstraction, fiercely practical developments, immediately and hardheadedly *useful* responses to the evolving world at the beginning of the century.

The founder of America's **pragmatic** school of thought was **Charles Sanders Peirce**, a philosopher of the 1870s, but his ideas were updated and made popular in 1906 by **William James**. William and his younger brother Henry, the novelist, came from a wealthy Boston family; their father, Henry James Sr., was a writer of 'mystical and amorphous philosophic tracts.'[9] William James's debt to Peirce was made plain in the title he gave to a series of lectures delivered in Boston in 1907: *Pragmatism: A New Name for Some Old Ways of Thinking*. The idea behind pragmatism was to develop a philosophy shorn of idealistic dogma and subject to the rigorous empirical standards being developed in the physical sciences. What James added to Peirce's ideas was the notion that philosophy should be accessible to everyone; it was a fact of life, he thought, that everyone liked to have what they called a philosophy, a way of seeing and understanding the world, and his lectures (eight of them) were intended to help.

James's approach signalled another great divide in twentieth-century philosophy, in addition to the rift between the continental school of Franz Brentano, Edmund Husserl, and Henri Bergson, and the analytic school of Bertrand Russell, Ludwig Wittgenstein, and what would become the Vienna Circle. Throughout the century, there were those philosophers who drew their concepts from ideal situations: they tried to fashion a worldview and a code of conduct in thought and behaviour that derived from a theoretical, 'clear' or 'pure' situation where equality, say, or freedom was assumed as a given, and a system constructed hypothetically around that. In the opposite camp were those philosophers who started from the world as it was, with all its untidiness, inequalities, and injustices. James was firmly in the latter camp.

He began by trying to explain this divide, proposing that there are two very different basic forms of 'intellectual temperament,' what he called the 'tough-' and 'tender-minded.' He did not actually say that he thought these temperaments were genetically endowed – 1907 was a bit early for anyone to use such a term – but his choice of the word *temperament* clearly hints at such a view. He thought that the people of one temperament invariably had a low opinion of the other and that a clash between the two was inevitable. In his first lecture he characterised them as follows:

Tender-minded	Tough-minded
Rationalistic (going by principle)	Empiricist (going by facts)
Optimistic	Pessimistic
Religious	Irreligious
Free-willist	Fatalistic
Dogmatic	Pluralistic
	Materialistic
	Sceptical

One of his main reasons for highlighting this division was to draw attention to how the world was changing: 'Never were as many men of a decidedly empiricist proclivity in existence as there are at the present day. Our children, one may say, are almost born scientific.'[10]

Nevertheless, this did not make James a scientific atheist; in fact it led him to pragmatism (he, after all, had published an important book *Varieties of Religious Experience* in 1902).[11] He thought that philosophy should above all be practical, and here he acknowledged his debt to Peirce. Beliefs, Peirce had said, 'are really rules for action.' James elaborated on this theme, concluding that 'the whole function of philosophy ought to be to find out what definite difference it will make to you and me, at definite instants of our life, if this world-formula or that world-formula be the true one. ... A pragmatist turns his back resolutely and once for all upon a lot of inveterate habits dear to professional philosophers. He turns away from abstraction and insufficiency, from verbal solutions, from bad *a priori* reasons, from fixed principles, closed systems, and pretended absolutes and origins. He turns towards concreteness and adequacy, towards facts, towards action, and towards power.'[12] Metaphysics, which James regarded as primitive, was too attached to the big words – 'God,' 'Matter,' 'the Absolute.' But these, he said, were only worth dwelling on insofar as they had what he called 'practical cash value.' What *difference* did they make to the conduct of life? Whatever it is that makes a practical difference to the way we lead our lives, James was prepared to call 'truth.' Truth was/is not absolute, he said. There are many truths, and they are only true so long as they are practically useful. That truth is beautiful doesn't make it eternal. This is why truth is good: by definition, it makes a practical difference. James used his approach to confront a number of metaphysical problems, of which we need consider only one to show how his arguments worked: Is there such a thing as the soul, and what is its relationship to consciousness? Philosophers in the past had proposed a 'soul-substance' to account for certain kinds of intuitive experience, James wrote, such as the feeling that one has lived before within a different identity. But if you take away consciousness, is it practical to hang on to 'soul'? Can a soul be said to exist without consciousness? No, he said. Therefore, why bother to concern oneself with it? James was a convinced Darwinist, evolution he thought was essentially a pragmatic approach to the universe; that's what adaptations – species – are.[13]

America's third pragmatic philosopher, after Peirce and James, was **John Dewey**. A professor in Chicago, Dewey boasted a Vermont drawl, rimless eyeglasses, and a complete lack of fashion sense. In some ways he was the most successful pragmatist of all. Like James he believed that everyone has his own philosophy, his own set of beliefs, and that such philosophy should help people to lead happier and more productive lives. His own life was particularly productive: through newspaper articles, popular books, and a number of debates conducted with other philosophers, such as Bertrand Russell or Arthur Lovejoy, author of *The Great Chain of Being*, Dewey became known to the general public as few philosophers are.[14] Like James, Dewey was a convinced Darwinist, someone who believed that science and the scientific approach needed to be incorporated into other areas of life. In particular, he believed that the discoveries of science should be adapted to the education of children. For Dewey, the start of the twentieth century was an age of 'democracy, science and industrialism,' and this, he argued, had profound consequences for education. At that time, attitudes to children were changing fast. In 1909 the Swedish feminist **Ellen Key** published her book *The Century of the Child*, which reflected the general view that the child had been rediscovered – rediscovered in the sense that there was a new joy in the possibilities of childhood and in the realisation that children were different from adults and from one another.[15] This seems no more than common sense to us, but in the nineteenth century, before the victory over a heavy rate of child mortality, when families were much larger and many children died, there was not – there could not be – the same investment in children, in time, in education, in *emotion*, as there was later. Dewey saw that this had significant consequences for teaching. Hitherto schooling, even in America, which was in general more indulgent to children than Europe, had been dominated by the rigid authority of the teacher, who had a concept of what an educated person should be and whose main aim was to convey to his or her pupils the idea that knowledge was the 'contemplation of fixed verities.'[16]

Dewey was one of the leaders of a movement that changed such thinking, in two directions. The traditional idea of education, he saw, stemmed from a leisured and aristocratic society, the type of society that was disappearing fast in the European democracies and had never existed in America. Education now had to meet the needs of democracy. Second, and no less important, education had to reflect the fact that children were very different from one another in abilities and interests. For children to make the best contribution to society they were capable of, education should be less about 'drumming in' hard facts that the teacher thought necessary and more about drawing out what the individual child was capable of. In other words, pragmatism applied to education.

Dewey's enthusiasm for science was reflected in the name he gave to the 'Laboratory School' that he set up in 1896.[17] Motivated partly by the ideas of **Johann Pestalozzi**, a pious Swiss educator, and the German philosopher **Friedrich Fröbel**, and by the child psychologist **G. Stanley Hall**, the institution operated on the principle that for each child there were negative and positive consequences of individuality. In the first place, the child's natural abilities set limits to what it was capable of. More positively, the interests and qualities within the child had to be discovered in order to see where 'growth' was possible. Growth was an important concept for the 'child-centred' apostles of the 'new education' at the beginning of the century. Dewey believed that since antiquity society had been divided into leisured and aristocratic classes, the custodians of knowledge, and the working classes, engaged in work and practical knowledge. This separation, he believed, was fatal, especially in a democracy. Education along class lines must be rejected, and inherited notions of learning discarded as unsuited to democracy, industrialism, and the age of science.[18]

The ideas of Dewey, along with those of Freud, were undoubtedly influential in attaching far more importance to childhood than before. The notion of personal growth and the drawing back of traditional, authoritarian conceptions of what knowledge is and what education should seek to do were liberating ideas for many people. In America, with its many immigrant groups and wide geographical spread, the new education helped to create many individualists. At the same time, the ideas of the 'growth movement' always risked being taken too far, with children left to their own devices too much. In some schools where teachers believed that 'no child should ever know failure' examinations and grades were abolished.[19] This lack of structure ultimately backfired, producing children who were more conformist precisely because they lacked hard knowledge or the independent judgement that the occasional failure helped to teach them. Liberating children from parental 'domination' was, without question, a form of freedom. But later in the century it would bring its own set of problems.

It is a cliché to describe the university as an ivory tower, a retreat from the hurly-burly of what many people like to call the 'real world,' where professors (James at Harvard, Dewey at Chicago, or Bergson at the Collège de France) can spend their hours contemplating fundamental philosophical concerns. It therefore makes a nice irony to consider next a very pragmatic idea, which was introduced at Harvard in 1908. This was the **Harvard Graduate School of Business Administration**. Note that it was a *graduate* school. Training for a life/career in business had been provided by other American universities since the 1880s, but always as undergraduate study. The Harvard school actually began as an idea for an administrative college, training diplomats and civil servants. However, a stock market panic of 1907 showed a need for better-trained businessmen.

The Graduate School of Business Administration opened in October 1908 with fifty-nine candidates for the new degree of Master of Business Administration (M.B.A.).[20] At the time there was conflict not only over what was taught but *how* it was to be taught. Accountancy, transportation, insurance, and banking were covered by other institutions, so Harvard evolved its own definition of business: 'Business is making things to sell, at a profit, decently.' Two basic activities were identified by this definition: manufacturing, the act of production; and merchandising or marketing, the act of distribution. Since there were no readily available textbooks on these matters, however, businessmen and their firms were spotlighted by the professors, thus evolving what would become Harvard's famous system of case studies. In addition to manufacturing and distribution, a course was also offered for the study of **Frederick Winslow Taylor**'s *Principles of Scientific Management*.[21] Taylor, an engineer by training, embraced the view, typified by a speech that President Theodore Roosevelt had made in the White House, that many aspects of American life were inefficient, a form of waste. For Taylor, the management of companies needed to be put on a more 'scientific' basis – he was intent on showing that management was a science, and to illustrate his case he had investigated, and improved, efficiency in a large number of companies. For example, research had discovered, he said, that the average man shifts far more coal or sand (or whatever substance) with a shovel that holds 21 pounds rather than, say, 24 pounds or 18 pounds. With the heavier shovel, the man gets tired more quickly from the weight. With the lighter shovel he gets tired more quickly from having to work faster. With a 21-pound shovel, the man can keep going longer, with fewer breaks. Taylor devised new strategies for many businesses, resulting, he said, in higher wages for the workers and higher profits for the company. In the case of pig-iron handling, for example, workers increased their wages from $1.15 a day to $1.85, an increase of 60 percent, while average production went up from 12.5 tons a day to 47 tons, an increase of nearly 400 percent. As a result, he said, everyone was satisfied.[22] The final elements of the Harvard curriculum were research, by the faculty, shoe retailing being the first business looked into, and employment experience, when the students spent time with firms during the long vacation. Both elements proved successful. Business education at Harvard thus became a mixture of case study, as was practised in the law department, and a 'clinical' approach, as was pursued in the medical school, with research thrown in. The approach eventually became famous, with many imitators. The 59 candidates for M.B.A. in 1908 grew to 872 by the time of the next stock market crash, in 1929, and included graduates from fourteen foreign countries. The school's publication, the *Harvard Business Review*, rolled off the presses for the first time in 1922, its editorial aim being to demonstrate the relation between fundamental economic theory and the everyday experience and problems of the executive in business, the ultimate exercise in pragmatism.[23]

What was happening at Harvard, in other business schools, and in business itself was one aspect of what Richard Hofstadter has identified as 'the practical culture' of America. To business, he added farming, the American labor movement (a much more practical, less ideological form of socialism than the labor movements of Europe), the tradition of the self-made man, and even religion.[24] Hofstadter wisely points out that Christianity in many parts of the United States is entirely practical in nature. He takes as his text a quote of theologian **Reinhald Niebuhr**, that a strain in American theology 'tends to define religion in terms of adjustment to divine reality for the sake of gaining power rather than in terms of revelation which subjects the recipient to the criticism of that which is revealed.'[25] And he also emphasises how many theological movements use 'spiritual technology' to achieve their ends: 'One . . . writer tells us that . . . "the body is . . . a receiving set for the catching of messages from the Broadcasting Station of God" and that "the greatest of Engineers . . . is your silent partner." '[26] In the practical culture it is only natural for even God to be a businessman.

The intersection in New York's Manhattan of Broadway and Twenty-third Street has always been a busy crossroads. Broadway cuts through the cross street at a sharp angle, forming on the north side a small triangle of land quite distinctive from the monumental rectangular 'blocks' so typical of New York. In 1903 the architect Daniel Burnham used this unusual sliver of ground to create what became an icon of the city, a building as distinctive and as beautiful now as it was on the day it opened. The narrow wedge structure became known – affectionately – as the **Flatiron** Building, on account of its shape (its sharp point was rounded). But shape was not the only reason for its fame: the Flatiron was 285 feet – twenty-one storeys – high, and New York's first skyscraper.[27]

Buildings are the most candid form of art, and the skyscraper is the most pragmatic response to the huge, crowded cities that were formed in the late nineteenth century, where space was at a premium, particularly in Manhattan, which is built on a narrow slice of an island.[28] Completely new, always striking, on occasions beautiful, there is no image that symbolised the early twentieth century like the skyscraper. Some will dispute that the Flatiron was the first such building. In the nineteenth century there *were* buildings twelve, fifteen, or even nineteen storeys high. George Post's Pulitzer Building on Park Row, built in 1892, was one of them, but the Flatiron Building was the first to rule the skyline. It immediately became a focus for artists and photographers. **Edward Steichen**, one of the great early American photographers, who with **Alfred Stieglitz** ran one of New York's first modern art galleries (and introduced Cézanne to America), portrayed the Flatiron Building as rising out of the misty haze, almost a part of the natural landscape. His photographs of it showed diminutive, horse-drawn carriages making their way along the streets, with gaslights giving the image the feel almost of an impressionist painting of Paris.[29] The Flatiron created downdraughts that lifted the skirts of women going by, so that youths would linger around the building to watch the flapping petticoats.[30]

The skyscraper, which was to find its full expression in New York, was actually conceived in Chicago.[31] The history of this conception is an absorbing story with its own tragic hero, **Louis Henry Sullivan** (1856–1924). Sullivan was born in Boston, the son of a musically gifted mother of German-Swiss-French stock and a father, Patrick, who taught dance. Louis, who fancied himself as a poet and wrote a lot of bad verse, grew up loathing the chaotic architecture of his home city, but studied the subject not far away, across the Charles River at MIT.[32] A round-faced man with brown eyes, Sullivan had acquired an imposing self-confidence even by his student days, revealed in his dapper suits, the pearl studs in his shirts, the silver-topped walking cane that he was never without. He travelled around Europe, listening to Wagner as well as looking at buildings, then worked briefly in Philadelphia and the Chicago office of William Le Baron Jenney, often cited as the father of the skyscraper for introducing a steel skeleton and elevators in his Home Insurance Building (Chicago, 1883–5).[33] Yet it is doubtful whether this building – squat by later standards – really qualifies as a skyscraper. In Sullivan's view the chief property of a skyscraper was that it 'must be tall, every inch of it tall. The force and power of altitude must be in it. It must be every inch a proud and soaring thing, rising in sheer exaltation that from top to bottom it is a unit without a single dissenting line.'[34]

In 1876 Chicago was still in a sense a frontier town. Staying at the Palmer House Hotel, Rudyard Kipling found it 'a gilded rabbit warren ... full of people talking about money and spitting,' but it offered fantastic architectural possibilities in the years following the great fire of 1871, which had devastated the city core.[35] By 1880 Sullivan had joined the office of Dankmar Adler and a year later became a full partner. It was this partnership that launched his reputation, and soon he was a leading figure in the Chicago school of architecture.

Though Chicago became known as the birthplace of the skyscraper, the notion of building very high structures is of indeterminable antiquity. The intellectual breakthrough was the realisation that a tall building need not rely on masonry for its support.*

The metal-frame building was the answer: the frame, iron in the earlier examples, steel later on, is bolted (later riveted for speedier construction) together to steel plates, like shelves, which constitute the floors of each storey. On this structure curtain walls could be, as it were, hung. The wall is thus a cladding of the building, rather than truly weight bearing. Most of the structural problems regarding skyscrapers were solved very early on. Therefore, as much of the debate at the turn of the century was about the aesthetics of design as about engineering. Sullivan passionately joined the debate in favour of a modern architecture, rather than pastiches and sentimental memorials to the old orders. His famous dictum, 'Form ever follows function,' became a rallying cry for modernism, already mentioned in connection with the work of Adolf Loos in Vienna.[36]

Sullivan's early masterpiece was the **Wainwright Building** in Saint Louis.

---

* The elevator also played its part. This was first used commercially in 1889 in the Demarest Building in New York, fitted by Otis Brothers & Co., using the principle of a drum driven by an electric motor through a 'worm gear reduction.' The earliest elevators were limited to a height of about 150 feet, ten storeys or so, because more rope could not be wound upon the drum.

This, again, was not a really high structure, only ten storeys of brick and terra-cotta, but Sullivan grasped that intervention by the architect could 'add' to a building's height.[37] As one architectural historian wrote, the Wainwright is 'not merely tall; it is about *being* tall – it is tall architecturally even more than it is physically.'[38] If the Wainwright Building was where Sullivan found his voice, where he tamed verticality and showed how it could be controlled, his finest building is generally thought to be the **Carson Pirie Scott** department store, also in Chicago, finished in 1903–4. Once again this is not a skyscraper as such – it is twelve storeys high, and there is more emphasis on the horizontal lines than the vertical. But it was in this building above all others that Sullivan displayed his great originality in creating a new kind of decoration for buildings, with its 'streamlined majesty,' 'curvilinear ornament' and 'sensuous webbing.'[39] The ground floor of Carson Pirie Scott shows the Americanisation of the art nouveau designs Sullivan had seen in Paris: a Metro station turned into a department store.[40]

Frank Lloyd Wright was also experimenting with urban structures. Judging by the photographs – which is all that remains since the edifice was torn down in 1950 – his Larkin Building in Buffalo, on the Canadian border, completed in 1904, was at once exhilarating, menacing, and ominous.[41] (John Larkin built the Empire State Building in New York, the first to have more than 100 floors.) An immense office space enclosed by 'a simple cliff of brick,' its furnishings symmetrical down to the last detail and filled with clerks at work on their long desks, it looks more like a setting for automatons than, as Wright himself said, 'one great official family at work in day-lit, clean and airy quarters, day-lit and officered from a central court.'[42] It was a work with many 'firsts' that are now found worldwide. It was air-conditioned and fully fireproofed; the furniture – including desks and chairs and filing cabinets – was made of steel and magnesite; its doors were glass, the windows double-glazed. Wright was fascinated by materials and the machines that made them in a way that Sullivan was not. He built for the 'machine age,' for standardisation. He became very interested also in the properties of ferro-concrete, a completely new building material that revolutionised design. Steel was pioneered in Britain as early as 1851 in the Crystal Palace, a precursor of the steel-and-glass building, and reinforced concrete (*béton armé*) was invented in France in the same year, by François Hennebique. But it was only in the United States, with the building of skyscrapers, that these materials were exploited to the full. In 1956 Wright proposed a *mile-high* skyscraper for Chicago.[43]

Further down the eastern seaboard of the United States, 685 miles away to be exact, lies **Kill Devil Hill**, near the ocean banks of North Carolina. In 1903 it was as desolate as Manhattan was crowded. A blustery place, with strong winds gusting in from the sea, it was conspicuous by the absence of the umbrella pine trees that populate so much of the state. This was why it had been chosen for an experiment that was to be carried out on 17 December that year – one of the most exciting ventures of the century, destined to have an enormous impact on the lives of many people. The skyscraper was one way of leaving the ground; this was another, and far more radical.

At about half past ten that morning, four men from the nearby lifesaving station and a boy of seventeen stood on the hill, gazed down to the field which lay alongside, and waited. A pre-arranged signal, a yellow flag, had been hoisted nearby, at the village of Kitty Hawk, to alert the local coastguards and others that something unusual might be about to happen. If what was supposed to occur did occur, the men and the boy were there to serve as witnesses. To say that the sea wind was fresh was putting it mildly. Every so often **the Wright brothers – Wilbur and Orville**, the object of the observers' attention – would disappear into their shed so they could cup their freezing fingers over the stove and get some feeling back into them.[44]

Earlier that morning, Orville and Wilbur had tossed a coin to see who would be the first to try the experiment, and Orville had won. Like his brother, he was dressed in a three-piece suit, right down to a starched white collar and tie. To the observers, Orville appeared reluctant to start the experiment. At last he shook hands with his brother, and then, according to one bystander, 'We couldn't help notice how they held on to each other's hand, sort o' like they hated to let go; like two folks parting who weren't sure they'd ever see each other again.'[45] Just after the half-hour, Orville finally let go of Wilbur, walked across to the machine, stepped on to the bottom wing, and lay flat, wedging himself into a hip cradle. Immediately he grasped the controls of a weird contraption that, to observers in the field, seemed to consist of wires, wooden struts, and huge, linen-covered wings. This entire mechanism was mounted on to a fragile-looking wooden monorail, pointing into the wind. A little trolley, with a cross-beam nailed to it, was affixed to the monorail, and the elaborate construction of wood, wires and linen squatted on that. The trolley travelled on two specially adapted bicycle hubs.

Orville studied his instruments. There was an anemometer fixed to the strut nearest him. This was connected to a rotating cylinder that recorded the distance the contraption would travel. A second instrument was a stopwatch, so they would be able to calculate the speed of travel. Third was an engine revolution counter, giving a record of propeller turns. That would show how efficient the contraption was and how much fuel it used, and also help calculate the distance travelled through the air.[46] While the contraption was held back by a wire, its engine – a four-cylinder, eight-to-twelve-horsepower gasoline motor, lying on its side – was opened up to full throttle. The engine power was transmitted by chains in tubes and was connected to two airscrews, or propellers, mounted on the wooden struts between the two layers of linen. The wind, gusting at times to thirty miles per hour, howled between the struts and wires. The brothers knew they were taking a risk, having abandoned their safety policy of test-flying all their machines as gliders before they tried powered flight. But it was too late to turn back now. Wilbur stood by the right wingtip and shouted to the witnesses 'not to look sad, but to laugh and hollo and clap [their] hands and try to cheer Orville up when he started.'[47] As best they could, amid the howling of the wind and the distant roar of the ocean, the onlookers cheered and shouted.

With the engine turning over at full throttle, the restraining wire was suddenly slipped, and the contraption, known to her inventors as *Flyer*, trundled forward. The machine gathered speed along the monorail. Wilbur Wright ran alongside *Flyer* for part of the way, but could not keep up as it achieved a speed of thirty miles per hour, lifted from the trolley and rose into the air. Wilbur, together with the startled witnesses, watched as the *Flyer* careered through space for a while before sweeping down and ploughing into the soft sand. Because of the wind speed, *Flyer* had covered 600 feet of air space, but 120 over the ground. 'This flight only lasted twelve seconds,' Orville wrote later, 'but it was, nevertheless, the first in the history of the world in which a machine carrying a man had raised itself by its own power into the air in full flight, had sailed forward without reduction of speed, and had finally landed at a point as high as that from which it had started.' Later that day Wilbur, who was a better pilot than Orville, managed a 'journey' of 852 feet, lasting 59 seconds. The brothers had made their point: their flights were *powered*, *sustained*, and *controlled*, the three notions that define proper heavier-than-air flight in a powered aircraft.[48]

Men had dreamed of flying from the earliest times. Persian legends had their kings borne aloft by flocks of birds, and Leonardo da Vinci conceived designs for both a parachute and a helicopter.[49] Several times in history ballooning has verged on a mania. In the nineteenth century, however, countless inventors had either killed themselves or made fools of themselves attempting to fly contraptions that, as often as not, refused to budge.[50] The Wright brothers were different. Practical to a fault, they flew only four years after becoming interested in the problem.

It was Wilbur who wrote to the Smithsonian Institution in Washington, D.C., on 30 May 1899 to ask for advice on books to read about flying, describing himself as 'an enthusiast but not a crank.'[51] Born in 1867, thus just thirty-two at the time, Wilbur was four years older than Orville. Though they were always a true brother-brother team, Wilbur usually took the lead, especially in the early years. The sons of a United Brethren minister (and later a bishop) in Dayton, Ohio, the Wright brothers were brought up to be resourceful, pertinacious, and methodical. Both had good brains and a mechanical aptitude. They had been printers and bicycle manufacturers and repairers. It was the bicycle business that gave them a living and provided modest funds for their aviation; they were never financed by anyone.[52] Their interest in flying was kindled in the 1890s, but it appears that it was not until **Otto Lilienthal**, the great German pioneer of gliding, was killed in 1896 that they actually did anything about their new passion. (Lilienthal's last words were, 'Sacrifices must be made.')[53]

The Wrights received a reply from the Smithsonian rather sooner than they would now, just three days after Wilbur had written to them: records show that the reading list was despatched on 2 June 1899. The brothers set about studying the problem of flight in their usual methodical way. They immediately grasped that it wasn't enough to read books and watch birds – they had to get up into the air themselves. Therefore they started their practical researches by building a glider. It was ready by September 1900, and they took it to Kitty Hawk, North Carolina, the nearest place to their home that had constant and satisfactory winds. In all, they built three gliders between 1900 and 1902, a sound commercial move that enabled them to perfect wing shape and to develop the rear rudder, another of their contributions to aeronautical technology.[54] In fact, they made such good progress that by the beginning of 1903 they thought they were ready to try powered flight. As a source of power, there was only one option: the internal combustion engine. This had been invented in the late 1880s, yet by 1903 the brothers could find no engine light enough to fit onto an aircraft. They had no choice but to design their own. On 23 September 1903, they set off for Kitty Hawk with their new aircraft in crates. Because of unanticipated delays – broken propeller shafts and repeated weather problems (rain, storms, biting winds) – they were not ready to fly until 11 December. But then the wind wasn't right until the fourteenth. A coin was tossed to see who was to make the first flight, and Wilbur won. On this first occasion, the *Flyer* climbed too steeply, stalled, and crashed into the sand. On the seventeenth, after Orville's triumph, the landings were much gentler, enabling three more flights to be made that day.[55] It was a truly historic moment, and given the flying revolution that we now take so much for granted, one might have expected the Wrights' triumph to be front-page news. Far from it. There had been so many crackpot schemes that newspapers and the public were thoroughly sceptical about flying machines. In 1904, even though the Wrights made 105 flights, they spent only forty-five minutes in the air and made only two five-minute flights. The U.S. government turned down three offers of an aircraft from the Wrights without making any effort to verify the brothers' claims. In 1906 no airplanes were constructed, and neither Wilbur nor Orville left the ground even once. In 1907 they tried to sell their invention in Britain, France, and Germany. All attempts failed. It was not until 1908 that the U.S. War Department at last accepted a bid from the Wrights; in the same year, a contract was signed for the formation of a French company.[56] It had taken four and a half years to sell this revolutionary concept.

The principles of flight could have been discovered in Europe. But the Wright brothers were raised in that practical culture described by Richard Hofstadter, which played a part in their success. In a similar vein a group of painters later called the Ashcan school, on account of their down-to-earth subject matter, shared a similar pragmatic and reportorial approach to their art. Whereas the cubists, Fauves, and abstractionists concerned themselves with theories of beauty or the fundamentals of reality and matter, the **Ashcan school** painted the new landscape around them in vivid detail, accurately portraying what was often an ugly world. Their vision (they didn't really share a style) was laid out at a groundbreaking exhibition at the Macbeth Gallery in New York.[57]

The leader of the Ashcan school was **Robert Henri** (1865–1929), descended from French Huguenots who had escaped to Holland during the Catholic massacres of the late sixteenth century.[58] Worldly, a little wild, Henri, who visited Paris in 1888, became a natural magnet for other artists in Philadelphia, many of whom worked for the local press: **John Sloan, William Glackens, George Luks**.[59] Hard-drinking, poker playing, they had the newspaperman's eye for detail and a sympathy – sometimes a sentimentality – for the underdog. They met so often they called themselves Henri's Stock Company.[60] Henri later moved to the New York School of Art, where he taught George Bellows, Stuart Davis, Edward Hopper, Rockwell Kent, Man Ray, and Leon Trotsky. His influence was huge, and his approach embodied the view that the American people should 'learn the means of expressing themselves in their own time and in their own land.'[61]

The most typical Ashcan school art was produced by John Sloan (1871–1951), George Luks (1867–1933), and George Bellows (1882–1925). An illustrator for the *Masses*, a left-wing periodical of social commentary that included John Reed among its contributors, Sloan sought what he called 'bits of joy' in New York life, colour plucked from the grim days of the working class: a few moments of rest on a ferry, a girl stretching at the window of a tenement, another woman smelling the washing on the line – all the myriad ways that ordinary people seek to blunt, or even warm, the sharp, cold life at the bottom of the pile.[62]

George Luks and George Bellows, an anarchist, were harsher, less sentimental.[63] Luks painted New York crowds, the teeming congestion in its streets and neighbourhoods. Both he and Bellows frequently represented the boxing and wrestling matches that were such a feature of working-class life and so typical of the raw, naked struggle among the immigrant communities. Here was life on the edge in every way. Although prize fighting was illegal in New York in the 1900s, it nonetheless continued. Bellows's painting *Both Members of This Club*, originally entitled *A Nigger and a White Man*, reflected the concern that many had at the time about the rise of the blacks within sports: 'If the Negro could beat the white, what did that say about the Master Race?'[64] Bellows, probably the most talented painter of the school, also followed the building of Penn Station, the construction of which, by McKim, Mead and White, meant boring a tunnel halfway under Manhattan and the demolition of four entire city blocks between Thirty-first and Thirty-third Streets. For years there was a huge crater in the centre of New York, occupied by steam shovels and other industrial appliances, flames and smoke and hundreds of workmen. Bellows transformed these grimy details into things of beauty.[65]

The achievement of the Ashcan School was to pinpoint and report the raw side of New York immigrant life. Although at times these artists fixed on fleeting beauty with a generally uncritical eye, their main aim was to show people at the bottom of the heap, not so much suffering, but making the most of what they had. Henri also taught a number of painters who would, in time, become leading American abstractionists.[66]

At the end of 1903, in the same week that the Wright brothers made their first flight, and just two blocks from the Flatiron Building, the first celluloid print of *The Great Train Robbery* was readied in the offices of Edison Kinetograph, on Twenty-third Street. Thomas Edison was one of a handful of people in the United States, France, Germany, and Britain who had developed silent movies in the mid-1890s.

Between then and 1903 there had been hundreds of staged fictional films, though none had been as long as *The Great Train Robbery*, which lasted for all of six minutes. There had been chase movies before, too, many produced in Britain right at the end of the nineteenth century. But they used one camera to tell a simple story simply. *The Great Train Robbery*, directed and edited by **Edwin Porter**, was much more sophisticated and ambitious than anything that had gone before. The main reason for this was the way Porter told the story. Since its inception in France in 1895, when the Lumière brothers had given the first public demonstration of moving pictures, film had explored many different locations, to set itself apart from theatre. Cameras had been mounted on trains, outside the windows of ordinary homes, looking in, even underwater. But in *The Great Train Robbery*, in itself an ordinary robbery followed by a chase, Porter in fact told *two* stories, which he intercut. That's what made it so special. The telegraph operator is attacked and tied up, the robbery takes place, and the bandits escape. At intervals, however, the operator is shown struggling free and summoning law enforcement. Later in the film the two narratives come together as the posse chase after the bandits.[67] We take such 'parallel editing' – intercutting between related narratives – for granted now. At the time, however, people were fascinated as to whether film could throw light on the stream of consciousness, Bergson's notions of time, or Husserl's phenomenology. More practical souls were exercised because parallel editing added immeasurably to the psychological tension in the film, and it couldn't be done in the theatre.[68] In late 1903 the film played in every cinema in New York, all ten of them. It was also responsible for Adolph Zukor and Marcus Loew leaving their fur business and buying small theatres exclusively dedicated to showing movies. Because they generally charged a nickel for entry, they became known as 'nickelodeons.' Both William Fox and Sam Warner were fascinated enough by Porter's *Robbery* to buy their own movie theatres, though before long they each moved into production, creating the studios that bore their names.[69]

Porter's success was built on by another man who instinctively grasped that the intimate nature of film, as compared with the theatre, would change the relationship between audience and actor. It was this insight that gave rise to the idea of the movie star. **David Wark (D. W.) Griffith** was a lean man with grey eyes and a hooked nose. He appeared taller than he was on account of the high-laced hook shoes he wore, which had loops above their heels for pulling them on – his trouser bottoms invariably rode up on the loops. His collar was too big, his string tie too loose, and he liked to wear a large hat when large hats were no longer the fashion. He looked a mess, but according to many, he 'was touched by genius.' He was the son of a Confederate Kentucky colonel, 'Roaring Jake' Griffith, the only man in the army who, so it was said, could shout to a soldier five miles away.[70] Griffith had begun life as an actor but transferred to movies by selling story synopses (these were silent movies, so no scripts were necessary). When he was thirty-two he joined an early film outfit, the Biograph Company in Manhattan, and had been there about a year when **Mary Pickford** walked in. Born in Toronto in 1893, she was sixteen. Originally christened Gladys Smith, she was a precocious if delicate child. After her father was killed in a paddle-steamer accident, her mother, in reduced circumstances, had been forced to let the master bedroom of their home to a theatrical couple; the husband was a stage manager at a local theatre. This turned into Gladys's opportunity, for he persuaded Charlotte Smith to let her two daughters appear as extras. Gladys soon found she had talent and liked the life. By the time she was seven, she had moved to New York where, at $15 a week, the pay was better. She was now the major breadwinner of the family.[71]

In an age when the movies were as young as she, theatre life in New York was much more widespread. In 1901–2, for example, there were no fewer than 314 plays running on or off Broadway, and it was not hard for someone with Gladys's talent to find work. By the time she was twelve, her earnings were $40 a week. When she was fourteen she went on tour with a comedy, *The Warrens of Virginia*, and while she was in Chicago she saw her first film. She immediately grasped the possibilities of the new medium, and using her recently created and less harsh stage name Mary Pickford, she applied to several studios. Her first efforts failed, but her mother pushed her into applying for work at the Biograph. At first Griffith thought Mary Pickford was 'too little and too fat' for the movies. But he was impressed by her looks and her curls and asked her out for dinner; she refused.[72] It was only when he asked her to walk across the studio and chat with actors she hadn't met that he decided she might have screen appeal. In those days, movies were short and inexpensive to make. There was no such thing as a makeup assistant, and actors wore their own clothes (though by 1909 there had been some experimentation with lighting techniques). A director might make two or three pictures a week, usually on location in New York. In 1909, for example, Griffith made 142 pictures.[73]

After an initial reluctance, Griffith gave Pickford the lead in *The Violin-Maker of Cremona* in 1909.[74] A buzz went round the studio, and when it was first screened in the Biograph projection room, the entire studio turned up to watch. Pickford went on to play the lead in twenty-six more films before the year was out.

But Mary Pickford's name was not yet known. Her first review in the *New York Dramatic Mirror* of 21 August 1909 read, 'This delicious little comedy introduced again an ingenue whose work in Biograph pictures is attracting attention.' Mary Pickford was not named because all the actors in Griffith's movies were, to begin with, anonymous. But Griffith was aware, as this review suggests, that Pickford was attracting a following, and he raised her wages quietly from $40 to $100 a week, an unheard-of figure for a repertory actor at that time.[75] She was still only sixteen.

Three of the great innovations in filmmaking occurred in Griffith's studio. The first change came in the way movies were staged. Griffith began to direct actors to come on camera, not from right or left as they did in the theatre, but from *behind* the camera and exit toward it. They could therefore be seen in long range, medium range, and even close-up in the same shot. The close-up was vital in shifting the emphasis in movies to the looks of the actor as much as his or her talent. The second revolution occurred when Griffith hired another director. This allowed him to break out of two-day films and plan bigger projects, telling more complex stories. The third revolution built on the first and was arguably the most important.[76] Florence Lawrence, who was marketed as the 'Biograph Girl' before Mary, left for another company. Her contract with the new studio contained an unprecedented clause: anonymity was out; instead she would be billed under her own name, as the 'star' of her pictures. Details about this innovation quickly leaked all over the fledgling movie industry, with the result that it was not Lawrence who took the best advantage of the change she had wrought. Griffith was forced to accept a similar contract with Mary Pickford, and as 1909 gave way to 1910, she prepared to become the world's first movie star.[77]

A vast country, teeming with immigrants who did not share a common heritage, America was a natural home for the airplane and the mass-market movie, every bit as much as the skyscraper. The Ashcan school recorded the poverty that most immigrants endured when they arrived in the country, but it also epitomised the optimism with which most of the emigrés regarded their new home. The huge oceans on either side of the Americas helped guarantee that the United States was isolated from many of the irrational and hateful dogmas and idealisms of Europe which these immigrants were escaping. Instead of the grand, all-embracing ideas of Freud, Hofmannsthal, or Brentano, the mystical notions of Kandinsky, or the vague theories of Bergson, Americans preferred more practical, more limited ideas that worked, relishing the difference and isolation from Europe. That pragmatic isolation would never go away entirely. It was, in some ways, America's most precious asset.

6

$$E = mc^2, \; \supset / \equiv / \mathbf{v} + C_7H_{38}O_{43}$$

Pragmatism was an American philosophy, but it was grounded in empiricism, a much older notion, spawned in Europe. Although figures such as Nietzsche, Bergson, and Husserl became famous in the early years of the century, with their wide-ranging monistic and dogmatic theories of explanation (as William James would have put it), there were many scientists who simply ignored what they had to say and went their own way. It is a mark of the division of thought throughout the century that even as philosophers tried to adapt to science, science ploughed on, hardly looking over its shoulder, scarcely bothered by what the philosophers had to offer, indifferent alike to criticism and praise. Nowhere was this more apparent than in the last half of the first decade, when the difficult groundwork was completed in several hard sciences. ('Hard' here has two senses: first, intellectually difficult; second, concerning hard matters, the material basis of phenomena.) In stark contrast to Nietzsche and the like, these men concentrated their experimentation, and resulting theories, on very restricted aspects of the observable universe. That did not prevent their results having a much wider relevance, once they were accepted, which they soon were.

The best example of this more restricted approach took place in Manchester, England, on the evening of 7 March 1911. We know about the event thanks to **James Chadwick**, who was a student then but later became a famous physicist. A meeting was held at the Manchester Literary and Philosophical Society, where the audience was made up mainly of municipal worthies – intelligent people but scarcely specialists. These evenings usually consisted of two or three talks on diverse subjects, and that of 7 March was no exception. A local fruit importer spoke first, giving an account of how he had been surprised to discover a rare snake mixed in with a load of Jamaican bananas. The next talk was delivered by **Ernest Rutherford**, professor of physics at Manchester University, who introduced those present to what is certainly one of the most influential ideas of the entire century – the basic structure of the atom. How many of the group understood Rutherford is hard to say. He told his audience that the atom was made up of 'a central electrical charge concentrated at a point and surrounded by a uniform spherical distribution of opposite electricity equal in amount.' It sounds dry, but to Rutherford's colleagues and students present, it was the most exciting news they had ever heard. James Chadwick later said that he remembered the meeting all his life. It was, he wrote, 'a most shattering performance to us, young boys that we were. ... We realised that this was obviously the truth, this was it.'[1]

Such confidence in Rutherford's revolutionary ideas had not always been so evident. In the late 1890s Rutherford had developed the ideas of the French physicist **Henri Becquerel**. In turn, Becquerel had built on **Wilhelm Conrad Röntgen**'s discovery of **X rays**, which we encountered in chapter three. Intrigued by these mysterious rays that were given off from fluorescing glass, Becquerel, who, like his father and grandfather, was professor of physics at the Musée d'Histoire Naturelle in Paris, decided to investigate other substances that 'fluoresced.' Becquerel's classic experiment occurred by accident, when he sprinkled some uranyl potassium sulphate on a sheet of photographic paper and left it locked in a drawer for a few days. When he looked, he found the image of the salt on the paper. There had been no naturally occurring light to activate the paper, so the change must have been wrought by the uranium salt. Becquerel had discovered naturally occurring **radioactivity**.[2]

It was this result that attracted the attention of Ernest Rutherford. Raised in New Zealand, Rutherford was a stocky character with a weatherbeaten face who loved to bellow the words to hymns whenever he got the chance, a cigarette hanging from his lips. 'Onward Christian Soldiers' was a particular favourite. After he arrived in Cambridge in October 1895, he quickly began work on a series of experiments designed to elaborate Becquerel's results.[3] There were three naturally radioactive substances – uranium, radium, and thorium – and Rutherford and his assistant Frederick Soddy pinned their attentions on thorium, which gave off a radioactive gas. When they analysed the gas, however, Rutherford and Soddy were shocked to discover that it was completely inert – in other words, it wasn't thorium. How could that be? Soddy later described the excitement of those times in a memoir. He and Rutherford gradually realised that their results 'conveyed the tremendous and inevitable conclusion that the element thorium was spontaneously transmuting itself into [the chemically inert] argon gas!' This was the first of Rutherford's many important experiments: what he and Soddy had discovered was the spontaneous decomposition of the radioactive elements, a modern form of alchemy. The implications were momentous.[4]

This wasn't all. Rutherford also observed that when uranium or thorium decayed, they gave off two types of radiation. The weaker of the two he called 'alpha' radiation, later experiments showing that '**alpha particles**' were in fact helium atoms and therefore positively charged. The stronger '**beta radiation**', on the other hand, consisted of electrons with a negative charge. The electrons, Rutherford said, were 'similar in all respects to cathode rays.' So exciting were these results that in 1908 Rutherford was awarded the Nobel Prize at age thirty seven, by which time he had moved from Cambridge, first to Canada and then back to Britain, to Manchester, as professor of physics.[5] By now he was devoting all his energies to the alpha particle. He reasoned that because it was so much larger than the beta electron (the electron had almost no mass), it was far more likely to interact with matter, and that interaction would obviously be crucial to further understanding. If only he could think up the right experiments, the alpha might even tell him something about the structure of the atom. 'I was brought up to look at the atom as a nice hard fellow, red or grey in colour, according to taste,' he said.[6] That view had begun to change while he was in Canada, where he had shown that alpha particles sprayed through a narrow slit and projected in a beam could be deflected by a magnetic field. All these experiments were carried out with very basic equipment – that was the beauty of Rutherford's approach. But it was a refinement of this equipment that produced the next major breakthrough. In one of the many experiments he tried, he covered the slit with a very thin sheet of mica, a mineral that splits fairly naturally into slivers. The piece Rutherford placed over the slit in his experiment was so thin – about three-thousandths of an inch – that in theory at least alpha particles should have passed through it. They did, but not in quite the way Rutherford had expected. When the results of the spraying were 'collected' on photographic paper, the edges of the image appeared fuzzy. Rutherford could think of only one explanation for that: some of the particles were being deflected. That much was clear, but it was the *size* of the deflection that excited Rutherford. From his experiments with magnetic fields, he knew that powerful forces were needed to induce even small deflections. Yet his photographic paper showed that some alpha particles were being knocked off course by as much as two degrees. Only one thing could explain that. As Rutherford himself was to put it, 'the atoms of matter must be the seat of very intense electrical forces.'[7]

Science is not always quite the straight line it likes to think it is, and this result of Rutherford's, though surprising, did not automatically lead to further insights. Instead, for a time Rutherford and his new assistant, Ernest Marsden, went doggedly on, studying the behaviour of alpha particles, spraying them on to foils of different material – gold, silver, or aluminium.[8] Nothing notable was observed. But then Rutherford had an idea. He arrived at the laboratory one morning and 'wondered aloud' to Marsden whether (with the deflection result still in his mind) it might be an idea to bombard the metal foils with particles sprayed *at an angle*. The most obvious angle to start with was 45 degrees, which is what Marsden did, using foil made of gold. This simple experiment 'shook physics to its foundations.' It was 'a new view of nature ... the discovery of a new layer of reality, a new dimension of the universe.'[9] Sprayed at an angle of 45 degrees, the alpha particles did not pass *through* the gold foil – instead they were bounced back by 90 degrees onto the zinc sulphide screen. 'I remember well reporting the result to Rutherford,' Marsden wrote in a memoir, 'when I met him on the steps leading to his private room, and the joy with which I told him.'[10] Rutherford was quick to grasp what Marsden had already worked out: for such a deflection to occur, a massive amount of energy must be locked up somewhere in the equipment used in their simple experiment.

But for a while Rutherford remained mystified. 'It was quite the most incredible event that has ever happened to me in my life,' he wrote in his autobiography. 'It was almost as incredible as if you fired a 15-inch shell at a piece of tissue paper and it came back and hit you. On consideration I realised that this scattering backwards must be the result of a single collision, and when I made calculations I saw that it was impossible to get anything of that order of magnitude unless you took a system in which the greatest part of the mass of the atom was concentrated in a minute **nucleus**.'[11] In fact, he brooded for months before feeling confident he was right. One reason was because he was slowly coming to terms with the fact that the idea of the atom he had grown up with – J. J. Thomson's notion that it was a miniature plum pudding, with electrons dotted about like raisins – would no longer do.[12] Gradually he became convinced that another model entirely was far more likely. He made an analogy with the heavens: the nucleus of the atom was orbited by electrons just as planets went round the stars.

As a theory, the planetary model was elegant, much more so than the 'plum pudding' version. But was it correct? To test his theory, Rutherford suspended a large magnet from the ceiling of his laboratory. Directly underneath, on a table, he fixed another magnet. When the pendulum magnet was swung over the table at a 45-degree angle and when the magnets were matched in polarity, the swinging magnet bounced through 90 degrees just as the alpha particles did when they hit the gold foil. His theory had passed the first test, and atomic physics had now become nuclear physics.[13]

For many people, particle physics has been the greatest intellectual adventure of the century. But in some respects there have been two sides to it. One side is exemplified by Rutherford, who was brilliantly adept at thinking up often very simple experiments to prove or disprove the latest advance in theory. The other project has been *theoretical* physics, which involved the imaginative use of already existing information to be reorganised so as to advance knowledge. Of course, experimental physics and theoretical physics are intimately related; sooner or later, theories have to be tested. Nonetheless, within the discipline of physics overall, theoretical physics is recognised as an activity in its own right, and for many perfectly respectable physicists theoretical work is all they do. Often the experimental verification of theories in physics cannot be tested for years, because the technology to do so doesn't exist.

The most famous theoretical physicist in history, indeed one of the most famous figures of the century, was developing his theories at more or less the same time that Rutherford was conducting his experiments. **Albert Einstein** arrived on the intellectual stage with a bang. Of all the scientific journals in the world, the single most sought-after collector's item by far is the *Annalen der Physik*, volume XVII, for 1905, for in that year Einstein published not one but three papers in the journal, causing 1905 to be dubbed the annus mirabilis of science. These three papers were: the first experimental verification of Max Planck's quantum theory; Einstein's examination of Brownian motion, which proved the existence of molecules; and the special theory of relativity with its famous equation, **$E=mc^2$**.

Einstein was born in Ulm, between Stuttgart and Munich, on 14 March 1879, in the valley of the Danube near the slopes that lead to the Swabian

Alps. Hermann, his father, was an electrical engineer. Though the birth was straightforward, Einstein's mother Pauline received a shock when she first saw her son: his head was large and so oddly shaped, she was convinced he was deformed.[14] In fact there was nothing wrong with the infant, though he did have an unusually large head. According to family legend, Einstein was not especially happy at elementary school, nor was he particularly clever.[15] He later said that he was slow in learning to talk because he was 'waiting' until he could deliver fully formed sentences. In fact, the family legend was exaggerated. Research into Einstein's early life shows that at school he always came top, or next to top, in both mathematics and Latin. But he did find enjoyment in his own company and developed a particular fascination with his building blocks. When he was five, his father gave him a compass. This so excited him, he said, that he 'trembled and grew cold.'[16]

Though Einstein was not an only child, he was fairly solitary by nature and independent, a trait that was encouraged by his parents' habit of encouraging self-reliance in their children at a very early age. Albert, for instance, was only three or four when he was given the responsibility of running errands, alone in the busy streets of Munich.[17] The Einsteins encouraged their children to develop their own reading, and while studying math at school, Albert was discovering Kant and Darwin for himself at home – very advanced for a child.[18] This did, however, help transform him from being a quiet child into a much more 'difficult' and rebellious adolescent. His character was only part of the problem here. He hated the autocratic approach used in his school, as he hated the autocratic side of Germany in general. This showed itself politically, in Germany as in Vienna, in a crude nationalism and a vicious anti-Semitism. Uncomfortable in such a psychological climate, Einstein argued incessantly with his fellow pupils and teachers, to the point where he was expelled, though he was thinking of leaving anyway. Aged sixteen he moved with his parents to Milan, attended university in Zurich at nineteen, though later he found a job as a patent officer in Bern. And so, half educated and half-in and half-out of academic life, he began in 1901 to publish scientific papers. His first, on the nature of liquid surfaces, was, in the words of one expert, 'just plain wrong.' More papers followed in 1903 and 1904. They were interesting but still lacked something – Einstein did not, after all, have access to the latest scientific literature and either repeated or misunderstood other people's work. However, one of his specialities was statistical techniques, which stood him in good stead later on. More important, the fact that he was out of the mainstream of science may have helped his originality, which flourished unexpectedly in 1905. One says unexpectedly, so far as Einstein was concerned, but in fact, at the end of the nineteenth century many other mathematicians and physicists – Ludwig Boltzmann, Ernst Mach, and Jules-Henri Poincaré among them – were inclining towards something similar. Relativity, when it came, both was and was not a total surprise.[19]

Einstein's three great papers of that marvellous year were published in March, on quantum theory, in May, on **Brownian motion**, and in June, on the special theory of relativity. Quantum physics, as we have seen, was itself new, the brainchild of the German physicist Max Planck. Planck argued that light is a form of electromagnetic radiation, made up of small packets or bundles – what he called quanta. Though his original paper caused little stir when it was read to the Berlin Physics Society in December 1900, other scientists soon realised that Planck must be right: his idea explained so much, including the observation that the chemical world is made up of discrete units – the elements. Discrete elements implied fundamental units of matter that were themselves discrete. Einstein paid Planck the compliment of thinking through other implications of his theory, and came to agree that light really does exist in discrete units – **photons**. One of the reasons why scientists other than Einstein had difficulty accepting this idea of quanta was that for years experiments had shown that light possesses the qualities of a wave. In the first of his papers Einstein, showing early the openness of mind for which physics would become celebrated as the decades passed, therefore made the hitherto unthinkable suggestion that light was *both*, a wave at some times and a particle at others. This idea took some time to be accepted, or even understood, except among physicists, who realised that Einstein's insight fitted the available facts. In time the **wave-particle duality**, as it became known, formed the basis of quantum mechanics in the 1920s. (If you are confused by this, and have difficulty visualising something that is both a particle and a wave, you are in good company. We are dealing here with qualities that are essentially mathematical, and all visual analogies will be inadequate. Niels Bohr, arguably one of the century's top two physicists, said that anyone who wasn't made 'dizzy' by the very idea of what later physicists called 'quantum weirdness' had lost the plot.)

Two months after his paper on quantum theory, Einstein published his second great work, on Brownian motion.[20] Most people are familiar with this phenomenon from their school days: when suspended in water and inspected under the microscope, small grains of pollen, no more than a hundredth of a millimetre in size, jerk or zigzag backward and forward. Einstein's idea was that this 'dance' was due to the pollen being bombarded by molecules of water hitting them at random. If he was right, Einstein said, and molecules were bombarding the pollen at random, then some of the grains should not remain stationary, their movement cancelled out by being bombarded from all sides, but should move at a certain pace through the water. Here his knowledge of statistics paid off, for his complex calculations were borne out by experiment. This was generally regarded as the first proof that molecules exist.

But it was Einstein's third paper that year, the one on the special theory of relativity, published in June, that would make him famous. It was this theory which led to his conclusion that $E=mc^2$. It is not easy to explain the **special theory of relativity** (the general theory came later) because it deals with extreme – but fundamental – circumstances in the universe, where common sense breaks down. However, a thought experiment might help.[21] Imagine you are standing at a railway station when a train hurtles through from left to right. At the precise moment that someone else on the train passes you, a light on the train, in the middle of a carriage, is switched on. Now, assuming the train is transparent, so you can see inside, you, as the observer on the platform, will see that by the time the light beam reaches the back of the carriage, the carriage will have moved forward. In other words, that light beam has travelled slightly less than half the length of the carriage. However, the person inside the train will see the light beam hitting the back of the carriage at the same time as it hits the front of the carriage, because to that person it has travelled exactly half the length of the carriage. Thus the time the light beam takes to reach the back of the carriage is different for the two observers. But it is the same light beam in each case, travelling at the same speed. The discrepancy, Einstein said, can only be explained by assuming that the perception is relative to the observer and that, because the speed of light is constant, time must change according to circumstance.

The idea that time can slow down or speed up is very strange, but that is exactly what Einstein was suggesting. A second thought experiment, suggested by Michael White and John Gribbin, Einstein's biographers, may help. Imagine a pencil with a light upon it, casting a shadow on a tabletop. The pencil, which exists in three dimensions, casts a shadow, which exists in two, on the tabletop. As the pencil is twisted in the light, or if the light is moved around the pencil, the shadow grows or shrinks. Einstein said in effect that objects essentially have four dimensions in addition to the three we are all familiar with – they occupy space-time, as it is now called, in that the same object lasts over time.[22] And so if you play with a four-dimensional object the way we played with the pencil, then you can shrink and extend time, the way the pencil's shadow was shortened and extended. When we say 'play' here, we are talking about some hefty tinkering; in Einstein's theory, objects are required to move at or near the speed of light before his effects are shown. But when they do, Einstein said, time really does change. His most famous prediction was that clocks would move more slowly when travelling at high speeds. This anti-commonsense notion was actually borne out by experiment many years later. Although there might be no immediate practical benefit from his ideas, physics was transformed.[23]

Chemistry was transformed, too, at much the same time, and arguably with much more benefit for mankind, though the man who effected that transformation did not achieve anything like the fame of Einstein. In fact, when the scientist concerned revealed his breakthrough to the press, his name was left off the headlines. Instead, the *New York Times* ran what must count as one of the strangest headlines ever: 'HERE'S TO $C_7H_{38}O_{43}$.'[24] That formula gave the chemical composition for plastic, probably the most widely used substance in the world today. Modern life – from airplanes to telephones to television to computers – would be unthinkable without it. The man behind the discovery was **Leo Hendrik Baekeland**.

Baekeland was Belgian, but by 1907, when he announced his breakthrough, he had lived in America for nearly twenty years. He was an individualistic and self-confident man, and plastic was by no means the first of his inventions, which included a photosensitive paper called Velox, which he sold to the Eastman Company for \$750,000 (about \$40 million now) and the Townsend

Cell, which successfully electrolysed brine to produce caustic soda, crucial for the manufacture of soap and other products.[25]

The search for a synthetic plastic was hardly new. Natural plastics had been used for centuries: along the Nile, the Egyptians varnished their sarcophagi with resin; jewellery of amber was a favourite of the Greeks; bone, shell, ivory, and rubber were all used. In the nineteenth century shellac was developed and found many applications, such as with phonograph records and electrical insulation. In 1865 Alexander Parkes introduced the Royal Society of Arts in London to Parkesine, the first of a series of plastics produced by trying to modify nitrocellulose.[26] More successful was **celluloid**, camphor gum mixed with pyroxyline pulp and made solvent by heating, especially as the basis for false teeth. In fact, the invention of celluloid brought combs, cuffs, and collars within reach of social groups that had hitherto been unable to afford such luxuries. There were, however, some disturbing problems with celluloid, notably its flammability. In 1875 a *New York Times* editorial summed up the problem with the alarming headline 'Explosive Teeth.'[27]

The most popular avenue of research in the 1890s and 1900s was the admixture of phenol and formaldehyde. Chemists had tried heating every combination imaginable to a variety of temperatures, throwing in all manner of other compounds. The result was always the same: a gummy mixture that was never quite good enough to produce commercially. These gums earned the dubious honour of being labelled by chemists as the 'awkward resins.'[28] It was the very awkwardness of these substances that piqued Baekeland's interest.[29] In 1904 he hired an assistant, Nathaniel Thurlow, who was familiar with the chemistry of phenol, and they began to look for a pattern among the disarray of results. Thurlow made some headway, but the breakthrough didn't come until 18 June 1907. On that day, while his assistant was away, Baekeland took over, starting a new laboratory notebook. Four days later he applied for a patent for a substance he at first called '**Bakalite**.'[30] It was a remarkably swift discovery.

Reconstructions made from the meticulous notebooks Baekeland kept show that he had soaked pieces of wood in a solution of phenol and formaldehyde in equal parts, and heated it subsequently to 140–150°C. What he found was that after a day, although the surface of the wood was not hard, a small amount of gum had oozed out that was very hard. He asked himself whether this might have been caused by the formaldehyde evaporating before it could react with the phenol.[31] To confirm this he repeated the process but varied the mixtures, the temperature, the pressure, and the drying procedure. In doing so, he found no fewer than four substances, which he designated A, B, C, and D. Some were more rubbery than others; some were softened by heating, others by boiling in phenol. But it was mixture D that excited him.[32] This variant, he found, was 'insoluble in all solvents, does not soften. I call it Bakalite and it is obtained by heating A or B or C in closed vessels.'[33] Over the next four days Baekeland hardly slept, and he scribbled more than thirty-three pages of notes. During that time he confirmed that in order to get D, products A, B, and C needed to be heated well above 100°C, and that the heating had to be carried out in sealed vessels, so that the reaction could take place under pressure. Wherever it appeared, however, substance D was described as 'a nice smooth ivory-like mass.'[34] The Bakalite patents were filed on 13 July 1907. Baekeland immediately conceived all sorts of uses for his new product – insulation, moulding materials, a new linoleum, tiles that would keep warm in winter. In fact, the first objects to be made out of Bakalite were billiard balls, which were on sale by the end of that year. They were not a great success, though, as the balls were too heavy and not elastic enough. Then, in January 1908, a representative of the Loando Company from Boonton, New Jersey, visited Baekeland, interested in using Bakelite, as it was now called, to make precision bobbin ends that could not be made satisfactorily from rubber asbestos compounds.[35] From then on, the account book, kept by Baekeland's wife to begin with (although they were already millionaires), shows a slow increase in sales of Bakelite in the course of 1908, with two more firms listed as customers. In 1909, however, sales rose dramatically. One event that helps explain this is a lecture Baekeland gave on the first Friday in February that year to the New York section of the American Chemical Society at its building on the corner of Fourteenth Street and Fifth Avenue.[36] It was a little bit like a rerun of the Manchester meeting where Rutherford outlined the structure of the atom, for the meeting didn't begin until after dinner, and Baekeland's talk was the third item on the agenda. He told the meeting that substance D was a polymerised oxy-benzyl-methylene-glycol-anhydride, or $n(C_7H_{38}O_{43})$. It was past 10:00 P.M. by the time he had finished showing his various samples, demonstrating the qualities of Bakelite, but even so the assembled chemists gave him a standing ovation. Like James Chadwick attending Rutherford's talk, they realised they had been present at something important. For his part, Baekeland was so excited he couldn't sleep afterward and stayed up in his study at home, writing a ten-page account of the meeting. Next day three New York papers carried reports of the meeting, which is when the famous headline appeared.[37]

The first **plastic** (in the sense in which the word is normally used) arrived exactly on cue to benefit several other changes then taking place in the world. The electrical industry was growing fast, as was the automotive industry.[38] Both urgently needed insulating materials. The use of electric lighting and telephone services was also spreading, and the phonograph had proved more popular than anticipated. In the spring of 1910 a prospectus was drafted for the establishment of a Bakelite company, which opened its offices in New York six months later on 5 October.[39] Unlike the Wright brothers' airplane, in commercial terms Bakelite was an immediate success.

Bakelite evolved into plastic, without which computers, as we know them today, would probably not exist. At the same time that this 'hardware' aspect of the modern world was in the process of formation, important elements of the 'software' were also gestating, in particular the exploration of the logical basis for mathematics. The pioneers here were **Bertrand Russell** and **Alfred North Whitehead**.

Russell – slight and precise, a finely boned man, 'an aristocratic sparrow' –

is shown in Augustus John's portrait to have had piercingly sceptical eyes, quizzical eyebrows, and a fastidious mouth. The godson of the philosopher John Stuart Mill, he was born halfway through the reign of Queen Victoria, in 1872, and died nearly a century later, by which time, for him as for many others, nuclear weapons were the greatest threat to mankind. He once wrote that 'the search for knowledge, unbearable pity for suffering and a longing for love' were the three passions that had governed his life. 'I have found it worth living,' he concluded, 'and would gladly live it again if the chance were offered me.'[40]

One can see why. John Stuart Mill was not his only famous connection – T. S. Eliot, Lytton Strachey, G. E. Moore, Joseph Conrad, D. H. Lawrence, Ludwig Wittgenstein, and Katherine Mansfield were just some of his circle. Russell stood several times for Parliament (but was never elected), championed Soviet Russia, won the Nobel Prize for Literature in 1950, and appeared (sometimes to his irritation) as a character in at least six works of fiction, including books by Roy Campbell, T. S. Eliot, Aldous Huxley, D. H. Lawrence, and Siegfried Sassoon. When Russell died in 1970 at the age of ninety-seven there were more than sixty of his books still in print.[41]

But of all his books the most original was the massive tome that appeared first in 1910, entitled, after a similar work by Isaac Newton, *Principia Mathematica*. This book is one of the least-read works of the century. In the first place it is about mathematics, not everyone's favourite reading. Second, it is inordinately long – three volumes, running to more than 2,000 pages. But it was the third reason which ensured that this book – which indirectly led to the birth of the computer – was read by only a very few people: it consists mostly of a tightly knit argument conducted not in everyday language but by means of a specially invented set of symbols. Thus 'not' is represented by a curved bar; a boldface $\mathbf{v}$ stands for 'or'; a square dot means 'and,' while other logical relationships are shown by devices such as a U on its side ($\supset$) for 'implies,' and a three-barred equals sign ($\equiv$) for 'is equivalent to.' The book was ten years in the making, and its aim was nothing less than to explain the logical foundations of mathematics.

Such a feat clearly required an extraordinary author. Russell's education was unusual from the start. He was given a private tutor who had the distinction of being agnostic; as if that were not adventurous enough, this tutor also introduced his charge first to Euclid, then, in his early teens, to Marx. In December 1889, at the age of seventeen, Russell went to Cambridge. It was an obvious choice, for the only passion that had been observed in the young man was for mathematics, and Cambridge excelled in that discipline. Russell loved the certainty and clarity of math. He found it as 'moving' as poetry, romantic love, or the glories of nature. He liked the fact that the subject was totally uncontaminated by human feelings. 'I like mathematics,' he wrote, 'because it is *not* human & has nothing particular to do with this planet or with the whole accidental universe – because, like Spinoza's God, it won't love us in return.' He called Leibniz and Spinoza his 'ancestors.'[42]

At Cambridge, Russell attended Trinity College, where he sat for a scholarship. Here he enjoyed good fortune, for his examiner was Alfred North Whitehead. Just twenty-nine, Whitehead was a kindly man (he was known in Cambridge as 'cherub'), already showing signs of the forgetfulness for which he later became notorious. No less passionate about mathematics than Russell, he displayed his emotion in a somewhat irregular way. In the scholarship examination, Russell came second; a young man named Bushell gained higher marks. Despite this, Whitehead convinced himself that Russell was the abler man – and so burned all of the examination answers, and his own marks, before meeting the other examiners. Then he recommended Russell.[43] Whitehead was pleased to act as mentor for the young freshman, but Russell also fell under the spell of G. E. Moore, the philosopher. Moore, regarded as 'very beautiful' by his contemporaries, was not as witty as Russell but instead a patient and highly impressive debater, a mixture, as Russell once described him, of 'Newton and Satan rolled into one.' The meeting between these two men was hailed by one scholar as a 'landmark in the development of modern ethical philosophy.'[44]

Russell graduated as a 'wrangler,' as first-class mathematics degrees are known at Cambridge, but if this makes his success sound effortless, that is misleading. Russell's finals so exhausted him (as had happened with Einstein) that afterward he sold all his mathematical books and turned with relief to philosophy.[45] He said later he saw philosophy as a sort of no-man's-land between science and theology. In Cambridge he developed wide interests (one reason he found his finals tiring was because he left his revision so late, doing other things). Politics was one of those interests, the socialism of Karl Marx in particular. That interest, plus a visit to Germany, led to his first book, *German Social Democracy*. This was followed by a book on his 'ancestor' Leibniz, after which he returned to his degree subject and began to write *The Principles of Mathematics*.

Russell's aim in *Principles* was to advance the view, relatively unfashionable for the time, that mathematics was based on logic and 'derivable from a number of fundamental principles which were themselves logical.'[46] He planned to set out his own philosophy of logic in the first volume and then in the second explain in detail the mathematical consequences. The first volume was well received, but Russell had hit a snag, or as it came to be called, a paradox of logic. In *Principles* he was particularly concerned with 'classes.' To use his own example, all teaspoons belong to the class of teaspoons. However, the class of teaspoons is not itself a teaspoon and therefore does not belong to the class. That much is straightforward. But then Russell took the argument one step further: take the class of all classes that do not belong to themselves – this might include the class of elephants, which is not an elephant, or the class of doors, which is not a door. Does the class of all classes that do not belong to themselves belong to itself? Whether you answer yes or no, you encounter a contradiction.[47] Neither Russell nor Whitehead, his mentor, could see a way around this, and Russell let publication of *Principles* go ahead without tackling the paradox. 'Then, and only then,' writes one of his biographers, 'did there take place an event which gives the story of mathematics one of its moments of high drama.' In the 1890s Russell had read *Begriffsschrift* ('Concept-Script'), by the German mathematician **Gottlob Frege**, but had failed to understand it. Late in 1900 he bought the first volume of the same author's *Grundgesetze der Arithmetik* (Fundamental Laws of Arithmetic) and realised to his shame and horror that Frege had anticipated the paradox, and also failed to find a solution. Despite these problems, when *Principles* appeared in 1903 – all 500 pages of it – the book was the first comprehensive treatise on the logical foundation of mathematics to be written in English.[48]

The manuscript for *Principles* was finished on the last day of 1900. In the final weeks, as Russell began to think about the second volume, he became aware that Whitehead, his former examiner and now his close friend and colleague, was working on the second volume of *his* book *Universal Algebra*. In conversation, it soon became clear that they were both interested in the same problems, so they decided to collaborate. No one knows exactly when this began, because Russell's memory later in his life was a good deal less than perfect, and Whitehead's papers were destroyed by his widow, Evelyn. Her behaviour was not as unthinking or shocking as it may appear. There are strong grounds for believing that Russell had fallen in love with the wife of his collaborator, after his marriage to Alys Pearsall Smith collapsed in 1900.[49]

The collaboration between Russell and Whitehead was a monumental affair. As well as tackling the very foundations of mathematics, they were building on the work of Giuseppe Peano, professor of mathematics at Turin University, who had recently composed a new set of symbols designed to extend existing algebra and explore a greater range of logical relationships than had hitherto been specifiable. In 1900 Whitehead thought the project with Russell would take a year.[50] In fact, it took ten. Whitehead, by general consent, was the cleverer mathematician; he thought up the structure of the book and designed most of the symbols. But it was Russell who spent between seven and ten hours a day, six days a week, working on it.[51] Indeed, the mental wear and tear was on occasions dangerous. 'At the time,' Russell wrote later, 'I often wondered whether I should ever come out at the other end of the tunnel in which I seemed to be. ... I used to stand on the footbridge at Kennington, near Oxford, watching the trains go by, and determining that tomorrow I would place myself under one of them. But when the morrow came I always found myself hoping that perhaps "**Principia Mathematica**" would be finished some day.'[52] Even on Christmas Day 1907, he worked seven and a half hours on the book. Throughout the decade, the work dominated both men's lives, with the Russells and the Whiteheads visiting each other so the men could discuss progress, each staying as a paying guest in the other's house. Along the way, in 1906, Russell finally solved the paradox with his theory of types. This was in fact a logico-philosophical rather than a purely logical solution. There are two ways of knowing the world, Russell said: acquaintance (spoons) and description (the class of spoons), a sort of secondhand knowledge. From this, it follows that a description about a description is of a higher order than the description it is about. On this analysis, the paradox simply disappears.[53]

Slowly the manuscript was compiled. By May 1908 it had grown to 'about

6,000 or 8,000 pages.'[54] In October, Russell wrote to a friend that he expected it to be ready for publication in another year. 'It will be a very big book,' he said, and 'no one will read it.'[55] On another occasion he wrote, 'Every time I went for a walk I used to be afraid that the house would catch fire and the manuscript get burnt up.'[56] By the summer of 1909 they were on the last lap, and in the autumn Whitehead began negotiations for publication. 'Land in sight at last,' he wrote, announcing that he was seeing the Syndics of the Cambridge University Press (the authors carried the manuscript to the printers on a four-wheeled cart). The optimism was premature. Not only was the book very long (the final manuscript was 4,500 pages, almost the same size as Newton's book of the same title), but the alphabet of **symbolic logic** in which it was half written was unavailable in any existing printing font. Worse, when the Syndics considered the market for the book, they came to the conclusion that it would lose money – around £600. The press agreed to meet 50 percent of the loss, but said they could publish the book only if the Royal Society put up the other £300. In the event, the Royal Society agreed to only £200, and so Russell and Whitehead between them provided the balance. 'We thus earned minus £50 each by ten years' work,' Russell commented. 'This beats "Paradise Lost." '[57]

Volume 1 of *Principia Mathematica* appeared in December 1910, volume 2 in 1912, volume 3 in 1913. General reviews were flattering, the *Spectator* concluding that the book marked 'an epoch in the history of speculative thought' in the attempt to make mathematics 'more solid' than the universe itself.[58] However, only 320 copies had been sold by the end of 1911. The reaction of colleagues both at home and abroad was awe rather than enthusiasm. The theory of logic explored in volume 1 is still a live issue among philosophers, but the rest of the book, with its hundreds of pages of formal proofs (page 86 proves that $1+1=2$), is rarely consulted. 'I used to know of only six people who had read the later parts of the book,' Russell wrote in the 1950s. 'Three of these were Poles, subsequently (I believe) liquidated by Hitler. The other three were Texans, subsequently successfully assimilated.'[59]

Nevertheless, Russell and Whitehead had discovered something important: that most mathematics – if not all of it – could be derived from a number of axioms logically related to each other. This boost for mathematical logic may have been their most important legacy, inspiring such figures as Alan Turing and John von Neumann, mathematicians who in the 1930s and 1940s conceived the early computers. It is in this sense that Russell and Whitehead are the grandfathers of software.[60]

In 1905 in the British medical periodical the *Lancet*, E. H. Starling, professor of physiology at University College, London, introduced a new word into the medical vocabulary, one that would completely change the way we think about our bodies. That word was **hormone**. Professor Starling was only one of many doctors then interested in a new branch of medicine concerned with 'messenger substances.' Doctors had been observing these substances for decades, and countless experiments had confirmed that although the body's ductless glands –

the thyroid in the front of the neck, the pituitary at the base of the brain, and the adrenals in the lower back – manufactured their own juices, they had no apparent means to transport these substances to other parts of the body. Only gradually did the physiology become clear. For example, at Guy's Hospital in London in 1855, Thomas Addison observed that patients who died of a wasting illness now known as Addison's Disease had adrenal glands that were diseased or had been destroyed.[61] Later Daniel Vulpian, a Frenchman, discovered that the central section of the adrenal gland stained a particular colour when iodine or ferric chloride was injected into it; and he also showed that a substance that produced the same colour reaction was present in blood that drained away from the gland. Later still, in 1890, two doctors from Lisbon had the ostensibly brutal idea of placing half of a sheep's thyroid gland under the skin of a woman whose own gland was deficient. They found that her condition improved rapidly. Reading the Lisbon report, a British physician in Newcastle-upon-Tyne, George Murray, noticed that the woman began her improvement as early as the day after the operation and concluded that this was too soon for blood vessels to have grown, connecting the transplanted gland. Murray therefore concluded that the substance secreted by the gland must have been absorbed directly into the patient's bloodstream. Preparing a solution by crushing the gland, he found that it worked almost as well as the sheep's thyroid for people suffering from thyroid deficiency.[62]

The evidence suggested that messenger substances were being secreted by the body's ductless glands. Various laboratories, including the Pasteur Institute in New York and the medical school of University College in London, began experimenting with extracts from glands. The most important of these trials was conducted by George Oliver and E. A. Sharpy-Shafer at University College, London, in 1895, during which they found that the 'juice' obtained by crushing adrenal glands made blood pressure go up. Since patients suffering from Addison's disease were prone to have low blood pressure, this confirmed a link between the gland and the heart. This messenger substance was named adrenaline. John Abel, at Johns Hopkins University in Baltimore, was the first person to identify its chemical structure. He announced his breakthrough in June 1903 in a two-page article in the *American Journal of Physiology*. The chemistry of adrenaline was surprisingly straightforward; hence the brevity of the article. It comprised only a small number of molecules, each consisting of just twenty-two atoms.[63] It took a while for the way **adrenaline** worked to be fully understood and for the correct dosages for patients to be worked out. But adrenaline's discovery came not a moment too soon. As the century wore on, and thanks to the stresses of modern life, more and more people became prone to heart disease and blood pressure problems.

At the beginning of the twentieth century people's health was still dominated by a 'savage trinity' of diseases that disfigured the developed world: tuberculosis, alcoholism, and **syphilis**, all of which proved intractable to treatment for many years. TB lent itself to drama and fiction. It afflicted the young as well as the old, the well-off and the poor, and it was for the most part a slow, lingering death – as consumption it features in *La Bohème*, *Death in Venice*, and *The Magic Mountain*. Anton Chekhov, Katherine Mansfield, and Franz Kafka all died of the disease. Alcoholism and syphilis posed acute problems because they were not simply constellations of symptoms to be treated but the charged centre of conflicting beliefs, attitudes, and myths that had as much to do with morals as medicine. Syphilis, in particular, was caught in this moral maze.[64]

The fear and moral disapproval surrounding syphilis a century ago mingled so much that despite the extent of the problem, it was scarcely talked about. Writing in the *Journal of the American Medical Association* in October 1906, for example, one author expressed the view that 'it is a greater violation of the proprieties of public life publicly to mention venereal disease than privately to contract it.'[65] In the same year, when Edward Bok, editor of the *Ladies' Journal*, published a series of articles on venereal diseases, the magazine's circulation slumped overnight by 75,000. Dentists were sometimes blamed for spreading the disease, as was the barber's razor and wet nurses. Some argued it had been brought back from the newly discovered Americas in the sixteenth century; in France a strong strand of anticlericalism blamed 'holy water.'[66] Prostitution didn't help keep track of the disease either, nor Victorian medical ethics that prevented doctors from telling one fiancée anything about the other's infections unless the sufferer allowed it. On top of it all, no one knew whether syphilis was hereditary or congenital. Warnings about syphilis sometimes verged on the hysterical. *Vénus*, a 'physiological novel,' appeared in 1901, the same year as a play called *Les Avariés* (The Rotting or Damaged Ones), by Eugène Brieux, a well-known playwright.[67] Each night, before the curtain went up at the Théâtre Antoine in Paris, the stage manager addressed the audience: 'Ladies and Gentlemen, the author and director are pleased to inform you that this play is a study of the relationship between syphilis and marriage. It contains no cause for scandal, no unpleasant scenes, not a single obscene word, and it can be understood by all, if we acknowledge that women need have absolutely no need to be foolish and ignorant in order to be virtuous.'[68] Nonetheless, *Les Avariés* was quickly banned by the censor, causing dismay and amazement in the editorials of medical journals, which complained that blatantly licentious plays were being shown in café concerts all across Paris with 'complete impunity'.[69]

Following the first international conference for the prevention of syphilis and venereal diseases in Brussels in 1899, **Dr Alfred Fournier** established the medical speciality of syphilology, using epidemiological and statistical techniques to underline the fact that the disease affected not just the demimonde but all levels of society, that women caught it earlier than men, and that it was 'overwhelming' among girls whose poor background had forced them into prostitution. As a result of Fournier's work, journals were established that specialised in syphilis, and this paved the way for clinical research, which before long produced results. On 3 March 1905 in Berlin, **Fritz Schaudinn**, a zoologist, noticed under the microscope 'a very small spirochaete, mobile and very difficult to study' in a blood sample taken from a syphilitic. A week later Schaudinn and **Eric Achille Hoffmann**, a bacteriologist, observed the same spirochaete in samples taken from different parts of the body of a patient who only later developed roseolae, the purple patches that disfigure the skin of syphilitics.[70] Difficult as it was to study, because it was so small, the spirochaete was clearly the syphilis microbe, and it was labelled *Treponema* (it resembled a twisted thread) *pallidum* (a reference to its pale colour). The invention of the ultramicroscope in 1906 meant that the spirochaete was now easier to experiment on than Schaudinn had predicted, and before the year was out a diagnostic staining test had been identified by August Wassermann. This meant that syphilis could now be identified early, which helped prevent its spread. But a cure was still needed.[71]

The man who found it was **Paul Ehrlich** (1854–1915). Born in Strehlen, Upper Silesia, he had an intimate experience of infectious diseases: while studying tuberculosis as a young doctor, he had contracted the illness and been forced to convalesce in Egypt.[72] As so often happens in science, Ehrlich's initial contribution was to make deductions from observations available to everyone. He observed that, as one bacillus after another was discovered, associated with different diseases, the cells that had been infected also varied in their response to staining techniques. Clearly, the biochemistry of these cells was affected according to the bacillus that had been introduced. It was this deduction that gave Ehrlich the idea of the antitoxin – what he called the 'magic bullet' – a special substance secreted by the body to counteract invasions. Ehrlich had in effect discovered the principle of both antibiotics and the human immune response.[73] He went on to identify what antitoxins he could, manufacture them, and employ them in patients via the principle of inoculation. Besides syphilis he continued to work on tuberculosis and diphtheria, and in 1908 he was awarded the Nobel Prize for his work on immunity.[74]

By 1907 Ehrlich had produced no fewer than 606 different substances or 'magic bullets' designed to counteract a variety of diseases. Most of them worked no magic at all, but '**Preparation 606**,' as it was known in Ehrlich's laboratory, was eventually found to be effective in the treatment of syphilis. This was the hydrochloride of dioxydiaminoarsenobenzene, in other words an arsenic-based salt. Though it had severe toxic side effects, arsenic was a traditional remedy for syphilis, and doctors had for some time been experimenting with different compounds with an arsenic base. Ehrlich's assistant was given the job of assessing the efficacy of 606, and reported that it had no effect whatsoever on syphilis-infected animals. Preparation 606 therefore was discarded. Shortly afterward the assistant who had worked on 606, a relatively junior but fully trained doctor, was dismissed from the laboratory, and in the spring of 1909 a Japanese colleague of Ehrlich, Professor Kitasato of Tokyo, sent a pupil to Europe to study with him. **Dr Sachachiro Hata** was interested in syphilis and familiar with Ehrlich's concept of 'magic bullets.'[75] Although Ehrlich had by this stage moved on from experimenting with Preparation 606, he gave Hata the salt to try out again. Why? Was the verdict of his former (dismissed) assistant still rankling two years later? Whatever the reason, Hata was given a substance that had been already studied and discarded. A few weeks

later he presented Ehrlich with his laboratory book, saying, 'Only first trials –
only preliminary general view.'[76]

Ehrlich leafed through the pages and nodded. 'Very nice . . . very nice.' Then
he came across the final experiment Hata had conducted only a few days before.
With a touch of surprise in his voice he read out loud from what Hata had
written: 'Believe 606 *very* effacious.' Ehrlich frowned and looked up. 'No,
surely not? *Wieso denn . . . wieso denn?* It was all minutely tested by Dr R. and
he found nothing – *nothing!*'

Hata didn't even blink. 'I found *that.*'

Ehrlich thought for a moment. As a pupil of Professor Kitasato, Hata wouldn't
come all the way from Japan and then lie about his results. Then Ehrlich
remembered that Dr R had been dismissed for not adhering to strict scientific
practice. Could it be that, thanks to Dr R, they had missed something? Ehrlich
turned to Hata and urged him to repeat the experiments. Over the next few
weeks Ehrlich's study, always untidy, became clogged with files and other
documents showing the results of Hata's experiments. There were bar charts,
tables of figures, diagrams, but most convincing were the photographs of
chickens, mice, and rabbits, all of which had been deliberately infected with
syphilis to begin with and, after being given Preparation 606, showed progressive
healing. The photographs didn't lie but, to be on the safe side, Ehrlich and
Hata sent Preparation 606 to several other labs later in the year to see if different
researchers would get the same results. Boxes of this particular magic bullet
were sent to colleagues in Saint Petersburg, Sicily, and Magdeburg. At the
Congress for Internal Medicine held at Wiesbaden on 19 April 1910, Ehrlich
delivered the first public paper on his research, but by then it had evolved one
crucial stage further. He told the congress that in October 1909 twenty-four
human syphilitics had been successfully treated with Preparation 606. Ehrlich
called his magic bullet **Salvarsen**, which had the chemical name of asphen-
amine.[77]

The discovery of Salvarsen was not only a hugely significant medical breakthrough but also produced a social change that would in years to come influence
the way we think in more ways than one. For example, one aspect of the
intellectual history of the century that has been inadequately explored is the
link between syphilis and psychoanalysis. As a result of syphilis, as we have seen,
the fear and guilt surrounding illicit sex was much greater at the beginning of
the century than it is now, and helped account for the climate in which
Freudianism could grow and thrive. Freud himself acknowledged this. In his
Three Essays on the Theory of Sexuality, published in 1905, he wrote, 'In more
than half of the severe cases of hysteria, obsessional neurosis, etc., which I have
treated, I have observed that the patient's father suffered from syphilis which
had been recognised and treated before marriage. . . . I should like to make it
perfectly clear that the children who later became neurotic bore no physical
signs of hereditary syphilis. . . . Though I am far from wishing to assert that
descent from syphilitic parents is an invariable or necessary etiological condition
of a neuropathic constitution, I believe that the coincidences which I have
observed are neither accidental nor unimportant.'[78]

This paragraph appears to have been forgotten in later years, but it is crucial. The chronic fear of syphilis in those who didn't have it, and the chronic guilt in those who did, created in the turn-of-the-century Western world a psychological landscape ready to spawn what came to be called depth psychology. The notion of germs, spirochaetes, and bacilli was not all that dissimilar from the idea of electrons and atoms, which were not pathogenic but couldn't be seen either. Together, this hidden side of nature made the psychoanalytic concept of the unconscious acceptable. The advances made by the sciences in the nineteenth century, together with the decline in support for organised religion, helped to produce a climate where 'a scientific mysticism' met the needs of many people. This was scientism reaching its apogee. Syphilis played its part.

One should not try too hard to fit all these scientists and their theories into one mould. It is, however, noticeable that one characteristic does link most of these figures: with the possible exception of Russell, each was fairly solitary. Einstein, Rutherford, Ehrlich, and Baekeland, early in their careers, ploughed their own furrow – not for them the Café Griensteidl or the Moulin de la Galette. Getting their work across to people, whether at conferences or in professional journals, was what counted. This was – and would remain – a significant difference between scientific 'culture' and the arts, and may well have contributed to the animosity toward science felt by many people as the decades went by. The self-sufficiency of science, the self-absorption of scientists, the sheer *difficulty* of so much science, made it inaccessible in a way that the arts weren't. In the arts, the concept of the avant-garde, though controversial, became familiar and stabilised: what the avant-garde liked one year, the bourgeoisie would buy the next. But new ideas in science were different; very few of the bourgeoisie would ever fully comprehend the minutiae of science. Hard science and, later, weird science, were hard and/or weird in a way that the arts were not.

For non-specialists, the inaccessibility of science didn't matter, or it didn't matter very much, for the technology that was the product of difficult science worked, conferring a continuing authority on physics, medicine, and even mathematics. As will be seen, the main effect of the developments in hard science were to reinforce two distinct streams in the intellectual life of the century. Scientists ploughed on, in search of more and more fundamental answers to the empirical problems around them. The arts and the humanities responded to these fundamental discoveries where they could, but the raw and awkward truth is that the traffic was almost entirely one-way. Science informed art, not the other way round. By the end of the first decade, this was already clear. In later decades, the issue of whether science constitutes a special kind of knowledge, more firmly based than other kinds, would become a major preoccupation of philosophy.

LADDERS OF BLOOD

On the morning of Monday, 31 May 1909, in the lecture theatre of the Charity Organization Society building, not far from Astor Place in New York City, three pickled brains were displayed on a wooden bench. One of the brains belonged to an ape, another was the brain of a white person, and the third was a Negro brain. The brains were the subject of a lecture given by Dr Burt Wilder, a neurologist from Cornell University. Professor Wilder, after presenting a variety of charts and photographs and reporting on measurements said to be relevant to the 'alleged prefrontal deficiency in the Negro brain,' reassured the multiracial audience that the latest science had found no difference between white and black brains.[1]

The occasion of this talk – which seems so dated and yet so modern – was in some ways historic. It was the opening morning of a three-day 'National Negro Conference,' the very first move in an attempt to create a permanent organisation to work for civil rights for American blacks. The conference was the brainchild of Mary Ovington, a white social worker, and had been nearly two years in the making. It had been conceived after she had read an account by William Walling of a race riot that had devastated Springfield, Illinois, in the summer of 1908. The trouble that flared in Springfield on the night of 14 August signalled that America's race problem was no longer confined to the South, no longer, as Walling wrote, 'a raw and bloody drama played out behind a magnolia curtain.' The spark that ignited the riot was the alleged rape of a white woman, the wife of a railway worker, by a well-spoken black man. (The railroads were a sensitive area at the time. Some southern states had 'Jim Crow' carriages: as the trains crossed the state line, arriving from the North, blacks were forced to move from interracial carriages to the blacks-only variety.) As news of the alleged rape spread that night, there were two lynchings, six fatal shootings, eighty injuries, more than $200,000 worth of damage. Two thousand African Americans fled the city before the National Guard restored order.[2]

William Walling's article on the riot, 'Race War in the North,' did not appear in the *Independent* for another three weeks. But when it did, it was much more than a dispassionate report. Although he reconstructed the riot and its immediate cause in exhaustive detail, it was the passion of Walling's rhetoric that moved Mary Ovington. He showed how little had changed in attitudes towards blacks since the Civil War; he exposed the bigotry of certain governors in southern states, and tried to explain why racial troubles were now spreading north. Reading Walling's polemic, Mary Ovington was appalled. She contacted him and suggested they start some sort of organisation. Together they rounded up other white sympathisers, meeting first in Walling's apartment and then, when the group got too big, at the Liberal Club on East Nineteenth Street. When they mounted the first National Negro Conference, on that warm May day, in 1909, just over one thousand attended. Blacks were a distinct minority.

After the morning session of science, both races headed for lunch at the Union Square Hotel close by, 'so as to get to know each other.' Even though nearly half a century had elapsed since the Civil War, integrated meals were unusual even in large northern towns, and participants ran the risk of being jeered at, or worse. On that occasion, however, lunch went smoothly, and duly fortified, the lunchers walked back over to the conference centre. That afternoon, the main speaker was one of the black minority, a small, bearded, aloof academic from Fisk and Harvard Universities, called **William Edward Burghardt Du Bois**.

W. E. B. Du Bois was often described, especially by his critics, as arrogant, cold and supercilious.[3] That afternoon he was all of these, but it didn't matter. This was the first time many white people came face to face with a far more relevant characteristic of Du Bois: his intellect. He did not say so explicitly, but in his talk he conveyed the impression that the subject of that morning's lectures – whether whites were more intelligent than blacks – was a matter of secondary importance. Using the rather precise prose of the academic, he said he appreciated that white people were concerned about the deplorable housing, employment, health, and morals of blacks, but that they 'mistook effects for causes.' More important, he said, was the fact that black people had sacrificed their own self-respect because they had failed to gain the vote, without which the 'new slavery' could never be abolished. He had one simple but all-important message: economic power – and therefore self-fulfilment – would only come for the Negro once political power had been achieved.[4]

By 1909 Du Bois was a formidable public speaker; he had a mastery of detail and a controlled passion. But by the time of the conference he was undergoing a profound change, in the process of turning from an academic into a politician – and an activist. The reason for Du Bois's change of heart is instructive. Following the American Civil War, the Reconstruction movement had taken hold in the South, intent on turning back the clock, rebuilding the former Confederate states with de facto, if not de jure, segregation. Even as late as the turn of the century, several states were still trying to disenfranchise blacks, and even in the North many whites treated blacks as an inferior people. Far from advancing since the Civil War, the fortunes of blacks had actually regressed. The situation was not helped by the theories and practices of the first prominent black leader, a former slave from Alabama, **Booker T. Washington**. He took the view that the best form of race relations was accommodation with the whites, accepting that change would come eventually, and that any other approach risked a white backlash. Washington therefore spread the notion that blacks 'should be a labour force, not a political force,' and it was on this basis that his Tuskegee Institute was founded, in Alabama, near Montgomery, its aim being to train blacks in the industrial skills mainly needed on southern farms. Whites found this such a reassuring philosophy that they poured money into the **Tuskegee Institute**, and Washington's reputation and influence grew to the point where, by the early years of the twentieth century, few federal black appointments were made without Theodore Roosevelt, in the White House, canvassing his advice.[5]

Washington and Du Bois could not have been more different. Born in 1868, three years after the Civil War ended, the son of northern blacks, and with a little French and Dutch blood in the background, Du Bois grew up in Great Barrington, Massachusetts, which he described as a 'boy's paradise' of hills and rivers. He shone at school and did not encounter discrimination until he was about twelve, when one of his classmates refused to exchange visiting cards with him and he felt shut off, as he said, by a 'vast veil.'[6] In some respects, that veil was never lifted. But Du Bois was enough of a prodigy to outshine the white boys in school at Great Barrington, and to earn a scholarship to Fisk University, a black college founded after the Civil War by the American Missionary Association in Nashville, Tennessee. From Fisk he went to Harvard, where he studied sociology under William James and George Santayana. After graduation he had difficulty finding a job at first, but following a stint at teaching he was invited to make a sociological study of the blacks in a slum area in Philadelphia. It was just what he needed to set him off on the first phase of his career. Over the next few years Du Bois produced a series of sociological surveys – *The Philadelphia Negro*, *The Negro in Business*, *The College-Bred Negro*, *Economic Cooperation among Negro Americans*, *The Negro Artisan*, *The Negro Church*, and eventually, in the spring of 1903, *Souls of Black Folk*. James Weldon Johnson, proprietor of the first black newspaper in America, an opera composer, lawyer, and the son of a man who had been free before the Civil War, described this book as having 'a greater effect upon and within the Negro race in America than any other single book published in this country since *Uncle Tom's Cabin*.'[7]

Souls of Black Folk summed up Du Bois's sociological research and thinking of the previous decade, which not only confirmed the growing dis-enfranchisement and disillusion of American blacks but proved beyond doubt the brutal economic effects of discrimination in housing, health, and employ-ment. The message of his surveys was so stark, and showed such a deterioration in the overall picture, that Du Bois became convinced that Booker T. Wash-ington's approach actually did more harm than good. In *Souls*, Du Bois rounded on Washington. It was a risky thing to do, and relations between the two leaders quickly turned sour. Their falling-out was heightened by the fact that Washington had the power, the money, and the ear of President Roosevelt. But Du Bois had his intellect and his studies, his evidence, which gave him an unshakeable conviction that higher education must become the goal of the 'talented tenth' of American blacks who would be the leaders of the race in the future.[8] This was threatening to whites, but Du Bois simply didn't accept the Washington 'softly, softly' approach. Whites would only change if forced to do so.

For a time Du Bois thought it was more important to argue the cause against whites than to fight his own color. But that changed in July 1905 when, with feelings between the rival camps running high, he and twenty-nine others met secretly at Fort Erie in Ontario to found what became known as the '**Niagara movement**.'[9] Niagara was the first open black protest movement, and altogether more combative than anything Washington had ever contemplated. It was intended to be a nationwide outfit with funds to fight for civil and legal rights both in general and in individual cases. It had committees to cover health, education, and economic issues, press and public opinion, and an anti-lynching fund. When he heard about it, Washington was incensed. Niagara went against everything he stood for, and from that moment he plotted its downfall. He was a formidable opponent, not without his own propaganda skills, and he pitched this battle for the souls of black folk as between the 'soreheads,' as the protesters were referred to, and the 'responsible leaders' of the race. Washington's campaign scared away white support for Niagara, and its membership never reached four figures. Indeed, the Niagara movement would be completely forgotten now if it hadn't been for a curious coincidence. The last annual meeting of the movement, attended by just twenty-nine people, was adjourned in Oberlin, Ohio, on 2 September 1908. The future looked bleak and was not helped by the riot that had recently taken place in Springfield. But the very next day, William Walling's article on the riot was published in the *Independent*, and Mary Ovington took up the torch.[10]

The conference Ovington and Walling organised, after its shaky start discussing brains, did not fizzle out – far from it. The first National Negro Conference (NNC) elected a Committee of Forty, also known as the National Committee for the Advancement of the Negro. Although predominantly staffed by whites, this committee turned its back on Booker T. Washington, and from that moment his influence began to wane. For the first twelve months, the activities of the NNC were mainly administrative and organisational – putting finance and a nationwide structure in place. By the time they met again in May 1910, they were ready to combat prejudice in an organised way.[11]

Not before time. Lynchings were still running at an average of ninety-two a year. Roosevelt had made a show of appointing a handful of blacks to federal positions, but William Howard Taft, inaugurated as president in 1909, 'slowed the trickle to a few drops,' insisting that he could not alienate the South as his predecessor had done by 'uncongenial black appointments.'[12] It was therefore no surprise that the theme of the second conference was 'disenfranchisement and its effects upon the Negro,' mainly the work of Du Bois. The battle, the argument, was being carried *to* the whites. To this end, the conference adopted a report worked out by a Preliminary Committee on Organisation. This allowed for a National Committee of One Hundred, as well as a thirty-person executive committee, fifteen to come from New York and fifteen from elsewhere.[13] Most important of all, funds had been raised for there to be five full-time, paid officers – a national president, a chairman of the Executive Committee, a treasurer and his assistant, and a director of publications and research. All of these officeholders were white, except the last – W. E. B. Du Bois.[14]

At this second meeting delegates decided they were unhappy with the word *Negro*, feeling that their organisation should campaign on behalf of all people with dark skin. As a result, the name of the organisation was changed, and the National Negro Conference became the **National Association for the Advancement of Colored People** (NAACP).[15] Its exact form and approach owed more to Du Bois than to any other single person, and this aloof black intellectual stood poised to make his impact, not just on the American nation but worldwide.

There were good practical and tactical reasons why Du Bois should have ignored the biological arguments linked to America's race problem. But that didn't mean that the idea of a biological ladder, with whites above blacks, would go away: social Darwinism was continuing to flourish. One of the crudest efflorescences of this idea had been displayed at the World's Fair in Saint Louis, Missouri, in 1903, lasting for six months. The Saint Louis World's Fair was the most ambitious gathering of intellectuals the new world had ever seen. In fact, it was the largest fair ever held, then or since.[16]

It had begun life as The Louisiana Purchase Exhibition, held to commemorate the hundredth anniversary of President Jefferson's purchase of the state from the French in 1803, which had opened up the Mississippi and helped turn the inland port of Saint Louis into America's fourth most populous city after New York, Chicago, and Philadelphia. The fair had both highbrow and lowbrow aspects. There was, for instance, an International Congress of Arts and Sciences, which took place in late September. (It was depicted as 'a Niagara of scientific talent,' though literature also featured.) Among the participants were John B. Watson, the founder of behaviourism, Woodrow Wilson, the new president of Princeton, the anthropologist Franz Boas, the historian James Bryce, the economist and sociologist Max Weber, Ernest Rutherford and Henri Poincaré in physics, Hugo de Vries and T. H. Morgan in genetics. Although they were not there themselves, the brand-new work of Freud, Planck, and Frege was discussed. Perhaps more notable for some was the presence of Scott Joplin, the king of ragtime, and of the ice cream cone, invented for the fair.[17]

Also at the fair was an exhibition showing 'the development of man.' This had been planned to show the triumph of the 'Western' (i.e., European) races. It was a remarkable display, comprising the largest agglomeration of the world's non-Western peoples ever assembled: Inuit from the Arctic, Patagonians from the near-Antarctic, Zulu from South Africa, a Philippine Negrito described as 'the missing link,' and no fewer than fifty-one different tribes of Indians, as native Americans were then called. These 'exhibits' were on show all day, every day, and the gathering was not considered demeaning or politically incorrect by the whites attending the fair. However, the bad taste (as we would see it) did not stop there. Saint Louis, because of the World's Fair, had been chosen to host the 1904 Olympic Games. Using this context as inspiration, an alternative 'Games' labelled the 'Anthropology Days' was organised as part of the fair.

Here all the various members of the great ethnic exhibition were required to pit themselves against each other in a contest organised by whites who seemed to think that this would be a way of demonstrating the differing 'fitness' of the races of mankind. A Crow Indian won the mile, a Sioux the high jump, and a Moro from the Philippines the javelin.[18]

Social Darwinist ideas were particularly virulent in the United States. In 1907, Indiana introduced sterilisation laws for rapists and imbeciles in prison. But similar, if less drastic, ideas existed elsewhere. In 1912 the **International Eugenics Conference** in London adopted a resolution calling for greater government interference in the area of breeding. This wasn't enough for the Frenchman **Charles Richet**, who in his book *Sélection humaine* (1912) openly argued for all newborn infants with hereditary defects to be killed. After infancy Richet thought castration was the best policy but, giving way to horrified public opinion, he advocated instead the prevention of marriage between people suffering from a whole range of 'defects' – tuberculosis, rickets, epilepsy, syphilis (he obviously hadn't heard of Salvarsen), 'individuals who were too short or too weak,' criminals, and 'people who were unable to read, write or count.'[19] Major **Leonard Darwin**, Charles Darwin's son and from 1911 to 1928 president of the British Eugenics Education Society, didn't go quite this far, but he advocated that 'superior' people should be encouraged to breed more and 'inferior' people encouraged to reproduce less.[20] In America, eugenics remained a strong social movement until the 1920s, the Indiana sterilisation laws not being repealed until 1931. In Britain the Eugenics Education Society remained in business until the 1920s. The story in Germany is a separate matter.

Paul Ehrlich had not allowed his studies of syphilis to be affected by the prevailing social views of the time, but the same cannot be said of many geneticists. In the early stages of the history of the subject, a number of reputable scientists, worried by what they perceived as the growth of alcoholism, disease, and criminality in the cities, which they interpreted as degeneration of the racial stock, lent their names to the eugenic societies and their work, if only for a while. The American geneticist **Charles B. Davenport** produced a classical paper, still quoted today, proving that Huntington's chorea, a progressive nervous disorder, was inherited via a Mendelian dominant trait. He was right. At much the same time, however, he campaigned for eugenic sterilisation laws and, later, for immigration to the United States to be restricted on racial and other biological/genetic grounds. This led him so much astray that his later work was devoted to trying to show that a susceptibility to violent outbursts was the result of a single dominant gene. One can't 'force' science like that.[21]

Another geneticist affiliated to the eugenics movement for a short time was **T. H. Morgan**. He and his co-workers made the next major advance in genetics after Hugo de Vries's rediscovery of Mendel in 1900. In 1910, the same year that America's eugenic society was founded, Morgan published the first results of his experiments on the fruit fly, *Drosophila melanogaster*. This may not sound much, but the simplicity of the fruit fly, and its rapid breeding time, meant that in years to come, and thanks to Morgan, *Drosophila* became the staple research tool of genetics. Morgan's **'fly room'** at Columbia University in New York became famous.[22] Since de Vries's rediscovery of Mendel's laws in 1900, the basic mechanism of heredity had been confirmed many times. However, Mendel's approach, and de Vries's, was statistical, centring on that 3 : 1 ratio in the variability of offspring. The more that ratio was confirmed, the more people realised there had to be a physical, biological, and cytological grounding for the mechanism identified by Mendel and de Vries. There was one structure that immediately suggested itself. For about fifty years, biologists had been observing under the microscope a certain characteristic behaviour of cells undergoing reproduction. They saw a number of minute threads forming part of the nuclei of cells, which separated out during reproduction. As early as 1882, Walther Flemming recorded that, if stained with dye, the threads turned a deeper colour than the rest of the cell.[23] This reaction led to speculation that the threads were composed of a special substance, labelled chromatin, because it coloured the threads. These threads were soon called **chromosomes**, but it was nine years before H. Henking, in 1891, made the next crucial observation, that during meiosis (cell division) in the insect *Pyrrhocoris*, half the spermatozoa received eleven chromosomes while the other half received not only these eleven but an additional body that responded strongly to staining. Henking could not be sure that this extra body was a chromosome at all, so he simply called it 'X.' It never crossed his mind that, because half received it and half didn't, the '**X body**' might determine what sex an insect was, but others soon drew this conclusion.[24] After Henking's observation, it was confirmed that the same chromosomes appear in the same configuration in successive generations, and Walter Sutton showed in 1902 that during reproduction similar chromosomes come together, then separate. In other words, chromosomes behaved in exactly the way Mendel's laws suggested.[25] Nonetheless, this was only inferential – circumstantial – evidence, and so in 1908 T. H. Morgan embarked on an ambitious program of animal breeding designed to put the issue beyond doubt. At first he tried rats and mice, but their generations were too long, and the animals often became ill. So he began work on the common fruit fly, *Drosophila melanogaster*. This tiny creature is scarcely exotic, nor is it as closely related to man. But it does have the advantage of a simple and convenient lifestyle: 'To begin with it can thrive in old milk bottles, it suffers few diseases and it conveniently produces a new generation every couple of weeks.'[26] Unlike the twenty-odd pairs of chromosomes that most mammals have, *Drosophila* has four. That also made experimentation simpler.

The fruit fly may have been an unromantic specimen, but scientifically it turned out to be perfect, especially after Morgan noticed that a single white-eyed male suddenly occurred among thousands of normal red-eyed flies. This sudden mutation was something worth getting to the bottom of. Over the next few months, Morgan and his team mated thousands and thousands of flies in their laboratory at Columbia University in New York. (This is how the 'fly room' got its name.) The sheer bulk of Morgan's results enabled him to conclude that mutations formed in fruit flies at a steady pace. By 1912, more than twenty recessive mutants had been discovered, including one they called 'rudimentary wings' and another that produced 'yellow body colour.' But that wasn't all. The mutations only ever occurred in one sex, males or females, never in both. This observation, that mutations are always sex-linked, was significant because it supported the idea of *particulate* inheritance. The only *physical* difference between the cells of the male fruit fly and the female lay in the 'X body'. It followed, therefore, that the X body *was* a chromosome, that it determined the sex of the adult fly, and that the various mutations observed in the fly room were also carried on this body.[27]

Morgan published a paper on *Drosophila* as early as July 1910 in *Science*, but the full force of his argument was made in 1915 in *The Mechanism of Mendelian Inheritance*, the first book to air the concept of the 'gene.'[28] For Morgan and his colleagues the gene was to be understood 'as a particular segment of the chromosome, which influenced growth in a definite way and therefore governed a specific character in the adult organism'. Morgan argued that the gene was self-replicating, transmitted unchanged from parent to offspring, mutation being the only way new genes could arise, producing new characteristics. Most importantly, mutation was a random, accidental process that could not be affected in any way by the needs of the organism. According to this argument, the inheritance of acquired characteristics was logically impossible. This was Morgan's basic idea. It promoted a great deal of laboratory research elsewhere, especially across the United States. But in other long-established fields (like palaeontology), scientists were loath to give up non-Mendelian and even non-Darwinian ideas until the modern synthesis was formed in the 1940s (see below, chapter 20).[29] There were of course complications. For example, Morgan conceded that a single adult characteristic can be controlled by more than one gene, while at the same time a single gene can affect several traits. Also important was the position of a gene on the chromosome, since its effects could occasionally be modified by neighbouring genes.

Genetics had come a long way in fifteen years, and not just empirically, but philosophically too. In some senses the gene was a more potent fundamental particle than either the electron or the atom, since it was far more directly linked to man's humanity. The accidental and uncontrollable nature of mutation as the sole mechanism for evolutionary change, under the 'indifferent control of natural selection,' was considered by critics – philosophers and religious authorities – as a bleak imposition of banal forces without meaning, yet another low point in man's descent from the high ground he had occupied when religious views had ruled the world. For the most part, Morgan did not get involved in these philosophical debates. Being an empiricist, he realised that genetics was more complicated than most eugenicists believed, and that no useful purpose could be achieved by the crude control techniques favoured by the social Darwinist zealots. Around 1914 he left the eugenics movement. He was also aware that recent results from anthropology did not support the easy certainties of the race biologists, in particular the work of a colleague whose office was only a few blocks from Columbia University on the Upper West Side of New York, at the American Museum of Natural History, located at Seventy-ninth Street and

Central Park West. This man's observations and arguments were to prove just as influential as Morgan's.

Franz Boas was born in Minden in northwestern Germany in 1858. Originally a physicist-geographer, he became an anthropologist as a result of his interest in Eskimos. He moved to America to write for *Science* magazine, then transferred to the American Museum of Natural History in New York as a curator. Small, dark-haired, with a very high forehead, Boas had a relaxed, agreeable manner. At the turn of the century he studied several groups of native Americans, examining the art of the Indians of the north Pacific Coast and the secret societies of the Kwakiutl Indians, near Vancouver. Following the fashion of the time for craniometry, he also became interested in the development of children and devised a range of physical measurements in what he called the 'Cephalic Index.'[30] The wide diversity of Boas's work and his indefatigable research made him famous, and with Sir James Frazer, author of *The Golden Bough*, he helped establish anthropology as a respected field of study. As a consequence he was called upon to record the native American population for the U.S. Census in 1900 and asked to undertake research for the Dillingham Commission of the U.S. Senate. This report, published in 1910, was the result of various unformed eugenic worries among politicians – that America was attracting too many immigrants of the 'wrong sort,' that the 'melting pot' approach might not always work, and that the descendants of immigrants might, for reasons of race, culture, or intelligence, be unable or unwilling to assimilate.[31] This is a not unfamiliar argument, even today, but in 1910 the fears of the restrictionists were rather odd, considered from this end of the century. Their anxieties centred upon the physical dimensions of immigrants, specifically that they were 'degenerate' stock. Boas was asked to make a biometric assessment of a sample of immigrant parents and children, an impertinence as controversial then as it would be scandalous now. With the new science of genetics making waves, many were convinced that physical type was determined solely by heredity. Boas showed that in fact immigrants assimilated rapidly, taking barely one or at most two generations to fall in line with the host population on almost any measure you care to name. As Boas, himself an immigrant, sharply pointed out, newcomers do not subject themselves to the traumas of emigration, an arduous and long journey, merely to stand out in their new country. Most want a quiet life and prosperity.[32]

Despite Boas's contribution, the *Dillingham Commission Report* – eighteen volumes of it – concluded that immigrants from Mediterranean regions were 'biologically inferior' to other immigrants. The report did not, however, recommend the exclusion of 'degenerate races,' concentrating its fire instead on 'degenerate individuals' who were to be identified by a test of reading and writing.★[33]

Given the commission's conclusions, the second book Boas published that year took on added significance. **The Mind of Primitive Man** soon became a

★ Passed into law over the president's veto in 1917.

classic of social science: it was well known in Britain, and the German version
was later burned by the Nazis. Boas was not so much an imaginative anthropologist as a measurer and statistician. Like Morgan he was an empiricist and a
researcher, concerned to make anthropology as 'hard' a science as possible and
intent on studying 'objective' things, like height, weight, and head size. He had
also travelled, got to know several different races or ethnic groups, and was
highly conscious that, for most Americans at least, their contact with other
races was limited to the American Negro.

Boas's book begins, 'Proud of his wonderful achievements, civilised man
looks down upon the humbler members of mankind. He has conquered the
forces of nature and compelled them to serve him.'[34] This statement was
something of a lure, designed to lull the reader into complacency. For Boas
then set out to question – all but eradicate – the difference between 'civilised'
and 'primitive' man. In nearly three hundred pages, he gently built argument
upon argument, fact upon fact, turning the conventional 'wisdoms' of the day
upside-down. For example, psychometric studies had compared the brains of
Baltimore blacks with Baltimore whites and found differences in brain structure,
in the relative size of the frontal and orbital lobes and the corpus callosum. Boas
showed that there were equally great differences between the northern French
and the French from central France. He conceded that the dimensions of the
Negro skull were closer to those of apes than were the skulls of the 'higher
races,' but argued that the white races were closer to apes because they were
hairier than the Negro races, and had lips and limb proportions that were closer
to other primates than were the corresponding Negroid features. He accepted
that the average capacity of the skulls of Europeans was 1560 cc, of African
Negroes 1405 cc, and of 'Negroes of the Pacific' 1460 cc. But he pointed out
that the average cranial capacity of several hundred murderers had turned out
to be 1580 cc.[35] He showed that the 'primitive' races were quite capable of
nonimpulsive, controlled behaviour when it suited their purposes; that their
languages were just as highly developed, once you understood the languages
properly; that the Eskimos, for example, had many more words for snow than
anyone else – for the obvious reason that it mattered more to them. He
dismissed the idea that because some languages did not have numerals above
ten, as was true of certain native American tribes, this did not mean that
members of those tribes could not count above ten in English once they had
been taught to speak it.[36]

An important feature of Boas's book was its impressive references. Anthropological, agricultural, botanical, linguistic, and geological evidence was used,
often from German and French language journals beyond the reach of his
critics. In his final chapter, 'Race Problems in the United States,' he surveyed
Lucca and Naples in Italy, Spain and Germany east of the Elbe, all of which
had experienced large amounts of immigration and race mixing and had scarcely
suffered physical, mental, or moral degeneration.[37] He argued that many of the
so-called differences between the various races were in fact ephemeral. Quoting
from his own research on the children of immigrants in the United States, he
explained how within two generations at the most they began to conform,

even in physical dimensions, to those around them, already arrived. He ended by calling for studies to be made about how immigrants and Negroes had adapted to life in America, how they differed as a result of their experiences from their counterparts in Europe or Africa or China who had not migrated. He said it was time to stop concentrating on studies that emphasised often imaginary or ephemeral differences. 'The similarity of fundamental customs and beliefs the world over, without regard to race and environment, is so general that race [appears] ... irrelevant,' he wrote, and expressed the hope that anthropological findings would 'teach us a greater tolerance of forms of civilisation different from our own.'[38]

Boas's book was a tour-de-force. He became very influential, leading anthropologists and the rest of us away from unilinear evolutionary theory and race theory and toward cultural history. His emphasis on cultural history helped to fashion what may be the single most important advance in the twentieth century in the realm of pure ideas: relativism. Before World War I, however, his was the only voice advancing such views. It was another twenty years before his students, Margaret Mead and Ruth Benedict in particular, took up the banner.

At the same time that Boas was studying the Kwakiutl Indians and the Eskimos, archaeologists were also making advances in understanding the history of native Americans. The thrust was that native Americans had a much more interesting culture and past than the race biologists had been willing to admit. This came to a head with the discoveries of **Hiram Bingham**, an historian with links to Yale.[39]

Born in Honolulu in 1875, Bingham came from a family of missionaries who had translated the Bible into some of the world's most remote languages (such as Hawaiian). A graduate of Yale, with a Ph.D. from Harvard, he was a prehistorian with a love of travel, adventure, exotic destinations. This appetite led him in 1909 to Peru, where he met the celebrated historian of Lima, Carlos Romero, who while drinking coca tea with Bingham on the verandah of his house showed him the writings of Father de la Calancha, which fired Bingham's imagination by describing to him the lost Inca city of Vilcabamba.[40] Although some of the larger ancient cities of pre-Columbian America had been recorded in detail by the Spanish conquerors, it was not until the work of the German scholar Eduard Seler in the late 1880s and 1890s that systematic study of the region was begun. Romero kept Bingham enthralled with his account of how **Vilcabamba** – the lost capital of **Manco Inca**, the last great Inca king – had obsessed archaeologists, historians, and treasure hunters for generations.

It was, most certainly, a colourful tale. Manco Inca had taken power in the early sixteenth century when he was barely nineteen. Despite his youth, he proved a courageous and cunning opponent. As the Spanish, under the Pizarro brothers, made advances into the Inca lands, Manco Inca gave ground and retreated to more inaccessible hideouts, finally reaching Vilcabamba. The crunch came in 1539 when Gonzalo Pizarro led three hundred of 'the most distinguished captains and fighting men' in what was by sixteenth-century standards a massive assault. The Spaniards went as far as they could on horseback (horses had become extinct in America before the Spanish arrived).[41] When they could go no farther as a mounted force, they left their animals with a guard and advanced on foot. Crossing the Urumbamba River, they wound their way up the valley of the Vilcabamba to a pass beyond Vitcos. By now, the jungle was so dense as to be all but impassable, and the Spaniards were growing nervous. Suddenly they encountered two new bridges over some mountain streams. The bridges were inviting, but their newness should have made Pizarro suspicious: it didn't, and they were caught in an ambush. Boulders cascaded down on them, to be followed by a hail of arrows. Thirty-six Spaniards were killed, and Gonzalo Pizarro withdrew. But only temporarily. Ten days later, with a still bigger party, the Spaniards negotiated the bridges, reached Vilcabamba, and sacked it. By then, however, Manco Inca had moved on. He was eventually betrayed by Spaniards whose lives he had spared because they had promised to help him in the fight against Pizarro, but not before his cunning and courage had earned him the respect of the Spaniards.[42] Manco Inca's legend had grown over the intervening centuries, as had the mystery surrounding Vilcabamba. In fact, the city assumed even greater significance later in the sixteenth century after silver was discovered there. Then, in the seventeenth century, after the mines had been exhausted, it was reclaimed by the jungle. Several attempts were made in the nineteenth century to find the lost city, but they all failed.

Bingham could not resist Romero's story. When he returned to Yale, he persuaded the millionaire banker Edward Harkness, who was a member of the board of the Metropolitan Museum in New York, a friend of Henry Clay Frick and John Rockefeller, and a collector of Peruvian artefacts, to fund an expedition. In the summer of 1911 Bingham's expedition set out and enjoyed a measure of good fortune, not unlike that of Arthur Evans at Knossos. In 1911 the Urumbamba Valley was being opened up anyway, due to the great Amazonian rubber boom. (Malaya had not yet replaced South America as the chief source of the world's rubber.)[43] Bingham assembled his crew at Cuzco, 350 miles southeast of Lima and the ancient centre of the Inca Empire. The mule train started out in July, down the new Urumbamba road. A few days out from Cuzco, Bingham's luck struck. The mule train was camped between the new road and the Urumbamba River.[44] The noise of the mules and the smell of cooking (or the other way around) attracted the attention of a certain Melchor Arteaga, who lived alone nearby in a run-down shack. Chatting to members of Bingham's crew and learning what their aim was, Arteaga mentioned that there were some ruins on the top of a hill that lay across the river. He had been there 'once before.'[45] Daunted by the denseness of the jungle and the steepness of the canyon, no one felt inclined to check out Arteaga's tip – no one, that is, except Bingham himself. Feeling it was his duty to follow all leads, he set out with Arteaga on the morning of 24 July, having persuaded one other person, a Peruvian sergeant named Carrasco, to accompany them.[46] They crossed the roaring rapids of the Urumbamba using a makeshift bridge of logs linking the boulders. Bingham was so terrified that he crawled across on all fours. On the far side they found a path through the forest, but it was so steep at times that, again, they were forced to crawl. In this manner they climbed two thousand feet above the river, where they stopped for lunch. To Bingham's surprise, he found they were not alone; up here there were two 'Indians' who had made themselves a farm. What was doubly surprising was that the farm was formed from a series of terraces – and the terraces were clearly very old.[47]

Finishing lunch, Bingham was of two minds. The terraces were interesting, but no more than that. An afternoon of yet more climbing was not an attractive proposition. On the other hand, he had come all this way, so he decided to go on. Before he had gone very far, he realised he had made the right decision. Just around the side of a hill, he came upon a magnificent flight of stone terraces – a hundred of them – rising for nearly a thousand feet up the hillside.[48] As he took in the sight, he realised that the terraces had been roughly cleared, but beyond them the deep jungle resumed, and anything might be hidden there. Forgetting his tiredness, he swiftly scaled the terraces – and there, at the top, half hidden among the lush green trees and the spiky undergrowth, he saw ruin after ruin. With mounting excitement, he identified a holy cave and a three-sided temple made of granite ashlars – huge stones carved into smooth squares or rectangles, which fitted together with the precision and beauty of the best buildings in Cuzco. In Bingham's own words, 'We walked along a path to a clearing where the Indians had planted a small vegetable garden. Suddenly we found ourselves standing in front of the ruins of two of the finest and most interesting structures in ancient America. Made of beautiful white granite, the walls contained blocks of Cyclopean size, higher than a man. The sight held me spellbound. . . . Each building had only three walls and was entirely open on one side. The principal temple had walls 12 feet high which were lined with exquisitely made niches, five high up at each end, and seven on the back. There were seven courses of ashlars in the end walls. Under the seven rear niches was a rectangular block 14 feet long, possibly a sacrificial altar, but more probably a throne for the mummies of departed Incas, brought out to be worshipped. The building did not look as though it had ever had a roof. The top course of beautifully smooth ashlars was left uncovered so that the sun could be welcomed here by priests and mummies. I could scarcely believe my senses as I examined the larger blocks in the lower course and estimated that they must weigh from ten to fifteen tons each. Would anyone believe what I had found? Fortunately . . . I had a good camera and the sun was shining.'[49]

One of the temples he inspected on that first day contained three huge windows – much too large to serve any useful purpose. The windows jogged his memory, and he recalled an account, written in 1620, about how the first Inca, Manco the Great, had ordered 'works to be executed at the place of his birth, consisting of a masonry wall with three windows.' 'Was that what I had found? If it was, then this was not the capital of the last Inca but the birthplace of the first. It did not occur to me that it might be both.' On his very first attempt, Hiram Bingham had located **Machu Picchu**, what would become the most famous ruin in South America.[50]

Though Bingham returned in 1912 and 1915 to make further surveys and discoveries, it was Machu Picchu that claimed the world's attention. The city that emerged from the careful excavations had a beauty that was all its own.[51] This was partly because so many of the buildings were constructed from interlocking Inca masonry, and partly because the town was remarkably well preserved, intact to the roofline. Then there was the fact of the city's unity – house groups surrounded by tidy agricultural terraces, and an integrated network of paths and stairways, hundreds of them. This made it easy for everyday life in Inca times to be imagined. The location of Machu Picchu was also extraordinary: after the jungle had been cleared, the remoteness on a narrow ridge surrounded by a hairpin canyon many feet below was even more apparent. An exquisite civilisation had been isolated in a savage jungle.[52]

Bingham was convinced that Machu Picchu was Vilcabamba. One reason he thought this was because he had discovered, beyond the city, no fewer than 135 skeletons, most of them female and many with skulls that had been trepanned, though none in the town itself. Bingham deduced that the trepanned skulls belonged to foreign warriors who had not been allowed inside what was clearly a holy city. (Not everyone agrees with this interpretation.) A second exciting and strange discovery added to this picture: a hollow tube was found which Bingham believed had been used for inhalation. He thought the tube had probably formed part of an elaborate religious ceremony and that the substance inhaled was probably a narcotic such as the yellow seed of the local huilca tree. By extension, therefore, this one tube could be used to explain the name Vilcabamba: plain (bamba) of Huilca. Bingham's final argument for identifying the site as Vilcabamba was based on the sheer size of Machu Picchu. Its roughly one hundred houses made it the most important ruin in the area, and ancient Spanish sources had described Vilcabamba as the largest city in the province – therefore it seemed only common sensical that when Manco Inca sought refuge from Pizarro's cavalry he would have fallen back to this well-defended place.[53] These arguments seemed incontrovertible. Machu Picchu was duly identified as Vilcabamba, and for half a century the majority of archaeological and historical scholars accepted that the city was indeed the last refuge of Manco Inca, the site of his wife's terrible torture and death.[54]

Bingham was later proved wrong. But at the time, his discoveries, like Boas's and Morgan's, acted as a careful corrective to the excesses of the race biologists who were determined to jump to the conclusion that, following Darwin, the races of the world could be grouped together on a simple evolutionary tree. The very strangeness of the Incas, the brilliance of their art and buildings, the fantastic achievement of their road network, stretching over 19,000 miles and superior in some ways to the European roads of the same period, showed the flaws in the glib certainties of race biology. For those willing to listen to the evidence in various fields, evolution was a much more complex process than the social Darwinists allowed.

There was no denying the fact that the idea of evolution was growing more popular, however, or that the work of Du Bois, Morgan, Boas, and Bingham did hang together in a general way, providing new evidence for the links between animals and man, and between various racial groups across the world. The fact that social Darwinism was itself so popular showed how powerful the idea of evolution was. Moreover, in 1914 it received a massive boost from an entirely new direction. Geology was beginning to offer a startling new understanding of how the world itself had evolved.

Alfred Wegener was a German meteorologist. His *Die Entstehung der Kontinente und Ozeane* (The Origin of Continents and Oceans) was not particularly original. His idea in the book that the six continents of the world had begun life as one supercontinent had been aired earlier by an American, F. B. Taylor, in 1908. But Wegener collected much more evidence, and more impressive evidence, to support this claim than anyone else had done before. He set out his ideas at a meeting of the German Geological Association at Frankfurt-am-Main in January 1912.[55] In fact, with the benefit of hindsight one might ask why scientists had not reached Wegener's conclusion sooner. By the end of the nineteenth century it was obvious that to make sense of the natural world, and its distribution around the globe, some sort of intellectual explanation was needed. The evidence of that distribution consisted mostly of fossils and the peculiar spread of related types of rocks. Darwin's *On the Origin of Species* had stimulated an interest in fossils because it was realised that if they could be dated, they could throw light on the development of life in bygone epochs and maybe even on the origin of life itself. At the same time, quite a lot was known about rocks and the way one type had separated from another as the earth had formed, condensing from a mass of gas to a liquid to a solid. The central problem lay in the spread of some types of rocks across the globe and their links to fossils. For example, there is a mountain range that runs from Norway to north Britain and that should cross in Ireland with other ridges that run through north Germany and southern Britain. In fact, it looked to Wegener as though the crossover actually occurs near the coast of north America, as if the two seaboards of the north Atlantic were once contiguous.[56] Similarly, plant and animal fossils are spread about the earth in a way that can only be explained if there were once land connections between areas that are now widely separated by vast oceans.[57] The phrase used by nineteenth-century scientists was 'land bridges,' convenient devices that were believed to stretch across the waters to link, for example, Africa to South America, or Europe to North America. But if these land bridges had never existed, where had they gone to? What had provided the energy by which the bridges had arisen and disappeared? What happened to the seawaters?

Wegener's answer was bold. There were no land bridges, he said. Instead, the six continents as they now exist – Africa, Australia, North and South America, Eurasia, and Antarctica – were once one huge continent, one enormous land mass which he called **Pangaea** (from the Greek for *all* and *earth*). The continents had arrived at their present positions by 'drifting,' in effect floating like huge icebergs. His theory also explained midcontinent mountain ridges, formed by ancient colliding land masses.[58] It was an idea that took some getting used to. How could entire continents 'float'? And on what? And if the continents had moved, what enormous force had moved them? By Wegener's time the earth's essential structure was known. Geologists had used analysis of earthquake waves to deduce that the earth consisted of a crust, a mantle, an outer core, and an inner core. The first basic discovery was that all the continents of the earth are made of one form of rock, granite – or a granular igneous rock (formed under intense heat) – made up of feldspar and quartz. Around the granite continents may be found a different form of rock – basalt, much denser and harder. Basalt exists in two forms, solid and molten (we know this because lava from volcanic eruptions is semi-molten basalt). This suggests that the relation between the outer structures and the inner structures of the earth was clearly related to how the planet formed as a cooling mass of gas that became liquid and then solid.

The huge granite blocks that form the continents are believed to be about 50 kilometres (30 miles) thick, but below that, for about 3,000 kilometres (1,900 miles), the earth possesses the properties of an 'elastic solid,' or semi-molten basalt. And below that, to the centre of the earth (the radius of which is about 6,000 kilometres – nearly 4,000 miles), there is liquid iron.* Millions of years ago, of course, when the earth was much hotter than it is today, the basalt would have been less solid, and the overall situation of the continents would have resembled more closely the idea of icebergs floating in the oceans. On this view, the drifting of the continents becomes much more conceivable.

Wegener's theory was tested when he and others began to work out how the actual land masses would have been pieced together. The continents do not of course consist only of the land that we see above sea level at the present time. Sea levels have risen and fallen throughout geological time, as ice ages have lowered the water table and warmer times raised them, so that the continental shelves – those areas of land currently below water but relatively shallow, before the contours fall off sharply by thousands of feet – are just as likely to make the 'fit.' Various unusual geological features fall into place when this massive jigsaw is pieced together. For example, deposits from glaciation of permo-carboniferous age (i.e., ancient forests, which were formed 200 million years ago and are now coalfields) exist in identical forms on the west coast of South Africa and the east coast of Argentina and Uruguay. Areas of similar Jurassic and Cretaceous rocks (roughly 100–200 million years old) exist around Niger in West Africa and around Recife in Brazil, exactly opposite, across the South Atlantic. And a geosyncline (a depression in the earth's surface) that extends across southern Africa also strikes through mid-Argentina, aligning neatly. Finally, there is the distribution of the distinctive *Glossopteris* flora, similar fossils of which exist in both South Africa and other faraway southern continents, like South America and Antarctica. Wind is unlikely to account for this dispersal, since the seeds of *Glossopteris* were far too bulky to have been spread

---

* Both the pressure of the rock, and its age, mean it is molten. As matter is condensed, its temperature rises – witness the world's deepest gold mine, the Robinson Deep, in South Africa. The walls are so hot that a half-million-dollar air-conditioning plant (at 1960 prices) had to be installed to prevent the miners from being roasted alive. Studies show, in fact, that the below-ground temperature reached 100° centigrade, the boiling point of water, at about 7,200 feet.

in that way. Here too, only continental drift can account for the existence of this plant in widely separated places.

How long was Pangaea in existence, and when and why did the breakup occur? What kept it going? These are the final questions in what is surely one of the most breathtaking ideas of the century. (It took some time to catch on: in 1939, geology textbooks were still treating continental drift as 'a hypothesis only.' Also see chapter 31, below.)[59]

The theory of **continental drift** coincided with the other major advance made in geology in the early years of the century. This related to the age of the earth. In 1650, James Ussher, archbishop of Armagh in Ireland, using the genealogies given in the Bible, had calculated that the earth was created at 9:00 A.M. on 26 October 4004 B.C.* It became clear in the following centuries, using fossil evidence, that the earth must be at least 300 million years old; later it was put at 500 million. In the late nineteenth century William Thomson, Lord Kelvin (1824–1907), using ideas about the earth's cooling, proposed that the crust formed between 20 million and 98 million years ago. All such calculations were overtaken by the discovery of radioactivity and radioactive decay. In 1907 Bertram Boltwood realised that he could calculate the age of rocks by measuring the relative constituents of uranium and lead, which is the final decay product, and relating it to the half-life of uranium. The oldest substances on earth, to date, are some zircon crystals from Australia dated in 1983 to 4.2 billion years old; the current best estimate of the age of the earth is 4.5 billion years.[60]

The age of the oceans has also been calculated. Geologists have taken as their starting point the assumption that the world's oceans initially consisted entirely of fresh water, but gradually accumulated salts washed off the continents by the world's rivers. By calculating how much salt is deposited in the oceans each year, and dividing that into the overall salinity of the world's body of seawater, a figure for the time such salination has taken can be deduced. The best answer at the moment is between 100 and 200 million years.[61]

In trying to set biology to one side in his understanding of the Negro position in the United States, Du Bois grasped immediately what some people took decades to learn: that change for the Negro could only come through political action that would earn for a black skin the same privileges as a white one. He nevertheless underestimated (and he was not alone) the ways in which different forms of knowledge would throw up results that, if not actually haphazard, were not entirely linear either, and which from the start began to flesh out Darwin's theory of evolution. Throughout the twentieth century, the idea of evolution would have a scientific life and a popular life, and the two were not always identical. What people thought about evolution was as important as what evolution really was. This difference was especially important in the United States, with its unique ethnic/biological/social mix, a nation of immi-

* In some geology departments in modern universities, the twenty-sixth of October is still celebrated − ironically − as the earth's birthday.

grants so different from almost every other country in the world. The role of genes in history, the brainpower of the different races, as evolved, would never go away as the decades passed.

The slow pace of evolution, operating over geological time, and typified by the new realisation of the great age of the earth, contributed to the idea that human nature, like fossils, was set in stone. The predominantly unvarying nature of genes added to that sense of continuity, and the discovery of sophisticated civilisations that had once been important but had collapsed encouraged the idea that earlier peoples, however colourful and inventive, had not become extinct without deserving to. And so, while physics undermined conventional notions of reality, the biological sciences, including archaeology, anthropology, and geology, all started to come together, even more so in the popular mind than in the specialist scientific mind. The ideas of linear evolution and of racial differences went together. It was to prove a catastrophic conjunction.

8

VOLCANO

Every so often history gives us a time to savour, a truly defining moment that stands out for all time. 1913 was such a moment. It was as if Clio, the muse of history, was playing tricks with mankind. With the world on the brink of the abyss, with World War I just months away, with its terrible, unprecedented human wastage, with the Russian Revolution not much further off, dividing the world in a way it hadn't been divided before, Clio gave us what was, in creative terms, arguably the most fecund – and explosive – year of the century. As **Robert Frost** wrote in *A Boy's Will*, his first collection of poems, also published that year:

> The light of heaven falls whole and white . . .
> The light for ever is morning light.[1]

Towards the end of 1912 Gertrude Stein, the American writer living in Paris, received a rambling but breathless letter from Mabel Dodge, an old friend: 'There is an exhibition coming on the 15 Feb to 15 March, which is the most important public event that has ever come off since the signing of the Declaration of Independence, & it is of the same nature. Arthur Davies is the President of a group of men here who felt the American people ought to be given a chance to see what the modern artists have been doing in Europe, America & England of late years. . . . This will be a *scream!*'[2]

In comparing what became known as the **Armory Show** to the Declaration of Independence, Mabel Dodge was (one hopes) being ironic. Nonetheless, she was not wholly wrong. One contemporary American press clipping said, 'The Armory Show was an eruption only different from a volcano's in that it was made by man.' The show opened on the evening of 17 February 1913. Four thousand people thronged eighteen temporary galleries bounded by the shell of the New York Armory on Park Avenue and Sixty-fifth Street. The stark ceiling was masked by yellow tenting, and potted pine trees sweetened the air. The proceedings were opened by John Quinn, a lawyer and distinguished patron of contemporary art, who numbered Henri Matisse, Pablo Picasso, André Derain, W. B. Yeats, Ezra Pound, and James Joyce among his friends.[3] In his speech Quinn said, 'This exhibition will be epoch-making in the history of American art. Tonight will be the red-letter night in the history not only of American art but of all modern art.'[4]

The Armory Show was, as Mabel Dodge had told Gertrude Stein, the brainchild of **Arthur Davies**, a rather tame painter who specialised in 'unicorns and medieval maidens.' Davies had hijacked an idea by four artists of the Pastellists Society, who had begun informal discussions about an exhibition, to be held at the Armory, showing the latest developments in American art. Davies was well acquainted with three wealthy New York wives – Gertrude Vanderbilt Whitney, Lillie P. Bliss, and Mrs Cornelius J. Sullivan. These women agreed to finance the show, and Davies, together with the artist **Walt Kuhn** and **Walter Pach**, an American painter and critic living in Paris, set off for Europe to find the most radical pictures the Continent had to offer.

The Armory Show was in fact the third great exhibition of the prewar years to introduce the revolutionary painting being produced in Paris to other countries. The first had taken place in London in 1910 at the Grafton Galleries. *Manet and the Post-Impressionists* was put together by the critic **Roger Fry**, assisted by the artist **Clive Bell**. Fry's show began with Edouard Manet (the last 'old masterly' painter, yet the first of the moderns), then leapt to Paul Cézanne, Vincent Van Gogh, and Paul Gauguin without, as the critic John Rewald has said, 'wasting time' on the other impressionists. In Fry's eyes, Cézanne, Van Gogh, and Gauguin, at that point virtually unknown in Britain, were the immediate precursors of modern art. Fry was determined to show the differences between the impressionists and the **Post-impressionists**, who for him were the greater artists. He felt that the aim of the Post-impressionists was to capture 'the emotional significance of the world that the Impressionists merely recorded.'[5] Cézanne was the pivotal figure: the way he broke down his still lifes and landscapes into a patchwork of coloured lozenges, as if they were the building blocks of reality, was for Fry a precursor of cubism and abstraction. Several Parisian dealers lent to the London show, as did Paul Cassirer of Berlin. The exhibition received its share of criticism, but Fry felt encouraged enough to hold a second show two years later.

This second effort was overshadowed by the German **Sonderbund**, which opened on 25 May 1912, in Cologne. This was another volcano – in John Rewald's words, a 'truly staggering exhibition.' Unlike the London shows, it took for granted that people were already familiar with nineteenth-century painting and hence felt free to concentrate on the most recent movements in modern art. The Sonderbund was deliberately arranged to provoke: the rooms devoted to Cézanne were next to those displaying Van Gogh, Picasso was next to Gauguin. The exhibition also featured Pierre Bonnard, André Derain, Erich Heckel, Aleksey von Jawlensky, Paul Klee, Henri Matisse, Edvard Munch, Emil Nolde, Max Pechstein, Egon Schiele, Paul Signac, Maurice de Vlaminck and Edouard Vuillard. Of the 108 paintings in the show, a third had German owners; of the twenty-eight Cézannes, seventeen belonged to Germans. They were clearly more at home with the new painting than either the British or the Americans.[6] When Arthur Davies received the catalogue for the Sonderbund, he was so startled that he urged Walt Kuhn to go to Cologne immediately.

Kuhn's trip brought him into contact with much more than the Sonderbund. He met Munch and persuaded him to participate in the Armory; he went to Holland in pursuit of Van Goghs; in Paris all the talk was of cubism at the Salon d'Automne and of the futurist exhibition held that year at the Bernheim-Jeune Gallery. Kuhn ended his trip in London, where he was able to raid Fry's second exhibition, which was still on.[7]

The morning after Quinn's opening speech, the attack from the press began – and didn't let up for weeks. The cubist room attracted most laughs, and was soon rechristened the Chamber of Horrors. One painting in particular was singled out for ridicule: **Marcel Duchamp**'s *Nude Descending a Staircase*. Duchamp was already in the news for 'creating' that year the first 'readymade,' a work called simply *Bicycle Wheel*. Duchamp's *Nude* was described as 'a lot of disused golf clubs and bags,' 'an orderly heap of broken violins,' and 'an explosion in a shingle factory.' Parodies proliferated: for example, *Food Descending a Staircase*.[8]

But the show also received serious critical attention. Among the New York newspapers, the *Tribune*, the *Mail*, the *World*, and the *Times* disliked the show. They all applauded the aim of the Association of American Painters and Sculptors to present new art but found the actual pictures and sculptures difficult. Only the *Baltimore Sun* and the *Chicago Tribune* liked what they saw. With critical reception weighted roughly five to two against it, and popular hilarity on a scale rarely seen, the show might have been a commercial disaster, but it was nothing of the kind. As many as ten thousand people a day streamed through the Armory, and despite the negative reviews, or perhaps because of them, the show was taken up by New York society and became a *succès d'estime*. Mrs Astor went every day after breakfast.[9]

After New York the Armory Show travelled to Chicago and Boston, and in all 174 works were sold. In the wake of the show a number of new galleries opened up, mainly in New York. Despite the scandal surrounding the new modern art exhibitions, there were plenty of people who found something fresh, welcome, and even wonderful in the new images, and they began collecting.[10]

Ironically, resistance to the newest art was most vicious in Paris, which at the same time prided itself on being the capital of the avant-garde. In practice, what was new one minute was accepted as the norm soon after. By 1913, impressionism – which had once been scandalous – was the new orthodoxy in painting; in music the controversy surrounding Wagner had long been forgotten, and his lush chords dominated the concert halls; and in literature the late-nineteenth-century symbolism of Stephane Mallarmé, Arthur Rimbaud, and Jules Laforgue, once the enfants terribles of the Parisian cultural scene, were now approved by the arbiters of taste, people such as Anatole France.

Cubism, however, had still not been generally accepted. Two days after the Armory Show closed in New York, **Guillaume Apollinaire**'s publishers announced the almost simultaneous release of his two most influential books, *Les Peintres cubistes* and *Alcools*. Apollinaire was born illegitimate in Rome in

1880 to a woman of minor Polish nobility who was seeking political refuge at the papal court. By 1913 he was already notorious: he had just been in jail, accused on no evidence whatsoever of having stolen Leonardo da Vinci's *Mona Lisa* from the Louvre. After the painting was found, he was released, and made the most of the scandal by producing a book that drew attention to the work of his friend, Pablo Picasso (who the police thought also had had a hand in the theft of the *Mona Lisa*), Georges Braque, Robert Delaunay, and a new painter no one had yet heard of, Piet Mondrian. When he was working on the proofs of his book, Apollinaire introduced a famous fourfold organisation of cubism – **scientific, physical, orphic**, and **instinctive** cubism.[11] This was too much for most people, and his approach never caught on. Elsewhere in the book, however, he wrote sympathetically about what the cubists were trying to achieve, which helped to get them accepted. His argument was that we should soon get bored with nature unless artists continually renewed our experience of it.[12]

Brought up on the Côte d'Azur, Apollinaire appealed to Picasso and the *bande à Picasso* (Max Jacob, André Salmon, later Jean Cocteau) for his 'candid, voluble, sensuous' nature. After he moved to Paris to pursue a career as a writer, he gradually earned the title 'impresario of the avant-garde' for his ability to bring together painters, musicians, and writers and to present their works in an exciting way. 1913 was a great year for him. Within a month of *Les Peintres cubistes* appearing, in April, Apollinaire produced a much more controversial work, *Alcools* (Liquors), a collection of what he called art poetry, which centred on one long piece of verse, entitled 'Zone.'[13] 'Zone' was in many ways the poetic equivalent of Arnold Schoenberg's music or Frank Lloyd Wright's buildings. Everything about it was new, very little recognisable to traditionalists. Traditional typography and verse forms were bypassed. So far as punctuation was concerned, 'The rhythm and division of the lines form a natural punctuation; no other is necessary.'[14] Apollinaire's imagery was thoroughly modern too: cityscapes, shorthand typists, aviators (French pilots were second only to the Wright brothers in the advances being made). The poem was set in various areas around Paris and in six other cities, including Amsterdam and Prague. It contained some very weird images – at one point the bridges of Paris make bleating sounds, being 'shepherded' by the Eiffel Tower.[15] 'Zone' was regarded as a literary breakthrough, and within a few short years, until Apollinaire died (in a 'flu epidemic), he was regarded as the leader of the modernist movement in poetry. This owed as much to his fiery reputation as to his writings.[16]

Cubism was the art form that most fired Apollinaire. For the Russian composer **Igor Stravinsky**, it was fauvism. He too was a volcano. In the words of the critic Harold Schonberg, Stravinsky's 1913 ballet produced the most famous *scandale* in the history of music.[17] *Le Sacre du printemps* (The Rite of Spring) premiered at the new Théâtre des Champs-Elysées on 29 May and overnight changed Paris. Paris, it should be said, was changing in other ways too. The gaslights were being replaced by electric streetlamps, the *pneumatique* by the telephone, and the last horse-drawn buses went out of service in 1913. For some, the change produced by Stravinsky was no less shocking than Rutherford's atom bouncing off gold foil.[18]

Born in Saint Petersburg on 17 June 1882, Stravinsky was just thirty-one in 1913. He had already been famous for three years, since the first night of his ballet *Firebird*, which had premiered in Paris in June 1910. Stravinsky owed a lot to his fellow Russian Serge Diaghilev, who had originally intended to become a composer himself. Discouraged by Nicolai Andreyevich Rimsky-Korsakov, who told him he had no talent, Diaghilev turned instead to art publishing, organising exhibitions, and then putting on music and ballet shows in Paris. Not unlike Apollinaire, he discovered his true talent as an impresario. Diaghilev's great passion was ballet; it enabled him to work with his three loves – music, dance and painting (for the scenery) – all at the same time.[19]

Stravinsky's father had been a singer with the Saint Petersburg opera.[20] Both Russian and foreign musicians were always in and out of the Stravinsky home, and Igor was constantly exposed to music. Despite this, he went to university as a law student, and it was only when he was introduced to Rimsky-Korsakov in 1900 and taken on as his pupil after showing some of his compositions that he switched. In 1908, the year Rimsky-Korsakov died, Stravinsky composed an orchestral work that he called *Fireworks*. Diaghilev heard it in Saint Petersburg, and the music stuck in his mind.[21] At that stage he had not formed the Ballets Russes, the company that was to make him and many others famous. However, having staged concerts and operas of Russian music in Paris, Diaghilev decided in 1909 to found a permanent company. In no time, he made the Ballets Russes a centre of the avant-garde. His composers who wrote for the Ballets Russes included Claude Debussy, Manuel de Falla, Sergei Prokofiev, and Maurice Ravel; Picasso and **Leon Bakst** designed the sets; and the principal dancers were **Vaslav Nijinsky**, Tamara Karsavina, and Léonide Massine. Later, Diaghilev teamed up with another Russian, George Balanchine.[22] Diaghilev decided that for the 1910 season in Paris he wanted a ballet on the Firebird legend, to be choreographed by the legendary Michel Fokine, the man who had done so much to modernise the Imperial Ballet. Initially, Diaghilev commissioned Anatol Liadov to write the music, but as the rehearsals approached, Liadov failed to deliver. Growing desperate, Diaghilev decided that he needed another composer, and one who could produce a score in double-quick time. He remembered *Fireworks* and got word to Stravinsky in Saint Petersburg. The composer immediately took the train for Paris to attend rehearsals.[23]

Diaghilev was astounded at what Stravinsky produced. *Fireworks* had been promising, but *Firebird* was far more exciting, and the night before the curtain went up, Diaghilev told Stravinsky it would make him famous. He was right. The music for the ballet was strongly Russian, and recognisably by a pupil of Rimsky-Korsakov, but it was much more original than the impresario had expected, with a dark, almost sinister opening.[24] Debussy, who was there on the opening night, picked out one of its essential qualities: 'It is not the docile servant of the dance.'[25] *Petrushka* came next in 1911. That too was heavily Russian, but at the same time Stravinsky was beginning to explore polytonality.

At one point two unrelated harmonies, in different keys, come together to create an electrifying effect that influenced several other composers such as Paul Hindemith. Not even Diaghilev had anticipated the success that *Petrushka* would bring Stravinsky.

The young composer was not the only Russian to fuel scandal at the Ballets Russes. The year before *Le Sacre du printemps* premiered in Paris, the dancer Vaslav Nijinsky had been the star of Debussy's *L'Après-midi d'un faune*. No less than Apollinaire, Debussy was a sybarite, a sensualist, and both his music and Nijinsky's dancing reflected this. Technically brilliant, Nijinsky nonetheless took ninety rehearsals for the ten-minute piece he had choreographed himself. He was attempting his own *Les Demoiselles d'Avignon*, a volcanic, iconoclastic work, to create a half-human, half-feral character, as disturbing as it was sensual. His creature, therefore, had not only the cold primitivism of Picasso's *Demoiselles* but also the expressive order (and disorder) of *Der Blaue Reiter*. Paris was set alight all over again.

Even though those who attended the premier of *Le Sacre* were used to the avant-garde and therefore were not exactly expecting a quiet night, *this* volcano put all others in the shade. *Le Sacre* is not mere folk lore: it is a powerful legend about the sacrifice of virgins in ancient Russia.[26] In the main scene the Chosen Virgin must dance herself to death, propelled by a terrible but irresistible rhythm. It was this that gave the ballet a primitive, archetypal quality. Like Debussy's *Après-midi*, it related back to the passions aroused by primitivism – blood history, sexuality, and the unconscious. Perhaps that 'primitive' quality is what the audience responded to on the opening night (the premiere was held on the anniversary of the opening of *L'Après-midi*, Diaghilev being very superstitious).[27] The trouble in the auditorium began barely three minutes into the performance, as the bassoon ended its opening phrase.[28] People hooted, whistled, and laughed. Soon the noise drowned out the music, though the conductor, Pierre Monteux, manfully kept going. The storm really broke when, in the 'Dances des adolescents', the young virgins appeared in braids and red dresses. The composer Camille Saint-Saëns left the theatre, but Maurice Ravel stood up and shouted 'Genius.' Stravinsky himself, sitting near the orchestra, also left in a rage, slamming the door behind him. He later said that he had never been so angry. He went backstage, where he found Diaghilev flicking the house lights on and off in an attempt to quell the noise. It didn't work. Stravinsky then held on to Nijinsky's coattails while the dancer stood on a chair in the wings shouting out the rhythm to the dancers 'like a coxswain.'[29] Men in the audience who disagreed as to the merits of the ballet challenged each other to duels.[30]

'Exactly what I wanted,' said Diaghilev to Stravinsky when they reached the restaurant after the performance. It was the sort of thing an impresario would say. Other people's reactions were, however, less predictable. 'Massacre du Printemps' said one paper the next morning – it became a stock joke.[31] For many people, *The Rite of Spring* was lumped in with cubist works as a form of barbarism resulting from the unwelcome presence of 'degenerate' foreigners in the French capital. (The cubists were known as *métèques*, damn foreigners, and foreign artists were often likened in cartoons and jokes to epileptics.)[32] The critic for *Le Figaro* didn't like the music, but he was concerned that he might be too old-fashioned and wondered whether, in years to come, the evening might turn out to have been a pivotal event.[33] He was right to be concerned, for despite the first-night scandal, *Le Sacre* quickly caught on: companies from all over requested permission to perform the ballet, and within months composers across the Western world were imitating or echoing Stravinsky's rhythms. For it was the rhythms of *Le Sacre* more than anything else that suggested such great barbarity: 'They entered the musical subconscious of every young composer.'

In August 1913 Albert Einstein was walking in the Swiss Alps with the widowed **Marie Curie**, the French physicist, and her daughters. Marie was in hiding from a scandal that had blown up after the wife of Paul Langevin, another physicist and friend of Jules-Henri Poincaré, had in a fit of pique published Marie's love letters to her husband. Einstein, then thirty-four, was a professor at the Federal Institute of Technology, the Eidgenössische Technische Hochschule, or ETH, in Zurich and much in demand for lectures and guest appearances. That summer, however, he was grappling with a problem that had first occurred to him in 1907. At one point in their walks, he turned to Marie Curie, gripped her arm, and said, 'You understand, what I need to know is exactly what happens to the passengers in an elevator when it falls into emptiness.'[34]

Following his special theory of relativity, published in 1905, Einstein had turned his ideas, if not on their head, then on their side. As we have seen, in his special theory of relativity, Einstein had carried out a thought experiment involving a train travelling through a station. (It was called the 'special' theory because it related only to bodies moving in relation to one another.) In that experiment, light had been travelling in the same direction as the train. But he had suspected since 1911 that gravity attracted light.[35] Now he imagined himself in an elevator falling down to earth in a vacuum and therefore accelerating, as every schoolchild knows, at 32 feet per second. However, without windows, and if the acceleration were constant, there would be no way of telling that the elevator was not stationary. Nor would the person in the elevator feel his or her own weight. This notion startled Einstein. He conceived of a thought experiment in which a beam of light struck the elevator not in the direction of movement but at right angles. Again he compared the view of the light beam seen by a person inside the elevator and one outside. As in the 1905 thought experiment, the person inside the elevator would see the light beam enter the box or structure at one level and hit the opposite wall at the same level. The observer outside, however, would see the light beam *bend* because, by the time it reached the other side of the elevator, the far wall would have moved on. Einstein concluded that if acceleration could curve the light beam, and since the acceleration was a result of gravity, then gravity must also be able to bend light. Einstein revealed his thinking on this subject in a lecture in Vienna later in the year, where it caused a sensation among physicists. The implications of

Einstein's **General Theory of Relativity** may be explained by a model, as the special theory was explained using a pencil twisting in the light, casting a longer and shorter shadow. Imagine a thin rubber sheet set out on frame, like a picture canvas, and laid horizontally. Roll a small marble or a ball bearing across the rubber sheet, and the marble will roll in a straight line. However, if you place a heavy ball, say a cannonball, in the centre of the frame, depressing the rubber sheet, the marble would then roll in a curve as it approaches this massive weight. In effect, this is what Einstein argued would happen to light when it approached large bodies like stars. There is a curvature in space–time, and light bends too.[36]

General relativity is a theory about gravity and, like special relativity, a theory about nature on the cosmic scale beyond everyday experience. J. J. Thomson was lukewarm about the idea, but Ernest Rutherford liked the theory so much that he said even if it wasn't true, it was a beautiful work of art.[37] Part of that beauty was that Einstein's theory could be tested. Certain deductions followed from the equations. One was that light should bend as it approaches large objects. Another was that the universe cannot be a static entity – it has to be either contracting or expanding. Einstein didn't like this idea – he thought the universe was static – and he invented a correction so he could continue to think so. He later described this correction as 'the biggest blunder of my career,' for, as we shall see, both predictions of the general theory were later supported by experimentation – and in the most dramatic circumstances. Rutherford had it right; relativity was a most beautiful theory.[38]

The other physicist who produced a major advance in scientific understanding in that summer of 1913 could not have been more different from Einstein. **Niels Henrik David Bohr** was a Dane and an exceptional athlete. He played soccer for Copenhagen University; he loved skiing, bicycling, and sailing. He was 'unbeatable' at table tennis, and undoubtedly one of the most brilliant men of the century. C. P. Snow described him as tall with 'an enormous, domed head,' with a long, heavy jaw and big hands. He had a shock of unruly, combed-back hair and spoke with a soft voice, 'not much above a whisper.' All his life, Bohr talked so quietly that people strained to hear him. Snow also found him to be 'a talker as hard to get to the point as Henry James in his later years.'[39]

This extraordinary man came from a civilised, scientific family – his father was a professor of physiology, his brother was a mathematician, and all were widely read in four languages, as well as in the work of the Danish philosopher Søren Kierkegaard. Bohr's early work was on the surface tension of water, but he then switched to radioactivity, which was the main reason that drew him to Rutherford, and England, in 1911. He studied first in Cambridge but moved to Manchester after he heard Rutherford speak at a dinner at the Cavendish Laboratory in Cambridge. At that time, although Rutherford's theory of the atom was widely accepted by physicists, there were serious problems with it, the most worrying of which was the predicted instability of the atom – no one could see why electrons didn't just collapse in on the nucleus. Shortly after Bohr arrived to work with Rutherford, he had a series of brilliant intuitions, the most important of which was that although the radioactive properties of matter originate in the atomic nucleus, chemical properties reflect primarily the number and distribution of electrons. At a stroke he had explained the link between physics and chemistry. The first sign of Bohr's momentous breakthrough came on 19 June 1912, when he explained in a letter to his brother Harald what he had discovered: 'It could be that I've found out a little bit about the structure of atoms ... perhaps a little piece of reality.' What he meant was that he had an idea how to make more sense of the electrons orbiting Rutherford's nucleus.[40] That summer Bohr returned to Denmark, got married, and taught at the University of Copenhagen throughout the autumn. He struggled on, writing to Rutherford on 4 November that he expected 'to be able to finish the paper [with his new ideas] in a few weeks.' He retreated to the country and wrote a very long article, which he finally divided into three shorter ones, since he had so many ideas to convey. He gave the papers a collective title – *On the Constitution of Atoms and Molecules*. Part 1 was mailed to Rutherford on 6 March 1913; parts 2 and 3 were finished before Christmas. Rutherford had judged his man correctly when he allowed Bohr to transfer to Cambridge. As Bohr's biographer has written, 'A revolution in understanding had taken place.'[41]

As we have seen, Rutherford's notion of the atom was inherently unstable. According to 'classical' theory, if an electron did not move in a straight line, it lost energy through radiation. But electrons went round the nucleus of the atom in orbits – such atoms should therefore either fly apart in all directions or collapse in on themselves in an explosion of light. Clearly, this did not happen: matter, made of atoms, is by and large very stable. Bohr's contribution was to put together a proposition and an observation.[42] He proposed '**stationary**' **states** in the atom. Rutherford found this difficult to accept at first, but Bohr insisted that there must be certain orbits electrons can occupy without flying off or collapsing into the nucleus and without radiating light.[43] He immeasurably strengthened this idea by adding to it an observation that had been known for years – that when light passes through a substance, each element gives off a characteristic spectrum of color and moreover one that is stable and discontinuous. In other words, it emits light of only particular wavelengths – the process known as spectroscopy. Bohr's brilliance was to realise that this spectroscopic effect existed because electrons going around the nucleus cannot occupy 'any old orbit' but only certain permissible orbits.[44] These orbits meant that the atom was stable. But the real importance of Bohr's breakthrough was in his unification of Rutherford, Planck, and Einstein, confirming the quantum – discrete – nature of reality, the stability of the atom, and the nature of the link between chemistry and physics. When Einstein was told of how the Danish theories matched the spectroscopies so clearly, he remarked, 'Then this is one of the greatest discoveries.'[45]

In his own country, Bohr was fêted and given his own Institute of Theoretical Physics in Copenhagen, which became a major centre for the subject in the years between the wars. Bohr's quiet, agreeable, reflective personality – when speaking he often paused for minutes on end while he sought the correct word – was an important factor in this process. But also relevant to the rise of the Copenhagen Institute was Denmark's position as a small, neutral country where, in the dark years of the century, physicists could meet away from the frenetic spotlight of the major European and North American centres.

For psychoanalysis, 1913 was the most significant year after 1900, when *The Interpretation of Dreams* was published. Freud published a new book, *Totem and Taboo*, in which he extended his theories about the individual to the Darwinian, anthropological world, which, he argued, determined the character of society. This was written partly in response to a work by Freud's former favourite disciple, **Carl Jung**, who had published *The Psychology of the Unconscious*, two years before, which marked the first serious division in psychoanalytic theory. Three major works of fiction, very different from one another but each showing the influence of Freudian ideas as they extended beyond the medical profession to society at large, also appeared.

Thomas Mann's great masterpiece *Buddenbrooks* was published in 1901, with the subtitle 'Decline of a Family.' Set in a north German, middle-class family (Mann was himself from Lübeck, the son of a prosperous corn merchant), the novel is bleak. Thomas Buddenbrook and his son Hanno die at relatively young ages (Thomas in his forties, Hanno in his teens) 'for no other very good reason than they have lost the will to live.'[46] The book is lively, and even funny, but behind it lies the spectre of Nietzsche, nihilism, and degeneracy.

Death in Venice, a novella published in 1913, is also about degeneracy, about instincts versus reason, and is an exploration of the author's unconscious in a far more brutally frank way than Mann had attempted or achieved before. Gustav von Aschenbach is a writer newly arrived in Venice to complete his masterpiece. He has the appearance, as well as the first name, of Gustav Mahler, whom Mann fiercely admired and who died on the eve of Mann's own arrival in Venice in 1911. No sooner has Aschenbach arrived than he chances upon a Polish family staying in the same hotel. He is struck by the dazzling beauty of the young son, Tadzio, dressed in an English sailor suit. The story follows the ageing Aschenbach's growing love for Tadzio; meanwhile he neglects his work, and his body succumbs to the cholera epidemic encroaching on Venice. Aschenbach fails to complete his work and he also fails to alert Tadzio's family to the epidemic so they might escape. The writer dies, never having spoken to his beloved.

Von Aschenbach, with his ridiculously quiffed hair, his rouge makeup, and his elaborate clothes, is intended by Mann to embody a once-great culture now deracinated and degenerate. He is also the artist himself.[47] In Mann's private diaries, published posthumously, he confirmed that even late in life he still fell romantically in love with young men, though his 1905 marriage to Katia Pringsheim seemed happy enough. In 1925 Mann admitted the direct influence of Freud on *Death in Venice*: 'The death wish is present in Aschenbach's consciousness though he's unaware of it.' As Ronald Hayman, Mann's biographer has stressed, *Ich* was frequently used by Mann in a Freudian way, to suggest an aspect or segment of the personality that asserts itself, often competing against instinct. (*Ich* was Freud's preferred usage; the Latin *ego* was an innovation of his English translator.)[48] The whole atmosphere of Venice represented in the book – dark, rotting back alleys, where 'unspeakable horrors' lurk unseen and unquantified – recalls Freud's primitive id, smouldering beneath the surface of the personality, ready to take advantage of any lapse by the ego. Some critics have speculated that the very length of time it took Mann to write this short work – several years – reflected the difficulty he had in admitting his own homosexuality.[49]

1913 was also the year in which **D. H. Lawrence**'s *Sons and Lovers* was published. Whether or not Lawrence was aware of psychoanalysis as early as 1905, when he wrote about infantile sexuality 'in terms almost as explicit as Freud's,' he was exposed to it from 1912 on, when he met Frieda Weekley. Frieda, born Baroness Frieda von Richthofen at Metz in Germany in 1879, had spent some time in analysis with her lover Otto Gross, a psychoanalyst.[50] His technique of treatment was an eclectic mix, combining the ideas of Freud and Nietzsche. *Sons and Lovers* tackled an overtly Freudian theme: the Oedipal. Of course, the Oedipal theme pre-dated Freud, as did its treatment in literature. But Lawrence's account of the Morel family – from the Nottinghamshire coalfields (Nottingham being Lawrence's own home county) – places the Oedipal conflict within the context of wider issues. The world inhabited by the Morels is changing, reflecting the transition from an agricultural past to an industrial future and war (Paul Morel actually predicts World War I).[51] Gertrude Morel, the mother in the family, is not without education or wisdom, a fact that sets her apart from her duller, working-class husband. She devotes all her energies to her sons, William and Paul, so that they may better themselves in this changing world. In the process, however, Paul, an artist, who also works in a factory, falls in love and tries to escape the family. Where before there had been conflict between wife and husband, it is now a tussle between mother and son. 'These sons are *urged* into life by their reciprocal love of their mother – urged on and on. But when they come to manhood, they can't love, because their mother is the strongest power in their lives, and holds them. ... As soon as the young men come into contact with women, there's a split. William gives his sex to a fribble, and his mother holds his soul.'[52] Just as Mann tried to break the taboo on homosexuality in *Death in Venice*, Lawrence talks freely of the link between sex and other aspects of life in *Sons and Lovers* and in particular the role of the mother in the family. But he doesn't stop there. As Helen and Carl Baron have said, socialist and modernist themes mingle in the book: low pay, unsafe conditions in the mines, strikes, the lack of facilities for childbirth, or the lack of schooling for children older than thirteen; the ripening ambition of women to obtain work and to agitate for votes; the unsettling effect of evolutionary theory on social and moral life; and the emergence of an interest in the unconscious.[53] In his art studies, Paul encounters the new theories about social Darwinism and gravity. Mann's story is about a world that is ending, Lawrence's about one world giving way to another. But both reflect the Freudian theme of the primacy of sex and the instinctual side of life, with the ideas of Nietzsche and social Darwinism in the background. In both, the unconscious plays a not altogether wholesome role. As Gustav Klimt and

Hugo von Hofmannsthal pointed out in fin-de-siècle Vienna, man ignores the instinctive life at his peril: whatever physics might say, biology is the everyday reality. Biology means sex, reproduction, and behind that evolution. *Death in Venice* is about the extinction of one kind of civilisation as a result of degeneracy. *Sons and Lovers* is less pessimistic, but both explore the Nietzschean tussle between the life-enhancing barbarians and the overrefined, more civilised, rational types. Lawrence saw science as a form of overrefinement. Paul Morel has a strong, instinctive life force, but the shadow of his mother is never absent.

Marcel Proust never admitted the influence of Freud or Darwin or Einstein on his work. But as the American critic Edmund Wilson has pointed out, Einstein, Freud and Proust, the first two Jewish, the latter half-Jewish, 'drew their strength from their marginality which heightened their powers of observance.' In November 1913 Proust published the first volume of his multi-volume work *A la recherche du temps perdu*, normally translated as *Remembrance of Things Past,* though many critics/scholars now prefer *In Search of Lost Time*, arguing that it better conveys Proust's idea that the novel has some of the qualities of science – the research element – and Proust's great emphasis on time, time being lost and recovered rather than just gone.

Proust was born in 1871 into a well-off family and never had to work. A brilliant child, he was educated at the Lycée Condorcet and at home, an arrangement that encouraged a close relationship with his mother, a neurotic woman. After she died in 1905, aged fifty seven, two years after her husband, her son withdrew from the world into a cork-lined room where he began to correspond with hundreds of friends and convert his meticulously detailed diaries into his masterpiece. *A la recherche du temps perdu* has been described as the literary equivalent of Einstein or Freud, though as the Proust scholar Harold March has pointed out, such comparisons are generally made by people unfamiliar with either Freud or Einstein. Proust once described his multi-volume work in an interview as 'a series of novels of the unconscious'. But not in a Freudian sense (there is no evidence that Proust ever read Freud, whose works were not translated into French until the novelist was near the end of his life). Proust 'realised' one idea to wonderful heights. This was the notion of involuntary memory, the idea that the sudden taste of a pastry, say, or the smell of some old back stairs, brings back not just events in the past but a whole constellation of experiences, vivid feelings and thoughts *about* that past. For many people, Proust's insight is transcendentally powerful, for others it is overstated (Proust has always divided the critics).

His real achievement is what he makes of this. He is able to evoke the intense emotions of childhood – for example, near the beginning of the book when he describes the narrator's desperate desire to be kissed by his mother before he goes to sleep. This shifting back and forth in time is what has led many people to argue that Proust was giving a response to Einstein's theories about time and relativity though there is no real evidence to link the novelist and the physicist any more than there is to link him with Freud. Again, as Harold March has said, we should really consider Proust on his own terms. Looked at in this way, *In Search of Lost Time* is a rich, gossipy picture of French aristocratic/

upper class life, a class that, as in Chekhov and Mann, was disappearing and vanished completely with World War I. Proust was used to this world – his letters constantly refer to Princess This, the Count of That, the Marquis of the Other.[54] His characters are beautifully drawn; Proust was gifted not only with wonderful powers of observation but with a mellifluous prose, writing in long, languid sentences interlaced with subordinate clauses, a dense foliage of words whose direction and meaning nonetheless always remains vivid and clear.

The first volume, published in 1913, *Du côté de chez Swann*, 'Swann's Way' (in the sense of Swann's area of town), comprised what would turn out to be about a third of the whole book. We slip in and out of the past, in and around Combray, learning the architecture, the layout of the streets, the view from this or that window, the flower borders and the walkways as much as we know the people. Among the characters are Swann himself, Odette, his lover and a prostitute, the Duchesse de Guermantes. Proust's characters are in some instances modelled on real people.[55] In sheer writing power, he is able to convey the joy of eating a madeleine, the erotic jealousy of a lover, the exquisite humiliation heaped on a victim of snobbery or anti-Semitism. Whether or not one feels the need to relate him to Bergson, Baudelaire or Zola, as others have done, his descriptions work *as writing*. It is enough.

Proust did not find it easy to publish his book. It was turned down by a number of publishers, including the writer André Gide at *Nouvelle Revue Française*, who thought Proust a snob and a literary amateur. For a while the forty-two-year-old would-be author panicked and considered publishing privately. But then Grasset accepted his book, and he now shamelessly lobbied to get it noticed. Proust did not win the Prix Goncourt as he had hoped, but a number of influential admirers wrote to offer their support, and even Gide had the grace to admit he had been wrong in rejecting the book and offered to publish future volumes. At that stage, in fact, only one other volume had been planned, but war broke out and publication was abandoned. For the time being, Proust had to content himself with his voluminous letters.

Since 1900 Freud had expended a great deal of time and energy extending the reach of the discipline he had founded; psychoanalytic societies now existed in six countries, and an International Association of Psychoanalysis had been formed in 1908. At the same time, the 'movement,' as Freud thought of it, had suffered its first defectors. **Alfred Adler**, along with Wilhelm Stekel, left in 1911, Adler because his own experiences gave him a very different view of the psychological forces that shape personality. Crippled by rickets as a child and suffering from pneumonia, he had been involved in a number of street accidents that made his injuries worse. Trained as an ophthalmologist, he became aware of patients who, suffering from some deficiency in their body, compensated by strengthening other faculties. Blind people, for example, as is well known, develop very acute hearing. A social Democrat and a Jew who had converted to Christianity, Adler tried hard to reconcile the Marxist doctrine of class struggle with his own ideas about psychic struggle. He formed the view that the libido is not a predominantly sexual force but inherently aggressive, the search for power becoming for him the mainspring of life and the 'inferiority complex' the directing force that gives lives their shape.[56] He resigned as spokesman of the Vienna Psychoanalytical Association because its rules stipulated that its aim was the propagation of Freud's views. Adler's brand of 'individual psychology' remained very popular for a number of years.

Freud's break with Carl Jung, which took place between the end of 1912 and the early part of 1914, was much more acrimonious than any of the other schisms because Freud, who was fifty-seven in 1913, saw Jung as his successor, the new leader of 'the movement.' The break came because although Jung had been devoted to Freud at first, he revised his views on two seminal Freudian concepts. Jung thought that the libido was not, as Freud insisted, solely a sexual instinct but more a matter of 'psychic energy' as a whole, a reconceptualisation that, among other things, vitiated the entire idea of childhood sexuality, not to mention the Oedipal relationship.[57] Second, and perhaps even more important, Jung argued that he had discovered the existence of the unconscious for himself, entirely independently of Freud. It had come about, he said, when he had been working at Burghölzli mental hospital in Zurich, where he had seen a 'regression' of the libido in schizophrenia and where he was treating a woman who had killed her favourite child.[58] Earlier in life the woman had fallen in love with a young man who, so she believed, was too rich and too socially superior ever to want to marry her, so she had turned to someone else. A few years later, however, a friend of the rich man had told the woman that he had in fact been inconsolable when she had spurned him. Not long after, she had been bathing her two young children and had allowed her daughter to suck the bath sponge even though she knew the water being used was infected. Worse, she gave her son a glass of infected water. Jung claimed that he had grasped for himself, without Freud's help, the central fact of the case – that the woman was acting from an unconscious desire to obliterate all traces of her present marriage to free herself for the man she really loved. The woman's daughter caught typhoid fever and died from the infected sponge. The mother's symptoms of depression, which appeared when she was told the truth about the wealthy man she had loved, turned worse after her daughter's death, to the point where she had to be sent to Burghölzli.

Jung did not at first question the diagnosis, 'dementia praecox.' The real story emerged only when he began to explore her dreams, which prompted him to give her the 'association test.' This test, which subsequently became very famous, was invented by a German doctor, **Wilhelm Wundt** (1832–1920). The principle is simple: the patient is shown a list of words and asked to respond to each one with the first word that comes into his/her head. The rationale is that in this way conscious control over the unconscious urges is weakened. Resurrecting the woman's case history via her dreams and the association test, Jung realised that the woman had, in effect, murdered her own daughter because of the unconscious urges within her. Controversially, he faced her with the truth. The result was remarkable: far from being untreatable, as the diagnostic label dementia praecox had implied, she recovered quickly and left hospital three weeks later. There was no relapse.

There is already something defiant about Jung's account of his discovery of the unconscious. Jung implies he was not so much a protégé of Freud's as moving in parallel, his equal. Soon after they met, when Jung attended the Wednesday Society in 1907, they became very close, and in 1909 they travelled to America together. Jung was overshadowed by Freud in America, but it was there that Jung realised his views were diverging from the founder's. As the years had passed, patient after patient had reported early experiences of incest, all of which made Freud lay even more emphasis on sexuality as the motor driving the unconscious. For Jung, however, sex was not fundamental – instead, it was itself a transformation from religion. Sex, for Jung, was one aspect of the religious impulse but not the only one. When he looked at the religions and myths of other races around the world, as he now began to do, he found that in Eastern religions the gods were depicted in temples as very erotic beings. For him, this frank sexuality was a symbol and one aspect of 'higher ideas.' Thus he began his famous examination of religion and mythology as 'representations' of the unconscious 'in other places and at other times.'

The rupture with Freud started in 1912, after they returned from America and Jung published the second part of *Symbols of Transformation*.[59] This extended paper, which appeared in the *Jahrbuch der Psychoanalyse*, was Jung's first public airing of what he called the '**collective unconscious**.' Jung concluded that at a deep level the unconscious was shared by everyone – it was part of the '**racial memory**.' Indeed, for Jung, that's what therapy *was*, getting in touch with the collective unconscious.[60] The more Jung explored religion, mythology, and philosophy, the further he departed from Freud and from the scientific approach. As J. A. C. Brown wrote, one 'gets much the same impression from reading Jung as might be obtained from reading the scriptures of the Hindus, Taoists, or Confucians; although well aware that many wise and true things are being said, [one] feels that they could have been said just as well without involving us in the psychological theories upon which they are supposedly based.'[61]

According to Jung, our psychological makeup is divided into three: consciousness, personal unconsciousness, and the collective unconscious. A common analogy is made with geology, where the conscious mind corresponds to that part of land above water. Below the water line, hidden from view, is the personal unconscious, and below that, linking the different landmasses, so to speak, is the 'racial unconscious' where, allegedly, members of the same race share deep psychological similarities. Deepest of all, equating to the earth's core, is the psychological heritage of all humanity, the irreducible fundamentals of human nature and of which we are only dimly aware. This was a bold, simple theory supported, Jung said, by three pieces of 'evidence.' First, he pointed to the 'extraordinary unanimity' of narratives and themes in the mythologies of different cultures. He also argued that 'in protracted analyses, any particular symbol might recur with disconcerting persistency but as analysis proceeded the symbol came to resemble the universal symbols seen in myths and legends.' Finally he claimed that the stories told in the delusions of mentally ill patients often resembled those in mythology.

The notion of archetypes, the theory that all people may be divided according to one or another basic (and inherited) psychological type, the best known being introvert and extrovert, was Jung's other popular idea. These terms relate only to the conscious level of the mind, of course; in typical psychoanalytic fashion, the truth is really the opposite – the extrovert temperament is in fact unconsciously introvert, and vice versa. It thus follows that for Jung psychoanalysis as treatment involved the interpretation of dreams and **free association** in order to put the patient into contact with his or her collective unconscious, a cathartic process. While Freud was sceptical of and on occasions hostile to organised religion, Jung regarded a religious outlook as helpful in therapy. Even Jung's supporters concede that this aspect of his theories is confused.[62]

Although Jung's very different system of understanding the unconscious had first come to the attention of fellow psychoanalysts in 1912, so that the breach was obvious within the profession, it was only with the release of *Symbols of Transformation* in book form in 1913 (published in English as *Psychology of the Unconscious*) that the split with Freud became public. After that there was no chance of a reconciliation: at the fourth International Psychoanalytic Congress, held in Munich in September 1913, Freud and his supporters sat at a separate table from Jung and his acolytes. When the meeting ended, 'we dispersed,' said Freud in a letter, 'without any desire to meet again.'[63] Freud, while troubled by this personal rift, which also had anti-Semitic overtones, was more concerned that Jung's version of psychoanalysis was threatening its status as a science.[64] Jung's concept of the collective unconscious, for example, clearly implied the inheritance of acquired characteristics, which had been discredited by Darwinism for some years. As Ronald Clark commented: 'In short, for the Freudian theory, which is hard enough to test but has some degree of support, Jung [had] substituted an untestable system which flies in the face of current genetics.'[65]

Freud, to be fair, had seen the split with Jung coming and, in 1912, had begun a work that expanded on his own earlier theories and, at the same time, discredited Jung's, trying to ground psychoanalysis in modern science. Finished in the spring of 1913 and published a few months later, this work was described by Freud as 'the most daring enterprise I have ever ventured.'[66] *Totem and Taboo* was an attempt to explore the very territory Jung was trying to make his own, the 'deep ancestral past' of mankind. Whereas Jung had concentrated on the universality of myths to explain the collective – or racial – unconscious, Freud turned to anthropology, in particular to Sir James Frazer's *The Golden Bough* and to Darwin's accounts of the behaviour of primate groupings. According to Freud (who said from the start that *Totem and Taboo* was speculation), primitive society was characterised by an unruly horde in which a despotic male dom- inated all the females, while other males, including his own offspring, were either killed or condemned to minor roles. From time to time the dominant male was attacked and eventually overthrown, a neat link to the Oedipus complex, the lynchpin of 'classical' Freudian theory. *Totem and Taboo* was intended to show how individual and group psychology were knitted together, how psychology was rooted in biology, in 'hard' science. Freud said these theories could be tested (unlike Jung's) by observing primate societies, from which man had evolved.

Freud's new book also 'explained' something nearer home, namely Jung's attempt to unseat Freud as the dominant male of the psychoanalytic 'horde.' A letter of Freud's, written in 1913 but published only after his death, admitted that 'annihilating' Jung was one of his motives in writing *Totem and Taboo*.[67] The book was not a success: Freud was not as up-to-date in his reading as he thought, and science, which he thought he was on top of, was in fact against him.[68] His book regarded evolution as a unilinear process, with various races around the world seen as stages on the way to 'white,' 'civilised' society, a view that was already dated, thanks to the work of Franz Boas. In the 1920s and 1930s anthropologists like Bronislaw Malinowski, Margaret Mead, and Ruth Benedict would produce more and more fieldwork confirming *Totem and Taboo* as scientifically worthless. In attempting to head off Jung, Freud had shot himself in the foot.[69]

Nevertheless, it sealed the breach between the two men (it should not be forgotten that Jung was not the only person Freud fell out with; he also broke with Breuer, Fliess, Adler, and Stekel).[70] Henceforth, Jung's work grew increasingly metaphysical, vague, and quasi-mystical, attracting a devoted but fringe following. Freud continued to marry individual psychology and group behaviour to produce a way of looking at the world that attempted to be more scientific than Jung's. Until 1913 the psychoanalytic movement had been one system of thought. Afterward, it was two.

Mabel Dodge, in her letter to Gertrude Stein, had been right. The explosion of talent in 1913 was volcanic. In addition to the ideas reported here, 1913 also saw the birth of the modern assembly line, at Henry Ford's factory in Detroit, and the appearance of Charlie Chaplin, the little man with baggy trousers, bowler hat, and a cunning cheekiness that embodied perfectly the eternal optimism of an immigrant nation. But it is necessary to be precise about what was happening in 1913. Many of the events of that annus mirabilis were a maturation, rather than a departure in a wholly new direction. Modern art had extended its reach across the Atlantic and found another home; Niels Bohr had built on Einstein and Ernest Rutherford, as Igor Stravinsky had built on Claude Debussy (if not on Arnold Schoenberg); psychoanalysis had conquered Mann and Lawrence and, to an extent, Proust; Jung had built on Freud (or he thought he had), Freud had extended his own ideas, and psychoanalysis, like modern art, had reached across to America; film had constructed its first immortal character as opposed to star. People like Guillaume Apollinaire, Stravinsky, Proust, and Mann were trying to merge together different strands of thought – physics, psychoanalysis, literature, painting – in order to approach new truths about the human condition. Nothing characterised these developments so much as their optimism. The mainstreams of thought, set in flow in the first months of the century, seemed to be safely consolidating.

One man sounded a warning, however, in that same year. In *A Boy's Will*, Robert Frost's voice was immediately distinct: images of the innocent, natural

world delivered in a gnarled, broken rhythm that reminds one of the tricks
nature plays, not least with time:

> Ah, when to the heart of man
>> Was it ever less than a treason
> To go with the drift of things,
>> To yield with a grace to reason.[71]

9
COUNTER-ATTACK

The outbreak of World War I took many highly intelligent people by surprise. On 29 June, Sigmund Freud was visited by the so-called Wolf Man, a rich young Russian who during treatment had remembered a childhood phobia of wolves. The assassination of Archduke Franz Ferdinand of Austro-Hungary and his wife had taken place in Sarajevo the day before. The conversation concerned the ending of the Wolf Man's treatment, one reason being that Freud wanted to take a holiday. The Wolf Man later wrote, 'How little one then suspected that the assassination ... would lead to World War I.'[1] In Britain, at the end of July, J. J. Thomson, who discovered the electron and soon after became president of the Royal Society, was one of the eminent men who signed a plea that 'war upon [Germany] in the interests of Serbia and Russia will be a sin against civilisation.'[2] Bertrand Russell did not fully grasp how imminent war was until, on 2 August, a Sunday, he was crossing Trinity Great Court in Cambridge and met the economist John Maynard Keynes, who was hurrying to borrow a motorcycle with which to travel to London. He confided to Russell he had been summoned by the government. Russell went to London himself the following day, where he was 'appalled' by the war spirit.[3] Pablo Picasso had been painting in Avignon and, fearing the closure of Daniel Henry Kahnweiler's gallery (Kahnweiler, Picasso's dealer, was German) and a slump in the market for his own works, he rushed to Paris a day or so before war was declared and withdrew all his money from his bank account – Henri Matisse later said it amounted to 100,000 gold francs. Thousands of French did the same, but the Spaniard was ahead of most of them and returned to Avignon with all his money, just in time to go to the station to say good-bye to Georges Braque and André Derain, who had been called up and were both impatient to fight.[4] Picasso said later that he never saw the other two men again. It wasn't true; what he meant was that Braque and Derain were never the same after the war.

World War I had a direct effect on many writers, artists, musicians, mathematicians, philosophers, and scientists. Among those killed were August Macke, the Blaue Reiter painter, shot as the German forces advanced into France; the sculptor and painter Henri Gaudier-Brzeska, who died in the French trenches near the English Channel; and the German expressionist painter Franz Marc at Verdun. Umberto Boccioni, the Italian futurist, died on

Italy's Austrian front, and the English poet Wilfred Owen was killed on the Sambre Canal a week before the Armistice.[5] Oskar Kokoschka and Guillaume Apollinaire were both wounded. Apollinaire went home to Paris with a hole in his head and died soon afterward. Bertrand Russell and others who campaigned against the war were sent to jail, or ostracised like Albert Einstein, or declared mad like Siegfried Sassoon.[6] Max Planck lost his son, Karl, as did the painter Käthe Kollwitz (she also lost her grandson in World War II). Virginia Woolf lost her friend Rupert Brooke, and three other British poets, Isaac Rosenberg, Julian Grenfell, and Charles Hamilton Sorley, were also killed. The mathematician and philosopher Lieutenant Ludwig Wittgenstein was interned in a 'Campo Concentramento' in northern Italy, from where he sent Bertrand Russell the manuscript of his recently completed work *Tractatus Logico-Philosophicus*.[7]

Many of the intellectual consequences of the war were much more indirect and took years to manifest themselves. The subject is vast, engrossing, easily worth the several books that have been devoted to it.[8] The sheer carnage, the military stalemate that so characterised the hostilities that took place between 1914 and 1918, and the lopsided nature of the armistice all became ingrained in the mentality of the age, and later ages. The Russian Revolution, which occurred in the middle of the war, brought about its own distorted political, military, and intellectual landscape, which would last for seventy years. This chapter will concentrate on ideas and intellectual happenings that were introduced during World War I and that can be understood as a direct response to the fighting.

Paul Fussell, in *The Great War in Modern Memory*, gives one of the most clear-eyed and harrowing accounts of World War I. He notes that the toll on human life even at the beginning of the war was so horrific that the height requirement for the British army was swiftly reduced from five feet eight in August 1914 to five feet five on 11 October.[9] By 5 November, after thirty thousand casualties in October, men had to be only five feet three to get in. Lord Kitchener, secretary of state for war, asked at the end of October for 300,000 volunteers. By early 1916 there were no longer enough volunteers to replace those that had already been killed or wounded, and Britain's first conscript army was installed, 'an event which could be said to mark the beginning of the modern world.'[10] General Douglas Haig, commander in chief of the British forces, and his staff devoted the first half of that year to devising a massive offensive.

World War I had begun as a conflict between Austro-Hungary and Serbia, following the assassination of the Archduke Franz Ferdinand. But Germany had allied itself with Austro-Hungary, forming the Central Powers, and Serbia had appealed to Russia. Germany mobilised in response, to be followed by Britain and France, which asked Germany to respect the neutrality of Belgium. In early August 1914 Russia invaded East Prussia on the same day that Germany occupied Luxembourg. Two days later, on 4 August, Germany declared war on France, and Britain declared war on Germany. Almost without meaning to, the world tumbled into a general conflict.

After six months' preparation, the Battle of the Somme got under way at seven-thirty on the morning of 1 July 1916. Previously, Haig had ordered the bombardment of the German trenches for a week, with a million and a half shells fired from 1,500 guns. This may well rank as the most unimaginative military manoeuvre of all time – it certainly lacked any element of surprise. As Fussell shows, 'by 7.31' the Germans had moved their guns out of the dugouts where they had successfully withstood the previous week's bombardment and set up on higher ground (the British had no idea how well dug in the Germans were). Out of the 110,000 British troops who attacked that morning along the thirteen-mile front of the Somme, no fewer than 60,000 were killed or wounded on the first day, *still* a record. 'Over 20,000 lay dead between the lines, and it was days before the wounded in No Man's Land stopped crying out.'[11] Lack of imagination was only one cause of the disaster. It may be too much to lay the blame on social Darwinist thinking, but the British General Staff did hold the view that the new conscripts were a low form of life (mainly from the Midlands), too simple and too animal to obey any but the most obvious instructions.[12] That is one reason why the attack was carried out in daylight and in a straight line, the staff feeling the men would be confused if they had to attack at night, or by zigzagging from cover to cover. Although the British by then had the tank, only thirty-two were used 'because the cavalry preferred horses.' The disaster of the Somme was almost paralleled by the attack on Vimy Ridge in April 1917. Part of the infamous Ypres Salient, this was a raised area of ground surrounded on three sides by German forces. The attack lasted five days, gained 7,000 yards, and cost 160,000 killed and wounded – more than twenty casualties for each yard of ground that was won.[13]

Passchendaele was supposed to be an attack aimed at the German submarine bases on the Belgian coast. Once again the ground was 'prepared' by artillery fire – 4 million shells over ten days. Amid heavy rain, the only effect was to churn up the mud into a quagmire that impeded the assault forces. Those who weren't killed by gun- or shell-fire died either from cold or literally drowned in the mud. British losses numbered 370,000. Throughout the war, some 7,000 officers and men were killed or wounded every day: this was called 'wastage.'[14] By the end of the war, half the British army was aged less than nineteen.[15] No wonder people talked about a 'lost generation.'

The most brutally direct effects of the war lay in medicine and psychology. Major developments were made in the understanding of cosmetic surgery and vitamins that would eventually lead to our current concern with a healthy diet. But the advances that were of the most immediate importance were in blood physiology, while the most contentious innovation was the IQ – **Intelligence Quotient** – test. The war also helped in the much greater acceptance afterwards of psychiatry, including psychoanalysis.*

---

* The hostilities also hastened man's understanding of flight, and introduced the tank. But the principles of the former were already understood, and the latter, though undeniably important, had little impact outside military affairs.

It has been estimated that of some 56 million men called to arms in World War I, around 26 million were casualties.[16] The nature of the injuries sustained was different from that of other wars insofar as high explosives were much more powerful and much more frequently used than before. This meant more wounds of torn rather than punctured flesh, and many more dismemberments, thanks to the machine gun's 'rapid rattle.' Gunshot wounds to the face were also much more common because of the exigencies of trench warfare; very often the head was the only target for riflemen and gunners in the opposing dugouts (steel helmets were not introduced until the end of 1915). This was also the first major conflict in which bombs and bullets rained down from the skies. As the war raged on, airmen began to fear fire most of all. Given all this, the unprecedented nature of the challenge to medical science is readily appreciated. Men were disfigured beyond recognition, and the modern science of **cosmetic surgery** evolved to meet this dreadful set of circumstances. Hippocrates rightly remarked that war is the proper school for surgeons.

Whether a wound disfigured a lot or a little, it was invariably accompanied by the loss of blood. A much greater understanding of blood was the second important medical advance of the war. Before 1914, **blood transfusion** was virtually unknown. By the end of hostilities, it was almost routine.[17] William Harvey had discovered the circulation of the blood in 1616, but it was not until 1907 that a doctor in Prague, **Jan Jansky**, showed that all human blood could be divided into **four groups**, O, A, B, and AB, distributed among European populations in fairly stable proportions.[18] This identification of blood groups showed why, in the past, so many transfusions hadn't worked, and patients had died. But there remained the problem of clotting: blood taken from a donor would clot in a matter of moments if it was not immediately transferred to a recipient.[19] The answer to this problem was also found in 1914, when two separate researchers in New York and Buenos Aires announced, quite independently of each other and almost at the same time, that a 0.2 percent solution of sodium citrate acted as an efficient anticoagulant and that it was virtually harmless to the patient.[20] **Richard Lewisohn**, the New York end of this duo, perfected the dosage, and two years later, in the killing fields of France, it had become a routine method for treating haemorrhage.[21] Kenneth Walker, who was one of the pioneers of blood transfusion, wrote in his memoirs, 'News of my arrival spread rapidly in the trenches and had an excellent effect on the morale of the raiding party. "There's a bloke arrived from G.H.Q. who pumps blood into you and brings you back to life even after you're dead," was very gratifying news for those who were about to gamble with their lives.'[22]

Mental testing, which led to the concept of the IQ, was a French idea, brainchild of the Nice-born psychologist **Alfred Binet**. At the beginning of the century Freudian psychology was by no means the only science of behaviour. The Italo-French school of craniometry and stigmata was also popular. This reflected the belief, championed by the Italian Cesare Lombroso and the Frenchman Paul Broca, that intelligence was linked to brain size and that personality – in particular personality defects, notably criminality – was related to facial or other bodily features, what Lombroso called 'stigmata.'

Binet, a professor at the Sorbonne, failed to confirm Broca's results. In 1904 he was asked by France's Minister of Public Education to carry out a study to develop a technique that would help identify those children in France's schools who were falling behind the others and who therefore needed some form of special education. Disillusioned with craniometry, Binet drew up a series of very short tasks associated with everyday life, such as counting coins or judging which of two faces was 'prettier.' He did not test for the obvious skills taught at school – math and reading for example – because the teachers already knew which children failed on those skills.[23] Throughout his studies, Binet was very practical, and he did not invest his tests with any mystical powers.[24] In fact, he went so far as to say that it didn't matter what the tests were, so long as there were a lot of them and they were as different from one another as could be. What he wanted to be able to do was arrive at a single score that gave a true reflection of a pupil's ability, irrespective of how good his or her school was and what kind of help he or she received at home.

Three versions of Binet's scale were published between 1905 and 1911, but it was the 1908 version that led to the concept of the so-called IQ.[25] His idea was to attach an age level to each task: by definition, at that age a normal child should be able to fulfil the task without error. Overall, therefore, the test produced a rounded 'mental age' of the child, which could be compared with his or her actual age. To begin with, Binet simply subtracted the 'mental age' from the chronological age to get a score. But this was a crude measure, in that a child who was two years behind, say, at age six, was more retarded than a child who was two years behind at eleven. Accordingly, in 1912 the German psychologist **W. Stern** suggested that mental age should be *divided* by chronological age, a calculation that produced the intelligence quotient.[26] It was never Binet's intention to use the IQ for normal children or adults; on the contrary, he was worried by any attempt to do so. However, by World War I, his idea had been taken to America and had completely changed character.

The first populariser of Binet's scales in America was **H. H. Goddard**, the contentious director of research at the Vineland Training School for Feeble-minded Girls and Boys in New Jersey.[27] Goddard was a much fiercer Darwinian than Binet, and after his innovations mental testing would never be the same again.[28] In those days, there were two technical terms employed in psychology that are not always used in the same way now. An '**idiot**' was someone who could not master full speech, so had difficulty following instructions, and was judged to have a mental age of not more than three. An '**imbecile**,' meanwhile, was someone who could not master written language and was considered to have a mental age somewhere between three and seven. Goddard's first innovation was to coin a new term – '**moron**,' from the Greek, meaning foolish – to denote the feebleminded individuals who were just below normal intelligence.[29] Between 1912 and the outbreak of war Goddard carried out a number of experiments in which he concluded, alarmingly – or absurdly – that between 50 and 80 percent of ordinary Americans had mental ages of eleven or less and were therefore morons. Goddard was alarmed because, for him, the moron was the chief threat to society. This was because idiots and imbeciles were obvious, could be locked up without too much public concern, and were in any case extremely unlikely to reproduce. On the other hand, for Goddard, morons could never be leaders or even really think for themselves; they were workers, drones who had to be told what to do. There were a lot of them, and most would reproduce to manufacture more of their own kind. Goddard's real worry was immigration, and in one extraordinary set of studies where he was allowed to test the immigrants then arriving at Ellis Island, he managed to show to his own satisfaction (and again, alarm) that as many as four-fifths of Hungarians, Italians, and Russians were 'moronic.'[30]

Goddard's approach was taken up by **Lewis Terman**, who amalgamated it with that of Charles Spearman, an English army officer who had studied under the famous German psychologist Wilhelm Wundt at Leipzig and fought in the Boer War. Until Spearman, most of the practitioners of the young science of psychology were interested in people at the extremes of the intelligence scale – the very dull or the very bright. But Spearman was interested in the tendency of those people who were good at one mental task to be good at others. In time this led him to the concept of intelligence as made up of a 'general' ability, or g, which he believed underlay many activities. On top of g, said Spearman, there were a number of specific abilities, such as mathematical, musical, and spatial ability. This became known as the two-factor theory of intelligence.[31]

By the outbreak of World War I, Terman had moved to California. There, attached to Stanford University, he refined the tests devised by Binet and his other predecessors, making the '**Stanford-Binet**' tests less a diagnosis of people in need of special education and more an examination of 'higher,' more complex cognitive functioning, ranging over a wider spread of abilities. Tasks included such things as size of vocabulary, orientation in space and time, ability to detect absurdities, knowledge of familiar things, and eye–hand coordination.[32] Under Terman, therefore, the IQ became a general concept that could be applied to anyone and everyone. Terman also had the idea to multiply Stern's calculation of the IQ (mental age divided by chronological age) by 100, to rule out the decimal point. By definition, therefore, an average IQ became 100, and it was this round figure that, as much as anything, caused 'IQ' to catch on in the public's imagination.

It was at this point that world events – and the psychologist Robert Yerkes – intervened.[33] Yerkes was nearly forty when the war started, and by some accounts a frustrated man.[34] He had been on the staff of the Harvard faculty since the beginning of the century, but it rankled with him that his discipline still wasn't accepted as a science. Often, for example, in universities psychology was part of the philosophy department. And so, with Europe already at war, and with America preparing to enter, Yerkes had his one big idea – that psychologists should use mental testing to help assess recruits.[35] It was not forgotten that the British had been shocked during the Boer War to find out how poorly their recruits rated on tests of physical health; the eugenicists had been complaining for years that the quality of American immigrants was declining; here was a chance to kill two birds with one stone – assess a huge number of people to gain some idea of what the average mental age really was and see how immigrants compared, so that they too might be best used in the coming war effort. Yerkes saw immediately that, in theory at least, the U.S. armed services could benefit enormously from psychological testing: it could not only weed out the weaker men but also identify those who would make the best commanders, operators of complex equipment, signals officers, and so forth. This ambitious goal required an extraordinary broadening of available intelligence testing technology in two ways – there would have to be group testing, and the tests would have to identify high flyers as well as the inadequate rump. Although the navy turned down Yerkes's initiative, the army adopted it – and never regretted it. He was made a colonel, and he would later proclaim that mental testing 'had helped to win the war.' This was, as we shall see, an exaggeration.[36]

It is not clear how much use the army made of Yerkes's tests. The long-term significance of the military involvement lay in the fact that, over the course of the war, Yerkes, Terman, and another colleague named **C. C. Brigham** carried out tests on no fewer than 1.75 million individuals.[37] When this unprecedented mass of material had been sifted (after the war), three main results emerged. The first was that the average mental age of recruits was thirteen. This sounds pretty surprising to us at this end of the century: a nation could scarcely hope to survive in the modern world if its average mental age really was thirteen. But in the eugenicist climate of the time, most people preferred the 'doom' scenario to the alternative view, that the tests were simply wrong. The second major result was that European immigrants could be graded by their country of origin, with (surprise, surprise) darker people from the southern and eastern parts of the continent scoring worse than those fairer souls from the north and west. Third, the Negro was at the bottom, with a mental age of ten and a half.[38]

Shortly after World War I, Terman collaborated with Yerkes to introduce the National Intelligence Tests, constructed on the army model and designed to measure the intelligence of groups of schoolchildren. The market had been primed by the army project's publicity, and intelligence testing soon became big business. With royalties from the sales of his tests, Terman became a wealthy as well as a prominent psychologist. And then, in the 1920s, when a fresh wave of xenophobia and the eugenic conscience hit America, the wartime IQ results came in very handy. They played their part in restricting immigration, with what results we shall see.[39]

The last medical beneficiary of World War I was psychoanalysis. After the assassination of the archduke in Sarajevo, Freud himself was at first optimistic about a quick and painless victory by the Central Powers. Gradually, however, like others he was forced to change his mind.[40] At that stage he had no idea that the war would affect the fortunes of psychoanalysis so much. For example, although America was one of the half-dozen or so foreign countries that had a psychoanalytic association, the discipline was still regarded in many quarters as a fringe medical speciality, on a level with faith healing or yoga. The situation was not much different in Britain. When *The Psychopathology of Everyday Life* was published in translation in Britain in the first winter of the war, the book was viciously attacked in the review pages of the *British Medical Journal*, where psychoanalysis was described as 'abounding nonsense' and 'a virulent pathogenic microbe.' At other times, British doctors referred slightingly to Freud's 'dirty doctrines.'[41]

What caused a change in the views of the medical profession was the fact that, on both sides in the war, a growing number of casualties were suffering from shell shock (or combat fatigue, or battle neurosis, to use the terms now favoured). There had been cases of men breaking down in earlier wars, but their numbers had been far fewer than those with physical injuries. What seemed to be crucially different this time was the character of hostilities – static trench warfare with heavy bombardment, and vast conscript armies which contained large numbers of men unsuited for war.[42] Psychiatrists quickly realised that in the huge civilian armies of World War I there were many men who would not normally have become soldiers, who were unfit for the strain, and that their 'civilian' neuroses would express themselves under the terror of bombardment. Doctors also learned to distinguish such men from those who had more resilient psychoses but through fatigue had come to the end of their tether. The intense scrutiny of the men on the stage in the theatre of war revealed to psychology much that would not have been made evident in years and years of peace. As Rawlings Rees noted, 'The considerable incidence of battle neurosis in the war of 1914–18 shook psychiatry, and medicine as a whole, not a little.' But it also helped make psychiatry respectable.[43] What had been the mysteries of a small group of men and women was now more widely seen as a valuable aid to restoring some normality to a generation that had gone almost insane with the horror of it all. An analysis of 1,043,653 British casualties revealed that neuroses accounted for 34 percent.[44]

Psychoanalysis was not the only method of treatment tried, and in its classical form it took too long to have an effect. But that wasn't the point. Both the Allied and Central powers found that officers were succumbing as well as enlisted men, in many cases highly trained and hitherto very brave men; these behaviours could not in any sense be called malingering. And such was the toll of men in the war that clinics well behind enemy lines, and even back home, became necessary so that soldiers could be treated, and then returned to the front.[45] Two episodes will show how the war helped bring psychoanalysis within the fold. The first occurred in February 1918, when Freud received a copy of a paper by **Ernst Simmel**, a German doctor who had been in a field hospital as a medical staff officer. He had used hypnosis to treat so-called malingerers but had also constructed a human dummy against which his patients could vent their repressed aggression. Simmel had found his method so successful that he had applied to the German Secretary of State for War for funds for a plan to set up a psychoanalytic clinic. Although the German government never took any action on this plan during wartime, they did send an observer to the International Congress of Psychoanalysis in 1918 in Budapest.[46] The second episode took place in 1920 when the Austrian government set up a commission to investigate the claims against **Julius von Wagner-Jauregg**, a professor of psychiatry in Vienna. Wagner-Jauregg was a very distinguished doctor who won the Nobel Prize in 1927 for his work on the virtual extinction of cretinism (mental retardation caused by thyroid deficiency) in Europe, by countering the lack of iodine in the diet. During the war Wagner-Jauregg had been responsible for the treatment of battle casualties, and in the aftermath of defeat there had been many complaints from troops about the brutality of some of his treatments, including electric-shock therapy. Freud was called before the commission, and his testimony, and Wagner-Jauregg's, were soon seen as a head-to-head tussle of rival theories. The commission decided that there was no case against Wagner-Jauregg, but the very fact that Freud had been called by a government-sponsored commission was one of the first signs of his more general acceptance. As Freud's biographer Ronald Clark says, the Freudian age dates from this moment.[47]

'At no other time in the twentieth century has verse formed the dominant literary form' as it did in World War I (at least in the English language), and there are those, such as Bernard Bergonzi, whose words these are, who argue that English poetry 'never got over the Great War.' To quote Francis Hope, 'In a not altogether rhetorical sense, all poetry written since 1918 is war poetry.'[48] In retrospect it is not difficult to see why this should have been so. Many of the young men who went to the front were well educated, which in those days included being familiar with English literature. Life at the front, being intense and uncertain, lent itself to the shorter, sharper, more compact structure of verse, war providing unusual and vivid images in abundance. And in the unhappy event of the poet's death, the elegiac nature of a slim volume had an undeniable romantic appeal. Many boys who went straight from the cricket field to the Somme or Passchendaele made poor poets, and the bookshops were crammed with verse that, in other circumstances, would never have been published. But amid these a few stood out, and of those a number are now household names.[49]

The poets writing during World War I can be divided into two groups. There were those early poets who wrote about the glory of war and were then killed. And there were those who, killed or not, lived long enough to witness the carnage and horror, the awful waste and stupidity that characterised so much of the 1914–18 war.[50] **Rupert Brooke** is the best known of the former group. It has been said of Brooke that he was prepared all his short life for the role of war poet/martyr. He was handsome, with striking blond hair; he was clever, somewhat theatrical, a product of the Cambridge milieu that, had he lived, would surely have drawn him to Bloomsbury. Frances Cornford wrote a short stanza about him while he was still at Cambridge:

A young Apollo, golden-haired,
Stands dreaming on the verge of strife,
Magnificently unprepared
For the long littleness of life.[51]

Before the war Brooke was one of the Georgian Poets who celebrated rural

England; their favoured techniques were unpretentious and blunt, if somewhat complacent.[52] In 1914 there had been no major war for a hundred years, since Waterloo in 1815; reacting to the unknown was therefore not easy. Many of Brooke's poems were written in the early weeks of the war when many people, on both sides, assumed that hostilities would be over very quickly. He saw brief action outside Antwerp in the autumn of 1914 but was never really in any danger. A number of his poems were published in an anthology called *New Numbers*. Little notice was taken of them until on Easter Sunday, 1915, the dean of St Paul's Cathedral quoted Brooke's 'The Soldier' in his sermon. As a result *The Times* of London reprinted the poem, which gave Brooke a much wider audience. A week later his death was reported. It wasn't a 'glamorous' death, for he had died from blood poisoning in the Aegean; he had not been killed in the fighting, but he had been on active service, on his way to Gallipoli, and the news turned him into a hero.[53]

Several people, including his fellow poet Ivor Gurney, have remarked that Brooke's poetry is less about war than about what the English felt – or wanted to feel – about the events of the early months of the war.[54] In other words, they tell us more about the popular state of mind in England than about Brooke's own experience of fighting in the war at the front. His most famous is 'The Soldier' (1914):

If I should die, think only this of me:
    That there's some corner of a foreign field
That is for ever England. There shall be
    In that rich earth a richer dust concealed;
A dust whom England bore, shaped, made aware,
    Gave, once, her flowers to love, her ways to roam,
A body of England's, breathing English air,
    Washed by the rivers, blest by suns of home.

Robert Graves, born in Wimbledon in 1895, was the son of the Irish poet Alfred Perceval Graves. While serving in France, he was wounded, lay unconscious on a stretcher in a converted German dressing station, and was given up for dead.[55] Graves was always interested in mythology, and his verse was curiously distant and uncomfortable. One of his poems describes the first corpse he had seen – a German dead on the trench wire whom, therefore, Graves couldn't bury. This was hardly propaganda poetry, and indeed many of Graves's stanzas rail against the stupidity and bureaucratic futility of the conflict. Most powerful perhaps is his reversal of many familiar myths:

One cruel backhand sabre-cut –
'I'm hit! I'm killed!' young David cries,
Throws blindly forward, chokes … and dies.
Steel-helmeted and grey and grim
Goliath straddles over him.[56]

This is antiheroic, deflating and bitter. Goliath isn't supposed to win. Graves himself suppressed his poetry of war, though *Poems about War* was reissued after his death in 1985.[57]

Unlike Brooke and Graves, **Isaac Rosenberg** did not come from a middle-class, public school background, nor had he grown up in the country. He was born into a poor Jewish family in Bristol and spent his childhood in London's East End, suffering indifferent health.[58] He left school at fourteen, and some wealthy friends who recognised his talents paid for him to attend the Slade School to learn painting, where he met David Bomberg, C. R. W. Nevinson, and Stanley Spencer.[59] He joined the army, he said, not for patriotic reasons but because his mother would benefit from the separation allowance. He found army life irksome and never rose above private. But never having been schooled in any poetic tradition, he approached the war in a particular way. He kept art and life separate and did not try to turn the war into metaphor; rather he grappled with the unusual images it offered to re-create the experience of war, which is a part of life and yet not part of most people's lives:

> The darkness crumbles away –
> It is the same old druid Time as ever.
> Only a live thing leaps my hand –
> A queer sardonic rat –
> As I pull the parapet's poppy
> To stick behind my ear.

And later,

> Poppies whose roots are in man's veins
> Drop, and are ever dropping;
> But mine in my ear is safe,
> Just a little white with the dust.
>
> – 'Break of Day in the Trenches,' 1916

Above all, you are *with* Rosenberg. The rat, skittering through no-man's-land with a freedom no man enjoys, the poppies, drawing life from the blood-sodden ground, are powerful as images, but it is the immediacy of the situation that is conveyed. As he said in a letter, his style was 'surely as simple as ordinary talk.'[60] Rosenberg's is an unflinching gaze, but it is also understated. The horror speaks for itself. This is perhaps why Rosenberg's verse has lost less of its power than other war poems as the years have gone by. He was killed on April Fool's Day, 1918.

Wilfred Owen is generally regarded as Rosenberg's only equal, and maybe even his superior. Born in Oswestry in Shropshire in 1893, into a religious, traditional family, Owen was twenty-one when war was declared.[61] After matriculating at London University, he became the pupil and lay assistant to a vicar in an Oxfordshire village, then obtained a post as a tutor in English at the Berlitz School of Languages in Bordeaux. In 1914, after war broke out, he witnessed the first French casualties arriving at the hospital in Bordeaux and wrote home to his mother vividly describing their wounds and his pity. In

October 1915 he was accepted for the Artists' Rifles (imagine a regiment with that name now) but was commissioned in the Manchester Regiment. He sailed to France on active service at the end of December 1916, attached to the Lancashire Fusiliers. By then, the situation at the front was in strong contrast to the image of the front being kept alive by government propaganda back home.

Owen's first tour of duty on the Somme was an overwhelming experience, as his letters make clear, and he went through a rapid and remarkable period of maturing. He was injured in March 1917 and invalided home via a series of hospitals, until he ended up in June in Craiglockhart Hospital outside Edinburgh, which, says his biographer, 'was the most considerable watershed in Wilfred's short life.'[62] This was the famous psychiatric hospital where **W. H. Rivers**, one of the medical staff, was making early studies, and cures, of shell shock. While at Craiglockhart, Owen met Edmund Blunden and Siegfried Sassoon, who both left a record of the encounter in their memoirs. Sassoon's *Siegfried's Journey* (not published until 1948) has this to say about their poetry: 'My trench sketches were like rockets, sent up to illuminate the darkness. They were the first of their kind, and could claim to be opportune. It was Owen who revealed how, out of realistic horror and scorn, poetry might be made.'[63] Owen went back to the front in September 1918, partly because he believed in that way he might argue more forcefully against the war. In October he won the Military Cross for his part in a successful attack on the Beaurevoir-Fonsomme line. It was during his final year that his best poems were composed. In 'Futility' (1918), Owen is light years away from Brooke and very far even from Rosenberg. He paints a savage picture of the soldier's world, a world very different from anything his readers back home would have ever encountered. His target is the destruction of youth, the slaughter, the maiming, the sense that it might go on for ever, while at the same time he discovers a language wherein the horror may be shown in a clear, beautiful, but always terrible way:

Move him into the sun —
Gently its touch awoke him once,
At home, whispering of fields unsown.
Always it woke him, even in France,
Until this morning and this snow.
If anything might rouse him now
The kind old sun will know.

Think how it wakes the seeds —
Woke, once, the clays of a cold star.
Are limbs, so dear-achieved, are sides,
Full-nerved — still warm — too hard to stir?
Was it for this the clay grew tall?
— O what made fatuous sunbeams toil
To break earth's sleep at all?

In poems like 'The Sentry' and 'Counter-Attack,' the physical conditions and the terror are locked into the words; carnage can occur at any moment.

> We'd found an old Boche dug out, and he knew,
> And gave us hell; for shell on frantic shell
> Lit full on top, but never quite burst through.
> Rain, guttering down in waterfalls of slime,
> Kept slush waist-high and rising hour by hour

For Owen the war can never be a metaphor for anything – it is too big, too horrific, to be anything other than itself. His poems need to be read for their cumulative effect. They are not rockets 'illuminating the darkness' (as Sassoon described his own work), but rather like heavy artillery shells, pitting the landscape with continual bombardment. The country has failed Owen; so has the church; so – he fears – has he failed himself. All that is left is the experience of war.[64]

> I have made fellowships –
>> Untold of happy lovers in old song.
>> For love is not the binding of fair lips
>> With the soft silk of eyes that look and long,
>
> By Joy, whose ribbon slips, –
>> But wound with war's hard wire whose stakes
>>> are strong;
>> Bound with the bandage of the arm that drips;
>> Knit in the webbing of the rifle-thong.
>
>> *–Apologia Pro Poemate Meo*, 1917

Owen saw himself, in Bernard Bergonzi's felicitous phrase, as both priest and victim. W. B. Yeats notoriously left him out of the *Oxford Book of Modern Verse* (1936) with the verdict that 'passive suffering was not a proper subject for poetry,' a spiteful remark that some critics have put down to jealousy. Owen's verse has certainly lasted. He was killed in action, trying to get his men across the Sambre Canal. It was 4 November 1918, and the war had less than a week to go.

The war in many ways changed incontrovertibly the way we think and what we think about. In 1975, in *The Great War and Modern Memory*, Paul Fussell, then a professor at Rutgers University in New Jersey and now at the University of Pennsylvania, explored some of these changes. After the war the idea of progress was reversed, for many a belief in God was no longer sustainable, and irony – a form of distance from feeling – 'entered the modern soul as a permanent resident.'[65] Fussell also dates what he calls 'the modern *versus* habit' to the war – that is, a dissolution of ambiguity as a thing to be valued, to be replaced instead by 'a sense of polarity' where the enemy is so wicked that his position is deemed a flaw or perversion, so that 'its total submission is called for.' He noted the heightened erotic sense of the British during the war, one aspect being the number of women who had lost lovers at the front and who came together afterward to form lesbian couples – a common sight in the 1920s and 1930s. In turn, this pattern may have contributed to a general view that female homosexuality was more unusual in its aetiology than is in fact the case. But it may have made lesbianism more acceptable as a result, being overlaid with sympathy and grief.

Building on the work of Fussell, Jay Winter, in *Sites of Memory, Sites of Mourning* (1995), made the point that the apocalyptic nature of the carnage and the unprecedented amount of bereavement that it caused drove many people away from the novelties of modernism – abstraction, *vers libre*, atonalism and the rest – and back to more traditional forms of expression.[66] War memorials in particular were realistic, simple, conservative. Even the arts produced by avant-gardists – Otto Dix, Max Beckmann, Stanley Spencer, and even Jean Cocteau and Pablo Picasso in their collaboration with Erik Satie on his modernist ballet *Parade* (1917) – fell back on traditional and even Christian images and themes as the only narratives and myths that could make sense of the overwhelming nature of 'a massive problem shared.'[67] In France, there was a resurgence of *images d'Epinal*, pietistic posters that had not been popular since the early nineteenth century, and a reappearance of apocalyptic, 'unmodern' literature, especially but not only in France: Henri Barbusse's *Le Feu* and Karl Kraus's *Last Days of Mankind* are two examples. Despite its being denounced by the Holy See, there was a huge increase in spiritualism as an attempt to talk to the dead. And this was not merely a fad among the less well educated. In France the Institut Métaphysique was headed by Charles Richet, Nobel Prize-winning physiologist, while in Britain the president of the Society for Psychical Research was Sir Oliver Lodge, professor of physics at Liverpool University and later principal of Birmingham University.[68] Winter included in his book 'spirit photographs' taken at the Remembrance Day ceremony in Whitehall in 1922, when the dead allegedly appeared to watch the proceedings. Abel Gance used a similar approach in one of the great postwar films, *J'accuse* (1919), in which the dead in a battlefield graveyard rise up with their bandages and crutches and walking sticks and return to their villages, to see if their sacrifices were worth it: 'The sight of the fallen so terrifies the townspeople that they immediately mend their ways, and the dead return to their graves, their mission fulfilled.'[69] They were easily satisfied.

But other responses – and perhaps the best – would take years to ripen. They would form part of the great literature of the 1920s, and even later.

All the developments and episodes discussed so far in this chapter were direct responses to war. In the case of **Ludwig Wittgenstein**, the work he produced *during* the war was not a response to the fighting itself. At the same time, had not Wittgenstein been exposed to the real possibility of death, it is unlikely that he would have produced *Tractatus Logico-Philosophicus* when he did, or that it would have had quite the tone that it did.

Wittgenstein enlisted on 7 August, the day after the Austrian declaration of war on Russia, and was assigned to an artillery regiment serving at Kraków on the eastern front.[70] He later suggested that he went to war in a romantic mood, saying that he felt the experience of facing death would, in some indefinable manner, improve him (Rupert Brooke said much the same). On the first sight of the opposing forces, he confided in a letter, 'Now I have the chance to be a decent human being, for I am standing eye to eye with death.'[71]

Wittgenstein was twenty-five when war broke out, one of eight children. His family was Jewish, wealthy, perfectly assimilated into Viennese society. Franz Grillparzer, the patriotic poet and dramatist, was a friend of Ludwig's father, and Johannes Brahms gave piano lessons to both his mother and his aunt. The Wittgensteins' musical evenings were well known in Vienna: Gustav Mahler and Bruno Walter were both regulars, and Brahms's Clarinet Quintet received its first performance there. Margarete Wittgenstein, Ludwig's sister, sat for Gustav Klimt, whose painting of her is full of gold, purple, and tumbling colours.[72] Ironically, Ludwig, now the best remembered of the Wittgensteins, was originally regarded by other family members as the dullest. Margarete had her beauty; Hans, one of the older brothers, began composing at the age of four, by which time he could play the piano and the violin; and Rudolf, another older brother, went to Berlin to be an actor. Had Hans not disappeared, sailing off Chesapeake Bay in 1903, and Rudolf not taken cyanide in a Berlin bar after buying the pianist a drink and requesting him to play a popular song, 'I Am Lost,' Ludwig might never have shone.[73] Both his brothers were tortured by the feeling that they had failed to live up to their father's stiff demands that they pursue successful business careers. Rudolf was also tormented by what he felt was a developing homosexuality.

Ludwig was as fond of music as the rest of the family, but he was also the most technical and practical minded. As a result, he wasn't sent to the grammar school in Vienna but to Realschule in Linz, a school chiefly known for the teaching of the history master, **Leopold Pötsch**, a rabid right-winger who regarded the Habsburg dynasty as 'degenerate.'[74] For him, loyalty to such an entity as the Habsburgs was absurd; instead he revered the more accessible *völkisch* nationalism of the Pan-German movement. There is no sign that Wittgenstein was ever attracted by Pötsch's theories, but a fellow pupil, with whom he overlapped for a few months, certainly was. His name was Adolf Hitler. After Linz, Wittgenstein went to Berlin, where he became interested in philosophy. He also developed a fascination with aeronautics, and his father, still anxious for one of his sons to have a lucrative career, suggested he go to Manchester University in England, where there was an excellent engineering department. Ludwig duly enrolled in the engineering course as planned. He also attended the seminars of Horace Lamb, the professor of mathematics. It was in one of his seminars that Wittgenstein was introduced by a fellow student to Bertrand Russell's *Principles of Mathematics*. This book, as we have seen earlier, showed that mathematics and logic are the same. For Wittgenstein, Russell's book was a revelation. He spent months studying *The Principles* and also Gottlob Frege's *Grundgesetze der Arithmetik* (Fundamental Laws of Arithmetic).[75] In the late summer of 1911 Wittgenstein travelled to Jena in Germany to visit Frege, a small man 'who bounced around the room when he talked,' who was impressed enough by the young Austrian to recommend that he study under Bertrand Russell at Cambridge.[76] Wittgenstein's approach to Russell coincided with the Englishman just having finished *Principia Mathematica*. The young Viennese arrived in Cambridge in 1911, and to begin with people's opinions of him were mixed. Nicknamed 'Witter-Gitter,' he was generally considered dull, with a laboured Germanic sense of humour. Like Arnold Schoenberg and Oskar Kokoschka he was an autodidact and didn't care what people thought of him.[77] But it soon got about that the pupil was rapidly overtaking the master, and when Russell arranged for Wittgenstein to be invited to join the Apostles, a highly secret and selective literary society dating back to 1820 and dominated at that time by Lytton Strachey and Maynard Keynes, 'Cambridge realised that it had another genius on its hands.'[78]

By 1914, after he had been in Cambridge for three years, Wittgenstein, or Luki as he was called, began to formulate his own theory of logic.[79] But then, in the long vacation, he went home to Vienna, war was declared, and he was trapped. What happened over the next few years was a complex interplay between Wittgenstein's ideas and the danger he was in at the front. Early on in the war he conceived what he called the picture theory of language – and it was this that was refined during the Austrian army's chaotic retreat under Russian attack. In 1916, however, Wittgenstein was transferred to the front as an ordinary soldier after the Russians attacked the Central Powers on their Baltic flank. He proved brave, asking to be assigned to the most dangerous place, the observation post on the front line, which guaranteed he would be a target. 'Was shot at,' his diary records on 29 April that year.[80] Despite all this, he wrote some philosophy in those months, until June at least, when Russia launched its long-planned Brusilov offensive and the fighting turned heavy. At this point Wittgenstein's diaries show him becoming more philosophical, even religious. At the end of July the Austrians were driven back yet again, this time into the Carpathian Mountains, in icy cold, rain, and fog.[81] Wittgenstein was shot at once more, recommended for the Austrian equivalent of the Victoria Cross (he was given a slightly lesser honour) and promoted three times, eventually to officer.[82] At officer school he revised his book in collaboration with a kindred spirit, Paul Engelmann, and then returned as a *Leutnant* on the Italian front.[83] He completed the book during a period of leave in 1918 after his uncle Paul had bumped into him at a railway station where Wittgenstein was contemplating suicide. The uncle persuaded his nephew to go with him to Hallein, where he lived.[84] There Wittgenstein finished the new version before returning to his unit. Before the manuscript was published, however, Wittgenstein was taken prisoner in Italy, with half a million other soldiers. While incarcerated in a concentration camp, he concluded that his book had solved all the outstanding problems of philosophy and that he would give up the discipline after the war and become a schoolteacher. He also decided to give away his fortune. He did both.

Few books can have had such a tortuous birth as the *Tractatus*. Wittgenstein had great difficulty finding a publisher, the first house he approached agreeing to take the book only if he paid for the printing and the paper himself.[85] Other publishers were equally cautious and his book did not appear in English until

1922.[86] But when it did appear, *Tractatus Logico-Philosophicus* created a sensation. Many people did not understand it; others thought it 'obviously defective', 'limited' and that it stated the obvious. Frank Ramsay, in the philosophical journal *Mind*, said, 'This is a most important book containing original ideas on a large range of topics, forming a coherent system ...'[87] Keynes wrote to Wittgenstein, 'Right or wrong, it dominates all fundamental discussions at Cambridge since it was written.'[88] In Vienna, it attracted the attention of the philosophers led by Moritz Schlick – a group that eventually evolved into the famous Vienna Circle of logical positivists.[89] As Ray Monk, Wittgenstein's biographer describes it, the book comprises a Theory of Logic, a Picture Theory of Propositions and a 'quasi-Schopenhauerian mysticism.' The argument of the book is that language corresponds to the world, as a picture or model corresponds to the world that it attempts to de*pict*. The book was written in an uncompromising style. 'The *truth* of the thoughts that are here communicated,' so runs the preface, 'seems to me unassailable and definitive.' Wittgenstein added that he had found the solution to the problems of philosophy 'on all essential points,' and concluded the preface, 'if I am not mistaken in this belief, then the second thing in which the value of this work consists is that it shows how little is achieved when these problems are solved.' The sentences in the book are simple, and numbered, remark 2.151 a refinement of 2.15, which cannot be understood without reference to the remarks in 2.1. Few of these remarks are qualified; instead each is advanced, as Russell once put it, 'as if it were a Czar's ukase.'[90] Frege, whose own work had inspired the *Tractatus*, died without ever understanding it.

It is perhaps easier to grasp what Wittgenstein was driving at in the *Tractatus* if we concentrate on the second half of his book. His major innovation was to realise that language has limitations, that there are certain things it cannot do and that these have logical and therefore philosophical consequences. For example, Wittgenstein argues that it is pointless to talk about value – simply because 'value is not part of the world'. It therefore follows that all judgements about moral and aesthetic matters cannot – ever – be meaningful uses of language. The same is true of philosophical generalisations that we make about the world as a whole. They are meaningless if they cannot be broken down into elementary sentences 'which really are pictures.' Instead, we have to lower our sights, says Wittgenstein, if we are to make sense. The world can only be spoken about by careful description of the individual facts of which it is comprised. In essence, this is what science tries to do. Logic he thought was essentially tautologous – different ways of saying the same thing, conveying 'no substantial information about the world.'

Wittgenstein has been unfairly criticised for starting a trend in philosophy – 'an obsession with word games.' He was in fact trying to make our use of language more precise, by emphasising what we can and cannot meaningfully talk about. The last words of the *Tractatus* have become famous: 'Whereof one cannot speak, thereof one must be silent.'[91] He meant that there is no point in talking about areas where words fail to correspond to reality. His career after this book was as remarkable as it had been during its compilation, for he fulfilled the sentiments of that last sentence in his own highly idiosyncratic way. He fell silent, becoming a schoolteacher in the Austrian countryside, and never published another book in his lifetime.[92]

During the war many artists and writers retreated to Zurich in neutral Switzerland. **James Joyce** wrote much of *Ulysses* by the lake; **Hans Arp**, **Frank Wedekind** and **Romain Rolland** were also there. They met in the cafés of Zurich, which for a time paralleled in importance the coffeehouses of Vienna at the turn of the century. The Café Odéon was most well known. For many of those in exile in Zurich, the war seemed to mark the end of the civilisation that had spawned them. It came after a period in which art had become a proliferation of 'isms,' when science had discredited both the notion of an immutable reality and the concept of a wholly rational and self-conscious man. In such a world, the Dadaists felt they had to transform radically the whole concept of art and the artist. The war exploded the idea of progress, which in turn killed the ambition to make durable, classic works for posterity.[93] One critic said the only option facing artists was silence or action.

   Among the regulars at the Café Odéon were **Franz Werfel**, **Aleksey Jawlensky**, and **Ernst Cassirer**, the philosopher. There was also a then-unknown German writer, a Catholic and an anarchist at the same time, named **Hugo Ball**, and his girlfriend, **Emmy Hennings**. Hennings was a journalist but also performed as a cabaret actress, accompanied by Ball on the piano. In February 1916 they had the idea to open a review or cabaret with a literary bent. It was ironically called the Cabaret Voltaire (ironic because Dada eschewed the very reason for which Voltaire was celebrated)[94] and opened on the Spiegelgasse, a steep and narrow alley where Lenin lived. Among the first to appear at Voltaire were two Romanians, the painter Marcel Janco and a young poet, Sami Rosenstock, who adopted the pen name of **Tristan Tzara**. The only Swiss among the early group was Sophie Taueber, Hans Arp's wife (he was from Alsace). Others included Walter Serner from Austria, Marcel Slodki from Ukraine, and Richard Hülsenbeck and Hans Richter from Germany. For a review, in June 1916 Ball produced a programme, and it was in his introduction to the performance that the word **Dada** was first used. Ball's own journal records the kinds of entertainment at Cabaret Voltaire: 'rowdy provocateurs, primitivist dance, cacophony and Cubist theatricals.'[95] Tzara always claimed to have found the word *Dada* in the Larousse dictionary, but whether the term ever had any intrinsic meaning, it soon acquired one, best summed up by Hans Richter.[96] He said it 'had some connection with the joyous Slavonic affirmative "Da, da," ... "yes, yes," to life.' In a time of war it lauded play as the most cherished human activity. 'Repelled by the slaughterhouses of the world war, we turned to art,' wrote Arp. 'We searched for an elementary art that would, we thought, save mankind from the furious madness of those times ... we wanted an anonymous and collective art.'[97] Dada was designed to rescue the sick mind that had brought mankind to catastrophe, and restore its health.[98] Dadaists questioned whether, in the light of scientific and political developments, art – in the broadest sense – was possible. They doubted whether reality could be represented, arguing that it was too elusive, according to science, and therefore dubious both morally and socially. If Dada valued anything, it was the freedom to experiment.[99]

Dada, no less than other modern movements, harboured a paradox. For though they doubted the moral or social usefulness of art, the Dadaists had little choice but to remain artists; in their attempt to restore the mind to health, they still supported the avant-garde idea of the explanatory and redemptive powers of art. The only difference was that, rather than follow any of the 'isms' they derided, they turned instead to childhood and chance in an attempt to recapture innocence, cleanliness, clarity – above all, as a way to probe the unconscious.

No one succeeded in this more than Hans Arp and **Kurt Schwitters**. Arp produced two types of image during the years 1916–20. There were his simple woodcuts, toylike jigsaws; like children he loved to paint clouds and leaves in straightforward, bright, immediate colours. At the same time he was open to chance, tearing off strips of paper that he dropped and fixed wherever they fell, creating random collages. Nonetheless, the work which Arp allowed into the public domain has a meditative quality, simple and stable.[100] Tristan Tzara did the same thing with poetry, where, allegedly, words were drawn at random from a bag and then tumbled into 'sentences.'[101] Kurt Schwitters (1887–1948) made collages too, but his approach was deceptively unrandom. Just as Marcel Duchamp converted ordinary objects like urinals and bicycle wheels into art by renaming them and exhibiting them in galleries, Schwitters found poetry in rubbish. A cubist at heart, he scavenged his native Hanover for anything dirty, peeling, stained, half-burnt, or torn. When these objects were put together by him, they were transformed into something else entirely that told a story and was beautiful.[102] Although his collages may appear to have been thrown together at random, the colors match, the edges of one piece of material align perfectly with another, the stain in a newspaper echoes a form elsewhere in the composition. For Schwitters these were '**Merz**' paintings, the name forming part of a newspaper advertisement for the Kommerz- und Privat-Bank, which he had used in an early collage. The detritus and flotsam in Schwitters's collages were for him a comment, both on the culture that leads to war, creating carnage, waste, and filth, and on the cities that were the powerhouse of that culture and yet the home of so much misery. If Edouard Manet, Charles Baudelaire, and the impressionists had celebrated the fleeting, teeming beauty of late-nineteenth-century cities, the environment that gave rise to modernism, Schwitters's collages were uncomfortable elegies to the end of an era, a new form of art that was simultaneously a form of relic, a condemnation of that world, and a memorial. It was this kind of ambiguity, or paradox, that the Dadaists embraced with relish.[103]

Towards the end of the war, Hugo Ball left Zurich for the Ticino, the Italian-speaking part of Switzerland, and the centre of gravity of Dada shifted to Germany. Hans Arp and Max Ernst, another collagist, went to Cologne, and Schwitters was in Hanover. But it was in Berlin that Dada changed, becoming far more political. Berlin, amid defeat, was a brutal place, ravaged by shortages, despoiled by misery everywhere, with politics bitterly divided, and with revolution in the wake of Russian events a very real possibility. In November 1918 there was a general socialist uprising, which failed, its leaders Karl Liebknecht and Rosa Luxemburg murdered. The uprising was a defining moment for, among others, Adolf Hitler, but also for the Dadaists.[104]

It was **Richard Hülsenbeck** who transported 'the Dada virus' to Berlin.[105] He published his Dada manifesto in April 1918, and a Dada club was established. Early members included **Raoul Hausmann, George Grosz, John Heartfield**, and **Hannah Höch**, who replaced collage with photomontage to attack the Prussian society that they all loathed. Dadaists were still being controversial and causing scandals: Johannes Baader invaded the Weimar Assembly, where he bombarded the delegates with leaflets and declared himself president of the state.[106] Dada was more collectivist in Berlin than in Zurich, and a more long-term campaign was that waged by the Dadaists against the German expressionists, such as Erich Heckel, Ernst Ludwig Kirchner, and Emil Nolde, who, they claimed, were no more than bourgeois German romantics.[107] George Grosz and Otto Dix were the fiercest critics among the painters, their most striking image being the wretched half-human forms of the war cripple. These deformed, grotesque individuals were painful reminders for those at home of the brutal madness of the war. Grosz, Dix, Höch and Heartfield were no less brutal in their depiction of figures with prostheses, who looked half-human and half-machine. These mutilated figures were gross metaphors for what would become the Weimar culture: corrupt, disfigured, with an element of the puppet, the old order still in command behind the scenes – but above all, a casualty of war.

No one excoriated this society more than Grosz in his masterpiece *Republican Automatons* (1920), where the landscape is forbidding, with skyscrapers that are bleak in a way that Giorgio de Chirico, before long, would make menacing. In the foreground the deformed figures, propped up by prostheses of absurd complexity and yet at the same time atavistically dressed in traditional bowler hat, stiff high collar, boiled shirt, *and* sporting their war medals, wave the German flag. It is, like all Grosz's pictures, a mordant image of virulent loathing, not just of the Prussians but also of the bourgeoisie for accepting an odious situation so glibly.[108] For Grosz, the evil had not ended with the war; indeed the fact that so little had changed, despite the horror and the mutilation, was what he railed against. 'In Grosz's Germany, everything and everybody is for sale [prostitutes were a favourite subject]. ... The world is owned by four breeds of pig: the capitalist, the officer, the priest and the hooker, whose other form is the socialite wife. It was no use objecting ... that there were some decent officers, or cultivated bankers. The rage and pain of Grosz's images simply swept such qualifications aside.'[109]

Tristan Tzara took the idea of Dada to Paris in 1920. **André Breton, Louis Aragon**, and **Philippe Soupault**, who together edited the modernist review *Littérature*, were sympathetic, being already influenced by Alfred Jarry's brand of symbolism and its love of absurdity.[110] They also enjoyed a tendency to shock. But unlike in Berlin, Dada in Paris took a particularly literary form, and by the end of

1920 there were at least six Dada magazines in existence and as many books, including Francis Picabia's *Pensées sans langage* (Thoughts without Language) and Paul Eluard's *Les Nécessités de la vie et les conséquences des rêves* (The Necessities of Life and the Consequences of Dreams). The magazines and books were reinforced by salons and soirées in which the main aim was to promise the public something scandalous and then disappoint them, forcing the bourgeoisie to confront its own futility, 'to look over into an abyss of nothing.'[111] It was this assault on the public, this fascination with risk, this 'surefootedness on the brink of chaos,' that linked Paris, Berlin, and Zurich Dada.[112]

Unique to Paris Dada was **automatic writing**, a psychoanalytic technique where the writer allowed himself to become 'a recording machine,' listening for the '**unconscious murmur**.' André Breton thought that a deeper level of reality could be realised through automatic writing, 'that analogical sequences of thought' were released in this way, and he published a short essay in 1924 about the deeper meaning of our conscious thoughts.[113] Called *Manifeste du Surréalisme*, it had an enormous influence on artistic/cultural life in the 1920s and 1930s. Even though surrealism did not flower until the mid-1920s, Breton maintained that it was 'a function of war.'[114]

Across from the Austrian front line, where Wittgenstein was writing and rewriting the *Tractatus*, on the Russian side several artists were recording hostilities. **Marc Chagall** drew wounded soldiers. **Natalya Goncharova** published a series of lithographs, *Mystical Images of War*, in which ancient Russian icons appeared under attack from enemy aircraft. **Kasimir Malevich** produced a series of propaganda posters ridiculing German forces. But the immediate and crude intellectual consequence of the war for Russia was that it cut off the Russian art community from Paris.

Before World War I the Russian artistic presence in Paris was extensive. Futurism, begun by the Italian poet Filippo Marinetti, in 1909, had been taken up by **Mikhail Larionov** and Natalya Goncharova in 1914. Its two central ideas were first, that machinery had created a new kind of humanity, in so doing offering freedom from historical constraints; and second, that operating by confrontation was the only way to shake people out of their bourgeois complacencies. Although it didn't last long, the confrontational side of futurism was the precursor to that aspect of Dada, surrealism, and the 'happenings' of the 1960s. In Paris, Goncharova designed *Le Coq d'or* for Nicolai Rimsky-Korsakov, and Alexandre Benois worked for Serge Diaghilev's Ballets Russes. Guillaume Apollinaire reviewed the exhibition of paintings by Larionov and Goncharova at the Galérie Paul Guillaume in *Les Soirées de Paris*, concluding that 'a universal art is being created, an art in which painting, sculpture, poetry, music and even science in all its manifold aspects will be combined.' In the same year, 1914, there was an exhibition of Chagall in Paris, and several paintings by Malevich were on show at the Salon des Indépendants. Other Russian artists in Paris before the war included **Vladimir Tatlin**, **Lydia Popova**, Eliezer Lissitzky, Naum Gabo, and Anton Pevsner. Wealthy Russian bourgeois collectors like **Sergey Shchukin** and **Ivan Morozov** collected some of the best modern pictures the French school had to offer, making friends with Picasso, Braque, Matisse, and Gertrude and Leo Stein.[115] By the outbreak of war, Shchukin had collected 54 Picassos, 37 Matisses, 29 Gauguins, 26 Cézannes, and 19 Monets.[116]

For Russians, the ease of travel before 1914 meant that their art was both open to international modernistic influences and yet distinctively Russian. The works of Goncharova, Malevich, and Chagall combined recognisable themes from the Russian 'East' but also images from the modern 'West': Orthodox icons and frozen Siberian landscapes but also iron girders, machines, airplanes, the whole scientific palette. Russian art was not backward before the revolution. In fact, 'suprematism,' a form of geometrical abstraction born of Malevich's obsession with mathematics, appeared between the outbreak of war and revolution – yet another 'ism' to add to the profusion in Europe. But the explosion of revolution, coming in the middle of war, in October 1917, transformed painting and the other visual arts. Three artists and one commissar typified the revolution in Russian art: Malevich, Vladimir Tatlin, **Alexandr Rodchenko**, and **Anatoli Lunacharsky**.

Lunacharsky was a sensitive and idealistic writer of no fewer than thirty six books who was convinced that art was central to the revolution and the regeneration of Russian life and he had firm ideas about its role.[117] Now that the state was the only patron of art (the Shchukin collection was nationalised on 5 November 1918), Lunacharsky conceived the notion of a new form of art, **agitprop**, combining agitation and propaganda. For him art was a significant medium of change.[118] As commissar for education, an authority on music and theatre, Lunacharsky had Lenin's ear, and for a time several grandiose plans were considered – for example, a proposal to erect at well-known landmarks in Moscow a series of statues, monuments of great international revolutionaries of the past. Loosely interpreted, many of the 'revolutionaries' were French: Georges-Jacques Danton, Jean-Paul Marat, Voltaire, Zola, Cézanne.[119] The scheme, like so many others, failed simply for lack of resources: there was no shortage of artists in Russia, but there was of bronze.[120] Other agitprop schemes were realised, at least for a while. There were agitprop posters and street floats, agitprop trains, and agitprop boats on the Volga.[121] Lunacharsky also shook up the art schools, including the two most prestigious institutions, in Vitebsk, northwest of Smolensk, and Moscow. In 1918 the former was headed by Chagall, and Malevich and Lissitzky were members of its faculty; the latter, the Higher State Art Training School, or **Vkhutemas School**, in Moscow, was a sort of Bauhaus of Russia, 'the most advanced art college in the world, and the ideological centre of Russian Constructivism.'[122]

The early works of Kasimir Malevich (1878–1935) owe much to impressionism, but there are also strong echoes of Cézanne and Gauguin – bold, flat colour – and the Fauves, especially Matisse. Around 1912 Malevich's images began to break up into a form of cubism. But the peasants in the fields that dominate this period of his work are clearly Russian. From 1912 on Malevich's work changed again, growing simpler. He was always close to **Velimir Khlebnikov**, a poet and a mathematician, and Malevich's paintings have been

described as analogues to poetry, exploiting abstract, three-dimensional forms –
triangles, circles, rectangles, with little colour variation.[123] His shapes are less
solid than those of Braque or Picasso. Finally, Malevich changed again, to his
celebrated paintings of a black square on a white background and, in 1918, a
white square on a white background. As revolution was opening up elsewhere,
Malevich's work represented one kind of closure in painting, about as far as it
could be from representation. (A theoretician of art as well as a painter, he
entitled one essay 'The Objectless World.')[124] Malevich aimed to represent the
simplicity, clarity, and cleanliness that he felt was a characteristic of mathematics,
the beautiful simplicity of form, the essential shapes of nature, the abstract
reality that lay beneath even cubism. Malevich revolutionised painting in Russia,
pushing it to the limits of form, stripping it down to simple elements the way
physicists were stripping matter.

Malevich may have revolutionised painting, but **constructivism** was itself
part of the revolution, closest to it in image and aim. Lunacharsky was intent
on creating a people's art, 'an art of five kopeks,' as he put it, cheap and available
to everyone. Constructivism responded to the commissar's demands with
images that looked forward, that suggested endless movement and sought to
blur the boundaries between artist and artisan, engineer or architect. Airplane
wings, rivets, metal plates, set squares, these were the staple images of constructivism.[125] Vladimir Tatlin (1885–1953), the main force in constructivism,
was a sailor and a marine carpenter, but he was also an icon painter. Like
Kandinsky and Malevich, he wanted to create new forms, logical forms.[126] Like
Lunacharsky he wanted to create a proletarian art, a socialist art. He started to
use iron and glass, 'socialist materials' that everyone knew and was familiar
with, materials that were 'not proud.'[127] Tatlin's theories came together in 1919,
two years after the revolution, when he was asked to design a monument to
mark the Third Communist International, the association of revolutionary
Marxist parties of the world. The design he came up with – unveiled at the
Eighth Congress of the Soviets in Moscow in 1920 – was a **slanting tower**,
1,300 feet high, dwarfing even the Eiffel Tower, which was 'only' 1,000 feet.
The slanting tower was a piece of propaganda for the state and for Tatlin's
conception of the place of engineering in art (he was a very jealous man, keenly
competitive with Malevich).[128] Designed in three sections, each of which
rotated at a different speed, and built of glass and steel, Tatlin's tower was
regarded as the defining monument of constructivism, an endlessly dynamic
useful object, loaded with heavy symbolism. The banner that hung above the
model when it was unveiled read 'Engineers create new forms.' But of course,
a society that had no bronze for statues of Voltaire and Danton had no steel or
glass for Tatlin's tower either, and it never went beyond the model stage: 'It
remains the most influential non-existent object of the twentieth-century, and
one of the most paradoxical – an unworkable, probably unbuildable metaphor
of practicality.'[129] It was the perfect epitome of Malevich's objectless world.

The third of revolutionary Russia's artistic trinity was the painter Alexander
Rodchenko (1891–1956). Fired by the spirit of the revolution, he created his
own brand of futurism and agitprop. Beginning with a variety of constructions,

part architectural models, part sculpture, he turned to the stark realism of photography and the immediate impact of the poster.[130] He sought an art form that was, in the words of Robert Hughes, as 'arresting as a shout in the street':[131] 'The art of the future will not be the cosy decoration of family homes. It will be just as indispensable as 48-storey skyscrapers, mighty bridges, wireless [radio], aeronautics and submarines, which will be transformed into art.' With one of Russia's great modernist poets, **Vladimir Mayakovsky**, Rodchenko formed a partnership whose common workshop stamp read, 'Advertisement Constructors, Mayakovsky-Rodchenko.'[132] Their posters were advertisements for the new state. For Rodchenko, propaganda became great art.[133]

Rodchenko and Mayakovsky shared Tatlin's and Lunacharsky's ideas about proletarian art and about the reach of art. As true believers in the revolution, they thought that art should belong to everyone and even shared the commissar's view that the whole country, or at least the state, should be regarded as a work of art.[134] This may seem grandiose to the point of absurdity now; it was deadly serious then. For Rodchenko, photography was the most proletarian art: even more than typography or textile design (other interests of his), it was cheap, and could be repeated as often as the situation demanded. Here are some typical Rodchenko arguments:

Down with ART as bright PATCHES
    on the *undistinguished* life of the
    man of property.
Down with ART as a precious STONE
  midst the dark and filthy *life* of the
    pauper.
Down with art as a means of
    ESCAPING from LIFE which is
    not worth living.[135]

and:

Tell me, frankly, what ought to remain of Lenin:
an art bronze,
oil portraits,
etchings,
watercolours,
his secretary's diary, his friends' memoirs –

or a file of photographs taken of him at work and at rest,
archives of his books, writing pads, notebooks,
shorthand reports, films, phonograph records?
I don't think there's any choice.
Art has no place in modern life. . . . Every modern cultured
man must wage war against art, as against opium.

Don't lie.
Take photo after photo![136]

Taking this perfect constructivist material – modern, humble, real, influenced by his friend, the Russian film director **Dziga Vertov** – Rodchenko began a series of photomontages that used repetition, distortion, magnification and other techniques to interpret and reinterpret the revolution to the masses. For Rodchenko, even beer, a proletarian drink, could be revolutionary, an explosive force.

Even though they were created as art forms for the masses, suprematism and constructivism are now considered 'high art.' Their intended influence on the proletariat was ephemeral. With the grandiose schemes failing for lack of funds, it was difficult for the state to continue arguing that it was a work of art. In the 'new' modern Russia, art lost the argument that it was the most important aspect of life. The proletariat was more interested in food, jobs, housing, and beer.

It does not diminish the horror of World War I, or reduce our debt to those who gave their lives, to say that most of the responses considered here were positive. There seems to be something in human nature such that, even when it makes an art form, or a philosophy, out of pessimism, as Dada did, it is the art form or the philosophy that lasts, not the pessimism. Few would wish to argue which was the worst period of darkness in the twentieth century, the western front in 1914–18, Stalin's Russia, or Hitler's Reich, but something *can* be salvaged from 'the Great War'.

Civilisations and Their Discontents

One of the most influential postwar ideas in Europe was published in April 1918, in the middle of the Ludendorff offensive – what turned out to be the decisive event of the war in the West, when General Erich Ludendorff, Germany's supreme commander in Flanders, failed to pin the British against the north coast of France and Belgium and separate them from other forces, weakening himself in the process. **Oswald Spengler**, a schoolmaster living in Munich, wrote *Der Untergang des Abendlandes* (literally, The Sinking of the Evening Lands, translated into English as *The Decline of the West*) in 1914, using a title he had come up with in 1912. Despite all that had happened, he had changed hardly a word of his book, which he was to describe modestly ten years later as '*the* philosophy of our time.'[1]

Spengler was born in 1880 in Blankenburg, a hundred miles southwest of Berlin, the son of emotionally undemonstrative parents whose reserve forced on their son an isolation that seems to have been crucial to his formative years. This solitary individual grew up with a family of very Germanic giants: Richard Wagner, Ernst Haeckel, Henrik Ibsen, and Friedrich Nietzsche. It was Nietzsche's distinction between *Kultur* and *Zivilisation* that particularly impressed the teenage Spengler. In this context, *Kultur* may be said to be represented by Zarathustra, the solitary seer creating his own order out of the wilderness. *Zivilisation*, on the other hand, is represented, say, by the Venice of Thomas Mann's *Death in Venice*, glittering and sophisticated but degenerate, decaying, corrupt.[2] Another influence was the economist and sociologist Werner Sombart, who in 1911 had published an essay entitled 'Technology and Culture,' where he argued that the human dimension of life was irreconcilable with the mechanical, the exact reverse of the Futurist view. There was a link, Sombart said, between economic and political liberalism and the 'oozing flood of commercialism' that was beginning to drag down the Western world. Sombart went further and declared that there were two types in history, Heroes and Traders. These two types were typified at their extremes by, respectively, Germany – heroes – and the traders of Britain.

In 1903 Spengler failed his doctoral thesis. He managed to pass the following year, but in Germany's highly competitive system his first-time failure meant that the top academic echelon was closed to him. In 1905 he suffered a nervous breakdown and wasn't seen for a year. He was forced to teach in schools, rather than university, which he loathed, so he moved to Munich to become a full-time writer. Munich was then a colorful city very different from the highly academic centres such as Heidelberg and Göttingen. It was the city of Stefan George and his circle of poets, of Thomas Mann, just finishing *Death in Venice*, of the painters Franz Marc and Paul Klee.[3]

For Spengler the defining moment, which led directly to his book, occurred in 1911. It was the year he moved to Munich, when in May the German cruiser *Panther* sailed into the Moroccan port of Agadir in an attempt to stop a French takeover of the country. The face-off brought Europe to the edge of war, but in the end France and Britain prevailed by forcing Germany to back down. Many, especially in Munich, felt the humiliation keenly, none more so than Spengler.[4] He certainly saw Germany, and the German way of doing things, as directly opposed to the French and, even more, the British way. These two countries epitomised for him the rational science that had arisen since the Enlightenment, and for some reason Spengler saw the Agadir incident as signalling the end of that era. It was a time for heroes, not traders. He now set to work on what would be his life's project, his theme being how Germany would be *the* country, *the* culture, of the future. She might have lost the battle in Morocco, but a war was surely coming in which she, and her way of life, would be victorious. Spengler believed he was living at a turning point in history such as Nietzsche had talked of. The first title for his book was *Conservative and Liberal*, but one day he saw in the window of a Munich bookshop a volume entitled *The Decline of Antiquity* and at once he knew what he was going to call his book.[5]

The foreboding that Germany and all of Europe was on the verge of a major change was not of course confined to Spengler. Youth movements in France and Germany were calling for a 'rejuvenation' of their countries, as often as not in militaristic terms. Max Nordau's *Degeneration* was still very influential and, with no wholesale war for nearly a century, ideas about the ennobling effects of an honourable death were far from uncommon. Even Ludwig Wittgenstein shared this view, as we have seen.[6] Spengler drew on eight major world civilisations – the Babylonians, the Egyptians, the Chinese, the Indians, the pre-Columbian Mexicans, the classical or Graeco-Roman, the Western European, and the 'Magian,' a term of his own which included the Arabic, Judaic, and Byzantine – and explained how each went through an organic cycle of growth, maturity, and inevitable decline. One of his aims was to show that Western civilisation had no privileged position in the scheme of things: 'Each culture has its own new possibilities of self-expression which arise, ripen, decay and never return.'[7] For Spengler, *Zivilisation* was not the end product of social evolution, as rationalists regarded Western civilisation; instead it was *Kultur*'s old age. There was no science of history, no linear development, simply the repeated rise and fall of individual *Kulturs*. Moreover, the rise of a new *Kultur* depended on two things – the race and the *Geist* or spirit, 'the inwardly lived experience of the "we."' For Spengler, rational society and science were evidence only of a triumph of the indomitable Western will, which would collapse in the face of a stronger will, that of Germany. Germany's will was stronger because her sense of 'we' was stronger; the West was obsessed with matters 'outside' human nature, like materialistic science, whereas in Germany there was more feeling for the inner spirit. This is what counted.[8] Germany was like Rome, he said, and like Rome the Germans would reach London.[9]

The Decline was a great and immediate commercial success. Thomas Mann compared its effect on him to that of reading Schopenhauer for the first time.[10] Ludwig Wittgenstein was astounded by the book, but Max Weber described Spengler as a 'very ingenious and learned dilettante.' Elisabeth Förster-Nietzsche read the book and was so impressed that she arranged for Spengler to receive the Nietzsche Prize. This made Spengler a celebrity, and visitors were required to wait three days before he could see them.[11] He tried to persuade even the English to read Nietzsche.[12]

From the end of the war throughout 1919, Germany was in chaos and crisis. Central authority had collapsed, revolutionary ferment had been imported from Russia, and soldiers and sailors formed armed committees, called 'soviets.' Whole cities were 'governed' at gunpoint, like Soviet republics. Eventually, the Social Democrats, the left-wing party that installed the **Weimar Republic**, had to bring in their old foes the army to help restore order; this was achieved but involved considerable brutality – thousands were killed. Against this background, Spengler saw himself as the prophet of a nationalistic resurgence in Germany, concluding that only a top-down command economy could save her. He saw it as his role to rescue socialism from the Marxism of Russia and apply it in the 'more vital country' of Germany. A new political category was needed: he put Prussianism and Socialism together to come up with National Socialism. This would lead men to exchange the 'practical freedom' of America and England for an 'inner freedom,' 'which comes through discharging obligations to the organic whole.'[13] One of those impressed by this argument was Dietrich Eckart, who helped form the German Workers' Party (GWP), which adopted the symbol of the Pan-German Thule Society Eckart had previously belonged to. This symbol of 'Aryan vitalism,' the swastika, now took on a political significance for the first time. Alfred Rosenberg was also a fan of Spengler and joined the GWP in May 1919. Soon after, he brought in one of his friends just back from the front, a man called Adolf Hitler.

From 18 January 1919 the former belligerent nations met in Paris at a peace conference to reapportion those parts of the dismantled Habsburg and German Empires forfeited by defeat in war, and to discuss reparations. Six months later, on 28 June, Germany signed the treaty in what seemed the perfect location: the **Hall of Mirrors**, at the **Palace of Versailles**, just outside the French capital.

Adjoining the Salon de la Guerre, the Galérie des Glaces is 243 feet in length, a great blaze of light, with a parade of seventeen huge windows overlooking the formal gardens designed in the late seventeenth century by André Le Nôtre. Halfway along the length of the hall three vast mirrors are set between marble pilasters, reflecting the gardens. Among this overwhelming splendour, in an historic moment captured by the British painter **Sir William Orpen**, the Allied leaders, diplomats, and soldiers convened. Opposite them, their faces away from the spectator, sat two German functionaries, there to sign the treaty. Orpen's picture perfectly captures the gravity of the moment.[14]

In one sense, Versailles stood for the continuity of European civilisation, the very embodiment of what Spengler hated and thought was dying. But this overlooked the fact that Versailles had been a museum since 1837. In 1919, the centre stage was held not by any of the royal families of Europe but by the politicians of the three main Allied and Associated powers. Orpen's picture focuses on Georges Clemenceau, greatly advanced in years, with his white walrus moustache and fringe of white hair, looking lugubrious. Next to him sits a very upright President Woodrow Wilson – the United States was an Associated Power – looking shrewd and confident. David Lloyd George, then at the height of his authority, sits on the other side of Clemenceau, his manner thoughtful and judicious. Noticeable by its absence is Bolshevik Russia, whose leaders believed the Allied Powers to be as doomed by the inevitable march of history as the Germans they had just defeated. A complete settlement, then, was an illusion at Versailles. In the eyes of many it was, rather, a punishment of the vanquished and a dividing of the spoils. For some present, it did not go unnoticed that the room where the treaty was signed was a hall of mirrors.

Barely was the treaty signed than it was exploded. In November 1919 *The Economic Consequences of the Peace* scuttled what public confidence there was in the settlement. Its author, **John Maynard Keynes**, was a brilliant intellectual, not only a theorist of economics, an original thinker in the philosophical tradition of John Stuart Mill, but a man of wit and a central figure in the famous Bloomsbury group. He was born into an academically distinguished family – his father was an academic in economics at Cambridge, and his mother attended Newnham Hall (though, like other women at Cambridge at that time, she was not allowed to graduate). As a schoolboy at Eton he achieved distinction with a wide variety of noteworthy essays and a certain fastidiousness of appearance, which derived from his habit of wearing a fresh boutonnière each morning.[15] His reputation preceded him to King's College, Cambridge, where he arrived as an undergraduate in 1902. After only one term he was invited to join the **Apostles** alongside Lytton Strachey, Leonard Woolf, G. Lowes Dickinson and E. M. Forster. He later welcomed into the society Bertrand Russell, G. E. Moore and Ludwig Wittgenstein. It was among these liberal and rationalist minds that Keynes developed his ideas about reasonableness and civilisation that underpinned his attack on the politics of the peace settlement in *The Economic Consequences*.

Before describing the main lines of Keynes's attack, it is worth noting the path he took between Cambridge and Versailles. Convinced from an early age that no one was ever as ugly as he – an impression not borne out by photographs and portraits, although he was clearly far from being physically robust – Keynes put great store in the intellectual life. He also possessed a sharpened appreciation for physical beauty. Among the many homosexual affairs of his that originated at Cambridge was one with Arthur Hobhouse, another Apostle. In 1905 he wrote to Hobhouse in terms that hint at the emotional delicacy at the centre of

Keynes's personality: 'Yes I have a clever head, a weak character, an affectionate disposition, and a repulsive appearance ... keep honest, and – if possible – like me. If you never come to love, yet I shall have your sympathy – and that I want as much, at least, as the other.'[16] His intellectual pursuits, however, were conducted with uncommon certainty. Passing the civil service examinations, Keynes took up an appointment at the India Office, not because he had any interest in India but because the India Office was one of the top departments of state.[17] The somewhat undemanding duties of the civil service allowed him time to pursue a fellowship dissertation for Cambridge. In 1909 he was elected a fellow of King's, and in 1911 he was appointed editor of the *Economic Journal*. Only twenty-eight years old, he was already an imposing figure in academic circles, which is where he might have remained but for the war.

Keynes's wartime life presents an ironic tension between the economic consequences of his expertise as a member of the wartime Treasury – in effect, negotiating the Allied loans that made possible Britain's continuance as a belligerent – and the convictions that he shared with conscientious objectors, including his close Bloomsbury friends and the pacifists of Lady Ottoline Morrell's circle. Indeed, he testified on behalf of his friends before the tribunals but, once the war was being waged, he told Lytton Strachey and Bertrand Russell, 'There is really no practical alternative.' And he was practical: one of his coups in the war was to see that there were certain war loans France would never repay to Britain. In 1917, when the Degas collection came up for sale in Paris after the painter's death, Keynes suggested that the British government should buy some of the impressionist and postimpressionist masterpieces and charge them to the French government. The plan was approved, and he travelled to Paris with the director of the National Gallery, both in disguise to escape the notice of journalists, and landed several bargains, including a Cézanne.[18]

Keynes attended the peace treaty talks in Versailles representing the chancellor of the exchequer. In effect, terms were dictated to Germany, which had to sue for peace in November 1918. The central question was whether the peace should produce reconciliation, reestablishing Germany as a democratic state in a newly conceived world order, or whether it should be punitive to the degree that Germany would be crippled, disabled from ever again making war. The interests of the Big Three did not coincide, and after months of negotiations it became clear that the proposals of the Armistice would not be implemented and that instead an enormous reparation would be exacted from Germany, in addition to confiscation of a considerable part of German territory and redistribution to the victors of her overseas empire.

Keynes was appalled. He resigned in 'misery and rage.' His liberal ideals, his view of human nature, and his refusal to concur with the Clemenceau view of German nature as endemically hostile, combined with a feeling of guilt over his noncombatant part in the war (as a Treasury official he was exempt from conscription), propelled him to write his book exposing the treaty. In it Keynes expounded his economic views, as well as analysing the treaty and its effects. Keynes thought that the equilibrium between the Old and New Worlds which the war had shattered should be reestablished. Investment of European surplus capital in the New World produced the food and goods needed for growing populations and increased standards of living. Thus markets must be freer, not curtailed, as the treaty was to do for Germany. Keynes's perspective was more that of a European than of a nationalist. Only in this way could the spectre of massive population growth, leading to further carnage, be tamed.[19] Civilisation, said Keynes, must be based on shared views of morality, of prudence, calculation, and foresight. The punitive impositions on Germany would produce only the opposite effect and impoverish Europe. Keynes believed that enlightened economists were best able to secure the conditions of civilisation, or at any rate prevent regression, not politicians. One of the most far-reaching aspects of the book was Keynes's argument, backed with figures and calculations, that there was no probability that Germany could repay, in either money or kind, the enormous reparations required over thirty years as envisaged by the Allies. According to Keynes's theory of probability, the changes in economic conditions simply cannot be forecast that far ahead, and he therefore urged much more modest reparations over a much shorter time. He could also see that the commission set up to force Germany to pay and to seize goods breached all the rules of free economic association in democratic nations. His arguments therefore became the basis of the pervasive opinion that Versailles inevitably gave rise to Hitler, who could not have taken control of Germany without the wide resentment against the treaty. It didn't matter that, following Keynes's book, reparations were in fact scaled down, or that no great proportion of those claimed were ever collected. It was enough that Germany thought itself to have been vengefully treated.

Keynes's arguments are disputable. From the outset of peace, there was a strong spirit of noncompliance with orders for demilitarisation among German armed forces. For example, they refused to surrender all the warplanes the Allies demanded, and production and research continued at a fast pace.[20] Did the enormous success of Keynes's book create attitudes that undermined the treaty's more fundamental provisions by putting such an emphasis upon what may have been a peripheral part of the treaty?[21] And was it instrumental in creating the climate for Western appeasement in the 1930s, an attitude on which the Nazis gambled? Such an argument forms the basis of a bitter attack on Keynes published in 1946, after Keynes's death and that of its author, **Etienne Mantoux**, who might be thought to have paid the supreme price exacted by Keynes's post-Versailles influence: he was killed in 1945 fighting the Germans. The grim title of Mantoux's book conveys the argument: *The Carthaginian Peace; or, The Economic Consequences of Mr Keynes*.[22]

What is not in dispute is Keynes's brilliant success, not only in terms of polemical argument but also in the literary skill of his acid portraits of the leaders. Of Clemenceau, Keynes wrote that he could not 'despise him or dislike him, but only take a different view as to the nature of civilised man, or indulge at least a different hope.' 'He had one illusion – France; and one disillusion – mankind, including Frenchmen and his colleagues not least.' Keynes takes the reader into Clemenceau's mind: 'The politics of power are inevitable, and there is nothing very new to learn about this war or the end it was fought for; England had destroyed, as in each preceding century, a trade rival; a mighty chapter had been closed in the secular struggle between the glories of Germany and France. Prudence required some measure of lip service to the "ideals" of foolish Americans and hypocritical Englishmen, but it would be stupid to believe that there is much room in the world, as it really is, for such affairs as the League of Nations, or any sense in the principle of self-determination except as an ingenious formula for rearranging the balance of power in one's own interest.'[23]

This striking passage leads on to the 'foolish' American. Woodrow Wilson had come dressed in all the wealth and power of mighty America: 'When President Wilson left Washington he enjoyed a prestige and a moral influence throughout the world unequalled in history.' Europe was dependent on the United States financially and for basic food supplies. Keynes had high hopes of a new world order flowing from New to Old. It was swiftly dashed. 'Never had a philosopher held such weapons wherewithal to bind the princes of this world. . . . His head and features were finely cut and exactly like his photographs. . . . But this blind and deaf Don Quixote was entering a cavern where the swift and glittering blade was in the hands of the adversary. . . . The President's slowness amongst the Europeans was noteworthy. He could not, all in a minute, take in what the rest were saying, size up the situation in a glance . . . and was liable, therefore, to defeat by the mere swiftness, apprehension, and agility of a Lloyd George.' In this terrible sterility, 'the President's faith withered and dried up.'

Among the intellectual consequences of the war and Versailles was the idea of a universal – i.e., worldwide – government. One school of thought contended that the Great War had mainly been stumbled into, that it was an avoidable catastrophe that would not have happened with better diplomacy. Other historians have argued that the 1914–18 war, like most if not all wars, had deeper, coherent causes. The answer provided by the Versailles Treaty was to set up a League of Nations, a victory in the first instance for President Wilson. The notion of international law and an international court had been articulated in the seventeenth century by Hugo Grotius, a Dutch thinker. The **League of Nations** was new in that it would provide a permanent arbitration body and a permanent organisation to enforce its judgements. The argument ran that if the Germans in 1914 had had to face a coalition of law-abiding nations, they would have been deterred from the onslaught on Belgium. The Big Three pictured the League very differently. For France a standing army would be to control Germany. Britain's leaders saw it as a conciliation body with no teeth. Only Wilson conceived of it as both a forum of arbitration and as an instrument of collective security. But the idea was dead in the water in the United States; the Senate simply refused to ratify an arrangement that took fundamental decisions away from its authority. It would take another war, and the development of atomic weapons, before the world was finally frightened into acting on an idea similar to the League of Nations.

★

Before World War I, Germany had held several concessions in Shandong, China. The Versailles Treaty did not return these to the Beijing government but left them in the hands of the Japanese. When this news was released, on 4 May 1919, some 3,000 students from Beida (Beijing University) and other Beijing institutions besieged the Tiananmen, the gateway to the palace. This led to a battle between students and police, a student strike, demonstrations across the country, a boycott of Japanese goods – and in the end the 'broadest demonstration of national feeling that China had ever seen.'[24] The most extra-ordinary aspect of this development – what became known as the **May 4 movement** – was that it was the work of both mature intellectuals and students. Infused by Western notions of democracy, and impressed by the advances of Western science, the leaders of the movement put these new ideas together in an anti-imperialist program. It was the first time the students had asserted their power in the new China, but it would not be the last. Many Chinese intellectuals had been to Japan to study. The main Western ideas they returned with related to personal expression and freedom, including sexual freedom, and this led them to oppose the traditional family organisation of China. Under Western influence they also turned to fiction as the most effective way to attack traditional China, often using first-person narratives written in the vernacular. Normal as this might seem to Westerners, it was very shocking in China.

The first of these new writers to make a name for himself was **Lu Xun**. His real name was Zhou Shuren or Chou Shu-jen, and, coming from a prosperous family (like many in the May 4 movement), he first studied Western medicine and science. One of his brothers translated Havelock Ellis's theories about sexuality into Chinese, and the other, a biologist and eugenicist, translated Darwin. In 1918, in the magazine *New Youth*, Lu Xun published a satire entitled 'The Diary of a Madman.' The 'Diary' was very critical of Chinese society, which he depicted as cannibalistic, devouring its brightest talents, with only the mad glimpsing the truth, and then as often as not in their dreams – a theme that would echo down the years, and not just in China. The problem with Chinese civilisation, Lu Xun wrote, was that it was 'a culture of serving one's masters, who are triumphant at the cost of the misery of the multitude.'[25]

The Versailles Treaty may have been the immediate stimulus for the May 4 movement, but a more general influence was the ideas that shaped Chinese society after 1911, when the Qing dynasty was replaced with a republic.[26] Those ideas – essentially, of a civil society – were not new in the West. But the Confucian heritage posed two difficulties for this transition in China. The first was the concept of individualism, which is of course such a bulwark in Western (and especially American) civil society. Chinese reformers like **Yan (or Yen) Fu**, who translated so many Western liberal classics (including John Stuart Mill's *On Liberty* and Herbert Spencer's *Study of Sociology*), nonetheless saw individualism only as a trait to be used in support of the state, not against it.[27] The second difficulty posed by the Confucian heritage was even more problematic. Though the Chinese developed something called the New Learning, which encompassed 'foreign matters' (i.e., modernisation), what in practice was taught may be summarised, in the words of Harvard historian John

Fairbanks, as 'Eastern ethics and Western science.'[28] The Chinese (and to an extent the Japanese) persisted in the belief that Western ideas – particularly science – were essentially technical or purely functional matters, a set of tools much shallower than, say, Eastern philosophy, which provided the 'substance' of education and knowledge. But the Chinese were fooling themselves. Their own brand of education was very thinly spread – male literacy in the late Qing period (i.e., up to 1911) was 30 to 45 percent for men and as low as 2 to 10 percent for women. As a measure of the educational backwardness of China at this time, such universities as existed were required to teach and examine many subjects – engineering, technology, and commerce – using English-language textbooks: Chinese words for specialist terms did not yet exist.[29]

In effect, China's educated elite had to undergo two revolutions. They had first to throw off Confucianism, and the social/educational structure that went with it. Then they had to throw off the awkward amalgam of 'Eastern ethics, Western science' that followed. In practice, those who achieved this did so only by going to the United States to study (provided for by a U.S. congressional bill in 1908). To a point this was effective, and in 1914 young Chinese scientists who had studied in America founded the Science Society. For a time, this society offered the only real chance for science in the Chinese/Confucian context.[30] Beijing University played its part when a number of scholars who had trained abroad attempted to cleanse China of Confucianism 'in the name of science and democracy.'[31] This process became known as the **New Learning** – or **New Culture** – movement.[32] Some idea of the magnitude of the task facing the movement can be had from the subject it chose for its first campaign: the Chinese writing system. This had been created around 200 B.C. and had hardly changed in the interim, with characters acquiring more and more meanings, which could only be deciphered according to context and by knowing the classical texts.[33] Not surprisingly (to Western minds) the new scholars worked to replace the classical language with everyday speech. (The size of the problem is underlined when one realises this was the step taken in Europe during the Renaissance, four hundred years before, when Latin was replaced by national vernaculars.)[34] Writing in the new vernacular, Lu Xun had turned his back on science (many in China, as elsewhere, blamed science for the horrors of World War I), believing he could have more impact as a novelist.[35] But science was integral to what was happening. For example, other leaders of the May 4 movement like **Fu Sinian** and **Luo Jialun** at Beida advocated in their journal *New Tide* (*Renaissance*) – one of eleven such periodicals started in the wake of May 4 – a Chinese 'enlightenment.'[36] By this they meant an individualism beyond family ties and a rational, scientific approach to problems. They put their theories into practice by setting up their own lecture society to reach as many people as possible.[37]

The May 4 movement was significant because it combined intellectual and political concerns more intimately than at other times. Traditionally China, unlike the West since the Enlightenment, had been divided into two classes only: the ruling elite and the masses. Following May 4, a growing bourgeoisie in China adopted Western attitudes and beliefs, calling for example for birth control and self-government in the regions. Such developments were bound to provoke political awareness.[38] Gradually the split between the more academic wing of the May 4 movement and its political phalanx widened. Emboldened by the success of Leninism in Russia, the political wing became a secret, exclusive, centralised party seeking power, modelled on the Bolsheviks. One intellectual of the May 4 movement who began by believing in reform but soon turned to violent revolution was the burly son of a Hunan grain merchant whose fundamental belief was eerily close to that of Spengler, and other Germans.[39] His name was Mao Zedong.

The old Vienna officially came to an end on 3 April 1919, when the Republic of Austria abolished titles of nobility, forbidding the use even of 'von' in legal documents. The peace left Austria a nation of only 7 million with a capital that was home to 2 million of them. On top of this overcrowding, the years that followed brought famine, inflation, a chronic lack of fuel, and a catastrophic epidemic of influenza. Housewives were forced to cut trees in the woods, and the university closed because its roof had not been repaired since 1914.[40] Coffee, historian William Johnston tells us, was made of barley, bread caused dysentery. Freud's daughter Sophie was killed by the epidemic, as was the painter Egon Schiele. It was into this world that Alban Berg introduced his opera *Wozzeck* (1917–21, premiered 1925), about the murderous rage of a soldier degraded by his army experiences. But morals were not eclipsed entirely. At one point an American company offered to provide food for the Austrian people and to take payment in the emperor's Gobelin tapestries: a public protest stopped the deal.[41] Other aspects of Vienna style went out with the 'von.' It had been customary, for example, for the doorman to ring once for a male visitor, twice for a female, three times for an archduke or cardinal. And tipping had been ubiquitous – even elevator operators and the cashiers in restaurants were tipped. After the terrible conditions imposed by the peace, all such behaviour was stopped, never to resume. There was a complete break with the past.[42] Hugo von Hofmannsthal, Freud, Karl Kraus, and Otto Neurath all stayed on in Vienna, but it wasn't the same as before. Food was so scarce that a team of British doctors investigating 'accessory food factors,' as vitamins were then called, was able to experiment on children, denying some the chance of a healthy life without any moral compunction.[43] Now that the apocalypse had come to pass, the gaiety of Vienna was entirely vanished.

In Budapest, the changes were even more revealing, and more telling. A group of brilliant scientists – physicists and mathematicians – were forced to look elsewhere for work and stimulation. These included **Edward Teller**, **Leo Szilard**, and **Eugene Wigner**, all Jews. Each would eventually go to Britain or the United States and work on the atomic bomb. A second group, of writers and artists, stayed on in Budapest, at least to begin with, having been forced home by the outbreak of war. The significance of this group lay in the fact that its character was shaped by both World War I and the Bolshevik revolution in Russia. For what happened in the **Sunday Circle**, or the Lukács Circle, as it was called, was the eclipse of ethics. This eclipse darkened the world longer than most.

The Budapest Sunday Circle was not formed until after war broke out, when a group of young intellectuals began to meet on Sunday afternoons to discuss various artistic and philosophical problems mainly to do with modernism. The group included **Karl Mannheim**, a sociologist, art historian **Arnold Hauser**, the writers **Béla Balázs** and **Anna Leznai**, and the musicians **Béla Bartók** and **Zoltán Kodály**, all formed around the critic and philosopher **George Lukács**. Like Teller and company, most of them had travelled widely and spoke German, French, and English as well as Hungarian. Although Lukács – a friend of Max Weber – was the central figure of the 'Sundays,' they met in Balázs's elegant, 'notorious,' hillside apartment.[44] For the most part the discussions were highly abstract, though relief was provided by the musicians – it was here, for example, that Bartók tried out his compositions. To begin with, the chief concern of this group was 'alienation'; like many people, the Sunday Circle members took the view that the war was the logical endpoint of the liberal society that had developed in the nineteenth century, producing industrial capitalism and bourgeois individualism. To Lukács and his friends, there was something sick, unreal, about that state of affairs. The forces of industrial capitalism had created a world where they felt ill at ease, where a *shared* culture was no longer part of the agenda, where the institutions of religion, art, science, and the state had ceased to have any communal meaning. Many of them were influenced in this by the lectures of George Simmel, 'the Manet of philosophy', in Berlin. Simmel made a distinction between 'objective' and 'subjective' culture. For him, objective culture was the best that had been thought, written, composed, and painted; a 'culture' was defined by how its members related to the canon of these works. In subjective culture, the individual seeks self-fulfilment and self-realisation through his or her own resources. Nothing need be shared. By the end of the nineteenth century, Simmel said, the classic example of this was the business culture; the collective 'pathology' arising from a myriad subjective cultures was alienation. For the Sunday Circle in Budapest the stabilising force of objective culture was a sine qua non. It was only through shared culture that the self could become known to others, and thus to itself. It was only by having a standpoint that was to be shared that one could recognise alienation in the first place. This solitude at the heart of modern capitalism came to dominate the discussions of the Sunday Circle as the war progressed and after the Bolshevik revolution they were led into radical politics. An added factor in their alienation was their Jewishness: in an era of growing anti-Semitism, they were bound to feel marginalised. Before the war they had been open to international movements – impressionism and aestheticism and to Paul Gauguin in particular, who, they felt, had found fulfilment away from the anti-Semitic business culture of Europe in far-off Tahiti. 'Tahiti healed Gauguin,' as Lukács wrote at one point.[45] He himself felt so marginalised in Hungary that he took to writing in German.

The Sunday Circle's fascination with the redemptive powers of art had some predictable consequences. For a time they flirted with mysticism and, as Mary Gluck describes it, in her history of the Sunday circle, turned against science.

(This was a problem for Mannheim; sociology was especially strong in Hungary and regarded itself as a science that would, eventually, explain the evolution of society.) The Sundays also embraced the erotic.[46] In *Bluebeard's Castle*, Béla Balázs described an erotic encounter between a man and a woman, his focus being what he saw as the inevitable sexual struggle between them. In Bartók's musical version of the story, Judith enters Prince Bluebeard's Castle as his bride. With increasing confidence, she explores the hidden layers – or chambers – of man's consciousness. To begin with she brings joy into the gloom. In the deeper recesses, however, there is a growing resistance. She is forced to become increasingly reckless and will not be dissuaded from opening the seventh, forbidden door. Total intimacy, implies Balázs, leads only to a 'final struggle' for power. And power is a chimera, bringing only 'renewed solitude.'[47]

Step by step, therefore, Lukács and the others came to the view that art could only ever have a limited role in human affairs, 'islands in a sea of fragmentation.'[48] This was – so far as art was concerned – the eclipse of meaning. And this cold comfort became the main message of the **Free School for Humanistic Studies**, which the Sunday Circle set up during the war years. The very existence of the Free School was itself instructive. It was no longer Sunday-afternoon discussions – but action.

Then came the Bolshevik revolution. Hitherto, Marxism had sounded too materialistic and scientistic for the Sunday Circle. But after so much darkness, and after Lukács's own journey through art, to the point where he had much reduced expectations and hopes of redemption in that direction, socialism began to seem to him and others in the group like the only option that offered a way forward: 'Like Kant, Lukács endorsed the primacy of ethics in politics.'[49] A sense of urgency was added by the emergence of an intransigent left wing throughout Europe, committed to ending the war without delay. In 1917 Lukács had written, 'Bolshevism is based on the metaphysical premise that out of evil, good can come, that it is possible to lie our way to the truth. [I am] incapable of sharing this faith.'[50] A few weeks later Lukács joined the Communist Party of Hungary. He gave his reasons in an article entitled 'Tactics and Ethics.' The central question hadn't changed: 'Was it justifiable to bring about socialism through terror, through the violation of individual rights,' in the interests of the majority? Could one lie one's way to power? Or were such tactics irredeemably opposed to the principles of socialism? Once incapable of sharing the faith, Lukács now concluded that terror *was* legitimate in the socialist context, 'and that therefore Bolshevism was a true embodiment of socialism.' Moreover, 'the class struggle – the basis of socialism – was a transcendental experience and the old rules no longer applied.'[51]

In short, this was the eclipse of ethics, the replacement of one set of principles by another. Lukács is important here because he openly admitted the change in himself, the justification of terror. Conrad had already foreseen such a change, Kafka was about to record its deep psychological effects on all concerned, and a whole generation of intellectuals, maybe two generations, would be compromised as Lukács was. At least he had the courage to entitle his paper

'Tactics and Ethics.' With him, the issue was out in the open, which it wouldn't always be.

By the end of 1919 the Sunday Circle was itself on the verge of eclipse. The police had it under surveillance and once went so far as to confiscate Balász's diaries, which were scrutinised for damaging admissions. The police had no luck, but the attention was too much for some of the Sundays. The Circle was reconvened in Vienna (on Mondays), but not for long, because the Hungarians were charged with using fake identities.[52] By then Lukács, its centre of gravity, had other things on his mind: he had become part of the Communist underground. In December 1919 Balázs gave this description: 'He presents the most heart-rending sight imaginable, deathly pale, hollow cheeked, impatient and sad. He is watched and followed, he goes around with a gun in his pocket. . . . There is a warrant out for his arrest in Budapest which would condemn him to death nine times over. . . . And here [in Vienna] he is active in hopeless conspiratorial party work, tracking down people who have absconded with party funds . . . in the meantime his philosophic genius remains repressed, like a stream forced underground which loosens and destroys the ground above.'[53] Vivid, but not wholly true. At the back of Lukács's mind, while he was otherwise engaged on futile conspiratorial work, he was conceiving what would become his most well known book, *History and Class Consciousness.*

The Vienna–Budapest (and Prague) axis did not disappear completely after World War I. The Vienna Circle of philosophers, led by Moritz Schlick, flourished in the 1920s, and Franz Kafka and Robert Musil produced their most important works. The society still produced thinkers such as Michael Polanyi, Friedrich von Hayek, Ludwig von Bertalanffy, Karl Popper, and Ernst Gombrich – but they came to prominence only after the rise of the Nazis caused them to flee to the West. Vienna as a buzzing intellectual centre did not survive the end of empire.

Between 1914 and 1918 all direct links between Great Britain and Germany had been cut off, as Wittgenstein discovered when he was unable to return to Cambridge after his holiday. But Holland, like Switzerland, remained neutral, and at the University of Leiden, in 1915, **W. de Sitter** was sent a copy of Einstein's paper on the general theory of relativity. An accomplished physicist, de Sitter was well connected and realised that as a Dutch neutral he was an important go-between. He therefore passed on a copy of Einstein's paper to **Arthur Eddington** in London.[54] Eddington was already a central figure in the British scientific establishment, despite having a 'mystical bent,' according to one of his biographers.[55] Born in Kendal in the Lake District in 1882, into a Quaker family of farmers, he was educated first at home and then at Trinity College, Cambridge, where he was senior wrangler and came into contact with J. J. Thomson and Ernest Rutherford. Fascinated by astronomy since he was a boy, he took up an appointment at the Royal Observatory in Greenwich from 1906, and in 1912 became secretary of the Royal Astronomical Society. His first important work was a massive and ambitious survey of the structure of the universe. This survey, combined with the work of other researchers and the development of more powerful telescopes, had revealed a great deal about the size, structure, and age of the heavens. Its main discovery, made in 1912, was that the brightness of so-called **Cepheid** stars pulsated in a regular way associated with their sizes. This helped establish real distances in the heavens and showed that our own galaxy has a diameter of about 100,000 light-years and that the sun, which had been thought to be at its centre, is in fact about 30,000 light-years excentric. The second important result of Cepheid research was the discovery that the spiral nebulae were in fact extragalactic objects, entire galaxies themselves, and very far away (the nearest, the Great Nebula in Andromeda, being 750,000 light-years away). This eventually provided a figure for the distance of the farthest objects, 500 million light-years away, and an age for the universe of between 10 and 20 billion years.[56]

Eddington had also been involved in ideas about the evolution of stars, based on work that showed them to consist of giants and dwarves. Giants are in general less dense than dwarves, which, according to Eddington's calculations, could be up to 20 million degrees Kelvin at their centre, with a density of one ton per cubic inch. But Eddington was also a keen traveller and had visited Brazil and Malta to study eclipses. His work and his academic standing thus made him the obvious choice when the Physical Society of London, during wartime, wanted someone to prepare a *Report on the Relativity Theory of Gravitation*.[57] This, which appeared in 1918, was the first complete account of general relativity to be published in English. Eddington had already received a copy of Einstein's 1915 paper from Holland, so he was well prepared, and his report attracted widespread attention, so much so that Sir Frank Dyson, the Astronomer Royal, offered an unusual opportunity to test Einstein's theory. On 29 May 1919, there was to be a total eclipse. This offered the chance to assess if, as Einstein predicted, light rays were bent as they passed near the sun. It says something for the Astronomer Royal's influence that, during the last full year of the war, Dyson obtained from the government a grant of £1,000 to mount not one but two expeditions, to Principe off the coast of West Africa and to Sobral, across the Atlantic, in Brazil.[58]

Eddington was given Principe, together with E. T. Cottingham. In the Astronomer Royal's study on the night before they left, Eddington, Cottingham, and Dyson sat up late calculating how far light would have to be deflected for Einstein's theory to be confirmed. At one point, Cottingham asked rhetorically what would happen if they found twice the expected value. Drily, Dyson replied, 'Then Eddington will go mad and you will have to come home alone!'[59] Eddington's own notebooks continue the account: 'We sailed early in March to Lisbon. At Funchal we saw [the other two astronomers] off to Brazil on March 16, but we had to remain until April 9 ... and got our first sight of Principe in the morning of April 23. ... We soon found we were in clover, everyone anxious to give every help we needed ... about May 16 we had no difficulty in getting the check photographs on three different nights. I had a good deal of work measuring these.' Then the weather changed. On the morning of 29 May, the day of the eclipse, the heavens opened, the downpour lasted for hours, and Eddington began to fear that their arduous journey was a waste of time. However, at one-thirty in the afternoon, by which time the partial phase of the eclipse had already begun, the clouds at last began to clear. 'I did not see the eclipse,' Eddington wrote later, 'being too busy changing plates, except for one glance to make sure it had begun and another half-way through to see how much cloud there was. We took sixteen photographs. They are all good of the sun, showing a very remarkable prominence; but the cloud has interfered with the star images. The last six photographs show a few images which I hope will give us what we need. ... June 3. We developed the photographs, 2 each night for 6 nights after the eclipse, and I spent the whole day measuring. The cloudy weather upset my plans. ... But the one plate that I measured gave a result agreeing with Einstein.' Eddington turned to his companion. 'Cottingham,' he said, 'you won't have to go home alone.'[60]

Eddington later described the experiment off West Africa as 'the greatest moment of my life.'[61] Einstein had set three tests for relativity, and now two of them had supported his ideas. Eddington wrote to Einstein immediately, giving him a complete account and a copy of his calculations. Einstein wrote back from Berlin on 15 December 1919, 'Lieber Herr Eddington, Above all I should like to congratulate you on the success of your difficult expedition. Considering the great interest you have taken in the theory of relativity even in earlier days I think I can assume that we are indebted primarily to your initiative for the fact that these expeditions could take place. I am amazed at the interest which my English colleagues have taken in the theory in spite of its difficulty.'[62]

Einstein was being disingenuous. The publicity given to Eddington's confirmation of relativity made Einstein the most famous scientist in the world. 'EINSTEIN THEORY TRIUMPHS' blazed the headline in the *New York Times*, and many other newspapers around the world treated the episode in the same way. The Royal Society convened a special session in London at which Frank Dyson gave a full account of the expeditions to Sobral and Principe.[63] **Alfred North Whitehead** was there, and in his book *Science and the Modern World*, though reluctant to commit himself to print, he relayed some of the excitement: 'The whole atmosphere of tense interest was exactly that of the Greek drama: we were the chorus commenting on the decree of destiny as disclosed in the development of a supreme incident. There was dramatic quality in the very staging: – the traditional ceremonial, and in the background the picture of Newton to remind us that the greatest of scientific generalisations was now, after more than two centuries, to receive its first modification. Nor was the personal interest wanting: a great adventure in thought had at length come safe to shore.'[64]

Relativity theory had not found universal acceptance when Einstein had first proposed it. Eddington's Principe observations were therefore the point at which many scientists were forced to concede that this exceedingly uncommon idea about the physical world was, in fact, true. Thought would never be the same again. Common sense very definitely had its limitations. And Eddington's, or rather Dyson's, timing was perfect. In more ways than one, the old world had been eclipsed.

THE ACQUISITIVE WASTELAND

Much of the thought of the 1920s, and almost all of the important literature, may be seen, unsurprisingly perhaps, as a response to World War I. Not so predictable was that so many authors should respond in the same way – by emphasising their break with the past through new *forms* of literature: novels, plays, and poems in which the way the story was told was as important as the story itself. It took a while for authors to digest what had happened in the war, to grasp what it signified, and what they felt about it. But then, in 1922, a year to rival 1913 as an annus mirabilis in thought, there was a flood of works that broke new ground: James Joyce's *Ulysses*; T. S. Eliot's *Waste Land*; Sinclair Lewis's *Babbitt*; Marcel Proust's ninth volume of *A la Recherche du Temps Perdu, Sodome et Gomorrhe II*; Virginia Woolf's first experimental novel, *Jacob's Room*; Rainer Maria Rilke's *Duino Elegies*; and Pirandello's *Henry IV*, all foundation stones for the architecture of the literature of the century.

What Joyce, Eliot, Lewis, and the others were criticising, among other things, was the society – and not only the war society – which capitalism had brought about, a society where value was placed on possessions, where life had become a race to acquire things, as opposed to knowledge, understanding, or virtue. In short, they were attacking the acquisitive society. This was in fact a new phrase, coined the year before by **R. H. Tawney** in a book that was too angry and too blunt to be considered great literature. Tawney was typical of a certain kind of figure in British society at the time (William Beveridge and George Orwell were others). Like them, Tawney came from an upper-class family and was educated at a public school (Rugby) and Balliol College, Oxford; but he was interested all his life in poverty and especially in inequality. After university, he decided, instead of going into the City, as many of his background would have done, to work at Toynbee Hall in London's East End (Beveridge, the founder of Britain's welfare state, was also there). The idea behind Toynbee Hall was to bring a university atmosphere and lifestyle to the working classes, and in general it had a profound effect on all who experienced it. It helped turn Tawney into the British socialist intellectual best in touch with the unions.[1] But it was the miners' strike in February 1919 that was to shape Tawney's subsequent career. Seeking to head off confrontation, the government established a Royal Commission on the Coal Mines, and Tawney was one of six men representing the labour side (another was Sidney Webb).[2] Millions of words of evidence were put before the commission, and Tawney read all of them. He was so moved by the accounts of danger, ill-health, and poverty that he wrote the first of the three books for which he is chiefly known. These were *The Acquisitive Society* (1921), *Religion and the Rise of Capitalism* (1926), and *Equality* (1931).

Tawney, a mild man whose bushy moustache made him appear avuncular, hated the brutalism of unbridled capitalism, particularly the waste and inequalities it produced. He served in the trenches in the war as an ordinary soldier, refusing a commission. He expected capitalism to break down afterward: he thought that it misjudged human nature, elevating production and the making of profit, which ought to be a means to certain ends, into ends in themselves. This had the effect, he argued, of encouraging the wrong instincts in people, by which he meant acquisitiveness. A very religious man, Tawney felt that acquisitiveness went against the grain – in particular, it sabotaged 'the instinct for service and solidarity' that is the basis for traditional civil society.[3] He thought that in the long run capitalism was incompatible with culture. Under capitalism, he wrote, culture became more private, less was shared, and this trend went against the common life of men – individuality inevitably promoted inequality. The very concept of culture therefore changed, becoming less and less an inner state of mind and more a function of one's possessions.[4] On top of that, Tawney also felt that capitalism was, at bottom, incompatible with democracy. He suspected that the inequalities endemic in capitalism – inequalities made more visible than ever by the acquisitive accumulation of consumer products – would ultimately threaten social cohesion. He saw his role, therefore, as helping to provide an important *moral* counterattack against capitalism for the many like himself who felt it had been at least partly responsible for war.[5]

But this wasn't Tawney's only role. He was an historian, and in his second book he looked at capitalism historically. The thesis of *Religion and the Rise of Capitalism* was that 'economic man,' the creature of classical economics, was by no means the universal figure in history he was supposed to be, that human nature was not necessarily shaped as classical liberals said it was. Tawney argued that the advent of capitalism was not inevitable, that its successes were relatively recent, and that in the process it had rendered extinct a whole range of behaviours and experiences and replaced them with its own. In particular capitalism had extinguished religion, though the church had to take some share of the blame insofar as it had abdicated its role as a moral leader.[6]

In retrospect, not all of Tawney's criticisms of capitalism ring true anymore.[7] Most obviously, and importantly, capitalism has not proved incompatible with democracy. But he was not wholly wrong; capitalism probably is inimical to what Tawney meant by culture – indeed, as we shall see, capitalism has changed what we *all* mean by culture; and it is arguable that capitalism has aided the change in morality we have seen during the century, though there have been other reasons as well.

★

Tawney's vision was bitter and specific. Not everyone was as savage about capitalism as he was, but as the 1920s wore on and reflection about World War I matured, an unease persisted. What characterised this unease, however, was that it concerned more than capitalism, extending to Western civilisation as a whole, in some senses an equivalent of Oswald Spengler's thesis that there was decay and ruin everywhere in the West. Without question the man who caught this mood best was both a banker — the archsymbol of capitalism — and a poet, the licensed saboteur.

T. S. Eliot was born in 1888, into a very religious Puritan family. He studied at Harvard, took a year off to study poetry in Paris, then returned to Harvard as a member of the faculty, teaching philosophy. Always interested in Indian philosophy and the links between philosophy and religion, he was infuriated when Harvard tried to separate the one from the other as different disciplines. In 1914 he transferred to Oxford, where he hoped to continue his philosophical studies. Shortly after, war broke out. In Europe, Eliot met two people who had an immense effect on him: **Ezra Pound** and **Vivien Haigh-Wood**. At the time they met, Pound was a much more worldly figure than Eliot, a good teacher and at that time a better poet. Vivien Haigh-Wood became Eliot's first wife. Initially happy, the marriage had turned into a disaster by the early 1920s: Vivien descended steadily into madness, and Eliot found the circumstances so trying that he himself sought psychiatric treatment in Switzerland.[8]

The puritanical world Eliot grew up in had been fiercely rational. In such a world science had been dominant in that it offered the promise of relief from injustice. Beatrice Webb had shared Eliot's early hopes when, in 1870, she said, 'It was by science, and by science alone, that all human misery would be ultimately swept away.'[9] And yet by 1918 the world insofar as Eliot was concerned was in ruins. For him, as for others, science had helped produce a war in which the weapons were more terrible than ever, in which the vast nineteenth-century cities were characterised as much by squalor as by the beauty the impressionists painted, where in fact the grinding narratives of Zola told a grimmer truth. Then there was the new physics that had helped remove more fundamental layers of certainty; there was Darwin undermining religion, and Freud sabotaging reason itself. A consolidated edition of Sir James Frazer's *The Golden Bough* was also published in 1922, the same year as *The Waste Land*, and this too hit hard at Eliot's world. It showed that the religions of so-called savages around the world were no less developed, complex, or sophisticated than Christianity. At a stroke the simple social Darwinian idea that Eliot's world was the current endpoint in the long evolutionary struggle, the 'highest' stage of man's development, was removed. Also subverted was the idea that there was anything special about Christianity itself. Harvard had been right after all to divorce philosophy and religion. In Max Weber's term, the West had entered a phase of *Entzauberung*, 'unmagicking' or disenchantment. At a material, intellectual, and spiritual level — in all senses — Eliot's world was laid waste.[10]

Eliot's response was a series of verses originally called *He Do the Police in Different Voices*, taken from Charles Dickens's *Our Mutual Friend*. Eliot was at the time working in the colonial and foreign branch of Lloyds Bank, 'fascinated by the science of money' and helping with the prewar debt position between Lloyds and Germany. He got up at five every morning to write before going into the bank, a routine so exhausting that in the autumn of 1921 he took a prolonged leave.[11] Pound's poem *Hugh Selwyn Mauberly*, published the year before, had a not dissimilar theme to *The Waste Land*. It explored the sterility, intellectual, artistic, and sexual, of the old world afflicted by war. In *Mauberly*, 1920, Pound described Britain as 'an old bitch, gone in the teeth.'[12] But *Mauberly* did not have either the vividly savage images of *He Do the Police*, nor its shockingly original form, and Pound, to his credit, immediately recognised this. We now know that he worked hard on Eliot's verses, pulling them into shape, making them coherent, and giving them the title *The Waste Land* (one of the criteria he used was whether the lines read well out loud).[13] Eliot dedicated the work to Pound, as *il miglior fabbro*, 'the better maker.'[14] His concern in this great poem is the sterility that he regards as the central fact of life in the postwar world, a dual sterility in both the spiritual and sexual spheres. But Eliot is not content just to pin down that sterility; he contrasts the postwar world with other worlds, other possibilities, in other places and at other times, which were fecund and creative and not at all doomed. And this is what gave *The Waste Land* its singular poetic architecture. As in Virginia Woolf's novels, Joyce's *Ulysses*, and Proust's *roman fleuve*, the form of Eliot's poem, though revolutionary, was integral to its message. According to Eliot's wife, the poem – partly autobiographical – was also partly inspired by Bertrand Russell.[15] Eliot juxtaposed images of dead trees, dead rats, and dead men – conjuring up the horrors of Verdun and the Somme – with references to ancient legends; scenes of sordid sex run into classical poetry; the demeaning anonymity of modern life is mingled with religious sentiments. It is this collision of different ideas that was so startling and original. Eliot was trying to show how far we have fallen, how far evolution is a process of *descent*.

The poem is divided into six parts: 'The Epigraph,' 'The Burial of the Dead,' 'A Game of Chess,' 'The Fire Sermon,' 'Death by Water,' and 'What the Thunder Said.' All the titles are evocative and all, on first acquaintance, obscure. There is a chorus of voices, sometimes individual, sometimes speaking in words borrowed from the classics of various cultures, sometimes heard via the incantations of the 'blind and thwarted' Tiresias.[16] At one moment we pay a visit to a tarot reader, at another we are in an East End pub at closing time, next there is a reference to a Greek legend, then a line or two in German. Until one gets used to it, the approach is baffling, quite unlike anything encountered elsewhere. Even stranger, the poem comes with notes and references, like an academic paper. These notes, however, repay inspection. For study of the myths introduces other civilisations, with different but coherent worldviews and a different set of values. And this is Eliot's point: if we are to turn our back on the acquisitive society, we have to be ready to *work*:

> At the violet hour, when the eyes and back
> Turn upward from the desk, when the human engine waits

Like a taxi throbbing waiting,
I Tiresias, though blind, throbbing between two lives,
Old man with wrinkled female breasts, can see
At the violet hour, the evening hour that strives
Homeward, and brings the sailor home from sea,
The typist home at teatime, clears her breakfast, lights
Her stove, and lays out food in tins.

It takes no time at all for the poem to veer between the heroic and the banal, knitting a sense of pathos and bathos, outlining an ordinary world on the edge of something finer, yet not really aware that it is.

There is a shadow under this red rock,
(Come in under the shadow of this red rock),
And I will show you something different from either
Your shadow at morning striding behind you
Or your shadow at morning rising to meet you;
I will show you fear in a handful of dust.
   Frisch weht der Wind
   Der Heimat zu
   Mein Irisch Kind
   Wo weilest du?[17]

The first two lines hint at Isaiah's prophecy of a Messiah who will be 'as rivers of water in a dry place, as the shadow of a great rock in a weary land' (Isaiah 32.2). The German comes direct from Wagner's opera *Tristan und Isolde*: 'Fresh blows the wind/Toward home/My Irish child/Where are you waiting?' The imagery is dense, its aim ambitious. *The Waste Land* cannot be understood on one reading or without 'research' or work. It has been compared (by Stephen Coote, among others) to an Old Master painting in which we have first to learn the iconography before we can understand fully what is being said. In order to appreciate his poem, the reader has to open himself or herself to other cultures, to attempt an escape from this sterile one. The first two 'confidential copies' of the poem were sent to John Quinn and Ezra Pound.[18]

Eliot, incidentally, did not share the vaguely Freudian view of most people at the time (and since) that art was an expression of the personality. On the contrary, for him it was 'an escape from personality.' He was no expressionist pouring his 'over-charged soul' into his work. *The Waste Land* is, instead, the result of detailed reflection, of craftsmanship as well as art, owing as much to the rewards of a good education as the disguised urges of the unconscious. Much later in the century, Eliot would publish considerably fiercer views about the role of culture, particularly 'high' culture in all our lives, and in less poetic terms. In turn, he himself would be accused of snobbery and worse. He was ultimately, like so many writers and artists of his day, concerned with 'degeneration' in cultural if not in individual or biological terms.

Frederick May, the critic and translator, has suggested that **Luigi Pirandello**'s highly innovative play *Six Characters in Search of an Author* is a dramatic analogue of *The Waste Land*: 'Each is a high poetic record of the disillusionment and spiritual desolation of its time, instinct with compassion and poignant with the sense of loss ... each has become in its own sphere at once the statement and the symbol of its age.'[19]

Born in Caos, near Girgenti (the modern Agrigento) in Sicily in 1867, in the middle of a cholera epidemic, Pirandello studied literature in Palermo, Rome and Bonn. He began publishing plays in 1889, but success did not arrive fully until 1921, by which time his wife had entered a nursing home for the insane. His two plays that will be considered here, *Six Characters in Search of an Author* (1921), and *Henry IV* (1922), are united in being concerned with the impossibility of describing, or even conceiving, reality. 'He dramatises the subconscious.' In the earlier title, six characters invade the rehearsal of a play, a play Pirandello had himself written a few years earlier, insisting that they are not actors, nor yet people, but characters who need an 'author' to arrange the story that is within them. As with Wittgenstein, Einstein, and Freud, Pirandello is drawing attention to the way words break down in describing reality. What is the difference – and the overlap – between character and personality, and can we ever hope to pin them down in art? Just as Eliot was trying to produce a new form of poetry, Pirandello was creating a new form of drama, where theatre itself comes under the spotlight as a form of truth-telling. The characters in his plays know the limits to their understanding, that truth is relative, and that their problem, like ours, is to realise themselves.

Six Characters created a scandal when it was first performed, in Rome, but a year later received a rapturous reception in Paris. *Henry IV* had a much better reception in Italy when it was premiered in Milan, and after that Pirandello's reputation was made. As did Eliot's, his wife descended into madness and Pirandello later formed a relationship with the Italian actress Marta Abba.[20] Unlike Eliot, whose art was forged despite his personal circumstances, Pirandello several times used madness as a dramatic device.[21] *Henry IV* tells the story of a man who, twenty years before, had fallen from his horse during a masquerade in which he was dressed as the German emperor Henry IV, and was knocked unconscious when he hit his head on the paving. In preparation for the masquerade, the man had read widely about the emperor and, on coming to, believed he was in fact Henry IV. To accommodate his illness his wealthy sister has placed him in a mediaeval castle surrounded by actors dressed as eleventh-century courtiers who enable him to live exactly as Henry IV did, though they move in and out of their roles, confusingly and at times hilariously (without warning, a costumed actor will suddenly light up a cigarette). Into this scene are introduced old friends, including Donna Matilda, still beautiful, her daughter Frida, and a doctor. Here Pirandello's mischief is at its highest, for we can never be sure whether Henry is still mad, or only playing a part.

Like the fool in earlier forms of theatre, Henry asks his fellow characters penetrating questions: 'Do you remember always being the same?' Therefore, we never quite know whether Henry is a tragic figure, and aware that he is. This would make him moving – and also sane. It would also make all the others in the play either fools or mad, or possibly both. But if Henry is fully sane, does it make sense for him to live on as he does? Everyone in the play, though real enough, is also desperate, living a lie.

The real tragedy occurs when the doctor, in order to 'treat' Henry by facing him with a shocking reality, provokes him into murder. In *Henry IV* no one really understands themselves completely, least of all the man of science who, so certain of himself and his methods, precipitates the greatest calamity. Devastated by the wasteland of his life, Henry had opted for a 'planned' madness, only to have that backfire on him too. Life, for Pirandello, was like a play within a play, a device he used many times: one can never be entirely sure who is acting and who is not. One cannot even be sure when one is acting oneself.

Wittgenstein's *Tractatus*, discussed in chapter 9, was actually published in the annus mirabilis of 1922. So too was *The Last Days of Mankind*, the great work of Wittgenstein's Viennese friend **Karl Kraus**. Kraus, who was Jewish, had been part of *Jung Wien* at the Café Griensteidl in the early years of the century, mixing with Hugo von Hofmannsthal, Arthur Schnitzler, Adolf Loos, and Arnold Schoenberg. He was a difficult and slightly deformed man, with a congenital abnormality in his shoulders that gave him a stoop. A satirist of almost unrivalled mordancy, he earned most of his considerable income from lectures and readings. At the same time, he published a magazine, *Die Fackel* (The Torch), three times a month, from 1899 until his death in 1936. This made him a lot of enemies but also earned him a wide following, which even extended to the troops on the front line in World War I. Punctilious to a degree, he was no less interested in language than his philosopher friend and was genuinely pained by solecisms, infelicitous turns of phrase, ungainly constructions. His aim, he once said, 'is to pin down the Age between quotation marks.'[22] Bitterly opposed to feminine emancipation, which he regarded as 'a hysterical response to sexual neurosis,' he hated the smugness and anti-Semitism of the Viennese press, together with the freewheeling freemasonry that, more than once, led him into the libel courts. Kraus was in effect doing in literature and society what Loos was doing in architecture, attacking the pompous, self-regarding self-satisfaction of the *ancien régime*. As he himself described his aim in *Die Fackel*: 'What has been laid down here is nothing else than a drainage system for the broad marshes of phraseology.'[23]

The Last Days of Mankind was written – usually late at night – during the summers of World War I and immediately afterward. On occasions Kraus escaped to Switzerland, to avoid the turmoil of Vienna and the attentions of the censor. His deformity had helped him avoid military service, which made him already suspect in the eyes of certain critics, but his opposition to the aims of the Central Powers earned him even more opprobrium. The play was his verdict on the war, and although certain passages appeared in *Die Fackel* in 1919 it wasn't completed until 1921, by which time Kraus had added much new material.[24] The play draws a cumulative strength from hundreds of small vignettes, all taken from newspaper reports and, therefore, not invented. Life at the front, in all its horror and absurdity, is juxtaposed (in a verbal equivalent of Kurt Schwitters's technique) with events back in Vienna, in all *their* absurdity and venality. Language is still the central element for Kraus (*Last Days* is essentially a play for voices rather than action). We witness the Kaiser's voice, that of the poet, the man at the front, Jewish dialects from Vienna, deliberately cheek-by-jowl with one another to throw each crime – of thought or action – into relief. The satirist's technique, of holding one phrase (or thought, or belief, or conviction) against its opposite, or reciprocal, is devastatingly effective, the more so as time passes.

The play has been rarely performed because of its length – ten hours – and Kraus himself claimed that it was intended only for performances on Mars because 'people on Earth could not bear the reality presented to them.'[25] At the end of the play, mankind destroys itself in a hail of fire, and the last lines, put into the mouth of God, are those attributed to the Kaiser at the start of the war: 'I did not want it.' Brecht's epitaph of Kraus was: 'As the epoch raised its hand to end its life, he was this hand.'[26]

The most overwhelming of the great books that appeared in 1922 was *Ulysses*, by **James Joyce**. On the surface, the form of Joyce's *Ulysses* could not be more different from *The Waste Land* or Virginia Woolf's *Jacob's Room*, which will be considered later. But there are similarities, and the authors were aware of them. *Ulysses* was also in part a response to the war – the last line reads: 'Trieste-Zurich-Paris, 1914–1921.' As Eliot does in *The Waste Land*, Joyce, as Eliot himself commented in a review, uses an ancient myth (in this case Homer) as 'a way of controlling, of ordering, of giving a shape and a significance to the immense panorama of futility and anarchy which is contemporary history.'[27]

Born in Dublin in 1882, Joyce was the oldest child in a family of ten. The family struggled financially but still managed to give James a good education at Jesuit schools and University College, Dublin. He then moved to Paris, where at first he thought he might be a doctor. Soon, though, he started to write. From 1905 he lived in Trieste with Nora Barnacle, a young woman from Galway who he had met on Nassau Street, Dublin, in 1904. *Chamber Music* was published in 1907, and *Dubliners*, a series of short stories, in 1914. On the outbreak of war, Joyce was obliged to move to neutral Zurich (Ireland was then ruled by Great Britain), though he considered Prague as an alternative.[28] During hostilities, he published *A Portrait of the Artist as a Young Man*, but it was *Ulysses* that brought him international fame. Some chapters appeared first in 1919 in a London magazine, the *Egoist*. However, the printers and some subscribers took objection, and publication of subsequent chapters was discontinued. Joyce next turned to an avant-garde American magazine, the *Little Review*, which published other chapters of the book, but in February 1921 that magazine was found guilty of obscenity, and the editors were fined.[29] Finally Joyce approached a young bookseller in Paris, another American named **Sylvia Beach**, and her shop, Shakespeare & Co., published the book in its entirety on 2 February 1922. For the first edition, one thousand copies were printed.

There are two principal characters in *Ulysses*, though many of the countless minor ones are memorable too. **Stephen Dedalus** is a young artist going through a personal crisis (like Western civilisation he has dried up, lost his large ambitions and the will to create). **Leopold Bloom** – 'Poldy' to his wife, and modelled partly on Joyce's father and brother – is a much more down-to-earth character. Joyce (influenced by the theories of Otto Weininger) makes him Jewish and slightly effeminate, but it is his unpretentious yet wonderfully rich life, inner and outer, that makes him Ulysses.[30] For it is Joyce's point that the age of heroes is over.* He loathed the 'heroic abstractions' for which so many soldiers were sacrificed, 'the big words which make us so unhappy.'[31] The odyssey of *his* characters is not to negotiate the fearsome mythical world of the Greeks – instead, he gives us Bloom's entire day in Dublin on 16 June 1904.[32] We follow Bloom from the early preparation of his wife's breakfast, his presence at the funeral of a friend, encounters with newspaper acquaintances, racing aficionados, his shopping exploits, buying meat and soap, his drinking, a wonderfully erotic scene where he is on the beach near three young women and they are watching some fireworks, and a final encounter with the police on his way home late at night. We leave him gently climbing into bed next to his wife and trying not to wake her, when the book shifts perspective and gives us his wife Molly's completely unpunctuated view of Bloom.

It is one of the book's attractions that it changes style several times, from stream of consciousness, to question-and-answer, to a play that is also a dream, to more straightforward exchanges. There are some lovely jokes (Shakespeare is 'the chap that writes like Synge', 'My kingdom for a drink') and some hopelessly childish puns ('I beg your parsnips'); incredibly inventive language, teeming with allusions; endless lists of people and things and references to the latest developments in science. One point of the very great length of the book (933 pages) is to recreate a world in which the author slows life down for the reader, enabling him or her to relish the language, a language that never sleeps. In this way, Joyce draws attention to the richness of Dublin in 1904, where poetry, opera, Latin and liturgy are as much a part of everyday lower-middle-class life as are gambling, racing, minor cheating and the lacklustre lust of a middle-aged man for virtually every woman he meets.[33] 'If *Ulysses* isn't fit to read', said Joyce to his cousin, responding to criticism, 'life isn't fit to live.' Descriptions of food are never far away, each and every one mouthwatering ('Buck Mulligan slit a steaming scone in two and plastered butter over its smoking pith.'). Place names are left to hang, so we realise how improbable but very beautiful even proper names are: Malahide, Clonghowes, Castleconnel. Joyce revisits words, rearranges spelling and punctuation so that we see these words, and what they represent, anew: 'Whether these be sins or virtues old

* In fact, *Ulysses* is more deeply mythical than many readers realise, various parts being based on different areas of the body (the kidneys, the flesh); this was spelled out in *James Joyce's Ulysses*, published in collaboration with Stuart Gilbert in 1930. It is not necessary to know this for a rich and rewarding experience in reading the book.[36]

Nobodaddy will tell us at doomsday . . .', 'He smellsipped the cordial . . .', 'Her ample bedwarmed flesh . . .', 'Dynamitard'.[34]

In following Bloom the reader – like Dedalus – is exhilarated and liberated.[35] Bloom has no wish to be anything other than who he is, 'neither Faust nor Jesus'. Bloom inhabits an amazingly *generous* world, where people allow each other to be as they are, celebrating everyday life and giving a glimpse of what civilisation can evolve into: food, poetry, ritual, love, sex, drink, language. They can be found anywhere, Joyce is saying. They are what peace – inner and out – is.

T. S. Eliot wrote an essay about *Ulysses* in the *Dial* magazine in 1923, in which he confessed that the book for him had 'the importance of a scientific discovery,' and indeed part of Joyce's aim was to advance language, feeling it had dropped behind as science had expanded. He also liked the fact that Joyce had used what he called 'the mythical method.'[37] This, he believed, might be a way forward for literature, replacing the narrative method. But the most revealing difference between *Ulysses*, on the one hand, and *The Waste Land*, *Jacob's Room*, and *Henry IV* on the other, is that in the end Stephen Dedalus is redeemed. At the beginning of the book, he is in an intellectual and moral wasteland, bereft of ideas and hope. Bloom, however, shows himself throughout the book as capable of seeing the world through others' eyes, be it his wife Molly, who he knows intimately, or Dedalus, a relative stranger. This not only makes Bloom profoundly unprejudiced – in an anti-Semitic world – but it is, on Joyce's part, a wonderfully optimistic message, that connections are possible, that solitude and atomisation, alienation and ennui are not inevitable.

In 1922 Joyce's Irish colleague **W. B. Yeats** was named a senator in Ireland. Two years later he received the Nobel Prize for Literature. Yeats's fifty-seven-year career as a poet spanned many different periods, but his political engagement was of a piece with his artistic vision. An 1899 police report described him as 'more or less of a revolutionary,' and in 1916 he had published 'Easter 1916,' about the botched Irish nationalist uprising. This contained lines that, though they refer to the executed leaders of the uprising, could also serve, in the ending, as an epitaph for the entire century:

> We know their dream; enough
> To know they dreamed and are dead;
> And what if excess of love
> Bewildered them till they died?
> I write it out in a verse –
> MacDonagh and MacBride
> And Connolly and Pearse
> Now and in time to be,
> Wherever the green is worn,
> All changed, changed utterly:
> A terrible beauty is born.[38]

Yeats recognised that he had a religious temperament at a time when science had largely destroyed that option. He believed that life was ultimately tragic, and that it is largely determined by 'remote . . . unknowable realities.'[39] For him the consensus of life, its very structure, will defeat us, and the search for greatness, the most noble existential cause, must involve a stripping away of the 'mask': 'If mask and self could be unified, one would experience completeness of being.'[40] This was not exactly Freudianism but close and, as David Perkins has shown, it led Yeats to a complicated and highly personal system of iconography and symbols in which he pitched antitheses against one another: youth and age, body and soul, passion and wisdom, beast and man, creative violence and order, revelation and civilisation, time and eternity.[41]

Yeats's career is generally seen in four phases – before 1899, 1899–1914, 1914–28, and after 1928 – but it is his third phase that marks his highest achievement. This period includes *The Wild Swans at Coole* (1919), *Michael Robartes and the Dancer* (1921), *The Tower* (1928), and the prose work *A Vision* (1925). This latter book sets out Yeats's occult system of signs and symbols, which were partly the result of his 'discovery' that his wife had psychic powers and that spirits 'spoke through her' in automatic writing and trances.[42] In anyone else such an approach might have been merely embarrassing, but in Yeats the craftsmanship shines through to produce a poetic voice that is clear and distinctive, wholly autonomous, conveying 'the actual thoughts of a man at a passionate moment of life.'[43] Yeats the man is not at all like Bloom, but they are embarked on the same journey:

> The trees are in their autumn beauty,
> The woodland paths are dry,
> Under the October twilight the water
> Mirrors a still sky;
> Upon the brimming water among the stones
> Are nine-and-fifty swans . . .
>
> Unwearied still, lover by lover,
> They paddle in the cold
> Companionable streams or climb the air;
> Their hearts have not grown old;
> Passion or conquest, wander where they will,
> Attend upon them still.
>
>   – 'The Wild Swans at Coole,' 1919

Yeats was affected by the war and the wilderness that followed.

> Many ingenious lovely things are gone
> That seemed sheer miracle to the multitude . . .
>
> O but we dreamed to mend
> Whatever mischief seemed
> To afflict mankind, but now
> That winds of winter blow
>
>   – 'Nineteen Hundred and Nineteen,' 1919

But, like Bloom, he was really more interested in creating afresh from nature than lamenting what had gone.

> That is no country for old men. The young
> In one another's arms, birds in the trees,
> – Those dying generations – at their song,
> Those salmon-falls, the mackerel-crowded seas,
> Fish, flesh, or fowl, commend all summer long
> Whatever is begotten, born, and dies.
> Caught in that sensual music all neglect
> Monuments of unageing intellect.
>
> – 'Sailing to Byzantium,' 1928

Yeats had begun his career trying to put the legends of Ireland to poetic use. He never shared the modernist desire to portray the contemporary urban landscape; instead, as he grew older he recognised the central reality of 'desire in our solitude,' the passion of private matters, and that science had nothing worthwhile to say on the matter.[44] Greatness, as Bloom realised, lay in being wiser, more courageous, more full of insight, even in little ways, especially in little ways. Amid the wasteland, Yeats saw the poet's role as raising his game, in order to raise everybody's. His poetry was very different from Eliot's, but in this one aim they were united.

Bloom is, of course, a standing reproach for the citizens of the acquisitive society. He is not short of possessions, but he doesn't have much, or all that he might have, yet that doesn't bother him in the slightest. His inner life is what counts. Nor does he judge other people by what they have; he just wants to get inside their heads to see how it might be different from his own, to aid his experience of the world.

Four years after *Ulysses*, in 1926, **F. Scott Fitzgerald** published his novel *The Great Gatsby*, which, though a much more conventional work, addresses the same theme albeit from virtually the opposite direction. Whereas Leopold Bloom is a lower-middle-class Dubliner who triumphs over small-scale adversity by redemptive wit and low-level cunning, the characters in *Gatsby* are either very rich or want to be, and sail through life in such a way that hardly anything touches them, inhabiting an environment that breeds a moral and intellectual emptiness that constitutes its own form of wasteland.

The four main characters in the book are Jay Gatsby, Daisy and Tom Buchanan, and Nick Carraway, the narrator. The action takes place one summer on an island, West Egg, a cross between Nantucket, Martha's Vineyard, and Long Island, but within driving distance of Manhattan. Carraway, who has rented the house next to Gatsby by accident, is a relative of Daisy. To begin with, Gatsby, who shared some biographical details with Fitzgerald, the Buchanans, and Carraway lead relatively separate lives; then they are drawn together.[45] Gatsby is a mysterious figure. His home is always open for large, raucous, Jazz Age parties, but he himself is an enigmatic loner; no one really knows who he is, or how he made his money. He is often on the phone, long distance (when long distance was expensive and exotic). Gradually, however, Nick is drawn into Gatsby's orbit. In parallel with this he learns that Tom Buchanan is having an affair with a Myrtle Wilson whose husband owns a gas station where he often refuels on his way to and from Manhattan. Daisy, the original 'innocent,' a 1920s bright young thing, is blissfully unaware of this. The book is barely 170 pages long, and nothing is laboured. There is an early mention of '*The Rise of the Colored Empires*, by this man Goddard,' a reference to Lothrop Stoddard's eugenic tract *The Rising Tide of Colour*. This provokes a discussion by Tom about race: 'If we don't look out the white race will be – will be utterly submerged. It's all scientific stuff; it's been proved . . . it's up to us, who are the dominant race, to watch out or these other races will have control of things. . . . The idea is that we're Nordics . . . and we've produced all the things that go to make civilisation – oh, science and art, and all that. Do you see?'[46] The area where the fatal accident takes place, where Myrtle is killed, is known as the Valley of Ashes, based on Flushing Meadow, a swamp filled with garbage and ash. At other times, 'breeding' is a matter of exquisite fascination to the characters. But these points are lightly made, not forced on the reader.

Permeating all is the doubt that surrounds Gatsby. Dark rumours abound about the way he made his fortune – liquor, drugs, gambling. It soon transpires that Gatsby wants an introduction to Daisy and asks Nick, her relative, to arrange a meeting. When he does so, it turns out that Gatsby and Daisy already know each other and were in love before she married Tom. (Fitzgerald was worried that this was the weak point of the book: he had not explained adequately Gatsby's earlier relations with Daisy.)[47] They resume their affair. One afternoon a group of them go in two cars to Manhattan. In the city Tom accuses Gatsby and Daisy of being lovers. At Gatsby's instigation, Daisy confesses she has never loved Tom. Angered, Tom reveals he has been checking up on Gatsby: he did go to Oxford, as he claimed; he was decorated in the war. Like Nick, the reader warms to Gatsby. We also know by now that his real name is James Gatz, that he comes from a poor background, and that fortune smiled on him as a young man when he was able to do a millionaire a favour. But Tom has amassed evidence that Gatsby is in fact now involved in a number of unwholesome, even illegal schemes: bootlegging and dealing in stolen secur-ities. Before we can digest this, the confrontation breaks up, and the parties drive back to the island in two cars, Gatsby and Daisy in one, the rest in the other. We surmise that the confrontation will continue later. On the way, however, Gatsby's car kills Myrtle Wilson, Tom's lover, but doesn't stop. Tom, Nick, and the others, travelling well behind, arrive to find the police at the scene and Mr Wilson distraught. Mr Wilson has begun to suspect that his wife is being unfaithful but doesn't know who her lover is. He now suspects Gatsby, deciding his wife was killed to keep her quiet, so he goes to Gatsby's house, finds him in the pool, shoots him, and then turns the gun on himself. What Wilson doesn't know, and what Tom never finds out, is that Daisy was driving. This is kept from the police. Daisy, whose carelessness kills Myrtle, gets off scot-free. Tom's affair, which triggers all this tragedy, is never disclosed. Tom and Daisy disappear, leaving Carraway to arrange Gatsby's funeral. By now Gatsby's shady business deals have been confirmed, and no one attends.[48]

The last scene in the book takes place in New York, when Nick sees Tom on Fifth Avenue and refuses to shake hands. It is clear from this meeting that Tom still has no idea that Daisy was driving the car, but for Nick this innocence is irrelevant, even dangerous. It is what enchants and disfigures America: Gatsby betrays and is betrayed.[49] He feels that even if Tom is unaware that Daisy was driving, their behaviour is so despicable it really makes no difference to his judgement of them. He also has some harsh words to say about Daisy, that she smashed up things, and then 'retreated back' into her money. In attacking her, Nick is forsaking the blood link, disallying himself from the 'Nordics' who have 'produced civilisation.' What Tom and Daisy have left behind, despite their breeding, is catastrophe. The Buchanans – and others like them – sail through life in a moral vacuum, incapable of distinguishing the significant from the trivial, obsessed with the trappings of luxury. Everywhere you turn in *The Great Gatsby* is a wasteland: moral, spiritual, biological, even, in the Valley of Ashes, topographical.

James Joyce and Marcel Proust met in 1922, on 18 May, after the first night of Igor Stravinsky's *Renard*, at a party for Serge Diaghilev also attended by Pablo Picasso, who had designed the sets. Afterwards Proust gave Joyce a life home in a taxi, and during the journey the drunken Irishman told Proust he had never read a single word he had written. Proust was very offended and took himself off to the Ritz, where he had an agreement that he would always be fed, however late.[50]

Joyce's insult was unbecoming. After the delay in publication of other volumes of *A la recherche du temps perdu*, caused by war, Proust had published four titles in fairly rapid succession. *A l'ombre des jeunes filles en fleurs* (which won the Prix Goncourt) was published in 1919, *Le Côté de Guermantes* came out the year after, and both *Le Côté de Guermantes II* and *Sodome et Gomorrhe I* were released in May 1921. *Sodome et Gomorrhe II* was published in May 1922, the very month Proust and Joyce met. Three more volumes – *La Prisonnière*, *Albertine disparue*, and *Le temps retrouvé* – all came out after Proust died in 1922.

Despite the delay in publication, *Jeunes filles* and *Le Coté de Guermantes* take us back to Swann, the salons of Paris, the minutiae of aristocratic snobbishness, the problems associated with Swann's love for Gilberte and Odette. But with *Sodome et Gomorrhe* there is a change, and Proust fixes his gaze on one of the areas singled out by Eliot and Joyce: the landscape of sex in the modern world. However, unlike those two, who wrote about sex outside marriage, outside the church, casual and meaningless sex, Proust focused his attention on homosexuality. Proust, who was himself homosexual, had suffered a double tragedy during the war years when his driver and typist, Alfred Agostinelli, with whom he had fallen in love, left him for a woman and went to live in the south of France. A short while later, Agostinelli was killed in a flying accident, and for months Proust was inconsolable.[51] After this episode, homosexuality begins to make a more frank appearance in his work. Proust's view was that homosexuality was more widespread than generally realised, that many more men were homosexual than even they knew, and that it was a malady, a kind of nervous complaint that gave men female qualities (another echo of Otto Weininger). This changed dramatically Proust's narrative technique. It becomes apparent to the reader that a number of the male characters lead a double life. This makes their stiff, self-conscious grandeur and their snobbery more and more absurd, to the extent that *Sodome et Gomorrhe* finally becomes subversive of the social structure that dominates the earlier books. The most enviable life, he is showing us, is a low comedy based on deceit.

In fact, the comedy is far from funny for the participants.[52] The last books in the sequence are darker; the war makes an appearance, and there is a remarkable description of grief in *Albertine disparue*. Sex also continues to make its presence felt. But possibly the most poignant moment comes in the very last book, when the narrator steps on two uneven flagstones and an involuntary memory floods in on him, just as it did at the very start of the series. Proust does not bring us full circle, however. This time the narrator refuses to follow that path, preferring to keep his mind focused on the present. We are invited to think that this is a decisive change in Proust himself, a rejection of all that has gone before. He has kept the biggest surprise till the end, like the masterful storyteller that he is. But still, one cannot call it much of a climax, after so many volumes.[53]

At the time of his death, Proust's reputation was high. Now, however, some critics argue that his achievement no longer merits the enormous effort. For others, *A la recherche du temps perdu* is still one of the outstanding achievements of modern literature, 'the greatest exploration of a self by anyone, including Freud.'[54]

The first volume of Proust's novel, it will be recalled, had been turned down by among others **André Gide** at the *Nouvelle Revue Française* (*NRF*). The tables were soon turned, however. Gide apologised for his error, and in 1916 Proust migrated to NRF. At Proust's death, Gide's great novel *The Counterfeiters* was barely begun. He did in fact record a dream about Proust in his journal for 15 March 1923 (Proust had died the previous November). Gide was sitting in Proust's study and 'found himself' holding a string which was attached to two books on Proust's shelves. Gide pulled the string, and unwound a beautiful binding of Saint-Simon's *Memoirs*. Gide was inconsolable in the dream but did acknowledge later that his action may have been intentional.[55]

The Counterfeiters, which had been on the author's mind since 1914, is not really like *A la recherche du temps perdu*, but some similarities have been noted and are pertinent.[56] 'Gide's novel has its own Baron de Charlus, its band of adolescents, its preoccupation with the cities of the plain. In both works the chief character is writing a novel that turns out to be, more or less, the very novel we are reading. But the most important resemblance is, that each was written with the conscious intention of writing a great novel. Gide was attempting to rival Proust on his own ground. In the dream the element of jealousy in Gide's attitude to Proust is 'brought to a head, confessed, and reconciled.'[57] The novel, with its highly complex plot, is important for a number of reasons, one of which is that Gide also kept a journal in which he recorded his thoughts about composition. This journal is probably the most complete account of a major literary work in formation. The main lesson to be learned is how Gide progressively changed and winnowed away at his early ideas and cut out characters. His aim was to produce a book where there is no main character but a variety of different characters, all equally important, a little bit like the paintings of Picasso, where objects are 'seen' not from one predominant direction but from all directions at once. In his journal he also included some newspaper cuttings, one about a band of young men passing counterfeit coins, another about a school pupil who blew his brains out in class under pressure from his friends. Gide weaves these elements into a complex plot, which includes one character, Edouard, who is writing a novel called *The Counterfeiters*, and in which, in essence, everyone is a counterfeiter of sorts.[58] Edouard, as a writer, and the boys with the false money are the most obvious counterfeiters, but what most shocked readers was Gide's indictment of French middle-class life, riddled with illegitimacy and homosexuality while all the time counterfeiting an attitude of respectable propriety (and not so dissimilar in subject matter from the later volumes of Proust). The complexity of the plot has its point in that, as in real life, characters are at times unaware of the consequences of their own actions, unaware of the reasons for other people's actions, unaware even of when they are being truthful or counterfeiting. In such a milieu how can anything – especially art – be expected to work? (Here there is an overlap with Luigi Pirandello.) While it is obvious why some counterfeiting (such as passing false money) works, some episodes of life, such as a boy blowing his brains out, will always remain at some level a mystery, inexplicable. In such a world, what rules is one to live by? *The Counterfeiters* is perhaps the most realistic diagnosis of our times. The novel offers no prescription; it infers that none is really available. If our predicament is ultimately tragic, why don't more people commit suicide? That too is a mystery.

Gide was unusually interested in English literature: William Blake, Robert Browning, Charles Dickens. But he also knew the Bloomsbury set – Gide had studied English at Cambridge, the Bloomsbury outpost, in 1918. He met Clive Bell in Paris in 1919, stayed with Lady Ottoline Morrell in Garsington in 1920, carried on a lengthy correspondence with Roger Fry (both shared a love of Nicolas Poussin), and later served on an antifascist committee of intellectuals with **Virginia Woolf**.

As she was preparing her novel *Jacob's Room*, Virginia Woolf was only too well aware that what she was trying to do was also being attempted by other authors. In her diary for 26 September 1920, she wrote, 'I reflected how what I'm doing is probably being better done by Mr Joyce.'[59] T. S. Eliot, she knew, was in touch with James Joyce, for he kept her informed of what the Irishman was doing.

Virginia Woolf was born in 1882 into an extremely literary family (her father was founding editor of the *Dictionary of National Biography*, and his first wife was a daughter of William Makepeace Thackeray). Although she was denied the education given to her brothers, she still had the run of the family's considerable library and grew up much better read than most of her female contemporaries. She always wanted to be a writer and began with articles for the *Times Literary Supplement* (which had begun as a separate publication from its parent, the London *Times*, in 1902). But she didn't publish her first novel, *The Voyage Out*, until 1915, when she was thirty-three.[60]

It was with *Jacob's Room* that the sequence of experimental novels for which Woolf is most remembered was begun. The book tells the story of a young man, Jacob, and its central theme, as it follows his development through Cambridge, artistic and literary London, and a journey to Greece, is the description of a generation and class that led Britain into war.[61] It is a big idea; however, once again it is the form of the book which sets it apart. In her diary for early 1920 she had written, 'I figure that the approach will be entirely different this time; no scaffolding; scarcely a brick to be seen; all crepuscular, but the heart, the passion, humour, everything as bright as fire in the mist.'[62] *Jacob's Room* is an urban novel, dealing with the anonymity and fleeting experiences of city streets, the 'vast atomised masses scurrying across London's bridges', staring faces glimpsed through the windows of tea shops, either bored or bearing the marks of 'the desperate passions of small lives, never to be known.'[63] Like *Ulysses* and like Proust's work, the book consists of a stream of consciousness – erratic at times – viewed through interior monologues, moving backward and forward in time, sliding from one character to another without warning, changing viewpoint and attitude as fast and as fleetingly as any encounter in any major urban centre you care to name.[64] Nothing is settled in *Jacob's Room*. There isn't much plot in the conventional sense (Jacob's early promise is never fulfilled, characters remain unformed, people come and go; the author is as interested in marginal figures, like a flower seller on the street, as in those who are, in theory, more central to the action), and there is no conventional narrative. Characters are simply cut off, as in an impressionist painting. 'It is no use trying to sum people up,' says one of the figures, who could have stepped out of Gide, 'One must follow hints, not exactly what is said, nor yet entirely what is done.'[65] Woolf is describing, and making us feel, what life is like in vast cosmopolitan cities of the modern world. This fragmentation, this dissolution of the familiar categories – psychological as well as physical – is just as much the result of World War I, she is saying, as the military/political/economic changes that have been wrought, and is arguably more fundamental.

The effect of Sigmund Freud's psychological ideas on **André Breton** (1896–1966) was very direct. During World War I he stood duty as an hospital orderly at the Saint-Dizier psychiatric centre, treating victims of shell shock. And it was in Saint-Dizier that Breton first encountered the (psycho)analysis of dreams, in which – as he later put it – he did the 'groundwork' for **surrealism**. In particular, he remembered one patient who lived entirely in his own world. This man had been in the trenches but had become convinced he was invulnerable. He thought the whole world was 'a sham,' played by actors who used dummy bullets and stage props. So convinced was he of this vision that he would show himself during the fighting and gesture excitedly at the explosions. The miraculous inability of the enemy to kill him only reinforced his belief.[66]

It was the 'parallel world' created by this man that had such an effect on Breton. For him the patient's madness was in fact a rational response to a world that had gone mad, a view that was enormously influential for several decades in the middle of the century. Dreams, another parallel world, a route to the unconscious as Freud said, became for Breton the route to art. For him, art and the unconscious could form 'a new alliance,' realised through dreams, chance, coincidence, jokes – all the things Freud was investigating. This new reality Breton called *sur*-reality, a word he borrowed from Guillaume Apollinaire. In 1917 Picasso, Jean Cocteau, Erik Satie, and Léonide Massine had collaborated on a ballet, *Parade*, which the French poet had described as 'une espèce de *sur-réalisme*.'[67]

Surrealism owed more to what its practitioners *thought* Freud meant than to what he actually wrote. Few French and Spanish surrealists could read Freud's works, as they were still only available in German. (Psychoanalysis was not really popular in France until after World War II; in Britain the British Psychoanalytic Association was not formed until 1919.) Breton's ideas about dreams, about neurosis as a sort of 'ossified' form of permanent dreaming, would almost certainly have failed to find favour with Freud, or the surrealists' view that neurosis was 'interesting,' a sort of mystical, metaphysical state. It was in its way a twentieth-century form of romanticism, which subscribed to the argument that neurosis was a 'dark side' of the mind, the seat of dangerous new truths about ourselves.[68]

Though surrealism started as a movement of poets, led by Breton, Paul Eluard (1895–1952), and Louis Aragon (1897–1982), it was the painters who were to achieve lasting international fame. Four painters became particularly well known, and for three of them the wasteland was a common image.

Max Ernst was the first artist to join the surrealists (in 1921). He claimed to have hallucinated often as a child, so was predisposed to this approach.[69] His landscapes or objects are oddly familiar but subtly changed. Trees and cliffs, for example, may actually have the texture of the insides of the body's organs; or the backside of a beast is so vast, so out of scale, that it blocks the sun. Something dreadful has either just happened or seems about to. Ernst also painted apparently cheerful scenes but gave these works long and mysterious titles that suggest something sinister: *The Inquisitor: At 7:07 Justice Shall Be Made.*[70] For example, on the surface *Two Children Threatened by a Nightingale* is cheerfully colourful. The picture consists of a bird, a clock that resembles a cuckoo clock, a garden enclosed by a wall. But then we notice that the figures in the picture are running away after an episode not shown. And the picture is actually painted on a small door, or the lid of a box, with a handle attached. If the door is opened what will be revealed? The unknown is naturally menacing.

The most unsettling of the surrealists was **Giorgio de Chirico** (1888–1978), the 'painter of railway stations,' as Picasso dubbed him. An Italian of Greek descent, de Chirico was obsessed by the piazzas and arcades of north Italian towns: 'I had just come out of a long and painful intestinal illness. I was in a nearly morbid state of sensitivity. The whole world, down to the marble of the buildings and the fountains, seemed to me to be convalescent. . . . The autumn sun, warm and unloving, lit the statue and the church façade. Then I had the strange impression that I was looking at these things for the first time.'[71] These landscapes, these townscapes, are always depicted in the same way by de Chirico. The light is always the same (it is afternoon light, coming from the right or left, rather than from above); there are long, forbidding shadows; darkness is not far away.[72] Second, there are next to no people – these townscapes are deserted. Sometimes there is a tailor's mannequin, or a sculpture, figures that resemble people but are blind, deaf, dumb, insensate, echoing, as Robert Hughes has said, the famous lines of Eliot: 'These fragments have I shored against my ruins.' There are often humanlike shadows just around the corner. De Chirico's is a cold world; the mood is forbidding, with a feeling that this is perhaps the last day of all, that the universe is imploding, and the sun about to cease shining forever. Again something dreadful has either happened or is about to happen.[73]

At first sight, **Joan Miró** (1893–1983) was a much more cheerful, playful painter than the other two. He never joined the political wing of the surrealists: he didn't get involved in manifestos or campaigns.[74] But he did contribute to group shows, where his style contrasted strongly with the others. A Catalan by birth, he trained in Barcelona at a time when that city was a cosmopolitan capital, before it was cut off from the rest of Europe by the Spanish Civil War. He showed an early interest in cubism but turned against it; after a childhood spent on a farm, his interest in wildlife kept bubbling through.[75] This gave his paintings their biological lyricism, increasingly abstract as time went by. In *The Farm* 1921–2, he painted scores of animals in scientific detail, to produce a work that pleases both children and adults. (He carried dried grasses all the way from Barcelona to Paris to be sure he got the details right.) In his later Constellation series, the myriad forms echo earlier artists such as Hieronymus Bosch but are joyful, more and more abstract, set in a nebulous sky where the stars have biological rather than physico-chemical forms. Miró met the surrealists through the painter André Masson, who lived next door to him in Paris. He took part in the first surrealist group show in 1924. But he was less a painter of dread than of the survival of the childlike in adult life, the 'uncensored self,' another confused concept drawn from psychoanalysis.[76]

The wastelands of **Salvador Dalí** are famous. And they *are* wastelands: even where life appears, it corrupts and decays as soon as it blooms. After Picasso, Dalí is the most famous artist of the twentieth century, though this is not the same as saying he is the second best. It has more to do with his extraordinary technique, his profound fear of madness, and his personal appearance – his staring eyes and handlebar moustache, adapted from a Diego Velázquez portrait of Philip IV of Spain.[77] Discovering his facility with paint, Dalí found he was able to render crystal-clear landscapes that, given the themes he pursued, played with reality, again in the way dreams are supposed to do. He had the lyricism of Miró, the afternoon light of de Chirico, and Ernst's sense of dread derived from subtly changing familiar things. His images – cracked eggs ('Dalinian DNA'), soft watches, elongated breasts, dead trees in arid landscapes – are visually lubricious and disturbing to the mind.[78] They convey a world pullulating with life, but uncoordinated, as if the guiding principles, the very laws, of nature have broken down, as if biology is coming to an end and the Darwinian struggle has gone mad.

René Magritte (1898–1967) was never part of the salon of surrealists – he spent all his life in Brussels – but he shared their obsession with dread, adding too an almost Wittgensteinian fascination with language and the hold it has on meaning. In his classic paintings, Magritte took ordinary subjects – a bowler hat, a pipe, an apple, an umbrella – and made extraordinary things happen to them (he himself often wore a bowler).[79] For example, in The Human Condition (1934), a painting of a view through a window overlaps exactly with the same view, so that they fuse together and one cannot tell where the painting begins and ends. The world 'out there,' he is saying, is really a construction of the mind, an echo of Henri Bergson. In The Rape, also 1934, a naked female torso, framed in hair, forms a face, a prim yet at the same time wild face, casting doubt on the nature of primness itself, suggesting a raw sexuality that lies hidden. This image is seen against a flat, empty landscape, a purely psychoanalytic wasteland.[80]

The surrealists played with images – and the verb is pertinent; they were seriously suggesting that man could play himself out of trouble, for in play the unconscious was released. By the same token they brought eroticism to the surface, because repression of sexuality cut off man from his true nature. But above all, taking their lead from dreams and the unconscious, their work showed a deliberate rejection of reason. Their art sought to show that progress, if it were possible, was never a straight line, that nothing was predictable, and that the alternative to the banalities of the acquisitive society, now that religion was failing, was a new form of enchantment.

Ironically, the wasteland was a very fertile metaphor. What underlines all the works considered here is a sense of disenchantment with the world and with the joint forces of capitalism and science, which created the wasteland. These targets were well chosen. Capitalism and science were to prove the century's most enduring modes of thought and behaviour. And by no means everyone would find them disenchanting.

In the 1920s the eugenicists and scientific racists were especially persistent in America. One of their main texts was a book by C.C. Brigham called *A Study of American Intelligence*, which was published in 1923. Brigham, an assistant professor of psychology at Princeton University, was a disciple of Robert Yerkes, and in his book he relied on the material Yerkes had obtained during the war (Yerkes wrote the foreword for Brigham's book). Despite evidence that the longer immigrants were in the United States, the better they performed on IQ tests, Brigham's aim was to show that the southern and eastern peoples of Europe, and Negroes, were of inferior intelligence. In making his arguments he relied on the much earlier notions of such figures as Count Georges Vacher de Lapouge, who thought that Europe was divided into three racial types, according to the shape of their skulls. Given this, Brigham's conclusions were not surprising: 'The decline in intelligence [in America] is due to two factors, the change in the races migrating to this country, and to the additional factor of the sending of lower and lower representatives of each race. ... Running parallel with the movements of these European peoples, we have the most sinister development in the history of this continent, the importation of the negro. ... The decline of American intelligence will be more rapid than the decline of the intelligence of European national groups, owing to the presence here of the negro.'[1]

In such a context, the idea for a return to segregation was never far below the surface. Cornelia Cannon, noting that 89 percent of blacks had tested as 'morons,' wrote in the American periodical *Atlantic Monthly*, 'Emphasis must necessarily be laid on the development of the primary schools, on the training in activities, habits, occupations which do not demand the more evolved faculties. In the South particularly ... the education of the whites and colored in separate schools may have justification other than that created by race prejudice.'[2] Henry Fairfield Osborn, a trustee of Columbia University and president of the American Museum of Natural History, believed 'those tests were worth what the war cost, even in human life, if they served to show clearly to our people the lack of intelligence in our country, and the degrees of intelligence in different races who are coming to us, in a way which no one can say is the result of prejudice. ... We have learned once and for all that the negro is not like us.'[3]

The battles over biology did not stop with the victory the eugenicists achieved in getting the 1924 Immigration Restriction Act passed. The following year biology was back in the public eye in the notorious Scopes trial. As early as 1910 the Presbyterian General Assembly had drawn up a list of the 'Five Fundamentals' which they believed to be the basis of Christianity. These were: the miracles of Christ; the Virgin birth; the Resurrection; the Crucifixion, understood as atonement for mankind's sins; and the Bible as the directly inspired word of God. It was the latter that was the focus of **the Scopes trial**. The facts of the case were not in dispute.[4] John Scopes, of Dayton, Tennessee, had taught a biology class using as a textbook *Civic Biology* by George William Hunter, which had been adopted as a standard text by the State Textbook Commission in 1919. (It had actually been used in some schools since 1909, so it was in circulation for fifteen years before it was considered dangerous.)[5] The part of Hunter's book which Scopes had used reported evolution as a fact. This, the prosecution argued, was contrary to Tennessee law. Evolution was a theory that contradicted the Bible, and it should not be asserted as bald fact. The trial turned into a circus. The prosecution was led by **William Jennings Bryan**, three times a presidential nominee, a former secretary of state, and a man who told Seventh Day Adventists before the trial that it would determine whether evolution or Christianity survived. He also said, 'All the ills from which America suffers can be traced back to the teachings of evolution. It would be better to destroy every book ever written, and save just the first three verses of Genesis.'[6] The defence was led by a no less colourful person, **Clarence Darrow**, a skilled orator and a fabled criminal lawyer. While Bryan was determined to make the trial a contest of Darwin versus the Bible, Darrow's technique was to tie his adversary in knots, aided by eminent scientists and theologians who had arrived in Dayton determined to see that Bryan did not have his fundamentalist way. At one point, when Bryan insisted on testifying as an expert in biblical science, he proved unwilling or unable to answer questions about the age of the earth or of well-known archaeological sites. He defended himself by saying, 'I do not think about things I do not think about.' Darrow replied drily, 'Do you think about things you do think about?' In fact, Bryan won the case, but on a technicality. The judge kept the focus of the trial not on whether Darwin was right or wrong but on whether or not Scopes had taught evolution. And since Scopes admitted what he had done, the result was a foregone conclusion. He was given a fine of $100, which was then successfully appealed because the judge rather than the jury had set the fine. But that technicality apart, Bryan lost heavily. He was humiliated and mocked in the press, not just in America but around the world. He died five days after the trial ended.[7]

Religion, however, explained only part of the reaction to the Scopes trial. In his *Anti-Intellectualism in American Life*, Richard Hofstadter argues that particularly in the American South and Midwest, people used the Christianity/evolution struggle as a cipher for revolting against modernity. The rigid defence of Prohibition, then in force, was another side to this. Hofstadter quotes with some sympathy Hiram W. Evans, the imperial wizard of the Ku

Klux Klan, who, he says, summed up the major issue of the time 'as a struggle between "the great mass of Americans of the old pioneer stock" and the "intellectually mongrelised Liberals."' 'We are a movement,' Evans wrote, 'of the plain people, very weak in the matter of culture, intellectual support, and trained leadership. We are demanding, and we expect to win, a return of power into the hands of the everyday, not highly cultured, not overly intellectualised, but entirely unspoiled and not de-Americanised, average citizen of the old stock. . . . This is undoubtedly a weakness. It lays us open to the charge of being "hicks" and "rubes" and "drivers of second-hand Fords." We admit it.'[8] The words of the Klan wizard highlight the atmosphere in America at the time, so different from that in Europe, where in London and Paris modernism was flourishing.

America had ended the war transformed: she alone was stronger, unravaged. The prevailing American mood was still pragmatic, practical, independent of the great isms of the Old World. 'This is essentially a business country,' said Warren Harding in 1920, and he was echoed by Calvin Coolidge's even more famous words, uttered in 1922: 'The business of America is business.' All these different strands – anti-intellectualism, business, the suspicion of Europe, or at least her peoples – were brilliantly brought together in the novels of **Sinclair Lewis**, the best of which, *Babbitt*, appeared in that remarkable year, 1922.

It would be hard to imagine a character more different from Dedalus, or Tiresias, or Jacob or Swann, than **George F. Babbitt**. A realtor from **Zenith, Ohio**, a medium-size town in the American Midwest, Babbitt is hardworking, prosperous, and well liked by his fellow citizens. But Babbitt's success and popularity are just the beginning of his problems. Lewis was a fierce critic of the materialistic, acquisitive society that Oswald Spengler, R. H. Tawney, and T. S. Eliot so loathed. Eliot and Joyce had stressed the force of ancient myth as a way to approach the modern world, but as the twenties passed, Lewis dissected a number of modern American myths. Babbitt, like the 'heroes' of Lewis's other books, is, although he doesn't know it, a victim.

Born in 1885, Harry Sinclair Lewis was raised in the small Minnesota town of Sauk Center, which, he was to say later, was 'narrow-minded and socially provincial.' One of Lewis's central points in his books was that small-town America was nowhere near as friendly or as agreeable as popular mythology professed. For Lewis, small-town Americans were suspicious of anyone who did not share their views, or was different.[9] Lewis's own growing up was aided and eased by his stepmother, who came from Chicago – although not the most sophisticated place at the time, at least not a small town. His stepmother encouraged the young Harry to read 'foreign' books and to travel. He attended Oberlin Academy and then headed east to Yale. There he learned poetry and foreign languages and met people who had travelled even more than his stepmother. After Yale, he went to New York, where at the age of twenty-five he found work as a reader of manuscripts and as a press agent for a publisher. This introduced him to the reading tastes of the American public. He had a series of short stories published in the *Saturday Evening Post*. Each was slightly subversive of the American self-image, but the stories' length did not do full justice to what he wanted to say. It was only when he published his first novel, *Main Street*, which appeared in October 1920, that 'a new voice was loosed on the American ear'.[10] Published in late autumn, in time for the Christmas rush, *Main Street* was that rare phenomenon, a best-seller created by word of mouth. It was set in Gopher Prairie, a small town that, naturally enough, had a lot in common with Lewis's own Sauk Center. The inhabitants of Gopher, their prejudices and peccadilloes, were brilliantly observed, their foibles and their fables about themselves cleverly caught, so that the book proved as popular in middle America as it was among more sophisticated types who would not have been seen dead in 'the sticks.' The book was so popular that at times the publisher could not find enough paper to issue reprints. It even managed to cause a scandal back east when it was revealed that the Pulitzer Prize jury had voted for *Main Street* as winner but, unusually, the Columbia University trustees who administered the prize had overturned their decision and given the prize instead to Edith Wharton, for *The Age of Innocence*. Lewis didn't mind; or not much. He was a fan of Wharton and dedicated his next book, *Arrowsmith*, to her.[11]

In *Babbitt*, Lewis moved on, from small-town America to the medium-size midwestern city. This was in many ways a more typical target; Zenith, the city where the story is set, exhibited not only America's advantages but also its problems. By 1922 there had already been a number of novels about businessmen in America – for example, **Dean Howells's** *Rise of Silas Lapham* (1885) and **Theodore Dreiser's** *Financier* (1912). But none of them had the tragic structure of *Babbitt*. Lewis, with his passion for 'foreign' literature, took a leaf out of Emile Zola's book. The Frenchman had ridden the railways on the footplate and descended into the mines to research his great series of Rougon-Macquart novels in the last quarter of the nineteenth century. Likewise, Lewis travelled by train to visit several midwestern towns, lunching in the Rotary associations with realtors, mayors, chairmen of the chambers of commerce. Like Zola, he took copious notes, recording in his grey notebooks typical phrases and figures of speech, collecting suitable names for people and places. All this produced Babbitt, a man who lies 'at the very heart' of American materialist culture.[12] The central quality that Lewis gives Babbitt is his success, which for him entails three things: material comfort; popularity with his fellow citizens, who think like he does; and a sense of superiority over the less successful. Complacent without recognising his complacency, Babbitt lives by a code of Efficiency, Merchandising, and 'Goods' – things, material possessions. For Lewis, paralleling Eliot, these are false gods; in Babbitt's world, art and religion have been perverted, in the service, always, of business. The point at which Lewis makes this most clear is when one of the characters, called Chum Frink, delivers a speech to the '**Booster's Club**,' a sort of Rotary association. The theme of Chum's speech concerns why Zenith should have its own symphony orchestra: 'Culture has become as necessary an adornment and advertisement for a city to-day as pavements or bank-clearances. It's Culture, in theaters and art galleries and so on, that brings thousands of visitors. ... [So] I call on you brothers to whoop it up for Culture and A World-beating Symphony Orchestra!"[13]

The self-satisfaction is all but unbearable, and Lewis doesn't let it last. A shadow begins to form in this perfect world when Babbitt's closest friend kills his wife. There is no mystery about the death; and it is manslaughter, not murder. Even so, the friend is sent to prison. This set of events is thoroughly dislocating for Babbitt and provokes in him a number of changes. To the reader these are small changes, insignificant rebellions, but each time Babbitt tries to rebel, to lead what he thinks of as a more 'bohemian' life, he realises that he cannot do it: the life he has made is dominated by, depends on, conformity. There is a price to pay for success in America, and Lewis presents it as a kind of Faustian bargain where, for Babbitt and his kind, heaven and hell are the same place.

Lewis's indictment of materialism and the acquisitive society is no less effective than Tawney's, but his creation, certainly more memorable, is much less savage.[14] He made Babbitt's son Ted somewhat more reflective than his father, a hint, perhaps, that middle America might evolve. This slight optimism on Lewis's part may have been a clever move to aid the book's success. Upon its publication, on 14 September 1922, the word *Babbitt*, or *Babbittry*, immediately entered the vocabulary in America as shorthand for *conformism*. Even more strongly, *boosterism* came into widespread use to describe an all-too-familiar form of American self-promotion. Upton Sinclair thought the book 'a genuine American masterpiece,' while Virginia Woolf judged it 'the equal of any novel written in English in the present century.'[15] What sets Babbitt apart from the European literary figures being created at the same time is that he doesn't realise he is a tragic figure; he lacks the insight of classic figures in tragedy. For Lewis, this complacency, this incapacity for being saved, was middle America's besetting sin.[16]

As well as being a classic middle American, Babbitt was also a typical 'middle-brow,' a 1920s term coined to describe the culture espoused by the British Broadcasting Corporation (BBC). However, it applied a fortiori in America, where a whole raft of new media helped to create a new culture in the 1920s in which Babbitt and his booster friends could feel at home.

At this end of the century the electronic media – television in particular, but also radio – are generally regarded as more powerful than print media, with a much bigger audience. In the 1920s it was different. The principles of radio had been known since 1873, when James Clerk Maxwell, a Scot, and Heinrich Hertz, from Germany, carried out the first experiments. Guglielmo Marconi founded the first wireless telegraph company in 1900, and Reginald Fessenden delivered the first 'broadcast' (a new word) in 1906 from Pittsburgh. Radio didn't make real news, however, until 1912, when its use brought ships to the aid of the sinking *Titanic*. All belligerents in World War I had made widespread use of radio, as propaganda, and afterwards the medium seemed ready to take America by storm – radio seemed the natural vehicle to draw the vast country together. David Sarnoff, head of RCA, envisaged a future in which America might have a broadcasting system where profit was not the only criterion of excellence, in effect a public service system that would educate as well as entertain. Unfortunately, the business of America was business. The early 1920s saw a 'radio boom' in the United States, so much so that by 1924 there were no fewer than 1,105 stations. Many were tiny, and over half failed, with the result that radio in America was never very ambitious for itself; it was dominated from the start by advertising and the interests of advertisers. Indeed, at one time there were not enough wavelengths to go round, producing 'chaos in the ether.'[17]

As a consequence of this, new print media set the agenda for two generations, until the arrival of television. An added reason, in America at least, was a rapid expansion in education following World War I. By 1922, for example, the number of students enrolled on American campuses was almost double what it had been in 1918.[18] Sooner or later that change was bound to be reflected in a demand for new forms of media. Radio apart, four new entities appeared to meet that demand. These were *Reader's Digest, Time*, the Book-of-the-Month Club, and the *New Yorker*.

If war hadn't occurred, and infantry sergeant **DeWitt Wallace** had not been hit by shrapnel during the Meuse-Argonne offensive, he might never have had the 'leisure' to put into effect the idea he had been brooding upon for a new kind of magazine.[19] Wallace had gradually become convinced that most people were too busy to read everything that came their way. Too much was being published, and even important articles were often too wordy and could easily be reduced. So while he was convalescing in hospital in France, he started to clip articles from the many magazines that were sent through from the home front. After he was discharged and returned home to Saint Paul, Minnesota, he spent a few more months developing his idea, winnowing his cuttings down to thirty-one articles he thought had some long-term merit, and which he edited drastically. He had the articles set in a common typeface and laid out as a magazine, which he called *Reader's Digest*. He ordered a printing of 200 copies and sent them to a dozen or so New York publishers. Everyone said no.[20]

Wallace's battles to get *Reader's Digest* on a sound footing after its launch in 1922 make a fine American adventure story, with a happy ending, as do **Briton Hadden's** and **Henry Luce's** efforts with *Time*, which, though launched in March 1912, did not produce a profit until 1928. The **Book-of-the-Month-Club**, founded by the Canadian **Harry Scherman** in April 1926, had much the same uneven start, with the first books, Sylvia Townsend Warner's *Lolly Willowes*, T. S. Stribling's *Teeftallow*, and *The Heart of Emerson's Journals*, edited by Bliss Perry, being returned 'by the cartload.'[21] But Wallace's instincts had been right: the explosion of education in America after World War I changed the intellectual appetite of Americans, although not always in a direction universally approved. Those arguments were especially fierce in regard to the Book-of-the-Month Club, in particular the fact that a committee was deciding what people should read, which, it was said, threatened to 'standardise' the way Americans thought.[22] 'Standardisation' was worrying to many people in those days in many walks of life, mainly as a result of the 'Fordisation' of industry following the invention of the moving assembly line in 1913. Sinclair Lewis

had raised the issue in *Babbitt* and would do so again in 1926, when he turned
down the Pulitzer Prize for his novel *Arrowsmith*, believing it was absurd to
identify any book as 'the best.' What most people objected to was the mix of
books offered by the Book-of-the-Month Club; they claimed that this produced
a new way of thinking, chopping and changing between serious 'high culture'
and works that were 'mere entertainment.' This debate produced a new concept
and a new word, used in the mid-1920s for the first time: *middlebrow.* The
establishment of a professoriate in the early decades of the century also played
a role here, as did the expansion of the universities, before and after World War
I, which helped highlight the distinction between 'highbrow' and 'lowbrow.'
In the mid- and late 1920s, American magazines in particular kept returning to
discussions about middlebrow taste and the damage it was or wasn't doing to
young minds.

Sinclair Lewis might decry the very idea of trying to identify 'the best,' but he
was unable to stop the influence of his books on others. And he earned perhaps
a more enduring accolade than the Pulitzer Prize from academics – sociologists –
who, in the mid-1920s, found the phenomenon of Babbitt so fascinating that
they decided to study for themselves a middle-size town in middle America.

   Robert and Helen Lynd decided to study an ordinary American town, to
describe in full sociological and anthropological detail what life consisted of.
As Clark Wissler of the American Museum of Natural History put it in his
foreword to their book, *Middletown*, 'To most people, anthropology is a mass
of curious information about savages, and this is so far true, in that anthropology
deals with the less civilised.' Was that irony – or just cheek?[23] The fieldwork
for the study, financed by the Institute of Social and Religious Research, was
completed in 1925, some members of the team living in 'Middletown' for
eighteen months, others for five. The aim was to select a 'typical' town in the
Midwest, but with certain specific aspects so that the process of social change
could be looked at. A town of about 30,000 was chosen (there being 143 towns
between 25,000 and 50,000, according to the U.S. Census). The town chosen
was homogeneous, with only a small black population – the Lynds thought it
would be easier to study cultural change if it was not complicated by racial
change. They also specified that the town have a contemporary industrial
culture and a substantial artistic life, but they did not want a college town with
a transient student population. Finally, Middletown should have a temperate
climate. (The authors attached particular importance to this, quoting in a
footnote on the very first page of the book a remark of J. Russell Smith in his
North America: 'No man on whom the snow does not fall ever amounts to a
tinker's damn')[24] It later became known that the city they chose was Muncie,
Indiana, sixty miles northeast of Indianapolis.

   No one would call *Middletown* a work of great literature, but as sociology it
had the merit of being admirably clearheaded and sensible. The Lynds found
that life in this typical town fell into six simple categories: getting a living;
making a home; training the young; using leisure in various forms of play, art,
and so forth; engaging in religious practices; and engaging in community

activities. But it was the Lynds' analysis of their results, and the changes they observed, that made *Middletown* so fascinating. For example, where many observers – certainly in Europe – had traditionally divided society into three classes, upper, middle, and working, the Lynds detected only two in Middletown: the business class and the working class. They found that men and women were conservative – distrustful of change – in different ways. For instance, there was far more change, and more acceptance of change, in the workplace than in the home. Middletown, the Lynds concluded, employed 'in the main the psychology of the last century in training its children in the home and the psychology of the current century in persuading its citizens to buy articles from its stores.'[25] There were 400 types of job in Middletown, and class differences were apparent everywhere, even at six-thirty on the average morning.[26] 'As one prowls Middletown streets about six o'clock of a winter morning one notes two kinds of homes: the dark ones where people still sleep, and the ones with a light in the kitchen where the adults of the household may be seen moving about, starting the business of the day.' The working class, they found, began work between six-fifteen and seven-thirty, 'chiefly seven.' For the business class the range was seven-forty-five to nine, 'but chiefly eight-thirty.' Paradoxes abounded, as modernisation affected different aspects of life at different rates. For example, modern (mainly psychological) ideas 'may be observed in [Middletown's] courts of law to be commencing to regard individuals as not entirely responsible for their acts,' but not in the business world, where 'a man may get his living by operating a twentieth-century machine and at the same time hunt for a job under a *laisser-faire* individualism which dates back more than a century.' 'A mother may accept community responsibility for the education of her children but not for the care of their health.'[27]

In general, they found that Middletown learned new ways of behaving toward material things more rapidly than new habits addressed to persons and nonmaterial institutions. 'Bathrooms and electricity have pervaded the homes of the city more rapidly than innovations in the personal adjustments between husband and wife or between parents and children. The automobile has changed leisure-time life more drastically than have the literature courses taught the young, and tool-using vocational courses have appeared more rapidly in the school curriculum than changes in the arts courses. The development of the linotype and radio are changing the technique of winning political elections [more] than developments in the art of speechmaking or in Middletown's method of voting. The Y.M.C.A., built about a gymnasium, exhibits more change in Middletown's religious institutions than do the weekly sermons of its ministers.'[28] A classic area of personal life that had hardly changed at all, certainly since the 1890s, which the Lynds used as the basis for their comparison, was the 'demand for romantic love as the only valid basis for marriage. ... Middletown adults appear to regard romance in marriage as something which, like their religion, must be believed in to hold society together. Children are assured by their elders that "love" is an unanalysable mystery that "just happens." ... And yet, although theoretically this "thrill" is all-sufficient to insure per-

manent happiness, actually talks with mothers revealed constantly that, particularly among the business group, they were concerned with certain other
factors.' Chief among these was the ability to earn a living. And in fact the
Lynds found that Middletown was far more concerned with money in the 1920s
than it had been in 1890. In 1890 vicinage (the old word for neighbourhood) had
mattered most to people; by the 1920s financial and social status were much
more closely allied, aided by the automobile.[29]

Cars, movies, and the radio had completely changed leisure time. The passion
with which the car was received was extraordinary. Families in Middletown
told the Lynds that they would forgo clothes to buy a car. Many preferred to
own a car rather than a bathtub (and the Lynds did find homes where bathtubs
were absent but cars were not). Many said the car held the family together. On
the other hand, the 'Sunday drive' was hurting church attendance. But perhaps
the most succinct way of summing up life in Middletown, and the changes it
had undergone, came in the table the Lynds presented at the end of their book.
This was an analysis of the percentage news space that the local newspapers
devoted to various issues in 1890 and 1923:[30]

	1890	1923	% change
Cartoons	0.2	14.6	+7300
Women's news	0.5	3.4	+680
Sports	3.8	13.2	+347
Business	3.4	6.6	+94
Public affairs	9.1	15.7	+72
Science	2.0	1.0	−50
Accidents	5.4	1.9	−65
Agriculture	4.3	1.1	−74
Politics	17.3	1.2	−93

Certain issues we regard as modern were already developing. Sex education
was one; the increased role (and purchasing power) of youth was another (these
two matters not being entirely unrelated, of course). The Lynds also spent quite
a bit of time considering differences between the two classes in IQ. Middletown
had twelve schools; five drew their pupils from both working-class and business-
class parents, but the other seven were sufficiently segregated by class to allow
the Lynds to make a comparison. Tests on 387 first-grade (i.e., six-year-old)
children revealed the following picture:[31]

	% from business-class parents	% from working-class parents
Above average IQ 110–139	25.8	6.5
Average IQ 90–109	60.8	51.0
Below average IQ 70–89	13.4	36.2
Moron or imbecile IQ 25–69	00.0	6.3

The Lynds showed some awareness of the controversies surrounding intelligence testing (for example, by using the phrase 'intelligent test' in quotes) but nonetheless concluded that there were 'differences in the equipment with which, at any given time, children must grapple with their world.'

The Lynds had produced sociology, anthropology – and a new form of history. Their pictured lacked the passion and the wit of *Babbitt*, but Middletown was recognisably the same beast as Zenith. The book's defining discovery was that there were two classes, not three, in a typical American town. It was this which fuelled the social mobility that was to set America apart from Europe in the most fruitful way.

Babbitt's Middletown may have been typical America, intellectually, sociologically and statistically. But it wasn't the only America. Not everyone was in the 'digest' business, and not everyone was in a hurry or too busy to read, or needed others to make up his mind for him. These 'other' Americas could be identified by place: in particular Paris, Greenwich Village, and Harlem, black Harlem. Americans flocked to Paris in the 1920s: the dollar was strong, and modernism far from dead. Ernest Hemingway was there for a short time, as was F. Scott Fitzgerald. It was an American, Sylvia Beach, who published *Ulysses*. Despite such literary stars, the American influx into the French capital (and the French Riviera) was more a matter of social than intellectual history. Harlem and Greenwich Village were different.

When the British writer Sir Osbert Sitwell arrived in New York in 1926, he found that 'America was strenuously observing Prohibition by staying sempiternally [everlastingly] and gloriously drunk.' Love of liberty, he noted, 'made it almost a duty to drink more than was wise,' and it was not unusual, after a party, 'to see young men stacked in the hall ready for delivery at home by taxicab.'[32] But he had an even bigger surprise when, after an evening spent at Mrs Cornelius Vanderbilt's 'Fifth Avenue Chateau,' he was taken uptown, to **A'Lelia Walker's** establishment on 136th Street, in Harlem. The soirées of A'Lelia, the beneficiary of a fortune that stemmed from a formula to 'de-kink' Negro hair, were famous by this time. Her apartment was lavishly decorated, one room tented in the 'Parisian style of the Second Empire,' others being filled, inter alia, with a golden grand piano and a gold-plated organ, yet another dedicated as her personal chapel.[33] Here visiting grandees, as often as not from Europe, could mix with some of the most intellectually prominent blacks: W. E. B. Du Bois, **Langston Hughes, Charles Johnson, Paul Robeson, Alain Locke.** A'Lelia's was the home of what came to be called '**the new Negro,**' and hers was by no means the only establishment of its kind.[34] In the wake of the Great War, when American blacks in segregated units had fought with distinction, there was a period of optimism in race relations (on the East Coast, if not in the South), partly caused by and partly reflected in what became known as the Harlem Renaissance, a period of about a decade and a half when black American writers, actors, and musicians made their collective mark on the country's intellectual landscape and stamped one place, Harlem, with a vitality, a period of chic, never seen before or since.

The **Harlem Renaissance** began with the fusion of two bohemias, when the talents of Greenwich Village began at last to appreciate the abilities of black actors. In 1920 Charles Gilpin, a black actor, starred in Eugene O'Neill's *Emperor Jones*, establishing a vogue.[35] Du Bois had always argued that the way ahead for the Negro in America lay with its '**talented tenth**,' its elite, and the Harlem Renaissance was the perfect expression of this argument in action: for a decade or so there was a flowering of black stage stars who all shared the belief that arts and letters had the power to transform society. But the renaissance also had its political edge. Race riots in the South and Midwest helped produce the feeling that Harlem was a place of refuge. Black socialists published magazines like the *Messenger* ('The only magazine of scientific radicalism in the world published by Negroes').[36] And there was **Marcus Garvey**, 'a little sawed-off, hammered down black man' from Jamaica, whose Pan-African movement urged the return of all blacks to Africa, Liberia in particular. He was very much part of Harlem life until his arrest for mail fraud in 1923.[37]

But it was literature, theatre, music, poetry, and painting that held most people's hearts. Clubs sprang up everywhere, attracting jazz musicians like Jelly Roll Morton, Fats Waller, Edward Kennedy 'Duke' Ellington, Scott Joplin, and later, Fletcher Henderson. Nick La Rocca's Original Dixieland Jazz Band made the first jazz recording in New York in 1917, 'Dark Town Strutter's Ball.'[38] The renaissance threw up a raft of blacks – novelists, poets, sociologists, performers – whose very numbers conveyed an optimism about race even when their writings belied that optimism, people like **Claude McKay, Countee Cullen**, Langston Hughes, **Jean Toomer**, and **Jessie Fauset**. McKay's *Harlem Shadows*, for instance, portrayed Harlem as a lush tropical forest hiding (spiritual) decay and stagnation.[39] Jean Toomer's *Cane* was part poem, part essay, part novel, with an overall elegiac tone, lamenting the legacy of slavery, the 'racial twilight' in which blacks found themselves: they can't – won't – go back, and don't know the way forward.[40] Alain Locke was a sort of impresario, an Apollinaire of Harlem, whose *New Negro*, published in 1925, was an anthology of poetry and prose.[41] Charles Johnson was a sociologist who had studied under Robert Park at Chicago, who organised intellectual gatherings at the Civic Club, attended by Eugene O'Neill, Carl van Doren, and Albert Barnes, who spoke about African art. Johnson was also the editor of a new black magazine to put alongside Du Bois's *Crisis*. It was called *Opportunity*, its very name reflecting the optimism of the time.[42]

The high point and low point of the Harlem Renaissance is generally agreed to have been the publication in 1926 of *Nigger Heaven*, by **Carl Van Vechten**, described as 'Harlem's most enthusiastic and ubiquitous Nordic.' Van Vechten's novel is scarcely read now, though sales soared when it was first released by Alfred A. Knopf. Its theme was High Harlem, the Harlem that Van Vechten knew and adored but was, when it came down to it, an outsider in. He *thought* life in Harlem was perfect, that the blacks there were, as he put it, 'happy in their skin,' reflecting the current view that African Americans had a vitality that whites lacked, or were losing with the decadence of their civilisation. All that may have been acceptable, just; but Van Vechten was an outsider, and he made two unforgivable mistakes which vitiated his book: he ignored the problems that even sophisticated blacks knew had not gone away; and in his use of slang, and his comments about the 'black gait' and so forth, though he may have thought he was being 'anthropological,' he came across as condescending and embarrassing. *Nigger Heaven* was not at all ironic.[43]

The Harlem Renaissance barely survived the 1929 Wall Street debacle and the subsequent depression. Novels and poems continued to be put out, but the economic constraints caused a return to deeper segregation and a recrudescence of lynchings, and against such a background it was difficult to maintain the sense of optimism that had characterised the renaissance. Art, the arts, might have offered temporary respite from the realities of life, but as the 1930s matured, American blacks could no longer hide from the bleak truth: despite the renaissance, underneath it all nothing had changed.

The wider significance of the Harlem Renaissance was twofold: in the first place, that it occurred at all, at the very time that the scientific racists were introducing the Immigration Restriction Act and trying to prove that blacks were simply not capable of producing the sort of work that characterised the renaissance; and second, that once it was over, it was so comprehensively forgotten. That too was a measure of racism.*

In a sense, by the 1920s the great days of Greenwich Village were over. It was still a refuge for artists, and still home to scores of little literary magazines, some of which, like the *Masses* and the *Little Review,* enjoyed a period of success, and others, like the *New Republic* and the *Nation*, are still with us. The Provincetown Players and the Washington Square Players still performed there in season, including the early plays of O'Neill. But after the war the costume balls and more colourful excesses of bohemia now seemed far too frivolous. The spirit of the Village lived on, however, or perhaps it would be truer to say that it matured, in the 1920s, in a magazine that reflected the Village's values by flying in the face of *Time, Reader's Digest*, Middletown, and the rest. This was the *New Yorker.*

The fact that the *New Yorker* could follow this bold course owed everything to its editor, **Harold Ross**. In many respects Ross was an improbable editor – for a start, he wasn't a New Yorker. Born in Colorado, he was a 'poker-playing, hard-swearing' reporter who had earlier edited the *Stars and Stripes*, the U.S. Army's newspaper, published from Paris during the war years. That experience had given Ross a measure of sophistication and scepticism, and when he returned to New York he joined the circle of literary types who lunched at the famous Round Table at the Algonquin Hotel on Forty-Fourth Street. Ross became friendly with **Dorothy Parker, Robert Benchley**, Marc Connelly, Franklin P. Adams, and Edna Ferber. Less famous but more important for Ross's career was the poker game that some of the Round Table types took part in on Saturday evenings. It was over poker that Ross met **Raoul Fleischmann**, a

* The history of Harlem was not fully recovered until the 1980s, by such scholars as David Levering Lewis and George Hutchinson. My account is based chiefly on their work.

baking millionaire, who agreed to bankroll his idea for a satirical weekly.[44]

Like all the other publishing ventures started in the 1920s, the *New Yorker* did not prosper at first. Initially, sales of around 70,000 copies were anticipated, so when the first issue, appearing in February 1925, sold only 15,000, and the second dropped to 8,000, the future did not look good. Success only came, according to another legend, when a curious package arrived in the office, unsolicited. This was a series of articles, written by hand but extravagantly and expensively bound in leather. The author, it turned out, was a debutante, **Ellin Mackay**, who belonged to one of New York's society families. Making the most of this, Ross published one of the articles with the headline, 'Why We Go to Cabarets.' The thrust of the article, which was wittily written, was that New York nightlife was very different, and much more fun, than the stiff society affairs organised for her by Miss Mackay's parents. The knowing tone was exactly what Ross had in mind, and appealed to other writers: **E. B. White** joined the *New Yorker* in 1926, **James Thurber** a year later, followed by **John O'Hara, Ogden Nash**, and **S. J. Perelman**.[45]

But a dry wit and a knowing sophistication were not the only qualities of the *New Yorker*; there was a serious side, too, as reflected in particular in its profiles. *Time* sought to tell the news through people, successful people. The *New Yorker*, on the other hand, elevated the profile to, if not an art form, a high form of craft. In the subsequent years, a *New Yorker* reporter might spend five months on a single article: three months collecting information, a month writing and a month revising (all this before the fact checkers were called in). 'Everything from bank references to urinalysis was called for and the articles would run for pages.'[46] The *New Yorker* developed a devoted following, its high point being reached immediately after World War II, when it sold nearly 400,000 copies weekly. In the early 1940s, no fewer than four comedies based on *New Yorker* articles were playing on Broadway: *Mr and Mrs North, Pal Joey, Life with Father* and *My Sister Eileen*.[47]

The way radio developed in Britain reflected a real fear that it might have a bad influence on levels of information and taste, and there was a strong feeling, in the 'establishment,' that central guidance was needed. 'Chaos in the ether' was to be avoided at all costs.[48] To begin with, a few large companies were granted licences to broadcast experimentally. After that, a syndicate of firms which manufactured radio sets was founded, financed by the Post Office, which levied a 10-shilling (50 pence) fee payable by those who bought the sets. Adverts were dispensed with as 'vulgar and intrusive.'[49] This, the **British Broadcasting Company**, lasted for four years. After that, the **Corporation** came into being, granted a royal charter to protect it from political interference.

In the early days the notion of the BBC as a public service was very uncertain. All manner of forces were against it. For a start, the country's mood was volatile. Britain was still in financial straits, recovering from the war, and 1.5 million were unemployed. Lloyd George's coalition government was far from popular, and these overall conditions led to the general strike of 1926, which itself imperilled the BBC. A second factor was the press, which viewed the BBC as a threat, to such an extent that no news bulletins were allowed before 7:00 P.M. Third, no one had any idea what sort of material should be broadcast – audience research didn't begin until 1936, and 'listening in,' as it was called, was believed by many to be a fad that would soon pass.[50] Then there was the character of the Corporation's first director, a thirty-three-year-old Scottish engineer named **John Reith**. Reith, a high-minded Scottish Presbyterian, never doubted for a moment that radio should be far more than entertainment, that it should also educate and inform. As a result, the BBC gave its audience what Reith believed was needed rather than what the people wanted. Despite this high-handed and high-minded approach, the BBC proved popular. From a staff of 4 in the first year, it grew to employ 177 twelve months after that. In fact, the growth of radio actually outstripped that of television a generation or so later, as these figures show:[51]

No. of licences issued

Radio		TV	
1922	35,744	1947	14,560
1923	595,496	1948	45,564
1924	1,129,578	1949	126,567
1925	1,645,207	1950	343,882
1926	2,178,259	1951	763,941
	(+6094%)		(+5246%)

Note: Television sets were, comparatively speaking, much more expensive than radio sets. Even so, the numbers are revealing.

To be set against this crude measure of popularity, there was a crop of worries about the intellectual damage radio might do. 'Instead of solitary thought,' said the headmaster of Rugby School, 'people would listen in to what was said to millions of people, which could not be the best of things.'[52] Another worry was that radio would make people 'more passive,' producing 'all-alike girls.' Still others feared radio would keep husbands at home, adversely affecting pub attendance. In 1925 *Punch* magazine, referring to the new culture established by the BBC, labelled it as 'middlebrow.'[53]

Editorially speaking, the BBC's first test arrived in 1926 with the onset of the General Strike. Most newspapers were included in the strike, so for a time the BBC was virtually the only source of news. Reith responded by ordering five bulletins a day instead of the usual one. The accepted view now is that Reith complied more or less with what the government asked, in particular putting an optimistic gloss on government policy and actions. In his official history of the BBC, Professor Asa Briggs gives this example of an item broadcast during the strike: 'Anyone who is suffering from "strike depression" can do no better than to pay a visit to "RSVP" [a show] at the New Vaudeville Theatre.' Not everyone thought that Reith was a stool pigeon, however. Winston Churchill, then chancellor of the exchequer, actually thought the BBC should be taken over. He saw it as a rival to his own *British Gazette*, edited from his official address at 11 Downing Street.[54] Churchill failed, but people had seen the danger, and it was partly as a result of this tussle that the 'C' in BBC was changed in 1927 from Company to Corporation, protected by royal charter. The General Strike was therefore a watershed for the BBC in the realm of politics. Before the strike, politics (and other 'controversial' subjects) were avoided entirely, but the strike changed all that, and in 1929 *The Week in Parliament* was launched. Three years later, the corporation began its own news-gathering organisation.[55]

The historian J. H. Plumb has said that one of the great unsung achievements of the twentieth century has been the education of vast numbers of people. Government-funded schools and universities led the way here, but the various forms of new media, many of which started in the 1920s, have also played their part. The term *middlebrow* may be intended as an insult by some, but for millions, like the readers of *Time* or those listening in to the BBC, it was more a question of wising up than dumbing down.

13

HEROES' TWILIGHT

In February 1920 a horror film was released in Berlin that was, in the words of one critic, 'uncanny, demonic, cruel, "Gothic",' a Frankenstein-type story filled with bizarre lighting and dark, distorted sets.[1] Considered by many to be the first 'art film,' *The Cabinet of Dr Caligari* was a huge success, so popular in Paris that it played in the same theatre every day between 1920 and 1927.[2] But the film was more than a record breaker. As the historian of interwar Germany Peter Gay writes, 'With its nightmarish plot, its Expressionist sets, its murky atmosphere, *Caligari* continues to embody the Weimar spirit to posterity as palpably as Gropius's buildings, Kandinsky's abstractions, Grosz's cartoons, and Marlene Dietrich's legs ... But *Caligari*, decisive for the history of film, is also instructive for the history of Weimar. ... There was more at stake here than a strange script or novelties of lighting.'[3]

Following World War I, as we have seen, Germany was turned almost overnight into a republic. Berlin remained the capital but Weimar was chosen as the seat of the assembly after a constitutional conference had been held there to decide the form the new republic would take, because of its immaculate reputation (Goethe, Schiller), and because of worries that the violence in Berlin and Munich would escalate if either of those cities were selected. The **Weimar Republic** lasted for fourteen years until Hitler came to power in Germany in 1933, a tumultuous interregnum between disasters which astonishingly managed to produce a distinctive culture that was both brilliant and characterised by its own style of thought, the very antithesis of Middletown.

The period can be conveniently divided into three clear phases.[4] From the end of 1918 to 1924, 'with its revolution, civil war, foreign occupation, and fantastic inflation, [there] was a time of experimentation in the arts; Expressionism dominated politics as much as painting or the stage.'[5] This was followed, from 1924 to 1929, by a period of economic stability, a relief from political violence, and increasing prosperity reflected in the arts by the *Neue Sachlichkeit*, the 'new objectivity,' a movement whose aims were matter-of-factness, even sobriety. Finally, the period 1929 to 1933 saw a return to political violence, rising unemployment, and authoritarian government by decree; the arts were cowed into silence, and replaced by propagandistic *Kitsch*.[6]

★

Caligari was a collaboration between two men, **Hans Janowitz**, a Czech, and **Carl Meyer**, an Austrian, who had met in Berlin in 1919.[7] Their work was not only fiercely antiwar but also explored what expressionism could do in the cinema. The film features the mad Dr Caligari, a fairground vaudeville act who entertains with his somnambulist, Cesare. Outside the fair, however, there is a second string to the story, and it is far darker. Wherever Caligari goes, death is never far behind. Anyone who crosses him ends up dead. The story proper starts after Caligari kills two students – or thinks that he has. In fact, one survives, and it is this survivor, Francis, who begins to investigate. Nosing around, he discovers Cesare asleep in a box. But the killings continue, and when Francis returns to the sleeping Cesare, he realises this time that the 'person' motionless in the box is merely a dummy. It dawns on Francis, and the police, whose help he has now enlisted, that the sleepwalking Cesare is unconsciously obeying Caligari's instructions, killing on his behalf without understanding what he has done. Realising he has been discovered, Caligari flees into an insane asylum. But this is more than it seems, for Francis now finds out that Caligari is also the *director* of the asylum. Shocking as this is, there is no escape for Caligari, and when his double life is exposed, far from being cathartic, he loses all self-control and ends up in a straitjacket.[8]

This was the original story of Caligari, but before the film appeared it went through a drastic metamorphosis. Janowitz and Meyer had intended their story to be a powerful polemic against military obedience and assumed that when the script was accepted by **Erich Pommer**, one of the most successful producers of the day, he would not change it in any way.[9] However, Pommer and the director, **Robert Wiene**, actually turned the story inside out, rearranging it so that it is Francis and his girlfriend who are mad. The ideas of abduction and murder are now no more than *their* delusions, and the director of the asylum is in reality a benign doctor who cures Francis of his evil thoughts. Janowitz and Meyer were furious. Pommer's version of the story was the opposite of theirs. The criticism of blind obedience had disappeared and, even worse, authority was shown as kindly, even safe. It was a travesty.[10]

The irony was that Pommer's version was a great success, commercially and artistically, and film historians have often wondered whether the original version would have done as well. And perhaps there is a fundamental point here. Though the plot was changed, the style of telling the story was not – it was still expressionistic. Expressionism was a force, an impulse to revolution and change. But, like the psychoanalytic theory on which it was based, it was not fully worked out. The expressionist **Novembergruppe**, founded in December 1918, was a revolutionary alliance of all the artists who wanted to see change – **Emil Nolde, Walter Gropius, Bertolt Brecht, Kurt Weill, Alban Berg**, and **Paul Hindemith**. But revolution needed more than an engine; it needed direction. Expressionism never provided that. And perhaps in the end its lack of direction was one of those factors that enabled Adolf Hitler's rise to power. He hated expressionism as much as he hated anything.[11]

But it would be wrong to see Weimar as a temporary way station on the path to Hitler. It certainly did not see itself in that light, and it boasted many solid achievements. Not the least of these was the establishment of some very prestigious academic institutions, still centres of excellence even today. These included the **Psychoanalytic Institute in Berlin** – home to Franz Alexander, Karen Horney, Otto Fenichel, Melanie Klein, and Wilhelm Reich – and the **Deutsche Hochschule für Politik**, which had more than two thousand students by the last year of the republic: the teachers here included Sigmund Neumann, Franz Neumann, and Hajo Holborn. And then there was the **Warburg Institute of Art History**.

In 1920 the German philosopher Ernst Cassirer paid a visit to the Warburg art historical library in Hamburg. He had just been appointed to the chair in philosophy at the new university in Hamburg and knew that some of the scholars at the library shared his interests. He was shown around by Fritz Saxl, then in charge. The library was the fantastic fruit of a lifetime's collecting by **Aby Warburg**, a rich, scholarly, and 'intermittently psychotic individual' who, not unlike T. S. Eliot and James Joyce, was obsessed by classical antiquity and the extent to which its ideas and values could be perpetuated in the modern world.[12] The charm and value of the library was not just that Warburg had been able to afford thousands of rare volumes on many recondite topics, but the careful way he had put them together to illuminate one another: thus art, religion, and philosophy were mixed up with history, mathematics, and anthropology. For Warburg, following James Frazer, philosophy was inseparable from study of the 'primitive mind.' The Warburg Institute has been the home of many important art historical studies throughout the century, but it started in Weimar Germany, where among the papers published under its auspices were **Erwin Panofsky's** *Idea, Dürers 'Melancolia 1,' Hercules am Scheidewege* and **Percy Schramm's** *Kaiser, Rom und Renovatio*. Panofsky's way of reading paintings, his 'iconological method,' as it was called, would prove hugely influential after World War II.[13]

Europeans had been fascinated by the rise of the skyscraper in America, but it was difficult to adapt on the eastern side of the Atlantic: the old cities of France, Italy, and Germany were all in place, and too beautiful to allow the distortion that very tall buildings threatened.[14] But the new materials of the twentieth century, which helped the birth of the skyscraper, were very seductive and proved popular in Europe, especially steel, reinforced concrete, and sheet glass. The latter in particular transformed the appearance of buildings and the experience of being *inside* a structure. With its different colours, reflectivity, and transparency, glass was a flexible, expressive skin for buildings constructed in steel. In the end, glass and steel had a bigger effect on European architects than concrete did, and especially on three architects who worked together in the studio of the leading industrial designer in Germany, **Peter Behrens** (1868–1940). These were **Walter Gropius, Ludwig Mies van der Rohe**, and **Charles-Edouard Jeanneret**, better known as **Le Corbusier**. Each would make his mark, but the first was Gropius. It was Gropius who founded the **Bauhaus**.

It is not difficult to see why Gropius should have taken the lead. Influenced by Marx and by William Morris, he always believed, contrary to Adolf Loos, that craftsmanship was as important as 'higher' art. He had also learned from

Behrens, whose firm was one of the first to develop the modern 'design package,' providing AEG with a corporate style that they used for everything, from letterheads and arc lamps to the company's buildings themselves. Therefore, when the Grand Ducal Academy of Art, which was founded in the mid-eighteenth century, was merged with the Weimar Arts and Crafts School, established in 1902, he was an obvious choice as director. The fused structure was given the name Staatliche Bauhaus, with *Bauhaus* – literally, 'house for building' – chosen because it echoed the *Bauhütten*, mediaeval lodges where those constructing the great cathedrals were housed.[15]

The early years of the Bauhaus, in Weimar, were troubled. The government of Thuringia, where Weimar was located, was very right-wing, and the school's collectivist approach, the rebelliousness of its students, and the style of its first head teacher, Johannes Itten, a quarrelsome mystic-religious, proved very unpopular.[16] The school's budget was cut, forcing its removal to Dessau, which had a more congenial administration.[17] This change in location seems to have brought about a change in Gropius himself. He produced a second manifesto, in which he announced that the school would concern itself with practical questions of the modern world – mass housing, industrial design, typography, and the 'development of prototypes.' The obsession with wood was abandoned: Gropius's design for the school's new building was entirely of steel, glass, and concrete, to underline the school's partnership with industry. Inside the place, Gropius vowed, students and staff would explore a 'positive attitude to the living environment of vehicles and machines . . . avoiding all romantic embellishment and whimsy.'[18]

After a lost war and an enormous rise in inflation, there was no social priority of greater importance in Weimar Germany than mass housing. And so Bauhaus architects were among those who developed what became a familiar form of social housing, the *Siedlung* or 'settlement.' This was introduced to the world iein 1927, at the Stuttgart trade fair exhibition. Corbusier, Mies van der Rohe, Gropius, J. P. Oud, and Bruno Taut all designed buildings for the Weissenhof (White House) *Siedlung*, 'and twenty thousand people came every day to marvel at the flat roofs, white walls, strip windows and *pilotis* of what Rohe called "the great struggle for a new way of life." '[19] Although the *Siedlungen* were undoubtedly better than the nineteenth-century slums they were intended to replace, the lasting influence of the Bauhaus has been in the area of applied design.[20] The Bauhaus philosophy, 'that it is far harder to design a first-rate teapot than paint a second-rate picture,' has found wide acceptance – folding beds, built-in cupboards, stackable chairs and tables, designed with mass-production processes in mind and with an understanding of the buildings these objects were to be used in.[21]

The catastrophe of World War I, followed by the famine, unemployment, and inflation of the postwar years, for many people confirmed Marx's theory that capitalism would eventually collapse under the weight of its own 'insoluble contradictions'. However, it soon became clear that it wasn't communism that was appearing from the rubble, but fascism. Some Marxists were so disillusioned by this that they abandoned Marxism altogether. Others remained convinced of the theory, despite the evidence. But there was a third group, people in between, who wished to remain Marxists but felt that Marxist theory needed reconstructing if it was to remain credible. This group assembled in Frankfurt in the late 1920s and made a name for itself as **the Frankfurt School**, with its own institute in the city. Thanks to the Nazis, the institute didn't stay long, but the name stuck.[22]

The three best-known members of the Frankfurt School were **Theodor Adorno**, a man who 'seemed equally at home in philosophy, sociology and music,' **Max Horkheimer**, a philosopher and sociologist, less innovative than Adorno but perhaps more dependable, and the political theorist **Herbert Marcuse**, who in time would become the most famous of all. Horkheimer was the director of the institute. In addition to being a philosopher and sociologist, he was also a financial wizard, who brilliantly manipulated the investments of the institute, both in Germany and afterward in the United States. According to Marcuse, nothing that was written by the Frankfurt School was published without previous discussion with him. Adorno was the early star. According to Marcuse, 'When he talked it could be printed without any changes.' In addition there was Leo Lowenthal, the literary critic of the school, Franz Neumann, a legal philosopher, and Friedrich Pollock, who was one of those who argued – against Marx and to Lenin's fury – that there were no compelling internal reasons why capitalism should collapse.[23]

In its early years the school was known for its revival of the concept of **alienation**. This, a term originally coined by Georg Wilhelm Friedrich Hegel, was taken up and refined by Marx but, for half a century, from the 1870s, ignored by philosophers. 'According to Marx, "alienation" was a socio-economic concept.'[24] Basically, Marcuse said, alienation meant that under capitalism men and women could not, in their work, fulfil their own needs. The capitalist mode of production was at fault here, and alienation could only be abolished by radically changing this mode of production. The Frankfurt School, however, developed this idea so that it became above all a *psychological* entity, and one, moreover, that was not necessarily, or primarily, due to the capitalist mode of production. Alienation, for the Frankfurt School, was more a product of all of modern life. This view shaped the school's second and perhaps most enduring preoccupation: the attempted **marriage of Freudianism and Marxism**.[25] Marcuse took the lead to begin with, though Erich Fromm wrote several books on the subject later. Marcuse regarded Freudianism and Marxism as two sides of the same coin. According to him, Freud's unconscious primary drives, in particular the life instinct and the death instinct, are embedded within a social framework that determines how they show themselves. Freud had argued that repression necessarily increases with the progress of civilisation; therefore aggressiveness must be produced and released in ever greater quantities. And so, just as Marx had predicted that revolution was inevitable, a dislocation that capitalism must bring on itself, so, in Marcuse's hands, Freudianism produced a parallel, more personal backdrop to this scenario, accounting for a buildup of destructiveness – self-destruction and the destruction of others.[26]

The third contribution of the Frankfurt School was a more general analysis of social change and progress, the introduction of an interdisciplinary approach – sociology, psychology, philosophy – to examine what the school regarded as the vital question of the day: 'What precisely has gone wrong in Western civilisation, that at the very height of technical progress we see the negation of human progress: dehumanisation, brutalisation, revival of torture as a "normal" means of interrogation, the destructive development of nuclear energy, the poisoning of the biosphere, and so on? How has this happened?'[27] To try to answer this question, they looked back as far as the Enlightenment, and then traced events and ideas forward to the twentieth century. They claimed to discern a 'dialectic,' an interplay between progressive and repressive periods in the West. Moreover, each repressive period was usually greater than the one before, owing to the growth of technology under capitalism, to the point where, in the late 1920s, 'the incredible social wealth that had been assembled in Western civilisation, mainly as the achievement of Capitalism, was increasingly used for preventing rather than constructing a more decent and human society.'[28] The school saw fascism as a natural development in the long history of capitalism after the Enlightenment, and in the late 1920s earned the respect of colleagues with its predictions that fascism would grow. The Frankfurt School's scholarship most often took the form of close readings of original material, from which views uncontaminated by previous analyses were formed. This proved very creative in terms of the new understanding it produced, and the Frankfurt method became known as **critical theory**.[29] Adorno was also interested in aesthetics, and he had his own socialist view of the arts. He felt that there are insights and truths that can be expressed only in an artistic form, and that therefore the aesthetic experience is another variety of liberation, to put alongside the psychological and political, which should be available to as many people as possible.

The Psychoanalytic Institute, the Warburg Institute, the Deutsche Hochschule für Politik, and the Frankfurt School were all part of what Peter Gay has called 'the community of reason,' an attempt to bring the clear light of scientific rationality to communal problems and experiences. But not everyone felt that way.

One part of what became a campaign against the 'cold positivism' of science in Weimar Germany was led by the *Kreis* ('circle') of poets and writers that formed around **Stefan George**, 'king of a secret Germany.'[30] Born in 1868, George was already fifty-one when World War I ended. He was very widely read, in all the literatures of Europe, and his poems at times bordered on the precious, brimming over with an 'aesthetic of arrogant intuitionism.' Although led by a poet, the *Kreis* was more important for what it stood for than for what it actually produced. Most of its writers were biographers – which wasn't accidental. Their intention was to highlight 'great men,' especially those from more 'heroic' ages, men who had by their will changed the course of events. The most successful book was Ernst Kantorowicz's biography of the thirteenth-century emperor Frederick II.[31] For George and his circle, Weimar Germany was a distinctly unheroic age; science had no answer to such a predicament, and the task of the writer was to inspire others by means of his superior intuition.

George never had the influence that he expected because he was over-shadowed by a much greater poetic talent, **Rainer Maria Rilke**. Born René Maria Rilke in Prague in 1875 (he Germanised his name only in 1897), Rilke was educated at military school.[32] An inveterate traveller and something of a snob (or at least a collector of aristocratic friendships), his path crossed with those of Friedrich Nietzsche, Hugo von Hofmannsthal, Arthur Schnitzler, Paula Modersohn-Becker, Gerhart Hauptmann, Oskar Kokoschka, and Ellen Key (author of *The Century of the Child*; see chapter 5).[33] Early in his career, Rilke tried plays as well as biography and poetry, but it was the latter form that, as he grew older, distinguished him as a remarkable writer, influencing W. H. Auden, among others.[34] His reputation was transformed by *Five Cantos/August 1914*, which he wrote in response to World War I. Young German soldiers 'took his slim volumes with them to the front, and his were often the last words they read before they died. He therefore had the popularity of Rupert Brooke without the accompanying danger, becoming … "the idol of a generation without men." '[35] Rilke's most famous poems, the *Duino Elegies*, were published in 1923 during the Weimar years, their mystical, philosophical, 'oceanic' tone perfectly capturing the mood of the moment.[36] The ten elegies were in fact begun well before World War I, while Rilke was a guest at Duino Castle, south of Trieste on the Adriatic coast, where Dante was supposed to have stayed. The castle belonged to one of Rilke's many aristocratic friends, Princess Marie von Thurn und Taxis-Hohenlohe. But the bulk of the elegies were 'poured out' in a 'spiritual hurricane' in one week, between 7 and 14 February 1922.[37] Lyrical, metaphysical, and very concentrated, they have proved lastingly popular, no less in translation than in the original German. After he had finished his exhausting week that February, he wrote to a friend that the elegies 'had arrived' (it had been eleven years since he had started), as if he were the mouthpiece of some other, perhaps divine, voice. This is indeed how Rilke thought and, according to friends and observers, behaved. In the elegies Rilke wrestles with the meaning of life, the 'great land of grief,' casting his net over the fine arts, literary history, mythology, and the sciences, in particular biology, anthropology, and psychoanalysis.[38] The poems are peopled by angels, lovers, children, dogs, saints, and heroes, reflecting a very Germanic vision, but also by more down-to-earth creatures such as acrobats and the *saltimbanques* Rilke had seen in Picasso's early work. Rilke celebrates life, heaping original image upon original image (in a slightly uncomfortable rhythm that keeps the reader focused on the words), and yet juxtaposes the natural world with the mechanics of modernity. At the same time that he celebrates life, however, Rilke reminds us of its fragility, the elegiac quality arising from man's unique awareness among life forms of his approaching death. For E. M. Butler, Rilke's biographer, the poet's concept of 'radiant angels' was his truest poetical creation; not 'susceptible of rational interpretation … they stand like a liquid barrier of fire between man and his maker.'

Earliest triumphs, and high creation's favourites,
Mountain-ranges and dawn-red ridges,
Since all beginning, pollen of blossoming godhead,
Articulate light, avenues, stairways, thrones,
Spaces of being, shields of delight, tumults
Of stormily-rapturous feeling, and suddenly, singly,
Mirrors, drawing back within themselves
The beauty radiant from their countenance.[39]

Delivering a eulogy after Rilke's death, Stefan Zweig accorded him the accolade of *Dichter*.[40] For Rilke, the meaning of life, the sense that could be made of it, was to be found in language, in the ability to speak or 'say' truths, to transform machine-run civilisation into something more heroic, more spiritual, something more worthy of lovers and saints. Although at times an obscure poet, Rilke became a cult figure with an international following. Thousands of readers, mostly women, wrote to him, and when a collection of his replies was published, his cult received a further boost. There are those who see in the Rilke cult early signs of the *völkisch* nationalism that was to overtake Germany in the late 1920s and 1930s. In some ways, certainly, Rilke anticipates Heidegger's philosophy. But in fairness to the poet, he himself always saw the dangers of such a cult. Many of the young in Germany were confused because, as he put it, they 'understood the call of art as a call *to* art.'[41] This was an echo of the old problem identified by Hofmannsthal: What is the fate of those who cannot create? For Rilke, the cult of art was a form of retreat from life, by those who wanted to *be* artists rather than lead a life.[42] Rilke did not create the enthusiasm for spirituality in Weimar Germany; it was an old German obsession. But he did reinvigorate it. Peter Gay again: 'His magnificent gift for language paved the way to music rather than to logic.'[43]

Whereas Rilke shared with Hofmannsthal the belief that the artist can help shape the prevailing mentality of an age, **Thomas Mann** was more concerned, as Schnitzler had been, to describe that change as dramatically as possible. Mann's most famous novel was published in 1924. *The Magic Mountain* did extremely well (it was published in two volumes), selling fifty thousand copies in its first year. It is heavily laden with symbolism, and the English translation has succeeded in losing some of Mann's humour, not exactly a rich commodity in his work. But the symbolism is important, for as we shall see, it is a familiar one. *The Magic Mountain* is about the wasteland that caused, or at least preceded, *The Waste Land*. Set on the eve of World War I, it tells the story of **Hans Castorp**, 'a simple young man' who goes to a Swiss sanatorium to visit a cousin who has tuberculosis (a visit Alfred Einstein actually made, to deliver a lecture).[44] Expecting to stay only a short time, he catches the disease himself and is forced to remain in the clinic for seven years. During the course of the book he meets various members of staff, fellow patients, and visitors. Each of these represents a distinct point of view competing for the soul of Hans. The overall symbolism is pretty heavy-handed. The hospital is Europe, a stable, long-standing institution but filled with decay and corruption. Like the generals starting the war,

Hans expects his visit to the clinic to be short, over in no time.[45] Like them, he is surprised – appalled – to discover that his whole time frame has to be changed. Among the other characters there is the liberal Settembrini, anti-clerical, optimistic, above all rational. He is opposed by Naphta, eloquent but with a dark streak, the advocate of heroic passion and instinct, 'the apostle of irrationalism.'[46] Peeperkorn is in some ways a creature out of Rilke, a sensualist, a celebrant of life, whose words come tumbling out but expose him as having little to say. His body is like his mind: diseased and impotent.[47] Clawdia Chauchat, a Russian, has a different kind of innocence from Hans's. She is self-possessed but innocent of knowledge, particularly scientific knowledge. Hans assumes that by revealing all the scientific knowledge he has, he will possess her. They enjoy a brief affair, but Hans no more possesses her mind and soul than scientific facts equal wisdom.[48] Finally, there is the soldier Joachim, Hans's cousin, who is the least romantic of all of them, especially about war. When he is killed, we feel his loss like an amputation. Castorp is redeemed – but through a dream, the sort of dream Freud would have relished (but which in fact rarely exists in real life), full of symbolism leading to the conclusion that love is the master of all, that love is stronger than reason, that love alone can conquer the forces that are bringing death all around. Hans does not forsake reason entirely, but he realises that a life without passion is but half a life.[49] Unlike Rilke, whose aim was to transform experience into art, Mann's goal was to sum up the human condition (at least, the Western condition), in detail as well as in generalities, aware as Rilke was that a whole era was coming to an end. With compassion and an absence of mysticism, Mann grasped that heroes were not the answer. For Mann, modern man was self-conscious as never before. But was self-consciousness a form of reason? Or an instinct?

Over the last half of the nineteenth century and the first decades of the twentieth, Paris, Vienna, and briefly Zurich dominated the intellectual and cultural life of Europe. Now it was Berlin's turn. Viscount D'Abernon, the British ambassador to Berlin, described in his memoirs the period after 1925 as an 'epoch of splendour' in the city's cultural life.[50] Bertolt Brecht moved there; so did Heinrich Mann and Erich Kästner, after he had been fired from the Leipzig newspaper where he worked. Painters, journalists, and architects flocked to the city, but it was above all a place for performers. Alongside the city's 120 newspapers, there were forty theatres providing, according to one observer, 'unparalleled mental alertness.'[51] But it was also a golden age for political cabaret, art films, satirical songs, Erwin Piscator's experimental theatre, Franz Léhar operettas.

Among this concatenation of talent, this unparalleled mental alertness, three figures from the performing arts stand out: Arnold Schoenberg, **Alban Berg**, and Bertolt Brecht. Between 1915 and 1923 Schoenberg composed very little, but in 1923 he gave the world what one critic called '**a new way of musical organisation**.'[52] Two years before, in 1921, Schoenberg, embittered by years of hardship, had announced that he had 'discovered something which will assure the supremacy of German music for the next hundred years.'[53] This was what became known as 'serial music.' Schoenberg himself gave rise to the phrase when he wrote, 'I called this procedure "Method of composing with twelve tones which are related only with one another." '[54] 'Procedure' was an apt word for it, since serialism is not so much a style as a 'new grammar' for music. Atonalism, Schoenberg's earlier invention, was partly designed to eliminate the individual intellect from musical composition; serialism took that process further, minimalising the tendency of any note to prevail. Under this system a composition is made up of a series from the twelve notes of the chromatic scale, arranged in an order that is chosen for the purpose and varies from work to work. Normally, no note in the row or series is repeated, so that no single note is given more importance than any other, lest the music take on the feeling of a tonal centre, as in traditional music with a key. Schoenberg's tone series could be played in its original version, upside down (inversion), backward (retrograde) or even backward upside down (retrograde inversion). The point of this new music was that it was horizontal, or contrapuntal, rather than vertical, or harmonic.[55] Its melodic line was often jerky, with huge leaps in tone and gaps in rhythm. Instead of themes grouped harmonically and repeated, the music was divided into 'cells.' Repetition was by definition avoided. Huge variations were possible under the new system – including the use of voices and instruments in unusual registers. However, compositions always had a degree of harmonic coherence, 'since the fundamental interval pattern is always the same.'[56]

   The first completely serial work is generally held to be Schoenberg's Piano Suite (op. 25), performed in 1923. Both Berg and Anton von Webern enthusiastically adopted Schoenberg's new technique, and for many people Berg's two operas Wozzeck and Lulu have become the most familiar examples of, first, atonality, and second, serialism. Berg began to work on Wozzeck in 1918, although it was not premiered until 1925, in Berlin. Based on a short unfinished play by Georg Büchner, the action revolves around an inadequate, simple soldier who is preyed upon and betrayed by his mistress, his doctor, his captain, and his drum major; in some ways it is a musical version of George Grosz's savage pictures.[57] The soldier ends up committing both murder and suicide. Berg, a large, handsome man, had shed the influence of romanticism less well than Schoenberg or Webern (which is perhaps why his works are more popular), and Wozzeck is very rich in moods and forms – rhapsody, lullaby, a military march, rondo, each character vividly drawn.[58] The first night, with Erich Kleiber conducting, took place only after 'an unprecedented series of rehearsals,' but even so the opera created a furore.[59] It was labelled 'degenerate,' and the critic for Deutsche Zeitung wrote, 'As I was leaving the State Opera, I had the sensation of having been not in a public theatre but in an insane asylum. On the stage, in the orchestra, in the stalls – plain madness. ... We deal here, from a musical viewpoint, with a composer dangerous to the public welfare.'[60] But not everyone was affronted; some critics praised Berg's 'instinctive perception,' and other European opera houses clamoured to stage it. Lulu is in some ways the reverse of Wozzeck. Whereas the soldier was prey to those around him, Lulu is a predator, an amoral temptress 'who ruins all she touches.'[61]

Based on two dramas by **Frank Wedekind**, this serial opera also verges on atonality. Unfinished at Berg's death in 1935, it is full of bravura patches, elaborate coloratura, and confrontations between a heroine-turned-prostitute and her murderer. Lulu is the 'evangelist of a new century,' killed by the man who fears her.[62] It was the very embodiment of the Berlin that **Bertolt Brecht**, among others, was at home in.

Like Berg, Kurt Weill, and Paul Hindemith, Brecht was a member of the Novembergruppe, founded in 1918 and dedicated to disseminating a new art appropriate to a new age. Though the group broke up after 1924, when the second phase of life in the Weimar Republic began, the revolutionary spirit, as we have seen, survived. And it survived in style in Brecht. Born in Augsburg in 1898, though he liked to say he came from the Black Forest, Brecht was one of the first artists/writers/poets to grow up under the influence of film (and Chaplin, in particular). From an early age, he was always fascinated by America and American ideas – jazz and the work of Upton Sinclair were to be other influences later. Augsburg was about forty miles from Munich, and it was there that Brecht spent his formative years. Somewhat protected by his parents, Bertolt (christened Eugen, a name he later dropped) grew up as a self-confident and even 'ruthless' child, with the 'watchful eyes of a raccoon.'[63] Initially a poet, he was also an accomplished guitarist, with which talent, according to some (like Lion Feuchtwanger) he used to 'impose himself' on others, smelling 'unmistakably of revolution'.[64] He collaborated and formed friendships with Karl Kraus, Carl Zuckmayer, Erwin Piscator, Paul Hindemith, Kurt Weill, Gerhart and Elisabeth Hauptmann, and an actor who 'looked like a tadpole.' The latter's name was Peter Lorre. In his twenties, Brecht gravitated toward theatre, Marxism, and Berlin.[65]

Brecht's early works, like *Baal*, earned him a reputation among the avant-garde, but it was with *The Threepenny Opera* (titled *Die Dreigroschenoper* in German) that he first found real fame. This work was based on a 1728 ballad opera by John Gay, *The Beggar's Opera*, which had been revived in 1920 by Sir Nigel Playfair at the Lyric Theatre in London, where it ran for four years. Realising that it could be equally successful in Germany, Elisabeth Hauptmann translated it for Brecht.[66] He liked it, found a producer and a theatre, and removed himself to Le Lavandou, in the south of France near Saint Tropez, with the composer **Kurt Weill** to work on the show. John Gay's main aim had been to ridicule the pretensions of Italian grand opera, though he did also take the odd swipe at the prime minister of the day, Sir Robert Walpole, who was suspected of taking bribes and having a mistress. But Brecht's aim was more serious. He moved the action to Victorian times – nearer home – and made the show an attack on bourgeois respectability and its self-satisfied self-image. Here too the beggars masquerade as disabled, like the war cripples so vividly portrayed in George Grosz's paintings. Rehearsals were disastrous. Actresses walked out or suffered inexplicable illness. The stars objected to changes in the script and even to some of the moves they were directed to make. Songs about sex had to be removed because the actresses refused to sing them. And this was not the only way *Dreigroschenoper* resembled *Salomé*: rumours about the backstage dramas circulated in Berlin, together with the belief that the theatre owner was desperately searching for another show to stage as soon as Brecht's and Weill's had failed.[67]

The first night did not start well. For the first two songs the audience sat in unresponsive silence. There was a near-disaster when the barrel organ designed to accompany the first song refused to function and the actor was forced to sing the first stanza unaided (the orchestra rallied for the second verse). But the third song, the duet between Macheath and the Police Chief, Tiger Brown, reminiscing about their early days in India, was rapturously received.[68] The manager had specified that no encores would be sung that night, but the audience wouldn't let the show proceed without repeats and so he had to overrule himself. The opera's success was due in part to the fact that its avowed Marxism was muted. As Brecht's biographer Ronald Hayman put it, 'It was not wholly insulting to the bourgeoisie to expatiate on what it had in common with ruthless criminals; the arson and the throat-cutting are mentioned only casually and melodically, while the well-dressed entrepreneurs in the stalls could feel comfortably superior to the robber gang that aped the social pretensions of the *nouveaux-riches*.'[69] Another reason for the success was the fashion in Germany at the time for *Zeitoper*, opera with a contemporary relevance. Other examples in 1929–30 were Hindemith's *Neues von Tage* (Daily News), a story of newspaper rivalry; *Jonny spielt auf*, by Ernst Kreutz; Max Brandt's *Maschinist Hopkins*; and Schoenberg's *Von Heute auf Morgen*.[70] Brecht and Weill repeated their success with the *Rise and Fall of the City of Mahagonny* – like *The Threepenny Opera*, a parable of modern society. As Weill put it, 'Mahagonny, like Sodom and Gomorrah, falls on account of the crimes, the licentiousness and the general confusion of its inhabitants.'[71] Musically, the opera was popular because the bitter, commercialised sounds of jazz symbolised not the freedom of Africa or America but the corruption of capitalism. The idea of degeneration wasn't far away, either. Brecht's version of Marxism had convinced him that works of art were conditioned, like everything else, by the commercial network of theatres, newspapers, advertisers, and so on. *Mahagonny*, therefore, was designed so that 'some irrationality, unreality and frivolity should be introduced in the right places to assert a double meaning.'[72] It was also epic theatre, which for Brecht was central: 'The premise for dramatic theatre was that human nature could not be changed; epic theatre assumed not only that it could but that it was already changing.'[73]

Change there certainly was. Before the show opened, the Nazis demonstrated outside the theatre. The first night was disrupted by whistles from the balcony, then by fistfights in the aisles, with a riot soon spreading to the stage. For the second night police lined the walls, and the house lights were left on.[74] The Nazis took more and more interest in Brecht, but when he sued the film producer who had bought the rights to *Die Dreigroschenoper* because the producer wanted to make changes against the spirit of the contract, the Brownshirts had a dilemma: How could they take sides between a Marxist and a Jew? The brownshirts would not always be so impotent. In October 1929, when Weill attended one of their rallies out of mere curiosity, he was appalled to hear himself denounced 'as a danger to the country,' together with Albert Einstein and Thomas Mann. He left hurriedly, unrecognised.[75]

One man who hated Berlin – he called it Babylon – who hated all cities, who in fact elevated his hatred of city life to an entire philosophy, was **Martin Heidegger**. Born in southern Germany in 1889, he studied under Edmund Husserl before becoming himself a professional teacher of philosophy.[76] His deliberate provincialism, his traditional mode of dress – knickerbockers – and his hatred of city life all confirmed his philosophy for his impressionable students. In 1927, at the age of thirty-eight, he published his most important book, *Being and Time*. Despite the fame of Jean-Paul Sartre in the 1930s, 1940s and 1950s, Heidegger was – besides being earlier – a more profound existentialist.

Being and Time is an impenetrable book, 'barely decipherable,' in the words of one critic. Yet it became immensely popular.[77] For Heidegger the central fact of life is man's existence in the world, and we can only confront this central fact by *describing* it as exactly as possible. Western science and philosophy have all developed in the last three or four centuries so that 'the primary business of Western man has been the conquest of nature.' As a result, man regards nature as though he is the subject and nature the object. Philosophically, the nature of knowledge is the central dilemma: 'What do we know? How can we know that we know?' Ever since Descartes these questions have been paramount. For Heidegger, however, reason and intellect are 'hopelessly inadequate guides to the secret of being.' Indeed, at one point he went so far as to say that 'thinking is the mortal enemy of understanding.'[78] Heidegger believed that we are thrust into the world willy-nilly, and by the time we have got used to being here, we are facing death. Death, for Heidegger, is the second central fact of life, after being.[79] We can never experience our own death, he said, but we can fear it, and that fear is all-important: it gives meaning to our being. We must spend our time on earth creating ourselves, 'moving into an open, uncertain, as yet uncreated future.' One other element of Heidegger's thought is essential to understanding him. Heidegger saw science and technology as an expression of the will, a reflection of our determination to control nature. He thought, however, that there was a different side to man's nature, which is revealed above all in poetry. The central aspect of a poem, said Heidegger, was that 'it eludes the demands of our will'. 'The poet cannot will to write a poem, it just comes.'[80] This links him directly with Rilke. Furthermore, the same argument applies to readers: they must allow the poem to work its magic on them. This is a central factor in Heidegger's ideas – the split between the will and those aspects of life, the interior life, that are beyond, outside, the will, where the appropriate way to understanding is not so much thinking as submission. At one level this sounds a little bit like Eastern philosophies. And Heidegger certainly believed that the Western approach needed sceptical scrutiny, that science was becoming intent on mastery rather than understanding.[81] He argued, as the philosopher William Barrett has said, summing up Heidegger, that there may come a time 'when we should stop asserting ourselves and just submit, let be.' Heidegger quoted Friedrich Hölderlin: We are in the period of darkness between the gods that have vanished and the god that has not yet come, between Matthew Arnold's two worlds, 'one dead, the other powerless to be born.'[82]

This is, inevitably perhaps, a rather bloodless summary of Heidegger's thinking. What made it so immediately popular was that it gave respectability to the German obsession with death and unreason, with the rejection of urban rationalist civilisation, with, in effect, a hatred of Weimar itself. Moreover, it gave tacit approval to those *völkisch* movements then being spawned that appealed not to reason but to heroes, that called for submission in the service of an alternative will to science, to those who, in Peter Gay's striking phrase, 'thought with their blood.' Heidegger did not create the Nazis, or even the mood that led to the Nazis. But as the German theologian Paul Tillich, who was himself dismissed from his chair, was to write later, 'It is not without *some* justification that the names of Nietzsche and Heidegger are connected with the anti-moral movements of fascism and national socialism.' *Being and Time* was dedicated to Edmund Husserl, Heidegger's mentor, who was Jewish. When the book was reprinted during the Nazi era, the dedication was omitted.[83]

We last left **George Lukács** in chapter 10, in Vienna, in exile from Budapest, 'active in hopeless conspiratorial [Communist] Party work, tracking down people who have absconded with party funds.'[84] Throughout the 1920s Lukács's life remained difficult. In the early years he vied with Béla Kun for leadership of the Hungarian Party in exile – Kun had fled to Moscow. Lukács met Lenin in Moscow and Mann in Vienna, making enough of an impact on the latter for him to model the Communist Jesuit Naphta in *The Magic Mountain* partly on Lukács.[85] Most of the time, however, he lived in poverty, and in 1929 he stayed illegally in Hungary before going to Berlin and on to Moscow. He worked there at the Marx-Engels Institute, where Nikolai Ryazanov was editing the newly discovered manuscripts of the young Marx.[86]

Despite these difficulties, Lukács published in 1923 *History and Class Consciousness*, for which he was to become famous.[87] These nine essays were about both literature and politics. So far as literature was concerned, Lukács's theory was that, beginning with Miguel de Cervantes' *Don Quixote*, novelists have fallen predominantly into two groups, those who portray 'the incommensurability between self (or hero) and environment (or society),' as Cervantes, Friedrich von Schiller, and Honoré de Balzac did, as 'world fleeing,' or as in Gustave Flaubert, Ivan Sergeyevich Turgenev, or Lev Nikolayevich Tolstoy, in 'the romanticism of disillusionment,' involved in life but aware that man cannot be improved, as Joseph Conrad had said.[88] In other words, both approaches were essentially antipositive, antiprogressive. Lukács moved from literature to politics to argue that the different classes have different forms of consciousness. The bourgeoisie, while glorifying individualism and competition, respond in literature, and in life, to a stance that assumes that society is 'bound by immutable laws, as dehumanised as the natural laws of physics.'[89] In contrast, the proletariat seeks a new order of society, which acknowledges that human nature *can* change, that there can be a new synthesis between self and society. Lukács saw it as his role to explain this dichotomy to the bourgeoisie so they would understand the revolution, when it came. He thought the popularity of film lay in the fact that in movies things lost presence, and that people liked the illusion, to live 'without fate, without causes, without motives.'[90] He also argued that while Marxism explained these different class consciousnesses, after the revolution, with the new synthesis of self and society that he posited, Marxism would be superseded. He came to the conclusion, therefore, that 'communism should not be reified by its own builders.'[91]

Lukács was roundly condemned and ostracised for being a revisionist and anti-Leninist. He never really recovered, never counterattacked, and eventually admitted his 'error.' However, his analysis of Marxism, class-consciousness, and literature found an echo in Walter Benjamin's work in the 1930s, and was revived in modified form after World War II by Raymond Williams and others in the doctrine of cultural materialism (see chapters 26 and 40).

In 1924, the year after *History and Class Consciousness* was published, a group of philosophers and scientists in Vienna began to meet every Thursday. Originally organised as the Ernst Mach Society, in 1928 they changed their name to the *Wiener Kreis*, **the Vienna Circle**. Under this title they became what is arguably the most important philosophical movement of the century (and one, incidentally, directly opposed to Heidegger).

The guiding spirit of the circle was **Moritz Schlick** (1882–1936), Berlin-born who, like many members of the *Kreis*, had trained as a scientist, in his case as a physicist under Max Planck, from 1900–4. The twenty-odd members of the circle that Schlick put together included **Otto Neurath** from Vienna, a remarkable Jewish polymath; **Rudolf Carnap**, a mathematician who had been a pupil of Gottlob Frege at Jena; Philipp Frank, another physicist; Heinz Hartmann, a psychoanalyst; **Kurt Gödel**, a mathematician; and at times **Karl Popper**, who became an influential philosopher after World War II. Schlick's original label for the kind of philosophy that evolved in Vienna in the 1920s was *konsequenter Empirismus*, or consistent empiricism. However, after he visited America in 1929 and again in 1931–2, the term **logical positivism** emerged – and stuck.

The logical positivists made a spirited attack on metaphysics, against any suggestion that 'there might be a world beyond the ordinary world of science and common sense, the world revealed to us by our senses.'[92] For the logical positivists, any statement that wasn't empirically testable – verifiable – or a statement in logic or mathematics was nonsensical. And so vast areas of theology, aesthetics, and politics were dismissed. There was more to it than this, of course. As the British philosopher A. J. Ayer, himself an observer of the circle for a short time, described it, they were also against 'what we might call the German past,' the romantic and to them rather woolly thinking of Hegel and Nietzsche (though not Marx).[93] The American philosopher Sidney Hook, who travelled in Germany at the time, confirmed the split, that the more traditional German philosophers were hostile to science and saw it as their duty 'to advance the cause of religion, morality, freedom of the will, the *Volk* and the organic nation state.'[94] The aim of the Vienna Circle was to clarify and simplify philosophy, using techniques of logic and science. Under them, philosophy became the handmaiden of science and a 'second-order subject.' First-order subjects talk about the world (like physics and biology); second-order subjects talk about their talk about the world.[95] Wittgenstein's *Tractatus* was one of the main influences on the Vienna Circle, and he too had been interested in the role of language in experience, and was very critical of traditional metaphysics. In this way, as the Oxford philosopher Gilbert Ryle said, philosophy came to be regarded as 'talk about talk.'[96]

Neurath was perhaps the most talented of the circle. Though he trained as a mathematician, he also studied with Max Weber and wrote a book called *Anti Spengler* (1921). He was close to the Bauhaus people and developed a system of two thousand symbols (called isotypes) designed to help educate the illiterate (he would sign his own letters with an isotype of an elephant, happy or sad, as the case might be).[97] But this huge ebullient character was intensely serious and agreed with Wittgenstein that one should remain silent regarding metaphysics, because it is nonsense, while recognising 'that one is being silent about something that does not exist.'[98]

The self-conscious organisation of the Vienna Circle, and their enthusiasm for their new approach, was also a factor in their influence. It was as if they suddenly knew what philosophy was. Science describes the world, the only world there is, the world of things around us. All philosophy can do, therefore, is analyse and criticise the concepts and theories of science, so as to refine them, make them more accurate and useful. This is why the legacy of logical positivism is known as analytic philosophy.

In the same year that Moritz Schlick started the Vienna Circle, 1924, the year that *The Magic Mountain* appeared, **Robert Musil** began work in Vienna on his masterpiece, *The Man without Qualities*. If he had never written a book, Musil would still be worth remembering for describing Hitler in 1930 as 'the living unknown soldier.'[99] But his three-volume work, the first volume of which was published in the same year, is for some people the most important novel in German written this century, eclipsing anything Mann wrote. Rated by many as on a par with Joyce and Proust, it is still far less well known than *Ulysses, A la recherche du temps perdu,* or *The Magic Mountain.*

Born in Klagenfurt in 1880, Musil came from an upper-middle-class family, part of the Austrian 'mandinarate.' He trained in science and engineering and wrote a thesis on Ernst Mach. *The Man without Qualities* is set in 1913 in the mythical country of 'Kakania.' Kakania is clearly Austro-Hungary, the name referring to *Kaiserlich und Königlich,* or K.u.K, standing for the royal kingdom of Hungary and the imperial-royal domain of the Austrian crown lands.[100] The book, though daunting in length, is for many the most brilliant literary response to developments in other fields in the early twentieth century, one of a handful of creations that is incapable of over-interpretation. It is: post-Bergson, post-Einstein, post-Rutherford, post-Bohr, post-Freud, post-Husserl, post-Picasso, post-Proust, post-Gide, post-Joyce and above all post-Wittgenstein.

There are three intertwined themes which provide a loose narrative. First, there is the search by the main character, Ulrich von . . ., a Viennese intellectual in his early thirties, whose attempt to penetrate the meaning of modern life involves him in a project to understand the mind of a murderer. Second, there is Ulrich's relationship (and love affair) with his sister, who he had lost contact with in childhood. Third, the book is a social satire on Vienna on the eve of World War I.[101]

But the real theme of the book is what it means to be human in a scientific age. If all we can believe are our senses, if we can know ourselves only as scientists know us, if all generalisations and talk about value, ethics and aesthetics are meaningless, as Wittgenstein tells us, how are we to live? asks Musil. He accepts that the old categories in which men thought – the 'halfway house' ideas of racialism, or religion – are of no use any more, but with what are we to replace them? Ulrich's attempts to understand the mind of the murderer, Moosbrugger, recall Gide's arguments that some things are inexplicable. (Musil studied under the psychologist Carl Stumpf, as did Husserl, and so was not especially in thrall to Freud, believing that although there *was* an unconscious it was an unorganised 'Proustian' jumble of forgotten memories. He also researched his book in a scientific way, studying a real murderer in jail in Vienna.) At one point Ulrich notes that he is tall, with broad shoulders, that 'his chest cavity bulged like a spreading sail on a mast' but that on occasions he felt small and soft, like 'a jelly-fish floating in the water' when he read a book that moved him. In other words, no one description, no one characteristic or quality, fitted him. It is in this sense that he is a man without qualities: 'We no longer have any inner voices. We know too much these days; reason tyrannises our lives.'

Musil had hardly finished his massive work when he died, nearly destitute, in 1942, and the time it took for completion reflected his view that, in the wake of other developments, the novel had to change in the twentieth century. He thought that the traditional novel, as a way of telling stories, was dead. Instead, for him the modern novel was the natural home of metaphysics. Novels – his novel anyway – were a kind of thought experiment, on a par with Einstein's, or Picasso's, where a figure might be seen in profile and in full face at the same time. The two intertwined principles underlying experience, he believed, were violence and love, which is what links him to Joyce: science may be able to explain sex – but love? And love can be so exhausting that getting through today is all we can manage. Thinking about tomorrow – philosophy – is incommensurate with that. Musil wasn't anti-science, as so many others were. (Ulrich 'loved mathematics because of the kind of people who could not endure it.') But he thought novelists could help discover where science might lead us. For him the fundamental question was whether the soul could ever be replaced by logic. The search for objectivity and the search for meaning are irreconcilable.

Franz Kafka was also obsessed by what it means to be human, and by the battle between science and ethics. In 1923, when he was thirty-nine, he realised a long-cherished ambition to move from Prague to Berlin (he was educated in the German language and spoke it at home). But he was in Berlin less than a year before the tuberculosis in his throat forced him to transfer to a sanatorium near Vienna, where he died. He was forty-one.

Few details of Kafka's private life suggest how he came by his extraordinarily strange imagination. A slim, well-dressed man with a hint of the dandy about him, he had trained in law and worked in insurance successfully. The only clue to his inner unconventionality lay in the fact that he had three unsuccessful engagements, two of them to the same woman.[102] Just as Freud was ambivalent about Vienna, so Kafka felt much the same about Prague. 'This little mother has claws' is how he once described the city, and he was always intending to leave, but could never quite give up his well-paid job in insurance, not until 1922, when it was too late.[103] He often clashed with his father, and this may have had an effect on his writings, but as with all great art, the link between Kafka's books and his life is far from straightforward.

Kafka is best known for three works of fiction, *Metamorphosis* (1916), *The Trial* (1925; posthumous), and *The Castle* (1926; posthumous). But he also kept a diary for fourteen years and wrote copious letters. These reveal him to have been a deeply paradoxical and enigmatic man. He often claimed that his primary aim was independence, yet he lived in his parents' home until he left for Berlin; he was engaged to the same woman for five years, yet saw her fewer than a dozen times in that period; and he amused himself by imagining the most gruesome way he could die. He lived for writing and would work for months, collapsing in exhaustion afterward. Even so, he might jettison what he had done if he felt it was unworthy. He had relatively few correspondents, yet wrote to them often – very often, and very long letters. He wrote 90 letters to one woman in the two months after he met her, including several of between twenty and thirty pages, and to another he wrote 130 letters in five months. He wrote a famous forty-five-page typed letter to his father when he was thirty-six, explaining why he was still afraid of him, and another long letter to a prospective father-in-law, whom he had met only once, explaining that he was probably impotent.[104]

Although Kafka's novels are ostensibly about very different subjects, they have some striking similarities, so much so that the cumulative effect of Kafka's work is much more than the sum of its parts. *Metamorphosis* begins with one of the most famous opening lines in literature: 'As Gregor Sams awoke one morning from uneasy dreams he found himself transformed in his bed into a gigantic insect.' This might seem as if the plot had been given away right there and then, but in fact the book explores Gregor's response to his fantastic condition, and his relationship with his family and with his colleagues at work. If a man is turned into an insect, does this help him/us understand what it means to be human? In *The Trial*, Joseph K. (we never know his last name) is arrested and put on trial.[105] But neither he nor the reader ever knows the nature of his offence, or by what authority the court is constituted, and therefore he and we cannot know if the death sentence is warranted. Finally, in *The Castle* K. (again, that is all we are told) arrives in a village to take up an appointment as land surveyor at the castle that towers above the village and whose owner also owns all the houses there. However, K. finds that the castle authorities deny all knowledge of him, at least to begin with, and say he cannot even stay at the inn in the village. There then follows an extraordinary chain of events in which characters contradict themselves, vary unpredictably in their moods and attitudes to K., age virtually overnight, or lie – even K. himself is reduced to lying on occasions. Emissaries from the castle arrive in the village, but he never sees any sign of life in the castle itself, and never reaches it.[106]

An added difficulty with interpreting Kafka's work is that he never completed any of his three major novels, though we know from his notebooks what he intended at the time of his death. He also told his friend Max Brod what he planned for *The Castle*, his most realised work. Some critics argue that each of his ideas is an exploration of the inner workings of the mind of a mentally unstable individual, particularly *The Trial*, which on this reading becomes a sort of imaginative case history of someone with a persecution complex. In fact, one needn't go this far. All three stories show a man not in control of himself, or of his life. In each case he is swept along, caught up in forces where he cannot impose his will, where those forces – biological, psychological, logical – lead blindly. There is no development, no progress, as conventionally understood, and no optimism. The protagonist doesn't always win; in fact, he always loses. There are forces in Kafka's work, but no authority. It is bleak and chilling. Jewish, and Czech, an outsider at Weimar, Kafka nevertheless saw where that society was headed. There are similarities between Kafka and Heidegger in that Kafka's characters must submit to greater forces, forces they don't truly understand. He once said, 'I sometimes believe I understand the Fall of Man as no one else.'[107] Kafka parts company with Heidegger, however, in saying that not even submission brings satisfaction; indeed, satisfaction, or fulfilment, may not be possible in the modern world. This is what makes *The Castle* Kafka's masterpiece, for many people a latter-day *Divine Comedy*. W. H. Auden once said, 'Had one to name the author who comes nearest to bearing the same kind of relation to our age as Dante, Shakespeare or Goethe have to theirs, Kafka is the first one would think of.'[108]

In *The Castle* life in the village is dominated by the eponymous building. Its authority is unquestioned but also unexplained. The capriciousness of its bureaucracy is likewise unquestioned, but all attempts by K. to understand that capriciousness are nullified. Though obviously and perhaps too heavily allegorical of modern societies, with their faceless bureaucratic masses, verging on terror, their impersonality, marked by a pervading feeling of invasion (by science and machines) and of dehumanisation, Kafka's works both reflect and prophesy a world that was becoming a reality. *The Castle* was the culmination of Kafka's work, at least in the sense that the reader tries to understand the book as K. tries to understand the castle. In all his books, however, Kafka succeeds in *showing* the reader terror and the uncomfortable, alienated, disjunctive feelings that so characterise modern life. Eerily, he also prefigured the specific worlds that were soon to arrive: Stalin's Russia and Hitler's Germany.

In 1924, the year that tuberculosis killed Kafka, **Adolf Hitler** celebrated his thirty-fifth birthday – in prison. He was in Landsberg jail, west of the Bavarian capital, serving a five-year sentence for treason and his part in the Munich putsch. There were several other National Socialists in prison with him, and as well as being given minimum sentences, they had an easy time inside. There was plenty of good food, they were allowed out into the garden, Hitler was a favourite with the jailers, and on his birthday he received numerous parcels and bunches of flowers. He was putting on weight.[109]

The trial had been front-page news in every German newspaper for more than three weeks, and for the first time Hitler broke through to a national audience. Later, he was to claim that the trial and the publicity surrounding it were a turning point in his career. It was during his time in prison that Hitler wrote the first part of *Mein Kampf*. It is entirely possible that he might never have written anything had he not been sent to Landsberg. At the same time, as Alan Bullock has pointed out, the opportunity was invaluable. *Mein Kampf* helped Hitler establish himself as the leader of the National Socialists, helped him lay the foundation of the Hitler myth, and helped him clarify his ideas. Hitler instinctively grasped that a movement such as the one he planned needed a 'sacred text,' a bible.[110]

Whatever his other attributes, Hitler certainly thought of himself as a thinker, with a grasp of technical-military matters, of natural science, and above all of history. He was convinced that this grasp set him apart from other men, and in this he was not entirely wrong. We need to remember that he started adult life as an artist and an aspiring architect. He was transformed into the figure he became first by World War I and the ensuing peace, but also by the education he gave himself. Perhaps the most important thing to grasp about Hitler's intellectual development is that it was so far removed from that of most if not all the people we have been considering in this chapter. As even a cursory examination of *Mein Kampf* will show, this is because most of Hitler's ideas were nineteenth-century or turn-of-the-century ideas – the kind examined here in chapters 2 and 3 – and once they were formed, Hitler never changed them. The Führer's ideas, as revealed in his table talk during World War II, are directly traceable to his thinking as a young man.[111]

The historian George L. Mosse has disinterred the more distant intellectual origins of the Third Reich, on which this section is chiefly based.[112] He shows how an amalgam of *völkisch* mysticism and spirituality grew up in Germany in the nineteenth century, in part a response to the romantic movement and to the bewildering pace of industrialisation, and was also an aspect of German unification. While the *Volk* were coming together, forging one heroic Pan-German nation, the 'rootless Jew' was a convenient, negative comparison (though of course this was not at all fair: in Germany Jews could not be government officials or full professors until 1918). Mosse traces the influence of thinkers and writers, many completely forgotten now, who helped create this cast of mind – people like Paul Lagarde and Julius Langbehn, who stressed '**German intuition**' as a new creative force in the world, and Eugen Diederichs, who openly advocated 'a culturally grounded nation guided by the initiated elite,' by the revival of German legends, such as the Edda, which stressed

Germany's great antiquity and its links to Greece and Rome (great civilisations but also pagan). The point about all this was that it elevated the **Volk** almost to the level of a deity.[113] There were nineteenth-century German books such as that by Ludwig Woltmann, examining the art of the Renaissance, identifying 'Aryans' in positions of power and showing how much the Nordic type was admired.[114] Mosse also emphasises how social Darwinism threaded through society. In 1900, for example, Alfred Krupp, the wealthy industrialist and arms manufacturer, sponsored a public essay competition on the topic, 'What can we learn from the principles of Darwinism for application to inner political development and the laws of the state?'[115] Not surprisingly, the winner advocated that *all* aspects of the state, without exception, should be viewed and administered in social Darwinian terms. Mosse further describes the many German attempts at utopias – from 'Aryan' colonies in Paraguay and Mexico to nudist camps in Bavaria, which tried to put *völkisch* principles into effect. The craze for physical culture grew out of these utopias, and so too did the movement for rural boarding schools with a curriculum based on 'back to nature' and *Heimatkunde*, rendered as 'lore of the homeland,' emphasising Germanness, nature, and ancient peasant customs. As a boy, Hitler grew up in this milieu without realising that there was any alternative.[116]

In fact, Hitler never made any secret of this. Linz, where he was raised, was a semirural, middle-class town populated by German nationalists. The town authorities turned a blind eye to the gatherings of the banned '**Gothia**' or '**Wodan**' societies, with their Pan-German tendencies.[117] As a boy, Hitler belonged to these groups, but he also witnessed the intolerant nationalism of the town's adults, whose anti-Czech feelings boiled over so easily that they even took against the eminent violinist Jan Kubelik, who was scheduled to perform in Linz. These memories, all too evident in *Mein Kampf*, helped account for Hitler's attacks on the Habsburgs for the 'Slavisation' of the Austrians. In his book Hitler also insists that while at school in Linz he 'learned to understand and grasp the meaning of history.' 'To "Learn" history,' he explained, 'means to seek and find the forces which are the causes leading to those effects which we subsequently perceive as historical events.'[118] One of these forces, he felt (and this too he had picked up as a boy), was that Britain, France, and Russia were intent on encircling Germany, and he thereafter never rid himself of this view. Perhaps not surprisingly, for him history was invariably the work of great men – his heroes were Charlemagne, Rudolf von Habsburg, Frederick the Great, Peter the Great, Napoleon, Bismarck, and Wilhelm I. Hitler therefore was much more in the mould of Stefan George, or Rainer Maria Rilke, rather than Marx or Engels, for whom the history of class struggle was paramount. For Hitler, history was a catalogue of racial struggles, although the outcome always depended on great men: '[History] was the sum total of struggle and war, waged by each against all with no room for either mercy or humanity.'[119] He often quoted Helmut von Moltke, a nineteenth-century German general, who had argued that one should always use the most terrible weapons and tactics available because, by shortening hostilities, lives would be saved.

Hitler's biological thinking was an amalgam of Thomas R. Malthus, Charles Darwin, Joseph Arthur Gobineau, and William McDougall: 'Man has become great through struggle. ... Whatever goal man has reached is due to his originality plus his brutality. ... All life is bound up in three theses: struggle is the father of all things, virtue lies in blood, leadership is primary and decisive. ... He who wants to live must fight, and he who does not want to fight in this world where eternal struggle is the law of life has no right to exist.'[120] Malthus had argued that the world's population was outstripping the earth's capacity to provide for it. The result must be famine and war. Birth control and much-improved agriculture offered the only hope for Malthus, but for Hitler there was another answer: 'a predatory war of annihilation as a means to an end, an historically all-important act in response to natural law and necessity.' According to Werner Maser, one of Hitler's biographers, his brutal attitude to 'weaklings' was transplanted from the teachings of Alfred Ploetz, whose book, *Die Tüchtigkeit unserer Rasse und der Schutz der Schwachen* (The Efficiency of our Race and the Protection of the Weak), Hitler had read as a young man in Vienna before World War I. The following extract from Ploetz will show how his thinking had 'advanced' since the nineteenth century: 'Advocates of racial hygiene [the new phrase for eugenics] will have little objection to war since they see in it one of the means whereby the nations carry on their struggle for existence. ... In the course of the campaign it might be deemed advisable deliberately to muster inferior variants at points where the main need is for cannon fodder and where the individual's efficiency is of secondary importance.'[121]

Hitler's biologism was intimately linked to his understanding of history. He knew very little about prehistory but certainly regarded himself as something of a classicist. He was fond of saying that his 'natural home' was ancient Greece or Rome, and he had more than a passing acquaintance with Plato. Partly because of this, he considered the races of the East (the old 'Barbarians') as inferior. 'Retrogression' was a favourite idea of Hitler's, something he applied to the 'Habsburg brood,' who ruled in Vienna but for him were doomed to degeneracy. Similarly, organised religion, Catholicism in particular, was also doomed, owing to its antiscientific stance and its unfortunate interest in the poor ('weaklings'). For Hitler mankind was divided into three – creators of culture, bearers of culture, and destroyers of culture – and only the 'Aryans' were capable of creating culture.[122] The decline of culture was always due to the same reason: miscegenation. The Germanic tribes had replaced decadent cultures before – in ancient Rome – and could do so again with the decadent West. Here again, the influence of Linz can be detected. For one thing, it helps explain Hitler's affinity for Hegel. Hegel had argued that Europe was central in history, that Russia and the United States were peripheral. Landlocked Linz reinforced this view. 'Throughout his life Hitler remained an inland-orientated German, his imagination untouched by the sea. ... He was completely rooted within the cultural boundaries of the old Roman Empire.'[123] This attitude may just have been crucial, leading Hitler to fatally underestimate the resolve of that periphery – Britain, the United States, and Russia.

If Linz kept Hitler's thinking in the nineteenth century, Vienna taught him to hate. Werner Maser says, interestingly, that 'Hitler perhaps hated better than he loved.'[124] It was the Vienna Academy that twice rejected him and his efforts to become an art student and an architect. And it was in Vienna that Hitler first encountered widespread anti-Semitism. In *Mein Kampf* he argued that he did not come across many Jews or any anti-Semitism until he reached Vienna, and that anti-Semitism had a rational basis, 'the triumph of reason over feeling.' This is flatly contradicted by **August Kubizek**, Hitler's friend from his Vienna years (*Mein Kampf* is now known to be wrong on several biographical details). According to Kubizek, Adolf's father was not a broadminded cosmopolitan, as he is portrayed, but an out-and-out anti-Semite and a follower of Georg Ritter von Schönerer, the rabid nationalist we met in chapter 3. Kubizek also says that in 1904, when they first met and Hitler was fifteen and still at school, he was already 'distinctly anti-Semitic.'[125] Research has confirmed that there were fifteen Jews at Hitler's school, not one, as he says in *Mein Kampf*.

Whether or not Kubizek or Hitler is right about the anti-Semitism in Linz, Vienna, as we have seen, was a sump of vicious anti-Jewish feeling. For a start, Hitler early on encountered a series of pamphlets entitled *Ostara*, a periodical that was often stamped with a swastika on its cover.[126] Founded in 1905 by a wild racist who called himself George Lanz von Liebenfels, this journal at one point claimed a circulation of 100,000 copies. Its editorials revealed its stance openly: 'The *Ostara* is the first and only periodical devoted to investigating and cultivating heroic racial characteristics and the law of man in such a way that, by actually applying the discoveries of ethnology, we may through systematic eugenics ... preserve the heroic and noble race from destruction by socialist and feminist revolutionaries.' Lanz von Liebenfels was also the founder of the 'Order of the New Temple,' whose membership 'was restricted to fair-haired, blue-eyed men, all of whom were pledged to marry fair-haired, blue-eyed women.' Between 1928 and 1930 *Ostara* reprinted Liebenfels's 1908 tome *Theozoology; or, the Science of Sodom's Apelings and the Divine Electron: An Introduction to the Earliest and Most Recent World View and a Vindication of Royalty and the Nobility.* 'Sodom's apelings' was the appealing label given to dark-skinned 'inferior races,' whom Liebenfels regarded as 'God's bungled handiwork.'[127] But Hitler's anti-Semitism was also fanned by Georg Ritter von Schönerer, who in turn owed a debt to the German translation of Gobineau's *Essai sur l'inégalité des races humaines.* At the 1919 meeting of the Pan-German League, one of the League's specific aims was identified as combating 'the disruptive, subversive influence of the Jews – a racial question which has nothing to do with questions of religion.' As Werner Maser remarks, 'This manifesto thus marked the launch of biological antisemitism.'[128] Certainly, by the time Hitler came to write *Mein Kampf*, more than five years later, he referred to Jews as 'parasites,' 'bacilli,' 'germ-carriers,' and 'fungus.' From then on, from a National Socialist point of view, Jews were deprived of all human attributes.

It is doubtful that Hitler was as well read as his admirers claimed, but he did know some architecture, art, military history, general history, and technology, and also felt at home in music, biology, medicine, and the history of civilisation and religion.[129] He was often able to surprise his listeners with his detailed knowledge in a variety of fields. His doctor, for example, was once astonished to discover that the Führer fully grasped the effects of nicotine on the coronary vessels.[130] But Hitler was largely self-taught, which had significant consequences. He never had a teacher able to give him a systematic or comprehensive grounding in any field. He was never given any objective, outside viewpoint that might have had an effect on his judgement or on how he weighed evidence. Second, World War I, which began when Hitler was twenty-five, acted as a brake (and a break) in his education. Hitler's thoughts stopped developing in 1914; thereafter, he was by and large confined to the halfway house of ideas in Pan-Germany described in chapters 2 and 3. Hitler's achievement showed what could be wrought by a mixture of Rilke's mysticism, Heidegger's metaphysics, Werner Sombart's notion of heroes versus traders, and that hybrid cocktail of social Darwinism, Nietzschean pessimism, and the visceral anti-Semitism that has become all too familiar. It was a mix that could flourish only in a largely landlocked country obsessed with heroes. Traders, especially in maritime nations, or America, whose business was business, learned too much respect for other peoples in the very act of trading. It would be entirely fitting, though not often enough stressed, that Hitler's brand of thought was so comprehensively defeated by Western rationalism, so much the work of Jews.

We must be careful, however, not to pitch Hitler's thought too high. For a start, as Maser highlights, much of his later reading was done merely to confirm the views he already held. Second, in order to preserve a consistency in his position, he was required to do severe violence to the facts. For example, Hitler several times argued that Germany had abandoned its expansion toward the East 'six hundred years ago.' This had to do with his explanation of Germany's failure in the past, and its future needs. Yet both the Habsburgs and the Hohenzollerns had had a well established *Ostpolitik* – Poland, for instance, being partitioned three times. Above all there was Hitler's skill at drawing up his own version of history, convincing himself and others that he was right and academic opinion usually wrong. For example, whereas most scholars believed that Napoleon's downfall was the result of his Russian campaign, Hitler attributed it to his Corsican 'sense of family' and his 'want of taste' in accepting the imperial crown, which meant that he made 'common cause with degenerates.'[131]

In political terms, Hitler's accomplishments embraced the Third Reich, the Nazi Party, and, if they can be called accomplishments, World War II and the Holocaust. In the context of this book, however, he represents the final convulsions of the old metaphysics. Weimar was a place of both 'unparalleled mental alertness' and the dregs of nineteenth-century *völkisch* romanticism, where people 'thought with their blood.' That the Weimar culture which Hitler hated should be exported virtually en bloc in years to come was entirely apropos. Hitler's intellectual failings shaped the second half of the century every bit as much as did his military megalomania.

THE EVOLUTION OF EVOLUTION

Perhaps the greatest intellectual casualty of World War I was the idea of progress. Prior to 1914 there had been no major conflict for a hundred years, life expectancy in the West had increased dramatically, many diseases and child mortality had been conquered, Christianity had spread to vast areas of Africa and Asia. Not everyone agreed this was progress – Joseph Conrad had drawn attention to racism and imperialism, and Emile Zola to squalor. But for most people the nineteenth century had been an era of moral, material, and social progress. World War I overturned that at a stroke.

Or did it? Progress is a notoriously elusive concept. It is one thing to say that mankind has made no moral progress, that our capacity for cruelty and injustice has grown in parallel with our technological advances; but that there has been technological progress, few would doubt. As the war was ending, **J. B. Bury**, Regius Professor of Modern History at Cambridge, embarked on an inquiry into the idea of progress, to see how it had developed, how best it should be understood, and what lessons could be learned. *The Idea of Progress* was released in 1920, and it contained one very provocative – even subversive – thought.[1] Bury found that the idea of progress had itself progressed. In the first place, it was mainly a French idea, but until the French Revolution it had been pursued only on a casual basis. This was because in a predominantly religious society most people were concerned with their own salvation in a future life and because of this were (relatively speaking) less concerned with their lot in the current world. People had all sorts of ideas about the way the world was organised, for the most part intuitive. For example, Bernard de Fontenelle, the seventeenth-century French writer, did not believe any aesthetic progress was possible, arguing that literature had reached perfection with Cicero and Livy.[2] Marie Jean de Condorcet (1743–94), the French philosopher and mathematician, had argued that there had been ten periods of civilisation, whereas Auguste Comte (1798–1857) thought there had been three.[3] Jean-Jacques Rousseau (1712–78) had gone the other way, believing civilisation was actually a degenerate – i.e., retrogressive – process.[4] Bury unearthed two books published (in French) in the late eighteenth century, *The Year 2000* and *The Year 2440*, which predicted, among other things, that the perfect, progressive society would have no credit, only cash, and where historical and literary records of the past would have all been burned, history being regarded as 'the disgrace of humanity, every page ... crowded with crime and follies.'[5] Bury's second period ran from the French Revolution, 1789, to 1859, embracing the era of the first industrial revolution, which he found to be an almost wholly optimistic time when it was believed that science would transform society, easing poverty, reducing inequality, even doing God's work. Since 1859 and the publication of Darwin's *On the Origin of Species*, however, Bury thought that the very notion of progress had become more ambiguous: people were able to read both optimistic and pessimistic outcomes into the evolutionary algorithm.[6] He viewed the hardening of the idea of progress as the result of the decline of religious feeling, directing people's minds to the present world, not the next one; to scientific change, giving man greater control over nature, so that more change was possible; and to the growth of democracy, the formal political embodiment of the aim to promote freedom and equality. Sociology he saw as the science of progress, or the science designed to define it and measure the change.[7] He then added the thought that maybe the very idea of progress itself had something to do with the bloodiness of World War I. Progress implied that material and moral conditions would get better in the future, that there was such a thing as posterity, if sacrifices were made. Progress therefore became something worth dying for.[8]

The last chapter of Bury's book outlined how 'progress' had, in effect, evolved into the idea of evolution.[9] This was a pertinent philosophical change, as Bury realised, because evolution was nonteleological – had no political, or social, or religious significance. It theorised that there would be progress without specifying in what direction progress would take place. Moreover, the opposite – extinction – was always a possibility. In other words, the idea of progress was now mixed up with all the old concepts of social Darwinism, race theory, and degeneration.[10] It was a seductive idea, and one immediate practical consequence was that a whole range of disciplines – geology, zoology, botany, palaeontology, anthropology, linguistics – took on a historical dimension: all discoveries, whatever value they had in themselves, were henceforth analysed for the way they filled in our understanding of evolution – progress. In the 1920s in particular our understanding of the progress, evolution, of civilisation was pushed back much further.

T. S. Eliot, James Joyce and Adolf Hitler, so different in many ways, had one thing in common – a love of the classical world. In 1922, the very year that both Eliot and Joyce published their masterpieces and Hitler was asked to address the National Club in Berlin, which consisted mainly of army officers, senior civil servants, and captains of industry, an expedition was leaving London, bound for Egypt. Its aim was to search for the man who may have been the greatest king of all in classical times.

Before World War I there had been three elaborate excavations in the Valley of the Kings, about 300 miles south of Cairo. In each, the name **Tutankhamen** kept appearing: it was inscribed on a faïence cup, on some gold leaf, and on some clay seals.[11] Tutankhamen was therefore believed to have been an important personage, but most Egyptologists never imagined his remains would ever be found. Despite the fact that the Valley of the Kings had already been excavated so often, the British archaeologist **Howard Carter** and his sponsor, **Lord Carnarvon**, were determined to dig there. They had tried to do so for a number of years, and had been prevented by the war. But neither would give up. Carter, a slim man with dark eyes and a bushy moustache, was a meticulous scientist, patient and thorough, who had been excavating in the Middle East since 1899. After the Armistice, Carnarvon and he finally obtained a licence to excavate across the Nile from Karnak and Luxor.

Carter left London without Carnarvon. Nothing notable occurred until the morning of 4 November.[12] Then, as the sun began to bleach the surrounding slopes, one of his diggers scraped against a stone step cut into the rock. Excavated carefully, twelve steps were revealed, leading to a doorway that was sealed and plastered over.[13] 'This seemed too good to be true,' but, deciphering the seal, Carter was astonished to discover he had unearthed a royal necropolis. He was itching to break down the door, but as he rode his donkey back to camp that evening, having left guards at the site, he realised he must wait. Carnarvon was paying for the dig and should be there when any grand tomb was opened. Next day, Carter sent a telegram giving him the news and inviting him to come.[14]

Lord Carnarvon was a romantic figure – a great shot, a famous yachtsman who, at the age of twenty-three, had sailed around the world. He was also a passionate collector and the owner of the third automobile licensed in Britain. It was his love of speed that led, indirectly, to the Valley of the Kings. A car accident had permanently damaged his lungs, making England uncomfortable in wintertime. Exploring Egypt in search of a mild climate, he discovered archaeology.

Carnarvon arrived in Luxor on the twenty-third. Beyond the first door was a small chamber filled with rubble. When this was cleared away, they found a second door. A small hole was made, and everyone retreated, just in case there were any poisonous gases escaping. Then the hole was enlarged, and Carter shone the beam of his torch through the hole to explore the second chamber.

'Can you see anything?' Carnarvon was peremptory.

Carter didn't reply for a moment. When he did, his voice broke. 'Yes.' Another pause. 'Wonderful things.'[15]

He did not exaggerate. 'No archaeologist in history has ever seen by torchlight what Carter saw.'[16] When they finally entered the second chamber, the tomb was found to be packed with luxurious objects – a gilded throne, two golden couches, alabaster vases, exotic animal heads on the walls, and a golden snake.[17] Two royal statues faced each other, 'like sentinels,' wearing gold kilts and gold sandals on their feet. There were protective cobras on their heads, and they each held a mace in one hand, a staff in the other. As Carnarvon and Carter took in this amazing splendour, it dawned on them that there was something missing – there was no sarcophagus. Had it been stolen? It was only now that Carter realised there was a *third* door. Given what they had found already, the inner chamber promised to be even more spectacular. But Carter was a professional. Before the inner chamber could be opened up, he determined to make a proper archaeological study of the outer room, lest precious knowledge be lost. And so the antechamber, as it came to be called, was resealed (and of course heavily guarded) while Carter called in a number of experts from around the world to collaborate on an academic investigation. The inscriptions needed study, as did the seals, and even the remains of plants that had been found.[18]

The tomb was not reopened until 16 December. Inside were objects of the most astounding quality.[19] There was a wooden casket decorated with hunting scenes of a kind never yet seen in Egyptian art. There were three animal-sided couches that, Carter realised, had been seen illustrated in other excavations – in other words, this site was famous even in ancient Egypt.[20] And there were four chariots, completely covered in gold and so big that the axles had to be broken in two before they could be installed. No fewer than thirty-four heavy packing cases were filled with objects from the antechamber and put on a steam barge on the Nile, where they began the seven-day journey downriver to Cairo. Only when that had been done was the way clear to open the inner room. When Carter had cut a large enough hole, he shone his torch through it as before. 'He could see nothing but a shining wall. Shifting the flashlight this way and that, he was still unable to find its outer limits. Apparently it blocked off the whole entrance to the chamber beyond the door. Once more, Carter was seeing something never seen before, or since. He was looking at a wall of solid gold.' The door was dismantled, and it became clear that the gold wall was part of a shrine that occupied – all but filled – the third chamber. Measurements taken later would show that the shrine measured seventeen feet by eleven feet by nine feet high and was completely covered in gold except for inlaid panels of brilliant blue faïence, depicting magic symbols to protect the dead.[21] Carnarvon, Carter, and the workmen were speechless. To complete their astonishment, in the main shrine there was a room within a room. Inside the inner shrine was a third, and inside that a fourth.

Removing these layers took eighty-four days.[22] A special tackle had to be devised to lift the lid of the sarcophagus. And here the final drama was enacted. On the lid of the coffin was a golden effigy of the boy-ruler Tutankhamen: 'The gold glittered as brightly as if it had just come from the foundry.'[23] 'Never was there such a treasure as the king's head, his face made of gold, his brows and [eye]lids of lapis lazuli blue glass and the eyes contrasting in obsidian and aragonite.' Most moving of all were the remains of a small wreath of flowers, 'the last farewell offering of the widowed girl-queen to her husband.'[24] After all that, and perhaps inevitably, the body itself proved a disappointment. The boy-king had been so smothered in 'unguents and other oils' that, over the centuries, the chemicals had mixed to form a pitchy deposit and had invaded the swaddling clothes. Layers of jewels had been poured between the wrappings, which had reacted with the pitch, causing a spontaneous combustion that carbonised the remains and surrounding linen. Nonetheless, the age of the king could be fixed at nearer seventeen than eighteen.[25]

In life Tutankhamen was not an especially important pharaoh. But his treasures and sumptuous tomb stimulated public interest in archaeology as never before, more even than had the discoveries at Machu Picchu. The high drama of the excavation, however, concealed a mystery. If the ancient Egyptians buried a seventeen-year-old monarch with such style, what might they have done for older, more accomplished kings? If such tombs haven't been found – and they haven't – does this mean they have been lost to plunderers? And at what cost to knowledge? If they are still there, how might they change our understanding of the way civilisations evolve?

Much of the fascination in Middle Eastern archaeology, however, lay not in finding gold but in teasing out fact from myth. By the 1920s the biblical account of man's origins had been called into question time and again. While it was clear that some of the Bible was based on fact, it was no less obvious that the Scriptures were wildly inaccurate in many places. A natural area of investigation was the birth of writing, as the earliest record of the past. But here too there was a mystery.

The mystery arose from the complicated nature of cuneiform writing, a system of wedges cut in clay that existed in Mesopotamia, an area between the rivers of Tigris and Euphrates. Cuneiform was believed to have developed out of pictograph script, spreading in time throughout Mesopotamia. The problem arose from the fact that cuneiform was a mixture of pictographic, syllabic, and alphabetic scripts which could not have arisen, all by themselves, at one time and in one place. It followed that cuneiform must have evolved from an earlier entity – but what? And belonging to which people? Analysis of the language, the type of words that were common, the business transactions typically recorded, led philologists to the idea that cuneiform had not been invented by the Semitic Babylonians or Assyrians but by another people from the highlands to the east. This was pushing the 'evidence' further than it would go, but this theoretical group of ancestors had even been given a name. Because the earliest known rulers in the southern part of Mesopotamia had been called 'Kings of Sumer and Akkad,' they were called the **Sumerians**.[26]

It was against this background that a Frenchman, **Ernest de Sarzec**, excavated a mound at Telloh, near **Ur and Uruk**, north of modern Basra in Iraq, and found a statue of a hitherto unknown type.[27] This naturally sparked fresh interest in the 'Sumerians,' and other digs soon followed, carried out mainly by Americans and Germans. These unearthed among other things huge ziggurats, which confirmed that the ancient civilisation (then called Lagash) was sophisticated. The dating was provocative too: 'It seemed almost as if its beginnings coincided with the times described in Genesis. The Sumerians might well be the same people, it was thought, who populated the earth after the punitive deluge that wiped out all humankind but Noah and his kin.' These excavations revealed not only how early civilisations evolved but also how early man thought, which is why, in 1927, the British archaeologist **Leonard Woolley** began to dig in the biblical **Ur of Chaldea**, the alleged home of Abraham, founder of the Jews.

Woolley, born in 1880, was educated at Oxford. He was a friend and colleague of T. E. Lawrence ('Lawrence of Arabia'); together they excavated Carchemish, where the Euphrates flows from modern Turkey into Syria. In

World War I Woolley did intelligence work in Egypt but then spent two years as a prisoner of war in Turkey. He made three important discoveries at Ur: first, he found several royal tombs, including the grave of **Queen Shub-ad**, which contained almost as many gold and silver vessels as the tomb of Tutankhamen; second, he unearthed the so-called mosaic standard of Ur, which featured a cluster of chariots, showing that it was the Sumerians, at the end of the fourth millennium BC, who had introduced this device into warfare; and third, he discovered that the royal corpses in Ur were not alone.[28] Alongside the king and queen, in one chamber, lay a company of soldiers (copper helmets and spears were found next to their bones). In another chamber were the skeletons of nine ladies of the court, still wearing elaborate gold headdresses.[29] Not only were these very grisly practices, but more significant, *no text had ever hinted at this collective sacrifice.* Woolley therefore drew the conclusion that the sacrifice had taken place before writing had been invented to record such an event. In this way the sacrifices confirmed the Sumerians, at that stage, as the oldest civilisation in the world.

It was only after these astounding discoveries that Woolley reached the forty-feet level. And here he came upon nothing.[30] For more than eight feet there was just clay, completely free from shards and rubbish or artefacts of any kind. Now, for a deposit of clay eight feet thick to be laid down, a tremendous **flood** must at some time have inundated the land of Sumer. Was this, then, the deluge mentioned in the Bible?[31] Like all classical archaeologists, Woolley was familiar with the Middle Eastern legend of **Gilgamesh**, half-man, half-god, who endured many trials and adventures, including a massive flood ('the waters of death').[32] Were there other correspondences between the Sumerians and the early Bible? When he looked, Woolley found many of them. The most intriguing was the account in Genesis that between Adam and the Deluge there were ten 'mighty forefathers which were old.' The Sumerian literature also referred to their 'primal kings,' which were eight in number. Moreover, the Israelites boasted improbably long life spans. Adam, for example, who begot his first son at the age of 130, is said to have lived for 800 years. Woolley found that the life spans of the ancient Sumerians were supposed to have been even greater.[33] According to one account, the reigns of eight ancestral kings stretched over 241,200 years, an average of 30,400 years per king.[34] The central point was this: the more he looked, the more Woolley found that the Sumerians over-lapped with the early biblical account of Genesis, and that Sumer occupied a pivotal point in human development.[35] For example, they boasted the first schools and were the first to use gardens to provide shade. The first library was theirs, and they had the concept of the 'Resurrection' long before the Bible. Their law was impressive and in some respects surprisingly modern.[36] 'The astounding thing about this legal code from a modern point of view, is the way it is governed by a clear and consistent concept of guilt.'[37] The juristic approach was emphasised at all times, with a deliberate suppression of religious considerations. Vendettas, for example, were all but abolished in Sumer, the important point being that the state took over from the individual as the arbiter of justice. This justice was harsh but did its best to be objective. Medicine and mathematics were also highly regarded professions in Sumer, and the Sumerians appeared to have discovered the arch. Like we do, they polished apples before they ate them, and the idea that a black cat is unlucky comes from Sumer, as does the division of the clock face into twelve hours.[38] Sumer was, then, a missing link in the evolution of civilisation. From what Woolley was able to deduce, the Sumerians were non-Semitic, a dark-haired people who displaced two other Semitic peoples in the Mesopotamian delta.[39]

Though Woolley could go no further than this, more light was thrown on Hebrew origins, and on the evolution of writing, by discoveries made at **Ras Shamra**. Ras Shamra lies in northwestern Syria, near the Mediterranean bay of Alexandretta, at the angle between Syria and Asia Minor. Here, on a hill above a small harbour, was an ancient settlement excavated in 1929 by the French, led by **Claude Schaeffer**. They were able to construct a full chronology of the site, in which was embedded Ras Shamra's written records, dating to the fifteenth and fourteenth centuries BC. This showed the site to have been named Ugarit, and that it was occupied by a Semitic people of the Amorite-Canaanite class.[40] According to the Bible, this was the period when the Israelites were entering Palestine from the south and beginning to spread among Canaanites, kinsmen of the inhabitants of Ugarit. The library was discovered in a building that stood between the temples of Baal and Dagon. Belonging to the high priest, it consisted mainly of tablets with writing in a cuneiform style but adapted to an alphabetic script, comprising twenty-nine signs. This made it the earliest known alphabet.[41]

The contents of the texts proved to be legal works, price lists, medical and veterinary treatises, and a huge number of religious writings. These showed that Ugarit's supreme god was **El**, a very familiar name from the Old Testament as one of the names of the God of Israel. For example, in chapter 33, verse 20, of Genesis, Jacob erects his altar to 'El, the God of Israel.' In the Ras Shamra tablets, 'El is the king, the supreme judge, the father of years' and 'He reigns over all the other gods.'[42] The land of Canaan is referred to as 'the whole land of El.' El has a wife, Asherat, with whom he has a son, Baal. El is often represented as a bull, and in one text Crete is described as the abode of El. Thus there are overlaps not only between Ras Shamra and Sumeria, Assyrian and Cretan ideas, but also with Hebrew concepts. Many of the writings describe Baal's adventures – for example, his fights with Lotan, 'the sinuous serpent, the mighty one with seven heads,' which recalls the Hebrew Leviathan, and whose seven heads remind us of the beast with seven heads in Revelation and in Job.[43] In another set of writings, El gives Keret command of a huge army, called the 'army of the Negeb.' This is recognisable as the Negev Desert area in the extreme south of Palestine. Keret's orders were to conquer some invaders who are called Terachites, immediately identified as the descendants of Terah, the father of Abraham – in other words the Israelites, who were at that time (according to the then generally accepted chronology) occupying the desert during their forty years' wanderings.[44] The Ras Shamra/Ugarit texts contained other parallels with the Old Testament and provide a strong if not entirely clear link between the bull cults dated to circa 2,000–4,000 BC throughout the Middle East, and religions as we recognise them today.

The discoveries at Ras Shamra matter for two reasons. In the first place, in a country in which the existence of Palestine and then Israel highlights the differences between the Arabs and the Jews, Ras Shamra shows how Judaism grew out of – evolved from – Canaanite religion by a natural process that proves the ancient peoples of this small area, Canaanite and Israelite, to have been essentially the same. Second, the existence of writing – and an alphabet – so early, revolutionised thinking about the Bible. Until the excavation of Ugarit, the accepted view was that writing was unknown to the Hebrews before the ninth century BC and that it was unknown to the Greeks until the seventh. This implied that the Bible was handed down orally for several centuries, making its traditions unreliable and subject to embellishment. In fact, writing was half a millennium older than anyone thought.

In classical archaeology, and in palaeontology, the traditional method of dating is stratigraphy. As common sense suggests, deeper layers are older than the layers above them. However, this only gives a relative chronology, helping to distinguish later from earlier. For absolute dates, some independent evidence is needed, like a king list with written dates, or coins with the date stamped on them, or reference in writings to some heavenly event, like an eclipse, the date of which can be calculated back from modern astronomical knowledge. Such information can then be matched to stratigraphic levels. This is of course not entirely satisfactory. Sites can be damaged, deliberately or accidentally, by man or nature. Tombs can be reused. Archaeologists, palaeontologists, and historians are therefore always on the lookout for other dating methods. The twentieth century offered several answers in this area, and the first came in 1929.

In the notebooks of Leonardo da Vinci there is a brief paragraph to the effect that dry and wet years can be traced in tree rings. The same observation was repeated in 1837 by Charles Babbage – more famous as the man who designed the first mechanical calculators, ancestors of the computer – but he added the notion that tree rings might also be related to other forms of dating. No one took this up for generations, but then an American physicist and astronomer, **Dr Andrew Ellicott Douglass**, director of the University of Arizona's Steward Observatory, made a breakthrough. His research interest was the effect of sunspots on the climate of the earth, and like other astronomers and climatologists he knew that, crudely speaking, every eleven years or so, when sunspot activity is at its height, the earth is racked by storms and rain, one consequence of which is that there is well above average moisture for plants and trees.[45] In order to prove this link, Douglass needed to show that the pattern had been repeated far back into history. For such a project, the incomplete and occasional details about weather were woefully inadequate. It was then that Douglass remembered something he had noticed as a boy, an observation familiar to everyone brought up in the countryside. When a tree is sawn through and the top part carted away, leaving just the stump, we see row upon row of concentric rings. All woodmen, gardeners, and carpenters know, as part of the lore of their trade, that tree rings are annual rings. But what Douglass observed, which no one else had thought through, was that the rings are not of equal thickness. Some years there are narrow rings, other years the rings are broader. Could it be, Douglass wondered, that the broad rings represent 'fat years' (i.e., moist years), and the thin rings represent 'lean years' – in other words, dry years?[46]

It was a simple but inspired idea, not least because it could be tested fairly easily. Douglass set about comparing the outer rings of a newly cut tree with weather reports from recent years. To his satisfaction he discovered that his assumption fitted the facts. Next he moved further back. Some trees in Arizona where he lived were three hundred years old; if he followed the rings all the way into the pith of the trunk, he should be able to re-create climate fluctuations for his region in past centuries. Every eleven years, coinciding with sunspot activity, there had been a 'fat period,' several years of broad rings. Douglass had proved his point that sunspot activity and weather are related. But now he saw other uses for his new technique. In Arizona, most of the trees were pine and didn't go back earlier than 1450, just before the European invasion of America.[47] At first Douglass obtained samples of trees cut by the Spaniards in the early sixteenth century to construct their missions. During his research, Douglass wrote to a number of archaeologists in the American Southwest, asking for core samples of the wood on their sites. Earl Morris, working amid the Aztec ruins fifty miles north of Pueblo Bonito, a prehistoric site in New Mexico, and Neil Judd, excavating Pueblo Bonito itself, both sent samples.[48] These Aztec 'great houses' appeared to have been built at the same time, judging by their style and the objects excavated, but there had been no written calendar in North America, and so no one had been able to place an exact date on the pueblos. Some time after Douglass received his samples from Morris and Judd, he was able to thank them with a bombshell: 'You might be interested to know,' he said in a letter, 'that the latest beam in the ceiling of the Aztec ruins was cut just exactly nine years before the latest beam from Bonito.'[49]

A new science, **dendrochronology**, had been born, and Pueblo Bonito was the first classical problem it helped solve. Douglass's research had begun in 1913, but not until 1928–9 did he feel able to announce his findings to the world. At that point, by overlapping trees of different ages felled at different times, he had an unbroken sequence of rings in southwest America going back first to AD 1300, then to AD 700.[50] The sequence revealed that there had been a severe drought, which lasted from AD 1276 to 1299 and explained why there had been a vast migration at that time by Pueblo Indians, a puzzle which had baffled archaeologists for centuries.

These discoveries placed yet more of man's history on an evolutionary ladder, with ever more specific time frames. The evolution of writing, of religions, of law, and even of building all began to slot into place in the 1920s, making history and prehistory more and more comprehensible as one linked story. Even the familiar events of the Bible appeared to fit into the emerging sequence of events. Such a view had its dangers, of course. Order could be imposed where there may have been none, and complex processes could be over-simplified. Many people were fascinated by scientific discovery and found the new narrative satisfying, but others were disturbed by what they took to be further 'disenchantment' of the world, the removal of mystery. That was one reason why a very short book, published in 1931, had the impact that it did.

Herbert Butterfield was still only twenty-six when, as a young don at Peterhouse, Cambridge, he published *The Whig Interpretation of History* and made his reputation.[51] Controversial as it was, and although he was not really concerned with evolution as such, his argument concerned 'the friends and enemies of progress' and was nonetheless therefore a useful corrective to the emerging consensus. Butterfield exploded the teleological view of history – that it is essentially a straight line leading to the present. To Butterfield, the idea of 'progress' was suspect, as was the notion that in any conflict there were always the good guys who won and the bad guys who lost. The particular example he used was the way the Renaissance led to the Reformation and then on to the contemporary world. The prevailing view, what he called the Whig view, was to see a straight line from the essentially Catholic Renaissance to the Protestant Reformation to the modern world with all its freedoms, as a result of which many attributed to Luther the intention of promoting greater liberty.[52] Butterfield argued that this view assumed 'a false continuity in events': the Whig historian 'likes to imagine religious liberty issuing beautifully out of Protestantism when in reality it emerges painfully and grudgingly out of something quite different, out of the tragedy of the post-Reformation world.'[53]

The motive for this habit on the part of historians was, said Butterfield, contemporary politics – in its broadest sense. The present-day historian's enthusiasm for democracy or freedom of thought or the liberal tradition led him to conclude that people in the past were working toward these goals.[54] One consequence of this tendency, Butterfield thought, was that the Whig historian was overfond of making moral judgements on the past: 'For him the voice of posterity is the voice of God and the historian is the voice of posterity. And it is typical of him that he tends to regard himself as the judge when by his methods and his equipment he is fitted only to be the detective.'[55] This fashion for moral judgements leads the Whig historian into another mistake, that more evil is due to conscious sin than to unconscious error.[56] Butterfield was uneasy with such a stance. He offered the alternative view – that all history could do was approach its subjects in more and more detail, and with less and less abridgement. No moral judgements are necessary for him because it is impossible to get within the minds of people of bygone ages and because the great quarrels of history have not been between two parties of which one was 'good' and the other 'evil' but between opposing groups (not necessarily two in number) who had rival ideas about where they wanted events, and society, to go. To judge backward from the present imposes a modern mindset on events which cannot be understood in that way.[57]

Butterfield's ideas acted as a check on the growth of evolutionary thought, but only a check. As time went by, and more results came in, the evidence amassed for one story was overwhelming. Progress was a word less and less used, but evolution went from strength to strength, invading even history itself.

The discoveries of the 1920s pushed forward the idea that a complete history of mankind might one day be possible. This expanding vision was further fuelled by parallel developments in physics.

THE GOLDEN AGE OF PHYSICS

The period from 1919, when Ernest Rutherford first split the atom, to 1932, when his student James Chadwick discovered the neutron, was a golden decade for physics. Barely a year went by without some momentous breakthrough. At that stage, America was far from being the world leader in physics it has since become. All the seminal work of the golden decade was carried out in one of three places in Europe: the Cavendish Laboratory in Cambridge, England; Niels Bohr's Institute of Theoretical Physics in Copenhagen; and the old university town of Göttingen, near Marburg in Germany.

For Mark Oliphant, one of Rutherford's protégés in the 1920s, the main hallway of the Cavendish, where the director's office was, consisted of 'uncarpeted floor boards, dingy varnished pine doors and stained, plastered walls, indifferently lit by a skylight with dirty glass.'[1] For C. P. Snow, however, who also trained there and described the lab in his first novel, *The Search*, the paint and the varnish and the dirty glass went unremarked. 'I shall not easily forget those Wednesday meetings in the Cavendish. For me they were the essence of all the *personal* excitement in science; they were romantic, if you like, and not on the plane of the highest experience I was soon to know [of scientific discovery]; but week after week I went away through the raw nights, with east winds howling from the fens down the old streets, full of a glow that I had seen and heard and been close to the leaders of the greatest movement in the world.' Rutherford, who followed Maxwell as director of the Cavendish in 1919, evidently agreed. At a meeting of the British Association in 1923 he startled colleagues by suddenly shouting out, 'We are living in the heroic age of physics!'[2]

In some ways, Rutherford himself – now a rather florid man, with a moustache and a pipe that was always going out – embodied in his own person that heroic age. During World War I, particle physics had been on hold, more or less. Officially, Rutherford was working for the Admiralty, researching submarine detection. But he carried on research when his duties allowed. And in the last year of war, in April 1919, just as Arthur Eddington was preparing his trip to West Africa to test Einstein's predictions, Rutherford sent off a paper that, had he done nothing else, would earn him a place in history. Not that you would have known it from the paper's title: '**An Anomalous Effect in**

Nitrogen.' As was usual in Rutherford's experiments, the apparatus was simple to the point of being crude: a small glass tube inside a sealed brass box fitted at one end with a zinc-sulphide scintillation screen. The brass box was filled with nitrogen and then through the glass tube was passed a source of alpha particles – helium nuclei – given off by radon, the radioactive gas of radium. The excitement came when Rutherford inspected the activity on the zinc-sulphide screen: the scintillations were indistinguishable from those obtained from hydrogen. How could that be, since there was no hydrogen in the system? This led to the famously downbeat sentence in the fourth part of Rutherford's paper: 'From the results so far obtained it is difficult to avoid the conclusion that the long-range atoms arising from collision of [alpha] particles with nitrogen are not nitrogen atoms but probably atoms of hydrogen. . . . If this be the case, we must conclude that the nitrogen atom is disintegrated.' The newspapers were not so cautious. Sir Ernest Rutherford, they shouted, had *split the atom*.[3] He himself realised the importance of his work. His experiments had drawn him away, temporarily, from antisubmarine research. He defended himself to the overseers' committee: 'If, as I have reason to believe, I have disintegrated the nucleus of the atom, this is of greater significance than the war.'[4]

In a sense, Rutherford had finally achieved what the old alchemists had been aiming for, transmuting one element into another, nitrogen into oxygen and hydrogen. The mechanism whereby this artificial transmutation (the first ever) was achieved was clear: an alpha particle, a helium nucleus, has an atomic weight of 4. When it was bombarded on to a nitrogen atom, with an atomic weight of 14, it displaced a hydrogen nucleus (to which Rutherford soon gave the name **proton**). The arithmetic therefore became: $4+14-1=17$, the oxygen isotope, O^{17}.[5]

The significance of the discovery, apart from the philosophical one of the transmutability of nature, lay in the new way it enabled the nucleus to be studied. Rutherford and Chadwick immediately began to probe other light atoms to see if they behaved in the same way. It turned out that they did – boron, fluorine, sodium, aluminum, phosphorus, all had nuclei that could be probed: they were not just solid matter but had a structure. All this work on light elements took five years, but then there was a problem. The heavier elements were, by definition, characterised by outer shells of many electrons that constituted a much stronger electrical barrier and would need a stronger source of alpha particles if they were to be penetrated. For James Chadwick and his young colleagues at the Cavendish, the way ahead was clear – they needed to explore means of accelerating particles to higher velocities. Rutherford wasn't convinced, preferring simple experimental tools. But elsewhere, especially in America, physicists realised that one way ahead lay with particle accelerators.

Between 1924 and 1932, when Chadwick finally isolated the neutron, there were no breakthroughs in nuclear physics. **Quantum physics**, on the other hand, was an entirely different matter. Niels Bohr's Institute of Theoretical Physics opened in Copenhagen on 18 January 1921. The land had been given by the city, appropriately enough next to some soccer fields (Niels and his brother, Harald, were both excellent players).[6] The large house, on four floors, shaped like an 'L,' contained a lecture hall, library, and laboratories (strange for an institute of *theoretical* physics), as well as a table-tennis table, where Bohr also shone. 'His reactions were very fast and accurate,' says Otto Frisch, 'and he had tremendous will power and stamina. In a way those qualities characterised his scientific work as well.'[7] Bohr became a Danish hero a year later when he won the Nobel Prize. Even the king wanted to meet him. But in fact the year was dominated by something even more noteworthy – Bohr's final irrevocable linking of chemistry and physics. In 1922 Bohr showed how atomic structure was linked to the periodic table of elements drawn up by Dmitri Ivanovich Mendeléev, the nineteenth-century Russian chemist. In his first breakthrough, just before World War I, Bohr had explained how electrons orbit the nucleus only in certain formations, and how this helped explain the characteristic spectra of light emitted by the crystals of different substances. This idea of natural orbits also married atomic structure to Max Planck's notion of quanta. Bohr now went on to argue that successive orbital shells of electrons could contain only a precise number of electrons. He introduced the idea that elements that behave in a similar way chemically do so because they have a similar arrange-ment of electrons in their outer shells, which are the ones most used in chemical reactions. For example, he compared barium and radium, which are both alkaline earths but have very different atomic weights and occupy, respectively, the fifty-sixth and eighty-eighth place in the periodic table. Bohr explained this by showing that barium, atomic weight 137.34, has electron shells filled successively by 2, 8, 18, 18, 8, and 2 (=56) electrons. Radium, atomic weight 226, has on the other hand electron shells filled successively by 2, 8, 18, 32, 18, 8, and 2 (=88) electrons.[8] Besides explaining their position on the periodic table, the fact that the outer shell of each element has two electrons means barium and radium are chemically similar despite their considerable other differences. As Einstein said, 'This is the highest form of musicality in the sphere of thought.'[9]

During the 1920s the centre of gravity of physics – certainly of quantum physics – shifted to **Copenhagen**, largely because of Bohr. A big man in every sense, he was intent on expressing himself accurately, if painfully slowly, and forcing others to do so too. He was generous, avuncular, completely devoid of those instincts for rivalry that can so easily sour relations. But the success of Copenhagen also had to do with the fact that Denmark was a small country, neutral, where national rivalries of the Americans, British, French, Germans, Russians, and Italians could be forgotten. Among the sixty-three physicists of renown who studied at Copenhagen in the 1920s were Paul Dirac (British), Werner Heisenberg (German), and Lev Landau (Russian).[10]

There was also the Swiss-Austrian, Wolfgang Pauli. In 1924 Pauli was a pudgy twenty-three-year-old, prone to depression when scientific problems defeated him. One problem in particular had set him prowling the streets of the Danish capital. It was something that vexed Bohr too, and it arose from the fact that no one, just then, understood why all the electrons in orbit around the nucleus didn't just crowd in on the inner shell. This is what should have happened, with the electrons emitting energy in the form of light. What was known by now, however, was that each shell of electrons was arranged so that the inner shell always contains just one orbit, whereas the next shell out contains four. Pauli's contribution was to show that no orbit could contain more than two electrons. Once it had two, an orbit was 'full,' and other electrons were excluded, forced to the next orbit out.[11] This meant that the inner shell (one orbit) could not contain more than two electrons, and that the next shell out (four orbits) could not contain more than eight. This became known as **Pauli's exclusion principle**, and part of its beauty lay in the way it expanded Bohr's explanation of chemical behaviour.[12] Hydrogen, for example, with one electron in the first orbit, is chemically active. Helium, however, with two electrons in the first orbit (i.e., that orbit is 'full' or 'complete'), is virtually inert. To underline the point further, lithium, the third element, has two electrons in the inner shell and one in the next, and is chemically very active. Neon, however, which has ten electrons, two in the inner shell (filling it) and eight in the four outer orbits of the second shell (again filling those orbits), is also inert.[13] So together Bohr and Pauli had shown how the chemical properties of elements are determined not only by the number of electrons the atom possesses but also by the dispersal of those electrons through the orbital shells.

The next year, 1925, was the high point of the golden age, and the centre of activity moved for a time to Göttingen. Before World War I, British and American students regularly went to Germany to complete their studies, and Göttingen was a frequent stopping-off place. Moreover, it had held on to its prestige and status better than most in the Weimar years. Bohr gave a lecture there in 1922 and was taken to task by a young student who corrected a point in his argument. Bohr, being Bohr, hadn't minded. 'At the end of the discussion he came over to me and asked me to join him that afternoon on a walk over the Hain Mountain,' **Werner Heisenberg** wrote later. 'My real scientific career only began that afternoon.'[14] In fact it was more than a stroll, for Bohr invited the young Bavarian to Copenhagen. Heisenberg didn't feel ready to go for two years, but Bohr was just as welcoming after the delay, and they immediately set about tackling yet another problem of quantum theory, what Bohr called 'correspondence.'[15] This stemmed from the observation that, at low frequencies, quantum physics and classical physics came together. But how could that be? According to quantum theory, energy – like light – was emitted in tiny packets; according to classical physics, it was emitted continuously. Heisenberg returned to Göttingen enthused but also confused. And Heisenberg hated confusion as much as Pauli did. And so when, toward the end of May 1925, he suffered one of his many attacks of hay fever, he took two weeks' holiday in Heligoland, a narrow island off the German coast in the North Sea, where there was next to no pollen. An excellent pianist who could also recite huge tracts of Goethe, Heisenberg was very fit (he liked climbing), and he cleared his head with long walks and bracing dips in the sea.[16] The idea that came to Heisenberg in that cold, fresh environment was the first example of what came to be called quantum weirdness. Heisenberg took the view that we should stop trying to visualise what goes on inside an atom, as it is impossible to observe directly something so small.[17] All we can do is measure its properties. And so, if something is measured as continuous at one point, and discrete at another, that is the way of reality. If the two measurements exist, it makes no sense to say that they disagree: they are just measurements.

This was Heisenberg's central insight, but in a hectic three weeks he went further, developing a method of mathematics, known as matrix math, originating from an idea by David Hilbert, in which the measurements obtained are grouped in a two-dimensional table of numbers where two matrices can be multiplied together to give another matrix.[18] In Heisenberg's scheme, each atom would be represented by one matrix, each 'rule' by another. If one multiplied the 'sodium matrix' by the 'spectral line matrix,' the result should give the matrix of wavelengths of sodium's spectral lines. To Heisenberg's, and Bohr's, great satisfaction, it did; 'For the first time, atomic structure had a genuine, though very surprising, mathematical base.'[19] Heisenberg called his creation/discovery quantum mechanics.

The acceptance of Heisenberg's idea was made easier by a new theory of **Louis de Broglie** in Paris, also published in 1925. Both Planck and Einstein had argued that light, hitherto regarded as a wave, could sometimes behave as a particle. De Broglie reversed this idea, arguing that particles could sometimes behave like waves. No sooner had de Broglie broached this theory than experimentation proved him right.[20] The **wave-particle** duality of matter was the second weird notion of physics, but it caught on quickly. One reason was the work of yet another genius, the Austrian **Erwin Schrödinger**, who was disturbed by Heisenberg's idea and fascinated by de Broglie's. Schrödinger, who at thirty-nine was quite 'old' for a physicist, added the notion that the electron, in its orbit around the nucleus, is not like a planet but like a wave.[21] Moreover, this wave pattern determines the size of the orbit, because to form a complete circle the wave must conform to a whole number, not fractions (otherwise the wave would descend into chaos). In turn this determined the distance of the orbit from the nucleus. Schrödinger's work, set out in four long papers in *Annalen der Physik* in spring and summer 1926, was elegant and explained the position of Bohr's orbits. The mathematics that underlay his theory also proved to be much the same as Heisenberg's matrices, only simpler. Again knowledge was coming together.[22]

The final layer of weirdness came in 1927, again from Heisenberg. It was late February, and Bohr had gone off to Norway to ski. Heisenberg paced the streets of Copenhagen on his own. Late one evening, in his room high up in Bohr's institute, a remark of Einstein's stirred something deep in Heisenberg's brain: 'It is the theory which decides what we can observe.'[23] It was well after midnight, but he decided he needed some air, so he went out and trudged across the muddy soccer fields. As he walked, an idea began to germinate in his brain. Unlike the immensity of the heavens above, the world the quantum physicists dealt with was unimaginably small. Could it be, Heisenberg asked himself, that at the level of the atom there was a limit to what could be known? To identify the position of a particle, it must impact on a zinc-sulphide screen. But this alters its velocity, which means that it cannot be measured at the crucial moment. Conversely, when the velocity of a particle is measured by scattering gamma rays from it, say, it is knocked into a different path, and its exact position at the point of measurement is changed. Heisenberg's **uncertainty principle**, as it came to be called, posited that the exact position and precise velocity of an electron could not be determined at the same time.[24] This was disturbing both practically and philosophically, because it implied that in the subatomic world cause and effect could never be measured. The only way to understand electron behaviour was statistical, using the rules of probability. 'Even in principle,' Heisenberg said, 'we cannot know the present in all detail. For that reason everything observed is a selection from a plenitude of possibilities and a limitation on what is possible in the future.'[25]

Einstein, no less, was never very happy with the basic notion of quantum theory, that the subatomic world could only be understood statistically. It remained a bone of contention between him and Bohr until the end of his life. In 1926 he wrote a famous letter to the physicist Max Born in Göttingen. 'Quantum mechanics demands serious attention,' he wrote. 'But an inner voice tells me that this is not the true Jacob. The theory accomplishes a lot, but it does not bring us closer to the secrets of the Old One. In any case, I am convinced that He does not play dice.'[26]

For close on a decade, quantum mechanics had been making news. At the height of the golden age, German preeminence was shown by the fact that more papers on the subject were published in that language than in all others put together.[27] During that time, experimental particle physics had been stalled. It is difficult at this distance to say why, for in 1920 Ernest Rutherford had made an extraordinary prediction. Delivering the Bakerian lecture before the Royal Society of London, Rutherford gave an insider's account of his nitrogen experiment of the year before; but he also went on to speculate about future work.[28] He broached the possibility of a third major constituent of atoms in addition to electrons and protons. He even described some of the properties of this constituent, which, he said, would have 'zero nucleus charge.' 'Such an atom,' he argued, 'would have very novel properties. Its external [electrical] field would be practically zero, except very close to the nucleus, and in consequence it should be able to move freely through matter.' Though difficult to discover, he said, it would be well worth finding: 'it should readily enter the structure of atoms, and may either unite with the nucleus or be disintegrated by its intense field.' If this constituent did indeed exist, he said, he proposed calling it the neutron.[29]

Just as **James Chadwick** had been present in 1911, in Manchester, when Rutherford had revealed the structure of the atom, so he was in the audience for the Bakerian lecture. After all, he was Rutherford's right-hand man now. At the time, however, he did not really share his boss's enthusiasm for the neutron. The symmetry of the electron and the proton, negative and positive, seemed perfect, complete. Other physicists may never have read the Bakerian lecture – it was a stuffy affair – and so never have had their minds stimulated. Throughout the late 1920s, however, anomalies built up. One of the more intriguing was the relationship between atomic weight and atomic number. The atomic number was derived from the nucleus's electrical charge and a count of the protons. Thus helium's atomic number was 2, but its atomic weight was 4. For silver the equivalent numbers were 47 and 107, for uranium 92 and 235 or 238.[30] One popular theory was that there were additional protons in the nucleus, linked with electrons that neutralised them. But this only created another, theoretical anomaly: particles as small and as light as electrons could only be kept within the nucleus by enormous quantities of energy. That energy should show itself when the nucleus was bombarded and had its structure changed – and that never happened.[31] Much of the early 1920s was taken up by repeating the nitrogen transmutation experiment with other light elements, so Chadwick scarcely had time on his hands. However, when the anomalies showed no sign of being satisfactorily resolved, he came round to Rutherford's view. Something like a **neutron** must exist.

Chadwick was in physics by mistake.[32] A shy man, with a gruff exterior that concealed his innate kindness, he had wanted to be a mathematician but turned to physics after he stood in the wrong queue at Manchester University and was impressed by the physicist who interviewed him. He had studied in Berlin under Hans Geiger but failed to leave early enough when war loomed and was interned in Germany for the duration. By the 1920s he was anxious to be on his way in his career.[33] To begin with, the experimental search for the neutron went nowhere. Believing it to be a close union of proton and electron, Rutherford and Chadwick devised various ways of, as Richard Rhodes puts it, 'torturing' hydrogen. The next bit is complicated. First, between 1928 and 1930, a German physicist, **Walter Bothe**, studied the gamma radiation (an intense form of light) given off when light elements such as lithium and oxygen were bombarded by alpha particles. Curiously, he found intense radiation given off not only by boron, magnesium, and aluminum – as he had expected, because alpha particles disintegrated those elements (as Rutherford and Chadwick had shown) – but also by beryllium, which was not disintegrated by alpha particles.[34] Bothe's result was striking enough for Chadwick at Cambridge, and **Irène Curie**, daughter of Marie, and her husband **Frédéric Joliot** in Paris, to take up the German's approach. Both labs soon found anomalies of their own. H. C. Webster, a student of Chadwick, discovered in spring 1931 that 'the radiation [from beryllium] emitted in the same direction as the ... alpha particles was harder [more penetrating] than the radiation emitted in a backward direction.' This mattered because if the radiation was gamma rays – light – then it should spray equally in all directions, like the light that shines from a lightbulb. A *particle*, on the other hand, would behave differently. It might well be knocked forward in the direction of an incoming alpha.[35] Chadwick thought, 'Here's the neutron.'[36]

In December 1931 Irène Joliot-Curie announced to the French Academy of Sciences that she had repeated Bothe's experiments with beryllium radiation but had standardised the measurements. This enabled her to calculate that the energy of the radiation given off was *three times* the energy of the bombarding alphas. This order of magnitude clearly meant that the radiation wasn't gamma;

some other constituent must be involved. Unfortunately Irène Joliot-Curie had never read Rutherford's Bakerian lecture, and she took it for granted that the beryllium radiation was caused by protons. Barely two weeks later, in mid-January 1932, the Joliot-Curies published another paper. This time they announced that paraffin wax, when bombarded by beryllium radiation, emitted high-velocity protons.[37]

When Chadwick read this account in the *Comptes rendus*, the French physics journal, in his morning mail in early February, he realised there was something very wrong with this description and interpretation. Any physicist worth his salt knew that a proton was 1,836 times heavier than an electron: it was all but impossible for a proton to be dislodged by an electron. While Chadwick was reading the report, a colleague named Feather, who had read the same article and was eager to draw his attention to it, entered his room. Later that morning, at their daily progress meeting, Chadwick discussed the paper with Rutherford. 'As I told him about the Curie-Joliot observation and their views on it, I saw his growing amazement; and finally he burst out "I don't believe it." Such an impatient remark was utterly out of character, and in all my long association with him I recall no similar occasion. I mention it to emphasise the electrifying effect of the Curie-Joliot report. Of course, Rutherford agreed that one must believe the observations; the explanation was quite another matter.'[38] Chadwick lost no time in repeating the experiment. The first thing to excite him was that he found the beryllium radiation would pass unimpeded through a block of lead three-quarters of an inch thick. Next, he found that bombardment by the beryllium radiation knocked the protons out of some elements by up to 40 centimetres, fully 16 inches. Whatever the radiation was, it was huge – and in terms of electrical charge, it was neutral. Finally, Chadwick took away the paraffin sheet that the Joliot-Curies had used so as to see what happened when elements were bombarded directly by beryllium radiation. Using an oscilloscope to measure the radiation, he found first that beryllium radiation displaced protons whatever the element, and crucially, that the energies of the displaced protons were just too huge to have been produced by gamma rays. Chadwick had learned a thing or two from Rutherford by now, including a habit of understatement. In the paper, entitled 'Possible Existence of a Neutron,' which he rushed to *Nature*, he wrote, 'It is evident that we must either relinquish the application of the conservation of energy and momentum in these collisions or adopt another hypothesis about the nature of radiation.' Adding that his experiment appeared to be the first evidence of a particle with 'no net charge,' he concluded, 'We may suppose it to be the "neutron" discussed by Rutherford in his Bakerian lecture.'[39] The process observed was $^4\text{He} + {}^9\text{Be} \rightarrow {}^{12}\text{C} + n$, where n stands for neutron of mass number 1.[40]

The Joliot-Curies were much embarrassed by their failure to spot what was, for Rutherford and Chadwick, the obvious (though the French would make their own discoveries later). Chadwick, who had worked day and night for ten days to make sure he was first, actually announced his results initially to a meeting of the Kapitza Club at Cambridge, which had been inaugurated by Peter Kapitza, a young Russian physicist at the Cavendish. Appalled by the formal, hierarchical structure of Cambridge, Kapitza had started the club as a discussion forum where rank didn't matter. The club met on Wednesdays, and on the night when Chadwick, exhausted, announced that he had discovered the third basic constituent of matter, he delivered his address – very short – and then remarked tartly, 'Now I want to be chloroformed and put to bed for a fortnight.'[41] Chadwick was awarded the Nobel Prize for his discovery, the result of dogged detective work. The neutral electrical charge of the new particle would allow the nucleus to be probed in a far more intimate way. Other physicists were, in fact, already looking beyond his discovery – and in some cases they didn't like what they saw.

Physics was becoming the queen of sciences, a fundamental way to approach nature, with both practical and deeply philosophical implications. The transmutability of nature apart, its most philosophical aspect was its overlap with astronomy.

At this point we need to return – briefly – to Einstein. At the time he produced his theory of relativity, most scientists took it for granted that the universe was static. The nineteenth century had produced much new information about the stars, including ways to measure their temperatures and distances, but astronomers had not yet observed that heavenly bodies are clustered into galaxies, or that they were moving away from one another.[42] But relativity had a surprise for astronomers: Einstein's equations predicted that the universe must either be expanding or contracting. This was a wholly unexpected consequence, and so weird did it appear, even to Einstein himself, that he tinkered with his calculations to make his theoretical universe stand still. This correction he later called the biggest blunder of his career.[43]

Curiously, however, a number of scientists, while they accepted Einstein's theory of relativity and the calculations on which it was based, never accepted the cosmological constant, and the correction on which *it* was based. **Alexander Friedmann**, a young Russian scientist, was the first man to cause Einstein to think again ('cosmological constant' was actually his term). Friedmann's background was brutish. His mother had deserted his father – a cruel, arrogant man – taking the boy with her. Convicted of 'breaking conjugal fidelity,' she was sentenced by the imperial court to celibacy and forced to give up Alexander. He didn't see his mother again for nearly twenty years. Friedmann taught himself relativity, during which time he realised Einstein had made a mistake and that, cosmological constant or no, the universe must be either expanding or contracting.[44] He found this such an exciting idea that he dared to improve on Einstein's work, developing a mathematical model to underline his conviction, and sent it to the German. By the early 1920s, however, Arthur Eddington had confirmed some of Einstein's predictions, and the great man had become famous and was snowed under with letters: Friedmann's ideas were lost in the avalanche.[45] Undaunted, Friedmann tried to see Einstein in person, but that move also failed. It was only when Friedmann was given an introduction by a mutual colleague that Einstein finally got to grips with the Russian's ideas. As a result, Einstein began to have second thoughts about his cosmological constant – and its implications. But it wasn't Einstein who pushed Friedmann's ideas forward. A Belgian cosmologist, **Georges Lemaître**, and a number of others built on his ideas so that as the 1920s advanced, a fully realised geometric description of a homogeneous and expanding universe was fleshed out.[46]

A theory was one thing. But planets and stars and galaxies are not exactly small entities; they occupy vast spaces. Surely, if the universe really was expanding, it could be observed? One way to do this was by observation of what were then called 'spiral nebulae.' Nowadays we know that nebulae are distant galaxies, but then, with the telescopes of the time, they were simply indistinct smudges in the sky, beyond the solar system. No one knew whether they were gas or solid matter; and no one knew what size they were, or how far away. It was then discovered that the light emanating from spiral nebulae is shifted toward the red end of the spectrum. One way of illustrating the significance of this redshift is by analogy to the **Doppler effect**, after Christian Doppler, the Austrian physicist who first explained the observation in 1842. When a train or a motorbike comes toward us, its noise changes, and then, as it goes past and away, the noise changes a second time. The explanation is simple: as the train or bike approaches, the sound waves reach the observer closer and closer together – the intervals get shorter. As the train or bike recedes, the opposite effect occurs; the source of the noise is receding at all times, and so the interval between the sound waves gets longer and longer. Much the same happens with light: where the source of light is approaching, the light is shifted toward the blue end of the spectrum, while light where the source is receding is shifted toward the red end.

The first crucial tests were made in 1922, by Vesto Slipher at the Lowell Observatory in Flagstaff, Arizona.[47] The Lowell had originally been built in 1893 to investigate the 'canals' on Mars. In this case, Slipher anticipated finding redshifts on one side of the nebulae spirals (the part swirling away from the observer) and blueshifts on the other side (because the spiral was swirling toward earth). Instead, he found that all but four of the forty nebulae he examined produced only redshifts. Why was that? Almost certainly, the confusion arose because Slipher could not really be certain of exactly how far away the nebulae were. This made his correlation of redshift and distance problematic. But the results were nonetheless highly suggestive.[48]

Three years elapsed before the situation was finally clarified. Then, in 1929, **Edwin Hubble**, using the largest telescope of the day, the 100-inch reflector scope at Mount Wilson, near Los Angeles, managed to identify individual stars in the spiral arms of a number of nebulae, thereby confirming the suspicions of many astronomers that 'nebulae' were in fact entire galaxies. Hubble also located a number of 'Cepheid variable' stars. Cepheid variables – stars that vary in brightness in a regular way (periods that range from 1–50 days) – had been known since the late eighteenth century, but it was only in 1908 that Henrietta Leavitt, at Harvard, showed that there is a mathematical relationship between the average brightness of a star, its size, and its distance from earth.[49] Using the Cepheid variables that he could now see, Hubble was able to calculate how far away a score of nebulae were.[50] His next step was to correlate those distances with their corresponding redshifts. Altogether, Hubble collected information on twenty-four different galaxies, and the results of his observations and calculations were simple and sensational: he discovered a straightforward linear relationship. The farther away a galaxy was, the more its light was redshifted.[51] This became known as Hubble's law, and although his original observations were made on twenty-four galaxies, since 1929 the law has been proven to apply to thousands more.[52]

Once more then, one of Einstein's predictions had proved correct. His calculations, and Friedmann's, and Lemaître's, had been borne out by experiment: the universe was indeed expanding. For many people this took some getting used to. It involved implications about the origins of the universe, its character, the very meaning of time. The immediate impact of the idea of an expanding universe made Hubble, for a time, almost as famous as Einstein. Honours flowed in, including an honorary doctorate from Oxford, *Time* put him on its cover, and the observatory became a stopping-off place for famous visitors to Los Angeles: Aldous Huxley, Andrew Carnegie, and Anita Loos were among those given privileged tours. The Hubbles were taken up by Hollywood: the letters of Grace Hubble, Edwin's wife, written in the early thirties, talk of dinners with Helen Hayes, Ethel Barrymore, Douglas Fairbanks, Walter Lippmann, Igor Stravinsky, Frieda von Richthofen (D. H. Lawrence's widow), Harpo Marx and Charlie Chaplin.[53] Jealous colleagues pointed out that, far from being a Galileo or Copernicus of his day, Hubble was not all that astute an observer, and that since his findings had been anticipated by others, his contribution was limited. But Hubble did arduous spadework and produced enough accurate data so that sceptical colleagues could no longer scoff at the theory of an expanding universe. It was one of the most astonishing ideas of the century, and it was Hubble who put it beyond doubt.

At the same time that physics was helping explain massive phenomena like the universe, it was still making advances in other areas of the minuscule world, in particular the world of molecules, helping us to a better understanding of chemistry. The nineteenth century had seen the first golden age of chemistry, industrial chemistry in particular. Chemistry had largely been responsible for the rise of Germany, whose nineteenth-century strength Hitler was so concerned to recover. For example, in the years before World War I, Germany's production of sulphuric acid had gone from half that of Britain to 50 percent more; its production of chlorine by the modern electrolytic method was three times that of Britain; and its share of the world's dyestuffs market was an incredible 90 percent.

The greatest breakthrough in theoretical chemistry in the twentieth century was achieved by one man, **Linus Pauling**, whose idea about the nature of the chemical bond was as fundamental as the gene and the quantum because it showed how physics governed molecular structure and how that structure was related to the properties, and even the appearance, of the chemical elements. Pauling explained the logic of why some substances were yellow liquids, others white powders, still others red solids. The physicist Max Perutz's verdict was that Pauling's work transformed chemistry into 'something to be understood and not just memorised.'[54]

Born the son of a pharmacist, near Portland, Oregon, in 1901, Pauling was blessed with a healthy dose of self-confidence, which clearly helped his career. As a young graduate he spurned an offer from Harvard, preferring instead an institution that had started life as Throop Polytechnic but in 1922 was renamed the California Institute of Technology, or Caltech.[55] Partly because of Pauling, Caltech developed into a major centre of science, but when he arrived there were only three buildings, surrounded by thirty acres of weedy fields, scrub oak, and an old orange grove. Pauling initially wanted to work in a new technique that could show the relationship between the distinctively shaped crystals into which chemicals formed and the actual architecture of the molecules that made up the crystals. It had been found that if a beam of X rays was sprayed at a crystal, the beam would disperse in a particular way. Suddenly, a way of examining chemical structure was possible. X-ray crystallography, as it was called, was barely out of its infancy when Pauling got his Ph.D., but even so he quickly realised that neither his math nor his physics were anywhere near good enough to make the most of the new techniques. He decided to go to Europe in order to meet the great scientists of the day: Niels Bohr, Erwin Schrödinger, Werner Heisenberg, among others. As he wrote later, 'I had something of a shock when I went to Europe in 1926 and discovered that there were a good number of people around that I thought to be smarter than me.'[56]

So far as his own interest was concerned, the nature of the chemical bond, his visit to Zurich was the most profitable. There he came across two less famous Germans, **Walter Heitler** and **Fritz London**, who had developed an idea about how electrons and wave functions applied to chemical reactions.[57] At its simplest, imagine the following: Two hydrogen atoms are approaching one another. Each is comprised of one nucleus (a proton) and one electron. As the two atoms get closer and closer to each other, 'the electron of one would find itself drawn to the nucleus of the other, and vice versa. At a certain point, the electron of one would jump to the new atom, and the same would happen with the electron of the other atom.' They called this an 'electron exchange,' adding that this exchange would take place *billions* of times a second.[58] In a sense, the electrons would be 'homeless,' the exchange forming the 'cement' that held the two atoms together, 'setting up a chemical bond with a definite length.' Their theory put together the work of Pauli, Schrödinger, and Heisenberg; they also found that the 'exchange' determined the architecture of the molecule.[59] It was a very neat piece of work, but from Pauling's point of view there was one drawback about this idea: it wasn't his. If he were to make his name, he needed to push the idea forward. By the time Pauling returned to America from Europe, Caltech had made considerable progress. Negotiations were under way to build the world's biggest telescope at Mount Wilson, where Hubble would work. A jet propulsion lab was planned, and T. H. Morgan was about to arrive, to initiate a biology lab.[60] Pauling was determined to outshine them all. Throughout the early 1930s he released report after report, all part of the same project, and all having to do with the **chemical bond**. He succeeded magnificently in building on Heitler and London's work. His early experiments on carbon, the basic constituent of life, and then on silicates showed that the elements could be systematically grouped according to their electronic relationships. These became known as Pauling's rules. He showed that some bonds were weaker than others and that this helped explain chemical properties. Mica, for example, is a silicate that, as all chemists know, splits into thin, transparent sheets. Pauling was able to show that mica's crystals have strong bonds in two directions and a weak one in a third direction, exactly corresponding to observation. In a second instance, another silicate we all know as talc is characterised by weak bonds all around, so that it crumbles instead of splitting, and forms a powder.[61]

Pauling's work was almost as satisfying for others as it was for him.[62] Here at last was an atomic, electronic explanation of the observable properties of well-known substances. The century had begun with the discovery of fundamentals that applied to physics and biology. Now the same was happening in chemistry. Once more, knowledge was beginning to come together. During 1930–5, Pauling published a new paper on the bond every five weeks on average.[63] He was elected to the National Academy of Sciences in America at thirty-two, the youngest scientist ever to receive that honour.[64] For a time, he was so far out on his own that few other people could keep up. Einstein attended one lecture of his and admitted afterward that it was beyond him. Uniquely, Pauling's papers sent to the *Journal of the American Chemical Society* were published unrefereed because the editor could think of no one qualified to venture an opinion.[65] Even though Pauling was conscious of this, throughout the 1930s he was too busy producing original papers to write a book consolidating his research. Finally, in 1939 he published *The Nature of the Chemical Bond*. This revolutionised our understanding of chemistry and immediately became a standard text, translated into several languages.[66] It proved crucial to the discoveries of the molecular biologists after World War II.

The fresh data that the new physics was producing had very practical ramifications that arguably have changed our lives far more directly than was at first envisaged by scientists mainly interested in fundamental aspects of nature. Radio, in use for some time, moved into the home in the 1920s; television was first shown in August 1928. Another invention, using physics, revolutionised life in a completely different way: this was the **jet engine**, developed with great difficulty by the Englishman **Frank Whittle**.

Whittle was the working-class son of a mechanic who lived on a Coventry housing estate. As a boy he educated himself in Leamington Public Library, where he spent all his spare time devouring popular science books about aircraft – and turbines.[67] All his life Frank Whittle was obsessed with flight, but his background was hardly natural in those days for a university education, and so at the age of fifteen he applied to join the Royal Air Force as a technical apprentice. He failed. He passed the written examination but was blocked by the medical officer: Frank Whittle was only five feet tall. Rather than give up, he obtained a diet sheet and a list of exercises from a friendly PE teacher, and within a few months he had added three inches to his height and another three to his chest measurement. In some ways this was as impressive as anything else he did later in life. He was finally accepted as an apprentice in the RAF, and although he found the barrack-room life irksome, in his second year as a cadet at Cranwell, the RAF college – at the age of nineteen – he wrote a thesis on future developments in aircraft design. It was in this paper that Whittle began to sketch his ideas for the jet engine. Now in the Science Museum in London, the paper is written in junior handwriting, but it is clear and forthright.[68] His crucial calculation was that 'a 100mph wind against a machine travelling at 600mph at 120,000 feet would have less effect than a 20mph head wind against the same machine at 1,000 feet.' He concluded, 'Thus everything indicates that designers should aim at altitude.' He knew that propellers and petrol engines were inefficient at great heights, but he also knew that rocket propulsion was suitable only for space travel. This is where his old interest in turbines resurfaced; he was able to show that the efficiency of turbines increased at higher altitudes. An indication of Whittle's vision is apparent from the fact that he was contemplating an aircraft travelling at a speed of 500mph at 60,000 feet, while in 1926 the top speed of RAF fighters was 150 mph, and they couldn't fly much above 10,000 feet.

After Cranwell, Whittle transferred to Hornchurch in Essex to a fighter squadron, and then in 1929 moved on to the Central Flying School at Wittering in Sussex as a pupil instructor. All this time he had been doggedly worrying how to create a new kind of engine, most of the time working on an amalgam of a petrol engine and a fan of the kind used in turbines. While at Wittering, he suddenly saw that the solution was alarmingly simple. In fact, his idea was so simple his superiors didn't believe it. Whittle had grasped that a turbine would drive the compressor, 'making the principle of the jet engine essentially circular.'[69] Air sucked in by the compressor would be mixed with fuel and ignited. Ignition would expand the gas, which would flow through the blades of the turbine at such a high speed that not only would a jet stream be created, which would drive the aircraft forward, but the turning of the blades would also draw fresh air into the compressor, to begin the process all over again. If the compressor and the turbine were mounted on the same shaft, there was in effect only one moving part in a jet engine. It was not only far more powerful than a piston engine, which had many moving parts, but incomparably safer. Whittle was only twenty-two, and just as his height had done before, his age now acted against him. His idea was dismissed by the ministry in London. The rebuff hit him hard, and although he took out patents on his inventions, from 1929 to the mid-1930s, nothing happened. When the patents came up for renewal, he was still so poor he let them lapse.[70]

In the early 1930s, **Hans von Ohain**, a student of physics and aerodynamics at Göttingen University, had had much the same idea as Whittle. Von Ohain could not have been more different from the Englishman. He was aristocratic, well off, and over six feet tall. He also had a different attitude to the uses of his jet.[71] Spurning the government, he took his idea to the private planemaker **Ernst Heinkel**. Heinkel, who realised that high-speed air transport was much needed, took von Ohain seriously from the start. A meeting was called at his country residence, at Warnemünde on the Baltic coast, where the twenty-five-year-old Ohain was faced by some of Heinkel's leading aeronautical brains. Despite his youth, Ohain was offered a contract, which featured a royalty on all engines that might be sold. This contract, which had nothing to do with the air ministry, or the Luftwaffe, was signed in April 1936, seven years after Whittle wrote his paper.

Meanwhile in Britain Whittle's overall brilliance was by now so self-evident that two friends, convinced of Whittle's promise, met for dinner and decided to raise backing for a jet engine as a purely business venture. Whittle was still only twenty-eight, and many more experienced aeronautical engineers thought his engine would never fly. Nonetheless, with the aid of O. T. Falk and Partners, city bankers, a company called Power Jets was formed, and £20,000 raised.[72] Whittle was given shares in the company (no royalties), and the Air Ministry agreed to a 25 percent stake.

Power Jets was incorporated in March 1936. On the third of that month Britain's defence budget was increased from £122 million to £158 million, partly to pay for 250 more aircraft for the Fleet Air Arm for home defence. Four days later, German troops occupied the demilitarised zone of the Rhineland, thus violating the Treaty of Versailles. War suddenly became much more likely, a war in which air superiority might well prove crucial. All doubts about the theory of the jet engine were now put aside. From then on, it was simply a question of who could produce the first operational jet.

The intellectual overlap between physics and mathematics has always been considerable. As we have seen in the case of Heisenberg's matrices and Schrödinger's calculations, the advances made in physics in the golden age often involved the development of new forms of mathematics. By the end of the 1920s, the twenty-three outstanding math problems identified by David Hilbert at the Paris conference in 1900 (see chapter 1) had for the most part been settled, and mathematicians looked out on the world with optimism. Their confidence was more than just a technical matter; mathematics involved logic and therefore had philosophical implications. If math was complete, and internally consistent, as it appeared to be, that said something fundamental about the world.

But then, in September 1931, philosophers and mathematicians convened in Königsberg for a conference on the 'Theory of Knowledge in the Exact Sciences,' attended by, among others, Ludwig Wittgenstein, Rudolf Carnap, and Moritz Schlick. All were overshadowed, however, by a paper from a young mathematician from Brünn, whose revolutionary arguments were later published in a German scientific journal, in an article entitled 'On the Formally Undecidable Propositions of *Principia Mathematica* and Related Systems.'[73] The author was **Kurt Gödel**, a twenty-five-year-old mathematician at the University of Vienna, and this paper is now regarded as a milestone in the history of logic and mathematics. Gödel was an intermittent member of Schlick's Vienna Circle, which had stimulated his interest in the philosophical aspects of science.

In his 1931 paper he demolished Hilbert's aim of putting all mathematics on irrefutably sound foundations, with his theorem that tells us, no less firmly than Heisenberg's uncertainty principle, that there are some things we cannot know. No less importantly, he demolished Bertrand Russell's and Alfred North Whitehead's aim of deriving all mathematics from a single system of logic.[74]

There is no hiding the fact that Gödel's theorem is difficult. There are two elements that may be stated: one, that 'within any consistent formal system, there will be a sentence that can neither be proved true nor proved false'; and two, 'that the consistency of a formal system of arithmetic cannot be proved *within* that system'.[75] The simplest way to explain his idea makes use of the so-called Richard paradox, first put forward by the French mathematician Jules Richard in 1905.[76] In this system integers are given to a variety of definitions about mathematics. For example, the definition 'not divisible by any number except one and itself' (i.e., a prime number), might be given one integer, say 17. Another definition might be 'being equal to the product of an integer multiplied by that integer' (i.e., a perfect square), and given the integer 20. Now assume that these definitions are laid out in a list with the two above inserted as 17th and 20th. Notice two things about these definitions: 17, attached to the first statement, is itself a prime number, but 20, attached to the second statement, is not a perfect square. In Richardian mathematics, the above statement about prime numbers is not Richardian, whereas the statement about perfect squares is. Formally, the property of being Richardian involves 'not having the property designated by the defining expression with which an integer is correlated in the serially ordered set of definitions.' But of course this last statement is itself a mathematical definition and therefore belongs to the series and has its own integer, n. The question may now be put: Is n itself Richardian? Immediately the crucial contradiction appears. 'For n is Richardian if, and only if, it does *not* possess the property designated by the definition with which n is correlated; and it is easy to see that therefore n is Richardian if, and only if, n is not Richardian.'[77]

No analogy like this can do full justice to Gödel's theorem, but it at least conveys the paradox adequately. It is for some a depressing conclusion (and Gödel himself battled bouts of chronic depression. After living an ascetic personal life, he died in 1978, aged seventy-two, of 'malnutrition and inanition' brought about by personality disturbance).[78] Gödel had established that there were limits to math and to logic. The aim of Gottlob Frege, David Hilbert, and Russell to create a unitary deductive system in which all mathematical (and therefore all logical) truth could be deduced from a small number of axioms could not be realised. It was, in its way and as was hinted at above, a form of mathematical uncertainty principle – and it changed math for all time. Furthermore, as Roger Penrose has pointed out, Gödel's 'open-ended mathematical intuition is fundamentally incompatible with the existing structure of physics.'[79]

In some ways Gödel's discovery was the most fundamental and mysterious of all. He certainly had what most people would call a mystical side, and he thought we should trust [mathematical] intuition as much as other forms of

experience.[80] Added to the uncertainty principle, his theory described limits
to knowledge. Put alongside all the other advances and new avenues of thought,
which were then exploding in all directions, it injected a layer of doubt and
pessimism. Why should there be limits to our knowledge? And what did it
mean to know that such limits existed?

CIVILISATIONS AND THEIR DISCONTENTS

On 28 October 1929 the notorious stock market crash occurred on Wall Street, and U.S. loans to Europe were suspended. In the weeks and months that followed, and despite the misgivings of many, Allied troops prepared and then executed their departure from the Rhineland. In France, Georges Clemenceau died at the age of eighty-eight, while in Thuringia Wilhelm Frick was about to become the first member of the Nazi Party to be appointed minister in a state government. Benito Mussolini was clamouring for the revision of the Versailles Treaty, and in India Mohandâs Gandhi began his campaign of civil disobedience. In Britain in 1931 a National Government was formed to help balance the budget, while Japan abandoned the gold standard. There was a widespread feeling of crisis.

Sigmund Freud, then aged seventy-three, had far more personal reasons to feel pessimistic. In 1924 he had undergone two operations for cancer of the mouth. Part of his upper jaw had to be removed and replaced with a metal prosthesis, a procedure that could only be carried out using a local anaesthetic. After the operation he could chew and speak only with difficulty, but he still refused to stop smoking, which had probably been the cause of the cancer in the first place. Before he died in London in 1939, Freud underwent another two dozen operations, either to remove precancerous tissue or to have his prosthesis cleaned or renewed. During all this time he never stopped working.

In 1927 Freud had published *The Future of an Illusion*, which both explained away and yet amounted to an attack on organised religion. This was the second of three 'cultural' works by Freud (the first, *Totem and Taboo*, was discussed earlier: see above, page 141). At the end of 1929, as Wall Street was crashing, Freud delivered the third of these works, *Civilisation and Its Discontents*. There had been famine in Austria and attempted revolution and mega-inflation in Germany, and capitalism appeared to have collapsed in America. The devastation and moral degeneration of World War I was still a concern to many people, and Hitler was on the rise. Wherever you looked, Freud's title fitted the facts.[1]

In *Civilisation and Its Discontents*, Freud developed some of the ideas he had explored in *Totem and Taboo*, in particular that society − civilisation − evolves out of the need to curb the individual's unruly sexual and aggressive appetites. He now argued that civilisation, suppression, and neurosis are inescapably intertwined because the more civilisation there is, the more suppression is needed and, as a direct result, the more neurosis. Man, he said, cannot help but be more and more unhappy in civilisation, which explains why so many seek refuge in drink, drugs, tobacco, or religion. Given this basic predicament, it is the individual's 'psychical constitution' which determines how any individual adjusts. For example, 'The man who is predominantly erotic will give first preference to his emotional relationships with other people; the narcissistic man, who inclines to be self-sufficient, will seek his main satisfactions in his internal mental process.'[2] And so on. The point of his book, he said, was not to offer easy panaceas for the ills of society but to suggest that ethics – the rules by which men agree to live together – can benefit from psychoanalytic understanding, in particular, the psychoanalytic concept of the superego, or conscience.[3]

Freud's hopes were not to be fulfilled. The 1930s, especially in the German-speaking countries, were dominated more by a complete lack of conscience than any attempt to refine or understand it. Nevertheless, his book spawned a raft of others that, though very different from his, were all profoundly uneasy with Western capitalist society, whether the source of concern was economics, science and technology, race, or man's fundamental nature as revealed in his psychology. The early 1930s were dominated by theories and investigations exploring the discontents of Western civilisation.

The book closest to Freud's was published in 1933 by the former crown prince of psychoanalysis, now turned archrival. Carl Jung's argument in *Modern Man in Search of a Soul* was that 'modern' society had more in common with 'archaic,' primitive society than it did with what had gone immediately before – i.e., the previous phase of civilisation.[4] The modern world was a world where the ancient 'archetypes' revealed themselves more than they had done in the recent past. This explained modern man's obsession with his psyche and the collapse of religion. The modern condition was that man knew he was the culmination of evolution – science told him so – but also knew that 'tomorrow he will be surpassed,' which made life 'solitary, cold, and frightening.'[5] Further, psychoanalysis, by replacing the soul with the psyche (which Jung clearly thought had happened), only offered a palliative. Psychoanalysis, as a technique, could only be used on an individual basis; it could not become 'organised' and used to help millions at a time, like Catholicism, say. And so, the participation mystique, as the anthropologist Lucien Lévy-Bruhl called it, was a whole dimension of life closed to modern man. It set Western civilisation, a new civilisation, apart from the older Eastern societies.[6] This lack of a collective life, ceremonies of the whole as Hugo von Hofmannsthal called them, contributed to neurosis, and to general anxiety.[7]

For fifteen years, Karen Horney practised in Weimar Germany as an orthodox Freudian analyst, alongside Melanie Klein, Otto Fenichel, Franz Alexander, Karl Abraham and Wilhelm Reich at the Berlin Psychoanalytic Institute. Only after she moved to the United States, first as associate director of the Chicago Institute and then in New York, at the New School for Social Research and the New York Psychoanalytic Institute, did she find herself capable of offering criticism of the founder of the movement. Her book, *The Neurotic Personality of Our Time*, overlapped with both Freud and Jung but was also an attack on capitalistic society for the way it induced neurosis.[8]

Horney's chief criticism of Freud was his antifeminist bias (her early papers included 'The Dread of Women' and 'The Denial of the Vagina'). But she was also a Marxist and thought Freud too biological in outlook and 'deeply ignorant' of modern anthropology and sociology (she was right). Psychoanalysis had itself become split by this time into a right wing and a left wing. What may be characterised as the right wing concentrated on biological aspects, delving further and further into infantile experience. Melanie Klein, a German disciple of Freud who moved to Britain, was the leader of this approach. The left wing, which consisted in the main of Horney, Erich Fromm, and Harry Stack Sullivan, was instead more concerned with the individual's social and cultural background.[9]

Horney took the line that 'there is no such thing as a universal normal psychology.'[10] What is regarded as neurotic in one culture may be normal elsewhere, and vice versa. For her, however, two traits invariably characterised all neurotics. The first was 'rigidity in reaction,' and the second was 'a discrepancy between potentiality and achievement.' For example, a normal person by definition becomes suspicious of someone else only after that person has behaved badly toward them; the neurotic 'brings his or her suspicion with them at all times.' Horney didn't believe in the Oedipus complex either. She preferred the notion of 'basic anxiety,' which she attributed not to biology but to the conflicting forces of society, conflicts that act on an individual from childhood. Basic anxiety she characterised as a feeling of 'being small, insignificant, helpless, endangered, in a world that is out to abuse, cheat, attack, humiliate, betray, envy.'[11] Such anxiety is worse, she said, when parents fail to give their children warmth and affection. This usually occurs in families where the parents have their own unresolved neuroses, initiating a vicious circle. By definition, the neurotic personality has lost, or never had, 'the blissful certainty of being wanted.'[12] Such a child grows up with one of four rigid ways of approaching life, which interfere with achievement: the neurotic striving for affection; the neurotic striving for power; neurotic withdrawal; and neurotic submissiveness.[13]

The most contentious part of Horney's theory, for nonpsychoanalysts, was her blaming neurosis on the contradictions of, in particular, contemporary American life. She insisted that in America more than anywhere else there existed an inherent contradiction between competition and success on the one hand ('never give a sucker an even break') and good neighborliness on the other ('love your neighbour as yourself'); between the promotion of ambition by advertising ('keeping up with the Joneses') and the inability of the individual to satisfy these ambitions; between the creed of unfettered individualism and the ever more common curbs brought about by environmental concerns and more laws.[14] This modern world, despite its material advantages, foments the feeling in many individuals that they are 'isolated and helpless.'[15] Many would agree that they feel isolated and helpless, and maybe even neurotically so. But

Horney's theory never explains why some neurotics need affection, and others power, and why some become submissive. She herself denied that biology was responsible but never clarified what else might account for such large differences in behaviour.

Horney's feminism was new but not unique. The campaign to gain women the vote had exercised politicians in several countries prior to World War I, not least in Austria and Great Britain. Immediately after the war other matters had taken priority, both economically and psychologically, but as the 1920s passed, the status of women again became an issue.

One of the minor themes in Virginia Woolf's *Jacob's Room* is the easy effortlessness of the men who led Britain into war, and their casual treatment of women. Whereas all the men in the book have comfortable sets of rooms from which to embark on their fulfilling lives, the women always have to share, or are condemned to cold and draughty houses. This was a discrepancy Woolf was to take up in her most famous piece of nonfiction, *A Room of One's Own*, published in 1929. It appears that being turned away from an Oxbridge college library because she was a woman propelled her to write her feminist polemic. And it is certainly arguable that the greatest psychological revolution of the century has been in the female sensibility.[16]

By 1929 Virginia Woolf had published six novels. These included *Jacob's Room*, in the miracle year of 1922, *Mrs Dalloway* (1925), *To the Lighthouse* (1927), and *Orlando* in 1928. Her success, however, only seems to have made her more unsettled about the situation most female writers found themselves in. Her central argument in the 100-page essay was that 'a woman must have money and a room of her own if she is to write fiction.'[17] Her view, which was to be echoed by others in different ways later in the century, was that a writer 'is the product of his or her historical circumstances and that material conditions are crucially important' – not just to whether the books get written but to the psychological status of the writer, male or female. But women were the main focus of her attention, and she went on to show how, in Britain at least, until the Married Women's Property Acts of 1870 and 1882, a married woman's income legally belonged to her husband. There could be no freedom of the mind, she felt, without freedom of circumstance. This meant that prior to the end of the seventeenth century there were very few women writers, and those who did write often only dabbled in it. Woolf herself suffered, in that the boys in her own family went to boarding school and then to university, whereas she and the other girls were educated at home.[18] This brought several consequences. Female experience was underreported in fiction, and what experience was reported was inevitably distorted and/or restricted to certain kinds. For example, she felt that Jane Austen was not given access to the wider world that her talent demanded, with similar restrictions applying also to Elizabeth Barrett Browning: 'It cannot be doubted that the long years of seclusion had done her [Browning] irreparable damage as an artist.'[19]

Though she felt feminist anger, Woolf was very clear that such anger had no place in fiction, which should have larger ambitions for itself, and she criticised earlier writers, like Browning and Charlotte Brönte, for giving way to that anger. She then moved on to consider the ways in which the female mind might complement the male mind, in an effort to show what literature has lost by the barriers erected against women. For example, she considers Samuel Taylor Coleridge's notion of the androgynous mind, with male and female qualities coexisting in harmony, to be open to all possibilities. She makes no case for the superiority of either sex, but rather for the mind that allows both sympathies equal access. She actually wrote that it is 'fatal for anyone who writes to think of their sex.'[20] She herself described *A Room* as a trifle, but she also said she wrote it with ardour, and it has certainly been a huge success. One reason is the style. When the book was published, in October 1929, it was reviewed in the *Sunday Times* of London by Desmond MacCarthy, who described it as 'feminist propaganda' but added 'yet it resembles an almond-tree in blossom.'[21] Woolf's style is conversational, intimate. She manages to be both angry and above anger at the wrongs done to women writers, and would-be women writers, in the past. She devotes pages to the lunches she has eaten at Oxbridge colleges – where she says the food is much better in the women's colleges than the men's. And she makes it matter. Of course, Virginia Woolf's fiction should be read alongside *A Room of One's Own*. She did help emancipate women not only by her polemic but also by her example.

Psychoanalysts and novelists were not the only people analysing the short-comings of civilisations. Anthropologists, sociologists, philosophers, and report-ers were all obsessed by the same theme. The 1930s were an especially fruitful time for anthropology. This discipline not only offered implicit comparison with, and criticism of, the capitalist way of life, but provided examples of more or less successful alternatives.

Franz Boas still dominated anthropology. His 1911 book *The Mind of Primitive Man* made clear his loathing of nineteenth-century ideas that took for granted the inherent superiority of white Westerners. For Boas, anthropology 'could free a civilisation from its own prejudices.' The sooner data from other civ-ilisations could be gathered and assimilated into the general consciousness, the better. Boas's powerful and passionate advocacy had made anthropology seem a thrilling subject and an advance on the outmoded ethnocentrism of previous decades and the vague biologism of psychoanalysis. Two of Boas's students, **Margaret Mead** and **Ruth Benedict**, produced highly influential studies that further undermined biologism. Like Boas, Mead and Benedict were interested in the link between race, genetics (still an infant subject) and culture. Mead had a master's degree in psychology, but like many others she found anthropology more alluring and had been inspired by Ruth Benedict. Reticent to the point where her fellow students thought her depressed (they hated what they called her 'castor oil' faces), Ruth Benedict began to inspire respect. She and Mead eventually formed part of an influential international network of anthropologists and psychiatrists which also included Geoffrey Gorer, Gregory Bateson, Harry Stack Sullivan, Erik Erikson, and Meyer Fortes.

For Boas anthropology was, as Mead later put it, 'a giant rescue operation' to show the importance of culture.[22] Boas gave Margaret Mead the idea that made her famous while she was still in her twenties: he suggested she study adolescence in a non-Western society. It was a clever choice, for adolescence was arguably part of the pathology of Western culture. In fact, adolescence had been 'invented' only in 1905, in a study by the American psychologist G. Stanley Hall (a friend of Freud).[23] His *Adolescence: Its Psychology and Its Relations to Physiology, Anthropology, Sociology, Sex, Crime, Religion and Education* referred to over sixty studies of physical growth alone and portrayed adolescence 'as the period in which idealism flowered and rebellion against authority waxed strong, a period during which difficulties and conflicts were absolutely inevitable.'[24] In other words, it was psychologically crucial. Boas was sceptical of the idea that the problems of adolescence were purely or largely biological. He felt they must owe as much to culture as to genes.[25]

In September 1925 Margaret Mead spent several weeks in Pago Pago, capital of Tutuila, the chief island of American **Samoa** in the southwest Pacific Ocean.[26] She stayed at a hotel made famous by Somerset Maugham in his 1920 story 'Rain,'[27] learning the basics of the Samoan language before launching on her field study.[28] Mead told Boas that from her preliminary survey she proposed to spend her time on Ta'u, one of three small islands in the Manu'a group, about a hundred miles east of Pago Pago. This was 'the only island with villages where there are enough adolescents, which are at the same time primitive enough and where I can live with Americans. I can eat native food, but I can't live on it for six months; it is too starchy.'[29] A government steamer stopped at the islands every few weeks, but she thought that was too infrequent to spoil the island's status as an uncontaminated separate culture; the people of Ta'u were 'much more primitive and unspoiled than any other part of Samoa. . . . There are no white people on the island except the navy man in charge of the dispensary, his family, and two corpsmen.' The climate was far from perfect: year-round humidity of 80 percent, temperatures of 70–90 degrees, and 'furious rains' five times a day, which fell in 'drops the size of almonds.' Then the sun would come out, and everything on the island, including the people, would 'steam' until they were dry.[30]

Mead's account of her fieldwork, *Coming of Age in Samoa*, was phenomenally successful when it appeared in 1928. Her introduction to the book concluded with an account of what happened on the island after dark. In the moonlight, she wrote, 'men and maidens' would dance and 'detach themselves and wander away among the trees. Sometimes sleep will not descend upon the village until long past midnight; then at last there is only the mellow thunder of the reef and the whisper of lovers, as the village rests until dawn.'[31] She described 'horseplay' between young people, 'particularly prevalent in groups of young women, often taking the form of playfully snatching at the sex organs.' She said she was satisfied that, for these girls, adolescence 'represented no period of crisis or stress, but was instead an orderly development of a set of slowly maturing interests and activities. The girls' minds were perplexed by no conflicts, troubled by no philosophical queries, beset by no remote ambitions. . . . To live as a girl with as many lovers as possible and then to marry in one's own village, near one's own relatives and to have many children, these were uniform and satisfying ambitions.' Samoans, she insisted, had not the faintest idea of 'romantic love as it occurs in our civilisation, inextricably bound up with ideas of monogamy, exclusiveness, jealousy and undeviating fidelity.'[32] At the same time, the concept of celibacy was 'absolutely meaningless.'[33]

Samoa, or at least Ta'u, was an idyll. For Mead, the island existed only in 'pastel tones,' and she assumed that the picture was true for Samoa as a whole. In fact, this generalisation was inaccurate, for the main island had recently, in 1924, seen political problems and a killing. In Ta'u Mead was isolated and treated very well, the Samoans nicknaming her 'Makelita' after one of their dead queens. One of the reasons why *Coming of Age in Samoa* was so successful was that when her publisher, William Morrow, received the first draft of the manuscript, he suggested that she add two chapters explaining the relevance of her findings for Americans and American civilisation. In doing so, she stressed 'Papa Franz's' approach, emphasising the predominance of culture over that of biology. Adolescence didn't need to be turbulent: Freud, Horney, and the others were right – Western civilisation had a lot to answer for. The book was welcomed by the sexologist Havelock Ellis; by Bronislaw Malinowski, an anthropologist and the author of *The Sexual Life of Savages*; and by H. L. Mencken. Mead quickly became the most famous anthropologist in the world.[34]

She followed *Samoa* with two more field studies in the early 1930s, *Growing Up in New Guinea* (1930) and *Sex and Temperament in Three Primitive Societies* (1935). In these books, one critic remarked, Margaret Mead took a 'diabolical delight' in stressing how little difference there is between so-called civilised man and his more 'primitive' cousins. But that was unfair. Mead was not uncritical of primitive societies, and the whole thrust of her books was to draw attention to variation in cultures. In New Guinea, children might be allowed to play all day long, but, she said, 'alas for the theorists, their play is like that of young puppies or kittens. Unaided by the rich hints for play which children of other societies take from the admired adult traditions, they have a dull, uninteresting child life, romping good-humoredly until they are tired, then lying inert and breathless until rested sufficiently to romp again.'[35] In *Sex and Temperament*, in which she looked at the Arapesh, she found that warfare was 'practically unknown,' as was personal aggression. The Arapesh had little in the way of art and, what she foundest oddest of all, little differentiation between men and women, at least in terms of psychology.[36] Moving on from the Arapesh to the Mundugumor, on the Yua River, a tributary of the Sepik (also in New Guinea), she found a people that, she said, she loathed.[37] Only three years before, headhunting and cannibalism had been outlawed. Here she recorded that it was not uncommon to see the bodies of very small children floating, 'unwashed and unwanted,' down the river.[38] 'They are always throwing away infants here,' Mead wrote. Babies that were wanted, she said, were carried around in rigid baskets that they couldn't see out of and which didn't let in much light. The children were never cuddled or comforted when they cried, so that for Mead it was hardly surprising they should grow up feeling unloved or that Mundugumor society should be 'riddled with suspicion and distrust.' In the third society, the

Tchambuli, fifty miles up the Sepik River, the familiar roles of men and women in Western society were reversed. Women were the 'dominant, impersonal, managing partners,' and men were 'less responsible and emotionally dependent.'[39] Mead's conclusion, after this 'orgy of fieldwork,' was that 'human nature is almost unbelievably malleable, responding accurately and contrastingly to contrasting cultural conditions.'

Ruth Benedict's *Patterns of Culture*, published the same year as *Sex and Temperament in Three Primitive Societies*, might have been called 'Sex and Temperament, Economic Exchange, Religion, Food Production and Rivalry in Three Primitive Societies,' for the two books had much in common.[40] Benedict looked at the Zuni Indians of New Mexico (native Americans were called 'Indians' in those days, even by anthropologists), the Dobu of New Guinea, and the Kwakiutl, who lived on the Pacific coast of Alaska and Puget Sound. Here again large idiosyncrasies in culture were described. The Zuni were 'a people who value sobriety and inoffensiveness above all other virtues,' who placed great reliance on imitative magic: water was sprinkled on the ground to produce rain.[41] Children were whipped ceremonially from time to time 'to take off the bad happenings.'[42] Ownership of property – in particular the sacred fetishes – was in the matrilineal line, and the dominant aspect of Zuni life, religion apart, was its polite orderliness, with individuality lost within the group. The Dobu, in contrast, were 'lawless and treacherous'; 'the social forms which obtain in Dobu put a premium on ill-will and treachery and make of them the recognised virtues of their society.'[43] Faithfulness was not expected between husband and wife, broken marriages were 'excessively common,' and a special role was played by disease. If someone fell ill, it was because someone else willed it. Disease-charms were widely sold, and some individuals had a monopoly on certain diseases. In trade the highest value was put on cheating the other party. 'The Dobu, therefore, is dour, prudish and passionate, consumed with jealousy and suspicion and resentment. Every moment of prosperity he conceives himself to have wrung from a malicious world by a conflict in which he has worsted his opponent.'[44] Ecstatic dancing was the chief aspect of Kwakiutl religion, and inherited property – which even included areas of the sea, where halibut was found, for example – was the chief organisational basis of society. Immaterial things, like songs and myths, were forms of wealth, some of which could be gained by killing their possessors. The Kwakiutl year was divided into two, the summer, when wealth and social privileges were honoured, and winter, when a more egalitarian society prevailed.[45]

Benedict's chapters reporting on primitive societies were bracketed by polemical ones. Here her views clearly owe a huge debt to Boas. Her main theme aimed to show human nature as very malleable; that geographically separate societies may be integrated around different aspects of human nature, giving these societies a distinctive character. Some cultures, she said, were 'Dionysian,' organised around feeling, and others 'Apollonian,' organised around rationality.[46] And in a number of wide-ranging references she argued that Don Quixote, Babbitt, Middletown, D. H. Lawrence, the homosexuality in Plato, may all best be understood in an anthropological context, that is to say as normal variations in human nature that are fundamentally incommensurable. Societies must be understood on their own terms, not on some single scale (where, of course, 'we' – whites – always come out on top). In creating their own 'patterns of culture,' other societies, other civilisations, have avoided some of the problems Western civilisation faces, and created their own.[47]

It is almost impossible now to recover the excitement of anthropology in the 1920s and 1930s.[48] This was an era before mass air travel, mass tourism, or television, and the exploration of these 'primitive' societies, before they changed or were killed off, was one of the last great adventures of the world. The anthropologists were a small number of people who all knew each other (and in some cases married each other: Mead had three husbands, two of them anthropologists, and was for a time Benedict's lover). There was an element of the crusade in their work, to show that all cultures are relative, a message wrapped up in their social/political views (Mead believed in open marriage; Benedict, from a farming family, was self-educated).

Benedict's book was as successful as Mead's, selling hundreds of thousands of copies over the years, available not just in bookstores but in drugstores, too. Together these two students of Boas, using their own research but also his and that of Malinowski and Mead's husband, Reo Fortune, transformed the way we look at the world. Unconscious ethnocentrism, not to say sexual chauvinism, was much greater in the first half of the century than it is now, and their conclusions, presented scientifically, were vastly liberating. The aim of Boas, Benedict, and Mead was to put beyond doubt the major role played by culture in determining behaviour and to argue against the predominating place of biology. Their other aim – to show that societies can only be understood on their own terms – proved durable. Indeed, for a comparatively small science, anthropology has helped produce one of the biggest ideas of the century: relativism. Margaret Mead put this view well. In 1939, lying on her back, her legs propped against a chair ('the only posture,' she explained, 'for a pregnant woman'), she jotted down some thoughts for the foreword to *From the South Seas*, an anthology of her writing about Pacific societies. 'In 1939,' she noted prophetically, 'people are asking far deeper and more searching questions from the social sciences than was the case in 1925. . . . We are at a crossroads and must decide whether to go forward towards a more ordered heterogeneity, or make frightened retreat to some single standard which will waste nine-tenths of the potentialities of the human race in order that we may have a too dearly purchased security.'[49]

Sociologists were not tempted by exotic foreign lands. There was enough to do at home, trying to make sense of the quiddities thrown up by Western capitalism. Here, a key figure was **Robert E. Park**, professor of sociology at the University of Chicago and the man who more than anyone else helped give sociology a more scientific status. Chicago University was the third of the three great research universities established in America in the late nineteenth century, after Johns Hopkins and Clark. (It was these research universities that first made the Ph.D. a requirement for would-be scholars in the United States.) Chicago established four great schools of thought: philosophy, under John

Dewey, sociology, under Park, political science, under Charles Merriam, and economics, much later in the century, under Milton Friedman. Park's great achievement in sociology was to turn it from an essentially individual, observational activity into a much more empirically based discipline.[50]

The first noteworthy Chicago study was *The Polish Peasant in Europe and America*, now generally forgotten but regarded by sociologists as a landmark that blended empirical data and generalisation. W. I. Thomas and Florian Znaniecki spent several months in Poland, then followed thousands of Polish immigrants to America, examining the same people on both sides of the Atlantic. They gained access to private correspondence, archives from the Bureau of Immigration, and newspaper archives to produce as complete a picture as possible of the whole migration experience. That was followed by a series of Chicago studies which examined various 'discontents' of the age, or symptoms of it – *The Gang*, by Frederic Thrasher, in 1927; *The Ghetto*, by Louis Wirth, *Suicide*, by Ruth Shonle Cavan, and *The Strike*, by E. T. Hiller, all published in 1928; and *Organised Crime in Chicago*, by John Landesco, released in 1929. Much of this research was directly related to policy – helping Chicago reduce crime or suicide, or get the gangs off the streets. Park always worked with a local community committee to ensure his studies chimed with the real concerns of local people. But the importance of Chicago sociology, which exerted its greatest influence between 1918 and 1935, had as much to do with the development of survey techniques, nondirective interviewing, and attitude measurement, all of which were intended to produce more psychological ways of grouping people, going beyond the picture painted in bland government censuses.[51]

The most significant Chicago survey was an examination of the discontent that most maimed American civilisation (a rival even to the unemployment caused by the Great Depression): race. In 1931 **Charles Johnson** published *The Negro in American Civilisation* and for the first time froze a statistical picture of the black American against which his progress, or lack of it, could be measured.[52] Johnson was actually on the faculty of Fisk University when the book came out, but he had trained under Park and, in 1922, published *The Negro in Chicago* as one of the sociology department's series of studies.[53] Johnson, more than anyone else, helped create the Harlem Renaissance and believed that if the American Negro could not achieve equality or respect in any other way, he should exploit the arts. Throughout the 1920s, Johnson had edited the New York magazine for blacks, *Opportunity*, but toward the end of the decade he returned to academia. The subtitle of his new book was 'A Study of Negro Life and Race Relations in the Light of Social Research,' and the research element was its strong point. The book, the most thorough analysis of Negro status yet produced, compiled government records and reports, health and crime statistics, charts, tables, graphs, and lists. At that time, many blacks – called Negroes then – could remember slavery, and some had fought in the Civil War.

The statistics showed that the lives of blacks had improved. Illiteracy had been reduced among Negroes from 70 percent in 1880 to 22.9 percent in 1920.

But of course that compared very badly, still, with the white illiteracy rate of 4.1 percent in 1920.[54] The number of lynchings was down from 155 in 1892 to 57 in 1920 and 8 in 1928, the first time it had fallen to single figures. But eight lynchings a year was still a fearful statistic.[55] More enlightening, perhaps, was the revealing way in which prejudices had evolved. For example, it was widely assumed that there was so pronounced a susceptibility among Negroes to tuberculosis that expenditures for preventive or corrective measures were practically useless. At the same time, it was believed that Negroes had a corresponding immunity to such diseases as cancer, malaria, and diabetes, so that no special measures of relief were necessary. It did not go unnoticed among Negroes that the majority opinion always interpreted the evidence to the minorities' disadvantage.[56] What Johnson's survey also showed, however, and for the first time in a thorough way, was that many social factors, rather than race per se, predetermined health. In one survey of fifteen cities, including New York, Louisville and Memphis, the population density of Negroes was *never* less than that for whites, and on occasions four times as high.[57] Mortality rates for Negroes in fifteen states were always higher than for whites, and in some cases twice as high. What emerged from the statistics was a picture that would become familiar – Negroes were beginning to occupy the inner-city areas, where the houses were smaller, less well built, and had fewer amenities. Already there were differences in what was then called 'law observance.'[58] A survey of ten cities – Cleveland, Detroit, Baltimore, and others – showed Negroes as two to five times as likely to be arrested as whites, though they were three and a half times *less* likely to be sentenced to a year or more in prison. Whatever was being shown here, it wasn't a biological propensity on the part of Negroes to commit violence, as many whites argued.

W. E. B. Du Bois's chapter in Johnson's book repeated his argument that the supposed biological differences between the races must be ignored. Instead attention should be focused on the sociological statistics – now amply widened – which disclosed the effects of discrimination on the status of the Negro. The statistics were particularly useful, he said, in the realm of education. In 1931 there were 19,000 black college students compared with 1,000 in 1900, 2,000 black bachelors of arts compared with 150. Those figures nailed the view that Negroes could never benefit from education.[59] Du Bois never wavered from his position that the obsession with biological and psychological differences was a device for prejudiced whites to deny the very real sociological differences between races, for which they – the whites – were largely to blame. Herbert Miller, a sociologist from Ohio State University, felt that the tighter controls on immigration introduced in the 1920s had 'profoundly affected race relations by substituting the Negro for the European' as the object of discrimination.[60] The long-term message of *The Negro in American Civilisation* was not optimistic, confounding America's view of itself as a place where everything is possible.

Charles Johnson, the black, urban, sophisticated polymath and star of the Harlem Renaissance, could not have been more different from **William Faulkner**, a rural, white monomaniac (in the nicest sense) from the Deep South.

Between 1929 and 1936 Faulkner produced his four masterpieces, *The Sound and the Fury* (1929), *As I Lay Dying* (1930), *Light in August* (1932), and *Absalom, Absalom!* (1936), the last two of which specifically confront the issue of black and white.

Faulkner, who lived in Oxford, Mississippi, was obsessed by the South, its obsession with itself and with its history, what his biographer called 'the great discovery.'[61] For Faulkner the South's defeat in the Civil War had trapped it in the past. He realised that whereas most of America was an optimistic country without much of a past, and with immigrants forever reshaping the present, the South was a very different enclave, almost the opposite of the thrusting North and West Coast. Faulkner wanted to explain the South to itself, to re-create its past in an imaginative way, to describe the discontents of a civilisation that had been superseded but refused to let go. All his great books about the South concern proud dynastical families, the artificial, arbitrary settings in which barriers are forever being transgressed, in particular those of class, sex, and race. Families are either on the rise or on the wane, and in the background is shame, incest, and in the case of *Light in August* and *Absalom, Absalom!* miscegenation. These unions raise passions, violent passions, death and suicide, frustrating dynastic ambitions.

Most typical of Faulkner's approach is *Absalom, Absalom!* for in addition to its plot, this book, like *The Sound and the Fury* and *As I Lay Dying*, is notoriously difficult. Faulkner imposes strong demands on the reader – flashbacks in time, rapid alternation in viewpoint without warning, obscure references that are only explained later.[62] His aim is to *show* the reader the confusion of society, unhelped by the author's guiding hand. Just as his characters work on themselves to create their identities and fortunes, the reader must work out Faulkner's meaning.[63]

Absalom, Absalom! begins when Miss Rosa Coldfield summons Quentin Compson, a friend and amateur historian, and tells him a story about the rise and fall of Thomas Sutpen, the founder of a southern dynasty whose son, Henry, shot his friend Charles Bon, who he had fought with in the war, causing the demise of the dynasty.[64] What motive could Henry Sutpen have had for killing his best friend? Gradually Compson fills in the gaps in the story – using his imagination where facts are too sparse.[65] Eventually, the mystery is solved. Charles Bon was actually the fruit of an earlier union by Thomas Sutpen and a Negro (and therefore his eldest child). In Sutpen's refusal to recognise his eldest son, we see the 'great guilt' underlying the whole edifice of the dynasty, and by implication the South itself. Faulkner does not shirk the moral dilemmas, but his main aim was to describe the pain that is their consequence. While Charles Johnson catalogued the shortcomings of northern urban American society, Faulkner illuminated – with sympathy – that the South had its imperfections too.

If race was (still) America's abiding problem, in Europe and particularly in Britain it was class that divided people. Here, one man who did so much to publicise the great poverty associated with Britain's lower classes, especially in the 1930s following the great crash, was the writer and reporter **George Orwell**. It was no accident that Orwell was a reporter as well as a novelist, or that he should prefer reportage to bring home his message. The great age of reportage, as Eric Hobsbawm tells us, had only recently begun, in the 1920s, following the growth of new media, like *Time* and newsreels. The word *reportage* itself first appeared in French dictionaries in 1929, and in English in 1931. Many novelists of the time (Ernest Hemingway, Theodore Dreiser, Sinclair Lewis) were or had been or would become reporters.[66]

Orwell, born Eric Blair in the remote town of Motihari in Bengal, northwest of Calcutta, on 25 June 1903, received a conventional – that is to say, privileged – middle-class upbringing in Britain. He went to Saint Cyprian's school near Eastbourne, where Cyril Connolly was a friend and where he wet the bed, then was sent to Wellington and Eton.[67] After school he joined the Indian imperial police and served in Burma. Dissatisfied with his role in the imperial police, Blair cut short his time in Burma and began his career as a writer. 'Feeling tainted by his "success" as a young officer in the East, he wanted to shun anything that reminded him of the unjust system which he had served. "I felt that I had got to escape not merely from imperialism but from every form of man's dominion over man," he explained later. "Failure seemed to me to be the only virtue. Every suspicion of self-advancement, even to 'succeed' in life to the extent of making a few hundreds a year, seemed to me spiritually ugly, a species of bullying." '[68]

It is too simple to say that Blair's desire not to succeed was the direct result of his experience in Burma.[69] The idea had planted itself in his mind long before he became a police officer. Saint Cyprian's, says his biographer Michael Shelden, had prejudiced him against success very early in life by giving him such a corrupt view of merit. Winning was the only thing that mattered at the school, and one became a winner by 'being bigger, stronger, handsomer, richer, more popular, more elegant, more unscrupulous than other people' – in short, 'by getting the better of them in every way.' Later, he put it like this: 'Life was hierarchical and whatever happened was right. There were the strong, who deserved to win and always did win, and there were the weak, who deserved to lose and always did lose, everlastingly.'[70] He was made to feel that he was one of the weak, and that, whatever he did 'he would never be a winner. The one consolation for him was the knowledge that there was honour in losing. One could take pride in rejecting the wrong view of success . . . I could accept my failure and make the best of it.'[71] Of Orwell's four most famous books, two explored in reportorial fashion the weakest (and poorest) elements of society, the flotsam of the 1930s capitalist world. The other two, produced after World War II, explored the nature of power, success, and the way they so easily become abused.

After leaving the police, Blair stayed with his parents for a few months but in the autumn of 1927 found a small room in the Portobello Road, in west London. He tried his hand at fiction and began to explore the East End of the city, living cheek by jowl with tramps and beggars in order to understand how the poor lived, and to experience something of their suffering.[72] Having rejected 'every form of man's dominion over man,' he wanted 'to get right down among the oppressed, to be one of them and on their side against their tyrants.' Blair worried at his appearance on these visits. He acquired a shabby coat, black dungaree trousers, 'a faded scarf, and a rumpled cap'. He changed the way he spoke, anxious that his educated accent would give him away. He soon grew to know the seedy area around the West India docks, mixing with stevedores, merchant sailors, and unemployed labourers and sleeping at a common lodging house in Limehouse Causeway (paying nine pence a night). Being accepted in this way, he decided to go 'on the road' and for a while meandered through the outreaches of the East End, overnighting in dingy 'spikes' – the barracks of local workhouses. These sallies formed the backbone of *Down and Out in Paris and London*, which came out in 1933. Of course, Orwell was never really down and out; as Michael Shelden says, his tramping was something of a game, one that reflected his ambivalence toward his own background, his ambitions, and his future. But the game was not entirely frivolous. The best way he could help those who were less fortunate was to speak up for them, 'to remind the rest of the world that they existed, that they were human beings who deserved better and that their pain was real.'[73]

In 1929 Orwell went to Paris, to show that the misery wasn't confined to just one country. There he took a small room at a run-down hotel in the rue du Pot de Fer, a narrow, mean lane in the Latin Quarter. He described the walls of his room as thin; 'there was dirt everywhere in the building and bugs were a constant nuisance.'[74] He suffered a nervous breakdown.[75] There were more cheerful neighborhoods not far away, however, in one of which could be found the Ecole Normale Supérieure, where Jean-Paul Sartre was a star pupil and where Samuel Beckett was just beginning to teach. Further on was the place de la Contrescarpe, which Hemingway describes in *The Snows of Kilimanjaro*, affectionately sketching its mix of 'drunks, prostitutes, and respect-able working folk.'[76] Orwell says in the book that he was the victim of a theft that left him almost penniless.[77]

The book was published by **Victor Gollancz**, who had begun his company in 1929 with offices in Covent Garden. Gollancz was a driven man, a canny bargainer, and soon his business was thriving. He paid his authors small advances but spent much larger sums on advertising. He published all kinds of books, but politics was his first love, and he was a passionate socialist. Orwell's book was as much sociological as political, but it appealed to Gollancz 'as a powerful statement against social injustice.'[78] Published at the beginning of January 1933, it was an immediate success, widely praised in the press (by, among others, Compton Mackenzie). Orwell realised that no quick or glib remedy for poverty could possibly work. What he was after was a change in perception, so that poverty would no longer be regarded 'as a kind of shameful disease which infects people who are incapable of helping themselves.'[79] He emphasised the point that even many charity workers expected 'some show of contrition, as though poverty signified a sinful soul.' This attitude, he felt, and the continued existence of poverty were linked.

Down and Out was followed by three novels, *Burmese Days*, *A Clergyman's Daughter*, and *Keep the Aspidistra Flying*. Each of these examined an aspect of

British life and helped establish Orwell's reputation. In 1937 he returned to his reportorial/sociological writing with *The Road to Wigan Pier*, which arose out of his heightened political awareness, the rise of Hitler and Mussolini, and Orwell's growing conviction that 'Socialism is the only real enemy Fascism has to face.'[80] Gollancz had asked him to write a book about unemployment – the scourge of the 1930s since the great crash. It was hardly an original idea, and indeed Orwell had himself refused an almost identical proposal from the *News Chronicle* some months before.[81] But feeling that he had to be more politically engaged, he agreed. Starting in Coventry, he moved north to Manchester, where he boarded with a trade union official who suggested that Orwell visit Wigan.[82] He found lodgings over a tripe shop, sleeping in shifts, and in his room he found no sign that anyone had bothered to clean or dust 'in ages'; he was told by other lodgers 'that the supplies of tripe in the cellar were covered with black beetles'. One day he was 'disconcerted' to find a full chamberpot under the table at breakfast.[83] According to Shelden, he spent hours at the local library compiling statistics on the coal industry and on unemployment, but most of the time he spent travelling, inspecting housing conditions, the canals, and the mines, interviewing workers and unemployed. He later described Wigan as a 'dreadful place' and the mines as a 'pretty devastating experience.' He had to go to bed for a day to get over it.[84] 'He had not realised that a man of his height could not stand upright in the mine, that the walk from the shaft to the coal face could be up to three miles and that this cramped combination "was enough to put my legs out of action for four days." Yet this walk was only the beginning and end of the miner's work day. "At times my knees simply refused to lift me after I had knelt down." '[85]

Figures Orwell obtained in the library – available to anyone – established that miners suffered an appalling rate of accidents. In the previous eight years, nearly 8,000 men had been killed in the mines; one miner in six was injured. Death was so common in the mines it was almost routine: 'A shilling was deducted from the men's pay whenever a fellow-miner was killed – and the money contributed to a fund for the widow. But this deduction, or "stoppage," occurred with such grim regularity that the company used a rubber stamp marked "Death stoppage" to make the notation on the pay-checks.'[86] After two months in the north, Orwell was on the train home when he had one final shocking image of the cost exacted by the town's grim reality. He noticed a young woman standing at the back of her house, trying to unblock a pipe with a stick. 'She looked up as the train passed, and I was almost near enough to catch her eye. She had a round pale face, the usual exhausted face of the slum girl who is twenty-five and looks forty, thanks to miscarriages and drudgery; and it wore, for the second in which I saw it, the most desolate, hopeless expression I have ever seen. It struck me then that we are mistaken when we say that "It isn't the same for them as it would be for us," and that people bred in the slums can imagine nothing but the slums. ... She knew well enough what was happening to her – understood as well as I did how dreadful a destiny it was to be kneeling there in the bitter cold, on the slimy stones of a slum backyard, poking a stick up a foul drain-pipe.'[87]

Orwell had been made so angry by his experiences that he wrote the book in two parts. In the first he let the harsh facts speak for themselves. Part 2 was an emotional polemic against the capitalist system and in favour of socialism, and the publishers entertained some doubts about its merit.[88] Many critics found little sense of remedy in this section, its prose vague and overwrought. But the stark details of part 1 were undeniable, as shaming for Britain as Johnson's were for America. *The Road to Wigan Pier* caused a sensation.

Criticism of a very different aspect of civilisation came from the writer **Lewis Mumford**, part of a coterie who gathered around the photographer Alfred Stieglitz in New York. In the early 1920s Mumford had taught architecture at the New School for Social Research in Manhattan, and was then taken on as architecture correspondent for the *New Yorker*. His growing fame led to more lecturing at MIT, Columbia, and Stanford, which he published as a book, *Technics and Civilisation*, in 1934.[89] In this work he charted the evolution of technology. In the eotechnic phase, society was characterised by machines made of wood, and driven by water or wind power.[90] In the palaeotechnic phase, what most people called the first industrial revolution, the main form of energy was steam and the main material iron. The neotechnic age (the second industrial revolution) was characterised by electricity, aluminum, new alloys, and synthetic substances.[91]

For Mumford, technology was essentially driven by capitalism, which needed continued expansion, greater power, greater reach, faster speeds. He thought that dissatisfaction with capitalism arose because although the neotechnic age had arrived by the 1920s, social relations were stuck in the palaeotechnic era, where work was still alienating for the vast majority of people in the sense that they had no control over their lives. A neat phrasemaker ('Robbery is probably the greatest labour-saving device ever invented'), Mumford posed as a solution 'Basic Communism,' by which he didn't mean Soviet communism so much as the municipal organisation of work, just as there was the municipal organisation of parks, fire services and swimming pools.[92] Mumford's book was remarkable for being one of the first to draw attention to the damage capitalist enterprises were doing to the environment, and how consumerism was being led, and misled, by advertising. Like many others, he saw World War I as the culmination of a technological race that met the needs of capitalists and militarists alike, and he thought the only way forward lay in economic planning. Cannily, Mumford predicted that the industrial proletariat (Orwell's subject) would disappear as the old-style factories became obsolete, and he thought the neotechnic industries would be spread more evenly across countries (less congregated around ports or mines) and across the world. He forecast that Asia and Africa would become market and neotechnic forces in years ahead. He predicted that biology would replace physics as the most important and contentious science, and that population would become a major issue of the future. The immediate dangers for Americans, however, arose from a 'purposeless materialism' and an unthinking acceptance that unbridled capitalism was the only organising principle for modern life. In this basically optimistic book (there was a section on the beauty of machines), Mumford's criticisms of Western society were ahead of their time, which only makes them more impressive, for with the benefit of hindsight we can say that he was right far more than he was wrong.[93]

Four years later, Mumford published *The Culture of Cities*, which looked at the history of the city.[94] Beginning around 1,000 AD, when Mumford said the city revived after the Dark Ages, he defined cities according to the main collective dramas they played out. In mediaeval cities this was the market, the tournament, and the church's processionals. In the Baroque city, the court offered the best drama, and in the industrial city the station, the street, and the political meeting were what counted.[95] Mumford also distinguished six phases of city life: eopolis – village communities, domestication of animals; polis – an association of villages or blood groups, for defence; metropolis – the crucial change to a 'mother city,' with a surplus of regional products; megalopolis – beginning of decline, mechanisation, standardisation (a megalopolis was characterised by the lack of drama, replaced instead by routine); tyrannopolis – overexpansion, decadence, decline in numbers; nekropolis – war, famine, disease. The two last stages were predictions, but Mumford thought that megalopolis had already been reached in several cases, for example, New York.[96]

Mumford believed that the answer to the crisis of the alienation and poverty that characterised cities was to develop the regions (although he also considered garden cities). Here too Mumford was prescient; the last chapter of his book is almost wholly devoted to environmental and what we would now call 'quality of life' issues.

Despite his focus on the environment and the effects of technology on the quality of life, Mumford was not anti-science in the way that some others were. Even at the time that people like Freud and Mead and Johnson thought science could provide answers to society's ills, sceptics thought that every advantage of science was matched by a corresponding disadvantage. That was what gave it such a terrible beauty. Also, religion may have taken a battering at the hands of science, but it had not gone away, not by a long chalk. No doubt chronic unemployment had something to do with the scepticism toward science as a palliative, but as the 1930s progressed, religion reasserted itself.

The most extraordinary element in this reaffirmation of religion was a series of lectures given by **Ernest William Barnes**, the bishop of Birmingham, and published in 1933 as *Scientific Theory and Religion*.[97] Few readers, picking up a book by a bishop, would expect the first 400 pages to consist of a detailed discussion of advanced mathematics. Yet Ernest Barnes was a highly numerate scientist, a D.Sc., and a Fellow of the Royal Society. In his book he wanted to show that as a theologian he knew a great deal about modern science and was not afraid of it. He discussed all the recent developments in physics as well as the latest advances in geology, evolution, and mathematics. It was a tour de force. Barnes without exception endorsed the advances in particle physics, relativity, space-time, the new notions of an expanding universe, the findings of geology about the age of the earth and the record of life in the rocks. He was convinced of evolution.[98] At the same time, he dismissed various forms of mysticism and the paranormal. (Incidentally, despite its panoramic survey of recent twentieth-century science, it made not a single mention of Freud.)

So what would the bishop say about God? His argument was that there is a Universal Mind which inhabits all matter in the universe, and that the purpose of the universe is to evolve consciousness and conscience in order to produce goodness and, above all, beauty. His view on immortality was that there is no such thing as a 'soul,' and that the goodness and beauty that people create lives on after them. But he did also say that he personally believed in an afterlife.[99]

A copy of the book was sent to another eminent theologian, **William Ralph Inge**, dean of St Paul's and the man who had quoted Rupert Brooke's poems during his sermon on Easter Sunday, 1915. When he received Barnes's book, Inge was already correcting the proofs of a book of his own, *God and the Astronomers*, which was published later that same year, 1933. It too had started life as a series of lectures, in Inge's case the Warburg lectures, which he gave at Lincoln's Inn Chapel in London.[100] As well as being dean of St Paul's, Inge was a fellow of Jesus College, Cambridge, and Hertford College, Oxford, and well known as a lecturer, writer, and intellectual. His provocative views on contemporary topics had already been published as *Outspoken Essays*. *God and the Astronomers* tackled the second law of thermodynamics, entropy, and evolution. For Inge these fields were linked fundamentally because each was about time. The idea of a universe being created, expanding, contracting, and disappearing in a final Götterdämmerung, as he put it, was clearly worrying, since it raised the idea that there is no such thing as eternity.

The chief effect of evolution was to demote ideas in the past, arguing that more modern ideas had 'evolved' beyond them.[101] Inge therefore deliberately made widespread use of the ancient philosophers – mainly Greek – to support his arguments. His aim was to show how brilliant their minds were, in comparison to those of the present. He made several references to 'dysgenic' trends, to suggest that evolution did not always produce advances. And he confessed that his arguments were intuitive, insisting (much as the poets were doing in Weimar Germany) that the very existence of intuition was a mark of the divine, to which science had no real answer.[102] Like Henri Bergson, Inge acknowledged the existence of the *élan vital* and of an 'impassable gulf' between scientific knowledge and God's existence. Like Barnes, he took as evidence for God's existence the very concept of goodness and the mystical experiences of rapture that, as often as not, took place during prayer, which he said could not be explained by any science. He thought that civilisation, with its pressures and pace, was distancing us from such experiences. He hinted that God's existence might be similar to the phenomenon that scientists call 'emergent property,' the classic example here being molecules of water, which are not themselves liquid in the way that water is. In other words, this was a scientific metaphor to support the argument for God.[103] Inge, unlike Barnes, was unable to accept recent scientific advances: 'It is a strange notion that God reveals himself more clearly and more directly in inanimate nature than in the human mind or heart. . . . My conclusion is that the fate of the material universe is not a vital question for religion.'[104] Like Barnes, Inge made no reference to Freud.

A year after Barnes and Inge had their say, **Bertrand Russell** published a short but pithy book, *Religion and Science*. Russell's relationship with religion was complicated.[105] He had a number of friends who were religious (in particular Lady Ottoline Morrell), and he was both envious of and irritated by them. In a letter written in January 1912 he had said, 'What we *know* is that things come into our lives sometimes which are so immeasurably better than the things of everyday, that it *seems* as though they were sent from another world and could not come out of ourselves.'[106] But later he added, 'Yet I have another vision . . . in this vision, sorrow is the ultimate truth . . . we draw our breath in pain . . . thought is the gateway to despair.'[107]

In *Religion and Science*, Russell covered much the same ground as Barnes and Inge – the Copernican revolution, the new physics, evolution, cosmic purpose – but he also included an analysis of medicine, demonology, and miracles, and a chapter on determinism and mysticism.[108] Throughout most of the book, he showed the reader how science could explain more and more about the world. For a scientist, he was also surprisingly easy on mysticism, declaring that some of the psychic experiments he had heard about were 'convincing to a reasonable man.' In his two concluding chapters, on science and ethics, he wrote as a fierce logician, trying to prove that there is no such thing as objective beauty or goodness. He began with the statement, 'All Chinese are Buddhists.' Such a statement, he said, could be refuted 'by the production of a Chinese Christian.'[109] On the other hand, the statement 'I believe that all Chinese are Buddhists' cannot be refuted 'by any evidence from China [i.e., about Buddhists in China]', but only by evidence that 'I do not believe what I say.' If a philosopher says, 'Beauty is good,' it may mean one of two things: 'Would that everybody loved the beautiful' (which corresponds to 'All Chinese are Buddhists') or 'I wish that everybody loved the beautiful' (which corresponds to 'I believe that all Chinese are Buddhists'). 'The first of these statements makes no assertion but expresses a wish; since it affirms nothing, it is logically impossible that there should be evidence for or against it, or for it to possess either truth or falsehood. The second sentence, instead of being merely optative, does make a statement, but it is one about the philosopher's state of mind, and it could only be refuted by evidence that he does not have the wish that he says he has. This second sentence does not belong to ethics, but to psychology or biology. The first sentence, which does belong to ethics, expresses a desire for something, but asserts nothing.'[110]

Russell went on, 'I conclude that, while it is true that science cannot decide questions of value [Inge's argument], this is because they cannot be intellectually decided at all, and lie outside the realm of truth and falsehood. Whatever knowledge is attainable, must be attained by scientific methods; and what science cannot discover, mankind cannot know.'[111] Again, there was no reference to Freud.

A quite different line of attack on science came from Spain, from **José Ortega y Gasset's** *Revolt of the Masses*, published in 1930. Ortega was professor of philosophy at the University of Madrid, and his main thesis was that society was degenerating, owing to the growth of mass-man, the anonymous, alienated individual of mass society, this growth itself of course due in no small measure to scientific advances. For Ortega, true democracy occurred only when power was voted to a 'super minority.' What in fact was happening, he said, was 'hyper-democracy,' where average man, mediocre man, wanted power, loathed everyone not like himself and so promoted a society of 'homogenised ... blanks.' He blamed scientists in particular for the growth of specialisation, to the point where scientists were now 'learned ignoramuses,' who knew a lot about very little, focusing on their own small areas of interest at the expense of the wider picture. He said he had found such scientists 'self-satisfied,' examples of a very modern form of degeneration, which helped account for the growing absence of culture he saw encroaching all around him.

Ortega y Gasset was a sort of cultural social Darwinist, or Nietzschean. In *The Dehumanisation of Art*, he argued that it was 'the essential function of modern art to divide the public into two classes – those who can understand it and those who cannot.'[112] He thought that art was a means by which the elite, 'the privileged minority of the fine senses,' could recognise themselves and distinguish themselves from the 'drab mass of society,' who are the 'inert matter of the historical process.' He believed that the vulgar masses always wanted the man behind the poet and were rarely interested in any purely aesthetic sense (Eliot would have been sympathetic here). For Ortega y Gasset, science and mass society were equally inimical to 'fine' things.

With fascism on the rise in Germany and Italy, and the West in general beset by so many problems, people began to look to Soviet Russia to examine an alternative system of social organisation, to see whether the West could learn anything. Many Western intellectuals, such as George Bernard Shaw and Bertrand Russell, paid visits to Russia in the 1920s and '30s, but the most celebrated at the time was the journey by **Sidney and Beatrice Webb**, whose account of their visit, *Soviet Communism: A New Civilisation?* was published in 1935.

Well before the book appeared, the Webbs had a profound influence on British politics and society and were very well connected, with friends such as the Balfours, the Haldanes, the Dilkes, and the Shaws.[113] Sidney Webb became a cabinet minister in both interwar Labour governments, and the couple formed one of the most formidable intellectual partnerships ever (Sidney was once called 'the ablest man in England').[114] They founded the London School of Economics (LSE) in 1896, and the *New Statesman* in 1913, and were instrumental in the creation of the welfare state and in developing the Fabian Society, a socialist organisation that believed in the inevitability of gradual change. They were the authors, either singly or jointly, of nearly a hundred books and pamphlets, including *The Eight Hours Day, The Reform of the Poor Law, Socialism and Individualism, The Wages of Men and Women: Should They Be Equal?* and *The Decay of Capitalist Civilisation*. Committed socialists all their lives, the Webbs met when Beatrice wanted someone to help her study the co-op movement and a friend suggested Sidney. Lisanne Radice, the Webbs' biographer, makes the point that, on the whole, Sidney and Beatrice were more successful together, as organisers and theoreticians, than he was as a practical politician, in the cabinet. Their prolific writings and their uncompromising socialist views meant that few people were indifferent to them. Leonard Woolf liked them, but Virginia did not.[115]

The Webbs went to Russia in 1932, when they were both already in their mid-seventies. Beatrice instigated the visit, feeling that capitalism was in terminal decay and that Russia might just offer an alternative. In their books, the Webbs had always argued that, contrary to Marx, socialism could arrive gradually, without revolution; that through reason people could be convinced, and equality would evolve (this was the very essence of Fabianism). But with fascism on the rise, she and Sidney felt that if capitalism could be swept away, so too could Fabianism.[116] In these circumstances, Russian collective planning became more viable. At the end of 1930 Beatrice began reading Russian literature, her choice being assisted by the Soviet ambassador to London and his wife. Almost immediately Beatrice made a note in her diary: 'The Russian Communist government may still fail to attain its end in Russia, as it will certainly fail to conquer the world with a Russian brand of Communism, but its exploits exemplify the Mendelian view of sudden jumps in biological evolution as against the Spencerian vision of slow adjustment.' (The social Darwinist Herbert Spencer had been a close friend of Beatrice's father.) A year later, just before her trip, Beatrice wrote the words that were to be remembered by all her detractors: 'In the course of a decade, we shall know whether American capitalism or Russian communism yields the better life for the bulk of the people ... without doubt, we are on the side of Russia.'[117]

The Russia the Webbs set foot in in 1932 was near the end of the first Five-Year Plan, which Stalin had introduced in 1929 to force through rapid industrialisation and rural collectivisation. (Such plans were popular just then: Roosevelt introduced his New Deal in 1933, and in 1936 Germany brought in the four-year Schacht plan for abolishing unemployment by expanding public works). Stalin's 'plan' led directly to the extermination of a million kulaks, mass deportation and famine; it extended the grip of the OGPU, the secret police, a forerunner of the KGB, and vitiated the power of trade unions by the introduction of internal passports, which restricted people's movement. There were achievements – education improved and was available to more children, there were more jobs for women – but, as Lisanne Radice describes it, the first Five-Year Plan, 'stripped of its propaganda verbiage ... foreshadowed a profound extension of the scope of totalitarian power.'[118]

The Webbs, treated as important foreign guests, were kept well away from these aspects of Russia. They had a suite at the Astoria Hotel in Leningrad, so huge that Beatrice worried, 'We seem to be a new kind of royalty.' They saw a tractor plant at Stalingrad and a Komsomol conference. In Moscow they stayed in a guest house belonging to the Foreign Ministry, from where they were taken to schools, prisons, factories, and theatres. They went to Rostow, 150 miles northeast of Moscow, where they visited several collective farms. Dependent on interpreters for their interviews, the Webbs encountered only one failure, a motor plant that was not meeting its production targets, and the only statistics they managed to collect were provided by the government. Here were the founders of the LSE and the *New Statesman* accepting information from sources no self-respecting academic or journalist would dream of publishing without independent corroboration. They could have consulted Malcolm Muggeridge, the *Manchester Guardian*'s correspondent in Moscow, who was married to Beatrice's niece. But he was highly critical of the regime, and they took little notice of him. And yet, on their return, Beatrice wrote, 'The Soviet government . . . represents a new civilisation . . . with a new outlook on life – involving a new pattern of behaviour in the individual and his relation to the community – all of which I believe is destined to spread to many other countries in the course of the next hundred years.'[119]

In Lisanne Radice's words, *Soviet Communism: a new civilisation?* was 'monumental in conception, in scope, and in error of judgement.'[120] The Webbs truly believed that Soviet communism was superior to the West because ordinary individuals had more opportunity to partake in the running of the country. Stalin was not a dictator to them, but the secretary of 'a series of committees.' The Communist Party, they said, was dedicated to the removal of poverty, with party members enjoying 'no statutory privileges.' They thought OGPU did 'constructive work.' They changed the title of their book in later editions, first to *Is Soviet Communism a New Civilisation?* (1936), then *Soviet Communism: Dictatorship or Democracy?* (released later the same year) – suggesting a slight change of heart. But they were always reluctant to retract fully what they had written, even after the Stalinist show trials in the later 1930s. In 1937, the height of the terror, their book was republished as *Soviet Communism: a new civilisation* – i.e., without the question mark. On their forty-seventh wedding anniversary, in July 1939, Beatrice confided to her diary that *Soviet Communism* was 'the crowning achievement of Our Partnership.'[121] Dissatisfaction with the performance of capitalism led few people as far astray as it did the Webbs.

Russian communism was one alternative to capitalism. Another was beginning to reveal itself in Germany, with the rising confidence of the Nazis. During the Weimar years, as we have seen, there was a continual battle between the rationalists – the scientists and the academics – and the nationalists – the pan-Germans, who remained convinced that there was something special about Germany, her history, the instinctive superiority of her heroes. Oswald Spengler had stressed in *The Decline of the West* how Germany was different from France, the United States and Britain, and this view, which appealed to Hitler, gained ground among the Nazis as they edged closer to power. In 1928 this growing confidence produced a book which, almost certainly, would never have found a publisher in Paris, London, or New York.

The text was inflammatory enough, but the pictures were even more so. On one side of the page were reproductions of modern paintings by artists such as Amedeo Modigliani and Karl Schmidt-Rottluff, but on the other were photographs of deformed and diseased people – some with bulging eyes, others with Down's syndrome, still others who had been born cretinous. The author of the book was a well-known architect, **Paul Schultze-Naumburg**; its title was *Kunst und Rasse* (Art and Race); and its thesis, though grotesque, had a profound effect on National Socialism.[122] Schultze-Naumburg's theory was that the deformed and diseased people shown in his book were the prototypes for many of the paintings produced by modern – and in particular, expressionist – artists. Schultze-Naumburg said this art was *entartet* – degenerate. His approach appears to have been stimulated by a scientific project carried out a few years earlier in the university town of Heidelberg, which had become a centre for the study of art produced by schizophrenics as a means of gaining access to the central problems of mental illness. In 1922 psychiatrist Hans Prinzhorn had published his study *Bildnerei der Geisteskranken* (Image-making by the Mentally Ill), based on material he gathered by examining more than 5,000 works by 450 patients. The study, which demonstrated that the art of the insane exhibited certain qualities, received serious attention from critics well beyond the medical profession.[123]

Art and Race caught Hitler's attention because its brutal 'theory' suited his aims. From time to time he attacked modern art and modern artists, but like other leading Nazis, he was by temperament an anti-intellectual; for him, most great men of history had been doers, not thinkers. There was, however, one exception to this mould, a would-be intellectual who was even more of an outsider in German society than the other leading Nazis – **Alfred Rosenberg**.[124] Rosenberg was born beyond the frontiers of the Reich. His family came from Estonia, which until 1918 was one of Russia's Baltic provinces. There is some evidence (established after World War II) that Rosenberg's mother was Jewish, but at the time no suspicion ever arose, and he remained close to Hitler for longer than many of their early colleagues. As a boy he was fascinated by history, especially after he encountered the work of Houston Stewart Chamberlain.[125] Chamberlain was a renegade Englishman, an acolyte and relative by marriage of Wagner, who regarded European history 'as the struggle of the German people against the debilitating influences of Judaism and the Roman Catholic Church'. When Rosenberg came across Chamberlain's *Foundations of the Nineteenth Century* on a family holiday in 1909, he was transformed. The book provided the intellectual underpinning of his German nationalistic feelings. He now had a reason to hate the Jews every bit as much as his experiences in Estonia gave him reason to hate the Russians. Moving to Munich after the Armistice in 1918, he quickly joined the NSDAP and began writing vicious anti-Semitic pamphlets. His ability to write, his knowledge of Russia, and his facility with Russian all helped to make him the party's expert on the East; he also became editor of the *Völkischer Beobachter* (National Observer), the Nazi Party's newspaper. As the 1920s passed, Rosenberg, together with Martin Bormann and Heinrich Himmler, began to see the need for a Nazi ideology that went beyond *Mein Kampf*. So in 1930 he published what he believed provided the intellectual basis for National Socialism. In German its title was *Der Mythus des 20. Jahrhunderts*, usually translated into English as *The Myth of the Twentieth Century*.

Mythus is a rambling and inconsistent book, and consequently hard to summarise. (One example of how obscure it was: a contemporary admirer published a glossary of 850 terms that needed explaining.) It conducts a massive assault on Roman Catholicism as the main threat to German civilisation. The text stretches to more than 700 pages, with the history of Germany and German art making up more than 60 percent of the book.[126] The third section is entitled 'The Coming Reich'; other parts deal with 'racial hygiene,' education, and religion, with international affairs at the end. Rosenberg argues that Jesus was not Jewish and that his message had been perverted by Paul, who *was* Jewish, and that it was the Pauline/Roman version that had forged Christianity into its familiar mould by ignoring ideas of aristocracy and race and creating fake doctrines of original sin, the afterlife, and hell as an inferno, all of which beliefs, Rosenberg thought, were 'unhealthy.'

Rosenberg's aim – and at this distance his audacity is breathtaking – was to create a substitute faith for Germany. He advocated a 'religion of the blood' which, in effect, told Germans that they were members of a master race, with a 'race-soul.' Rosenberg appropriated famous German figures from the past, such as the painter Meister Eckhart and the religious leader Martin Luther, who had resisted Rome, though here again he only used those parts of the story that suited his purpose. He quoted the works of the Nazis' chief academic racialist, H. F. K. Guenther, who 'claimed to have established on a scientific basis the defining characteristics of the so-called Nordic-Aryan race'. As with Hitler and others before him, Rosenberg did his best to establish a connection to the ancient inhabitants of India, Greece, and Germany, and he brought in Rembrandt, Herder, Wagner, Frederick the Great, and Henry the Lion, to produce an entirely arbitrary but nonetheless heroic history specifically intended to root the NSDAP in the German past.

For Rosenberg, race – the religion of the blood – was the only force that could combat what he saw as the main engines of disintegration – individualism and universalism. 'The individualism of economic man,' the American ideal, he dismissed as 'a figment of the Jewish mind to lure men to their doom.'[127] At the same time he had to counter the universalism of Rome, and in creating his own new religion certain Christian symbols had to go, including the crucifix. If Germans and Germany were to be renewed after the chaos of military defeat, 'the Crucifix was too powerful a symbol to permit of change.' By the same token, 'The Holy Land for Germans,' Rosenberg wrote, 'is not Palestine. ... Our holy places are certain castles on the Rhine, the good earth of Lower Saxony and the Prussian fortress of Marienburg.' In some respects, the *Mythus* fell on fertile ground. The 'religion of the blood' fitted in well with new rituals, already developing among the faithful, whereby Nazis who had been killed early on in the 'struggle' were proclaimed 'martyrs' and were wrapped in flags that, once tainted with their blood, became 'blood flags' and were paraded as totems, used in ceremonies to dedicate other flags. (Another invented tradition was for party members to shout out 'Here' when the names of the dead were read out during roll call.) Hitler, however, seems to have had mixed feelings about the *Mythus*. He held on to the manuscript for six months after Rosenberg submitted it to him, and publication was not sanctioned until 15 September

1930, after the Nazi Party's sensational victory at the polls. Perhaps Hitler had put off approving the book until the party was strong enough to risk losing the support of Roman Catholics that would surely follow publication. The book sold 500,000 copies, but that means little, as all secondary schools and institutes of higher education were forced to buy copies.[128]

If Hitler did delay publication because of the effect *Mythus* might have on Catholics, he was being no more than realistic. The Vatican was incensed by its argument and, in 1934, placed it on the Index of Prohibited Books. **Cardinal Schulte**, archbishop of Cologne, set up a 'Defence Staff' of seven young priests, who worked round the clock to list the many errors in the text. These were published in a series of anonymous pamphlets printed simultaneously in five different cities to evade the Gestapo. The most brutal use of the book was as a tool to expose priests: Catholic Nazis were ordered to refer to the *Mythus* in the confessional, and then denounce any priest who was so duped into criticising the ideology of the NSDAP.[129] For a time it seems that Rosenberg truly believed that a new religion was coming into being – he told Hermann Göring as much in August 1939. Within a month, however, Germany was at war, and after that the impact of the *Mythus* was patchy. Rosenberg himself remained popular with Hitler, and when the war began, he was given his own unit, the Einsatzstab Reichsleiter Rosenberg, or ERR, charged with looting art.

Although they were incoherent and arbitrary, *Art and Race* and the *Mythus* were linked by the fact that each attacked the intellectual and cultural life of Germany. Whatever their shortcomings and failings, however crude and tendentious, they represented an attempt by the Nazis to focus on thought beyond the confines of party politics. In publicising such views, the Nazis now left no doubt as to what they thought was wrong with German civilisation.

With many people so worried about the direction civilisation was taking, with evidence for such a dire fate being adduced on all sides, it is perhaps not surprising that such a period, such a mood, produced one of the great works of literature of the century. One could argue that John Steinbeck was *the* chronicler of unemployment in the 1930s, that Christopher Isherwood's novels about Berlin served as an antidote to the sinister absurdities of the *Mythus*. But the worries and the bleak mood went far wider than unemployment and Germany, and this pessimism was clearly captured by someone else – by **Aldous Huxley**, in *Brave New World*.

Seven years younger than his brother Julian, the eminent biologist, Aldous Huxley was born in 1894.[130] His poor eyesight exempted him from service in World War I and he spent the time working on Lady Ottoline Morrell's farm near Oxford, where he met Lytton Strachey, T. S. Eliot, Mark Gertler, Middleton Murry, D. H. Lawrence, and Bertrand Russell. (Eliot said Huxley showed him some early verse, which he was 'unable to show any enthusiasm for.')[131] Very well read and deeply sceptical, Huxley had written four books by the early 1930s, including the novels *Crome Yellow* and *Antic Hay*.[132] *Brave New World*, published in 1932, is a dystopian novel, a pessimistic taste of the possible horrific consequences of twentieth-century thought. It is, at one level, science fiction. But *Brave New World* was also designed to be a cautionary tale; if Freud, in *Civilisation and Its Discontents*, explored the superego as the basis of a new ethics, Huxley described a new ethic itself – in which the new psychology was as much to blame as anything.[133]

Huxley's targets in the book are primarily biology, genetics, behavioural psychology, and mechanisation. *Brave New World* is set well into the future, in AF 632, AF standing for After Ford (which would make it around 2545 AD). Technology has moved on, and a technique known as Bokanovsky's Process enables one ovary in certain circumstances to produce sixteen thousand persons, perfect for Mendelian mathematics, the basis for a new society in which vast numbers of people are, even more than now, all the same. There are neo-Pavlovian infant-conditioning methods (books and flowers are linked with noxious shocks), and a 'sleep-teaching process by which infants acquire, among other things, the rudiments of class-consciousness.'[134] Sex is strictly controlled: women are allowed a pregnancy substitute, and there are bandolier-containers, known as Malthusian belts, which carry not bullets but contraceptives. Polygamy is the accepted norm, monogamy a disgrace. The family, and parenthood, are obsolete. It has become 'improper' to want to spend time alone, to fall in love, and to read books for pleasure. In a chilling echo of the *Mythus* (Huxley's book was published in the same year), the Christian cross has been abolished by the simple expedient of having its head removed to form the letter *T*, after the 'model T Ford.' Organised religion has been replaced by 'Solidarity Services.' The book solemnly informs us that this new world resulted from a nine-year war in which biological weapons wrought such devastation that 'a world-wide federation and foolproof control of the people were the only acceptable alternative.' Huxley is specific about the eugenics that help exercise this foolproof control, showing how eggs are graded (alphas to epsilons) and then immersed in 'a warm bouillon containing free-swimming spermatozoa.' We encounter half-familiar organisations such as the 'Central London Hatchery and Conditioning Centre.' Some of the characters – Mustapha Mond, the Resident Controller for Western Europe, and Bernard Marx and Lenina Crowne – remind us of what the new world has lost from the past and what the society has chosen to remember. Huxley is also careful to show that snobbery and jealousy still exist, as does loneliness, 'despite all attempts to eradicate such feelings.'[135]

This sounds heavy-handed in summary, but Huxley is a funny writer. His vision of the future is not wholly bad – the elite still enjoy life, as elites tend to do.[136] And it is this which links Huxley to Freud, where this chapter began. Freud's view was that a better understanding of the superego, by psychoanalysis, would ultimately lead to a better understanding of ethics, and more ethical behaviour. Huxley was more sceptical, and he had more in common with Russell. He thought there were no absolutes of good and bad, and that man must continually renew his political institutions in the light of new knowledge, to create the best society possible. The society of *Brave New World* may seem horrible to us, but it doesn't seem all that horrible to the people in the story, who know nothing else, just as the Dobu, or the Arapesh, or the Kwakiutl, know nothing else beyond their societies, and are happy enough. To get the world you want, Huxley affirms, you have to fight for it. And, by implication, if your world is collapsing, you aren't fighting hard enough. That was where, in 1932, he was most prescient of all, in suggesting that there was a fight coming.

INQUISITION

On 30 January 1933, Adolf Hitler became Chancellor of Germany. Barely six weeks later, on 11 March, he established the Reich Ministry for Popular Enlightenment and Propaganda, with Joseph Goebbels as minister.[1] This was a name straight out of *Brave New World*, and between them Hitler and Goebbels would soon wreak havoc on the cultural life of Germany on a scale never seen before. Their brutal actions did not come out of the blue. Hitler had always been very clear that when the Nazi Party formed a government, there would be 'accounts' to settle with a wide range of enemies. Foremost among those he singled out were artists. In 1930, in a letter to Goebbels, he assured the future minister that when the party came to power, it would not simply become a 'debating society' so far as art was concerned. The party's policies, laid out in the manifesto as early as 1920, called for 'a struggle' against the 'tendencies in the arts and literature which exercise a disintegrating influence on the life of the people.'[2]

The first **blacklist of artists** was published on 15 March. George Grosz, visiting the United States, was stripped of his German citizenship. The Bauhaus was closed. Max Liebermann (then aged eighty-eight) and Käthe Kollwitz (sixty-six), Paul Klee, Max Beckmann, Otto Dix and Oskar Schlemmer were all dismissed from their posts as teachers in art schools. So swift were these actions that the sackings had to be made legal retroactively by a law that wasn't passed until 7 April 1933.[3] In the same month the first exhibition defaming modern art – called **Chamber of Horrors** – was held in Nuremberg, then travelled to Dresden and Dessau.[4] A week before Hitler became chancellor, Ernst Barlach had been rash enough to describe him on radio as 'the lurking destroyer of others' and called National Socialism 'the secret death of mankind.'[5] Now, in retribution, the local Nazis called for the artist's Magdeburg Memorial to be removed from the cathedral there, and no sooner had this demand been voiced than the work was shipped to Berlin 'for storage.'[6] *Der Sturm*, the magazine that had done so much to promote modern art in Germany, was shut down, and so were *Die Aktion* and *Kunst und Kunstler* (Art and Artists). Herwarth Walden, publisher of *Der Sturm*, escaped to the Soviet Union, where he died in 1941.[7] The collagist John Heartfield fled to Prague.

In 1933 modern artists made several attempts to align themselves with the

Nazis, but Goebbels would have none of it, and the exhibitions were forced to close. For a time he and Rosenberg competed for the right to set policy in the cultural/intellectual sphere, but the propaganda minister was a superb organiser and sidelined his rival as soon as an official **Chamber for Arts and Culture** came into being under Goebbels's control. The powers of the chamber were formidable – each and every artist was forced to join a government-sponsored professional body, and unless artists registered, they were forbidden from exhibiting in museums or from receiving commissions. Goebbels also stipulated that there were to be no public exhibitions of art without official approval.[8] In a speech to the party's annual meeting in September 1934, Hitler emphasised 'two cultural dangers' that threatened National Socialism. On the one hand, there were the modernists, the 'spoilers of art' – identified specifically as 'the cubists, futurists and Dadaists.' What he and the German people wanted, he said, was a German art that was 'clear,' 'without contortion' and 'without ambiguity.' Art was not 'auxiliary to politics,' he said. It must become a 'functioning part' of the Nazi political program.[9] The speech was an important moment for those artists who had not yet been dismissed from their positions or had their art taken off display. Goebbels, who had shown some sympathy for people like Emil Nolde and Ernst Barlach, quickly hardened his opinions. Confiscations recommenced, and another raft of painters and sculptors was dismissed from teaching or museum positions. Hans Grundig was forbidden to paint. Books by or about modern artists also became targets. Copies of the catalogue of Klee's drawings, published in 1934, were seized even before they arrived in the shops. Two years later a catalogue of the works of Franz Marc was seized (Marc had been dead nearly twenty years), as was a volume of Barlach's drawings – labelled a danger to 'public safety, peace and order.' The book was later pulped by the Gestapo.[10] In May 1936 all artists registered with the Reichskammer had to prove their Aryan ancestry. In October 1936 the National Gallery in Berlin was instructed to close its modern art galleries, and in November Goebbels outlawed all 'unofficial art criticism.' From then on, only the *reporting* of art events was allowed.

Some artists tried to protest. Ernst Ludwig Kirchner, as he was forced out of the Prussian Academy, insisted that he was 'neither a Jew nor a social democrat'. 'For thirty years I have struggled for a new, strong, and true German art and will continue to do so for as long as I live.'[11] Max Pechstein could not believe what was happening to him, and reminded the Gestapo that he had fought for Germany on the western front in World War I, that one of his sons was a member of the SA, and another was in the Hitler Youth. Emil Nolde, an enthusiastic supporter of the party from the early 1920s, criticised the 'daubings' of some of his colleagues, whom he described as 'half-breeds, bastards and mulattoes' in his autobiography, *Years of Struggle*, published in 1934.[12] That year he wrote directly to Goebbels, insisting that his own art was 'vigorous, durable, ardent and German.' Goebbels wasn't listening; in June 1937, 1,052 of Nolde's works were confiscated.[13] Oskar Schlemmer stood up for artists when they were attacked by **Gottfried Benn** in *The New State and the Intellectuals*, which was a highly charged defence of the Nazis and an intemperate attack on their perceived enemies. Schlemmer's argument was that the artists identified by Benn as 'decadent' were nothing of the sort and that the real decadence lay in the 'second-raters' who were replacing their betters with, as he put it, 'kitsch.'[14]

Such protests went nowhere. Hitler's mind had been made up long ago, and he wasn't about to change it. Indeed, these artists were lucky not to have provoked reprisals. All that was left for them was to protest in their art. Otto Dix was one of those who led the way, portraying Hitler as 'Envy' in his 1933 picture *The Seven Deadly Sins*. (He meant, of course, that Hitler, the failed artist, envied real ones.) Max Beckmann caricatured the chancellor as a 'Verführer,' a seducer. When informed that he had been expelled from the Prussian Academy, Max Liebermann, the most popular living painter in pre-World War I Germany, remarked tartly, 'I couldn't possibly eat as much as I would like to puke.'[15]

Many artists eventually took the option of emigration and exile.[16] Kurt Schwitters went to Norway, Paul Klee to Switzerland, Lyonel Feininger to the United States, Max Beckmann to the Netherlands, Heinrich Campendonck to Belgium and then to Holland, Ludwig Meidner to England, and Max Liebermann to Palestine. Liebermann had loved Germany; it had been good to him before World War I, and he had met, and painted, some of its most illustrious figures. And yet, shortly before his death in 1935, he sadly concluded that there was only one choice for young German artists who were Jewish: 'There is no other salvation than emigration to Palestine, where they can grow up as free people and escape the dangers of remaining refugees.'[17]

For the most part, one would think that science – especially the 'hard' sciences of physics, chemistry, mathematics and geology – would be unaffected by political regimes. It is, after all, generally agreed that research into the fundamental building blocks of nature is as free from political overtones as intellectual work can be. But in Nazi Germany nothing could be taken for granted.

The persecution of Albert Einstein began early. He came under attack largely because of the international acclaim he received after Arthur Eddington's announcement, in November 1919, that he had obtained experimental confirmation for the predictions of general relativity theory. The venom came from both political and scientific extremists. He had some support – for example, the German ambassador in London in 1920 warned his Foreign Office privately in a report that 'Professor Einstein is just at this time a cultural factor of first rank. . . . We should not drive such a man out of Germany with whom we can carry on real cultural propaganda.' Yet two years later, following the political assassination of Walther Rathenau, the foreign minister, unconfirmed reports leaked out that Einstein was also on the list of intended victims.[18]

When the Nazis finally achieved power, ten years later, action was not long delayed. In January 1933 Einstein was away from Berlin on a visit to the United States. He was then fifty-four, and although he found his fame burdensome, preferring to bury himself in his work on general relativity theory and cosmology, he also realised that he couldn't altogether avoid being a public figure. So he made a point of announcing that he would not return to his positions at the university in Berlin and the Kaiser Wilhelm Gesellschaft as long as the

Nazis were in charge.[19] The Nazis repaid the compliment by freezing his bank account, searching his house for weapons allegedly hidden there by Communists, and publicly burning copies of a popular book of his on relativity. Later in the spring, the regime issued a catalogue of 'state enemies.' It had been carefully edited to show the most unflattering photographs of the Nazis' opponents, with a brief text underneath each one. Einstein's picture headed the list, and below his photograph was the text, 'Not yet hanged.'[20]

In September Einstein was in Oxford, shortly before he was scheduled to return to the teaching position he had at Caltech, the California Institute of Technology. It was by no means clear then where he would settle. He told a reporter that he felt he was European and that, whatever might happen in the short term, he would eventually return. Meanwhile, 'in a fit of absent mindedness,' he had accepted professorships in Spain, France, Belgium, and the Hebrew University in Jerusalem, and at the newly formed Institute for Advanced Study (IAS) at Princeton. In Britain there were plans to give him an appointment at Oxford, and a bill was before the House of Commons to give him the status of a naturalised citizen.[21] By the early 1930s, however, America was no longer a backwater in physics. It was beginning to generate its own Ph.D.s (1,300 in the 1920s), who were carrying on Einstein's work. Also, he liked America, and he needed no further inducements to leave after Hitler became chancellor. He didn't go to Caltech, however, but to Princeton. In 1929 the American educationalist Abraham Flexner had succeeded in raising money to build an **advanced research institute** at **Princeton, New Jersey**. Five million dollars had been pledged by Louis Bamberger and his sister Caroline Fuld, a successful business family from New Jersey.[22] The basic idea was to establish a centre for the advanced study of science where eminent figures could work in a peaceful and productive environment, free of any teaching burden. Flexner had stayed with Einstein at Caputh, his home, and there, as they walked by the lake, Einstein's enthusiasm for Princeton grew still more. They even got as far as talking money. Asked what he wished to be paid, Einstein hesitated: 'Three thousand dollars a year? Could I live on less?' 'You couldn't live on that,' Flexner said promptly, and suggested he should sort it out with Mrs Einstein. In no time, Flexner and Elsa had arrived at a figure of $16,000 per annum.[23] This was a notable coup for Flexner. When the news was released, at a stroke he had dramatically increased the profile of his project. Inside Germany, reactions were somewhat different. One newspaper ran the headline: 'GOOD NEWS FROM EINSTEIN – HE IS NOT COMING BACK.' Not everyone in America wanted Einstein. The National Patriotic Council complained he was a Bolshevik who espoused 'worthless theories.' The American Women's League also branded him a Communist, clamouring for the State Department to refuse Einstein an entry permit. They were ignored.[24] Einstein might be the most famous physicist to leave Germany, but he was by no means the only one. Roughly one hundred world-class colleagues found refuge in the United States between 1933 and 1941.[25]

For scientists only slightly less famous than Einstein, the attitude of the Nazis could pose serious problems, offering fewer chances of a safe haven abroad. **Karl von Frisch** was the first zoologist to discover 'the language of the bees,' by means of which bees informed other bees about food sources, through dances on the honeycomb. 'A round dance indicated a source of nectar, while a tail-wagging dance indicated pollen.' Von Frisch's experiments caught the imagination of the public, and his popular books were best-sellers. This cut little ice with the Nazis, who under the Civil Service Law of April 1933 still required Von Frisch to provide proof of his Aryan descent. The sticking point was his maternal grandmother, and it was possible, he admitted, that she was 'non-Aryan.' A virulent campaign was therefore conducted against von Frisch in the student newspaper at Munich University, and he survived only because there was in Germany an outbreak of nosema, a bee disease, killing several hundred thousand bee colonies in 1941. This seriously damaged fruit growing and dislocated agricultural ecology. At that stage Germany had to grow its own food, and the Reich government concluded that von Frisch was the best man to rescue the situation.[26]

According to recent research about 13 percent of biologists were dismissed between 1933 and the outbreak of war, four-fifths of them for 'racial' reasons. About three-quarters of those who lost their jobs emigrated, the expelled biologists on average proving considerably more successful than their colleagues who remained in Germany. The subject suffered most in two areas: the molecular genetics of bacteria, and phages (viruses that prey on bacteria). This had less to do with the quality of scientists who remained than with the fact that the scientific advances in these areas were chiefly made in the United States, and the normal dialogue between colleagues simply did not take place, neither in the late 1930s, nor throughout the war, nor for a considerable period afterward.[27]

In 1925 Walter Gropius and Laszlo Moholy-Nagy had moved the **Bauhaus** from Thuringia when the right-wing authorities there cut its budget, and transferred to Dessau. In the Saxony-Anhalt state elections of May 1932, however, the Nazis gained a majority, and their election manifesto included a demand for 'the cancellation of all expenditures for the Bauhaus' and ranted against 'Jewish Bauhaus culture.'[28] The new administration made good its promise, and in September the Bauhaus was closed. Bravely, Ludwig Mies van der Rohe moved on to the Steglitz suburb of Berlin, running the Bauhaus as a private school without state or municipal support. But money wasn't the real problem, and on 11 April 1933 the Bauhaus was surrounded by police and storm troopers. Students were detained, files seized, and the building sealed. Police guards prevented entry for months. When the Bauhaus had closed in Dessau, there had at least been protests in the press. Now, in Berlin, there was a press campaign *against* the Bauhaus, which was dismissed as a 'germ cell of Bolshevik subversion,' sponsored by the 'patrons and popes of the Arty German Empire of the Jewish nation.'[29] Attempts were made to reopen the school; the Nazis actually had a policy for this, called *Gleichschaltung* – assimilation into the status quo.[30] In the case of the Bauhaus, Mies was told that this would require the dismissal of, among others, Wassily Kandinsky. In the end, the differences between Mies and the Nazi authorities could not be reconciled, and the Bauhaus closed for good in Germany. It was more than just anti-Semitism. In trying to marry classical tradition to modern ideas, the Bauhaus stood for everything the Nazis loathed.

Those who went into exile included some of the most prominent Bauhaus teachers. Walter Gropius, Ludwig Mies van der Rohe, Josef Albers, Marcel Breuer and Laszlo Moholy-Nagy, all members of the inner circle, left Germany in either 1933–4 or 1937–8. Most went because their careers were stalled rather than because their lives were threatened, though the weaver Otti Berger was murdered at Auschwitz.[31] Gropius moved to Britain in 1934, but only after he had received official permission. In Britain he avoided any contact with the politically active German artists who were also there at the time (known as the Oskar-Kokoschka-Bund). When he was made professor at Harvard in 1937, the news received favourable coverage in the German papers.[32] In America Gropius soon became a highly respected authority on modernism, but he still eschewed politics. Art historians have been unable to trace any public statement of his about events in Nazi Germany – not even the *Entartete Kunst* (Degenerate Art) exhibition (see below), held in the very year of his appointment, and in which practically all of his Bauhaus artist colleagues and friends were vilely defamed.

The closure of the Warburg Institute in Hamburg actually preceded that of the Bauhaus. Aby Warburg died in 1929, but in 1931, fearing that a Jewish-founded institute would become a target for the Nazis if they came to power, his friends took the precaution of moving the books and the institute itself to the safety of Britain, to become the Graduate Art History Department of the University of London. Later in the 1930s, one of the Warburg's most illustrious disciples, Erwin Panofsky, who had written his famous study of perspective at the institute in Hamburg, also left Germany. He was dismissed in 1933, and he too was hired by Abraham Flexner at Princeton.

Most members of the Frankfurt Institute for Social Research were not only Jewish but openly Marxist. According to Martin Jay, in his history of the Institute, its endowment was moved out of Germany in 1931, to Holland, thanks to the foresight of the director, Max Horkheimer. Foreign branches of the school had already been set up, in Geneva, Paris, and London (the latter at the London School of Economics). Shortly after Hitler assumed power, Horkheimer left his house in the Kronberg suburb of Frankfurt and installed himself and his wife in a hotel near the main railway station. During February 1933 he gave up his classes on logic and turned instead to politics, especially the meaning of freedom. A month later, he quietly crossed the border into Switzerland, only days before the institute was closed down for 'tendencies hostile to the state.'[33] The building on Victoria-Allee was confiscated, as was the library of 60,000 volumes. A few days after he had escaped, Horkheimer was formally dismissed, together with Paul Tillich and Karl Mannheim. By then almost all the senior staff had fled. Horkheimer and his deputy, Friedrich Pollock, went to Geneva, and so did Erich Fromm. Offers of employment were received from France, initiated by Henri Bergson and Raymond Aron. Theodor Adorno meanwhile went to Merton College, Oxford, where he remained from 1934 to 1937. Sidney Webb, R. H. Tawney, Morris Ginsberg and Harold Laski all helped preserve the London branch until 1936. Geneva, however, gradually became less hospitable. According to Pollock, 'fascism also makes great progress in Switzerland.' He and Horkheimer made visits to London and New York to sound out the possibility of transferring there. They received a much more optimistic reception at Columbia University than from William Beveridge at the LSE, and so, by the middle of 1934, the Frankfurt Institute for Social Research was reconstituted in its new home at 429 West 117th Street. It remained there until 1950, during which time much of its more influential work was carried out. The combination of German analysis and U.S. empirical methods helped give sociology its postwar flavour.[34]

The migration of the philosophers of the Vienna Circle was perhaps less traumatic than with other academics. Thanks to the pragmatic tradition in America, not a few scholars there were very sympathetic to what the logical positivists were saying, and several of the circle crossed the Atlantic in the late 1920s or early 1930s to lecture and meet similar-minded colleagues. They were aided by a group known as Unity in Science, which consisted of philosophers and scientists searching for the constancies from one discipline to another. This international group held meetings all over Europe and North America. Then, in 1936, A. J. Ayer, the British philosopher, published *Language, Truth and Logic*, a brilliantly lucid account of logical positivism that popularised its ideas still more in America, making the members of the circle especially welcome on the other side of the ocean. Herbert Feigl was the first to go, to Iowa in 1931; Rudolf Carnap went to Chicago in 1936, taking Carl Hempel and Olaf Helmer with him. Hans Reichenbach followed, in 1938, establishing himself at UCLA. A little later, Kurt Gödel accepted a research position at the Institute of Advanced Studies at Princeton and so joined Einstein and Erwin Panofsky.[35]

The Nazis had always viewed psychoanalysis as a 'Jewish science.' Even so, it was a rude shock when, in October 1933, the discipline was banned from the Congress of Psychology in Leipzig. Psychoanalysts in Germany were forced to look elsewhere for work. For some Freud's hometown, Vienna, provided a refuge for a few years, but most went to the United States. American *psychologists* were not especially favourable to Freudian theory – William James and pragmatism were still influential. But the American Psychological Association did set up a Committee on Displaced Foreign Psychologists and by 1940 was in touch with 269 leading professionals (not all psychoanalysts), 134 of whom had already arrived in America: Karen Horney, Bruno Bettelheim, Else Frenkel-Brunswik, and David Rapaport among them.[36]

Freud was eighty-two and far from well when, in March 1938, Austria was declared part of the Reich. Several sets of friends feared for him, in particular Ernest Jones in London. Even President Roosevelt asked to be kept informed. William Bullitt, U.S. ambassador to Paris, was instructed to keep an eye on 'the

Freud situation,' and he ensured that staff from the consul general in Vienna showed 'a friendly interest' in the Freuds.[37] Ernest Jones hurried to Vienna, having taken soundings in Britain about the possibility of Freud settling in London, but when he arrived Jones found Freud unwilling to move. He was only persuaded by the fact that his children would have more of a future abroad.[38]

Before Freud could leave, his 'case' was referred as high as Himmler, and it seems it was only the close interest of President Roosevelt that guaranteed his ultimate safety, but not before Freud's daughter Anna was arrested and removed for a day's interrogation. The Nazis took care that Freud settled all his debts before leaving and sent through the exit visas one at a time, with Freud's own arriving last. Until that moment he worried that the family might be split up.[39] When his papers did at last arrive, the Gestapo also brought with them a document, which he was forced to sign, which affirmed that he had been properly treated. He signed, but added, 'I can heartily recommend the Gestapo to anyone.' He left, via the Orient Express, for Paris, before going on to London. A member of the American legation was instructed to go with him, to ensure Freud's safety.[40] In London, Freud stayed first at 39 Elsworthy Road in Hampstead. He was visited by Stefan Zweig, Salvador Dalí, Bronislaw Malinowski, Chaim Weizmann, and the secretaries of the Royal Society, who brought the society's Charter Book for him to sign, an honour previously bestowed only on the king.

Within a month of his arrival Freud began work on *Moses and Monotheism*, which he first conceived as an historical novel. In this book Freud claimed that the biblical Moses was an amalgam of two historical personages, an Egyptian and a Jew, and that the Egyptian, autocratic Moses had been murdered, a crime that lay at the root of Jewish guilt, which had been handed down. He thought the early Jews a barbarous people who worshipped a god of 'volcanoes and wildernesses,' and who, in their practice of circumcision, inspired in gentiles a fear of castration, the root cause of anti-Semitism.[41] It is difficult not to see the book as a reply to Hitler, almost a turning of the other cheek. The real significance of the book was its timing; Freud turned his back on Judaism (intellectually, if not emotionally) at Judaism's darkest hour. He was hinting that the Jews' separateness was psychologically profound, and partly their own fault. Freud didn't agree with the Führer that the Jews were evil, but he did admit they were flawed.[42] Many Jewish scholars implored him not to publish the book, on the grounds that it was historically inaccurate as much as because it would offend politico/religious sensibilities. But he went ahead.

It was not a fitting epitaph. At the end of 1938, and early 1939, new lumps appeared in Freud's mouth and throat. His Viennese doctor had obtained special permission to treat Freud without British qualifications, but there was little to be done. Freud died in September 1939, three weeks after war was declared.

As a young philosophy student of eighteen, **Hannah Arendt** arrived in Marburg in 1924 to study under Martin Heidegger, then arguably the most famous living philosopher in Europe and in the final process of completing his most important work, *Being and Time*, which appeared three years later. When Arendt first met Heidegger, he was thirty-five and married, with two young children. Born a Catholic and intended for the priesthood, he developed into an extremely charismatic lecturer – his classes were complicated and dazzling intellectual displays. Students found his approach bewitching, but those who couldn't cope with the intellectual fireworks often despaired. At least one committed suicide.

Arendt came from a very different background – an elegant, cosmopolitan, totally assimilated Jewish family in Königsberg. Both her father and grandfather had died when she was young, and her mother travelled a great deal, so the young Hannah constantly worried that she would not return. Then her mother remarried, to a man Hannah never warmed to; nor did she take to the two stepsisters she acquired as a result of this union. When she arrived in Marburg, she was therefore intense but emotionally insecure, very much in need of love, protection and guidance.[43] Marburg was then a small university town, staid, respectable, quiet. For a professor to risk his position in such an environment with one of his young students says a lot about the passions that Hannah's arrival had aroused in him. Two months after she started attending his classes, he invited her to his study to discuss his work. Within another two weeks they had become lovers. Heidegger was transformed by Hannah. She was totally different from the 'Teutonic Brunhildas' he was used to, and one of the brightest students he had ever known.[44] Instead of being a rather morose, even sullen man, he became much more outgoing, writing Hannah passionate poetry. For months they indulged in clandestine meetings with an elaborate code of lights in Heidegger's house to indicate when it was safe to meet, and where. Working on *Being and Time* was an intense emotional experience for both of them, and Hannah adored being part of such an important philosophical project. After the initial passion, both realised it would be better if Hannah left Marburg, and she transferred to Heidelberg, where she studied under Karl Jaspers, a friend of Heidegger. But Hannah and Heidegger continued to correspond, and to meet, sharing their love for Beethoven and Bach, Rainer Maria Rilke and Thomas Mann, with an abandon that neither had known before. They met in a series of small German or Swiss towns where Heidegger had devised excuses to visit.[45]

After she had finished her Ph.D., Hannah moved to Berlin and married a man whom, although he was Jewish, she did not love. For her, it was a survival device. He too was a philosopher, but not as dedicated as she, and he became a journalist. They moved in a left-wing circle, and among their close friends was the playwright Bertolt Brecht and the philosopher–social scientists from the Frankfurt School – Theodor Adorno, Herbert Marcuse, Erich Fromm. Hannah still corresponded with Heidegger. Then, in 1933, after the Nazis took power, Hannah and Heidegger's lives turned dramatically in different directions. He was made rector of Freiburg University, and rumours soon reached her that he was refusing to recommend Jews for positions and even turning his back on them. She wrote to him, and he replied immediately, 'furiously' denying the charge.[46] She put it out of her head. Her left-wing husband decided he should leave Germany for Paris. Soon after, at Heidegger's rectorial address, he made a very anti-Semitic and pro-Hitler speech, which was reported all over the world.[47] Hannah was deeply upset and very confused by Martin's behavior. To make matters worse, Bertolt Brecht was being persecuted as a Communist and forced to flee the country. He left behind most of his personal possessions, including his address book, which contained Hannah's name and phone number. She was arrested, and spent eight days in jail being interrogated. Her husband was already in Paris; Martin could have helped her; he didn't.[48]

As soon as Hannah was released from jail, she left Germany and settled in Paris. From then on her world and Heidegger's were about as different as could be. As a Jew in exile, homeless, careerless, cut off from her family and all that she had known, for Arendt the late 1930s and early 1940s were a desperately tragic time. She joined a Jewish organisation, Youth Aliyah, which trained students who wanted to move to the Holy Land. She visited Palestine but didn't like it and wasn't a Zionist. Yet she needed a job and wished to help her people.[49]

Heidegger's life was very different. He played a crucial role in Germany. As a philosopher, he gave his weight to the Third Reich, helping develop its thinking, which grounded Nazism in history and the German sense of self. In this he had the support of Goebbels and Himmler.[50] As an academic figure he played a leading role in the reorganisation of the universities, the chief 'policy' under this scheme being the removal of all Jews. Through Heidegger's agency both Edmund Husserl, the founder of phenomenology and his own professor, and Karl Jaspers, who had a Jewish wife, were forced out of their university posts. Hannah later wrote that 'Martin murdered Edmund.' When *Being and Time* was republished in 1937, the dedication to Husserl had been removed.[51] Heidegger allowed both himself and his philosophy to become part of the Nazi state ideological apparatus. He changed his thinking to extol war (this happened when his rectorial address was republished in 1937). He argued that the Nazis were not Nietzschean enough, not enough concerned with great men and struggle. He played a part in linking biology to history by drawing parallels between modern Germany and ancient Greece, in its obsession with sport and physical purity.

The encounter between Hannah Arendt and Martin Heidegger was revealing not just in itself but also for the way it showed that intellectuals were not only victims of Hitler's inquisition; they helped perpetrate it too.

This is an area of prewar and wartime activity that has only become crystal clear since the fall of the Berlin Wall in 1989, which made many more archives available to scholars. Among the scientists who are now known to have conducted unethical research (to put it no stronger) are **Konrad Lorenz**, who went on to win the Nobel Prize in 1973, Hans Nachtsheim, a member of the notorious Kaiser Wilhelm Institute for Anthropology and Human Genetics in Berlin, and Heinz Brucher at the Ahnenerbe Institute for Plant Genetics at Lannach.

Lorenz's most well known work before the war was in helping to found ethology, the comparative study of animal and human behaviour, where he discovered an activity he named 'imprinting.' In his most famous experiment he found that young goslings fixated on whatever image they first encountered at a certain stage of their development. With many of the birds it was Lorenz himself, and the photographs of the professor walking on campus, followed by a line of young birds, proved very popular in the media. Imprinting was theoretically important for showing a link between Gestalt and instinct. Lorenz had read Oswald Spengler's *Decline of the West* and was not unsympathetic to the Nazis.[52] In that climate, he began to conceive of imprinting as a disorder of the domestication of animals, and drew a parallel between that and civilisation in humans: in both cases, he thought, there was degeneration. In September 1940, at the instigation of the Party and over the objections of the faculty, he became professor and director of the Institute for Comparative Psychology at the University of Königsberg, a government-sponsored position, and from then until 1943 Lorenz's studies were all designed to reinforce Nazi ideology.[53] He claimed, for instance, that people could be classified into those of 'full value' (*vollwertig*) and those of 'inferior value' (*minderwertig*). Inferior people included the 'defective type' (*Ausfalltypus*), created by the evolutionary conditions of big cities, where breeding conditions paralleled the 'domesticated animal that can be bred in the dirtiest stable and with any sexual partner.' For Lorenz, any policy that reduced 'the ethically inferior,' or 'elements afflicted with defects,' was legitimate.[54]

The Kaiser Wilhelm Institute for Anthropology and Human Genetics (KWI) was founded in 1927 at Berlin-Dahlem, on the occasion of the Fifth International Congress for Genetics, held in the German capital. The institute, and the congress, were both designed to gain international recognition for the study of human inheritance in Germany because, like other scientists, its biologists had been boycotted by scholars from other countries after World War I.[55] The first director of the institute was **Eugen Fischer**, the leading German anthropologist, and he grouped around him a number of scientists who became infamous. They included Kurt Gottschaldt, who ran hereditary pathology; Wolfgang Abel, racial science; Fritz Lenz, racial hygiene; and **Hans Nachtsheim**, in charge of the department of experimental hereditary pathology. Nearly all the scientists at the KWI supported the racial-political goals of the Nazis and were involved in their practical implementation – for example, by drawing up expert opinions on 'racial membership' in connection with the Nuremberg laws. There were also extensive links between the institute's doctors and Josef Mengele in Auschwitz. The institute itself was dissolved by the Allies after the war.[56]

Nachtsheim studied epilepsy, which he suspected was caused by lack of oxygen to the brain. Since the very young react more overtly to oxygen deficiency than adults, it became 'necessary' to experiment on children aged five to six. In order to determine which of these children (if any) suffered from epilepsy, they were all forced to inhale an oxygen mixture that corresponded to a high altitude – say, 4,000 metres (roughly 13,000 feet). This was enough to kill some children, but if epilepsy did result, the children could be lawfully sterilised. These were not *völkisch* brutes carrying out such experiments, but educated men.[57]

★

Using newly opened archives in Berlin and Potsdam, Ute Deichmann has shown the full extent to which **Heinrich Himmler** (1900–45) largely shaped the goals of the science policy of the SS as well as the practical content of the scientific and medical research it initiated. He grew up in a strict Catholic home and, even as a child, took an interest in warfare and agriculture, notably animal and plant breeding. He also developed an early interest in alternative forms of medicine, in particular homeopathy. A superstitious man, he shared with Hitler a firm belief in the superior racial value of the Germanic people. It was Himmler's Institute for Practical Research in Military Science, within the framework of another SS branch, **Das Ahnenerbe** (Ancestral Heritage), which set about clarifying the 'Jewish question' anthropologically and biologically. Himmler played a decisive role in the establishment of Das Ahnenerbe in 1935 and was the first curator. A detailed analysis of SS research authorised by Das Ahnenerbe shows that Himmler's central concern was the study of the history of, threat to, and preservation of the Nordic race, 'the race he regarded as the bearer of the highest civilisation and culture.'[58]

At the Institute for Practical Research in Military Science, experiments were carried out on cooling, using inmates from Dachau. The ostensible reason for this research was to study the effects of recovery of humans who suffered frostbite, and to examine how well humans adapted to the cold. Some 8,300 inmates died during the course of these experiments. Second, were the experiments on yellow cross, otherwise known as mustard gas. So many people were killed in this experiment that after a while no more 'volunteers' could be found with the promise of being released afterward. August Hirt, who carried out these 'investigations', was allowed to murder 115 Jewish inmates of Auschwitz at his own discretion to establish 'a typology of Jewish skeletons.' (He committed suicide in 1945.)[59] No less brutal was the Ahnenerbe's Institute for Plant Genetics at Lannach, near Graz, and in particular the work of Heinz Brücher. Brücher had the distinction of having an entire commando unit at his disposal. During the German invasion of Russia, this unit stole Nikolai Vavilov's collection of seeds (see below, page 319). The aim here was to find hardy strains of wheat so as to be able to provide enough food for the German people in the ever-expanding Reich. Brücher and his unit also went on expeditions to areas like Tibet, carrying out ethnological as well as plant studies, which show that they were thinking far ahead, identifying remote areas where 'inferior' peoples would be forced to produce these foods, or else to make way for others who would.[60]

On 2 May 1938, Hitler signed his will. In it he ordered that, upon his death, his body was to be taken to Munich – to lie in state in the Feldherrnhalle and then to be buried nearby. More than any other place, even more than Linz, Munich was home to him. In *Mein Kampf*, Hitler had described the city as 'this metropolis of German art,' adding that 'one does not know German art if one has not seen Munich.' It was here that the climax of his quarrel with the artists took place in 1937.[61]

On 18 July that year, Hitler opened the House of German Art in Munich,

nearly 900 paintings and pieces of sculpture by such Nazi favourites as Arno
Breker, Josef Thorak and Adolf Ziegler. There were portraits of Hitler as well
as Hermann Hoyer's *In the Beginning Was the Word*, a nostalgic view of the
Führer consulting his 'colleagues' during the early days of the Nazi Party.[62] One
critic, mindful that speculative criticism was now outlawed, and only reporting
allowed, disguised his criticism in reportage: 'Every single painting on display
projected either soulful elevation or challenging heroism . . . the impression of
an intact life from which the stresses and problems of modern existence were
entirely absent – and there was one glaringly obvious omission – not a single
canvas depicted urban and industrial life.'[63]

On the day that the exhibition opened, Hitler delivered a ninety-minute
speech, a measure of the importance he attached to the occasion. During the
course of his remarks he reassured Germany that 'cultural collapse' had been
arrested and the vigorous classical-Teutonic tradition revived. He repeated
many of his by now well known views on modern art, which he depicted this
time as 'slime and ordure' heaped on Germany. But he had more to offer than
usual. Art was very different from fashion, he insisted: 'Every year something
new. One day Impressionism, then Futurism, Cubism, and maybe even
Dadaism.' No, he insisted, art 'is not founded on time, but only on peoples. It
is therefore imperative that the artist erect a monument not to a time but to his
people.'[64] Race – the blood – was all, Hitler said, and art must respect that.
Germany, he insisted, 'demands . . . an art that reflects our growing racial
unification and, thus, the portrayal of a well-rounded, total character.' What
did it mean to be German? It meant, he said, 'to be clear.' Other races might
have other aesthetic longings, but 'this deep, inner yearning for a German art
that expresses this law of clarity has always been alive in our people.' Art is for
the people, and the artists must present what the people see – 'not blue
meadows, green skies, sulphur-yellow clouds, and so on.' There can be no
place for 'pitiful unfortunates, who obviously suffer from some eye disease.'[65]
Warming to his theme, he promised to wage 'an unrelenting war of purification
against the last elements of putrefaction in our culture,' so that 'all these cliques
of chatterers, dilettantes and art forgers will be eliminated.'[66]

Of course, art criticism was not the only form of criticism outlawed in
Germany; speeches by the Führer were apt to get an easy ride, too. This time,
however, there *was* criticism of a sort, albeit in a heavily disguised way. For the
very next day, 19 July, in the Municipal Archaeological Institute, across town
in Munich, the exhibition *Entartete Kunst* (Degenerate Art) opened.[67] This was
a quite different show, almost an antishow. It displayed works by 112 German
and non-German artists. There were twenty-seven Noldes, eight Dixes, thirteen
Heckels, sixty-one Schmidt-Rottluffs, seventeen Klees, and thirty-two Kir-
chners, plus works by Gauguin, Picasso, and others. The paintings and sculptures
had been plundered from museums all over Germany.[68] This exhibition surely
ranks as the most infamous ever held. It not only broke new ground in its
theme – freely vilifying some of the greatest painters of the century – but it
also set new standards in the display of art. Even the Führer himself was taken
aback by the way in which some of the exhibits were presented. Paintings and

sculptures were juxtaposed at random making them appear bizarre and strange. Sarcastic labels, which ran around, over, and under the pictures, were designed to provoke ridicule. Ernst Ludwig Kirchner's *Peasants at Midday*, for example, was labelled, 'German Peasants as Seen by the Yids.' Max Ernst's *The Creation of Eve; or, The Fair Gardener* was labelled, 'An Insult to German Womanhood.' Ernst Barlach's statue *The Reunion*, which showed the recognition of Christ by Saint Thomas, was labelled, 'Two Monkeys in Nightshirts.'[69]

If Hitler and Ziegler thought they had killed off modern art, they were mistaken. Over the four months that *Entartete Kunst* remained in Munich, more than two million people visited the Archaeological Institute, far more than the thin crowds that attended the House of German Art.[70] This was small consolation for the artists, many of whom found the show heartbreaking. Emil Nolde wrote yet again to Goebbels, more than a trace of desperation in his demand that 'the defamation against me cease.' Max Beckmann was more realistic, and on the day the show opened, he took himself off into exile. Lyonel Feininger, born in New York of German parents but living in Europe since 1887, fell back on his American passport and sailed for the New World.

After it closed in Munich *Entartete Kunst* travelled to Berlin and a host of other German cities. Yet another retroactive law, the degenerate art law of May 1938, was passed, enabling the government to seize 'degenerate art' in museums without compensation. Some of the pictures were sold for derisory sums at a special auction held at the Fischer gallery in Lucerne; there were even some pictures that the Nazis decided were too offensive to exist – approximately 4,000 of these were simply burned in a huge bonfire, held on Kopernikerstrasse in Berlin in March 1938.[71] The exhibition was a one-off, mercifully, but the House of German Art became an annual fixture, at least until 1944. Here the sort of art that Hitler liked – pastoral scenes, military portraits, mountainscapes similar to those he himself had painted when he was younger – hardly changed from year to year.[72]

Hitler's assault on painters and sculptors has received more attention from historians, but his actions against musicians were no less severe. Here too there was an initial tussle between Goebbels and Rosenberg; the modernist repertoire was purged from early on in 1933, with 'degenerate' composers like Arnold Schoenberg, Kurt Weill, Hanns Eisler, and Ernst Toch, and conductors who included Otto Klemperer and Hermann Scherchen expelled. An *Entartete Musik* exhibition was held in Dusseldorf in May 1938. This was the brainchild of Adolf Ziegler, and a major feature was photographs of composers – Schoenberg, Stravinsky, Hindemith, Webern – who were considered to have a destructive influence on German music. Jazz was treated less harshly. Goebbels realised how popular it was with the masses and that its curtailment might lose the Nazis much sympathy, so it could be performed, provided it was German musicians who were playing. Opera, on the other hand, came under strict Nazi control, with the 'safer' works of Wagner, Verdi, Puccini, and Mozart dominating the repertoire as modernist works were discouraged or banned outright.[73]

If Alfred Rosenberg, on behalf of the Nazis, was to create a new National Socialist religion, as he hoped, then such religions as existed had to be destroyed.

More than anyone else, Protestant or Catholic, one man realised this and the dangers it posed: **Dietrich Bonhoeffer**. The son of a psychiatrist, Bonhoeffer was born in 1906 in Breslau, the brother in a set of nonidentical twins, the sixth and seventh in a family of eight. His father was one of the leaders of the opposition to Freud. He was taken aback when his son felt called to the church but, as a liberal, raised no objection.

Bonhoeffer had an academic bent and High Church leanings. Although he was a Protestant, he liked the confessional nature of Catholicism and was much influenced by Heidegger and existentialism, but in a negative sense. One of the most influential theologians of the century, he wrote his major books in the 1930s, during the Nazi era – *The Communion of Saints* (1930), *Act and Being* (1931), and *The Cost of Discipleship* (1937) – though *Ethics* (1940–4, never completed) and *Letters and Papers from Prison* (1942) – also became famous. As the second title hints, Bonhoeffer agreed with Heidegger that it was necessary to act in order to be, but he did not think that man was alone in this world or faced with the necessarily stark realities that Heidegger identified. It was clear to Bonhoeffer that community was the answer to the solitariness bemoaned by so many modern philosophers, and that the natural community was the church.[74] Community life was therefore, in theory at least, far more rewarding than atomised society, but it did involve certain sacrifices if it was to work. These sacrifices, he said, were exactly the same as those demanded by Christ, on behalf of God: obedience, discipline, even suffering on occasion.[75] And so the church, rather than God, became for Bonhoeffer the main focus of attention and thought. Operating within the church – as a body that had existed for centuries, since Jesus himself – teaches us how to behave; and this is where ethics fitted in. This community, of saints and others, teaches us how to think, how to advance theology: in this context we pray, a religious existential act by means of which we hope to become more like Christ.[76]

It was no accident that Bonhoeffer's emphasis on community, obedience, and discipline should become central theological issues at a time when the Nazis were coming to power, and stressing just these qualities. Bonhoeffer saw immediately the dangers that the Nazis posed, not just to society at large but specifically so far as the church was concerned. On 1 February 1933, the very day after Hitler took power, Bonhoeffer broadcast a contentious speech over Berlin radio. It was entitled 'The Younger Generation's Changed Views of the Concept of Führer,' and it was so directly confrontational that it was cut off before he had a chance to finish. In it he argued that modern society was so complex that a cult of youth was exactly what was *not* needed, that there was a false generation gap being created by the Hitler Youth movement, and that parents and youth needed to work together, so that the energies of youth could be tempered by the experience of age. He was in effect arguing that the Nazis had whipped up the fervour of the youth because mature adults could see through the bombastic and empty claims of Hitler and the other leaders.[77] This speech reflected Bonhoeffer's beliefs and attitude but, as Mary Bosanquet, his biographer, makes clear, it also highlighted his courage. From then on, he was one of those who repeatedly attacked efforts by the state to take over the church, and the functions of the church. The church, he said, was founded on confession, man's relation with God, not with the state. He showed further courage by opposing the 'Aryan' clause when it was introduced the following month, and arguing that it was a Christian duty to care for the Jews. This made him so unpopular with the authorities that in summer 1933 he accepted an invitation to become a pastor of a German parish in London. He stayed until April 1935, when he returned to take charge of a seminary at Finkelwalde. While there he published *The Cost of Discipleship* (1937), his first book to attract widespread attention.[78] One of its themes was a comparison of spiritual community and psychological manipulation. In other words, he was contrasting the ideas of the church and Rosenberg's notions in the *Mythus* and, by extension, Hitler's techniques in eliciting support. Finkelwalde was closed by Himmler in that same year, the seminarians sequestered, and later in the war sent to the front, where twenty-one died. Bonhoeffer was left untouched but not allowed to teach or publish. In the summer of 1939 he was invited to America by the theologian Reinhald Niebuhr, but no sooner had he arrived in New York, in June, than he realised his mistake and returned to Germany, taking one of the last ships before war broke out.[79]

Unable to take part in ordinary life, Bonhoeffer joined the underground. His brother-in-law worked in military intelligence under Admiral Canaris, and in 1940 Bonhoeffer was given the task of holding clandestine meetings with Allied contacts in neutral countries like Sweden and Switzerland, to see what the attitude would be if Hitler were assassinated.[80] Nothing came of these encounters, though the group around Canaris continued to work toward the first plot to kill the Führer, in Smolensk in 1943. This failed, as did the attempt in the summer of 1944, and in April 1945 Bonhoeffer was arrested and held in Tegel military prison in Berlin. From here he sent out letters and other writings, which were published in 1951 as *Letters and Papers from Prison*.[81] The Gestapo had never been absolutely sure how close Bonhoeffer was to the German underground, but after the second attempt on Hitler's life failed, on 20 July 1944, files were found at Zossen which confirmed the link between the *Abwehr* and the Allies. As a result Bonhoeffer was transferred to the Gestapo prison on Prinz-Albert-Strasse and then, in February 1945, sent to Buchenwald. It was a slow journey, with the Reich collapsing, and before he reached the camp, Bonhoeffer's party was overtaken by emissaries from Hitler. Trapped in his Bunker, the Führer was determined that no one involved in the plot to kill him should survive the war. Bonhoeffer received a court-martial during the night of 8–9 April and was hanged, naked, early the next morning.[82]

Hitler had devised a system to persecute and destroy millions, but Bonhoeffer's death was one of the last he ordered personally. He hated God even more than he hated artists.

In 1938 a young (twenty-year-old) Russian writer, or would-be writer, sent an account of his experiences in Kolyma, the vast, inaccessible region of Siberia that contained the worst camps of the Gulag, to the Union of Writers in Moscow. Or he thought he had. **Ivan Vasilievich Okunev**'s report, written in a simple school notebook, never went anywhere. It was kept by the KGB in his file until it was found in the early 1990s by Vitali Shentalinsky, a fellow writer and poet who, after years of trying, finally managed to persuade the Russian authorities to divulge the KGB's 'literary archive.' His tenacity was well rewarded.[83]

Okunev had been arrested and sent to the Gulag because he had allowed his (internal) passport to lapse. That is all. He was put to work in a mine, as a result of which, after several weeks, the sleeves of his coat became torn. One day the camp director announced that if anyone had any complaints, they should say so before that day's shift began. Okunev and another man explained about their sleeves, and two others said they needed new gloves. Everyone else was sent off to the mines, but the four who had raised their hands were sent instead to the punishment block. There they were sprayed with water for twenty minutes. As it was December, in Siberia, the temperature was fifty degrees below zero, and the water froze on Okunev and the others, so that the four men became united as one solid block of ice. They were cut apart with an axe, but since they couldn't walk – their clothes being frozen solid – they were kicked over and rolled in the snow back to the hut where they slept. As he fell, Okunev hit his face on the frozen ground and lost two teeth. At the hut, he was left to thaw out near the stove. Next morning, when he woke, his clothes were still wet and he had pneumonia, from which he took a month to recover. Two of the others who had formed the same block of ice with him didn't make it.[84]

Okunev was lucky, if you call surviving in such conditions lucky. It is now known that up to 1,500 writers perished under the Soviet system, mainly in the late 1930s. Many others were driven into exile. As **Robert Conquest** has pointed out, *The Penguin Book of Russian Verse*, published in 1962, shows that since the revolution, poets who lived in exile survived to an average age of seventy-two, whereas for those who remained in or returned to the Soviet Union, it was forty-five. Many scientists were also sent into exile, imprisoned, or shot. At the same time, Stalin realised that, in order to produce more food, more machinery, and as the 1930s wore on, better weapons, he needed scientists. Great pressure was therefore put on scientists to accede to Marxist ideology, even if that meant ignoring inconvenient results. Special camps were developed for scientists, called *sharashki*, where they were better fed than other prisoners, while forced to work on scientific problems.

This Russian inquisition did not arrive overnight. In summer 1918, when the civil war started, all non-Bolshevik publications were banned. However, with the New Economic Policy, unveiled in 1922, the Communist Party (as the Bolsheviks were now called) allowed a curious form of mixed economy, in which private entrepreneurs and co-operatives were established. As a result, several pre-revolutionary publishers re-emerged, but also more than a hundred literary cooperatives, some of which, like RAPP (the **Russian Association of Proletarian Writers**), became quite powerful. In literature the 1920s were an uneasy time. Several writers went into exile, but as yet there was no firm distinction between what was and was not acceptable as literature. The mind of the leadership was clearly on more pressing things than writing, though two new journals, *Krasnaya nov* (1921) and *Novy mir* (1925), were under the control of hard-line Marxists. Certain writers, like **Osip Mandelstam** and Nikolay Klyuev, already found it difficult to be published. In 1936, a decade later, no fewer than 108 newspapers and 162 periodicals were still being published, in the Russian language, *outside* the Soviet Union.[85]

Science had been 'nationalised' by the Bolsheviks in 1917, thus becoming the property of the state.[86] To begin with, according to Nikolai Krementsov, in his history of Stalinist science, many scientists had not objected because under the tsars Russian science, though expanding slowly, lagged well behind its counterparts in other European countries. For the Bolsheviks, science was expected to play an important role in a technocratic future, and during the civil war scientists were given several privileges, including enlarged food rations (*paiki*) and exemption from military service. In 1919 there was a special decree 'to improve the living conditions for scholars.' During the early 1920s international currency was made available for scientists to buy foreign equipment and to make specially sanctioned 'expeditions' abroad. In 1925 the Lenin Prize for scientific research was established. Scientists occupied places on the highest councils, and under their guidance numerous institutes were opened, such as the X Ray Institute, the Soil Institute, the Optical Institute, and the Institute of Experimental Biology, a large outfit that housed departments of cytology, genetics, eugenics, zoo-psychology, hydrology, histology, and embryology.[87] This modern approach was also reflected in the publication of the first *Great Soviet Encyclopedia*, and the period saw the great flowering of 'Soviet physics,' in particular the Leningrad Physico-Technical Laboratory, when relations with the West were good.[88] Science was no longer bourgeois.

In the mid-1920s, however, a change began to be seen in science in the language used. A new lexicon, and a new style – far more polemical – started to surface, even in the journals. Professional societies like the Society of Mathematician-Materialists and the Society of Marxist-Agrarians began to appear. Books with titles such as *Psychology, Reflexology and Marxism* (1925) were published, and the journal of the Communist Academy, *Under the Banner of Marxism*, carried a series of articles by accomplished scientists which nonetheless argued that the results of experiments had nothing to do with their interpretation. Specifically Communist universities were formed, along with an **Institute of Red Professors**, the aim of both being 'to create a new, Communist intelligentsia.'[89] In May 1928, at the Eighth Congress of the Union of Communist Youth – the Komsomol – Stalin indicated that he was ready for a new phase in Soviet life. In a speech he said, 'A new fortress stands before us. This fortress is called science, with its numerous fields of knowledge. We must seize this fortress at any cost. Young people must seize this fortress, if they want to be builders of a new life, if they want truly to replace the old guard. ... *A mass attack of the revolutionary youth on science is what we need now, comrades.*'[90]

A year later, what Stalin called *Velikii Perelom* (the Great Break, or Great Leap Forward) was launched. All private initiative was crushed, market forces removed, and the peasantry collectivised. On Stalin's orders, the state exercised from now on a total monopoly over resources and production. In science there was a period of 'sharpened class struggle,' in effect the first arrests, exiles, and show trials, but also the intervention of party cadres into agriculture. This was disastrous and led directly to the famines of 1931–3. Science was expanded (by about 50 percent) under the first Five-Year Plan, which was the main plank under the Great Break, but it was as much a political move as an intellectual one. Party activists took over all the new establishments and also infiltrated those that already existed, including the Academy of Sciences.[91] Even **Ivan Pavlov**, the great psychologist and a Nobel Prize winner in physiology, was shadowed continually (he was eighty), and the 'Great Proletarian Writer,' **Maxim Gorky**, a friend of Stalin, was put in charge of genetics and medical research.[92] Later, in July 1936, entire areas of psychology and pedagogy were abolished; the Academy of Sciences, originally a club for award-winning scholars, was forced to become the administrative head of more than a hundred laboratories, observatories, and other research institutions, though of course by then the academy was stuffed with 'red directors' at the highest levels. 'Cadres decide everything' was the official slogan: *Kadry reshaiut vse*. A circle of physicists-mathematicians-materialists was established. 'It sought to apply Marxist methodology to mathematics and physics.'[93] The *Nomenklatura* was a list of posts that could not be occupied (or, indeed, vacated) without permission of the appropriate party committee, the higher the post, the higher the committee that had to authorise appointment: the president of the Academy, for instance, had to be authorised by the Politburo.[94] At the same time, foreign contacts were discouraged; there was careful screening of scientists who applied to travel and of foreign scientists who wished to come to Russia. A special agency, Glavlit, censored all publications, even scientific ones, sometimes removing 'harmful' literature from libraries.[95]

By now, some scientists had learned to live with the system, liberally sprinkling the introductions to their publications with appropriate quotations from approved writers, like Marx, before getting on with the main business of the paper. Beginning in December 1930, Stalin charged the discipline of philosophy with the task of combating traditional notions and with developing Lenin's philosophy. This policy was launched through the Institute of Red Professors of Philosophy and Natural Sciences. The idea behind it was that science had a 'class nature' and must be made more 'proletarian.'[96] There was also a campaign to make science more 'practical.' Applied science was lauded over basic research. 'Militant' scientists criticised their less militant (but often more talented) colleagues and engaged them in public discussions where these colleagues were forced to admit previous 'errors.' By the mid-1930s, therefore, Soviet science had changed completely in character. It was now run by party bureaucrats and, insofar as this was possible, organised along lines in keeping with the tenets of Marxism and Leninism. Naturally, this led to absurdities.[97] The most notorious occurred in the discipline of genetics. Genetics had not existed in Russia prior to the revolution, but in the 1920s it began to flourish. In 1921 a Bureau of Eugenics was created, though in Russia this was predominantly concerned with plant breeding, and in 1922 one of T. H. Morgan's aides had visited Russia and brought valuable *Drosophila* stocks. Morgan, William Bateson, and Hugo de Vries were all elected foreign members of the Academy of Sciences in 1923 and 1924.[98]

In the latter half of the 1920s, however, the situation became more complex and sinister. In the immediate postrevolutionary climate in Russia, Darwinism was at first seen as aiding Marxism in creating a new socialist society. But genetics, besides explaining how societies evolve, inevitably drew attention to the fact that many characteristics are inherited. This was inconvenient for the Bolsheviks, and geneticists who espoused this view were suppressed in 1930, along with the Russian Eugenics Society. In the Soviet context, with the country's food problems, its vast expanses of land, and its inhospitable extremes of climate, genetics was potentially of enormous importance in developing strains of wheat, for example, that gave higher yields and/or grew on previously inhospitable land. The key figure here in the late 1920s and early 1930s was **Nikolai Vavilov**, one of the three scientists who had helped establish the science in the early 1920s, who was close to many foreign geneticists such as T. H. Morgan in the United States and C. D. Darlington in Great Britain. But this, of course, was a 'traditional' way of thinking. In the early 1930s a new name began to be heard in Russian genetics circles – **Trofim Lysenko**.[99]

Born in 1898 into a peasant family, Lysenko had no formal academic training, and in fact research was never his strong point; instead he became noted for a number of polemical papers about the role of genetics in Soviet society, in particular what genetics research *ought* to show. This was exactly what the party bosses wanted to hear – it was, after all, extremely 'practical' – and in 1934 Lysenko was appointed scientific chief of the Odessa Institute of Genetics and Breeding and 'elected' to membership of the Ukrainian Academy of Sciences.[100] Lysenko's doctrine, termed 'agrobiology,' was an amalgam of physiology, cytology, genetics, and evolutionary theory in which the new element was his concept of vernalisation. Vernalisation relates to the way plant seeds respond to the temperature of the seasons; Lysenko argued that if temperature could be manipulated, plants would 'think' that spring and summer had come early, and produce their harvest sooner rather than later. The question was – did it work? And second, with agriculture used as metaphor, vernalisation showed that what a plant produced was at least partly due to the way it was treated, and therefore not entirely due to its genetic component. To Marxists, this showed that surroundings – and by extension society, upbringing, education, in the human context – were as important, if not more important, than genetics. Throughout the early 1930s, in his *Bulletin of Vernalization* and in press campaigns organised for him by party friends, Lysenko conducted a noisy assault on his rivals.[101] This culminated in 1935, when Vavilov was dismissed from the presidency of the Lenin All-Union Academy of Agricultural Sciences, the most prestigious position in plant breeding and genetics, and replaced by a party hack. At the same time, Lysenko was appointed as a member of the same academy. The changing landscape was clear.[102]

Vavilov did not go without a fight, and the academy held a discussion of Lysenko's controversial views, at which it was outlined how unusual and unreliable they were.[103] Lysenko dismissed the very idea of the gene as a physical unit of heredity, claimed that Mendel was wrong and insisted that environmental conditions could directly influence the 'heredity' of organisms.[104] The scientists on Vavilov's side argued that the results of Lysenko's experiments were of dubious validity, had never been replicated or sustained by further experimentation, and flew in the face of research in other countries. The people on Lysenko's side, says Krementsov, accused their opponents of being 'fascists' and 'anti-Darwinists,' and pointed to the link between German biologists and the Nazis' ideas of a master race. At that stage, the academy actually seems to have been more favourable to Vavilov than to Lysenko, at least to the extent of not accepting the latter's results, and ordering more research. An International Genetics Conference was scheduled for Moscow in 1938, when Vavilov's allies felt sure that contact with foreign geneticists would kill off Lysenkoism for all time. Then came the Great Terror.

Nine leading geneticists were arrested and shot in 1937 (though in all eighty-three biologists were killed, and twenty-two physicists).[105] The geneticists' crime was to hold to a view that the gene was the unit of heredity and to be suspicious of Lysenko's officially approved notion of vernalisation. The institutes these geneticists headed either faded away or were taken over by acolytes of Lysenko. He himself assumed the role previously occupied by Vavilov, as president of the Lenin All-Union Academy of Agricultural Sciences, but he was also promoted further, to become a member of the USSR Supreme Soviet. Still, Lysenko did not have things all his own way. In 1939 Vavilov and other colleagues who had escaped the Terror, which ended in March that year, sent a joint six-page letter to **Andrei Zhdanov**, secretary of the Central Committee and of the Leningrad City Party, arguing for traditional genetics over Lysenkoism. (Zhdanov and his son were both chemists.)[106] They were fortified by the recent award of a Nobel Prize to T. H. Morgan, in 1933.[107] Their letter stressed the 'careerism' of Lysenko and his associates, the unreliability of his results, and the incompatibility of his ideas with both Darwinism and the international consensus in genetics. The letter received serious attention, and the Party Secretariat – which included Stalin – decided to let the philosophers judge. This meeting took place on 7–14 October 1939 at the Marx-Engels-Lenin Institute in Moscow. All four 'judges' were graduates of the Institute of Red Professors.

Fifty-three academics of one kind or another took part in the discussions. Formally, the dialogue, as identified by the philosophers in their invitation, was 'to define the Marxist-Leninist line of work in the field of genetics and breeding, which must mobilise all workers in this field in the general struggle for the development of socialist agriculture and the real development of the theory of Darwinism.' At one level the discussion was familiar. The Lysenkoists accused their opponents of work that was 'impractical' because it involved the fruit fly, whereas theirs used tomatoes, potatoes, and other useful plants and animals. The Lysenkoists no longer argued that the rival camp were 'fascists,' however. By October 1939 Russia had signed the Molotov-Ribbentrop nonaggression pact, and such a reference would have been highly inappropriate. For their part, the geneticists pointed to the unreliability of Lysenko's results, arguing that his hasty theoretical conclusions would simply lead to disaster for Soviet agriculture when they were found not to produce the predicted results. At another level, however, the debate was over Darwinism. By now, in Soviet Russia, Marxism and Darwinism had become blended.[108] The inevitability of biological evolution was assumed by Marxists to be paralleled in the sociological field, which naturally made the USSR the most highly 'evolved' society, the pinnacle that all others would reach eventually.

In their judgement, the philosophers found that Lysenko had transgressed some rules of Soviet bureaucracy, but they agreed with him that formal genetics was 'anti-Darwinian' and its methods 'impractical.' The **Leningrad Letter**, as it was called, had changed nothing. The lesser role of the formal geneticists was still allowed, but Lysenko had not been damaged and still occupied all the positions he had before the letter was written. Indeed, that position was soon consolidated; in the summer of 1940 Vavilov was arrested by the secret police as a British spy. What seems to have triggered this was his correspondence with the British geneticist C. D. Darlington, who arranged to have one of Vavilov's publications translated into English. It was not hard for the secret police to fabricate charges or secure a 'confession' about how Vavilov had provided the British with important details about Russian genetics research, which could have affected her ability to feed herself.[109]

Vavilov died in prison, and with him a huge part of Russian genetics. He was perhaps the most important scientist to succumb to the Great Terror, but genetics/agriculture was not the only discipline that was devastated: psychology and other areas of biology were also deeply affected. Vavilov was probably mourned more outside Russia than within, and is still remembered today as a great scientist. Lysenko remained where he was.[110]

On 20 June 1936 Maxim Gorky died at his *dacha*, in Gorki, just outside Moscow. He was, at the time of his death, probably the most well-known writer in Russia, a novelist, a playwright, and a poet, though he had first become famous as a short-story writer in the 1890s. He had participated in the 1905 revolution, joined the Bolsheviks, but from 1906 to 1913 had lived in Capri.[111] His novel *The Mother* (1906) is generally regarded as the pioneer of socialist realism; it was written in the United States while he was fund-raising for the Bolsheviks. A friend of Lenin, he was in favour of the 1917 revolution and afterward founded the newspaper *Novaya zhizm*. He left Russia again in the early 1920s, as a protest against the treatment of intellectuals, but Stalin persuaded him back in 1933.

To those who knew the sixty-two-year-old writer and his poor health, his death was not a surprise, but wild rumours immediately began to circulate. One version had it that he had been killed by Genrikh Yagoda, the bureaucrat in charge of the Writers' Union, because he intended to denounce Stalin to André Gide, the French author (and someone who had retracted his earlier enthusiasm for Soviet Russia). Another rumour had it that Gorky had been administered 'heart stimulants in large quantities,' including camphor, caffeine, and cardiosal. According to this version, the ultimate culprits were 'Rightists and Trotskyites' funded by foreign governments, intent on destabilising Russian society by the murder of public figures.[112] When Vitaly Shentalinsky was given access to the KGB literary archive in the 1990s, he found the Gorky file. This contained two versions of Gorky's own death, the 'official' one and the authentic one. What does seem at least theoretically possible is that the murder of Gorky's son in 1934 was designed to break the father, psychologically speaking. Even this is not entirely convincing because Gorky was not an enemy of the regime. As an old friend of Lenin, he may have felt he had to tread carefully where Stalin was concerned, and certainly, as time went by, a coldness developed between Stalin and Gorky. But as the KGB's file makes clear, Stalin visited the writer twice during his last illness. Gorky's death was natural.[113]

The rumours surrounding his death nevertheless underline the unhappy atmosphere in which writers and other artists, no less than scientists, lived. In the decade between the Great Break and World War II, literature in Russia went through three distinct phases, though this owed more to attempts by the authorities to coerce writers than to any aesthetic innovations. The first phase, from 1929 to 1932, saw the rise of proletarian writers, who followed Stalin rather than Lenin. This movement was led by RAPP, the Russian Association of Proletarian Writers, infiltrated by a new breed of author who began a campaign against the older literary types, who held to the view that the writer, like all intellectuals, should remain 'outside society, the better to be able to criticise it.' RAPP therefore attacked 'psychologism' on the grounds that a concern with individual motives for action was 'bourgeois.' RAPP also took exception to writing in which the peasants were portrayed in anything other than a flattering light.[114] The peasants were noble, not envious; and the kulaks warranted no sympathy. RAPP took part in the establishment of 'Writers' Brigades,' whose job it was to describe what the party bureaucrats were doing, collectivisation in particular. Osip Mandelstam, Boris Pasternak, and Vladimir Mayakovsky were all criticised by RAPP.[115] From 1932 to 1935 the pendulum swung back. Anyone with any sense could see that under the RAPP system, people with little or no talent were hounding much better writers into silence. The new approach granted authors special privileges – dachas, rest homes, sanitaria, foreign travel – but they were also required to join a new organisation: RAPP was abolished, to be replaced by the Writers' Union. This was more than just a union, however. It epitomised a compulsory orthodoxy: socialist realism. It was the introduction of this dogma that caused Gorky to be called home.

Socialist realism was a trinity. First, it was required to appeal to the newly educated masses and to be didactic, 'showing real events in their revolutionary context.'[116] Second, writing should not be 'too abstract', it had to be 'a guide to action,' and involve a 'celebratory' tone, since that made it 'worthy of the great epoch in socialism.' Third, socialist realism should show *Partiinost*, or 'party-mindedness,' an echo of 'Cadres decide everything' in the scientific field.[117] Gorky, for one, realised that great literature was unlikely to be produced under such circumstances. Certain ponderous projects, such as a vast history of the civil war, a history of factories, and a literature of the famine, were worth doing, but they were bound to be stolid, rather than imaginative.[118] Gorky's main aim, therefore, was to ensure that Soviet literature was not reduced to banal propaganda. The high point of socialist realism was the infamous **First Congress of Soviet Writers**, which met in the Hall of Columns in Moscow in 1935. For the congress, the hall was decorated with huge portraits of Shakespeare, Cervantes, Pushkin, and Tolstoy – none of these immortals, so it seemed, was bourgeois. Delegations of workers and peasants, carrying tools, trooped through the proceedings to remind Soviet delegates of their 'social responsibilities.'[119] Gorky gave an ambiguous address. He underlined his sympathies with the emerging talents of Russia, which the revolution had uncovered, and he went out of his way to criticise bureaucrats who, he said, could never know what it was like to be a writer. This barb was, however, directed as much at the bureaucracy of the Writers' Union itself as at other civil servants. He was implying that socialist realism had to be real, as well as socialist – the same point that Vavilov was fighting in biology. As it turned out, all the proposals the congress gave rise to were overtaken by the Great Terror. That same year a score of writers was shot in Ukraine, after the murder of Kirov. At the same time, libraries were told to remove the works of Trotsky, Zinoviev, and others. Most chilling of all, Stalin began to take a personal interest in literature. There were phone calls to individual writers, like Pasternak, verdicts on specific works (approval for *Quiet Flows the Don*, disapproval for Shostakovich's opera *Lady Macbeth of the Mtsensk District*). Stalin even read L. M. Leonov's *Russian Forest*, correcting it with a red pencil.[120]

Stalin's involvement with **Osip Mandelstam** was much more dramatic. Mandelstam's file was another of those discovered in the KGB archive by Vitaly Shentalinsky, and the most moving. Mandelstam was arrested twice, in 1934 and 1938. The second time he was seized while **Anna Akhmatova** was in his flat (she had just arrived from Leningrad).[121] Mandelstam was later interrogated by Nikolay Shivarov, in particular about certain poems he had written, including one about Stalin.

> *Question*: 'Do you recognise yourself guilty of composing works of a counter-revolutionary character?'
> *Answer*: 'I am the author of the following poem of a counter-revolutionary nature:
>
> We live without sensing the country beneath us,
> At ten paces, our speech has no sound
> And when there's the will to half-open our mouths
> The Kremlin crag-dweller bars the way.
> Fat fingers as oily as maggots,
> Words sure as forty-pound weights,
> With his leather-clad gleaming calves
> And his large laughing cockroach eyes.
>
> And around him a rabble of thin-necked bosses,
> He toys with the service of such semi-humans.
> They whistle, they meouw, and they whine:

He alone merely jabs with his finger and barks,
Tossing out decree after decree like horseshoes –
Right in the eye, in the face, the brow or the groin.
Not one shooting but swells his gang's pleasure,
And the broad breast of the Ossetian.'

There was also a poem about a terrible famine in Ukraine. As a result, Mandelstam was sent into exile for three years; it might have been worse had not Stalin taken a personal interest and told his captors to 'isolate but preserve' him.[122] Mandelstam was accused again in 1938, under the same law as before. 'This time the sentence was to "isolate" but not necessarily "preserve." '[123] Mandelstam, who had not been back from his first exile for very long, was already thin and emaciated, and the authorities, Stalin included, knew that he would never survive five years (for a second offence) in a camp. Sentence was passed in August; by December, in the transit camp, he had not even the strength to get up off his bed boards. He collapsed on 26 December and died the next day. The file says that a board was tied to his leg, with his number chalked on it. Then the corpse was thrown onto a cart and taken to a common grave. His wife Nadezhda only found out he had died on 5 February 1939, six weeks later, when a money order she had sent to him was returned 'because of the death of the addressee.'[124]

Isaac Babel, a celebrated short story writer whose best-known works include *Red Cavalry* (1926) and *Odessa Tales* (1927), an account of his civil war experience, was never a party member; he was also Jewish. Appalled at what was happening in Russia, he wrote little in the 1930s (and came under attack for it). Nonetheless, he was arrested in May 1939 and not seen again. Throughout the 1940s his wife was told periodically, 'He is alive, well and being held in the camps.'[125] In 1947 she was officially told that Isaac would be released in 1948. Not until March 1955 was she told that her husband had died 'while serving his sentence,' on 17 March 1941. Even that was wrong. The KGB file makes it clear he was shot on 27 January 1940.

The period 1937–8 became known among intellectuals as the era of *Yezhov-shchina* (Yezhov's misrule), after N. I. Yezhov, boss of the NKVD, the forerunner of the KGB. The term was originally coined by Boris Pasternak, who had always referred to *shigalyovshchina*, recalling Shigalyov in Fyodor Dostoyevsky's *The Possessed*, a book that features a dystopia in which denunciation and surveillance are paramount. Writers, artists, and scholars killed in the Great Terror included the philosopher Jan Sten, who had taught Stalin; Leopold Averbakh, Ivan Katayev, Alexander Chayanov, Boris Guber, Pavel Florensky, Klychkov Lelevich, Vladimir Kirshans, Ivan Mikhailovich Bespalov, Vsevelod Meyerhold, Benedikt Livshits, the historian of futurism, and Prince Dmitry Sviatopolk-Mirsky.[126] Estimates for the number of writers who died during the Terror range from 600 to 1,300 to 1,500. Even the lower figure was a third of the membership of the Writers' Union.[127]

The result of all this brutality, obsession with control, and paranoia was sterility. Socialist realism failed, though this was never admitted in Stalin's lifetime. The literature of the period – the history of factories, for example – is not read, if it is read at all, for pleasure or enlightenment, but only for its grim historical interest. What happened in literature was a parallel of what was happening in psychology, linguistics, philosophy, and biology. In retrospect, the best epitaph came from a real writer, Vladimir Mayakovsky. In an early futurist poem, one of the characters visits the hairdresser. When asked what he wants, he replies simply: 'Please, trim my ears.'[128]

COLD COMFORT

Despite what was happening in Germany, and in Soviet Russia, and despite the widespread unemployment on both sides of the Atlantic, new ideas, new works of art, could not be suppressed. In some ways the 1930s were surprisingly fertile.

At the time of the Wall Street crash in 1929 and the depression which followed, the cinema was overtaken by the introduction of sound.[1] The first film director to appreciate fully the implications of sound was the Frenchman **René Clair**. The first 'talkie' was *The Jazz Singer*, with Al Jolson, directed by Alan Crosland. That film was an example of what the film historian Arthur Knight calls the early 'tyranny of sound,' in which raw noise was used at every available opportunity simply because it was new. In early talkies you could hear picknickers crunching celery, in the place of written credits, actors were introduced by other actors wearing capes. Billboards advertised movies as the first '100% all-talking drama filmed outdoors,' or 'the first all-Negro all-talking picture.'[2]

Clair was much more subtle. To begin with, he was actually opposed to sound. Overcoming his reluctance, he chose to use dialogue and sound effects sparingly, most notably employing them *against* the images for heightened effect. He didn't *show* a door closing; instead, the audience heard it slam. The most dramatic instance of Clair's technique is a fight in *Sous les toits de Paris*, which happens in the dark near a railway line. The clatter and urgent rhythm of the passing trains – which we hear but do not see – adds to the muffled thuds and grunts of the shadowy fighters. Clair's invention was in essence a new filmic language, an allusive way of adding information, moods, and atmosphere that had been entirely absent hitherto.[3]

The psychological change that showed in the movies made in America in particular owed a lot to the depression, the election of Franklin D. Roosevelt, and his prompt introduction of New Deal relief measures in 1933, designed to stimulate economic revival. This brought a measure of optimism to the public mood, but the speed with which the president acted only underlined the urgency and depth of the problem. In Hollywood, as the depression lasted, the traditional comedies and even the vogue for musicals since the introduction of sound no longer seemed to be enough to help people cope with the grim reality of the early 1930s. Audiences still wanted to escape at the movies, but there was also a growing demand for realistic stories that addressed their problems.

Warner Brothers' *Little Caesar* was the first gritty drama in this vein to become big box office, the earliest successful gangster movie (it was based on the life of Al Capone). But Hollywood quickly followed it with a long string of similar films (fifty in 1931 alone) and equally sensational exposés, lifting the lid on rackets, political corruption, prison brutality, and bank failures. Among these were *The Big House* (1930), *The Front Page* (1931), *The Public Enemy* (1931), and *The Secret Six* (1931), each with a story that took the audience behind the headlines.[4] Some oversimplified, but by no means all. *I Am a Fugitive from a Chain Gang* (1932) was based on a true story and brought about real changes in the chain-gang system. Poverty was tackled head on in *Blonde Venus* (1932) and *Letty Lynton* (1932).[5] After Roosevelt's election, the mood changed again. The focus on social problems – slum housing, unemployment or the conditions of agricultural workers – remained, but films now conveyed the view that these matters needed to be addressed by democracy, that whether the actual story line had a happy or an unhappy ending there were systematic *political* faults in the country underlying the personal tragedies. The developing taste for 'biopics' likewise grew out of the same sensibility by showing the heroic struggle of successful individuals to overcome the odds. Biopics of Lincoln, Louis Pasteur, Marie Curie, and Paul Ehrlich all proved popular, though the best was probably *The Life of Emile Zola* (1937), which in Zola's classic defence of Captain Dreyfus offered a scathing attack on anti-Semitism, which was not only disfiguring Nazi Germany but prevalent in the United States as well.[6]

At the New York World's Fair in 1939, every conceivable kind of film – from travelogue to sales promotion – was on display, but what stood out was a very different way of filming the 1930s. This was the **British documentary**. In straightforward entertainment films, Britain was already far behind not only Hollywood but other European countries.[7] The documentary tradition, however, was a different matter. It owed its early virility to the **Empire Marketing Board Film Unit**, which was begun in 1929 as a propaganda outfit that devised posters and brochures to promote Britain's food supply from what was then still the Empire. A film unit was added after a gritty Scot, **John Grierson**, educated in America and much impressed by American advertising skills, persuaded Sir Stephen Tallents, who ran the board, that film could have a much wider effect than the written word.[8] Grierson's aim was to use the talents of major directors – people like **Eric von Stroheim** and **Serge Eisenstein** – to bring 'real life' to the screen, to convey the drama and heroism of real people, mainly working-class people which he believed was now possible with the invention of sound. For Grierson, the documentary was a new art form waiting to be born.[9] The early films, of fishermen, potters, or miners, in fact contained little drama and even less art. Then, in 1933, the Film Unit was moved, virtually intact, to the General Post Office, where it was to remain until the war.[10] In its new home the Film Unit produced a groundbreaking series of documentaries; the new art form that Grierson had yearned for was finally born. There was no one style. Basil Wright's touch in *Song of Ceylon* was allusive, gently intercutting 'the ageless ritual of tea-picking' with the harsher sounds of tea traders and the more prosaic sights of parts of the London Stock Exchange. Harry Watts's *Night Mail* was probably the most famous documentary of all for generations of British people (like the others, it was distributed by schools). It followed the nightly run of the mail train from London to Scotland, with a commentary by **W. H. Auden** and set to the music of **Benjamin Britten**. Auden was a perfect choice; his poem conveyed at once the lyrical rhythms of the train, its urgency, and the routine ordinariness of the operation, plus the effect that even an unexceptional letter can have on the lives of people:[11]

> And none will hear the postman's knock
> Without a quickening of the heart.
> For who can bear to feel himself forgotten?[12]

It would take a war for the British to see the propaganda value of film. By then, however, Germany had been living with propaganda for nearly a decade – Hitler moved in on the filmmakers as soon as he moved in on the artists. One of the first initiatives of **Joseph Goebbels**, when he was appointed propaganda minister, was to call together Germany's most prominent filmmakers and show them Eisenstein's *Potemkin*, his 1925 masterpiece that commemorated the revolution, and which was both a work of art and a piece of propaganda. 'Gentlemen,' Goebbels announced when the lights came on, 'that's an idea of what I want from you.'[13] The minister wasn't looking for obvious propaganda; he was clever and knew better. But the films he wanted must glorify the Reich: there was to be no argument about that. At the same time, he insisted that every cinema must include in its program a government-sponsored newsreel and, on occasions, a short documentary. By the outbreak of war, Goebbels's newsreels could be as long as forty minutes, but it was the documentaries that had most effect. Technically brilliant, they were masterminded by **Leni Riefenstahl**, an undistinguished actress in the Weimar years who had reinvented herself as a director and editor. Any summary of these films sounds boring – party meetings, Göring's new air force, the land army, the Olympic Games. It was the method of presentation, Riefenstahl's directorial skills, that made them memorable. The best was *Triumph of the Will* (1937), at three hours hardly short as Goebbels had stipulated, but then it was commissioned by the Führer himself as a record of the first party convention at Nuremberg. To judge by what was captured on camera – the parades, the oratory, the drilling of the troops, the vast numbers of people engrossed in sports or simply being fed – there were almost as many people behind the cameras as in front of them. In fact, sixteen cameras crews were involved.[14] When it was shown, after two years of editing, *Triumph of the Will* had a mesmerising effect on some people.[15] The endless torchlit parades, one speaker after another shouting into the microphone, the massive regularity of Brownshirts and Blackshirts absorbed in the rhetoric and then bellowing 'Sieg Heil' in unison, were hypnotic.[16]

Almost as clever was the film *Olympia*, which Goebbels ordered to be made about the 1936 Olympic Games, staged in Berlin. It was there that the modern Olympic Games emerged, thanks to the Nazis. The games had been restarted in 1896 in Athens, but it was not until the Los Angeles games in 1932 that Negroes first excelled. Germany won few medals, disappointing to all but the National Socialists, who had opposed participation in the games on the grounds that they were cosmopolitan, and 'racially inclusive.' This made it all the more dramatic, then, that the 1936 games were to be held in Germany.[17]

After taking power, the Nazis glorified sport as a noble ideal, a stabilising force in the modern state. Despite its racially inclusive nature, therefore, Hitler and Goebbels saw the 1936 games as a perfect way to show off the Third Reich, to display to the world its achievements and high ideals – and to teach its rivals a lesson. Jews had been excluded from sports clubs in Nazi Germany, which provoked an Olympic boycott in the United States. But that soon faded when the Germans assured everyone that all would be welcome. Hitler and Goebbels set about making the games a spectacle. Berlin streets were renamed after foreign athletes for the duration of the games, and the main stadium was erected specially for the occasion by **Albert Speer**, Hitler's architect. The Nazis initiated the 'torch run,' whereby a flaming torch was carried by a succession of runners from Greece to Berlin, arriving in time to open the games in style.[18]

For Leni Riefenstahl's film of the games, *Olympia*, she had the use of eighty cameramen and crew, and virtually unlimited state funds.[19] She shot 1.3 million feet of film and eventually produced, in 1938, a two-part, six-hour film with sound tracks in German, English, French, and Italian. As one critic wrote, 'Riefenstahl's film accepted and hardened all the synthetic myths about the modern Olympic Games. She intertwined symbols of Greek antiquity with motifs of industrial society's sports theater. She ennobled good losers, supreme winners, and dwelled on fine musculature, particularly that of Jesse Owens,' the Negro athlete from the United States who, to Hitler's extreme displeasure, won four gold medals.[20] 'Riefenstahl was the first cinematographer to use slow-motion filming and radical cutting to reveal the intensity of effort required for supreme athletic performance. Some of *Olympia*'s sections, most particularly the one dealing with platform diving, are unsurpassingly beautiful.'[21]*

After the war had started, Goebbels used all the powers at his command to make the most of propaganda. Cameramen accompanied the Stuka bombers and Panzer divisions as they knifed through Poland – but these documentaries were not only used for audiences back home. Specially edited versions were shown to government officials in Denmark, Holland, Belgium and Romania to underline 'the futility of resistance.'[22] Goebbels liked to say that 'pictures don't lie.' He must have kept his fingers crossed when he said it.

Stalin was not far behind Goebbels in his instinctive understanding of the link

---

* Until the Berlin Olympics, the events were mainly about individual prowess. However, journalists covering the games devised their own points system so that the relative performances of the different countries could be compared. This had never happened before, but became the basis for the system now in place at all Olympic Games. Under the system, Germany won most points in 1936, then came the United States, then Italy. The Japanese beat the British.

between film and propaganda. One of the aims of the first Five-Year Plan was to increase the amount of projection equipment throughout Russia. Between 1929 and 1932, the number of projectors trebled to 27,000, 'drastically altering the status of the film in the Soviet Union.'[23] What the party officials said they wanted from this new industry was 'socialist realism,' but it was really propaganda.

The tone was set in 1934 with *Chapayev*, directed by two brothers, **Sergei** and **Grigori Vassiliev**. This was a clever, funny, and romantic film about a Red guerrilla leader during Russia's civil war, an ordinary peasant who led his people to victory then became 'a well-disciplined Bolshevik.' At the same time it managed to be human by not hiding the hero's faults.[24] *Chapayev* became the model for most Russian films up to World War II. *We Are from Kronstadt* (1936), *Baltic Deputy* (1937), and the *Maxim* trilogy (1938–40) all featured revolutionary heroes who become good Bolsheviks.[25] In contrast, films about contemporary life were conspicuous by their absence and it is not hard to see why. 'Socialist realism,' as it is commonly understood, would have involved social criticism – a very dangerous enterprise in Stalinist Russia. One development that *was* allowed was the making of historical films, showing life in prerevolutionary Russia as not wholly bad. This idea had its roots in Stalin's growing belief, in the mid-1930s, that worldwide revolution would never happen and that Germany was emerging as the greatest threat to the Soviet Union. Directors were allowed to tell stories about Peter the Great, Ivan the Terrible, and others, so long as these figures had contributed to the unification of Russia.[26] Soon, however, nationalism was not enough to meet Stalin's propaganda needs. With the growing tension between Germany and Russia, films with an even stronger message were wanted. In *Alexander Nevsky* (1938), Serge Eisenstein argued that the eponymous hero had led the Russians to victory over the Teutonic knights of the thirteenth century, and they could repeat the feat if called upon to do so. At the end, Nevsky speaks directly to the camera: 'Those who come to us with sword in hand will perish by the sword.'[27] Other films were more explicit: *Soldiers of the Marshes* (1938) and *The Oppenheim Family* (1939) showed the harsh realities of Germany's anti-Semitism and the desperate conditions inside the concentration camps.[28] The trouble with propaganda, of course, is that it can never escape politics. When Molotov signed the Nazi-Soviet nonaggression pact in August 1939, all anti-German films were suddenly banned.

A different view of film was provided in 1936 in **Walter Benjamin**'s celebrated essay 'The Work of Art in the Age of Mechanical Reproduction,' published in the newly founded *Zeitschrift für Sozialforschung* (Journal for Social Research), put out by the exiled Frankfurt Institute. Benjamin, born in Berlin in 1892, the son of a Jewish auctioneer and art dealer, was a radical intellectual, a 'cultural Zionist' as he described himself (meaning he was an advocate of Jewish liberal values in European culture) who earned his living as a historian, philosopher, art and literary critic, and journalist.

Of a slightly mystical bent, Benjamin spent World War I in medical exile in Switzerland, afterward forming friendships with Hugo von Hofmannsthal, the sculptress Julia Cohn, Bertolt Brecht, and the founders of the Frankfurt School. In a series of essays and books – *Elective Affinities, The Origin of German Tragic Drama,* and 'The Politicisation of the Intelligentsia' – he compared and contrasted traditional and new art forms, anticipating in a general way the ideas of Raymond Williams, Andy Warhol, and Marshall McLuhan.[29] In the most celebrated, 'The Work of Art in the Age of Mechanical Reproduction,' written when he was already in exile, he advanced his theory of '**non-auratic**' art.[30] According to Benjamin, art from antiquity to the present has its origin in religion, and even secular work kept to itself an 'aura,' the possibility that it was a glimpse of the divine, however distant that glimpse might be. As Hofmannsthal, Rainer Maria Rilke, and José Ortega y Gasset had said, this implied a crucial difference between the artist and the non-artist, the intellectual and the proletariat. In the era of mechanical reproduction, however, and especially in film – a group rather than an individual activity – this tradition, and the distance between artists and nonartists, breaks down. Art can no longer appeal to the divine; there is a new freedom between the classes, no distinction between author and public, the latter ready to become the former if given the chance. For Benjamin the change is a good thing: in an age of mechanical reproduction the public are less an agglomeration of isolated souls, and film in particular, in offering mass entertainment, can address the psychological problems of society. As a result, social revolution might be possible without violence.[31] Benjamin's arguments, written by a liberal intellectual in exile, may be contrasted with Goebbels's. Both understood the political power of film. Goebbels appreciated its force as a political instrument in the short run; but Benjamin was one of the first to see that the very nature of art was changing, that part of its meaning was draining away. He had identified a phase in cultural evolution that would accelerate in the second half of the century.

In 1929 the Museum of Modern Art had opened in New York, its first exhibition devoted to Paul Cézanne, Paul Gauguin, Georges Seurat, and Vincent van Gogh. Arguably more influential, however, was an exhibition about architecture since 1920, held at the same museum in 1932. This was where the terms '**international style**' or '**international modern style**' were first coined. In New York at that time the new buildings attracting attention were the Chrysler headquarters (1930) and the Rockefeller Center (1931–9). Neither was in the international style, but it was the Manhattan designs that were the anachronisms. In the twentieth century, the international style would prove more influential than any other form of architecture. This was because it was more than just a style, but rather a whole way of conceiving buildings. Its aims were first clearly set out at the **International Congress of Modern Architecture** (CIAM), meeting during a cruise between Marseilles and Athens in 1933.[32] There, CIAM issued a dogmatic manifesto, known as the Athens Charter, which insisted on the importance of city planning, of 'functional zoning', and of high-rise, widely spaced apartment blocks. The moving spirit behind this approach was a forty-six-year-old Swiss, christened Charles-Edouard Jeanneret but known since 1920 as Le Corbusier. Walter Gropius,

Alvar Aalto (a Finn), Philip Johnson (the curator of the MoMA show, who coined the term International Style), and even Frank Lloyd Wright shared Le Corbusier's passion for new materials and clean straight lines in their search for a more democratic form of their art. But Le Corbusier was the most innovative, and the most combative.[33]

Le Corbusier studied art and architecture in Paris in the early years of the century, much influenced by John Ruskin and the social ideals of the Arts and Crafts Movement. He worked in Peter Behrens's office in Berlin in 1910–11 and was affected by Wright and by the Bauhaus, many of whose aims he shared, and who produced similar buildings.[34] After World War I, Le Corbusier's schemes for new architecture gradually became more radical. First came his '**Citrohan**' houses, a variation of Citroën, suggesting that houses were as up-to-date as cars. These houses abolished conventional walls and were raised on stilts or *piloti*.[35] In 1925, at the Exposition Internationale des Arts Décoratifs et Industriels, in Paris, he designed a stark white house with a tree growing out of it. The house was part of a *plan voisin* (neighbourhood plan) that envisaged demolishing much of central Paris and replacing it with eighteen huge sky-scrapers.[36] Le Corbusier's distinctive international style finally found expression in the Villa Savoye at Passy (1929–32) and in his Swiss pavilion at University City, near Paris (1930–32). These were both plain white rectangular slabs, raised off the ground.[37] Here, and in the Salvation Army Hostel, also in Paris (1929–33), Le Corbusier sought to achieve a simplicity and a purity, combining classical antiquity and modernity with the 'fundamentals' of new science.[38] He said he wanted to celebrate what he called 'the white world': precise materials, clarity of vision, space, and air, as against the 'brown world' of cluttered, closed, muddled design and thinking.[39] It was a noble aim, publicly acknowledged when he was given the commission to design the Pavillon des Temps Nouveaux for the Exposition Universelle held in Paris in 1937 (where Picasso's *Guernica* was shown).

Unfortunately, there were serious problems with Le Corbusier's approach. The available materials didn't do justice to his vision. Plain white surfaces soon stained, or cracked, or peeled. People didn't like living or working inside such buildings, especially minimalist apartment blocks.[40] The white world of the international movement would dominate the immediate post-World War II landscape, with its passion for planning. In many ways it was a disaster.

It is common now to speak of an 'Auden generation' of poets, which included Christopher Isherwood, Stephen Spender, Cecil Day Lewis, John Betjeman and, sometimes, Louis MacNeice. Not all of them spoke in an identical 'Audenesque' voice – nonetheless, **Audenesque** entered the language.

Born in 1907, **Wystan Hugh Auden** grew up in Birmingham (though he went to school in Norfolk), a middle-class boy fascinated by mythology and by the industrial landscape of the Midlands – railways, gasworks, the factories and machinery associated with the motor trade.[41] He went to Oxford to read biology, and although he soon changed to English, he always remained interested in science, and psychoanalysis especially. One of the reasons he changed to

English was because he already knew that he wanted to be a poet.[42] His first verse was published in 1928, by **Stephen Spender**, whom he met at Oxford, who had his own hand press. T. S. Eliot, by then an editor at Faber & Faber, had previously rejected one collection of Auden's poems, but the firm published a new set in 1930.[43] The collection showed that at twenty-three Auden had achieved a striking originality in both voice and technique. His background in the already decaying industrial heartland of Britain, and his interest in science and psychology, helped him to an original vocabulary, set in contemporary and realistic locations. At the same time he dislocated his syntax, juxtaposing images in deliberately jarring ways, reminiscent of the arrhythmia of machines. There was something familiar, almost ordinary, about the way many lines ended.

> The dogs are barking, the crops are growing,
> But nobody knows how the wind is blowing:
> Gosh, to look at we're no great catch;
> History seems to have struck a bad patch.[44]

Or:

> Brothers, who when the sirens roar
> From office, shop and factory pour
>   'Neath evening sky;
> By cops directed to the fug
> Of talkie-houses for a drug,
> Or down canals to find a hug
>   Until you die.[45]

Reading Auden is strangely calming, as though a 'stranger were making our acquaintance,' perhaps because, in the changing insecure world of the 1930s, his familiar, clear images were something to hold on to.[46] He was not averse to drawing his ideas from sociology and the sort of information gleaned from surveys carried out by Gallup, which started its polling activities in America in 1935 and opened an office in Britain a year later.[47] Auden's later poems, as Bernard Bergonzi has observed, had a more political edge, but it was really the new 'palette' he discovered that characterised the Auden style, appropriating the rhythms of jazz, Hollywood musicals, and popular songs (now infinitely more popular than hitherto because of the radio), and peppering his lines with references to film stars such Garbo or Dietrich.

> The soldier loves his rifle,
>   The scholar loves his books,
> The farmer loves his horses,
>   The film star loves her looks.
> There's love the whole world over
>   Wherever you may be;
> Some lose their rest for gay Mae West,
>   But you're my cup of tea.[48]

Auden was quickly imitated, but the quality and intensity of his own poetry fell off at the end of the 1930s, after one of his finest works, *Spain*. Auden was in Spain in January 1937, not to take part as a combatant in the civil war, as so many prominent intellectuals did, but to drive an ambulance for the Republican side, though that didn't happen. While there he came across the desperate infighting among the different Republican factions, and he was shocked by their cruelty to the priests. Despite these misgivings, he still thought a fascist victory needed to be prevented, and on his return to Britain he wrote *Spain*, which was completed in less than a month.[49] His main concern is liberalism, what it is and whether it can survive.

> All presented their lives.
> On that arid square, that fragment nipped off from hot
> Africa, soldered so crudely to inventive Europe;
> On that tableland scored by rivers,
>
> Our thoughts have bodies; the menacing shapes of our
> fever
> Are precise and alive.[50]

Among the lines, however, was the following:

> Today the deliberate increase in the chances of
> death,
> The conscious acceptance of guilt in the necessary
> murder.

George Orwell, who wrote his own account of the civil war, in which he himself fought, *Homage to Catalonia*, vehemently attacked Auden for this poem, saying that these lines could have been written only 'by a person to whom murder is at most a *word*.'[51] In fact, Auden was unhappy about the phrase and later changed it to 'the fact of murder.' He was subsequently attacked for being one of a group of intellectuals who favoured political murder and turned a collective blind eye to the terror in Russia.

Orwell didn't go that far. Like Auden, he feared a fascist victory in Spain and so felt obliged to fight. So did many others. In fact, the range of writers and other intellectuals who travelled to Spain to take part in the civil war was remarkable: from France, André Malraux, François Mauriac, Jacques Maritain, Antoine de Saint-Exupéry, Louis Aragon, and Paul Eluard; from Britain, besides Orwell and Auden, there was Stephen Spender, C. Day Lewis, and Herbert Read; from the United States, Ernest Hemingway, John Dos Passos, and Theodore Dreiser; from Russia, Ilya Ehrenburg and Michael Kol'tsov; from Chile, Pablo Neruda.[52] There was not yet the grand disillusion with the Soviet system that would come later, and many intellectuals were worried about the further extension of fascism beyond Germany and Italy (fascist parties existed in Finland, Portugal, and Britain, as well as elsewhere). They thought it was a 'just war.' A small number of writers supported Franco – George Santayana and Ezra Pound among them – because they thought he might impose a nationalistic and aristocratic social order, which would rescue culture from its inevitable decline; and there were a number of Roman Catholic writers who wanted a return to a Christian society. Some authors, after the senseless slaughter in the nationalist zone of Spain's own best poet, Federico García Lorca, also joined the fight. From among these writers the war generated several first-person accounts.[53] Most of the issues raised were overtaken by World War II and the Cold War that soon followed. But the Spanish Civil War generated at least two great novels that have lasting value, and one painting. These are André Malraux's *L'Espoir* (translated as *Days of Hope*), Ernest Hemingway's *For Whom the Bell Tolls*, and Pablo Picasso's *Guernica*.

André Malraux was involved in the war far more than most other intellectuals, and far more than as a writer. He was an accomplished pilot, and spent time obtaining tanks and airplanes for the Republicans and even travelled to the United States to raise funds (successfully). His novel *L'Espoir* followed the fortunes of the International Brigade, in particular the air squadron, from the beginning of the war, in Madrid, to Barcelona and Toledo, ending at the battle of Guadalajara in March 1937.[54] It is in part a combat diary and at other times an exploration of different philosophies as reflected in the experiences and attitudes of the various members of the brigade.[55] The underlying theme is that courage alone is not enough in war: victory will go to the side that best *organises* that courage. This was designed to be a two-edged message. *L'Espoir* was published while the war was still going on, so Malraux was speaking to his fellow combatants as well as to the world at large. While courage is clearly needed for a revolution, the author says, organisation raises entirely different issues, of discipline, rank, sacrifice. With one eye firmly on Lenin and Stalin, organisers par excellence, Malraux drew attention to the dangers inherent in revolution, reminding readers that organisation can be a weapon, and as with any weapon, in the wrong hands it is a calamity.

Ernest Hemingway's book is set later in the war, in the early summer of 1937, an important date because at that time a Republican defeat was beginning to seem likely. The plot centres on a group of Republican partisans, drawn from all over Spain, subsisting in a cave high among the pines of the Sierra del Guadaramas, one hundred kilometres southwest of Madrid, and behind fascist lines. Much more than in *L'Espoir*, Hemingway's book is a study of doom and betrayal, of a dawning awareness among some of the characters that the cause for which they are fighting cannot win and the beginning of an analysis of who and why that situation has come about. Hemingway's view was that the Spanish people had been betrayed, by the international powers who had not delivered on their promises, but also by Spain herself, by self-interest, factionalism, undisciplined individualism. Some of the power and poignancy of the novel arises from the fact that the American Robert Jordan realises that there is a stage in every war when the possibility of defeat appears, and yet that possibility cannot be admitted, and one has to go on killing. Where does that leave the liberal conscience?[56]

A month after the battle of Guadalajara, which formed a set piece in Malraux's novel, on 26 April 1937, forty-three Heinkels from the German

Luftwaffe attacked the tiny Spanish town of **Guernica** in the Basque region. One aircraft after another descended on the town in the afternoon light and strafed and bombed the defenceless roofs and churches and squares of an ancient and sacred place. By the time the attack was over, 1,600 of Guernica's 7,000 inhabitants had been killed, and 70 percent of the town destroyed. It was an amazing act of wanton cruelty. Prior to this, Pablo Picasso had been commissioned by the Spanish government to produce a canvas for the Spanish Pavilion at the Paris World's Fair later in 1939. He had procrastinated despite the fact that he hated Franco and, at the beginning of the year, had composed 'Dream and Lie of Franco,' a poem full of violent imagery, designed to ridicule the general, whom he presented as a loathsome, barely human hairy slug. Having dithered for months over the government commission, the attack on Guernica finally stimulated him into action. He started within weeks of the attack and completed the huge canvas, twenty-five feet by eleven feet, in a frenzy in only a month or so.[57] For the first time Picasso allowed himself an audience while he worked. Dora Maar, his companion, was always present, photographing the development of the composition; Paul Eluard was another member of this select group, together with Christian Zervos, André Malraux, Maurice Raynal, and Jean Cassou, watching him, sleeves rolled up, often talking about Goya, whose paintings had recorded the horrors of the Napoleonic wars.[58] The painting was a distillation of forty years of Picasso's art, deeply introspective and personal as well as having wider significance.[59] It shows a woman, bull, and horse as terrified companions in a black-and-white nightmare. The novelist Claude Roy, then a law student, saw *Guernica* at the Paris World's Fair and thought it was 'a message from another planet. Its violence dumbfounded me, it petrified me with an anxiety I have never experienced before.'[60] Herbert Read said, 'Art long ago ceased to be monumental, the age must have a sense of glory. The artist must have some faith in his fellow men, and some confidence in the civilisation to which he belongs. Such an attitude is not possible in the modern world.... The only logical monument would be some sort of negative monument. A monument to disillusion, to despair, to destruction. It was inevitable that the greatest artist of our time should be driven to this conclusion. Picasso's great fresco is a monument to destruction, a cry of outrage and horror amplified by the spirit of genius.'[61]

The painting is above all Picasso. The frantic, screaming woman, the horse, shrieking in pain, its eyeballs distended in agony, the sinister bull, all broken, disfigured by war and bereavement, are entirely in black and white, with traces of newsprint on the horse's torso. In his despair, Picasso is hinting that even his monument may prove no more permanent than a newspaper. As Robert Hughes has written, *Guernica* was the last great history painting.[62] It was also the last major painting that took its subject from politics 'with the intention of changing the way large numbers of people thought and felt about power.' By the end of World War II the role of 'war artist' would be rendered obsolete by war photography.[63] Early in the war, in the autumn of 1940, when Picasso was living in occupied Paris, the Nazis checked up on his assets. They visited the strongrooms in his bank and inventoried his paintings there. Then they visited his apartment. One of the officers noticed a photograph of *Guernica* lying on a table. The officer examined the photo and said, 'Did you do this?'

'No,' Picasso replied. 'You did.'[64]

Picasso was wrong about one thing, though. The images in *Guernica* have lasted, and still have resonance today. So does the Spanish Civil War. George Orwell, who fought with the Republican partisans in and around Barcelona and produced a splendid account, *Homage to Catalonia*, explained how the war seemed a catalyst to him: 'The Spanish Civil War and other events in 1936–7 turned the scale and thereafter I knew where I stood. Every line of serious work that I have written since 1936 has been written, directly or indirectly, *against* totalitarianism and *for* democratic socialism, as I understand it.'[65] In other words, Orwell knew what totalitarianism was like in 1936. It would take others decades to admit as much.

Homage to Catalonia not only conveys the horror of war, the cold, the lice, the pain (Orwell was shot in the neck), but also the boredom.[66] It was impossible to fight off the cold or the lice, but in a brief aside Orwell says that he staved off the boredom because he had brought with him, in his knapsack, 'a few Penguins.' This is one of the first references in print to a new literary phenomenon of the thirties: the paperback book.

Homage to Catalonia itself became a very popular Penguin, but the books available to Orwell in Spain were unlikely to have been particularly highbrow. **Penguin Books** had a difficult and rather undistinguished birth. The idea for the company arose from a weekend visit which **Allen Lane** made to Devon in the spring of 1934 to stay with Agatha Christie and her second husband, Max Mallowan, an archaeologist. Lane was then managing director of the Bodley Head, a London publisher. He very much enjoyed the weekend, finding his hosts in excellent spirits. (Christie used to say, 'An archaeologist is the best person to be married to – the older you get the more interested he is.') On the journey home, however, Lane found himself with nothing to read.[67] Changing trains at Exeter, he had an hour to wait, time to inspect the station's bookstalls. All he could find were magazines, cheap thrillers, and romances in dreary hard covers. The very next day, at the morning meeting with his two brothers, Dick and John, who were also directors of the Bodley Head, he said that he had had an idea for a new kind of book: reprints of quality fiction and nonfiction, but bound in cheerful paper covers which would mean they could be priced at sixpence, well below the price of normal hardcovers and the same as a packet of ten cigarettes. The idea did not go down well with the brothers. If the books were to sell for sixpence, they said, how could they hope to make a profit? Allen's answer was one word: Woolworth – though it might easily have been Ford, or Fordism. Because these paperbacks would be unimaginably cheap, he insisted, they would sell in unimaginably large quantities. Unit costs would be minimal and income maximised. Allen's enthusiasm gradually won over his brothers. There had been cheap books before, but none of them spawned the change in reading habits that Allen Lane brought about.[68] His first choice of name for the new series was Dolphin, part of the coat of arms of Bristol, Lane's hometown. It was already being used, and so was Porpoise. Penguin, however, was free. It proved far harder to sell the idea to the rest of the book trade than Lane had envisaged, and Penguin only became remotely commercial, says J. E. Morpurgo, Lane's biographer, after the wife of Woolworth's senior buyer happened to be present at one of the meetings and said she liked the range of titles for the first ten books, and the jacket design.[69] Her husband then placed a bulk order.

The first Penguins were a mixed bunch. Number one was André Maurois's *Ariel*, followed by Hemingway's *A Farewell to Arms*. Then came Eric Linklater's *Poet's Pub*, Susan Ertz's *Madame Claire*, Dorothy L. Sayers's *The Unpleasantness at the Bellona Club* and Agatha Christie's *The Mysterious Affair at Styles*. These were followed by Beverley Nichols's *Twenty-five*, E. H. Young's *William*, and Mary Webbs's *Gone to Earth*. At number ten was Compton Mackenzie's *Carnival*. It was a solid list, but it cannot be said to have broken new ground intellectually – sensible but safe, in the words of one friend.[70] It was, however, an immediate commercial success. Some of the sociological reasons given at the time for the impact made by Penguin were more plausible than others. For example, it was argued that during the depression books were a cheap form of escape; alternatively, that large private libraries were no longer possible, in the smaller houses that **J. B. Priestley** had written about in *English Journey*, an examination of the social changes in Britain in the 1930s.[71] But a better understanding of Penguin's success emerged from a study Lane was familiar with, since it had been published only two years before, in 1932, which had examined people's reading habits. This was **Q. D. Leavis**'s *Fiction and the Reading Public*. Queenie Leavis was the wife of **F. R. Leavis**, a controversial don and literary critic in the English department at Cambridge. 'English' was then a relatively new subject at that university. The department, formed shortly after World War I, was run by the professor there, Hector Munro Chadwick, and his colleagues I. A. Richards, William Empson, and the Leavises. They had two main interests: the belief that literature was man's noblest adventure, *the* attempt above all others to forge an ethical, moral, and therefore ultimately an enjoyable and satisfying life; and the corrupting influence on literature, and therefore on the mind, of commercial culture. In 1930 F. R. Leavis had produced *Mass Civilisation and Minority Culture*, in which he argued that the 'discerning appreciation' of art and literature always depends on a small minority and that 'fine living' stems crucially from the 'unprompted first-hand judgement' of this minority.[72] High culture was led by poetry.

In Cambridge, Richards and the Leavises were surrounded by scientists. Empson originally went to Cambridge to read mathematics, Kathleen Raine was there and read biology, and the leading student literary magazine was edited by a man better known as a scientist, Jacob Bronowski. There is no question but that they were affected by this. As Leavis's biographer tells us, poetry, for him, 'belonged to the "vast *corpus* of problems" that are addressed by subjective opinion, rather than scientific method or conventional rule of thumb: "The whole world, in brief, of abstract opinion and disputation about *matters of feeling*." Poetry invited subjectivity, so it was an eminently suitable *bait* for anyone

who wishes to trap current opinions and responses" '[73] (italics in original). Leavis
and Richards were interested in what 'ordinary' people (as opposed to critics)
thought about poetry, about specific poems, and carried out surveys (science
of sorts) to gauge reactions. Discussion of these 'protocols' introduced a new
interaction in the lecture room, which was also revolutionary for the time. It
was an attempt to be more objective, more scientific, as was *Fiction and the
Reading Public*, in which Q. D. Leavis described herself as a sort of anthropologist
looking at literature.

The focus of her attention was 'the best-seller' and why best-sellers are never
regarded as great literature. Her early chapters were based on a questionnaire
sent to best-selling authors, but were overshadowed by the rest of the book,
which was historical, describing the rise of the fiction-reading public in Britain.
Leavis noted how in Elizabethan times the most popular form of culture was
music; in the seventeenth and eighteenth centuries, the Puritan conscience
maintained a canon of literature that was designed to be uplifting, a reflection
of the fact that, at the least, the established church put 'a scholar and a gentleman
in every parish' who helped to lead taste. The changes that followed all stemmed
from one thing: the growth in and changes to journalism. In the late eighteenth
century, with the growth in popularity of periodicals like the *Tatler* and the
Spectator, the reading of fiction quadrupled. This change, Leavis says, was so
rapid that standards fell; novelists wrote more quickly to meet the expanding
demand, producing inferior works. Then, in the early nineteenth century, the
demand for novels written in serial form meant that novelists were forced to
write more quickly still, in instalments, where each instalment had to end in as
sensational a way as possible. Standards still fell further. Finally, at the end of
the nineteenth century, with the arrival of the rotary press and the modern
newspaper – and Lord Northcliffe and his *Daily Mail* in particular – standards
fell yet again under the rubric 'Give the public what it wants.' By stages, Leavis
said, the novel acquired a standing and then lost it; where once it had been a
highbrow exploration of man's essential ethical nature, it had since fallen a long
way, step by step, to become mere storytelling. By the end of her book, Leavis
had quite abandoned her anthropological stance and her scientific impartiality.
Fiction and the Reading Public ends up as an angry work, angry with Lord
Northcliffe in particular.[74]

The book did, however, offer some clues as to the success of Allen Lane and
Penguin Books. Several of the authors Leavis mentions – Hemingway, G. K.
Chesterton, Hilaire Belloc – were included in the early lists. Hemingway, she
said, glorified the 'regular man,' the figure set up by journalists in opposition
to the highbrow; Chesterton and Belloc used a prose that, though more polished
than journalism, was recognisably of that genre, carefully crafted to make no
intellectual demands on the reader.[75] This was not entirely fair on Lane. His
lists were a mix, and with some of his other titles he did try to raise people's
horizons. For example, the second ten Penguins were better than the first
ten: Norman Douglas's *South Wind*, W. H. Hudson's *Purple Land*, Dashiell
Hammett's *Thin Man*, Vita Sackville-West's *Edwardians*, and Samuel Butler's
Erewhon. In May 1937 Lane launched the Pelican imprint, and it was this range

of nonfiction books that may have brought him his greatest triumph.[76] It was the 1930s, and something was clearly wrong with Western capitalism, or the Western system.[77] **Pelican** actually started after Allen had been sent one of George Bernard Shaw's notorious postcards, in the summer of 1936. Shaw's message was that he liked the first Penguins, and he recommended Apsley Cherry-Garrard's *Worst Journey in the World* as a 'distinguished addition.' Lane had already dismissed that very title on the grounds that, at sixpence a book, it was far too long to make a profit. And so, when he replied to Shaw, he was careful to make no promises, but he did say that what he really wanted was Shaw's own *Intelligent Woman's Guide to Socialism, Capitalism and Sovietism*. Shaw simply replied: 'How much?'[78] With Shaw on board, H. G. Wells, Julian Huxley, G. D. H. Cole, and Leonard Woolley soon followed. As this list shows, Penguin moved into science immediately and took a predominantly left-of-centre view of the world. But by now, 1937, the world was turning darker, and to adjust, Lane introduced a third innovation: the **Penguin Special**.[79] The first was *Germany Puts the Clock Back*, which came out in November 1937, written by the opinionated American journalist Edgar Mowrer. The tone of the text was polemical, but also relevant to its success was the fact that the book had been quickly produced to address a specific predicament. This note of urgency was new, making Penguin Specials feel different from the traditional, leisured manner of the book trade. Before the outbreak of war, Penguin produced thirty-six specials, among them *Blackmail or War?*, *China Struggles for Unity*, *The Air Defence of Britain*, *Europe and the Czechs*, *Between Two Wars?*, *Our Food Problem*, and *Poland* (the latter released only two months before Hitler's invasion).[80]

Allen Lane, and Penguin, were often too left-wing for many. But commercially speaking, the great majority of titles were a success, selling on average 40,000 but with the political specials reaching six figures.[81] And in a way, Queenie Leavis had been confounded. There might not be much of a taste, by her standards, for serious fiction, but there was a healthy demand for serious *books*. It was, as no one needed to be reminded, a serious time.

Clive Bell, the artist, was in no doubt about the cleverest man he had ever met: John Maynard Keynes. Many people shared Bell's view, and it is not hard to see why. Keynes's Political Economy Club, which met in King's College, Cambridge, attracted the cleverest students and economists from all over the world. Nor did it hurt Keynes's reputation that he had made himself comfortably rich by a number of ventures in the City of London, a display of practical economics rare in an academic. Since publication of *The Economic Consequences of the Peace*, Keynes had been in an anomalous position. So far as the establishment was concerned, he was an outsider, but as part of the Bloomsbury group he was by no means invisible. He continued to correct politicians, criticising Winston Churchill, chancellor of the exchequer, in 1925 for the return to the gold standard at $4.86 to the pound, which in Keynes's view made it about 10 percent overvalued.[82] He also foresaw that as a result of the mines of the Ruhr being allowed back into production in 1924, coal prices would drop significantly, leading to the conditions in Britain which provoked the General Strike of 1926.[83]

Being right did not make Keynes popular. But he refused to hold his tongue. Following the Wall Street crash in 1929 and the depression that followed, when unemployment rose to nearly 25 percent in the United States and 33 percent in areas of Europe, and when no fewer than 9,000 banks failed in America, most economists at the time believed that the correct course of action was no action.[84] Conventional wisdom held that depressions were 'therapeutic,' that they 'squeezed out' the inefficiency and waste that had accumulated in a nation's economy like poison; to interfere with that natural economic homeopathy risked inflation. Keynes thought this was nonsense. Worse, given the hardship caused by mass unemployment, it was immoral nonsense. Traditional economists based their views of inaction on Say's law of markets, after Jean-Baptiste Say, the nineteenth-century French economist. Say's law maintained that the general overproduction of goods was impossible, as was general unemployment, because men produced goods only in order to enjoy the consumption of other goods. Every increase in investment was soon followed by an increase in demand. Savings were likewise used by the banks to fund loans for investments, so there was no real difference between spending and saving. Such unemployment as arose was temporary, soon rectified, or voluntary, when people took time off to enjoy their earnings.[85]

Keynes was not the only one to point out that in the 1930s the system had produced a situation in which unemployment was not only widespread but involuntary, and far from temporary. His radical observation was that people do not spend every increase in income they receive. They spend more, but they hold back some. This may not seem very significant, but Keynes saw that it had a domino effect whereby businessmen would not spend all their profits in investment: as a result the system outlined by Say would gradually slow down and, eventually, stop. This had three effects: first, that an economy depended as much on people's *perceptions* of what was about to happen as on what actually happened; second, that an economy could achieve stability with a significant measure of unemployment within it, with all the social damage that followed; and third, that investment was the key matter. This led to his crucial insight, that if private investment wasn't happening, the state should intervene, using government credits, and manipulation of interest rates, to create jobs. Whether these jobs were useful (building roads) or merely wasteful didn't really matter: they provided cash that would be spent in real ways, generating income for others, which would then be passed on.[86]

Keynes was still outside the heart of the British establishment, and it would need another war to bring him in from the cold. He had always been a 'practical visionary,' but others refused to recognise that.[87] Ironically, the first place Keynes's policies were tried was in Nazi Germany. From the moment he assumed office in 1933, Hitler behaved almost like the perfect Keynesian, building railways, roads, canals, and other public projects, while implementing strict exchange controls that prevented Germans sending their money abroad and forced them to buy domestic products. Unemployment was abolished inside two years, and prices and wages began to rise in tandem.[88] Germany, however, didn't count for many people. The horror of Hitler prevented them giving him credit for anything. In 1933, on a visit to Washington, Keynes tried to interest Franklin D. Roosevelt in his ideas, but the new president, preoccupied with his own New Deal, did not fully engage with Keynes, or Keynesianism. After this failure, Keynes decided to write a book in the hope of gaining a wider audience for his ideas. *The General Theory of Employment, Interest and Money* appeared in 1936. For some economists, it was sensational, and merited comparison with Adam Smith's *Wealth of Nations* (1776) and Marx's *Capital* of 1867. For others, Keynes's radicalism was every bit as odious as Marx's, and maybe more dangerous, because it stood a greater chance of working.[89] To begin with, the book had a bigger practical effect in America than in Britain. The universities there took up *The General Theory*, and then it spread to Washington. J. K. Galbraith remembers that 'on Thursday and Friday nights in the New Deal years the Federal Express out of Boston to Washington would be half-filled with Harvard faculty members, old and young. All were on the way to impart wisdom to the New Deal. After *The General Theory* was published, the wisdom that the younger economists sought to impart was that of Keynes.'[90]

In 1937, a few months after Keynes's book was published, it seemed that the depression was easing, and signs of recovery were at last showing themselves. Unemployment was still high, but production and prices were at least creeping up. No sooner had these green shoots begun to appear than the classical economists came out of hibernation, arguing that federal spending be cut and taxes raised, to balance the budget. Immediately, the recovery slowed, stopped, and then reversed itself. Gross national product (GNP) fell from $91 billion to $85 billion, and private investment halved.[91] It is not often that nature offers a natural laboratory to test hypotheses, but this time it did.[92] War was now not far away. When hostilities began in Europe, unemployment in the United States was still at 17 percent, and the depression was a decade old. World War II would remove unemployment from the American scene for generations and herald what has aptly been called the Age of Keynes.

The essence of the 1930s as a grey, menacing time is nowhere more contradicted than in the work – and words – of **Cole Porter**. Queenie Leavis and her husband might lament the influence of mass culture on the general quality of thought (and their pessimism would be echoed time and again in the years to follow), but once in a while, individuals of near-genius have produced popular art, and in music, Porter stands out. Although he continued to produce good work up to 1955 (in *Silk Stockings*), the 1930s were his decade.[93] Porter's oeuvre in the 1930s included 'Don't Fence Me In,' 'Night and Day,' 'Just One of Those Things,' 'In the Still of the Night,' 'I've Got You under My Skin,' 'You're the Top,' 'Begin the Beguine,' 'Easy to Love,' and 'I Get a Kick out of You':

I get no kick from champagne;
Mere alcohol doesn't thrill me at all.

So tell me why should it be true
That I get a kick out of you.

I get no kick in a plane.
Flying too high with some guy in the sky
Is my idea of nothing to do,
Yet I get a kick out of you.

Porter's work suffered when a horse fell on him in 1937, crushing both legs, and he became a semi-invalid, but until then his sophistication and cleverness were only part of his genius. His topical eye for detail was second to none, even Audenesque, according to Graham Greene.[94]

You're the purple light of a summer night in Spain
You're the National Gallery
You're Garbo's salary
You're cellophane!

And

In olden days a glimpse of stocking
Was looked on as something shocking,
Now heaven knows, anything goes![95]

Cellophane and stockings. They were, in fact, much more impressive than Garbo's salary.[96] The 1930s, even as Linus Pauling was discovering the nature of the chemical bond, were also the decade when Baekeland's discovery of plastic began to deliver its legacy in a proliferation of synthetic substances that hit the market one after another. The first **acetylene-based fabrics** were marketed in 1930, as was acrylic plastic, leading to **Perspex**, **Plexiglass**, and **Lucite**. **Cellophane** proper appeared wrapped around Camel cigarettes, also in 1930.[97] Neoprene synthetic rubber was available a year later, and polyamide synthetic fibres in 1935. Perlon, an early form of nylon, was introduced in Germany in 1938, and commercial polythene in 1939. In 1940 in America cellophane was voted the third 'most beautiful' word in the language (after 'mother' and 'memory'), a triumph of that other 'm' word, marketing. But it was the chemistry that mattered, and here **nylon** was the most instructive.[98]

Despite being on the losing side in World War I, Germany had maintained a strong base in industrial chemistry. In fact, because the Allied naval blockade had been so successful, Germany was forced to experiment with synthetic foods and products, keeping her ahead of her enemies. Beginning in 1925, with the formation of I. G. Farben Chemical Group, a team of talented organic chemists was brought together to carry out basic research in polymer chemistry, aiming to build specific molecules with specific properties.[99] This was categorised as fundamental research and so escaped the Allied sanctions against military products. The team synthesised a new polymer every day for a period of years. British and American industries were aware of this commercial threat, even though the politicians dismissed the military risk, so much so that in 1927 the

Du Pont Company of Wilmington, Delaware, increased the research budget of the chemical department from $20,000 a year to $25,000 a month.[100]

At the time it was believed that chemical substances were divided into two, those like sugar or salt whose molecules would pass through a fine membrane, and which were crystal; and those with larger molecules, like rubber or gelatin, which would not pass through such a membrane, classified as '**colloids**.' Colloids were conceived as a series of smaller molecules held together by a mysterious 'electrical' force. As Linus Pauling's experiments were showing, however, the chemical bond was basic, a part of physics: there was no 'mysterious' force. Once the mystery was removed, and the way molecules were linked together became clearer, the possibility of synthesising substances similar to, and maybe better than, rubber or gelatin became a practical option. In particular, there was a need for a silk substitute, silk being expensive and difficult to obtain from Japan, which was then at war with China. The fundamental breakthrough was the work of **Wallace Hume Carothers**, 'Doc,' who had been lured to Wilmington against a rival offer from Harvard with the promise of 'massive funds' for basic research. He began to build up ever larger chain molecules – polyesters – by using so-called difunctional molecules. In classical chemistry, alcohols react with acids to produce esters. In difunctional molecules, there are two acid or alcohol groups at each end of the molecule, not one, and Carothers discovered that such molecules 'are capable of reacting continually with each other to set off chain reactions,' which grow into longer and longer molecules.[101] As the 1930s progressed, Carothers built up molecules with molecular weights of 4,000, 5,000, and then 6,000 (sugar has a molecular weight of 342, haemoglobin 6,800, and rubber approximately 1,000,000). One of the properties to emerge was the ability to be drawn out as a long, fine, strong filament. To begin with, says Stephen Fenichell, in his history of plastic, these were too brittle, or too expensive, to be commercially useful. Then, in late March 1934, Carothers asked an assistant, **Donald Coffman**, to try to build a fibre from an ester not studied before. If any synthetic fibre were to be commercially viable, it needed the capacity to be 'cold drawn,' which showed how it would behave at normal temperatures. The standard test was to insert a cold glass rod into the mixture and pull it out. Coffman and Carothers found that the new polymer turned out to be tough, not at all brittle, and lustrous.

After this discovery, Du Pont went into frantic action to be the first to create a successful synthetic silk. The patent was filed on 28 April 1937, and the world was introduced to the new substance at Du Pont's 'Wonder World of Chemistry' at the New York World's Fair in 1939. Nylon – in the form of nylon stockings – stole the show. It was originally called fibre 66; hundreds of names had been tried, from Klis (silk backward) to nuray and wacara (imagine asking for 'a pair of wacaras, please'). Nylon was preferred because it sounded synthetic and couldn't be confused with anything else. After the fair demand for nylon built up; many stores restricted customers to two pairs each. There was a serious side to the nylon frenzy, however, which the *New York Times* pointed out: 'Usually a synthetic is a reproduction of something found in nature. ... This nylon is different. It has no chemical counterpart in nature. ... It is ... control over matter so perfect that men are no longer utterly dependent upon animals, plants and the crust of the earth for food, raiment and structural material.'[102]

In the depths of the depression, only twenty-eight of the eighty-six legitimate theatres on Broadway were open, but **Eugene O'Neill**'s *Mourning Becomes Electra* had sold out even its top-of-the-range six-dollar seats.[103] O'Neill had been confirmed as 'the great US playwright, the man with whom true American theatre really begins,' long before *Mourning*, which premiered on 26 October 1931.[104] Curiously, however, it was not until the other end of the decade, by which time O'Neill had turned fifty, that his two great masterpieces *The Iceman Cometh* and *Long Day's Journey into Night*, were written. The intervening years have become known as 'The Silence.'

More than for most artists, certain biographical details of O'Neill are crucial to understanding his work. When he was not yet fourteen, he found that his own birth had precipitated a morphine addiction in his mother. He also discovered that his parents blamed their first son, Jamie, for infecting their second son, Edmund, with measles, from which he had died, aged eighteen months. In 1902 Ella O'Neill, who was addicted to drugs, had run out of morphine and tried suicide; this set off in Eugene, then in adolescence, a period of binge drinking and self-destructive behaviour; he also began to hang around theatres (his father was an actor).[105] After an unsuccessful marriage, O'Neill attempted suicide himself, overdosing in a flophouse in 1911, after which he saw several psychiatrists; a year later his TB was diagnosed. In 1921 his father died tragically from cancer, his mother following in 1922; his brother Jamie died twelve months after that, from a stroke, which itself followed an alcoholic psychosis. He was forty-five. O'Neill had intended to study at Princeton, taking a science course. At university, however, he was greatly influenced by Nietzsche, adopting an approach to life that his biographer calls 'scientific mysticism.' He was eventually removed from the course because he attended so few classes. He began writing in 1912, as a journalist, but soon turned to plays.[106]

Autobiography apart, O'Neill's dramatic philosophy may be understood from this verdict on the United States.: America, he said, 'instead of being the most successful country in the world, is the greatest failure. It's the greatest failure because it was given everything, more than any other country. . . . Its main idea is that everlasting game of trying to possess your own soul by the possession of something outside it.'[107] Both *The Iceman Cometh* and *Long Day's Journey into Night* are very long, lasting several hours, and both are talking plays, with little action. The characters, and the audience, are trapped within the same room: here conversation is unavoidable. In *The Iceman*, the characters all wait in Harry Hope's saloon, where they drink and tell each other the same stories day in, day out, stories that are in fact pipe dreams, hopes and illusions that will never happen.[108] One man wants to get back into the police force, another to be re-elected as a politician, a third simply wants to go home. As time goes by, from one thing and another that is said, the audience realises that even these far-from-exceptional aims are, in the case of these characters, illusions – pipe dreams, in O'Neill's own words. Later it becomes clear that the characters are spending their time waiting, waiting for Hickey, a travelling salesman who, they believe, will make things happen, be their saviour (Hickey is the son of a preacher). But when Hickey finally appears, he punctures their dreams one by one. O'Neill is not making the glib point that reality is invariably cold. Instead he is saying there is no reality; there are no firm values, no ultimate meanings, and so all of us need our pipe dreams and illusions.[109] Hickey leads an 'honest' life; he works and tells himself the truth, or what he thinks of as the truth. But it turns out that he has killed his wife because he could not bear the way she 'simply' accepted the fact of his numerous, casual infidelities. We never know how she explained her life to herself, what illusions she had, and how she kept herself going. But, we realise, they *did* keep her going. The Iceman, of course, is death. It has often been remarked that the play could be called *Waiting for Hickey*, emphasising the similarities to Samuel Beckett's *Waiting for Godot*. Both, as we shall see, provided a chilling view of the world that followed the discoveries of Charles Darwin, T. H. Morgan, Edwin Hubble, and others.

Long Day's Journey is O'Neill's most autobiographical work, a 'play of old sorrow, written in tears and blood.'[110] The action takes place in one room, in four acts, at four times of the day: breakfast, lunch, dinner, and bedtime, when the members of the Tyrone family gather together. There are no great action scenes, but there are two events: Mary Tyrone returns to her dope addiction, and Edmund Tyrone (Edmund, remember, was O'Neill's brother who died) discovers he has TB. As the day wears on, the weather turns darker and foggier outside, and the house seems more and more isolated.[111] Various episodes are returned to time and again in the conversation, as characters reveal more about themselves and give their version of events recounted earlier by others. At the centre of the play is O'Neill's pessimistic view of life's 'strange determinism.' 'None of us can help the things life has done to us,' says Mary Tyrone. 'They're done before you realize it, and once they're done they make you do other things until at last everything comes between you and what you'd like to be, and you've lost your true self forever.'[112] Elsewhere, one brother says to the other, 'I love you much more than I hate you.' And then, right at the end, the three Tyrone men, Mary's husband and two sons, watch her enter the room in a deep dream, her own fog.[113] The men watch as she laments, 'That was in the winter of senior year. Then in the spring something happened to me. Yes, I remember. I fell in love with James Tyrone and was so happy for a time.' These are the last lines of the play and, as Normand Berlin has written, it is those three final words, 'for a time,' that are so heartbreaking (O'Neill's relatives hated the play).[114] For O'Neill, it was a mystery how one can be in love, and then not in love, and then be trapped for ever. In such devastating ways, O'Neill is saying, the past lives on in the present, and this is nothing science can say anything about.[115]

It is arguable whether the works of Orwell, Auden, or O'Neill best encapsulate the 1930s. The period was far from being the disaster, 'the low dishonest decade,' that Auden called it. Yet there is no escaping the fact that it was a journey toward the night, with the iceman waiting at the end. Whatever happened in the 1930s – and a lot did – it was cold comfort.

'Do you know that European birds have not half the melody of ours?' One kind of epitaph was set on the period by **Alfred Kazin**, the critic, who uses this quote from Abigail Adams to John Adams to open the last chapter of his *On Native Grounds*, published in New York in 1942. It was an apt enough sentence, for his argument in the book was that, between the Civil War and World War II, American literature had come of age, explained America to itself, and now, with Europe bent on self-destruction, it fell to America to maintain and evolve the Western tradition.[116]

But the book's other main message lay in its use of material, which was itself peculiarly American. Kazin's subtitle was 'An Interpretation of Modern American Prose Literature.' This meant of course that he left out poetry and drama (and therefore figures like Wallace Stevens and Eugene O'Neill) but did not mean that he confined himself, as a European critic might well have done, to fiction only. Instead Kazin included as literature: criticism, muckraking journalism, philosophy, and even photojournalism. His argument here was that American fiction was firmly rooted in pragmatic realism (unlike Virginia Woolf, say, or Kafka, or Thomas Mann or Aldous Huxley), and that its chief battle, its big theme, within this overall context, was with business and materialism. Discussing the novels of Theodore Dreiser, Sinclair Lewis, F. Scott Fitzgerald, Willa Cather, John Dos Passos, John Steinbeck, Ernest Hemingway, William Faulkner, and Thomas Wolfe alongside the writings of Thorsten Veblen, John Dewey, H. L. Mencken, and Edmund Wilson, Kazin first identified the various influential segments of the American psyche – pioneers, scholars, journalists/muckrakers, businessmen, and the leftovers of the feudal South. These competed, he said, to produce a literature that sometimes 'touches greatness' but is often 'half-sentimental, half-commercial.' His own analysis, as this comment reveals, was wholly unsentimental. He identified as peculiarly American the theme of 'perpetual salesmanship' highlighted by Sinclair Lewis, Van Wyck Brooks's complaint that the most energetic talents in America went into business and politics and not the arts or humanities, that several writers, like John Dos Passos in *USA*, 'feel that the victory of business in America has been a defeat for the spirit, and that this had all achieved a tragicomic climax' in the late 1930s, where education was 'only a training for a business civilisation, in politics only the good life of materialism.'[117] At the same time, Kazin noted the development of criticism, from liberal criticism in the 1920s to Marxist criticism to 'scientific criticism' in the early 1930s, with such books as Max Eastman's *The Literary Mind: Its Place in an Age of Science* (1931), in which the author argued that science would soon have the answer to 'every problem that arises' and that literature in effect 'had no place in such a world'.[118] Kazin also recorded the early rise of 'semiosis,' the understanding of language as a system of signs.

But Kazin, as that quote at the beginning of his last chapter showed, felt that since 1933 Europe had been closed and that now, in 1942, American literature, for all its faults and its love-hate affair with business, was 'the repository of

Western culture in a world overrun by fascism.'[119] This, he felt, was a profound shift, coinciding with a reawakening of America's own tradition. The stock market crash and the rise of fascism, which led many in Europe to question capitalism and to gravitate to Russia, had the effect in the United States of driving Americans back on themselves, to a moral transformation realised through nationalism as a coalescing force that, at the same time, would counteract the excesses of business, industrialisation, and science. For Kazin, this nationalism was not blind or parochial: it was a kind of conscience, which gave America dignity. Literature was only part of this society-wide trend, but Kazin thought that its role could only grow in the future. That was cold comfort too.

A parallel with Kazin's main thesis, albeit in a very different medium, can be found in what for some people is the greatest film ever made, released not long before *On Native Grounds* appeared. This was **Orson Welles**'s *Citizen Kane* (1941). Welles, born in 1915 in Kenosha, Wisconsin, was a prodigy, an innovative man of the theatre and radio by his mid-twenties, during which time he had staged a successful *Macbeth* with black actors, and startled the entire nation with his version of H. G. Wells's *War of the Worlds*, presented as a news program, which many people were panicked into believing was a real invasion from Mars. He was brought to Hollywood while he was still in his early twenties and given a virtually unique contract in which he was to write, direct, and star in his own movies.

Destined by his bulky frame to play 'big' characters (as he himself put it), he sought a subject for his first, much-publicised and much-awaited movie and hit on Kane, it seems, because his first wife, Virginia Nicholson, had married the nephew of Marion Davies, the film star who lived with William Randolph Hearst.[120] *Citizen Kane* was filmed in great secrecy, partly for publicity purposes and partly to prevent Hearst finding out, and some effort was made for legal reasons to distance the main character from the newspaper baron. But the fact remains that the film is about a media mogul who uses his power to help the theatrical career of his consort, while living in a palatial mansion peopled by an esoteric mix of friends and hangers-on. There was really no disguising who Kane was, and for a time, when filming had been completed, there was doubt as to whether the film would be released, RKO fearing a massive libel and invasion-of-privacy suit from Hearst. In the event Hearst did not sue, but some cinema chains did not carry or show the film for fear of him. Partly for that reason (and partly because, as impresario Sol Hurok said of the punters, 'If they don't want to come, nothing will stop them'), *Citizen Kane* was not a commercial success.

It was, however, a massive critical and artistic success. To begin with, it introduced technical innovations on a wide scale. This was partly the work of the cameraman, Gregg Toland, and of Linwood Dunn, in the special effects department.[121] In those days, special effects did not mean creating beings from outer space, but filming scenes more than once, so that, for example, all that greets the eye is in focus, thus providing an experience more akin to theatre – quite new in cinema. Welles also played scenes from beginning to end without intercuts and with the camera following the action. He himself, in the role of Kane, aged some fifty years – the makeup on the film was another major special effect. Other technical innovations were the introduction of a 'newsreel' into the film, to tell the life story of Kane. The film had its corny elements: at the beginning a reporter is set off on an 'investigation' to find the meaning of Kane's dying word, 'Rosebud.' But people were impressed.

When the film finally premiered, in three separate cities, the reviews were ecstatic: 'sensational' (*New York Times*); 'magnificent' (*New York Herald Tribune*); 'masterpiece' (*New York World-Telegram*); 'unfettered intelligence' (*New York Post*); 'Something new has come to the movie world at last' (the *New Yorker*).[122] The more partisan right-wing press accused Welles of mounting a Communist attack on Hearst, and this is where the link to Kazin's thesis comes in. For *Kane* was an attack on big business, but not so much a political attack, such as a regular Communist might have made, but a psychological attack. *Kane* shows that, for all a man's possessions, for all his power, his vast acres and thousands of sculptures that populate those acres, he may lack – as does Kane – an emotional core, and remain lonely and unloved. This was scarcely a new message, as Kazin had shown, but in America at the end of the 1930s, it was no less powerful for all that, especially in the way that Welles told it. The enigma that has remained (Jorge Luis Borges called Kane a labyrinth without a centre) is whether Welles meant the film to have a cold centre too.[123] He once said that personality was unknowable ('Throw away all biographies'), and it is at least possible that another aim of the film was to show this unknowability in Kane. In general, though, the verdict of his critics is that this aspect of the film was a failure, rather than an intentional device.

Riches, for Welles, as for Kane – as indeed for Hearst – were cold comfort. The rest of Welles's career was really a coda to his early flowering and the magnificence of *Kane*. The film had closed everywhere by the end of the year, before Kazin's book appeared. After that, it was for Welles – albeit very slowly - downhill all the way.

HITLER'S GIFT

A famous photograph exists, taken on the occasion of an exhibition, *Artists in Exile*, at the Pierre Matisse Gallery in New York in March 1942. Pierre Matisse, the son of the painter, Henri Matisse, had been a successful dealer in Manhattan since the early 1930s, but there had been no show like this one. Pictured in the photograph, all dressed 'respectably' in suits or tweed jackets, are: (front row) Matta, Ossip Zadkine, Yves Tanguy, Max Ernst, Marc Chagall, Fernand Léger; (back row) André Breton, Piet Mondrian, André Masson, Amédée Ozenfant, Jacques Lipchitz, Pavel Tchelitchev, Kurt Seligmann, and Eugene Berman. Such a range and quality of artistic talent can seldom, if ever, have been gathered together in one room, and critics felt the same about the art on display. *American Mercury* headlined its review of the show 'Hitler's Gift to America.'[1]

Between January 1933 and December 1941, 104,098 German and Austrian refugees arrived in America, of whom 7,622 were academics and another 1,500 were artists, journalists specialising in cultural matters, or other intellectuals. The trickle that began in 1933 swelled after Kristallnacht in 1938, but it never reached a flood. By then it had been made difficult for many to leave, and anti-Semitism, and anti-immigrant feeling generally in America, meant that many were turned away. The United States had operated a quota system since 1924, limiting immigration to 165,000, with each Caucasian nation represented in the 1890 census restricted to 2 percent of their numbers at that time. The quotas for Austrian and German immigrants actually remained unfilled throughout the 1930s and 1940s, a little-known statistic of shame for the United States among its many acts of humanitarianism.

Other artists and academics fled to Amsterdam, London, or Paris. In the French capital Max Ernst, Otto Freundlich, and Gert Wollheim formed the Collective of German Artists, and then later the Free League of Artists, which held a counter-exhibition to the Nazi *Entartete Kunst* (Degenerate Art) show in Munich. In Amsterdam Max Beckmann, Eugen Spiro, Heinrich Campendonck, and the Bauhaus architect Hajo Rose formed a close-knit group, for which Paul Citroën's private art school served as a focus. In London such artists as John Heartfield, Kurt Schwitters, Ludwig Meidner, and Oskar Kokoschka were the most well known in an intellectual community of exiles that was about two hundred strong, organised into the Free German League of

Culture by the Artists' Refugee Committee, the New English Arts Club, and the Royal Academy. The league's most potent gesture was its *Exhibition of Twentieth-Century German Art* held in the New Burlington Galleries in 1938. The title was deliberately bland, so as not to offend the government, then embarked on its policy of appeasing Hitler. When war broke out, Heartfield and Schwitters were interned as enemy aliens.[2] In Germany itself, artists such as Otto Dix, Willi Baumeister, and Oskar Schlemmer retreated into what they called 'inner exile.' Dix hid away at Lake Constance, where he painted landscapes; that, he said, was 'tantamount to emigration.'[3] Karl Schmidt-Rottluff and Erich Heckel removed themselves to obscure hamlets, hoping to escape attention. Ernst Ludwig Kirchner was so depressed by the whole business that he took his life.

But it was the emigration to the United States that was most important and significant, and not only because of the numbers involved. As a result of that intellectual migration, the landscape of twentieth-century thought was changed dramatically. It was probably the greatest transfer of its kind ever seen.

After Hitler's inquisition had become plain for all to see, emergency committees were set up in Belgium, Britain, Denmark, France, Holland, Sweden, and Switzerland, of which two may be singled out. In Britain the Academic Assistance Council (AAC) was formed by the heads of British universities, under Sir William Beveridge of the LSE. By November 1938 it had placed 524 persons in academic positions in 36 countries, 161 in the United States. Many members of British universities taxed their own salaries between 2 and 3 percent to raise money, and there were American academics who, hearing of this, sent equivalent proportions across the Atlantic. In this way the AAC raised some £30,000. (It was not finally disbanded until 1966, continuing to support academics in other countries who were persecuted for political or racial reasons.) A group of refugee German scholars established the **Emergency Society of German Scholars Abroad**. This sought to place colleagues in employment where it could, but it also produced a detailed list of 1,500 names of Germans dismissed from their academic posts, which proved very useful for other societies as the years passed. The Emergency Society also took advantage of the fact that in Turkey, in spring 1933, Ataturk reorganised the University of Istanbul, as part of his drive to Westernise the country. German scholars (among them Paul Hindemith) were taken on under this scheme and a similar one, in 1935, when the Istanbul law school was upgraded to a university. These scholars even established their own academic journal, since it was so difficult for them to publish either back home or in Britain or in the United States. The journal carried papers on anything from dermatology to Sanskrit. Its issues are collectors' items now.[4]

The German journal in Turkey only lasted for eighteen issues. A more enduring gift from Hitler was a very different periodical, *Mathematical Reviews*. The first issue of this new journal went largely unremarked when it appeared – most people had other things on their minds in 1939. But, in its own quiet way, the appearance of *MR*, as mathematicians soon began calling it, was both dramatic and significant. Until that time, the most important mathematical periodical, which abstracted articles from all over the world, in dozens of languages, was the *Zentralblatt für Mathematik und ihre Grenzgebiete*, launched in 1931, by Springer Verlag in Berlin. Thanks partly to the golden age of physics, but also to the work of Gottlob Frege, David Hilbert, Bertrand Russell, and Kurt Gödel, mathematics was proliferating, and a comprehensive abstracting journal helped people keep in touch.[5] In 1933–4, however, a problem loomed: the journal's editor, **Otto Neugebauer**, a faculty member in Richard Courant's famous department at Göttingen, was politically suspect. In 1934, he escaped to Denmark. He remained a board member of the *Zentralblatt* until 1938, but then the Italian mathematician **Tullio Levi-Civita**, who was a fellow board member and Jewish, was dismissed. Neugebauer resigned in sympathy, together with several members of the international advisory board. At the end of the year the Russian involvement on the board was also terminated, and refugee mathematicians were even banned as reviewers. An article in *Science* reported that articles by Jews now went unabstracted in the *Zentralblatt*.

American mathematicians watched the situation with dismay and alarm. At first they considered buying the title, but the Berlin company wouldn't sell. Springer did, however, make a counter-suggestion, offering two editorial boards, which would have produced different versions of the journal, one for the United States, Britain, the Commonwealth, and the Soviet Union, the other for Germany and nearby countries. American mathematicians were so incensed by this insult that in May 1939 they voted to establish their own journal.[6]

As early as April 1933 officials at the Rockefeller Foundation began to consider how they might help individual scholars. Funds were found for an emergency committee, which started work in May. This committee had to move carefully, for the depression was still hurting, and jobs were scarce. The first task was to assess the size of the problem. In October 1933, **Edward R. Murrow**, vice chairman of the emergency committee, calculated that upward of 2,000 scholars, out of a total of 27,000, had been dropped from 240 institutions. That was a lot of people, and wholesale immigration not only risked displacing American scholars but might trigger anti-Semitism. A form of words was needed that would confine the numbers who were encouraged to cross the Atlantic and in the end the emergency committee decided that its policy would be 'to help scholarship, rather than relieve suffering.' Thus they concentrated on older scholars, whose achievements were already acknowledged. The most well known beneficiary was Richard Courant from Göttingen.[7]

The two mathematicians who did most to help their German-speaking colleagues were **Oswald Veblen** (1880–1960) and **R. G. D. Richardson** (1878–1949). The former, a nephew of Thorstein Veblen, the great social theorist, was a research fellow at the Institute for Advanced Study (IAS) in Princeton, while Richardson was chairman of the mathematics department at Brown University and secretary of the American Mathematical Society. With the aid of the society, which formally joined the emergency committee, fifty-one mathematicians were brought to America before the outbreak of the war in Europe in 1939; and by the end of the war the total migration was just under 150. Every scholar, whatever his or her age, found work. Put alongside the six million Jews who perished in the gas ovens, 150 doesn't sound much; yet there were more mathematicians helped than any other professional group. Today, out of the top eight world-class mathematics institutes, the United States has three. Germany has none.[8]

In addition to the artists, musicians, and mathematicians who were brought to America, there were 113 senior biologists and 107 world-class physicists whose decisive influence on the outcome of the war we shall meet in chapter 22. Scholars were also helped by a special provision in the U.S. immigration law, created by the State Department in 1940, which allowed for 'emergency visitor' visas, available to imperilled refugees 'whose intellectual or cultural achievements or political activities were of interest to the United States.' Max Reinhardt, the theatre director, Stefan Zweig, the writer, and Roman Jakobson, the linguist, all entered the United States on emergency visas.[9]

Of all the various schemes to help refugees whose work was deemed important in the intellectual sphere, none was so extraordinary, or so effective, as the **Emergency Rescue Committee** (ERC) organised by the American Friends of German Freedom. The Friends had been formed in America by the ousted German socialist leader Paul Hagen (also known as Karl Frank), to raise money for anti-Nazi work. In June 1940, three days after France signed the armistice with Germany, with its notorious 'surrender on demand' clause, the committee's members held a lunch to consider what now needed to be done to help threatened individuals in the new, much more dangerous situation.[10] The ERC was the result, and $3,000 was raised immediately. The aim, broached at the lunch, was to prepare a list of important intellectuals – scholars, writers, artists, musicians – who were at risk and would be eligible for special visa status. One of the committee's members, **Varian Fry**, was chosen to go to France, to find as many threatened intellectuals as he could and help them to safety.

Fry, a slight, bespectacled Harvard graduate, had been in Germany in 1935 and seen at first hand what the Nazi pogroms were like. He spoke German and French and was familiar with the work of their living writers and painters. At that time, with anti-Semitism running high in America, his first move was to visit Eleanor Roosevelt in the White House, soliciting her support. The first lady promised to help, but to judge by the behaviour of the State Department subsequently, her husband did not share her views. Fry arrived in Marseilles in August 1940 with $3,000 in his pocket and a list of two hundred names that he had memorised, judging it too dangerous to carry written lists. These names had been collected in an ad hoc way. Thomas Mann had provided the names of German writers at risk, Jacques Maritain a list of French writers, Jan Masaryk the Czechs. Alvin Johnson, president of the New School of Social Research, submitted names of academics, and Alfred Barr, director of MoMA, supplied the names of artists. To begin with, many of those Fry had been sent to help – especially the artists – didn't want to leave. Pablo Picasso, Henri Matisse, Marc

Chagall, and Jacques Lipchitz all refused to emigrate (Chagall asked if there were 'any cows' in America). Amedeo Modigliani wanted to leave but wouldn't do anything illegal. Fry's offer was also turned down by Pablo Casals, André Gide, and André Malraux.[11]

Fry soon came to understand that not all the people on his list were in mortal danger. The Jews were, as well as the more outspoken, long-standing political opponents of Nazism. At the same time, it became clear that if many of the very famous, non-Jewish 'degenerate' artists were protected by their celebrity in Vichy France, there were far more lesser-known figures who *were* in real danger. Without referring back to New York, therefore, Fry changed the policy of the ERC and set about helping as many people as he could who fell within the ambit of the special visa law, whether they were on his list or not.[12] He installed the Centre Américain de Secours, a 'front' organisation on the rue Grignan in Marseilles, which dispensed routine aid to refugees – small amounts of money, help with documentation or in communicating with the United States. Meanwhile he set up his own clandestine network, using several members of the French underground, which transported selected refugees out of France into Portugal, where, with a visa, they could sail for America. He found a 'safe house,' the Villa Air Bel, just north of Marseilles, and there he equipped his refugees with false documents and local guides who could lead them via obscure and arduous pathways across the Pyrenees to freedom. The best-known figures who escaped in this dramatic fashion included André Breton, Marc Chagall, Max Ernst, Lion Feuchtwanger, Konrad Heiden (who had written a critical biography of Hitler), Heinrich Mann, Alma Mahler-Werfel, André Masson, Franz Werfel, and the Cuban painter Wilfredo Lam. In all, Fry helped around two thousand individuals, ten times the number he had been sent out to look for.[13]

Until Pearl Harbor (by which time Fry was home), the American public was largely indifferent to the plight of European refugees, and positively hostile to Jewish ones. The State Department was itself staffed by many anti-Semites in senior positions, not excluding the assistant secretary of state himself, Breckinridge Long, who hated what Fry was doing. Fry was constantly harassed by the U.S. Consul in Marseilles as a matter of departmental policy; almost certainly, the consul had a hand in Fry's arrest in September 1941, and his brief imprisonment by the Vichy authorities.[14] Despite this, between 1933 and 1941 several thousand scientists, mathematicians, writers, painters, and musicians crossed the Atlantic, many of them to remain in America permanently. Alvin Johnson, at the **New School for Social Research** in New York, took ninety scholars to create a University in Exile, where the faculty included Hannah Arendt, Erich Fromm, Otto Klemperer, Claude Lévi-Strauss, Erwin Piscator, and Wilhelm Reich. Most of these scholars he had either met or corresponded with in editing the groundbreaking *Encyclopedia of the Social Sciences*.[15] Later, after the fall of France, he also created another exilic institute, the **Ecole Libre des Hautes Etudes**. Laszlo Moholy-Nagy recreated a New Bauhaus in Chicago, and other former colleagues initiated something similar in what became **Black Mountain College**. Located at 2,400 feet, in the wooded hills and streams of

North Carolina, this was a place where architecture, design, and painting were taught alongside biology, music, and psychoanalysis. At one time or another its faculty included Joseph Albers, Willem de Kooning, Ossip Zadkine, Lyonel Feininger, and Amédée Ozenfant. Although the college was in the South, Negroes were represented among both faculty and students. After the war the college was home to a prominent school of poets and it remained in existence until the 1950s.[16] The Frankfurt Institute at Colombia University and Erwin Panofsky's Institute of Fine Arts at New York University were also started and staffed by exiles. Hitler's gift turned out to be incalculable.

The *Artists in Exile* exhibition at the Pierre Matisse Gallery in 1942, and others like it, introduced Americans to the work of important European artists. But it was only the beginning of a two-way process. Several painters who showed at Matisse never felt comfortable in America and returned to Europe as soon as they could; others adapted and stayed; none could fail to respond to the apocalyptic events they had been through.

Beckmann, Kandinsky, Schwitters, Kokoschka, and the surrealists hit back directly at fascism and the departure from liberalism, reason, and modernity that it represented. Chagall and Lipchitz interpreted events more personally, exploring the changing essence of Jewishness. **Fernand Léger** and **Piet Mondrian** looked forward, and around them, at their new country. Léger himself admitted that though he was struck by the great skyscraper canyons of cities like New York, what impressed him most about America, and helped account for its great vitality and 'electric intensity,' was the clash and complementarity of a huge country, with 'vast natural resources and immense mechanical forces.'[17] The colour in his paintings became bolder and brighter, yet simpler, whereas his black lines became starker, less part of the three-dimensional effect. Léger's American paintings are like intimate, mysterious billboards. Piet Mondrian's late paintings (he died in 1944, aged seventy-two) are probably the most accessible abstract paintings of all time. Electric, vivid, flickering lattices, *New York City; New York City 1*, *Victory Boogie-Woogie* and *Broadway Boogie-Woogie* shimmer with movement and excitement, Manhattan grids seen from the air or the tops of skyscrapers, capturing the angular, anonymous beauty of this new world, abstract and expressionistic at the same time, emphasising how, in the New World, the old categories break down.[18]

Other exhibitions were mounted during wartime, mainly in New York, showing the work of European artists living in America. *War and the Artist* was mounted in 1943, and *Salon de la Libération* in 1944. What counted here was less the way America affected the emigrés and more the way the emigrés affected a group of young American artists who were anxious to see everything the Europeans could produce. Their names were Willem de Kooning, Robert Motherwell, and Jackson Pollock.

One of Hitler's greatest gifts to the new world was Arnold Schoenberg. Once the Nazis took power, there was never much doubt that the composer would have to leave. Although he had converted from Judaism to Christianity early in life, that never made any impression with the authorities, and in 1933 he reverted to being a Jew. In the same year he was blacklisted as a 'cultural Bolshevik' and dismissed from his Berlin professorship. He moved first to Paris, where for a while he was penniless and stranded. Then, out of the blue, he received an invitation to teach at a small private conservatory in Boston, founded and directed by the cellist Joseph Malkin. Schoenberg accepted immediately, arriving in America in October.

America, however, was not quite ready for Schoenberg, and he found the early months hard going. The winter was harsh, his English was poor, there weren't many students, and his work was too difficult for conductors. As soon as he could, he transferred to Los Angeles, where at least the weather was better. He remained in Los Angeles until his death in 1951, his reputation steadily spreading. A year or so after he moved to Los Angeles, Schoenberg was appointed professor of music at the University of Southern California; in 1936 he accepted a similar position at UCLA. He never lost sight of what he was trying to do in music, and he successfully resisted the blandishments of Hollywood: when MGM inquired if he would like to write for films, he put them off by quoting so high a price ($50,000) that they melted away as quickly as they had appeared.[19]

The first music he wrote in America was a light piece for a student orchestra, but then came the **Violin Concerto (op. 36)**. Not only was this his American debut, it was also his first concerto. Rich and passionate, it was – for Schoenberg – fairly conventional in form, though it demanded phenomenally difficult finger work from the violinist. Schoenberg continued to think of himself as a conservative, in search of a new harmony, never quite (in his own mind) finding it.

Twenty years younger than Schoenberg, **Paul Hindemith** was not Jewish – in fact, he was of 'pure' German stock. But he was also devoid of any nationalistic or ethnic feelings, and the string trio he helped to make famous contained a Jew, a tie he saw no reason to break. That was one black mark against him. Another was that as a teacher at the Berlin Hochschule from 1927 to 1934 he had become known as a high-profile German composer. He had a fervent following at the time, not least among music critics at certain influential newspapers and the conductor Wilhelm Furtwängler. But Goebbels was unimpressed, and Hindemith too was branded a 'cultural Bolshevik.' After a stint in Turkey, he went to America in 1937. Béla Bartók, Darius Milhaud and Igor Stravinsky all followed to the United States. Many of the virtuoso performers, being frequent travellers as a matter of course, were already familiar with America, and America with them. Artur Rubinstein, Hans von Bülow, Fritz Kreisler, Efrem Zimbalist and Mischa Elman all settled in America in the late 1930s.[20]

The only rival to New York as a base for exiles in wartime was, as Schoenberg found out, Los Angeles, where the roster of famous names living in close proximity (close in Los Angeles terms) was remarkable. Apart from Schoenberg, it included Thomas Mann, Bertolt Brecht, Lion Feuchtwanger, Theodor

Adorno, Max Horkheimer, Otto Klemperer, Fritz Lang, Artur Rubinstein, Franz and Alma Werfel, Bruno Walter, Peter Lorre, Sergei Rachmaninoff, Heinrich Mann, Igor Stravinsky, Man Ray, and Jean Renoir.[21] The historian Lawrence Weschler has gone so far as to prepare an 'alternative' Hollywood map, displaying the addresses of intellectuals and scholars, as opposed to the more conventional map showing the homes of movie stars – worth doing, but in today's world it could never have the same appeal.[22] Arnold Schoenberg's widow used to amuse her guests by taking them outside when the tour bus came round. It would stop outside the Schoenberg house, from where the voice of the tour guide could be clearly heard, over the loudspeaker. As the tourists peered across the garden and into the house, the guide would say: 'And on the left you can see the house where Shirley Temple lived in the days when she was filming.'[23]

When he was at Harvard, Varian Fry had edited an undergraduate literary magazine with a friend and classmate named **Lincoln Kirstein**. Like Fry, Kirstein later in life went to Europe and helped bring a piece of Old World culture to America. In Kirstein's case, however, the emigration had nothing to do with the war, anti-Semitism, or Hitler. In addition to his literary interests, Kirstein was a balletomane: he thought America needed a boost in the realm of modern dance, and that only one man could fit the bill.

Kirstein was very tall, very wealthy, and very precocious. Born into a Jewish family in Rochester, New York, he started collecting art when he was ten, saw his first ballet (Pavlova) when he was twelve, published a play – set in Tibet – when he was barely fourteen, and in that same year summered in London, where he met the Bloomsbury set, encountering Lytton Strachey, John Maynard Keynes, E. M. Forster, and the Sitwells. But it was ballet that was to make the difference in Kirstein's life.[24] He had been fascinated by the dance ever since he was nine, when his parents had refused to allow him to see Diaghilev's company perform *Scheherezade* in Boston. Then, as a young man of twenty-two, visiting Venice, he had chanced on a funeral in an Orthodox church. An exotic barge of black and gold was moored to the church steps, waiting to take the body to Sant'Erasmus, the Isle of the Dead on the lagoon. Inside the church, beyond the mourners, Kirstein saw a bier, 'blanketed with heaped-up flowers, below a great iconostasis of burnished bronze.'[25] Some of the faces that came out into the sunlight after the service was over he thought he recognised, though he couldn't be sure. Three days later, according to Bernard Taper, his biographer, he chanced upon a copy of the London *Times*, and discovered that the church he had slipped into was San Giorgio dei Greci, and that the funeral was that of none other than Serge Diaghilev.

The following year Kirstein graduated from Harvard, at which point his father took him to one side and said, 'Look here, I'm going to leave you a lot of money. Do you want it now or when I die?' Kirstein took it there and then: he was still in his early twenties, but his early passion for ballet had matured into a specific ambition. Ballet in America should not have to rely on 'itinerant Russians,' or itinerants of any kind. Kirstein's mission in life was to bring ballet to America, to make it an indigenous art form.[26] The musicals of the early 1930s, newly transferred to film, were showing all America that its people could dance, but dance in a certain way. For Kirstein, ballet was the highest form of dance, and he instinctively felt that this was an area where, given the chance, America would shine.

Kirstein had tried ballet himself, taking lessons in New York from **Mikhail Fokine**, the great Russian choreographer.[27] He helped **Romola Nijinska** with her biography of her husband, and he studied ballet history. None of this satisfied him; but his study of the history of the dance showed him that ballet had only been successfully transplanted into new countries three or four times in the three hundred years since the first company had been chartered by the king of France. That made Kirstein determined, and in 1933, when many artistic refugees were beginning to stream to America, he travelled to Europe. He started in Paris, where, he later said, he behaved 'like a groupie.'[28] That was where **George Balanchine** was, and Balanchine, Kirstein knew, was the best choreographer alive. Everyone he met agreed on Balanchine's stature – but their enthusiasm went little further than that. One problem was Balanchine's ill health; Romola Nijinsky told Kirstein she thought the choreographer 'would be dead within three years'; apparently a clairvoyant had even named the exact date. Then there was his temperamental nature and his legendary lack of taste in certain areas, like clothes (he wore a string tie). Kirstein refused to be put off. All genuinely creative people were difficult, he himself had enough taste for two people, and as to Balanchine's health ... well, as he confided to his diary, 'Much can be accomplished in three years.'[29] But with all this to-ing and fro-ing he didn't meet the choreographer himself in Paris, and he was forced to follow him to London, where the company was playing next. When they finally met in Kirstein's hotel, Kirstein, speaking in French, broached the reason why he had come to Europe.[30] It made for an incongruous encounter. Kirstein was tall, rich, and earnest; Balanchine was slight, penniless, and congenitally distrustful of solemnity (he liked to say 'ballet is like coffee, it smells better than it tastes').[31] Kirstein had prepared his speech and was as articulate as he was passionate, praising Balanchine's choreography, extolling the spirit of America, promising that the Russian could, in the not-too-distant future, have his own company and his own theatre. When he had the chance, Balanchine remarked that he would dearly love to go to a country that had produced Ginger Rogers. It took Kirstein a moment to realise that this was the choreographer's way of saying yes.[32]

Balanchine reached Manhattan in October that year. It was a bleak time for such a radical venture. The depression was at its deepest, and the arts were expected to be relevant, or at least not to add to people's troubles by being costly and apparently wasteful. It had been Kirstein's intention to set up the company in a quiet backwater in Connecticut, where Balanchine could begin training the dancers. Balanchine would have none of it. He was a city man through and through, equally at home in Saint Petersburg, Paris, and London. He had never heard of the small town Kirstein had in mind and said he would rather return to Europe than 'lose myself in this Hartford place.'[33] Kirstein therefore found a classroom in an old building on Madison Avenue at Fifty-ninth Street. The School of American Ballet opened on 1 January 1934. Twenty-five were accepted as pupils, all but three females. The young Americans were in for a shock. Normally, dance directors never laid a finger on their students, but Balanchine was forever 'whacking, pushing, tugging, touching, poking.'[34] In this way he made them do things they had never thought possible.

Balanchine's first ballet in the New World, performed on 10 June 1934, was *Serenade*, which immediately became a classic.[35] As an instinctive showman he realised that to work, and work well, his first ballet had to be about dance itself and about America. He needed to show American audiences that for all its classical heritage, ballet is an ever-changing, contemporary, relevant art, not a static thing, not just *Giselle* or *The Nutcracker*. So Balanchine improvised. 'The first evening he worked on it, seventeen young women were present, so he choreographed the opening scene for seventeen. At one point, a woman fell down and cried – that became a step. On another evening several dancers were late, so that too became part of the ballet.'[36] The story within the story in *Serenade* is about how young, inexperienced dancers achieve mastery of their craft, and how, in a wider sense, they are refined and dignified in the process. He was showing the ennobling powers of art, and why therefore it was necessary to have a ballet company in the first place.[37] For Edward Denby, the ballet critic, the crux of *Serenade* was the 'sweetness' of the bond between all the young dancers. Americans, Denby felt, were not like Russians, who had ballet in their very bones. Americans came from a more individualistic, rational, less emotional culture, with less of a shared heritage. Feeling could, therefore, be created by membership of the company instead. This, Denby said, was the basis for Balanchine's controversial approach – which he always stuck to – that in modern dance the company is more important than any individual dancer; that there should be no stars.[38]

Serenade was initially performed before a private, 'invitation only' audience. The lawn where the stage was erected 'never recovered from the shock.'[39] The first public performances were given in a two-week season at the Adelphi Theater, beginning 1 March 1935. The company, which comprised twenty-six dancers from the school plus two guest artists – Tamara Geva (Balanchine's first wife) and Paul Haakon – was called American Ballet.[40] The ballets danced included *Serenade*, *Reminiscences*, and *Transcendence*. Kirstein was naturally thrilled that his venture across the Atlantic had paid off so handsomely and so soon. On the first night, however, Balanchine was more circumspect, and he was right. Acceptance would take a while. The following day, in the *New York Times*, the paper's dance critic, John Martin, singled out Balanchine as 'precious and decadent,' an example of the kind of 'Riviera aesthetics' that America could do without (a crack at Scott Fitzgerald and Bertolt Brecht). The best thing for American Ballet, he advised, would be to jettison Balanchine, 'with his international notions,' and replace him with 'a good American dance man.' But this was ballet, not musicals, and mercifully no one listened.

One measure of Hitler's gift arrived in the form of the Benjamin Franklin lectures at the University of Pennsylvania, delivered in the spring of 1952, in which all the speakers were exiles. Franz Neumann spoke on the social sciences, Henri Peyre on the study of literature, Erwin Panofsky on the history of art, Wolfgang Kohler on scientists, and Paul Tillich entitled his talk 'The Conquest of Theological Provincialism.' His use of the word *conquest* was optimistic, but he ended by posing a question that remains remarkably vivid even today: 'Will America remain what it has become to us [exiles], a country in which people from every country can overcome their spiritual provincialism? One can be both a world power politically and a provincial people spiritually.'[41]

Britain declared war on Germany on a Sunday, the morning of 3 September 1939. It was a balmy day in Berlin. William Shirer, the American newspaperman who later wrote a vivid history of the rise and fall of the Third Reich, reported that the city streets were calm, but the faces of Berliners registered 'astonishment, depression.' Before lunch he had drinks at the Adlon Hotel with about a dozen members of the British embassy. 'They seemed completely unmoved by events. They talked about *dogs* and such stuff.'

Others were required to show a greater sense of urgency. The very next day, Monday 4 September, **Alan Turing** reported to the **Government Code and Cipher School at Bletchley Park** in Buckinghamshire.[1] Bletchley town was an unlovely part of England, not far from the mud and dust of the county's famous brickfields. It did, however, have one advantage: it was equidistant from London, Cambridge, and Oxford, the heart of intellectual Britain, and at Bletchley station the railway from London to the north crossed the local line that linked Oxford with Cambridge. North of the station, on an insignificant rise, stood Bletchley Park. In the early years of war, Bletchley's population was swollen by two very different kinds of stranger. One kind was children, hundreds of them, evacuated from East London mainly, a precaution against the bombing that became known as the Blitz. The second kind was people like Turing, though it was never explained to the locals who these people actually were and what they were doing.[2] Life at Bletchley Park was so secret that the locals took against these 'do-nothings' and asked their local MP to table a question in Parliament. He was firmly dissuaded from doing so.[3] Turing, a shy, unsophisticated man with dark hair that lay very flat on his head, found a room over a pub, the Crown, in a village about three miles away. Even though he helped in the bar when he could, the landlady made no secret of the fact that she didn't see why an able-bodied young man like Turing shouldn't be in the army.

In a sense, Bletchley Park had already been at war for a year when Turing arrived. In 1938 a young Polish engineer called **Robert Lewinski** had slipped into the British embassy in Warsaw and told the chief of military intelligence there that he had worked in Germany in a factory which made code-signalling machines. He also said he had a near-photographic memory, and could remember the details of the machine, the **Enigma**. The British believed him and smuggled Lewinski to Paris, where he was indeed able to help build a machine.[4] This was the first break the British had in the secret war of codes. They knew that Enigma was used to send orders to military commanders both on land and at sea. But this was the first chance anyone had had to see it close up.

It turned out that the machine was extremely simple, but its codes were virtually unbreakable.[5] In essence it looked like a typewriter with parts added on. The person sending the message simply typed what he or she had to say, in plain German, having first set a special key to one of a number of pointers. A series of rotor arms then scrambled the message as it was sent. At the other end, a similar machine received the message and, provided it was set to the same key, the message was automatically decoded. All personnel operating the machines were issued with a booklet indicating which key setting was to be used on which day. The rotors enabled billions of permutations. Since the key was changed three times a day, with the Germans transmitting thousands of messages in any twenty-four-hour period, the British were faced with a seemingly impossible task. The story of how the Enigma was cracked was a close secret for many years, and certainly one of the most dramatic intellectual adventures of the century. It also had highly pertinent long-term consequences – not only for the course of World War II but for the development of computers.

Turing was a key player here. Born in 1912, he had a father who worked in the Indian civil service, and the boy was sent to boarding school, where he suffered considerable psychological damage. His experience at school brought on a stutter and induced in him an eccentricity that probably contributed to his suicide some years later. He discovered in traumatic circumstances that he was homosexual, falling in love with another pupil who died from tuberculosis. Yet Turing's brilliance at mathematics shone through, and in October 1931 he took up a scholarship at King's College, Cambridge. This was the Cambridge of John Maynard Keynes, Arthur Eddington, James Chadwick, the Leavises, and George Hardy, another brilliant mathematician, so that intellectually at least Turing felt comfortable. His arrival in Cambridge also coincided with publication of Kurt Gödel's famous theorem: it was an exciting time in mathematics, and with so much ferment in Germany, people like Erwin Schrödinger, Max Born, and Richard Courant, from Göttingen, all passed through.[6] Turing duly graduated with distinction as a wrangler, was elected to a fellowship at King's, and immediately set about trying to take maths beyond Gödel. The specific problem he set himself was this: What *was* a computable number, and how was it calculated? To Turing, calculation was so logical, so straightforward, so independent of psychology, that it could even be followed by a machine. He therefore set about trying to describe what properties such a machine would have.

His solution had distinct echoes of Gödel's theorem. Turing theorised first a machine that could find the number of 'factors' in an integer – that is, the prime numbers it is divisible by. In his account of Turing, Paul Strathern quotes a familiar example as follows:[7]

$$180 \div 2 = 90$$
$$90 \div 2 = 45$$
$$45 \div 3 = 15$$
$$15 \div 3 = 5$$
$$5 \div 5 = 1$$

Thus $180 = 2^2 \times 3^2 \times 5$.

Turing believed that it would not be long before a machine was devised to follow these rules. He next assumed that a machine could be invented (as it now has) that could follow the rules of chess. Third, Turing conceived what he called a universal machine, a device that could perform *all* calculations. Finally (and this is where the echo of Gödel is most strong), he added the following idea: assume that the universal machine responds to a list of integers corresponding to certain types of calculation. For example, 1 might mean 'finding factors,' 2 might mean 'finding square roots,' 3 might mean 'following the rules of chess,' and so on. What would happen, Turing now asked, if the universal machine was fed a number that corresponded to itself? How could it follow an instruction to behave as it was already doing?[8] His point was that such a machine could not exist *even in theory*, and therefore, he implied, a calculation of that type was simply not computable. There were/are no rules that explain how you can prove, or disprove, something in mathematics, using mathematics itself. Turing published his paper in 1936 in the *Proceedings of the London Mathematical Society*, though publication was delayed because, as in Pauling's case with the chemical bond, there was no one judged competent to referee Turing's work. Entitled 'On Computable Numbers,' the paper sparked as much attention as Gödel's 'catastrophe' had done.[9] Turing's idea was important mathematically, for it helped define what computation was. But it was also important for the fact that it envisaged a kind of machine – now called a Turing machine – that was a precursor, albeit a theoretical precursor, to the computer.

Turing spent the mid-1930s at Princeton, where he completed his Ph.D. The mathematics department there was in the same building as the recently established Institute for Advanced Study (IAS), and so he joined some of the most famous brains of the day: Einstein, Gödel, Courant, Hardy, and a man he became particularly friendly with, the Austro-Hungarian mathematician **Johann von Neumann**. Whereas Einstein, Gödel, and Turing were solitary figures, eccentric and unstylish, von Neumann was much more worldly, a sophisticate who missed the cafés and the dash of his native Vienna.[10] Despite their differences, however, von Neumann was the man who most appreciated Turing's brilliance – he invited the Englishman to join him at the IAS after he had finished his Ph.D. Though Turing was flattered, and although he liked America, finding it a more congenial environment for a homosexual, he nonetheless returned to Britain.[11] Here he came across another brilliant eccentric, Ludwig Wittgenstein, who had reappeared in Cambridge after many years absence. Wittgenstein's lectures were open only to a select few, the philosopher/mathematician having lost none of his bizarre habits. Turing, like the others in the seminar, was provided with a deck chair in an otherwise bare room. The subject of the seminars was the philosophical basis of mathematics; by all accounts, Turing knew little philosophy, but he had the edge when it came to mathematics, and there were several pointed exchanges.[12]

In the middle of these battles the real war broke out, and Turing was summoned to Bletchley. There, his encounter with the military brass was almost comical: anyone less suited to army life would be hard to find. To the soldiers in uniform, Turing was positively weird. He hardly ever shaved, his trousers were held up using a tie as a belt, his stutter was as bad as ever, and he kept highly irregular hours. The only distinction that he recognised between people was intellectual ability, so he would dismiss even senior officers whom he regarded as fools and spend time instead playing chess with the lower ranks if they showed ability. Since his return from America, he was much more at home with his homosexuality, and at Bletchley often made open advances – this, at a time when homosexuality in Britain was an imprisonable offence.[13] But cracking Enigma was an intellectual problem of a kind where he shone, so he was tolerated.[14] The basic difficulty was that Turing and all the others working with him had to search through thousands of intercepted messages, looking for any regularities, and then try to understand them. Turing immediately saw that in theory at least this was a problem for a Turing machine. His response was to build an electromagnetic device capable of high-speed calculation that could accept scrambled Enigma messages and search for any regularities.[15] This machine was given the name **Colossus**. The first Colossus (ten versions eventually became operational) was not built until December 1943.[16] Details of the machine were kept secret for many years, but it is now known to have had 1,500 valves and, in later versions, 2,400 vacuum tubes computing in 'binary' (i.e., all information was contained in 'bits,' various arrangements of either 0 or 1).[17] It is in this sense that Colossus is now regarded as the forerunner of the electromagnetic digital computer. Colossus was slightly taller than the size of a man, and photographs show that it occupied the entire wall of a small room in Hut F at Bletchley. It was a major advance in technology, able to scan 25,000 characters a second.[18] Despite this, there was no sudden breakthrough with Enigma, and in 1943 the Atlantic convoys bringing precious food and supplies from North America were being sunk by German U-boats in worrying numbers. At the darkest time, Britain had barely enough food to last a week. However, by dogged improvements to Colossus, the time it took to crack the coded messages was reduced from several days to hours, then minutes. Finally, Bletchley's code breakers were able to locate the whereabouts of every German U-boat in the Atlantic, and shipping losses were reduced considerably. The Germans became suspicious but never imagined that Enigma had been cracked, an expensive mistake.[19]

Turing's work was regarded as so important that he was sent to America to share it with Britain's ally.[20] On that visit he again met Von Neumann, who had also begun to convert the ideas from 'On Computable Numbers' into practice.[21] This was to result in **ENIAC** (the Electronic Numerical Integrator and Calculator), built at the University of Pennsylvania. Bigger even than Colossus, this had some 19,000 valves and would in time have a direct influence on the development of computers.[22] But ENIAC was not fully operational until after the war and benefited from the teething problems of Colossus.[23] There is no question that Colossus helped win the war – or at least helped Britain avoid defeat. The 'do-nothings' at Bletchley had proved their worth. At the end of hostilities, Turing was sent to Germany as part of a small contingent of scientists and mathematicians assigned to investigate German progress in the realm of communications.[24] Already, news was beginning to leak out about Colossus, not so much details about the machine itself as that Bletchley had housed 'a great secret.' In fact, Enigma/Colossus did not break upon the world for decades, by which time computers had become a fixture of everyday life. Turing did not live to see this; he committed suicide in 1954.

In a survey conducted well after the war was over, a group of senior British servicemen and scientists was asked what they thought were the most important scientific contributions to the outcome of the war. Those surveyed included: Lord Hankey, secretary of the Committee of Imperial Defence; Admiral Sir William Tennant, who commanded the Mulberry harbour organisation during the Normandy landings; Field Marshal Lord Slim, commander of the Four-teenth Army in Burma; Marshal of the Royal Air Force Sir John Slessor, commander-in-chief of RAF Coastal Command during the critical period of the U-boat war; Sir John Cockcroft, a nuclear physicist responsible for radar development; Professor P. M. S. Blackett, a physicist and member of the famous Tizard Committee (which oversaw the development of radar), and later one of the developers of operational research; and Professor R. V. Jones, physicist and wartime director of scientific intelligence in the Air Ministry. This group concluded that there were six important developments or devices that 'arose or grew to stature because of the war.' These were: atomic energy, radar, rocket propulsion, jet propulsion, automation, and operational research (there was, of course, no mention of Bletchley or Enigma). Atomic energy is considered separately in chapter 22; of the others, by far the most intellectually radical idea was radar.[25]

Radar was an American name for a British invention. During the war, the fundamental notion came to have a great number of applications, from antisubmarine warfare to direction finding, but its most romantic role was in the Battle of Britain in 1940, when the advantage it provided to the British aircrews may just have made all the difference between victory and defeat. As early as 1928, one of the physicists at the Signals School in Portsmouth, England, took out a patent for a device that could detect ships by radio waves. Few of his superior officers believed in the need for such a piece of equipment, and the patent lapsed. Six years later, in June 1934, with the threat of German rearmament becoming clearer, the director of scientific research at the Air Ministry ordered a survey of what the ministry was doing about air defence. Collecting all fifty-three files bearing on the subject, the responsible bureaucrat saw 'no hope in any of them.'[26] It was the bleak picture revealed in this survey that led directly to the establishment of the **Tizard Committee**, a subcommittee of the Committee of Imperial Defence. Sir Henry Tizard was an Oxford

chemist, an energetic civilian, and it was his committee, formally known as the
Scientific Survey of Air Defence, that pushed radar research to the point where
it would make a fundamental contribution not only to Britain's fate in World
War II, but also to aircraft safety.

Three observations came together in the development of radar. Ever since
Heinrich Hertz had first shown that radio waves were related to light waves, in
1885, it had been understood that certain substances, like metal sheets, reflected
these waves. In the 1920s a vast electrified layer had been discovered high in
the atmosphere, which also acted as a reflector of radio waves (originally called
the Heaviside Layer, after the scientist who made the discovery, it later became
known as the **ionosphere**). Third, it was known from experiments with
prototype television sets, carried out in the late 1920s, that aircraft interfered
with transmission. Only in 1935 were these observations put together, but even
then radar emerged almost by accident. It happened because **Sir Robert
Watson-Watt**, in the radio department of the National Physical Laboratory in
Middlesex, was researching a 'death ray.' He had the bloodthirsty idea that an
electromagnetic beam might be created of sufficient energy to melt the thin
metal skin of an aircraft and kill the crew inside. Calculations proved that this
futuristic idea was a pipe dream. However, Watson-Watt's assistant, A. F.
Wilkins, the man doing the arithmetic, also realised that it might be practicable
to use such a beam to detect the presence of aircraft: the beam would be re-
radiated, bounced back toward the transmitting source in an 'echo.'[27] Wilkins's
ideas were put to the test on 26 February 1935 near the Daventry broadcasting
station in the Midlands. Tizard's committee, closeted in a caravan, saw that the
presence of an aircraft (though not, at that stage, its exact location) could indeed
be detected at a distance of about eight miles. The next steps took place on the
remote East Anglian coast. Masts some seventy feet high were erected, and with
their aid, aircraft up to forty miles away could be tracked. By now the Tizard
Committee realised that ultimate success depended on a reduction of the wavelength of the radio beams. In those days wavelengths were measured in metres,
and it was not thought practicable to create wavelengths of less than 50 centimetres
(20 inches). But then **John Randall** and **Mark Oliphant** at Birmingham University came up with an idea they called a **cavity magnetron**, essentially a glass
tube with halfpennies at each end, fixed with sealing wax. The air was sucked out,
creating a vacuum; an electromagnet provided a magnetic field, and a loop of
wire was threaded into one of the cavities 'in the hope that it would extract high-
frequency power' (i.e., generating shorter waves). It did.[28]

It was now 21 February 1940.[29] Anticipating success, a chain of coastal radar
stations, stretching from Ventnor on the Isle of Wight to the Firth of Tay in
Scotland, had been begun, which meant that once the cavity magnetron had
proved itself, radar stations could monitor enemy aircraft even as they were
getting into formation in France and Belgium. The British were even able to
gauge the rough strength of the enemy formations, their height, and their
speed, and it was this 'which enabled the famous "few," Britain's fighter pilots,
to intercept the enemy with such success.'[30]

★

May 1940 was for Britain and its close European allies the darkest hour of the war. On the tenth of the month German forces invaded Holland, Belgium, and Luxembourg, followed by the surrender of the Dutch and Belgian armies, with King Leopold III being taken prisoner. On the twenty-sixth, the evacuation of 300,000 British and French troops trapped in northeast France was begun at Dunkirk. Oswald Mosley and 750 other British fascists were interned. Neville Chamberlain resigned as prime minister, to be replaced by Winston Churchill.

Though the war dominated everyone's thoughts, on Saturday, 25 May, two scientists in Oxford's University Pathology Department conducted the first experiments in a series that would lead to 'the most optimistic medical breakthrough of the century'. **Ernst Chain** was the son of a Russo-German industrial chemist, and an exile from Nazi Germany; **N. G. Heatley** was a British doctor. On that Saturday, they injected streptococci bacteria into mice and then administered some of the mice with penicillin. After that, Chain went home, but Heatley stayed in the lab until 3:30 the next morning. By then every single untreated mouse had died – but *all* of the treated mice were alive. When Chain returned to the pathology lab on Sunday morning, and saw what Heatley had seen, he is reported to have started dancing.[31]

The age of **antibiotics** had taken a while to arrive. The word *antibiotic* itself first entered the English language at the turn of the century. Doctors were aware that bodies have their own defences – up to a point – and since 1870 it had been known that some *Penicillium* moulds acted against bacteria. But until the 1920s, most medical attempts to combat microbial infection had largely failed – quinine worked for malaria, and the 'arsenicals' worked for syphilis, but these apart, there was a general rule that 'chemicals' in therapy did as much damage to the patient as to the microbe. This is why the view took hold that the best way forward was some device to take advantage of the body's own defences, the old principle of homeopathy. A leading centre of this approach was Saint Mary's Hospital in Paddington, in London, where one of the doctors was **Alexander Fleming**. To begin with, Fleming worked on the Salvarsen trials in Britain (see chapter 6). However, he dropped into the lab in Paddington one day in the summer of 1928, having been away for a couple of weeks on holiday, and having left a number of cultures in the lab to grow in dishes.[32] He noticed that one culture, *Penicillium*, appeared to have killed the bacteria in the surrounding region.[33] Over the following weeks, various colleagues tried the mould on themselves – on their eye infections, for example – but Fleming failed to capitalise on this early success. Who knows what Fleming would or would not have done, but for a very different man?

Howard Walter Florey (later Lord Florey, PRS; 1898–1968) was born in Australia but came to Britain in 1922 as a Rhodes scholar. He worked in Cambridge under Sir Charles Sherrington, moving on to Sheffield, then Oxford. In the 1930s his main interest was in the development of spermicidal substances that would form the basis of vaginal contraceptive gels. Besides the practical importance of the gels, their theoretical significance lay in the fact that they embodied 'selective toxicity' – the spermatozoa were killed without the walls of the vagina being damaged.[34] At Oxford, Florey recruited E. B. (later Sir Ernst) Chain (1906–1979). Chain had a Ph.D. in chemistry from the Friedrich-Wilhelm University in Berlin. Being Jewish, he had been forced to leave Germany, also relinquishing his post as the distinguished music critic of a Berlin newspaper, yet another example of the 'inferior' form of life that Hitler considered the Jews. Chain and Florey concentrated on three antibiotica – *Bacillus subtilis, Pseudomonas pyocyanea*, and *Penicillium notatum*. After developing a method to freeze-dry the mould (penicillin was highly unstable at ordinary temperatures), they began their all-important experiments with mice.

Encouraged by the remarkable results mentioned above, Florey and Chain arranged to repeat the experiment using human subjects. Although they obtained enough penicillin to start trials, and although the results were impressive, the experiment was nonetheless spoiled by the death of at least one patient because Florey, in wartime, could not procure enough antibiotics to continue the study.[35] Clearly this was unacceptable, even if the shortage was understandable in the circumstances, so Florey and Heatley left for America. Florey called in on funding agencies and pharmaceutical companies, while Heatley spent several weeks at the U.S. Department of Agriculture's North Regional Research Laboratory in Peoria, Illinois, where they were expert at culturing microorganisms. Unfortunately, Florey didn't get the funds he sought, and Heatley, though he found himself in the company of excellent scientists, also found them anti-British and isolationist. The result was that penicillin became an American product (the pharmaceutical companies took Florey's results but did their own clinical trials). For many, penicillin has always been an American invention.[36] Without the help of the U.S. pharmaceutical companies, penicillin would no doubt not have had the impact it did (or have been so cheap so early), but the award of the Nobel Prize in 1945 to Fleming, Florey, and Chain showed that the intellectual achievement belonged to the British-Australians and the Russo-German Jew Chain.

Montignac, a small town in the Dordogne region of France, about thirty miles southeast of Périgueux, straddles the Vézère River where it has carved a narrow gorge through the limestone. On the morning of 12 September 1940, just after the Blitz had begun in London and with France already sundered into the occupied and unoccupied zones, five boys left town looking for birds and rabbits to shoot. They headed toward a wooded hill where they knew there were birch, hazel, and the small oaks that characterised the region. They saw rabbits aplenty, but no pheasant or partridge.[37]

They moved slowly and silently so as not to disturb the wildlife. Shortly before midday they came to a shallow depression, caused some decades before when a large fir tree had been toppled in a storm. This was known to the locals as the 'Donkey Dip' because a donkey had once strayed into the area, broken its leg, and had to be put down. Passing the Dip, the boys moved on; the trees grew denser here, and they hoped for some birds. However, one of the boys had brought a dog, Robot, a mongrel with a dark patch over one eye. Suddenly, he was nowhere to be seen (this part of the account is now disputed – see references).[38] The boys were all fond of Robot and began calling for him. When he didn't respond, they turned back, calling and whistling. Eventually, as they returned to the vicinity of the Dip, they heard the dog's barks, but they were strangely muffled. They then realised that Robot must have fallen through a hole in the floor of the forest; there were caves all over the area, so that wasn't too much of a surprise. Sure enough, the barking led them to a small hole, through which they dropped a stone. Listening carefully, they were surprised it took so long to fall, and then they heard it crack on other stones, then plop into water.[39] Breaking branches off the birch and beech trees, they hacked at the hole until the smallest of the boys could scramble down. He had taken some matches, and with their aid he soon found the dog. But that was not all he found. By the light of the matches he could see that, below ground, the narrow passage that Robot had fallen through opened out into a large hall about sixty feet long and thirty feet wide. Impressed, he called to the others to come and see. Grumbling about the birds they were missing, the others joined him. One of the things that immediately caught their eye was the rock formation in the ceiling of the cave. They were later to say that these 'resembled nothing so much as rocky clouds, tortured into fantastic shapes by centuries of underground streams coming and going with the storms'. Alongside the rocks, however, was something even more surprising: strange paintings of animals, in red, yellow, and black. There were horses, deer, stags, and huge bulls. The deer had delicate, finely rendered antlers; the bulls were stippled, some of them, and up to their knees in grass. Still others seemed to be stampeding across the ceiling.[40]

The matches soon gave out, and darkness returned. The boys walked back to the village but told no one what they had discovered. Over the following few days, leaving the village at ten-minute intervals so as not to attract attention and using a makeshift torch, they explored every nook and cranny in the cave.[41] Discussing the matter among themselves, they decided to call in the local schoolteacher, M. Léon Laval. At first he suspected a practical joke. Once he saw the cave for himself, however, his attitude changed completely. In a matter of only a few days, the caves at Lascaux were visited by none other than the **Abbé Breuil**, an eminent archaeologist. Breuil, a French Catholic priest, was until World War II the most important student of cave art. He had visited even the most inaccessible sites, usually on muleback. Arrested as a spy in Portugal in World War I, he had carried on his research regardless, under armed guard, until he was cleared of all charges.[42] At Montignac Breuil was impressed by what he saw. There was no question that the Lascaux paintings were genuine, and very old. Breuil said that the cave the boys had found was bettered only by **Altamira** in Spain.

When it occurred, the discovery of **Lascaux** was the most sensational find of its kind this century.[43] **Prehistoric art** had first been identified as such in 1879 at Altamira, a cave hidden in the folds of the Cantabrian Mountains in northern Spain. There was a personal sadness associated with this discovery, for the man who made it, Don Marcelino de Sautuola, a Spanish aristocrat and amateur archaeologist, died without ever convincing his professional colleagues that what he had found in Altamira was genuine. No one could believe that such vivid, modern-looking, *fresh* images were old. By the time Robot fell through that hole in Lascaux, however, too many other sites had been found for them all to be hoaxes.[44] In fact, there had been so many discoveries of cave art by the time of World War II that two things could be said with certainty. First, many of the caves with art in them were concentrated in the mountains of northern Spain and around the rivers of central France. Since then, prehistoric art has been found all over the world, but this preponderance in southern France and northern Spain still exists, and has never been satisfactorily explained. The second point relates to dating. Lascaux fitted into a sequence of prehistoric art in which simple drawings, apparently of vulvas, begin to occur around 30,000–35,000 years ago; then came simple outline drawings, 26,000–21,000 years ago; then more painted, three-dimensional figures, after 18,000 years ago. This 'creative explosion' has also been paired with the development of stone tools, beginning about 31,000 years ago, and the widespread distribution of the so-called **Venus figurines**, big-breasted, big-buttocked carvings of females found all over Europe and Russia and dating to 28,000–26,000 years ago. Archaeologists believed at the time Lascaux was discovered that this 'explosion' was associated in some way with the emergence of a new species of man, the Cro-Magnon people (after the area of France where they were found), formally known as *Homo sapiens sapiens*, and which replaced the more archaic *Homo sapiens* and the Neanderthals. Related discoveries suggested that these peoples were coming together in larger numbers than ever before, a crucial development from which everything else (such as civilisation) followed.[45] Breuil's view, shared by others, was that the Venus figurines were fertility goddesses and the cave paintings primitive forms of 'sympathetic magic.'[46] In other words, early man believed he could improve his kill rate in the hunt by 'capturing' the animals he wanted on the walls of what would be a sacred place, and making offerings to them. After the war, at another French site known as **Trois Frères**, a painting of a figure was discovered that appears to show a human wearing a bison skin and a mask with antlers. Was this '**sorcerer**' (as he became known), a primitive form of shaman? If so, it would support the idea of sympathetic magic. One final mystery remains: this explosion of creative activity appears to have died out about 10,000 years ago. Again, no one knows why.

Halfway across the world, much rarer evidence relating to man's remote past became a direct casualty of hostilities. China and Japan had been at war since 1937. The Japanese had invaded Java at the end of February 1941 and were advancing through Burma. In June, they attacked the U.S. Aleutian chain – China was being encircled. Among these great affairs of state, a few old bones counted for not very much. But in fact the hominid fossils from the cave of **Zhoukoudien** were just about as important as any anthropological/archaeological relic could be.

Until World War II, such evidence as existed for early man had been found mainly in Europe and Asia. The most famous were the bones and skulls unearthed in 1856 in a small cave in the steep side of the Neander Valley (Neander Thal), through which the river Düssel reaches the Rhine. Found in sediments dating to 200,000 to 400,000 years old, these remains raised the possibility that Neanderthal man was our ancestor. More modern-looking skulls had been found at Cro-Magnon ('Big Cliff') in the valley of the Vézère River in France, suggesting that modern man had lived side by side with Neanderthals.[47] And the anatomical details of Raymond Dart's discovery, in South Africa in 1925, of *Australipithecus africanus*, 'the man-ape of South Africa,' implied that the find spot, a place called Taung, near Johannesburg, was where the apes had first left the trees and walked upright. But more discoveries had been made in Asia, in China and Java, associated with fire and crude stone artefacts. It was believed at that stage that most of the characteristics that made the early hominids human first appeared in Asia, which made the bones found at Zhoukoudien so significant.

Chinese academics raised the possibility of sending these precious objects to the United States for safety. Throughout most of 1941, however, the custodians of the bones dithered, and the decision to export them was not made until shortly before the attack on Pearl Harbor in December that year.[48] Barely twenty-four hours after the attack, the Japanese in Beijing searched the fossils' repository. They found only casts. That did not mean, however, that the fossils were safe. What appears to have happened is that they were packed in a couple of footlockers and put in the care of a platoon of U.S. Marines headed for the port of Tientsin. The plan was for the fossils to be loaded on board the SS *President Harrison*, bound for home. Unfortunately, the *Harrison* was sunk on her way to the port, and the fossils vanished. They have never been found.

The Zhoukoudien fossils were vital because they helped clarify the theory of evolution, which at the outbreak of war was in a state of chaos. Throughout the 1930s, the attention of palaeontologists had continued to focus on Zhoukoudien, in China, rather than Java or Africa for the simple reason that spectacular discoveries continued to be made there. In 1939, for example, Franz Weidenreich reported that of the forty or so individuals found in the Zhoukoudien caves (fifteen of whom were children), not one was a complete skeleton. In fact, the great preponderance were skulls, and smashed skulls at that. Weidenreich's conclusion was dramatic: these individuals had been killed – and eaten. The remains were an early ritualistic killing, a primitive religion in which the murderers had eaten the brains of their victims in order to obtain their power. Striking as these observations were, evolutionary theory and its relation to known fossils was still incoherent and unsatisfactory.[49]

The incoherence was removed by four theoretical books, all published between 1937 and 1944, and thanks to these four authors several nineteenth-century notions were finally laid to rest. Between them, these studies created what is now known as '**the evolutionary synthesis**,' which produced our modern understanding of how evolution actually works. In chronological order, these books were: *Genetics and the Origin of Species*, by Theodosius Dobzhansky (1937); *Evolution: The Modern Synthesis*, by Julian Huxley (1942); *Systematics and the Origin of Species*, by Ernst Mayr (also 1942); and *Tempo and Mode in Evolution*, by George Gaylord Simpson (1944). The essential problem they all sought to deal with was this:[50] Following the publication of Charles

Darwin's *On the Origin of Species* in 1859, two of his theories were accepted relatively quickly, but two others were not. The idea of evolution itself – that species change – was readily grasped, as was the idea of 'branching evolution,' that all species are descended from a common ancestor. What was not accepted so easily was the idea of gradual change, or of natural selection as an engine of change. In addition, Darwin, in spite of the title of his book, had failed to provide an account of speciation, how new species arise. This made for three major areas of disagreement.

The main arguments may be described as follows. First, many biologists believed in 'saltation' – that evolution proceeded not gradually but in large jumps; only in this way, they thought, could the great differences between species be accounted for.[51] If evolution proceeded gradually, why wasn't this reflected in the fossil record; why weren't 'halfway' species ever found? Second, there was the notion of 'orthogenesis,' that the direction of evolution was somehow preordained, that organisms somehow had a final destiny toward which they were evolving. And third, there was a widespread belief in 'soft' inheritance, better known as the inheritance of acquired characteristics, or Lamarckism. Julian Huxley, grandson of T. H. Huxley, 'Darwin's bulldog,' and the brother of Aldous, author of *Brave New World*, was the first to use the word *synthesis*, but he was really the least original of the four. What the others did between them was to bring together the latest developments in genetics, cytology, embryology, palaeontology, systematics, and population studies to show how the new discoveries fitted together under the umbrella of Darwinism.

Ernst Mayr, a German emigré who had been at the Museum of Natural History in New York since 1931, directed attention away from individuals and toward populations. He argued that the traditional view, that species consist of large numbers of individuals and that each conforms to a basic archetype, was wrong. Instead, species consist of populations, clusters of unique individuals where there is no ideal type.[52] For example, the human races around the world are different, but also alike in certain respects; above all, they can interbreed. Mayr advanced the view that, in mammals at least, major geographical boundaries – like mountains or seas – are needed for speciation to occur, for then different populations become separated and begin developing along separate lines. Again as an example, this could be happening with different races, and may have been happening for several thousand years – but it is a gradual process, and the races are still nowhere near being 'isolated genetic packages,' which is the definition of a species. Dobzhansky, a Russian who had escaped to New York just before Stalin's Great Break in 1928 to work with T. H. Morgan, covered broadly the same area but looked more closely at genetics and palaeontology. He was able to show that the spread of different fossilised species around the world was directly related to ancient geological and geographical events. Dobzhansky also argued that the similarity of Peking Man and Java Man implied a greater simplicity in man's descent, suggesting there had been fewer, rather than a greater number of, ancestors. He believed it was highly unlikely that more than one hominid form occupied the earth at a time, as compared with the prewar view that there may have been several.[53] Simpson, Mayr's colleague at the American Museum of Natural History, looked at the pace of evolutionary change and the rates of mutation. He was able to confirm that the known rates of mutation in genes produced sufficient variation sufficiently often to account for the diversity we see on earth. Classical Darwinism was thus reinforced, and all the lingering theories of saltation, Lamarckianism, and orthogenesis were killed off. Such theories were finally laid to rest (in the West anyway) at a symposium at Princeton in 1947. After this, biologists with an interest in evolution usually referred to themselves as 'neo-Darwinists.'

What Is Life? published in 1944 by **Erwin Schrödinger**, was not part of the evolutionary synthesis, but it played an equally important part in pushing biology forward. Schrödinger, born in Vienna in 1887, had worked as a physicist at the university there after graduating, then in Zurich, Jena, and Breslau before succeeding Max Planck as professor of theoretical physics in Berlin. He had been awarded the 1933 Nobel Prize for his part (along with Werner Heisenberg and Paul Dirac) in the quantum mechanics revolution considered in chapter 15, 'The Golden Age of Physics.' In the same year that he had won the Nobel, Schrödinger had left Germany in disgust at the Nazi regime. He had been elected a fellow of Magdalen College, Oxford, and taught in Belgium, but in October 1939 he moved on to Dublin, since in Britain he would have been forced to contend with his 'enemy alien' status.

An added attraction of Dublin was its brand-new Institute for Advanced Studies, modelled on the IAS at Princeton and the brainchild of Eamon de Valera ('Dev'), the Irish taoiseach, or prime minister. Schrödinger agreed to give the statutory public lectures for 1943 and took as his theme an attempted marriage between physics and biology, especially as it related to the most fundamental aspects of life itself and heredity. The lectures were described as 'semi-popular,' but in fact they were by no means easy for a general audience, containing a certain amount of mathematics and physics. Despite this, the lectures were so well attended that all three, originally given on Fridays in February, had to be repeated on Mondays.[54] Even *Time* magazine reported the excitement in Dublin.

In the lectures, Schrödinger attempted two things. He considered how a physicist might define life. The answer he gave was that a life system was one that took order from order, 'drinking orderliness from a suitable environment.'[55] Such a procedure, he said, could not be accommodated by the second law of thermodynamics, with its implications for entropy, and so he forecast that although life processes would eventually be explicable by physics, they would be new laws of physics, unknown at that time. Perhaps more interesting, and certainly more influential, was his other argument. This was to look at the hereditary structure, the chromosome, from the point of view of the physicist. It was in this regard that Schrödinger's lectures (and later his book) could be said to be semipopular. In 1943 most biologists were unaware of both quantum physics and the latest development on the chemical bond. (Schrödinger had been in Zurich when Fritz London and Walter Heitler discovered the bond;

no reference is made in *What Is Life?* to Linus Pauling.) Schrödinger showed that, from the physics already known, the gene must be 'an aperiodic crystal,' that is, 'a regular array of repeating units in which the individual units are not all the same.'[56] In other words, it was a structure half-familiar already to science. He explained that the behaviour of individual atoms could be known only statistically; therefore, for genes to act with the very great precision and stability that they did, they must be a minimum size, with a minimum number of atoms. Again using the latest physics, he also showed that the dimensions of individual genes along the chromosome could therefore be calculated (the figure he gave was 300 Å, or Angstrom units), and from that both the number of atoms in each gene and the amount of energy needed to create mutations could be worked out. The rate of mutation, he said, corresponded well with these calculations, as did the discrete character of mutations themselves, which recalled the nature of quantum physics, where intermediate energy levels do not exist.

All this was new for most biologists in 1943, but Schrödinger went further, to infer that the gene must consist of a long, highly stable molecule that contains a code. He compared this code to the Morse code, in the sense that even a small number of basic units would provide great diversity.[57] Schrödinger was thus the first person to use the term *code*, and it was this, and the fact that physics had something to say about biology, that attracted the attention of biologists and made his lectures and subsequent book so influential.[58] On the basis of his reasoning, Schrödinger concluded that the gene must be 'a large protein molecule, in which every atom, every radical, every heterocyclic ring, plays an individual role.'[59] The chromosome, he said, is a message written in code. Ironically, just as Schrödinger's basic contribution was the application of the new physics to biology, so he himself was unaware that, at the very time his lectures were delivered, Oswald Thomas Avery, across the Atlantic at the Rockefeller Institute for Medical Research in New York, was discovering that 'the transforming principle' at the heart of the gene was not a protein but deoxyribonucleic acid, or DNA.[60]

When he came to convert his lectures into a book, Schrödinger added an epilogue. Even as a young man, he had been interested in Vedanta, the Hindu doctrine, and in the epilogue he considered the question – central to Hindu thought – that the personal self is identical with the 'all-comprehending universal self.' He admitted that this was both 'ludicrous and blasphemous' in Christian thought but still believed the idea was worth advancing. This was enough to cause the Catholic Dublin publishing house that was considering releasing the lectures in print to turn its back on Schrödinger, even though the text had already been set in type. The title was released instead by Cambridge University Press a year later, in 1944.

Despite the epilogue, the book proved very influential; it is probably the most important work of biology written by a physicist. Timing also had something to do with the book's influence: not a few physicists were turned off their own subject by the development of the atomic bomb. At any rate, among those who read *What Is Life?* and were excited by its arguments were

Francis Crick, James Watson, and Maurice Wilkins. What they did with Schrödinger's ideas is considered in a later chapter.

Intellectually speaking, the most significant consequence of World War II was that science came of age. The power of physics, chemistry, and the other disciplines had been appreciated before, of course. But radar, Colossus, and the atomic bomb, not to mention a host of lesser discoveries – like operational research, new methods of psychological assessment, magnetic tape, and the first helicopters – directly affected the outcome of the war, much more so than the scientific innovations (such as the IQ test) in World War I. Science was itself now a – or perhaps *the* – colossus in affairs. Partly as a result of that, whereas the earlier war had been followed by an era of pessimism, World War II, despite the enormous shadow of the atomic bomb, was followed by the opposite mood, an optimistic belief that science could be harnessed for the benefit of all. In time this gave rise to the idea of The Great Society.

NO WAY BACK

It was perhaps only natural that a war in which very different regimes were pitched against one another should bring about a reassessment of the way men govern themselves. Alongside the scientists and generals and code breakers trying to outwit the enemy, others devoted their energies to the no less fundamental and only marginally less urgent matter of the rival merits of fascism, communism, capitalism, liberalism, socialism, and democracy. This brought about one of the more unusual coincidences of the century, when a quartet of books was published during the war by exiles from that old dual monarchy, Austria and Hungary, looking forward to the type of society man should aim for after hostilities ceased. Whatever their other differences, these books had one thing in common to recommend them: thanks to the wartime paper rationing, they were all mercifully short.

The first of these, *Capitalism, Socialism and Democracy*, by Joseph Schumpeter, appeared in 1942, but for reasons that will become apparent, it suits us to consider first **Karl Mannheim**'s *Diagnosis of Our Time*, which appeared a year later.[1] Mannheim was a member of the Sunday Circle who had gathered around George Lukács in Budapest during World War I, and included Arnold Hauser and Béla Bartók. Mannheim had left Hungary in 1919, studied at Heidelberg, and attended Martin Heidegger's lectures at Marburg. He was professor of sociology at Frankfurt from 1929 to 1933, a close colleague of Theodor Adorno, Max Horkheimer and the others, but after Hitler took power, he moved to London, teaching at the LSE and the Institute of Education. He also became editor of the International Library of Sociology and Social Reconstruction, a large series of books published by George Routledge and whose authors included Harold Lasswell, professor of political science at Chicago, E. F. Schumacher, Raymond Firth, Erich Fromm, and Edward Shils.

Mannheim took a '**planned society**' completely for granted. For him the old capitalism, which had produced the stock market crash and the depression, was dead. 'All of us know by now that from this war there is no way back to a laissez-faire order of society, that war as such is the maker of a silent revolution by preparing the road to a new type of planned order.'[2] At the same time he was equally disillusioned with Stalinism and fascism. Instead, according to him, the new society after the war, what he called the Great Society, could be achieved only by a form of planning that did not destroy freedom, as had happened in the totalitarian countries, but which took account of the latest developments in psychology and sociology, in particular psychoanalysis. Mannheim believed that society was ill – hence 'Diagnosis' in his title. For him the Great Society was one where individual freedoms were maintained, but informed by an awareness of how societies operated and how modern, complex, technological societies differed from peasant, agricultural communities. He therefore concentrated on two aspects of contemporary society: youth and education, on the one hand, and religion on the other. Whereas the Hitler Youth had become a force of conservatism, Mannheim believed youth was naturally progressive if educated properly.[3] He thought pupils should grow up with an awareness of the sociological variations in society, and the causes of them, and that they should also be made aware of psychology, the genesis of neurosis, how this affects society, and what role it might play in the alleviation of social problems. He concentrated the last half of his book on religion because he saw that at bottom the crisis facing the Western democracies was a crisis of values, that the old class order was breaking down but was yet to be replaced by anything else systematic or productive. While he saw the church as part of the problem, he believed that religion was still, with education, the best way to instil values, but that organised religion had to be modernised – again, with theology being reinforced by sociology and psychology. Mannheim was thus *for* planning, in economics, education, and religion, but by this he did not imply coercion or central control. He simply thought that postwar society would be much more informed about itself than prewar society.[4] He did acknowledge that socialism had a tendency to centralise power and degenerate into mere control mechanisms, but he was a great Anglophile who thought that Britain's 'unphilosophical and practically-minded citizens' would see off would-be dictators.

Joseph Schumpeter had little time for sociology or psychology. For him, insofar as they existed at all, they were subordinate to economics. In his wartime book *Capitalism, Socialism and Democracy,* he sought to change thinking about economics no less than John Maynard Keynes had done.[5] Schumpeter was firmly opposed to Keynes, and to Marx as well, and it is not hard to see why. Born in Austria in 1883, the same year as Keynes, he was educated at the Theresianum, an exclusive school reserved for the sons of the aristocracy.[6] Schumpeter was there by virtue of the fact that his mother had remarried a general after his father, an undistinguished man, had died. As a result of his 'elevation,' Schumpeter was always rather self-consciously aristocratic; he would appear at university meetings in riding habit and inform anyone who was listening that he had three ambitions in life – to be a great lover, a great horseman, and a great economist. After university in Vienna (during its glorious period, covered earlier in this book), he became economic adviser to a princess in Egypt, returning to a professorship in Austria after he had published his first book. After World War I he was invited to become finance minister in the newly formed centre-socialist government, and though he worked out a plan to stabilise the currency, he soon resigned and became president of a private bank. In the debacle after Versailles the bank failed. Eventually, Schumpeter made his way to Harvard, 'where his manner and his cloak quickly made him into a campus figure.'[7] All his life he believed in elites, 'an aristocracy of talent.'

Schumpeter's main thesis was that the **capitalist system is essentially static**: for employers and employees as well as for customers, the system settles down with no profit in it, and there is no wealth for investment. Workers receive just enough for their labour, based on the cost of producing and selling goods. Profit, by implication, can only come from innovation, which for a limited time cuts the cost of production (until competitors catch up) and allows a surplus to be used for further investment. Two things followed from this. First, capitalists themselves are not the motivating force of capitalism, but instead entrepreneurs who invent new techniques or machinery by means of which goods are produced more cheaply. Schumpeter did not think that entre-preneurship could be taught, or inherited; it was, he believed, an essentially 'bourgeois' activity. What he meant by this was that, in any urban environment, people would have ideas for innovation, but who had those ideas, when and where they had them, and what they did with them was unpredictable. Bour-geois people acted not out of any theory or philosophy but for pragmatic self-interest. This flatly contradicted Marx's analysis. The second element of Schumpeter's outlook was that profit, as generated by entrepreneurs, was temporary.[8] Whatever innovation was introduced would be followed up by others in that sector of industry or commerce, and a new stability would eventually be achieved. This meant that for Schumpeter capitalism was inevitably characterised by cycles of boom and stagnation.[9] As a result, his view of the 1930s was diametrically opposite to Keynes's. Schumpeter thought that the depression was to an extent inevitable, a cold, realistic *douche*. By wartime he had developed doubts that capitalism could survive. He thought that, as a basically bourgeois activity, it would lead to increasing bureaucratisation, a world for 'men in lounge suits' rather than buccaneers. In other words, it contained the seeds of its own ultimate failure; it was an economic success but not a sociological success.[10] Moreover, in em-bodying a competitive world, capitalism bred in people an almost endemic critical approach that in the end would be turned on itself. At the same time (1942), he thought socialism could work, though for him socialism was a benign, bureaucratic, planned economy rather than full-blooded Marxism or Stalinism.[11]

If Mannheim took planning for granted in the postwar world, and if Schum-peter was lukewarm about it, the third Austro-Hungarian, **Friedrich von Hayek**, was downright hostile. Born in 1899, Hayek came from a family of scientists, distantly related to the Wittgensteins. He took two doctorates at the University of Vienna, becoming professor of economics at the LSE in 1931, and acquired British citizenship in 1938. He too loathed Stalinism and fascism equally, but he was much less convinced than the others that the same cen-tralising and totalitarian tendencies that existed in Russia and Germany couldn't extend eventually to Britain and even America. In *The Road to Serfdom* (1944), also published by George Routledge, he set out his opposition to planning and linked freedom firmly to the market, which, he thought, helped produce a 'spontaneous social order.' He was critical of Mannheim, regarded Keynesian economics as 'an experiment' that, in 1944, had yet to be proved, and reminded his readers that democracy was not an end in itself but 'essentially a means, a utilitarian device for safeguarding internal peace and individual freedom.'[12] He acknowledged that the market was less than perfect, that one shouldn't make a fetish of it, but again reminded his readers that the rule of law had grown up at the same time as the market, and in part as a response to its shortcomings: the two were intertwined achievements of the Enlightenment.[13] His reply to Mannheim's point about the importance of having greater sociological know-ledge was that markets are 'blind,' producing effects that no one can predict, and that that is part of their point, part of their contribution to freedom, the 'invisible hand' as it has been called. For him, therefore, planning was not only wrong in principle but impractical. Von Hayek then went on to produce three reasons why, under planning, 'the worst get on top.' The first was that the more highly educated people are always those who can see through arguments and don't join the group or agree to any hierarchy of values. Second, the centraliser finds it easier to appeal to the gullible and docile; and third, it is always easier for a group of people to agree on a negative program – on the hatred of foreigners or a different class, say – than on a positive one. He attacked historians like E. H. Carr who aimed to present history as a science (as indeed did Marx), with a certain inevitability about it, and he attacked science itself, in the person of C. H. Waddington, author of *The Scientific Attitude*, which had predicted that the scientific approach would soon be applied to politics.[14] For Hayek, science in that sense was a form of planning. Among the weaknesses of capitalism, he conceded that the tendency to monopoly needed to be watched, and guarded against, but he saw a greater practical threat from the monopolies of the labour unions under socialism.

As the war was ending, a fourth Austro-Hungarian released *The Open Society and Its Enemies*.[15] This was **Karl Popper**. Popper's career had an unusual trajectory. Born in Vienna in 1902, he did not enjoy good health as a young man, and in 1917 a lengthy illness kept him away from school. He flirted with socialism, but Freud and Adler were deeper influences, and he attended Einstein's lectures in Vienna. He completed his Ph.D. in philosophy in 1928, then worked as a social worker with children abandoned after World War I, and as a teacher. He came into contact with the Vienna Circle, especially Herbert Feigl and Rudolf Carnap, and was encouraged to write. His first books, *The Two Fundamental Problems of the Theory of Knowledge* and *Logik der Forschung* (The Logic of Scientific Discovery), attracted enough attention for him to be invited to Britain in the mid-1930s for two long lecture tours. By then the mass emigration of Jewish intellectuals had begun, and when, in 1936, Moritz Schlick was assassinated by a Nazi student, Popper, who had Jewish blood, accepted an invitation to teach at the University of Canterbury in New Zealand. He arrived there in 1937 and spent most of World War II in the calm and relative isolation of his new home. It was in the Southern Hemisphere that he produced his next two books, *The Poverty of Historicism* and *The Open Society* *and Its Enemies*, many of the arguments of the former title being included in *The Open Society*.[16] Popper shared many of the views of his fellow Viennese exile Friedrich von Hayek, but he did not confine himself to economics, ranging far more widely.

The immediate spur to *The Open Society* was the news of the Anschluss, the annexation of Austria by Germany in 1938. The longer-term inspiration arose from the 'pleasant sensation' Popper felt on arriving for the first time in England, 'a country with old liberal traditions,' as compared with a country threatened with National Socialism, which for him was much more like the original closed society, the primitive tribe or feudal arrangement, where power and ideas are concentrated in the hands and minds of a few, or even one, the king or leader: 'It was as if the windows had been suddenly opened.' Popper, like the logical positivists of the Vienna Circle, was profoundly affected by the scientific method, which he extended to politics. For him, there were two important ramifications. One was that political solutions were like scientific ones – they 'can never be more than provisional and are always open to improvement.' This is what he meant by the poverty of historicism, the search for deep lessons from a study of history, which would provide the 'iron laws' by which society should be governed.[17] Popper thought there was no such thing as history, only historical interpretation. Second, he thought that the social sciences, if they were to be useful, 'must be capable of making prophecies.' But if that were the case, again historicism would work, and human agency, or responsibility, would be reduced and perhaps eliminated. This, he thought, was nonsense. He ruled out the very possibility that there could be 'theoretical history' as there was theoretical physics.[18]

This led Popper to the most famous passage in his book, the attack on Plato, Hegel, and Marx. (The book was originally going to be called *False Prophets: Plato, Hegel, Marx*.) Popper thought that Plato might well have been the greatest philosopher who ever lived but that he was a reactionary, who put the interests of the state above everything, including the interpretation of justice. For example, according to Plato, the guardians of the state, who are supposed to be philosophers, are allowed the right to lie and cheat, 'to deceive enemies or fellow-citizens in the interests of the state.'[19] Popper was attacked for his dismissal of Plato, but he clearly saw him as an opportunist and as the precursor of Hegel, whose dogmatic dialectical arguments had led, he felt, to an identification of the good with what prevails, and the conclusion that 'might is right.'[20] Popper thought that this was simply a mischaracterisation of dialectic. In reality, he said, it was merely a version of trial and error, as in the scientific method, and Hegel's idea that thesis generates antithesis was wrong – romantic but wrong: thesis, he said, generates modifications as much as it generates the opposite to itself. By the same token, Marx was a false prophet because he insisted on holistic change in society, which Popper thought had to be wrong simply because it was unscientific – it couldn't be tested. He himself preferred piecemeal change, so that each new element introduced could be tested to see whether it was an improvement on the earlier arrangement.[21] Popper was not against the aims of Marxism, pointing out, for example, that much of the program outlined in the *Communist Manifesto* had actually been achieved by Western societies. But that was his point: this had been achieved piecemeal, without violence.[22]

Popper shared with Hayek a belief that the state should be kept to a minimum, its basic raison d'être being to ensure justice, that the strong did not bully the weak. He disagreed with Mannheim, believing that planning would lead to more closure in society, simply because planning involved a historicist approach, a holistic approach, a utopian approach, all of which went against the scientific method of trial and error.[23] This led Popper to consider democracy as the only viable possibility because it was the only form of government that embodied the scientific, trial-and-error method and allowed society to modify its politics in the light of experience, and to change government without bloodshed.[24] Like Hayek's writings, Popper's ideas may not seem so original today, for the very reason that we take them so much for granted. But at the time, with totalitarianism in full flood, with the stock market crash and the depression still fresh in the mind, with World War I not so far in the past as it is now, many people took the view that history did have a hidden structure (Popper specifically attacks Oswald Spengler's *Decline of the West* thesis as 'pointless'), that it had a cyclical nature, particularly in the economic sphere, that there was something inevitable about either communism or fascism. Popper believed that ideas matter in human life, in society, that they can have power in changing the world, that political philosophy needs to take account of these new ideas to continually reinvent society.

The coincidence of these four books by Austro-Hungarian emigrés was remarkable but, on reflection, perhaps not so surprising. There was a war on, a war being fought for ideas and ideals as much as for territory. These emigrés had each seen totalitarianism and dictatorship at close hand and realised that even when the war with Germany and Japan ended, the conflict with Stalinism would continue.

When he completed *Christianity and the Social Order* in 1941, **William Temple** was archbishop of York.[25] By the time the book appeared, in early 1942, published as a Penguin Special, Temple was archbishop of Canterbury and head of the Church of England. Leaders of the church do not often publish tracts of a social scientific, still less a political, nature, and the book's high-profile author helped ensure its success: it was reprinted twice in 1942 and soon sold well over 150,000 copies. Temple's book perfectly illustrates one aspect of the intellectual climate in the war years.

The main part of the book was rather general. Temple took some time justifying the church's right to 'interfere' (his word) in social questions that inevitably had political consequences, and there was an historical chapter where he described the church's earlier interventions, and in which he revealed himself as extremely knowledgeable about economics, providing an original and entertaining interpretation of what the biblical authorities had to say on that score.[26] He tried to sketch out some '**Christian Social Principles,**' discussing such matters as fellowship in the workplace, God's purpose, and the nature of freedom. But it was really the appendix to Temple's book that comprised its main attraction. Temple thought it wrong for the Established Church to put out an 'official' view on what ought to be done once the war was over, and so in the body of the book he kept his remarks very broad. In the appendix, on the other hand, he set out his own very specific agenda.

To begin with, he agreed with Mannheim over planning. Right at the beginning of the appendix, Temple writes, 'No one doubts that in the post-war world our economic life must be "planned" in a way and to an extent that Mr Gladstone (for example) would have regarded, and condemned, as socialistic.'[27] Temple had concluded the main part of his book by outlining six fundamental principles on the basis of which a Christian society should be governed; he now set about describing how they could be brought about. His first principle was that everyone should be housed with decency, and for this he wanted a Regional Commissioner of Housing with power to say whether land should be used for that purpose.[28] Draconian powers were to be given to these commissioners, who were to prevent speculation in land. The second principle was that every child should have the opportunity of education to the years of maturity, and for this Temple wanted the school-leaving age to be raised from fourteen to eighteen. The third principle concerned an adequate income for everyone, and here he advocated straight Keynesianism, with a certain number of public works being maintained, 'from which private enterprise should be excluded,' and which could be expanded or contracted according to need. Fourth, all citizens should have a say in the conduct of the business or industry where they worked; Temple advocated a return to the mediaeval guilds with workers, management, and capital represented on the boards of all major undertakings. Fifth, all citizens needed adequate leisure to enjoy family life and give them dignity; Temple therefore recommended a five-day week with 'staggered' time off to help enterprises cope; he also proposed holidays with pay.[29] Last, he advocated freedom of worship, of speech, and of assembly.

This last provision was by far the most unexceptional. As for the others, Temple was anxious to make it plain that he was not anti-business and went out of his way to say that 'profit' was not a dirty word. He also underlined his awareness that planning could lead to a loss of freedom, but he thought that certain freedoms were hardly worth having. For example, he quoted figures which showed that 'three-quarters of the businesses which are started go into liquidation within three years. Frankly, it would seem to be a gain all round that there should be less inducement to start these precarious businesses, of which the extinction must cause inconvenience and may cause real distress.' He thought that a percentage of profits should be used for a 'wage-equalisation fund,' and he looked forward to a time whereby the capital accumulated by one generation was made to 'wither' away over the next two or three generations by death duties. For Temple, money was 'primarily an intermediary.' The prime necessities of life, he said, were air, sunshine, land, and water.[30] No one claimed to own the first two, and he made it plain that in his view the same principle should apply to the others.

The huge sales of Temple's book reflected the wide interest in planning and social justice that lay behind the more immediate contingencies of war. The scars of the stock market crash, the depression, and the events of the 1930s ran deep. How deep may be judged from the fact that although 'planning' was anathema in some quarters, for others it wasn't strong enough. Many people in Britain and America, for example, had a sneaking respect for the way Hitler had helped eliminate unemployment. After the experience of depression, the lack of a job seemed for some more important than political freedom, and so totalitarian planning – or central direction – was perhaps a risk worth taking. This attitude, as was mentioned earlier, also transferred to Stalin's 'planning,' which, because Russia just then was an ally, never received in wartime the critical scrutiny it deserved. It was against this intellectual background that there appeared a document that had a greater impact in Britain than any other in the twentieth century.

Late on the evening of 30 November 1942 queues began to form outside the London headquarters of His Majesty's Stationery Office in Holborn, Kingsway. This was, to say the least, an unusual occurrence. Government publications are rarely best-sellers. But, when HMSO opened the following morning, its offices were besieged. Sixty thousand copies of the report being released that day were sold out straight away, at 2 shillings (24 old pence, now 10 pence) a time, four times the cost of a Penguin paperback, and by the end of the year sales reached 100,000. Nor could it be said that the report was Christmas-present material – its title was positively off-putting: *Social Insurance and Allied Services*. And yet, in one form or another, this report eventually sold 600,000 copies, making it the best-selling government report until Lord Denning's inquiry into the Profumo sex and spying scandal twenty years later.[31] Why all the fuss? *Social Insurance and Allied Services* became better known as the Beveridge Report, and it created the modern welfare state in Britain, stimulating a whole climate of opinion in the postwar world. The frenzy that attended its publication was as important an indicator of a shift in public sensibility as was the report itself.

The idea of a **welfare state** was not new. In Germany in the 1880s Bismarck had obtained provision for accident, sickness, old age, and disability insurance. Austria and Hungary had followed suit. In 1910 and 1911, following agitation by the Webbs, Bernard Shaw, H. G. Wells, and other Fabians, Lloyd George, then chancellor in a Liberal British government, introduced legislation that provided for unemployment and an old age pension insurance. At Cambridge, in the 1920s, the economist Arthur Pigou held that, so long as total production was not reduced, the redistribution of wealth – a welfare economy – was entirely feasible, the first real break with 'classical economics.' In America in the 1930s, in the wake of Roosevelt's New Deal and in light of Keynes's theories, John Connor, Richard Ely, and Robert La Folette conceived the Wisconsin Plan, which provided for statewide unemployment compensation, with rudimentary federal provision for the old, needy, and dependent children following in 1935.[32] But the Beveridge Report was comprehensive and produced in wartime, thus benefiting from and helping to provoke a countrywide change in attitude.[33]

The report came about inadvertently, when in June 1941 **Sir William Beveridge** was asked by Arthur Greenwood, Labour minister for reconstruction in the wartime coalition, to chair an interdepartmental committee on the coordination of social insurance. Beveridge was being asked merely to patch up part of Britain's social machinery but, deeply disappointed (he wanted a more active wartime role), he quickly rethought the situation and saw its radical and far-reaching possibilities.[34]

Beveridge was a remarkable and well-connected man, and his connections were to play a part in what he achieved. Born the son of a British judge in India in 1879, into a household supported by twenty-six servants, he was educated at Charterhouse and Balliol College, Oxford, where he read mathematics and classics. At Balliol, like Tawney, he fell under the influence of the master, Edward Caird, who used to urge his newly minted graduates 'to go and discover why, with so much wealth in Britain, there continues to be so much poverty and how poverty can be cured.' Like Tawney, Beveridge went to Toynbee Hall, where, he said later, he learned the meaning of poverty 'and saw the consequence of unemployment.'[35] In 1907 he visited Germany to inspect the post-Bismarck system of compulsory social insurance for pensions and sickness, and on his return several articles he wrote in the *Morning Post* about German arrangements came to the attention of Winston Churchill, who invited him to join the Board of Trade as a full-time civil servant. Beveridge therefore played a key role in the Liberal government's 1911 legislation, which introduced old-age pensions, labour exchanges, and a statutory insurance scheme against unemployment. Churchill himself was so taken with social reform that he declared liberalism to be 'the cause of the left-out millions.'[36] After World War I, Beveridge became director of the LSE, transforming it into a powerhouse for the social sciences. By World War II he was back in Oxford, as Master of University College. His long career had brought him many connections: Tawney was his brother-in-law, Clement Attlee and Hugh Dalton had been hired by him at the LSE, and were now in Parliament and the government. He knew Churchill, Keynes, and Seebohm Rowntree, whose alarming picture of poverty in York in 1899 had been partly responsible for the 1911 legislation and whose follow-up study, in 1936, was to help shape Beveridge's own document.[37] His assistant at Oxford, Harold Wilson, would be a future prime minister of Britain.[38]

A month after his meeting with Greenwood, in July 1941, Beveridge presented a paper to the committee he chaired, 'Social Insurance – General Considerations,' in which there was no mention of patchwork. 'The time has now come,' Beveridge wrote, 'to consider social insurance as a whole, as a contribution to a better new world after the war. How would one plan social insurance now if one had a clear field ... without being hampered by vested interests of any kind?'[39] Over the ensuing months, in the darkest days of the war, Beveridge's committee took 127 pieces of written evidence, and held more than 50 sessions where oral evidence was taken from witnesses. But, as Nicholas Timmins reveals in his history of the welfare state, 'only one piece of written evidence had been received by December 1941 when Beveridge circulated a paper entitled 'Heads of a Scheme' which contained the essence of the final report a year later.'[40] This influential report was essentially the work of one man.

His paper envisaged two things. There were to be a national health service, children's allowances, and unemployment benefits; and benefits were to be paid at a flat rate, high enough to live on, with contributions to come from the individual, his employer and the state. Beveridge was totally opposed to means tests or sliding scales, since he knew they would create more problems than they solved, not least the bureaucracy needed for administering a more complex system. He was familiar with all the arguments that benefits set too high would stop people from seeking work, but he was also sympathetic to the recent research of Rowntree, which had shown that low wages in large families were the primary cause of poverty.[41] This was not what the government had asked for, and Beveridge knew it. But he now began pulling strings with his many connections, calling in favours – in broadcasting, the press, Whitehall – all designed to set up a climate of anticipation ahead of the publication of his report, so that it would be an intellectual-political event of the first importance.

In terms of impact, Beveridge succeeded in everything he set out to achieve. Besides those sensational sales figures in Britain, the report had a notable reception abroad. The Ministry of Information got behind it, and details were broadcast by the BBC from dawn on 1 December in twenty-two languages. All troops received copies of the report, and it sold so well in the United States that the Treasury made a $5,000 profit. Bundles of the report were parachuted into France and other parts of Nazi-occupied Europe, and two even made their way to Hitler's bunker in Berlin, where they were found at the end of the war, together with commentaries, marked 'Secret.' One commentary assessed the plans as 'a consistent system ... of remarkable simplicity ... superior to the current German social insurance in almost all points.'[42]

There were two reasons for the report's impact. Beveridge's title may have been dry, but his text certainly was not. This was no governmentese, no civil servant's deadpan delivery. 'A revolutionary moment in the world's history,' he wrote, 'is a time for revolutions, not patching.' War was 'abolishing landmarks of every kind,' he said, and so 'offered the chance of real change,' for 'the purpose of victory is to live in a better world than the old world.' His principle line of attack, he said, was on Want – that was what security of income, social security, was all about. 'But ... Want is only one of five giants on the road of reconstruction, and in some ways the easiest to attack. The others are Disease, Ignorance, Squalor and Idleness. ... The State should offer security for service and contribution. The State in organising security should not stifle incentive, opportunity, responsibility; in establishing a national minimum, it should leave room and encouragement for voluntary action by each individual to provide more than the minimum for himself and his family.'[43] But that minimum should be given 'as of right and without means test, so that individuals may build freely upon it. ... [This] is one part of an attack upon five giant evils: upon the physical Want with which it is directly concerned, upon Disease which often causes that Want and brings many other troubles in its train, upon Ignorance which no democracy can afford among its citizens, upon Squalor . . . and upon Idleness which destroys wealth and corrupts men.'[44]

Few people in those dark days expected a government report to be moving, still less exalting, but Beveridge seems to have grasped instinctively that *because* the days were so bleak, because the threat at the moment came so obviously from outside, that now was the time to spark a change in attitude, a change in feeling toward the dangers within British society, dangers that, despite all that had happened, were still there. From his vantage point, Beveridge knew better than most how little Britain had changed in the twentieth century.[45] As Beveridge well knew, after the Great War, Britain's share of international trade had shrunk, spoiled still further by Churchill's insistence on a return to the gold standard at too high a rate, bringing about sizeable cuts in public spending and a return of social divisions in Britain (67 percent unemployment in Jarrow, 3 percent in High Wycombe).[46] As R.A. Butler, the Conservative creator of the 1944 Education Act, itself the result of the Beveridge plan, wrote later, 'It was realised with deepening awareness that the "two nations" still existed in England a century after Disraeli had used the phrase.'[47] The success of Beveridge's plan, as he himself acknowledged, also owed something to Keynes, but the social and intellectual change that hit Britain, and other countries, was deeper than just economics. Mass Observation, the poll organisation run by W. H. Auden's friend Charles Madge, found in 1941 that 16 percent said the war had changed their political views. In August 1942, four months before the Beveridge Report, one in three had changed their political views.[48] More than anything, the Beveridge Report offered hope at a time when that commodity was in short supply.[49] A month before, Rommel had retreated in North Africa, British forces had retaken Tobruk, and Eisenhower had landed in Morocco. To celebrate, Churchill had ordered church bells to be rung in Britain for the first time since war was declared (they had been kept in reserve, to signify invasion).

Despite the Great Terror in Russia, Stalin's regime continued to benefit from its status as a crucial ally. In November 1943 Churchill, Roosevelt, and the Russian dictator met in Tehran to discuss the last phase of the war, in particular the invasion of France. At that meeting Churchill presented Stalin with a sword of honor for the people of Stalingrad. Not everyone thought the Soviet leader a suitable recipient for the honor, among them, as we have seen, Friedrich von Hayek and Karl Popper. But the extent to which Stalin was appeased in the middle of war is shown by **George Orwell**'s experiences in trying to get another slim volume published.

Subtitled 'A Fairy Story,' *Animal Farm* is about a revolution that goes wrong and loses its innocence when the animals in Mr Jones's farm, stimulated to rebellion by an old Middle White boar, Major, take over the farm and expel Mr Jones and his wife. The allegory is hardly subtle. Old Major, when he addresses the other animals before he dies, refers to them as Comrades. The rebellion itself is dignified by its leaders (among them the young boar Napoleon) with the name Animalism, and Orwell, although he'd had the idea in 1937, while fighting in Spain, never made any secret of the fact that his satire was directed at Stalin and his apparatchiks. He wrote the book at the end of 1943 and the beginning of 1944, important months when the Russians finally turned back the Germans, 'and the road to Stalingrad became the road to Berlin.'[50] The revolution on the farm is soon corrupted: the pigs, looking after their own, gradually take over; a litter of puppies is conditioned to grow up as a vicious Gestapo-like Praetorian guard; the original commandments of Animalism, painted on the barn wall, are secretly amended in the dead of night ('All animals are equal/*but some are more equal than others*'); and finally the pigs start to walk on two legs, after months when the main slogan has been 'Two-legs bad! Four-legs-good!'

The book appeared in August 1945, the same month that the United States dropped atomic bombs on Hiroshima and Nagasaki, and the delay between completion and release is partly explained by the difficulties Orwell experienced in getting the book published. Victor Gollancz was only one of the publishers who turned *Animal Farm* down – at Faber & Faber, T. S. Eliot did too.[51] As a Christian, Eliot was no friend of communism, and he needed no convincing of Orwell's abilities. However, in rejecting the book, he wrote, 'We have no conviction ... that this is the right point of view from which to criticise the political situation at the present time.'[52] Four publishers rejected the book, and Orwell began to grow angry at the self-censorship he saw in these decisions. He considered publishing the book himself, but then Warburgs took it on, though not immediately, owing to the paper shortage.[53] Perhaps the further delay was just as well. When the book finally appeared, the war had just ended, but the terror of the atomic bomb had recently arrived, and following the Potsdam conference in July, the postwar – Cold War – world was emerging. The evidence of the Nazi concentration camps was becoming known, with its bleak confirmation of what man was capable of doing to man.

Animal Farm was no more a fairy story than Stalin was a political role model. Though he might have had sociopolitical aims very similar to those of William Temple, Orwell was more realistic and, like von Hayek and Popper, grasped that though the battle against Hitler had been won, the battle against Stalin was far from over, and so far as twentieth-century thought and ideas were concerned, was much more important. A whole mode of thought – the liberal imagination – was called into question by Stalinism, collectivism, and planning.

Many of the Nazi and Japanese wartime atrocities were not fully revealed until hostilities had ended. They set the seal on six grim years. And yet, for the optimistic, there was another silver lining amid the gloom. Almost all the major belligerents in the war, including the remoter areas of the British Empire, such as Australia and New Zealand, had achieved full employment. The curse of the 1930s had been wiped out. In America, where the depression had begun and hit hardest, unemployment by 1944 had shrunk to 1.2 percent.[54] Except among his grudging opponents, this was regarded as a triumph for Keynes's ideas. Wartime governments had everywhere run huge public expenditure programs – weapons manufacture – which consisted entirely of waste (unlike roads, say, which lasted and went on being useful), and combined this with vast deficits.

The U.S. national debt, $49 billion in 1941, escalated to $259 billion in 1945.[55]

Keynes had been fifty-six at the outbreak of World War II, and although he had made his name in the first war, his role was actually more crucial in the second. Within two months of the outbreak of hostilities, he produced three articles for *The Times* of London, rapidly reprinted as a pamphlet entitled *How to Pay for the War*. (These actually appeared in the German press first, owing to a leak from a lecture he gave.)[56] Keynes's ideas this time had two crucial elements. He saw immediately that the problem was not, at root, one of money but of raw materials: wars are won or lost by the physical resources capable of being turned rapidly into ships, guns, shells, and so forth. These raw materials are capable of being measured and therefore controlled.[57] Keynes also saw that the difference between a peacetime economy and a war economy was that in peace workers spend any extra income on the goods they have themselves worked to produce; in war, extra output – beyond what the workers need to live on – goes to the government. Keynes's second insight was that war offers the opportunity to stimulate social change, that the 'equality of effort' needed in national emergency could be channelled into financial measures that would not only reflect that equality of effort but help ensure greater equality after the war was over. And that, in turn, if widely publicised, would help efficiency. After Winston Churchill became prime minister, and despite the hostility to his ideas by the Beaverbrook press, Keynes was taken on as one of his two economic advisers (Lord Catto was the other).[58] Keynes lost no time in putting his ideas into effect. Not all of them became law, but his influence was profound: 'The British Treasury fought the Second World War according to Keynesian principles.'[59]

In the United States the situation was similar. There was an early recognition in some influential quarters that wartime was a classic Keynesian situation, and a team of seven economists from Harvard and Tufts argued for a vigorous expansion of the public sector so that, as in Britain, the opportunity could be taken to introduce various measures designed to increase equality after the war.[60] The National Resources Planning Board (with planning in its name, be it noted) set down nine principles in a 'New Bill of Rights' that sounded suspiciously like William Temple's Six Christian Principles, and magazines like the *New Republic* made such declarations as, 'It had better be recognised at the very start that the old ideal of laissez-faire is no longer possible. ... Some sort of planning and control there will have to be, *to an increasing degree.*'[61] In America, as in Britain, the Keynesians didn't win everything; traditional business interests successfully resisted many of the more socially equitable ideas. But the achievement of World War II, coming after the gloom of the 1930s, was that governments in most of the Western democracies – Britain, the United States, Canada, New Zealand, Australia, Sweden, and South Africa – all accepted that preserving high levels of employment was a national priority, and it was Keynes and his ideas that had brought about both the knowledge as to how to do this and the recognition that governments should embrace such responsibility.[62]

If Keynes had won the day in regard to the regulation of domestic economics, his experiences were to be less happy in dealing with the problems facing international trade. This was the issue addressed by the famous conference at **Bretton Woods** in the summer of 1944.[63] Around 750 people attended this conference, in the White Mountains in New Hampshire, which gave birth to the **World Bank** and the **International Monetary Fund** – both part of Keynes's key vision, though their powers were much diluted by the American team. Keynes understood that two problems faced the postwar world, 'only one of which was new.' The old problem was to prevent a return to the competitive currency devaluations of the 1930s, which had had the overall effect of reducing international trade and adding to the effects of the depression. The new problem was that the postwar world would be divided into two: debtor nations (such as Britain) and creditor nations (most obviously the United States). So long as this huge imbalance existed, the recovery of international trade would be hampered, affecting everyone. Keynes, who was in brilliant form at the conference, clearly grasped that a system of international currency and an international bank were needed, so as to extend the principles of domestic economics into the international field.[64] The chief point of the international bank was that it could extend credit and make loans (provided by creditor countries) in such a way that debtor countries could change their currency ratios without provoking tit-for-tat reprisals from others. The plan also removed the world from the gold standard.[65] Keynes didn't have everything his own way, and the plan eventually adopted was as much the work of **Harry Dexter White**, in the U.S. Treasury, as it was of Keynes.[66] But the intellectual climate in which these problems were thrashed out at Bretton Woods was that created by Keynes in the interwar years. It was not planning as such – Keynes, as we have seen, was a great believer in markets – but he saw that world trade was interlinked, that the greatest prosperity for the greatest number could be achieved only by recognising that wealth needs customers as well as manu-facturers and that they are one and the same people. Keynes taught the world that capitalism works on cooperation almost as much as on competition.

The end of World War II was the high point of Keynesian economics. People thought of Keynes as 'a magician.'[67] Many wanted to see his principles enshrined in law, and to a limited extent they were. Others took a more Popperian view: if economics had any pretence to science, Keynes's ideas would be modified as time went by, which is in fact what happened. Keynes had brought about an amazing change in intellectual viewpoint (not just in wartime, but over a lifetime of writings), and although he would be much criticised in later years, and his theories modified, the attitude we have to unemployment now – that it is to an extent under the control of government – is thanks to him. But he was just one individual. The end of the war, and despite Keynes, brought with it a widespread fear of a rapid return to the dismal performance of the 1930s.[68] Only economists like W. S. Woytinsky saw that there would be a boom, that people had been starved of consumer goods, that labourers and technicians, who had spent the war working overtime, had had no chance to spend the extra, that massive numbers of soldiers had years of pay saved up, that huge amounts of war bonds had been bought, which would now be redeemed, and that the technological advances made in wartime in regard to military equipment could now be rapidly turned to peacetime products. (Woytinsky calculated that there was $250 billion waiting to be spent.)[69] In practice, once the world settled down, the situation would meet no one's expectations: there was no return to the high unemployment levels of the 1930s, though in America unemployment was never as low as it had been in wartime. Instead, in the United States, it fluctuated between 4 and 7 percent – 'high enough to be disturbing, but not high enough to alarm the prosperous majority.'[70] This split-level society puzzled economists for years, not least because it had not been predicted by Keynes.

In America, although the Keynesian economists of Harvard and Tufts wanted to promote a more equal society after the war, the main problem was not poverty as such, for the country was enjoying more or less full employment. No, in America, the war merely highlighted the United States' traditional problem when it came to equality: race. Many blacks fought in Europe and the Pacific, and if they were expected to risk their lives equally with whites, why shouldn't they enjoy equality afterward?

The document that would have as profound an impact on American society as Beveridge's did on Britain was released just as the war was turning firmly in the Allies' favour, in January 1944. It was a massive work, six years in preparation, entitled *An American Dilemma: The Negro Problem and Modern Democracy.*[71] The report's author, **Gunnar Myrdal** (1898–1987), was a Swede, and he had been chosen in 1937 by Frederick Keppel, president of the Carnegie Foundation, who paid for the study, because Sweden was assumed to have no tradition of imperialism. The report comprised 1,000 pages, 250 pages of notes, and ten appendices. Unlike Beveridge's one-man band, Myrdal had many assistants from Chicago, Howard, Yale, Fisk, Columbia, and other universities, and in his preface he listed scores of distinguished thinkers he had consulted, among others: Ruth Benedict, Franz Boas, Otto Klineberg, Robert Linton, Ashley Montagu, Robert Park, Edward Shils.[72]

Since the 1920s of Lothrop Stoddard and Madison Grant, the world of 'racial science' and eugenics had shifted predominantly to Europe with the Nazi rise to power in Germany and the campaigns of Trofim Lysenko in Soviet Russia. Britain and America had seen a revulsion against the simpleminded and easy truths of earlier authors, and doubts were even being thrown over race as a scientific concept. In 1939, in *The Negro Family in the United States*, E. Franklin Frazier, professor of sociology at Howard University, who had started his researches in Chicago in the early 1930s, chronicled the general disorganisation of the Negro family.[73] He argued that this went all the way back to slavery, when many couples had been separated at the whim of their owners, and to emancipation, which introduced sudden change, further destroying stability. The drift to the towns hadn't helped, he said, because it had contributed to the stereotype of the Negro as 'feckless, promiscuous, prone to crime and delinquency.' Frazier admitted that there was some truth to these stereotypes but disputed the causes.

Myrdal went much further than Frazier. While accepting that America had certain institutions that were an advance on those in Europe, that it was a more rational and optimistic country, he nonetheless concluded that even these advanced institutions were too weak to cope with the special set of circumstances that prevailed in the United States. The dilemma, he said, was entirely the responsibility of the whites.[74] The American Negro's lifestyle, every aspect of his being, was conditioned, a secondary reaction to the white world, the most important result of which was that blacks had been isolated from the law and the various institutions of the Republic, including in particular politics.[75]

Myrdal's solution was every bit as contentious as his analysis. Congress, he judged, was unwilling and/or incapable of righting these wrongs.[76] Something more was needed, and that 'something,' he felt, could be provided only by the courts. These, he said, should be used, and seen to be used, as a way to enforce legislation that had been on the statute books for years, designed to improve the condition of blacks, and to bring home to whites that the times were changing. Like Beveridge and Mannheim, Myrdal realised that after the war there would be no going back. And so the neutral Swede told America – just as it was rescuing democracy from dictatorship across the world – that at home it was unremittingly racist. It was not a popular verdict, at least among whites. Myrdal's conclusions were even described as 'sinister.'[77] On the other hand, in the long run there were two significant reactions to Myrdal's thesis. One was the use of the courts in exactly the way that he called for, culminating in what Ivan Hannaford described as 'the most important single Supreme Court decision in American history,' *Brown* v. *Board of Education of Topeka* (1954) in which the Court unanimously ruled that segregated schools violated the Fourteenth Amendment guaranteeing equal protection under the law, and were thus unconstitutional. This played a vital part in the civil rights movement of the 1950s and 1960s.

The other reaction to Myrdal was more personal. It was expressed first by Ralph Ellison, the black musician and novelist, who wrote a review of *An American Dilemma* that contained these words: 'It does not occur to Myrdal that many of the [Negro/black] cultural manifestations which he considers merely reflective might also embody a *rejection* of what he considers "high values." '[78] In some respects, that rejection of 'high values' (and not only by blacks) was the most important intellectual issue of the second half of the twentieth century.

LIGHT IN AUGUST

If there was a single moment when an atomic bomb moved out of the realm of theory and became a practical option, then it occurred one night in early 1940, in Birmingham, England. The Blitz was in full spate, there were blackouts every night, when no lights were allowed, and at times **Otto Frisch** and **Rudolf Peierls** must have wondered whether they had made the right decision in emigrating to Britain.

Frisch was Lise Meitner's nephew, and while she had gone into exile in Sweden in 1938, after the Anschluss, he had remained in Copenhagen with Niels Bohr. As war approached, Frisch grew more and more apprehensive. Should the Nazis invade Denmark, he might well be sent to the camps, however valuable he was as a scientist. Frisch was also an accomplished pianist, and his chief consolation was in being able to play. But then, in the summer of 1939, Mark Oliphant, joint inventor of the cavity magnetometer, who by now had become professor of physics at Birmingham, invited Frisch to Britain, ostensibly for discussions about physics. (After Rutherford's death in 1937 at the age of fifty-six, from an infection following an operation, many from the Cavendish team had dispersed.) Frisch packed a couple of bags, as one would do for a weekend away. Once in England, however, Oliphant made it clear to Frisch he could stay if he wished; the professor had made no elaborate plans, but he could read the situation as well as anyone, and he realised that physical safety was what counted above all else. While Frisch was in Birmingham, war was declared, so he just stayed. All his possessions, including his beloved piano, were lost.[1]

Peierls was already in Birmingham, and had been for some time. A wealthy Berliner, he was one of the many brilliant physicists who had trained with Arnold Sommerfeld in Munich. Peierls had been in Britain in 1933, in Cambridge on a Rockefeller fellowship, when the purge of the German universities had begun. He could afford to stay away, so he did. He would become a naturalised citizen in Britain in February 1940, but for five months, from 3 September 1939 onward, he and Frisch were technically enemy aliens. They got round this 'inconvenience' in their conversations with Oliphant by pretending that they were only discussing theoretical problems.[2]

Until Frisch joined Peierls in Birmingham, the chief argument against an atomic bomb had been the amount of uranium needed to '**go critical,**' start a

chain reaction and cause an explosion. Estimates had varied hugely, from thirteen to forty-four tons and even to a hundred tons. Had this been true, it would have made the bomb far too heavy to be transported by aircraft and in any case would have taken as long as six years to assemble, by which time the war would surely have been long over. It was Frisch and Peierls, walking through the blacked-out streets of Birmingham, who first grasped that the previous calculations had been wildly inaccurate.[3] Frisch worked out that, in fact, not much more than a kilogram of material was needed. Peierls's reckoning confirmed how explosive the bomb was: this meant calculating the available time before the expanding material separated enough to stop the chain reaction proceeding. The figure Peierls came up with was about four millionths of a second, during which there would be eighty neutron generations (i.e., 1 would produce 2 would produce $4 \rightarrow 8 \rightarrow 16 \rightarrow 32$ ... and so on). Peierls worked out that eighty generations would give temperatures as hot as the interior of the sun and 'pressures greater than the centre of the earth where iron flows as a liquid.'[4] A kilogram of uranium, which is a heavy metal, is about the size of a golf ball – surprisingly little. Frisch and Peierls rechecked their calculations, and did them again, with the same results. And so, as rare as U_{235} is in nature (in the proportions $1 : 139$ of U_{238}), they dared to hope that enough material might be separated out – for a bomb and a trial bomb – in a matter of months rather than years. They took their calculations to Oliphant. He, like them, recognised immediately that a threshold had been crossed. He had them prepare a report – just three pages – and took it personally to Henry Tizard in London.[5] Oliphant's foresight, in offering sanctuary to Frisch, had been repaid more quickly than he could ever have imagined.

Since 1932, when James Chadwick identified the neutron, atomic physics had been primarily devoted to obtaining two things: a deeper understanding of radioactivity, and a clearer picture of the structure of the atomic nucleus. In 1933 the Joliot-Curies, in France, had finally produced important work that won them the Nobel Prize. By bombarding medium-weight elements with alpha particles from polonium, they had found a way of making matter artificially radioactive. In other words, they could now transmute elements into other elements almost at will. As Rutherford had foreseen, the crucial particle here was the neutron, which interacted with the nucleus, forcing it to give up some of its energy in radioactive decay.

Also in 1933 the Italian physicist **Enrico Fermi** had burst on the scene with his theory of **beta decay** (despite *Nature* turning down one of his papers).[6] This too related to the way the nucleus gave up energy in the form of electrons, and it was in this theory that Fermi introduced the idea of the 'weak interaction.' This was a new type of force, bringing the number of basic forces known in nature to four: gravity and electromagnetism, operating at great distances, and the strong and weak forces, operating at the subatomic level. Although theoretical, Fermi's paper was based on extensive research, which led him to show that although lighter elements, when bombarded, were transmuted to still lighter elements by the emission of either a proton or an alpha particle, heavier elements acted in the opposite way. That is to say, their stronger electrical barriers *captured* the incoming neutron, making them heavier. However, being now unstable, they decayed to an element with one more unit of atomic number. This raised a fascinating possibility. Uranium was the heaviest element known in nature, the top of the periodic table, with an atomic number of 92. If *it* was bombarded with neutrons and captured one, it should produce a heavier isotope: U_{238} should become U_{239}. This should then decay to an element that was entirely new, never before seen on earth, with the atomic number 93.[7]

It would take a while to produce what would be called '**transuranic**' elements, but when they did arrive, Fermi was awarded the 1938 Nobel Prize. The day that Fermi heard he had been awarded the ultimate honour was exciting in more ways than one. First there was a telephone call early in the morning; it was the local operator, to say they had been told to expect a call that evening at six o'clock, from Stockholm. Suspecting he had won the coveted award, Fermi and his family spent the day barely able to concentrate, and when the phone rang promptly at six, Fermi rushed to answer it. But it wasn't Stockholm; it was a friend, asking them what they thought of the news.[8] The Fermis had been so anxious about the phone call that they had forgotten to switch on the radio. Now they did. A friend later described what they heard: 'Hard, emphatic, pitiless, the commentator's voice read the . . . set of racial laws. The laws issued that day limited the activities and the civil status of the Jews [in Italy]. Their children were excluded from the public schools. Jewish teachers were dismissed. Jewish lawyers, physicians and other professionals could practise for Jewish clients only. Many Jewish firms were dissolved. . . . Jews were to be deprived of full citizenship rights, and their passports would be withdrawn.'[9]

Laura Fermi was Jewish.

That was not the only news. The evening before, in Germany itself, anti-Semitism had boiled over: mobs had torched synagogues across the country, pulled Jewish families into the streets, and beaten them. Jewish businesses and stores had been destroyed in their thousands, and so much glass had been shattered that the evening became infamous as Kristallnacht.

Eventually the call from Stockholm came through. Enrico had been awarded the Nobel Prize, 'for your discovery of new radioactive substances belonging to the entire race of elements and for the discovery you made in the course of this work of the selective power of slow neutrons.' Was that reference fortuitous? Or was it Swedish irony?

Until that moment, although some physicists talked about 'nuclear energy,' most of them didn't really think it would ever happen. Physics was endlessly fascinating, but as a fundamental explanation of nature rather than anything else. Ernest Rutherford gave a public lecture in 1933 in which he specifically said that, exciting as the recent discoveries were, 'the world was not to expect practical application, nothing like a new source of energy, such as once had been hoped for from the forces in the atom.'[10]

But in Berlin **Otto Hahn** spotted something available to any physicist but missed. The more common isotope of uranium, U_{238}, is made up of 92 protons and 146 neutrons in its nucleus. If neutron bombardment were to create new, transuranic elements, they would have not only different weights but different chemical properties.[11] He therefore set out to look for these new properties, always keeping in mind that if the neutrons were not being captured, but were chipping particles *out* of the nucleus, he ought to find radium. A uranium atom that lost two alpha particles (helium nuclei, atomic weight four for each) would become radium, R_{230}. He didn't find radium, and he didn't find any new elements, either. What he did find, time and again when he repeated the experiments, was barium. Barium was much lighter: 56 protons and 82 neutrons, giving a total of 138, well below uranium's 238. It made no sense. Puzzled, Hahn shared his results with **Lise Meitner**. Hahn and Meitner had always been very close, and he had helped protect her throughout the 1930s, because she was Jewish. She was kept employed because, technically speaking, she was Austrian, and therefore, technically speaking, the racial laws didn't apply to her. After the Anschluss, however, in March 1938, when Austria became part of Germany, Meitner could no longer be protected, and she was forced to escape to Göteborg in Sweden. Hahn wrote to her just before Christmas 1938 describing his unusual results.[12]

As luck would have it, Meitner was visited that Christmas by her nephew Otto Frisch, then with Bohr in Copenhagen. The pair were very pleased to see each other – both were in exile – and they went *lang-laufing* in the nearby woods, which were covered in snow. Meitner told her nephew about Hahn's letter, and they turned the barium problem over in their minds as they walked between the trees.[13] They began to consider radical explanations for Hahn's puzzling observation, in particular a theory of Bohr's that the nucleus of an atom was like a drop of water, which is held together by the attraction that the molecules have for each other, just as the nucleus is held together by the nuclear force of *its* constituents. Until then, as mentioned earlier, physicists had considered that when the nucleus was bombarded, it was so stable that at most the odd particle could be chipped off.[14] Now, huddled on a fallen tree in the Göteborg woods, Meitner and Frisch began to wonder whether the nucleus of uranium was like a drop of water in other ways, too.[15] In particular they allowed the possibility that instead of being chipped away at by neutrons, a nucleus could in certain circumstances be cleaved in two. They had been in the woods, skiing and talking, for three hours. They were cold. Nonetheless, they did the calculations there and then before turning for home. What the arithmetic showed was that if the uranium atom *did* split, as they thought it might, it could produce barium (56 protons) and krypton (36) – 56+36=92. They were right, and when Frisch told Bohr, he saw it straight away. 'Oh, what idiots we have all been,' he cried. 'This is just as it must be.'[16] But that wasn't all. As the news sank in around the world, people realised that, as the nucleus split apart, it released energy, as heat. If that energy was in the form of neutrons, and in sufficient quantity, then a chain reaction, and a bomb, might indeed be possible. Possible, but not easy. Uranium is very stable, with a half-life of 4.5 billion years; as Richard Rorty dryly remarks, if it was apt to give off energy that sparked chain reactions, few physics labs would have been around to tell the tale. It was Bohr who grasped the essential truth – that U_{238}, the common isotope, was stable, but U_{235}, the much less common form, was susceptible to **nuclear fission** (the brand–new term for what Hahn had observed and Meitner and Frisch had been the first to understand). Bring two quantities of U_{235} together to form a critical mass, and you had a bomb. But how much U_{235} was needed?

The pitiful irony of this predicament was that it was still only early 1939. Hitler's aggression was growing, sensible people could see war coming, but the world was, technically, still at peace. The Hahn/Meitner/Frisch results were published openly in *Nature*, and thus read by physicists in Nazi Germany, in Soviet Russia, and in Japan, as well as in Britain, France, Italy, and the United States.[17] Three problems now faced the physicists. How likely was a chain reaction? This could be judged only by finding out what energy was given off when fission occurred. How could U_{235} be separated from U_{238}? And how long would it take? This third question involved the biggest drama. For even after war broke out in Europe, in September 1939, and the race for the bomb took on a sharper urgency, America, with the greatest resources, and now the home of many of the exiles, was a nonbelligerent. How could she be persuaded to act? In the summer of 1939 a handful of British physicists recommended that the government acquire the uranium in the Belgian Congo, if only to stop others.[18] In America the three Hungarian refugees Leo Szilard, Eugene Wigner, and Edward Teller had the same idea and went to see Einstein, who knew the queen of Belgium, to ask her to set the ball rolling.[19] In the end they decided to approach Roosevelt instead, judging that Einstein was so famous, he would be listened to.[20] However, an intermediary was used, who took six weeks to get in to see the president. Even then, nothing happened. It was only after Frisch and Peierl's calculations, and the three-page paper they wrote as a result, that movement began. By that stage the Joliot-Curies had produced another vital paper – showing that each bombardment of a U_{235} atom released, on average, 3.5 neutrons. That was nearly twice what Peierls had originally thought.[21]

The Frisch-Peierls memorandum was considered by a small subcommittee brought into being by Henry Tizard, which met for the first time in the offices of the Royal Society in April 1940. This committee came to the conclusion that the chances of making a bomb in time to have an impact on the war were good, and from then on the development of an atomic bomb became British policy. The job of persuading the Americans to join in fell to Mark Oliphant, Frisch and Peierls's professor at Birmingham. Strapped by war, Britain did not have the funds for such a project, and any location, however secret, might be bombed.[22] In America, a 'Uranium Committee' had been established, whose chairman was Vannevar Bush, a dual-doctorate engineer from MIT. Oliphant and John Cockroft travelled to America and persuaded Bush to convey some of the urgency they felt to Roosevelt. Roosevelt would not commit the United States to build a bomb, but he did agree to explore whether a bomb could be built. Without informing Congress, he found the necessary money 'from a special source available for such an unusual purpose.'[23]

★

While Bush set to work to check on the British findings, Niels Bohr in Copenhagen received a visit from his former pupil, the creator of the Uncertainty Principle, Werner Heisenberg. Denmark had been invaded in April 1940. Bohr had refused a guarantee by the American embassy of safe passage to the United States and instead did what he could to protect more junior scholars who were Jewish. After much talk, Bohr and Heisenberg went for a walk through the brewery district of Copenhagen, near the Carlsberg factories. Heisenberg was one of those in charge of the German bomb project in Leipzig, and on that walk he raised the prospect of the military applications of atomic energy.[24] He knew that Bohr had just been in America, and Bohr knew that he knew. At the meeting Heisenberg also passed to Bohr a diagram of the reactor he was planning to build – and this is what makes the meeting so puzzling and dramatic in retrospect. Was Heisenberg letting Bohr know how far the Germans had got, because he hated the Nazis? Or was he, as Bohr subsequently felt, using the diagram as a lure, to get Bohr to talk, so he would tell Heisenberg how far America and Britain had progressed? The real reason for this encounter has never been established, though its drama has not diminished as the years have passed.[25]

The National Academy of Sciences report, produced as a result of Bush's October conversation with the president, was ready in a matter of weeks and was considered at a meeting chaired by Bush in Washington on Saturday, 6 December 1941. The report concluded that a bomb was possible and should be pursued. By this stage, American scientists had managed to produce two **'transuranic' elements**, called neptunium and plutonium (because they were the next heavenly bodies beyond Uranus in the night sky), and which were by definition unstable. Plutonium in particular looked promising as an alternative source of chain-reaction neutrons to U_{235}. Bush's committee also decided which outfits in America would pursue the different methods of isotope separation – electromagnetic or by centrifuge. Once that was settled, the meeting broke up around lunchtime, the various participants agreeing to meet again in two weeks. The very next morning the Japanese attacked Pearl Harbor, and America, like Britain, was now at war. As Richard Rhodes put it, the lack of urgency in the United States was no longer a problem.[26]

The early months of 1942 were spent trying to calculate which method of U_{235} separation would work best, and in the summer a special study session of theoretical physicists, now known as the **Manhattan Project**, was called at Berkeley. The results of the deliberations showed that much more uranium would be needed than previous calculations had suggested, but that the bomb would also be far more powerful. Bush realised that university physics departments in big cities were no longer enough. A secret, isolated location, dedicated to the manufacture of an actual bomb, was needed.

When Colonel **Leslie Groves**, commander of the Corps of Engineers, was offered the job of finding the site, he was standing in a corridor of the House of Representatives Office Building in Washington, D.C. He exploded. The job offer meant staying in Washington, there was a war on, he'd only ever had 'desk' commands, and he wanted some foreign travel.[27] When he found that as part of the package he was to be promoted to brigadier, his attitude started to change. He quickly saw that if a bomb *was* produced, and it did decide the war, here was a chance for him to play a far more important role than in any assignment overseas. Accepting the challenge, he immediately went off on a tour of the project's laboratories. When he returned to Washington, he singled out Major **John Dudley** as the man to find what was at first called Site Y. Dudley's instructions were very specific: the site had to accommodate 265 people; it should be west of the Mississippi, and at least 200 miles from the Mexican or Canadian border; it should have some buildings already, and be in a natural bowl. Dudley came up with, first, Oak City, Utah. Too many people needed evicting. Then he produced Jemez Spring, New Mexico, but its canyon was too confining. Farther up the canyon, however, on the top of the mesa, was a boys' school on a piece of land that looked ideal. It was called **Los Alamos.**[28]

As the first moves to convert Los Alamos were being made, Enrico Fermi was taking the initial step toward the nuclear age in a disused squash court in Chicago (he had emigrated in 1938). By now, no one had any doubt that a bomb could be made, but it was still necessary to confirm Leo Szilard's original idea of a nuclear chain reaction. Throughout November 1942, therefore, Fermi assembled what he called a 'pile' in the squash court. This consisted of six tons of uranium, fifty tons of uranium oxide, and four hundred tons of graphite blocks. The material was built up in an approximate sphere shape in fifty-seven layers and in all was about twenty-four feet wide and nearly as high. This virtually filled the squash court, and Fermi and his colleagues had to use the viewing gallery as their office.

The day of the experiment, 2 December, was bitterly cold, below zero.[29] That morning the first news had been received about 2 million Jews who had perished in Europe, with millions more in danger. Fermi and his colleagues gathered in the gallery of the squash court, wearing their grey lab coats, 'now black with graphite.'[30] The gallery was filled with machines to measure the neutron emission and devices to drop safety rods into the pile in case of emergency (these rods would rapidly absorb neutrons and kill the reactions). The crucial part of the experiment began around ten as, one by one, the cadmium absorption rods were pulled out, six inches at a time. With each movement, the clicking of the neutron records increased and then levelled off, in sync and exactly on cue. This went on all through the morning and early afternoon, with a short break for lunch. Just after a quarter to four Fermi ordered the rods pulled out enough for the pile to go critical. This time the clicks on the neutron counter did not level off but rose in pitch to a roar, at which point Fermi switched to a chart recorder. Even then they had to keep changing the scale of the recorder, to accommodate the increasing intensity of the neutrons. At 3:53 P.M., Fermi ordered the rods put back in: the pile had been self-sustaining for more than four minutes. He raised his hand and said, 'The pile has gone critical.'[31]

Intellectually, the central job of Los Alamos was to work on three processes designed to produce enough fissile material for a bomb.[32] Two of these concerned uranium, one plutonium. The first uranium method was known as gaseous diffusion. Metal uranium reacts with fluorine to produce a gas, uranium hexafluoride. This is composed of two kinds of molecule, one with U_{238} and another with U_{235}. The heavier molecule, U_{238}, is slightly slower than its half-sister, so when it is passed through a filter, U_{235} tends to go first, and gas on the far side of the filter is richer in that isotope. When the process is repeated (several thousand times), the mixture is even richer; repeat it often enough, and the 90 percent level the Los Alamos people needed is obtained. It was an arduous process, but it worked. The other method involved stripping uranium atoms of their electrons in a vacuum and then giving them an electrical charge that made them susceptible to outside fields. These were then passed in a beam that curved within an electrical field so that the heavy isotope would take a wider course than the lighter form, and become separated. In plutonium production, the more common isotope, U_{238}, was bombarded with neutrons, to create a new, transuranic element, plutonium-239, which did indeed prove fissile, as the theoreticians had predicted.[33]

At its height, 50,000 people were employed at Los Alamos on the Manhattan Project, and it was costing $2 billion a year, the largest research project in history.[34] The aim was to produce one uranium and one plutonium bomb by late summer 1945.

In early 1943 Niels Bohr received a visit from a captain in the Danish army. They took tea and then retired to Bohr's greenhouse, which they thought more secure. The captain said he had a message from the British via the underground, to say that Bohr would shortly receive some keys. Minute holes had been drilled in these keys, in which had been hidden a microdot, and the holes then filled in with fresh metal. He could find the microdot by slowly filing the keys at a certain point: 'The message can then be extracted or floated out on to a microslide.'[35] The captain offered the army's help with the technical parts, and when the keys arrived, the message was from James Chadwick, inviting Bohr to England to work 'on scientific matters.' Bohr guessed what that meant, but as a patriot he didn't immediately take up the offer. The Danes had managed to do a deal with the Nazis, so that in return for providing the Reich with food, Danish Jews would go unmolested. Though the arrangement worked for a while, strikes and sabotage were growing, especially after the German surrender at Stalingrad, when many people sensed that the course of the war was decisively changing. Finally, sabotage became so bad in Denmark that on 29 August 1943 the Nazis reoccupied the country, immediately arresting certain prominent Jews. Bohr was warned that he was on the list of those to be arrested, and at the end of September, with the help of the underground, he escaped, taking a small boat through the minefields of the Öresund and flying from

Sweden to Scotland. He soon moved on from Britain to Los Alamos. There, although he took an interest in technical matters and made suggestions, his very presence was what mattered, giving the younger scientists a boost: he was a symbol for those scientists who felt that the weapon they were building was so terrible that all attempts should be made to avoid using it; that the enemy should be shown what it was capable of and given the chance to surrender. There were those who went further, who said that the technical information should be shared, that the moral authority this would bring would ensure there would never be an arms race. A plan was therefore mounted for Bohr to see Roosevelt to put forward this view. Bohr got as far as Felix Frankfurter, the president's aide, who spent an hour and a half discussing the matter with Roosevelt. Bohr was told that the president was sympathetic but wanted the Dane to see Churchill first. So Bohr recrossed the Atlantic, where the British prime minister kept him waiting for several weeks. When they finally did meet, it was a disaster. Churchill cut short the meeting and in effect told Bohr to stop meddling in politics. Bohr said later that Churchill treated him like a schoolboy.[36]

Churchill was understandably worried (and he was frantically planning the Normandy invasions at the time). How could they know that the Germans, or the Japanese, or the Russians were not ahead of them? With the benefit of hindsight, no one was anywhere near the Allies on this matter.[37] In Germany, **Fritz Houtermans** had concentrated since about 1939 on making element 94, and the Germans – in the 'U-PROJECT,' as it was called – had thus neglected isotope separation. Bohr had been given that diagram of a heavy-water reactor and, drawing their own conclusions, the British had bombed the **Vemork** factory in Norway, the only establishment that manufactured such a product.[38] But that had been rebuilt. Fresh attempts to blow it up were unsuccessful, and so a different plan was called for when, via the underground, it was learned that the heavy water was to be transferred to Germany in late February 1944. According to intelligence, the water was to be taken by train to Tinnsjö, then across the sea by ferry. On the twentieth of the month, a team of Norwegian commandos blew up the ferry, the *Hydro*, with the loss of twenty-six of the fifty-three people on board. At the same time, thirty-nine drums containing 162 gallons of heavy water went to the bottom of the sea. The Germans later conceded that 'the main factor in our failure to achieve a self-sustaining atomic reactor before the war ended' was due to their inability to increase their stocks of heavy water, thanks to the attacks on Vemork and the *Hydro*.[39] This was almost certainly the most significant of the many underground acts of sabotage during the war.

The Japanese never really got to grips with the problem. Their scientists had looked at the possibility, but the special naval committee set up to oversee the research had concluded that a bomb would need a hundred tons of uranium, half the Japanese output of copper and, most daunting of all, consume 10 percent of the county's electricity supply. The physicists turned their attention instead to the development of radar. The Russians were more canny. Two of their scientists had published a paper in *Physical Review* in June 1940, making certain new observations about uranium.[40] This paper brought no response from American physicists, the Russians thus concluding (and maybe this was the real point of their paper) that the lack of a follow-through implied that the Western Allies were already embarked on their own bomb project, which was secret. The Russians also noticed what the Germans and the Japanese must have noticed as well, that the famous physicists of the West were no longer submitting original papers to the scientific journals; obviously they were busy doing something else. In 1939, therefore, the Russians started looking hard at a bomb, though work was stopped after Hitler invaded (again radar research and mine detection occupied the physicists, while the labs and materials were moved east, for safety). After Stalingrad, the program was resuscitated, and scientists were recalled from forward units. What was called 'Laboratory Number Two,' located in a derelict farm on the Moscow River, was the Russian equivalent of Los Alamos. But the lab only ever housed about twenty-five scientists and conducted mainly theoretical work regarding the chain reaction and isotope separation. The Russians were on the right lines, but years behind – for the time being.[41]

On 12 April 1945 President Roosevelt died of a massive cerebral haemorrhage. Within twenty-four hours his successor, Harry Truman, had been told about the atomic bomb.[42] Inside a month, on 8 May, the war in Europe was at an end. But the Japanese hung on, and Truman, a newcomer to office, was faced with the prospect of being the man to issue the instruction to use the awesome weapon. By V-E Day, the target researchers for the atomic bombs had selected Hiroshima and Nagasaki, the delivery system had been perfected, the crews chosen, and the aeronautical procedure for actually dropping the mechanism tried out and improved. Critical amounts of plutonium and uranium became available after 31 May, and a test explosion was set for 05.50 hours on 16 July in the desert at Alamogordo, near the Rio Grande, the border with Mexico, in an area known locally as Jornada del Muerto, 'the journey of death'.[43]

The test explosion went exactly according to plan. **Robert Oppenheimer**, the scientific director of Los Alamos, watched with his brother Frank as the clouds turned 'brilliant purple' and the echo of the explosion went on, and on, and on.[44] The scientists were still split among themselves as to whether the Russians should be told, whether the Japanese should be warned, and whether the first bomb should be dropped in the sea nearby. In the end total secrecy was maintained, one important reason for doing so being the fear that the Japanese might move thousands of captured American servicemen into any potential target area as a deterrent.[45]

The U_{235} bomb was dropped on Hiroshima shortly before 9:00 A.M. local time, on 6 August. In the time it took for the bomb to fall, the *Enola Gay*, the plane it had been carried in, was eleven and a half miles away.[46] Even so, the light of the explosion filled the cockpit, and the aircraft's frame 'crackled and crinkled' with the blast.[47] The plutonium version fell on Nagasaki three days later. Six days after that the emperor announced Japan's surrender. In that sense, the bombs worked.

The world reacted with relief that the war was over and with horror at the

means used to achieve that result. It was the end of one era and the beginning
of another, and for once there was no exaggeration in those words. In physics
it was a terrible culmination of the greatest intellectual adventure in what has
traditionally been called 'the beautiful science.' But a culmination is just that:
physics would never again be quite so heroic, but it wasn't over.

Four long years of fighting the Japanese had given the rest of the world,
especially the Americans, an abiding reason for being interested in the enemy,
who – with their *kamikaze* pilots, their seemingly gratuitous and baffling cruelty,
and their unswerving devotion to the emperor – seemed so different from
Westerners. By 1944 many of these differences had become obvious, so much
so that it was felt important in the military hierarchy in America to commission
a study of the Japanese in order fully to understand what the nation was – and
was not – capable of, how it might react and behave in certain circumstances.
(In particular, of course – though no one was allowed to say this – the military
authorities wanted to know how Japan would behave when faced with an
atomic bomb, should one be prepared. By then it was already clear that many
Japanese soldiers and units fought to the bitter end, even against overwhelming
odds, rather than surrender, as Allied or German troops would do in similar
circumstances. Would the Japanese surrender in the face of one or more atomic
bombs? If they didn't, how many were the Allies prepared to explode to bring
about surrender? How many was it *safe* to explode?)

In June 1944 the anthropologist Ruth Benedict, who had spent the previous
months in the Foreign Morale Division of the Office of War Information, was
given the task of exploring Japanese culture and psychology.[48] She was known
for her fieldwork, and of course that was out of the question in this case. She
got round the problem as best she could by interviewing as many Japanese as
possible, Japanese who had emigrated to America before the war and Japanese
prisoners of war. She also studied captured propaganda films, regular movies,
novels, and the few other political or sociological books that had been published
about Japan in English. As it happened, her study wasn't completed until 1946,
but when it did appear, published as *The Chrysanthemum and the Sword*, and
despite the fact that it was aimed at policy makers, it created a sensation.[49]
There were still half a million American servicemen in Japan, as part of the
occupying force, and this once terrifying enemy had accepted the foreign troops
with a gentleness and courtesy that was as widespread as it was surprising. The
Japanese were no less baffling in peacetime than they had been in war, and this
helped account for the reception of Benedict's book, which became much
better known than her earlier fieldwork studies.[50]

Benedict set herself the task of explaining the paradox of the Japanese, 'a
people who could be so polite and yet insolent, so rigid and yet so adaptable
to innovations, so submissive and yet so difficult to control from above, so loyal
and yet so capable of treachery, so disciplined and yet occasionally insubordinate,
so ready to die by the sword and yet so concerned with the beauty of the
chrysanthemum.'[51] Her greatest contribution was to identify Japanese life as a
system of interlocking obligations, from which all else stemmed. In Japanese

society, she found, there is a strict hierarchy of such obligations, each with its associated way of behaving. *On* is the name for the obligations one receives from the world around – from the emperor, from one's parents, from one's teacher, all contacts in the course of a lifetime.[52] These obligations impose on the individual a series of reciprocal duties: *chu* is the duty to the emperor, *ko* to one's parents – and these are subsets of *Gimu*, debts that can only ever be repaid partially but for which there is no time limit. In contrast, there is *Giri*, debts regarded as having to be repaid 'with mathematical equivalence to the favour received' and to which there are time limits. There is *Giri*-to-the-world, for example (aunts, uncles), and *Giri*-to-one's-name, clearing one's reputation of insult or the imputation of failure. Benedict explained that in Japanese psychology there is no sense of sin, as Westerners would understand the concept, which means that drama in life comes instead from dilemmas over conflicting obligations. Japanese society is based not on guilt but on shame, and from this much else derives.[53] For example, failure is much more personally traumatic in Japanese society than in Western society, being felt as an insult, with the result that great attempts are made to avoid competition. In school, records are kept not of performance but of attendance only. Insults suffered at school can be harboured for years and may not be 'repaid' until adult life and even though the 'recipient' is never aware that 'repayment' is being made. Children are allowed great freedom until about nine, Benedict says, much more so than in the West, but at around that age they begin to enter the adult world of obligations. One result, she says, is that they never forget this golden period, and this accounts for many of the problems among Japanese – heaven is something they have lost before they are even aware of it.[54] Another crucial aspect of Japanese psychology is that the absence of guilt means that they can consciously and carefully enjoy the pleasures of life. Benedict explored these – in particular, baths, food, alcohol, and sex. Each of these, she found, was pursued assiduously by the Japanese, without the attendant frustrations and guilt of the Westerner. Food, for example, is consumed in huge, long meals, each course very small, savoured endlessly, and the appearance of the food is as important as the taste. Alcohol, rarely consumed with food, often results in intoxication, but again without any feelings of remorse. Since marriages are arranged, husbands feel free to visit geishas and prostitutes. Sex outside marriage is not available to women in quite the same way, but Benedict reports that masturbation is available to the wife; here too no guilt attaches, and she found that Japanese wives often had elaborate collections of antique devices to aid masturbation. More important than any of these pleasures, per se, however, was the more widespread Japanese attitude that these aspects of life were minor. The earthly pleasures were there to be enjoyed, savoured even, but what was central for the Japanese was the interlocking system of obligations, mostly involving the family, such obligations to be met with a firm self-discipline.[55]

Benedict's study quickly established itself as a classic at a time when such international cross-cultural comparisons were thin on the ground (a situation very different from now). It was thorough, jargon-free, and did not smack of intellectualism: the generals liked it.[56] It certainly helped explain what many in the occupying forces had found: that despite the ferocity with which the Japanese had fought the war, the Americans could travel the length and breadth of the country without weapons, being welcomed wherever they went. The important point, as Benedict discovered, was that the Japanese had been allowed to keep their emperor, and he had given the order for surrender. Though there was shame attached to military defeat, the obligations of *chu* meant that the emperor's order was accepted without question. It also enabled the conquered people the freedom to emulate those who had conquered them – this, too, was a natural consequence of the Japanese psychology.[57] There were no hints in Benedict's study of the remarkable commercial success that the Japanese would later enjoy, but with hindsight, they were there. Under Japanese ways of thinking, as Benedict concluded, militarism was 'a light that failed,' and therefore Japan had now to earn respect in the world by 'a New Art and New Culture.'[58] That involved emulating her victor, the United States.

PART THREE

SARTRE TO THE SEA OF TRANQUILITY

The New Human Condition
and The Great Society

PARIS IN THE YEAR ZERO

In October 1945, following his first visit to the United States, which had impressed him, at least temporarily, with its vitality and abundance, the French philosopher **Jean-Paul Sartre** returned to a very different Paris. After the years of war and occupation, the city was wrecked, emotionally more so than physically (because the Germans had spared it), and the contrast with America was stark. Sartre's first task on his return was to deliver a lecture at the university entitled 'Existentialism is a Humanism.' To his consternation, so many people turned up for the lecture that all the seats were occupied, and he himself couldn't get in. The lecture started an hour late. Once begun, 'he spoke for two hours without stopping, without notes, and without taking his hands out of his pockets,' and the occasion became famous.[1] It became famous not only for the virtuosity of its delivery but because it was the first public admission by Sartre of a change in his philosophy. Much influenced by what had happened in Vichy France and the ultimate victory of the Allies, Sartre's **existentialism**, which before the war had been an essentially pessimistic doctrine, now became an idea 'based on optimism and action.'[2] Sartre's new ideas, he said, would be 'the new creed' for 'the Europeans of 1945.' Sartre was one of the most influential thinkers in the immediate postwar world, and his new attitude, as Arthur Herman makes plain in his study of cultural pessimism, was directly related to his experiences in the war. 'The war really divided my life in two,' Sartre said. Speaking of his time in the Resistance, he described how he had lost his sense of isolation: 'I suddenly understood that I was a social being ... I became aware of the weight of the world and my ties with all the others and their ties with me.'[3]

Born in Poitiers in 1905, Sartre grew up in comfortable surroundings with sophisticated and enlightened parents who exposed their son to the best in art, literature, and music (his grandfather was Albert Schweitzer's uncle).[4] He attended the Lycée Henri IV, one of the most fashionable schools in Paris, and then went on to the Ecole Normale Supérieure. Initially he intended to become a poet, Baudelaire being a particular hero of his, but he soon came under the influence of Marcel Proust and, most important, of Henri Bergson. 'In Bergson,' he said, 'I immediately found a description of my own psychic life.' It was as if 'the truth had come down from heaven.'[5] Other influences were Edmund

Husserl and Martin Heidegger, Sartre's attention being drawn to the Germans in the early 1930s by **Raymond Aron**, a fellow pupil at the same *lycée*. Aron was at the time more knowledgeable than Sartre, having just returned from studying with Husserl in Berlin. It was Husserl's theory that much of the formal structure of traditional philosophy is nonsense, that true knowledge comes from 'our immediate intuition of things as they are', and that truth can best be grasped in 'boundary situations' – sudden, extreme moments, as when someone steps off the pavement in front of an oncoming car. Husserl called these moments of 'unmediated existence,' when one is forced to 'choose and act,' when life is 'most real.'[6]

Sartre followed Aron to Berlin in 1933, apparently ignoring Hitler's rise.[7] In addition to the influence of Husserl, Heidegger, and Bergson, Sartre also took advantage of the intellectual climate created in Paris in the 1930s by a seminar at the Sorbonne organised by a Russian emigré named **Alexandre Kojève**. This introduced a whole generation of French intellectuals – Aron, **Maurice Merleau-Ponty**, Georges Bataille, **Jacques Lacan**, and André Breton – to Nietzsche and to Hegel's ideas of history as progress.[8] Kojève's argument was that Western civilisation and its associated democracy had triumphed over every alternative (ironic in view of what was happening then in Germany and Russia) and that everyone, eventually, including the presently downtrodden working classes, would be 'bourgeoisified.' Sartre, however, drew different conclusions – being far more pessimistic in the 1930s than his Russian teacher. In one of his most famous phrases, he described man as 'condemned to be free.' For Sartre, following Heidegger much more than Kojève, man was alone in the world and gradually being overtaken by materialism, industrialisation, standardisation, *Americanisation* (Heidegger, remember, had been influenced by Oswald Spengler). Life in such a darkening world, according to Sartre, was **'absurd'** (another famous coinage of his). This absurdity, a form of emptiness, Sartre added, produced in man a sense of **'nausea,'** a new version of alienation and a word he used as the title for a novel he published in 1938, *La Nausée*. One of the protagonists of the novel suffers this complaint, living in a provincial bourgeois world where life drags on with 'a sort of sweetish sickness' – *Madame Bovary* in modern dress.[9] Most people, says Sartre, prefer to be free but are not: they live in 'bad faith.' This was essentially Heidegger's idea of authenticity/inauthenticity, but Sartre, owing to the fact that he used more accessible language and wrote novels and, later, plays, became much more well known as an existentialist.[10] Although he became more optimistic after the war, both phases of his thinking are linked by a distaste – one might almost say a hatred – for the bourgeois life. He loved to raise the spectre of the surly waiter, whose surliness – *La Nausée* – existed because he hated being a waiter and really wanted to be an artist, an actor, knowing that every moment spent waiting was spent in 'bad faith.'[11] Freedom could only be found by breaking away from this sort of existence.

Intellectual life in Paris experienced a resurgence in 1944, precisely because the city had been occupied. Many books had been banned, theatres censored, magazines closed; even conversation had been guarded. As in the other occupied countries of Eastern Europe and in Holland and Belgium, the Einsatzstab Reichsleiter Rosenberg (ERR), a special task force under Alfred Rosenberg, whose job it was to confiscate both private and public art collections, had descended on France. The paper shortage had ensured that books, newspapers, magazines, theatre programs, school notebooks, and artists' materials were in short supply. Sartre apart, this was the age of André Gide, Albert Camus, Louis Aragon, Lautréamont, of Federico García Lorca and Luis Buñuel, and all the formerly banned American authors – Ernest Hemingway, John Steinbeck, Thornton Wilder, Damon Runyon.[12] Nineteen-forty-four also became known as the year of 'Ritzkrieg': though the world was still at war, Paris had been liberated and was inundated with visitors. Hemingway visited Sylvia Beach – her famous bookshop, Shakespeare & Co. (which had published James Joyce's *Ulysses*) had closed down, but she had survived the camps. Lee Miller, of *Vogue*, hurried to resume her acquaintance with Pablo Picasso, Jean Cocteau, and Paul Eluard. Other visitors of that time included Marlene Dietrich, William Shirer, William Saroyan, Martha Gellhorn, A. J. Ayer, and George Orwell. The change in feeling was so marked, the feeling of renewal so complete, that Simone de Beauvoir talked about 'Paris in the Year Zero.'[13]

For someone like Sartre, the *épuration*, the purge of collaborators, was also, if not exactly joyful, at the least a satisfying display of justice. Maurice Chevalier and Charles Trenet were blacklisted, for having sung on the German-run Radio-Paris. Georges Simenon was placed under house arrest for three months for allowing some of his Maigret books to be made into films by the Germans. The painters André Derain, Dunoyer de Segonzac, Kees van Dongen, and Maurice Vlaminck (who had gone into hiding at the liberation) were all ordered to paint a major work for the state as a punishment for accepting a sponsored tour of Germany during the war; and the publisher Bernard Grasset was locked up in Fresnes prison for paying too much heed to the 'Otto List,' the works proscribed by the Germans, named after Otto Abetz, the German ambassador in Paris.[14] More serious was the fate of authors such as Louis-Ferdinand Céline, Charles Maurras, and Robert Brasillach, who had been close to the Vichy administration. Some were put on trial and convicted as traitors, some fled abroad, others committed suicide. The most notorious was the writer Brasillach, an 'exultant fascist' who had become editor of the virulently anti-Semitic *Je suis partout* ('I am everywhere', but nicknamed *Je suis parti*, 'I have left'). He was executed by firing squad in February 1945.[15] Sacha Guitry, the dramatist and actor, a sort of French Noël Coward, was arrested and asked why he had agreed to meet Göring. He replied, 'Out of curiosity.' Serge Lifar, Serge Diaghilev's protégé and the Vichy-appointed director of the Paris Opéra, was initially banned for life from the French stage, but this was later commuted to a year's suspension.[16]

Sartre, who had been in the army, interned in Germany and a member of the resistance, saw the postwar world as his moment, and he wanted to carve out a new role for the intellectual and the writer. His aim, as a philosopher, was still the creation of *l'homme revolté*, the rebel, whose aim was the overthrow of the bourgeoisie; but to this he now added an attack on analytic reason which he described as 'the official doctrine of bourgeois democracy.' Sartre had been struck, in wartime, by the way man's sense of isolation had disappeared, and he now felt that existentialism should be adapted to this insight – that action, choice, was the solution to man's predicament. Philosophy, existentialism, became for him – in a sense – a form of guerrilla war in which individuals, who are both isolated souls and yet part of a joint campaign, find their being. With Simone de Beauvoir and Maurice Merleau-Ponty, Sartre (as editor in chief) founded a new political, philosophical, and literary journal called *Les Temps modernes* (Modern Times), the motto for which was, 'Man is total: totally committed and totally free.'[17] This group in effect joined the long line of thinkers – Bergson, Spengler, Heidegger – who felt that positivism, science, analytic reason, and capitalism were creating a materialistic, rational but crass world that denuded man of a vital life force. In time this would lead Sartre to an equally crass anti-Americanism (as it had Spengler and Heidegger before him), but to begin with he declared in his *Existentialism* (1947) that 'man is only a situation,' one of his most important phrases. Man, he said, had 'a distant purpose,' to *realise* himself, to make choices in order to *be*. In doing so, he had to liberate himself from bourgeois rationality.[18] There is no doubt that Sartre was a gifted phrase maker, the first soundbite philosopher, and his ideas appealed to many in the postwar world, especially his belief that the best way to achieve an existential existence, the best way to be 'authentic,' as Heidegger would have put it, was to be *against* things. The critic, he said, has a fuller life than the acquiescer. (He even refused, in later life, the award of the Nobel Prize.)[19] It was this approach that led him in 1948 to found the Revolutionary Democratic Association, which tried to lead intellectuals and others away from the obsession that was already dominating their lives: the Cold War.[20]

Sartre was a Marxist – 'It is not my fault if reality is Marxist,' is how he put it. But in one important regard he was overtaken by the other member of the trinity that founded *Les Temps modernes*, Maurice Merleau-Ponty. Merleau-Ponty had also attended Kojève's seminar in the 1930s, and he too had been influenced by Husserl and Heidegger. After the war, however, he pushed the 'anti' doctrine much further than Sartre. In *Humanism and Terror*, published in 1948, Merleau-Ponty welded Sartre and Stalin in the ultimate existential argument.[21] His central point was that the Cold War was a classic '**boundary situation**,' which required 'fundamental decisions from men where the risk is total.' Successful revolutions, he claimed, had not shed as much blood as the capitalist empires, and therefore the former was preferable to the latter and had 'a humanistic future.' On this analysis, Stalinism, for all its faults, was a more 'honest' form of violence than that which underlay liberal capitalism. Stalinism acknowledged its violence, Merleau-Ponty said, whereas the Western empires did not. In this respect at least, Stalinism was to be preferred.[22]

Existentialism, Sartre, and Merleau-Ponty were, therefore, the conceptual fathers of much of the intellectual climate of the postwar years, particularly in France, but elsewhere in Europe as well. When people like Arthur Koestler – whose *Darkness at Noon*, exposing Stalinist atrocities, sold 250,000 copies in France alone – took them to task, they were denounced as liars.[23] Then Sartre *et al.* fell back on such arguments as that the Soviets covered up because they were ashamed of *their* violence, whereas in Western capitalist democracies violence was implicit and openly condoned. Sartre and Merleau-Ponty were one factor in France having the most powerful Communist Party outside the Soviet bloc (in 1952 *Les Temps modernes* became a party publication in all but name), and their influence did not really dissolve until after the student rebellions of 1968. Their stance also led to a philosophical hatred of America, which had never been entirely absent from European thought but now took on an unprecedented virulence. In 1954 Sartre visited Russia and returned declaring that 'there is total freedom of criticism in the USSR.'[24] He knew that wasn't true but felt it was more important to be anti-America than critical of the Soviet Union. This attitude persisted, in Sartre as in others, and showed itself in the philosopher's espousal of other Marxist anti-American causes: Tito's Yugoslavia, Castro's Cuba, Mao's China, and Ho Chi Minh's Vietnam. Nearer home, of course, he was a natural leader for the protests against France's battle with Algeria in the mid-1950s, where Sartre supported the FLN rebels. It was this support that led to his friendship with the man who would carry his thinking one important stage further: Frantz Fanon.[25]

France, more than most countries, lays great store by its intellectuals. Streets are named after philosophers and even minor writers. Nowhere is this more true than in Paris, and the period after World War II was the golden age of intellectuals. During the occupation the intellectual resistance had been led by the Comité National des Ecrivains, its mouthpiece being *Les Lettres françaises*. After the liberation the editorship was taken over by **Louis Aragon**, 'a former surrealist now turned Stalinist.' His first act was to publish a list of 156 writers, artists, theatre people, and academics who had collaborated and for whom the journal called for 'just punishment.'[26]

Nowadays, the image of the French intellectual is invariably of someone wearing a black turtleneck sweater and smoking a harsh cigarette, a Gauloise, say, or a Gitane. This certainly owes something to Sartre, who like everyone in those days smoked a great deal, and always carried scraps of paper in his pockets.[27] The various groups of intellectuals each had their favourite cafés. Sartre and de Beauvoir used the Flore at the corner of the boulevard Saint-Germain and the rue Saint-Benôit.[28] Sartre arrived for breakfast (two cognacs) and then sat at a table upstairs and wrote for three hours. De Beauvoir did the same but at a separate table. After lunch they went back upstairs for another three hours. The proprietor at first didn't recognise them, but after Sartre became famous he received so many telephone calls at the café that a line was installed solely for his use. The Brasserie Lipp, opposite, was shunned for a while because its Alsatian dishes had been favoured by the Germans (though Gide had eaten there). Picasso and Dora Maar used Le Catalan in the rue des

Grands Augustins, the Communists used the Bonaparte on the north side of the *place*, and musicians preferred the Royal Saint-Germain, opposite the Deux Magots, Sartre's second choice.[29] But in any event, the existential life of 'disenchanted nonchalance' took place only between the boulevard Saint-Michel in the east, the rue des Saint-Pères in the west, the *quais* along the Seine in the north, and the rue Vaugirard in the south; this was 'la cathédrale de Sartre.'[30] In those days, too, many writers, artists and musicians did not live in apartments but took rooms in cheap hotels – one reason why they made so much use of café life. The only late-night café in those days was Le Tabou in the rue Dauphine, frequented by Sartre, Merleau-Ponty, **Juliette Gréco**, the *diseuse* (a form of singing almost like speaking), and Albert Camus. In 1947 Bernard Lucas persuaded the owners of Le Tabou to rent him their cellar, a tubelike room in which he installed a bar, a gramophone, and a piano. Le Tabou took off immediately, and from then on, Saint-Germain and *la famille Sartre* were tourist attractions.[31]

Few tourists, however, read *Les Temps modernes*, the journal that had been started in 1945, funded by Gaston Gallimard and with Sartre, de Beauvoir, Camus, Merleau-Ponty, Raymond Queneau, and Raymond Aron on the board. Simone de Beauvoir saw *Les Temps modernes* as the showpiece of what she called the 'Sartrean ideal,' and it was certainly intended to be the flagship of an era of intellectual change. Paris at the time was resurgent intellectually, not just in regard to philosophy and existentialism. In the theatre, Jean Anouilh's *Antigone* and Sartre's own *Huis clos* had appeared in 1944, Camus's *Caligula* a year later, the same year as Giraudoux's *Madwoman of Chaillot*. Sartre's *Men without Shadows* appeared in 1946. Eugène Ionesco and Samuel Beckett, influenced by Luigi Pirandello, were waiting in the wings.

Exciting as all this was, the climate of *les intellos* in Paris soon turned sour thanks to one issue that dominated everything else: Stalinism.[32] France, as we have seen, had a strong Communist Party, but after the centralisation of Yugoslavia, in the manner of the USSR, the Communist takeover in Czechoslovakia, and the death of its foreign minister, Jan Masaryk, many in France found it impossible to continue their membership of the PCF, or were expelled when they expressed their revulsion. A number of disastrous strikes in France also drove a wedge between French intellectuals and workers, a relationship that was in fact never as strong as the intellectuals pretended. Two things followed. In one, Sartre and his 'famille' joined in 1947 the Rassemblement Démocratique Révolutionnaire, a party created to found a movement independent of the USSR and the United States.[33] The Kremlin took this seriously, fearing that Sartre's 'philosophy of decadence,' as they called existentialism, could become a 'third force,' especially among the young. Andrei Zhdanov, we now know, saw to it that Sartre was attacked on several fronts, in particular at a peace conference in Wrocław, Poland, in August 1948, where Picasso too was vilified.[34] Sartre later changed his tune on Stalinist Russia, arguing that whatever wrongs had been committed had been carried out for the greater good. This tortuous form of reasoning became ever more necessary as the 1940s wore on and more and more evidence was revealed about Stalin's atrocities. But Sartre's continuing hatred of American materialism kept him more in the Soviet camp than anywhere else. This position received a massive setback in 1947, however, with the publication of *I Chose Freedom*, by **Victor Kravchenko**, a Russian engineer who had defected from a Soviet trade mission to the United States in 1944. This book turned into a runaway success and was translated into a score of languages.[35] Russian-authored, it was the earliest first-person description of Stalin's labour camps, his persecution of the kulaks, and his forced collectivisations.[36]

In France, due to the strength of the Communist Party, no major publishing house would touch the book (echoes of Orwell's *Animal Farm* in Britain). But when it did appear, it sold 400,000 copies and won the Prix Sainte-Beuve. The book was attacked by the Communist Party, and *Les Lettres françaises* published an article by one Sim Thomas, allegedly a former OSS officer, who claimed that the book had been authored by American intelligence agents rather than Kravchenko, who was a compulsive liar and an alcoholic.[37] Kravchenko, who by then had settled in the United States, sued for libel. The trial was held in January 1949 amid massive publicity. *Les Lettres françaises* had obtained witnesses from Russia, with NKVD help, including Kravchenko's former wife, **Zinaïda Gorlova**, with whom, he said, he had witnessed many atrocities. Since Gorlova's father was still in a prison camp, her evidence was naturally tainted several times over. Despite this, faced by her ex-husband in the witness box, she physically deteriorated, losing weight almost overnight and becoming 'unkempt and listless'. She was eventually taken to Orly airport, where a Soviet military aircraft was waiting to fly her back to Moscow. 'Sim Thomas' was never produced; he did not exist. The most impressive witness for Kravchenko was **Margarete Buber-Neumann**, the widow of the prewar leader of the German Communist Party, Heinz Neumann. After Hitler achieved power, the Neumanns had fled to Soviet Russia but had been sent to the labour camps because of 'political deviationism.'[38] After the Molotov-Ribbentrop nonaggression pact, in 1940, they had been shipped back to Germany and the camp at Ravensbrück. So Margarete Buber-Neumann had been in camps on both sides of what became the Iron Curtain: what reason had she to lie?

The verdict was announced on 4 April, the same day that the North Atlantic Alliance was signed. Kravchenko had won. He received only minimal damages, but that wasn't the point. Many intellectuals resigned from the party that year, and soon even Albert Camus would follow.[39] Sartre and de Beauvoir did not resign, however. For them, all revolutions have their 'terrible majesty.'[40] For them, the hatred of American materialism outweighed everything else.

After the war, Paris seemed set to resume its position as the world capital of intellectual and creative life, the City of Light that it had always been. Breton and Duchamp were back from America, mixing again with Cocteau. This was the era of Anouilh's *Colombe*, Gide's *Journals* and his Nobel Prize, Malraux's *Voices of Silence*, Alain Robbe-Grillet's *Les Gommes*; it was again, after an interlude, the city of Edith Piaf, Sidney Bechet, and Maurice Chevalier, of

Matisse's Jazz series, of major works by the *Annales* school of historians, which are considered in a later chapter, of the new mathematics of 'Nikolas Bourbaki,' of Frantz Fanon's *Black Skin, White Masks*, and of Jacques Tati's *Mr. Hulot's Holiday*. Coco Chanel was still alive, and Christian Dior had just started. In serious music it was the time of **Olivier Messiaen**. This composer was splendidly individualistic. Far from being an existentialist, he was a theological writer, 'dedicated to the task of reconciling human imperfection and Divine Glory through the medium of Art.' Messiaen detested most aspects of modern life, preferring the ancient grand civilisations of Assyria and Sumer. Much influenced by Debussy and the Russian composers, his own works sought to create timeless, contemplative moods, and although he tried serialism, his works frequently employed repetition on a large scale and, his particular innovation, the transcription of birdsong. In the decade and a half after the war, Messiaen used adventurous techniques (including new ways of dividing up the piano keyboard), birdsong, and Eastern music to forge a new religious spirit in music: *Turangalîla* (Hindu for 'love song'), 1946–1948; *Livre d'Orgue*, 1951; *Réveil des Oiseaux*, 1953. Messiaen's opposition to existentialism was underlined by his pupil Pierre Boulez, who described his music as closer to the Oriental philosophy of 'being' rather than the Western idea of 'becoming.'[41]

And yet, despite all this, the 1950s would witness a slow decline in Paris, as the city was overtaken by New York and, to a lesser extent, by London. It would be eclipsed further in the student rebellions of the late 1960s. This was as true of painting as of philosophy and literature. Alberto Giacometti produced some of his greatest, gauntest, figures in postwar Paris, the epitome for many people of existential man; and Jean Dubuffet painted his childlike but at the same time very sophisticated pictures of intellectuals and animals (cows mainly), grotesque and gentle at the same time, revealing mixed feelings about the earnestness with which the postwar Parisian philosophical and literary scene regarded itself. Lesser School of Paris artists like Bernard Buffet, René Mathieu, Anton Tapiès, and Jean Atlan all sold embarrassingly well in France, much better than their British or North American contemporaries. But the hardships of war caused a marked shortsightedness among dealers and artists alike, leading to speculation and a collapse in prices in 1962. Contemporary painting in France has never really recovered. In reality de Beauvoir had got it back-to-front when she said that Paris was in the year zero, being reborn. It was yet another instance of a sunset being mistaken for a dawn. The decade after the end of World War II was the last great shining moment for the City of Light. Existentialism had been invigorated and was popular in France because it was in part a child of the Resistance, and therefore represented the way the French, or at least French intellectuals, liked to think of themselves. Sartre apart, Paris's final glory was delivered by four men, three of whom were French by adoption and not native-born, and a third who loathed most of what Paris stood for. These were **Albert Camus**, **Jean Genet**, **Samuel Beckett**, and **Eugène Ionesco**.

Camus, a *pied-noir* born in Algeria, was raised in poverty and never lost his sympathy for the poor and oppressed. Briefly a Marxist, he edited the Resistance newspaper *Combat* during the war. Like Sartre, he too became obsessed with man's 'absurd' condition in an indifferent universe, and his own career was an attempt to show how that situation could (or should) be met. In 1942 he produced *The Myth of Sisyphus*, a philosophical tract that first appeared in the underground press. His argument was that man must recognise two things: that all he can rely upon is himself, and what goes on inside his head; and that the universe *is* indifferent, even hostile, that life is a struggle, that we are all like Sisyphus, pushing a stone uphill, and that if we stop, it will roll back down again.[42] This may seem – may indeed be – futile, but it is all there is. He moved on, to publish *The Plague* in 1947. This novel, a much more accessible read, starts with an outbreak of bubonic plague in an Algerian city, Oran. There is no overt philosophising in the book; instead, Camus explores the way a series of characters – Dr Rieux, his mother, or Tarrou – react to the terrible news, and deal with the situation as it develops.[43] Camus's main objective is to show what community does, and does not, mean, what man can hope for and what he cannot – the book is in fact a sensitive description of isolation. And that of course is the plague that afflicts us. In this there are echoes of Dietrich Bonhoeffer and his ideas of community, but also of Hugo von Hofmannsthal; after all, Camus has created a work of art out of absurdity and isolation. Does that redeem him? Camus received the Nobel Prize for Literature in 1957 but was killed in a car crash three years later.

Jean Genet – Saint Genet in Sartre's biography – introduced himself one day in 1944 to the philosopher and his consort as they sat at the Café Flore. He had a shaven head and a broken nose, 'but his eyes knew how to smile, and his mouth could express the astonishment of childhood.'[44] His appearance owed not a little to his upbringing in reformatories, prisons, and brothels, where he had been a male prostitute. Genet's future reputation would lie in his brilliance with words and his provocative plots, but he was of interest to the existentialists because as an aggressive homosexual and a criminal he occupied two prisons (psychological as well as physical), and in living on the edge, in boundary situations, he at least stood the chance of being more alive, more authentic, than others. He was also of interest to de Beauvoir because, being homosexual and having been forced to play 'female' roles in prison (on one occasion he was a 'bride' in a prison ménage), Genet's views about sex and gender were quite unlike anyone else's. Genet certainly lived life to the full in his way, even going so far as to desecrate a church to see what God would do about it. 'And the miracle happened. There was no miracle. God had been debunked. God was hollow.'[45]

In a series of novels and plays Genet regaled his public with life as it really was among the 'queers' and criminals he knew, the vicious sexual hierarchies within prisons, the baroque sexual practices and inverted codes of behaviour (calling someone 'a cocksucker' was enough to get one murdered).[46] But Genet instinctively grasped that low life, on the edge of violence, the boundary situation par excellence, evoked not only a prurient interest on the part of the bourgeois but deeper feelings too. It opened a longing for something, whether

it was latent masochism or latent homosexuality or a sneaking lust for violence –
whatever it was, the very popularity of Genet's work showed up the inadequacies of bourgeois life much more than any analysis by Sartre or the others.
Our Lady of the Flowers (1946) was written while Genet was in Mettray prison
and details the petty but all-important victories and defeats in a closed world
of natural and unnatural homosexuals. *The Maids* (1948) is ostensibly about two
maids who conspire to murder their mistress; however, Genet's insistence that
all the roles are played by young men underlines the play's real agenda, the
nature of sexuality and its relation to our bodies. By the same token, in *The
Blacks* (1958) his requirement that some of the white roles be played by blacks,
and that one white person must always be in the audience for any performance,
further underlined Genet's point that life is about feeling (even if that feeling
is shame or embarrassment) rather than 'just' about thought.[47] As an erstwhile
criminal, he knew what Sartre didn't appear to grasp: that a rebel is not
necessarily a revolutionary, and that the difference between them is, at times,
critical.

   Samuel Beckett's most important creative period overlapped with those of
Camus and Genet, and in this time he completed *Waiting for Godot, Endgame,*
and *Krapp's Last Tape.* It should be noted, however, that both *Endgame* and
Krapp's Last Tape received their world premieres in London. By then, Paris was
slipping. Born in 1906, Beckett was the son of well-to-do Protestants who lived
at Foxrock, near Dublin. As Isaiah Berlin watched the October Revolution in
Petrograd, so Beckett watched the Easter Rebellion from the hills outside the
Irish capital.[48] He attended Trinity College, Dublin, like James Joyce, and after
a spell at teaching he travelled all over Europe.[49] He met the author of *Ulysses*
in Paris, becoming a friend and helping defend the older man's later work
(Joyce was writing *Finnegans Wake*).[50] Beckett settled first in London, however,
after his father died and left him an annuity. In 1934 he began analysis at the
Tavistock Clinic, with Wilfred Bion, by which time he was writing short
stories, poems, and criticism.[51] In 1937 he moved back to Paris, where he
eventually had his novel *Murphy* published, by Routledge, after it had been
rejected by forty-two houses. During the war he distinguished himself in the
resistance, winning two medals. But he also spent a long time in hiding (with
the novelist Nathalie Sarraute) in Vichy France, which, as several critics have
remarked, gave him an extended experience in waiting. (When he came back,
Nancy Cunard thought he had the look of 'an Aztec eagle about him.')[52]
Beckett was by now thoroughly immersed in French culture – he was an expert
on Proust, mixed in the circle around *Transition* magazine, imbibed the work
of the symbolist poets, and could not help but be affected by Sartre's existentialism. All of Beckett's major plays were written in French and then translated
back into English, mostly by him but occasionally with help.[53] As the critic
Andrew Kennedy has said, this experience with 'language pains' surely helped
his writing.

   Beckett wrote his most famous work, *Waiting for Godot*, in less than four
months, starting in early October 1948 and finishing the following January.
It was, however, another four years before it was performed, at the Théâtre

de Babylone in Paris. Despite mixed reviews, and his friends having to 'corral' people into attending, it was worth the wait, for *Godot* has become one of the most discussed plays of the century, loved and loathed in equal measure, at least to begin with, though as time has gone by its stature has, if anything, grown.[54] It is a spare, sparse play; its two main characters (there are five in all) occupy a stage that is bare save for a solitary tree.[55] The two central figures are usually referred to as literary tramps, and they are often cast wearing bowler hats, though the stage directions do not call for this. The play is notable for its long periods of silence, its repetitions of dialogue (when dialogue occurs), its lurches between metaphysical speculation and banal cliché, the near-repetitions of the action, such as it is, in the two halves of the play, and the final nonappearance of the eponymous Godot. In its unique form, its references to itself, and the demands it makes on the audience, it is one of the last throws of modernism. It was cleverly summed up by one critic, who wrote, 'Nothing happens, twice!'[56] This is true enough on the surface, but a travesty nonetheless. As with all the masterpieces of modernism, *Godot*'s form is integral to the play, and to the experience of the work; no summary can hope to do it justice. It is a post-*Waste Land* play, a post-O'Neill play, post-Joyce, post-Sartre, post-Proust, post-Freud, post-Heisenberg, and post-Rutherford. You can find as many twentieth-century influences as you care to look for – which is where its richness lies. Vladimir and Estragon, the two tramps, are waiting for Godot. We don't know why they are waiting, where they are waiting, how long they have been waiting, or how long they expect to wait. The act of waiting, the silences and the repetitions, conspire to bring the question of time to the fore – and of course in bewildering and intriguing the audience, who must also wait through these silences and repetitions, *Godot* provides an experience to be had nowhere else, causing the audience to think. (The play's French title is *En attendant Godot*; 'attending,' as in paying attention to, amplifies waiting.) In some respects, *Godot* is the reverse of Proust's *A la recherche du temps perdu*. Proust made something out of nothing; Beckett is making nothing out of something, but the result is the same, to force the audience to consider what nothing and something are, and how they differ (and recalls Wolfgang Pauli's question from the 1920s – why is there something rather than nothing?).[57]

Both acts are interrupted by the arrival, first, of Lucky and Pozzo, and of the Boy. The first two are a sort of vaudeville act, the former deaf and the latter dumb.[58] The Boy is a messenger from Mr Godot, but he has no message, recalling Kafka's *Castle*. There is much else, of course – a lot of cursing, a hat-passing routine, comic miming, problems with boots and bodily functions. But the play is essentially about emptiness, silence, and meaning. One is reminded of the physicists' analogous scale when illustrating the atom – that the nucleus (which nonetheless has most of the mass), is no more than a grain of sand at the centre of an electron shell-structure the size of an opera house. This is not only bleak, Beckett is saying; communication is not only fatuous, futile, and absurd, but it is also comic. All we are left with is either cliché or speculation so removed from any reality that we can never know if it has any meaning – shades of Wittgenstein. Though Beckett loved Chaplin, his message is the very opposite; there is nothing heroic about Vladimir or Estragon, their comedy evokes no identification on our part. It is, it is intended to be, terrifying. Beckett is breaking down all categories. Vladimir and Estragon occupy space-time; in the early French editions Pozzo and Lucky are described as 'les comiques staliniens'; the play is about humanity – the universe – running down, losing energy, cooling; the characters have, as the existentialists said, been thrown into the world without purpose or essence, only feeling.[59] They must wait, with patience, because they have no idea what will come, or even *if* it will come, save death of course. Vladimir and Estragon do stay together, the play's one positive, optimistic note, till they reach the superb culmination – as an example of the playwright's art, it can hardly be bettered. Vladimir cries, 'We have kept our appointment, and that's an end to that. We are not saints, but we have kept our appointment. How many people can boast as much?'

The important point with Beckett, as with O'Neill and Eliot, is to experience the work. For he was no cynic, and the only satisfactory way to conclude writing about him is to quote him. His endings are better than anyone else's. The end of *Godot* reads as follows:

> Vladimir: Well, shall we go?
> Estragon: Yes, let's go.
> [*They do not move.*]

Or we can end by quoting Beckett's letter to fellow playwright Harold Pinter: 'If you insist on finding form [for my plays] I'll describe it for you. I was in hospital once. There was a man in another ward, dying of throat cancer. In the silences I could hear his screams continually. That's the kind of form my work has.'

For Beckett at midcentury, the speculations of Sartre were pointless; they were simply statements of the obvious. Science had produced a cold, empty, dark world in which, as more details were grasped, the bigger picture drained away, if only because words were no longer enough to account for what we know, or think we know. Dignity has almost disappeared in *Godot*, and humour survives ironically only by grim effort, and uncertainly at best. Comforting though it is, Beckett can see no point to dignity. As for humour . . . well, the best that can be said is – it helps the waiting.

Beckett and Genet both came from outside the French mainland, but it was Paris that provided the stage for their triumphs. The position of the third great playwright of those years, **Eugène Ionesco**, was slightly different. Ionesco was of Romanian background, grew up in France, spent several years in Romania, during the Soviet occupation, and then returned to Paris, where his first play, *The Bald Prima Donna*, was produced in 1950. Others followed in rapid succession, including *The Chairs* (1955), *The Stroller in the Air* (1956), *How to*

Get Rid of It (1958), *The Killer* (1959) and *Rhinoceros* (1959). One of the biographies of Beckett was given the subtitle 'The Last Modernist,' but the title could have applied equally to Ionesco, for he was in some ways the perfect amalgam of Wittgenstein, Karl Kraus, Freud, Alfred Jarry, Kafka, Heidegger, and the Dada/surrealists. Ionesco admitted that many of his ideas for plays came from his dreams.[60] His main aim, he said, certainly in his earlier plays, was to convey the astonishment he felt simply at existing, at why there is something rather than nothing. Not far behind came his concern with language, his dissatisfaction at our reliance on cliché and, more profoundly, the sheer inadequacy of language when portraying reality. Not far behind this came his obsession with psychology, in particular the new group psychology of the modern world of mass civilisation in great cities, how that affected our ideas of solitude and what separated humanity from animality.

In *The Bald Prima Donna* it is as if the figures in a de Chirico landscape are speaking, virtual automatons who show no emotion, whose words come out in a monotone.[61] Ionesco's purpose here is to show the magic of genuine language, to draw our attention to what it is and how it is produced. In *The Stroller in the Air*, one of his plays based on a dream (of flying), the main character can see, from his vantage point, into the lives of others. This one-way sharing, however, which offers great comic possibilities, is in the end tragic, for as a result of his unique vantage point the stroller experiences a greater solitude than anyone else. In *The Chairs*, chairs are brought on to the stage at a rapid pace, to create a situation that words simply fail to describe, and the audience therefore has to work out the situation for itself, find its own words. Finally, in *Rhinoceros*, the characters gradually metamorphose into animals, exchanging an individual human psychology for something more 'primitive,' more group-centred, all the time provoking us to ask how great this divide really is.[62]

Ionesco was very attuned to the achievements of science, the psychology of Freud and Jung in particular, but biology too. It instilled in him his own brand of pessimism. 'I wonder if art hasn't reached a dead-end,' he said in 1970. 'If indeed in its present form, it hasn't already reached its end. Once, writers and poets were venerated as seers and prophets. They had a certain intuition, a sharper sensitivity than their contemporaries, better still, they discovered things and their imaginations went beyond the discoveries even of science itself, to things science would only establish twenty-five or fifty years later. In the relation to the psychology in his time, Proust was a precursor. ... But for some time now, science and the psychology of the subconscious have been making enormous progress, whereas the empirical revelations of writers have been making very little. In these conditions, can literature still be considered as a means to knowledge?' And he added, 'Telstar [the television satellite] in itself is an amazing achievement. But it's used to bring us a play by Terence Rattigan. Similarly, the cinema is more interesting as an achievement than the films that are shown in its theatres.'[63]

These observations by Ionesco were no less timely than his plays. Paris in the 1950s saw the last great throw of modernism, the last time high culture could be said to dominate any major civilisation. As we shall see in chapters 25 and 26, a seismic change in the structure of intellectual life was beginning to make itself felt.

'La famille Sartre' was the name given to the group of writers and intellectuals around the philosopher/novelist/playwright. This was not without irony, certainly so far as his chief companion, **Simone de Beauvoir**, was concerned, for by the late 1940s their ménage was fairly complicated. The couple had met in 1929, at the Lycée Janson de Sailly, where de Beauvoir took courses to become a trainee teacher (together with Maurice Merleau-Ponty and Claude Lévi-Strauss). She easily attracted attention to herself by virtue of her exceptional cleverness, so that she was eventually accepted into the elite intellectual *bande* at the school, led by Sartre. This began the long-term and somewhat unusual relationship between these two – unusual in that no sooner had they begun their affair than Sartre told de Beauvoir that he was not attracted to her in bed. This was less than flattering, but she adjusted to the situation and always considered herself his main companion, even to the extent of helping him to procure other lovers, as well as acting as his chief spokesperson after he developed his theory of existentialism.[1] For his part, Sartre was generous, supporting de Beauvoir financially (as he did several others) when his early novels and plays proved successful. There was no secret about their relationship, and de Beauvoir did not lack admirers. She became the object of a powerful lesbian passion from the writer Violette le Duc.[2]

Sartre and de Beauvoir were always irked by the fact that the world viewed them as existentialists – and only as existentialists. But on occasion it paid off. In spring 1947, de Beauvoir left France for America for a coast-to-coast lecture tour where she was billed as 'France's No. 2 existentialist.' While in Chicago she met **Nelson Algren**, a writer who insisted on showing her what he called 'the real America' beyond the obvious tourist traps. They became lovers immediately (they only had two days together), and she had, she later admitted, achieved her 'first complete orgasm' (at the age of thirty-nine).[3] With him, she said, she learned 'how truly passionate love could be between men and women.' Despite her dislike of America (a feeling she shared with Sartre), she considered not returning to France. As it was, when she did return, it was as a different woman. Until then she had been rather frumpy (Sartre called her 'Castor,' meaning Beaver, and others called her La Grande Sartreuse). But she was not unattractive, and the experience with Algren reinforced that. At that stage nothing she had written could be called memorable (articles in *Les Temps modernes* and *All Men Are Mortal*), but she returned to France with something different in mind that had nothing to do with existentialism. The idea wasn't original to her; it had first been suggested for her by Colette Audry, a long-standing friend who had taught at the same school as de Beauvoir, in Rouen.[4] Audry was always threatening to write the book herself but knew her friend would do a better job.[5] Audry's idea was a book that investigated the situation of women in the postwar world, and after years of prevarication de Beauvoir seems to have been precipitated into the project by two factors. One was her visit to America, which had shown her the similarities – and very great differences – between women in the United States and in Europe, especially France. The second reason was her experience with Algren, which highlighted her own curious position vis-à-vis Sartre. She was in a stable relationship; they were viewed by all their friends and colleagues as 'a couple' ('La Grande Sartreuse' was very revealing); yet they weren't married, didn't have sex, and she was supported by him financially. This 'marginal' position, which distanced her from the situation 'normal' women found themselves in, gave de Beauvoir a vantage point that, she felt, would help her write about her sex with objectivity and sympathy. 'One day I wanted to explain myself to myself. I began to reflect all about myself and it struck me with a sort of surprise that the first thing I had to say was "I am a woman." ' At the same time, she was reflecting something more general: 1947 was the year women got the vote in France, and her book appeared at almost exactly the time Alfred Kinsey produced his first report on sex in the human male. No doubt the war had something to do with the changed conditions between men and women. De Beauvoir began her research in October 1946 and finished in June 1949, spending four months in America in 1947.[6] She then went back to *la famille Sartre*, the work a one-off, at a distance from her other offerings and, in a sense, from her. Years later a critic said that she understood the feminine condition because she herself had escaped it, and she agreed with him.[7]

De Beauvoir relied on her own experience, supported by wide reading, and she also carried out a series of interviews with total strangers. The book is in two parts – the French edition was published in two volumes. Book 1, called *Facts and Myths*, provides an historical overview of women and is itself divided into three. In 'Destiny,' the female predicament is examined from a biological, psychoanalytic, and historical standpoint. In the historical section women are described, for example, in the Middle Ages, in primitive societies, and in the Enlightenment, and she closes the section with an account of present-day women. In the section on myth she examines the treatment of women in five (male) authors: Henri de Montherlant, D. H. Lawrence, Paul Claudel, André Breton, and Stendhal. She did not like Lawrence, believing his stories to be 'tedious,' though she conceded that 'he writes the simple truth about love.' On the other hand, she felt that Stendhal was 'the greatest French novelist.' The second volume, or book 2, is called *Women's Life Today* and explores childhood, adolescence, maturity, and old age.[8] She writes of love, sex, marriage, lesbianism. She made use of her impressive gallery of friends and acquaintances, spending several mornings with Lévi-Strauss discussing anthropology and with Jacques Lacan learning about psychoanalysis.[9] Algren's influence is as much evident in the book as Sartre's. It was the American who had suggested she also look at black women in a prejudicial society and introduced her not only to black Americans but to the literature on race, including Gunnar Myrdal's *An American Dilemma*. Initially she thought of calling her book *The Other Sex*; the title used, *The Second Sex*, was suggested by Jacques-Laurent Bost, one of the *premiers disciples* of Sartre, during an evening's drinking in a Left Bank café.[10]

When *The Second Sex* appeared, there were those critics (as there are always those critics) who complained that she didn't say anything new. But there were many more who felt she had put her finger on something that other people, other women, were working out for themselves at that time, and moreover that, in doing her research, she had provided them with ammunition: 'She had provided a generation of women with a voice.'[11] The book was translated into English very early, thanks to **Blanche Knopf** – wife of the publisher Alfred – whose attention had been drawn to the book by the Gallimard family when she was on a visit to Paris. Conscious of the great interest among American students in the bohemia of the Left Bank at the time, both Blanche and Alfred believed the book was bound to be a sound commercial proposition. They were right. When the book was released in America in February 1953, it was by and large well received, though there were several reviewers – Stevie Smith and Charles Rollo among them – who didn't like her tone, who thought she 'carried the feminist grievance too far.'[12] The most interesting reaction, however, was that of the editors of the *Saturday Review of Literature*, who believed the book's theme was too large for one reviewer and so commissioned six, among them the psychiatrist Karl Menninger, Margaret Mead, and another anthropologist, Ashley Montagu. Mead found the book's central argument – that society has wasted women's gifts – a sound one but added that de Beauvoir had violated every canon of science in her partisan selection of material. Above all, however, de Beauvoir's book was taken seriously, which meant that the issues it raised were considered soberly, something that had not always happened. De Beauvoir's strange idea that women represented 'the other' in society caught on, and would much infuse the feminine movement in years to come. Brendan Gill, in a review entitled 'No More Eve' in the *New Yorker*, summed up his reaction in a way others have noted: 'What we are faced with is more than a work of scholarship; it is a work of art, with the salt of recklessness that makes art sting.'[13]

When Blanche Knopf had first come across *The Second Sex*, on her visit to Paris, her appetite had been whetted on being told that it read like 'a cross between Havelock Ellis and the Kinsey Report.'[14] Havelock Ellis was old news; *Studies in the Psychology of Sex*, begun in 1897, had ceased publication as long ago as 1928, and he had been dead since 1939. The Kinsey Report, however, was new. Like *The Second Sex*, *Sexual Behaviour in the Human Male* reflected a changed, postwar world.

The generation that came back from World War II settled down almost immediately. They took opportunities to be educated, they got married – and then proceeded to have more children than their parents' generation: this was the baby boom. But they had seen life; they knew its attractions, and they knew its shadows. Living in close proximity to others, often in conditions of great danger, they had known intimacy as few people experience it. So they were particularly aware that there was a marked gap between the way people were supposed to behave and the way they did behave. And perhaps this gap was greatest in one area: sex. Of course, sex happened before World War II, but it wasn't talked about to anywhere near the same extent. When the Lynds carried out their study of Middletown in the 1920s, they had looked at marriage and dating, but not at sex per se. In fact, though, they had chronicled the one important social change that was to alter behaviour in this regard in the 1930s more than anything else: the motor car. The car took adolescents out of the home and away from parental supervision. It took adolescents to meeting places with their friends, as often as not the movie houses where Hollywood was selling the idea of romance. Most important of all, the car provided an alternative venue, a private area where intimate behaviour could take place. All of which meant that, by the late 1940s, behaviour had changed, but public perceptions of that behaviour had not kept up. It is this which mainly accounts for the unprecedented reception of a dry, 804-page academic report that appeared in 1948 under the title *Sexual Behaviour in the Human Male*. The author was a professor of zoology at the University of Indiana (not so far from Muncie).[15] The medical publisher who released the book printed an initial run of 5,000 copies but before long realised his error.[16] Nearly a quarter of a million copies were eventually sold, and the book spent twenty-seven weeks on the *New York Times* best-seller list. **Alfred Kinsey**, the professor of zoology, became famous and appeared on the cover of *Time* magazine.[17]

The scientific tone of the book clearly helped. Its elaborate charts and graphs, methodological discussions of the interviewing process, and consideration of the validity of 'the data', set it apart from pornography and allowed people to discuss sex in detail without appearing prurient or salacious. Moreover, Kinsey was an unlikely figure to spark such controversy. He had built his reputation on the study of wasps. His interest in human sexuality had begun when he had taught a course on marriage and the family in the late 1930s. He found students hungry for 'accurate, unbiased information about sex,' and indeed, being a scientist, Kinsey was dismayed by the dearth of 'reliable, non-moralistic data' concerning human sexual behaviour.[18] He therefore began to amass his own statistics by recording the sexual practices of students. He subsequently put together a small team of researchers and trained them in interviewing techniques, which meant they could explore a subject's sex life in about two hours. Over ten years he collected material on 18,000 men and women.[19]

In their study *Sexuality in America*, John d'Emilio and Estelle Freedman say, 'Behind the scientific prose of *Sexual Behaviour in the Human Male* lay the most elaborate description of the sexual habits of ordinary white Americans (or anyone, for that matter) ever assembled. In great detail, Kinsey tabulated the frequency and incidence of masturbation, premarital petting and coitus, marital intercourse, extramarital sex, homosexuality, and animal contacts. Avoiding as far as possible the moralistic tone he disliked in other works, Kinsey adopted a "count-and-catalogue" stance: how many respondents had done what, how many times and at what ages. His findings proved shocking to traditional moralists.[20] His study of the male revealed, for example, that masturbation and heterosexual petting were 'nearly universal, that almost nine out of ten men had sex before marriage, that half had affairs, and that over a third of adult males had had at least one homosexual experience.' Virtually all males had established a regular sexual outlet by the age of fifteen, and 'fully 95 per cent had violated the law at least once on the way to orgasm.'[21] A second volume in the series, *Sexual Behaviour in the Human Female*, was published in 1953 and caused a similar storm. Although the figures for women were lower (and less shocking) than for men, six out of ten had engaged in masturbation, half had had sex before marriage, and a quarter had had affairs.[22] Taken together, Kinsey's statistics pointed to a vast hidden world of sexual experience sharply at odds with publicly espoused norms. The reports became cultural landmarks.[23] But perhaps the most interesting reaction was the public's. In general there was no shock/horror reaction from middle America. Instead, opinion polls suggested that a large majority of the public approved of scientific research on sexuality and were eager to learn more. Undoubtedly, the revelation of a wide divergence between ideals and actual behaviour alleviated the anxiety of many individuals as to whether their own private behaviour set them apart from others.

In retrospect three of Kinsey's findings were to have sustained social, psychological, and intellectual effects, for good or ill. The first was the finding that many – most, when considering males and females – indulged in extramarital affairs. A decade after the studies were published, as we shall see, people began to act on this finding: whereas hitherto people had just had affairs, now they didn't stop there, and divorced in order to remarry. The second was the finding that there was a 'distinct and steady increase in the number of females reaching orgasm in their marital coitus.'[24] Looking at the age of the women in his sample, Kinsey found that most of the women born at the end of the nineteenth century had never reached orgasm (remember Simone de Beauvoir not achieving it until she was thirty-nine), whereas among those born in the 1920s, most of them 'always achieved it [orgasm] during coitus.' Although Kinsey was unwilling to equate female orgasm with a happy sex life, publication of his findings, and the solid size of his sample, clearly encouraged more women who were not achieving orgasm to seek to do so. This was by no means the only concern of the women's movement, which gathered pace in the decade following Kinsey, but it was a contributing element. The third important finding that proved of lasting significance was that which showed a much higher proportion of homosexual activity than had been anticipated – a third of adult men, it will be recalled, reported such experiences.[25] Here again Kinsey's report seems to have shown a large number of people that the behaviour they thought set them apart – made them odd and unusual – was in fact far more common than they could ever have known.[26] In doing so, the Kinsey reports not only allayed anxieties but may have encouraged more of such behaviour in the years that followed.

Kinsey's immediate successor was a balding, well-tanned obstetrician-gynae-cologist based at the Washington University Medical School in Saint Louis, Missouri, named **William Howell Masters**, born in Cleveland, Ohio, the son of well-to-do parents. Bill Masters's approach to sex research was very different from Kinsey's. Whereas Kinsey was interested in survey research, Masters was above all a biologist, a doctor, interested in the physiology of the orgasm and orgasmic dysfunction in particular, in order to discover how sexual physiology might affect infertile couples and what could be done to help them.[27]

Masters had been interested in sex research since 1941, when he had worked with Dr George Washington Corner at the Carnegie Institute of Experimental Embryology in Baltimore. Corner, the mentor of Alfred Kinsey as well of Masters, later discovered progesterone, one of the two female sex hormones.[28] Masters carefully prepared himself for his career in sex research – he knew that he was playing with fire and needed to be, as it were, 'above suspicion' professionally before he even began in the area. Throughout the 1940s he collected degrees and academic qualifications and published solid research on steroid replacement and the correct dosages for men and women. He also got married. In 1953, after both Kinsey reports had been published, he finally approached the board of trustees at his own university to request that he might study human sexual behaviour. The university was not enthusiastic, but Kinsey had established a precedent, and on grounds of academic freedom Masters was given the go-ahead a year afterward. Early on, he established that there were few books to which he could turn, and so he was soon back before the university chancellor requesting permission to mount his own first piece of research, a one-year study of prostitutes (as people who knew something about sex). Again he was given the go-ahead, but only on condition that he worked with a review board that consisted of the local commissioner of police, the head of the local Catholic archdiocese, and the publisher of the local news-paper.[29] Gaining their approval, Masters spent eighteen months working with both male and female prostitutes in brothels in the Midwest, West Coast, Canada, and Mexico, investigating a variety of sexual experiences, 'including all known variations of intercourse, oral sex, anal sex, and an assortment of fetishes.'[30] He asked the prostitutes how their sex organs behaved during intercourse, and what they had observed about orgasm. In the next phase of his research, and in the greatest secrecy, Masters opened a three-room clinic on the top floor of a maternity hospital associated with the university. Apart from the office, the two back rooms were separated by a one-way mirror through which, in time, Masters filmed 382 women and 312 men having sex, providing footage of 10,000 orgasms.[31]

As his researches continued, Masters realised he needed a female partner, the better to understand female sexual physiology and to ask the right questions. And so, in January 1957 he was joined by **Virginia Johnson**, a singer who had no degree, which Masters believed might help her frame different questions from himself. She became just as dedicated to 'the cause' as he was, and together they devised many new pieces of equipment with which to further their research; for example, there was one for measuring blood-volume changes in the penis, and 'a nine-inch-long clear Lucite phallus with a ray of cold light emanating from its glans so that the camera lens inside the shaft' could inspect the vaginal walls for what that might reveal about female orgasm. At that stage the main mystery of sex was felt to be the difference in women – promulgated by Freud, among others – between the clitoral and the vaginal orgasm.[32] Kinsey had come out against such a distinction, and Masters and Johnson, too, never found any evidence for Freud's theory. One of their first findings, however, was confirmation that whereas the penis was capable of only one orgasm at a time, with a refractory period in between, the clitoris was capable of repeated climax. This was an important development, 'on an almost Copernican scale,' in John Heidenry's words, for it had consequences both for female psychology (sexual fulfilment was no longer modelled on that of the male) and in sex therapy.[33] In a highly contentious area, Masters and Johnson's most controversial innovation was the employment of surrogates. Prostitutes were used at first – they were available and experienced – but this provoked objections from senior figures in the university, and so they advertised for female volunteers from among students.

As they developed their studies, and techniques of therapy, some of the early results were published in professional journals such as *Obstetrics and Gynecology* – they planned a large book later on. In November 1964, however, the secrecy they had maintained for a decade was blown away when they were attacked in the pages of *Commentary* by Leslie Farber, a psychoanalyst who sniggered in print, questioning their motives.[34] Their response was to bring forward publication of *Human Sexual Response* to April 1966. The book was deliberately written in a nonsensationalist, even leaden, prose, but that proved no barrier; the first printing sold out in a week, and eventually sales topped 300,000.[35] Fortunately for them, the *Journal of the American Medical Association* pronounced their work worthwhile, and so most of the mainstream press treated their findings with respect. The long-term importance of Masters and Johnson, coming on top of Kinsey, was that they brought out into the open the discussion of sexual matters, throwing light on to an area where there had been darkness and ignorance before. Many people objected to this change on principle, but not those who had suffered some form of sexual dysfunction and misery for years. Masters and Johnson found, for example, that some 80 percent of couples who sought treatment for sexual dysfunction responded almost immediately, and although there were relapses, the progress of many was excellent. They also found that secondary impotence in men – caused by alcohol, fatigue, or tension – was easily treated, and that one of the effects of pornography was to give people exaggerated expectations of what they might achieve in the sex act. Far from *being* pornography, *Human Sexual Response* put pornography in its place.

The Second Sex, the Kinsey reports, and *Human Sexual Response* all helped change attitudes. But they were themselves also the fruit of an attitude change that was already taking place. In Britain, this change was particularly acute owing to the war. During wartime, for example, there was in Britain a marked rise in illegitimate births, from 11.8 percent in 1942 to 14.9 percent in 1945.[36] At the same time, a shortage of rubber meant that sheaths (as condoms were then called) and caps were both in short supply and substandard. Simultaneously, the main problem in the Family Planning Association was *sub*fertility. There was so much concern that, in 1943, Prime Minister Winston Churchill broadcast to the nation about the need to encourage 'our people ... by every means to have large families.' This worry eventually led to the appointment, in 1944, of a **Royal Commission on Population**. This did not report until 1949, by which time concerns – and behaviour – had changed. The commission found, for example, that in fact, after falling continuously for half a century, family size in Britain had been comparatively stable for twenty years, at about 2.2 children per married couple, which meant a slow population increase over time.[37] But it also became clear to the commission that although central government did not seem concerned about birth control (there was no provision in the new National Health Service, for example, for family planning clinics), the population at large, especially women, did take the matter very seriously indeed; they well understood the link between numbers of children and the standard of living, and they had accordingly extended their knowledge of contraception. This was yet another area of sexual behaviour where there had been many private initiatives, though no one was aware of the wider picture. In particular, the commission concluded that 'the permeation of the small family system through nearly all classes had to be regarded as a fundamental adjustment to modern conditions in which the most significant feature was the gradual acceptance of control over the size of one's family, particularly by means of contraception, as a normal part of personal responsibility.'[38]

Artificial contraception was an issue that split the church. The Anglican Church voted to approve it in 1918, but the Roman Catholic Church has not done so yet. So it is an especially poignant fact that **Dr John Rock**, the chief of obstetrics and gynaecology at Harvard Medical School and the man who, in 1944, became the first scientist to fertilise a human egg in a test tube and was one of the first to freeze a human sperm for up to a year without impairing its potency, was a Catholic. His initial aim was to effect the opposite of contraception, and help infertile women conceive.[39] Rock believed that administering the female hormones progesterone and oestrogen might stimulate conception but also stabilise the menstrual cycle, enabling all religious couples to use the theologically sound 'rhythm method.'[40] Unfortunately the action of these hormones was only partly understood – progesterone, for example, worked because it inhibited ovulation, but exactly how was not clear. But what Rock did notice was that when he administered progesterone to a number of so-called infertile women, although the progesterone didn't appear to work at first, a substantial number became pregnant as soon as the treatment was stopped.[41] Enlisting the aid of **Dr Gregory Pincus**, a Harvard biologist also interested in infertility, he eventually established that a combination of oestrogen and progesterone suppressed gonadotrophic activity and consequently prevented ovulation. Conception therefore could be prevented by taking the chemicals on the right days, so that the normal process of menstruation was interfered with. In 1956 the first clinical trials were organised by Rock and Pincus among two hundred women in Puerto Rico, since birth control was still unlawful in Massachusetts.[42] When the nature of his work became known, there were attempts to have Rock excommunicated, but in 1957 the Food and Drug Administration in the United States approved the Rock-Pincus pill for treating women with menstrual disorders. Another trial followed, this time with a sample of nearly nine hundred women, the results of which were so promising that on 10 May 1960 the FDA sanctioned the use of **Enovid**, a birth-control pill manufactured by G. D. Searle & Co. in Chicago.[43] The development rated two inches in the *New York Times*, but it was enough: by the end of 1961 some 400,000 American women were taking the pill, and that number doubled the next year and the year after that. By 1966 six million American women were on the pill, and the same number across the rest of the world.[44] Some idea of the immediate success of the pill can be had from the British statistics. (Britain had a long tradition of family planning, with well-informed and proselytising volunteers, a residue of the benign end of the eugenics movement in the early years of the century. This made its statistics excellent.) In 1960, in Family Planning Association clinics, 97.5 percent of new birth control clients were advised to use the cap (the pill wasn't available in Britain until 1961); by 1975, 58 percent were advised to use the pill.[45] What the research into sexual statistics showed above all was that public perceptions of intimate behaviour were, by and large, wrong, outdated. People had been changing privately, silently, in countless small ways that nonetheless added up to a sexual revolution. This is why de Beauvoir, Kinsey, and Masters and Johnson had sold so well; there was the thrill of recognition among the hundreds of thousands who bought their books.

Publishers and writers could read the signs, too. The 1950s saw several works of literature that were far franker about sexual matters than ever before. These titles included Vladimir Nabokov's *Lolita* (1953), J. P. Donleavy's *Ginger Man* and Françoise Sagan's *Bonjour Tristesse* (both 1955), William Burroughs's *Naked Lunch* (1959), and Allen Ginsberg's 1956 poem *Howl. Howl* and D. H. Lawrence's *Lady Chatterley's Lover*, the latter available in France since 1929, both became the subject of celebrated obscenity trials, in the United Kingdom and in the United States, in 1959; both eventually escaped censorship on the grounds that they had redeeming artistic merit. Curiously, Nabokov's *Lolita* avoided the courthouse, perhaps because he did not use such explicit obscenities as the other authors did. But in some ways his theme, the love of a middle-aged man for an underage 'nymphet,' was the most 'perverse' of all.

But then **Nabokov** was an extraordinary man. Born in Saint Petersburg into an aristocratic family who had lost everything in the revolution, he was educated at Cambridge, then lived in Germany and France until he settled in America in 1941. As well as writing equally vividly in Russian and English, he was a passionate chess player and a recognised authority on butterflies.[46] *Lolita* is by turns funny, sad, pathetic. It is a story as much about age as sex, about the sorrow that comes with knowledge, the difference between biological sex and psychological sex, about the difference between sex and love and passion and about how love can be a wound, imprisoning rather than liberating. Lolita is the butterfly, beautiful, delicate, with a primitive life force that an older man can only envy, but she is also vulgar, a far from idealised figure.[47] The middle-aged 'hero' loses her, of course, just as he loses everything, including his self-respect. Although Lolita realises what is happening to her, it is far from clear what, if anything, rubs off. Has the warmth in him created the coldness in her; or has it made no difference? In *Lolita* the sexes are as far apart as can be.

The final report of these years built on the earlier investigations and events to produce a definite advance. This was **Betty Friedan**'s *Feminine Mystique*, which appeared in 1963. After graduating from Smith College, Friedan (née Goldstein) lived in Greenwich Village in New York, working as a reporter. In 1947 she married Carl Friedan, moving soon after to the suburbs, where Betty became a full-time mother, ferrying her children to school each day. She liked mother-hood well enough, but she also wanted a career and again took up journalism. Or she tried to. Her fifteenth college reunion came round in 1957, and she decided to write an article about it for *McCall's* magazine, using a questionnaire she had devised as the basis for the information.[48] The questions she asked chiefly concerned her classmates' reactions to being a woman and the way their sex, or gender, had affected their lives. She found that 'an overwhelming number of women felt unfulfilled and isolated, envying their husbands who had other lives, friends, colleagues, and challenges away from home.'

But *McCall's* turned her article down: 'The male editor said it couldn't be true.' She took it back and submitted the same piece to *Ladies' Home Journal*. They rewrote the article so it said the opposite of what she meant. Next she tried *Redbook*. There the editor told her agent, 'Betty has gone off her rocker.'[49] He thought only 'neurotic' women would identify with what she was saying. Belatedly, Friedan realised that what she had written 'threatened the very raison d'être of the women's magazine world,' and she then decided to expand what she had discovered about women into a book.[50] To begin with this had the title *The Togetherness Woman*, later changed to *The Feminine Mystique*. By the feminine mystique, Friedan meant the general assumption that women liked being housewives and mothers at home, having no interest in wider social, political, or intellectual matters, nor feeling a need for a career. She was surprised to find that it had not always been so, that the very magazines that had turned down her articles had, until World War II, printed very different material. 'In 1939 the heroines of women's magazine stories were not always young, but in a certain sense they were younger than their fictional counterparts today. . . . The majority of heroines in the four major women's magazines (then *Ladies' Home Journal*, *McCall's*, *Good Housekeeping*, and *Women's Home Companion*) were career women. . . . And the spirit, courage, independence, determination – the strength of char-acter they showed in their work as nurses, teachers, artists, actresses, copywriters, saleswomen – were part of their charm. There was a definite aura that their individuality was something to be admired, not unattractive to men, that men were drawn to them as much for their spirit and character as for their looks.'[51]

The war had changed all that, she felt. Going away to war had been supremely fulfilling for a whole generation of men, but they had returned to the 'little women' waiting at home, often raising a family deliberately conceived before the man went away. These men returned to good jobs or, via the GI bill, good educational opportunities, and a new pattern had been set, not helped by the flight to the suburbs, which had only made women's isolation more acute. By 1960, however, Friedan said that women's frustration was boiling over; anger and neuroses were at an unprecedented level, if the results of the questionnaire she had sent out were to be believed. But part of the problem was that it had no name; that's where her book came in. The problem with no name became *The Feminine Mystique.*

Friedan's attack was wide-ranging and extensively researched, her anger (for the book was a polemical but calmly marshalled thesis) directed not just at women's magazines and Madison Avenue, for portraying women as members of a 'comfortable concentration camp,' surrounded by the latest washing machines, vacuum cleaners, and other labour-saving devices, but also at Freud, Margaret Mead, and the universities for making women try to conform to some stereo-typical ideal.[52] Freud's theory of penis envy, she thought, was an outmoded way of trying to say that women were inferior, and there was no credible evidence for it. She argued that Mead's anthropological studies, although describing differences between women of differing cultures, still offered an ideal of womanhood that was essentially passive, again conforming to stereotypes. She made the telling point that Mead's own life – a career, two husbands, a lesbian lover, an open marriage – was completely at variance with what she described in her writings, and a much better model for the modern Western woman.[53] But Friedan's study was also one of the first popular works to draw attention to the all-important nuts-and-bolts of womanhood. She explored how many women got married in their teens, as a result of which their careers and intellectual lives went nowhere; she wondered how many supported their husbands in a 'qualification' – she ironically called it the Ph.T. (putting husband through [college]).[54] And she was one of the first to draw attention to the fact that, as a result of these demanding circumstances, it was always the mother who ended up battering and abusing her children.

Friedan's book hit a nerve, not just in its mammoth sales, but also in that it helped spark the President's Commission on the Status of Women. This commission's report, when it appeared in 1965, detailed the discriminatory wages women were earning (half the average for men) and the declining ratio of women in professional and executive jobs. When the report was buried in the Washington bureaucracy, a group of women decided they had to take things into their own hands. Betty Friedan was one of those who met in Washington to create what someone at the meeting called 'an NAACP for women.'[55] The acronym eventually became NOW, the **National Organization of Women**. The modern feminist movement had begun.[56]

THE NEW HUMAN CONDITION

Part of the message of the Kinsey reports, and of Betty Friedan's investigation, was that Western society was changing in the wake of war, and in some fairly fundamental ways. America was in the forefront here, but the changes applied in other countries as well, if less strongly. Before the war, anthropology had been the social science that, thanks to Franz Boas, Ruth Benedict, and Margaret Mead, most caught the imagination, certainly so far as the general public was concerned. Now, however, the changes within Western society came under the spotlight from the other social sciences, in particular sociology, psychology, and economics.

The first of these investigations to make an impact was *The Lonely Crowd*, published in 1950 by the Harvard sociologist **David Riesman** (who later moved to Stanford). Riesman began by stressing what sociology had to offer over and above anthropology. Compared with sociology, he said, anthropology was 'poor.' That is to say, it was not a big discipline, and many of its field studies were little more than one-man (or one-woman) expeditions, because funds were unavailable for more ambitious projects. As a result, fieldwork in anthropology was amateurish and, more important, 'inclined to holistic over-generalisation from a general paucity of data.' By contrast, public opinion surveys – the bread-and-butter material of sociologists, which had become more plentiful since the inception of **Gallup** in the mid-1930s and their widespread use during World War II to gauge public feeling, aided by advances in statistics for the manipulation of data – were rich both in quantitative terms, in the level of detail they amassed, and in the representativeness of their samples. In addition to survey data, Riesman also added the study of such things as advertisements, dreams, children's games, and child-rearing practices, all of which, he claimed, had now become 'the stuff of history.' He and his colleagues therefore felt able to deliver verdicts on the national character of Americans with a certainty that anthropologists could not match. (He was later to regret his overconfident tone, especially when he was forced to retract some of his generalisations.)[1]

Riesman was a pupil of Erich Fromm, and therefore indirectly in the tradition of the Frankfurt School. Like them, his ideas owed a lot to Freud, and to Max Weber, insofar as *The Lonely Crowd* was an attempt to relate individual psychology, and that of the family, to whole societies. His argument was twofold. In the first place, he claimed that as societies develop, they go through three phases relating to changes in population. In older societies, where there is a stable population at fairly low levels, people are 'tradition-directed.' In the second phase, populations show a rapid increase in size, and individuals become 'inner-directed.' In the third phase, populations level off at a much higher level, where the people are 'other-directed.' The second part of his argument described how the factors that shape character change as these other developments take place. In particular, he saw a decline in the influence and authority of parents and home life, and a rise in the influence of the mass media and the peer group, especially as it concerned the lives of young people.[2]

By the middle of the twentieth century, Riesman said, countries such as India, Egypt, and China remained tradition-directed. These locations are in many areas sparsely populated, death rates are high, and very often the people are nonliterate. Here life is governed by patterns and an etiquette of relationships that have existed for generations. Youth is regarded as an obvious period of apprenticeship, and admission to adult society is marked by initiation ceremonies that are formal and which everyone must go through. These ceremonies bring on added privilege but also added responsibility. The 'Three Rs' of this world are ritual, routine, and religion, with 'Little energy ... directed towards finding new solutions to age-old problems.'[3] Riesman did not devote any space to how tradition-oriented societies develop or evolve, but he saw the next phase as clearly marked and predicated upon a rapid increase in population, which creates a change in the relatively stable ratio of births to deaths, which in turn becomes both the cause and consequence of other social changes. It is this imbalance that puts pressure on society's customary ways of coping. The new society is characterised by increased personal mobility, by the rapid accumulation of capital, and by an almost constant expansion. Such a society (for example, the Renaissance or the Reformation), Riesman says, breeds character types 'who can manage to live socially without strict and self-evident tradition-direction.' The concept of 'inner-direction' covers a wide range of individuals, but all share the experience that the values that govern their lives and behaviour are implanted early in life by their elders, leading to a distinct individualism marked by a consistency within the individual from one situation to another. Inner-directed people are aware of tradition, or rather traditions, but each individual may come from a different tradition to which he or she owes allegiance. It is as if, says Riesman, each person has his own 'internal gyroscope.' The classic inner-directed society is Victorian Britain.[4]

As the birth rate begins to follow the death rate down, populations start to stabilise again, but at higher levels than before. Fewer people work on the land, more are in the cities, there is more abundance and leisure, societies are centralised and bureaucratised, and increasingly, '*other people* are the problem, not the material environment.'[5] People mix more widely and become more sensitive to each other. This society creates the other-directed person. Riesman thought that the other-directed type was most common and most at home in twentieth-century America, which lacked a feudal past, and especially in American cities, where people were literate, educated, and well provided for in the necessities of life.[6] Amid the new abundance, he thought that parental discipline suffered, because in the new, smaller, more biologically stable families it was needed less, and this had two consequences. First, the peer group becomes as important as, if not more important than, the family as a socialising influence – the peer group meaning other children the same age as the child in question. Second, the children in society become a marketing category; they are targeted by both the manufacturers of children's products and the media that help sell these products. It is this need for direction from, and the approval of, others that creates a modern form of conformity in which the chief area of sensitivity is wanting to be liked by other people – i.e., to be popular.[7] This new other-directed group, he said, is more interested in its own psychological development than in work for personal gain, or the greater good of all; it does not want to be esteemed but loved; and its most important aim is to 'relate' to others.

Riesman went on to qualify and expand this picture, devoting chapters to the changing role of parents, teachers, the print media, the electronic media, the role of economics, and the changing character of work. He thought that the changes he had observed and described had implications for privacy and for politics, and that whatever character type an individual was, there were three fates available – adjustment, anomie, and autonomy.[8] Later he recanted some of his claims, conceding he had overstated the change that had come over America. But in one thing he was surely right: his observation that Americans were concerned above all with 'relationships' foreshadowed the obsession later in the century with all manner of psychologies specifically designed to help in this area of life.

The Lonely Crowd was released in the same year that Senator Joseph McCarthy announced to the Women's Republican Club in Wheeling, West Virginia, that 'I hold in my hand' a list of Communist agents in the State Department. Until that point, McCarthy had been an undistinguished Midwestern politician with a drinking problem.[9] But his specific allegations now sparked a 'moral panic' in America, as it was described, in which 151 actors, writers, musicians, and radio and TV entertainers were accused of Communist affiliations, and the U.S. attorney general issued a list of 179 'Totalitarian, Fascist, Communist, subversive and other organisations.'* While McCarthy and the U.S. attorney general were worrying about Communists and 'subversives,' others were just as distressed about the whole moral panic itself and what that said about America. In fact, many people – especially refugee scholars from Europe – were by now worried that America itself had the potential to become fascist. It was thinking of this kind that underlay a particular psychological investigation that overlapped with *The Lonely Crowd* and appeared at more or less the same time.

The Authoritarian Personality had been conceived as early as 1939 as part of a

---

* Names included Leonard Bernstein, Lee J. Cobb, Aaron Copland, José Ferrer, Lillian Hellman, Langston Hughes, Burl Ives, Gypsy Rose Lee, Arthur Miller, Zero Mostel, Dorothy Parker, Artie Shaw, Irwin Shaw, William L. Shirer, Sam Wanamaker, and Orson Welles.

joint project, with the Berkeley Public Opinion Study and the American Jewish Committee, to investigate anti-Semitism.[10] The idea was for a questionnaire survey to explore whether a psychological profile of the 'potential fascist character' could be identified. It was the first time that the critical school of Frankfurt had used a quantitative approach, and the results of their **'F' (for fascist) scale** 'seemed to warrant alarm.'[11] 'Anti-Semitism turned out to be ... the visible edge of a dysfunctional personality revealed in the many "ethno-centric" and "conventional" attitudes of the general American population, as well as of a disquietingly submissive attitude towards authority of all kinds.'[12] This is where the link to Riesman came in: these potential fascists were 'other-directed,' normal, conventional Americans. *The Authoritarian Personality* therefore concluded with a warning that fascism rather than communism was the chief threat facing America in the postwar world, that fascism was finding 'a new home' on the western side of the Atlantic, and that bourgeois America and its great cities were now 'the dark heart of modern civilisation.'[13] The book's other conclusion was that the Holocaust was not simply the result of Nazi thinking, and its specific theories about degeneration, but that the rationality of Western capitalist civilisation itself was responsible. Theodor Adorno, the exile from Frankfurt and the main author of the report, found that whereas left-wing types were emotionally more stable, usually happier than their conservative counterparts, capitalism tended to throw up dysfunctional personalities, highly authoritarian anti-Semites who linked reason to power. For them, the pogrom was the ultimate expression of this power.[14] If *The Lonely Crowd* may be seen as an early effort to combine public opinion survey material with social psychology and sociology to understand whole nations, a rational – if not entirely suc-cessful – project to assimilate new forms of knowledge, *The Authoritarian Personality* is best understood as a late throw of the Germanic tradition of Freud and Spengler, yet another overarching attempt to denigrate the Western/Atlantic alliance of rationalism, science, and democracy. It was an arresting thesis, especially when read against the backdrop of the McCarthy shenanigans. But in fact it was immediately attacked by fellow social scientists, who systematically and ruthlessly disassembled its findings. By then, however, the unsubstantiated phrase 'the authoritarian personality' had caught on.

A better picture of totalitarianism, both as to its origins and its possible expression in the postwar world (especially America), was given by **Hannah Arendt**. She had been in New York since 1941, after she escaped from France. In Manhattan she had lived in poverty for a time, learned English, and begun to write, moving among the intellectuals of the *Partisan Review* milieu. At various times she was a professor at Princeton, Chicago, and the University of California as well as being a regular contributor to the *New Yorker*. She finally settled at the New School for Social Research in New York, where she taught until she died in 1975.[15] As home to the University in Exile, for emigré European intellectuals fleeing fascism in the 1930s, one aim of the New School was to develop an amalgam of European and American thought. Arendt made a name for herself with three influential – and highly controversial – books: *The Origins of Totalitarianism* (1951), *The Human Condition* (1958), and *Eichmann* *in Jerusalem* (1963).[16] She began *The Origins of Totalitarianism* after the war ended, and it took several years.[17] Her main aim was to explain why so 'unimportant' a matter in world politics as 'the Jewish question,' or anti-Semitism, could become the 'catalytic agent for, first, the Nazi movement, then a world war, and finally the establishment of the death factories.'[18] Her answer was that mass society led to isolation and loneliness – the lonely crowd of Riesman's title. In such a condition, she realised, normal political life deteriorated, fascism and communism drew their remarkable strength, offering a form of politics that provided people with a public life: uniforms, denoting belonging; specific ranks, recognised and respected by others; massed rallies, the experience of participation.[19] That was the positive side. At the same time, 'loneliness' she identified as 'the common ground for terror, the essence of totalitarian government.'[20] And this is where the controversy started, for although she equated Stalinism with Nazism and left many thinking that there was therefore no alternative to the emerging American way of life, she still implied that the 'massification' of society was 'a step towards totalitarianism', towards 'radical evil,' a key phrase, and that 'the new mass society in the West was in danger of converging with the totalitarian East.'[21]

In *The Human Condition* Arendt tried to offer some solutions for the problems she had identified in her earlier book.[22] The essential difficulty with modern society, she felt, was that modern man felt alienated politically (as opposed to psychologically). The ordinary individual did not have access to the inside information that the political elite had, there was bureaucracy everywhere, one man, one vote didn't mean that much, and such predicaments were all much more important now because, with the growth of huge corporations, individuals had less control over their work; there was less craftwork to offer satisfaction, and less control over income. Man was left alone but knew he couldn't act, live, alone.[23] Her solution, as Elisabeth Young-Bruehl, her biographer, has said, was ahead of its time; Arendt thought that society would evolve what she called the personalisation of politics – what we now call single-issue politics (the environment, feminism, genetically modified foods).[24] In this way, she said, people could become as informed as the experts, they could attempt to control their own lives, and they could have an effect. Arendt was right about the personalisation of politics: later in the century it would become an important element in collective life.

Like Hannah Arendt, **Erich Fromm** was German and Jewish. A member of the Frankfurt School, he had emigrated with the other members of the school in 1934 and sailed for America, continuing as an affiliate of the Frankfurt Institute for Social Research, attached to Columbia University. Fromm's family had been very religious; he himself had helped found an academy of Jewish thought (with Martin Buber), and this had translated, in Frankfurt, into a project to study the formation of class-consciousness, an exploration – one of the first of its kind – into the links between psychology and politics. On the basis of more than one thousand replies to a questionnaire he sent out, Fromm found that people could not be grouped, as he had expected, into 'revolutionary' workers and 'nonrevolutionary' bourgeois. Not only were some workers conservative, and some bourgeois revolutionary, but very left-wing workers often confessed to 'strikingly non-revolutionary, authoritarian attitudes' in many areas normally regarded as nonpolitical, such as child-rearing and women's fashion.[25] It was this, as much as anything, that convinced Fromm and the others of the Frankfurt School that Marxism needed to be modified in the light of Freud.

Fromm's 1920s work was not translated into English until the 1980s, so it never had the impact that perhaps it deserved. But it shows that he had the same sort of interests as Riesman, Adorno, and Arendt. He went considerably further, in fact, with his 1955 book, *The Sane Society.*[26] Instead of just looking at the shortcomings of mass society, he examined the much more extreme idea as to whether an entire society can be considered unhealthy. To many, Fromm's central notion was so presumptuous as to be meaningless. But he tackled it head-on. He admitted to begin with that his book was an amalgam of Tawney's *The Acquisitive Society* (which, he reminded readers, had originally been called *The Sickness of an Acquisitive Society*) and Freud's *Civilisation and Its Discontents.* Fromm started with the by-now familiar statistics, that America and other Protestant countries, like Denmark, Norway, and Sweden, had higher rates of suicide, murder, violence, and drug and alcohol abuse than other areas of the world.[27] So he thought that on any measure these societies were sicker than most. The rest of his argument was a mixture of psychoanalysis, economics, sociology, and politics. The central reality, he said, was that 'whereas in the nineteenth century God was dead, in the twentieth century man is dead.'[28] The problem with capitalism, for all its strengths, and itself the result of so many freedoms, was that it had terrible consequences for mankind. In a neat phrase he said that 'work can be defined as the performance of acts which cannot yet be performed by machines.' He was putting in modern garb a familiar argument that twentieth-century work, for most people, was dehumanising, boring, and meaningless, and provoked in them a wide array of problems. Words like *anomie* and *alienation* were resurrected, but the significance of Fromm's critique lay in his claim that the constricting experience of modern work was directly related to mental health. Mass society, he wrote, turned man into a commodity; 'his value as a person lies in his saleability, not his human qualities of love, reason, or his artistic capacities.'[29] Near the end of his book Fromm stressed the role of love, which he regarded as an 'art form,' because, he said, one of the casualties of super-capitalism, as he called it, was 'man's relationship to his fellow men.' Alienating work had consequences for friendship, fairness, and trust. Riesman had said that the young were more concerned about relationships and popularity, but Fromm worried that people were becoming indifferent to others; and if everyone was a commodity, they were no different from things.[30] He made it clear that he had scoured the literature, collecting accounts of how people's lives were drying up, losing interest in the arts, say, as work became all-engrossing. For Fromm, the aim was the recovery not so much of man's sanity as his dignity, the theme of Arthur Miller's 1949 play *Death of a Salesman,* to which he made pointed reference.[31] Fromm, for all his psychoanalytic approach and his diagnosis of the postwar world as an insane

society, offered no psychological remedies. Instead, he faced frankly the fact
that the character of work had to change, that the social arrangements of the
factory, or office, and participation in management decision making, needed
to be revamped if the harsh psychological damage that he saw all around him
was to be removed.

One of the main entities responsible for the condition Fromm was describing
was the vast corporation, or 'organisation,' and this was a matter taken up
specifically in **W. H. Whyte**'s *Organisation Man*, published the following year.
This was a much sharper, more provocative book than Fromm's, though the
overlap in subject matter was considerable.[32] Whyte's book was better written
(he was a journalist on *Fortune*) and more pointed, and what he provided was
a telling and not overly sympathetic account of the life and culture of 'other-
directed' people in postwar America. Whyte considered that vast organisations
both attracted and bred a certain type of individual, that there was a certain
kind of psychology most suited to corporate or organisational life. First and
foremost, he saw in the organisation a decline of the Protestant ethic, in the
sense that there was a marked drop in individualism and adventurousness.[33]
People knew that the way to get on in an organisation was to be part of a
group, to be popular, to avoid 'rocking the boat.' Organisation man, Whyte
says, is a conservative (with a small 'c'), and above all works for somebody else,
not himself.[34] Whyte saw this as a significant crossover point in American
history. The main motives inside corporations, he said, were 'belongingness'
and 'togetherness.' Whyte's subsidiary points were no less revealing. There had
recently been an historic change in the U.S. educational system, and he
produced a chart of education courses that described those changes clearly.
Between 1939–46 and 1954–5, whereas enrolments in fundamental courses
(the humanities, the physical sciences) had declined, subscriptions to practical
courses (engineering, education, agriculture) had increased.[35] He thought this
was regrettable because it represented a narrowing factor in life; people not
only knew less, they would only mix with fellow students with the same
interests, and therefore go on knowing less, leading a narrower life.[36] Whyte
went on to attack the personnel industry and the concept of 'personality' and
personality testing, which, he felt, further promoted the conforming and
conservative types. What he most objected to were the psychoanalytic interpretations of personality tests, which he thought were little better than astrology.
He saved his final attack for an assault on suburbia, which he saw as 'the branch
office' of the organisation and a complete extension of its group psychology.
With little maps of suburban developments, he showed how social life was
extremely constricted, being neighborhood-based (a rash of bridge parties, fish
picnics, Valentine costume parties), and underlined his central argument that
Organisation Man led his life in a regime he characterised as a 'benign tyranny.'[37]
Under this tyranny, people must be 'outgoing,' by far the most important
quality. They sacrifice their privacy and their idiosyncrasies and replace them
with an enjoyable but unreflective lifestyle that moves from group activity to
group activity and goes nowhere because one in three of such families will in
any case move within a year, most likely to a similar community hundreds of

miles away. Whyte recognised that, as Riesman had said of other-directed people, Organisation Man was tolerant, without avarice, and not entirely unaware that there *are* other forms of existence. His cage was gilded, but it was still a cage.

Whyte didn't like the changes he saw happening, but he was candid about them rather than angry. The same could not be said for **C. Wright Mills**. Mills liked to describe himself as 'an academic outlaw.'[38] As a native Texan, he fitted this image easily, aided by the huge motorcycle that he rode, but Mills wasn't joking, or not much. Trained as a sociologist, who had taught in Washington during the war and been exposed to the new social survey techniques that had come into being in the late 1930s and matured in wartime, Mills had recognised from these surveys that American society (and, to an extent, that of other Western countries) was changing – and he hated that fact. Unlike David Riesman or Whyte, however, he was not content merely to describe sociological change; he saw himself as a combatant in a new fight, where it was his job to point out the dangers overtaking America. This forced him up against many of his academic colleagues, who thought he had overstepped the mark. It was in this sense that he was an outlaw.

Born in 1916, Wright had taught at the University of Maryland in wartime, and it was while he was in Washington that he had been drawn into the work carried out by **Paul Lazersfeld** at Columbia University's Bureau of Applied Social Research, which did a lot of surveys for the government. Lazersfeld's essentially statistical approach to evidence had grown rapidly as war-related interest in practical social research was reflected in government spending.[39] This wartime experience had two consequences for Mills. It gave him greater awareness of the changes overtaking American society, and he had acquired a long-lasting belief that sociology should be practical, that it should strive not just to understand the way societies worked but to provide the common man with the basis for informed decisions. This was essentially the same idea that Karl Mannheim was having in London at much the same time. After the war Mills moved to New York, where he mixed with a group of other intellectuals who included Philip Rahv, Dwight Macdonald, and Irving Howe, who were connected to the *Partisan Review,* and Daniel Bell, editor of the *New Leader.*[40] At Columbia he got to know Robert Lynd, famous for his study *Middletown,* though Lynd's star was then on the wane. Between 1948 and 1959 Mills wrote a clutch of books that hung together with a rare intellectual consistency. The late 1940s and early 1950s, thanks to the GI bill, saw a flood of students into higher education. This raised standards in general and in turn produced a new kind of society with more jobs, more interesting kinds of job, and more specialities being professionalised. Mills saw it as his role to describe these new situations and to offer a critique.

Mills's books were published in the following order: *The New Men of Power* (1948), *White Collar* (1951), *The Power Elite* (1956), and *The Sociological Imagination* (1959). All reflected his view that, in essence, labor had ceased to be the great question in society: 'The end of the labor question in domestic politics was accompanied by the transformation of Russia from ally to enemy and the rise of the Communist threat. The end of utopia was also the end of ideology as the labor movement shifted from social movement to interest group. The defining political issue became totalitarianism versus freedom, rather than capitalism versus socialism.' He felt that the automobile had made suburban living possible, with the housewife as the centerpiece, 'a specialist in consumption and in nurturing a spirit of togetherness in the family.'[41] The home, and the private sphere, rather than the workplace and the union hall, had become the center of attention. He believed the 1930s, with so much government intervention because of the depression, was the crossover point. He was also the first to consider 'celebrities' as a group.[42] The result of all this, he said, was that the formerly 'ruggedly individualist' American citizens had become 'the masses,' 'conformist creatures of habit rather than free-thinking activists.'[43] Whereas in *Organisation Man* Whyte had found his interest in the middle orders of corporations, in *The New Men of Power* Mills concentrated on the leaders, arguing that there had appeared a new type of labor leader – he was now the head of a large bureaucratic organisation, part of a new power elite, part of the mainstream. In *White Collar*, his theme was the transformation of the American middle class, which he characterised as 'rootless and amorphous, a group whose status and power did not rest on anything tangible . . . truly a class in the middle, uncertain of itself,' essentially anomic and prone to take the tranquillisers then coming into existence.[44] 'The white collar people slipped quietly into modern society. Whatever history they have is a history without events; whatever common interests they have do not lead to unity; whatever future they have will not be of their own making.'[45] 'The idea born in the nineteenth century and nurtured throughout the 1930s, that the working class would be the bearers of a new, more progressive society,' was laid to rest, Mills concluded. In a section on mentalities, he introduced the subversive idea that the white-collar classes were in fact not so much the new middle classes as the new working classes.[46]

This reconceptualisation of American society culminated in 1956 in *The Power Elite*, a phrase and a thesis that many of the student revolutionaries of the 1960s would find congenial. Here Mills built on Max Weber's ideas (he had helped translate Weber into English), seeing 'the cohesiveness of modern society as a new form of domination, a social system in which power was more diffuse and less visible than in early forms of social order. Rather than the direct power exerted by the factory owner over his employees and the autocratic ruler over his subjects, modern power had become bureaucratised and thus less easy to locate and recognise. . . . The new face of power in mass society was a corporate one, an interlocking hierarchical system.'[47] In traditional America, Mills wrote, 'the family, the school and the church were the main institutions around which social order congealed. In modern America, these had been replaced by the corporation, the state, and the army, each embedded in a technology, a system of interlocking processes.'[48]

The Sociological Imagination, Mills's last book, took as its title another clever phrase designed to encapsulate a new way of looking at the world, and at experiences, to help the modern individual 'understand his own experience and gauge his own fate ... by locating himself within his own period, [so] that he can know his own chances in life ... by becoming aware of all those individuals in his circumstances' (again, reminiscent of Mannheim).[49] Like Hannah Arendt, Mills realised that as the old categories had broken down, so the nature of politics had changed; individual identities, as members of groups, had also collapsed and no longer applied; it was therefore, to him at least, part of the task of sociology to create a new pragmatism, to convert 'personal troubles into public issues, and public issues into the terms of their human meaning for a variety of individuals.'[50] Mills's vision was invigorating, based as it was not on his prejudices, or not only on his prejudices, but on survey material. His analysis complemented others', and his enthusiasm for using knowledge for practical purposes prefigured the more direct involvement of many academics – especially sociologists – in politics in the decades to follow. Mills was a kind of Sartrean *homme revolté* in the academy, a role he relished and which others, without the same success, tried to emulate.[51]

A different version of the change coming over American society, and by implication other Western societies, was provided by the economist **John Kenneth Galbraith**. Galbraith, a six-foot-five academic from Harvard and Princeton who had been in charge of wartime price control and director of the U.S. Strategic Bombing Survey, detected a major shift in economic sensibility in the wake of World War II and the advent of mass society. In the views he propounded, he was following – unwittingly perhaps – Karl Popper's idea that truth is only ever temporary, in the scientific sense: that is, until it is modified by later experience.

For Galbraith, the discipline of economics, the so-called 'dismal science,' had been born in poverty. For the vast span of history, he said, man has been bound by massive privation for the majority, and great inequality, with a few immensely rich individuals. Moreover, there could be no change in this picture, for the basic economic fact of life was that an increase in one man's wages inevitably meant a decrease in another man's profits: 'Such was the legacy of ideas in the great central tradition of economic thought. Behind the façade of hope and optimism, there remained the haunting fear of poverty, inequality and insecurity.'[52] This central vision of gloom was further refined by two glosses, one from the right, the other from the left. The social Darwinists said that competition and in some cases failure was quite normal – that was evolution working itself out. The Marxists argued that privation, insecurity, and inequality would increase to the point of revolution that would bring everything tumbling down. For Galbraith, productivity, inequality, and insecurity were the 'ancient preoccupations' of economics.[53] But, he argued, we were now living in an *Affluent Society* (the title of his book), and in such a world the ancient preoccupations had changed in two important respects. In the wake of World War II and the 'great Keynesian prosperity' it had brought about, especially in the United States, inequality had shown no tendency to get violently worse.[54] Therefore the Marxist prediction of a downward spiral to revolution did not appear to be on the cards. Second, the reason for this change, and something that, he said, had been insufficiently appreciated, was the extent to which modern business firms had inured themselves to economic insecurity. This had been achieved by various means, not all of them entirely ethical in the short run, such as cartels, tariffs, quotas, or price-fixing by law, all of which ameliorated the rawer effects of capitalist competition. But the long-term effect had been profound, Galbraith maintained. It had, for the first time in history (and admittedly only for the Western democracies), removed economic insecurity from the heart of human concerns. No one, any more, lived dangerously. 'The riskiness of modern corporate life is, in fact, the harmless conceit of the modern corporate executive, and that is why it is vigorously proclaimed.'[55]

This profound change in human psychology, Galbraith said, helped explain much modern behaviour – and here there were echoes of Riesman, though Galbraith never mentioned him by name. With the overwhelming sense of economic insecurity gone from people's lives, and with the truce on inequality, 'we are left with a concern only for the production of goods.' Only by higher levels of production, and productivity, can levels of income be maintained and improved. There is no paradox that the goods being produced are no longer essential to survival (in that sense they are peripheral), for in an 'other-directed' society, when keeping up with the Joneses comes to be an important social goal, it does not matter that goods are not essential to life – 'the desire to get superior goods takes on a life of its own.'[56]

For Galbraith there are four significant consequences of this. One is that advertising takes on a new importance. With goods inessential to life, the want has to be created: 'the production of goods creates the wants that the goods are presumed to satisfy,' so that advertising comes to be an integral aspect of the production process.[57] Advertising is thus a child to, and a father of, mass culture. Second, the increased production – and consumption – of goods can only be achieved by the deliberate creation of more debt (in a telling coincidence, credit cards were introduced in the same year that Galbraith's book was published). Second, in such a system there will always be a tendency to inflation, even in peace (in the past inflation had generally been associated with wars). For Galbraith, this is systemic, arising from the very fact that the producers of goods must also create the wants for those same goods, if they are to be bought. In an expanding economy, firms will always be operating at or near their capacity, and therefore always building new plants, which require capital investment. In a competitive system, successful firms will need to pay the highest wages – which must be paid before the returns on capital investment are brought in. There is, therefore, always an upward pressure on inflation in the consumer society. Third, and as a result of this, public services – paid for by the government because no market can exist in these areas – will always lag behind private, market-driven goods.[58] Galbraith both observes and predicts that public services will always be the poor relation in the affluent society, and that public service workers will be among the least well off. His last point is that with the arrival of the product-driven society there also arrives the age of the businessman –

'more precisely, perhaps, the important business executive.' So long as inequality was a matter of serious concern, says Galbraith, the tycoon had at best an ambiguous position: 'He performed a function of obvious urgency. But he was also regularly accused of taking too much for his services. As concern for inequality has declined, this reaction has disappeared.'

Having set out his description of modern mass society, Galbraith went on to make his famous distinction between private affluence and public squalor, showing how it is the obsession with private goods that helps create the poor public services, with overcrowded schools, under-strength police forces, dirty streets, inadequate transport. 'These deficiencies are not in new or novel services but in old and established ones,' he says, because only in private goods can advertising – that is, the creation of wants – work. It makes no sense to advertise roads, or schools, or police forces. He concludes, therefore, that the truce on inequality should be replaced with a concern for the *balance* between private affluence and public squalor. Inflation only makes that imbalance worse, and things are at their very worst in local, as opposed to central, government areas (the local police are always underfunded compared to the FBI, for instance).[59]

Galbraith's solutions to the problems of the affluent society were twofold. One was taken up widely. This was the local sales tax.[60] If consumer goods are the prime success story of modern society, and at the same time a cause of the problem, as Galbraith maintained, there is a certain justice in making them part of the solution too. His second solution was more radical, more unusual psychologically, and cannot be said to have been acted upon in any serious way as yet, though it may come. Galbraith noted that many people in the affluent society took large salaries not because they needed them but because it was a way of keeping score, a reflection of prestige. Such people actually enjoyed working; it was no longer a way to avoid economic insecurity but intellectually satisfying in itself. He thought what was needed was a new leisure class. In fact, he thought it was growing naturally, but he wanted it to be a matter of policy to encourage further growth. His point was that the **New Class**, as he called it, with initial capitals, would have a different system of morality. Better educated, with a greater concern for the arts and literature, having made enough money in the earlier parts of their careers, members of this New Class would retreat from work, changing the value attached to production and helping redress the social balance between private affluence and public squalor, maybe even devoting the latter part of their careers to public service.[61]

The Affluent Society may have sparked other books, but many were in preparation at the end of the 1950s, born of similar observations. For example, in *The Stages of Economic Growth*, completed in March 1959 and published a year later W. W. Rostow produced a book that in some ways showed affinities with both Galbraith and Riesman. Rostow, an economist at MIT who had spent a great deal of time in Britain, mainly but not only at Cambridge, agreed with Riesman that the modern world had developed through stages, from traditional societies to the age of high mass consumption. He echoed Galbraith in regarding economic growth as the engine not just of material change but of political, social, and intellectual change as well. He even thought that the stages of economic growth had a hand – but only a hand – in wars.[62]

For Rostow, societies fell into five stages. In the beginning, the pre-Newtonian world, there is the traditional society. This included the dynasties in China, the civilisations of the Middle East and the Mediterranean, the world of mediaeval Europe. What they shared was a ceiling on their productivity. They were capable of change, but slowly. At some point, he said, traditional societies broke out of their situation, mainly because the early days of modern science came along, with new techniques enabling individuals 'to enjoy the blessings and choices opened up by the march of compound interest.'[63] In this stage, the precondition for takeoff, several things happened, the most important being the emergence of an effective, centralised nation state, the lateral expansion of trade across the world, and the appearance of banks for mobilising capital. Sometimes this change was promoted by the intrusion of a more advanced society. What Rostow called 'The Take-Off' he regarded as the 'great watershed of modern life in modern society.'[64] This required two things; a surge in technology, but also a group of individuals, organised politically, 'prepared to regard the modernisation of the economy as serious, high-order political business.' During the takeoff the rate of effective investment and savings more than doubles, say from 5 percent to 10 percent and above. The classic example of this stage is the great railway booms. Some sixty years after the takeoff begins, Rostow says, the fourth stage, maturity, is reached.[65] Here there is a shift from, say, the coal, iron, and heavy engineering industries of the railway phase to machine tools, chemicals, and electrical equipment. Rostow produced a number of tables that illustrate his approach. Here, two of the more interesting have been amalgamated:[66]

Country	Takeoff	Maturity
United Kingdom	1783–1802	1850
United States	1843–60	1900
Germany	1850–73	1910
France	1830–60	1910
Sweden	1868–90	1930
Japan	1878–1900	1940
Russia	1890–1914	1950
Canada	1896–1914	1950

Speculating on the sixty-year gap between takeoff and maturity, Rostow puts this down to the time needed for the arithmetic of compound interest to take effect and/or for three generations of individuals to live under a regime where growth is the normal condition. In the fifth stage, the age of high mass consumption, there is a shift to durable consumer goods – cars, refrigerators, other electrically powered household gadgets.[67] There is also the emergence of a welfare state.[68] But *The Stages* was a book of its time in other senses than that it followed Galbraith. This was the height of the Cold War (the Berlin Wall would go up in the following year, with the Cuban missile crisis a year after that), and the arms race was at its height, the space race beginning in earnest. Rostow clearly saw his stages as an alternative, and better, analysis of social and economic change than Marxism, and he considered the stages of growth to be partly related to war. Rostow observed three kinds of war: colonial wars, regional wars, and the mass wars of the twentieth century.[69] Wars tended to occur, he said, when societies, or countries, were changing from one stage of growth to another – war both satisfied and encouraged the energies being unleashed at these times. Conversely, countries that were stagnating, as France and Britain were after World War II, became targets of aggression for expanding powers. His most important point, certainly in the context of the times when his book appeared, but still of great interest, was that the shift into high mass consumption was the best hope for peace[70] – not only because it created very satisfied societies who would not want to make war, but also because they had more to lose in an era of weapons of mass destruction. He noted that the USSR spent far too much on defence to allow its citizens to profit properly from consumer goods, and he hoped its citizens would one day realise how these two facts were related and prevail on their governments to change.[71] Rostow's analysis and predictions were borne out – but not until more than a quarter of a century had elapsed.

Rostow's view was therefore fundamentally optimistic, more optimistic certainly than Galbraith's. Other critics' were much less so. One of Galbraith's main points in the analytic section of his book was the relatively new importance of advertising, in creating the wants that the private consumer goods were intended to satisfy. Almost simultaneously with his book, an American journalist-turned-social-critic published three volumes that took a huge swipe at the advertising industry, expanding and amplifying Galbraith's argument, examining the 'intersection of power, money and writing.' **Vance Packard** called his trilogy *The Hidden Persuaders* (1957), *The Status Seekers* (1959), and *The Waste Makers* (1960). All of them reached the number-one slot in the *New York Times* best-seller list, in the process transforming Packard's own fortunes. He had lost his job just before Christmas 1956 when the magazine he wrote for, *Collier's*, folded.[72] In early 1957 he had taken his first unemployment cheque but already had a manuscript with the publishers. This manuscript had an odd life. In the autumn of 1954 *Reader's Digest* magazine had given Packard an assignment, which he later said 'they apparently had lying around,' on the new psychological techniques then being used in advertising. Packard researched the article, wrote it, but then learned that the *Digest* had 'recently broken its long-standing tradition and decided to begin carrying advertisements. Subsequently, he was paid for his article, but it never appeared, and he was outraged when he learned there was a connection between the decision not to publish his piece and the magazine's acceptance of advertising, the subject of his attack.'[73] He thus turned the article into a book.

The main target of Packard's attack was the relatively new technique of **motivational research** (MR), which relied on intensive interviewing, psychoanalytic theory, and qualitative analysis, and in which sex often figured prominently. As Galbraith had emphasised, many people did not question advertising – they thought it important in helping fuel the demand on which mass society's prosperity was based. In 1956 the prominent MR advocate **Ernest Dichter** had announced, 'Horatio Alger is dead. We do not any longer really believe that hard work and savings are the only desirable things in life; yet they remain subconscious criteria of our feeling of morality.' For Dichter, consumption had to be linked to pleasure, consumers had to be shown that it was 'moral' to enjoy life. This should be reflected in advertising.[74]

Packard's main aim in *The Hidden Persuaders* was to show – via a catalogue of case histories – that American consumers were little more than 'mindless zombies' manipulated by the new psychological techniques. In one revealing case, for example, he quoted a marketing study by Dichter himself.[75] Headed 'Mistress versus Wife,' this was carried out for the Chrysler Corporation and explored why men bought sedans even though they preferred sporty models. The report argued that men were drawn into automobile showrooms by the flashy, sporty types in the window, but actually *bought* less flashy cars, 'just as he once married a plain girl.' 'Dichter urged the auto maker to develop a hardtop, a car that combined the practical aspects men sought in a wife with the sense of adventure they imagined they would find in a mistress.'[76] Packard believed that MR techniques were antidemocratic, appealing to the irrational, mind-moulding on a grand scale. Such techniques applied to politics could take us nearer to the world of *1984* and *Animal Farm* and, Packard thought, following Riesman, that the 'other-directed' types of mass society were most at risk. Advertising not only helped along the consumer society, it stopped people achieving autonomy.

Packard's second book, *The Status Seekers*, was less original, attacking the way advertising used status and people's fears over loss of status to sell goods.[77] His more substantial point was that, just then in America, there was much debate over whether the country was really less class-ridden than Europe, or had its own system, based more on material acquisitions rather than heredity. (This also was an issue that Galbraith had raised.) Packard advanced the view that business was essentially hypocritical in its stance. On the one hand, it claimed that the wider availability of the consumer products it was selling made America less divided; on the other, one of its major methods of selling used exactly these differences in status – and anxiety over those differences – as a device for promoting the sales of goods. His third book, *The Waste Makers*, used as its starting point a 1957 paper by a Princeton undergraduate, William Zabel, on planned obsolescence, in other words the deliberate manipulation of taste so that goods would seem out of date – and therefore be replaced – long before they were physically exhausted.[78] This last book was probably Packard's most overstated case; even so, analysis of his correspondence showed that many people were already disenchanted by the underlying nature of mass consumer society but felt so atomised they didn't know what to do about it. As he himself was to put it later, the people who wrote to him were members of 'The Lonely Crowd.'[79]

Naturally, the business community didn't relish these attacks; as an editorial in *Life* put it, 'Some of our recent books have been scaring the pizazz out of us with the notion of the Lonely Crowd ... bossed by a Power Elite ... flim-flammed by hidden persuaders and emasculated into a neuter drone called the Organisational Man.'[80]

One general notion underpinned and linked these various ideas. It was that, as a result of changes in the workplace and the creation of mass society, and as a direct consequence of World War II and the events leading up to it, a new socio-politico-psychology, a new human condition, was abroad. The traditional sources from which people took their identity had changed, bringing new possibilities but also new problems. Riesman, Mills, Galbraith, and the others had each chipped away, sculpting part of the picture, but it was left to another man to sum it all up, to describe this change of epoch in the language it deserved.

Daniel Bell was born in the Lower East Side of New York City in 1919 and grew up in the garment district in a family that had migrated from Bialystok, between Poland and Russia (the family name was Bolotsky). Bell was raised in such poverty, he says, that there was 'never any doubt' that he would become a sociologist, in order to explain what he saw to himself. At the City College of New York he joined a reading group that included Melvin J. Lasky, Irving Kristol, Nathan Glazer, and Irving Howe, all well-known sociologists and social critics. Some were Trotskyists, though most later changed their beliefs and formed the backbone of the neoconservative movement. Bell also worked as a journalist, editing the *New Leader*, then at *Fortune* with Whyte, but he also had a stint at the end of the war as a sociologist at the University of Chicago, with David Riesman, and moonlighted as a sociology lecturer at Columbia from 1952–1956. He later joined Columbia full time before moving on to Harvard, in 1965 founding *The Public Interest* with Irving Kristol as a place to rehearse the great public debates.[81] It was while he was moonlighting at Columbia that he produced the work for which he first became known to the world outside sociology. This was *The End of Ideology*.

In 1955 Bell attended the **Congress for Cultural Freedom** in Milan, where several notable liberal and conservative intellectuals addressed a theme set by Raymond Aron, 'The End of the Ideological Age?' Among those present, according to Malcolm Waters, in his assessment of Bell, were Edward Shils, Karl Polanyi, Hannah Arendt, Anthony Crosland, Richard Crossman, Hugh Gaitskell, Max Beloff, J. K. Galbraith, José Ortega y Gassett, Sidney Hook, and Seymour Martin Lipset. Bell's contribution was a lecture on America as a mass society. The 'End of Ideology' debate – which would recur in several forms during the rest of the century – was seen originally by Aron as a good thing because he thought that ideologies prevent the building of a progressive state. In particular, Aron identified nationalism, liberalism, and Marxist socialism as the three dominant ideologies that, he said, were crumbling: nationalism because states were weakening as they became interdependent, liberalism because it could offer no 'sense of community or focus for commitment,' and

Marxism because it was false.[82] Bell's contribution was to argue that this whole process had gone further, faster, in the United States. For him, ideology was not only a set of governing ideas but ideas that were 'infused with passion,' and sought 'to transform the whole way of life.' Ideologies therefore take on some of the characteristics of a secular religion but can never replace real religion because they do not address the great existential questions, particularly death. For Bell, ideologies had worked throughout the nineteenth century and the earlier years of the twentieth because they helped offer moral guidance and represented real differences between the various interest groups and classes in society. But those differences had been eroded over the years, thanks to the emergence of the welfare state, the violent oppression carried out by socialist regimes against their populations, and the emergence of new stoic and existential philosophies that replaced the romantic ideas of the perfectibility of human nature.[83] Mass society, for Bell and for the United States at least, was a society of abundance and optimism where traditional differences were minimised and a consensus of views had emerged. The blood, sweat, and tears had gone out of politics.[84]

Bell wasn't seeking a prescription, merely attempting to describe what he saw as an epochal change in society, where its members were no longer governed by dominant ideas. Like Fromm or Mills he was identifying a new form of life coming into being. We are now apt to take that society for granted, especially if we are too young to have known anything else.

Few if any of these writers were associated intimately with any political party, but the majority were, for a time at least, of the left rather than of the right. The equality of effort demanded from all sections of society in wartime had a powerful significance that was much more than symbolic. This was reflected not only in the creation and provisions of the welfare state but in all the analyses of mass society, which accepted implicitly that all individuals had an equal right to the rewards that life had to offer. This equality was also part of the new human condition.

But was that justified? **Michael Young**, a British educationalist, an arch innovator, and a friend and colleague of Daniel Bell, produced a satire in 1958 that poked fun at some of these cherished assumptions.[85] *The Rise of the Meritocracy* was ostensibly set in 2034 and was cast as an 'official' report written in response to certain 'disturbances' that, to begin with, are not specified.[86] The essence of the satire is that the hereditary principle in life has been abolished, to be replaced by one of merit (IQ+Effort=Merit), with the 'aristocracy' replaced by a 'meritocracy.' Interestingly, Young found it very difficult to publish the book – it was turned down by eleven publishers.[87] One suggested that it would only be worth publishing if it were rewritten as a satire like *Animal Farm* (as if *that* had been easy to publish). Young did rewrite the book as a satire, but even so the publisher still declined to take it on. Young was also criticised for coining a term, meritocracy, that had both a Greek and a Latin root. In the end the book was published by a friend at Thames & Hudson, but only as an act of friendship – whereupon *The Rise* promptly sold several hundred thousand copies.[88]

The book is divided into two sections. 'The Rise of the Elite' is essentially an optimistic gloss on the way high-IQ people have been let loose in the corridors of power; the second section, 'The Decline of the Lower Classes,' is a gleeful picture of the way such social engineering is almost bound to backfire. Young doesn't take sides; he merely fires both barrels of the argument as to what would happen if we really did espouse wholeheartedly the mantra 'equality of opportunity.' His chief point is that such an approach would be bound to lead to eugenic nonsenses and monstrosities, that the new lower classes – by definition stupid – would have no leadership worth the name, and that the new IQ-rich upper classes would soon devise ways to keep themselves in power. Here he 'reveals' that society in 2034 has discovered ways of predicting the IQ of an infant at three months; the result is predictable – a black market in babies in which the stupid children of high IQ parents are swapped, along with large 'dowries,' for high-IQ children of stupid parents.[89] It is this practice that, when exposed in the newspapers, gives rise to the 'disturbances,' an incoherent rising by a leaderless, stupid mob, which has no chance of success.

Young's argument overlaps with Bell's, and others, insofar as he is saying that the new human condition risks being a passionless, cold, boring block of bureaucracy in which tyranny takes not the form of fascism or communism or socialism but benevolent bureaucratisation.[90] Scientism is a factor here, too, he says. You can measure IQ, maybe, but you can never measure good parenting or put a numerical value on being an artist, say, or a corporate CEO. And maybe any attempt to try only creates more problems than it solves.

Young had pushed Bell's and Riesman's and Mills's reasoning to its limits, its logical conclusion. Man's identity was no longer politically determined; and he was no longer an existential being. His identity was psychological, biological, predetermined at birth. If we weren't careful, the end of ideology meant the end of our humanity.

In November 1948 the Nobel Prize for Literature was awarded to T. S. Eliot. For him it was a year of awards – the previous January he had been given the Order of Merit by King George VI. Interviewed by a reporter in Princeton after the announcement from Stockholm, Eliot was asked for what the Nobel had been awarded. He said he assumed it was 'for the entire *corpus*.' 'When did you publish *that*?' replied the reporter.[1]

Between *The Waste Land* and the prize, Eliot had built an unequalled reputation for his hard, clear poetic voice, with its bleak vision of the emptiness and banality running through modern life. He had also written a number of carefully crafted and well-received plays peopled with mainly pessimistic characters, who had lost their way in a world that was exhausted. By 1948 Eliot was extremely conscious of the fact that his own work was, as his biographer Peter Ackroyd put it, 'one of the more brightly chiselled achievements of a culture that was dying,' and that partly explains why, in the same month that he travelled to Stockholm to meet the Swedish king and receive his prize, he also published his last substantial prose book.[2] *Notes Towards the Definition of Culture* is not his best book, but it interests us here because of its timing and the fact that it was the first of a small number of works on both sides of the Atlantic that, in the aftermath of war, formed the last attempt to define and preserve the traditional 'high' culture, which Eliot and others felt to be mortally threatened.[3]

As we saw in chapter 11, *The Waste Land*, besides its grim vision of the post-World War I landscape, had been constructed in a form that was frankly high culture – fiercely elitist and deliberately difficult, with elaborate references to the classics of the past. In the post-World War II environment, Eliot clearly felt that a somewhat different form of attack, or defence, was needed – in effect, a balder statement of his views, plain speaking that did not risk being misunderstood or overlooked. *Notes* begins by sketching out various meanings of the term 'culture' – as in its anthropological sense ('primitive culture'), its biological sense (bacterial culture, agriculture), and in its more usual sense of referring to someone who is learned, civil, familiar with the arts, who has an easy ability to manipulate abstract ideas.[4] He discusses the overlap between these ideas before concentrating on his preferred subject, by which he means that, to him, culture is a way of life. Here he advances the paragraph that was to become famous: 'The term *culture* . . . includes all the characteristic activities and interests of a people; Derby Day, Henley Regatta, Cowes, the twelfth of August, a cup final, the dog races, the pin table, the dart board, Wensleydale cheese, boiled cabbage cut into sections, beetroot in vinegar, 19th-century Gothic churches and the music of Elgar. The reader can make his own list.'[5]

But if this list seems ecumenical, Eliot soon makes it clear that he distinguishes many *levels* in such a culture. He is not blind to the fact that producers of culture – artists, say – need not necessarily have high intellectual gifts themselves.[6] But for him, culture can only thrive with an elite, a cultural elite, and cannot exist without religion, his point being that religion brings with it a shared set of beliefs to hold a way of life together – Eliot is convinced therefore that democracy and egalitarianism invariably threaten culture. Although he often refers to 'mass society,' his main target is the breakdown of the family and family life. For it is through the family, he says, that culture is transmitted.[7] He ends by discussing the unity of European culture and the relation of culture to politics.[8] The overall unity of European culture, he argues, is important because – like religion – it offers a shared context, a way for the individual cultures within Europe to keep themselves alive, taking in what is new and recognising what is familiar. He quotes Alfred North Whitehead from *Science and the Modern World* (1925): 'Men require from their neighbours something sufficiently akin to be understood, something sufficiently different to provoke attention, and something great enough to command admiration.'[9] But perhaps the most important point of culture, Eliot says, lies in its impact on politics. The power elite needs a cultural elite, he argues, because the cultural elite is the best antidote, provides the best critics for the power brokers in any society, and that criticism pushes the culture forward, prevents it stagnating and decaying.[10] He therefore thinks that there are bound to be classes in society, that class is a good thing, though he wants there to be plenty of movement between classes, and he recognises that the chief barrier to the ideal situation is the family, which quite naturally tries to buy privilege for its offspring. He views it as obvious that cultures have evolved, that some cultures are higher than others, but does not see this as cause for concern or, be it said, as an excuse for racism (though he himself was later to be accused of anti-Semitism).[11] For Eliot, within any one culture, the higher, more evolved levels positively influence the lower levels by their greater knowledge of, and use of, *scepticism*. For Eliot, that is what knowledge is for, and its chief contribution to happiness and the common good.

In Britain Eliot was joined by F. R. Leavis. Much influenced by Eliot, Leavis, it will be recalled from chapter 18, was born and educated in Cambridge. Being a conscientious objector, he spent World War I as a stretcher bearer. Afterward he returned to Cambridge as an academic. On his arrival he found no separate English faculty, but he, his wife Queenie, and a small number of critics (rather than novelists or poets or dramatists) set about transforming English studies into what Leavis was later to call 'the centre of human consciousness.' All his life Leavis evinced a high moral seriousness because he believed, quite simply, that that was the best way to realise 'the possibilities of life.' He thought that writers – poets especially but novelists too – were 'more alive' than anyone else, and that it was the responsibility of the university teacher and critic to show why some writers were greater than others. 'English was the route to other disciplines.'[12]

Early in his career, in the 1930s, Leavis extended the English syllabus to include assessments of advertisements, journalism, and commercial fiction, 'in order to help people resist conditioning by what we now call the "media."' However, in 1948 he published *The Great Tradition* and in 1952 *The Common Pursuit*.[13] Note the words 'Tradition' and 'Common,' meaning shared. Leavis believed passionately that there is a common human nature but that we each have to discover it for ourselves – as had the authors he concentrated on in his two books: Henry James, D. H. Lawrence, George Eliot, Joseph Conrad, Jane Austen, Charles Dickens. No less important, he felt that in judging serious literature there was the golden – the transcendent – opportunity to exercise judgement 'which is both "personal" and yet more than personal.'[14] This transcendental experience was what literature, and criticism, were for, and why literature is the central point of human consciousness, the poet 'the point at which the growth of the mind shows itself.' Leavis's literary criticism was the most visible example of Eliot's high-level scepticism at work.[15]

From New York Eliot and Leavis found kindred spirits in **Lionel Trilling** and **Henry Commager**. In *The Liberal Imagination* Trilling, a Jewish professor at Columbia University, was concerned, like Eliot, with the 'atomising' effects of mass society, or with what David Riesman was to call 'The Lonely Crowd.'[16] But Trilling's main point was to warn against a new danger to intellectual life that he perceived. In the preface to his book he concentrated on '**liberalism**' which, he said, was not just the dominant intellectual tradition in the postwar world but, in effect, the only one: 'For it is the plain fact that nowadays there are no conservative or reactionary ideas in general circulation.' Leaving aside whether this particular claim was true (and Eliot, for one, would have disagreed), Trilling's main interest was the effect of this new situation on literature. In particular, he foresaw a coarsening of experience. This came about, he said, because in liberal democracies certain dominant ideas spring up, find popular approval, and in consequence put ideas about human nature into a series of straitjackets. He drew his readers' attention to some of these straitjackets – Freudian psychoanalysis was one, sociology another, and Sartrean philosophy a third.[17] He wasn't against these ideas – in fact, he was very positive about Freud and psychoanalysis in general. But he insisted that it was – and is – the job of great literature to go beyond any one vision, to point up the shortcomings of each attempt to provide an all-enveloping account of human experience, and he clearly thought that in an atomised, democratised mass society, this view of literature is apt to get lost. As mass society moves toward consensus and conformity (as was happening at that time, especially in America with the McCarthy hearings), it is the job of literature, Trilling wrote, to be something else entirely. He dwelt in particular on the fact that some of the greatest writers of the twentieth century – he quoted Pound, Yeats, Proust, Joyce, Lawrence, and Gide – were far from being liberal democrats, that their very strength was drawn from being in the opposing camp. That, for Trilling, was at the root of the matter. For him, the job of the critic was to identify the consensus in order that artists might know what to kick *against*.[18]

Henry Steele Commager's *American Mind: An Interpretation of American Thought and Character since the 1880s* was also published in 1950, the same year as Trilling's book.[19] Ostensibly, Commager took a different line, in that he tried to pin down what it was that separated American thought from its European counterpart. The organisation of Commager's book was itself a guide to his thinking. It concentrated neither on the 'great men' of the period, in the sense of monarchs (which of course America did not have), nor on politicians (politics occupy chapters 15 and 16 out of a total of 20), nor on the vast mass of people and their lives (the Lynds' *Middletown* is mentioned, but their statistical approach is eschewed entirely). Instead, Commager concentrated his fire on the great individuals who had shone during the period – in philosophy, religion, literature, history, law, and what he saw as the new sciences of economics and sociology.[20] Running through his entire argument, however, and clarifying his approach, was an account of how Darwin and the theory of evolution had affected American intellectual life. After the more literal applications of the late nineteenth century, as exercised through the influence of Herbert Spencer (and discussed in chapter 3 of this book), Commager thought Darwinism had been taken on board by the American mind in the form of a pragmatic individualism. Americans, he implied, accepted that society moved forward through the achievements of outstanding individuals, that recognition of these individuals and their achievements was the responsibility of historians such as himself, that it was the role of literature to make the case both for tradition and for change, to help the debate along, and that it was also the writer's, or the academic's, job to recognise that individualism had its pathological side, which had to be kept in check and recognised for what it was.[21] He thought, for instance, that a number of writers (Jack London and Theodore Dreiser are discussed) took Darwinian determinism too far, and that the proliferation of religious sects in America was in some senses a pathological turning away from individualism (Reinhold Niebuhr was to make much the same point), as was the more general 'cult of the irrational,' which he saw as a revolt against scientific determinism. For him, the greatest success in America was the pragmatic evolution of the law, which recognised that society was not, and could not be, a static system but should change, and be made to change.[22] In other words, whereas Eliot saw the scepticism of the higher cultural elite as the chief antidote to the would-be excesses of politicians, Commager thought that the American legal system was the most considerable achievement of a post-Darwinian pragmatic society.

These four views shared a belief in reason, in the idea of progress, and in the role of serious literature to help cultures explain themselves to themselves. They even agreed, broadly, on what serious literature – high culture – *was*.

Barely was the ink dry on the pages of these books, however, than they were challenged. *Challenged* is perhaps too weak a word, for the view they represented was in fact assaulted and attacked and bombarded from all sides at once. The attack came from anthropology, from history, and from other literatures; the bombardment was mounted by sociology, science, music, and television; the assault was launched even from inside Leavis's own English department at Cambridge. The campaign is still going on and forms one of the main intellectual arteries of the last half of the twentieth century. It is one of the background factors that helps account for the rise of the individual. The initial and underlying motor for this change was powered by the advent of mass society, in particularly the psychological and sociological changes foreseen and described by David Riesman, C. Wright Mills, John Kenneth Galbraith, and Daniel Bell. But a motor provides energy, not direction. Although Riesman and the others helped to explain the way people were changing in general, as a result of mass society, specific direction for that change still had to be provided. The rest of this chapter introduces the main figures responsible for change, beginning with the neatest example.

No one could have predicted that when he stood up to recite his poem *Howl* in San Francisco in October 1955, **Allen Ginsberg** would spark an entire alternative 'Beat' culture, but on a closer reading of the man himself, some signs were there. Ginsberg had studied English literature at Columbia University under Lionel Trilling, whose defence of American liberalism he had found both 'inspiring and off-putting.' And while he composed *Howl*, Ginsberg worked as a freelance market researcher – and therefore knew as well as anyone what conventional attitudes and behaviour patterns were. If he could be sure what the norm was, he knew how to be different.[23]

Also, Ginsberg had for some time been moving in a world very different from Trilling's. Born in Paterson, New Jersey, the son of a poet and teacher, in the 1940s he had met both **William Burroughs Jr.** and Jack Kerouac in a New York apartment where they were all 'sitting out' World War II.[24] Burroughs Jr, much older, came from a wealthy Protestant Saint Louis family and had studied literature at Harvard and medicine in Vienna before falling among thieves – literally – around Times Square in Midtown Manhattan and the bohemian community of Greenwich Village. These two aspects of Burroughs, educated snob and lowlife deviant, fascinated Ginsberg. Like the older man, Ginsberg suffered from the feeling that he was outside the main drift of American society, a feeling that was intensified when he studied under Trilling.[25] Disliking the formalism of Trilling, Ginsberg was one of those who developed an alternative form of writing, the main characteristics of which were spontaneity and self-expression.[26] Ginsberg's style verged on the primitive, and was aimed at subverting what he felt was an almost official culture based on middle-class notions of propriety and success, an aspect of society now more visible than ever thanks to the commercials on the new television. Still, the evening when *Howl* received its first performance was hardly propitious. When Ginsberg got to his feet in that upstairs room in San Francisco, about a hundred other people present could see that he was nervous and that he had drunk a good deal.[27] He had, according to one who was there, a 'small, intense voice, but the alcohol and the emotional intensity of the poem quickly took over, and he was soon swaying to its powerful rhythm, chanting like a Jewish cantor, sustaining his long breath length, savouring the outrageous language.'[28] Among the others present was his old New York companion, Jean-Louis – Jack – Kerouac, who cheered at the end of each line, yelling 'Go! Go!' Soon others joined in. The chorus swelled as Ginsberg lathered himself into a trancelike state. The words Ginsberg opened with that night were to become famous, as did the occasion itself:

> I saw the best minds of my generation destroyed by madness,
>   starving hysterical naked,
> dragging themselves through the negro streets at dawn looking
>   for an angry fix,
> angelheaded hipsters burning for ancient heavenly connection
>   to the
> starry dynamo in the machinery of night

Kenneth Rexroth, a critic and key figure in what was to become known as the San Francisco poetry renaissance, said later that *Howl* made Ginsberg famous 'from bridge to bridge,' meaning from the Triboro in New York to the Golden Gate.[29] But this overlooks the real significance of Ginsberg's poem. What mattered most was its form and the mode of delivery. *Howl* was primitive not just in its title and the metaphors it employed but in the fact that it referred back to 'pre-modern oral traditions,' in which performance counted as much as any specific meaning to the words. In doing this, Ginsberg was helping to 'shift the meaning of culture from its civilising and rationalising connotations to the more communal notion of collective experience'.[30] This was a deliberate move by Ginsberg. From the first, he actively sought out the mass media – *Time, Life,* and other magazines – to promote his ideas, rather than the intellectual reviews; he was a market researcher, after all. He also popularised his work through the expanded paperback book trade – the publisher of *Howl* was **Lawrence Ferlinghetti**, owner of **City Lights**, the first paperback bookstore in the United States.[31] (In those days, paperbacks were still seen as an alternative, potentially radical form of information distribution.) And it was after *Howl* was picked up by the mass media that the Beat culture was transformed into an alternative way of life. The Beat culture would come to have three important ingredients: an alternative view of what culture was, an alternative view of experience (mediated through drugs), and its own frontier mentality, as epitomised by the road culture. Ironically, these were all intended to convey greater individualism and in that sense were slap in the middle of the American tradition. But the Beats saw themselves as radicals. The most evocative example of the road culture, and the other defining icon of the Beats, was Jack Kerouac's 1957 book *On the Road*.

Kerouac, born **Jean-Louis Lebris de Kerouac** in Lowell, Massachusetts, on 12 March 1922, did not have a background propitious for a writer. His parents were French-speaking immigrants from Quebec in Canada, so that English was not his first language. In 1939 he entered Columbia University, but on a football scholarship.[32] It was his meeting with Ginsberg and Burroughs that made him want to be a writer, but even so he was thirty-five before his most famous book (his second) was published.[33] The reception of Kerouac's book was partly helped by the fact that, two weeks before, Ginsberg's *Howl and Other Poems* had been the subject of a celebrated obscenity trial in San Francisco that had not yet been decided (the judge eventually concluded that the poems had 're-deeming social importance'). So 'Beat' was on everyone's lips. Kerouac explained to countless interviewers who wanted to know what Beat meant that it was partly inspired by a Times Square hustler 'to describe a state of exalted exhaustion' and was partly linked in Kerouac's mind to a Catholic beatific vision.[34] In the course of these interviews it was revealed that Kerouac had written the book in one frenzied three-week spell, using typing paper stuck together in a continuous ribbon so as to prevent the need to stop work in the middle of a thought. Though many critics found this technique absorbing, even charming, Truman Capote was moved to remark, 'That isn't writing; it's typing.'[35]

Like everything else Kerouac wrote, *On the Road* was strongly auto-biographical. He liked to say he had spent seven years on the road, researching the book, moving with a vague restlessness from town to town and drug to drug in search of experience.[36] It also included the characters and experiences of his friends, especially Neal Cassady – called Dean Moriarty in the book – who wrote wild, exuberant letters to Kerouac and Ginsberg detailing his 'sexual and chemical exploits.'[37] It was this sense of rootless, chaotic, yet essentially sympathetic energy of the 'courage-teachers' that Kerouac sought to re-create in his book, it being his deliberate aim to do for the 1950s what the F. Scott Fitzgerald novels had done for the 1920s and the Hemingway books for the 1930s and 1940s. (He was not keen on their writing styles but was anxious to emulate their experience as observers of a key sensibility.) In a flat, deliberately casual prose, the book did all the stock things people say about radical ventures – it challenged 'the complacency of a prosperous America' and brought out clearly, for example, the role of pop music (bebop and jazz) for the young.[38] But most of all it gave us the road book, which would lead to the road movie. 'The road' became the symbol of an alternative way of life, rootless but not aimless, mobile but with a sense of place, materially poor but generous and spiritually abundant, intellectually and morally adventurous rather than phys-ically so. With Kerouac, travel became part of the new culture.[39]

The Beat culture's turning away from Trilling, Commager, and the others was every bit as deliberate as Eliot's highbrow imagery in his poetry. The highly original use of a vernacular shared by the drug, biker, and Greyhound bus subculture, the 'strategic avoidance' of anything complex or difficult, and the transfer into an 'alternative' consciousness as mediated by chemicals were in all respects assiduously subversive.[40] But not all the alternatives to traditional high culture in the 1950s were as self-conscious. That certainly applied to one of the most powerful: pop music.

No matter how far back in time we can date popular music, its expression was always constrained by the technology available for its dissemination. In the days of sheet music, live bands, and dance halls, and then of radio, its impact was relatively limited. There was an elite, an in-group who decided what music was printed, which bands were invited to perform, either in the dance halls or on radio. It was only with the invention of the long-playing record, by the Columbia Record Company in 1948, and the first 'single,' introduced by RCA a year later, that the music world as we know it took off. After that, anyone with a gramophone in their home could play the music of their choice whenever they pleased. Listening to music was transformed. At the same time, the new generation of 'other-directed' youth arrived on the scene perfectly primed to take advantage of this new cultural form.

It is usually agreed that pop music emerged in 1954 or 1955 when black R & B (rhythm and blues) music broke out of its commercial ghetto (it was known before World War II as 'race music'). Not only did black singers enjoy a success among white audiences, but many white musicians copied the black styles. Much has been written about the actual beginnings, but the one generally agreed upon has **Leo Mintz**, a Cleveland record store owner, approaching **Alan Freed**, a disc jockey at the WJW station in Cleveland, Ohio, and telling him that suddenly white teenagers were 'eagerly buying up all the black R & B records they could get.' Freed paid a visit to Mintz's store and later described what he saw: 'I heard the tenor saxophones of Red Prysock and Big Al Sears. I heard the blues-singing, piano-playing Ivory Joe Hunter. I wondered. I wondered for about a week. Then I went to the station manager and talked him into permitting me to follow my classical program with a rock 'n' roll party.'[41] Freed always claimed that he invented the term *rock 'n' roll*, though insiders say it was around in black music well before 1954, black slang for sexual intercourse.[42] But whether he discovered R & B, or rock 'n' roll, Freed was certainly the first to push it on air; he shouted at the records, rather like Kerouac yelling 'Go!' at Ginsberg's first performance of *Howl*.[43]

Freed's renaming of R & B was shrewd. Repackaged, it was no longer race music, and white stations could play it. Record companies soon caught on, one response being to issue white (and usually sanitised) versions of black songs. For instance, some regard 'Sh-Boom,' by the Chords, as the very first rock 'n' roll number.[44] No sooner had it hit the airwaves, however, than Mercury Records released the Crew Cuts' sanitised 'cover' version, which entered the Top Ten in a week. Soon, white performers like Bill Haley and Elvis Presley were imitating black music and outdoing them in terms of commercial success.[45] Films like *The Blackboard Jungle* and TV programs like *American Bandstand* further popularised the music, which above all provided a cohesive and instantly recognisable force for teenagers everywhere.[46] For the sociologically minded, early pop/rock songs reflected Riesman's theories very neatly – for example, Paul Anka's 'Lonely Boy' (1959), the Videls' 'Mr Lonely' (1960), Roy Orbison's 'Only the Lonely' (1960), and Brenda Lee's 'All Alone Am I' (1962), although loneliness, one assumes, had existed before sociology. A crucial aspect of the rock business, incidentally, and often overlooked, was the hit chart. In the new transient conformist communities that W. H. Whyte had poked fun at, statistics were important, to show people what others were doing, and to allow them to do the same.[47] But the most significant thing about the advent of rock/pop was that it was yet another nail in the coffin of high culture. The words that went with the music – fashion, the 'altered consciousness' induced by drugs, love, and above all sex – became the anthems of the generation. The sounds of rock drowned out everything else, and the culture of young people would never be the same again.

It was no accident that pop developed as a result of the white middle classes adopting black music, or a version of it. As the 1950s wore on, black self-consciousness was rising. American blacks had fought in the war, shared the risks equally with whites. Quite naturally they wanted their fair share of the prosperity that followed, and as it became clear in the 1950s that that wasn't happening, especially in the South, where segregation was still humiliatingly obvious, the black temper began to simmer. After the U.S. Supreme Court ruling on 17 May 1954 that racial segregation in schools was unconstitutional, thereby repudiating the 'separate but equal' doctrine that had prevailed until then, it was only a matter of time (in fact, eighteen months) until Rosa Parks, a black American, was arrested for sitting at the front of the bus in a section reserved for whites, in Montgomery, Alabama. The civil rights movement, which was to tear America apart, may be said to have begun that day. Internationally, there were parallel developments, as former colonies that had also fought in World War II negotiated their independence and with it a rising self-consciousness. (India achieved independence in 1947, Libya in 1951, Ghana in 1957, Nigeria in 1960.) The result was that black writing flourished in the 1950s.

In the United States we have already seen what the Harlem Renaissance had accomplished in the 1920s. The career of **Richard Wright** spanned the war, his two most important books appearing at either end of the conflict, *Native Son* in 1940, and *Black Boy* in 1945. Beautifully written, Wright's books agonisingly describe what was then a slowly changing world. A protégé of Wright's found this even harder to take.

Ralph Ellison had wanted to be a musician since he was eight years old, when his mother had bought him a cornet. But he 'blundered into writing' after attending Booker T. Washington's Tuskegee Institute in 1933 and discovering in the library there T. S. Eliot's *Waste Land*.[48] Inspired jointly by his friendship with Wright and by Hemingway's reports from the Spanish Civil War in the *New York Times*, Ellison eventually produced *Invisible Man* in 1952. In this large book, the hero (unnamed) passes through all the stages of modern American black history: 'a Deep South childhood; a Negro college supported by northern philanthropy; factory work in the North; exposure to the frenzy of sophisticated Negro city life in Harlem; a "back-to-Africa" movement; a Communist-type outfit known as "The Brotherhood"; and even a "hipster" episode.'[49] Yet each of these regurgitates him: the invisible man fits in nowhere. Ellison, despite his earlier criticism of Gunnar Myrdal, had little positive to offer beyond this bleak criticism of all the possibilities that face the black man. And he himself fell strangely silent after this novel, becoming not a little invisible himself. It was left to the third of the American Negro writers to really get under the skin of the whites, and he only did it when he was thrown by force of circumstance into the fire.

Born in 1924, one of ten children, James Arthur Jones grew up in crushing poverty and never knew his father. He took his stepfather's name when his mother married David Baldwin some years later. That stepfather was a preacher of 'incendiary' sermons, with an 'ingrained' hatred of whites, so that by the time he was fourteen **James Baldwin** had acquired both characteristics.[50] But his preaching and his moralising had revealed him to have a talent for writing, and he had been introduced to the *New Leader* (where C. Wright Mills got his break) by Philip Rahv. Because he was homosexual as well as black, Baldwin took a leaf out of Richard Wright's book and became an exile in Paris, where he wrote his first works. These were firmly in the tradition of American pragmatic realism, influenced by Henry James and John Dos Passos. Baldwin defined his role then as being 'white America's inside-eye on the closed families and locked churches of Harlem, the discreet observer of homosexual scenes in Paris, above all the sensitive recorder of the human heart in conflict with itself.'[51] He made a name for himself with *Go Tell It on the Mountain* (1953) and *Giovanni's Room* (1956), but it was with the emergence of the civil rights movement in the later 1950s that his life took on new and more urgent dimensions. Returning to the United States from France in July 1957, in September he was commissioned by *Harper's* magazine to cover the struggle for integration in Little Rock, Arkansas, and Charlotte, North Carolina. On 5 September that year, Governor Orval Faubus of Arkansas had attempted to prevent the admission of black pupils to a school in Little Rock, whereupon President Eisenhower sent in federal troops to enforce integration and protect the children.

The experience changed Baldwin: 'From being a black writer carving out a career in a white world, Baldwin was becoming black.'[52] No longer a mere observer, he conquered his fear of the South (as he himself put it) in the pages of *Harper's*, his anger and his honesty laid bare for the white readers to accept or reject. The message he conveyed, in painful, raw language, was this: 'They [the students in the sit-ins and freedom marches] are not the first Negroes to face mobs: they are merely the first Negroes to frighten the mob more than the mob frightens them.'[53] Two of Baldwin's essays were reprinted as a book, *The Fire Next Time*, which attracted a great deal of attention as he eloquently discovered a language for the Negro experience and explained to whites the virulent anger inside blacks. 'For the horrors of the American Negro's life there has been almost no language. ... I realised what tremendous things were happening and that I did have a role to play. I can't be happy here, but I can work here.'[54] The anger of the blacks was out of the bag and could never be put back.

Elsewhere, black writing was also making advances, though in Britain the novels of **Colin MacInnes** (*Absolute Beginners*, 1959, and *Mr Love and Mr Justice*, 1960) were more astute observations on the way of life of West Indians in London, who had been arriving since 1948 to work in the capital's transport system, than arguments with any direct social or political point.[55] In France, the concept of *négritude* had been coined before World War II but had only entered general usage since 1945. Its main theme was a glorification of the

African past, often stressing black emotion and intuition as opposed to Hellenic reason and logic. Its main exponents were **Léopold Senghor**, president of Senegal, Aimé Césaire, and Frantz Fanon. Fanon, a psychiatrist from Martinique who worked in Algeria, is considered in chapter 30 (page 526). *Négritude* was a somewhat precious word that made the process it described sound safer than it did in the hands of, say, Baldwin or Ellison. But its central message, like theirs, was that black culture, black life, was every bit as rich, as meaningful, and yes, as satisfying as any other, that art that was original, moving, and worth sharing, could be made out of the black experience.

In fact, *négritude* was a European label for something that was happening in francophone Africa.[56] And what was happening was much tougher and more profound than the word made it appear. This process – decolonisation – was an inevitable by-product of World War II. Not only were the colonial powers now too enfeebled to maintain their hold on their possessions, having relied on colonial manpower to help them fight their wars, they were under strong moral pressure to relinquish their political hold. These developments were naturally accompanied by parallel intellectual changes.

The first modern realistic novel to be published in West Africa was Cyprian Ekwensi's *People of the City* (1954), although it was the publication in 1951 of Amos Tutuola's *Palm-Wine Drinkard* that made the Western metropolitan countries aware of the new literary developments occurring in Africa.[57] Above all, however, **Chinua Achebe**'s novel *Things Fall Apart*, published in 1958, was the archetypal African novel. It described a situation – the falling apart of a traditional African society as a result of the arrival of the white man – in vivid terms that contained beautiful English. It was recognisably sophisticated yet set in an unmistakable non-Western landscape – non-Western emotionally and non-Western geographically. And it was all woven into a superb tragedy.[58]

Achebe's mother tongue was Ibo, but he learned English as a boy and in 1953 became one of the first students to graduate, in English literature, from University College, Ibadan. Besides Achebe's profound sympathy for the imperfections of his characters, the beauty of his approach is his realisation – revealed in his title – that all societies, all civilisations, contain the seeds of their destruction, so that the arrival of the white man in his story is not so much the cause as the catalyst to speed along what was happening anyway. Okonkwo, the hero of the novel, a member of the Igbo culture, is a respected elder of his village, a macho man, a successful farmer and wrestler, but at odds with his son, a far gentler soul.[59] The reader is drawn into the rhythms of the village, Umofia, so successfully that even the Western reader accepts that the 'barbaric' customs of the society have good reason. Indeed, we are given a crystal-clear picture of a society that is stable, rich, 'complex, and fundamentally humane' – that is *thought out*. When Okonkwo breaks the rules of the village, we accept that this must mean seven years in exile. When the hostage he has raised in his family – whose existence and love for Okonkwo we have come to accept – is murdered, and when Okonkwo himself delivers one of the blows, we accept even this, in itself a remarkable achievement of Achebe's. And when the white man arrives, we too are as baffled by his behaviour as are the villagers of Umofia.

But Achebe, much as he loathed colonialism, was not intent on merely white-man-bashing. He drew attention to the shortcomings of Umofia society – its stasis, its inability to change, the ways in which its own outcasts or misfits might well be drawn to Christianity (Okonkwo is himself unchanged, which is part of his tragedy). *Things Fall Apart* is a profoundly affecting work, beautifully constructed.[60] In Onkokwo and Umofia, Achebe created a character and a society of universal significance.

A second Nigerian, **Wole Soyinka**, a poet and playwright, published his first work, *The Lion and the Jewel*, a year after Achebe's, in 1958. This was a play in verse, a comedy, also set in an African village, which enjoyed a great success. Soyinka was a more 'anthropological' writer than Achebe, using Yoruba myths to great effect (he even made an academic study of them). Anthropology was itself one of several academic disciplines that helped reshape what was regarded as 'culture,' and here **Claude Lévi-Strauss** was the most influential figure, with two works published in 1955. Born in Belgium in 1908, Lévi-Strauss grew up near Versailles and became a student at the University of Paris. After graduating, he did fieldwork in Brazil while he was professor of sociology at the University of São Paulo. Further fieldwork followed, in Cuba, but Lévi-Strauss returned to France in 1939 for military service. In 1941 he arrived as a refugee at the New School for Social Research in New York, and after the war he was French cultural attaché to the United States. Eventually, he would be appointed to the Chair of Social Anthropology at the Collège de France, in 1959, but by then he had begun his remarkable series of publications. These fell into three kinds. There were his studies in kinship, examining the way familial relationships were understood among many different (but mainly Amerindian) tribes; there were his studies of mythologies, exploring what they reveal about the way people very different on the surface think about things; and third, there was a sort of autobiographical/philosophical/travelogue, *Tristes Tropiques*, published in 1955.[61]

Lévi-Strauss's theories were very complex and not helped by his own style, which was far from easy and on more than one occasion defeated his translators. He is, therefore, an author very difficult to do justice to in a book of this kind. Nevertheless we may say that, his studies of kinship apart, Lévi-Strauss's work has two main elements. In his paper 'The Structural Study of Myth,' published in the *Journal of American Folklore* in 1955, the same year as *Tristes Tropiques* appeared, and later developed in his four-volume *Mythologiques*, Lévi-Strauss examined hundreds of myths around the world. Though trained in anthropology, he came to this work, he said, with 'three mistresses' – geology, Marx, and Freud.[62] The Freudian element in his work is much more obvious than the Marxian, or the geology, but what he appears to have meant is that, like Marx and Freud, he was seeking to find the universal structures that underlie human experience; like the historians of the *Annales* school (chapter 31), he saw the broad sweeps of history as more important than more proximate events.[63]

All mythologies, Lévi-Strauss said, share a universal, inbuilt logic. Any corpus of mythological tales, he observed, contains a recurrent harping on elementary themes – incest, fratricide, patricide, cannibalism. Myth was 'a kind of collective dream,' an 'instrument of darkness' capable of being decoded.[64] In all, in what became four volumes, he examined 813 different stories with an extraordinary ingenuity that many, especially his Anglo-Saxon critics such as Edmund Leach, have refused to accept. He observes for instance that across the world, where figures from myth are born of the earth rather than from woman, they are given either very unusual names or some deformity such as a clubfoot to signify the fact.[65] At other times myths concern themselves with 'overrated' kin relationships (incest) or 'underrated' relationships (fratricide/parricide). Other myths concern themselves with the preparation of food (cooked/raw), whether there is sound or silence, whether people are dressed or undressed. It was Lévi-Strauss's claim, essentially, that if myth could be understood, it would explain how early man first came to decipher the world and would therefore represent the fundamental, unconscious structure of the mind. His approach, which came as a revelation for many people, also had one important secondary effect. He himself said explicitly that on the basis of his inquiries, there is really no difference between the 'primitive' mind and the 'developed' mind, that so-called savages are just as sophisticated in their storytelling, just as removed from the truly primitive, as we are ourselves.[66]

Earlier in the century, as we have seen, Margaret Mead and Ruth Benedict's work had been important in showing how different peoples around the world differ in various aspects of their behaviour (such as sex).[67] Conversely, the thrust of Lévi-Strauss's work was to show how, at root, myths reveal the essential similarity, the basic concordance, of human nature and beliefs right across the globe. This was an immensely influential view in the second half of the twentieth century, not only helping to undermine the validity of evolved high culture put forward by Eliot, Trilling, et alia, but promoting the idea of 'local knowledge,' the notion that cultural expression is valid even though it applies only to specific locations, whose reading of that expression may be much more diverse and complex – richer – than is evident to outsiders. In this, Lévi-Strauss and Chinua Achebe were saying the same thing.

This development in anthropology was aided by a parallel change in its sister discipline, archaeology. In 1959 Basil Davidson published Old Africa Rediscovered, a detailed account of the 'Dark Continent's' distant past. A later year, Oxford University Press released its magisterial History of African Music. Both these works will be properly considered in chapter 31, where we examine new concepts in historical thinking.[68] But they belong here too, for running through the work of Ellison, Baldwin, MacInnes, Achebe, Lévi-Strauss, and Basil Davidson was the experience of being black in a non-black world. Responses differed, but what they shared was a growing awareness that the art, history, language, and very experience of being black had been deliberately devalued, or rendered invisible, in the past. That history, that language, that experience, needed to be urgently reclaimed, and given a shape and a voice. It was a different alternative culture to that of the Beats, but it was no less rich, varied, or valid. Here was a common pursuit that had its own great tradition.

Britain in the 1950s did not yet have a large black population. Black immigrants had been arriving since 1948, their lives chronicled now and then by writers such as Colin MacInnes, as was referred to above. The first Commonwealth Immigrants Act, restricting admission from the 'New' Commonwealth (i.e., predominantly black countries), was not passed until 1961. Until that point, then, there was little threat to the traditional British culture from race. Instead, the 'alternative' found its strength in an equivalent social divide that for many created almost as much passion: class.

In 1955 a small coterie of like-minded serious souls got behind an idea to establish a theatre in London that would endeavour to do something new: find fresh plays from completely new sources, in an effort to revitalise contemporary drama and search out a new audience. They named the venture the **English Stage Company** and bought the lease of a small theatre known as the **Royal Court** in Sloane Square in Chelsea. The theatre turned out to be ideal. Set in the heart of bourgeois London, its program was revolutionary.[69] The first artistic director was **George Devine** who had trained in Oxford and in France, and he brought in as his deputy **Tony Richardson**, twenty-seven, who had been working for the BBC. Devine had experience, Richardson had the flair. In fact, says Oliver Neville in his account of the early days of the ESC, it was the solid Devine who spotted the first piece of flair. While launching the company, he had paid for an ad in *The Stage*, the theatrical weekly, soliciting new plays on contemporary themes, and among the seven hundred manuscripts that arrived 'almost by return of post' was one by a playright named **John Osborne**, which was called *Look Back in Anger*.[70] Devine was much taken by the 'abrasive' language that he grasped instinctively would play well on stage. He discovered that the writer was an out-of-work actor, a man who was in many ways typical of a certain post-war figure in Britain. The 1944 Education Act (brought in as a result of the Beveridge Report) had raised the school-leaving age and initiated the modern system of primary, secondary and tertiary schools; it had also provided funds to help lower-class students attend acting schools. But in drab post-war England, there were now more students than jobs. Osborne was one of these over-trained types and so was Jimmy Porter, the 'hero' of his play.[71]

'Hero' deserves inverted commas because it was one of the hallmarks of *Look Back in Anger* that its lower-middle-class protagonist, while attacking everything around him, also attacked himself. Jimmy Porter is, in this sense, a direct cousin of Okonkwo, 'driven by [a] furious energy directed towards a void.'[72] The structure of *Look Back in Anger* has been frequently criticised as falling apart at the end, where Jimmy and his middle-class wife retreated into their private fantasy world of cuddly toys.[73] Despite this, the play was a great success and marked the beginning of a time when, as one critic put it, plays 'would no longer be concerned with middle class heroes, or set in country houses.'[74] Its title helped give rise to the phrase '**angry young men**,' which, together with '**Kitchen Sink Drama**,' described a number of plays and novels that, in the mid- to late-1920s in Great Britain, drew attention to the experiences of working-class men (they were usually men).[75] So it is in this sense that the trend typified by Osborne fits in with the rest of the reconceptualisation of culture, with which we are concerned. In reality, in Osborne's play, just as in Bernard

Kops's *Hamlet of Stepney Green* (1957), John Arden's *Waters of Babylon* (1957) and *Live Like Pigs* (1958), Arnold Wesker's *Chicken Soup with Barley* (1958) and *Roots* (1959), together with a raft of novels – John Braine's *Room at the Top* (1957), Alan Sillitoe's *Saturday Night, Sunday Morning* (1958), and David Storey's *This Sporting Life* (1960) – the main characters were working-class 'heroes,' or antiheroes as they came to be called. These antiheroes are all aggressive, all *escaping* from their lower-class backgrounds because of their educational or other skills, but unsure where they are headed. Although each of these authors could see the shortcomings of lower-class society, no less than other kinds, their work lent a legitimacy to lower-class experience and provided another alternative to traditional cultural forms. In Eliot's terms, these works were profoundly sceptical.

A somewhat similar change was overtaking poetry. On 1 October 1954 an anonymous article appeared in the *Spectator* entitled 'In the **Movement**.' This, actually the work of the magazine's literary editor, **J. D. Scott**, identified a new grouping in British literature, a covey of novelists and poets who 'admired Leavis, Empson, Orwell and Graves,' were 'bored by the despair of the forties ... extremely impatient of poetic sensibility ... and ... sceptical, robust, ironic.'[76] The *Spectator* article identified five authors, but after D. J. Enright had published *Poets of the 1950s* in 1955, and Robert Conquest's *New Lines* had appeared a year later, nine poets and novelists came to be regarded as comprising what was by then known as the Movement: Kingsley Amis, Robert Conquest, Donald Davie, Enright himself, Thom Gunn, Christopher Holloway, Elisabeth Jennings, **Philip Larkin**, and John Wain. One anthologist, perhaps going a shade over the top, described the Movement as 'the greatest rupture in cultural tradition since the eighteenth century.' Its core texts included Wain's novel, *Hurry On Down* (1953), and Amis's *Lucky Jim* (1954), and its prevailing tone was 'middlebrow scepticism' and 'ironical commonsense.'[77]

The most typical poet of the Movement, the man who characterised its approach to life and literature most cleanly, was Larkin (1922–85). He grew up in Coventry, not too far from Auden's Birmingham, and after Oxford began a career as a university librarian (Leicester, 1946–50; Belfast, 1950–55; Hull, 1955–85) mainly because, as it seems, he needed a regular job. He wrote two early novels, but it was as a poet that he became famous. Larkin liked to say that poetry chose him, rather than the other way around. His poetic voice, as revealed in his first mature collection, *The Less Deceived*, which appeared in 1955, was 'sceptical, plain-speaking, unshowy,' and above all modest, fortified by common sense. It wasn't angry, like Osborne's plays, but Larkin's rejection of old literature, of tradition, lofty ideas, psychoanalysis – the 'common myth-kitty' as he put it – do echo the down-to-earth qualities of 'kitchen-sink' drama, even if the volume control is turned down.[78] One of his most famous poems was 'Church Going,' with the lines

> I take off
> My cycle-clips in awkward reverence

which immediately convey Larkin's 'intimate sincerity,' not to mention a certain comic awareness. For Larkin, man 'has a hunger for meaning but for the most part is not quite sure he is up to the task; the world exists without question – there's nothing philosophical about it; what's philosophical is that man can't do anything about that fact – he is a "helpless bystander"; his feelings have no meaning and therefore no place. Why therefore do we have them? That is the struggle.' He observes

> the hail
> Of occurrence clobber life out
> To a shape no one sees

Larkin verges on the sentimental purposely, in order to draw attention to the very shortcomings of sentimentality, only too aware that that is all many people have. His is a world of disenchantment and defeat ('two can live as stupidly as one' is his verdict on marriage), a 'passive realism whose diminished aim in life is not to feel grand passion but to prevent himself from ever hurting.' It is the message of someone who is aware of just enough science for it to pain and depress him, but who sees through existentialism, and all the other 'big' words, come to that. This is why Larkin's stature has grown; his view may not be heroic, but it is perfectly tenable. As Blake Morrison has pointed out, Larkin was regarded as a minor poet for decades, but at the end of the century, 'Larkin now seems to dominate the history of English poetry in the second half of the century much as Eliot dominated the first.'[79]

Overlapping with the angry young men, and the Movement, or at least with the world they attempted to describe, was **Richard Hoggart**'s highly original *Uses of Literacy*. Published a year after *Look Back in Anger* was first staged, in 1957, Hoggart was, with Raymond Williams, Stuart Hall, and E. P. Thompson, one of the founders of the school of thought (and now academic discipline) known as cultural studies. Born in Leeds in 1918 and educated at the university there, Hoggart saw action in World War II in North Africa and Italy. Military experience had a marked experience on him, as it did on Williams. After the war Hoggart worked alongside Larkin, in his case as a tutor in literature in the Department of Adult Education at the University of Hull, and while there published his first full-length critical work, *Auden*. But it was in *The Uses of Literacy* that all his experience, his working-class background, his army life, his teaching in the adult education department of a provincial university, came together. It was as if he had found a vocabulary for a side of life that, hitherto, had lacked one.[80]

Hoggart was trained in the traditional methods of practical literary criticism as devised by I. A. Richards (see chapter 18), and the 'Great Tradition' of F. R. Leavis, but his actual experience led him in a very different direction. He moved against Leavis rather as Ginsberg had moved against Lionel Trilling.[81] Instead of following in the Cambridge tradition, he brought Richards's methods to bear on the culture he himself knew – from the singing in working men's clubs to weekly family magazines, from commercial popular songs to the films that ordinary people flocked to time and again. Like an anthropologist he described and analysed the customs he had grown up not even questioning, such as washing the car on a Sunday morning, or scrubbing the front step. His book did two things. It first described in detail the working-class culture, in particular its language – in the books, magazines, songs, and games it employed. In doing so, it showed, second, how rich this culture was, how much more there was to it than its critics alleged. Like Osborne, Hoggart wasn't blind to its shortcomings, or to the fact that, overall, British society deprived people born into the working class of the chance to escape it. But Hoggart's aim was more description and analysis than any nakedly political intent. Many responded to Hoggart and Osborne alike. A legitimacy, a voice, was suddenly given to an aspect of affairs that hitherto had been overlooked. Here was another fine tradition.[82]

Hoggart led naturally to **Raymond Williams**. Like Hoggart, Williams had served in the war, though most of his life had been spent in the English Department at Cambridge, where he could not help but be aware of Leavis. Williams was more of a theoretician than Hoggart and a less compelling observer, but he was equally convincing in argument. In a series of books, beginning with *Culture and Society* in 1958, Williams made plain and put into context what had been implicit in the narrow scope of Hoggart's work.[83] This was in effect a new aesthetic. Williams's basic idea was that a work of art – a painting, a novel, a poem, a film – does not exist without a context. Even a work with wide applicability, 'a universal icon,' has an intellectual, social, and above all a political background. This was Williams's main argument, that the imagination cannot avoid a relation with power, that the form art takes and our attitudes toward it are themselves a form of politics. Not necessarily party politics but the acknowledgement of this relationship – culture and power – is the ultimate form of self-awareness. In *Culture and Society*, having first considered Eliot, Richards, and Leavis, all as authors who consider 'culture' as having different levels and where only an educated minority can really benefit from and contribute toward the highest level, Williams proceeds to a chapter headed 'Marxism and Culture.' In Marxist theory, Williams reminds us, the determining fact of life is the means of production and distribution, and so the progress of culture, like everything else, is dependent upon the material conditions for the production of that culture. Culture therefore cannot help but reflect the social makeup of society, and on such an analysis it is only natural that those at the top should not want change. On this view, then, Eliot and Leavis are merely reflecting the social circumstances of their time, and in so doing are exhibiting a conspicuous lack of self-awareness.[84]

Several things follow from this (oversimplified) account of Williams's arguments. One is that there is no one criterion by which to judge an artist, or a work of art. Elites, as viewed by Eliot or Leavis, are merely one segment of the population with their own special interests. Instead, Williams advises us to trust our own experience as to whether an artist or his work is relevant, the point being that all viewpoints may be equally relevant or valid. In this sense, though Williams himself was steeped in what most people would recognise as high culture, he was attacking that very tradition. Williams's theories also imply that, in developing new ideas, artists are breaking new ground not only aesthetically but politically as well. It was this conjoining of art and politics that would lead in time to what is sometimes known as the Cultural Left.

Two final assaults on the Eliot-Leavis-Trilling-Commager canon came from history and from science. The historical challenge was led first by the French *Annales* school, and second by the British school of Marxist historians. The achievements of their approach will be discussed more fully in chapter 31, but for now it is enough to say that these historians drew attention to the fact that 'history' happens to 'ordinary' people as well as to kings and generals and prime ministers, that such history as that pertaining to entire peasant villages, as reconstructed from, say, birth, marriage, and death records, can be just as gripping and important as the chronicles of major battles and treaties, that life moves forward and acquires meaning by other ways than war or politics. In so doing, history joined other disciplines in drawing attention to the world of the 'lower orders,' revealing how rich their lives could be. What Hoggart had done for the working class of twentieth-century Britain, the *Annales* school did, for example, for the peasants of fifteenth-century Languedoc or Montaillou. The British Marxist historians – Rodney Hilton, Christopher Hill, Eric Hobsbawm, and E. P. Thompson among others – also concentrated on the lives of 'ordinary' people: peasants, the lower ranks of the clergy, and in Thompson's classic work, the English working classes. The thrust of all these studies was that the lower orders were an important element in history and that they knew they were, acting rationally in their own interests, not mere fodder for their social superiors.

History, anthropology, archaeology, even the discipline of English itself in Williams's hands and, quite separately, in Achebe's, Baldwin's, Ginsberg's, Hoggart's, and Osborne's works, all conspired in the mid- to late 1950s to pull the rug out from under the traditional ideas of what high culture was. New writing, new discoveries, were everywhere. The idea that a limited number of 'great books' could provide the backbone, the core, of a civilisation seemed increasingly untenable, remote from reality. In material terms, America was now vastly more prosperous than Europe; why should its people look to European authors? Former colonies were exalted by their newfound histories; what need did they have of any other? There were answers to these questions – good answers – but for a time no one seemed interested. And then came an unexpected blow from a quite different direction.

The most frontal attack on Eliot–Leavis *et alia* may be precisely dated and located. The setting was Cambridge, England, and the time a little after five o'clock on the afternoon of 7 May 1959. That was when a 'bulky, shambling figure approached the lectern at the western end of the Senate House,' a white stone building in the centre of the city.[85] The room, in an ornately plastered neoclassical building, was packed with senior academics, students, and a number of distinguished guests, assembled for one of Cambridge's 'showpiece public occasions,' the annual Rede lecture. That year the speaker was Sir Charles

Snow, later to be Lord Snow but universally known by his initials, as **C. P. Snow**. 'By the time he sat down over an hour later,' as Stefan Collini tells the story, 'Snow had done at least three things: he had launched a phrase, perhaps even a concept, on an unstoppably successful international career; he had formulated a question ... which any reflective observer of modern societies needs to address; and he had started a controversy which was to be remarkable for its scope, its duration, and, at least at times, its intensity.'[86] The title of Snow's lecture was '**The Two Cultures** and the Scientific Revolution,' and the two cultures he identified were those of 'the literary intellectuals' and of the natural scientists, 'between whom he claimed to find a profound mutual suspicion and incomprehension, which in turn, he said, had damaging consequences for the prospects of applying technology to the world's problems.'[87]

Snow had chosen his moment. Cambridge was Britain's foremost scientific institution, but it was also the home of F. R. Leavis (and Raymond Williams), as we have seen, one of the country's foremost advocates of traditional literary culture. And Snow was himself a Cambridge man, who had worked in the Cavendish Laboratory under Ernest Rutherford (though he was an undergraduate at Leicester). His scientific career had suffered a setback in 1932 when, after announcing that he had discovered how to produce vitamin A by artificial methods, he was forced to recant because his calculations proved faulty.[88] He never did scientific research again after that but instead became a government scientific adviser and a novelist, with a multivolume series, 'Strangers and Brothers', about the decision-making processes in a series of closed communities (such as professional societies or Cambridge colleges). These were much derided by advocates of 'high' literature who found, or affected to find, his style stilted and pompous. Snow thus both bridged – and yet did not bridge – the two cultures about which he had such strong views.

Snow's central point applied across the world, he said, and the reaction to his lecture certainly justified that claim. But it was also true that it applied more than anywhere in Britain, where it was thrown into its starkest contrast. Literary intellectuals, said Snow, controlled the reins of power both in government and in the higher social circles, which meant that only people with, say, a knowledge of the classics, history, and/or English literature were felt to be educated. Such people did not know much – or often any – science; they rarely thought it important or interesting and as often as not left it out of the equation when discussing policy in government, or regarded it as boring socially. He thought this form of ignorance was disgraceful, dangerous, and when applied to government, that it failed the country. At the same time, he thought scientists culpable in often being ill-educated in the humanities, apt to dismiss literature as invalid subjectivism with nothing to teach *them*.

Reading Snow's lecture, one is struck by the many sharp observations he makes along the way. For example, he finds scientists more optimistic than the literary intellectuals, that they tend to come from poorer homes (both in Britain and 'probably' in the United States). He found literary intellectuals vainer than scientists, in effect 'tone-deaf' to the other culture, whereas at least scientists knew what they were ignorant of.[89] He also found the literary intellectuals jealous of their scientific colleagues: 'No young scientist of any talent would feel that he isn't wanted or that his work is ridiculous, as did the hero of *Lucky Jim*, and in fact some of the disgruntlement of [Kingsley] Amis and his associates is the disgruntlement of the under-employed arts graduate.'[90] Many literary intellectuals, he concluded, were natural Luddites. But it was the description of the two cultures, and the immense gap in between, that was his main point, supported by his argument that the world was then entering a scientific revolution.[91] This he separated from the industrial revolution in the following way. The industrial revolution had been about the introduction of machinery, the creation of factories and then cities, which had changed human experience profoundly. The scientific revolution, he said, dated from 'when atomic particles were first made industrial use of. I believe the industrial society of electronics, atomic energy, automation, is in cardinal respects different in kind from any that has gone before, and will change the world much more.' He surveyed science education in Britain, the United States, Russia, France, and Scandinavia and found Britain most wanting (he thought the Russians had it about right but was uncertain of what they had produced).[92] He concluded by arguing that the proper administration of science, which could only come about when the literary intellectuals became familiar with these alien disciplines and dropped their prejudices, would help solve the overriding problems of rich and poor countries that bedevilled the planet.[93]

Snow's lecture provoked an immense reaction. It was discussed in many languages Snow could not speak, so he never knew what was being said (in, for example, Hungary, Japan, Poland). Many of the comments agreed with him, more or less, but from two sources came withering – and in one case very personal – criticism. This latter was none other than F. R. Leavis, who published a lecture he had given on Snow as an article in the *Spectator*. Leavis attacked Snow on two grounds. At the more serious level, he argued that the methods of literature related to the individual quite differently from the methods of science, 'because the language of literature was in some sense the language of the individual – not in an obvious sense but at least in a *more* obvious sense than the language of science.' 'For Leavis, neither the physical universe nor the discourse of its notation was possessed by observers in the way in which literature could be possessed by its readers; or by its writers – because he would claim that literature and literary culture was constructed not from words learned but from intercourse.'[94] At the same time, however, Leavis also mounted a personal attack on Snow himself. So personal was Leavis's venom that both the *Spectator* and the publishers Chatto & Windus, who reprinted the article in an anthology, approached Snow to see if he would sue. He did not, but it is difficult to see how he could not have been hurt.[95] Leavis began, 'If confidence in oneself as a master-mind, qualified by capacity, insight, and knowledge to pronounce authoritatively on the frightening problems of our civilisation, is genius, then there can be no doubt about Sir Charles Snow's. He has no hesitations.' When Leavis delivered the lecture, a pause followed this sentence. Then he went on: 'Yet Snow is, in fact, portentously ignorant.'[96]

Nonetheless, the most cogent criticism came not from Leavis but from Lionel

Trilling in New York. He put down Leavis, both for his bad manners and for being so personal, and because he had come to the defence of modern writers that, hitherto, he had no time for. At the same time, Trilling thought Snow had absurdly overstated his case. It was impossible, he said, to characterise a vast number of writers in what he described as a 'cavalier' way. Science might hang together logically or conceptually, but not literature. The activities that comprise 'literature' are too varied to be compared with science in so simple a fashion.[97] But was that true? Whatever Trilling might say, the 'two cultures' debate is still going on in some quarters – Snow's lecture was reprinted in 1997 with a long introduction by Stefan Collini detailing its many ramifications all over the world, and in 1999 the BBC held a public debate entitled 'The Two Cultures 40 Years On.' It is now obvious at least that Snow was right about the importance of the electronic/information revolution. And Snow himself is remembered more for his lecture than for his novels.[98] As will be argued in the conclusion, the end of the twentieth century sees us living in what might be termed a 'crossover culture,' where popular (but quite difficult) science books sell almost as well as novels and rather better than books of literary criticism. People *are* becoming more scientifically literate. Whether or not one agrees wholeheartedly with Snow, it is difficult not to feel that, like Riesman, he had put his finger on something.

And so, piece by piece, book by book, play by play, song by song, discipline by discipline, the traditional canon began to crumble, or be undermined. For some this change had a liberating effect; for others it was profoundly unsettling, producing a sense of loss. Others, more realistic perhaps, took the changes in their stride. Knowing more science, or being familiar with the works of, say, Chinua Achebe, James Baldwin, or John Osborne, did not necessarily mean throwing traditional works out of the window. But undoubtedly, from the 1950s on, the sense of a common pursuit, a great tradition shared among people who regarded themselves as well educated and cultured, began to break down. Indeed, the very idea of high culture was regarded in many quarters with suspicion. The words 'high culture' themselves were often now written embedded (if not yet embalmed) in quotation marks, as if this were an idea not to be trusted or taken seriously. This attitude was fundamental to the new aesthetic which, in the later decades of the century, would become known as post-modernism.

Despite the viciousness of Leavis's attack on Snow, there was one especially powerful argument he didn't use, presumably because he was unaware of it, but which, in the 1950s, would grow increasingly important. Snow had emphasised the success of the scientific approach – empirical, coldly rational, self-modifying. Paradoxically, at the very time Snow and Leavis were trading blows, evidence was accumulating that the 'culture' of science was not quite the way Snow portrayed it, that it was actually a far more 'human' activity than appeared from a mere reading of what appeared in scientific journals. This new view of science, to which we now turn, would also help shape the so-called postmodern condition.

FORCES OF NATURE

By insisting that science was a 'culture' just as much as serious literature was, C. P. Snow was emphasising both the intellectual parity of the two activities and, at the same time, their differences. Perhaps the most important difference was the scientific *method* – the process of empirical observation, rational deduction, and continuous modification in the light of experience. On this basis, scientists were depicted as the most rational of beings, unhindered in their activities by such personal considerations as rivalry, ambition, or ideology. Only the evidence counted. Such a view was supported by the scientific papers published in professional journals. The written style was invariably impersonal to the point of anonymity, with a near-universal formal structure: statement of the problem; review of the literature; method; results; conclusion. In the journals, science proceeded by orderly steps, one at a time.

There was only one problem with this view: it wasn't true. It wasn't close to true. Scientists knew this, but for a variety of reasons, one of which was the insecurity Snow highlighted, it was rarely if ever broadcast. The first person to draw attention to the real nature of science was yet another Austro-Hungarian emigré, **Michael Polanyi**, who had studied medicine and physical chemistry in Budapest and at the Kaiser Wilhelm Institute in Berlin before World War II. By the end of the hostilities, however, Polanyi was professor of sociology at Manchester University (his brother Karl was an economist at Columbia). In his 1946 Riddell lectures, at the University of Durham, published as *Science, Faith and Society*, Michael Polanyi advanced two fundamental points about science that would come to form a central plank in the late–twentieth-century sensibility.[1] He first said that much of science stems from guesswork and intuition and that although, in theory, science is continually modifiable, in practice it doesn't work out like that: 'The part played by new observations and experiment in the process of discovery is usually over-estimated.'[2] 'It is not so much new facts that advance science but new interpretations of known facts, or the discovery of new mechanisms or systems that account for known facts.' Moreover, advances 'often have the character of a gestalt, as when people suddenly "see" something that had been meaningless before.'[3] His point was that scientists actually behave far more intuitively than they think, and that, rather than being absolutely neutral or disengaged in their research, they start with a conscience, a scientific conscience. This conscience operates in more than one way. It guides the scientist in choosing a path of discovery, but it also guides him in accepting which results are 'true' and which are not, or need further study. This conscience, in both senses, is a fundamental motivating force for the scientist.

Polanyi, unlike others perhaps, saw science as a natural outgrowth of religious society, and he reminded his readers that some of the founders of the Christian church – like Saint Augustine – were very interested in science. For Polanyi, science was inextricably linked to freedom and to an atomised society; only in such an environment could men make up their own minds as true independents. But for him, this was an outgrowth of monotheistic religion, Christianity in particular, which gave the world the idea, the tradition, of 'transcendent truth,' beyond any one individual, truth that is 'out there,' waiting to be found. He examined the structure of science, observing for example that few fellows of the Royal Society ever objected that any of their colleagues were unworthy, and that few injustices were done, in that no one was left out of the society who was worthy of inclusion. Science, and fairness, are linked.

Polanyi saw the *tradition* of science, the search for objective, transcendent truth, as at base a Christian idea, though of course much developed – evolved – beyond the times when there was only revealed religion. The development of science, and the scientific method, he felt, had had an effect on toleration in society, and on freedom, every bit as important as its actual findings. In fact, Polanyi saw an eventual return to God; for him, the development of science, and the scientific way of thinking and working, was merely the latest stage in fulfilling God's purpose, as man makes moral progress. The fact that scientists operate so much from intuition and according to their consciences only underlines his point.[4]

George Orwell disagreed. He believed science to be coldly rational, and no one detested or feared this cold rationalism more than he did. Both *Animal Farm* and *Nineteen Eighty-Four* are ostensibly political novels. When the latter was published in 1948, it was no less contentious than Orwell's earlier book and was again interpreted by conservatives as an attack on the totalitarian nature of socialism by a former socialist who had seen the light. But this is not how the author saw it himself. As much as anything, it was a pessimistic attack on science. Orwell was pessimistic partly because he was ill with TB, and partly because the postwar world of 1948 was still very grim in Britain: the meat ration (two chops a week) was not always available, bread and potatoes were still rationed, soap was coarse, razor blades were blunt, elevators didn't work, and according to Julian Symons, Victory gin gave you 'the sensation of being hit on the head with a rubber club.'[5] But Orwell never stopped being a socialist, and he knew that if it was to develop and succeed, it would have to take on the fact of Stalinism's brutality and totalitarian nature. And so, among the ideas that Orwell attacks in *Nineteen Eighty-Four*, for example, is the central argument of *The Managerial Revolution* by James Burnham, that a 'managerial class' – chief among whom were scientists, technicians, administrators, and bureaucrats – was gradually taking over the running of society in all countries, and that terms like *socialist* and *capitalist* had less and less meaning.[6] But the real power of the book was Orwell's uncanny ability to evoke and predict totalitarian society, with its scientific and mock-scientific certainties. The book opens with the now-famous line, 'It was a bright cold day in April, and the clocks were striking thirteen.' The clocks do not (yet) strike thirteen, but Orwell's quasi-scientific ideas about Thought Police, Newspeak, and memory holes (a sort of shredder whereby the past is consigned to oblivion) are already chillingly familiar. Phrases like 'Big brother is watching you' have passed into the language partly because the technology now exists to make this possible.

Orwell's timing for *Nineteen Eighty-Four* could not have been better. The year in which the book was published, 1948, saw the beginning of the Berlin blockade, when Stalin cut off electricity to the western zones of the divided city, and all access by road and rail from West Germany. The threat of Stalinism was thus made plain for all to see. The blockade lasted nearly a year, until May 1949, but its effects were more permanent because the whole episode concentrated the minds of the Western powers, who now realised that the Cold War was here to stay. But Orwell's timing was also good because *Nineteen Eighty-Four* coincided exactly with a very different set of events taking place on the intellectual front inside Russia which showed, just as much as the Berlin blockade, what Stalinism was all about. This was the Lysenko affair.

We have already seen, in chapter 17, how in the 1930s Soviet biology was split between traditional geneticists, who supported Western ideas – Darwin, Mendelian laws of inheritance, Morgan's work on the chromosome and the gene – and those who followed the claims of **Trofim Lysenko**, who embraced the Lamarckian idea of the inheritance of acquired characteristics.[7] During and immediately after World War II the situation inside Russia changed substantially. War concentrates the mind wonderfully, and thanks to the requirements of a highly mechanised and highly technical war, the Russian leadership needed scientists as it had never needed them before. As a result, science inside Russia was rapidly reorganised, with scientists rather than party commissars being placed in charge of key committees. Everything from geology to medicine was revamped in this way, and in several cases leading scientists were elevated to the rank of general. Brought in from the cold after the inquisition of the 1930s, scientists were given priority housing, allowed to eat in the special restaurants otherwise reserved for party apparatchiks and to use the special hospitals and sanitaria that had hitherto been the prerogative only of high party officials. The Council of Ministers even passed a resolution that provided for the building of dachas for academicians. More welcome still was the abolition of strict control over science by party philosophers that had been in place since the mid-1930s.

The war was particularly beneficial for genetics in Russia because, from 1941 on, Soviet Russia was an ally in particular of the United States and Great Britain. As a direct result of this alliance, the scientific barriers erected by Stalinism in the 1930s were dismantled. Soviet scientists were allowed to travel again, to visit American and British laboratories; foreign scientists (for example, Henry Dale, J. B. S. Haldane, and Ernest Lawrence) were again elected to

Russian academies, and foreign journals were once more permitted inside the Soviet Union.[8] Many of the Russian geneticists who opposed Lysenko took this opportunity to enlist the aid of Western colleagues – especially British and American biologists, and Russian emigrés in the United States, people like Theodosius Dobzhansky. They were further aided by the development of the 'evolutionary synthesis' (see chapter 20), which linked genetics and Darwinism and therefore put intellectual pressure on Michurin and Lysenko. Mendelian and Morgan-style experimentation and theory were reinstated, and thousands of boxes of *Drosophila* were imported into Russia in the immediate postwar years. As a direct result of all this activity, Lysenko found his formerly strong position under threat, and there was even an attempt to remove him from his position as a member of the praesidium of the Academy of Sciences.[9] Letters of complaint were sent to Stalin, and for a while the Soviet leadership, hitherto very much in Lysenko's camp, stood back from the debate. But only for a while.

The start of the Cold War proper was signalled in spring 1946 by Winston Churchill's 'Iron Curtain' speech in Fulton, Missouri, but the confrontation really began in March 1947 with the announcement of the 'Truman Doctrine,' with aid to Greece and Turkey designed specifically to counteract the influence of communism. Shortly afterwards, Communists were expelled from the coalition governments in France and Italy. In Russia, one of the consequences was a new, strident ideological campaign that became known as *zhdanovshchina*, after **Andrei Zhdanov**, a member of the Politburo, who announced a series of resolutions laying down what was and was not politically correct in the media. At first writers and artists were cautioned against 'servility and slavishness before Western culture,' but at the end of 1946 an Academy of Social Sciences was created in Moscow under Agitprop control, and in the spring of 1947 *zhdanovshchina* was extended to philosophy. By the summer, science was included. At the same time, party ideologists resumed their control as authorities over science. Russian scientists who had gone abroad and not returned were now attacked publicly, the election of eminent Western scholars to Russian academies was stopped, and several academic journals were closed, especially those published in foreign languages. So far as science was concerned, Stalinist Russia had come full circle. As the pendulum swung back his way, Lysenko began to reassert his influence. His main initiative was to help organise a major public debate at VASKhNIL, the Lenin All-Union Academy of Agricultural Sciences, on the subject of 'the struggle for existence.' By putting Darwin centre stage, it was Lysenko's intention to highlight not only the division between 'Mendelian-Morganists' and 'Michurinists' but to extend that division from the narrow field of genetics to the whole of biology, a naked power play. The central issue in the debate was between those who, like Lysenko, denied that there was competition *within* species, only that *inter*specific competition existed, and those traditionalists who argued that there was competition throughout all spheres of life. Marx, it will be remembered, had admired Darwin, and had conceived history as a dialectic, a struggle. By Lysenko's time, however, the official doctrine of Stalinism was that men are equal, that in a socialist society cooperation – and not competition – is what counts, and that differences between people (i.e., within the species) are not hereditary but solely produced by the environment. The debate was therefore designed to smoke out which scientists were in which camp.[10]

For some reason Stalin had always warmed to Lysenko. It seems the premier had pronounced views of his own on evolution, which were clearly Lamarckian. One reason for this may have been because Lamarck's views were felt to accord more closely with Marxism. A more pressing reason may have been that the Michurinist/Lysenkoist approach fitted with Stalin's rapidly developing views about the Cold War and the need to denounce everything Western. At any rate, he gave Lysenko a special consignment of 'branching wheat' to test his theories, and in return the 'scientist' kept Stalin regularly informed about the battle between the Michurinists and the Mendelians. And so, when this issue finally reached the Lenin All-Union Academy meeting in August 1948, Stalin took Lysenko's line, even going so far as to annotate conference documents with his own comments.[11]

The conference itself was a carefully staged victory for Lysenko. Following his opening address, five days were devoted to a discussion. However, his opponents were not allowed to speak for the first half of the meeting, and overall only eight of the fifty-six speakers were allowed to criticise him.[12] At the end, not only did the conference ratify Lysenko's approach, but he revealed he had the support of the Central Committee, which meant, in effect, that he had Stalin's full endorsement for total control, over not just genetics but all of Soviet biology. The VASKhNIL meeting was also followed by a sustained campaign in *Pravda*. Normally, the newspaper consisted of four pages; that summer for nine days the paper produced six-page editions with an inordinate amount of space devoted to biology.[13] A colour film about Michurin was commissioned, with music by Shostakovich. It is difficult to exaggerate the intellectual importance of these events. Recent research, published by Nikolai Krementsov, has revealed that Stalin spent part of the first week of August 1948 editing Lysenko's address; this was at exactly the time he was meeting with the ambassadors of France, Britain, and the United States for prolonged con-sultations on the Berlin crisis. After the conference, at the premier's instigation, great efforts were made to export Michurinist biology to newborn socialist countries such as Bulgaria, Poland, Czechoslovakia, and Romania. Biology, more than any other realm of science, concerns the very stuff of human nature, for which Marx had set down certain laws. Biology was therefore more of a potential threat to Marxist thought than any other science. The Lysenko version of genetics offered the Soviet leadership the best hope for producing a science that posed no threat to Marxism, and at the same time set Soviet Russia apart from the West. With the Iron Curtain firmly in place and communications between Russian scientists and their Western colleagues cut to a minimum, the path was set for what has rightly been called the death of Russian genetics. For the USSR it was a disaster.

The personal rivalry, political manoeuvring, self-deception, and sheer cussedness that disfigured Soviet genetics for so long is of course the very antithesis of the way science prefers itself portrayed. It is true that the Lysenko affair may be the very worst example of political interference in an important scientific venture, and for that reason the lessons it offers are limited. In the West there was nothing strictly comparable but even so, in the 1950s, there were other very significant advances made in science which, on examination, were shown to be the fruits of anything but calm, reflective, disinterested reason. On the contrary, these advances also resulted from bitter rivalry, overweening ambition, luck, and in some cases downright cheating.

Take first the jealous nature of **William Shockley**. That, as much as anything, was to account for his massive input into twentieth-century intellectual history. That input may be said to have begun on Tuesday, 23 December 1947, just after seven o'clock in the morning, when Shockley parked his MG convertible in the parking lot of Bell Telephone Laboratories in Murray Hill, New Jersey, about twenty miles from Manhattan.[14] Shockley, a thin man without much hair, took the stairs to his office on the third floor of the lab. He was on edge. Later in the day, he and two colleagues were scheduled to reveal a new device they had invented to the head of Bell Labs, where they worked. Shockley was tense because although he was the nominal head of his little group of three, it had actually been the other two, **John Bardeen** and **Walter Brattain**, who had made the breakthrough. Shockley had been leapfrogged.[15] During the morning it started to snow. Ralph Bown, the research director of Bell, wasn't deterred however, and stopped by after lunch. Shockley, Bardeen, and Brattain brought out their device, a small triangle of plastic with a piece of gold foil attached, fixed in place by a small spring made from a paper clip.[16] Their contraption was encased in another piece of plastic, transparent this time, and shaped like a capital C. 'Brattain fingered his moustache and looked out at the snow. The baseball diamond below the lab window was beginning to disappear. The tops of the trees on the Wachtung Mountains in the distance were also lost as the low cloud closed in. He leaned across the lab bench and switched on the equipment. It took no time at all to warm up, and the oscilloscope to which it was connected immediately showed a luminous spot that raced across the screen.'[17] Brattain now wired the device to a microphone and a set of headphones, which he passed to Bown. Quietly, Brattain spoke a few words into the microphone – and Bown shot him a sharp glance. Brattain had only whispered, but what Bown heard was anything but a whisper, and that was the point of the device. The input had been amplified. The device they had built, an arrangement of germanium, gold foil, and a paper clip, was able to boost an electrical signal almost a hundredfold.[18]

Six months later, on 30 June 1948, Bown faced the press at the Bell Headquarters on West Street in Manhattan, overlooking the Hudson River. He held up the small piece of new technology. 'We have called it the **Transistor**,' he explained, 'because it is a resistor or semiconductor device which can amplify electrical signals as they are transferred through it.'[19] Bown had high hopes for the new device; at that time the amplifiers used in telephones were clumsy and unreliable, and the vacuum tubes that performed the same function in radios were bulky, broke easily, and were very slow in warming up.[20] The press, or at least the *New York Times*, did not share this enthusiasm, and its report was buried in an inside section. It was at this point that Shockley's jealousy paid off. Anxious to make his own contribution, he kept worrying about the uses to which the transistor might be put. Looking at the world around him, the mass-society world of standardisation, he grasped that if the transistor were to be manufactured in bulk, it needed to be simpler and stronger.

The transistor was in fact a development of two inventions made much earlier in the century. In 1906 **Lee de Forest** had stumbled across the fact that an electrified wire mesh, placed in the path of a stream of electrons in a vacuum tube, could 'amplify' the flow at the outgoing end.[21] This natural amplification was the most important aspect of what came to be called the electronics revolution, but de Forest's discovery was built on by solid-state physics. This was due to a better grasp of electricity, itself the result of advances in particle physics. A solid structure will conduct electricity if the electron in its outer shell is 'free' – i.e., that shell isn't 'full' (this goes back to Pauli's exclusion principle and Linus Pauling's research on the chemical bond and how it affected reactivity). Copper conducts electricity because there is only one electron in its outer shell, whereas sulphur, for example, which does not carry electricity at all, has all its electrons tightly bound to their nuclei. Sulphur, therefore, is an insulator.[22] But not all elements are this simple. 'Semiconductors' (silicon, say, or germanium) are forms of matter in which there are a few free electrons but not many. Whereas copper has one free electron for each atom, silicon has a free electron for every *thousand* atoms. It was subsequently discovered that such semiconductors have unusual and very useful properties, the most important being that they can conduct (and amplify) under certain conditions, and insulate under others. It was Shockley, smarting from being beaten to the punch by Bardeen and Brattain, who put all this together and in 1950 produced the first, simple, strong, semiconductor transistor, capable of being mass-produced.[23] It consisted of a sliver of silicon and germanium with three wires attached. In conversation this device was referred to as a 'chip.'[24]

Shockley's timing was perfect. Long-playing records and 'singles' had recently been introduced to the market, with great success, and the pop music business was taking off. In 1954, the very year Alan Freed started playing R & B on his shows, a Dallas company called Texas Instruments began to manufacture chip-transistors for the new portable radios that had just gone on sale, which were cheap (less than $50) and therefore ideal for playing pop all day long. For reasons that have never been adequately explained, TI gave up this market, which was instead taken over by a Japanese firm no one had ever heard of, Sony.[25] By then Shockley had fallen out with first one, then the other erstwhile colleague. Bardeen had stormed out of the lab in 1951, unable to cope with Shockley's intense rivalry, and Brattain, likewise unable to stomach his former boss, had himself reassigned to a different section of Bell Labs. When the three of them gathered in Stockholm in 1956 to receive the Nobel Prize for Physics, the atmosphere was icy, and it was the last time they would be in the same room together.[26] Shockley had himself left Bell by that time, forsaking the snow of

New Jersey for the sunshine of California, in particular a pleasant valley of apricot orchards south of San Francisco. There he opened the Shockley Semiconductor Laboratory.[27] To begin with, it was a small venture, but in time the apricots would be replaced by more laboratories. In conversation the area was referred to as **Silicon Valley.**

Shockley, Bardeen, and Brattain fought among themselves. With the discovery of **DNA**, the long-chain molecule that governs reproduction, the rivalry was between three separate groups of researchers, on different continents, some of whom never met. But feelings ran just as high as between Shockley and his colleagues, and this was an important factor in what happened.

The first the public knew about this episode came on 25 April 1953, in *Nature*, in a 900-word paper entitled 'Molecular Structure of Nucleic Acids.' The paper followed the familiar, ordered layout of *Nature* articles. But although it was the paper that created the science of molecular biology, and although it also helped kill off Lysenkoism, it was the culmination of an intense two-year drama in which, if science really were the careful, ordered world it is supposed to be, the wrong side won.

Among the personalities, **Francis Crick** stands out. Born in Northampton in 1916, the son of a shoemaker, Crick graduated from London University and worked at the Admiralty during World War II, designing mines. It was only in 1946, when he attended a lecture by Linus Pauling, that his interest in chemical research was kindled. He was also influenced by Erwin Schrödinger's *What Is Life?* and its suggestion that quantum mechanics might be applied to genetics. In 1949 he was taken on by the Cambridge Medical Research Council Unit at the Cavendish Laboratory, where he soon became known for his loud laugh (which forced some people to leave the room) and his habit of firing off theories on this or that at the drop of a hat.[28] In 1951 an American joined the lab. **James Dewey Watson** was a tall Chicagoan, twelve years younger than Crick but extremely self-confident, a child prodigy who had also read Schrödinger's *What Is Life?* while he was a zoology student at the University of Chicago, which influenced him toward microbiology. As science historian Paul Strathern tells the story, on a visit to Europe Watson had met a New Zealander, **Maurice Wilkins**, at a scientific congress in Naples. Wilkins, then based at King's College in London, had worked on the Manhattan Project in World War II but became disillusioned and turned to biology. The British Medical Research Council had a biophysics unit at King's, which Wilkins then ran. One of his specialities was X-ray diffraction pictures of DNA, and in Naples he generously showed Watson some of the results.[29] It was this coincidence that shaped Watson's life. There and then he seems to have decided that he would devote himself to discovering the structure of DNA. He knew there was a Nobel Prize in it, that molecular biology could not move ahead without such an advance, but that once the advance was made, the way would be open for genetic engineering, a whole new era in human experience. He arranged a transfer to the Cavendish. A few days after his twenty-third birthday Watson arrived in Cambridge.[30]

What Watson didn't know was that the Cavendish had 'a gentleman's agreement' with King's. The Cambridge laboratory was studying the structure of protein, in particular haemoglobin, while London was studying DNA. That was only one of the problems. Although Watson hit it off immediately with Crick, and both shared an amazing self-confidence, that was virtually all they had in common. Crick was weak in biology, Watson in chemistry.[31] Neither had any experience at all of X-ray diffraction, the technique developed by the leader of the lab, **Lawrence Bragg**, to determine atomic structure.[32] None of this deterred them. The structure of DNA fascinated both men so much that virtually all their waking hours were spent discussing it. As well as being self-confident, Watson and Crick were highly competitive. Their main rivals came from King's, where Maurice Wilkins had recently hired the twenty-nine-year-old **Rosalind Franklin** ('Rosy,' though never to her face).[33] Described as the 'wilful daughter' of a cultured banking family, she had just completed four years X-ray diffraction work in Paris and was one of the world's top experts. When Franklin was hired by Wilkins she thought she was to be his equal and that she would be in charge of the X-ray diffraction work. Wilkins, on the other hand, thought that she was coming as his assistant. The misunderstanding did not make for a happy ship.[34]

Despite this, Franklin made good progress and in the autumn of 1951 decided to give a seminar at King's to make known her findings. Remembering Watson's interest in the subject, from their meeting in Naples, Wilkins invited the Cambridge man. At this seminar, Watson learned from Franklin that DNA almost certainly had a helical structure, each helix having a phosphate-sugar backbone, with attached bases: adenine, guanine, thymine, or cytosine. After the seminar, Watson took Franklin for a Chinese dinner in Soho. There the conversation turned away from DNA to how miserable she was at King's. Wilkins, she said, was reserved, polite, but cold. In turn, this made Franklin on edge herself, a form of behaviour she couldn't avoid but detested. At dinner Watson was outwardly sympathetic, but he returned to Cambridge convinced that the Wilkins–Franklin relationship would never deliver the goods.[35]

The Watson–Crick relationship meanwhile flourished, and this too was not unrelated to what happened subsequently. Because they were so different, in age, cultural, and scientific background, there was precious little rivalry. And because they were so conscious of their great ignorance on so many subjects relevant to their inquiry (they kept Pauling's *Nature of the Chemical Bond* by their side, as a bible), they could slap down each other's ideas without feelings being hurt. It was light-years away from the Wilkins–Franklin ménage, and in the long run that may have been crucial.

In the short run there was disaster. In December 1951, Watson and Crick thought they had an answer to the puzzle, and invited Wilkins and Franklin for a day in Cambridge, to show them the model they had built: a triple-helix structure with the bases on the outside. Franklin savaged them, curtly grumbling that their model didn't fit any of her crystallography evidence, either for the helical structure or the position of the bases, which she said were on the *inside*. Nor did their model take any account of the fact that in nature DNA existed in association with water, which had a marked effect on its structure.[36] She was genuinely appalled at their neglect of her research and complained that her day in Cambridge was a complete waste of time.[37] For once, Watson and Crick's ebullient self-confidence let them down, even more so when word of the debacle reached the ears of their boss. Bragg called Crick into his office and put him firmly in his place. Crick, and by implication Watson, was accused of breaking the gentleman's agreement, of endangering the lab's funding by doing so. They were expressly forbidden from continuing to work on the DNA problem.[38]

So far as Bragg was concerned, that was the end of the matter. But he had misjudged his men. Crick did stop work on DNA, but as he told colleagues, no one could stop him *thinking* about it. Watson, for his part, continued work in secret, under cover of another project on the structure of the tobacco mosaic virus, which showed certain similarities with genes.[39] A new factor entered the situation when, in the autumn of 1952, **Peter Pauling**, Linus's son, arrived at the Cavendish to do postgraduate research. He attracted a lot of beautiful women, much to Watson's satisfaction, but more to the point, he was constantly in touch with his father and told his new colleagues that Linus was putting together a model for DNA.[40] Watson and Crick were devastated, but when an advance copy of the paper arrived, they immediately saw that it had a fatal flaw.[41] It described a triple-helix structure, with the bases on the outside – much like their own model that had been savaged by Franklin – and Pauling had left out the ionisation, meaning his structure would not hold together but fall apart.[42] Watson and Crick realised it would only be a matter of time before Pauling himself realised his error, and they estimated they had six weeks to get in first.[43] They took a risk, broke cover, and told Bragg what they were doing. This time he didn't object: there was no gentleman's agreement so far as Linus Pauling was concerned.

So began the most intense six weeks Watson or Crick had ever lived through. They now had permission to build more models (models were especially necessary in a three-dimensional world) and had developed their thinking about the way the four bases – **adenine, guanine, thymine**, and **cytosine** – were related to each other. They knew by now that adenine and guanine were attracted, as were thymine and cytosine. And, from Franklin's latest crystallography, they also had far better pictures of DNA, giving much more accurate measures of its dimensions. This made for better model building. The final breakthrough came when Watson realised they could have been making a simple error by using the wrong isomeric form of the bases. Each base came in two forms – *enol* and *keto* – and all the evidence so far had pointed to the *enol* form as being the correct one to use. But what if the *keto* form were tried?[44] As soon as he followed this hunch, Watson immediately saw that the bases fitted together on the inside, to form the perfect **double-helix** structure. Even more important, when the two strands separated in reproduction, the mutual attraction of adenine to guanine, and of thymine to cytosine, meant that the new double helix was identical to the old one – the biological information contained in the genes was passed on unchanged, as it had to be if the structure was to explain heredity.[45] They announced the new structure to their colleagues on 7 March 1953, and six weeks later their paper appeared in *Nature*. Wilkins, says Strathern, was charitable toward Watson and Crick, calling them a couple of 'old rogues.' Franklin instantly accepted their model.[46] Not everyone was as emollient. They were called 'unscrupulous' and told they did not deserve the sole credit for what they had discovered.[47] In fact, the drama was not yet over. In 1962 the Nobel Prize for Medicine was awarded jointly to Watson, Crick, and Wilkins, and in the same year the prize for chemistry went to the head of the Cavendish X-ray diffraction unit, Max Perutz and his assistant, John Kendrew. Rosalind Franklin got nothing. She died of cancer in 1958, at the age of thirty-seven.[48]

Years later Watson wrote an entertaining and revealing book about the whole saga, on which this account is partly based. Some of his success as an author lay in his openness about the scientific process, which made him and his colleagues seem far more human than had hitherto been the case. For most people up until then, science books were textbooks, thick as bricks and just as dry. Partly this was a tradition, a convention that what counted in science was the results, not how the participants achieved them. Another reason, of course, in the case of certain sciences at least, was the Cold War, which kept many crucial advances secret, at least for a while. In fact the Cold War, which succeeded in making scientists into faceless bureaucrats, along the lines Orwell had laid into in *Nineteen Eighty-Four*, also sparked a bitter rivalry between scientists on either side of the divide, very different from the cooperative international mood in physics in the early part of the century. The most secret discipline was in fact physics itself and its penumbra of activities. And it was here that the rivalry was keenest. Archival research carried out in Russia since perestroika has, for example, identified one great scientist who, owing to secrecy, was virtually unknown hitherto, not only in the West but in his own country, and who was almost entirely obsessed with rivalry. He was more or less single-handedly responsible for Soviet Russia's greatest scientific success, but his strengths were also his weaknesses, and his competitiveness led to his crucial failures.[49]

On Friday, 4 October 1957, the world was astounded to learn that Soviet Russia had launched an orbiting satellite. *Sputnik I* measured only twenty-three inches across and didn't *do* much as it circled the earth at three hundred miles a minute. But that wasn't the point: its very existence up there, overflying America four times during the first day, was a symbol of the Cold War rivalry that so preoccupied the postwar world and in which, for a time at least, the Russians seemed to be ahead.[50] Receiving the story in the late afternoon, next morning the *New York Times* took the unusual step of printing a three-decker headline, in half-inch capitals, running the whole way across the front page:

> SOVIET FIRES EARTH SATELLITE INTO SPACE;
> IT IS CIRCLING THE GLOBE AT 18,000 MPH;
> SPHERE TRACKED IN 4 CROSSINGS OVER U. S.[51]

Only then did Nikita Khrushchev, the Russian leader, realise what an opportunity *Sputnik*'s launch provided for some Cold War propaganda. The next day's *Pravda* was quite different from the day before, which had recorded the launch of *Sputnik* in just half a column. 'World's First Artificial Satellite of Earth Created in Soviet Nation,' ran the headline, and it too stretched all the way across page one. The paper also published the congratulations that poured in, not only from what would soon come to be called satellite states of the USSR, but from scientists and engineers in the West.[52]

Sputnik was news partly because it showed that space travel was possible, and that Russia might win the race to colonise the heavens – with all the psychological and material advantages that implied – but also because, in order to reach orbit, the satellite must have been launched at a speed of at least 8,000 metres per second and with an accuracy which meant the Russians had solved several technological problems associated with rocket technology. And it was rocket technology that lay at the heart of the Cold War arms race; both Russia and the United States were then trying their hardest to develop intercontinental ballistic missiles (ICBMs) that could carry nuclear warheads vast distances between continents. The launch of *Sputnik* meant the Russians had a rocket with enough power and accuracy to deliver hydrogen bombs on to American soil.[53]

After dropping behind in the arms race during World War II, the Soviet Union quickly caught up between 1945 and 1949, thanks to a small coterie of 'atomic spies,' including Julius and Ethel Rosenberg, Morton Sobell, David Greenglass, Harvey Gold, and Klaus Fuchs. But the *delivery* of atomic weapons was a different matter, and here, since the advent of perestroika, several investigations have been made of what was going on behind the scenes in the Russian scientific community. By far the most interesting is James Harford's biography of **Sergei Pavlovich Korolev**.[54] Korolev, who led an extraordinary life, may fairly be described as the father of both Russia's ICBM system and its space program.[55] Born in 1907 near Kiev, in Ukraine, into an old Cossack family, Sergei Pavlovich grew up obsessed with manmade flight. This led to an interest in rocket and jet propulsion in the 1930s. (It has also become clear since perestroika that the USSR had a spy in Wernher von Braun's team, and that Korolev and his colleagues – not to mention Stalin, Beria, and Molotov – were kept up-to-date with German progress.) But Korolev's smooth ride up the Soviet system came to an abrupt end in June 1937, when he was arrested in the purges and deported to the gulag, accused of 'subversion in a new field of technology.' He was given no trial but beaten until he 'confessed.'[56] He spent some of his time at the notorious camp in the Kolyma area of far-eastern Siberia, later made famous by Aleksandr Solzhenitsyn in *The Gulag Archipelago*.[57] Robert Conquest, in *The Great Terror*, says that Kolyma 'had a death rate of up to 30 per cent [per year],' but Korolev survived, and because so many people interceded on his behalf, he was eventually moved to a *sharashka*, a penal institution not as severe as the gulag, where scientists and engineers were made to work on practical projects for the good of the state.[58] Korolev was employed in a *sharashka* run by **Andrei Tupolev**, another famous aircraft designer.[59]

During the early 1940s the Tu-2 light bomber and the Ilyushin-2 attack aircraft were designed in the Tupolev *sharashka*, and had notable records later in the war. Korolev was released in the summer of 1944, but it was not until 1957 – the year *Sputnik* was launched – that he obtained complete exoneration for his alleged 'subversion.'[60]

Photographs of Korolev show a tough, round-faced bear of a man, and do nothing to dispel the idea that he was a force of nature, with a temper that terrified even senior colleagues. After the war he adroitly picked the brains of Germany's rocket scientists, whom Russia had captured, and it was the same story after the explosion of the first atomic bomb, and the leaking of atomic secrets to the Russians. It was Korolev who spotted that the *delivery* of weapons of mass destruction was every bit as important as the weapons themselves. Rockets were needed that could travel thousands of miles with great accuracy. Korolev also realised that this was an area where two birds could be killed with one stone. A rocket that could carry a nuclear warhead all the way from Moscow to Washington would need enough power to send a satellite into orbit.

There were sound scientific reasons for exploring space, but from the information recently published about Korolev, it is clear that a major ingredient in *his* motivation was to beat the Americans.[61] This was very popular with Stalin, who met Korolev several times, especially in 1947. Here was another field, like genetics, where Soviet science could be different from, and better than, its Western counterpart.[62] It was a climate where the idea of science as a cool, rational, reflective, *disinterested* activity went out the window. By the early 1950s Korolev was the single most important driving force behind the Russian rocket/space program, and according to James Harford his moods fluctuated wildly depending on progress. He had a German trophy car commandeered after the war, which he drove at high speeds around Moscow and the surrounding countryside to get the aggression out of his system. He took all failures of the project personally and obsessively combed the open American technical literature for clues as to how the Americans might be progressing.[63] In the rush to be first, mistakes were made, and the first five tests of what was called in Russia the R-7 rocket were complete failures. But at last, on 21 August 1957, an R-7 flew the 7,000 kilometres to the Kamchatka Peninsula in eastern Siberia.[64]

In July 1955 the Eisenhower administration had announced that the United States intended to launch a satellite called Vanguard as part of the International Geophysical Year, which was due to run from 1957 to 1958. Following this announcement, Korolev recruited several new scientists and began to build his own satellite. Recent accounts make it clear that Korolev was intensely aware of how important the project was historically – he just *had* to be first – and once R-7 had proved itself, he turned up the heat. Within a month of the first R-7 reaching Kamchatka, *Sputnik* lifted off its launchpad in Baikonur. The launch not only made headline news in the world's media but gave a severe jolt to aeronautical professionals in the West.[65] The Americans responded almost immediately, bringing forward by several months the launch of their own satellite, to December 1957. This too was scarcely the mark of cool, rational scientists – and it showed. In the full glare of the television cameras, the American satellite got only a few feet off the ground before it fell back to earth and exploded in flames. 'OH, WHAT A FLOPNIK!' crowed *Pravda*. 'KAPUTNIK!' said another newspaper; 'STAYPUTNIK,' a third.[66]

Realising the coup Korolev had produced, Khrushchev called him to the Kremlin and instructed him to provide something even more spectacular to celebrate the fortieth anniversary of the revolution.[67] Korolev's response was *Sputnik 2*, launched a month after *Sputnik 1* – with **Laika**, a mongrel dog, aboard. As a piece of theatre it could not be faulted, but as science it left a lot to be desired. After refusing to separate from its booster, *Sputnik 2*'s thermal control system failed, the satellite overheated – and Laika was roasted. Animal rights groups protested, but the Russians dismissed the complaints, arguing that Laika had been 'a martyr to a noble cause.'[68] And in any case, *Sputnik 2* was soon followed by *Sputnik 3*.[69] This was intended as the most sophisticated and productive of all the satellites, equipped with sensitive measuring devices to assess a whole range of atmospheric and cosmological phenomena. Korolev's immediate motive was to heap further humiliation on the United States – but he came another cropper. During tests for the satellite, a crucial tape recorder failed to work. To have rectified it thoroughly would have delayed the launch, and the man responsible, Alexei Bogomolov, 'did not want to be considered a loser in the company of winners.' He argued that the failure was due to electrical interference in the test room and that such interference wouldn't exist in space. No one else was taken in – except the one man who counted, Korolev.[70] The tape recorder duly failed in flight. Nothing sensational occurred – there was no spectacular explosion – but crucial information was not recorded. As a result, it was the Americans, whose *Explorer 3* had finally been launched on 26 March 1958, who observed a belt of massive radiation around the earth that became known as the Van Allen belts, after James Van Allen, who designed the instruments that *did* record the phenomenon.[71] And so, after the initial space flight, with all that implied, the first major scientific discovery was made not by Korolev but by the late-arriving Americans. Korolev's personality was responsible for both his successes and his failures.[72]

Nineteen-fifty-eight was the first full year of the space age, with twenty-two launch attempts, though only five were successful. Korolev went on securing 'firsts,' including unmanned landings on the moon and Venus, and in April 1961 Yuri Gagarin became the first human being to orbit the earth. When Korolev died, in January 1966, he was buried in the wall of the Kremlin, a supreme honour. But his identity was always kept secret while he was alive; it is only recently he has received his full due.

Character was certainly crucial to the fifth great scientific advance that took place in the 1950s. Neither can one rule out the role of luck. For the fact is that **Mary** and **Louis Leakey**, archaeologists and palaeontologists, had been excavating in Africa, in Kenya and Tanganyika (later Tanzania) since the 1930s without finding anything especially significant. In particular, they had dug at

Olduvai Gorge, a 300-foot-deep, thirty-mile-long chasm cut into the Serengeti Plain, part of the so-called Rift Valley that runs north-south through the eastern half of Africa and is generally held to be the border between two massive tectonic plates.[73] For scientists, the Olduvai Gorge had been of interest ever since it had first been discovered in 1911, when a German entomologist named Wilhelm Kattwinkel almost fell into it as he chased butterflies.[74] Climbing down into the gorge, which cuts through many layers of sediments, he discovered innumerable fossil bones lying around, and these caused a stir when he got them back to Germany because they included parts of an extinct horse. Later expeditions found sections of a modern human skeleton, and this led some scientists to the conclusion that Olduvai was a perfect place for the study of extinct forms of life, including – perhaps – ancestors of mankind.

It says a lot for the Leakeys' strength of character that they dug at Olduvai from the early 1930s until 1959 without making the earth-shattering discovery they always hoped for.[75] Until that time, as was mentioned in earlier chapters, it was believed early man originated in Asia. Born in Kenya to a missionary family, Louis had found his first fossils at the age of twelve and had never stopped from then on. His quixotic character involved to begin with a somewhat lackadaisical approach to scientific evidence, which ensured that he was never offered a formal academic position.[76] In the prewar moral climate Leakey's career was not helped either by an acrimonious divorce from his first wife, which put paid to his chances of an academic position in straitlaced Cambridge.[77] Another factor was his activity as a British spy at the time of Kenya's independence movement in the late 1940s and early 1950s, culminating in his appearance to give evidence in court against Jomo Kenyatta, the leader of the independence party, and later the country's first president.[78] (Kenyatta never seems to have borne a grudge.) Finally, there was Leakey's fondness for a succession of young women. There was nothing one-dimensional about Leakey, and his character was central to his discoveries and to what he made of them.

During the 1930s, until most excavation was halted because of the war, the Leakeys had dug at Olduvai more years than not. Their most notable achievement was to find a massive collection of early manmade tools. Louis and his second wife Mary were the first to realise that flint tools were not going to be found in that part of Africa, as they had been found all over Europe, say, because in East Africa generally, flint is lacking. They did, however, find 'pebble tools' – basalt and quartzite especially – in abundance.[79] This convinced Leakey that he had found a 'living floor,' a sort of prehistoric living room where early man made tools in order to eat the carcasses of the several extinct species that by now had been discovered in or near Olduvai. After the war, neither he nor Mary revisited Olduvai until 1951, in the wake of the Kenyatta trial, but they dug there through most of the 1950s. Throughout the decade they found thousands of hand axes and, associated with them, fossilised bones of many extinct mammals: pigs, buffalos, antelopes, several of them much bigger than today's varieties, evoking a romantic image of an Africa inhabited by huge, primitive animals. They renamed this living floor 'the Slaughter House.'[80] At that stage, according to Virginia Morrell, the Leakeys' biographer, they thought that the lowest bed in the gorge dated to about 400,000 years ago and that the highest bed was 15,000 years old. Louis had lost none of his enthusiasm, despite having reached middle age without finding any humans in more than twenty years of searching. In 1953 he got so carried away by his digging that he spent too long in the African sun and suffered such a severe case of sunstroke that his hair 'turned from brown to white, literally overnight.'[81] The Leakeys were kept going by the occasional find of hominid teeth (being so hard, teeth tend to survive better than other parts of the human body), so Louis remained convinced that one day the all-important skull would turn up.

On the morning of 17 July 1959, Louis awoke with a slight fever. Mary insisted he stay in camp. They had recently discovered the skull of an extinct giraffe, so there was plenty to do.[82] Mary drove off in the Land Rover, alone except for her two dogs, Sally and Victoria. That morning she searched a site in Bed I, the lowest and oldest, known as FLK (for Frieda Leakey's *Korongo*, Frieda Leakey being Louis's first wife and *korongo* being Swahili for gully). Around eleven o'clock, with the heat becoming uncomfortable, Mary chanced on a sliver of bone that 'was not lying loose on the surface but projecting from beneath. It seemed to be part of a skull. ... It had a hominid look, but the bones seemed enormously thick – too thick, surely,' as she wrote later in her autobiography.[83] Dusting off the topsoil, she observed 'two large teeth set in the curve of a jaw.' At last, after decades. There could be no doubt: it was a hominid skull.[84] She jumped back into the Land Rover with the two dogs and rushed back to camp, shouting 'I've got him! I've got him!' as she arrived. Excitedly, she explained her find to Louis. He, as he put it later, became 'magically well' in moments.[85]

When Louis saw the skull, he could immediately see from the teeth that it wasn't an early form of *Homo* but probably australopithecine, that is, more apelike. But as they cleared away the surrounding soil, the skull revealed itself as enormous, with a strong jaw, a flat face, and huge zygomatic arches – or cheekbones – to which great chewing muscles would have been attached. More important, it was the third australopithecine skull the Leakeys had found in association with a hoard of tools. Louis had always explained this by assuming that the australopithecines were the victims of *Homo* killers, who then feasted on the more primitive form of ancestor. But now Louis began to change his mind – and to ask himself if it wasn't the australopithecines who had made the tools. Tool making had always been regarded as the hallmark of humanity – and now, perhaps, humanity should stretch back to the australopithecines.

Before long, however, Louis convinced himself that the new skull was actually midway between australopithecines and modern *Homo sapiens* and so he called the new find *Zinjanthropus boisei* – *Zinj* being the ancient Arabic word for the coast of East Africa, *anthropos* denoting the fossil's humanlike qualities, and *boisei* after Charles Boise, the American who had funded so many of their expeditions.[86] Because he was so complete, so old and so strange, Zinj made the Leakeys famous. The discovery was front-page news across the world, and Louis became the star of conferences in Europe, North America, and Africa. At these conferences, Leakey's interpretation of Zinj met some resistance from other scholars who thought that Leakey's new skull, despite its great size, was not all that different from other australopithecines found elsewhere. Time would prove these critics right and Leakey wrong. But while Leakey was arguing his case with others about what the huge, flat skull meant, two scientists elsewhere produced a completely unexpected twist on the whole matter. A year after the discovery of Zinj, Leakey wrote an article for the *National Geographic* magazine, 'Finding the World's Earliest Man,' in which he put *Zinjanthropus* at 600,000 years old.[87] As it turned out, he was way off.

Until the middle of the century, the main dating technique for fossils was the traditional archaeological device of stratigraphy, analysing sedimentation layers. Using this technique, Leakey calculated that Olduvai dated from the early Pleistocene, generally believed to be the time when the giant animals such as the mammoth lived on earth alongside man, extending from 600,000 years ago until around 10,000 years ago. Since 1947, a new method of dating, the **carbon-14** technique, had been introduced. C^{14} dating depends on the fact that plants take out of the air carbon dioxide, a small proportion of which is radioactive, having been bombarded by cosmic rays from space. Photosynthesis converts this CO_2 into radioactive plant tissue, which is maintained as a constant proportion until the plant (or the organism that has eaten the plant) dies, when radioactive carbon uptake is stopped. Radioactive carbon is known to have a half-life of roughly 5,700 years, and so, if the proportion of radioactive carbon in an ancient object is compared with the proportion of radioactive carbon in contemporary objects, it is possible to calculate how long has elapsed since that organism's death. With its relatively short half-life, however, C^{14} is only useful for artefacts up to roughly 40,000 years old. Shortly after Leakey's *National Geographic* article appeared, two geophysicists from the University of California at Berkeley, Jack Evernden and Garniss Curtis, announced that they had dated some volcanic ash from Bed I of Olduvai – where Zinj had been found – using the **potassium-argon (K/Ar)** method. In principle, this method is analogous to C^{14} dating but uses the rate at which the unstable radioactive potassium isotope potassium-40 (K^{40}) decays to stable argon-40 (Ar^{40}). This can be compared with the known abundance of K^{40} in natural potassium, and an object's age calculated from the half-life. Because the half-life of K^{40} is about 1.3 billion years, this method is much more suitable for geological material.[88]

Using the new method, the Berkeley geophysicists came up with the startling news that Bed I at Olduvai was not 600,000 but 1.75 *million* years old.[89] This was a revelation, the very first clue that early man was, much, much older than anyone suspected. This, as much as the actual discovery of *Zinj*, made Olduvai Gorge famous. In the years that followed, many more skulls and skeletons of early hominids would be found in East Africa, sparking bitter controversy about how, and when, early man developed. But the 'bone rush' in the Rift Valley really dates from the fantastic publicity surrounding the discovery of *Zinj* and its great antiquity. This eventually produced the breathtakingly audacious idea – almost exactly one hundred years after Darwin – that man originated in Africa and then spread out to populate the globe.

★

Each of these episodes was important in itself, albeit in very different ways, and transformed our understanding of the natural world. But besides the advances in knowledge that at least four of them share and to which we shall return (Lysenko was eventually overthrown in the mid-1960s), they all have in common that they show science to be an untidy, emotional, obsessive, all-too-human activity. Far from being a calm, reflective, solely rational enterprise, carried out by dispassionate scientists only interested in the truth, science is revealed as not so very different from other walks of life. If this seems an unexceptional thing to say now, at the end of the century, that is a measure of how views have changed since these advances were made, in the 1940s and 1950s. Early on in that same decade, Claude Lévi-Strauss had expressed the general feeling of the time: 'Philosophers cannot insulate themselves against science,' he said. 'Not only has it enlarged and transformed our vision of life and the universe enormously: it has also revolutionised the rules by which the intellect operates.'[90] This mindset was underlined by Karl Popper in *The Logic of Scientific Discovery*, published in English in 1959, in which he set out his view that the scientist encounters the world – nature – essentially as a stranger, and that what sets the scientific enterprise apart from everything else is that it only entertains knowledge or experience that is capable of falsification. For Popper this is what distinguished science from religion, say, or metaphysics: revelation, or faith, or intuition have no part, at least no central role; rather, knowledge increases incrementally, but that knowledge is never 'finished' in the sense that anything is 'knowable' as true for all time.[91] But Popper, like Lévi-Strauss, focused only on the rationalism of science, the logic by which it attempted – and often managed – to move forward. The whole penumbra of activities – the context, the rivalry, the ambition and hidden agendas of the participants in these dramas (for dramas they often were) – were left out of the account, as somehow inappropriate and irrelevant, sideshows to the main event. At the time no one thought this odd. Michael Polanyi, as we have seen, had raised doubts back in 1946, but it was left to a historian of science rather than a philosopher to produce the book that changed for all time how science was perceived. This was **Thomas Kuhn**, whose *Structure of Scientific Revolutions* appeared in 1962.

Kuhn, a physicist turned historian of science at MIT, was interested in the way major changes in science come about. He was developing his ideas in the 1950s and so did not use the examples just given, but instead looked at much earlier episodes from history, such as the Copernican revolution, the discovery of oxygen, the discovery of X rays, and Einstein's ideas about relativity. Kuhn's chief argument was that science consists mainly of relatively stable periods, when nothing much of interest goes on and scientists working within a particular 'paradigm' conduct experiments that flesh out this or that aspect of the paradigm. In this mode, scientists are not especially sceptical people – rather, they are in a sort of mental straitjacket as laid down by the paradigm or theory they are following. Amid this set of circumstances, however, Kuhn observed that a number of anomalies will occur. To begin with, there is an attempt to incorporate the anomalies into the prevailing paradigm, and these will be more or less successful. Sooner or later, however, the anomalies grow so great that a crisis looms within whatever branch of science it may be – and then one or more scientists will develop a totally new paradigm that better explains the anomalies. A scientific revolution will have taken place.[92] Kuhn also noted that science is often a collaborative exercise; in the discovery of oxygen, for example it is actually very difficult to say precisely whether Joseph Priestley or Antoine-Laurent Lavoisier was primarily responsible: without the work of either, oxygen would not have been understood in exactly the way it was. Kuhn also observed that revolutions in science are often initiated by young people or those on the edge of the discipline, not fully trained – and therefore not fully schooled – in a particular way of thought. He therefore stressed the sociology and social psychology of science as a factor in both the advancement of knowledge and the reception of new knowledge by other scientists. Echoing an observation of Max Planck, Kuhn found that the bulk of scientists never change their minds – a new theory wins because adherents of the old theory simply die out, and the new theory is favoured by the new generation.[93] In fact, Kuhn makes it clear several times that he sees scientific revolutions as a form of evolution, with the better – 'fitter' – ideas surviving while the less successful become extinct. The view that science is more ordered than is in fact the case, Kuhn said, is aided by the scientific textbook.[94] Other disciplines use textbooks, but it is in science that they are most popular, reflecting the fact that many young scientists get their information predigested (and therefore repackaged), rather than by reading the original literature. So, very often scientists do not – or did not then – learn about discoveries at first hand, as someone interested in literature reads the original books themselves, as well as reading textbooks of literary criticism. (In this, Kuhn was echoing one of F. R. Leavis's main criticisms of C. P. Snow.)

Much was made of Kuhn's book, especially by nonscientists and antis-cientists, so it is necessary to emphasise that he was *not* seeking to pull the rug out from under the feet of science. Kuhn always maintained that science produced, as Lévi-Strauss said, a special kind of knowledge, a knowledge that worked in a distinctive way and very well.[95] Some of the uses to which his book was put would not have met with his approval. Kuhn's legacy is a reconceptualisation of science, not so much a culture, as Snow said, but a tradition in which many scientists serve their apprenticeship, which pre-determines the types of question science finds interesting, and the way it seeks answers to problems. Thus the scientific tradition is nowhere near as rational as is generally thought. Not all scientists find this view convincing, and obviously there is much scope for disagreement as to what is or is not a paradigm, and what is and is not normal science. But for historians of science, and many in the humanities, Kuhn's work has been very liberating, allowing scientific knowledge to be regarded as somehow more tentative than before.

MIND MINUS METAPHYSICS

At the end of 1959 the film director **Alfred Hitchcock** was producing a movie in absolute secrecy. Around the lot at Revue Studios, part of Universal Pictures in Los Angeles, the film was known on the clapper board and in company designation by its codename, 'Wimpy.' When it was ready, Hitchcock wrote to film critics in the press, begging them not to give away the ending and announcing at the same time that no member of the public would be allowed in to the film after it had started.

Psycho was a screen 'first' in many different ways. Hitherto Hitchcock had directed top-quality murder stories, set in exotic locations and usually made in Technicolor. In deliberate contrast, *Psycho* was cheap in appearance, filmed in black and white, and focused on an area of sleaze.¹ There were unprecedented scenes of violence. Most arresting of all, however, was the treatment of madness. The film was actually based on the real-life case of Ed Gein, a 'cannibalistic Wisconsin killer' whose terrible deeds also inspired *The Texas Chain Saw Massacre* and *Deranged*. In *Psycho*, Hitchcock – fashionably enough – pinpointed the source of Norman Bates's homicidal mania in his narrow and inadequate family and sexual history.²

The film starred Anthony Perkins and Janet Leigh, both of whom worked for Hitchcock for well below their usual fee in order to gain experience with a master storyteller (Leigh's character was actually killed off halfway through the film, another innovation). The film is rich in visual symbolism meant to signify madness, schizophrenia in particular. Apart from the gothic setting in a gingerbread-house motel on a stormy night, each of the characters has something to hide – whether it is an illicit affair, stolen cash, a concealed identity, or an undiscovered murder. Mirrors are widely used to alter images, which are elsewhere sliced in two to suggest the reversal of reality and the cutting, split world of the violently insane.³ Anthony Perkins, who pretends he is in thrall to his mother when in reality he has killed her long ago, spends his time 'stuffing birds' (nightbirds, like owls, which also watch him). All this tension builds to what became the most famous scene in the film, the senseless slashing of Janet Leigh in the shower, where 'the knife functions as a penis, penetrating the body in a symbolic rape' and the audience watches – horrified and enthralled – as blood gurgles down the drain of the shower.⁴ *Psycho* is in fact a brilliant example of a device that would become much debased as time passed – the manipulation of the cinema audience so that, to an extent, it understands, or at least experiences, the conflicting emotions wrapped up in a schizophrenic personality. Hitchcock is at his most cunning when he has the murderer, Perkins/Bates, dispose of Janet Leigh's body by sinking it in a car in a swamp. As the car is disappearing in the mud, it suddenly stops. Involuntarily, the audience wills the car to disappear – and for a moment is complicit in the crime.[5]

The film received a critical pasting when it was released, partly because the critics hated being dictated to over what they could and could not reveal. 'I remember the terrible panning we got when *Psycho* opened,' Hitchcock said. 'It was a critical disaster.' But the public felt otherwise, and although the movie cost only $800,000 to make, Hitchcock alone eventually recouped more than $20 million. In no time the movie became a cult. 'My films went from being failures to masterpieces without ever being successes,' said Hitchcock.[6]

Attempts to understand the mentally ill as if their sickness is a maladaptation, a pathology of logic or philosophy rather than a physical disease, has a long history and is at the root of the psychoanalytic school of psychiatry. In the same year as Hitchcock's film, a psychoanalytic book appeared in Britain that also achieved cult status quickly. Its author was a young psychiatrist from Glasgow in Scotland who described himself as an existentialist and went on to become a fashionable poet. This idiosyncratic career path was mirrored in his theories about mental illness. In *The Divided Self*, **Ronald D. Laing** applied Sartre's existentialism to frankly psychotic schizophrenics in an attempt to understand why they went mad. Laing was one of the leaders of a school of thought (David Cooper and Aaron Esterson were others) which argued that schizophrenia was not an organic illness, despite evidence even then that it was grouped in families and therefore to some extent inherited, but represented a patient's private response to the environment in which he or she was raised. Laing and his colleagues believed in an entity they labelled the 'schizophrenogenic' – or schizophrenia-producing – family. In *The Divided Self* and subsequent books, Laing argued that investigation of the backgrounds of schizophrenics showed that they had several things in common, the chief of which was a family, in particular a mother, who behaved in such a way that the person's sense of self became separated from his or her sense of body, that life was a series of 'games' which threatened to engulf the patient.[7]

The efficacy of Laing's theories, and their success or otherwise in generating treatment, will be returned to in just a moment, but Laing was important in more than the merely clinical sense: insofar as his approach represented an attempt to align existential philosophy with Freudian psychology, his theories were part of an important crossover that took place between about 1948 and the mid-1960s. This period saw the death of metaphysics as it had been understood in the nineteenth century. It was philosophers who laid it to rest, and ironically, one of the chief culprits was the Waynflete Professor of Metaphysical Philosophy at Oxford University, **Gilbert Ryle**. In *The Concept of Mind*, published in 1949, Ryle delivered a withering attack on the traditional,

Cartesian concept of duality, which claimed an essential difference between mental and physical events.[8] Using a careful analysis of language, Ryle gave what he himself conceded was a largely behaviourist view of man. There is no inner life, Ryle said, in the sense that a 'mind' exists independently of our actions, thoughts, and behaviours. When we 'itch' to do something, we don't really itch in the sense that we itch if a mosquito bites us; when we 'see' things 'in our mind's eye,' we don't see them in the way that we see a green leaf. This is all a sloppy use of language, he says, and most of his book is devoted to going beyond this sloppiness. To be conscious, to have a sense of self, is not a by-product of the mind; it *is* the mind in action. The mind does not, as it were, 'overhear' us having our thoughts; having the thoughts *is* the mind in action.[9] In short, there is no ghost in the machine – only the machine. Ryle examined the will, imagination, intellect, and emotions in this way, demolishing at every turn the traditional Cartesian duality, ending with a short chapter on psychology and behaviourism. He took psychology to be more like medicine – an agglomeration of loosely connected inquiries and techniques – than a proper science as generally understood.[10] In the end, Ryle's book was more important for the way it killed off the old Cartesian duality than for anything it did for psychology.

While Ryle was developing his ideas in Oxford, **Ludwig Wittgenstein** was pursuing a more or less parallel course in Cambridge. After he had published *Tractatus Logico-Philosophicus* in 1921, Wittgenstein abandoned philosophy for a decade, but he returned in 1929 to Cambridge, where at first he proceeded to dismantle the philosophy of the *Tractatus*, influential though that had been, and replace it with a view that was in some respects diametrically opposite. Throughout the 1930s and the 1940s he published nothing, feeling 'estranged' from contemporary Western civilisation, preferring to exert his influence through teaching (the 'deck-chair' seminars that Turing had attended).[11] Wittgenstein's second masterpiece, *Philosophical Investigations*, was published in 1953, after his death from cancer in 1951, aged sixty-two.[12] His new view took Ryle's ideas much further. Essentially, Wittgenstein thought that many philosophical problems are false problems, mainly because we are misled by language. All around us, says P. M. S. Hacker, who wrote a four-volume commentary on *Philosophical Investigations*, are grammatical similarities that mask profound logical differences. 'Philosophical questions are frequently not so much questions in search of an answer as questions in search of a sense. "Philosophy is a struggle against the bewitchment of our understanding by means of language." ' For example, 'the verb "to exist" looks no different from such verbs as "to eat" or "to drink" but while it makes sense to ask how many people in College don't eat meat or drink wine, it makes no sense to ask how many people in College don't exist.'[13]

This is not just a language game.[14] Wittgenstein's fundamental idea was that philosophy exists not to solve problems but to make the problems disappear, just as a knot in a piece of string disappears when it is unravelled. Put another way, 'Problems are solved, not by giving new information, but by [re]arranging what we have always known.'[15] The way forward, for Wittgenstein, was to rearrange the entire language.[16] No man could do that on his own, and Wittgenstein started by concentrating, as Ryle had done, on the mind-body duality. He went further in linking with it what he called the brain-body duality. Both dualities, he said, were misconceptions. Consciousness was misconceived, he said, when it was 'compared with a self-scanning mechanism in the brain.'[17] He took as his example pain. To begin with, he explains that one does not 'have' a pain in the sense that one has a penny. 'A pain cannot go round the world, like a penny can, independent of anyone owning it.' Equally, we do not look to see whether we are groaning before reporting that we have a pain – in that sense, the groan is part of the pain.[18] Wittgenstein next argued that the 'inner' life, 'introspection,' and the privacy of experience have also been misconceived. The pain that one person has *is* the same that another person has, just as two books can have covers coloured in the same red. Red does not exist in the abstract, and neither does pain.[19] On inspection, Wittgenstein is saying, all the so-called mental things we do, do not need 'mind': 'To make up one's mind is to decide, and to be in two minds about something is to be undecided. . . . There is such a thing as introspection but it is not a form of inner perception . . . it is the calling up of memories; of imagined possible situations, and of the feelings that one would have if . . .'[20] 'I want to win' is not a description of a state of mind but a *manifestation* of it.[21] Talk of 'inner' and 'outer' in regard to 'mental' life is, for Wittgenstein, only metaphor. We may say that toothache is physical pain and that grief is mental. But grief is not painful in the sense that toothache is; it does not 'hurt' as toothache hurts.[22] For Wittgenstein, we do not need the concept of mind, and we need to be very careful about the way we think about 'brain.' It is the *person* who feels pain, hope, disappointment, not his brain.

Philosophical Investigations was more successful in some areas than in others. But by Wittgenstein's own criteria, it made some problems disappear, the problem of mind being one of them. His was one of the books that helped move attention toward consciousness, which Wittgenstein did not successfully explain, and which dominated the attentions of philosophers and scientists at the end of the century.

The consequences of *Philosophical Investigations* for Freudian psychoanalysis have never been worked through, but Wittgenstein's idea of 'inner' and 'outer' as merely metaphor to a large extent vitiates Freud's central ideas. The **attack on Freud** was growing anyway in the late 1950s and has been chronicled by Martin Gross. Although the interwar years had been the high point of the Freudian age, the first statistical doubts over the efficacy of psychoanalytic treatment occurred as early as the 1920s, when a study of 472 patients from the clinic of the Berlin Psychoanalytic Institute revealed that only 40 percent could be regarded as cured. Subsequent studies in the 1940s at the London Clinic, the Chicago Institute for Psychoanalysis, and the Menninger Clinic in Kansas likewise revealed an average 'cure rate' of 44 percent. A series of studies throughout the 1950s showed with some consistency that 'a patient has approximately a 50–50 chance of getting off the couch in somewhat better mental condition than when he first lay down on it.'[23] Most damaging of all, however, was the study carried out in the mid-1950s by the Central Fact-Gathering

Committee of the American Psychoanalytic Association (the APsaA), chaired by Dr Harry Weinstock. His committee collected evidence on 1,269 psychoanalytic cases treated by members of the APsaA. The report, on the largest sample to date, was eagerly awaited, but in December 1957 the association decided against publication, noting that the 'controversial publicity on such material cannot be of benefit in any way.'[24] Mimeographed copies of the report then began to circulate confidentially in the therapeutic community, and gossip about the results preoccupied the psychiatric profession until the APsaA finally consented to release the findings – *a decade later*. Then the reason for the delay became clear. The 'controversial material' showed that, of those originally accepted for treatment, barely one in six were cured. This was damning enough, being the profession's own report; but it wasn't just the effectiveness of psychoanalysis that came under threat; so did Freud's basic theories. His idea that we are all a little bisexual was challenged, and so was the very existence of the Oedipus complex and infantile sexuality. For example, penile erection in infants had been regarded by psychoanalysts as firm evidence of infantile sexuality, but H. M. Halverson observed nine infants for ten days each – and found that seven of them had an erection at least once a day.[25] 'Rather than being a sign of pleasure, the erections tended to show that the child was uncomfortable. In 85 percent of cases, the erection was accompanied by crying, restlessness, or the stiff stretching of legs. Only when the erection subsided did the children become relaxed.' Halverson concluded that the erection was the result of abdominal pressure on the bladder, 'serving a simple bodily, rather than a Freudian, need.' Likewise, sleep research shows that the forgetting of dreams – which according to psychoanalysis are repressed – can be explained more simply. We dream at a certain stage of sleep, now known as **REM sleep**, for the rapid eye movements that occur at this time. If the patient is woken during REM sleep, he or she can easily remember dreams, but grows very irritated if woken too often, indicating that REM sleep is necessary for well-being. After REM sleep, however, later in the sleep cycle, if that person is wakened, remembrance of dreams is much harder, and there is much less irritation. Dreams are naturally evanescent.[26] Finally, there was the growth in the 1950s of anti-Freudian anthropological evidence. According to Freudian theory, the breast-feeding of infants is important, helping to establish the basic psychological bond between mother and child, which is of course itself part of the infant's psychosexual development. In 1956, however, the anthropologist Ralph Linton reported on the women of the Marquesas Islands, 'who seldom nurse their babies because of the importance of breasts in their culture.' The Marquesan infant is simply laid on a stone and casually fed a mixture of coconut milk and breadfruit.[27] Nonetheless, the Marquesan children grew up without any special problems, their relationships with their mothers unimpaired.

Beginning in the 1950s, Freud and Jung came in for increasingly severe criticism, for being unscientific, and for using evidence only when it suited them.

Not that other forms of psychology were immune to criticism. In the same year that Wittgenstein's posthumous *Philosophical Investigations* appeared, **Burrhus F.**

Skinner, professor of psychology at Harvard University, published the first of his controversial works. Raised in the small Pennsylvania town of Susquehanna, Fred Skinner at first wanted to be a writer and studied English at Hamilton College, where Robert Frost told him that he was capable of 'real niceties of observation.' Skinner never developed as a writer, however, because 'he found he had nothing to say.' And he gave up the saxophone because it seemed to him to be 'the wrong instrument for a psychologist.'[28] Abandoning his plan to be a writer, he studied psychology at Harvard, so successfully that in 1945 he became a professor.

Skinner's *Science and Human Behavior* overlapped more than a little with Ryle and Wittgenstein.[29] Like them, Skinner regarded 'mind' as a metaphysical anachronism and concentrated on behavior as the object of the scientist's attention. And like them he regarded language as an at-times-misleading representation of reality, it being the scientist's job, as well as the philosopher's, to clarify its usage. In Skinner's case he took as his starting point a series of experiments, mainly on pigeons and rats, which showed that if their environment was strictly controlled, especially in regard to the administration of rewards and punishments, their behavior could be altered considerably and in predictable ways. This demonstration of rapid learning, Skinner thought, was both philosophically and socially important. He accepted that instinct accounted for a sizeable proportion of human conduct but his aim, in *Science and Human Behavior*, was to offer a simple, rational explanation for the rest of the behavioral repertoire, which he believed could be done, using the principles of reinforcement. In essence Skinner sought to show that the vast majority of behaviors, including beliefs, certain mental illnesses, and even 'love' in some circumstances, could be understood in terms of an individual's history, the extent to which his or her behavior had been rewarded or punished in the past. For example, 'You ought to take an umbrella' may be taken to mean: 'You will be reinforced for taking an umbrella.' 'A more explicit translation would contain at least three statements: (1) Keeping dry is reinforcing to you; (2) carrying an umbrella keeps you dry in the rain; and (3) it is going to rain. ... The "ought" is aversive, and the individual addressed may feel guilty if he does not then take an umbrella.'[30] On this reading of behavior, Skinner saw alcoholism, for example, as a bad habit acquired because an individual may have found the effects of alcohol rewarding, in that it relaxed him in social situations where otherwise he may have been ill at ease. He objected to Freud because he thought psychoanalysis's concern with 'depth' psychology was wrongheaded; its self-declared aim was to discover 'inner and otherwise unobservable conflicts, repressions, and springs of action. The behavior of the organism was often regarded as a relatively unimportant by-product of a furious struggle taking place beneath the surface of the mind.'[31] Whereas for Freud neurotic behavior was the symptom of the root cause, for Skinner neurotic behavior was the object of the inquiry – stamp out the neurotic behavior, and by definition the neurosis has gone. One case that Skinner considers in detail is that of two brothers who compete for the affection of their parents. As a result one brother behaves aggressively toward his sibling and is punished, either by the brother or the parents. Assume this happens repeatedly, to the point where the anxiety associated with such an event generates guilt in the 'aggressive' brother, leading to self-control. In this sense, says Skinner, the brother 'represses' his aggression. 'The repression is successful if the behavior is so effectively displaced that it seldom reaches the incipient state at which it generates anxiety. It is unsuccessful if anxiety is frequently generated.' He then goes on to consider other possible consequences and their psychoanalytic explanations. As a result of *reaction formation* the brother may engage in social work, or some expression of 'brotherly love'; he may *sublimate* his aggression by, say, joining the army or working in an abattoir; he may *displace* his aggression by 'accidentally' injuring someone else; he may *identify* with prizefighters. For Skinner, however, we do not need to invent deep-seated neuroses to explain these behaviors. 'The dynamisms are not the clever machinations of an aggressive impulse struggling to escape from the restraining censorship of the individual or of society, but the resolution of complex sets of variables. Therapy does not consist of releasing a trouble-making impulse but of introducing variables which compensate for or correct a history which has produced objectionable behavior. Pent-up emotion is not the cause of disordered behavior; it is part of it. Not being able to recall an early memory does not produce neurotic symptoms; it is itself an example of ineffective behavior.'[32] In this first book, Skinner's aim was to explain behavior, and he ended by considering the many controlling institutions in modern society – governments and laws, organised religion, schools, psychotherapy, economics and money – his point being that many systems of rewards and punishments are already in place and, more or less, working. Later on, in the 1960s and 1970s, his theories enjoyed a vogue, and in many clinics 'behavior therapy' was adopted. In these establishments, symptoms were treated without recourse to any so-called underlying problem. For example, a man who felt he was dirty and suffered from a compulsive desire to collect towels was no longer treated for his inner belief that he was 'dirty' and so needed to wash a great deal, but simply rewarded (with food) on those days when he didn't collect towels. Skinner's theories were also followed in the development of teaching machines, later incorporated into computer-aided instruction, whereby pupils follow their own course of instruction, at their own pace, depending on rewards given for correct answers.

Skinner's approach to behavior, his understanding of what man is, was looked upon by many as revolutionary at the time, and he was even equated to Darwin.[33] His method linked Ryle and Wittgenstein to psychology. He maintained, for example, that consciousness is a 'social product' that emerges from the human interactions within a verbal community. But verbal behavior, or rather *Verbal Behavior*, published in 1957, was to be his undoing.[34] Like Ryle and Wittgenstein, Skinner understood that if his theory about man was to be convincing, it needed to explain language, and this he set about doing in the 1957 book. His main point was that our social communities 'select' and fine-tune our verbal utterances, what we 'choose' to say, by a process of social reinforcement, and this system, over a lifetime, determines the form of speech we use. In turn this same system of reinforcement of our verbal behavior helps shape our other behaviors – our 'character' – and the way that we understand ourselves, our consciousness. Skinner argued that there are categories of speech acts that may be grouped according to their relationship to surrounding contingencies. For example, 'mands' are classes of speech behavior that are followed by characteristic consequences, whereas 'tacts' are speech acts socially reinforced when emitted in the presence of an object or event.[35] Essentially, under this system, man is seen as the 'host' of behaviors affected by the outside, rather than as autonomous. This is very different from the Freudian view, or more traditional metaphysical versions of man, that something comes from within. Unfortunately, from Skinner's point of view, his radical ideas suffered a withering attack in a celebrated – notorious – review of his book in the journal *Language* in 1959, by **Noam Chomsky**. Chomsky, thirty-one in 1959, was born in Pennsylvania, the son of a Hebrew scholar who interested his son in language. Chomsky's own book, *Syntactic Structures*, was also published in 1957, the same year as Skinner's, but it was the review in *Language* and in particular its vitriolic tone that drew attention to the young author and initiated what came to be called the Chomskyan revolution in psychology.[36]

Chomsky, by then a professor at MIT, just two stops on the subway from Harvard, argued that there are inside the brain universal, innate, grammatical structures; in other words, that the 'wiring' of the brain somehow governs the grammar of languages. He based much of his view on studies of children in different countries that showed that whatever their form of upbringing, they tended to develop their language skills in the same order and at the same pace everywhere. His point was that young children learn to speak spontaneously without any real training, and that the language they learn is governed by where they grow up. Moreover, they are very creative with language, using at a young age sentences that are entirely new to them and that cannot have been related to experience. Such sentences cannot therefore have been learned in the way that Skinner and others said.[37] Chomsky argued that there is a basic structure to language, that this structure has two levels, surface structure and deep structure, and that different languages are more similar in their deep structure than in their surface structure. For example, when we learn a foreign language, we are learning the surface structure. This learning is in fact only possible because the deep structure is much the same. German or Dutch speakers may put the verb at the end of a sentence, which English or French speakers do not, but German, Dutch, French, and English *have* verbs, which exist in all languages in equivalent relationship to nouns, adjectives, and so on.[38] Chomsky's arguments were revolutionary not only because they went against the behaviorist orthodoxy but because they appeared to suggest that there is some sort of *structure* in the brain that is inherited and that, moreover, the brain is prewired in some way that, at least in part, determines how humans experience the world.

The Chomsky–Skinner affair was as personal as Snow–Leavis. Skinner apparently never finished reading the review, believing the other man had completely – and perhaps deliberately – misunderstood him. And he never replied.[39] One consequence of this, however, was that Chomsky's review became more widely known, and agreed with, than Skinner's original book, and as a result Skinner's influence has been blunted. In fact, he never denied that a lot of behavior is instinctive; but he was interested in how it was modified and could, if necessary, be modified still further. His views have always found a small but influential following.

Whatever the effects of Chomsky's attack on Skinner, it offered no support for Freud or psychoanalysis. Although conventional Freudian analysis remained popular in a few isolated areas, like Manhattan, several other well-known scientists, while not abandoning Freudian concepts entirely, began to adapt and extend them in more empirically grounded ways. One of the most influential was **John Bowlby.**

In 1948 the Social Commission of the United Nations decided to make a study of the needs of homeless children: in the aftermath of war it was realised that in several countries large numbers of children lacked fully formed families as a result of the men killed in the fighting. The World Health Organization (WHO) offered to provide an investigation into the mental health aspects of the problem. Dr Bowlby was a British psychiatrist and psychoanalyst who had helped select army officers during the war. He took up a temporary appointment with the WHO in January 1950, and during the late winter and early spring of that year he visited France, Holland, Sweden, Switzerland, Great Britain, and the United States of America, holding discussions with workers involved in child care and child guidance. These discussions led to the publication, in 1951, of *Maternal Care and Mental Health*, a famous report that hit a popular nerve and brought about a wholesale change in the way we think about childhood.[40]

It was this report that first confirmed for many people the crucial nature of the early months of an infant's life, when in particular the quality of mothering was revealed as all-important to the subsequent psychological development of a child. Bowlby's book introduced the key phrase **maternal deprivation** to describe the source of a general pathology of development in children, the effects of which were found to be widespread. The very young infant who went without proper mothering was found to be 'listless, quiet, unhappy, and unresponsive to a smile or a coo,' and later to be less intelligent, bordering in some cases on the defective.[41] No less important, Bowlby drew attention to a large number of studies which showed that victims of maternal deprivation failed to develop the ability to hold relationships with others, or to feel guilty about their failure. Such children either 'craved affection' or were 'affect-less.' Bowlby went on to show that studies in Spain during the civil war, in America, and among a sample of Copenhagen prostitutes all confirmed that delinquent groups were comprised of individuals who, more than their counterparts, were likely to have come from broken homes where, by definition, there had been widespread maternal deprivation.[42] The thrust of this research had two consequences. On the positive side, Bowlby's research put beyond doubt the idea that even a bad home is better for a child than a good institution. It was then the practice in many countries for illegitimate or unwanted children to be cared for in institutions where standards of nutrition, cleanliness, and medical matters could be closely monitored. But it became clear that such an environment was not enough, that something was lacking which affected mental health, rather in the way that vitamins had been discovered to be lacking in the artificial diets created for neglected children in the great cities of the nineteenth century. And so, following publication of the WHO report, countries began to change their approach to neglected children: adoptions were favoured over fostering, children with long-term illnesses were not separated from their parents when they went to hospital, and mothers sent to prison were allowed to take their young babies with them. At work, maternity leave was extended to include not just the delivery but the all-important early months of the child's life. There was in general a much greater sensitivity to the nature of the mother–child bond.[43]

Less straightforward was the link the WHO report found between a disrupted early family life and later delinquency and/or inadequacy. This was doubly important because children from such 'broken' families also proved in many cases to be problem parents themselves, thus establishing what was at first called 'serial deprivation' and later the 'cycle of deprivation.' Not all deprived children became delinquent; and not all delinquent children came from broken homes (though the great majority did). The exact nature of this link assumed greater intellectual prominence later on, but in the 1950s the discovery of the relationship between broken homes and delinquency, mediated via maternal deprivation, offered hope for the amelioration of social problems that disfigured postwar society in many Western countries.

The great significance of Bowlby's report was the way it took an essentially Freudian concept – the bond between mother and child – and examined it scientifically, using objective measures of behavior to understand what was going on, rather than concentrating on the inner workings of 'the mind.' As a psychoanalyst, Freud's work had led Bowlby to focus on the mother-child bond, and to discover its vital practical significance, but *Maternal Care and Mental Health* has only one reference to Freud, and none at all to the unconscious, the ego, id, or superego. In fact, Bowlby was as much influenced by his observations of behavior among animals, including a series of studies carried out in the 1930s in Nazi Germany. So Bowlby's work was yet another instance of 'mind' being eschewed in favour of behavior. The fact that he was a psychoanalyst himself only underlined the inadequacy of traditional Freudian concepts.

Interest in the child as a psychological entity had been spasmodically entertained since the 1850s. The *Journal of Educational Psychology* was founded in the United States in 1910, and the Yale Psycho-Clinic, which opened a year later, was among the first to study babies systematically. But it was in Vienna, in the wake of World War I, that child psychology really began in earnest, due partly to the prevailing Freudian atmosphere, now much more 'respectable' than before, and partly to the straitened circumstances of the country, which affected children particularly badly. By 1926 there were forty different agencies in Vienna concerned with child development.

The man who was probably the greatest child psychologist of the century was influenced less by Freud than by Jung. **Jean Piaget** was born in Neufchâtel, Switzerland, in 1896. He was brilliant even as a boy, publishing his first scientific paper when he was ten, and by fifteen he had a Europe-wide reputation for a series of reports on molluscs. He studied psychiatry under both Eugen Bleuler (who coined the term *schizophrenia*) and Carl Jung, then worked with Théodore Simon at the Sorbonne.[44] Simon had collaborated with Alfred Binet on intelligence tests, and in Paris Piaget was given the task of trying out a new test devised in England by Cyril Burt. This test had questions of the following kind: 'Jane is fairer than Sue; Sue is fairer than Ellen; who is fairer, Jane or Ellen?'[45] Burt was interested in intelligence in general, but Piaget took something rather different from this test, an idea that was to make him far more famous and influential than Burt ever was. Piaget's central idea had two aspects. First he claimed that children are, in effect, *tabulae rasae*, with no inbuilt logical – i.e., intellectual – capabilities; rather, these are learned as they grow up. Second, a child goes through a series of stages in his or her development, as he or she grasps various logical relations and then applies them to the practicalities of life. These theories of Piaget arose from a massive series of experiments carried out at the International Centre of Genetic Epistemology which Piaget founded in Geneva in 1955. (**Genetic epistemology** is concerned with the nature and origins of human knowledge.)[46] Here there is space for just one experiment. At six months a baby is adept at reaching for things, lifting them up, and dropping them. However, if an object is placed under a cushion, the baby loses interest even if the object is still within reach. Piaget claimed, controversially, that this is because the six-month-old child has no conception that unseen objects continue to exist. By roughly nine months, the child no longer has this difficulty.[47]

Over the years, Piaget described meticulously the infant's growing repertoire of abilities in a series of experiments that were close to being games.[48] Although their ingenuity is not in doubt, critics found some of his interpretations difficult to accept, chiefly that at birth the child has no logic whatsoever and must literally 'battle with the world' to learn the various concepts needed to live a successful life.[49] Many critics thought he had done no more than observe a maturational process, as the child's brain developed according to the 'wiring' set down at birth and based, as Chomsky had said, on the infant's heredity. For these critics, logic 'was the engine of development, not the product,' as Piaget said it was.[50] In later years the battle between nature and nurture, and their effects on behaviour, would grow more heated, but the significance of Piaget was that he aligned himself with Skinner and Bowlby in regarding *behavior* as central to the psychologist's concern, and showing how the first few years of life are all-important to later development. Once again, with Piaget the concept of mind took a back seat.

One other development in the 1950s helped discredit the traditional concept of mind: medical drugs that influenced the workings of the brain. As the century wore on, one 'mental' condition after another had turned out to have a physical basis: cretinism, general paralysis of the insane, pellagra (nervous disorder caused by niacin deficiency) – all had been explained in biochemical or physiological terms and, more important, shown themselves as amenable to medication.[51]

Until about 1950 the 'hard core' of insanity – schizophrenia and the manic-depressive psychoses – lacked any physical basis. Beginning in the 1950s, however, even these illnesses began to come within the scope of science, three avenues of inquiry joining together to form one coherent view.[52] From the study of nerve cells and the substances that governed the transmission of the nerve impulse from one cell to another, specific chemicals were isolated. This implied that modification of these chemicals could perhaps help in treatment by either speeding up or inhibiting transmission. The **antihistamines** developed in the 1940s as remedies for motion sickness were found to have the side effect of making people drowsy – i.e., they exerted an effect on the brain. Third, it was discovered that the Indian plant *Rauwolfia serpentina*, extracts of which were used in the West for treatment of high blood pressure, was also used in India to control 'overexcitement and mania.'[53] The Indian drug acted like the anti-histamines, the most active substance being promethazine, commercially known as Phenergan. Experimenting with variants of promethazine, the Frenchman Henri Laborit hit on a substance that became known as chlorpromazine, which produced a remarkable state of 'inactivity or indifference' in excited or agitated patients.[54] Chlorpromazine was thus the first **tranquiliser**.

Tranquilisers appeared to work by inhibiting neurotransmitter substances, like acetylcholine or noradrenaline. It was natural to ask what effect might be achieved by substances that worked in the opposite way – might they, for instance, help relieve depression? At the time the only effective treatment for chronic depression was **electroconvulsive therapy**. ECT, which many viewed as brutal despite the fact that it often worked, was based on a supposed antagonism between epilepsy and schizophrenia: induction of artificial fits was believed to help. In fact, the first breakthrough arose accidentally. Administering the new antituberculosis drug, isoniazid, doctors found there was a marked improvement in the well-being of the patients. Their appetites returned; they put on weight and they cheered up. Psychiatrists quickly discovered that isoniazid and related compounds were fairly similar to neurotransmitters, in particular the amines found in the brain.[55] These amines, it was already known, were decomposed by a substance called monoamine oxidase; so did isoniazid achieve its effect by inhibiting monoamine oxidase, preventing it from decomposing the neurotransmitters? The monoamine oxidase inhibitors, though they worked well enough in relieving depression, had too many toxic side effects to be lasting as a family of drugs. Shortly afterward, however, another relative of chlorpromazine, imipramine, was found to be effective as an **antidepressant**, as well as increasing people's desire for social contact.[56] This entered widespread use as Tofranil.

All these substances reinforced the view that the 'mind' was amenable to chemical treatment. During the 1950s and early 1960s, many tranquilisers and antidepressants came into use. Not all were effective with all patients; each had side effects. But whatever their shortcomings, and despite the difficulties and complexities that remain, even to this day, these two categories of drugs, besides relieving an enormous amount of suffering, pose profound questions about human nature. They confirm that psychological moods are the result of chemical states within the brain, and therefore throw into serious doubt the traditional metaphysical concept of mind.

In trying to be an amalgam of Freud and Sartre, of psychoanalysis and exist-entialism, R. D. Laing's ideas were going against the grain then becoming established in psychiatry. Why then, when it is debatable whether Laing's approach ever cured anyone, did he become a cult figure?

In the context of the times, Laing and colleagues such as David Cooper in Britain and Herbert Marcuse in America focused their attention on the *personal* liberation of individuals in a mass society, as opposed to the earlier Marxist idea of liberation of an entire *class* through revolution. Gregory Bateson, Marcuse, and Laing all argued that man lived in conflict with mass society, that society and the unconscious were constantly at war, the schizophrenic simply the most visible victim in this war.[57] The intolerable pressures put on modern families led to the famous '**double bind**,' in which all-powerful parents tell a child one thing but do another, with the result that children grow up in perpetual conflict. Essentially, Laing and the others were saying that society is mad and the schizophrenic response is no more or less than a rational reaction to that complex, confusing world, if only the private logic of the schizophrenic can be unravelled. For Laing, families were 'power units' on top of whatever else they might be, and it is liberation from this power structure that is part of the function of psychiatry. This led to experiments in specially created clinics where even the power structure between psychiatrist and patient was abolished.

Laing became a cult figure in the early 1960s, not only because of his radical approach to schizophrenia (**anti-psychiatry**, and **radical psychiatry** became popular terms), but also because of his approach to experience.[58] From about 1960, Laing was a fairly frequent user of the so-called mind-altering drugs, including LSD. Like others, he believed that the '**alternative consciousness**' they provided could be clinically useful in the liberation of false consciousness created by schizophrenogenic families, and for a time he persuaded the British Home Office to give him a licence to experiment (in his offices in Wimpole Street, London) with LSD, which was then manufactured commercially in Czechoslovakia.[59] As the 1960s progressed, Laing and Cooper were taken up by the New Left. The linking of psychiatry and politics seemed new, radical, in Britain but went back to the teachings of the Frankfurt School and its original attempts to marry Marx and Freud. This is one reason why the Laing cult was overshadowed by the Marcuse cult in America.

Herbert Marcuse, sixty-two in 1960, had been part of the Frankfurt School and, like Hannah Arendt, studied under Martin Heidegger and Edmund Husserl. With Max Horkheimer and Theodor Adorno he had emigrated to the United States following Hitler's rise to power, but unlike them, he did not return once the war was over. He put his linguistic skills at the disposal of wartime intelligence and remained in government service some time after

1945.[60] As an erstwhile Marxist, Marcuse's mind was radically changed by Hitler, Stalin, and World War II. Afterward he was motivated, he said, by three things: that Marxism had not predicted the rise of Nazism, the emergence out of capitalist society of an irrational, barbaric movement; the effects of technology on society, especially Fordism and Taylorism; and the fact that prosperous America still contained many hidden and uncomfortable assumptions and contradictions.[61] Marcuse's attempt at a rapprochement of Freud and Marx was more sophisticated than either Erich Fromm's or Laing's. He felt that Marxism, as an account of the human condition, failed because it took no measure of individual psychology. In *Eros and Civilisation* (1955) and *One-Dimensional Man* (1964), Marcuse examined the conformist mass society around him, where high-technology material goods were both the epitome of scientific rationalism and the means by which conformity in thought and behavior was maintained, and he offered a new emphasis on aesthetics and sensuality in human life.[62] For him, the most worthwhile response to mass society on the part of the individual was **negation** (an echo of Sartre's *l'homme revolté*). The United States was one-dimensional because there were no longer any permissible alternative ways to think or behave. His was, he said, a 'diagnosis of domination'. Life moved 'forward' by means of 'progress,' thanks to reason and 'the rigidity' of science.[63] This was, he said, a stifling totality that had to be countered with imagination, art, nature, 'negative thought,' all put together in **'a great refusal.'**[64] The already disastrous results in recent decades of very conformist societies, the new psychologies of mass society and affluence, what were perceived as the dehumanising effects of positivist science and philosophy – all combined, for Marcuse, into 'a criminally limited' one-dimensional world.[65] For many, Laing and Marcuse went together because the former's schizophrenics were the natural endpoint of the one-dimensional society, the reject-victims of a dehumanising world where the price of nonconformity carried the risk of madness. This had uncomfortable echoes of Thomas Mann and Franz Kafka, looking back even to the speeches of Hitler, who had threatened with imprisonment the artists who painted in ways he thought 'degenerate.' In the early 1960s the baby-boom generation was reaching university age. The universities were expanding fast, and on campus the notions of Laing, Marcuse, and others, though quite at variance with the clinical evidence, nonetheless proved irresistible. Riesman had found that it was a characteristic of the 'other-directed' personality that it hated its own conformist image. The popularity of Laing and Marcuse underlines that. And so the stage was set for personal, rather than political change. The 1960s were ready to begin.

MANHATTAN TRANSFER

On 11 May 1960, at six-thirty in the evening, Richard Klement got down as usual from the bus which brought him home from work at the Mercedes-Benz factory in the Suarez suburb of Buenos Aires. A moment later he was seized by three men and in less than a minute forced into a waiting car, which took him to a rented house in another suburb. Asked who he was, he replied instantly, 'Ich bin Adolf Eichmann,' adding, 'I know I am in the hands of the Israelis.' The Israeli Secret Service had had 'Klement' under surveillance for some time, the culmination of a determined effort on the part of the new nation that the crimes of World War II would not be forgotten or forgiven. After his capture, Eichmann was kept secretly in Buenos Aires for nine days until he could be secretly flown to Jerusalem on an El Al airliner. On 23 May, Prime Minister David Ben-Gurion announced to cheers in the Jerusalem parliament that Eichmann had arrived on Israeli soil that morning. Eleven months later, Eichman was brought to trial in the District Court of Jerusalem, accused on fifteen counts that, 'together with others,' he had committed crimes against the Jewish people, and against humanity.[1]

Among the scores of people covering the trial was Hannah Arendt, who was there on behalf of the *New Yorker* magazine and whose articles, published later as a book, caused a storm of controversy.[2] The offence arose from the subtitle of her account, 'A Report on the **Banality of Evil**,' a phrase that became famous. Her central argument was that although Eichmann had done monstrous things, or been present when monstrous things had been done to the Jews, he was not himself a monster in the accepted sense of the word. She maintained that no court in Israel – nor *any* court – had ever had to deal with someone like Eichmann. His was a crime that was on no statute book. In particular, Arendt was fascinated by Eichmann's conscience. It wasn't true to say that he didn't have one: handed a copy of *Lolita* to read in his cell during the trial, he handed it back unfinished. 'Das ist aber ein sehr unerfreuliches Buch,' he told his guard; 'Quite an unwholesome book.'[3] But Arendt reported that throughout the trial, although Eichmann calmly admitted what he had done, and although he knew somewhere inside him that what had been done had been wrong, he did not *feel* guilty. He said he had moved in a world where no one questioned the final solution, where no one had ever condemned him. He had obeyed orders; that was all there was to it. 'The postwar notion of open disobedience was a fairy tale: "Under the circumstances such behaviour was impossible. Nobody acted that way." It was "unthinkable." '[4] Some atrocities he helped to commit were done to advance his career.

Arendt caused offence on two grounds.[5] She highlighted that many Jews had gone to their deaths without rebellion, not willingly exactly but in acquiescence; and many of her critics felt that in denying that Eichmann was a monster, she was diminishing and demeaning the significance of the Holocaust. This second criticism was far from the truth. If anything, Arendt's picture of Eichmann, consoling himself with clichés, querying why the trial was being prolonged – because the Israelis already had enough evidence to hang him several times over – only made what Eichmann had done more horrendous. But she wrote as she found, reporting that he went to the gallows with great dignity, after drinking half a bottle of red wine (leaving the other half) and refusing the help of a Protestant minister. Even there, however, he was still mouthing platitudes. The 'grotesque silliness' of his last words, Arendt said, proved more than ever the 'word-and-thought-defying *banality of evil*.'[6]

Despite the immediate response to Arendt's report, her book is now a classic.[7] At this distance her analysis, correct in an important way, is easier to accept. One aspect of Arendt's report went unremarked, however, though it was not insignificant. It was written in English, for the *New Yorker*. Like many intellectual emigrés, Arendt had not returned to Germany after the war, at least not to live. The mass emigration of intellectual talent in the 1930s, the bulk of which entered the United States, infused and transformed all aspects of American life in the postwar world, and had become very clear by the early 1960s, when *Eichmann in Jerusalem* appeared. It coloured everything from music to mathematics, and from chemistry to choreography, but it was all-important in three areas: psychoanalysis, physics, and art.

After some early hesitation, America proved a more hospitable host to psychoanalytic ideas than, say, Britain, France, or Italy. Psychoanalytic institutes were founded in the 1930s in New York, Boston, and Chicago. At that time American psychiatry was less organically oriented than its European counterparts, and Americans were traditionally more indulgent toward their children, as referred to earlier. This made them more open to ideas linking childhood experience and adult character.

Assistance to refugee analysts was organised very early in the United States, and although numbers were not large in real terms (about 190, according to one estimate), the people helped were extremely influential. Karen Horney, Erich Fromm, and Herbert Marcuse have already been mentioned, but other well known analyst-emigrés included Franz Alexander, Helene Deutsch, Karl Abraham, Ernst Simmel, Otto Fenichel, Theodor Reik, and Hanns Sachs, one of the 'Seven Rings,' early colleagues of Freud pledged to develop and defend psychoanalysis, and given a ring by him to symbolise that dedication.[8] The reception of psychoanalysis was further aided by the psychiatric problems that came to light in America in World War II. According to official figures, in the period 1942–5 some 1,850,000 men were rejected for military service for psychiatric reasons, 38 percent of all rejections. As of 31 December 1946, 54 percent of all patients in veterans' hospitals were being treated for neuro-psychiatric disorders.

The other two most influential emigré psychoanalysts in America after World War II were **Erik Erikson** and **Bruno Bettelheim**. Erikson was Freud's last pupil in Vienna. Despite his Danish name, he was a north German, who arrived in America in 1938 when he was barely twenty-one and worked in a mental hospital in Boston. Trained as a lay therapist (America was also less bothered by the absence of medical degrees for psychoanalysts than Europe was), Erikson gradually developed his theory, in *Childhood and Society* (1950), that adolescents go through an '**identity crisis**' and that how they deal with this is what matters, determining their adult character, rather than any Freudian experience in childhood.[9] Erikson's idea proved extremely popular in the 1950s and 1960s, with the advent of the first really affluent adolescent 'other-directed' generation. So too did his idea that whereas hysteria may have been the central neurosis in Freud's Vienna, in postwar America it was narcissism, by which he meant a profound concern with one's own psychological development, especially in a world where religion was, for many people, effectively dead.[10] Bruno Bettelheim was another lay analyst, who began life as an aesthetician and arrived in America from Vienna, via a concentration camp. The account he derived from those experiences, *Individual and Mass Behavior in Extreme Situations*, was so vivid that General Eisenhower made it required reading for members of the military government in Europe.[11] After the war, Bettelheim became well known for his technique for helping autistic children, described in his book *The Empty Fortress*.[12] The two works were related because Bettelheim had seen people reduced to an 'autistic' state in the camps, and felt that children could therefore be helped by treatment that, in effect, sought to reverse the experience.[13] Bettelheim claimed up to 80 percent success with his method, though doubt was cast on his methods later in the century.[14]

In America, psychoanalysis became a much more optimistic set of doctrines than it had been in Europe. It embodied the view that there were moves individuals could make to help themselves, to rectify what was wrong with their psychological station in life. This was very different from the European view, that sociological class had as much to do with one's position in society, and that individuals were less able to change their situation without more widespread societal change.

Two matters divided physicists in the wake of World War II. There was first the development of the **hydrogen bomb**. The Manhattan Project had been a collaborative venture, with scientists from Britain, Denmark, Italy, and else-where joining the Americans. But it was undoubtedly *led* by Americans, and almost entirely paid for by them. Given that, and the fact that Germany was occupied and Britain, France, Austria, and Italy were wrecked by six years of war, fought on their soil, it was no surprise that the United States should assume the lead in this branch of research. Göttingen was denuded; Copenhagen had been forced to give up its place as a centre for international scholars; and in Cambridge, England, the Cavendish population had been dispersed and was changing emphasis toward molecular biology, a very fruitful manoeuvre. In the years after the war, four nuclear scientists who migrated to America were awarded the Nobel Prize, adding immeasurably to the prestige of American science: Felix Bloch in 1952, Emilio Segrè in 1959, and Maria Mayer and Eugene Wigner in 1963. The Atomic Energy Act of 1954 established its own prize, quickly renamed after its first winner, Enrico Fermi, and that too was won by five emigrés before 1963: Fermi, John von Neumann, Eugene Wigner, Hans Bethe, and Edward Teller. Alongside three native American winners – Ernest Lawrence, Glenn Seaborg, and Robert Oppenheimer – these prize-winners emphasised the progress in physics in the United States.

Many of these men (and a few women) were prominent in the 'movement of atomic scientists,' whose aim was to shape public thinking about the atomic age, and which issued its own *Bulletin of the Atomic Scientists*, for discussion of these issues. The *Bulletin* had a celebrated logo, a clock set at a few minutes to midnight, the hands being moved forward and back, according to how near the editors thought the world was to apocalypse. Scientists such as Oppenheimer, Fermi, and Bethe left the Manhattan Project after the war, preferring not to work on arms during peacetime. Edward Teller, however, had been interested in a hydrogen bomb ever since Fermi had posed a question over lunch in 1942: Once an atomic bomb was developed, could the explosion be used to initiate something similar to the thermonuclear reactions going on inside the sun? The news, in September 1949, that Russia had successfully exploded an atomic bomb caused a lot of soul-searching among certain physicists. The Atomic Energy Commission decided to ask its advisory committee, chaired by Oppenheimer, for an opinion. That committee unanimously decided that the United States should not take the initiative, but feelings ran high, summed up best by Fermi, whose view had changed over time. He thought that the new bomb should be outlawed before it was born – and yet he conceded, in the Cold War atmosphere then prevailing, that no such agreement would be possible; 'Failing that, one should with considerable regret go ahead.'[15] The agonising continued, but in January 1950 Klaus Fuchs in England confessed that while working at Los Alamos he had passed information to Communist agents. Four days after the confession, President Truman took the decision away from the scientists and gave the go-ahead for an American H-bomb project.

The essence of the hydrogen bomb was that when an atomic bomb exploded in association with deuterium, or tritium, it would produce temperatures never seen on earth, which would fuse two deuterium nuclei together and simultaneously release binding energy in vast amounts. Early calculations had shown that such a device could produce an explosion equivalent to 100 million tons of TNT and cause damage across 3,000 square miles. (For comparison, the amount of explosives used in World War II was about 3 million tons.)[16] The world's first thermonuclear device – a hydrogen bomb – was tested on 1 November 1952, on the small Pacific island of **Elugelab**. Observers forty miles away saw millions of gallons of seawater turned to steam, appearing as a giant bubble, and the fireball expanded to three miles across. When the explosion was over, the entire island of Elugelab had disappeared, vaporised. The bomb had delivered the equivalent of 10.4 million tons of TNT, *one thousand* times more violent than the bomb dropped on Hiroshima. Edward Teller sent a telegram to a colleague, using code: 'It's a boy.' His metaphor was not lacking in unconscious irony. The Soviet Union exploded its own device nine months later.[17]

But after World War II ended, most physicists were anxious to get back to 'normal' work. Quite what normal work was now was settled at two big physics conferences, one at **Shelter Island**, off the coast of Long Island, near New York, in June 1947, and the other at Rochester, upstate New York, in 1956.

The high point of the Shelter Island conference was a report by Willis Lamb that presented evidence of small variations in the energy of hydrogen atoms that should not exist if Paul Dirac's equations linking relativity and quantum mechanics were absolutely correct. This 'Lamb shift' produced a revised mathematical account, quantum electro-dynamics (QED), which scientists congratulate themselves on as being the 'most accurate theory in physics.'[18] In the same year as the conference, mathematically and physically trained cosmologists and astronomers began studying cosmic rays arriving on Earth from the universe and discovered new subatomic particles that did not behave exactly as predicted – for example, they did not decay into other particles as fast as they should have done. This anomaly gave rise to the next phase of particle physics, which has dominated the last half of the century, an amalgam of physics, maths, chemistry, astronomy, and – strange as it may seem – history. Its two achievements are an understanding of how the universe formed, how and in which order the elements came into being; and a systematic classification of particles even more basic than electrons, protons, and neutrons.

The study of elementary particles quickly leads back in time, to the very beginning of the universe. The 'Big Bang' theory of the origin of the universe began in the 1920s, with the work of Georges Lemaître and Edwin Hubble. Following the Shelter Island conference, in 1948, two Austrian emigrés in Britain, **Herman Bondi** and **Thomas Gold**, together with **Fred Hoyle**, a professor at Cambridge, advanced a rival **'steady state'** theory, which envisaged matter being quietly formed throughout the universe, in localised 'energetic events.' This was never taken seriously by more than a few scientists, especially as in the same year **George Gamow**, a Russian who had defected to the United States in the 1930s, presented new calculations showing how nuclear interactions taking place in the early moments of the fireball that created the expanding universe could have converted hydrogen into helium, explaining the proportions of these elements in very old stars. Gamow also said that there should be evidence of the initial explosion in the form of background radiation, at a low level of intensity, to be picked up wherever one looked for it in the universe.[19]

Gamow's theories, especially his chapter on 'The Private Life of Stars,' helped initiate a massive interest among physicists in **'nucleosynthesis,'** the ways in which the heavier elements are built up from hydrogen, the lightest element, and the role played by the various forms of elementary particles. This is where the study of cosmic rays came in. Almost none of the new particles discovered since World War II exists naturally on earth, and they could only be studied by accelerating naturally occurring particles to make them collide with others, in particle accelerators and cyclotrons. These were very large, very expensive pieces of equipment, and this too was one reason why 'Big Science' flourished most in America – not only was it ahead intellectually, but America more than elsewhere had the appetite and the wherewithal to fund such ambition. Hundreds of particles were discovered in the decade following the Shelter Island conference, but three stand out. The particles that did not behave as they should have done under the earlier theories were christened '**strange**' by **Murray Gell-Mann** at Caltech in 1953 (the first example of a fashion for whimsical names for entities in physics).[20] It was various aspects of strangeness that came under scrutiny at the second physics conference in Rochester in 1956. These notions of strangeness were brought together by Gell-Mann in 1961 into a classification scheme for particles, reminiscent of the periodic table, and which he called, maintaining the whimsy, '**The Eight-Fold Way.**' The Eight-Fold Way was based on mathematics rather than observation, and in 1962 mathematics led Gell-Mann (and almost simultaneously, George Zweig) to introduce the concept of the '**quark,**' a particle more elementary still than electrons, and from which all known matter is made. (Zweig called them 'aces' but 'quark' stuck. Their existence was not confirmed experimentally until 1977.) Quarks came in six varieties, and were given entirely arbitrary names such as 'up,' 'down,' or 'charmed.'[21] They had electrical charges that were fractions – plus or minus one-third or two-thirds of the charge on an electron – and it was this fragmentary charge that was so significant, further reducing the building blocks of nature. We now know that all matter is made up of two kinds of particle: '**baryons**' – protons and neutrons, fairly heavy particles, which are divisible into quarks; and '**leptons,**' the other basic family, much lighter, consisting of electrons, muons, the tun particle and neutrinos, which are *not* broken down into quarks.[22] A proton, for example, is comprised of two up quarks and one down quark, whereas a neutron is made up of two down quarks and one up. All this may be confusing to nonphysicists, but keep in mind that the elementary particles that exist naturally on Earth are exactly as they were in 1932: the electron, the proton, and the neutron. All the rest are found only either in cosmic rays arriving from space or in the artificial circumstances of particle accelerators.[23]

It was the main aim of physicists to amalgamate all these discoveries into a grand synthesis that would have two elements. It would explain the evolution of the universe, describe the creation of the elements and their distribution among the planets and stars, and explain the creation of carbon, which had made life possible. Second, it would explain the fundamental forces that enable matter to form in the way that it forms. God apart, it would in effect explain everything.

One day in the middle of 1960, Leonard Kessler, a children's-book illustrator, ran into **Andy Warhol** – a classmate from college – coming out of an art-supply store in New York. Warhol was carrying brushes, tubes of paint, and some raw canvases. Kessler stared at him. 'Andy! What are you doing?'

'I'm starting pop art,' Warhol replied.

All Kessler could think of to say was, 'Why?'

'Because I hate abstract expressionism. I hate it!'[24]

Do art movements really start at such specific moments? Maybe **pop art** did. As we shall see, it totally transformed not only art but also the role of the artist, a metamorphosis that in itself epitomises late-twentieth-century thought as much as anything else. But if Andy Warhol hated the **abstract expressionists**, it was because he was jealous of the success that they enjoyed in 1960. As Paris had faded, New York had become the new home of the avant-garde. Warhol would help change ideas about the avant-garde too.

The exhibition *Artists in Exile* at the Pierre Matisse Gallery in 1943, when Fernard Léger, Piet Mondrian, Marc Chagall, Max Ernst, André Breton, André Masson, and so many other European artists had shown their work, had had a big impact on American artists.[25] It would be wrong to say that this exhibition changed the course of American painting, but it certainly accelerated a process that was happening anyway. The painters who came to be called the abstract expressionists (the term was not coined until the late 1940s) all began work in the 1930s and shared one thing: Jackson Pollock, Mark Rothko, Arshile Gorky, Clyfford Still, and Robert Motherwell were fascinated by psychoanalysis and its implications for art. In their case it was Jungian analysis that attracted their interest (Pollock was in Jungian analysis for two years), in particular the theory of archetypes and the collective unconscious. This made them assiduous followers (but also critics) of surrealism. Forged in the years of depression, in a world that by and large neglected the artist, many of the abstract expressionists experienced great poverty. This helped foster a second characteristic – the view that the artist is a social rebel whose main enemy is the culture of the masses, so much of which (radio, talking pictures, *Time*, and other magazines) was new in the 1930s. The abstract expressionists were, in other words, natural recruits to the avant-garde.[26]

Between the Armory Show and World War II, America had received a steady flow of exhibitions of European art, thanks mainly to Alfred Barr at the Museum of Modern Art in New York. It was Barr who had organised the show of Cézanne, Van Gogh, Seurat, and Gauguin in 1929, when MoMA had opened.[27] He had a hand in the International Modern show at MoMA in 1934, and the Bauhaus show in 1937. But it was only between 1935 and 1945 that psychoanalytic thought, and in particular its relation to art, was explored in any detail in America, due to the influx of European psychoanalysts, as referred to above. Psychoanalysis was, for example, a central ingredient in the ballets of **Martha Graham** and **Merce Cunningham**, who in such works as *Dark Meadow* and *Deaths and Entrances* combined primitive (Native American) myths with Jungian themes. The first art exhibitions to really explore psychoanalysis also took place in wartime. Jackson Pollock's show in November 1943, at Peggy Guggenheim's gallery, started the trend, soon followed by Arshile Gorky's exhibition at Julien Levy's gallery in March 1945, for which André Breton wrote the foreword.[28] But the abstract expressionists were important for far more than the fact that theirs was the first avant-garde movement to be influential in America. The critics **Isaac Rosenfeld** and **Theodore Solotaroff** drew attention to something they described as a 'seismic change' in art: as a result of the depression and the war, they said, artists had moved 'from Marx to Freud.' The underlying ethic of art was no longer 'Change the world,' but 'Adjust yourself to it.'[29]

And this is what made the abstract expressionists so pivotal. They might see themselves as an avant-garde (they certainly did so up until the end of the war), and some of them, like **Willem de Kooning**, would always resist the blandishments of patrons and dealers, and paint what they wanted, how they wanted. But that was the point: what artists wanted to produce had changed. The criticisms in their art were personal now, psychological, directed inward rather than outward toward the society around them, echoing Paul Klee's remark in 1915, 'The more fearful the world becomes, the more art becomes abstract.' It is in some ways extraordinary that at the very time the Cold War was beginning – when two atomic bombs had been dropped and the hydrogen bomb tested, when the world was at risk as never before – art should turn in on itself, avoid sociology, ignore politics, and concentrate instead on an aspect of self – the unconscious – that by definition we cannot know, or can know only indirectly, with great difficulty and in piecemeal fashion. This is the important subject of Diana Crane's *Transformation of the Avant-Garde*, in which she chronicles not only the rise of the New York art market (90 galleries in 1949, 197 in 1965) but also the changing status and self-conception of artists. The modernist avant-garde saw itself as a form of rebellion, using among other things the new techniques and understanding of science to disturb and provoke the bourgeois, and in so doing change a whole class of society. By the 1960s, however, as the critic Harold Rosenberg noted, 'Instead of being . . . an act of rebellion, despair or self-indulgence, art is being normalised as a professional activity within society.'[30] Clyfford Still put it more pungently: 'I'm not interested in illustrating my time. . . . Our age – it is of science – of mechanism – of power and death. I see no point in adding to its mammoth arrogance the compliment of graphic homage.'[31] As a result the abstract expressionists would be criticised time and again for their lack of explicit meaning or any social implications, the beginning of a long-term change.

The ultimate example of this was pop art, which both **Clement Greenberg** and the Frankfurt School critics saw as essentially inimical to the traditional function of avant-garde art. Few pop artists experienced poverty the way the abstract expressionists had. Frank Stella had had a (fairly) successful father, Joseph, and Andy Warhol himself, though he came from an immigrant family, was earning $50,000 a year by the mid-1950s from his work in advertising. What did Warhol – or any of them – have to rebel against?[32] The crucial characteristic of pop art was its *celebration*, rather than criticism, of popular culture and of the middle-class lifestyle. All the pop artists – Robert Rauschenberg, Jasper Johns, James Rosenquist, Claes Oldenburg, Roy Lichtenstein, and Warhol – responded to the images of mass culture, advertising, comic books, and television, but the early 1960s were above all Warhol's moment. As Robert Hughes has written, Warhol did more than any other painter 'to turn the art world into the art business.'[33] For a few years, before he grew bored with himself, his art (or should one say his works?) managed to be both subversive and celebratory of mass culture. Warhol grasped that the essence of popular culture – the audiovisual culture rather than the world of books – was repetition rather than novelty. He loved the banal, the unchanging images produced by machines, though he was also the heir to Marcel Duchamp in that he realised that certain objects, like an electric chair or a can of soup, change their meaning when presented 'as art.' This new aesthetic was summed up by the artist Jedd Garet when he said, 'I don't feel a responsibility to have a vision. I don't think that is quite valid. When I read artists' writings of the past, especially before the two wars, I find it very amusing and I laugh at these things: the spirituality, the changing of the culture. It is possible to change the culture but I don't think art is the right place to try and make an important change except visually. ... Art just can't be that world-shattering in this day and age. ... Whatever kind of visual statement you make has first to pass through fashion design and furniture design until it becomes mass-produced; finally, a gas pump might look a little different because of a painting you did. But that's not for the artist to worry about.... Everybody is re-evaluating those strict notions about what makes up high art. Fashion entering into art and vice versa is really quite a wonderful development. Fashion and art have become much closer. It's not a bad thing.'[34]

From pop art onward, though it started with abstract expressionism, artists no longer proposed – or saw it as their task to propose – 'alternative visions.' They had instead become part of the 'competing lifestyles and ideologies' that made up the contemporary, other-directed, affluent society. It was thus entirely fitting that when Warhol was gunned down in his 'Factory' on Union Square in 1968 by a feminist actress, and survived after being pronounced clinically dead, the price of his paintings shot up from an average of $200 to $15,000. From that moment, the price of art was as important as its content.

Also characteristic of the arts in America at that time, and Manhattan in particular, was the overlap and links between different forms: art, poetry, dance and music. According to David Lehman the very idea of the avant-garde had itself transferred to America and not just in painting: the title of his book on the New York school of poets, which flourished in the early 1950s, was *The Last Avant-Garde*.[35] Aside from their poetry, which travelled an experimental road between the *ancien régime* of Eliot *et alia* and the new culture of the Beats, John Ashbery, Frank O'Hara, Kenneth Koch, and James Schuyler were all very friendly with the abstract expressionist painters De Kooning, Jane Freilicher, Fairfield Porter, and Larry Rivers. Ashbery was also influenced by the composer **John Cage**. In turn, Cage later worked with painters **Robert Rauschenberg** and **Jasper Johns**, and with the choreographer **Merce Cunningham**.

By the middle of the century two main themes could be discerned in serious music. One was the loss of commitment to tonality, and the other was the general failure of twelve-tone serialism to gain widespread acceptance.[36] Tonality did continue, notably in the works of Sergei Prokofiev and Benjamin Britten (whose *Peter Grimes*, 1945, even prefigured the 'antiheroes' of the angry young men of the 1950s). But after World War II, composers in most countries outside the Soviet Union were trying to work out the implications 'of the two great contrasted principles which had emerged during and after World War I: "rational" serialism and "irrational" Dadaism.' To that was added an exploration of the new musical technology: tape recording, electronic synthesis, computer techniques.[37] No one reflected these influences more than John Cage.

Born in Los Angeles in 1912, Cage studied under Schoenberg between 1935 and 1937, though rational serialism was by no means the only influence on him: he also studied under Henry Cowell, who introduced him to Zen, Buddhist, and Tantric ideas of the East. Cage met Merce Cunningham at a dance class in Seattle in 1938, and they worked together from 1942, when Cunningham formed his own company. Both were invited to Black Mountain College summer school in North Carolina in 1948 and again in 1952, where they met Robert Rauschenberg. Painter and composer influenced each other: Rauschenberg admitted that Cage's ideas about the everyday in art had an impact on his images, and Cage said that Rauschenberg's white paintings, which he saw at Black Mountain in 1952, gave him courage to present his 'silent' piece, *4' 33"*, for piano in the same year (see below). In 1954 Rauschenberg became artistic adviser to Cunningham's dance company.[38]

Cage was the experimentalist par excellence, exploring new sound sources and rhythmic structures (*Imaginary Landscape No. 1* was scored for two variable-speed gramophone turntables, muted piano, and cymbals), and in particular indeterminacy. It was this concern with chance that linked him back to Dada, across to the surrealist Theatre of the Absurd and, later, as we shall see, to Cunningham. Cage also anticipated postmodern ideas by trying to break down (as Walter Benjamin had foreseen) the barrier between artist and spectator. Cage did not believe the artist should be privileged in any way and sought, in pieces such as *Musiccircus* (1968), to act merely as an initiator of events, leaving the spectator to do much of the work, where the gulf between musical notation and performance was deliberately wide.[39] The 'archetypal' experimental composition was the aforementioned *4' 33"* (1952), a three-movement piece for piano where, however, not a note is played. In fact Cage's instructions make clear that the piece may be 'performed' on any instrument for any amount of time. The aim, beyond being a parody and a joke at the expense of the ordinary concert, is to have the audience listen to the ambient sounds of the world around them, and reflect upon that world for a bearably short amount of time.

The overlap with Cunningham is plain. Born in 1919 in Centralia in Washington State, Cunningham had been a soloist with the Martha Graham Dance Company but became dissatisfied with the emotional and narrative content and began to seek out a way to present movement as itself. Since 1951, Cunningham had paralleled Cage by introducing the element of chance into dance. Coin tossing and dice throwing or clues from the *I Ching* were used to select the order and arrangement of steps, though these steps were themselves made up of partial body movements, which Cunningham broke down like no one before him. This approach developed in the 1960s, in works such as *Story* and *Events*, where Cunningham would decide only moments before the production which parts of the dance would be performed that night, though even then it was left to the individual dancers to decide at certain points in the performance which of several courses to follow.[40]

Two other aspects of these works were notable. In the first, Cage or some other composer provided the music, and Rauschenberg, Johns, Warhol, or other artists would provide the settings. Usually, however, these three elements – dance, music, and set – did not come together until the day before the premiere. Cunningham did not know what Cage was producing, and neither of them knew what, say, Rauschenberg was providing. A second aspect was that, despite one of Cunningham's better-known works being given the title *Story*, this was strongly ironic. Cunningham did not feel that ballets had to tell a story – they were really 'events.' He intended spectators to make up their own interpretations of what was happening.[41] Like Cage's emphasis on silence as part of music, so Cunningham emphasised that stillness was part of dance. In some cases, notices in the wings instructed certain dancers to stay offstage for a specified amount of time. Costumes and lighting changed from night to night, as did some sets, with objects being moved around or taken away completely.

That said, the style of Cunningham's actual choreography is light, suggestive. In the words of the critic Sally Banes, it conveys a 'lightness, elasticity ... [an] agile, cool, lucid, analytic intelligence.'[42] Just as the music, dance, and settings were to be comprehended in their own right, so each of Cunningham's steps is presented so as to be whole and complete in itself, and not simply part of a sequence. Cunningham also shared with Jacques Tati a compositional approach where the most interesting action is not always going on in the front of the stage at the centre. It can take place anywhere, and equally interesting things may be taking place at the same time on different parts of the stage. It is up to the spectator to respond as he or she wishes.

Cunningham was even more influenced by Marcel Duchamp, and his questioning of what art is, what an artist is, and what the relationship with the spectator is. This showed most clearly in *Walkaround Time* (1968), which had decor by Jasper Johns based on *The Bride Stripped Bare by her Bachelors, Even* and with music by David Behrman entitled *... for nearly an hour*, based on Duchamp's *To Be Looked at (from the Other Side of the Glass) with One Eye, Close to, for Almost an Hour*. This piece was Johns's idea. He and Cunningham were at Duchamp's house one evening, and when Johns put the idea to him, the Frenchman answered, 'But who would do all the work?'[43] Johns said he would, and Duchamp, relieved, gave permission, adding that the pieces should be moved around during the performance to emulate the paintings.[44] The dance is characterised by people running in place, small groups moving in syncopated jerkiness, like machines, straining in slow motion, and making minuscule movements that can be easily missed. *Walkaround Time* has a 'machine-like grace' that made it more popular than *Story*.[45]

With Martha Graham and **Twyla Tharp**, Cunningham has been one of the most influential choreographers in the final decades of the century. This influence has been direct on people like Jim Self, though others, such as Yvonne Rainer, have rebelled against his aleatory approach.

Cunningham, Cage, the abstract expressionists, and the pop artists were all concerned with the form of art rather than its meaning, or content. This distinction was the subject of a famous essay by the novelist and critic **Susan Sontag**, writing in 1964 in the *Evergreen Review.* In 'Against Interpretation,' she argued that the legacy of Freud and Marx, and much of modernism, had been to overload works of art with meaning, content, interpretation. Art – whether it was painting, poetry, drama, or the novel – could no longer be enjoyed for what it was, she said, for the qualities of form or style that it showed, for its numinous, luminous, or 'auratic' quality, as Benjamin might have put it. Instead, all art was put within a 'shadow world' of meaning, and this impoverished it and us. She discerned a countermovement: 'Interpretation, based on the highly dubious theory that a work of art is composed of items of content, violates art. It makes art into an article for use, for arrangement into a mental scheme of categories. ... The flight from interpretation seems particularly a feature of modern painting. Abstract painting is the attempt to have, in the ordinary sense, no content; since there is no content, there can be no interpretation. Pop art works by the opposite means to the same result; using a content so blatant, so "what it is," it, too, ends by being uninterpretable.'[46] She wanted to put silence back into poetry and the magic back into words: 'Interpretation takes the sensory experience of the work of art for granted. ... What is important now is to recover our senses. ... In place of a hermeneutics we need an erotics of art.'[47]

Sontag's warning was timely. Cage and Cunningham were in some respects the last of the modernists. In the postmodern age that followed, interpretation ran riot.

EQUALITY, FREEDOM, AND JUSTICE IN
THE GREAT SOCIETY

In the spring of 1964, just weeks after the assassination of John F. Kennedy, his successor as president, Lyndon Johnson, delivered a speech on the campus of the University of Michigan, at Ann Arbor. That day he outlined a massive program for social regeneration in America. The program, he said, would recognise the existence and the persistence of poverty and its links to the country's enduring civil rights problem; it would acknowledge the growing concern for the environment, and it would attempt to meet the demands of the burgeoning women's liberation movement. Having reassured his listeners that economic growth in America appeared sustained, with affluence a fact of life for many people, he went on to concede that Americans were not only interested in material benefit for themselves 'but in the prospects for human fulfilment for *all* citizens.'' Johnson, an experienced politician, understood that Kennedy's killing had sent a shockwave across America, had been a catalyst that made the early 1960s a defining moment in history. He realised that to meet such a moment, he needed to act with imagination and vision. **The Great Society** was his answer.

Whatever judgements are made about the success or otherwise of Johnson's idea, he was right to recognise the moment, for the 1960s saw a collective shift in several areas of thought. Often characterised as a 'frivolous' decade of fashion frippery, musical 'intoxication,' sexual licence, and a narcotics-induced nihilism, the decade was in fact the time when, outside war, more people in the West than ever before faced up to – or were faced with – the most fundamental dilemmas of human existence: freedom, justice, and equality, what they meant and how they could be achieved. Before examining what Johnson *did*, it is necessary first to examine the context of his Michigan speech, which went back further, and ranged far wider, than the assassination of one man in Dallas on 22 November 1963.

On 17 August 1961, East German workers had begun building the Berlin Wall, a near-impregnable barrier sealing off West Berlin and preventing the escape of East Germans to the West. This followed an initiative by Nikita Khrushchev, of the USSR, to President Kennedy of the United States, that a German peace conference be held to conclude a treaty and establish Berlin as a free city, the

Soviet leader proposing simultaneously that talks be held about a ban on nuclear tests. Although talks about a test ban had begun in June, they had broken down a month later. The construction of the Berlin Wall thus marked the low point of the Cold War, and provided an enduring symbol of the great divide between East and West. Relations soured still more in January of the following year, when the three-power conference (United States, U.K. and USSR) on nuclear test bans collapsed after 353 meetings. And then, in October 1962 the Cuban missile crisis flared, after Russia agreed to provide Fidel Castro – who had seized power in Cuba in 1959 after a prolonged insurrection – with arms, including missiles. President Kennedy installed a blockade around Cuba, and the world waited anxiously as Soviet ships approached the island. The crisis lasted for six days until, on 28 October, Khrushchev announced that he had ordered the withdrawal of all 'offensive' weapons from Cuba. It was the closest the world had come to nuclear war.

In 1961 communism stretched beyond Russia to East Germany and seven East European states, to the Balkan countries of Yugoslavia and Albania, to China, North Korea, and North Vietnam, to Angola in Africa, Cuba in the Americas, with a major Soviet or local Communist Party presence in Italy, Chile, Egypt, and Mozambique. The Soviet Union was providing arms, education, and training to several other countries, such as Syria, the Congo, and India. The world had never before been so extensively polarised into two rival systems, the centralised, state-centred and state-led Communist economies on the one hand and the free-market economies of the West on the other. Against such a background, it is perhaps no surprise that books began to appear examining the very notion of freedom at its most fundamental. Communism involved coercion, to put it mildly. But it was being successful, even if it wasn't being popular.

One of the central tenets of Friedrich von Hayek's *Road to Serfdom*, published in 1944, was that there is in life a 'spontaneous social order,' which has grown up over the years and generations, that things are as they are for a reason, and that attempts to interfere with this spontaneous order are almost certainly doomed to failure. In 1960, at the height of the Cold War, Hayek published *The Constitution of Liberty*, in which he extended his argument beyond planning, the focus of his earlier book, to the moral sphere.[2] His starting point was that the values by which we organise and operate our lives have evolved in just the same way that our intelligence has. It follows from this, he says, that liberty – the rules of justice – 'is bound to take priority over any specific claim to welfare' simply because liberty and justice *create* that very welfare: 'If individuals are to be free to use their own knowledge and resources to best advantage, they must do so in a context of known and predictable rules governed by law.' Individual liberty, Hayek said, 'is a creature of the law and does not exist outside any civil society.' Laws, therefore, must be as universal as possible in their application, and abstract – that is, based on general, and generally accepted, concepts rather than on individual cases.[3] He adds two further important points: that liberty is intimately linked to property rights, and that the concept of 'social justice,' which would become very much a vogue in the following years, and which certainly underpinned the Great Society, was and is a myth. For Hayek, the freedom to live as one wishes on one's own private property, always supposing of course that one does not, in so doing, interfere with the rights of others, was the ultimate good. Being evolved, law is for Hayek 'part of the natural history of mankind; it emerges directly from men's dealings with each other, is coeval with society, and therefore, and crucially, antedates the emergence of the state. For these reasons it is not the creation of any governmental authority and it is certainly not the command of any sovereign.'[4] Hayek was therefore against socialism, in particular the Soviet variety, on very fundamental grounds: the government – the state – organised the law, and had no second chamber, which Hayek thought was the natural antidote in the realm of law. Nor did Soviet communism allow any private property, by which the general principles of liberty translated into something practical that everyone could understand; and because it was centrally directed, there was no scope for law to evolve, to maintain the greatest liberty for the greatest number. Socialism, in short, was an interference in the natural evolution of law. Finally, and most controversially at the time, Hayek thought that the concept of 'social justice' was the most powerful threat to law conceived in recent years. Social justice, said Hayek, 'attributes the character of justice or injustice to the whole pattern of social life, with all its component rewards and losses, rather than to the conduct of its component individuals, and in doing this it inverts the original and authentic sense of liberty, in which it is properly attributed only to individual actions.'[5] In other words, the law must treat men anonymously in order to treat them truly equally; if they are not treated individually, serious inequities result. What is more, he argued, modern notions of 'distributive' justice, as he called it, involve some notion of 'need' or 'merit' as criteria for the 'just' distribution in society.[6] He observes that 'not all needs are commensurate with each other,' as for example a medical need involving the relief of pain and another regarding the preservation of life when there is competition for scarce resources.[7] Other needs are not satiable. It follows from this, he says, that there is 'no rational principle available to settle the conflict'; this 'infects' the lives of citizens 'with uncertainty and dependency on unforeseeable bureaucratic interventions.'[8] Hayek's view was – and remains – influential, though there were two main criticisms. One concerned spontaneous order. Why should spontaneous order occur? Why not spontaneous *dis*order? How can we be sure that what has evolved is invariably the best? And isn't spontaneous order, the fruit of evolution, a form of panglossianism, an assumption that we live in the best of all possible worlds, and that we can do little to improve things?

Constitution of Liberty is primarily a work about law and justice. Economics and politics, though not absent, are in the background. In 1950 Hayek had left Britain when he was appointed professor of social and moral sciences and a member of the Committee on Social Thought at the University of Chicago. It was a colleague in Chicago who took up where Hayek had left off, reflecting a similar view but adding an economic dimension to the debate. In *Capitalism and Freedom* (1962), **Milton Friedman** advanced the then relatively unpopular view that the meaning of *liberalism* had been changed in the twentieth century, corrupted from its original nineteenth-century meaning – of economic liberalism, a belief in free trade and free markets – and converted instead to mean a belief in equality brought about by well-meaning central government.[9] His first aim was to regain for liberalism its original meaning, and his second was to argue that true freedom could only be brought about by a return to a true market economy, that real freedom could only exist when man was economically free.[10] This view was more contentious then that it is now because, in 1962, Keynesian economics were still in the ascendant. In fact, Friedman's arguments went much further than traditional economic interests in markets. Besides arguing that the depression had been brought about *not* by the Crash, but by economic mismanagement by the U.S. government in the wake of the Crash, Friedman argued that health, schooling, and racial discrimination could be helped by a return to free market economics. Health, he thought, was hampered by the monopoloy which physicians had over the training and licensing of fellow doctors. This had the effect, he said, of keeping down the supply of medical practitioners, which helped their earning power and acted to the disadvantage of patients. He outlined many 'medical' duties that could be carried out by technicians – were they allowed to exist – who could be paid much less than highly trained doctors.[11] With schools, Friedman's ideas distinguished, first, a **'neighborhood effect'** in education. That is to say, to an extent we all benefit from the fact that all of us are educated in a certain way – in the basic skills of citizenship, without which no society can function. Friedman thought that this type of schooling should be provided centrally but that all other forms of education, and in particular vocational courses (dentistry, hairdressing, carpentry) should be paid for.[12] Even basic citizenship education, he thought, should operate on a voucher system, whereby parents could exchange their vouchers for schooling for their children at the schools of their choice. He thought this would exert an influence on schools, through teachers, in that the vouchers would reward good teachers and ought to be transferred into income for them.[13] Regarding racial discrimination, Friedman took the long-term view, arguing that throughout history capitalism and free markets had been the friend of minority groups, whether those groups were blacks, Jews, or Protestants in predominantly Catholic countries. He therefore thought that, given time, free markets would help emancipate America's blacks.[14] He argued that legislation for integration was no more and no less ethical than legislation for segregation.

One of the criticism of Friedman's arguments was that they lacked the sense of urgency that was undoubtedly present in Johnson's speech in Michigan. Kennedy's assassination had an effect here, as did the rioting and stand-offs between blacks and law-enforcement agencies that flared throughout the 1960s. There was also the relentless aggressiveness of communism in the background. But in 1964 there was another factor: the 'rediscovery' of poverty in America, of squalor amid abundance, and its link to something that all Americans could see for themselves – the disfiguring decline of its cities, especially the inner areas. While Hayek's and Friedman's books, controversial as they were, were calm and reflective in tone, two very different works published at the same time were much more polemical, and as a result had an immediate impact. *The Death and Life of Great American Cities*, by **Jane Jacobs**, was ironic and argumentative. *The Other America: Poverty in the United States*, by **Michael Harrington**, was downright angry.[15]

The Other America must count as one of the most successful polemics ever written, if judged by its ability to provoke political acts. Published in 1961, it was taken up by the *New Yorker*, where it was summarised under the title 'Our Invisible Poor.' By the end of the following year, President Kennedy was asking for specific proposals as to what might be done about poverty in the country.[16] Harrington's style was combative, but he was careful not to overstate his case. He admitted, for example, that in absolute terms poverty in the third world was probably worse than in North America. And he granted that though the affluent society helped breed 'spiritual emptiness and alienation ... yet a man would be a fool to prefer hunger to satiety, and the material gains at least open up the possibility of a rich and full existence.'[17] But he added that the third world had one advantage – everyone was in the same boat, and they were all pulling together to fight their way out. In America, on the other hand, there was 'a culture of poverty,' 'an under-developed nation' within the affluent society, hidden, invisible, and much more widespread than anyone had hitherto thought. He claimed that as many as 50 million people, a quarter of the nation, were poor.[18] This sparked a subsidiary debate as to what the criteria should be for drawing the poverty line, and whether poverty in America was increasing, decreasing, or static. But Harrington was more concerned to show that, despite the size of the poor, middle America was blind to its plight. This was partly because poverty occurred in remote areas – among migrant workers on farms, in remote islands or pockets of the country such as the Appalachian Mountains, or in black ghettoes where the white middle classes never went.[19] Here he succeeded in shocking America into realising the problem it was ignoring in its own backyard. He also argued that there was a 'culture of poverty' – that the lack of work, the poor housing, the ill health, high crime and divorce rates, all went together. The cause of poverty was not simply lack of money but systemic changes in the capitalist system that caused, say, the failure of the mines (as in the Appalachians) or of the farms (as in areas of California). It followed from this that the poor were not primarily to blame for their plight, and that the remedy lay not with individual action on their part but with the government. Harrington himself thought that the key lay in better housing, where the federal government should take the lead. His book was, therefore, addressed to the 'affluent blind,' and his searing descriptions of specific instances of the culture of poverty were deliberately designed to remove the indifference and blindness. How far he succeeded may be judged from the fact that his phrases '**the culture of poverty**' and '**the cycle of deprivation**' became part of the language, and in Johnson's State of the Union address, in January 1964, four months before his Great Society speech, he announced a thirteen-point program that would wage 'unconditional war on poverty . . . a domestic enemy which threatens the strength of our nation and the welfare of our people.'[20]

The Death and Life of Great American Cities, which appeared in the same year as Harrington's polemic, had an impact that was almost as immediate.[21] Curiously, however, although many people did and do agree with her, the long-term impact of her book has not been what Jacobs hoped. *Death and Life* is probably the most sensible book ever written about cities. It is, first, an attack on Ebenezer Howard and his idea of garden cities (a contradiction in terms for Jacobs), on Lewis Mumford and his stages of city life ('morbid' and 'biased'), and above all on Le Corbusier, whose ideas for a 'Radiant City' she blames for much of the great **'Blight of Dullness'** that she saw around her.[22] She began by stressing that the basic component of the city is the street, and in particular the sidewalk (pavement, in British usage). Sidewalks and streets are safer if busy, she points out; they are communities in themselves, entirely natural communities, peopled by inhabitants who know each other, as well as by strangers. They are places where children can learn and be assimilated into adult life (she observes that 'street' gangs usually congregate in parks or schools). Streets stay busy, and safe, all day long if, and only if, they are home to diverse interests – i.e., they are occupied not just by offices or shops but by a *mix*, which includes a residential element.[23] She argues that parks and schools are far more 'fickle' than streets – there is no telling whether a park will become a skid row or a hangout for perverts (her word), or which school will work and which won't.[24] She thinks 'neighborhood' is a sentimental concept but hardly a real one. Apart from streets, cities should be divided into districts, but these should be natural districts, corresponding to the way the city is divided up in the minds of most residents. The purpose of a district is political, not psychological or personal. A district is there to fight the battles that streets are too small and too weak to fight – she quotes the case of drug peddlers moving into one street. It is the district that prevails on the police to move into a street in force for a limited period until the problem is dispersed. Districts, she says, should never be more than a mile and a half from end to end.[25]

The essence of the street, and the sidewalk in particular, where people meet and talk, is that it enables people to control their own privacy, an important aspect of freedom. She believed that people are less than straightforward about privacy, hiding behind the convenient phrase 'mind your own business.' This reflects the importance of gossip – people can gossip all they like, but often pretend they don't, or don't approve. In this way they can retreat into their own private world, their own 'business,' whenever they want without loss of face. This is psychologically very important, she says, and may be all-important for keeping cities alive. Only when these psychological needs are met – a cross between privacy and community, which is a city speciality – are people content, and content to stay put.[26]

Jacobs also identified what she called 'border vacuums' – railway tracks, freeways, stretches of water, huge parks like Central Park in New York. These, she said, contribute their own share of blight to a city and should be recognised by planners as 'a mixed blessing'; they need special devices to reduce their impact. For example, huge parks might have carousels or cafés on their perimeters to make them less daunting and encourage usage. She thought that old buildings must be preserved, partly because of their aesthetic value and because they provide breaks in the dull monotony of many cityscapes, but also because old buildings have a different economy to new buildings. Theatres go into new buildings, for example, but the studios and workshops that service theatres usually don't – they can't afford new buildings, but they can afford old buildings that paid for themselves a long time ago. Supermarkets occupy new buildings, but not bookshops. She thought that a city does not begin to be a city until it has 100,000 inhabitants. Only then will it have enough diversity, which is the essence of cities, and only then will it have a large enough population for the inhabitants to find enough friends (say thirty or so people) with like interests.[27] Understanding these dynamics, she said, helps keep cities alive. Finance, of course, is important, and here cities can help themselves. Jacobs felt that too often the financing of real estate is left to professional (i.e., private) companies, so that in the end the needs of finance determine the type of real estate that is mortgaged, rather than the other way round.[28] Provided her four cardinal principles were adhered to, she said, she felt certain that the blight of city centres could be halted, and 'unslumming' be made to work. These four principles were: every district must serve more than one, and preferably more than two, primary functions (business, commerce, residential), and these different functions must produce a different daily schedule among people; city blocks should be short – 'opportunities to turn corners must be frequent'; there must be a 'close-grained' mingling of structures of very different age; and the concentration of people must be sufficiently dense for what purposes they may be there.[29] Hers was an optimistic book, resplendent with common sense that, however, no one else had pointed out before. What she didn't explore, not in any detail, was the racial dimension. She made a few references to segregation and 'Negro slums,' but other than that she wrote strictly as an architect/town planner.

The issues raised by Harrington and Jacobs were both referred to by President Johnson. There is no question, however, that the main urgency that propelled him to his Great Society speech, apart from the 'deep background' of the Cold War, was race, especially the situation of American blacks. By 1966 a whole decade had elapsed since the landmark decision of the Supreme Court in 1954, in *Brown* v. *Board of Education of Topeka*, that racial segregation in schools was unconstitutional, repudiating the doctrine of 'separate but equal.' As Johnson realised, in the intervening years the basic statistics of black life were dispiriting. In 1963 there were more blacks in America in *de facto* segregated schools than there had been in 1952. There were more black unemployed than in 1954. More significant still, the median income of blacks had slipped from 57 percent that of whites in 1954 to 54 percent. Against this background, Milton Friedman's arguments about the long-term beneficial effects of capitalism on race relations looked thin, and in 1963, as Johnson recognised, action was needed to avert trouble.

Among the blacks themselves there was, as might be expected, a range of opinions as to the way forward. Some were in more of a hurry than others; some felt violence was necessary; others felt nonviolence ultimately had more impact. In March 1963 there had been riots in Birmingham, Alabama, when an economic boycott of downtown businesses had turned ugly following a decision by the commissioner for public safety, Eugene 'Bull' Connor, to have the police surround a church and prevent people from leaving. Among those arrested in the wake of these events (on Good Friday) was **Martin Luther King**, a thirty-four-year-old preacher from Atlanta who had made a name for himself by rousing, rhetorical speeches advocating nonviolence. While he was in solitary confinement, King had been denounced by a group of white clerics. His response was '**Letter from a Birmingham Jail**,' nineteen pages scribbled and scrawled on envelopes, lavatory rolls, and the margins of newspaper articles, smuggled out of the jail by his supporters. It set out in vivid and eloquent detail why the people of Birmingham (i.e., the whites) had 'left the Negro community with no alternative' but to take the course of civil disobedience and 'nonviolent tension' in pursuit of their aims.[30] 'Birmingham is probably the most thoroughly segregated city in the United States. ... There have been more unsolved bombings of Negro homes and churches in Birmingham than in any other city in the nation. ... We had no alternative to prepare for our direct action. ... The nations of Asia and Africa are moving at jet-like speed toward gaining political independence, but we still creep at horse-and-buggy pace toward gaining a cup of coffee at a lunch counter.'[31]

After his release from the Birmingham jail, King achieved the peak of his fame, and he was chosen as the main speaker for an historic march on Washington that summer, designed deliberately by a variety of black leaders to become a turning point in the civil rights campaign. The march was to be massive, so massive that although it was to be peaceful, it would nonetheless convey an implicit threat that if America didn't change, didn't do something – and soon – about desegregation, then ... The threat was left deliberately vague. About a quarter of a million people descended on Washington on 28 August 1963, between a quarter and a third of them white. The marchers were relatively good-natured, and kept in line by a team of black New York policemen who had volunteered as marshals. The entertainment was second to none: Joan Baez, Bob Dylan, Peter, Paul and Mary, and Mahalia Jackson, with a number of other celebrities showing up to lend support: Marlon Brando, Harry Belafonte, Josephine Baker, James Baldwin, Lena Horne, Sammy Davis Junior. But what everyone remembered about that day was the speech by King. In recent speeches he had used a phrase he had found to be effective -'**I have a dream**' – and on this occasion he lavished extra special care on his delivery.[32] Just as some men's face is their fortune, in King's case it was his voice. A very distinctive baritone, its dominant characteristic was a slight quiver. Combined with a rhetorical strength, this quiver made King's voice both strong and yet vulnerable, exactly matching the developing mood and political situation of ordinary American blacks. But it also had a universal appeal that whites could identify with too. For many, King's speech that day would prove to be the most memorable part of the civil rights campaign, or at least the part they chose to remember. 'Five score years ago,' he began, announcing his near-biblical tone, 'a great American, in whose symbolic shadow we stand, signed the Emancipation Proclamation.' With his first sentence he had hit his theme and rooted it in American history. 'But one hundred years later, we must face the tragic fact that the Negro is still not free. ... There will be neither rest nor tranquility in America until the Negro is granted his citizenship rights.' And then he opened out, saying that he had a dream that one day his four little children would be judged 'not by their colour but by their character.'[33] Even today, the recording of King's speech has the power to move.

King lived through and helped bring about turbulent times (Vietnam was a second factor). Between November 1955, when Rosa Parks, a black American, was arrested for sitting at the front of the bus in Montgomery, Alabama (blacks had traditionally only been allowed in the back of the bus), and 1973, when Los Angeles elected its first black mayor, an enormous social, political, and legislative revolution took place. That revolution was most visible in the United States, but it extended to other countries, in Europe, Africa, and the Far East, as this list, by no means exhaustive, indicates:

1958: Disturbances in Little Rock, Arkansas, when the state governor tries to prevent the admission of black pupils to a school.

1960: The Civil Rights Act is passed, empowering blacks to sue if denied their voting rights.

1961: The Congress for Racial Equality (CORE) organises 'freedom rides' to enforce bus desegregation.

1962: The Committee on Equal Employment Opportunity forms, chaired by Vice President Johnson. James Meredith, a black student, gains admission to the University of Mississippi, Oxford, under federal guard. The British Commonwealth Immigrants Act limits the rights of admission to Britain of certain Commonwealth immigrants.

1963: The March on Washington. Equal-pay law for men and women in the United States is enacted.

1964: The Civil Rights Act in the United States forbids discrimination in work, restaurants, unions, and public accommodation. The Economic Opportunity and Food Stamps Acts are passed, and the U.S. Survey of Educational Opportunity carried out.

1965: Great Society initiatives includes Head Start programs to support education for the poor and minorities; Medicaid and Medicare to provide medicine for the poor and elderly; urban development schemes; and other welfare benefits. Women are accepted as judges.

1966: NOW, the National Organization for Women, is founded, along with the Black Panthers, a black paramilitary outfit that calls for 'Black Power.'

Under the U.S. Child Nutrition Act, federal funds provide food for poor children. British Supplementary Benefit assists the sick, disabled, unemployed, and widows. Inner cities are rebuilt.

1967: Thurgood Marshall becomes the first black man appointed to the U.S. Supreme Court. Race riots in seventy American cities accelerate 'white flight' to the suburbs. Colorado is the first U.S. state to allow abortion. Homosexuality is legalised in Britain. In the United States, a report of the Commission on Civil Rights concludes that racial integration needs to be accelerated to reverse the underachievement of African-American children. Educational Priority Areas are created in Britain to combat inequality. Abortion becomes lawful in the U.K.

1968: The Urban Institute is founded. The Kerner Report on the previous year's race riots warns that the United States is becoming 'two societies, one black, one white, separate and unequal.' President Johnson announces 'affirmative action,' under which all government contractors must give 'preferential treatment' to African Americans and other minorities. Racial discrimination in the sale and renting of houses is outlawed. Shirley Chisholm is elected the first black congresswoman. The Immigration and Nationality Act replaces quota system with skill requirements. Hispanic workers protest against their treatment in the United States. The Race Relations Act in the U.K. makes racial discrimination illegal.

1969: Supreme Court nominees are withdrawn on grounds of their 'racism and incompetence.' Black Panthers are killed in a police raid in Chicago. Land begins to be returned to Native Americans. The United States ends censorship.

1970: Civil rights for women; in federal contracts companies must employ a quota of women. The Equal Pay Act is passed in the U.K. Divorce is made legal in Italy. The first desegregated classes are held in the United States.

1971: Bussing introduced to ensure a 'racial balance' in some U.S. schools. Switzerland accepts female suffrage. Slum primary schools in the U.K. are cleared. Medicare is implemented in Canada. The first women are ordained as priests (by the Anglican bishop of Hong Kong).

1972: Andrew Young becomes the first African American elected from the South to Congress since Reconstruction. Indians march on Washington, D.C. First woman governor of the New York Stock Exchange.

1973: In the United States abortion is made legal. The first black mayor of Los Angeles is elected.[34]

The change didn't end there, of course (the following year saw the first Hispanic and women governors of U.S. states, and the first female bishops). But the years of turbulence were over (which was also related to the ending of the war in Vietnam, and the economic downturn following the oil crisis in 1973 – see chapter 33 below). Not that all the change was in one direction, toward greater freedom for minority groups, women, and homosexuals. An alternative list reads as follows:

1964: Bantu Laws amendment, designed to limit the settlement of Africans to peripheral areas, is introduced in South Africa.

1966: Apartheid is extended to South West Africa (Namibia).

1967: Resettlement villages are accelerated in South Africa.

1968: *Humanae Vitae*, papal encyclical, prohibits use of artificial contraceptives by Roman Catholics.

1969: The Stonewall police raid on a homosexual club in New York results in several days of violence after the club is set on fire while police are inside. Anti-egalitarian 'Black papers' are published in Britain. Arthur Jensen, in the *Harvard Educational Review*, argues that African Americans score consistently less well on IQ tests than do whites.

1970: In South Africa all black Africans are consigned to one or other of the 'Bantu homelands.' Several books about race are banned in South Africa.

1971: South African Bantu areas are brought under control of central government.

1972: South Africa abolishes coloured representatives on municipal councils.

Throughout the late 1950s and 1960s, the growing illiberalism of South African society and the violence associated with the advance of the blacks in America, were increasingly seen as part of the same malaise – the same dilemma, as Myrdal had called it – circumstances that combined to produce some sharp thinking about race. Though these authors might match King in rhetoric, they rarely matched him in Christian feeling.

One of the authors James Baldwin had read when he was in Paris was **Frantz Fanon**, a black psychiatrist born in the French West Indian island of Martinique in 1925. After training in psychiatry in Paris, Fanon was assigned to a hospital in the North African colony of Algeria during the rising against the French. The experience appalled him; he took the Algerians' side and wrote a number of books in which, like Baldwin in the southern states of America, he became a spokesman for those suffering oppression. In *A Dying Colonialism* (1959) and *Black Skin, White Masks* (1960), originally published in French, Fanon proved himself an articulate critic of the last days of imperialism, and his activities for the FLN (National Liberation Front), including an address to the First Congress of Negro Writers in 1956, drew the attention of the French police.[35] Later that year he was forced to leave Algeria for Tunisia, where he continued to be one of the editors of *El Moudjahid*, an anticolonial magazine. His most poignant book was *The Wretched of the Earth* (1961), conceived at the time Fanon was diagnosed as suffering from leukaemia, and which consumed his final strength.[36]

Fanon was a more polemical writer than Baldwin, and a less gifted phrasemaker. But like the American his works are designed to worry whites and convince blacks that the battle – against racism and colonialism – can be won. Where *The Wretched of the Earth* was different was in Fanon's use of his experiences as a psychiatrist. Fanon was intent on showing fellow blacks that the alienation they felt as a result of colonialism *was* a result of colonialism, and not some natural inferiority inherent in the black race. In support of his argument he reported a number of psychiatric reactions he had seen in his clinic and which, he said, were directly related to the guerrilla war of independence then being waged inside the country. In one case an Algerian taxi driver and member of the FLN had developed impotence after his wife had been beaten and raped by a French soldier during interrogation. In another, two young Algerians, aged thirteen and fourteen, had killed their European playmate. As the thirteen-year-old put it, 'We weren't a bit cross with him. ... One day we decided to kill him, because the Europeans want to kill all the Arabs. We can't kill big people. But we could kill ones like him, because he was the same age as us.'[37] Fanon had many stories of disturbances in young people, and especially among the victims of torture. He pointed out that torture victims could be divided into two – 'those who know something' and 'those who know nothing.' He said he never saw those who knew something as patients (they never got ill; they had in a sense 'earned' their torture), but among those who knew nothing, there were all sorts of symptoms, usually related to the type of torture – indiscriminate, mass attack with truncheons or cigarette burns; electricity; and the so-called 'truth serum.' Victims of electric torture, for example, would develop an electricity phobia and become unable to touch an electric switch.[38]

Fanon's aim, like R. D. Laing's, was to show that mental illness was an extreme but essentially rational response to an intolerable situation, but he was also answering what he saw as oversimple arguments by European scientists and social scientists regarding 'the African mind' and African culture. In the mid-1950s, the World Health Organisation had commissioned a survey by a Scottish psychiatrist, Dr J. C. Carothers, on 'Normal and Pathological Psychology of the African.' Carothers had worked in Kenya and been medical officer in command of prisons there. His survey had concluded, 'The African makes very little use of his frontal lobes. All the particularities of African psychiatry can be put down to frontal laziness.' Carothers actually put forward the idea that the 'normal' African is like a 'lobotomised European.'[39] Fanon countered dismissively, arguing that Carothers had missed the point. At that stage, he said, African culture (like black American culture, like Baldwin's writing) was the *struggle* to be free; the fight – violence itself – was the shared culture of the Algerians, and took most of their creative energy. Like King, they had become 'creative extremists.' Fanon did not live to see peace restored to an autonomous Algeria. He had been too busy completing his book to seek treatment for his leukaemia, and although he was taken to Washington in late 1961, the disease was too far advanced. He died a few weeks after his book was published, aged thirty-six.

Polemical writing, like Fanon's, was exactly the sustenance blacks needed in the 1960s, and in America, after James Baldwin changed his stance in a series of novels, *Another Country* (1962), *Blues for Mister Charlie* (1964), and *Going to Meet the Man* (1965), his place was taken by **Eldridge Cleaver**. Born in Little Rock, Arkansas, in 1935, Cleaver liked to describe himself as having been 'educated in the Negro ghetto of Los Angeles and at the California state prisons of San Quentin, Folsom and Soledad.' Though ironic, this was also true, as

Cleaver had read widely in jail (he had been convicted of marijuana possession) and met several other inmates who nurtured his rebellious instincts. He eventually became minister of information in the Black Panther Party, an African-American paramilitary organisation. His first book, *Soul on Ice*, released the same year that King was assassinated, was a wide-ranging attack on Baldwin. 'There is in James Baldwin's work,' wrote Cleaver, 'the most gruelling, agonizing, total hatred of the blacks, particularly of himself, and the most shameful, fanatical, fawning, sycophantic love of the whites that one can find in the writings of any black American writer of note in our time.'[40] For Cleaver, as with Fanon, the situation facing African Americans was too urgent to allow the luxury of becoming an artist in any wider sense; the problem was so all-enveloping that to turn one's back on it, or place it in a wider context, as Baldwin attempted to do from time to time, was for Cleaver an avoidance akin to race crime. Three themes are interlaced in *Soul on Ice*, which was written in prison. One is the everyday brutality of whites toward blacks, highlighted by prison routine. Two, Cleaver's thoughts on international race politics, white myths about race, Africa, black history, black food, black music, showing how to build a countervailing and sustaining myth. And three, Cleaver's progressive thoughts about sex between the races, from the first essay, where he confesses that for him, as a young man, he found white women more attractive than black, to the last essay, a far more lyrical, near-mystical paean of praise to 'Black Beauty' – 'Let me drink from the river of your love at its source.'[41] Pointed as his criticisms of Baldwin were at the time, the latter's works have survived in better shape than Cleaver's essays.

Maya Angelou's books are very different. Her message is that blacks are already free – not in the political sense, maybe, but in every other sense. It is her isolation of the political from the rest that is her more important, and contentious, point. In *I Know Why the Caged Bird Sings*, the first of her five-part autobiography, published in 1969, Angelou records her life until she has her first baby at the age of sixteen.[42] We are treated to the richness of black life in Stamps, Arkansas, not a million miles from Little Rock, Cleaver's birthplace and the scene of so much racial violence. Angelou re-creates brilliantly her childhood world 'of starched aprons, butter-yellow piqué dresses, peanut patties, and games of mumbledypeg, with bathwater steaming on the cooking stove.' When bad things happen, tears course down her cheeks 'like warm milk.'[43] But there is more to this soft-focus world than scoops of corn thrown to the chickens. Although her father is absent for much of the time, the emotional and intellectual life of the family left behind – mother, son, and daughter – is not much impoverished. William Shakespeare 'was my first white love' in a world where Kipling and Thackeray jostle with Langston Hughes, James Weldon Johnson, and W. E. B. Du Bois.[44] Maya, or Marguerite as she then was, has a genuine affection for her brother Bailey and her mother, a strong, upright, beautiful woman who is not cowed by the system. As the children grow up, the adult world of work and discrimination encroaches on their idyll – for example, in the form of the dentist who would rather stick his hand in a dog's mouth than a 'nigger's.'[45] But this is not presented as tragedy. Maya and her mother retain their interest in the world, keep control of it, and keep thinking. Their lives remain rich, whatever changes fate has in store. Of course Angelou hates the system of discrimination, but her books emphasise that life is made up of two kinds of freedom: one big political freedom, and countless little freedoms that come from education, strength of character, humour, dignity, and thought. At one point her mother is asked, 'You all right, momma?' 'Aw,' she replies, 'they tell me the whitefolks still in the lead.'[46]

I Know Why the Caged Bird Sings fits as easily into a canon of works written by female authors as it does one written by blacks. Women's emancipation, though not involving violence on anything like the same scale as the civil rights movement, offered several parallels throughout the 1960s. The decade saw major changes in almost all areas of sexual liberation. In 1966 the **Kinsey Institute** had begun its important early study of **homosexuality**, which found that 4 percent of males and 2 percent of females were predominantly or exclusively homosexual, and that no fewer than 37 percent of men reported at least one homosexual experience.[47] In the same year, William Howell Masters and Virginia Johnson's *Human Sexual Inadequacy* showed that about half of all marriages suffered from one sexual problem or another (inability to maintain an erection or premature ejaculation in men, inability to achieve orgasm in women).[48] A year after, in 1967, modern mass-market, hard-core pornography began to appear, produced by Scandinavian magazine publishers. In that year too Hugh Hefner, the publisher of *Playboy*, then selling 4 million copies a month, made the cover of *Time*.[49] On 3 November 1968, Al Goldstein launched *Screw*, the self-proclaimed aim of which was to become the *Consumer Reports* of the 'sexual netherworld.' A year later Philip Roth published *Portnoy's Complaint*, exploring the 'agony and ecstasy' of male masturbation, and *Oh! Calcutta!* was produced in London and off Broadway, with full-frontal nudity and explicitly sexual dialogue. Nineteen-seventy saw the first pubic hair to be shown in a commercial magazine, *Penthouse*. In 1970 the President's Commission on Obscenity and Pornography reported that there was no substantial basis for the belief that exposure to erotica caused sex crimes. Some kind of closure was achieved in this area in 1973, when the U.S. Supreme Court voted seven to two to legalise abortion, and in the same year, the American Psychiatric Association removed homosexuality from its diagnostic manual, declaring that gays and lesbians did not suffer from a mental disorder.

Whereas the publishing/pornography revolution, and gay liberation, were chiefly about sexual freedom (many states in the United States still outlawed homosexuality), the women's liberation movement was about far more than the new sexual awareness of women. Though that was important, the change in women's thinking about themselves, set in motion after World War II by Simone de Beauvoir and developed by Betty Friedan, was much more fundamental and far-reaching. In 1970, slap in the middle of the sexual revo-lution, three books appeared almost simultaneously, each of which took an uncompromising look at the relationship between the sexes.

Germaine Greer was an Australian who had settled in England as a graduate student and had drawn attention to herself in *Suck* magazine, decrying the missionary position (she thought women were more in control and had more pleasure if they sat on men during intercourse). Her book *The Female Eunuch* did not neglect women's economic condition, though only one of the thirty chapters is devoted to work. Rather, it drew its force from Greer's unflinching comparison of the way women, love, and marriage are presented in literature, both serious and popular, and in everyday currency, as compared with the way things really are. 'Freud,' she writes, 'is the father of psychoanalysis. It had no mother.'[50] From Jane Austen to Lord Byron to *Women's Weekly*, Greer is withering in her criticisms of how men are presented as dominant, socially superior, older, richer, and taller than their women. (Greer is very tall herself.) In what is perhaps her most original contribution, she demolishes love and romance (both given their own chapters) as chimeras, totally divorced (an apt verb) from the much bleaker reality. In fact, she says, 'Women have very little idea of how much men hate them.' A chapter headed 'Misery' recounts the amount of medication women take, the paraphernalia of sexual aids, leading to the resentment that she argues many women feel at being saddled with such things.[51] Her diagnosis is unstinting, and her solution demands nothing else than a radical reassessment by women not just of their economic and psychological position vis-à-vis men but, more revolutionary still, a fundamental reappraisal of what love and romance really are. Greer has the grace to admit that she has not herself entirely shed the romantic notions she was brought up with, but makes it plain she suspects they are entirely – *entirely* – without foundation. As with all true liberation, this view is both bleak and exhilarating.

Juliet Mitchell's *Women's Estate* was hardly exhilarating.[52] A fellow immigrant to Britain from the Antipodes, this time from New Zealand, Mitchell also studied English at a British university, though she subsequently transferred to psychoanalysis. Mitchell's account was Marxist, claiming that although socialist countries are not very nice to women, social*ism* does not require the subjugation of women as capitalism does, with its ideology of 'the nuclear family,' which succeeds only in keeping women in their place, acquiring consumer goods and breeding 'little consumers.'[53] Mitchell went on to argue that women need to undergo two revolutions, the political and the personal, and here she took the black experience as a guide but also psychoanalysis.[54] At the same time that women regrouped politically, she said, they also needed to raise their level of self-consciousness as the blacks had done, especially as in America. Women, she insisted, have been taught by capitalism and by Freud that they are the repositories of feelings, but in fact there is no limit to their experience. She favoured small groups of six to twenty-four women joining in '**consciousness-raising**' sessions, taking a leaf out of the book of the Chinese revolutionaries' practice of 'speaking bitterness.'[55] Together with her survey of what has been achieved by women in other countries around the world, Mitchell's aim was to bring about a situation where women did not feel alone in their predicament, and to spread the psychoanalytically inspired function: 'Speaking the unspoken is, of course, also the purpose of serious psychoanalytic work.'[56]

Kate Millett's *Sexual Politics* was, like Greer's book, essentially an examination of literary texts, equally erudite, equally readable, and even more thorough.[57] As her title implied, the focus of her interest was the power inherent in the relations between the sexes, though she queried whether it really is 'inherent.' She had herself been molested when she was thirteen and held on to her secret for a decade until, in a women's group, she found that almost all the other members had gone through similar experiences. This had fired her up. In her book, after brief excursions into sociological, biological, anthropological, and even mythological explanations for gender differences, she reverted to late-eighteenth- and early-nineteenth-century England, to John Stuart Mill, John Ruskin, William Wordsworth, and Alfred Lord Tennyson, moving on to Friedrich Engels's and Thorsten Veblen's theories of the family, its relation to the state, private property, and revolutionary theory. Domestication, prostitution, and sexuality are discussed, in Christina Bronte, Thomas Hardy, and Oscar Wilde (*Salomé*), where Millett found some grounds for hope before the 'counterrevolution' of Nazism, Stalinism, and Freudianism. Few would need convincing that Nazism and Stalinism were bad for women, but by including Freudianism along with these two, Millett's argument succeeded on shock value alone, as did her call to abolish the family. Millett's full ire was reserved, however, for three writers – D. H. Lawrence, Henry Miller, and Norman Mailer, who she compares and contrasts with a fourth, Jean Genet. In his novels D. H. Lawrence, she says, 'manipulates' women, Miller has only 'contempt' for them, and Mailer 'wrestles' them.[58] The force of her argument lies both in her close textual reading of the books and in the way she shows how certain themes run through several works by each author (patriarchy and employment in Lawrence, for example, murder in Mailer). In contrasting Genet with the other three, her aim was to show that the idea of femininity can exist in man, and she approved of his linking sexual and racial roles.[59] Ultimately, Millett was concerned about virility per se, the part it plays in *Realpolitik* as well as sexual politics. Perhaps most valuably, she pointed out that 'alienation' was no longer a vague word used by philosophers and psychologists; it had been revised and refined into a number of specific grievances felt by women, blacks, students, and the poor. That refinement was in itself an advance.[60]

This line of thinking culminated in the work of two women, **Andrea Dworkin** and **Shere Hite**. Dworkin, who described herself as 'an overweight ugly duckling,' had a father who was a teacher and instilled in her a love for ideas, but in 1969 she married a fellow left-wing radical who turned out to be a 'vicious rapist' and frequently beat her to the point of unconsciousness.[61] Eventually finding enough courage to leave him, she became a writer, taking up where Millett left off. In 1974 she published *Women Hating* and addressed a New York 'speak out' organised by the National Organization of Women, giving her talk the title 'Renouncing Sexual "Equality."' She was given a ten-minute ovation, and many of the eleven hundred women in the audience were left 'crying and shaking.' Dworkin concentrated on pornography, which she argued was motivated by a hatred of women, and she countered by developing a radical man-hating ideology. She herself set an example of what she saw as the only way out for women: she lived in a sexless open nonmarriage with a male homosexual.[62]

The Hite Report appeared in 1976. Born Shirley Gregory in Saint Joseph, Missouri, Shere Hite kept the name of her husband, whom she divorced after a brief marriage. Intending to pursue a master's degree in cultural history at Columbia University, Hite quit early and turned to a variety of jobs to survive. A Pre-Raphaelite redhead, she worked as a model and posed nude for both *Playboy* magazine and *Oui*. The real change in her life took place when she was asked to pose for an ad for Olivetti, the Italian typewriter company, where the photograph showed a secretary in front of a typewriter, with the legend, 'The typewriter that's so smart she doesn't have to be.' After posing for the ad, Hite read in a newspaper that a women's group planned to picket the company. She joined in, and soon after embroiled herself in the women's movement. One of the things she learned from this, which drew her particular attention, was that the medical profession at the time regarded a woman who could not achieve orgasm through intercourse as having 'a medical problem.' Over the next few years, she amassed enough funds to send out 100,000 questionnaires to women to see how they really felt about orgasm. She received over three thousand replies. When her *Report* appeared, it was a revelation.[63] Her most important finding was that most women did not orgasm as a result of vaginal penetration; moreover, they found that this unrealistic expectation placed a great psychological burden on women (and on men). This was not the same as saying that women did not enjoy intercourse, rather that what they enjoyed was the intimacy and the touching. Second, she found that these same women achieved orgasm fairly quickly when masturbating, but that there was a strong taboo against women touching themselves. *The Hite Report* made Shere Hite a millionaire virtually overnight, as its findings hit a chord in women, who found its message liberating, if only because so many women discovered that their own situation, predicament, problem – call it what you will – was not unique to them but, statistically speaking at least, 'normal.' Its findings carried the implication that women were much like men in sexual behaviour.[64] Hite's statistics turned out to be a form of emancipation, a practical response to one aspect of 'alienation.' There was a certain amount of cynicism in Shere Hite's work – a compendium of statistics on orgasm and masturbation was bound to be a commercial success. Even so, the report marked the end of a phase in women's liberation, reflecting a view that genuine independence, sexual as well as economic, was available for those women who wanted it.

Not everyone was happy with this wholesale change. A 1963 report, *Beyond the Melting Pot*, by Nathan Glazer (a junior colleague of David Riesman on *The Lonely Crowd*) and Patrick Moynihan, unveiled 'middle America,' which they described as a 'unifying state of mind,' 'characterised by opposition to civil rights, the peace movement, the student movement, "welfare intellectuals" and so on."[65] It was against this background that President Johnson sought to launch his great experiment. He set out his agenda in a series of speeches where 'the Great Society' became as familiar as Martin Luther King's 'Dream': Medicare for the old, educational assistance for the young, tax rebates for business, a higher minimum wage for labour, subsidies for farmers, vocational training for the unskilled, food for the hungry, housing for the homeless, poverty grants for the poor, clean highways for commuters, legal protection for blacks, improved schooling for Indians, higher benefits for the unemployed, pensions for the retired, fair labelling for consumers. Countless task forces were set up, as often as not with academics at their head. Legislation was hurried through, Johnson insisting that the Great Society would fill all the hopes and more of the New Deal.

It was, perhaps, the greatest experiment in social engineering outside the Communist world.[66] Between 1965 and 1968, when Johnson declined to stand for reelection, when the war in Vietnam was beginning to divide the nation and its cost to have a marked effect on the economy, some five hundred social programs were created, some more successful than others. (Johnson's biographer, Doris Kearns, concluded that medicare and voting rights succeeded admirably, for example, model cities less well, and community action was 'self-defeating.') But the real battle, which would last for years – and is, to an extent, still with us – was the fight over education, the idea that blacks and other disadvantaged minorities should be given access to better schooling, that equality of educational opportunity was what counted above all in a society where to be free meant to be free of ignorance, where democratic attitudes of fairness and individualism meant that men and women ought to be given a fair start in life but after that they were on their own, to make of their life what they could. These ideas spawned thousands of socio-psychological studies in the 1960s and afterward, exploring the effects of a person's economic, social, and racial background on a variety of factors, by far the most controversial of which was the IQ. Despite repeated criticisms over the years, that it did not measure what it purported to, that it was biased in favour of middle-class white children and against almost everyone else, the IQ continued to be used widely both as a research tool and in schools and the workplace.

The first major study of the issues the Great Society was meant to help rectify was *Equality of Educational Opportunity*, by **James Coleman** and others, released through the U.S. Government Printing Office in Washington, D.C., in 1966.[67] The Coleman Report, the most thorough study of naturally desegregated schools in America, concluded that the socioeconomic level of a student's school had more effect on his or her achievement than any other measurable factor except the socioeconomic level of his or her home. In other words, blacks were better off in desegregated schools because, in general, those schools were more likely to be middle class. Blacks were not better off if desegregation merely meant they transferred to schools where the whites were as poor as the blacks. British thinking had followed American ideas and created, in the mid-1960s, what were called Educational Priority Areas; as their name implied, these aimed to boost disadvantaged groups in disadvantaged socio-economic areas. However, in one study, *All Our Future*, by **J. W. B. Douglas** and others and published in 1968, their conclusion was that the gap between middle-class and working-class pupils had not been reduced, in any appreciable way, by such social engineering.[68]

The real controversy started at the end of the 1960s with an article in the *Harvard Educational Review* by **Arthur Jensen**, a psychologist from the University of California at Berkeley, headed 'How Much Can We Boost IQ and Scholastic Achievement?' This long review article (there was no new research material, merely a reanalysis of studies already published) began, 'Compensatory education has been tried and apparently it has failed.' Jensen argued that as much as 80 percent of the variance in IQ is due to genes and that therefore the approximately 15 percent difference between the average IQ score of whites and blacks was due mainly to hereditary racial differences in intelligence. It followed, Jensen said, that no program of social action could equalise the social status of blacks and whites, 'and that blacks ought better to be educated for the more mechanical tasks to which their genes predisposed them.'[69] At times it must have seemed to blacks as if there had been no progress since Du Bois's day.

Less contentious than Jensen but far more influential in the long run was the study carried out by **Christopher Jencks**, professor of sociology at Harvard, and seven colleagues.[70] Jencks, another student of David Riesman, had always been interested in the limits of schooling, about which he had written a book in the early 1960s. After the Coleman Report was published, **Daniel Moynihan** and **Thomas Pettigrew** initiated a seminar at Harvard to reanalyse the data. Moynihan, Johnson's assistant secretary of labor, had produced his own *Moynihan Report* in March 1965, which argued that half the black population suffered from 'social pathology.' Pettigrew was a black psychologist. Jencks and others joined the seminar, which grew over the years into the Center for Educational Policy Research, of which Jencks's book *Inequality* was the first important result.

It is no exaggeration to say that the findings of *Inequality* shocked and infuriated a great many people on both sides of the Atlantic. The main results of the Harvard inquiries, which included a massive chapter examining the effects of cognitive skills on advancement in life, its relation to school and to race, among other variables, was that genes and IQ 'have relatively little effect on economic success', 'school quality has little effect on achievement or on economic success'; and therefore 'educational reform cannot bring about economic or social equality.' More particularly, the study concluded, 'We cannot blame economic inequality primarily on genetic differences in men's capacity for abstract reasoning, since there is nearly as much economic inequality among men with equal test scores as among men in general. We cannot blame economic inequality primarily on the fact that parents pass along their disadvantages to their children, since there is nearly as much inequality among men whose parents had the same economic status as among men in general. We cannot blame economic inequality on differences between schools, since differences between schools seem to have very little effect on any measurable attribute of those who attend them. . . . Economic success seems to depend on varieties of luck and on-the-job competence that are only moderately related to family background, schooling, or scores on standardised tests. The definition of competence varies greatly from one job to another, but it seems in most cases to depend more on personality than on technical skills. This makes it hard to imagine a strategy for equalising competence. A strategy for equalising luck is even harder to conceive.'[71]

The impact of *Inequality* undoubtedly stemmed from the sheer quantity of data handled by the Harvard team, and the rigorous mathematical analysis discussed in detail in a series of long notes at the end of each chapter and in three appendices, on IQ, intergenerational mobility, and statistics. Jensen was put in his place, for instance, the Harvard study finding that the heritability of IQ was somewhere between 25 and 45 percent rather than 80 percent, though they took care to add that admitting a genetic component in IQ did not make one a racist.[72] They commented, 'It seems to be symbolically important to establish the proposition that blacks can do as well on standardised tests as whites. But if either blacks or whites conclude that racial equality is primarily a matter of equalising reading scores, they are fooling themselves . . . blacks and whites with equal test scores still have very unequal occupational statuses and incomes.'[73] Regarding desegregation, the Harvard team concluded that if applied across the board, it would reduce the 15-point gap in IQ between whites and blacks to maybe 12 or 13 points. While this is not trivial, they acknowledged, 'it would certainly not have much effect on the overall pattern of racial inequality in America.' They then added, 'The case for or against desegregation should not be argued in terms of academic achievement. If we want a segregated society, we should have segregated schools. If we want a desegregated society we should have desegregated schools.' Only political and economic change will bring about greater equality, they said. 'This is what other countries usually call socialism.'[74]

Given the news about freedom and equality then coming out of socialist countries, such as Russia and China, it is perhaps not surprising that this final message of the Harvard team did not catch on. On the other hand, their notion that schools could not bring about the equality that the blacks wanted was heeded, and the leaders of the civil rights movement began to concentrate their fire on segregation and discrimination in the workplace, which, it was now agreed, had a greater effect on economic inequality than schooling.

Traditional school came under a very different kind of attack in *Deschooling Society* by **Ivan Illich**, a Viennese who had studied at the Gregorian University in Rome and as an assistant pastor in an Irish-Puerto Rico parish in New York City. His main aim was to develop educational institutions for poor Latin American countries (he also worked in Mexico), and he argued that schools, far from liberating students from ignorance and teaching them to make the most of their capabilities, were actually, by 1971, merely boring, bourgeois 'processing factories,' organised anonymously, producing 'victims for the consumer society.'[75] Teachers, he said, were custodians, moralists, and therapists, rather than conveyors of information that taught people how to make their lives more meaningful. Illich therefore argued for the complete abolition of schools and their replacement by four 'networks.' What he had in mind, for example, was that children should learn about farming and geography and botany *on* the land, or about flight at airports, or economics in factories. Second, he called for 'skill exchanges,' whereby children would go to 'skill models,' say, guitar players or dancers or politicians, to learn those subjects that they really felt interested in. Third, he advocated 'peer matching,' essentially clubs of people interested in the same subject – fishing, motorcycles, Greek – who would compare progress and criticise each other.[76] Fourth, he said there was a need for professional educators, in effect people experienced in the first three networks outlined above, who could advise parents where to send their children. But teachers as such, and schools as such, would be abolished. *Deschooling Society* was an unusual book in that its prognosis was as detailed as its diagnosis. It formed part of the intellectual thrust that became known as the counterculture, but it had little real impact on schools.

The Great Society lost its chief navigator, and therefore its way, in March 1968 when President Johnson announced that he would not seek reelection. One reason for this was the war in Vietnam. In 1968 America had nearly half a million troops in Asia, 25,000 of whom were being killed annually. Before he left office, Johnson announced his policy of 'affirmative action,' under which all government contractors had to give preferential treatment to African Americans and other minorities. He was being optimistic: 1968 descended into violence and conflict on all fronts.

On 8 February, three black students were killed in Orangeburg, South Carolina, when they attempted to desegregate a bowling alley. On 4 April Martin Luther King was shot and killed in Memphis, and for a week there was rioting and looting in several U.S. cities in protest. In June Robert Kennedy was shot and killed in California. The Miss America contest in the United States was disrupted by feminists. But America was not alone. In Britain a new race relations act was deemed necessary. In July in Czechoslovakia, the USSR refused to withdraw its troops after Warsaw Pact exercises – this followed moves by the Czech government toward greater press freedom, the removal of censorship, easier religious assembly, and other liberal reforms. This was also the year of widespread student rebellion, rebellion against the war in Vietnam, against racial and sexual discrimination, and against rigid tuition policies in universities around the world – in the United States, Britain, Germany (where there was an attempt on the life of the student leader Rudi Dutschke), in Italy, but above all in France, where students co-operated with workers who occupied factories and campuses, barricaded the streets of major cities, forcing several changes in government policy, including a rise of 33 percent in the minimum wage.

The student rebellions were one aspect of a social phenomenon that had a number of intellectual consequences. The social phenomenon was the 'baby boom,' a jump in the number of births during and immediately after World

War II. This meant that, beginning in the late 1950s, coinciding with the arrival of the affluent society (and, it should be said, widespread availability of television), there occurred a highly visible, and much more numerous than hitherto, generation of students. In 1963, following the Robbins Report on higher education in Britain, the government doubled the number of universities (from twenty-three to forty-six) almost overnight. Books such as Daniel Bell's *End of Ideology* and Herbert Marcuse's *One-Dimensional Man*, alongside dis-illusionment with traditional left-wing politics after Stalin's death and the increased publicity given to his atrocities, not to mention the brutal Soviet invasion of Hungary in 1956, had come together to create the New Left (with initial capitals) around 1960. The essence of the New Left, which was a force in several countries, was a fresh concern with Marx's concept of alienation. For the New Left, politics was more personal, more psychological; its proponents argued that *involvement* was the best way to counter alienation, and that such self-conscious groups as students, women and blacks were better agents of radical change than the working classes. The Campaign for (Unilateral) Nuclear Disarmament, an early focus of involvement, received a massive boost at the time of the Cuban missile crisis. But the civil rights and women's liberation movements were soon joined to the Cold War as a focus for radical engagement. The demonstrations and rebellions of 1968 were the culmination of this process. Similarly, the Woodstock music festival in 1969 illustrated the other stream of 1960s student thought – personal liberation not through politics but through new psychologies, sex, new music, and drugs, a cocktail of experiences that became known as the 'counterculture.'

One man who distilled these issues in his writings and provided a thread running through the decade was an American figure who, in some ways, was to the last half of the twentieth century what George Orwell had been to the first: **Norman Mailer**. Like Orwell, Mailer was a reporter and a novelist who had seen action in war. Throughout the 1960s he produced a series of books – *An American Dream* (1965), *Cannibals and Christians* (1967), *The Armies of the Night* (1968), *Miami and the Siege of Chicago* (1968), *Why Are We in Vietnam?* (1969) – that chronicle, as their very titles reveal, a violent decade. In *An American Dream*, Steve Rojack, the central character (*hero* is very definitely not the word), is a much-decorated war veteran, a congressman, and at the time the story starts, a television personality with his own show – everything an American could hope to be.[77] Yet in the first few pages he strangles his wife, creeps along the corridor to have (violent) sex with the maid, then throws his wife out of the apartment window from a great height in the hope that she will be so mangled by the fall and the traffic that any evidence of the strangling will be destroyed. He is unsuccessful in this, but never punished, for strings are pulled, on others' behalf as well as his own. He loses his TV show but in the course of the three days that the novel lasts, two other people – one a woman, one a black man – endure a much worse fate, being killed as a result of Rojack's activities. What runs through the novel is the fact that nothing that happens to Rojack really touches him; he is a complete narcissist. This, says Mailer, is what

America has come to. Another 1960s book was Henry Steele Commager's *Was America a Mistake?* Mailer certainly thinks Steve Rojack was.[78]

The Armies of the Night carries the subheading, 'History as a Novel/The Novel as History.' Ostensibly, the main part of the book tells the inside story of the March on the Pentagon on 21 October 1967, by up to 75,000 people demonstrating against the Vietnam War.[79] Mailer's account is a novel only in the sense that he refers to himself throughout in the third person and takes the reader backstage – backstage both in the organisation for the march and backstage of Mailer as well. The other 'characters' in the novel are real people – Robert Lowell, Noam Chomsky, and Dr Spock among them. Mailer describes his various forms of jealousy, of Lowell for instance; his own embarrassing performance at a lecture the night before the march; his love for his wife. So what he offers is an early example of what would later be called radical chic; it is taken as read that the book-buying public will be interested in what a celebrity gets up to behind the scenes of a political event; readers will automatically understand that celebrities are now part of the picture in any political movement and will follow the story more easily if they have someone to identify with, especially someone with a confessional tone. In the course of the story the marchers are attacked; Mailer (along with about a thousand other demonstrators) is arrested and spends a night in the cells, as a result of which he misses a party in New York. This being a novel, in one chapter Mailer is able to give an account of the Vietnam War and why he thinks America's involvement is wrong. The second, shorter section, 'The Novel as History: The Battle of the Pentagon,' gives a more general account of the same events, including many quotes from newspapers. In this section, Mailer shows how the newspapers often get things wrong, but he also shows how they expand and fill out what he has to say in the first part. Mailer is using the march as an example of several trends in contemporary American life and thought: how violence is very near the surface; how the media and 'image' matter as much as substantive events; how the press are one of the armies of the night as well as indispensable bringers of light; above all how no one method of truth-telling is enough.[80] His fundamental point, what links *The Armies of the Night* with *An American Dream*, and what finally lays to rest the pattern of thought prevalent in the 1950s, may be described in this way: Mailer was an *anti*existentialist. For him, violence – boundary situations – actually *dulls* thinking; people stop listening to each other. Thinking is the most intense, the most creative, form of living but, surrounded by violence, views become polarised, frozen. Vietnam was freezing thought in America.

The 1960s had begun with a significant increase in tension in the Cold War. The later years of the decade saw yet another round of events that reflected the very different attitudes to freedom, equality, and justice in Communist countries.

On 10 November 1965 a young literary critic in Shanghai, named **Yao Wenyuan**, writing in *Literary Current*, attacked a play, *Hai Jui Dismissed from Office*, which had been written four years before by Wu Han, deputy mayor of

Beijing. The play was about an honest Ming dynasty official who took exception to the emperor's land policy and was punished simply for being so forthright. Though it was set many years in the past, Mao Zedong took the play as an attack on himself and used it as an excuse to introduce change on a massive scale. What became known as the Cultural Revolution had two aspects: it was a major political move by Mao, but it also had an important, devastating impact on the artists, intellectuals, and academics of China, who suffered extraordinary deprivations of freedom of thought and action.

Mao's own wife, **Jiang Qing**, was appointed 'cultural adviser' to the army, and it was this move that proved decisive. Surrounding herself with young activists, she first took on what she called the 'scholar-tyrants' who used 'abstruse language' to silence the class struggle. Worse, she said that the universities kept themselves free of this dialectic by emphasising the 'fallacy that everyone is "equal before the truth."' [81] Although she had difficulty at first (the *People's Daily* refused to publish her early pronouncements), by the end of May 1966 Jiang had enlisted the aid of a new phenomenon – 'Hung Wei Ping,' the **Red Guards**. These were essentially high school and university students, and their main aim was to attack the 'spectacle wearers,' as teachers and other academics were called. They took to the streets in gangs, marching first on Tsinghua University and then on others, attacking the university authorities. [82] Later, street violence broke out, the Red Guards seizing anyone whose hair or clothes they didn't like. Shops and restaurants were ordered to change their displays or menus that betrayed any Western bias. Neon signs were destroyed, and huge street bonfires were held, burning 'forbidden goods' such as jazz records, works of art, and dresses. Coffee bars, theatres, and circuses were closed down, weddings forbidden, even holding hands and kite flying. One female star of the Peking opera recounted how she went into exile in the countryside, where she would go to a remote area of the forest every day to exercise her voice where no one else would hear; she also buried her costumes and makeup until after the Cultural Revolution was over. Paul Johnson's depressing account of the disaster continues: 'Libraries were closed, books burned.' In one well-known instance – the Peking Research Institute of Non-Ferrous Metals – only four scientists had the courage to use the library during the entire period. [83] Jiang Qing wallowed in her role, addressing countless mass rallies where she denounced, in turn, 'jazz, rock 'n' roll, striptease, impressionism, fauvism,' and every other 'ism' of modern art, plus capitalism itself, which she said destroyed art. She was against specialisation. [84] By the second half of 1966 virtually every important cultural institution in China was under army control. On 12 December that year many 'public enemies,' who included playwrights, actors, film and theatre directors, poets, and composers, were marched to the Workers' Stadium before 10,000 people, each with a wooden placard around his or her neck. Later Jiang took over TV and radio stations and confiscated equipment, scripts, scores, and film, reediting the latter and reissuing them in revised versions. She ordered composers to write works that were then played to 'the masses' and changed afterward according to what the masses wanted. In the ballet she banned 'orchid fingers' and upturned palms, demanding instead that the dancers used clenched fists and violent movements to confirm their 'hatred of the landlord class.'[85] The attacks on the universities and artists bred violence, and in the universities private armies were set up. Among the better known were the 'East Is Red' commune at the Peking Geological Institute; the 'Sky Faction' of the Aeronautical Institute was another.[86] In many scientific institutions professors were sent out into the countryside to make greater practical use of their findings, with peasants. At the Genetics Institute in Peking (there had been no genetics institute in China before 1949), the theories of Lysenko hung on even later than in Russia, thanks in part to the Red Guards. Perhaps the most extraordinary notion bred by the Cultural Revolution was that traffic lights should be changed. The Red Guards were worried that red, the revolutionary colour, should be for change, for forward progress – in other words for 'Go' rather than 'Stop.' Zhou Enlai killed the idea with a joke about red being better seen in fog, and therefore the safest colour. But the Cultural Revolution was no joke.[87] Before it ended, as many as 400,000 had been killed. The effect on China's traditional culture was devastating, and in this respect strongly reminiscent of Stalin's inquisition.

Not that the intellectual inquisition in Russia had died out with Stalin. It wasn't as widespread as in the 1930s, but it was no less vicious.[88] The first details about the dark side of Russian psychiatric hospitals had been released to the West in 1965, with the publication of Valery Tarsis's *Ward 7*, after which a number of psychiatrists in Europe and North America made it their business to investigate Soviet practices. But it was the forced hospitalisation of Zhores Medvedev on 29 May 1970 at Kaluga Psychiatric Hospital, just south of Moscow, that drew the attention of the world to what was being done in the name of psychiatry.

A Question of Madness, which was written by **Zhores Medvedev** and his brother, **Roy**, a professional historian, reads like a Kafka novel. Early on in 1970, the manuscript of a book that Zhores had written was seized by the KGB in the course of a raid on the flat of a friend. Zhores was not especially worried when he found out that the KGB had seized the book – which was unfinished and not at all secret – but he did begin to grow anxious when he was asked to attend Kaluga Psychiatric Hospital to discuss the behavior of his son, who was then giving the Medvedevs some cause for concern, going through an 'awkward' or 'hippie' phase. As soon as he arrived at the hospital, Zhores was locked in the waiting room. When, through a window, he saw his son leave, Zhores realised that he was the chief object of concern to the authorities. On that occasion he picked the lock and escaped, but a week later he received a visit at home by three policemen and two doctors.[89] From their conversations, it became clear that Medvedev had caused offence with a book he had written, originally called *Biology and the Cult of Personality* but later changed to *The Rise and Fall of T. D. Lysenko*, in which he had exposed the shameful history of Soviet genetics. This book had appeared in the West in 1969, published by Columbia University Press, while Lysenko was still alive (he died in 1976). Zhores was forcibly removed to Kaluga, where both the hospital psychiatrists and a commission sent out by the central authorities tried to make out that he was an incipient schizophrenic, about to become a danger to himself and others.[90] The authorities had, however, reckoned without Zhores's relatives and friends. For a start, his brother Roy was an identical twin. Schizophrenia is known to be (partly) inherited, and so, strictly speaking, if Zhores showed signs of the illness, so too should Roy. This clearly wasn't true. Many academicians complained to the authorities that they had known Zhores for many years, and he had never shown any abnormal symptoms. **Peter Kapitsa, Andrei Sakharov,** and **Aleksandr Solzhenitsyn** all rallied to Zhores's support, and as a result the matter received wide publicity in the West.[91] But it was still nearly three weeks before he was released, and during that time, as the account the Medvedevs jointly wrote shows, the netherworld of psychiatry was exposed. Various psychiatrists claimed that Zhores showed 'heightened nervousness,' 'deviation from the norm,' was 'ill-adapted to the environment,' suffered a 'hypochondriac delusional condition,' and had 'an exaggerated opinion of himself.' When questioned by family relatives, these psychiatrists claimed that only experienced doctors could detect the 'early stages' of mental illness.[92] Other psychiatrists were brought in as part of a 'special commission' to consider the case, including Professor Andrei Snezhnevsky, Professor Daniel Lunts, and Dr Georgy Morozov, head of the **Serbsky Institute of Forensic Psychiatry,** which would be revealed as the worst of the Soviet psychiatric institutions involved in psychiatric-political terror. Despite this, Zhores's friends succeeded in forcing his release on 17 June and having him reinstated to the Lenin Agricultural Academy as a senior research fellow, to work on amino acids. In this instance there was a happy ending, but later research showed that between 1965 and 1975 there were 210 'fully authenticated' cases of psychiatric terror and fourteen institutions devoted to the incarceration of alleged psychiatric cases who were in fact political prisoners.[93]

Chilling as they were, the special psychiatric hospitals in Russia only dealt with, at most, hundreds of people. In comparison, the world revealed by **Aleksandr Solzhenitsyn** concerned perhaps 66 million people and, together with the Holocaust against the Jews, must rank as the greatest horror story of human history.

   The Gulag Archipelago is a massive, three-volume work, completed in 1969 but not published in English until 1974, 1975, and 1976. Solzhenitsyn's previous books, particularly One Day in the Life of Ivan Denisovich (1962) and Cancer Ward (1968) had made him well known in the West.[94] Born an orphan in the Caucasus in December 1918 (his father had died in a shooting accident six months before), in an area where there was a lot of White Russian resistance to the Bolsheviks, Solzhenitsyn grew up in the early 1930s, as the Communist Party strengthened its grip on the country after Stalin's Great Break.[95] Despite poverty and hardship he shone at school, and then at university, in physics, math, and Marxism-Leninism.[96] He had a 'good' war (he was promoted to captain and won four medals) but was arrested by secret agents in early 1945. His letters had been intercepted and read: among his 'crimes' was a letter referring to Stalin as 'the man with the moustache,' and photographs of Nicholas II and of

Trotsky were found among his belongings. Convicted as a 'socially dangerous' person, he was moved from prison to prison and then to Novy Ierusalim, New Jerusalem, a corrective labour camp, and to Marfino, a scientific *sharashka* that at least had a library. By 1955 he was living in a mud hut in Kol Terek; this was exile rather than imprisonment, and it was here that he contracted, and was successfully treated for, cancer. These experiences became his first masterpiece, *Cancer Ward*, not published in English until 1968.

He arrived back in Moscow in June 1956, after an absence of more than eleven years, aged not quite thirty-eight. Over the next few years, while he was teaching outside Moscow, he wrote a novel initially entitled *Sh-854* after the *sharashka* he had been in. It was very shocking. The story concerned the ordinary, everyday life in one camp over a twenty-four-hour period. The shock lay in the fact that the camp life – the conditions described – are regarded by the inhabitants as normal and permanent. The psychology of the camp, so different from the outside world, is taken for granted, as are the entirely arbitrary reasons why people arrived there. Solzhenitsyn sent the manuscript to friends at *Novy mir*, the literary magazine – and what happened then has been told many times.[97] Everyone who read the manuscript was shocked and moved by it; everyone at the magazine wanted to see the book published – but what would Khrushchev say? In 1956 he had made an encouraging (but secret) speech at the Party Congress, hinting at greater liberalisation now that Stalin was dead. By coincidence, friends got the manuscript to the Soviet leader at a time when he was entertaining Robert Frost, the American poet. Khrushchev gave the go-ahead, and *Sh-854* was published in English in 1963, to world acclaim, as *One Day in the Life of Ivan Denisovich*.[98] This marked a high spot in Solzhenitsyn's life, and for a few years – a very few years – he was lionised in Russia. But then, in the mid-1960s, Khrushchev clamped down on the liberalisation he had himself started, and Solzhenitsyn lost the Lenin Prize he should have won because one member of the committee, the director of the Komsomol, alleged that he had surrendered to the Germans in the war and had been convicted of (an unspecified) criminal offence. Both allegations were untrue, but they showed the strength of feeling against Solzhenitsyn, and all that he stood for.

From 1965 he began to work on his history of the camps, which would become *The Gulag Archipelago*. Since his disillusion with Marxism he had returned to 'some sort of Christian faith.'[99] But Russia was changing again; Khrushchev had fallen from power, and in September 1965 the KGB raided the flat of some of Solzhenitsyn's friends and seized all three copies of the manuscript of another book, *The First Circle*. This described four days in the life of a mathematician in a *sharashka* outside Moscow, and is clearly a self-portrait. Now began a very tense time: Solzhenitsyn went into hiding and found it difficult to have his writings published. Publication of *The First Circle* and *Cancer Ward* in the West brought him greater fame, but led to a more open conflict with the Soviet authorities. This conflict culminated in 1970, when he was awarded the Nobel Prize for Literature, but the authorities made it clear that, if he went to Sweden to collect the prize, he would not be allowed back.[100]

And so, by the time *The Gulag Archipelago* appeared, Solzhenitsyn's life had taken on an epic dimension.

The new project was a massive exercise, as it had to be.[101] The gulag *was* overwhelming, a massive intervention in so many millions of lives that only an equally vast project could do justice to what was indeed 'the greatest horror story of human history.' Besides the eight years he spent in the camps, it took Solzhenitsyn nine years – April 1958 to February 1967 – to compile the book.[102] Parts of the story had escaped before, but Solzhenitsyn's aim was to present such a mass of material that no one would ever again doubt the gross and grotesque abuses of freedom in Soviet Russia. Eighteen hundred pages long, it is all but overwhelming – but, as a literary work as well as a record, that was Solzhenitsyn's aim.

The book first appeared in the West in Paris, on 28 December 1973. At the end of January 1974 the BBC World Service and its German counterpart began broadcasting excerpts from *The Gulag* in Russian. In the same week the German version of the book was published, and smuggled copies of the Russian version began to appear in Moscow: they were passed from hand to hand, 'each reader being allowed a maximum of 24 hours to read the entire volume.'[103] On 12 February Solzhenitsyn was arrested. At 8:30 A.M. on Wednesday the fourteenth, the Bonn government was informed that Russia wanted to expel Solzhenitsyn and asked if the Germans would accept him. Willy Brandt, the German chancellor, was at that moment chairing a session of the cabinet. Interrupted, he immediately agreed to Russia's request. *The Gulag* was published in the U.K. and the United States later that spring. Worldwide, by 1976, according to *Publisher's Weekly*, the first volume sold 8 to 10 million copies (2.5 million in the United States, a million-plus in Germany, and just under that in the U.K., France, and Japan). All together, Solzhenitsyn's books have sold 30 million copies.[104]

Gulag – GUlag in Russian – stands for *Glavnoye Upravleniye Lagerei* (Chief Administration of the Labour Camps). Throughout his long book, Solzhenitsyn is unsparing in the detail. From the techniques of arrest to the horrors of interrogation, from the 'ships' of the archipelago (the red-painted cattle trains that transported the prisoners) to the maps of the 202 detention camps, from the treatment of corpses to the salaries of the guards, nothing is omitted.[105] He tells us how the 'red cows,' the cattle trucks, were prepared, with holes carved out of the floor for drainage but steel sheets nailed down all around, so no one could escape.[106] We learn the name of the individual – Naftaly Aronovich Frenkel, a Turkish Jew, born near Constantinople – who first thought up the gulag.[107] We learn the death rates of the various camps and are given an unsparing list of thirty-one techniques of punishment during interrogation. These include a machine for squeezing fingernails, or 'bridling,' in which a towel is inserted between the prisoner's jaws like a bridle and then pulled back over his shoulders and tied to his heels, with his spine bent. The prisoner was then left for several days without food or water, sometimes having first been given a salt-water douche in the throat.[108]

But as Michael Scammell, Solzhenitsyn's biographer, says, the book is not just a series of lists or statistics. Solzhenitsyn re-creates a whole world, an entire culture. His tone is ironic, not self-pitying, as he gives us the jokes and the jargon of life in the camps – camps that, he tells us, varied widely, from prospecting camps to railroad-building camps, transit camps to collective labour camps, island camps to juvenile camps. He shows that people were sent to the camps for absurd reasons. Irina Tuchinskaya, for example, was charged with having 'prayed in church for the death of Stalin,' others for showing friendliness to the United States, or a negative attitude toward state loans. Then there is the jargon. A *dokhodyaga* is a man on his last legs, a 'goner'; *katorga* was hard labour; everything that was constructed in camps was, they said, built with 'fart power'; *nasedha* was 'stool pigeon,' and reality was deliberately reversed so that the worst camps were referred to as the most privileged.[109] However, as horror is piled on horror, as page after page passes – as the weeks and months pass for those in the gulag (and this is Solzhenitsyn's intention) – the reader gradually comes to realise that although countless millions have been murdered, the human spirit has not been killed, that hope and a black sense of humour can keep those who survive alive, not thriving exactly, but *thinking*. In one of the last chapters, describing a revolt in the Kengir camp that lasted for forty days, the reader feels like cheering, that reason and sanity and goodness can prevail, even though in the end the revolt is brutally put down, as we know it will be.[110] So the book, though nearly choking with bleak horrors, is not in the end an entirely bleak document, as Solzhenitsyn intended. It is a warning to all of us of what it means to lose freedom, but it is a warning to tyrants as well, that they can never hope to win in the end. The reader comes away chastened – very chastened – but not despairing. As W. L. Webb said, reviewing the book in the *Guardian*, 'To live now and not to know this work is to be a kind of historical fool missing a crucial part of the consciousness of the age.'[111]

The un-freedoms in the Communist world, described by Solzhenitsyn and the Medvedevs, or those which took place in the Cultural Revolution in China, were far worse than anything that occurred in the West. Their extent, the vast number of their victims, underlined the fragility of freedom, equality, and justice everywhere. And, just as the 1960s had opened with Hayek's and Friedman's examinations of freedom, so the decade closed with other philosophers addressing the same issues, after years of turbulence in the name of civil rights.

In his 1969 book *Four Essays on Liberty*, **Isaiah Berlin** built on Hayek's notion that, in order to be free, man needs an area of private life where he is accountable to no one, where he can he left alone, free of constraint. Born in 1909 in Riga, part of the Russian empire, Berlin had moved to Russia when he was six. In 1921 his family had moved to Britain, where he was educated at Oxford, becoming a fellow of All Souls and subsequently professor of social and political theory and founding president of Wolfson College. In his essays, Berlin made three points, the first that liberty is just that: freedom.[112] In a famous sentence, he wrote, 'Everything is what it is: liberty is liberty, not equality or fairness or justice or culture, or human happiness or a quiet conscience.'[113] Berlin was at pains to point out that one man's freedom may conflict with another's; they may indeed be irreconcilable. His second and third points were that there is an important distinction between what he called **'negative' freedom** and **'positive'** freedom. Negative freedom is, on this analysis, 'a certain minimum area of personal freedom which on no account must be violated; for if it is overstepped, the individual will find himself in an area too narrow for even that minimum development of his natural faculties which alone makes it possible to pursue, and even to conceive, the various ends which men hold good or right or sacred. It follows that a frontier must be drawn between the area of private life and that of public authority. ... Without adequate conditions for the use of freedom, what is the value of freedom?'[114] Berlin argued that this doctrine of negative freedom is relatively modern – it does not occur in antiquity – but that the desire not to be impinged upon, to be left to oneself, 'has been a mark of high civilisation.' Negative freedom is important for Berlin not merely because of what it stands for but also because it is a simple notion, and therefore something men of goodwill can agree upon.

Positive freedom, on the other hand, is much more complex.[115] This, he says, concerns all those issues that centre around the desire of the individual 'to be his own master.' This concept therefore involves issues of government, of reason, of social identity (race, tribe, church), of genuine autonomy. If the only true method of attaining freedom in this sense is the use of critical reason, then all those matters that affect critical reason – history, psychology, science, for example – must come into play. And, as Berlin says, 'all conflict, and con- sequently all tragedy, is due solely to the clash of reason with the irrational or the insufficiently rational'; insofar as man is a social being, what he is, is, to some degree, what others think and feel him to be. It is this, he says – this failure on the part of many to be recognised for what they wish to feel themselves to be – that was 'the heart of the great cry' at that time on the part of certain nations, classes, professions, and races.[116] This is akin to freedom, he says, and it may be no less passionately needed, but it is not freedom itself. Berlin's aim in saying all this is to underline that there can be no 'final solution' (his words), no final harmony 'in which all riddles are solved, all contradictions reconciled,' no single formula 'whereby all the diverse ends of man can be harmoniously realised.' Human goals are many, he says, not all commensurable, some in perpetual rivalry. This is the human condition, the background against which we must understand freedom, which can only be achieved by par- ticipation in the political system. Freedom will always be difficult to attain, so we must be crystal-clear about what it is.[117]

Both Raymond Aron, in *Progress and Disillusion* (1968), and Herbert Marcuse, in *An Essay on Liberation* (1969), believed the 1960s to have been a crucial decade, since they had revealed science and technology as real threats to freedom, not just in the form of weapons and weapons research, which had linked so many universities to the military, but also because the civil rights movement, the women's liberation movement, and the sexual revolution in general had been underpinned by a psychological transformation.[118] For them the whole idea of freedom had been extended; in the third world in particular the traditional Marxist classes still needed to be freed; the influx of Western consumer goods – aided by widespread television – was exploiting a new raft of people. At the same time, in the developed Western democracies, people – especially the young – were experiencing a new form of freedom, a personal liberation, insight into their own character as afforded by the new psychologies. Marcuse in particular looked forward to a new 'aesthetic' in politics, where art and the creative act would allow people greater fulfilment, producing in the process what he called 'prettier' societies, more beautiful countries. It was at last appropriate, he said, to speak of utopias.

An entirely different idea of freedom – what it is and what its fate is likely to be – came from **Marshall McLuhan**. Born in Edmonton, Alberta, in Canada in 1911, he took a Ph.D. in Cambridge in 1943, working with F. R. Leavis and I. A. Richards, founder of the New Criticism, which gave him an intellectual confidence from which stemmed his great originality. McLuhan's chief interest was the effect of the new 'electric' media on man's self-consciousness and behaviour, but he also thought this had important consequences for freedom. McLuhan's notion of the individual, and his relation to society as a whole, was quite unlike anyone else's.

For him there have been three all-important watersheds in history: the invention of the alphabet, the invention of the book, and the invention of the telegraph, the first of the electric media, though he also thought the arrival of television was another epochal event. McLuhan's writing style was allusive, aphoristic, showing great learning but also obscure at times, meaning he was not always easy to understand. Essentially, he thought the alphabet had destroyed the world of tribal man. Tribal man was characterised by an oral culture in which all of the senses were in balance, though this world was predominantly auditory; 'no man knew appreciably more than another.'[119] 'Tribal cultures even today simply cannot comprehend the concept of the individual or of the separate and independent citizen,' he wrote. Into this world, he said, the phonetic alphabet 'fell like a bombshell.' The components of the alphabet, unlike pictographs and hieroglyphics, were essentially meaningless and abstract; they 'diminished the role of the senses of hearing and touch and taste and smell' while promoting the visual. As a result whole man became fragmented man. 'Only alphabetic cultures have succeeded in mastering connected linear sequences as a means of social and psychic organisation.'[120] He thought that tribal man was much less homogeneous than 'civilised' man and that the arrival of the book accelerated this process, leading to nationalism, the Reformation, 'the assembly line and its offspring the Industrial Revolution, the whole concept of causality, Cartesian and Newtonian concepts of the universe, perspective in art, narrative chronology in literature and a psychological mode of introspection that greatly intensified the tendencies towards indivi dualism.'[121] But with the arrival of the electric media, McLuhan thought that this process was now going into reverse, and that we would see a revival of tribal man.

The ideas for which McLuhan became famous (or notorious, depending on your viewpoint) were '**The medium is the message**' and his division of media into 'hot' and 'cool.' By the former phrase he meant two things. One, as described above, that the media determine much else in life; and two, that we all share assumptions about media, that the way 'stories,' or 'news,' are reported is as important as the actual content of these events. In other words, content is only part of the story: attitudes and emotions too are carried by electric media, and it was in this sense of a collective experience that he meant a return to tribalisation.[122]

A photograph is high-definition, requiring very little work by the viewer to complete its message, and it is therefore 'hot.'[123] A cartoon, on the other hand, requires the viewer to complete the information conveyed and is therefore 'cool.' Radio is hot, TV cool. Lectures are hot, seminars are cool. In television culture, political leaders become more like tribal chieftains than traditional politicians: they perform emotional and social functions, where supporters/ followers can feel part of a collectivity, rather than offer intellectual leadership, thinking for their followers.[124]

For McLuhan all this completely changed the notion of freedom: 'The open society, the visual offspring of phonetic literacy, is irrelevant to today's retribalised youth; and the closed society, the product of speech, drum and ear technologies, is thus being reborn. ... Literate man is alienated, impoverished man; retribalised man can learn a far richer and more fulfilling life ... with a deep emotional awareness of his complete interdependence with all humanity. The old "individualistic" print society was one where the individual was "free" only to be alienated and disassociated, a rootless outsider bereft of tribal dreams; our new electronic environment compels commitment and participation, and fulfils man's psychic and social needs at profound levels.'[125] McLuhan, who knew how to turn familiar categories on their head, foresaw a time when, for example, Italy might choose to reduce television watching by five hours a day in order to promote newspaper reading in an election campaign, or that Venezuela might lay on extra TV in order to cool down political tensions.[126] For McLuhan, the idea of a 'public' consisting of 'a differentiated agglomerate of fragmented individuals, all dissimilar but all capable of acting in basically the same way like cogs in a production line' was less preferable than a mass society 'in which personal diversity is encouraged while at the same time everybody reacts and interacts simultaneously to every stimulus.'[127]

This appears to change the very notion of autonomous individuals, but then McLuhan did predict, in this new world, the demise of large cities, the imminent obsolescence of the motor car and the stock exchange, and that the concept of the job would be replaced by that of the role. In many ways, though he was strikingly original, McLuhan was (so far) wrong.

A very similar message came from France, in **Guy Debord's** *The Society of the Spectacle*, published there in 1967 but not translated into English until much later. Debord saw the spectacle – mainly the television-dominated society, but also sports, rock concerts, stage-managed politics – as the chief product of modern society. The spectacle, he said, comprised basically the 'uninterrupted monologue of self-praise' of the ruling order and the passivity of the rest: 'Spectators are linked only by a one-way relationship to the very center that maintains their isolation from one another. ... The spectator feels at home nowhere, for the spectacle is everywhere. ... The spectacle's function in society is the concrete manufacture of alienation. ... The spectacle corresponds to the historical moment at which the commodity completes its colonization of social life ... commodities are now *all* that there is to see; the world we see is the world of the commodity.' For Debord, far from being a form of freedom, the society of the spectacle was the final form of alienation, final because people think they are enjoying themselves but are in reality passive spectators. His book contained a long historical section, on Hegel, Marx, and George Lukács, with Debord arguing essentially that the spectacle was the final banalising triumph of capitalism. (One of his 'texts' was Shakespeare's *Henry IV,* part 1: 'O, gentlemen, the time of life is short! ... *And if we live, we live to tread on kings.*') In later editions he said that he thought Daniel Boorstin, the much-respected Librarian of Congress, who had published *The Image* in 1972, had got it wrong, because he had regarded commodities as 'consumed' in an authentic private life, whereas Debord argued that even the consumption of individual commodities in the theatre of advertising is itself a spectacle, which negates the very idea of 'society' as it has historically been known. So for Debord, the society of the spectacle represents the final failure of man's progress toward ever greater self-consciousness. Man is not only impoverished, enslaved, and his life negated; capitalism, in the society of the spectacle, has deluded him into thinking he *is* free.[128]

In Isaiah Berlin's terms, positive freedom was not as basic as the negative variety. For **John Rawls**, professor of philosophy at Harvard, justice comes before liberty, if only by a short head. In *A Theory of Justice,* completed in 1971 and published a year later, Rawls produced what fellow philosopher Robert Nozick called the most significant work of political philosophy since John Stuart Mill. Rawls argued that a just society will in fact guarantee more liberties for the greatest number of its members and that therefore it is crucial to know what justice is and how it might be attained. Specifically arguing against the utilitarian tradition (actions are right because they are useful), he tried to replace the social contracts of Locke, Rousseau, and Kant with something 'more rational.' This led him to the view that justice is 'the first virtue of social institutions, as truth is of systems of thought,' and that justice is best understood as 'fairness.' It was Rawls's way of achieving fairness that was to bring him so much attention. To achieve this, he proposed an **'original position'** and a **'veil of ignorance.'**[129]

In the original position, the individuals drawing up the contract, the rules by which their society will be governed, are assumed to be rational but ignorant. They do not know whether they are rich or poor, old or young, healthy or infirm; they do not know which god they follow, if any; they have no idea what race they are, how intelligent or stupid, or whatever other gifts they may have or lack. In the original position, no one knows his place in society – and so the principles of justice 'are chosen behind a veil of ignorance.'[130] For Rawls, whatever social institutions are chosen in this way, those engaged in choosing them 'can say to one another that they are co-operating on terms to which they would agree if they were free and equal persons whose relations with one another were fair'; 'a society satisfying the principles of justice as fairness comes as close as a society can to being a voluntary scheme, for it meets the principles which free and equal persons would assent to under circumstances that are fair. In this sense its members are autonomous and the obligations they recognise self-imposed.' Rawls further argues that, from this premise of the original position and the veil of ignorance, there are two principles of justice, and in this order: (1) each person is to have an equal right to the most extensive basic liberty compatible with a similar liberty for others; and (2) social and economic inequalities are to be arranged so that they are both (a) reasonably expected to be to everyone's advantage, and (b) attached to positions and offices open to all.[131] Putting it another way, Rawls writes, 'All social values – liberty and opportunity, income and wealth, and the bases of self-respect – are to be distributed equally unless an unequal distribution of any, or all, of these values is to everyone's advantage.' His later arguments proved more controversial. For example, he discusses self-respect as a 'good,' something a rational man in a just society is naturally entitled to. He discusses envy and the place of shame. All this leads him into direct conflict with, say, Hayek, in that Rawls firmly believes there *is* such an entity as social justice; in Isaiah Berlin's terms, there is not enough positive freedom for certain groups, since those groups are not being treated as they would be treated by rational men in the original position behind a veil of ignorance. And because the first principle of justice (justice seen as fairness) takes priority over the second principle, the basic liberties of the disadvantaged come before inequalities of wealth or income however much those inequalities benefit everyone. In other words, even supposing that blacks were better off under white rule than they would be under mixed rule (say), it is still wrong (unjust, unfair) if blacks' liberty is more proscribed than whites. Equality of liberty comes first.

This led Rawls into perhaps the most controversial section of all in his book, 'The Justification of Civil Disobedience.'[132] Here he argues that civil disobedience is justifiable if the majority, perhaps in the form of a political party representing the majority, refuses to grant the minority equal liberties. Attempts to have the laws changed should first be tried, he says; civil disobedience should always be a last resort and take into account the likelihood of other minorities acting in a civilly disobedient way, in which case there might be a threat to overall order, in which case there could be a risk of an overall loss of liberty, in which case civil disobedience is not justified. But these are technicalities. By arguing that self-respect is a natural good, sought and expected by rational men in a free and fair society, Rawls legitimated the idea of social justice that had suffered so much at the hands of Hayek.

Rawls *assumed* the original position and the veil of ignorance in order to arrive at the principles of a just – fair – society. His colleague at Harvard, **Robert Nozick**, took him to task for this. More grounded in the tradition of Hayek, Nozick preferred to start from a consideration of the way things *are*, the way society actually *is* organised, rather than by positing some perfect world, as Rawls had done.* In *Anarchy, State, and Utopia*, published in 1974, partly as a reply to Rawls's book, Nozick argued that all 'patterned' justice, such as affirmative action, was morally wrong, violating more individual rights than was permissible and thereby doing more harm than good as measured by the number of people helped.[133] Nozick pointed out what he felt were a number of logical flaws in Rawls's arguments, but his most important substantive point was to offer the concept of 'entitlement' in any social situation.[134] In Rawls's original position, the individuals contributing to the rules of society, behind the veil of ignorance, have no idea of their own attributes – their wealth, status, intelligence, and so on. But in real life, says Nozick, this can never happen, and Rawls's position is, to that extent, inadequate. No less important logically, people do have talents that vary, and which they are born with. This is inequality, if you like, but it is a special kind, insofar as one person having more of something (say, intelligence) does not in and of itself mean that every other person in that society has less and is worse off. One person having more of a natural talent does not *deprive* anyone else of that talent. And so, for a society to coerce its members so that the disparity in talent, and what flows from it, is removed, is wrong, says Nozick. And it is all the more wrong when, as is often the case, that extra talent is used by the person possessing it to benefit society. Nozick uses a number of deliberately absurd examples to point up what he sees as the shortcomings in Rawls's arguments. For instance, he compares the provision of medical care with barbering. In medical care, it is generally argued that 'need' is the key factor in providing medical care, over and above any ability to pay. Should this therefore apply to barbering ? Should barbering services be provided first and foremost to people who are in need of a shave? In another case, assume a woman has four suitors. Do we allow her to choose who to marry, or do we put it to the vote among the suitors? Does it make sense to say that the successful suitor is the one who 'needs' the woman more than the others? In giving these examples, Nozick's aim is to show that Rawls's theoretical version of the way man arranges his affairs is far too simple; and second, to stress that many areas of life are properly left to the actions and decisions of individuals, freely exercising their naturally endowed talents, because these talents neither impinge on anyone else nor on the overall performance of society. All of which leads him to the view that only a minimal state, performing the basic functions of protection, can be morally justified.[135]

A few hundred yards from the philosophy department at Harvard is the psychology building, William James Hall, named in honour of the great pragmatist. From there, in 1972, at the same time that Rawls and Nozick were in residence, **B. F. Skinner** produced his remarkable book about liberty, *Beyond Freedom and Dignity*. Skinner wrote as a psychologist, not as a philosopher, and indeed he made it pretty clear that he thought a lot of traditional philosophical ideas were wrong.[136] Yet his book was deeply philosophical in the sense that he

---

* This, it will be observed, has been an important distinction between philosophers throughout the century: those who start from an ideal original position, and those who accept the world as it is.

was concerned not with equality and its relation with freedom, but with the fundamental idea of freedom in the first place. As a scientist and a biologist, Skinner saw human nature as the product of evolution (and therefore to a large extent genetically grounded) and as an adaptation to the environment. For Skinner, there was only one way to change (and by implication to improve) man, and that was to change the environment. His second point was to argue that, at base, true freedom did not, and does not, exist. Man's nature is the result of his history – evolution – in collaboration with his environment. Therefore, man is by definition subject to a certain amount of control. For Skinner, freedom is merely the state in which man does not *feel* the control that is exerted over him.[137] Yet freedom does not primarily apply to feelings, he says, but to behavior. In other words, freedom is the lack of aversive stimuli in the environment, and what we call the feeling of freedom is in reality only the result of this absence. These aversive stimuli will be different for different people, with different histories, but in the concluding chapters of his book he attempted to sketch out a design for a culture where the aversive stimuli are kept to a minimum.[138] Skinner wanted to see mankind develop a technology of behavior which recognises that man's nature, man's collective nature as a vast number of individuals, is developed as a result of contingencies – rewards and punishments – acting on our genetic makeup. For Skinner, there is no autonomous man, or rather we should recognise the limits to our autonomy, if we wish to be truly free, in the sense of being at ease with our true nature.

Control and punishment, Skinner says, are necessary aspects of the environment where people live together in society, but they should not be understood as bad things, only as ways to achieve the maximum freedom for the maximum number of people – freedom understood as a lack of aversive stimuli. (This was written near the height of the period of student rebellion.) By creating a better environment, we shall get better men. He therefore criticises notions such as 'a spiritual crisis' (among students), the drug problem, and the gambling problem. These problems come not from within human nature, like a homunculus, but rather from the mismanagement of the control of society: 'Autonomous man is a device used to explain what we cannot explain in any other way. He has been constructed from our ignorance, and as our understanding increases, the very stuff of which he is composed vanishes. Science does not dehumanize man, it dehomunculises him.'[139]

Skinner's ideas have not proved anywhere near as influential as Rawls's or Nozick's or Hayek's. In part that is because he did little to show how freedom might be improved. But the main reason was that, in the 1960s, in the American context of civil rights, freedom and justice were assumed by most people to be the same thing.

The 'long 1960s,' ending around 1973, was not at all the frivolous decade it is so often painted. A good claim can be made for saying that it was the most important postwar period, the most pivotal, when man's basic condition – the nature of his very freedom – came under threat and under scrutiny, for the reason that his psychology, his self-awareness, was changing. The shift from a class-based sociology to an individual psychology, the rise of new groups to identify with (race, gender, students) changed not only self-awareness but politics, as Hannah Arendt had predicted. Much of what happens in the rest of this book, much of the thought in the last quarter of the century, can only be understood in that light.

Between September and November 1965, the United States National Science Foundation vessel *Eltanin* was cruising on the edges of the Pacific-Antarctic Ocean, collecting routine data about the seabed. This ship was essentially a laboratory belonging to the Lamont-Doherty Geological Observatory, part of New York's Columbia University. Oceanography had received a boost in World War II because of the need to understand U-boats and their environment, and since then with the arrival of deepwater nuclear submarines. The Lamont Institute was one of the most active outfits in this area.[1]

On that 1965 voyage, *Eltanin* zigzagged back and forth over a deep-sea geological formation known as the **Pacific-Antarctic Ridge**, located at 51 degrees latitude south. Special equipment measured the magnetic qualities of the rocks on the seabed. It had been known for a time that the magnetism of rocks for some reason reversed itself regularly, every million years or so, and that this pattern told geologists a great deal about the history of the earth's surface. The scientist in charge of legs 19, 20, and 21 of the *Eltanin*'s journey that time was **Walter Pitman III**, a Columbia-trained graduate student. While on board ship, he was too busy to do more than double-check that the instruments were working properly, but as soon as he got back to Lamont, he laid out his charts to see what they showed. What he had in front of him was a series of black-and-white stripes. These recorded the magnetic anomalies over a stretch of ocean floor. Each time the magnetic anomaly changed direction, the recording device changed from black to white to black, and so on. What was immediately obvious, that November day, was that one particular printout, which recorded the progress of *Eltanin* from 500 kilometres east of the Pacific-Antarctic Ridge to 500 kilometres west, was *completely symmetrical* around the ridge.[2] That symmetry could be explained in only one way: the rocks either side of the ridge had been formed at exactly the same time as each other and 'occupied the position they did because they had originated at the ridge and then spread out to occupy the seabed. In other words, the seabed was formed by rocks emerging from the depths of the earth, then spread out across the seafloor – and pushing the continents apart. This was a confirmation at last of continental drift, achieved by **seafloor spreading**.'[3]

It will be recalled that continental drift was proposed by Alfred Wegener in

1915 as a way to explain the distribution of the landmasses of the world and the pattern of life forms. *He* took the theory for granted, based on the evidence he had collected, but many geologists, especially in the United States, were not convinced. They were 'fixists,' who believed that the continents were rigid and immobile. In fact, geology was divided for years, at least until the war. But with the advent of nuclear submarines the U.S. Navy in particular needed far more information about the Pacific Ocean, the area of water that lay between it and its main enemy, Russia. The basic result to come out of this study was that the magnetic anomalies under the Pacific were shaped like enormous 'planks' in roughly parallel lines, running predominantly north-south, each one 15–25 kilometres wide and hundreds of kilometres long. This produced a tantalising piece of arithmetic: divide 25 kilometres by 1 million (the number of years after which, on average, the earth's polarity changes), and you get 2.5 centimetres. Did that mean the Pacific was expanding at that rate each year?[4]

There was other evidence to support the mobilists. In 1953 the French seismologist **Jean Pierre Rothé** produced a map at a meeting of the Royal Society in London that recorded earthquake epicentres for the Atlantic and Indian Oceans.[5] This was remarkably consistent, showing many earthquakes associated with the midocean ridges. Moreover, the further the volcanoes were from the ridges, the older they were, and the less active. Yet another spinoff from the war was the analysis of the seismic shocks sent shuddering across the globe by atomic bomb explosions. These produced the surprising calculation that the ocean floor was barely four miles thick, whereas the continents were twenty miles thick. Just a year before the *Eltanin*'s voyage, **Sir Edward Crisp Bullard**, a British geophysicist, had reconstructed the Atlantic Ocean margins, using the latest underwater soundings, which enabled 1,000-metre depth contours to be used, rather than sea-level contours. At that depth, the fit between the continents was even more complete.[6] Despite these various pieces of evidence, it wasn't until *Eltanin*'s symmetrical picture came ashore that the 'fixists' were finally defeated.

Capitalising on this, in 1968 **William Jason Morgan**, from Princeton, put forward an even more extreme '**mobilist**' view. His idea was that the continents were formed from a series of global, or '**tectonic**,' plates, slowly inching their way across the surface of the earth. He proposed that the movement of these plates – each one about 100 kilometres thick – together accounts for the bulk of seismic activity on earth. His controversial idea soon received support when a number of '**deep trenches**' were discovered, labelled **subduction zones**, up to 700 kilometres deep, in the floor of the Pacific Ocean. It was here that the sea floor was absorbed back into the underlying mantle (one of these trenches ran from Japan to the Kamchatka Peninsula in Russia, a distance of 1,800 kilometres).[7]

Continental drift and the wanderings of tectonic plates (many geophysicists prefer the term *blocks*) were initially of geological interest only. But geology is a form of history. One of the achievements of twentieth-century science has been to make accessible more and more remote areas of the past. Although these discoveries have arrived piecemeal, they have proved consistent – romantically consistent – in helping to provide the basis for one story, one narrative, culminating in mankind. This is perhaps the crowning achievement of twentieth-century thought.

In the same year as the *Eltanin*'s crucial voyage, twenty-seven scientists from six nations met at a conference at Stanford University in California to consider how America had been populated. These were members of the International Quaternary Association – geologists, palaeontologists, geographers, and ethnographers interested in the most recent of the four basic geological periods – and the papers presented to the conference all concerned a single theme: the **Bering land bridge**. Although Christopher Columbus famously 'discovered' America in 1492, and whether or not you accept that he was beaten to it by the Vikings in the Middle Ages – as many scholars believe – equally clearly there were 'native' populations throughout the New World who had arrived there thousands of years before. Around 1959, as we have seen, palaeontologists were beginning to accept the view that *Homo sapiens* had evolved first hundreds of thousands of years ago in the Rift Valley of East Africa. Work on tectonic plates had shown that this valley was the edge of just such a plate, perhaps accounting, for some unknown reason, for why mankind should have emerged there. Since that time, unless man evolved separately in different parts of the world, he must have spread out across the earth in an order that, in theory at least, can be followed. The farthest large pieces of land from East Africa are Australia, Antarctica, and the Americas. To get to the Americas, early man would either have had to navigate huge distances across the oceans, in enough boats to create the numbers needed for propagation at the destination (which they could not have known about in advance), or crossed the narrow (56-mile) gap between Siberia and Alaska. It was this possibility that the Stanford conference was called to consider.

The idea was not new, but the conference was presented with archaeological and geological evidence that for the first time fleshed out a hitherto vague picture. It appeared that man crossed the land bridge in **three waves**, the first two being between 40,000 and 20,000 years ago, and the third between 13,000 and 12,000 years ago.[8] The basic long-term context of the migrations was determined by the ice ages, which locked enormous amounts of water in the glaciers at the poles, reducing sea levels by up to a hundred metres, more than 300 feet (the Bering Straits are 24 fathoms deep, or nearly 150 feet). The idea of three migrations came initially from an analysis of artefacts and burial techniques, later from an analysis of art, language, and genes. Calculations by **C. Vance Haynes** in Denver, made the year after the conference, suggested that a tribe of only thirty mammoth hunters (say) could increase in 500 years to as many as 12,500 people in perhaps 425 tribes. The **Clovis** hunters, who comprised the third wave, distributed their characteristic spearheads (first found at Clovis, New Mexico, near the Texan border) all over the continent. According to Haynes they would have needed to migrate only four miles to the south each year to reach Mexico in 500 years. Thus the geological and ethnographical evidence for early man in America fits together very well. It also fits neatly into the 'one story' narrative.

The recovery of the American past was matched by developments in Africa. Here, the seminal work was **Basil Davidson's** book *Old Africa Rediscovered*, first published in 1959, which proved so popular that by the early 1960s it had gone into several editions.[9] The book followed an explosion of scholarship in African studies, with Davidson pulling the picture together. His achievement was to show that the 'dark continent' was not so dark after all; that it had its own considerable history, which a number of well-known Western historians had denied, and that several more or less sophisticated civilisations had existed in Africa from 2000 BC onward.

Davidson surveyed all of Africa, from Egypt and Libya in the north to Ghana, Mali, and Benin in the west, the coast of Zanj (or Zinj) in the east, and the south-central area around what was then Rhodesia (now Zimbabwe). He covered the appearance of 'Negro' peoples, about 3,000–5,000 BC, according to an analysis of some 800 skulls discovered at a site from predynastic Egypt, and the evidence of early migrations – for example, from the Nile area to West Africa ('The Forty Day Road'). He described the Kush culture, emerging from the decadence of imperial Egypt, the enormous slag heaps of Meroe ('the Birmingham of Africa'), about a hundred miles from modern Khartoum. Besides the palaces and temples, only a fraction of which has been excavated, the slag heaps are evidence of Meroe's enormous iron-smelting capability, on which its great wealth was based.[10] Having described the great coastal civilisations of Benin, Kilwa, Brava, Zanzibar, and Mombasa, Davidson's most remarkable chapters concern the great inland civilisations of Songhay, Jebel Uri, Engaruka, Zimbabwe, and Mapungubwe, mainly because such places, remote from foreign influence, most closely represent the African achievement, uncomplicated by international trade and the ideas such trade brings with it. **Engaruka**, on the borders of Kenya and Tanganyika, as it then was (now Tanzania), had been first discovered by a district officer in 1935 but excavated later by Louis Leakey. He found the main city to consist of nearly seven thousand houses, supporting a population, he thought, of at least thirty to forty thousand. The houses were well built, with terraces and engravings that he thought were 'clan marks.'[11] Three hundred miles from the coast, Engaruka was well defended on a steep escarpment of the Rift Valley and, Leakey felt, dated to the seventeenth century. There were stone structures he took to be irrigation channels and evidence of solitary burials. Later excavation showed that the city was surrounded by eight thousand acres that were once under grain, producing a surplus that was traded via roads to the north and south – villages of up to a hundred houses were grouped along these roads. Iron-using techniques spread south through this area of Africa from about 500 AD.

Great Zimbabwe is a vast group of stone ruins a few miles off the main road which links what is now Harare (Salisbury when Davidson published his book) and Johannesburg in South Africa, featuring an 'acropolis' and an elliptical 'temple.' All the buildings are made of local granite, flat, bricklike stones chopped from 'leaves' of exfoliated rock. The form of the defensive work, terraced battlements, shares some features with the building at Jebel Uri several hundred miles away, raising the possibility of commerce and the exchange of ideas over large distances. Zimbabwe and Mapungubwe both lie near the centre of a vast mining area – gold, copper, iron, and tin – which stretches as far north as Zambia and the Belgian Congo (now Zaïre) and as far south as Pretoria and Johannesburg in the Transvaal. Some scholars believe that Zimbabwe is as old as 2000 BC, with the main period of inhabitation between 600 and 1600 AD.[12]

Mapungubwe is less well known than Zimbabwe and even more mysterious. It is found on a small table mountain about 200 miles to the south, just across the Limpopo River. It was regarded as 'a place of fear' by the locals, but when it was finally visited (via a narrow 'chimney,' found to have holes cut in it opposite one another, so that a ladder could be built into the walls), the top of the flat mountain proved to contain thousands of tons of soil imported from the surrounding countryside, clearly evidence of a crop-growing civilisation. But what most attracted the attention of the people who found the site were the gold artefacts they discovered – and the skeletons.[13] One skeleton (twenty-three were unearthed) was covered in gold bangles. Analysis of the skeletons showed an absence of Negroid features; they were, rather, 'pre-Negro.' The burial practices were Bantu, but the skeletons were partly Hottentot and partly similar to those found on the coast. They buried their own dead and their cattle, evidence of religion.

Davidson took care to emphasise that much remained to be discovered in Africa. But he achieved his aim, adding to the contributions of Chinua Achebe, Wole Soyinka, and others, who were showing that Africa had a voice and a history. More, Davidson was helping flesh out the greater history of mankind across the globe – his book also explored the way stone tools and metal technology spread. The history of Africa, like history elsewhere, was shaped by larger forces than mere individuals.[14]

The extent of those larger forces of history – economic, sociological, geo-graphical, and climatological – rather than the actions of significant individuals, has been the main shift in history as an academic discipline throughout most of the century. And within this overall paradigm the two most prolific schools of thought have been the French *Annales* historians and the British Marxists.

The 1960s saw the publication of three enormously influential books from the so-called *Annales* school of French historians. These were: *Centuries of Childhood*, by Philippe Ariès (1960); *The Peasants of Languedoc*, by Emmanuel le Roy Ladurie (1966); and *The Structures of Everyday Life*, by Fernand Braudel (1967), the first volume of his massive, three-part *Civilisation and Capitalism*. The 1960s were in fact the third great flowering of the *Annales* school – the first had been in the 1920s and the second in the 1940s.

Of the three authors **Fernand Braudel** was by far the most senior. He was older and was a close colleague of the two founders of the *Annales* school, **Lucien Febvre** and **Marc Bloch**. These two men came together at the University of Strasbourg in the 1920s, where they founded a new academic journal, the *Annales d'histoire économique et social*. As its name implied, *Annales* from the first sought to concentrate on the social and economic context of events rather than the deeds of 'great men,' but what set it apart was the imagination that Febvre and Bloch brought to their writing, especially after they both returned to Paris in the mid-1930s.[15]

Bloch (a resistance hero in World War II) wrote two books for which he is remembered today, *The Royal Touch* and *Feudal Society*. *The Royal Touch* was concerned with the belief, prevalent in both England and France from the Middle Ages to the Enlightenment, that kings – by the mere act of touching – could cure scrofula, a skin disease known as 'the king's evil.'[16] But Bloch's study ranged much further than this curious belief; it drew on contemporaneous ideas in sociology, psychology, and anthropology in search of a context for what Bloch called the *mentalité* of the period. In *Feudal Society*, published on the eve of World War II, he attempted to re-create the historical *psychology* of feudal times, something that was completely novel.[17] For example, he explored the mediaeval sense of time, better described perhaps as an 'indifference' to time, or as a lack of interest in the exact measurement of time. In the same way, Febvre's *Rabelais* explored the *mentalité* of the sixteenth-century world. By an analysis of letters and other writings, the author was able to show, for example, that when Rabelais was denounced as an atheist, his critics didn't mean what we would mean today.[18] In the early sixteenth century, *atheist* had no precise meaning, simply because it was inconceivable for anyone to be an atheist as we would recognise the term. It was, instead, as Peter Burke confirms in his history of the *Annales* school, a general smear word. Febvre also explored time, showing for example that someone like Rabelais would not have known the year in which he was born, and that time was experienced not in a precise way, as measured by clocks, but rather by 'the length of an Ave Maria' or 'the flight of the woodcocks.'[19] It was the ability of Bloch and Febvre to get 'inside the heads' of individuals remotely removed in time that readers found exciting. This *felt* much more like history than the mere train of events that many historians wrote about. And it applied even more to Braudel, for he took the *Annales* approach much further with his first book, *The Mediterranean*, which appeared in 1949 and created a bigger stir.[20]

This book was conceived and written in extremely unusual circumstances. It had begun as a diplomatic history in the early 1920s. Then in 1935–7 Braudel accepted an appointment to teach at the University of São Paolo, and on the voyage back he met Febvre, who 'adopted him as *un enfant de la maison*.'[21] But Braudel didn't get round to writing the book until he was a prisoner of war in a camp near Lübeck. He lacked notes, but he had a near-photographic memory, and he drafted *The Mediterranean* in longhand in exercise books, which he posted to Febvre.

The Mediterranean is 1,200 pages long and divided into three very different sections. In the first part, Braudel treats his readers to 300 pages on the geography of the Mediterranean – the mountains and rivers, the weather, the islands and the seas, the coastlines and the routes that traders and travellers would have taken in the past. This leads to a discussion of the various cultures in different geographical circumstances – mountain peoples, coastal dwellers, islanders.[22] Braudel's aim here is to show the importance of what he called *la longue durée* – that the history of anywhere is, first and foremost, determined by where it is and how it is laid out. The second part of the book he called 'Collective Destinies and General Trends,' and here the focus of his attention was on states, economic systems, entire civilisations – less permanent than the physical geography, but still more durable than the lives and careers of individuals.[23] His gaze now centres on change that occurs over generations or centuries, shifts that individuals are barely aware of. Exploring the rise of both the Spanish and the Turkish Empires, for example, he shows how their growth was related to the size and shape of the Mediterranean (long from west to east, narrow from north to south); he also showed why they gradually came to resemble each other – because communications were long and arduous, because the land and the available technology supported similar population densities.[24] And finally, there is the level of events and characters on the historical stage. While Braudel acknowledges that people differ in character, he thinks those differences account for less than traditional historians claim. Instead, he argues that an understanding of how people in the past viewed their world can help explain a lot of their behaviour. One example he makes much of is Philip II's notorious slowness in reacting to events. This was not just due to his personality, says Braudel. During Philip's reign Spain was financially exhausted (thanks again to geographical factors), and communications were slow – it could take two months to travel from one end of the Mediterranean to the other. Philip's deliberation was born as much of Spain's economic and geographic situation as anything.[25]

Whereas Bloch's books, and Febvre's, had created a sensation among historians, *The Mediterranean* broke out of its academic fold and became known well beyond France. He himself was very ambitious for this to happen.[26] People found the new type of information it contained every bit as fascinating as the doings of monarchs and prime ministers. For his part, Febvre invited his *enfant de la maison* (now turned fifty) to join him in an even more massive collaborative venture. This was a complete history of Europe, stretching over four hundred years, from 1400 to 1800, exploring how the mediaeval world became the modern world, and using the new techniques. Febvre said he would tackle 'thought and belief,' and Braudel could write about material life. The project hadn't gone very far when Febvre died in 1956, but Braudel carried on, with the book eventually taking almost as long to complete as did his earlier work. The first volume of *Civilisation et capitalisme*, known in English as *The Structures of Everyday Life*, appeared in 1967; the last in 1979.[27]

Here again Braudel's conception was threefold – production at the base, distribution, and consumption at the top. (This was Marx-like, rather than specifically Marxist.) In the realm of production, for example, Braudel explored the relationship of wheat and maize and rice to the civilisations of the world. Rice, he found, 'brought high populations and [therefore] strict social discipline to the regions where they prospered' in Asia.[28] On the other hand, maize, 'a crop that demands little effort,' allowed the native Americans much free time to construct their huge pyramids for which these civilisations have become famous.[29] He thought that a crucial factor in Europe's success was its relatively small size, plus the efficiency of grain, and the climate.[30] The fact that so much of life was indoors fostered the development of furniture, which brought about the development of tools; the poorer weather meant that fewer days could be worked, but mouths still had to be fed, making labour in Europe relatively expensive. This led to a greater need for labour-saving devices, which, on top of the development of tools, contributed to the scientific and industrial revolution. The second volume, *The Wheels of Commerce*, and the third, *Perspective of the World*, followed the rise of capitalism. Braudel's central point was that geography governed raw materials, the creation of cities (the markets) and trade routes. There was in other words a certain inevitability about the way civilisations developed, which made Europe, rather than Asia, Africa, or America the cradle of both capitalism and science.[31]

Braudel's influence lay not just in his books but in the inspiration he offered to others (he died in 1985). Since World War II, the *Annales* school has spawned a very successful series of investigations, among them *The Peasants of Languedoc; Montaillou; Centuries of Childhood; The Hour of Our Death; The Coming of the Book; The Identity of France; The Great Cat Massacre; Catholicism from Luther to Voltaire; The Birth of Purgatory;* and *The Triumph of the Bourgeoisie*. Emmanuel le Roy Ladurie was widely regarded as Braudel's most brilliant pupil.[32] He too was interested in *la longue durée*, and in *The Peasants of Languedoc* and *Montaillou* he sought to recreate the *méntalité* of mediaeval Europe. Montaillou, situated in the Ariège region of southwest France, was in an area that had been 'home' to a number of nonconformists during the Cathar heresy of the fourteenth century. These heretics were captured and interrogated by the local bishop, and a written record of these interrogations has survived. This register was used by **Ladurie**, who interpreted it in the light of more recent advances in anthropology, sociology, and psychology.[33] Among the names on the register of interrogations, twenty-five came from one village, Montaillou, and for many readers Ladurie brought these individuals back to life. The first part of his book deals with the material aspects of village life – the structure of the houses, the layout of the streets, where the church was.[34] This was done with wit and imagination – Ladurie shows, for instance, that the stones were so uneven that there were always holes in the walls so that families could listen to their neighbours: privacy was unknown in Montaillou. But it is in the second part of the book, 'An Archaeology of Montaillou: From Body Language to Myth,' that the real excitement lies. Here we are introduced, for example, to Pierre Maury, a gentle shepherd, but also politically conscious, to Pierre Clergue, the obnoxious priest, too big for his boots and the seducer of Béatrice des Planissoles, impressionable, headstrong, and all too eager to grow up.[35]

The *Annales* school has proved very influential. Its attraction for many people lies in the imaginative use of new kinds of evidence, science added to a humanity that provides a technique to bridge the gap across the centuries, in such a way that we can really understand what happened in the past, and how people thought. The very idea of recreating *mentalités*, the psychology of bygone ages, is ambitious, but for many people by far the most intriguing use of history, the closest to time travel we have ever had. A second reason why the *Annales* form of history has proved popular is its interest in 'ordinary' people and everyday life, rather than in kings and parliaments, or generals and armies. This shift of interest, very marked during the century, reflected the greater literacy achieved in Western countries at the end of the nineteenth century; poorer readers naturally wanted to read about people like themselves. It was also yet another fruit of World War I – that disaster affected the lives of ordinary people far more profoundly than it affected the generals or the leaders. Finally, the shifts in history writing formed part of a general trend: with the growth of mass society, of new media and popular forms of entertainment, the worlds of 'ordinary' people were a focus of interest everywhere.

But in some quarters there was a more specific reason, and this found an outlet particularly in Britain, in the work of a small but very influential group of Marxist historians. The **British Marxist historians** were less original than their French counterparts but more coherent in their aim, which was essentially to rewrite British history from the end of the Middle Ages to the beginning of the twentieth century, 'from the bottom up' (a favoured phrase, which soon became hackneyed). Most of its seminal works were produced in or near the 1960s: *Puritanism and Revolution: Studies in Interpretation of the English Revolution of the Seventeenth Century*, by Christopher Hill (1958); *Primitive Rebels*, by Eric Hobsbawm (1959); *The Age of Revolution*, by Hobsbawm (1962); *Studies in the Development of Capitalism* (1963), by Maurice Dobb; *The Making of the English Working Classes* (1964) by E. P. Thompson ('the pivotal work of British Marxists,'[36] 'probably the most important work of social history written since the Second World War'); *Labouring Men*, by Hobsbawm (1964); *Intellectual Origins of the English Revolution*, by Hill (1965); *A Medieval Society: The West Midlands at the End of the Thirteenth Century*, by Rodney Hilton (1966); *Reformation to Industrial Revolution: A Social and Economic History of Britain, 1530–1780*, by Hill (1967); *The Decline of Serfdom in Medieval England*, by Hilton (1969); *Bandits*, by Hobsbawm (1969); *God's Englishman: Oliver Cromwell and the English Revolution*, by Hill (1970); and *Bond Men Made Free: Medieval Movements and the English Rising of 1381*, by Hilton (1973). Three men stand out in this history of the lower orders, Rodney Hilton, Christopher Hill, and E. P. Thompson. The issues they focus on are the way feudal society changed to capitalist society, and the struggle which produced the working class.

Rodney Hilton, professor of history at Birmingham University, was like the others a member of the British Communist Party until the events in Hungary in 1956. His main interest was in the precursors of the working class – the peasants – and besides his own books on the subject, he was instrumental in the founding of two journals in the 1960s, the *Journal of Peasant Studies* in Britain and *Peasant Studies* in the United States.[37] Hilton's aim was to show that peasants were not a passive class in Britain in the Middle Ages; they did not just accept their status but were continually trying to improve it. There was, Hilton argued, constant struggle, as the peasants tried to gain more land for themselves or have their rents reduced or abolished.[38] This was no 'golden time,' to use Harvey Kaye's words, in his survey of the British group, when everyone was in his place and satisfied with it; instead there was always a form of peasant 'class-consciousness' that contributed to the eventual decline of the feudal-seigneurial regime in England.[39] This was a form of social evolution, Hilton's point being that this struggle gave rise to agrarian capitalism, out of which industrial capitalism would emerge.[40]

The next stage in the evolution was examined by **Christopher Hill**, fellow and Tutor of Balliol from 1938, who devoted himself to the study of the English revolution. His argument was that just as the peasants had struggled to obtain greater power in mediaeval times, so the English revolution, traditionally presented as a constitutional, religious, and political revolution, was in fact the culmination of a class struggle in which capitalist merchants and farmers sought to seize power from the feudal aristocracy and monarchy. In other words, the motivation for the revolution was primarily economic.[41] He put it this way: 'The English revolution of 1640–60 was a great social movement like the French Revolution of 1789. The state power protecting an old order that was essentially feudal was violently overthrown, power passed into the hands of a new class [the bourgeoisie], and so the freer development of capitalism was made possible. ... Furthermore, the Civil War was a class war, in which the despotism of Charles I was defended by the reactionary forces of the established Church and conservative landlords. Parliament beat the king because it could appeal to the enthusiastic support of the trading and industrial classes in town and countryside, to the yeomen and progressive gentry, and to wider masses of the population whenever they were able by free discussion to understand what the struggle was really about.'[42] He added that the revolution also took some of its colour from recent developments in science and technology, very practical concerns that could in time be converted into new commercial outlets.

Like Hilton and Hill, **E. P. Thompson** also left the British Communist Party in 1956. Like them he remained convinced that English history was determined mainly by class struggle. In a long book, *The Making of the English Working Class,* one of his aims was to 'rescue' the working classes from 'the enormous condescension of posterity' and render visible such neglected people as weavers and artisans. In the process, he redefined the working classes as essentially a matter of *experience.* It was the experience – between 1790 and 1830 – of a declining and weakening position in the world. This, he said, was the essence of the industrial revolution for the working class in England – the loss of common rights by the landless, the increasing poverty of many trades brought about by the deliberate manipulation of employment to make it more precarious.[43] Part of the attraction in Thompson's book lies in the fact that it is so vividly written and humane, but it was original in a social Darwinian sense too. Before 1790 the English working classes existed in many disparate forms; the experience of oppression and the progressive loss of rights, far from resulting in their extinction, proved to be a major unifying (and therefore strengthening) force.[44]

The final element in this 'Great Leap Forward' of historical studies came in 1973 from the archaeologist **Colin Renfrew**, in Britain. Like the *Annales* school and the Marxists, and like archaeologists everywhere, he had an interest in *la longue durée*. But, again like the French and British historians, his main aim was less an obsession with dating as such as with a new understanding of history. Then professor at Southampton University, and now at Cambridge, his book was entitled *Before Civilisation: The Radiocarbon Revolution and Prehistoric Europe.*[45] The title sold the book short, however, for Renfrew gave an admirably clear, albeit brief history of the development of dating in archaeology in the twentieth century and how – and this was his main point – it has changed our grasp of the past, not just in terms of chronology but in the way we conceive of man's early development.

He began with a crisp statement of the problem, so far as archaeology was concerned. Various early-twentieth-century studies by geologists in Switzerland and Sweden had confirmed that the last ice age endured for 600,000 years and ended 10,000 years ago. The problem with ancient human history therefore stemmed from the fact that the written records stretched back no further than about 3,000 BC. What had happened between the end of the ice age and the birth of writing? Renfrew's main aim was to take a look at archaeology in the wake of the various new dating mechanisms – dendrochronology, radiocarbon dating, and potassium-argon dating. Radiocarbon dating was first devised by **Willard F. Libby** in New York in 1949 (he won the Nobel Prize for Chemistry in 1960 for his innovation), but his insight was added to significantly by the *American Journal of Science*, which in 1959 started an annual radiocarbon supplement, which quickly became an independent publication, *Radiocarbon*. This provided an easily accessible forum for all the revised dates that were then being obtained from across the world. It was perhaps the biggest intrusion of science into a subject that had hitherto been regarded as an art or a humanity.

Before Civilisation had two core arguments. First, it revised the timing of the way the earth was populated. For example, from about 1960 on it was known that Australia had been occupied by man as early as 4000 BC and maybe even as early as 17000 BC. Maize, it was established, was gathered systematically in Mexico by 5000 BC, and well before 3000 BC it was showing signs of domestication. The real significance of these dates was not just that they were earlier than anyone had thought hitherto, but that they killed off the vague theories then current that Meso-America had only developed civilisation after it had been imported, in some indefinable way, from Europe. The Americas had been cut off from the rest of the world since 12000–13000 BC, in effect the last ice age, and had developed all the hallmarks of civilisation – farming, building, metallurgy, religion – entirely separately.[46]

This revision of chronology, and what it meant, was the second element in Renfrew's book, and here he concentrated on the area he knew best, Europe and the classical world of the Middle East. In the traditional view, the civilisations of the Middle East – Sumer and Egypt, for example – were the mother civilisations, the first great collective achievements of mankind, giving rise to the Minoans on Crete, and the classical world of the Aegean: Athens, Mycenae, Troy. From there, civilisation had spread farther north, to the Balkans and then

Germany and Britain, and west to Italy and then France, and the Iberian peninsula. But after the C^{14} revolution, there was suddenly a serious problem with this model.[47] On the new dating, the huge megalithic sites of the Atlantic seaboard, in Spain and Portugal, in Brittany and Britain, and in Denmark, were either contemporaneous with the civilisations of the Aegean or actually preceded them. This wasn't just a question of an isolated date here and there but of many hundreds of revised datings, consistent with each other, and which in some cases put the Atlantic megaliths up to a thousand years earlier than Aegean cultures. The traditional model, for Egypt, the Middle East, and the Aegean, still held. But there was, as Renfrew put it, a sort of archaeological 'fault line' around the Aegean. Beyond that, a new model was needed.

The model he came up with started from a rejection of the old idea of '**diffusion**' – that there had been an area of mother civilisations in the Middle East from which ideas of farming, metallurgy, and, say, the domestication of plants and animals had started, and then spread to all other areas as people migrated. It seemed clear to Renfrew that up and down the Atlantic coasts of Europe, there had developed a series of chiefdoms, a level of social organisation midway between hunter-gatherers and full-blown civilisation as represented in Egypt, Sumer, or Crete, which had kings, elaborate palaces, a highly stratified society. The sovereign areas of the chiefdoms were smaller (six on the Isle of Arran in Scotland, for example), and they were centred around large tombs and occasionally religious/astronomical sites, such as Stonehenge.[48] Associated with these chiefdoms were a rudimentary social stratification and early trade. Sufficient numbers were needed to build the impressive stone works, funerary religious monuments around which the clans cohered. The megaliths were always found associated with arable land, suggesting that chiefdoms were a natural stage in the evolution of society: when man settled with the first domesticated crops, chiefdoms and megaliths soon followed.[49]

Renfrew's analysis, now generally accepted, concentrated on sites in Britain, Spain, and the Balkans, which illustrated his argument. But it was his general thrust that counted: although early man had no doubt spread out to populate the globe from an initial point (maybe East Africa), civilisation, culture – call it what you will – had not developed in one place and then spread in the same way; civilisations had grown up in different times at different places of their own accord.[50] This had two important long-term intellectual consequences, quite apart from killing off any idea (which still lingered then) that American civilisations had been seeded by European ones (such as the 'lost tribe' of Israel). First, it demolished the idea that, culturally speaking, the history of man is one story; all cultures across the world were *sui generis* and did not owe their being to a mother culture, the ancestor of all. Combined with the findings of anthropologists, this made all cultures equally potent and equally original, and therefore the 'classical' world was no longer the ultimate source.

At a deeper level, as Renfrew specifically pointed out, the discoveries of the new archaeology showed the dangers of succumbing too easily to Darwinian thinking.[51] The old diffusionist theory was a form of evolution, but a form of evolution so general as to be almost meaningless. It suggested that civilisations developed in an unbroken, single sequence. The new C^{14} and tree-ring evidence showed that that simply wasn't true. The new view wasn't any less 'evolutionary,' but it was very different. It was, above all, a cautionary tale.

HEAVEN AND EARTH

The words *historic moment* have been heavily overused in this century. But if any moment outside war can be described as truly historic it surely occurred at twenty seconds after 3:56 AM BST on Monday, 21 July 1969, when **Neil Armstrong** stepped off the ladder that led from the 'Eagle,' the landing module of the *Apollo 11* spacecraft, and on to the surface of the Moon, making him the first person to arrive on a celestial body outside Earth. As he did so, he spoke the words that have since become famous: 'That's one small step for man – one giant leap for mankind.'[1]

For the benefit of scientists back at Mission Control in Houston, he then went on in a more down-to-earth, scientifically informative way: 'The surface is fine and powdery, I can ... I can pick it up loosely with my toe. It does adhere in fine layers like powdered charcoal to the sole and sides of my boots. I can only go in a small fraction of an inch. Maybe an eighth of an inch, but I can see the footprints of my boots and the treads in the fine sandy particles. ... There seems to be no difficulty in moving around, as we suspected. ... We're essentially on a level place here – very level place here.'[2] If the greatest intellectual achievement of the first half of the twentieth century was undeniable – the conception and construction of the atomic bomb – the achievements of the second half were more diverse, including the isolation and understanding of DNA, and the computer. But space travel and the moon landing certainly count as one of the greatest achievements of the century.

After the Russians had sprung their surprise in 1957, stealing a march on the United States with the launch of *Sputnik 1*, they had built on their lead, putting the first animal in space, the first man (**Yuri Gagarin**, in 1961), and the first woman (**Valentina Tereshkova**, in 1963). The United States responded with something close to panic. President Kennedy called a frenzied meeting at the White House four days after the Bay of Pigs disaster (when 1,500 Cuban exiles, trained in America by the U.S. military, invaded the island, only to be killed or captured). Railing that 'something must be done,' he had shouted at Lyndon Johnson, the vice president, ordering him to find out if 'we have a chance of beating the Soviets by putting a laboratory in space, or by a trip round the Moon, or by a rocket to land on the Moon, or by a rocket to go to the Moon and back with a man?'[3] The Americans finally put **John Glenn** into orbit on

20 February 1962 (Alan Shephard made a fifteen-minute nonorbital flight in May 1961). But then the Americans began to catch up, thanks to Kennedy's commitment to the Apollo program with its aim to land a manned spacecraft on the Moon 'before this decade is out.'[4] Begun in 1963 (though NASA had been created in 1958), America spent up to $5 billion a year on space over the next ten years. That sum gives an idea of the size of the project, which involved, among other things, building a reliable spaceship bigger than a railway engine, designing and manufacturing a rocket heavier than a destroyer, and inventing several new materials.[5] The project benefited from the brains of 400,000 people from 150 universities and 20,000 firms. We already know from Korolev that rocket technology lay at the heart of the space program and the biggest U.S. rocket, *Saturn 5*, weighed 2,700 tons, roughly the same as 350 London buses. Developed under Wernher von Braun, another German emigré, *Saturn* was 364 feet high, had 2 million working parts, 2.5 million solder joints, and 41 separate engines for guidance purposes, and carried in all 11,400,000 gallons of fuel – liquid nitrogen, oxygen, hydrogen, and helium, some of it stored at minus 221 degrees centigrade to keep it liquid.[6] The oxygen alone filled the equivalent of fifty-four railway container tanks.[7] The Moonship contained the cone-shaped command module, the only part to come back to earth and which therefore needed to withstand the enormous temperatures on re-entry to the atmosphere (caused by friction at such high speeds).[8] One of the main engineering problems was to keep the cryogenic fuels cool enough. The tanks eventually designed were so airtight that, were ice cubes to have been installed inside, they would not have melted for nine years. The exit hatch of the module needed 150 new tools to be invented. Some bolts had to be locked in place by two men using a five-foot wrench.[9]

No one really knew how conditions in space would affect the men.[10] Great care was taken therefore with psychological selection and training. They were taught to be tolerant and careful (always avoiding acute angles where they might snag their suits), and they were given massages every day. The crews that advanced were those that had worked together in harmony for more than a year. Interestingly enough, over the years both the Americans and the Russians came up with a near-identical profile of the ideal astronaut: they should not be too old, no later than their late thirties, or too tall, no taller than five-eleven or six feet; they should be qualified jet and test pilots with degrees in engineering.[11] Finally, there was the reconnaissance of the Moon itself. Quite apart from the prospect in time of colonising space and its minerals, there were sound scientific reasons for studying the Moon close up. Since it lacked an atmosphere, the Moon was in some senses in pristine condition, 'a priceless antique,' as one scientist called it, in much the same condition as it had been when the universe, or the solar system at least, had first evolved. Examination of the rocks would also help decide how the Moon formed – whether it was once part of Earth, or broke off with Earth from the sun after collision with an asteroid, or was formed by very hot gas cooling.[12] Both American and Soviet space probes got progressively closer to the Moon, sending back better and better photographs until objects as small as five feet wide could be distinguished. Five areas were originally chosen for landing, then narrowed to one, the Sea of Tranquillity, actually a flat plain free of craters.[13]

The biggest disaster of the American program took place in 1967, when a spaceship caught fire on the launchpad at Cape Kennedy after liquid oxygen somehow ignited, killing all three men inside. The world never knew how many Russian astronauts perished because of the greater secrecy surrounding their program, but distress messages picked up by radio hams around the globe suggested that at least eight were in trouble between 1962 and 1967.[14] The greatest drama before the moon landing itself was the December 1968 flight of *Apollo 8* around the Moon, which involved going behind the Moon to the dark side, which no one had ever seen, and meant that the crew would be out of radio contact with Mission Control for about half an hour. If the 'burn' of the engines was too strong, it might veer off into deep space; if it was too weak, it might crash into the Moon, on the far side, never to be heard from again.[15] The pope sent a message of goodwill, as did a number of Russian space scientists, acknowledging implicitly at this point that the Americans were now decisively ahead.

At 9:59 AM on Christmas Eve, *Apollo 8* swung behind the Moon. Mission Control in Houston, and the rest of the world, waited. Ten minutes of silence passed; twenty; thirty. At 10:39 AM Frank Borman's voice could be heard reporting data from his instruments. *Apollo 8* was exactly on schedule and, as Peter Fairley narrates the episode in his history of the Apollo Project, after a journey of a quarter of a million miles, it had arrived on its trajectory within half a mile of the one planned.[16]

The scene was set for *Apollo 11*. Edward 'Buzz' Aldrin Jr. joined Neil Armstrong on the surface of the Moon, where they placed a plaque, and a flag, planted some seeds, and collected rock samples with specially designed tools that avoided them having to bend. Then it was back into the 'Lunar Bug,' rendezvous with Michael Collins in the command module, and the return journey, splashing down near Johnston Island in the Pacific, where they were met by the USS *Hornet* with President Richard Nixon on board. The men had returned safely to Earth, and the space age had begun.[17]

The landing on the Moon was, however, in some ways a climax rather than a debut. Crewed flights to the Moon continued until 1972, but then stopped. As the 1970s wore on, space probes went deeper into the heavens – Venus, Mars, Mercury, Jupiter, the Sun, Saturn, with *Pioneer 10*, launched in 1972, becoming the first manmade object to leave the solar system, which it did in 1983. Actual landings were considered less necessary after the first flush of excitement, and both the Americans and Russians concentrated on longer flights in orbit, to enable scientists to carry out experiments in space: in 1973, in the United States *Skylab*, astronauts spent eighty-four days aboard. The first stage of the space age may be said to have matured around 1980. In that year, *Intelsat 5* was launched, capable of relaying thousands of telephone calls and two television channels. And in the following year the *Columbia*, the first reusable space shuttle, was launched. In just ten years space travel had gone from being exotic to being almost mundane.

★

The space race naturally stimulated interest in the heavens in general, a happy coincidence, as the 1960s had in any case seen some very important advances in our understanding of the universe, even without the advantages conferred by satellite technology. In the first half of the century, apart from the development of the atomic bomb and relativity, the main achievement of physics was its unification with chemistry (as epitomised in the work of Linus Pauling). After the war, the discovery of yet more fundamental particles, especially quarks, brought about an equivalent unification, between physics and astronomy. The result of this consilience, as it would be called, was a much more complete explanation of how the heavens – the universe – began and evolved. It was, for those who do not find the reference blasphemous, an alternative Genesis.

Quarks, as we have seen, were originally proposed by Murray Gell-Mann and George Zweig, almost simultaneously in 1962. It is important to grasp that quarks do not exist in isolation in nature (at least on Earth), but the significance of the quark (and certain other particles isolated in the 1960s and 1970s but which we need not describe here) is that it helps explain conditions in the early moments of the universe, just after the Big Bang. The idea that the universe began at a finite moment in the past was accepted by most physicists, and many others, since Hubble's discovery of the red shift in 1929, but the 1960s saw renewed interest in the topic, partly as a result of Gell-Mann's theories about the quark but also because of an accidental discovery made at the Bell Telephone Laboratories in New Jersey, in 1965.

Since 1964, the Bell Labs had been in possession of a new kind of telescope. An antenna located on Crawford Hill at Holmdel communicated with the skies via the *Echo* satellite. This meant the telescope was able to 'see' into space without the distorting interference of the atmosphere, and that far more of the skies were accessible. As their first experiment, the scientists in charge of the telescope, **Arno Penzias** and **Robert Wilson**, decided to study the radio waves being emitted from our own galaxy. This was essentially baseline research, the idea being that once they knew what pattern of radio waves *we* were emitting, it would be easier to study similar waves coming from elsewhere. Except that it wasn't that simple. Wherever they looked in the sky, Penzias and Wilson found a persistent source of interference – like static. At first they had thought there was something wrong with their instruments. A pair of pigeons were nesting in the antenna, with the predictable result that there were droppings everywhere. The birds were captured and sent to another part of the Bell complex. They came back. This time, according to Steven Weinberg's account published later, they were dealt with 'by more decisive means.'[18] With the antenna cleaned up, the 'static' was reduced, but only minimally, and it still appeared from all directions. Penzias discussed his mystery with another radio astronomer at MIT, **Bernard Burke**. Burke recalled that a colleague of his, Ken Turner of the Carnegie Institute of Technology, had mentioned a talk he had heard at Johns Hopkins University in Baltimore given by a young theorist from Princeton, **P. J. E. Peebles**, which might bear on the 'static' mystery. Peebles's speciality was the early universe. This was a relatively new discipline and still very speculative. As we saw in chapter 29, in the 1940s an emigré from

Ukraine, George Gamow, had begun to think about applying the new particle physics to the conditions that must have existed at the time of the Big Bang. He started with 'primordial hydrogen,' which, he said, would have been partly converted into helium, though the amount produced would have depended on the temperature of the Big Bang. He also said that the hot radiation corresponding to the enormous fireball would have thinned out and cooled as the universe expanded. He went on to argue that this radiation 'should still exist, in a highly "red-shifted" form, as radio waves.'[19] This idea of 'relict radiation' was taken up by others, some of whom calculated that such radiation should now have a temperature of 5 K (i.e., 5 degrees above absolute zero). Curiously, with physics and astronomy only just beginning to come together, no physicist appeared to be aware that even then radio astronomy was far enough ahead to answer that question. So the experiment was never done. And when radio astronomers at Princeton, under **Robert Dicke**, began examining the skies for radiation, they never looked for the coolest kinds, not being aware of their significance. It was a classic case of the right hand not knowing what the left was doing. When Peebles, a Canadian from Winnipeg, started his Ph.D. at Princeton in the late 1950s, he worked under Robert Dicke. Gamow's theories had been forgotten but, more to the point, Dicke himself seems to have forgotten his own earlier work.[20] The result was that Peebles unknowingly repeated all the experiments and theorising of those who had gone before. He arrived at the same conclusion, that the universe should now be filled with 'a sea of background radiation' with a temperature of only a few K. Dicke, who either still failed to remember his earlier experiments or didn't realise their significance, liked Peebles's reasoning enough to suggest that they build a small radio telescope to look for the background radiation.

At this point, with the Princeton experiment ready to start, Penzias called Peebles and Dicke, an exchange that became famous in physics. Comparing what Dicke and Peebles knew about the evolution of background noise and the observations of Penzias and Wilson, the two teams decided to publish in tandem a pair of papers in which Penzias and Wilson would describe their observations while Dicke and Peebles gave the cosmological interpretation – that this was indeed the **radiation left over from the Big Bang**. Within science, this created almost as huge a sensation as the confirmation of the Big Bang itself.[21] It was this duo of papers published in the *Astrophysical Journal* that caused most scientists to finally accept the Big Bang theory – not unlike the acceptance of continental drift after *Eltanin*'s sweep across the Pacific-Antarctic Ridge.[22] In 1978, Penzias and Wilson received the Nobel Prize.

Long before then, there had been a synthesis, bringing together what was known about the behaviour of elementary particles, nuclear reactions, and Einstein's theories of relativity to produce a detailed theory about the origin and evolution of the universe. The most famous summing up of these complex ideas was **Steven Weinberg's** book *The First Three Minutes*, published in 1977 and on which my account is chiefly based. The first thing that may be said about the 'singularity,' as physicists call Time Zero, is that technically all the laws of physics break down. Therefore, we cannot know exactly what happened at the moment of the Big Bang, only nanoseconds later (a nanosecond is a millionth of a second). Steven Weinberg gives the following chronology, which for ease of digestion to the layperson is set out here as a table.

After 0.0001 (10^{-4}) seconds:
  This, the original 'moment of creation,' occurred 15 billion years ago. The temperature of the universe at this near-original moment was 10^{12} K, or 1,000 billion degrees (written out, that is 1,000,000,000,000 degrees). The density of the universe at this stage was 10^{14} – 100,000,000,000,000 – grams per cubic centimetre (the density of water is 1 gram per cubic centimetre). Photons and particles were interchangeable at this point.

After 0.01 (10^{-2}) seconds:
  The temperature was 100 billion K.

After 0.1 seconds:
  The temperature was 30 billion K.

After 13.8 seconds:
  The temperature was 3 billion K, and nuclei of deuterium were beginning to form. These consisted of one proton and one neutron, but they would have soon been knocked apart by collisions with other particles.

After 3 minutes, 2 seconds:
  The temperature was 1 billion K (about seventy times as hot as the sun is now). Nuclei of deuterium and helium formed.

After 4 minutes:
  The universe consisted of 25 percent helium and the rest 'lone' protons, hydrogen nuclei.

After 300,000 years:
  The temperature was 6,000 K (roughly the same as the surface of the sun), when photons would be too weak to knock electrons off atoms. At this point the Big Bang could be said to be over. The universe expands 'relatively quietly,' cooling all the while.

After 1 million years:
  Stars and galaxies begin to form, when nucleosynthesis takes place and the heavy elements are formed, which will give rise to the Sun and Earth.[23]

At this point the whole process becomes more accessible to experimentation, because particle accelerators allowed physicists to reproduce some of the conditions inside stars. These show that the building blocks of the elements are hydrogen, helium, and alpha particles, which are helium-4 nuclei. These are added to existing nuclei, so that the elements build up in steps of 4 atomic mass units: 'Two helium-4 nuclei, for example, become beryllium-8, three helium-4 nuclei become carbon-12, which just happens to be stable. This is important: each carbon-12 nucleus contains slightly less mass than three alpha particles which went to make it up. Therefore energy is released, in line with Einstein's famous equation, $E=mc^2$, releasing energy to produce more reactions and more elements. The building continued, in stars: oxygen-16, neon-20, magnesium-24, and eventually silicon-28.' 'The ultimate step,' as Weinberg describes it, 'occurs when pairs of silicon-28 nuclei combine to form iron-56 and related elements such as nickel-56 and cobalt-56. These are the most stable of all.' Liquid iron, remember, is the core of the earth. This narrative of the early universe was brilliant science but also a great work of the imagination, the second evolutionary synthesis of the century.[24] It was more even than that, for although imagination of a high order was required, it also needed to conform to the evidence (such evidence as there was, anyway). As an intellectual exercise it was on a par with the ideas of Copernicus, Galileo, and Darwin.[25]

But background radiation was not the only form of radio waves from deep space discovered in the 1960s. Astronomers had observed many other kinds of radio activity unconnected with optical stars or galaxies. Then, in 1963, the Moon passed in front of one of those sources, number 273 in the *Third Cambridge Catalogue of the Heavens* and therefore known as 3C 273. Astronomers carefully tracked the exact moment when the edge of the Moon cut off the radio noise from 3C 273 – pinpointing the source in this way enabled them to identify the objects as 'star-like,' but they also found that the source had a very large redshift, meaning it was well outside our Milky Way galaxy. It was subsequently shown that these 'quasi-stellar' objects, or **quasars**, form the heart of distant galaxies that are so far away that such light as reaches us left them when the universe was very young, more than 10 billion years ago. What brightness there is, however, suggests that their energy emanates from an area roughly one light day across, more or less the dimensions of the solar system. Calculations show that quasars must therefore radiate 'about 1,000 times as much energy as all the stars in the Milky Way put together.' In 1967 **John Wheeler**, an American physicist who had studied in Copenhagen and worked on the Manhattan Project, revived the eighteenth-century theory of **black holes** as the best explanation for quasars. Black holes had been regarded as mathematical curiosities until relativity theory suggested they must actually exist. A black hole is an area where matter is so dense, and gravity so strong, that nothing, not even light, can escape: 'The energy we hear as radio noise comes from masses of material being swallowed at a fantastic rate.'[26]

Pulsars were another form of astronomical object detected by radio waves. They were discovered – accidentally, like background radiation – in 1967 by **Jocelyn Burnell**, a radio astronomer in Cambridge. She was using a radio telescope to study quasars when she stumbled on a completely unknown radio source. The pulses were extremely precise – so precise that at first the Cambridge astronomers thought they might be signals from a distant civilisation. But the discovery of many more showed they must be a natural phenomenon. The pulsing was so rapid that two things suggested themselves: the sources were small, and they were spinning. Only a small object spinning fast could produce such pulses, rather like a very rapid lighthouse beam coming round every so often. The small size of the pulsars told astronomers that they must be either white dwarfs, stars with the mass of the sun? packed into the size of the earth, or **neutron stars**, with the mass of the sun 'packed into a sphere less than ten kilometres across.'[27] When it was shown that white dwarfs could not rotate fast enough to produce such pulses without falling apart, scientists finally had to accept that neutron stars exist.[28] These superdense stars, midway between white dwarfs and black holes, have a solid crust of iron above a fluid inner core made of neutrons and, possibly, quarks. The density of neutron stars has been calculated by physicist John Gribbin as 1 million billion times greater than water, meaning that each cubic centimetre of such a star would weigh 100 million tons.[29] The significance of pulsars being identified as neutron stars was that it more or less completed the sequence of **stellar evolution**. Stars form as cooling gas; as they contract they get hotter, so hot eventually that nuclear reactions take place; this is known as the '**main sequence**' of stars. After that, depending on their size and when a crucial temperature is reached, quantum processes trigger a slight expansion that is also fairly stable – and the star is now a **red giant**. Toward the end of its life, a star sheds its outer layers, leaving a dense core in which all nuclear reactions have stopped – it is now a **white dwarf** and will cool for millions of years, eventually becoming a black dwarf, unless it is very large, in which case it ends as a dramatic supernova explosion, when it shines very brightly, very briefly, scattering heavy elements into space, out of which other heavenly bodies form and without which life could not exist.[30] It is these supernovae explosions that give rise to neutron stars and, in some cases, black holes. And so, the marriage of physics and astronomy – quasars and quarks, pulsars and particles, relativity, the formation of the elements, the lives of stars – was all synthesised into one consistent, coherent, story.[31]

Once one gets over the breathtaking numbers involved in anything to do with the universe, and accepts the sheer weirdness not only of particles but of heavenly bodies, one cannot escape the fact of how inhospitable much of the universe is – very hot, very cold, very radioactive, unimaginably dense. No life as we can conceive it could ever exist in these vast reaches of space. The heavens were as awesome as they had ever been, ever since man's observation of the sun and the stars began. But heaven was no longer heaven, if by that was meant the same thing as paradise.

When the crew of *Apollo 8* returned from their dangerous mission around the Moon, at the end of 1968, they gave a broadcast in which they treated earthlings to readings from the Bible. 'And the Earth was without form, and void,' read Frank Borman, quoting from Genesis.[32] 'And darkness was upon the face of the deep,' continued Bill Anders. This did not please everyone, and the American television networks were swamped with calls from viewers complaining about the intrusion of religion at such a time. But you didn't have to be much of a philosopher to see that the revolution in the study of the heavens, and the theories being propounded as a result of so many observations, both before the advent of satellites and since, could not be easily reconciled with many traditional religious ideas. Not only had man evolved; so had the very heavens themselves. The modern sciences of astrophysics and cosmology were not the only aspects of the modern world to bring about changes in religious belief, not by a long way. But they were not irrelevant either.

So far as the major religions of the world were concerned, there were three important developments after the end of World War II. Two of these concerned

Christianity, and the third involved the religions of the East, especially India. (So far as Judaism and Islam were concerned, their problems were mainly political, arising from the creation of the state of Israel in 1948.) The upsurge of interest on the part of Westerners in the religions of the East is considered in the next chapter. Here we shall examine the two main areas of thought that taxed Christianity.

These may be simply put: the continuing discoveries of science, in particular archaeological discoveries in the Middle East, what some people called the Holy Land, and existentialism. In 1947, a year before Israel was founded, there occurred the most spectacular excavation of archaeological material since the unearthing of Tutankhamen's tomb in 1922. This was the discovery of the so-called **Dead Sea Scrolls** at **Qumran**, which were first found in a cave by an Arab boy, Muhammad Adh-Dhib, chasing a wayward goat as it scampered up a rock face overlooking the inland sea. There were fewer parallels with the boys who discovered Lascaux than at first seemed because events in the wake of Muhammad's find involved far darker dealings. The area was highly unstable politically, and local traders and even religious leaders held back on the truth, hiding documents by burying them in soil so unsuitable that many were destroyed. It took months for the full picture to emerge, so that by the time trained archaeologists were able to visit the cave where Muhammad had first stumbled across the jars containing the scrolls, much of the context had been destroyed.[33]

Even so, the significance of the scrolls could not be minimised. Until that point, the last word on biblical archaeology had been **F. G. Kenyon's** *The Bible and Archaeology*, published in 1940. Broadly speaking, this argued that the thrust of science had been to confirm the biblical account, in particular that Jericho, as the Bible said had existed from around 2000 BC to 1400 BC and then been destroyed. The significance of the scrolls was more profound. They had belonged to an early sect that had existed in Palestine from perhaps 135 BC to shortly before the destruction of Jerusalem in AD 70.[34] The scrolls contained early texts from parts of the Bible, including Isaiah. At that stage, scholars were divided on how the Bible had been put together, and many thought there had been a fight in the early centuries as to what should be included and what left out. In other words, in this scenario the Bible too had evolved. But the Qumran texts showed that the Old Testament at least was already, in the first century AD, more or less written as we know it. A second and even more incendiary significance of the Qumran texts was that, as research showed, they had belonged to a very ascetic sect known as the **Essenes**, who had a Teacher of Righteousness and called themselves either the Sons of Zadok or the Children of Light.[35] Jesus wasn't referred to in the Qumran texts, and there were some marked differences between their lifestyle and his. But the existence of this extremist sect, at the very time Jesus is supposed to have lived, threw a great deal of light on the emergence of Christianity. Many of the events referred to in the Qumran documents are either exactly as described in the Bible, or thinly disguised allegories. The prospect was held out, therefore, that Jesus was a similar figure, beginning his career as the leader of just such a Jewish sect.[36]

The very authority and plausibility of this overall historical context, as expanded by recent scholarship, was most threatening to Christianity. On 12 August 1950 Pope Pius XII issued *Humani Generis*, an encyclical designed specifically to counter 'extreme non-Christian philosophies of evolutionism, existentialism, and historicism as contributing to the spread of error.'[37] Not that the encyclical was entirely defensive: the document called on Catholic philosophers and theologians to study these other philosophies 'for the purpose of combating them,' conceding that 'each of these philosophies contains a certain amount of truth.'[38] The encyclical condemned all attempts to 'empty the Genesis accounts in the Old Testament,' took the view that evolution was not as yet a proven fact, and insisted that polygenism (the idea that man evolved more than once, in several places across the earth) could not be taught (i.e., accepted), 'for it is not yet apparent how polygenism is to be reconciled with the traditional teaching of the Church on original sin.'[39] The encyclical turned existential thinking on its head, blaming Heidegger, Sartre, and the others for the gloom and anxiety that many people felt.

More lively, more original – and certainly more readable – resistance to existentialism, evolutionism, and historicism came not from the Vatican but from independent theologians who, in some cases, were themselves at loggerheads with Rome. **Paul Tillich**, for example, was a pre-eminent religious existentialist. Born in August 1886 in a small village near Brandenburg, he studied theology in Berlin, Tübingen, and Halle and was ordained in 1912. He was a chaplain in the German army in World War I and afterward, in the mid-1920s, was professor of theology at Marburg, where he came under the influence of Heidegger. In 1929 he moved to Frankfurt, where he became professor of philosophy and came into contact with the Frankfurt School.[40] His books, especially *Systematic Theology* (2 volumes, 1953 and 1957) and *The Courage to Be* (1952), had an enormous impact. A great believer in the aims of socialism, including many aspects of Marxism, Tillich was instantly dismissed when the Nazis came to power. Fortunately, Reinhald Niebuhr happened to be in Germany that summer and invited him to the Union Theological Seminary in New York.

Tillich mounted a complete rethink of Christian theology, starting from commonsense propositions – at its most basic, the fact that there *is* something, rather than nothing; that many people sense the existence of God; that there is sin (he thought Freud's libido was the modern manifestation of the driving force of sin); and that atonement for our sins is a way of approaching God.[41] Tillich thought that these feelings or thoughts were so natural that they needed no complicated explanation; in fact, he thought they were forms of reason just as much as scientific or analytic reason – he spoke of 'ecstatic reason' and 'depth of reason': 'The depth of reason is the expression of something that is not reason, but which precedes reason and is manifest through it.' He appears to be saying, in other words, that intuition is a form of reason, and evidence of the divine. Ecstatic reason was like revelation, 'numinous astonishment,' which conveyed the feeling of being 'in the grip of a mystery, yet elated with awe.'[42] The Bible and the church had existed for centuries; this needed no explanation either; it merely reflected the reality of God. Tillich followed Heidegger in believing that one had to create one's life, to create something out of nothing, as God had done, using the unique phenomenon of Christ as a guide, showing the difference between the self that existed, and the self in essence, and in doing so remove man from 'the anxiety of non-being,' which he thought was the central predicament.

When he revisited Europe after World War II, Tillich summed up his impression of the theological scene in this way: 'When you come to Europe today, it is not as it was before, with **Karl Barth** in the centre of discussion; it is now **Rudolf Bultmann** who is in the centre.'[43] In the twenty years after the war, Bultmann's '**demythologising**' made a remarkable impact on theology, an impact comparable to that made by Barth after World War I. Barth's view was that man's nature does not change, that there is no moral progress, and that the central fact of life is sin, evil. He rebelled against the beliefs of modernity that man was improving. The calamity of World War I gave great credibility and popularity to Barth's views, and in the grim years between the wars his approach became known as '**Crisis Theology**.' Man was in perpetual crisis, according to Barth, on account of his sinful nature. The only way to salvation was to *earn* the love of God, part of which was a literal belief in the Holy Bible. This new orthodoxy proved very helpful for some people as an antidote to the pseudo-religions in Nazi Germany.

Bultmann took a crucially different attitude to the Bible. He was very aware that throughout the nineteenth century, and in the first decades of the twentieth, archaeologists and some theologians had sought evidence in the Holy Lands for the events recorded in the Old and New Testaments. (One high point in this campaign had been Albert Schweitzer's *Quest for the Historical Jesus*, published in 1906.) Rather than express 'caution' about these matters, as *Humani Generis* had done, Bultmann argued that it was time to call a halt to this search. It had been futile from the start and could not hope to settle the matter one way or the other. He argued instead that the New Testament should be 'demythologised,' a term that became famous. Science had made much progress, he said, one effect of which was to suggest most strongly that the miracles of the Bible – the Resurrection, even the Crucifixion – may never have taken place *as historical events*. Bultmann knew that much of the information about Jesus in the Bible had been handed down from the Midrash, Jewish commentary and legend. He therefore concluded that the Bible could only be understood theologically. There may have been an historical Jesus, but the details of his life mattered less than that he was an example of *kerygma*, 'the proclamation of the decisive act of God in Christ.'[44] When people have faith, said Bultmann, they can enter a period of 'grace,' when they may receive 'revelations' from God. Bultmann also adapted several ideas from existentialism, but Heidegger's variety, not Sartre's (Bultmann was German). According to Heidegger, all understanding involves interpretation, and in order to *be* a Christian, one had to *decide* (an existential act) to follow that route (that's what faith meant), using the Bible as a guide.[45] Bultmann acknowledged that history posed a problem for this analysis: Why did the crucial events in Christianity take place where they did so long ago? His answer was that history should be looked upon less in a scientific way, or even in the cyclical way that some Eastern religions did, but existentially, with a meaning fashioned by each faithful individual for himself and herself. Bultmann was not advocating an 'anything goes' philosophy – a great deal of time and effort was spent with critics discussing what, in the New Testament, could and could not be demythologised.[46] Faith, he was saying, cannot be achieved by studying the history of religion, or history per se, nor by scientific investigation. Religious *experience* was what counted, and *kerygma* could be achieved only by reading the Bible in the 'demythologised' way he suggested. His final contentious point was that Christianity *was* a special religion in the world. For him, Christianity, the existence of Christ as an act of God on earth, 'has an inescapably definitive character.' He thought that at the turn of the century, 'when it seemed as if Western culture was on its way to becoming the first world-culture, it ... seemed also that Christianity was on its way to attaining a definitive status for all men.' But of course that didn't happen, and by the 1950s it appeared 'likely that for a long time yet different religions will need to live together on the earth.'[47] This was close to saying that religions evolve, with Christianity being the most advanced.[48]

If Bultmann was the most original and uncompromising theologian in his response to existentialism and historicism, **Teilhard de Chardin** fulfilled an equivalent role in regard to evolution. Marie-Joseph-Pierre Teilhard de Chardin was born on 1 May 1881, the fourth of eleven children, seven of whom died. He went to a school run by Jesuits, where he proved himself very bright but besotted by rocks more than lessons. He became a Jesuit novitiate at Aix in 1890 and took his first vows in 1901.[49] But his obsession with rocks turned into a passion for geology, palaeontology – and evolution. In his one person Teilhard de Chardin combined the great battle between religion and science, between Genesis and Darwin. His religious duties took him to China in the 1920s, 1930s, 1940s, where he excavated at Choukoutien. He met Davidson Black and Wen Chung-Pei, two of the discoverers of Peking Man and Peking Man culture. He became friendly with the Abbé Breuil, who introduced him to many of the caves and cave paintings of northern Spain, and with George Gaylord Simpson and Julian Huxley, two of the scholars who helped devise the evolutionary synthesis, and with Joseph Needham, whose seven-volume *Science and Civilisation in China* began publication in 1954. He knew and corresponded with Margaret Mead. This background was especially significant because Teilhard's chosen field, the emergence of man, the birth of humanity, profoundly affected his theology. His gifts put him in the position of reconciling as no one else could the church and the sciences, especially the science of evolution.

For Teilhard, the ideas of Darwin showed that the world had moved out of the static cosmos that applied in the days of Plato and the other Greeks, into a dynamic universe that was evolving. In consequence, religions evolved too, and man's very discovery of evolution showed that, in unearthing the roots of his own humanity, he was making spiritual progress. The supreme event in the universe was the incarnation of Christ, which Teilhard accepted as a fact. The event of Christ, he said, as a self-evidently nonevolutionary event – the only one in the history of the universe – showed its importance; and Christ's true nature, as revealed in the Scriptures, therefore served the purpose of showing what man was evolving *toward*.[50] Evolution, he believed, was a divine matter because it not only pointed backward but, allied with the event of Christ, showed us the path to come. Although Teilhard himself did not make a great deal out of it, and claimed indignantly that he was not a racist, he said clearly that 'there are some races that act as the spearhead of evolution, and others that have reached a dead end.'[51]

All his life, Teilhard planned a major work of religious and scientific synthesis, to be called *The Phenomenon of Man*. This was completed in the early 1940s, but as a Jesuit and a priest, he had first to submit the book to the Vatican. The book was never actually refused publication, but he was asked several times to revise it, and it remained unpublished at his death in 1955.[52] When it finally did appear, it became clear that for Teilhard evolution is the source of sin, 'for there can be no evolution without groping, without the intervention of chance; consequently, checks and mistakes are always possible.'[53] The very fact that the Incarnation of Christ took place was evidence, he said, that man had reached a certain stage in evolution, so that he could properly appreciate what the event meant. Teilhard believed that there would be further evolution, religious as well as biological, that there would be a higher form of consciousness, a sort of group consciousness, and in this he acknowledged an affinity for Jung's views about the racial unconscious (at the same time deriding Freud's theories). Chardin was turned down for a professorship at the Collège de France (the Abbé Breuil's old chair), but he was elected to the Institute of France.

But the church was not only concerned with theology; it was a pastoral organisation as well. It was rethinking the church's pastoral work that most concerned the other influential postwar religious thinker, **Reinhald Niebuhr**. Significantly, since pastoral work is essentially more practical, more pragmatic, than theological matters, Niebuhr was American. He came from the Midwest of America and did his early pastoral work in the capital of the motor trade, Detroit. In *The Godly and the Ungodly* (1958), he set out to rescue postwar America from what he saw as a fruitless pietism, redefining Christianity in the process, and to reaffirm the areas of life that science could never touch.[54] The chapters in his book reveal Niebuhr's anxieties: 'Pious and Secular America,' 'Frustration in Mid-Century,' 'Higher Education in America,' 'Liberty and Equality,' plus chapters on the Negro and on anti-Semitism. Niebuhr thought that America was still, in some ways, a naive country, sentimental even. He acknowledged that naïveté had certain strengths, but on the downside he also felt that America's many sectarian churches still had a frontier mentality, a form of pietism that took them away from the world rather than toward it. He saw it as his job to lead by example, to mix religion with the social and political life of America. This was how Christians showed love, he said, how they could find meaning in the world. He thought higher education was partly to blame, that the courses offered in American universities were too standardised, too inward-looking, to breed truly sophisticated students and were a cause of the intolerance that he explored in his chapters on blacks and Jews. He made it plain that pious Americans labelled everything they didn't like 'Godless,' and this did no one any good.[55]

He identified '**three mysteries**,' which, he said, remained, and would always remain. These were the mysteries of creation, of freedom, and of sin. Science might push back the moment of creation further and further, he said, but there would always be a mystery beyond any point science could reach. Freedom and sin were linked. 'The mystery of the evil in man does not easily yield to rational explanations because the evil is the corruption of a good, namely, man's freedom.'[56] He did not hold out the hope of revelation in regard to any of these mysteries. He thought that America's obsession with business was actually a curtailment of freedom, and that true freedom, the true triumph over evil, came from social and political engagement with one's fellow men, in a religious spirit. Niebuhr's analysis was an early sign of the greater engagement with sociopolitical matters that would overtake the church in the following decades, though Niebuhr, as his calm prose demonstrated, was no radical.[57]

Catholics were – in theory at least – moved by the same spirit. On 11 October 1962, 2,381 cardinals, bishops and abbots gathered in Rome for a huge conference designed to reinvigorate the Catholic Church, involve it in the great social issues of the day, and stimulate a religious revival. The conference, the **Second Vatican Ecumenical Council**, had been called back in 1959 by the then-new pope, **Angelo Giuseppe Roncalli**, who had taken the name John XXIII. Elected only on the eleventh ballot, when he was within a month of his seventy-seventh birthday, Roncalli was seen as a stopgap pope. But this short, dumpy man surprised everyone. His natural, down-to-earth manner was perfectly attuned to the mood of the times, and as the first pope of the television age, he quickly achieved a world-wide popularity no pope had had before.

Great things were expected from Vatican II, as it was called, though in more traditional quarters there was surprise that the council had been called in the first place: Vatican I had been held ninety-two years before, when its most important decision was that the pope was infallible on theological matters – for such purists there was no need of another council. Questionnaires were sent out to all the bishops and abbots of the church, inviting them to Rome and soliciting their early views on a number of matters that it was proposed to discuss. By the time the council began, one thousand aides had been added, at least a hundred official observers from other religions, and several hundred press. It was by far the largest gathering of its kind in the twentieth century.[58]

As part of the preparations, the pope's staff in Rome prepared an agenda of sixty-nine items, later boiled down to nineteen, and then thirteen. For each of these a schema was drafted, a discussion document setting out the ideas of the pope and his immediate aides. Shortly before the council began, on 15 May 1961, the pope issued an encyclical, *Mater et Magistra*, outlining how the church could become more involved in the social problems facing mankind. As more than one observer noted, neither the encyclical nor the council came too soon; as the French Dominican Yves Congar wrote, in 1961 'one man out of every four is Chinese, two men out of every three are starving, one man out of every three lives under Communism, and one Christian out of every two is not Catholic.'[59] In practice, the council was far from being an unqualified success. The first session, which began on 11 October 1962, lasted until 8 December the same year, the bishops spending two to three hours in discussion every morning. The pope issued a second encyclical, *Pacem in Terris*, the following April, which specifically addressed issues of peace in the Cold War. Sadly, Pope John died on 3 June that year, but his successor, **Giovanni Battista Montini**, Paul VI, kept to the same schedule, and three more sessions of the council took place in the autumn of 1963, 1964, and 1965.

During that time, for close observers (and the world *was* watching), the Catholic Church attempted to modernise itself. But although Catholicism emerged stronger in many ways, Rome revealed itself as virtually incapable of change. Depending on the observer, the church had dragged itself out of the Middle Ages and moved ahead either to the seventeenth, eighteenth, or nineteenth century. But no one thought it had modernised itself. One problem was the style of debate.[60] On most issues there was a 'progressive' wing and a 'reactionary' wing. This was only to be expected, but too often open discussion, and dissension, was cut short by papal fiat, with matters referred to a small papal commission that would meet later, behind closed doors. Teaching was kept firmly in the hands of the bishops, with the laity specifically excluded, and in discussions of ecumenism with Protestants and the Eastern Orthodox forms of Christianity, it was made clear that Catholicism was primary. The liturgy *was* allowed to shift from Latin to the vernacular, and some historical mistakes were admitted, but against that the church's implacable opposition to birth control was, in the words of Paul Blanshard, who attended all four sessions of the council as an observer, 'the greatest single defeat for intelligence.'[61] On such matters as biblical scholarship, the status of Mary, and women within the church, Catholicism showed itself as unwilling to change and as driven by Rome. Perhaps expectations had been raised too high by calling a council in the first place: in itself that seemed to promise greater democracy. America was now a much greater power in the world, and in the church, and Rome's way of conducting itself did not sit well with attitudes on the other side of the Atlantic.[62] Quite what effect Vatican II had on the *numbers* of Catholics around the world is unclear; but in the years that followed the rates for divorce continued to rise, even in Catholic countries, and women took their own decisions, in private, so far as birth control was concerned. In that sense, Vatican II was a missed opportunity.

For many people, the most beautiful image of the twentieth century was not produced by Picasso, or Jackson Pollock, or the architects of the Bauhaus, or the cameramen of Hollywood. It was a photograph, a simple piece of reportage, but one that was nevertheless wholly original. It was a photograph of Earth itself, taken from space. This picture, showing the planet to be slightly blue, owing to the amount of water in the atmosphere, was affecting because it showed the world as others might see us – as one place, relatively small and, above all, finite. It was that latter fact that so many found moving. Our arrival on the Moon marked the point when we realised that the world's population could not go on expanding for ever, that Earth's resources are limited. It was no accident that the ecology movement developed in parallel with the space race, or that it culminated at the time when space travel became a fact.

The ecological movement began in the middle of the nineteenth century. The original word, *oekologie*, was coined by the German Ernst Haeckel, and was deliberately related to *oekonomie*, using as a root the Greek *oekos*, 'household unit.' There has always been a close link between ecology and economy, and much of the enthusiasm for ecology was shown by German economic thinkers in the early part of the century (it formed a plank of National Socialist thinking).[63] But whether that thinking was in Germany, Britain, or the United States (the three countries where it received most attention), before the 1960s it was more a branch of thought that set the countryside – nature, peasant life – against urbanity. This was reflected in the writings of not only Haeckel but the British planners (Ebenezer Howard's garden cities and the Fabians), the Woodcraft Folk, and such writers as D. H. Lawrence, Henry Williamson, and J. R. Tolkien.[64] In Germany Heinrich Himmler experimented, grotesquely, with organic farms, but it was not until the 1960s that the modern worries came together, and when they did, they had three roots. One was the population boom stimulated by World War II and only now becoming visible; a second was the wasteful and inhuman planning processes created in many instances by the welfare state, which involved the wholesale destruction of towns and cities; and third, the space race, after which it became common to refer to the planet as 'spaceship Earth.'

When President Johnson made his Great Society speech in Michigan in the spring of 1964, he referred to the impoverished environment as one of his reasons for acting. Partly, he had in mind the destruction of the cities, and the 'Great Blight of Dullness' that Jane Jacobs had railed against. But he was also provoked by the writings of another woman who did much to stir the world's conscience with a passionate exposé of the pesticide industry and the damage commercial greed was doing to the countryside – plants, animals, and humans. The exposé was called *Silent Spring*, and its author was Rachel Carson.[65]

Rachel Carson was not unknown to the American public in 1962, when her book appeared. A biologist by training, she had worked for many years for the U.S. Fish and Wildlife Service, which had been created in 1940. As early as 1951 she published *The Sea Around Us*, which had been serialised in the *New Yorker*, a Book-of-the-Month Club alternative choice and top of the *New York Times* best-seller list for months. But that book was not so much a polemic as a straightforward account of the oceans, showing how one form of life was dependent on others, to produce a balance in nature that was all-important to both its continued existence and its beauty.[66]

Silent Spring was very different. As Linda Lear, her biographer, reminds us, it was an angry book, though the anger was kept in check. As the 1950s passed, Carson, as a scientist, had gradually amassed evidence – from journals and from colleagues – about the damage pesticides were doing to the environment. The 1950s were years of economic expansion, when many of the scientific advances of wartime were put to peaceful use. It was also a period when the Cold War was growing in intensity, and would culminate at the very time *Silent Spring* appeared. There was a tragic personal dimension behind the book. At about the time *The Sea around Us* appeared, Carson had been operated on for breast cancer. While she was researching and writing *Silent Spring*, she was suffering from a duodenal ulcer and rheumatoid arthritis (she was fifty-three in 1960), and her cancer had reappeared, requiring another operation and radiotherapy. Large chunks of the book were written in bed.[67]

By the late 1950s, it was clear to those who wished to hear the evidence that with the passage of time, many pollutants that formed part of daily life had toxic side effects. The most worrying, because it directly affects humans, was tobacco. Tobacco had been smoked in the West for three hundred years, but the link between cigarette smoking and lung cancer was not fully aired until 1950, when two reports, one in the *British Medical Journal* and the other in the *Journal of the American Medical Association*, both showed that 'smoking is a factor, and an important factor, in the production of carcinoma of the lung.'[68] This result was surprising: the British doctors doing the experiment thought that other environmental factors – automobile exhaust and/or the tarring of roads – were responsible for the rise in lung cancer cases that had been seen in the twentieth century. But no sooner had the British and American results appeared than they were confirmed in the same year in Germany and Holland.

From the evidence that Carson was collecting, it was becoming clear to her that some pesticides were far more toxic than tobacco. The most notorious was DDT, introduced to great effect in 1945 but now, after more than a decade, implicated not just in the deaths of birds, insects, and plants but also in cancerous deaths in humans. An especially vivid example explored by Carson was Clear Lake in California.[69] Here DDD, a variant of DDT, had been introduced in 1949 to rid the lake of a certain species of gnat that plagued fishermen and holidaymakers. It was administered carefully, as was thought: the concentration was 1 part in 70 million. Five years later, however, the gnat was back, and the concentration increased to 1 in 50 million. Birds began to die. The association wasn't understood at first, however, and in 1957 more DDD was used on the lake. When more birds and then fish began to die, an investigation was begun – which showed that certain species of grebe had concentrations of 1,600 parts in a million, and the fish as much as 2,500 in a million. Only then was it realised that some animals accumulate concentrations of chemicals, to lethal limits.[70] But it wasn't just the unanticipated build-up of chemicals that so alarmed Carson; each case was different, and often human agency was involved. In the case of aminotriazole, a herbicide, it had been officially sanctioned for use on cranberry bogs, but only *after* the berries had been harvested. This particular sequence mattered because laboratory studies had shown that aminotriazole caused cancer of the thyroid in rats. When it emerged that some growers sprayed the cranberries before they were harvested, the herbicide could not be blamed directly.[71] This is why, when *Silent Spring* appeared in 1962, again serialised in the *New Yorker*, the book created such a furore. For Carson not only explored the science of pesticides, showing that they were far more toxic than most people realised, but that industry guidelines, sometimes woefully inadequate in the first place, were often flouted indiscriminately. She revealed when and where specific individuals had died, and named companies whose pesticides were responsible, in some cases accusing them of greed, of putting profit before adequate care for wildlife and humans.[72] Like *The Sea Around Us,* *Silent Spring* shot to the top of the best-seller lists, helped by the thalidomide scandal, which erupted at that time, when it was shown that certain chemicals taken (for sedation or sleeplessness) by mothers in the early stages of pregnancy could result in deformed offspring.[73] Carson had the satisfaction of seeing President Kennedy call a special meeting of his scientific advisory committee to discuss the implications of her book before she died, in April 1964.[74] But her true legacy came five years later. In 1969, the U.S. Congress passed the National Environmental Policy Act, which required an environmental impact statement for each governmental decision. In the same year the use of DDT as a pesticide was effectively banned, and in 1970 the Environmental Protection Agency was established in the United States, and the Clean Air Amendment Act was passed. In 1972 the United States passed the Water Pollution Control Act, the Coastal Zone Management Act, and the Noise Control Act, with the Endangered Species Act approved in 1973.

By then, thirty-nine nations had met in Rome in 1969 to discuss pollution. Their report, *The Limits to Growth,* concluded that 'the hour was late,' that at some stage in the next one hundred years the limit to growth would be reached, that the earth's finite resources would be exhausted and there would be a 'catastrophic' decline in population and industrial capacity.[75] Attempts to meet this looming problem should begin immediately. In the same year **Barbara Ward** and **René Dubos** presented a report to the United Nations World Conference on the Human Environment which, as its title, *Only One Earth,* showed, had much the same message.[76] Nineteen-seventy saw the founding of the 'Bauernkongress' in Germany, and in 1973 ecology candidates first stood for election in France and Britain. These events coincided with the Yom Kippur War in 1973, as a result of which the OPEC cartel of oil-producing nations raised oil prices sharply, an oil crisis that forced gasoline rationing in several countries, the first time such measures had been necessary since World War II. It was this, as much as anything, that highlighted not just the finite nature of the earth's resources, but that such limits to growth had political consequences.

Charles Reich, an academic who taught at both Yale and Berkeley, claimed that the environmental revolution was more than just that; it was a true turning point in history, a pivot when human nature changed. In *The Greening of America* (1970), he argued that there existed, in America at any rate, three types of consciousness: 'Consciousness I is the traditional outlook of the American farmer, small businessman and worker who is trying to get ahead. Consciousness II represents the values of an organisational society. Consciousness III is the new generation. ... One was formed in the nineteenth century, the second in the first half of this century, the third is just emerging.'[77]

Beyond this division, Reich's idea was a very clever piece of synthesis: he related many works of popular culture to his arguments, explaining why particular songs or films or books had the power and popularity they did. He paid relatively little attention to Consciousness I but had great fun debunking Consciousness II, where his argument essentially followed on from Herbert Marcuse, in *One-Dimensional Man*, and W. H. Whyte's *Organisation Man*. Since the mid-1950s, Reich said, that world had deteriorated; in addition to vast organisations, we now had the 'corporate state,' with widespread, anonymous, and in many cases seemingly arbitrary power. He argued that the works of Raymond Chandler, such as *The Big Sleep* or *Farewell My Lovely*, owed their appeal to their picture of a world in which no one could be trusted, where one could only survive by living on one's wits. James Jones's *From Here to Eternity* pitted a young man against a vast, anonymous organisation (in this case the army), as did Philip Roth's *Portnoy's Complaint*. The appeal of *Casablanca*, he said, lay in the fact that 'Humphrey Bogart plays a man who could still change fate by taking action. Perhaps *Casablanca* was the last moment when most Americans believed that.'[78]

Reich showed how a large number of popular works took aim at one or other aspect of Consciousness II society, and tried to move on. In Stanley Kubrick's *2001: A Space Odyssey*, a space traveller is in what appears to be a hotel or motel room, expensive and plastic but lacking entirely anything he can do anything *with*, 'no work, nothing that asks a reaction.'[79] 'Almost every portrayal of a man at work [in American films] shows him doing something that is clearly outside of modern industrial society [i.e., the corporate state]. He may be a cowboy, a pioneer settler, a private detective, a gangster, an adventure figure like James Bond, or a star reporter. But no films attempt to confer satisfaction and significance upon the ordinary man's labour. By contrast, the novels of George Eliot, Hardy, Dickens, Howells, Garland and Melville deal with ordinary working lives, given larger meaning through art. Our artists, our advertisers and our leaders have not taught us how to work in our world.'[80] The beginning of Consciousness III he took to be J. D. Salinger's *Catcher in the Rye* (1951) but to gather force with the music and words of Bob Dylan, Cream, the Rolling Stones, and Crosby, Stills and Nash. Dylan's 'It's All Right Ma (I'm only Bleeding),' Reich said, was a far more powerful, and considerably earlier, social critique of police brutality than any number of sociological treatises. 'Eleanor Rigby' and 'Strawberry Fields Forever' said more about alienation more succinctly than any psychologist's offering. The same argument, he said, applied to such works as 'Draft Morning' by the Byrds, *Tommy* by the Who, or 'I Feel Free' by Cream. He thought that the drug culture, the mystical sounds of Procul Harum, and even bell-bottom trousers came together in a new idea of community (the bell-bottoms, he said, somewhat fancifully, left the ankles free, an invitation to dance). The works of authors like Ken Kesey, who wrote *One Flew over the Cuckoo's Nest* (1979) about a revolt in a mental hospital, embodied the new consciousness, Reich said, and even Tom Wolfe, who in *The Kandy-Kolored Tangerine-Flake Streamline Baby* (1965) was critical of many aspects of the new consciousness, at least conceded that subcultures like stock-car racing and surfing showed people *choosing* their own alternative lifestyles, rather than simply accepting what was given them, as their parents had.

This all came together, Reich said, in **the 'green' movement**. Opposition to the Vietnam War was an added factor, but even there the underlying force behind the war was corporate America and technology; napalm destroyed the environment and the enemy almost equally. And so, along with a fear for the environment, a realisation that resources were finite, and a rejection of the corporate state, went an avoidance where possible of the technology that represented Consciousness II. People, Reich said, were beginning to choose to bake their own bread, or to buy only bread that was baked in environmentally friendly ways, using organically grown materials. He was in fact describing what came to be called the counterculture, which is explored in more detail in the next chapter. He wasn't naïve; he did not think Consciousness II, corporate America, would just roll over and surrender, but he did believe there would be a growth of environment-conscious communes, green political parties, and a return to 'vocations,' as opposed to careers, with people devoting their lives to preserving areas of the world from the depredations of Consciousness II corporations.

A related argument came from the economist **Fritz Schumacher** in two books, *Small Is Beautiful* (1973) and *A Guide for the Perplexed* published in 1977, the year he died.[81] Born in Bonn in 1911, into a family of diplomats and academics, Schumacher was given a very cosmopolitan education by his parents, who sent him to the LSE in London and to Oxford. A close friend of Adam von Trott, executed for his part in the attempt on Hitler's life in July 1944, Schumacher was working in London in the late 1930s and spent the war in Britain, overcoming his enemy alien status. After the war, he became very friendly with Nicholas Kaldor and Thomas Balogh, economic advisers to Prime Minister Harold Wilson in the 1960s, and was appointed to a senior position on the National Coal Board (NCB). Very much his own man, Schumacher early on saw that the resources of the earth were finite, and that something needed to be done. For many years, however, he was not taken seriously because, in being his own man, he took positions that others regarded as outlandish or even as evidence of instability. He was a convinced believer in unidentified flying objects, flirted with Buddhism, and though he had rejected religion as a younger man, was received into the Catholic Church in 1971, at the age of sixty.[82]

Schumacher had spent his life travelling the world, especially to the poorer parts, such as Peru, Burma, and India. Gradually, as his religious feelings grew, as the environmental crisis around him deepened, and as he realised that the vast corporations of the West could not hope to offer solutions that would counter the poverty of so many third-world countries, he developed an alternative view. 1971 was for him a turning point. He had recently become president of the Soil Association in Britain (he was an avid gardener), he had been received into the church, and he had resigned from the NCB. He set about writing the book he had always wanted to write, provisionally called 'The Homecomers,' because his argument was that the world was reaching a crisis point. The central reality, as he saw it, was that the affluence of the West was 'an abnormality which "the signs of the times" showed was coming to an end.' The inflation that had started to plague Western societies was one such sign. The party was over, said Schumacher, but 'Whose party was it anyhow? That of a small minority of countries and, inside those countries, that of a minority of people.'[83] This minority kept itself in power, as was to be expected, but the corporations did little to help the chronic poverty seen in the rest of the world. These countries could not go from their underdeveloped state to a sophisticated state overnight. What was needed, he said, was a number of small steps, which were manageable by the people on the ground – and here he introduced his concept of **intermediate technology**. There had been an Intermediate Technology Development Group in Britain since the mid-1960s, trying to develop technologies that were more efficient than traditional ones, in India, say, or South America, but far less complex than their counterparts in the developed West. (A classic example of this would be the clockwork radio, which works by being wound up rather than with batteries, which may not be obtainable in remote areas, or may weather badly.) By 'The Homecomers' he meant that people would in the future return to their homes from factories, go back to simpler technologies simply because they were more human and humane. The publishers didn't like the title and Anthony Blond came up with *Small Is Beautiful*, at the same time keeping Schumacher's subtitle: 'Economics – as if People Mattered.' The book was published to a scattering of reviews, but it soon took off as word of mouth spread, and it became a cult from Germany to Japan.[84] Schumacher had hit a nerve; his main focus was the third world, but it was clear that many people loathed the big corporations as much as he did and longed for a different way of life. Until his death in 1977, Schumacher was a world figure, fêted by state governors in America, entertained at the White House by President Carter, welcomed in India as a 'practical Gandhi.' His underlying argument was that there *is* room on earth for everyone, provided that world affairs are managed properly. That management, however, was not an economic question but a moral one, which is why for him economics and religion went together, and why they were the most important disciplines.[85] Schumacher's arguments illustrated Reich's Consciousness III at its most practical level.

Anxieties about the human influence on our planet accelerated throughout the 1970s, aided by a scare in Italy in 1976 when a massive cloud of dioxin gas escaped from a pesticide plant near Seveso, killing domestic and farm animals in the surrounding region. In 1978 the United States banned CFCs as spray propellants in order to reduce damage to the ozone layer, which normally filtered out ultraviolet radiation from the sun. This damage, it was believed, was causing global warming through the 'greenhouse effect.' In 1980 the World Climate Research Program was launched, a study specifically intended to explore human influence on climate and to predict what changes could be expected.

No one has been to the Moon for more than a quarter of a century. We have lost the universal sense of optimism in science that the Apollo program represented.

PART FOUR

THE COUNTER-CULTURE TO KOSOVO

The View from Nowhere, The View from Everywhere

33

A NEW SENSIBILITY

On Saturday, 6 October 1973, on the fast of Yom Kippur, the Day of Atonement, the holiest day in the Jewish calendar, a surprise attack on Israel was launched from Syria in the north and Egypt in the south. For forty-eight hours the very existence of Israel appeared threatened. Its 'Bar-Lev' line in Sinai was broken, and many of its military aircraft were destroyed on the ground by Arab missiles. Only the rapid response of the United States, sending more than $2 billion worth of arms inside two days, enabled Israel eventually to recoup its losses and then hit back and gain ground. When a ceasefire was declared on 24 October, Israeli forces were close enough to Damascus to shell it, and had a bridgehead on the Western bank of the Suez Canal.

But the Yom Kippur War, as it came to be called, was more than just a war. It was a catalyst that led directly and immediately to an event that Henry Kissinger, at the time the U.S. secretary of state, called 'one of the pivotal events in the history of this century.' In the very middle of the war, on 16 October, the Arab and several non-Arab oil-producing nations cut oil production, and raised prices 70 percent. Two days before Christmas, they raised them again, this time by 128 percent. Crude oil prices thus quadrupled in less than a year.[1] No country was immune from this 'oil crisis.' Many poorer countries of Africa and Asia were devastated. In the West, petrol rationing was introduced in places like Holland for a while, and lines at gas stations became familiar everywhere. And it introduced a phenomenon not anticipated by Keynes – stagflation. Before the Yom Kippur War the average rate of growth in the developed West was 5.2 percent, comfortably ahead of the average rate of price increases of 4.1 percent. After the oil shock, growth was reduced to zero or even minus, but inflation rose to 10 or 12 percent.[2]

The oil crisis, in historian Paul Johnson's words, was 'by far the most destructive economic event since 1945.' But the oil-producing nations' decision to raise prices and limit production were not only the result of the war or the fact that – in the end – they were defeated and lost territory as a result of America coming to Israel's aid. The world's economic structure was changing anyway, if less obviously. Ironically, the year of rebellion, 1968, when black and student violence had peaked in America, was also the time of the United States' greatest economic influence. In that year, American production was more than a third of the world total – 34 percent. But, like many success stories, this one hid within it incipient problems. Ever since 1949 the Communist Chinese had been worried that America might, in crisis, block any dollars they earned. They had therefore always kept their dollars in Paris. Over the years others had followed suit, and a market in 'Eurodollars' had grown up. In turn this spawned a Eurocredit and Eurobond market beyond the control of Washington, or anybody, helping to make money more volatile than it had ever been. Alongside this were two additional factors. One was the ecological sentiment that the earth was a finite resource, which translated into a steady rise in commodity prices. Second was a specific instance of this: from 1970, America's own oil production peaked and then began to decline. In 1960 she had imported 10 percent of her oil; by 1973 that figure was 36 percent.[3] A major shift was taking place in the very nature of developed societies. It had been gathering pace, and visibility, throughout the 1960s, but it took a war to bring it home to everyone.

One of the first to reflect on this change, in his usual elegant way, was the economist **J. K. Galbraith**. In 1967 he released *The New Industrial State*, in which he described a new business-economic order that, he maintained, drastically changed the nature of traditional capitalism. His starting point was that the nature of the large business enterprise had altered fundamentally by the 1960s as compared with the start of the century.[4] Whereas the likes of Ford, Rockefeller, Mellon, Carnegie, and Guggenheim had been entrepreneurs, taking huge risks to launch the companies that bore their names, by the time these companies had matured, they had changed character in two fundamental ways. In the first place, they were no longer run by one man, who was both a leader and a shareholder, but by managers – Galbraith actually called them the technostructure, for reasons that will become apparent – who owned a minority of shares. One important result of this, says Galbraith, is that the shareholders nowadays have only nominal control over the company that, in theory, they own, and this has significant psychological consequences for democracy. Second, mature companies, mass-producing expensive and complex products, in fact have very little interest in risk or competition. On the contrary they require political and economic stability so that demand, and the growth in demand, can (within certain limits) be predicted. The most important effect of this, Galbraith argued, is that mature corporations actually *prefer* planning in an economy. In traditional conservatism, planning smacks of socialism, Marxism, and worse, but in the modern world mature corporations, who operate in an oligopolistic situation, which to Galbraith is but a modified monopoly, cannot do without it.[5]

Everything else in the new industrial state, says Galbraith, stems from these two facts. Demand is regulated, as Keynes showed, partly by the fiscal policy of governments – which presupposes a symbiotic relationship between the state and the corporation – and by devices such as advertising (which, Galbraith believes, has had an incalculably 'dire' effect on the truthfulness of modern society, to the point where we no longer notice how routinely dishonest we are). An additional characteristic of modern industrial society, Galbraith says, is that more and more important decisions depend on information possessed by more than one individual. Technology has a great deal to do with this. One consequence is a new kind of specialism: people who have no special skills in the traditional sense but instead have a new skill – knowing how to evaluate information. Thus information becomes important in itself, and people who can handle information constitute an '**insider class**,' the managers or techno-structure, alongside an '**outsider class**,' the shareholders.[6] Galbraith clearly thought this distinction was more important than, in practice, it turned out to be (though for a while, in the 1980s, 'insider trading' was a scandal that contaminated business life on both sides of the Atlantic). One effect of all this, he said, was to change the business experience. Instead of being rugged, individualistic, competitive, and risk-taking, executive life became highly secure; when Galbraith wrote his book, recent studies had shown that in America three-quarters of executives surveyed had been with their company more than twenty years. Affluence plays a part, says Galbraith, because the further a man is from the breadline – the more affluent he is – the more his desires may be manipulated, and the bigger the role of advertising, and here it was fortunate that the rise of radio and then television coincided with the maturation of corporations and the rise of affluence.[7]

But Galbraith's aim was not simply to describe the new arrangement, important though that was. With an appropriate sense of mischief, he observed how the technostructure, the management of the mature corporations, presents itself. Far from telling the truth about the new state of play, where in fact the corporations rule the roost, the technostructure pays lip service to the idea that the 'consumer is king.' The real truth, that the corporation has pretty near total control over prices and a good grasp on the control of demand, goes by the board.[8] Galbraith's next point was that the nature of unemployment was changing – indeed, in a sense, it was starting to lose any meaning; 'More and more, the figures on unemployment enumerate those who are currently un-employable by the industrial system.'[9] This has a domino effect among the un-ions, who lose power, and the educational and scientific 'estates,' which gain it. Galbraith was undoubtedly correct in his analysis of the relative powers of the unions, the education services, and the scientists; where he was wrong was that he expected the latter two estates to acquire a political force, as the unions had been hitherto. This didn't happen. He also thought that scientists working for private companies would become a voice in society. That didn't happen either.

After a swipe at the defence industry, examining how the Cold War actually helped economies in a Keynesian sense (though traditional conservatives denied it), Galbraith suddenly changed tack completely and considered what he called the '**aesthetic experience**.' The world of artists, he says, is quite unlike that of the technostructure: 'Artists do not come in teams.' Athens, Venice, Agra, and Samarkand are quite unlike Nagoya, Düsseldorf, Dagenham, or Detroit and always will be. He saw it as the role of artists to attack and criticise the technostructure; there is, he says, an inevitable struggle: 'Aesthetic achievement is beyond the reach of the industrial system and, in substantial measure, in conflict with it. There would be little need to stress the conflict were it not part of the litany of the industrial system that none exists.'[10] Galbraith felt that aesthetic goals would ultimately prevail over industrial ones.

But the main argument of *The New Industrial State* was that traditional capitalism had changed out of all recognition and that traditional capitalists lied about that change, pretending it just hadn't happened. At the time his book went to press, Galbraith said, Boeing 'sells 65 percent of its output to the government; General Dynamics sells a like percentage; Raytheon ... sells 70 percent; Lockheed ... sells 81 percent; and Republican Aviation ... sells 100 percent.'[11] 'The future of the industrial system is not discussed partly because of the power it exercises over belief. It has succeeded, tacitly, in excluding the notion that it is a transitory, which would be to say that it is a somehow imperfect, phenomenon. ... Among the least enchanting words in the business lexicon are planning, government control, state support and socialism. To consider the likelihood of these in the future would be to bring home the appalling extent to which they are already a fact. And it would not be ignored that these grievous things have arrived, at a minimum with the acquiescence and, at a maximum, on the demand, of the system itself.' And finally: 'There is no natural presumption in favour of the market; given the growth of the industrial system the presumption is, if anything, the reverse. And to rely on the market where planning is required is to invite a nasty mess.'[12] Galbraith's was a spirited attack, making some uncomfortable points about the way capitalism had developed, and presented itself. He foresaw the increased role of science, the overwhelming importance of information and the changing nature of unemployment and the skills that would be needed in the future.

What Galbraith missed, **Daniel Bell** brought centre stage. In his study of Bell, Malcolm Waters describes how in 1973 both men featured in a list compiled by the sociologist Charles Kadushin, who had carried out a survey to discover which individuals were regarded as America's intellectual elite. Among the top ten were Noam Chomsky, J. K. Galbraith himself, Norman Mailer, and Susan Sontag, with Hannah Arendt and David Riesman further down, and W. H. Auden and Marshall McLuhan even lower. There was just one sociologist in the top ten: Daniel Bell.

Bell's *End of Ideology* was covered in chapter 25, on the new psychology of affluence. In 1975 and again a year later, he came up with two more 'big ideas.' The first was summed up in the title of his book *The Coming of Post-Industrial Society*. For Bell, life is divided into three '**realms**' – nature, technology, and society – that determine the basics of experience. History is also divided into three. Pre-industrial society may be seen as 'a game against nature,' the attempt to extract resources from the natural environment, where the main activities are hunting, foraging, farming, fishing, mining, and forestry.[13] Industrial society is 'a game against fabricated nature,' centring on human–machine relationships, and with economic activity focusing on the 'manufacturing and processing of tangible goods,' the central occupations being the semiskilled factory worker and engineer.[14] A post-industrial society is a 'game between persons,' 'in which an "intellectual technology," based on information, rises alongside machine technology.'[15] Post-industrial society centres around industries from three sectors – transportation and utilities; finance and capital exchange; health, education, research, public administration, and leisure. Among all these, says Bell, scientists 'are at the core': 'Given that the generation of information is the key problem and that science is the most important source of information, the organisation of the institutions of science, the universities and research institutes is the central problem in the post-industrial society. The strength of nations is given in their scientific capacity.'[16] As a result the character of work has changed, focusing now on relationships between people rather than between people and objects; 'the expansion of the service sector provides a basis for the economic independence of women that was not previously available'; the post-industrial society is meritocratic; there is a change in scarcity – 'scarcity of goods disappears in favour of scarcity of information and time.' Finally Bell identifies something he labels a **situs**, a 'vertical order of society as opposed to a horizontal one,' such as classes. Bell identifies four functional situses (scientific, technological, administrative, and cultural) and five institutional ones (business, government, university/research, social welfare, and military), a division that would be eerily paralleled in the organisation of e-mail (see chapter 42). Besides the situses, however, Bell identifies a 'knowledge class' (of, mainly, scientists). He points out, for example, that whereas only about a quarter of first degrees in the United States are in science, more than half the doctorates are in natural science and mathematics.[17] This knowledge class is crucial to the success of the post-industrial society, but Bell remained uncertain as to whether it would ever act *as a class* in the Marxist sense, because it would probably never have enough independence to undermine capitalism.★

A further factor of significance, Bell says, is that intellectual property is owned not individually but communally. This means that politics become more, and not less, important, because the planning, which maximises scientific output, requires national rather than regional or local organisation. 'Politics therefore becomes the "cockpit" of the post-industrial society, the visible hand that coordinates where the market can no longer be effective.'[18]

Bell's third 'big idea,' published a year later, in 1976, was *The Cultural Contradictions of Capitalism*. This too had three themes bound together by the thesis that contemporary society is dominated by irreconcilable contradictions. These were: (1) the tension between the asceticism of capitalism (as defined by Max Weber) and the acquisitiveness of later forms of capitalism; (2) the tensions between bourgeois society and modernism – modernism, through the avant-garde, was always attacking bourgeois society (the rejection of the past, the commitment to ceaseless change, and the idea that nothing is sacred); and (3) the separation of law from morality, 'especially since the market has become the arbiter of all economic and even social relations (as in corporate obligations to employees) and the priority of the legal rights of ownership and property over all other claims, even of a moral nature, has been renewed.'[19]

Put another way, for Bell there is a contradiction between the drive for efficiency in modern capitalism and the drive for self-realisation in modern

---

★ The economist Robert Solow made essentially the same observation in his work on growth theory.

culture. Culture is all-important for Bell, first, because art has taken upon itself (under the mantle of modernism) constantly to seek 'innovative forms and sensations,' and second, because culture is now no longer a source of authoritative morality but 'a producer of new and titillating sensations.'[20] For Bell, modernism was ending by 1930 and exhausted by 1960. 'Society and art have come together in the market so that aesthetic aura and conceptions of high culture have disappeared.' But the endless quest for novelty was taken up by the mass media, which themselves largely took form in the 1920s and adopted the same quest of feeding new images to people, unsettling traditional conventions and 'highlighting . . . aberrant and quirky behaviour.'[21] Along the way, the traditional sociological categories of age, gender, class, and religion became less reliable guides to behaviour – 'lifestyle, value-choice and aesthetic preference have become more idiosyncratic and personal.'[22] The result, says Bell, is chaos and disunity. In the past, most cultures and societies were unified – classical culture in the pursuit of virtue, Christian society unified around divinely ordained hierarchies, and early industrial culture unified around 'work, order, and rationalisation.' In contemporary society, however, there has been a massive dislocation. While the techno-economic side of things is still ruled by 'efficiency, rationality, orderliness, and discipline . . . the culture is governed by immediate gratification of the senses and the emotions and the indulgence of the undisciplined self.' The contradictions, for Bell, imply a major change in the way we live, but this has to do not only with capitalism: 'The exhaustion of modernism, the aridity of Communist life, the tedium of the unrestrained self, and the meaninglessness of monolithic political chants, all indicate that a long era is coming to a slow close.' There is a heavy price to pay for modernism: 'Modernity is individualism, the effort of individuals to remake themselves, and, where necessary, to remake society in order to allow design and choice. . . . It implies the rejection of any "naturally" ascribed or divinely ordained order, of external authority, and of collective authority in favour of the self as the sole point of reference for action.'[23] 'Under modernity there can be no question about the moral authority of the self. The only question is that of how the self is to be fulfilled – by hedonism, by acquisitiveness, by faith, by the privatisation of morality or by sensationalism.'[24] Technology, of course, had something to do with this change, in particular the automobile. 'The closed car became the *cabinet particulier* of the middle class, the place where adventurous young people could shed their sexual inhibitions and break the old taboos.'[25] Advertising also played its part, 'emphasising prodigality rather than frugality, or lavish display rather than asceticism.' Financial services helped, so that debt, once a source of shame, became a component of the lifestyle.[26]

Perhaps Bell's most profound point is that modern culture emphasises experience, with the audience placed central. There is no longer any sense in which the audience engages in a dialogue with the artist or the work of art. And because the appeal is to the emotions, once the experience is over, it is over. There is no dialogue to be continued inside the head of the members of the audience. For Bell, this means that the modern society, in effect, has no culture.

★

Theodore Roszak disagreed. For him, and countless others, the changes described by Galbraith and Bell had provoked a shift in the very nature of culture, so much so that they needed a new term, the **counter-culture**.

One way of looking at the counter-culture is to regard it as one of the 'soft landings' of the New Left that formed in several Western countries in the late 1950s and early 1960s, brought about, as we have seen, by disillusionment with the Soviet Union and the horrors of Stalinism, and especially the brutal Soviet invasion of Hungary in 1956. But the other important influence was the discovery of some early writings of Marx, the so-called Economic and Philo-sophical Manuscripts, written in 1844 but published only in 1932. These new papers did not catch on generally until after World War II and the 1950s when neo-Marxists, as they were called, were trying to develop a more humanist form of Marxism. In the United States there was an additional factor; there the birth of the New Left is usually traced to the Port Huron Statement, a manifesto issued in 1962 by Students for a Democratic Society (SDS), which read in part, 'We regard *men* as infinitely precious and possessed of unfulfilled capacities for reason, freedom and love. ... We oppose the depersonalisation that reduces human beings to the status of things. ... Loneliness, estrangement, isolation describe the vast distance between man and man today. These dominant tendencies cannot be overcome by better personnel management, nor by improved gadgets, but only when a love of man overcomes the idolatrous worship of things by man.'[27] The concept of alienation underpinned the counter-culture, which like the Beats, another progenitor, rejected the main concepts of mass society. Other influences were C. Wright Mills, in *The Power Elite*, and David Riesman, in *The Lonely Crowd*. Very rapidly a whole 'alternative' set of media was created to disseminate its ideas – newspapers (such as the San Francisco-based *Journal for the Protection of All Beings*), films, plays, music, and the *Whole Earth Catalogue*, which taught how to live off the land and avoid engagement with 'mainstream' society. These ideas were set down by Roszak, a professor of history at California State University, in 1970, in *The Making of a Counter Culture*.[28]

Roszak makes it clear that the counter-culture is a *youth* revolt and, as much as anything, is opposed to the reductionism of science and technology. Youth, *especially* educated youth, Roszak said, loathed the direction in which 'tech-nocratic' society was headed, and the form of its protest was to mount an alternative lifestyle. It was a living embodiment of the cultural contradictions of capitalism. For Roszak, the counter-culture had five elements: a variety of alternative psychologies; Eastern (mystical) philosophy; drugs; revolutionary sociology; rock music. Together, these were supposed to provide a viable basis for an alternative way of life to technocratic society, in the form of communes of one sort or another, which also helped counter the alienation of 'normal' life. Aspects of this counter-culture included free universities, free clinics, 'food conspiracies' (to help the poor), an underground press, 'tribal' families. 'Everything,' says Roszak, 'was called into question: family, work, education, success, child-rearing, male–female relations, sexuality, urbanism, science, tech-nology, progress. The means of wealth, the meaning of love, the meaning of life – all became issues in need of examination. What is "culture"? Who decides what "excellence" is? Or "knowledge," or "reason"?[29]

After an opening chapter criticising reductionist science, and the way it produced a 'one-dimensional' society, deeply unsatisfying to many people (he records in loving detail the numbers of British students turning away from science courses at university), Roszak addressed the main agenda of the counter-culture, 'the subversion of the scientific world view, with its entrenched commitment to an egocentric and cerebral mode of consciousness. . . . In its place, there must be a new culture in which the non-intellective capacities of the personality – those capacities that take fire from visionary splendour and the experience of human communion – become the arbiters of the good, the true, and the beautiful.'[30] In essence, Roszak says, class consciousness gives way 'as a generative principle' to *consciousness* consciousness.[31] 'One can discern,' he argues, 'a continuum of thought and experience among the young which links together the New Left sociology of Mills, the Freudian Marxism of Herbert Marcuse, the Gestalt-therapy anarchism of **Paul Goodman**, the apocalyptic body mysticism of **Norman Brown**, the Zen-based psychotherapy of **Alan Watts**, and finally **Timothy Leary's** . . . narcissism, wherein the world and its woes may shrink at last to the size of a mote in one's private psychedelic void. As we move along the continuum, we find sociology giving way to psychology, political collectivities yielding to the person, conscious and articulate behaviour falling away before the forces of the non-intellective deep.'[32] All this, he says, amounts to an intellectual rejection of the Great Society.

Roszak's first stop, having set the scene, is Marcuse and Brown, whose significance lies in their claim that alienation is a psychological condition, not a sociological one, as Freud said. Liberation is personal, not political, and therefore resolution is to be found in changing society by creating first a set of individuals who are different – liberated in, say, a sexual sense, or freed from the 'performance principle,' i.e., having to perform in a certain prescribed way (at work, for example). Whereas Marx thought that the 'immiserisation' of man came when he was confined by poverty, Marcuse argued that psychological immiserisation came at the time of maximum affluence, with people governed by acquisitiveness and 'subtle technological repression.' Roszak makes room for one sociologist, Paul Goodman, whose main skill was an 'inexhaustible capacity to imagine new social possibilities.'[33] Goodman's role in the counter-culture was to imagine some practical 'alternative' solutions and institutions that might replace those dominating the technocratic society. Among these were the free universities and 'general strikes for peace.' But above all there was Goodman's idea of Gestalt therapy, the basic idea of which was that people should be treated as a whole, not just by their symptoms. This meant accepting that certain forces in society are irreconcilable and that, for example, violence may be necessary to resolve a situation rather than burying one's feelings of anger and guilt. In Gestalt therapy you do not talk out your feelings, you act them out.

Abraham Maslow, another psychologist, was also part of the counter-culture. In *The Psychology of Science* (1966), taking his cue from Michael Polanyi's

Personal Knowledge (1959) and Thomas Kuhn's *Structure of Scientific Revolutions* (1962), Maslow put forward the view that there is no such thing as objectivity, even in the physical sciences.[34] The 'discovery' of order is really an imposition of order on an untidy world and corresponds more to the scientist finding 'beauty' in, say, tidiness rather than to any real order 'out there' in an objective sense. The imposition of order undervalues subjective experience, which is as real as anything we know. There are, say Maslow and Roszak, other ways to know the world that have just as much subjective impact – and that is an objective fact. In discussing psychedelic drugs, Roszak was careful to place marijuana and LSD, in particular, in what he saw as a legitimate tradition of William James, Havelock Ellis, and Aldous Huxley (in *The Doors of Perception*), all of whom studied hallucinogenic substances – nitrous oxide and peyote, for example – in a search for 'non-intellective powers.' But he concentrated on marijuana and the experiments on LSD by the Harvard professor Timothy Leary. Roszak was not entirely convinced by Leary (who was eventually dismissed from Harvard) and his claims of a 'psychedelic revolution' (that if you change the prevailing mode of consciousness, you change the world), but he was convinced that hallucinogenics offered emotional release and liberation in a difficult world and were no less damaging than the enormous numbers of tranquilisers and antidepressants then being prescribed for the middle classes, often the parents of the children who comprised the 'drug generation.'[35]

In his chapter on religion Roszak introduced Alan Watts. Watts began teaching at the School of Asian Studies in Berkeley after leaving his position as an Anglican counselor at Northwestern University. Aged fifty-five in 1970, he had been a child prodigy in his chosen field, Buddhist studies, and the author of seven books on Zen and mystical religion. Zen was the first of the Eastern mystical religions to catch on in the West, a fact Roszak put down to its vulnerability to 'adolescentisation.'[36] By this he meant its commitment to a 'wise silence, which contrasts so strongly with the preachiness of Christianity' and which, he said, appealed strongly to a generation raised on wall-to-wall television and a philosophy that 'the medium is the message.' Watts was himself highly critical of the way Zen was used, sometimes by pop stars, as little more than the latest fashion accessory, but the fascination with Zen led to an interest in other Eastern religions – Sufism, Buddhism, Hinduism, and then on to primitive shamanism, theosophy, even kabbala, the *I Ching*, and, perhaps inevitably, the *Kama Sutra*.

Zen received a massive boost from an entirely separate book, **Robert Pirsig**'s *Zen and the Art of Motorcycle Maintenance* (1974).[37] This was a road book. Pirsig took his young son and some friends on a vacation through the backroads of America – as the book opens, they are biking between Minneapolis and the Dakotas. The text alternates between lyrical passages of life on the road – the sheer walls of canyons, the soft beds of pine needles that the bikers sleep on, the smell of rain – and rhetorical discussions of philosophy. Pirsig's main target is what he calls the Church of Reason. He moves between Eastern mystics, Zen Buddhism, and classical Greek philosophers in particular. For him the motorcycle maintenance manual shows the typical dead hand of reason: meticulously accurate, dull, and before you can use it you need to know everything about bikes. Opposed to that is the 'feel' that a true mechanic has for machines. Pirsig's most original ideas are new ways to conceive experience: rhetoric, quality, and 'stuckness.' Reason does not have to be a dialectic, he says. Rhetoric carries with it the idea that knowledge is never neutral but always has value and therefore leads somewhere. Quality is a difficult entity to describe, but as Pirsig uses the idea, he says that we recognise quality in art, say, or literature, or in a machine, and that such recognition is *unthinking*. 'Stuckness' is being immersed in a line of thought with an inability to shake free. The form of Pirsig's book, itself rhetorical, was designed to show his appreciation of the quality of nature, and the way he had come unstuck in his own thinking.

'What the counter-culture offers us, then,' concluded Roszak, 'is a remarkable defection from the long-standing tradition of sceptical, secular intellectuality which has served as the prime vehicle for three hundred years of scientific and technical work in the West. Almost overnight (and astonishingly, with no great debate on the point) a significant portion of the younger generation has opted out of that tradition, rather as if to provide an emergency balance to the gross distortions of our technological society.'[38]

Although it has long since disappeared as the entity described by Roszak, the counter-culture was not a complete dead end. Besides its input into the green movement and feminism, many of the psychotherapies that flowered under the counter-culture bordered on the religious: Erhard Seminar Training (est), Insight, primal therapy, rebirthing, Arica, bioenergetics, and Silva Mind Control were more than therapies, offering group experiences and ritual similar to church. All of them involved some form of body manipulation – rapid, chaotic breathing to build tension, shouting or screaming as a form of release. Often, such activities ended in group sex. Equally often, these therapy-religions had quite a complex set of ideas behind them, but it was rarely necessary for the ordinary members to be familiar with that: there was always a clerisy on hand to help. What mattered was the experience of tension and its release.[39]

Judged by the numbers who still followed the mainstream belief systems, the new therapy-religions were small beer; they never comprised more than a few hundred thousands. Their significance lay in the fact that people turned to them because life had become 'so fragmented that they [found] it increasingly difficult to draw on their public roles for a satisfying and fulfilling sense of identity.'[40] This is why the historian of religion Steve Bruce called these new movements 'self-religions,' because they elevated the self, if not to centre stage, then at least to far more importance than the traditional mainstream faiths: each individual had his or her turn at the centre.

One man who was fascinated by this idea, and ran with it in a series of brilliantly witty essays, was the American journalist **Tom Wolfe**. Wolfe was the inventor (in the 1960s) of something that came to be called the New Journalism (the capitals are customary). This was Wolfe's attempt to get beyond the 'pale beige tone' of most reporting, and to do so he employed many tricks and devices from fiction in an effort to get inside the minds of the people being written about; far from being mere neutral reporting, his journalism was enriched (victims would say distorted) by a point of view. Essentially a comic, even a manic writer, Wolfe's main aim was to chronicle the fragmentation and diversity of (American) culture, which has evolved its own, often bizarre art forms, lifestyles, and status rituals.[41] *The Electric Kool-Aid Acid Test* (1968) included an hilarious account of a journey across America in a psychedelic-painted bus with a crew of 'acid-heads,' complete with vernacular conversation and punctuation. *Radical Chic* (1970) was about the svelte sophisticates of New York, conductor Leonard Bernstein in particular, entertaining the Black Panthers ('I've never met a Panther – this is a first for me!') and conducting an auction to help their cause where the bidders included Otto Preminger, Harry Belafonte, and Barbara Walters. *Mau-Mauing the Flak Catchers* (also 1970) chronicled the way black recipients of welfare hopelessly outwit the various functionaries whose job it is to see that the system is not abused.[42] But it was in *The Me Decade* (1976) that Wolfe took up where Daniel Bell, Theodore Roszak, and Steve Bruce left off.[43] For Wolfe actually attended some of the sessions of these self-religions, and he wasn't taken in for a moment – or at least that's how he saw it. He called them Lemon-Sessions, and 'Lemon-Session Central' was the Esalen Institute, a lodge perched on a cliff overlooking the Pacific in Big Sur, California; but Wolfe made it clear that he included Arica, Synanon, and Primal Scream therapy in this pantheon. Although many people wondered what the appeal was of spending days on end in the close company of people who were complete strangers, Wolfe knew: 'The appeal was simple enough. It is summed up in the notion: "Let's talk about *Me*."' Wolfe saw the obsession with the self as a natural (but unwholesome) development of the counter-culture, a follow-on of the campaign for personal liberation that went with the sexual revolution, experiments with drugs, and the new psychologies. It was, said Wolfe, the natural corollary of alienation (Marx), anomie (Durkheim), mass man (Ortega y Gasset), and the lonely crowd (Riesman). But then he added, in his usual style, 'This [alienated] victim of modern times has always been a most appealing figure to intellectuals, artists, and architects. The poor devil so obviously needs *us* to be his Engineers of the Soul, to use a term popular in the Soviet Union in the 1920s. ... But once the dreary little bastards started getting money in the 1940s, they did an astonishing thing – they took their money and ran! They did something only aristocrats (and intellectuals and artists) were supposed to do – they discovered and started doting on *Me*!'[44]

Wolfe identified the Me decade, but it was **Christopher Lasch**, a psychoanalyst and professor at the University of Rochester in New York State, who went further than anyone else had done on the theme of '**the Me decades**' and what would shortly be known as '**the Me generation**.' In *The Culture of Narcissism* (1979) Lasch's thesis was that the whole development of American society (and by implication other Western societies to a greater or lesser extent) had, since World War II, brought about the development of the narcissistic personality, so much so that it now dominated the entire culture. His book was a mixture of social criticism and psychoanalysis, and his starting point was not so very different from Daniel Bell's.[45] The subtitle of Lasch's book was

'American Life in an Age of Diminishing Expectations,' and it began, 'Defeat in Vietnam, economic stagnation, and the impending exhaustion of natural resources have produced a mood of pessimism in higher circles, which spreads through the rest of society as people lose faith in their leaders.'[46] Liberalism, once the only game in town when Lionel Trilling was alive, was now 'intellectually bankrupt. ... The sciences it has fostered, once confident of their ability to dispel the darkness of the ages, no longer provide satisfactory explanations of the phenomena they profess to elucidate. Neoclassical economic theory cannot explain the coexistence of unemployment and inflation; sociology retreats from the attempt to outline a general theory of modern society; academic psychology retreats from the challenge of Freud into the measurement of trivia. ... In the humanities, demoralisation has reached the point of a general admission that humanistic study has nothing to contribute to an understanding of the modern world.'[47] Against this background, Lasch said, economic man had given way to psychological man, 'the final product of bourgeois individualism.' Lasch didn't like this psychological man. Having set the scene, he waded into all aspects of a society that he thought had been affected by the essentially narcissistic personality of our time – work, advertising, sport, the schools, the courts, old age, and the relations between the sexes.

His first target was **the awareness movement**. 'Having no hope of improving their lives in any of the ways that matter, people have convinced themselves that what matters is psychic improvement: getting in touch with their feelings, eating health food, taking lessons in ballet or belly-dancing, immersing themselves in the wisdom of the East, jogging, learning how to "relate," overcoming their "fear of pleasure." '[48] Echoing Steve Bruce, Lasch argues that we have entered a period of 'therapeutic sensibility': therapy, he says, has established itself as the successor to rugged individualism and to religion, though he prefers to characterise it as an antireligion.[49] He further argues that eventually this approach will serve as a replacement for politics. Norman Mailer's *Advertisements for Myself*, Philip Roth's *Portnoy's Complaint*, and Norman Podhoretz's *Making It* are all examples of the self-absorption on the part of the middle and upper-middle classes, designed to insulate them against the horrors of poverty, racism, and injustice all around them. The **new narcissism** means that people are more interested in personal change than political change, and encounter groups, T-groups, and other forms of awareness training have, in effect, helped to abolish a meaningful inner private life – the private has become public in 'an ideology of intimacy.' This makes people less individualistic, less genuinely creative, and far more fad- and fashion-conscious. It follows, says Lasch, that lasting friendships, love affairs, and successful marriages are much harder to achieve, in turn thrusting people back on themselves, when the whole cycle recommences. He goes on to identify different aspects of the narcissistic society – the creation of celebrities who are 'famous for being famous,' the degradation of sport to commercialised entertainment rather than heroic effort, the permissiveness in schools and courts, which put the needs of 'personal development' above the more old-fashioned virtues of knowledge acquisition and punishment (and thus treat the young gently rather than inculcating the rugged individualism that was once the tradition). In this context he also raises an issue that assumed greater relevance as the years went by, namely the attack on elites and the judgements they arrive at (as for example in the canon of books to be studied in schools). 'Two contributors to a Carnegie Commission report on education condemn the idea that "there are certain works that should be familiar to all educated men" as inherently an "elitist notion." Such criticisms often appear in company with the contention that academic life should reflect the variety and turmoil of modern society instead of attempting to criticise and thus transcend this confusion.'[50]

But, and here we get to the nub of Lasch's criticism, he argued that the awareness movement had failed, and failed completely. It failed because, in so many words, it had produced a false consciousness. The emancipation that it supposedly brought about was in fact no emancipation at all, but merely a more sophisticated and more subtle form of control. The new awareness still involved old tricks to keep power and control in the hands of – generally speaking – those who had it before. The feminist movement may have brought about greater freedom for many women, but the cost was a huge rise in one-parent families, overwhelmingly a mother and child, which in turn put greater pressures on women and on the children, in many cases breeding a 'revulsion' against close personal relationships that made loving friendships more difficult and promoted a dependence on the self. One-parent families are often narcissistic families. In business, too, greater discussion and worker participation led in most cases only to talking shops, which may have made the management more liked but did not substantially change anything else. 'The popularisation of therapeutic modes of thought discredits authority, especially in the home and the classroom, while leaving domination uncriticised. Therapeutic forms of social control, by softening or eliminating the adversary relation between subordinates and superiors, make it more and more difficult for citizens to defend themselves against the state or for workers to resist the demands of the corporation. As the ideas of guilt and innocence lose their moral and even legal meaning, those in power no longer enforce their rules by means of the authoritative edict of judges, magistrates, teachers, and preachers. Society no longer expects authorities to articulate a clearly reasoned, elaborately justified code of law and morality; nor does it expect the young to internalise the moral standards of the community. It demands only conformity to the conventions of everyday intercourse, sanctioned by psychiatric definitions of normal behaviour.'[51]

Modern (i.e., late 1970s) man, says Lasch, is imprisoned in his self-awareness; he 'longs for the lost innocence of spontaneous feeling. Unable to express emotion without calculating its effects on others, he doubts the authenticity of its expression in others and therefore derives little comfort from audience reactions to his own performance.'[52] Both Lasch and Tom Wolfe therefore concur in finding the awareness movement, the obsession with self, and the therapeutic sensibility not only unsatisfying but largely a sham.

Roszak, Wolfe, and Lasch all drew attention to the way in which, for many, the private, confessional, anonymous nature of traditional religions was giving way to the public, intimate, narcissistic nature of the awareness movement. Another way of putting this is to say that one set of beliefs, one kind of faith, was giving way to another. Not entirely coincidentally, three books in the early 1970s by well-known historians examined analogous times in the past.

Described by Christopher Hill as one of the most original books on English history, **Keith Thomas**'s *Religion and the Decline of Magic* (1971) showed that although the psychological atmosphere of sixteenth- and seventeenth-century England was very different from the California or Paris of the late 1960s and early 1970s, nevertheless the parallels were there to be seen in the overlap between rival systems of belief, the link to societal change and political radicalism.[53] Thomas explains that magic in the sixteenth and seventeenth centuries is to be understood as being on a par with, say, drink and gambling as a way of coping with the high number of uncertainties of life, and in particular with uncertainties in the medical sphere. Organised religion itself used many magical practices to enforce its way of life. Miracles were regularly reported until the Reformation.[54] In 1591 John Allyn, an Oxford recusant, was said to possess a quantity of Christ's blood, 'which he sold at twenty pounds a drop.'[55] One of the reasons why the Reformation succeeded was because sceptics no longer believed in the 'magic' surrounding Mass, whereby the Host was turned into the body of Christ and the wine into his blood.[56] Thus Protestantism represented itself as a direct attempt to take magic out of religion.

Sects proliferated because their leaders continued to promote supernatural solutions to earthly problems that the Reformation sternly resisted. (One of these, incidentally, was the interpretation of dreams – Thomas Hill's *Most Pleasaunte Art of the Interpretation of Dreames*).[57] Many women expected to see their future husband in their sleep. With the outbreak of civil war, the number of people claiming to be the Messiah shot up – William Franklin, a London ropemaker who made this claim, appointed disciples in the role of destroying angel, healing angel, and John the Baptist; his activities attracted 'multitudes of persons' before he was forced to recant before Winchester Assizes in 1650.[58] Thomas believed that the chaos of the age, helped along by technological advance (in particular gunpowder, the printing press, and the mariner's compass), was helping to create these sects, the avowed aims of which were only part of the attraction. Satisfaction was achieved by many participants simply by taking part in some symbolic, ritualistic action, irrespective of its purpose.[59] There were many names for these magicians – cunning men, wise women, conjurers, sorcerers, witches – offering a variety of services from finding lost goods to healing and fortune-telling. Each had a method that involved an intimidating ritual.[60]

But perhaps the strongest parallel was in astrology, which at that time was the only other system that attempted to explain why individuals differed from one another, or to account for physical characteristics, aptitudes, and temperament.[61] Even Sir Isaac Newton produced *The Chronology of Ancient Kingdoms Amended* in 1728, which attempted to use astronomical data to reconstruct the otherwise lost chronology of the ancient world, the aim being to explain why various peoples had the character, manners, and laws that they did.[62] The appeal of astrology was intended to be intellectual, to provide a coherent and comprehensive system of thought, its other avowed aim being to help men resolve personal problems and to 'take their own decisions.'[63] Again, a number of celebrated figures with an interest in astrology are known to have had sectarian or radical associations – Anabaptists, Ranters, Quakers, and Shakers all included. According to Thomas the existence of rebellious feeling (in the political sense) led to prophecies, wish-fulfilment in effect, and these fuelled supernatural speculation.[64] Technological change also had an effect on the idea of progress. This may have arisen from the crafts, where knowledge was cumulative; but it was not until the sixteenth century that the 'modern' idea that 'newest is the best' established itself, and only then after a protracted battle between 'ancients' and 'moderns.' It rubbed off with sects, people imagining that, even in religion, newest might be best. For Thomas, magic arises at the weak point of the social structure of the time, whether that be social injustice, physical suffering, or unrequited offences. Ultimately, however, magic was a 'collection of miscellaneous recipes' rather than a comprehensive body of doctrine, such as Christianity, which was far more fulfilling overall. The century after the Reformation was a transitional period when magic continued because it offered something for those who found the Protestant notion of self-help too arduous.[65] The changes are to be seen as a result of the shifting aspirations of people: as the development of insurance took the threat out of everyday setbacks, and medicine made genuine advances, magic contracted. Today it still clings on in astrology, horoscopes, and fortune-tellers.

The World Turned Upside Down, published by **Christopher Hill** in 1972, overlapped with Thomas's book.[66] Hill considers the years in Britain immediately after the civil war, a time when, as in the 1960s and early 1970s, radical political ideas and new religious sects proliferated. Again, without making too much of the parallels, certain similarities may be noted, in particular the left-wing nature of the political ideas; the fact that these new religious notions internalised the spiritual, making God less 'out there' or 'up there,' more of a personal matter; and thirdly pacifism. Hill even went so far as to use the words 'counter-culture' once or twice. It was, he said, a period of 'glorious flux and intellectual excitement,' powered by large numbers of 'masterless men,' no longer tied to feudal masters – itinerant merchants, peddlers, craftsmen, vagabonds who, beholden to no one, no longer fitted into hierarchical society and therefore provided the backbone of the new sects: Anabaptists, Levellers, Ranters, Quakers, and Muggletonians.[67]

Hill discovered several new patterns of thought. One was a belief in the spirit of Christianity – mastering sin – rather than following the letter of the Bible; another was the stirrings of science and a mood of scepticism towards many of the traditional claims of Christianity. There were also many Communist ideas and constitutional criticisms, all of a left-wing kind, as we would recognise them. Property laws were attacked, and squatters appeared (also typical of the 1960s and early 1970s).[68] Church services were run along more democratic lines. Members of the congregation were invited to publicly comment on, and criticise, sermons (several 'riots and tumults' being the result). With the collapse in traditional beliefs, in hell and heaven in particular, a widespread despair set in, and people talked far more freely than hitherto of suicide (a mortal sin in Catholicism). Many flitted from sect to sect. Hill noted a taste for nakedness and a general attitude to the insane that was a mixture of awe and fear: the mad were routinely regarded as prophets. A number of new schools and universities were started. The change in the status of women was also considerable, as evidenced not only in the higher rate of divorce but in the greater role they had in the sects (compared with the established church), with some sects, like the Quakers, abolishing the vow in the marriage ceremony of the wife to obey the husband, and others, like the Ranters, ceasing to regard sex outside marriage as sinful.[69] Indeed, the Ranters' views at times resembled Marcuse's: 'The world exists for man, and all men are equal. There is no after-life: all that matters is here and now. ... In the grave there is no remembrance of either sorrow or joy. ... Nothing is evil that does not harm our fellow men. ... Swearing i'this light, gloriously, and "wanton kisses," may help to liberate us from the repressive ethic which our masters are trying to impose on us.'[70] Hill agreed with Thomas that this was a time when the idea of novelty, originality, 'ceased to be shocking and became in a sense desirable.' This was an all-important advance, not only because the acceptance of novelty hastened change, but because it drove man back in on himself, to see 'what light was inside and how it could be made to shine.'

A further shift took place in the nineteenth century, and it was this change that was described and analysed by **Owen Chadwick** in *The Secularisation of the European Mind in the Nineteenth Century* (1975).[71] Chadwick divided his book into two. In part 1, 'The Social Problem,' he considered the effects of economic liberation, Karl Marx's materialism, and a general anticlericalism. This 'unsettlement' was also the result of new machines, new cities, massive transfers of population. In part 2, 'The Intellectual Problem,' he looked at the impact of science on men's minds, at the effects of new historical (including archaeological) researches and Comtean philosophy, and at the ethics that developed out of these and other changes. Certain trends, he says, are clear, like churchgoing statistics. There was a downward turn in France, Germany, and England in the 1880s; the larger the town, the smaller the proportion of people who attended church on a Sunday; a cheap press enabled more atheistic literature to be published. But Chadwick's more original point is that as the nineteenth century wore on, the very idea of secularisation itself changed. To begin with it could be described as anticlericalism, and a fairly aggressive anticlericalism at that.[72] With the passage of time, however, Christianity, while undoubtedly weaker, adapted to the new forms of knowledge, so that by the end of the century the secular world was in effect a separate realm from that of the faith. There were still areas of life, or experience, like mourning or providence, which were left to religion, but in general the heat and fury had gone out of the debate; the agnostics and atheists went their own way, following Marx, Darwin, or the radical historians; the religious moved in and out of science, accepting what they wished to accept.[73] The secular world thought it understood religion, as a phase or stage on the way to a fully secular society, and the religious denied that science and history could ever address the matter of faith. Despite the title, Chadwick's book was in fact a chronicle of the tenacious hold that religion had on many people, the need for a spiritual mystery at the heart of existence.

The works of Galbraith, Bell, Roszak, and Lasch, on the one hand, and of Thomas, Christopher Hill, and Chadwick on the other, complement each other. Two things stand out from the historical studies, as a prerogative for a change of sensibility: new modes of communication (which help change self-awareness), and new forms of knowledge, scientific knowledge especially, which threaten old explanations.

Galbraith and Bell recognised this. No sooner had their analyses appeared than the most important of their predictions were confirmed. In the spring of 1975, two young men quit their regular positions, one as a computer programmer at Honeywell in Boston, the other as a student at Harvard, and started their own company, writing software for the new generation of small computers that had just been announced. A few months later, in 1976, a young microbiologist in San Francisco was approached by an equally young venture capitalist, and they too launched their own company, this one to synthesise a specific protein from strands of DNA. The first two were called **Paul Allen** and **Bill Gates** and they named their company **Microsoft**. The second two were **Herbert Boyer** and **Robert Swanson**, and because neither Boyer & Swanson nor Swanson & Boyer was acceptable, they named their company **Genentech**. As the last quarter of the century arrived, the new information technology and the new biotechnology were spawned in tandem. The world was about to be turned upside down again.

34
GENETIC SAFARI

In 1973 the Nobel Prize for Medicine and Physiology was given to three men. Two of them had been on different sides of the Nazi divide in pre-World War II Germany. Karl von Frisch had suffered at the hands of Nazi students because he was never able to prove that he was not one-eighth Jewish. He survived only because he was a world authority on bees at a time when Germany was suffering a virus that threatened its bee population, and badly needed his help to bolster food production. Konrad Lorenz, on the other hand, had fully subscribed to the prevailing Nazi ideology about 'degeneration' among the German-Jewish population and willingly taken part in various highly dubious experiments, particularly in Poland. He was captured as a Russian prisoner toward the end of the war and not released until 1948. Subsequently he apologised for his prewar and wartime activities, apologies that were accepted by colleagues, the most important of whom was the third in the trio to share the Nobel Prize in 1973. This was **Nikolaas Tinbergen**, a Dutchman, who spent the war in a hostage camp in danger of being shot in reprisal for the activities of the Dutch underground. If Tinbergen accepted Lorenz's apologies, he must have been convinced they were genuine.[1] The award of the prize was recognition for a relatively new discipline in which each man had been a founding father: ethology, the study of animal behaviour with a strong comparative element. Ethologists are interested in animal behaviour for what that might reveal about instinct, and what, if anything, separates man from all other forms of life.

Tinbergen's classic work, carried out since the war (and after he had moved from Leiden to Oxford), elaborated on Lorenz's ideas of 'fixed action patterns' and 'innate releasing mechanisms' (IRMs). Experimenting with the male three-spined stickleback, Tinbergen showed the crucial importance of why at times the fish stood on its head to display its red belly to the female: this stimulated a mating response. Similarly, he showed the significance of the red spot on a herring gull's bill: it elicited begging from a chick.[2] It was later shown that such IRMs were more complicated, but the elegance of Tinbergen's experiments caught the imagination of scientists and public alike. John Bowlby's research on maternal attachment drew inspiration from this ethological work, which also helped stimulate a great burst of fieldwork with animals phylogenetically closer to man than the insects, birds, and fish examined by the three Nobel Prize winners. This fieldwork concentrated on mammals and primates.

Since 1959, when Mary Leakey discovered Zinj, the Leakeys had made several other significant discoveries at Olduvai Gorge in Tanzania. The most important of these was that three hominids had existed at the same time – *Australopithecus boisei*, *Homo erectus* (Louis now conceded that Zinj was actually an especially large form of Peking Man), and a new find, dating from the early 1960s, which they had named *Homo habilis*, 'Handy Man,' because he was found associated with slightly more advanced stone tools. Mary Leakey, in her scientific volume entitled *Olduvai Gorge*, analysed 37,000 Olduvai artefacts, including twenty hominid remains, 20,000 animal remains, and many stone tools.[3] All this revealed Olduvai as an early, primitive culture with *Homo erectus* giving way to *Homo habilis* with more refined but still very primitive tools, and many species of extinct animals (such as hippos).

An American author and playwright, **Robert Ardrey**, drew yet more attention to Olduvai, and Africa. In a series of books, *African Genesis* (1961), *The Territorial Imperative* (1967), and *The Social Contract* (1970), Ardrey did much to familiarise the idea that all animals – from lions and baboons to lizards and jackdaws – had territories, which varied in size from a few feet for lizards to a hundred miles for wolf packs, and which they would go to extreme lengths to defend. He also drew attention to the rankings in animal societies and to the idea that there are a wide variety of sexual arrangements, even among primates, which, he thought, effectively demolished Freud's ideas ('Freud lived too soon,' Ardrey wrote). In popularising the idea that man originated in Africa, Ardrey also emphasised his own belief that *Homo sapiens* is emotionally a wild animal who is domesticating himself only with difficulty. He thought that man was originally an ape of the forest, who was defeated by the other great apes and forced into the bush: *Australopithecus robustus*, a vegetarian, evolved into *A. africanus*, a carnivore, who then, as *Homo sapiens* (or even earlier), evolved the use of tools – for which Ardrey preferred the word *weapons*. For Ardrey, mankind could only survive and prosper so long as he never forgot he was at heart a wild animal.[4] The fieldwork that lay at the heart of Ardrey's book helped establish the idea that, contrary to the view prevailing before the war, humanity originated not in Asia but in Africa and that, by and large, it emerged only once, somewhere along the Rift Valley, rather than several times in different places. A sense of urgency was added to this reorientation, because ethological research, besides showing that animals could be studied in the wild, also confirmed that in many cases numbers were dwindling. Ethology, therefore, became a contributor to the ecological movement.

Far and away the most influential people in persuading the wider public that ethology was valuable were three extraordinary women in Africa, whose imaginative and brave forays into the bush proved remarkably successful. These were **Joy Adamson**, who worked with lions in Kenya, **Jane Goodall**, who investigated chimpanzees at Gombe Stream, in Tanzania, and **Dian Fossey**, who spent several years working with gorillas in Uganda.

The Adamsons – Joy and George – were old Africa hands since before World War II, and friends of the Leakeys (**George Adamson** had been a locust control officer and a gold prospector in Kenya since 1929). Joy, of Austrian birth, was 'an often-married, egotistical, wilful and at times unstable woman of great energy and originality.'[5] In 1956, near where they lived, a lion had attacked and eaten a local boy. With others, George Adamson set off in pursuit of the man-eater, which by custom had to be killed because, having been 'rewarded,' it would certainly return. A female lion was found, and duly shot. However, three very young cubs were discovered nearby, still with film over their eyes, and were raised by the Adamsons. Two were eventually acquired by a zoo, while the Adamsons kept the other, the 'plainest,' named after an equally plain relative, **Elsa**.[6] Thus began the Adamsons' observations of lion behavior. These were hardly systematic in, say, a laboratory sense, but the closeness of the relationship between human and animal was nonetheless new and enabled certain insights into mammal behavior that would otherwise not have been made. For example, 'Elsa's most remarkable demonstration of understanding and restraint occurred when she knocked over a buffalo in the Ura and was efficiently drowning it. While her blood was still up, Nuru, a Muslim, rushed down to cut the animal's throat before it died so that he and other Africans could eat some of the meat. For a second Elsa turned on him, but suddenly realised he had come to share, not steal, her kill.'[7]

In 1958, for a variety of reasons, one of which was Elsa's growing strength and uncontrollability (she had at one stage taken Joy's head in her mouth), the lioness was reintroduced into the wild. This, a dangerous exercise for her, was completed successfully, but on several occasions thereafter she reappeared, accompanied by her new family, and for the most part behaved in a docile, friendly manner. It was now that Joy Adamson conceived the series of three books that were to make her famous: *Born Free* (1959), *Living Free* (1960), and *Forever Free* (1961).[8] The many photographs of apparently friendly lions had just as much impact as the text, if not more so, helping the book to sell more than five million copies in a dozen languages, not to mention a major movie and several documentaries. Joy had originally taken on the cubs because they were 'orphans,' and, in the 1950s, maternal deprivation in humans was, as we have seen, an important issue in the wake of war. Throughout the 1960s, 1970s, and 1980s, Joy and/or George continued to live close to lions, exploring in an informal but unique way their real nature. They were criticised for 'ruining' lions, making them less lionlike because they were friendly to humans, but between them the Adamsons were able to show that, fierce and wild as lions undoubtedly are, their violence is not completely programmed, by no means 100 percent instinctive; they at least appear to be capable of affection or respect or familiarity, and the needs of their stomach are not always paramount. Ted Hughes, Britain's poet laureate, had this to say in reviewing *Born Free*: 'That a lioness, one of the great moody aggressors, should be brought to display such qualities as Elsa's, is a step not so much in the education of lions as in the civilisation of man'*[9]

Jane Goodall, like Dian Fossey after her, was a protégé of Louis Leakey. Apart from his other talents, Leakey was a great womaniser, who had affairs with a number of female assistants. Goodall had approached Leakey as early as 1959, the year of Zinj, begging to work for/with him. When he met her, Leakey noted that Goodall was very knowledgeable about animals, and so was born a project that had been simmering at the back of his mind for some time. He knew of a community of chimpanzees at **Gombe Stream**, near Kigoma on the shore of Lake Tanganyika. Leakey's thinking was simple: Africa had a very rich ape population; man had evolved from the apes; and so the more we discovered about them, the more we were likely to understand how mankind – humanity – had evolved. Leakey thought Goodall suitable because, while she was knowledgeable, she wasn't too academic, and her mind wasn't 'cluttered by theory.' Not that there *was* much theory at the time – ethology was a new subject – but Goodall loved her assignment, and both her official reports and her popular account, *In the Shadow of Man*, published in 1971, managed to be both scientifically important and moving at the same time.[10]

Goodall found that it took the chimpanzees some months to accept her, but once they did she was able to get close enough to observe their behavior in the wild *and* to distinguish one chimpanzee from another. This simple insight proved extremely important. She was later criticised by other, more academically grounded scientists for giving her chimps names – David Greybeard, Flo, Flint, Flame, Goliath – instead of more neutral numbers, and for reading motives into chimp actions, but these were lame criticisms when set against the richness of her material.[11] Her first significant observation occurred when she saw a chimp insert a thin stick into a termite mound in order to catch termites that attached themselves to the stick – the chimp then raised the stick to its lips. Now here was a chimp using a tool, hitherto understood to be the hallmark of humanity. As the months passed, the social/communal life of these primates also began to reveal itself. Most notable was the hierarchy of males and the occasional displays of aggression that brought about this ranking, which by and large determined sexual privilege in the troupe, but not necessarily priority in food gathering. But Goodall also recorded that much of the aggressive displays were just that – displays – and that once the less dominant male had made deferential or submissive gestures, the dominant animal would pat his rival in what appeared to be a gesture of reassurance. Goodall also observed mother–offspring behavior, the importance of social grooming (picking unwanted matter out of each other's fur), and what appeared to be familial feeling. Young chimpanzees who for some reason lost their mothers shrivelled physically and/or became nervous – what we would call neurotic; and brothers, though they often fought with or were indifferent to each other, sometimes ran to one another for comfort and reassurance. Controversially, she thought that chimps

* Joy was stabbed to death in 1980 by an assistant who claimed he hadn't been paid. George was shot in an ambush by Somali poacher/farmers in 1989.

had a rudimentary sense of self and that children learned much behavior from their mothers. In one celebrated instance, she observed a mother with diarrhoea wipe herself with a handful of leaves; immediately, her two-year-old infant did the same although his bottom was clean.[12]

Dian Fossey's *Gorillas in the Mist* related her observations and experiences on the Rwanda/Zaire/Uganda border in the 1970s, and concerned one species of mountain gorilla, *Gorilla gorilla berengei*. While much more impressive physically than the chimpanzee, this primate was and remains the most threatened in terms of numbers. Rwanda is one of the most densely populated African countries, and the gorilla population had by then been falling by an average of 3 percent a year for more than twenty years, to the point where not much more than 250 were left. Fossey's work was therefore as much a part of ecology as biology.[13]

Fossey documented in shocking detail the vicious work of poachers, who sometimes kidnapped animals for zoos and sometimes killed them, cutting off their heads and hands in a primitive ritual. This aspect of her book, when it was published in 1983, shocked the world, stimulating action to conserve the dwindling numbers of an animal that, despite its fierce appearance and 'King Kong' reputation, the other part of her argument showed to be unfairly maligned. Fossey found that she was able to habituate herself to at least some of the gorilla groups near her research station, Karisoke, in the volcanic Parc des Virungas. The crucial element here was that she had learned what she called '**belch vocalisations**,' a soft, deep, purr, '*naoom, naoom*,' which resembled a stomach rumbling. These sounds, she found, which express contentment in gorillas, announced her presence and set the animals at ease to the point where, eventually, she could sit among them, exchanging sounds and observing close up. She found that gorillas had a family structure much closer to that of humans than did chimpanzees. They lived in relatively stable groups of about ten individuals. 'A typical group contains: one silverback, a sexually mature male over the age of fifteen years, who is the group's undisputed leader and weighs roughly 375 pounds, or about twice the size of a female; one blackback, a sexually immature male between eight and thirteen years weighing some 250 pounds; three to four sexually mature females over eight years, each about 200 pounds, who are ordinarily bonded to the dominant silverback for life; and, lastly, from three to six immature members, those under eight years. .... The prolonged period of association of the young with their parents, peers, and siblings offers the gorilla a unique and secure type of familial organisation bonded by strong kin ties. As the male and female offspring approach sexual maturity they often leave their natal groups. The dispersal of mating individuals is perhaps an evolved pattern to reduce the effects of inbreeding, though it seems that maturing individuals are more likely to migrate when there are no breeding opportunities within the group into which they are born.'[14]

Fossey found that different gorillas had very different characters, and that they used some seven different sounds – including alarm calls, pig-grunts when travelling, rebuttals to other sounds, and disciplinary enforcements between adults and young. Unfortunately, Dian Fossey was unable to further her studies;

at the end of 1985 she too, like the Adamsons, was murdered. Her black tracker and her white research assistant were both accused, though the charges against her tracker were dropped. Fearful of not receiving a fair trial, the white assistant fled the country, later to be convicted in his absence.[15] In the short run, Fossey's battle against poaching was more important than her ethological observations, as her death shows. But only in the short run. For example, her sensitive description of the gorilla Icarus's response to the death of another, Marchesa, raised profound questions about gorilla 'grief' and the nonhuman understanding of death. In many ways, the evolutionary psychology of gorillas is even more enlightening than that of chimpanzees.

George Schaller, director of the Wildlife Conservation Division of the New York Zoological Society, made it his life's work to study some of the ecologically threatened large animals of the world, in the hope that this would contribute to their survival. In a long career, he spent time studying pandas, tigers, deer, and gorillas but his most celebrated study, published in 1972, was *The Serengeti Lion*.[16] This book, which also included sections on the cheetah, leopard, wild dogs, and hyenas, took up where the Adamsons left off, in that Schaller was much more systematic and scientific in his approach – he counted the number of lions, the times of the day they hunted, the number of times they copulated, and the number of trees they marked out as their territory.[17] While this did not make his book an enthralling read, his overall picture of the delicate balance in Africa between predator and prey had a marked effect on the ecological movement. He showed that far from harming other wildlife (as was then thought), predators were actually good influences, weeding out the weaker vessels among their prey, keeping the herds healthy and alert. He also made the point that although lions were not as close to man as chimpanzees or gorillas were in phylogenetic terms, they were quite close in ecological terms to, say, *Australopithecus*. He argued that lions' hunting techniques were far more likely to resemble early man's, and his own studies, he said, showed that lions could hunt efficiently in prides without any sophisticated vocalisation or language. He did not therefore think that language in man necessarily evolved to cope with hunting, as other scholars believed.[18]

The final study in this great scientific safari on the Kenya/Tanzania/Uganda border was **Ian Douglas-Hamilton**'s investigation of elephants. A student of Nikolaas Tinbergen at Oxford, Douglas-Hamilton had originally wanted to study lions but was told that George Schaller had got there first. Douglas-Hamilton's study, published as *Among the Elephants* in 1975, was a cross between the Adamson-Goodall-Fossey approach and Schaller's more distanced research, mainly because elephants are far harder to habituate to in the wild.[19] He observed that elephants keep to family and kinship units and appear to show affection to other family members, which extends to a characteristic trunk-to-mouth gesture. Although he would never have been so anthropomorphic as to say this was 'kissing,' it is hard to know how else to describe it. Several family units make up kinship units. At times of abundant food supply, after the rains, elephants come together in massive 200-strong herds, whereas in drought they break up into smaller family groupings. Elephants show an extraordinary amount of interest in dead elephants – offspring will remain alongside the body of a dead mother for days, and a herd will sometimes dismember the carcass of an erstwhile colleague. Douglas-Hamilton's research meticulously catalogued which elephant stood next to which, and showed that there were clearly long-term 'friendships.'[20] As with the other big mammals of Africa, Douglas-Hamilton observed great individuality among elephants.

Much farther north than Olduvai, but still part of the Rift, the great valley splits into two: one part of the Y extends northeast into the Gulf of Aden, whereas the other heads northwest along the Red Sea. The area between the two arms of the Y is known as the Afar Triangle and is part of Ethiopia.

To begin with, the sites in Afar had been excavated by the Leakeys, especially Louis's son, Richard. They had dug there by invitation of Emperor Haile Selassie, who was himself interested in the origins of humankind and, on a state visit to Kenya in 1966, had met Louis Leakey and encouraged him to come north. Early digs consolidated the picture emerging farther south but were overshadowed by a discovery made by a rival French-American team. The guiding spirit of this team was **Maurice Taieb**, a geologist, who made the Afar Triangle his speciality (it was geologically unique). He called in a palaeontologist he had met elsewhere in Ethiopia, **Don Johanson**, a graduate student at Chicago University. Taieb had found an area, named Hadar, which he regarded as very fruitful – it was several thousand square kilometres in size and very rich in fossils. An expedition society was formed, which initially had the Leakeys as members. What happened on that expedition, and subsequently, became one of the most controversial incidents in palaeontology.

In November 1974, four miles from his camp, Johanson spotted a fragment of an arm bone sticking out of a slope. At first he thought it belonged to a monkey, but 'it lacked the monkey's distinguishing bony flange.'[21] His eye fell on another bony fragment higher up the slope – then a lower jaw, ribs, some vertebrae. He had in fact found the most complete hominid skeleton yet discovered, about 40 percent of the entire structure, and from the shape of the pelvic bone, almost certainly female. That night, back at camp, the team celebrated with beer and roast goat, and Johanson played the Beatles song 'Lucy in the Sky with Diamonds' over and over again. Famously, and unscientifically, the skeleton, officially recorded as **AL 288–1**, became known as '**Lucy**.'[22] The unparalleled importance of Lucy at the time was the fact that her anatomy indicated she had walked upright and could be precisely dated as being between 3.1 and 3.2 million years old. Her skull was not complete, but there was enough of it for Johanson to say that it was in the ape-size range. Her molar teeth were human-like, but the front molars were not bicuspids like ours.

Haile Selassie was overthrown in September 1974 in a coup which resulted in a Marxist military dictatorship in Ethiopia. This made work difficult, but Johanson managed to return and in 1975 made a yet further extraordinary discovery: a 'first family' of thirteen individuals – males, females, adults, juveniles, children, some two hundred fossils at one site, Site 333, as it became known. And in the following year, 1976, together with the French archaeologist

Hélène Roche, he found simple basalt tools, dating back to 2.5 million years. This all meant a complete revision of humankind's origins. Tool-making was much older than anyone imagined, as was upright walking. And it was clearly something indulged in first by *Australopithecus*, not the *Homo* genus.

Further finds in Hadar were prevented by another deterioration in the political situation in Ethiopia (another military coup in Addis Ababa). During this interregnum the southern end of the Rift Valley came back into the spotlight. In the mid-1970s **Mary Leakey** had been working in **Laetoli**, a site thirty miles from Olduvai, an area of sandstone gullies that cut into a plateau, very different from the gorge. She had been going there for many years and had recently found two jaws dating from 3.6 to 3.8 million years ago. In the last week of July 1976 she was joined by four other scientists, among them Andrew Hill and Kay Behrensmeyer. The newcomers, in high spirits, were all taken on a tour of the site on the morning after their arrival, and an elephant-dung fight broke out. Ducking into a flat gully to look for ammunition, Hill and Behrensmeyer came across a hard layer of volcanic ash – in which, as they suddenly noticed, there were elephant footprints. They dropped to their knees for a closer look, and then called the others. These were not fresh prints, but fossilised, and scattered near the elephant tracks were those of buffalos, giraffes, and birds. There were even a few ancient raindrops. What must have happened was that a spurt of volcanic ash, given off by a nearby mountain, had settled and then been rained upon, turning it into a form of cement. While this 'cement' was wet, animals walked across it, then another layer of ash was deposited on the top. Over the centuries, the top layer had weathered away to reveal the fossil footprints. It was an unusual find, but Mary Leakey told everyone to look out for **hominid footprints** – that would certainly make news. They searched all through August, but not until one day in September did they find some prints that looked hominid, with signs of a big toe. There were two sets, one much larger than the other, and they stretched for eighteen feet across the ancient 'cement.' In February 1978 Mary Leakey felt confident enough to announce the discovery. What was especially interesting was that the volcanic ash was dated to 3.7 million years ago, slightly earlier than the Ethiopian sites. From the pattern of indentations, some experts thought that whoever this hominid was, he did not walk upright all the time. So was this the period when man first *began* to walk upright?[23]

The answer did not come from Mary Leakey. The Laetoli bones and jaws had been given to **Tim White**, an American palaeontologist, whose job it was to describe them meticulously. However, White, a difficult man, fell out with both Richard and Mary Leakey. Worse, from the Leakey point of view, he subsequently teamed up with Don Johanson, and this pair proceeded to examine and analyse all the fossils from Laetoli and Hadar, all those aged between 3 and 4 million years old. They revealed their conclusions in 1979 in *Science*, claiming that what they had was a single species of hominid that was different to, and the ancestor of, many others.[24] This species, which they named *Australopithecus afarensis*, they said was fully bipedal and showed marked sexual dimorphism (the males were much bigger than the females), though even the males were no more than four feet six. Their brains were in the chimpanzee range and their faces pronounced, like the apes; their teeth were halfway between those of the apes and humans. Most controversially, Johanson and White claimed that *A. afarensis* was the ancestor of both the *Australopithecus* and the *Homo* genus, 'which therefore must have diverged some time after three million years ago.'[25]

At the beginning it had been Johanson and White's intention to include Mary Leakey as a coauthor, but Mary was unhappy with the label *Australopithecus* being attached to the fossils she had discovered. The convention in science is for the discoverer to have the first 'say' in publishing the fossils that he or she finds, and to name them. After that, of course, other scientists are free to agree or disagree. By including Mary's discoveries in their paper, Johanson and White were not only breaking with tradition; they knew by then that they were specifically going against her own interpretation. But they were anxious to claim for *A. afarensis* the title of common ancestor of almost all known hominid fossils and so went ahead anyway. This caused a bitter feud that has never healed.[26]

Beyond the personal dimension, however, *A. afarensis* has provoked much rethinking.[27] At the time it was given its name, the predominant view was that bipedalism and tool using were related: early man walked on two feet so as to free his hands for tools. But according to Johanson and White, early man was bipedal at least half a million years before tool using came in. The latest thinking puts bipedalism alongside a period of drying in Africa, when the forest retreated and open savannah grasslands spread. In such an environment, upright walking would have offered certain selective advantages – upright early man would have been faster, his body would have cooled more quickly, and he could have roamed over greater distances, with his hands free to carry food home, or back to his offspring. So although the bitterness was personally unpleasant, it did provoke useful new ideas about man's origins.[28]

Since the discovery of the helical structure of DNA in 1953, the next theoretical advance had come in 1961, when Francis Crick and **Sidney Brenner** in Cambridge had shown that the amino acids that make up the proteins of life are actually coded by a triplet of base pairs on DNA strands. That is, of the four bases – (a)denine, (c)ytosine, (g)uanine, and (t)hymine – three, in certain arrangements, such as CGT or ATG, code for specific acids. But more practical advances involved two ways of manipulating DNA that proved integral to the process of what became known as genetic engineering. The first was cloning, the second, gene sequencing.

In November 1972 **Stanley Cohen** heard a lecture in Hawaii delivered by **Herbert Boyer**, a microbiologist from the University of California at San Francisco. Boyer's lecture was about certain substances known as '**restriction enzymes**.' These were substances which, when they came across a certain pattern of DNA bases, cut them in two. For example, every time they came across a T(hymine) followed by an A(denine), one restriction enzyme (of which there are several) would sever the DNA at that point. However, as Boyer told the meeting, restriction enzymes did more than this. When they cut, they did not form a blunt end, with both strands of the double helix stopping at the same point; instead they formed jagged or steplike ends, one part jutting out, slightly longer than the other. Because of this, the ends were what scientists labelled 'sticky,' in that the flaps attracted complementary bases.[29] At the time he attended Boyer's lecture, Cohen was himself working on plasmids, microscopic loops of DNA that lurk outside a bacterium's chromosome and reproduce independently. As Cohen took in what Boyer was saying, he saw an immediate – and revolutionary – link to his own work. Because plasmids were loops, if they were cut with one of Boyer's restriction enzymes, they would become like broken rings, the two broken ends being mirror images of each other. Therefore, strips of DNA from other animals, and it didn't matter which (a lion, say, or an insect), if inserted into the bacterium with 'split rings,' would be taken up. The significance of Cohen's idea lay in the fact that the plasmid replicated itself many times in each cell, and the bacterium divided *every twenty minutes*. With this form of replication and division, more than a million copies of the spliced DNA could be created *within a day*.[30]

After the lecture, Cohen sought out Boyer. As Walter Bodmer and Robin McKie tell the story, in their history of the genome project, the two micro-biologists adjourned to a delicatessen near Waikiki Beach and, over corned beef sandwiches, agreed on a collaboration that bore its first fruits in the *Proceedings of the National Academy of Sciences* in November 1973, when they announced the first report of successful cloning. From now on there was enough DNA to experiment with.[31]

The next step – important both practically and theoretically – was to explore the *sequence* of bases in the DNA molecule. Sequencing was necessary because if biologists were to discover which genes governed which aspects of func-tioning, the exact order needed to be understood. **Fred Sanger** in Cambridge, England, and **Walter Gilbert** in Cambridge, Massachusetts (Harvard), both discovered methods of doing this, and both received a Nobel Prize for their efforts. But Sanger's method was identified first and is the more widely used.* Earlier, Sanger had developed a way of identifying the amino acids that make up proteins, and this had earned him his first Nobel, when he discovered the structure of insulin. But that method was far too slow to work with DNA, which is a very long molecule. Moreover, it is made up from only four subunits (A, C, G, and T), so long sequences would need to be understood before they could be related to properties. His breakthrough was the creative use of chemicals called dideoxy, otherwise known as chain terminators.'[32] These are in fact imperfect forms of adenine, cytosine, guanine, and thymine; when mixed with DNA polymerase, the DNA-copying enzyme, they form sequences, but incompletely – in fact they stop, are terminated, at either A, C, G, or T.[33] As a result they form DNA of varying lengths, each time stopping at the same base. Imagine, for the sake of argument, a strip of DNA that reads: CGTAGCATCGCTGAG. This, treated with adenine (A) terminators would

---

* Sanger won a *second* Nobel prize for this discovery, which meant that he joined a select band of individuals, double Nobel winners, which includes Marie Curie, John Bardeen and Linus Pauling.

produce strips in which growth stops at positions 4, 7, and 15, whereas the thymine (T) terminator would produce strips where growth stops at 3, 8, and 12, and so on. The technique actually to separate out these different strands consisted of placing the DNA in a tray of special gel, in which an electrical field had been applied to opposite ends. DNA, being negatively charged, is attracted to the positive pole, with the smaller fragments pulled faster than the larger ones, meaning that the strands eventually separate out, according to size. The DNA is then stained, and the sequence can be read. The technique was announced in *Nature* on 24 February 1977, and it was from that moment, coming on top of the cloning experiments, that genetic engineering may be said to have begun.[34]

Just over a year later, on 24 August 1978, Genentech, founded by Boyer and a young venture capitalist called Robert Swanson, announced that it had produced human insulin by this method – gene sequencing and cloning – and that it had concluded a deal with Eli Lilly, the pharmaceutical giant, for the mass manufacture of the substance. Two years later, in October 1980, when Genentech offered 1,100,000 of its shares for sale to the public, another phase in the microbiological revolution was born: offered at $35 a share, the stock immediately jumped to $89, and Boyer, who had invested just $500 in the company in early 1974, saw the value of his 925,000 shares leap to more than $80 million. No physicist was ever worth so much.[35]

Compared with the electron and other fundamental particles, the gene had taken some time to be isolated and broken down into its component parts. But as with physics, the experimental and theoretical work went in tandem.

Beginning in the 1970s a new form of literature began to appear. It grew out of Robert Ardrey's works but was more ambitious. These were books of biology but with a distinct philosophical edge. However, they were not written by journalists, or dramatists, as Ardrey was, or Gordon Rattray Taylor was, in *The Biological Time Bomb*, or by scientific popularisers, as Desmond Morris was essentially in *The Naked Ape*, but by the leading scientists themselves. These books each contained a fair amount of complex biology, but they had wider ambitions too.

The first appeared in 1970 in French and a year later in English. Its author was **Jacques Monod**, part of a three-man team that had won the Nobel Prize in 1965 for uncovering the mechanism by which genetic material synthesises protein. In *Chance and Necessity*, Monod sought to use the latest biology, since Watson and Crick's discovery of the double helix, to define life, and in considering what life itself is, went on to consider the implications that might have for ethics, politics, and philosophy. The book is almost certainly more impressive now, at the end of the century and with the benefit of hindsight, than when it was first published (it was republished by Penguin in 1997). This is because Monod's thinking foreshadowed many of the ideas promulgated by biologists and philosophers who are now much better known than Monod, authors like E. O. Wilson, Stephen J. Gould, Richard Dawkins, and Daniel Dennett.

Although a biologist, Monod's underlying insight was that life is essentially a physical and even mathematical phenomenon. His initial purpose was to show how entities in the universe can 'transcend' the laws of that universe while nevertheless obeying them. Or, as he put it, evolution does not confer 'the obligation' to exist but it does confer 'the right' to exist. For Monod, two of the great intellectual successes of the twentieth century, the free market and the transistor, share an important characteristic with life itself: amplification. The rules allow for the constituent parts to spontaneously – naturally – produce *more* of whatever system they are part of. On this reasoning there is nothing in principle unique about life.

In the technical part of his book, Monod showed how proteins and nucleic acids, the two components which all life is made from, *spontaneously* adopt certain three-dimensional forms, and that it is these three-dimensional forms which predetermine so much else. It is this spontaneous assembly that, for Monod, is the most important element of life. These substances, he says, are characterised by physical – and therefore mathematical – properties. 'Great thinkers, Einstein among them, have often ... wondered at the fact that mathematical entities created by man can so faithfully represent nature even though they owe nothing to experience.' Again Monod implies that there is nothing especially 'wonder'-ful about this – life is just as much about mathematics and physics as it is about biology. (This foreshadowed work we shall be considering in the last chapter.)

He went on to argue that evolution can only take place at all because of the ability of nucleic acids to reproduce themselves *exactly*, and this therefore means that only accident can produce mutations. In that sense, the universe was and is accidental (statistical and, therefore, again mathematical). This too, he felt, had profound implications. To begin with, evolution did not apply only to living things: adaptation is another expression of time, no less than another function of the second law of thermodynamics. Living things, as isolated, self-contained energetic systems, seem to operate against entropy, except that it is inconceivable for evolution – being a function of time – to go backwards. This implies that life, being an essentially physical phenomenon, is *temporary*: different life forms will battle against each other until a greater disorder takes over again.

No less controversially, but a good deal less apocalyptically, and anticipating the work of E. O. Wilson, Richard Dawkins, and others, Monod felt that ideas, culture, and language are survival devices, that there is survival value in myth (he avoided use of the term *religion*), but that they will in time be replaced. (He thought Christianity and Judaism more 'primitive' religions in this sense than, say, Hinduism, and implied that the latter would outlast Judaeo-Christianity.) And he felt that the scientific approach, as epitomised in the theory of evolution, which is a 'blind' process, not leading to any teleological conclusion, is the most 'objective' view of the world, in that it does not involve any one set of individuals having greater access to the truth than any other group. In this sense he thought that science disproves and replaces such ideas as animism, Bergson's vitalism, and above all Marxism, which presents itself as a scientific theory of the history of society. Monod therefore saw science not simply as a way of approaching the world, but as an *ethical* stance, from which other institutions of society could only benefit.

Not that he was blind to the problems such an attitude brought with it. 'Modern societies, woven together by science, living from its products, have become as dependent upon it as an addict on his drug. They owe their material wherewithal to this fundamental ethic upon which knowledge is based, and their moral weakness to those value-systems, devastated by knowledge itself, to which they still try to refer. The contradiction is deadly. It is what is digging the pit we see opening under our feet. The ethic of knowledge that created the modern world is the only ethic compatible with it, the only one capable, once understood and accepted, of guiding its evolution.'[36]

Monod's vision was broad, his tone tentative, as befitted someone new to philosophy, feeling his way and not trained in the discipline. His vision of 'objective knowledge' largely ignored the work of Thomas Kuhn and would come under sustained attack from philosophers in the years that followed. But not all the biologists who came after Monod were as humble. Two other books published in the mid-1970s were much more aggressive in making the link between genes, social organisation, and human nature.

In *Sociobiology: The New Synthesis* (1975), the Harvard zoologist **Edward O. Wilson** intended to show the extent to which social behavior – in all animals, including man – is governed by biology, by genes.[37] Widely read in every field of biology, and a world authority on insects, Wilson demonstrated that all manner of social behavior in insects, birds, fish, and mammals could be accounted for either by the requirements of the organism's relationship to its environment or to some strictly biological factor – such as smell – which was clearly determined by genetics. He showed how territoriality, for example, was related to food requirements, and how population was related not only to food availability but to sexual behavior, itself in turn linked to dominance patterns. He surveyed the copious evidence for birdsong, which showed that birds inherit a 'skeleton' of their songs but are able to learn a limited 'dialect' if moved.[38] He showed the importance of bombykol, a chemical substance that, in the male silkworm, stimulates the search for females, making the silkworm, according to Wilson, little more than 'a sexual guided missile.'[39] As little as one molecule of the substance is enough to set the silkworm off, he says, which shows how evolution might happen: a minute change in either bombykol or the receptor structure – equally fragile – could be enough to provoke a population of individuals sexually isolated from the parental stock. Wilson surveyed many of the works referred to earlier in this chapter – on gorillas, chimpanzees, lions, and elephants – as well as the studies of *Australopithecus*, and produced at the end of his book very contentious tables claiming to show how human societies, and human behavior, evolved. This produced a hierarchy with countries like the United States, Britain, and India at the top, Hawaii and New Guinea in the middle, and aborigines and Eskimos at the bottom.[40]

Wilson's arguments were rejected by critics as oversimple, racist (he was from America's South), and philosophically dubious; they called into question the entire concept of free will. A more technical area of controversy, but very important philosophically, related to his discussion of altruism and group selection. If evolution operated in the classical way (upon individuals), critics asked, how did altruism arise, in which one individual put another's interests before its own? How did group selection take place at all? And here the second book published in the mid-1970s provided a clearer answer. Perhaps surprisingly for nonbiologists, *The Selfish Gene* contained a fair amount of elementary mathematics.[41]

Its author, **Richard Dawkins** from Oxford, imagined in one of his crucial passages a bird population made up entirely of either hawks or doves. Hawks always fight but doves always back down. Now enters the mathematics. Dawkins attaches relative – and entirely arbitrary – values to various encounters. For example, the winner in a contest between two doves scores 50 but pays a penalty of −10 for a long staring ritual, meaning he scores 40 in all. The loser is penalised −10 for also wasting time and staring. On average therefore any one dove, in competition with other doves, can expect to win half and lose half of his contests, so the average payoff is half the difference between +40 and −10, which is +15.[42] Now assume a mutant, a hawk. He never backs down so wins every fight, at 50 a time. He enjoys a big advantage over doves whose average payoff is only +15. In such a world, there is a clear advantage to being a hawk. But now hawk genes spread through the population, and soon all fights will be hawk fights, where the winner scores +50 but the loser is so seriously injured that he scores −100. If a hawk wins half and loses half of his fights, the average payoff of all hawk fights is halfway between +50 and −100, which is −25. If, amid such society, a dove mutant arises, he loses all his fights but never gets injured, so his average payoff is 0. This may not sound much, but it beats −25. Dove genes should now start to spread. Looking at the arithmetic in this way, communities of birds would eventually arrive at an **evolutionary stable strategy** (ESS) in which 5/12 are doves and 7/12 are hawks. When this point is reached, the payoff for hawks and doves is the same, and selection does not favour either of them. The point about this admittedly simple example is to show that a group of birds can take on a certain character while selection is taking place on an individual level.

Now Dawkins moves on to a slightly more complex example. This time he asks us to assume that he is an animal who has found a clump of eight mushrooms – food. To these he attaches a value of +6 units each (again, these units are entirely arbitrary). He writes, 'The mushrooms are so big I could eat only three of them. Should I inform anybody else about my find, by giving the "food call"? Who is within earshot? Brother B (his relatedness to me is $\frac{1}{2}$ [i.e., he shares half my genes]), cousin C (relatedness to me = $\frac{1}{8}$), and D (no particular relation: his relatedness to me is some small number which can be treated as zero for practical purposes). The net benefit score to me if I keep quiet about my find will be +6 for each of the three mushrooms I eat, that is +18 in all. My net benefit score if I give the food call needs a bit of figuring. The eight mushrooms will be shared equally between the four of us. The payoff to me from the two that I shall eat will be the full +6 units each, that is +12 in all. But I shall also get some payoff when my brother and cousin eat their two mushrooms each, because of our shared genes. The actual score comes to $(1 \times 12) + (\frac{1}{2} \times 12) + (\frac{1}{8} \times 12) + (0 \times 12) = 19\frac{1}{2}$. The corresponding net benefit for the selfish behavior was $+18$: it is a close-run thing, but the verdict is clear. I should give the food call; altruism on my part would in this case pay my selfish genes.'[43] Dawkins's overriding point is that we must think of the central unit of evolution and natural selection as the gene: the gene, the replicating unit, is 'concerned' to see itself survive and thrive, and once we understand this, everything else falls into place: kinship patterns and behaviour in insects, birds, mammals, and humans are explained; altruism becomes sensible, as do the relations of non-kin groups (such as races) to one another.

Dawkins's argument, eloquently made, and Wilson's, together sparked a resurgence in Darwinian thinking that characterised the last quarter of the century. One remaining aspect of Dawkins's and Wilson's arguments is the link to Tom Wolfe, Christopher Lasch, John Rawls, and economics. They are yet another example of the way knowledge began to come together toward the end of the century. Wolfe's book *The Me Decades*, Lasch's *Culture of Narcissim*, and *The Selfish Gene* all reflect an emphasis on individuation and selfishness. They were quite different books, with ostensibly different aims, but the selfish theme common to them all was remarkable. The link to John Rawls's *Theory of Justice* is that his 'original position' and 'veil of ignorance' describe what is essentially the very opposite of the position the selfish gene is in: no one knows their inheritance, and only by not knowing, Rawls is saying, can we ever hope to arrive at a true system of fairness, a way of living life together with selfishness taken out. In Rawls's original position there are by definition no hawks or doves and no relatives. Rawls's system is all too well aware of Dawkins-type arguments and seeks to circumvent them. Daniel Bell had drawn attention to the cultural contradictions of capitalism; Rawls's ideas threw up some contradictions of Darwinism. Dawkins's ideas also show certain similarities to the market system. This arises partly from the way he attaches values to the outcomes of behavior, but though simplifications, these outcomes – gains and losses – are real enough. The situation of hawks and doves, for example, is mirrored to an extent in price-fixing agreements in humans. It is in the best interests of garage owners (say) to fix the price of petrol at one (relatively) stable price; in that way all garage owners benefit. However, the temptation always exists for a wayward 'hawk' to drop his prices for a very heavy quick profit. Of course, other garage owners would soon follow suit, until the situation again stabilised and, perhaps, price fixing is reestablished. Many democracies have laws against this sort of behavior being carried too far, but that does not cancel out the fact that in some respects evolution shares a lot of features with market economics.

On the evening of 31 January 1977, at twenty minutes past eight, regardless of who had arrived and who had not, the doors to the Galérie Beaubourg, the Centre Nationale d'Art et de Culture Georges Pompidou, in central Paris, were shut tight. President Valéry Giscard d'Estaing was about to make a speech, declaring the Pompidou Centre open, and there was no escape. Men in black tie and women in long dresses were milling on all floors, many searching for a drink – in vain, as the president, for reasons best known to himself, had decreed that there should be no refreshments. When he delivered his speech, Giscard began with a tribute to Pompidou (a former president of the French republic, who had nursed the project), snubbed Jacques Chirac, the mayor of Paris and the man in charge of the department that had commissioned the centre, and made no mention of either the building's designers or builders. Forced to listen to the president without a glass in their hands, the assembled guests wondered whether the omissions in his speech meant that he didn't like the building.[1]

Many didn't. Many thought that the **Pompidou Centre** was then, and still is, the ugliest building ever constructed. Whatever the truth of this, its importance cannot be doubted.[2] In the first place, it was intended not just as a gallery or museum but as an arts complex and library, designed to help Paris regain its place as a capital of the arts, a title it had lost since World War II and the rise of New York. Second, the centre was important architecturally because what-ever its appearance, it undoubtedly marked a robust attempt to get away from the modernist aesthetic that had predominated since the war. And third, it was important because the centre also housed IRCAM, the Institut de Recherche et de Coordination Acoustique/Musique, which was intended to become a world centre for experimental music. The directorship of IRCAM was offered to Pierre Boulez, to lure him back from America.[3]

But the significance of 'Beaubourg' was predominantly architectural. Its designers were **Renzo Piano**, an Italian, and **Richard Rogers**, from London, two of the most high-tech-minded men of the times, while the jury who selected them included **Philip Johnson**, **Jørn Utzon**, and **Oscar Niemayer**, respectively American, Danish, and Brazilian architects who had between them been responsible for some of the most famous buildings constructed since the war. Philip Johnson represented the mainstream of architecture as it had followed on from the Bauhaus – Walter Gropius, Mies van der Rohe, and Le Corbusier. In the thirty years between 1945 and 1975 most Western architecture was dominated, functionally, by two matters: the corporate building and mass housing. Following the International style (a term coined by Philip Johnson himself), architecture had devised solutions mainly by means of straight lines and flat planes, in buildings that were often either wholly black (as with Mies van der Rohe's Seagram's Tower in Manhattan) or, more usually, wholly white (as with countless housing projects). Despite heroic attempts to escape the tyranny of the straight line (zigzags, diamonds, lozenges, most notably successful in the building boom of new universities in the 1960s), too often modern architecture had resulted in what Jane Jacobs famously called the 'great blight of dullness,' or what the critic Reyner Banham labelled the 'new brutalism.' The problem, as identified by the Italian critics Manfredo Tafuri and Francesco Dal Co, was an 'obsessive preoccupation with restoring meaningful depth to a repertory of inherited forms that are devoid of meaning in themselves.'[4] The South Bank Complex in London (the ensemble that houses the National Theatre) and the Torre Velasca, near the Duomo in Milan, are good examples of these massive buildings, which feel almost menacing.

Niemayer and Utzon were notable for at least trying to break away from this tradition. Niemayer trained with Le Corbusier and became famous for his curved, shell-like concrete roofs, most notably in the new Brazilian capital, and vistas reminiscent of Giorgio de Chirico. Jørn Utzon designed many housing projects, but his most famous building was **Sydney Opera House**, in Australia, which, in its white, billowing roofs, sought to recapture the line of the sailing ships that had first discovered Australia not so very long before. Here too, though undoubtedly popular with the public (and without question strikingly original), the Opera House was perhaps too much of a one-off in function and location (on the waterfront, where it could be easily seen) to be widely imitated. Nevertheless, for all their faults, Niemayer and Utzon had tried hard to get away from the conventional architectural wisdom epitomised by Johnson, and in theory that made the Beaubourg jury a good one, the more so as it also included **Wilhelm Sandberg**, curator of the **Stedelijk** (Modern Art) **Museum** in Amsterdam, and widely regarded as the most important museum curator of the century (though Alfred Barr would surely run him close). They considered 681 valid submissions, which they reduced first to 100, then to 60, and finally to project 493 (all drawings were considered anonymously): Messrs Piano, Rogers, Franchini, architects; and Ove Arup and Partners, consultant engineers (who had worked both on the South Bank Complex and Sydney Opera House).[5]

Renzo Piano, born in 1937, was Genoese and did not consider himself only an architect but also an industrial designer – Olivetti were one of his clients. Richard Rogers was born in England in 1933 but came from a family that was mainly Italian – his cousin, Ernesto Rogers, taught Piano in Milan. A Fulbright scholar, Rogers had studied at the Architectural Association School of Architecture in London, and then at Yale, where he met his one-time partner Norman Foster, and Philip Johnson. Piano and Rogers' winning design had two main features. It did not use up all the space, an area of Paris of about seven acres that had been cleared many years before. Instead, a rectangle was left free in front of the main building for a piazza, not simply for tourists but for street theatre – jugglers, fire eaters, acrobats, and so on. A more controversial feature of the building was that its 'innards,' essentially the parts of a building that are usually hidden, such as the ducts for air conditioning, plumbing, and elevator motor rooms, were on the outside, and brightly painted, turned into a prominent design feature. One reason for this was flexibility: the building was expected to develop in the future, and the existing machinery might need to be changed.[6] Another reason was to avoid the idea that yet another 'monument' was being erected in Paris. By exposing those elements that would normally be hidden, the 'industrial' aspects of the centre were highlighted, making the building more urban.

An escalator also snaked up the building, on the outside, covered in a glass tube. This feature especially appealed to Philip Johnson.[7] The Pompidou Centre was only a shoebox festooned in ducts; yet it looked like nothing that had gone before, and certainly not like an international modern building. Like it or loathe it, Pompidou was different, and a mould was broken. It didn't inspire many copies, but it was a catalyst for change.

IRCAM was part of the specifications for the Pompidou Centre. The brief was to make it the world's pre-eminent centre for musical technology, with special studios that had absolutely no echo, advanced computers, and acoustical research laboratories, plus a hall for performances that would seat up to 500. This centre, which became known as 'Petit Beaubourg,' was originally conceived on five underground levels, with a glass roof, a library, and, in the words of Nathan Silver, the Pompidou's historian, 'studios for musical researchers from all over the world.'[8] It was cut back after Giscard became president, but even so it was enough to entice Boulez home.

Pierre Boulez was born in 1925. He was one of a handful of composers – **Karlheinz Stockhausen**, Milton Babbitt, and John Cage were others – who dominated musical innovation in the years after World War II. In the 1950s, as we have seen, serious composers had followed three main directions – serialism, electronics, and the vagaries of chance in composition. Boulez, Stockhausen, and Jean Barraqué had all been pupils of Olivier Messiaen. He, it will be recalled from chapter 23, had tried to write down the notes of birdsong, believing that all forms of sound could be made into music, and this was one important way he exerted an influence on his pupils. Stockhausen in particular was impressed by the music of Africa, Japan (where he worked in 1966), and South America, but Boulez too, in *Le Marteau sans maître* (The Hammer Unmastered), 1952–4, scored for vibraphone and xylorimba, used the rhythms of black African music. Serialism in the late compositions of Anton von Webern, who had died in 1945, was just as influential, however. Boulez described these compositions as 'the threshold,' and Stockhausen agreed, as did Milton Babbitt in America. In Europe the centre of this approach was the Kranichstein Institute in Darmstadt, where in the summer months composers and students met to discuss the latest advances. Stockhausen was a regular.[9]

Boulez was perhaps the most intellectual in a field that was, more than most, dominated by theory. For him, serialism was a search for 'an objective art of sound.' He saw himself as a scientist, an architect, or an engineer of sound as much as a composer. In a paper entitled 'Technology and the Composer,' he lamented the conservative tendencies in music which had, as he saw it, prohibited the development of new musical instruments, and this is why, according to the critic Paul Griffiths, he thought Messiaen's approach, electronic music, and the computer so important to the advance of his art form.[10] As one of his most famous compositions, *Structures*, shows, he was also concerned with 'structure,' which, he wrote, was 'the key word of our time.' In his writings Boulez made frequent references to Claude Lévi-Strauss, the Bauhaus, Ferdinand Braudel, and Picasso, each of which was a model. He had frequent meetings, some of them in public, with Jacques Lacan and Roland Barthes (see below). In a celebrated remark, he said that it was not enough to add a moustache to the *Mona Lisa*: 'It should simply be destroyed.' In order to do this, he rigorously pursued new forms of sound, where 'research' and mathematical patterns were not out of place.[11] Both Boulez and Cage used charts of numbers in setting up rhythmic structures.

Electronic music, including the electronic manipulation of natural sounds, metallic and aqueous (*musique concrète*), provided yet another avenue to explore, one that offered both new structures and a seemingly scientific element that was popular with this small group. New notations were devised, and new instruments, in particular Robert Moog's synthesiser, which arrived on the market in 1964, bringing with it a huge variety of new electronically generated sounds. Babbitt and Stockhausen both wrote a great deal of electronic music, and the latter even had a spherical auditorium (for maximum effect) built for him at the 1970 Osaka exhibition.

Chance in music was described by Paul Griffith as the equivalent of Jackson Pollock's drip paintings in art, and the swaying 'mobiles' of Alexander Calder in sculpture.[12] In America John Cage was the leading exponent; in Europe, chance arrived at Darmstadt in 1957, with Stockhausen's *Klavierstück XI* and Boulez's Piano Sonata no. 3. In Stockhausen's composition the musician was presented with a single sheet of paper containing nineteen fragments that could be played in any order. Boulez's work was less extreme: the piece was fully notated, but the musician was forced to make a choice of direction at various points.[13]

Boulez epitomised the radical character of these postwar composers, even to the extent that he questioned everything to do with music – the nature of concerts, the organisation of orchestras, the architecture of concert halls, above all the limitations imposed by existing instruments. It was this that led to the idea of IRCAM. John Cage had tried something similar in Los Angeles in the early 1950s, but Boulez didn't float the idea until May 1968, a revolutionary moment in France.[14] He was ambitious in his aims (he once said, 'What I want to do is to change people's whole mentality'). It is true, too, that Boulez, more than anyone else of his generation, more even than Stockhausen, saw himself in a sense in a 'Braudelian' way, as part of *la longue durée*, as a stage in the evolution of music. This was why he wanted IRCAM to make music more 'rational' (his word) in its search for creativity, in its employment of machines, like the '4X,' which was capable of 'generating' music.[15] In May 1977, in the *Times Literary Supplement*, Boulez set out his views. 'Collaboration between scientists and musicians – to stick to those two generic terms which naturally include a large number of more specialised categories – is therefore a necessity that, seen from the outside, does not appear to be inevitable. An immediate reaction might be that musical invention can have no need of a corresponding technology; many representatives of the scientific world see nothing wrong with this and justify their apprehensions by the fact that artistic creation is specifically the domain of intuition, of the irrational. They doubt whether this utopian marriage of fire and water would be likely to produce anything valid. If mystery is involved, it should remain a mystery: any investigation, any search for a meeting point is easily taken to be sacrilege. Uncertain just what it is that musicians are demanding from them, and what possible terrain there might be for joint efforts, many scientists opt out in advance, seeing only the absurdity of the situation.'[16] But, he goes on, 'In the end, musical invention will have somehow to learn the language of technology, and even to appropriate it. ... A virtual understanding of contemporary technology ought to form part of the musician's invention; otherwise, scientists, technicians and musicians will rub shoulders and even help one another, but their activities will only be marginal one to the other. Our grand design today, therefore, is to prepare the way for their integration and, through an increasingly pertinent dialogue, to reach a common language that would take account of the imperatives of musical invention and the priorities of technology. ... Future experiments, in all probability, will be set up in accordance with this permanent dialogue. Will there be many of us to undertake it?'[17]

The French Connection, William Friedkin's 1971 film about the Mafia and drug running into America, wasn't really about France (some of the villains are French-speaking Canadians), but the film's title did catch on as a description of something that was notable in philosophy, psychology, linguistics, and epistemology just as much as in historiography, anthropology, and music. This was a marked divergence between French thought and Anglo-Saxon thought that proved fruitful and controversial in equal measure. In the United States, Britain, and the rest of the English-speaking world, the Darwinian metanarrative was in the ascendant. But in France in particular the late 1960s, 1970s, and 1980s also saw a resurgence of the other two great nineteenth-century metanarratives: Freudianism and Marxism. It was not always easy to distinguish between these theories, for several authors embraced both, and some – generally French but also German – wrote in such a difficult and paradoxical style that, especially after translation, their language was extremely dense and often obscure. In the sections that follow, I have relied on the accessible commentaries quoted, in addition to the works themselves, in an effort to circumvent their obscurity. These French thinkers do represent a definite trend.

Jacques Lacan was a psychoanalyst in the Freudian tradition, who developed in highly idiosyncratic ways. Born in 1901 in Paris, in the 1930s Lacan attended Alexandre Kojève's seminars on Hegel and Heidegger, along with Raymond Aron, Jean-Paul Sartre, André Breton, Maurice Merleau-Ponty, and Raymond Queneau (see chapter 23). Psychoanalysis was not taken up as quickly in France as in the US, and so it wasn't until Lacan began giving his public seminars in 1953, which lasted for twenty-six years, that psychoanalysis in France was taken seriously. His seminars were intellectually fashionable, with 800 people crammed into a room designed for 650, with many prominent intellectuals and writers in the audience. Although the Marxist philosopher Louis Althusser thought enough of Lacan to invite him in 1963 to transfer his seminar to the Ecole Normale Supérieure, Lacan was forced to resign from the Société Psychoanalytique de Paris (SPP) and was expelled from the International Psychoanalytic Association, because of his 'eclectic' methods. After May 1968, the Department of Psychoanalysis at Vincennes (part of Paris University) was reorganised as Le Champ Freudien with Lacan as scientific director. Here was the mix of Freudianism and Marxism in action.[18]

Lacan's first book, *Ecrits* (Writings), published in 1966, contained major revisions of Freudianism, including the idea that there is no such thing as the ego.[19] But the aspect of Lacan's theory that was to provoke widespread attention, and lead on from Ludwig Wittgenstein and R. D. Laing, was his attention to language.[20] Like Laing, Lacan believed that going mad was a rational response to an intolerable situation; like Wittgenstein, he believed that words are imprecise, meaning both more and less than they appear to mean to either the speaker or the hearer, and that it was the job of the psychoanalyst to understand this question of meaning, as revealed through language, in the light of the unconscious. Lacan did not offer a cure, as such; for him psychoanalysis was a technique for listening to, and questioning, 'desire.' In essence, the language revealed in psychoanalytic sessions was the language of the unconscious uncovering, 'in tortured form,' desire. The unconscious, says Lacan, is not a private region inside us. It is instead the underlying and unknown pattern of our relations with one another, as mediated by language. Influenced by surrealism, and by the linguistic theories of Ferdinand de Saussure, Lacan became fascinated by the devices of language. For him there are 'four modes of discourse' – those of the master, the university, the hysteric, and the psychoanalyst, though they are rarely seen in pure form, the categories existing only for the purpose of analysis. A final important concept of Lacan was that there is no such thing as the whole truth, and it is pointless waiting until that point has been reached. Lacan liked to say that the patient terminates his psychoanalytic treatment when he realises that it can go on for ever. This is the use of language in the achievement of meaning: language brings home to the patient the true nature – the true meaning – of his situation. This is one reason, say Lacan's followers, why his own writing style is so dense and, as we would normally describe it, obscure. The reader has to 'recover' his own meaning from the words, just as a poet does in composing a poem (though presumably a poet's recovered meanings are more generally accessible than the patient's).[21] This is of course an oversimplification of Lacan's theories. Toward the end of his life he even introduced mathematical symbols into his work, though this does not seem to have made his ideas much clearer for most people, and certainly not for his considerable number of critics, who believe Lacan to have been eccentric, confused, and very wrong. Not least among the criticisms is that despite a long career in Paris, in which he made repeated attempts to synthesise Freud with Hegel, Spinoza, Heidegger, and the existentialism of Sartre, he nevertheless ignored the most elementary developments in biology and medicine. Lacan's enduring legacy, if there is one, was to be one of the founding fathers of 'deconstruction,' the idea that there is no intrinsic meaning in language, that the speaker means more and less than he or she knows, and that the listener/hearer must play his or her part. This is why his ideas lived on for a time not just in psychology but in philosophy, linguistics, literary criticism, and even in film and politics.

Among psychiatrists, none was so political and influential as **Michel Foucault**. His career was as interesting as his ideas. Born in Poitiers, in October 1926, Paul-Michel Foucault trained at the Ecole Normale Supérieure. One of *les grandes écoles*, the ENS was especially grand, all-resident, its graduates known as *normaliens*, supplying universities with teachers. There Foucault came under the friendship, protection, and patronage of Louis Althusser, a slender man with 'a fragile, almost melancholy beauty.' Far from well, often in analysis and even electroshock treatment, Althusser had a huge reputation as a grand theorist.[22] Foucault failed his early exams – to general consternation – but after he developed an interest in psychiatry, especially in the early years of the profession and its growth, his career blossomed. The success of his books brought him into touch with very many of the luminaries of French intellectual culture: Claude Lévi-Strauss, Roland Barthes, Ferdinand Braudel, Alain Robbe-Grillet, Jacques Derrida, and Emmanuel le Roy Ladurie. Following the events of 1968, he was elected to the chair in philosophy at the new University of Vincennes.[23] The University of Vincennes, officially known as the Vincennes Experimental University Centre, 'was the offspring of May 1968 and Edgar Faure,' the French minister for education. 'It was resolutely interdisciplinary, introduced novel courses on cinema, semiotics and psychoanalysis, and was the first French university to open its doors to candidates who did not have the *baccalauréat*.' 'It therefore succeeded in attracting (for a time) many wage earners and people outside the normal university recruitment pool.' 'The atmosphere . . . was like a noisy beehive.'[24] This aspect of Foucault's career, plus his well-publicised use of drugs, his involvement with the anti-Vietnam protests, his part in the campaign for prison reform, and his role in the gay liberation movement, show him as a typical central figure in the counter-culture. Yet at the same time, in April 1970 Foucault was elected to the Collège de France, a major plank in the French establishment, to a chair in the History of Systems of Thought, specially created around him. This reflected the very substantial body of work Foucault had amassed by that stage.[25]

Foucault shared with Lacan and Laing the belief that mental illness was a social construct – it was what psychiatrists, psychologists, and doctors said it was, rather than an entity in itself. In particular, he argued that modern societies control and discipline their populations by delegating to the practitioners of the human sciences the authority to make these decisions.[26] These sciences of man, he said, 'have subverted the classical order of political rule based on sovereignty and rights and replaced them with a new regime of power exercised through the stipulation of norms for human behavior'. As Mark Philp has put it, we now know, or think we know, what 'the normal child' is, what 'a stable mind' is, a 'good citizen,' or the 'perfect wife.' In describing normality, these sciences and their practitioners define deviation. These laws – 'the laws of speech, of economic rationality, of social behavior' – define who we are. For Foucault, this idea, of 'man as a universal category, containing within it a "law of being,"' is ... an invention of the Enlightenment' and both mistaken and unstable. The aim of his books was to aid the destruction of this idea and to argue that there is no 'single, cohesive human condition.' Foucault's work hung together with a rare consistency. His most important books examine the history of institutions: *Madness and Civilisation: A History of Insanity in the Age of Reason* (1964); *The Archaeology of Knowledge* (1969); *The Order of Things: An Archaeology of the Human Sciences* (1971); *The Birth of the Clinic: An Archaeology of Medical Perception* (1972); *Discipline and Punish: The Birth of the Prison* (1975); *The History of Sexuality* (1976).

But Foucault was not just writing a history, of psychiatry, penology, economics, biology, or philology, as the case may be. He was seeking to show how the way knowledge is organised reflects the power structures within a society and how the definition of the normal man, or mind, or body, is as much a political construct as one that reflects 'the truth.'[27] 'We are subject to the production of truth through power,' Foucault wrote. It is the human sciences, he says, that have given us the conception of a society as an organism 'which legitimately regulates its population and seeks out signs of disease, disturbance and deviation so that they can be treated and returned to normal functioning under the watchful eye of one or other policing systems.' Again, as Philp has emphasised, these are revealingly known as 'disciplines.' Foucault calls his books 'archaeologies' rather than histories because, as Lacan saw meaning as a 'recovering activity,' Foucault saw his work too as an excavation that not only describes the processes of the past but goes beyond, to recreate 'buried' knowledge. There was something of *l'homme revolté* about Foucault; he believed that man could only exist if he showed a 'recalcitrance' towards the normative pressures of the human sciences, and that there is no coherent or constant human 'condition' or 'nature,' no rational course to history, no 'gradual triumph of human rationality over nature.' There is struggle, but it is 'patternless.' His final argument in this vein was to show that bourgeois, humanistic man had 'passed.' Liberal humanism, he said, was shown up as a sham, disintegrating as it revealed itself as an instrument of class power and the socially privileged.[28] The individual subject, with a conscience and reason, is out-of-date in the modern state, intellectually, morally, and psychologically deconstructed.

Foucault's last important book was an investigation of the history of sexuality, in which he argued that but for rape and sex with children, there should be no restraint on behavior. This was entirely in line with the rest of his oeuvre, but for him it had the unfortunate consequence that the development of gay bars and bathhouses, of which he positively approved (he adored California and went there a lot), were probably responsible for the fact that he died, in June 1984, of an AIDS-related illness.

From psychiatry to psychology: **Jean Piaget**, the Swiss psychologist, was not only interested in child development and the systematic growth of intelligent behavior. Later in life, his interests widened, and using the ideas of Foucault and Lacan, he became a leading advocate of a mode of thought known as structuralism. Piaget's arguments also drew on the work of Noam Chomsky, but structuralism was really a concept developed in continental, and especially francophone, Europe, largely in ignorance of more empirical findings in the Anglo-Saxon, English-speaking world. This is one reason why many people outside France do not find it easy to say what structuralism is. Piaget's *Structuralism* (1971) was one of the clearer expositions.[29] Just as Foucault used the word *archaeology*, rather than history, to imply that he was uncovering something – structures – that already existed, so Piaget implied that there are '**mental structures**' that exist midway between genes and behavior. One of his starting points, for example, is that 'algebra is not "contained" in the behavior of bacteria or viruses' and – by implication, because they are relatively similar – in genes.[30] The capacity to act in a mathematical way, either as a bacterium (in dividing), or in the human, by adding and subtracting, is according to Piaget only partially inherited. Part of the ability arises from mental structures built up as the organism develops and encounters the world. For Piaget the organisation of grammar was a perfect example of a mental structure, in that it was partly inherited and partly 'achieved,' in the sense that Lacan thought patients achieved meaning in analysis.

'If asked to "locate" these structures,' Piaget writes, 'we would assign them a place somewhere midway between the nervous system and conscious behavior' (wherever that might be).[31] To add to the confusion, Piaget does not claim that these structures actually exist physically in the organism; structures are theoretic, deductive, a process. In his book he ranges widely, from the mathematical ideas of Ludwig von Bertalanffy, to the economics of Keynes, to Freud, to the sociology of Talcott Parsons. His main concern, however, is with mental structures, some of which, he believes, are formed unconsciously and which it is the job of the psychologist to uncover. Piaget's aim, in drawing attention to these mental structures, was to show that human experience could not be understood either through the study of observable behavior or through physiological processes, that 'something else' was needed.[32] Piaget, more than most of his continental counterparts, was aware of the contemporary advances being made in evolutionary biology and psychology: no one could accuse him of not doing the work. But his writings were still highly abstract and left a lot to be desired in the minds of his Anglo-Saxon critics.[33] So Piaget regarded the perfect life as an achieved structure, within biological limits but creatively individual also. The mind develops or matures, and the process cannot be hurried. One's understanding of life, as one grows up, is mediated by a knowledge of mathematics and language – two essentially logical systems of thought – which help us handle the world, and in turn help organise that world. For Piaget, the extent to which we develop our own mental constructs, and the success with which they match the world, affects our happiness and adjustment. The unconscious may be seen as essentially a disturbance in the system that it is the job of the psychoanalyst to resolve.

After the success of structuralism, it was no doubt inevitable that there should be a backlash. **Jacques Derrida**, the man who mounted that attack, was Algerian and Jewish. In 1962, at independence, the Jews in Algeria left *en masse*: France suddenly had the largest Jewish population on the continent, west of Russia.

Derrida began with a specific attack on a specific body of work. In France in the 1960s, Claude Lévi-Strauss was not merely an anthropologist – he had the status of a philosopher, a guru, someone whose structuralist views had extended well beyond anthropology to embrace psychology, philosophy, history, literary criticism, and even architecture.[34] We also have Lévi-Strauss to thank for the new term 'human sciences,' *sciences* of the human, which he claimed had left behind the 'metaphysical preoccupations of traditional philosophy' and were offering a more reliable perspective on the human condition. As a result, the traditional role of philosophy as 'the privileged point of synthesis of human knowledge' seemed increasingly vitiated: 'The human sciences had no need of this kind of philosophy and could think for themselves.'[35] Among the people to come under attack were Jean-Paul Sartre and the linguist Ferdinand de Saussure. Lévi-Strauss belittled the 'subjectivist bias' of existentialism, arguing that a philosophy based on personal experience 'can never tell us anything essential either about society or humanity.' Being an anthropologist, he also attacked the ethnocentric nature of much European thought, saying it was too culture-bound to be truly universal.

Derrida took Lévi-Strauss to task – for being imprisoned within his own viewpoint in a much more fundamental way. In *Tristes Tropiques*, Lévi-Strauss's autobiographical account of how and why he chose anthropology and his early fieldwork in Brazil, he had explored the link between writing and secret knowledge in primitive tribes like the Nambikwara.[36] This led Lévi-Strauss to the generalisation that 'for thousands of years,' writing had been the privilege of a powerful elite, associated with caste and class differentiation, that 'its primary function' was to 'enslave and subordinate.' Between the invention of writing and the advent of science, Lévi-Strauss said, there was 'no accretion of knowledge, just fluctuations up and down.'[37]

Derrida advanced a related but even more fundamental point. Throughout history, he said, writing was treated with less respect than oral speech, as somehow less reliable, less authoritative, less authentic.[38] Combined with its 'controlling' aspects, this makes writing 'alienating,' doing 'violence' to experience. Derrida, like Lacan and Foucault especially, was struck by the 'inexactitudes and imprecisions and contradictions of words,' and he thought these shortcomings philosophically important. Going further into Lévi-Strauss's text, he highlighted logical inconsistencies in the arguments, concepts that were limited or inappropriate. The Nambikwara, Derrida says, have all sorts of 'decorations' that, in a less ethnocentric person, might be called 'writing.' These include calabashes, genealogical trees, sketches in the soil, and so on, all of which undoubtedly have *meaning*. Lévi-Strauss's writing can never catch these meanings, says Derrida. He has his own agenda in writing his memoir, and this leads him, more or less successfully, to write what he does. Even then, however, he makes mistakes: he contradicts himself, he makes things appear more black and white than they are, many words only describe part of the things they refer to. Again, all common sense. But again Derrida is not content with that. For him, this failure of complete representation is as important as it is inevitable. For Derrida, as with Lacan, Foucault, and Piaget, language is the most important mental construct there is, something that (perhaps) sets man apart from other organisms, the basic tool of thought and therefore essential – presumably – to reason (though also of corruption).[39] For Derrida, once we doubt language, 'doubt that it accurately represents reality, once we are conscious that all individuals are ethnocentric, inconsistent, incoherent to a point, oversimplifiers ... then we have a new concept of man.' Consciousness is no longer what it appears to be, nor reason, nor meaning, nor – even – intentionality.[40] Derrida questions whether any single utterance by an individual can have one meaning even for that person. To an extent, words mean both more and less than they appear to, either to the person producing them or someone hearing or reading them.

This gap, or 'adjournment' in meaning, he labelled *the différance* and it led on to the process Derrida called '**deconstruction**,' which for many years proved inordinately popular, notorious even. As Christopher Johnson says, in his commentary on Derrida's ideas, deconstruction was an important ingredient in the postmodern argument or sensibility, enabling as many readings of a text as there are readers.[41] Derrida wasn't being entirely arbitrary or perverse here. He meant to say (in itself dangerous) not only that people's utterances have unconscious elements, but also that the words themselves have a history that is greater than any one person's experience of those words, and so anything anyone says is almost bound to mean more than that person means. This too is no more than extended common sense. Where Derrida grows controversial, or non-commonsensical, is when he argues that the nature of language robs even the speaker of any authority over the meaning of what he or she says or writes.[42] Instead, that 'meaning resides in the structure of language itself: we think only in signs, and signs have only an arbitrary relationship to what they signify.'[43] For Derrida, this undermines the very notion of philosophy as we (think we) understand it. For him there can be no progress in human affairs, no sense in which there is an accumulation of knowledge 'where what we know today is "better," more complete, than what was known yesterday.' It is simply that old vocabularies are seen as dead, but 'that too is a meaning that could change.' On this account even *philosophy* is an imprecise, incoherent, and therefore hardly useful word.

For Derrida, the chief aspect of the human condition is its 'undecided' quality, where we keep giving meanings to our experience but can never be sure that those meanings are the 'true' ones, and that in any case 'truth' itself is an unhelpful concept, which itself keeps changing.[44] 'Truth is plural.' There is no progress, there is no one truth that, 'if we read enough, or live life enough, we can finally grasp: everything is undecided and always will be.' We can never know exactly what we mean by anything, and others will never understand us exactly as we wish to be understood, or think that we are being understood. *That* (maybe) is the postmodern form of anomie.

Like Derrida, **Louis Althusser** was born in Algeria. Like Derrida, says Susan James, he was more Marxist than Marx, believing that not even the great revolutionary was 'altogether aware of the significance of his own work.' This led Althusser to question the view that the world of ideology and the empirical world are related. For example, 'the empirical data about the horrors of the gulag do not necessarily lead one to turn against Stalin or the USSR.' For Althusser, thinking along the same lines as Derrida, empirical data do not carry with them any one meaning; therefore one can (and Althusser did) remain loyal to, say, Stalin and the ideology of communism despite disparate events that happened inside the territory under Stalin's control. Althusser also took the view that history is overdetermined: so many factors contribute to one event, or phenomenon – be they economic, social, cultural, or political – that it is impossible to specify causes: 'There is, in other words, no such thing as a capability of determining the cause of a historical event. Therefore one can decide for oneself what is at work in history, which decision then constitutes one's ideology. Just as economic determinism cannot be proved, it cannot be disproved either. The theory of history is something the individual works out for himself; necessarily so, since it does not admit of empirical and rational demonstration.'[45] In any case, Althusser says, individuals are so much the creation of the social structures they inhabit that their intentions are to be regarded as consequences, rather than causes, of social practice.[46] More often than not, all societies – and especially capitalist societies – have what he calls Ideological State Apparatuses: the family, the media, schools, and churches, for example, which propagate and receive ideas, so much so that we are not really self-conscious agents. 'We acquire our identity as a result of the actions of these apparatuses.'[47] In Marxist terms, the key to Althusser is the relative autonomy of the superstructure, and he replaced the false consciousness of class, which Marx had made so much of, and 'substituted the false consciousness of ideology and individual identity, the aim being to shake people out of their ideological smugness and create a situation where change could be entertained.'[48] Unfortunately, his published ideas stopped in 1980 after he murdered his wife and was declared unfit to stand trial.

With their scepticism about language, especially as it relates to knowledge and its links with power in the search for meaning, structuralism and deconstruction are the kin of cultural studies, as outlined by Raymond Williams, with Marx looming large in the background. Taken together they amount to a criticism of both capitalist/materialist society and the forms of knowledge produced by the natural sciences.

The most frontal attack on the sciences also came from the continent, by **Jürgen Habermas**. Habermas is the latest major philosopher in the tradition of the Frankfurt School, the school of Horkheimer, Benjamin, Adorno, and Marcuse, and like theirs his aim was a modern synthesis of Marx and Freud. Habermas accepted that the social conditions that obtained when Marx was alive have changed markedly and that, for example, the working class long ago became 'integrated into capitalist society, and is no longer a revolutionary force.'[49] Anthony Giddens has drawn attention to the fact that Habermas shared with Adorno the view that Soviet society was a 'deformed' version of a socialist society. There are, Habermas said, two things wrong with regarding the study of human social life as a science on a par with the natural sciences. In the first place there is a tendency in modern intellectual culture to overestimate the role of science 'as the only valid kind of knowledge that we can have about either the natural or the social world.'[50] Second, science 'produces a mistaken view of what human beings are like as capable, reasoning actors who know a great deal about why they act as they do.' There cannot be 'iron laws' about people, says Habermas, criticising Marx as much as natural scientists. Otherwise there would be no such thing as humans. Instead, he says, humans have self-reflection or reflexivity, intentions and reasons for what they do. No amount of natural science can ever explain this. His more original point was that knowledge, for him, is emancipatory: 'The more human beings understand about the springs of their own behaviour, and the social institutions in which that behaviour is involved, the more they are likely to be able to escape from constraints to which previously they were subject.'[51] A classic case of this, says Habermas, occurs in psychoanalysis. The task of the analyst is to interpret the feelings of the patient, and when this is successful, the patient gains a greater measure of rational control over his or her behaviour – meanings and intentions change, entities that cannot be represented by the natural sciences.[52] He envisages an emancipated society in which all individuals control their own destinies 'through a heightened understanding of the circumstances in which they live.'[53] In fact, says Habermas, there is no single mould into which all knowledge can fit. Instead it takes three different forms – and here he produced his famous three-part argument, summed up in the following table which I have taken from Giddens:[54]

Aspects of Human Society	Knowledge-Constitutive interest	Type of Study
Labour	Prediction and control	Empirical-analytic sciences
Interaction	Understanding of meaning	Historical-hermeneutic disciplines
Domination (power)	Emancipation	Critical theory

The 'hard sciences' occupy the top row, activities like psychoanalysis and philosophy occupy the middle row, and critical theory, which we can now see really includes all the thinkers of this chapter, occupies the bottom row. As Foucault, Derrida, and the others would all agree, the understanding of the link between knowledge and power is the most emancipatory level of thinking.

What the French thinkers (and Habermas) produced was essentially a post-modern form of Marxism. Some of the authors seem reluctant to abandon Marx, others are keen to update him, but no one seems willing to jettison him entirely. It is not so much his economic determinism or class-based motivations that are retained as his idea of 'false consciousness,' expressed through the idea that knowledge, and reason, must always be forged or mediated by the power relations of any society – that knowledge, hermeneutics, and understanding always serve a purpose. Just as Kant said there is no pure reason, so, we are told from the Continent, there is no pure knowledge, and understanding this is emancipatory. While it would not be true to say that these writers are anti-scientific (Piaget, Foucault, and Habermas in particular are too well-informed to be so crude), there is among them a feeling that science is by no means the only form of knowledge worth having, that it is seriously inadequate to explain much, if not most, of what we know. These authors do not exactly ignore evolution, but they show little awareness of how their theories fit – or do not fit – into the proliferation of genetic and ethological studies. It is also noticeable that almost all of them accept, and enlist as support, evidence from psy-choanalysis. There is, for anglophone readers, something rather unreal about this late continental focus on Freud, as many critics have pointed out. Finally, there is also a feeling that Foucault, Lacan, and Derrida have done little more than elevate small-scale observations, the undoubted misuses of criminals or the insane in the past, or in Lacan's case vagaries in the use of language, into entire edifices of philosophy. Ultimately, the answer here must lie in how convincing others find their arguments. None has found universal acceptance.

At the same time, the ways in which they have subverted the idea that there is a general canon, or one way of looking at man, and telling his story, has undoubtedly had an effect. If nothing else, they have introduced a scepticism that Eliot and Trilling would have approved. In 1969, in a special issue of *Yale French Studies*, structuralism crossed the Atlantic. Postmodernist thought had a big influence on philosophy in America, as we shall see.

Roland Barthes is generally considered a poststructuralist critic. Born in 1915 in Cherbourg, the son of a naval lieutenant, he grew up with a lung illness that made his childhood painful and solitary. This made him unfit for service in World War II, during which he began his career as a literature teacher. A homosexual, Barthes suffered the early death of a lover (from TB), and the amount of illness in his life even led him to begin work on a medical degree. But during his time in the sanatorium, when he did a lot of reading, he became interested in Marxism and for a time was on the edge of Sartre's milieu. After the war he took up appointments in both Bucharest (then of course a Marxist country) and Alexandria in Egypt. He returned to a job in the cultural affairs section of the French Foreign Office. The enforced solitude, and the travel to very different countries, meant that Barthes added to his interest in literature a fascination with language, which was to make his name. Beginning in 1953, Barthes embarked on a series of short books, essays mainly, that drew attention to language in a way that gradually grew in influence until, by the 1970s, it was the prevailing orthodoxy in literary studies.[55]

The Barthes phenomenon was partly a result of the late arrival of Freudianism into France, as represented by Lacan. There was also a sense in which Barthes was a French equivalent of Raymond Williams in Cambridge. Barthes's argument was that there was more to modern culture than met the eye, that modern men and women were surrounded by all manner of signs and symbols that told them as much about the modern world as traditional writing forms. In *Mythologies* (1957, but not translated into English until 1972), Barthes focused his gaze on specific aspects of the contemporary world, and it was his choice of subject, as much as the content of his short essays, that attracted attention.[56] He was in essence pointing to certain aspects of contemporary culture and saying that we should not just let these phenomena pass us by without inspection or reflection. For example, he had one essay on margarine, another on steak and chips, another on soap powders and detergents. He was after the 'capillary meanings' of these phenomena. This is how he began his essay 'Plastic': 'Despite having names of Greek shepherds (Polystyrene, Polyvinyl, Polyethylene), plastic, the products of which have just been gathered in an exhibition, is in essence the stuff of alchemy . . . as its everyday name indicates, it is ubiquity made visible . . . it is less a thing than the trace of a movement. . . . But the price to be paid for this success is that plastic, sublimated as movement, hardly exists as a substance. . . . In the hierarchy of the major poetic substances, it figures as a disgraced material, lost between the effusiveness of rubber and the flat hardness of metal. . . . What best reveals it for what it is is the sound it gives, at once hollow and flat; its noise is its undoing, as are its colours, for it seems capable of retaining only the most chemical-looking ones. Of yellow, red and green, it keeps only the aggressive quality.'[57]

Barthes's Marxism gave him, like Sartre, a hatred of the bourgeoisie, and his very success in the analysis of the signs and symbols of everyday modern life (semiology, as it came to be called) turned him against the scientific stance of the structuralists. Fortified by Lacan's ideas about the unconscious, Barthes came down firmly on the side of humanistic interpretation, of literature, film, music. His most celebrated essay was '**The Death of the Author**,' published in 1968, though again not translated into English until the 1970s.[58] This echoed the so-called New Criticism, in the 1940s in America in particular, where the dominant idea was 'the intentional fallacy.' As refined by Barthes, this view holds that the intentions of an author of a text do not matter in interpreting that text. We all read a new piece of work having read a whole range of works earlier on, which have given words particular meanings that differ subtly from one person to another. An author, therefore, simply cannot predict what meaning his work will have for others. In *The Pleasures of the Text* (1975), Barthes wrote, 'On the stage of the text, no footlights: there is not, behind the text, someone active (the writer) and out front someone passive (the reader); there is not a subject and an object.'[59] 'The pleasure of the text is that moment when my body pursues its own ideas.'[60] Like Raymond Williams, Barthes was aware that all writing, all creation, is bound by the cultural context of its production, and he wanted to help people break out of those constraints, so that reading, far from being a passive act, could be more active and, in the end, more enjoyable. He was given a rather bad press in the Anglo-Saxon countries, although he became very influential nonetheless. At this distance, his views seem less exceptional than they did.* But he was such a vivid writer, with a gift for phrasemaking and acute observation, that he cannot be dismissed so easily by Anglo-Saxons.[61] He wanted to show the possibilities within language, so that it would be liberating rather than constricting. A particularly good example of this would occur a few years later, when Susan Sontag explored the metaphors of illness, in particular cancer and AIDS.

Among the systems of signs and symbols that Barthes drew attention to, a special place was reserved for film (Garbo, Eisenstein, Mankiewicz's *Julius Caesar*), and here there was an irony, an important one. For the first three decades after World War II, Hollywood was actually not as important as it is now: the most interesting creative innovations were going on elsewhere – and they were *structural*. Second, and herein lies another irony, the European film business, and the French film industry in particular, which was the most creative of all, was building the idea of the director (rather than the writer, or the actor or the cameraman) as *author*.

Hollywood went through several changes after the war. Box-office earnings for 1946, the first full year of peace, were the highest in U.S. film history and, allowing for inflation, may be a record that still stands. But then Hollywood's fortunes started to wane; attendances shrank steadily, so that in the decade between 1946 and 1957, 4,000 cinemas closed. One reason was changing lifestyles, with more people moving to the suburbs, and the arrival of television. There was a revival in the 1960s, as Hollywood adjusted to TV, but it was not long-lived, and between 1962 and 1969 five of the eight major studios changed hands, losing some $500 million (more than $4 billion now) in the process. Hollywood recovered throughout the 1970s, and a new generation of 'movie brat' directors led the industry forward. They owed a great deal to the idea of the film director as *auteur*, which matured in Europe.

The idea itself had been around, of course, since the beginning of movies. But it was the French, in the immediate postwar period, who revived it and popularised it, publicising the battle between various critics as to who should take preeminence in being credited with the success of a film – the screenwriter or the director. In 1951 **Jacques Doniol-Valcroze** founded the monthly magazine *Cahiers du cinéma*, which followed the line that films 'belonged' to their directors.[62] Among the critics who aired their views in *Cahiers* were Eric

---

* Barthes's views contain at least one contradiction. If an author's intentions mean little, how can Barthes's own views mean anything?

Rohmer, Claude Chabrol, Jean-Luc Godard, and François Truffaut. In a famous article Truffaut distinguished between 'stager' films, in which the director merely 'stages' a movie, written by a screenwriter, and proper *auteurs*, whom he identified as Jean Renoir, Robert Bresson, Jean Cocteau, Jacques Tati and Max Ophüls. This French emphasis on the *auteur* helped lead to the golden age of the art cinema, which occupied the 1950s, 1960s, and early 1970s.

Robert Bresson's early postwar films had a religious, or spiritual, quality, but he turned pessimistic as he grew older, focusing on the everyday problems of young people.[63] *Une Femme douce* (A Gentle Woman, 1969) is an allegory: a film about a simple woman who commits suicide without explanation. Her grief-stricken husband recalls their life together, its ups and downs, his late realisation of how much he loves her. Don't let life go by default, Bresson is saying, catch it before it's too late. *Le Diable probablement* (The Devil Probably, 1977) is one of Bresson's most minimal films. Again the main character commits suicide, but the 'star' of this film is Bresson's technique, the mystery and unease with which he invests the film, and which cause the viewer to come away questioning his/her own life.

The stumbling career of **Jacques Tati** bears some resemblance to that of his character Mr Hulot: he went bankrupt and lost control of his film prints.[64] Tati's films included *Holiday* (1949) and *Mr. Hulot's Holiday* (1953), but his best known are *Mon Oncle* (1958) and *Playtime* (1967). Mr. Hulot returns to the screen in the earlier of the two, again played by Tati himself, an ungainly soul who staggers through life in his raincoat and umbrella. In this manner he happens upon the family of his sister and brother-in-law, whose house is full of gadgets. Tati wrings joke after joke from these so-called labour-saving devices, which in fact only delay the forward movement, and enjoyment, of life.[65] Hulot forms a friendship with his nephew, Gerald, a sensitive boy who is quite out of sympathy with all that is going on around him. Tati's innovativeness is shown in the unusual shots, where sometimes more than one gag is going on at the same time, and his clever staging. In one famous scene, Hulot's in-laws walk back and forth in front of their round bedroom windows in such a way as to make us think that the house itself is rolling its eyes. *Playtime* (1967) was as innovative as anything Tati or Bresson had done. There is no main character in the film, and hardly any plot. Like *Mon Oncle* it is a satire on the shiny steel and gadgetry of the modern world. In a typical Tati scene, many visual elements are introduced, but they are not at all obvious – the viewer has to work at noticing them. But, once you learn to look, you see – and what Tati is seeing is the ordinary world around us, plotless, which we must incorporate into our own understanding. The parallel with Barthes, and Derrida, is intentional. It is also more fun.

The new wave or 'Left Bank' directors in France all derived from the influence of *Cahiers du cinéma*, itself reflecting the youth culture that flourished as a result of the baby boom. There was also a maturing of film culture: beginning in the late 1950s, international film festivals, with the emphasis on the international, proliferated: San Francisco and London, in 1957; Moscow two years later; Adelaide and New York in 1963; Chicago and Panama in 1965;

Brisbane a year after that; San Antonio, Texas, and Shiraz, Iran, in 1967. Cannes, the French mother of them all, had begun in 1939. Abandoned when Hitler invaded Poland, it was resumed in 1946.

The new wave directors were characterised by their technical innovations brought about chiefly by lighter cameras which allowed greater variety in shots – more unusual close-ups, unexpected angles, long sequences from far away. But their main achievement was a new directness, a sense of immediacy that was almost documentary in its feel. This golden age was responsible for such classics as *Les Quatre Cents Coups* (Truffaut, 1959), *Hiroshima mon amour* (Alain Resnais, 1959), *A bout de souffle* (Godard, 1960), *Zazie dans le métro* (Louis Malle, 1960), *L'Année dernière à Marienbad* (Resnais, 1961), *Jules et Jim* (Truffaut, 1962), *Cléo de 5 à 7* (Agnès Varda, 1962), *La Peau douce* (Truffaut, 1964), *Bande à part* (Godard, 1964), *Les Parapluies de Cherbourg* (Jacques Demy, 1964), *Alphaville* (Godard, 1965), *Fahrenheit 451* (Truffaut, 1966), *Deux ou Trois Choses que je sais d'elle* (Godard, 1967), *Ma nuit chez Maud* (Eric Rohmer, 1967), *La Nuit américaine* (Truffaut, 1973, translated into English as *Day for Night*).[66]

Most famous of the technical innovations was **Truffaut's 'jump-cut,'** removing frames from the middle of sequences to create a jarring effect that indicated the passage of time (a short time in particular) but also emphasised a change in emotion. There was a widespread use of the freeze frame, most notably in the final scene of *Les Quatre Cents Coups*, where the boy, at the edge of the sea, turns to face the audience. This often left the ending of a film open in a way that, combined with the nervous quality introduced by the jump-cuts, sometimes caused these films to be labelled 'existentialist' or 'deconstructionist,' leaving the audience to make what it could of what the director had offered.[67] The ideas of Sartre and the other existentialists certainly did influence the writers at *Cahiers*, as did *la longue durée* notions of Braudel, seen especially in Bresson's work. In return, this free reading introduced by the *nouvelle vague* stimulated Roland Barthes's celebrated thoughts about the death of the author.[68]

The film guides say that *Hiroshima mon amour* is as important to film history as *Citizen Kane*. As with all great films, *Hiroshima* is a seamless combination of story and form. Based on a script by Marguerite Duras, the film explores a two-day love affair in Hiroshima between a married French actress and a married Japanese architect. With Hiroshima so closely associated with death, the woman cannot help but recall an earlier affair which she had with a young German soldier whom she had loved during the occupation of France, and who had been killed on the day her town was liberated. For loving the enemy she had been imprisoned in a cellar by her family, and ostracised. In Hiroshima she relives the pain at the same time that she loves the architect. The combination of tender, deliquescent lovemaking and brutal war footage matched her mood exactly.[69]

Les Quatre Cents Coups is generally regarded as the best film ever made about youth. It was the first of a series of five films, culminating in *Love on the Run* (1979). 'The Four Hundred Blows' (a French expression for getting up to no good, doing things to excess, derived from a more literal meaning as the most punishment anyone can bear) tells the story of Antoine Doinel at age twelve.

Left alone by his parents, he gets into trouble, runs away, and is finally consigned to an observation centre for delinquents. Truffaut's point is that Antoine is neither very bad nor very good, simply a child, swept this way and that by forces he doesn't understand. The film is intended to show a freedom – geographical, intellectual, artistic – that the boy glimpses but only half realises is there before it is all swept away. Never really happy at school (we are shown others who *are* unthinkingly happy), the boy enters adulthood already tainted. The famous freeze-frame that ends the film is usually described as ambiguous, but *Les Quatre Cents Coups* is without doubt a sad film about what might have been.[70]

A bout de souffle (*Breathless*) has been described as the film equivalent of *Le Sacre du printemps* or *Ulysses*; **Godard**'s first masterpiece, it changed everything in film. Ostensibly about the final days of a petty (but dangerous) criminal, who initiates a manhunt after he guns down a policeman, the film follows the movements of a man (played by Jean-Paul Belmondo) who models himself on Bogart and the characters he has often seen in Hollywood B gangster movies.[71] He meets and falls in love with an American student (Jean Seberg), whose limited French underlines his own circumscribed world and personality. Their very different views on life, discussed in the pauses between frantic action, gave the film a depth that sets it apart from the B movies it both reveres and derides. Michel Poiccard, the Belmondo character, knows only too well the failures of life that Antoine Doinel is just waking up to. It too is about what might have been.[72]

L'Année dernière à Marienbad, directed by **Alain Resnais**, scripted by Alain Robbe-Grillet, is a sort of *nouveau roman* on screen. It concerns the attempts by X to convince A that they met last year in Marienbad, a resort hotel, where she promised (or may have promised) to run away with him this year. We never know whether the earlier meeting ever took place, whether A is ambiguous because her husband is near, or even whether the 'recollections' of X are in fact premonitions set in the future. That this plot seems improbable when written down isn't the point; the point is that Resnais, with the aid of some superb settings, beautifully shot, keeps the audience puzzled but interested all the way through. The most famous shot takes place in the huge formal garden, where the figures cast shadows but the towering bushes do not.[73]

Jules et Jim is 'a shrine to lovers who have known obsession and been destroyed by it,' a story about two friends, writers, and the woman they meet, who first has a child by one of them and then falls in love with the other.[74] Considered Truffaut's masterpiece, it is also Jeanne Moreau's triumph, playing Catherine. She is so convincing as the wilful third member of the friendship that when she drives into the Seine because Jules and Jim have not included her in a discussion of a Strindberg play, it seems entirely natural.

Deux ou Trois Choses que je sais d'elle (*Two or Three Things I Know about Her*) was described by the critic James Pallot as 'arguably the greatest film made by arguably the most important director to emerge since World War II.'[75] The plot is not strong, and not especially original: it features a housewife who works part-time as a prostitute. It is a notoriously difficult film, dense with images, with endless references to Marx, Wittgenstein, Braudel, structuralism, all related to film, how we watch film, and – a theme underlying all Godard's and Truffaut's work – the place of film in how we lead our lives. *Two or Three Things I Know about Her* is also regarded as a 'Barthian film,' creating and reflecting on 'mythologies' of the world, using signs in old and new ways, to show how they influence our thought and behavior.[76] This was an important ingredient of the French film renaissance, that it was willing to be associated with other areas of contemporary thought, that it saw itself as part of that collective activity: the fact that Godard's masterpiece was so difficult meant that he put intellectual content first, entertainment value second. And that was the point. In the third quarter of the century, so far as film was concerned, traditional Hollywood values took a back seat.

In 1980 **Peter Brook's Centre International de Créations Théâtricales**, in Paris, was given the New York Drama Critics' Circle Award. It was no more than he deserved. In many ways, Brook's relation to theatre was analogous to Boulez's in music. Each was very much his own man, who ploughed his own creative furrow, very international in outlook, very experimental. In the CICT Brook brought a 'research' element to theatre, much as Boulez was doing at IRCAM in music.[77]

Born in London in 1925, to parents of Russian descent, Brook left school at sixteen and, in wartime, worked briefly for the Crown Film Unit, before being persuaded by his parents that a university education would, after all, be desirable. At Magdalen College, Oxford, he began directing plays, after which he transferred to Birmingham Repertory Company ('Birmingham Rep'). In an age before television, this was a very popular form of theatre, almost unknown now: new productions were mounted every two weeks or so, a mix of new plays and classics, so that repertory companies with a stable of familiar actors played an important part in intellectual life, especially in provincial cities outside London. When the Royal Shakespeare Company was formed in 1961 in Stratford-upon-Avon and Brook was invited to take part, he became much better known. At Birmingham he introduced Arthur Miller and Jean Anouilh to Britain, and coaxed classic Shakespearean performances out of John Gielgud (*Measure for Measure*) and Laurence Olivier (*Titus Andronicus*).[78] But it was his sparse rendering of *King Lear*, with Paul Scofield in the title role, in 1962 that is generally regarded as the turning point in his career. Peter Hall, who helped found both the Royal Shakespeare Company and the National Theatre in Britain, asked Brook to join him, and Brook made it a condition of his employment that he could have 'an independent unit of research.'

Brook and his colleagues spent part of 1965 behind locked doors, and when they presented the fruits of their experiments to the public, they called their performances the '**Theatre of Cruelty**,' as homage to **Antonin Artaud**.[79] Cruelty was used in a special sense here: Brook himself once said, in his *Manifesto for the Sixties*, 'We need to look to Shakespeare. Everything remarkable in Brecht, Beckett, Artaud is in Shakespeare. For an idea to stick, it is not enough to state it: it must be burnt into our memories. *Hamlet* is such an idea.'[80]

The most celebrated production in the 'Theatre of Cruelty' season was Brook's direction of Peter Weiss's *Marat/Sade*. The full title of the play explains the plot: *The Persecution and Assassination of Jean-Paul Marat as performed by the Inmates of the Asylum of Charenton under the Direction of the Marquis de Sade*. Weiss himself described the play as Marxist, but this was not especially important for Brook. Instead he concentrated on the intensity of experience that can be conveyed in theatre (one of Brook's aims, as he admitted, was in helping the theatre to overcome the onslaught of television, which was the medium's driving force in the middle years of the century). For Brook the greatest technique for adding intensity in the theatre is the use of verse, in particular Shakespeare's verse, which helps actors, directors, and audience concentrate on what is important. But he realised that a twentieth-century technique was also needed, and for him *the* invention was Brecht's, 'what has been uncouthly labelled "alienation." Alienation is the art of placing an action at a distance so that it can be judged objectively and so that it can be seen in relation to the world – or rather, worlds – around it.' *Marat/Sade* showed Brook's technique at work. When they began rehearsals, he asked the actors to improvise madness. This resulted in such cliché-ridden eye-rolling and frothing at the mouth that he took the company off to a mental hospital to see for themselves. 'As a result, I received for the first time the true shocks that come from direct contact with the physically atrocious conditions of inmates in mental hospitals, in geriatric wards, and, subsequently, in prisons – images of real life for which pictures on film are no substitute. Crime, madness, political violence were there, tapping on the window, pushing open the door. There is no way. It was not enough to remain in the second room, on the other side of the threshold. A different involvement was needed.'[81]

It was after another Shakespearean success, *A Midsummer Night's Dream*, at the RSC in 1970 (and after he had directed a number of productions in France), that Brook was offered the financial assistance he needed to set up the Centre International de Recherche Théâtrale (CIRT), which later became CICT. Brook's aim in doing this was to move away from the constraints of the commercial theatre, which he regarded as a compromise on what he wanted to do.[82] Brook attracted actors from all over the world, interested as he was not only in acting but in the production and reception of theatre and in ways in which research might add to the intensity of the experience.

Orghast (1971) was a recasting of the Prometheus myth, performed at Persepolis and written in a new language devised by the British poet Ted Hughes (who was appointed poet laureate in 1984). This play in particular explored the way the *delivery* of the lines – many of which were sung in incantation – affected their reception. Hughes was also using the ideas of Noam Chomsky, and the deep structure of language, in the new forms he invented.[83]

Moving into the abandoned Bouffes du Nord theatre, built in 1874 but empty since 1952, Brook embarked on an ambitious – unparalleled – experiment, which had two strands. One was to use theatre in an attempt to find a

common, global language, by peopling his productions with actors from different traditions – South American, Japanese, European – but also by exposing
these experimental productions to *audiences* of different traditions, to see how
they were received and understood. This meant that Brook tackled some
improbable ideas – for example, *Conference of the Birds* (1979), based on a
twelfth-century Sufi poem, a comic but painful allegory about a group of birds
that sets off on a perilous journey in search of a legendary bird, the Simourg,
their hidden king.[84] Of course, the journey becomes a stripping away of
each bird's/person's façades and defences, which Brook used as a basis for
improvisation. A second improbable production was *The Mahabharata* (1985),
the Sanskrit epic poem, fifteen times the length of the Bible. This six-hour
production, which reduces the epic to the 'core narrative' of two warring
families, was researched in India, which Brook referred to fondly in his memoirs:
'Perhaps India is the last place where every period of history can coexist, where
the ugliness of neon lighting can illuminate ceremonies that have not changed
in ritual form nor in outer clothing since the origin of the Hindu faith.'[85] He
took his leading actors to India to spend time in holy places, so that they might
at least half-appreciate the Vedic world they were about to portray. (The various
versions of the script took ten years to pare down.) The third of Brook's great
non-Western innovations was *The Ik*, a play about famine in Africa. This looked
back to the books of the anthropologist Colin Turnbull, who had discovered a
series of extraordinary tribes around the world, and forward to the economics
of Amartya Sen, the Indian who would win the Nobel Prize in 1998 for his
theories, expressed throughout the 1980s, about the way famines developed.
To complement these non-Western ideas, Brook also took his company on
three great tours – to Iran, to Africa, and to the United States, not simply to
entertain and inform but also to study the reaction to his productions of the
audiences in these very different places. The tours were intended as a test for
Brook's ideas about a global language for theatre, and to see how the company
might evolve if it wasn't driven by commercial constraints.[86]

But Peter Brook was not only experimental, and not only French-oriented –
he thought theatre in Britain 'very vital.'[87] The CICT continued to produce
Shakespeare and Chekhov, and he himself produced several mainstream films
and opera: *Lord of the Flies* (1963), based on William Golding's novel about a
group of young boys marooned on a desert island, who soon 'return' to
savagery; *Meetings with Remarkable Men* (1979), based on the spiritualist George
Ivanovich Gurdjieff's autobiography; and *La Tragédie de Carmen* (1983). He also
produced a seminal book on the theatre, in which he described four categories
of theatrical experience – deadly, holy, rough, and immediate. At the end of
his memoirs, Brook said that 'at its origin, theatre was an act of healing, of
healing the city. According to the action of fundamental, entropic forces, no
city can avoid an inevitable process of fragmentation. But when the population
assembles together in a special place under special conditions to partake in a
mystery, the scattered limbs are drawn together, and a momentary healing
reunites the larger body, in which each member, re-membered, finds it place
... Hunger, violence, gratuitous cruelty, rape, crime – these are constant

companions in the present time. Theatre can penetrate into the darkest zones of terror and despair for one reason only: to be able to affirm, neither before nor after but at the very same moment, that light is present in darkness. Progress may have become an empty concept, but evolution is not, and although evolution can take millions of years, the theatre can free us from this time frame.'[88]

Later on, Brook made a play from Oliver Sacks's book *The Man Who Mistook His Wife for a Hat*, describing a number of neurological oddities. And that surely underlines Brook's great significance in the postwar world. His attempts to go beyond the narrow confines of nationality, to discover the humanity in science, to use scientific techniques in the production of great art, show an unusual vision on his part as to where the *healing* in modern society is necessary.[89] Brook, though he himself might eschew the term, is also an existentialist. To return to his memoirs: 'I have witnessed no miracles, but I have seen that remarkable men and women do exist, remarkable because of the degree to which they have worked on themselves in their lives.'[90] That applies exactly to Peter Brook. In particular, and perhaps uniquely, he showed how it was possible to bestride both the francophone and anglophone culture of his time.

36

DOING WELL, AND DOING GOOD

In 1944, in *An American Dilemma*, Gunnar Myrdal had forecast that in order to advance the situation of Negroes, the U.S. courts would need to take up the cause. In the years between the mid-1950s and the mid-1970s, that is exactly what happened, but then a reaction set in. President Richard Nixon and Vice President Spiro Agnew queried whether the Supreme Court in the United States was not taking decisions that (a) were more properly the work of government, making political decisions masquerading as legal ones, and (b) were flagrantly disregarding the views of the 'silent majority,' sowing social tension by always putting the minorities first.

President Nixon and Vice President Agnew left office in disgrace, and so their personal involvement in these debates was compromised. But the issues were real enough. They were addressed, in detail, by **Ronald Dworkin**, a professor at New York University's law school, in *Taking Rights Seriously*, which was published in 1977.[1] Dworkin's book was an examination of the way law evolves and itself an example of that evolution. It combined thinking about law, moral philosophy, linguistic philosophy, politics, and political economy, and took into account recent developments in civil rights, women's liberation, homosexual emancipation, and the theories and arguments of Ludwig Wittgenstein, Herbert Marcuse, Willard van Orman Quine, and even R. D. Laing. But Dworkin's main aim was to clarify certain legal concepts in the wake of the civil rights movement, Rawls's theory of justice, and Berlin's notions of negative and positive freedom. In doing so, Dworkin offered a defence of civil disobedience, a legal justification for reverse discrimination, and argued, most fundamentally, that there is no right to general liberty, when by *liberty* is meant *licence*. Instead, Dworkin argued, the most basic right (insofar as the phrase makes sense) is of the individual against the state, and this is best understood as having the right to be treated as an equal with everyone else. For Dworkin, in other words, equality before the law came ahead of everything else, and above all other meanings of freedom.

Besides considering the great social/legal problems of the day, Dworkin grounded his work in the all-important question of how, in a democracy, the rights of the majority, the minorities, and the state can be maintained. As with *Rawls* v. *Nozick*, in an earlier exchange, he compared utilitarian notions (that such-and-such a law would benefit the greatest number) in favour of the ideal (that fairness be seen as the greatest common good). He was suspicious of Isaiah Berlin's notion of 'negative' freedom as the basic form.[2] Berlin, it will be recalled, defined negative freedom as the right to be left alone, unconstrained, whereas positive freedom was the right to be appreciated as whatever kind of person one wished to be appreciated as. Dworkin thought that provided one had equality before the law, this distinction of Berlin's turned out to be false, and that therefore, in a sense, law preceded politics. (This was reminiscent of Friedrich von Hayek's view that the 'spontaneous' way that man has worked out his system of laws precedes any political party.) On Dworkin's analysis, equality before the law precluded a general right to property, which Hayek and Berlin thought was a sine qua non of freedom. Dworkin arrived at his view because, as the title of his books suggests, he thought that rights are serious in a modern society, and that without taking rights seriously, the law cannot be serious.[3] (His book was also a specific reply to Vice President Agnew, who in a speech had argued that 'rights are divisive,' that liberals' concern for individuals' rights 'was a headwind blowing in the face of the ship of state,' not so very different from President Nixon's comments about the silent majority.) As Dworkin put it at the end of his central chapter, 'If we want our laws and legal institutions to provide the ground rules within which these [social and political] issues will be contested then these ground rules must not be the conqueror's law that the dominant class imposes on the weaker, as Marx supposed the law of a capitalist society must be. The bulk of the law – that part which defines and implements social, economic, and foreign policy – cannot be neutral. It must state, in its greatest part, the majority's view of the common good. The institution of rights is therefore crucial, because it represents the majority's promise to the minorities that their dignity and equality will be respected. ... The Government will not re-establish respect for law without giving the law some claim to respect. It cannot do that if it neglects the one feature that distinguishes law from ordered brutality. If the Government does not take rights seriously, then it does not take the law seriously either.'[4] Dworkin's conclusion that in the modern age, the post-1960s age, the right to be treated equally – by government – was a prerequisite of all freedoms, was congenial to most liberals.

An alternative view of the legacy of the 1960s and 1970s, the freedom and the equalities the period had produced, and a very different picture of the laws that had been passed, came from two conservative economists in Chicago. Milton and Rose Friedman took equality before the law for granted but thought that freedom could only be guaranteed if there were economic freedom, if men and women were 'free to choose' – the title of their 1980 book – the way they earned their living, the price they paid for the goods they wished to buy, and the wage they were willing to pay anyone who worked for them.[5] **Milton Friedman** had advanced very similar views two decades before, in *Capitalism and Freedom*, published in 1962, and considered in chapter 30 (see page 519). He and his wife returned to the subject, they said, because they were anxious that during the interim, 'big government' had grown to the point where its mushrooming legal infrastructure, much of it concerned with 'rights,' seriously interfered with people's lives, because unemployment and inflation were growing to unacceptable levels in the West, and because they felt that, as they put it, 'the tide [was] turning,' that people were growing tired and sceptical of the 'liberal' approach to economics and government in the West, and looking for a new direction.[6]

Free to Choose, as its authors were at pains to point out, was a much more practical and concrete book than *Capitalism and Freedom*. The Friedmans had specific targets and specific villains in their view of the world. They began by reanalysing the 1929 stock market crash and the ensuing depression. Their aim was to counter the view that these two events amounted to the collapse of capitalism, and that the capitalist system was responsible for the failure of so many banks and the longest-running depression the world has ever known. They argued that there had been mismanagement of specific banks, specifically the Bank of the United States, which closed its doors on 11 December 1930, the largest financial institution ever to fail in the history of the U.S. Although a rescue plan had been drawn up for this bank, anti-Semitism on the part of 'leading members of the banking community' in New York was at least partly responsible for the plan not being put into effect. The Bank of the United States was run by Jews, servicing mainly the Jewish community, and the rescue plan envisaged it merging with another Jewish bank. But, according to Friedman (himself Jewish), this could not be stomached 'in an industry that, more than almost any other, has been the preserve of the well-born and the well placed.'[7] This sociological – rather than economic – failure was followed by others; by Britain abandoning the gold standard in 1931, by mismanagement of the Federal Reserve System's response to the various crises, and by the interregnum between Herbert Hoover's presidency and Franklin Roosevelt's in 1933, when neither man would take any action in the economic sphere for a period of three months. On the Friedman analysis, therefore, the great crash and the depression were more the result of technical mismanagement than anything fundamental to capitalism per se.

The crash/depression was important, however, because it was followed so soon by world war, when the intellectual climate changed: people saw – or thought they could see – that cooperation worked, rather than competition; the idea of a welfare state caught on in wartime and set the tone for government between 1945 and, say, 1980. But, and this was the main point of the Friedmans' book, 'New Deal liberalism,' as they called it, and Keynesianism, didn't work (though they were relatively easy on Keynes: even President Nixon had declared, 'We are all Keynesians now'). They looked at schools, at the unions, at consumer protection, and at inflation, and found that in all cases free-market capitalism not only produced a more efficient society but created greater freedom, greater equality, and more public benefit overall: 'Nowhere is the gap between rich and poor wider, nowhere are the rich richer and the poor poorer, than in those societies that do not permit the free market to operate. That is true of mediaeval societies like Europe, India before independence, and much of modern South America, where inherited status determines position. It is equally true of centrally planned societies, like Russia or China or India since independence, where access to government determines position. It is true even where central planning was introduced, as in all three of these countries, in the name of equality.[8] Even in the Western democracies, the Friedmans said, echoing something first observed by Irving Kristol, a 'new class' has arisen – government bureaucrats or academics whose research is supported by government funds, who are privileged but preach equality. 'They remind us very much of the old, if unfair, saw about the Quakers: "They came to the New World to do good, and ended up doing well." '[9]

The Friedmans gave many instances of how capitalism promotes freedom, equality, and the wider spread of benefits. In attacking the unions they did not confine themselves to the 'labour' unions but focused on middle-class unions as well, such as the doctors, and quoted the case of the introduction of 'paramedics' in one district of California. This had been vigorously opposed by doctors – ostensibly because only properly trained medical personnel could cope, but really because they wanted to limit entry to the profession, to keep up their salaries. In fact, the number of people surviving cardiac arrest rose in the first six months after the introduction of paramedics from 1 percent to 23 percent. In the case of consumer rights, the Friedmans claimed that in America there was far too much government legislation interfering with the free market, one result being a 'drug lag': the United States had dropped behind countries like Great Britain in the introduction of new drugs – they specifically referred to beta-blockers. The introduction of new drugs to the market, for example, had fallen, they said, by about 50 percent since 1962, mainly because the cost of testing their effects on the consumer had risen disproportionately. The Friedmans considered that government response to exposés like Rachel Carson's had been too enthusiastic; 'all the movements of the past two decades – the consumer movement, the ecology movement, the back-to-the-land movement, the hippie movement, the organic-food movement, the protect-the-wilderness movement, the zero-population-growth movement, the "small is beautiful" movement, the anti-nuclear movement – have had one thing in common. All have been anti-growth. They have been opposed to new developments, to industrial innovation, to the increased use of natural resources.'[10] It was time to shout enough is enough, that the forces for control, for 'rights', had gone too far. At the end of their book, however, the Friedmans said they thought a change was coming, that many people wanted 'big government' rolled back. In particular, they pointed to the election of Margaret Thatcher in Britain in 1979, on a promise 'to haul back the frontiers of the state,' and to the revolt in America against the government monopoly of the postal service. They ended by calling for an amendment to the U.S. Constitution, for what would in effect be an **Economic Bill of Rights** that would force the government to limit federal spending.

Why this change in public mood? The main reason, alluded to in an earlier chapter, was that following the oil crisis in 1973–74, the long stagnation in the living standards of the West produced a growing dissatisfaction. As the economist

Paul Krugman of MIT described it, the 'magic' of the Western economies, their ever-higher standards of living, went away in 1973. It took time for these trends to emerge, but as they did, certain academic economists, notably Martin Feldstein of Harvard, began to document the negative effects of taxation and government expenditure on investment and savings.[11] Friedman actually predicted that there would come a time of stagnation – zero growth – combined with inflation, which according to classical economics couldn't happen. Paul Samuelson gave this phenomenon its name, 'stagflation,' but it was Friedman, rightly, who received the Nobel Prize for the insight. Where Friedman and Feldstein led, others soon followed, and by the late 1970s there emerged a hard core of 'supply-side' economists who rejected Keynesianism and believed that a sharp reduction in taxation, meaning that more money would be 'supplied' to the economy, would produce such a surge in growth that there was no need to worry about expenditure. These ideas were behind the election of Margaret Thatcher in the United Kingdom in 1979, and of Ronald Reagan as president of the United States a year later. In the United States the Reagan years were marked by massive budget deficits, which were still being paid for in the 1990s, but also by a striking rally on Wall Street, which faltered between 1987 and 1992 but then recovered. In Britain, in addition to a similar rise in the stock market, there was also an important series of policy initiatives, known as privatisation, in which mainly public utilities were returned to private hands.[12] In social, economic, and political terms, privatisation was a huge success, transforming ungainly and outdated businesses into modern, efficient corporations where, in some cases at least, real costs to the consumer fell. The idea of privatisation was widely exported – to Western Europe, Eastern Europe, Asia, and Africa.

Nevertheless, despite all that was happening on the stock markets, the growth performance of the major Western economies remained unimpressive, certainly in comparison with pre-1973 levels. At the same time there was a major jump in the inequalities of wealth distribution. In the 1980s growth and inequality were the two main theoretical issues that concerned economists, much more so than politicians, Western politicians anyway.

Traditionally, three reasons are given for the slowdown in growth after the oil crisis. The first is technological. MIT's Robert Solow was the first economist to show exactly how this worked (he won a Nobel Prize in 1987). In his view, productivity growth comes from technological innovation – what is known in economics now as the Solow Residual.[13] Many technological breakthroughs matured in World War II, and in the period of peace and stability that followed, these innovations came to fruition as products. However, all of these high-tech goods – the jet, television, washing machines, long-playing records, portable radios, the car – once they had achieved saturation, and once they had developed to a certain point of sophistication, could no longer add further innovation worth the name, and by around 1970 the advances in technology were slowing down. Paul Krugman, in his history of economics, underlines this point with a reference to the Boeing 747 jet. Still, in 2000 AD, the backbone of many airlines, this first came into service in 1969. The second reason for the slowdown in growth was sociological. In the 1960s the baby-boom generation reached maturity. During that same decade many of the assumptions of capitalism came under attack, and several commentators observed a decline thereafter in educational standards. As Krugman wrote, 'The expansion of the underclass has been a significant drag on US growth. ... There is a plausible case to be made that social problems – the loss of economic drive among the children of the middle class, the declining standards of education, the rise of the underclass – played a significant role in the productivity slowdown. This story is very different from the technological explanation; yet it has in common with that story a fatalistic feel. ... [It] would seem to suggest that we should learn to live with slow productivity growth, not demand that the government somehow turn it around.'[14] The third explanation is political. This is the Friedman argument that government policies were responsible for the slow growth and that only a reduction in taxes and a rolling back of regulations would free up the forces needed for growth to recur. Of these three, the last, because it was overtly political, was the most amenable to change. The Thatcher government and the Reagan administration both sought to follow monetarist and supply-side policies. Feldstein was himself taken into the Reagan White House.

Ironically, however, again as Paul Krugman makes clear, 1980 was actually the high point of conservative economics, and since then ideas have moved on once more, concentrating on the more fundamental forces behind growth and inequality.[15] The two dominant centres of economic thinking, certainly in the United States, have been Chicago and Cambridge, Massachusetts – home to Harvard and MIT. Whereas Chicago was associated primarily with conservative economics, Cambridge, in the form of Feldstein, Galbraith, Samuelson, Solow, Krugman, and Sen (now at Cambridge, England), embraces both worldviews.

After his discovery of the '**residual**' named after him, Robert Solow's interest in understanding growth, its relation to welfare, work, and unemployment, is perhaps the best example of what currently concerns theoretical economists involved with macroeconomics (as opposed to the economics of specific, closed systems). The ideas of Solow and others, fashioned in the 1950s and 1960s, coalesced into **Old Growth Theory**.[16] This said essentially that growth was fuelled by technological innovation, that no one could predict when such innovation would arise, and that the gain produced would be temporary, in the sense that there would be a rise in prosperity but it would level off after a while. This idea was refined by **Kenneth Arrow** at Stanford, who showed that there was a further gain to be made – of about 30 percent – because workers learned on the job: they became more skilled, enabling them to complete tasks faster, and with fewer workers needed. This meant that prosperity lasted longer, but even here diminishing returns applied, and growth levelled off.[17]

New Growth Theory, which emerged in the 1980s, pioneered by **Robert Lucas** at Chicago but added to by Solow himself, argued that on the contrary, substantial investment by government and private initiative can ensure *sustained* growth because, apart from anything else, it results in a more educated and better motivated workforce, who realise the importance of innovation.[18] This idea was remarkable for two reasons. In the first place Lucas came from conservative Chicago, yet was making a case for *more* government intervention and expenditure. Second, it marked the coming together of sociology, social psychology, and economics: a final recognition of David Riesman's argument in *The Lonely Crowd*, which had shown that 'other-directed' people loved innovation. It is too soon to say whether New Growth Theory will turn out to be right.[19] The explosion of computer technology and biotechnology in the 1990s, the ease with which new ideas have been accepted, certainly suggests that it might do. Which makes it all the more curious that Margaret Thatcher railed so much against the universities while she was in power. Universities are one of the main ways governments can help fuel technological innovation and therefore stimulate growth.

Milton and Rose Friedman, and the Chicago school in general, based their theories on what they called the key insight of the Scotsman Adam Smith, 'the father of modern economics,' who wrote *The Wealth of Nations* in 1776. 'Adam Smith's key insight was that both parties to an exchange can benefit and that, so long as cooperation is strictly voluntary, no exchange will take place unless both parties do benefit.'[20] Free-market economics, therefore, not only work: they have an ethical base.

There was, however, a rival strand of economic thinking that did not share the Friedmans' faith in the open market system. There was little space in *Free to Choose* for a consideration of poverty, which the Friedmans thought in any case would be drastically reduced if their system were allowed full rein. But many other economists were worried about economic inequality, the more so after John Rawls and Ronald Dworkin had written their books. The man who came to represent these other economists was an Indian but Oxford- and Cambridge-trained academic, **Amartya Sen**. In a prolific series of papers and books Sen, who later held joint appointments at Harvard and Cambridge, attempted to move economics away from what he saw as the narrow interests of the Friedmans and the monetarists. One area he promoted was 'welfare economics,' in effect economics that looked beyond the operation of the market to scrutinise the institution of poverty and the concept of 'need.' Many of Sen's articles were highly technical mathematical exercises, as he attempted to measure poverty and different types of need. A classic Sen problem, for example, would be trying to calculate who was worse off, someone with more income but with a chronic health problem, for which treatment had to be regularly sought and paid for, or someone with less income but better health.

Sen's first achievement was the development of various technical measures which enabled governments to calculate how many poor people there were within their jurisdiction, and what exactly the level of need was in various categories. These were no mean accomplishments, but he himself called them 'engineering problems,' with 'nuts and bolts' solutions. Here too economics and sociology came together. Of wider relevance were two other ideas that contributed equally to his winning the Nobel Prize for Economics in 1998. The first of these was his marriage of economics and ethics. Sen took as a starting point a non sequitur that he had observed: many people who were not poor were nevertheless interested in the problem of poverty, and its removal, not because they thought it was more efficient to remove it, but because it was wrong. In other words, individuals often behaved ethically, without putting their own self-interest first. This, he noted, went against not only the ideas of economists like the Friedmans but also those of some evolutionary thinkers, like Edward O. Wilson and Richard Dawkins. In his book *On Ethics and Economics* (1987), Sen quoted the well-known Prisoners' Dilemma game, which Dawkins also made so much of in *The Selfish Gene*. Sen noted that, while cooperation might be preferable in the evolutionary context, in the industrial or commercial setting the selfish strategy is theoretically what pays any single person, seen from that person's point of view. In practice, however, various cooperative strategies are invariably adopted, because people have notions of other people's rights, as well as their own; they have a sense of community, which they want to continue. In other words, people *do* have a general ethical view of life that is not purely selfish. He thought these findings had implications for the economic organisation of society, taxation structure, financial assistance to the poor and the recognition of social needs.[21]

But the work of Sen's that really caught the world's imagination was *Poverty and Famines*, his 1981 report for the World Employment Programme of the International Labour Organisation, written when he was professor of political economy at Oxford and a Fellow of All Souls.[22] The subtitle of Sen's book was 'An Essay on Entitlement and Deprivation,' which brings us back to Dworkin's concept of rights. In his report, Sen examined four major famines – the Great Bengal Famine in 1943, when about 1.5 million people starved to death; the Ethiopian famines of 1972–74 (more than 100,000 deaths); the 1973 drought and famine in the Sahel (100,000 dead); and the 1974 flood and famine in Bangladesh (figures vary from 26,000 to 100,000 dead). His most important finding was that, in each case, in the areas most affected, there was no significant decline in the availability of food (FAD, for 'food availability decline' in the jargon); in fact, in many cases, and in many of the regions where famine was occurring, food production, and food production per capita, actually rose (e.g., in Ethiopia, barley, maize, and sorghum production was above normal in six out of fourteen provinces).[23] Instead, what Sen found typically happened in a famine was that a natural disaster, like a flood or a drought, (a) made people *think* there would be a shortage of food, and (b) at the same time affected the ability of certain sectors of the population – peasants, labourers, agricultural workers – to earn money. Possessors of food hoard what they have, and so the price rises at the very time large segments of the population suffer a substantial fall in income. Either the floods mean there is no work to be had on the land, or drought causes the poor to be evicted from where they are living, because they can't grow enough to earn enough to pay the rent. But the chief factor is, as Sen phrases it, a fall in 'entitlement': they have less and less to exchange for food. It is a failure of the market system, which operates on what people think is happening, or soon will happen. But, objectively, in terms of the aggregate food availability, the market is wrong. Sen's analysis was startling, partly because, as he himself said, it was counterintuitive, going against the grain of common sense, but also because it showed how the market could make a bad situation worse. Apart from helping governments understand in a practical way how famines develop, and therefore might be avoided or the effects mitigated, his empirical results highlighted some special limitations of the free-market philosophy and its ethical base. Famines might be a special case, but they affect a lot of people.

In his economic history of the last quarter of the century, *Peddling Prosperity*, the MIT economist, Paul Krugman charts the rise of right-wing economics and then describes its declining influence in the 1980s, devoting the last third of his book to the revival of Keynesianism (albeit in new clothes) in the late 1980s and 1990s.[24] Krugman's account described the failure of such right-wing doctrines as 'business-cycle' theory, and the drag on the U.S. economy brought about by the huge budget deficits, the result of Ronald Reagan's various monetarist policies. He similarly took to task the ideas of more recent, more liberal economic thinkers such as Lester Thurow, in *Zero-Sum Society* (1980), and notions of 'strategic trade' put forward by the Canadian economist James Brander and his Australian coauthor, Barbara Spencer. Strategic trade views countries as similar to companies – corporations – who seek to 'place' their economy in a strategic position vis-à-vis other economies. This view held sway in the Clinton White House, at least for a while – until Larry Summers became economic secretary in May 1999 – but it was misplaced, argues Krugman, for countries are not companies and do not necessarily need to compete to survive and prosper, and such apparently intelligent thinking is in any case doomed to failure because, as most research in the 1980s and 1990s showed, people behave not in a perfectly rational way, as classical economists always claimed, but in a 'near-rational' way, thinking mainly of the short term and using only such information as comes their way *easily*. For Krugman, this recent insight is an advance because it means that individual decisions, each one taken sensibly, can have disastrous collective consequences (in short, this is why recessions occur). Krugman therefore allies himself with the new Keynesians who believe that some government intervention in macroeconomic matters is essential to exert an influence on invention/inflation/unemployment/international trade. But Krugman's conclusion, in the mid-1990s, was that the two main economic problems still outstanding were slow growth and productivity on the one hand, and rising poverty on the other: 'Everything else is either of secondary importance, or a non-issue.'[25] This brings us to a familiar name: J. K. Galbraith.

Among professional economists, J. K. Galbraith is sometimes disparaged as a fellow professional much less influential with his colleagues than among the general public. That is to do him a disservice. In his many books he has used his 'insider' status as a trained economist to make some uncomfortable observations on the changing nature of society and the role economics plays in that change. Despite the fact that Galbraith was born in 1908, this characteristic showed no sign of flagging in the last decade of the century, and in 1992 he published *The Culture of Contentment* and, four years later, *The Good Society: The Human Agenda*, his eighteenth and twentieth books. (There was another, *Name Dropping*, a memoir, in 1999.)

The Culture of Contentment is a deliberate misnomer. Galbraith is using irony here, irony little short of sarcasm.[26] What he really means is the culture of smugness. His argument is that until the mid-1970s, round about the oil crisis, the Western democracies accepted the idea of a mixed economy, and with that went economic social progress. Since then, however, a prominent class has emerged, materially comfortable and even very rich, which, far from trying to help the less fortunate, has developed a whole infrastructure – politically and intellectually – to marginalise and even demonise them. Aspects of this include tax reductions to the better off and welfare cuts to the worse off, small, 'manageable wars' to maintain the unifying force of a common enemy, the idea of 'unmitigated laissez-faire as the embodiment of freedom,' and a desire for a cutback in government. The more important collective end result of all this, Galbraith says, is a blindness and a deafness among the 'contented' to the growing problems of society. While they are content to spend, or have spent in their name, trillions of dollars to defeat relatively minor enemy figures (Gaddafi, Noriega, Milosevic), they are extremely unwilling to spend money on the underclass nearer home. In a startling paragraph, he quotes figures to show that 'the number of Americans living below the poverty line increased by 28 per cent in just ten years, from 24.5 million in 1978 to 32 million in 1988. By then, nearly one in five children was born in poverty in the United States, more than twice as high a proportion as in Canada or Germany.'[27]

Galbraith reserves special ire for **Charles Murray**. Murray, a Bradley Fellow at the American Enterprise Institute, a right-wing think tank in Washington, D.C., produced a controversial but well-documented book in 1984 called *Losing Ground*.[28] This examined American social policy from 1950 to 1980 and took the position that, in fact, in the 1950s the situation of blacks in America was fast improving, that many of the statistics that showed they were discriminated against actually showed no such thing, rather that they were poor, that a minority of blacks pulled ahead of the rest as the 1960s and 1970s passed, while the bulk remained behind, and that by and large the social initiatives of the Great Society not only failed but made things worse because they were, in essence, fake, offering fake incentives, fake curricula in schools, fake diplomas in colleges, which changed nothing. Murray allied himself with what he called 'the popular wisdom,' rather than the wisdom of intellectuals or social scientists. This popular wisdom had three core premises: people respond to incentives and disincentives – sticks and carrots work; people are not inherently hardworking or moral – in the absence of countervailing influences, people will avoid work and be amoral; people must be held responsible for their actions – whether they *are* responsible in some ultimate philosophical or biochemical sense cannot be the issue if society is to function.[29] His charts, showing for instance that black entry into the labour force was increasing steadily in the 1955–80 period, or that black wages were rising, or that entry of black children into schools increased, went against the grain of the prevailing (expert) wisdom of the time, as did his analysis of illegitimate births, which showed that there was partly an element of 'poor' behaviour in the figures and partly an element of 'race.'[30] But overall his message was that the situation in America in the 1950s, though not perfect, was fast improving and ought to have been left alone, to get even better, whereas the Great Society intervention had actually made things worse.

For Galbraith, Murray's aim was clear: he wanted to get the poor off the federal budget and tax system and 'off the consciences of the comfortable.'[31] He confirmed this theme in *The Good Society* (1996). Galbraith could never be an 'angry' writer; he is almost Chekhovian in his restraint. But in *The Good Society*, no less than in *The Culture of Contentment*, his contempt for his opponents is there, albeit in polite disguise. The significance of *The Good Society*, and what links many of the ideas considered in this chapter, is that it is a book by an economist in which economics is presented as the *servant* of the people, not the engine.[32] Galbraith's agenda for the good society is unashamedly left of centre; he regards the right-wing orthodoxies, or would-be orthodoxies, of 1975–90, say, as a mere dead end, a blind alley. It is now time to get back to the real agenda, which is to re-create the high-growth, low-unemployment, low-inflation societies of the post–World War II era, not for the sake of it, but because they were more civilised times, producing social and moral progress before a mini–Dark Age of selfishness, greed, and sanctimony.[33] It is far from certain that Galbraith was listened to as much as he would have liked, or as much as he would have been earlier. Poverty, particularly poverty in the United States, remained a 'hidden' issue in the last years of the century, seemingly incapable of moving or shaking the contented classes.

The issue of race was more complicated. It was hardly an 'invisible' matter, and at a certain level – in the media, among professional politicians, in literature – the advances made by blacks and other minorities were there for all to see. And yet in a mass society, the mass media are very inconsistent in the picture they paint. In mass society the more profound truths are often revealed in less compelling and less entertaining forms, in particular through statistics. It is in this context that Andrew Hacker's *Two Nations: Black and White, Separate, Hostile, Unequal*, published in 1992, was so shattering.[34] It returns us not only to the beginning of this chapter, and the discussion of rights, but to the civil rights movement, to Gunnar Myrdal, Charles Johnson, and W. E. B. Du Bois. Hacker's message was that some things in America have not changed.

A professor of political science at Queen's College in New York City, **Andrew Hacker** probably understands the U.S. Census figures better than anyone outside the government; and he lets the figures lead his argument. He has been analysing America's social and racial statistics for a number of years, and is no firebrand but a reserved, even astringent academic, not given to hyperbole or rhetorical flourishes. He publishes his startling (and stark) conclusions mainly in the *New York Review of Books*, but *Two Nations* was more chilling than anything in the *Review.* His argument was so shocking that Hacker and his editors apparently felt the need to wrap his central chapters behind several 'softer' introductory chapters that put his figures into context, exploring and seeking to explain racism and the fact of being black in an anecdotal way, to prepare the reader for what was to come. The argument was in two parts. The figures showed not only that America was still deeply divided, after decades – a century – of effort, but that in many ways the situation had *deteriorated* since Myrdal's day, and despite what had been achieved by the civil rights movement. Dip into Hacker's book at almost any page, and his results are disturbing.

HOUSEHOLDS HEADED BY WOMEN

Year	Black	White	Black Multiple
1950	17.2%	5.3%	3.2
1960	24.4%	7.3%	3.3
1970	34.5%	9.6%	3.6
1980	45.9%	13.2%	3.5
1993	58.4%	18.7%	3.1

In other words, the situation in 1993 was, relatively speaking, no better than in 1950.[35]

'The real problem in our time,' wrote Hacker, 'is that more and more black infants are being born to mothers who are immature and poor. Compared with white women – most of whom are older and more comfortably off – black women are twice as likely to have anemic conditions during pregnancy, twice as likely to have no prenatal care, and twice as likely to give birth to low-weight babies. Twice as many of their children develop serious health problems, including asthma, deafness, retardation, and learning disabilities, as well as conditions stemming from their own use of drugs and alcohol during pregnancy.'[36] 'Measured in economic terms, the last two decades have not been auspicious ones for Americans of any race. Between 1970 and 1992, the median income for white families, computed in constant dollars, rose from $34,773 to $38,909, an increase of 11.9 percent. During this time black family income in fact went down a few dollars, from $21,330 to $21,161. In relative terms, black incomes fell from $613 to $544 for each $1,000 received by whites.'[37]

Despite a large chapter on crime, Hacker's figures on school desegregation were more startling. In the early 1990s, 63.2 percent of all black children, nearly two out of three, were still in segregated schools. In some states the percentage of blacks in segregated schools was as high as 84 percent. Hacker's conclusion was sombre: 'In allocating responsibility, the response should be clear. It is white America that has made being black so disconsolate an estate. Legal slavery may be in the past, but segregation and subordination have been allowed to persist. Even today, America imposes a stigma on every black child at birth. . . . A huge racial chasm remains, and there are few signs that the coming century will see it closed. A century and a quarter after slavery, white America continues to ask of its black citizens an extra patience and perseverance that whites have never required of themselves. So the question for white Americans is essentially moral: is it right to impose on members of an entire race a lesser start in life and then to expect from them a degree of resolution that has never been demanded from your own race?[38]

The oil crisis of 1973–74 surely proved Friedrich von Hayek and Milton Friedman right in at least one respect. Economic freedom, if not the most basic of freedoms, as Ronald Dworkin argues, is still pretty fundamental. Since the oil crisis, and the economic transformation it ignited, many areas of life in the West – politics, psychology, moral philosophy, and sociology – have been refashioned. The works of Galbraith, Sen, and Hacker, or more accurately the *failure* of these works to stimulate, say, the kind of popular (as opposed to academic) debate that Michael Harrington's *Other America* provoked in the early 1960s, is perhaps the defining element of the current public mood. Individualism and individuality are now so prized that they have tipped over into selfishness. The middle classes are too busy doing well to do good.[39]

THE WAGES OF REPRESSION

When Dr. Michael Gottlieb, from the University of California at Los Angeles, arrived in Washington in the second week of September 1981 for a conference at the National Institutes of Health (NIH), he was optimistic that the American medical authorities were finally taking seriously a new illness that, he feared, could soon reach epidemic proportions. The NIH is the world's biggest and most powerful medical organisation. Housed on a campus of more than 300 acres, ten miles northwest of Washington, D.C. in the Bethesda hills, it had by the end of the century an annual budget of $13 billion and housed among other things the National Institute for Allergy and Infectious Diseases, the National Heart, Lung and Blood Institute, and the National Cancer Institute.

The conference Gottlieb was attending had been called at the NCI to explore the outbreak of a rash of cases of a rare skin cancer in the United States, known as **Kaposi's sarcoma**.[1] One of the other doctors at the conference was Linda Laubenstein, a blood specialist at New York University. She had first seen KS in a patient of hers in September 1979, when it had presented as a generalised rash on a man's skin, associated with enlarged lymph nodes. At that point she had never heard of KS, and had looked it up after the cancer had been diagnosed by a dermatologist. This particular type of illness had originally been discovered in 1871 among Mediterranean and Jewish men, and in the century that followed, between 500 and 800 cases had been reported. It was also seen among the Bantu in Africa. It usually struck men in their forties and fifties and was generally benign; the lesions were painless, and the victims usually died much later of something else. But, as Laubenstein and Gottlieb now knew, KS in America was much more vicious: 120 cases had already been reported, often in association with a rare, parasitical form of pneumonia – pneumocystis – and in 90 percent of cases the patients were gay men.[2] An added, and worrying, complication was that these patients also suffered from a strange deficiency to their immune system – the antibodies in their blood simply refused to fight the different infections that came along, so that the men died from whatever illnesses they contracted while their bodies were already weakened by cancer.

Gottlieb was astonished by the Bethesda conference. He had arrived amid rumours that the NIH was at last going to fund a research program into this new disease. The Center for Disease Control, headquartered in Atlanta, Georgia, had been trying to trace where the outbreak had started and how it had spread, but the CDC was just the 'shock troops' in the fight against disease; it was now time for more fundamental research. So Gottlieb sat in quiet amazement as he and the others were lectured on KS and its treatment in Africa, as if there was no awareness at NIH that the disease had arrived in America, and in a far more virulent form than across the Atlantic. He left the meeting bewildered and depressed and returned to Los Angeles to work on a paper he was planning for the *New England Journal of Medicine* on the links between KS and *Pneumocystis carinii* that he was observing. But then he found that the *Journal* was not 'overly enthusiastic' about publishing his article, and kept sending it back for one amendment after another. All this prevarication raised in Gottlieb's mind the feeling that, among the powers-that-be, in the medical world at least, the new outbreak of disease was getting less attention than he thought it deserved, and that this was so because the great preponderance of victims were homosexual.[3]

It would be another year before this set of symptons acquired a name. First it was GRID, standing for Gay-Related Immune Deficiency, then ACIDS, for Acquired Community Immune Deficiency Syndrome, and finally, in mid-1982, AIDS, for **Acquired Immune Deficiency Syndrome**. The right name for the disease was the least of the problems. In March of the following year the *New York Native*, a Manhattan gay newspaper, ran the headline, '1,112 AND COUNTING.' That was the number of homosexual men who had died from the disease.[4] But AIDS was significant for two reasons over and above the sheer numbers it cut down, tragic though that was. It was important because it encompassed the two great strands of research that, apart from psychiatric drugs, had dominated medical thinking in the postwar period; and, second, because the people who were cut down by AIDS were, disproportionately, involved in artistic and intellectual life.

The two strands of inquiry dominating medical thought after 1945 were the biochemistry of the immunological system, and the nature of cancer. After the first reports in the early 1950s about the links between smoking and cancer, it was soon observed that there was almost as intimate a link between smoking and heart disease. Coronary thrombosis – heart attack – was found to be much more common in smokers than in nonsmokers, especially among men, and this provoked two approaches in medical research. In heart disease the crucial factor was blood pressure, and this deviated from the norm mainly due to two reasons. Insofar as smoking damaged the lungs, and made them less efficient at absorbing oxygen from the air, each breath sent correspondingly less oxygen into the body's system, causing the heart to work harder to achieve the same effect. Over time this imposed an added burden on the muscle of the heart, which eventually gave out. In such cases blood pressure was low, but high blood pressure was a problem also, this time because it was found that foods high in animal fats caused deposits of cholesterol to be laid down in the blood vessels, narrowing them and, in extreme cases, blocking them entirely. This also put pressure on the heart, and on the blood vessels themselves, because the same volume of blood was being compressed through less space. In extremes this could damage the muscle of the heart and/or rupture the walls of the blood vessels, including those of the brain, in a cerebral haemorrhage, or stroke. Doctors responded by trying to devise drugs that either raised or lowered blood pressure, in part by 'thinning' the blood and, where the heart had been irreparably damaged, replacing it in its entirety.

Before World War II there were in effect no drugs that would lower blood pressure. By 1970 there were no fewer than four families of drugs in wide use, of which the best-known were the so-called **beta-blockers**. These drugs grew out of a line of research that dated back to the 1930s, in which it had been found that acetylcholine, the transmitter substance that played a part in nerve impulses (see chapter 28, page 501), also exerted an influence on nervous structures that governed the heart and the blood vessels.[5] In the nerve pathways that eventually lead to the coronary system, a substance similar to adrenaline was released, and it was this which controlled the action of the heart and blood vessels. So began a search for some way of interfering with – blocking – this action. In 1948 **Raymond Ahlquist**, at the University of Georgia, found that the nerves involved in this mechanism consisted of two types, which he arbitrarily named alpha and beta, because they responded to different substances. The beta receptors, as Ahlquist called them, stimulated both the rate and the force of the heartbeat, which gave **James Black**, a British doctor, the idea to see if blocking the action of adrenaline might help reduce activity.[6] The first substance he identified, promethalol, was effective but was soon shown to produce tumours in mice and was withdrawn. Its replacement, propranolol had no such drawbacks and became the first of many 'beta-blockers.' They were in fact subsequently found to have a wide range of uses: besides lowering blood pressure, they prevented heart irregularities and helped patients survive after a heart attack.[7]

Heart transplants were a more radical form of intervention for heart disease, but as doctors watched developments in molecular biology, the option grew more attractive as it was realised that, at some point, cloning might become a possibility. The central intellectual problem with transplants, apart from the difficult surgery involved and the ethical problems of obtaining donor organs from newly deceased individuals, was immunological: the organs were in effect foreign bodies introduced into a person's physiological system, and therefore rejected as an intruder.

The research on immunosuppressants grew out of cancer research, in particular leukaemia, which is a tumour of the lymphocytes, the white blood cells that rapidly reproduce to fight off foreign bodies in disease.[8] After the war, and even before the structure of DNA had been identified, its role in reproduction suggested it might have a role in cancer research (cancer itself being the rapid reproduction of malignant cells). Early studies showed that a particular type of purine (such as adenine and guanine) and pyridamines (cytosine and thymine) did affect the growth of cells. In 1951 a substance known as 6-Mercaptopurine (6-MP) was found to cause remission for a time in certain leukemias. The good news never lasted, but the action of 6-MP was potent enough for its role in immunosuppression to be tested. The crucial experiments were carried out in the late 1950s at the New England Medical Center, where **Robert Schwartz** and **William Dameshek** decided to try two drugs used for leukaemia – methotrexate and 6-MP – on the immune response of rabbits. As Miles Weatherall tells the story in his history of modern medicine, this major breakthrough turned on chance. Schwartz wrote to **Lederle** Laboratories for samples of methotrexate, and to **Burroughs Wellcome** for 6-MP.[9] He never heard from Lederle, but Burroughs Wellcome sent him generous amounts of 6-MP. He therefore went ahead with this and found within weeks that it was indeed a powerful suppresser of the immune response. It was subsequently found that methotrexate had no effect on rabbits, so as Schwartz himself remarked, if the response of the two companies had been reversed, this particular avenue of inquiry would have been a dry run, and the great breakthrough would never have happened.[10] **Dr. Christian Barnard**, in South Africa, performed the world's first heart transplant between humans in December 1967, with the patient surviving for eighteen days; a year later, in Barnard's second heart-transplant operation, the patient lived for seventy-four days. A nerve transplant followed in Germany in 1970, and by 1978 immunosuppressant drugs were being sold commercially for use in transplant surgery. In 1984, at Loma Linda University Medical Center in California, the heart of a baboon was inserted into a two-week-old girl. She survived for only twenty days, but new prospects of 'organ farming' had been opened up.[11]

By the time the AIDS epidemic appeared, therefore, a lot was already known about the human body's immunological system, including a link between immune suppression and cancer. In 1978 **Robert Gallo**, a research doctor at the National Cancer Institute in Bethesda, discovered a new form of virus, known as a retrovirus, that caused leukaemia.[12] He had been looking at viruses because by then it was known that leukaemia in cats – feline leukaemia, a major cause of death in cats – was caused by a virus that knocked out the cats' immune system. Japanese researchers had studied T cell leukaemia (T cells being the recently discovered white blood cells that are the key components of the immune system) but it was Gallo who identified the human T cell leukaemia virus, or HTLV, a major practical and theoretical breakthrough. Following this approach, in February 1983, Professor **Luc Montagnier**, at the Pasteur Institute in Paris, announced that he was sure he had discovered a new virus that was cytopathic, meaning it killed certain kinds of cell, including T lymphocytes. It operated like the feline leukaemia virus, which caused cancer but also knocked out the immune system – exactly the way AIDS behaved. Montagnier did not feel that 'his' virus was the leukaemia virus – it behaved in somewhat different ways and therefore had different genetic properties. This view strengthened when he heard from a colleague about a certain category of virus, known as lentiviruses, based on the Latin *lentus*, 'slow'.[13] Lentiviruses lie dormant in cells before bursting into action. That is what seemed to happen with the AIDS virus, unlike the leukaemia virus. Montagnier therefore called the virus LAV, for lymphadenopathy-associated virus, because he had taken it from the lymph nodes of his patients.[14]

★

Intellectually speaking, there are five strands to current cancer research.[15] Viruses are one; the others are the environment, genes, personality (reacting with the environment), and auto-immunology, the idea that the body contains the potential for cancerous growth but is prevented from doing so by the immune system until old age, when the auto-immune system breaks down. There is no question that isolated breakthroughs have been made in cancer research, as with Gallo's viral discoveries and that of the link between tobacco and cancer, but the bleak truths about the disease were broadcast in 1993 by **Harold Varmus**, 1989 Nobel Prize winner in physiology and head of NIH in Washington, D.C., and **Robert Weinberg** of MIT in their book *Genes and the Biology of Cancer.*[16] They concluded that tobacco accounts for 30 percent of all cancer deaths in the United States, diet a further 35 percent, and that no other factor accounts for more than 7 percent. Once the smoking-related tumours are subtracted from the overall figures, however, the incidence and death rates for the great majority of cancers have either remained level or declined.[17] Varmus and Weinberg therefore attach relatively little importance to the environment in causing cancer and concentrate instead on its biology – **viruses and genes**. The latest research shows that there are proto-oncogenes, mutations formed by viruses that bring about abnormal growth, and tumour-suppresser genes that, when missing, fail to prevent abnormal growth. Though this may represent an intellectual triumph of sorts, even Varmus and Weinberg admit that this has not yet been translated into effective treatment. In fact, 'incidence and mortality have changed very little in the past few decades.'[18] This failure has become an intellectual issue in itself – the tendency for government and cancer institutes to say that cancer can be cured (which is true, up to a point), versus the independent voice of the medical journals, which underline from time to time that, with a few exceptions, incidence and survival rates have not changed, or that most of the improvements occurred years ago (also true).

Such debate, bitter at times, has often made cancer seem far more dreadful than other diseases, and it was this that provoked **Susan Sontag**, herself recovering from cancer, to write the first of two celebrated essays on illness. Her chief argument in *Illness as Metaphor* (1978) is that disease in general and cancer in particular in the late twentieth century is used as a metaphor for all sorts of political, military, and other processes, which demonise the illness and, more to the point, separate the sufferer from her/his family, friends, and life.[19] In many combative passages, she compares cancer now to TB a few generations ago. Illness, she says, 'is the night-side of life, a more onerous citizenship.'[20] There is, or is supposed to be, something uniquely frightening about cancer, so that even today, in France and Italy, it is still the rule for doctors to communicate a cancer diagnosis to the patient's family, not the patient. Since getting cancer jeopardises one's love life, chances of promotion, or even a job, people learn to be secretive. In literature, she points out, TB represents disintegration – it is 'a disease of liquids' – whereas cancer symbolises degeneration, 'the body tissue turning to something hard ... a demonic pregnancy.'[21] TB affects the lungs, the 'spiritual' part of the body, whereas 'cancer is notorious for attacking parts of the body (colon, bladder, rectum, breast, cervix, prostate, testicles) that are embarrassing to acknowledge.' Having a tumour generally arouses some feelings of shame, but 'in the hierarchy of the body's organs, lung cancer is felt to be less shameful than rectal cancer.'[22] The most striking similarity between TB and cancer, she says, is that both are diseases of passion – TB a sign of inward burning, romantic agony, whereas cancer 'is now imagined to be the wages of repression.' Surveying a wide range of literature from *Wings of the Dove* to *The Immoralist* to *The Magic Mountain* to *Long Day's Journey into Night* to *Death in Venice*, she finds the transformation of TB, a dreadful disease, into something romantic as 'preposterous,' a distortion as she sees it, and one that she does not want repeated with cancer.

Illness as Metaphor, provoked by Susan Sontag's own experience, was described by *Newsweek* as 'one of the most liberating books of our time.' In *AIDS and its Metaphors*, published a decade later, in 1989, Sontag returned to the attack.[23] AIDS she saw as one of the most 'meaning-laden' of diseases, and her aim was to 'retire' some of the many metaphors it had acquired. Sontag wanted – fiercely – to combat the aspect of punishment that was attaching to AIDS, to challenge 'a predictable mix of superstition and resignation [that] is leading some people with AIDS to refuse antiviral chemotherapy.'[24] She reserved special venom for those like the Christian right who argued that AIDS was a retribution for the sins and indulgences, the 'moral laxity and turpitude,' of the 1960s, and for those above all who were homosexual, understood as in some way abnormal. This *Kulturkampf*, she said, went beyond America. In France, where she lived part of the time, one right-wing politician had dismissed certain opponents as *sidatique* ('AIDS-ish'), or as suffering from 'mental AIDS.' But, she asked, could not AIDS better be understood as a capitalist-type disease of the consumer society in which 'appetite is *supposed* to be immoderate. ... Given the imperatives about consumption and the virtually unquestioned value attached to the expression of self, how could sexuality *not* have come to be, for some, a consumer option: an exercise of liberty, of increased mobility, of the pushing back of limits? Hardly an invention of the male homosexual subculture, recreational, risk-free sexuality is an inevitable reinvention of the culture of capitalism.'[25] She thought that the metaphors of AIDS have diminished us all. They have, for example, helped introduce the sad form of relationship that appeared in the late 1980s – telephone sex, which had the merit, if that is the word, of being safe. We have been further diminished by the widespread campaigns for the use of condoms and clean needles, which, she said, are 'felt to be tantamount to condoning and abetting illicit sex, illegal chemicals.'[26] It was time to understand illness, cancer, and AIDS for what they are: diseases of the body, with no moral or social or literary layers of meaning.

Other factors helped account for the changing perception of AIDS. Also relevant was the nature and quality of the victims themselves. When the *Hollywood Reporter* ran an item of news in its issue of 23 July 1985 saying that the handsome film actor Rock Hudson was suffering from AIDS, the illness finally received the publicity that in truth, given its killing power, it deserved.[27] But in addition to being the first AIDS victim most people had heard of, Hudson was also significant in being an actor. Over the following years, the arts and the humanities lost hundreds of bright lights as, despite the isolation of the virus responsible, AIDS took its toll: Michel Foucault, philosopher, June 1984, aged fifty-seven; Erik Bruhn, ballet dancer, 1986, aged fifty-eight; Bruce Chatwin, travel writer, January 1989, aged forty-eight; Robert Mapplethorpe, photographer, March 1989, aged forty-two; Keith Haring, graffiti artist, February 1990, aged thirty-one; Halston, fashion designer, March 1990, aged fifty-seven; Tony Richardson, film director, November 1991, aged sixty-three; Anthony Perkins, actor, September 1992, aged sixty; Denholm Elliott, actor, October 1992, aged seventy; Rudolf Nureyev, the most famous dancer of his day, who had defected from Russia in 1961, who had been director of the Paris Opera Ballet and danced for every leading company in the world, in January 1993, aged fifty-four. No disease this century has produced such carnage in the intellectual and artistic field.[28]

Carnage of a different kind took place in the psychiatric ward. On 29 March 1983 Dr John Rosen surrendered his medical licence in Harrisburg, Pennsylvania. He did this in order to avoid being tried by the State Board of Medical Education and Licensure of the Department of State of Pennsylvania, which was preparing to accuse him of sixty-seven violations of the Pennsylvania Medical Practices Act and thirty-five violations of the rules and regulations of the Medical Board.[29] Some of the abuses Rosen had subjected his patients to were horrific, none more so than in the case of Janet Katkow, who was taken to see him by her parents (the following details are taken from court documents, part of the public record). On their very first meeting, in front of her parents, Rosen asked Katkow if she had enjoyed her first sexual experience. She did not reply. When she expressed the wish to return to her home in the mountains of Colorado, he immediately made a 'deep interpretation' and explained that the snow-capped mountains were 'the next best thing' to 'a breast filled with mother's milk.' 'Defendant then told Plaintiff's mother that he had something better for Plaintiff to suck on and he simultaneously patted his groin with one hand.'[30] For the next seven years, Rosen forced Katkow to suck his penis during therapy. These sessions were invariably followed by vomiting on her part, which, he explained, was her throwing up her mother's milk. Another patient of Rosen's, Claudia Ehrman, who was treated by two of his assistants, was found dead in her room on 26 December 1979, having been heavily beaten, it emerged, by the assistants as part of therapy, in 'an attempt to force her to speak to them.'

An account of Dr. Rosen's extraordinary theories and practices, known in the psychiatric profession since 1959 as 'direct analysis,' and which culminated in the 102 charges against him being dropped, in exchange for his licence, forms the central chapter of **Jeffrey Masson's** book *Against Therapy*, published in 1988. Masson had himself trained as a psychoanalyst and was briefly projects director of the Sigmund Freud Archives, but he came to the conclusion that there was something very wrong with *psycho*therapy, whatever its genealogy. Masson's was an attack on psychoanalysis from a direction not seen before – that it was *by definition* corrupt and thereby irreconcilably flawed.

Masson began his book by going back to Freud himself and reexamining the very first patient, Dora. Masson's argument was that Freud had his own problems that he brought to the sessions with Dora, that they interfered with his interpretation of her condition, that she understood him every bit as well as he understood her, and that Freud 'simply ignored her needs in the service of his own, which was to find more evidence for the correctness of his psychological theories.'[31] In other words, psychoanalysis was flawed from the very beginning. From there, Masson moved forward, examining **Sandor Ferenczi's secret diary** (not published until 1985, although he had died in 1933), which showed that he too had had his doubts about the therapeutic relationship, to the point of even considering a variant, namely 'mutual analysis,' in which the patient analyses the therapist at the same time that the therapist is analysing the patient. He also looked at Jung's involvement with the Nazis, his anti-Semitism, and his mysticism, once again finding that Jung, like Freud, was an authoritarian, reading his own thoughts into whatever stories his patients told him, on the basis that the therapist is healthy, devoid of neuroses, and the patient is in this sense unclean. Masson also looked at the newer therapies, of Carl Rogers, for example, at Fritz Perls's Gestalt therapy, and the work of Rollo May, Abraham Maslow, and Milton Erickson.[32] Everywhere he found a great deal of authoritarianism and, more perniciously, a great concern with sex, especially sex *within* the therapeutic relationship. For Masson, it was clear that with many therapists the therapeutic situation served their needs as much as, or more than, the needs of the so-called patients, and for this reason he thought that therapy per se was impossible, that this is why the figures showing the inefficacy of psychoanalysis had to be right.

Much wittier than Masson's attack was **Ernest Gellner**'s in *The Psychoanalytic Movement* (1985), which must rank as one of the greatest intellectual putdowns of the century.[33] Gellner, born in Paris in 1925 and educated in Prague and England, became professor of both philosophy and sociology at the London School of Economics, then William Wye Professor of Social Anthropology at Cambridge. The subtitle of his book was 'The Cunning of Unreason,' and nothing in psychoanalysis – no non sequitur, no inconsistency, no piece of sloppy reasoning or logical laxity, no hypocrisy – was allowed to escape. His chief target is the unconscious, which he says is the new version of original sin.[34] Its official principle, he says, in only one of many wonderful belittlings, is 'Softlee Softlee Catchee Unconscious.' It is as if, he says, there is an Unconscious Secrets Act; the unconscious is not merely hidden from consciousness but seeks actively to remain so.[35] 'Neither intelligence nor conscious honesty nor theoretical learning in any way increase the prospects of by-passing and surmounting the counter-intelligence ploys of the Unconscious.'[36] By some strange set of events, however, Freud was able to break down this seemingly impregnable barrier and passed on the secret to others in a secular Apostolic Succession. But, asks Gellner, if the unconscious is so clever, why didn't it see Freud coming and disguise itself even more thoroughly? Gellner's aim was not simply to return to the statistical arguments against cure by psychoanalysis, however, but to debunk it. He quoted the Nobel Prize winner Friedrich von

Hayek: 'I believe men will look back on our age as an age of superstition, chiefly connected with the names of Karl Marx and Sigmund Freud.'[37] Yet Gellner really had no need of help from others. 'The Unconscious,' he wrote, 'is like some low hostelry just across the border, where all the thieves and smugglers indulge themselves with abandon, free of the need for camouflage and disguise which they prudently adopt, for fear of the authorities, when they are *this* side of the frontier ... [The unconscious] is like meeting all one's friends, enemies and acquaintances, but at the carnival and in fancy dress: one may be a bit surprised at what they get up to but there are few ... surprises as to personnel.'[38]

Freud was not alone in being debunked. At the end of January in 1983 the *New York Times* ran a front-page story headed: 'NEW SAMOA BOOK CHALLENGES MARGARET MEAD'S CONCLUSIONS.' The book was the work of the New Zealand-born Australian anthropologist **Derek Freeman**, who had been working in Samoa since 1940, mostly in an area some 130 miles from Ta'u, the village where Mead had done her fieldwork. His conclusion was that Mead had completely misunderstood Samoan society, and by implication drawn the wrong conclusions. The Samoans, said Freeman, were just as troublesome as people anywhere, and they had 'resented the way they were portrayed in *Coming of Age*,' as simple, playful people for whom sex was largely a game and whose nature was very different from that of people in other cultures.[39]

The *New York Times* story ran for 47 column inches, occupying almost a page inside the paper, and ignited a furious debate. Harvard University Press brought forward publication of Freeman's book *Margaret Mead and Samoa: The Making and Unmaking of an Anthropological Myth*, and he was invited on to television programs all over America. Several scientific seminars were held to consider his findings, the most important of which was a meeting of the American Anthropological Association.[40] Here Freeman's motivation was called into question. It was noted that he had hitherto been an obscure academic, working in Samoa by his own admission since 1940. Could he not have presented his arguments before, while Mead was alive to defend herself? He replied that he had put his preliminary doubts to her, and she had acknowledged certain shortcomings in her data, but it was not until 1981, when he was granted permission to examine Samoan court records, that he could conclude that Western Samoa was just as violent a society as elsewhere.[41] Other anthropologists doubted Freeman's account at this point; *they* had had no problem getting access to court records many years before. A bigger issue, however, was where Freeman's revelations, if revelations they were, left Franz Boas's ideas that culture, and not nature, is more important in determining behaviour patterns. Freeman was not himself a biological determinist, but there is no question that if he was right, his revision of Mead's findings provided support for a less 'cultural' understanding of human nature. The issue was never satisfactorily resolved, but Mead, like Freud, now has a definite shadow over her seminal work (no one doubts that many of her other findings were real).

In 1997 Roy Porter published *The Greatest Benefit to Mankind: A medical history of humanity from antiquity to the present.* In his chapter on clinical science, Porter quotes Sir David Weatherall, Regius Professor of Medicine at Oxford. Weatherall, as Porter reports, asked of modern medicine this question: How are we doing? and reached a surprisingly sombre conclusion. 'We seem to have reached an impasse in our understanding of the major killers of Western society, particularly heart and vascular disease, cancer and the chronic illnesses whose victims fill our hospitals. ... Although we have learned more and more about the minutiae of how these diseases make patients sick, we have made little headway in determining why they arise in the first place.'[42]

Weatherall's scepticism is realistic; his argument is well made. Triumphalism in science is unscientific. The same goes for the revisions of Freud, Jung, and Mead. The irony – and absurdity – of having a therapeutic sensibility when the therapies don't work cannot escape anyone. Porter's own conclusion, after his masterly survey of medicine, was hardly less pessimistic than Weatherall's: 'The root of the trouble is structural. It is endemic to a system in which an expanding medical establishment, faced with a healthier population, is driven to medicalising normal events like menopause, converting risks into diseases, and treating trivial complaints with fancy procedures. Doctors and "consumers" are becoming locked within a fantasy that *everyone* has *something* wrong with them, everyone and everything can be cured.'[43] This is one explanation, of course, for why the 'cure rates' for psychoanalysis are so dismal. Many who seek analysis have nothing wrong with them in the first place.

In 1979 the U.S. space probe *Pioneer 11* reached Saturn and travelled through its surrounding rings, which were found to be made of ice-covered rocks. The business use of personal computers was vastly expanded after the first software for spreadsheets was introduced. In the same year the Phillips Company launched its LaserVision video disc system, and Matsushita brought out its pocket-size flat-screen TV set. Physicists at Hamburg observed **gluons** – elementary particles that carry the strong nuclear force that holds quarks together. Science and technology were continuing to make impressive advances, though there was one blot on the landscape – almost literally, in the form of a major accident at the Three Mile Island nuclear power station in Pennsylvania, which lost its water buffer through operator error, allowing the escape of a small amount of radioactive material, with the reactor itself undergoing a partial meltdown. No one was injured, but everyone was chastened.

Although science was, far more often than not, offering material advance and intellectual excitement for those who wanted it, by 1979 there were also many countervailing voices. This was not simply antiscience in the old-fashioned sense, of the creationists, say, or the religious fundamentalists. By the end of the 1970s the critique of science, the scientific method, and science as a system of knowledge had become a central plank in postmodern thinking. *The Postmodern Condition*, by **Jean-François Lyotard**, was the first in a whole raft of books that began to question the very status of science. It is important to give the subtitle of Lyotard's book, 'A Report on Knowledge', for he was a French academic, at the Institut Polytechnique de Philosophie of the Université de Paris VIII (at Vincennes), who was commissioned by the Conseil des Universités of the government of Quebec to prepare an investigation.[1] Though a philosopher, Lyotard had begun adult life in postwar Paris as a left-wing political journalist. Later, while completing his academic qualifications in philosophy, he had developed an interest in psychoanalysis, trying to marry Freud and Marx, as so many colleagues were doing, and in the arts. His early writing he had grouped into the 'The Libidinal,' 'The Pagan,' and 'The Intractable.'[2] The first category clearly carried psychoanalytic overtones, but beyond that the use of *the libidinal* was meant to imply that, as he viewed the world, motivating sources were personal, individual, and even unconscious, rather than overtly political, or deriving from some particular metanarrative. Similarly, in using the term *pagan*, Lyotard intended to imply not so much false gods as alternative gods, and many different varieties, that one's interests in life could be satisfying and rewarding even when they had nothing to do with the official, or most popular 'truths.' By *intractable* he meant that some areas of study, of experience, are simply too complex or too random ever to be predicted or understood.

In *The Postmodern Condition*, however, Lyotard's specific target was science as a form of knowledge. He wanted to know in what important ways scientific knowledge differs from other forms of knowledge, and what effects the success of scientific knowledge is having on us, as individuals and as a society. 'Simplifying to the extreme,' he begins, 'I define *postmodern* as incredulity toward metanarratives.'[3] He goes on to compare different kinds of knowledge – for example, that contained in a fairy story, that produced by the law, and that produced by science. For many scientists, as Lyotard concedes, scientific knowledge is the only form of knowledge there is, but if so, how then do we understand fairy stories and laws? The most important form of knowledge that isn't scientific – in the sense that most scientists would accept the term – is, he says, knowledge about the self. The self, Lyotard says, has a history, is in part at least a narrative, and like no other. It is, therefore, unavailable to science, which produces knowledge that is essentially abstract in character.

In an historical excursion Lyotard explains how, in his view, the traditional scientific approach originated at the University of Berlin in the nineteenth century; he argues that science has essentially been a child of universities ever since, and therefore has usually been paid for by governments. This is important to Lyotard as the central fact in the sociology of (scientific) knowledge, what Nietzsche called 'the paranoia of reason,' though Lyotard prefers the 'tyranny of the experts.' This is why a certain kind of knowledge (such as, 'The earth revolves around the sun') came to have a higher status than others (such as, 'The minimum wage should be set at x dollars'). After 150 years of state-run science, we find it much easier to prove the former than the latter.[4] Is that because of the science we have pursued, or because the latter statement is intractable, incapable of proof? If there are certain categories of problem, or experience, or simple ways of talking that are intractable *in principle*, where does that leave science? Where does that leave the universities and the optimism (in those who possess it) that, given time, science can solve all our problems? Much influenced by Werner Heisenberg, Kurt Gödel, and Thomas Kuhn, Lyotard was impressed by the new ideas being broached in the late 1970s and 1980s, in particular catastrophe theory, chaos theory, and the problems posed by incomplete information, 'fracta': 'It is changing the meaning of the word *knowledge*. ... It is producing not the known but the unknown.'[5] Lyotard adds that many areas of life are language games – we manipulate language in relation to experience, but that relation is incomplete, complex, and in any case it is only one of the things we are doing when we use language. Perhaps the very notion of self is, in a sense, a game.

Lyotard's conclusion was not antiscience. But he argued that other forms of knowledge (including speculation, which scientists have been known to go in for) have their place, that science can never hope to provide anything like a complete answer to the philosophical problems that face us (or that we think face us). Science derives its power, its legitimacy, from its technological successes, and rightly so. But science can only go so far; there are many areas of life that will always remain intractable to science in principle. Of these the most important is the self.

Like Lyotard, **Richard Rorty** of Princeton is a philosopher fascinated by the status of scientific knowledge. This led him to write two books, *Philosophy and the Mirror of Nature* (1980) and *Objectivity, Relativism, and Truth* (1991), in which he offered a radical reinterpretation of what philosophy can ever hope to be.[6]

In *Philosophy and the Mirror of Nature* Rorty accepts that science has proved amazingly successful in producing a certain kind of knowledge. And he agrees, with Rudolf Carnap, that science has correctly destroyed a certain kind of speculation – traditional metaphysics. He agrees with Lyotard that scientific knowledge is not the only form of knowledge there is (he uses literary criticism and politics as other forms). His main point, however, is to try to prevent philosophy from becoming a mere adjunct of science. 'Some day,' thanks to science, he says, we shall be able 'in principle to predict every movement of a person's body (including those of his larynx and his writing hand) by reference to microstructures within his body.' But even when we can do this, says Rorty, we shall still not be able to predict what these people will say and/or mean. He says this with some confidence because it is his argument that people, persons, continually 'remake' themselves 'as we read more, earn more, and write more.' People are constantly 'edifying' themselves and in the process becoming different persons. It is in this sense that Rorty synthesises – for example – Freud, Sartre, and Wittgenstein. Freud (like Marx) realised that people could change when their self-consciousness changed, a change that could be brought about by words; this concept of a changing self was central to Sartre's existential notion of 'becoming' and to Lacan's idea of 'success' in treatment; and Wittgenstein's focus on the central aspect of language, and that metaphysics is a 'disease' of language, underpins Rorty's reevaluation of what philosophy is.[7]

For Rorty, the central mistake of philosophers has been twofold – to see philosophy as an extension of science, to try to speak in a scientific language, and to see philosophy as a *system*, which offers a more or less complete explanation or understanding of the world. Rorty, on the other hand, sees philosophy as an activity attempting to reach areas of human experience that science will never be able to conquer. Philosophy should be 'edifying' in the following sense: 'The attempt to edify [ourselves] ... may ... consist in the "poetic" activity of thinking up such new aims, new words, or new disciplines, followed by, so to speak, the inverse of hermeneutics: the attempt to reinterpret our familiar surroundings in the unfamiliar terms of our new inventions. ... [T]he activity is ... edifying without being constructive – at least if "constructive" means the sort of co-operation in the accomplishment of research programs which takes place in normal discourse. For edifying discourse is *supposed* to be abnormal, to take us out of our old selves by the power of strangeness, to aid us in becoming new beings.'[8] But, says Rorty, 'on the periphery of the history of modern philosophy, one finds figures who, without forming a "tradition," resemble each other in their distrust of the notion that man's essence is to be a knower of essences. Goethe, Kierkegaard, Santayana, William James, Dewey, the later Wittgenstein, the later Heidegger, are figures of this sort. They are often accused of relativism or cynicism. They are often dubious about progress, and especially about the latest claim that such-and-such a discipline has at last made the nature of human knowledge so clear that reason will now spread throughout the rest of human activity'.[9] 'These writers have kept alive the suggestion that, even when we have justified true belief about everything we want to know, we may have no more than conformity to the norms of the day. They have kept alive the historicist sense that this century's "superstition" was the last century's triumph of reason, as well as the relativist sense that the latest vocabulary, borrowed from the latest scientific achievement, may not express privileged representations of essences, but be just another of the potential infinity of vocabularies in which the world can be described. ... The mainstream philosophers are the philosophers I shall call "systematic," and the peripheral ones are those I shall call "edifying." These peripheral, pragmatic philosophers are skeptical primarily *about systematic philosophy* [italics in original], about the whole project of universal commensuration. In our time, Dewey, Wittgenstein, and Heidegger are the great edifying, peripheral, thinkers.'[10]

For Rorty, philosophy is in a way a parasitic activity, a guerrilla mode of thought, achieving its aims piecemeal and as a result of what is happening in other disciplines. John Dewey, Wittgenstein, and Kuhn were 'debunkers,' and Rorty is perhaps the greatest debunker of all when he likens philosophy to no more than 'conversation' (the last section of his book is called 'Philosophy in the Conversation of Mankind'). 'If we see knowing as not having an essence, to be described by scientists or philosophers, but rather as a right, by current standards, to believe, then we are well on the way to seeing *conversation* [italics in original] as the ultimate context within which knowledge is to be understood. ... The fact that we can continue the conversation Plato began without discussing the topics Plato wanted discussed, illustrates the difference between treating philosophy as a voice in a conversation and treating it as a subject.'[11]

In *Objectivity, Relativism, and Truth*, Rorty's two main areas of exploration are the objectivity of science and the relation of philosophy to politics.[12] Objectivity – the sense that there is something 'out there,' irrespective of who is doing the thinking, or observing – he sees as a doomed notion. The idea that 'green' or 'gravity' exist in some way different from the way 'justice' exists is a misconception, and merely reflects that more people agree on what 'green' is than what 'justice' is.[13] As Rorty puts it, there is more 'solidarity' in the practice. Think of the first person in early antiquity who first used the word *green* (in whatever language was spoken then); that person had to have a concept of green. The concept, and the word, have worked. But that is mere pragmatism. Think of the word *gravity*. This is an entity, whatever it is, that is still imperfectly understood. When and if it is ever understood, that word may prove to be inadequate, like *phlogiston* and *ether* in the past, and fall into disuse. In the end, Rorty thinks that the difference between truth and opinion is a matter of degree only, a question of solidarity, and we mislead ourselves if we think that there is some sense in which things are true for all time, and all cultures.

In his earlier book, one of Rorty's aims was to diminish our ambitions for what philosophy is, to make it more a 'conversation' than a system of thought. In the later book he did the same for reason. Reason, he says, is not an unalterable set of rules for thinking, which corresponds to reality 'out there.' Instead, it is much more like what we mean when we say something or someone is 'reasonable,' 'methodical,' or 'sane.' 'It names a set of moral virtues: tolerance, respect for the opinions of those around one, willingness to listen, reliance on persuasion rather than force. ... When so construed, the distinction between the rational and the irrational has nothing in particular to do with the difference between the arts and the sciences. On this construction, to be rational is simply to discuss any topic – religious, literary, or scientific – in a way which eschews dogmatism, defensiveness, and righteous indignation.'[14] 'On this view there is no reason to praise scientists for being more "objective" or "logical" or "methodical" or "devoted to the truth" than other people. But there is plenty of reason to praise the institutions they have developed and within which they work, and to use these as models for the rest of culture. For these institutions give concreteness and detail to the idea of "unforced agreement." Reference to such institutions fleshes out the idea of "a free and open encounter" – the sort of encounter in which truth cannot fail to win. On this view, to say that truth will win in such an encounter is not to make a metaphysical claim about the connection between human reason and the nature of things. It is merely to say that the best way to find out what to believe is to listen to as many suggestions and arguments as you can.'[15] As a pragmatist, Rorty admires the sciences for the qualities listed above, but it does not follow that he wants the rest of society to be organised in the same way: 'One consequence of [the pragmatic] view is the suggestion that perhaps "the human sciences" *should* [italics in original] look quite different from the natural sciences. This suggestion is not based on epistemological or metaphysical considerations which show that inquiry into societies must be different from inquiry into things. Instead, it is based on the observation that natural scientists are interested primarily in predicting and controlling the behaviour of things, and that prediction and control may not be what we want from our sociologists and our literary critics.'[16] There are no 'different worlds,' and all forms of inquiry – from physics to poetry – are equally legitimate.

Rorty's main aim when discussing politics is to argue that a political system does not need a concept of human nature in order to function. Indeed, Rorty says that this development is crucial to the existence of the bourgeois liberal democracies. He makes it clear that he believes the bourgeois liberal democracies to be the best form of government, and here he differs from many other postmodern scholars. He agrees with Lyotard, Jürgen Habermas, and other postmodernists that metanarratives are unhelpful and misleading, but he takes this farther, arguing that the very success of the American Constitution, and of the parliamentary democracies, stems from their tolerance, and that almost by definition this means that metanarratives about human nature have been eschewed. Rorty follows Dewey in arguing that the 'disenchantment' of the world, as for example in the loss of religion, has enabled personal liberation to replace it. As a result, history is made up of countless personal narratives rather than one great narrative. This is much the same as saying that the postmodern sensibility is one endpoint of bourgeois liberal democracy.

On this score, Rorty is somewhat at odds with a figure like Clifford Geertz, whom we shall come to shortly. Geertz, an anthropologist, cultural historian, and philosopher, put forward the argument in several books in the 1970s and 1980s that – to simplify for the moment – we can only ever have 'local knowledge,' knowledge grounded in space and time, that other cultures and societies need to be understood in their terms rather than ours. While agreeing with Geertz up to a point, Rorty clearly believes that a bourgeois liberal democracy has something other societies don't, if only because 'its sense of its own moral worth is founded on its tolerance of diversity. ... Among the enemies it diabolizes are the people who attempt to diminish this capacity, the vicious ethnocentrists.'[17] Rorty emphasises that the very anthropologists, of which Geertz is such a distinguished example, are part of bourgeois liberal democracy, *and that is the point*. Their actions have drawn to 'our' attention the existence of certain people who were 'outside' before. This is an example, he says, of the principal moral division in a liberal democracy, epitomised by 'the agents of love' and 'the agents of justice.'* The agents of love include ethnographers, historians, novelists, muckraking journalists, specialists in particularity rather than specialists in universality like theologians or, yes, the old idea of philosophers. In leaving to one side any overriding conception of human nature, liberal democracies have helped the 'forgetting' of philosophy as traditionally understood, i.e., as a *system* of thought: 'The *défaillance* of modernity strikes me as little more than the loss of ... faith in our ability to come up with a single set of criteria which everybody in all times and places can accept, invent a single language-game which can somehow take over all the jobs previously done by all the language-games ever played. But the loss of this theoretical goal merely shows that one of the less important sideshows of Western civilisation – metaphysics – is in the process of closing down. This failure to find a single grand commensurating discourse, in which to write a universal translation manual (thereby doing away with the need to constantly learn new languages) does nothing to cast doubt on the possibility (as opposed to the difficulty) of peaceful social progress. In particular, the failure of metaphysics does not hinder us from making a useful distinction between persuasion and force. We can see the pre-literate native as being persuaded rather than forced to become cosmopolitan just insofar as, having learned to play the language-games of Europe, he decides to abandon the ones he played earlier – without being threatened with loss of food, shelter, or *Lebensraum* if he makes the opposite decision.'[18]

---

* This terminology recalls exactly the title of Colin MacInnes's 1958 novel, *Mr Love and Justice*.

Although he doesn't develop the point, Rorty uses the words *défaillance* and *progress*. One translation of *défaillance* is extinction. Rorty is, therefore, marrying postmodernism to evolutionary theory, and in two ways. He and other philosophers are concerned partly with whether the nature of science, and the knowledge it produces, is in any sense different in kind from other forms of knowledge, whether and to what extent science itself may be regarded as an example of cultural evolution; and partly with whether postmodernism itself is an 'evolved' concept.

Thomas Nagel, professor of philosophy and law at New York University, likes to give his books arresting titles: *Mortal Questions, What Does It All Mean? The View from Nowhere, The Last Word.* Nagel stands out because, in a postmodern world, he considers traditional problems of philosophy. He uses new, clear language, but they *are* the old problems he is considering. He even uses words like *mind* without hesitation.

In *Mortal Questions* (1979) and *The View from Nowhere* (1986) Nagel's main focus is the objectivity-subjectivity divide, how it relates to the concept of the self, and to consciousness.[19] Nagel is one of those philosophers, like Robert Nozick and unlike John Rawls, who takes the world as he find it: 'I believe one should trust problems over solutions, intuition over arguments, and pluralistic discord over systematic harmony. Simplicity and elegance are never reasons to think that a philosophical theory is true: on the contrary, they are usually grounds for thinking it is false.'[20] Nagel's view is that there *are* such things as mental states, the most important of which is experience of the world. He doubts whether the physical sciences will ever be able to explain what experience of the world is, or the sense of self, and asks therefore whether we can ever have a concept of 'reality' that is anywhere near complete. Aren't we better off accepting these limitations, and shouldn't we just get on trying to understand experience and subjectivity in other ways? There is no law that says that philosophy shouldn't be useful. But Nagel shares with Lyotard, Rorty, and others a fascination with what science has done to us, in the sense of whether the knowledge that science produces is or is not some kind of special knowledge, more 'objective' than other kinds. His approach might be termed 'taking intuition seriously.' 'Objectivity of whatever kind is not the test of reality. It is just one way of understanding reality,'[21] he writes; and, 'The difference between mental and physical is far greater than the difference between electrical and mechanical.'[22] Just as the world of physics, and the way we understand objectivity, was changed by James Clerk Maxwell and Albert Einstein, so, Nagel believes, we may one day have a psychological Maxwell and Einstein who will change our understanding of reality in equally fundamental ways, though at the moment we are nowhere near it. Not only is Nagel dismissive of the kind of objectivity provided by physics, he is also sceptical of the claims of evolutionary theory. Darwinian theory 'may explain why creatures with vision or reason survive, but it does not explain how vision or reasoning are possible. These require not diachronic [historical] but timeless explanation. ... The possibility of minds capable of forming progressively more objective conceptions of reality is not something the theory of natural selection can attempt to explain, since it doesn't explain possibilities at all, but only selection among them.'[23]

Nagel does not have an alternative explanation to, say, evolutionary theory, but he says he doesn't need one to cast doubt on the grand claims that are being made for evolution. That is Nagel's charm and, maybe, his force: he is not afraid to tell us what he doesn't know, or even that some of his views may be absurd. His aim is to use language, and reason, to think in ways that haven't been done before. In his view, his intuition (as well as his powers of observation) tell him that the world is a big, complex place. Any one solution is extremely likely to be wrong, and it is intellectually lazy not to explore all possibilities. 'The capacity to imagine new forms of hidden order, and to understand new conceptions created by others, seems to be innate. Just as matter can be arranged to embody a conscious, thinking organism, so some of these organisms can rearrange themselves to embody more and more thorough and objective mental representations of the world that contains them, and this possibility too must exist in advance.'[24] Nagel describes this view as rational but anti-empiricist.[25] Since agreement is only possible to us through language, Nagel says, echoing Wittgenstein, there may well be things about our world – in fact, there probably are – that we cannot conceive. We are almost certainly limited by our biological capacity in this respect. In time that may change, but this also should change our view of what objectivity and reality are. 'Realism is most compelling when we are forced to recognise the existence of something which we cannot describe or know fully, because it lies beyond the reach of language, proof, evidence, or empirical understanding.'[26] So for Nagel we may some day be able to conceive what things were like before the Big Bang.[27]

For Nagel, ethics are just as objective as anything science has to offer, and the subjective experience of the world easily the most fascinating 'problem,' which science is nowhere near answering. The objective fact of our subjective lives is a conundrum that we don't even have the language or the right approach for. Empirical science as we know it is nowhere near an answer. Nagel's books are difficult, in the sense that one feels he is on the edge of language all the time, questioning our assumptions, throwing up new possibilities, rearranging (as Wittgenstein counselled) the familiar in new and exciting ways. One is reminded of Lionel Trilling's hope for fiction, that it would/should seek to remain outside any consensus and continually suggest new – and hitherto unimaginable – possibilities. And so Nagel is difficult, but exhilarating.

Clifford Geertz, at the Institute for Advanced Study, in Princeton, New Jersey, shares *very firmly* with other postmodernists like Lyotard the view that the world is 'a various place' and that we must confront this uncomfortable truth if we are to have any hope of understanding the 'conditions' by which we live. In two books, *The Interpretation of Cultures* (1973) and, even more, *Local Knowledge* (1983), he detailed his view that subjectivity is *the* phenomenon for anthropologists (and others in the human sciences) to tackle.[28] 'The basic unity of mankind,' according to Geertz, is an empty phrase if we do not take on board that drawing a 'line between what is natural, universal, and constant in man and what is conventional, local and variable [is] extraordinarily difficult. In fact, it suggests that to draw such a line is to falsify the human situation, or at least to misrender it seriously.'[29] The hunt for universals began with the Enlightenment, says Geertz, and that aim directed most Western thought, and has been a paradigm of Western science, and the Western notion of 'truth,' ever since. Pursuing fieldwork in Java, Bali, and Morocco, Geertz has dedicated his entire career to changing that view, to distinguishing between the '**thin**' and '**thick**' interpretations of cultures around the world, where 'thick' means to try to understand the signs and symbols and customs of another culture in its own terms, by assuming not, as Lévi-Strauss did, that all human experience across the globe can be reduced to structures, but instead that other cultures are just as 'deep' as our own, just as well thought out and rich in meaning, but perhaps 'strange,' not easily fitted into our own way of thinking.[30]

Geertz's starting point is palaeontology. It is wrong in his view to assume that the brain of *Homo sapiens* evolved biologically and that cultural evolution followed. Surely, he argues, there would have been a period of overlap. As man developed fire and tools, his brain would have still been evolving – and have evolved to take into account fire and tools. This evolution may well have been slightly different in different parts of the world, so that to talk of one human nature, even biologically speaking, may be misleading. Geertz's own anthropology therefore involves the meticulous description of certain alien practices among non-Western peoples, where the examples are chosen precisely because they appear strange to 'us.' He chooses, for example, a Balinese cockfight (where people gamble with their status in a way literally unthinkable in the West); the way the Balinese give names to people; Renaissance painters in Italy (a sort of historical anthropology, this); and certain aspects of North African law, tribal practices overlaid with Islam.[31] In each case his aim is not to show that these processes can be understood as 'primitive' versions of customs and practices that exist in the West, but as practices rich in themselves, with no exact counterpart in the West. The Balinese, for example, have five different ways of naming people; some of these are rarely used, but among those that are, are names that convey, all at the same time, the region one is from, the respect one is held in, and one's relation to certain significant others. In another example, he shows how a Balinese man, whose wife has left him, tries to take (Balinese) law into his own hands, but ends up in a near-psychotic state since his actions cause him to be rejected by his society.[32] These matters cannot be compared to their Western equivalents, says Geertz, because there *are* no Western equivalents. That is the point.

Cultural resources are, therefore, not so much accessory to thought as 'ingredient' to it. For Geertz, an analysis of a Balinese cockfight can be as rich and rewarding about Bali thought and society as, say, an analysis of *King Lear* or *The Waste Land* are about Western thought and society. For him, the old division between sociology and psychology – whereby the sociology of geographically remote societies differed, but the psychology stayed the same – has now broken down.[33] Geertz's own summing up of his work is that 'every people has its own sort of depth.'[34] 'Thinking is a matter of the intentional

manipulation of cultural forms, and outdoor activities like ploughing or peddling are as good examples of it as closet experiences like wishing or regret-
ting,'[35] he writes; and, 'The hallmark of modern consciousness ... is its
enormous multiplicity. For our time and forward, the image of a general
orientation, perspective, *Weltanschauung*, growing out of humanistic studies (or,
for that matter, out of scientific ones) and shaping the direction of culture is a
chimera. ... Agreement on the foundations of scholarly authority, old books
and older manners, has disappeared. ... The concept of a "new humanism,"
of forging some general "the best that is being thought and said" ideology
and working it into the curriculum, [is] not merely implausible but utopian
altogether. Possibly, indeed, a bit worrisome.'[36] Geertz does not see this as a
recipe for anarchy; for him, once we accept the 'depth of the differences'
between peoples and traditions, we can begin to study them and construct a
vocabulary in which to publicly formulate them. Life will in future be made
up of a variety of vivid vernaculars, rather than 'forceless generalities.' This is
the way the 'conversation of mankind' will continue.[37]

The main contribution of the philosopher **Hilary Putnam**, from Harvard, was
an examination of the impact of science on our notions of reason and rationality.
Putnam's argument is that what we call ' "truth" depends both on what there
is (the way things are) and on the contribution of the thinker ... there is a
human contribution, a conceptual contribution, to what we call "truth."
Scientific theories are not simply dictated to us by the facts.'[38] This view had
important implications, in Putnam's mind, for he felt that by now, the end of
the twentieth century, the 'scientific method' had become a very 'fuzzy' thing,
an idea that for him had peaked in the seventeenth century and had been
gradually dissolving since, making the logical positivists of the Vienna Circle
anachronistic. By this he meant the idea that science, and therefore reason,
could only apply to directly observable and neutral 'facts,' which led to easily
falsifiable theories. Many modern scientific theories, he pointed out, were by
no means easily falsifiable – evolution being a case in point.[39] He therefore
agreed with Rorty that 'reason' ought to mean what most of us mean by it,
how a reasonable person behaves in his/her approach to the world. But Putnam
went further in arguing that there is much less distinction between facts and
values than traditional scientists, or philosophers of science, allow. He agreed
with Kuhn and Polanyi that science often proceeds by some sort of intuitive or
inductive logic, because not all possible experiments are ever tried, merely the
most plausible, 'plausible' itself being derived from some 'reasonable' idea we
have of what we should do next. Arising from this, Putnam argued that certain
statements, traditionally taken to be values, or prejudices (in the widest sense),
are also facts just as much as the facts produced by science. Two examples he
gives are that Hitler was a bad man and that poetry is better than pushpin. In
the case of pushpin, for example, Jeremy Bentham said in the eighteenth
century that expressing a preference for poetry over the game is a mere
prejudice, subjective – an argument much loved by the relativists, who believe
that the subjective life of one person, and even more so of one culture, cannot

be fruitfully or meaningfully compared to that of another. Putnam's refutation was not anthropological but philosophical, because the argument gave credence to 'prejudice' as a mental entity while denying it to, say, 'enlarged sensibilities,' 'enlarged repertoires of meaning and metaphor,' 'self-realization,' and so on: 'The idea that values are not part of the Furniture of the World and the idea that "value judgements" are expressions of "prejudice" are two sides of the same coin.'[40] Value judgements, Putnam is saying, can be rationally supported, and it is time to get away once and for all from the idea that scientific facts are the only facts worthy of the name. 'Even the distinction between "classical" physics and quantum mechanics, with their rival views of the world, is itself observer-dependent.' 'The harm that the old picture of science does is this: if there is a realm of absolute fact that scientists are gradually accumulating, then everything else appears as non-knowledge.'

Willard van Orman Quine, another Harvard philosopher, took a very different line, while still retaining the importance of science, and the scientific method, for philosophy. In a series of books, *From a Logical Point of View* (1953), *Word and Object* (1960), *Roots of Reference* (1974), *Theories and Things* (1981), *Quiddities* (1987), and *From Stimulus to Science* (1995), Quine set out his view that philosophy is continuous with science, even part of science, and that there are essentially two aspects to reality: physical objects, which exist externally and independently of us, and abstract objects, notably mathematics. Quine is a dedicated materialist, holding that 'there is no change without a change in the distribution of microphysical properties over space.'[41] This approach, he says, enables him to eschew dualism, for 'mental' events are 'manifested' by behaviour. In other words, the understanding of mental events will ultimately be neurological, whether or not we ever reach such understanding. Mathematics, on Quine's formulation, has a twofold importance.[42] First, the existence and efficiency of numbers in helping describe and understand the universe is fundamental, the more so as numbers exist only as an abstract concept. Second, there is the idea of sets, the way some entities group together to form higher-order superentities, which imply similarity and difference. This, for Quine, relates number to words and words to sentences, the building blocks of experience. In zoology, for instance, living organisms have evolved into different genera and families – what does that mean philosophically? Are there genuine families and genera in nature, or are they a figment of our brains, based on our understanding of similarities, differences, and the relative importance of those similarities/differences? What goes on in the brain, at the microphysical level, when we think or talk about such matters? How closely do words, *can* words, correspond to what is 'out there,' and what does that mean for the microphysical processes in the brain?[43] When words that mean similar (but not identical) things in different languages are translated, what does that involve for microphysical properties in the brain? Quine is an unusually difficult philosopher to paraphrase, because many of his writings are highly technical, using mathematical notation, but broadly speaking he may be seen in the tradition of Bertrand Russell, the logical positivists, and B. F. Skinner, in that for him philosophy is not a discipline as Rorty or Nagel would have it, beyond science, but is a part of science, an extension that, although it asks questions that scientists themselves might not ask, nevertheless talks about them in ways that scientists *would* recognise.

Whose Justice? Which Rationality? (1988), by **Alasdair MacIntyre**, is perhaps the most subversive postmodern book yet, uniting as it does the work of Michel Foucault, Roland Barthes, Geertz, Rawls, and Dworkin in a most original fashion.[44] MacIntyre looked at notions of reason, and rationality, and their effects on ideas of justice, in earlier societies – in classical Greece, classical Rome, Saint Thomas Aquinas's teaching at the University of Paris in the thirteenth century, the Scottish enlightenment in the seventeenth and eighteenth centuries – and in modern liberal times. He looked at their arguments, as developed in political, philosophical, legal, and literary works, but also at their language and how it did, or did not, conform to modern notions. Rhetoric in Athens, for example, was regarded as the high point of reason, and its aim was to spur to action; it was not thought proper, therefore, to refer to rival points of view, to weigh both sides of the argument before deciding. Reasoning, as we would understand it, was kept to a discussion of means toward an end, not about the end and the justice of that end, which was understood implicitly to be shared by all. Only people who possessed the virtues were felt to be capable of reason in Athens, says MacIntyre, and this concept was even given a special name, *boulesis*, 'rational wish.' In this context, the rational person in Athens acted 'immediately and necessarily upon affirming his reasons for action ... very much at odds with our characteristically modern ways of envisaging a rational agent.'[45]

Saint Thomas Aquinas believed, along with all Christians, that everyone had the potentiality to act in a reasoned way, which would lead to a moral life, but that only education in a certain order – logic, mathematics, physics – could bring about full realisation of those potentialities. There was, for him, no difference between being rational and being moral. The Scottish enlightenment, on the other hand, turned back to an emphasis on the passions, David Hume distinguishing between the calm passions and the violent passions, which take priority over reason. 'Truth in itself according to Hume ... is not an object of desire. But how then are we to explain the pursuit of truth in philosophy? Hume's answer is that the pleasure of philosophy and of intellectual inquiry more generally "consists chiefly in the action of the mind, and the exercise of the genius and understanding in the discovery or comprehension of any truth." Philosophy, so it turns out, is like the hunting of woodcocks or plovers; in both activities the passion finds its satisfaction in the pleasures of the chase.' For Hume, then, reason cannot motivate us.[46] 'And the passions, which do motivate us, are themselves neither reasonable nor unreasonable. ... Passions are thus incapable of truth or falsity.'[47] Hume himself said, 'Reason is, and ought only to be, the slave of the passions and can never pretend to any other office than to serve and obey them.'[48]

In the modern liberal society, on the other hand, MacIntyre tells us there is a rival concept of reason and of justice, based on different assumptions, namely that people are individuals and nothing more: 'In Aristotelian practical reasoning it is the individual *qua* citizen who reasons; in Thomistic practical reasoning it is the individual *qua* enquirer into his or her good and the good of his or her community; in Humean practical reasoning it is the individual *qua* propertied or unpropertied participant in a society of a particular kind of mutuality and reciprocity; but in the practical reasoning of liberal modernity it is the individual *qua* individual who reasons.'[49] MacIntyre's conclusion is that our concepts of reasoning (and justice) are just one tradition among several. He offers no concept of evolution in these matters, and neither Darwin nor Richard Dawkins is mentioned in his book. Instead, MacIntyre thinks we continue to deform our relationship with the past by coarse translations of the classics (even when done by some scholars), which do not treat ancient words to their ancient meanings but instead offer crude modern near-equivalents. Quoting Barthes, he says that to understand the past, we need to include all the signs and other semiological clues that the ancients themselves would have had, to arrive at what Clifford Geertz (who *is* referred to in MacIntyre's book) would call a 'thick description' of their conceptions of reason and justice. The result of the liberal conception of reason, he says, has some consequences that might be seen as disappointing: 'What the student is in consequence generally confronted with ... is an apparent inconclusiveness in all argument outside the natural sciences, an inconclusiveness which seems to abandon him or her to his or her pre-rational preferences. So the student characteristically emerges from a liberal education with a set of skills, a set of preferences, and little else, someone whose education has been as much a process of deprivation as of enrichment.'[50]

The title of **David Harvey**'s book *The Condition of Postmodernity* is strikingly similar to Lyotard's *Postmodern Condition*. First published in 1980, it was reissued in 1989 in a much revised version, taking into account the many developments in postmodernism during that decade.[51] Contrasting postmodernity with modernity, Harvey begins by quoting an editorial in the architectural magazine *Precis 6*: 'Generally perceived as positivistic, technocentric, and rationalistic, universal modernism has been identified with the belief in linear progress, absolute truths, the rational planning of ideal social orders, and the standardisation of knowledge and production. Postmodernism, by way of contrast, privileges "heterogeneity and differences as liberative forces in the redefinition of cultural discourse." Fragmentation, indeterminacy, and intense distrust of all universal or 'totalising' discourses (to use the favoured phrase) are the hallmark of postmodernist thought. The rediscovery of pragmatism in philosophy (e.g., Rorty, 1979), the shift of ideas about the philosophy of science wrought by Kuhn (1962) and Feyerabend (1975), Foucault's emphasis on discontinuity and difference in history and his privileging of "polymorphous correlations in place of simple or complex causality," new developments in mathematics emphasising indeterminacy (catastrophe and chaos theory, fractal geometry), the re-emergence of concern in ethics, politics and anthropology for the validity and dignity of "the other," all indicate a widespread and profound shift in "the structure of feeling." What all these examples have in common is a rejection of 'meta-narratives' (large-scale theoretical interpretations purportedly of universal application).'[52] Harvey moves beyond this summing-up, however, to make four contributions of his own. In the first place, he describes postmodernism in architecture (the form, probably, where most people encounter it); most valuably, he looks at the political and economic conditions that brought about postmodernism and sustain it; he looks at the effect of postmodernism on our conceptions of space and time (he is a geographer, after all); and he offers a *critique* of postmodernism, something that was badly needed.

In the field of architecture and urban design, Harvey tells us that postmodernism signifies a break with the modernist idea that planning and development should focus on 'large-scale, metropolitan-wide, technologically rational and efficient urban *plans*, backed by absolutely no-frills architecture (the austere "functionalist" surfaces of "international style" modernism). Postmodernism cultivates, instead, a conception of the urban fabric as necessarily fragmented, a "palimpsest" of past forms superimposed upon each other, and a "collage" of current uses, many of which may be ephemeral.' Harvey put the beginning of postmodernism in architecture as early as 1961, with Jane Jacobs's *Death and Life of Great American Cities* (see chapter 30), one of the 'most influential anti-modernist tracts' with its concept of 'the great blight of dullness' brought on by the international style, which was too static for cities, where *processes* are of the essence.[53] Cities, Jacobs argued, need organised complexity, one important ingredient of which, typically absent in the international style, is diversity. Postmodernism in architecture, in the city, Harvey says, essentially meets the new economic, social, and political conditions prevalent since about 1973, the time of the oil crisis and when the major reserve currencies left the gold standard. A whole series of trends, he says, favoured a more diverse, fragmented, intimate yet anonymous society, essentially composed of much smaller units of diverse character. For Harvey the twentieth century can be conveniently divided into the Fordist years – broadly speaking 1913 to 1973 – and the years of 'flexible accumulation.' Fordism, which included the ideas enshrined in Frederick Winslow Taylor's *Principles of Scientific Management* (1911), was for Harvey a whole way of life, bringing mass production, standardisation of product, and mass consumption:[54] 'The progress of Fordism internationally meant the formation of global mass markets and the absorption of the mass of the world's population, outside the communist world, into the global dynamics of a new kind of capitalism.'[55] Politically, it rested on notions of mass economic democracy welded together through a balance of special-interest forces.[56] The restructuring of oil prices, coming on top of war, brought about a major recession, which helped catalyse the breakup of Fordism, and the 'regime of accumulation' began.[57]

The adjustment to this new reality, according to Harvey, had two main elements. Flexible accumulation 'is marked by a direct confrontation with the rigidities of Fordism. It rests on flexibility with respect to labour processes, labour markets, products and patterns of consumption. It is characterised by the emergence of entirely new sectors of production, new ways of providing financial services, new markets, and, above all, greatly intensified rates of commercial, technological, and organisational innovation.'[58] Second, there has been a further round of space-time compression, emphasising the ephemeral, the transient, the always-changing. 'The relatively stable aesthetic of Fordist modernism has given way to all the ferment, instability, and fleeting qualities of a postmodernist aesthetic that celebrates difference, ephemerality, spectacle, fashion, and the commodification of cultural forms.'[59] This whole approach, for Harvey, culminated in the 1985 exhibition at the Pompidou Centre in Paris, which had Lyotard as one of its consultants. It was called *The Immaterial.*

Harvey, as was said earlier, was not uncritical of postmodernism. Elements of nihilism are encouraged, he believes, and there is a return to narrow and sectarian politics 'in which respect for others gets mutilated in the fires of competition between the fragments.'[60] Travel, even imaginary travel, need not broaden the mind, but only confirms prejudices. Above all, he asks, how can we advance if knowledge and meaning are reduced 'to a rubble of signifiers'?[61] His verdict on the postmodern condition was not wholly flattering: 'confidence in the association between scientific and moral judgements has collapsed, aesthetics has triumphed over ethics as a prime focus of social and intellectual concern, images dominate narratives, ephemerality and fragmentation take precedence over eternal truths and unified politics, and explanations have shifted from the realm of material and political-economic groundings towards a consideration of autonomous cultural and political practices.'[62]

'THE BEST IDEA, EVER'

Narborough is a small village about ten miles south of Leicester, in the British East Midlands. Late on the evening of 21 November 1983 a fifteen-year-old girl, Lynda Mann, was sexually assaulted and strangled, her body left in a field not too far from her home. A manhunt was launched, but the investigation revealed nothing. Interest in the case died down until the summer of 1986, when on 2 August the body of another fifteen-year-old, Dawn Ashworth, was discovered in a thicket of blackthorn bushes, also near Narborough. She too had been strangled, after being sexually assaulted.

The manhunt this time soon produced a suspect, Richard Buckland, a porter in a nearby hospital.[1] He was arrested exactly one week after Dawn's body was found, following his confession. The similarities in the victims' ages, the method of killing, and the proximity to Narborough naturally made the police wonder whether Richard Buckland might also be responsible for the death of Lynda Mann, and with this in mind they called upon the services of a scientist who had just developed a new technique, which had become known to police and public alike as 'genetic fingerprinting.'[2] This advance was the brainchild of Professor **Alec Jeffreys** of Leicester University. Like so many scientific discoveries, Jeffreys's breakthrough came in the course of his investigation of something else – he was looking to identify the myoglobin gene, which governs the tissues that carry oxygen from the blood to the muscles. Jeffreys was in fact using the myoglobin gene to look for 'markers,' characteristic formations of DNA that would identify, say, certain families and would help scientists see how populations varied genetically from village to village, and country to country. What Jeffreys found was that on this gene one section of DNA was repeated over and over again. He soon found that the same observation – repeated sections – was being made in other experiments, investigating other chromosomes. What he realised, and no one else did, was that there seemed to be a widespread weakness in DNA that caused this pointless duplication to take place. As Walter Bodmer and Robin McKie describe it, the process is analogous to a stutterer who repeatedly stammers over the same letter. Moreover, this weakness *differed from person to person.* The crucial repeated segment was about fifteen base pairs long, and Jeffreys set about identifying it in such a way that it could be seen by eye with the aid of just a microscope. He first froze the DNA, then thawed it, which broke down the membranes of the red blood cells, but not those of the white cells that contain DNA. With the remains of the red blood cells washed away, an enzyme called proteinase K was added, exploding the white cells and freeing the DNA coils. These were then treated with another enzyme, known as HinfI, which separates out the ribbons of DNA that contain the repeated sequences. Finally, by a process known as electrophoresis, the DNA fragments were sorted into bands of different length and transferred to nylon sheets, where radioactive or luminescent techniques obtained images unique to individuals.[3]

Jeffreys was called in to try this technique with Richard Buckland. He was sent samples of semen taken from the bodies of both Lynda Mann and Dawn Ashworth, together with a few cubic centimetres of Buckland's blood. Jeffreys later described the episode as one of the tensest moments of his life. Until that point he had used his technique simply to test whether immigrants who came to Britain and were admitted on the basis of a law that allowed entry only to close relatives of those already living in the country really were as close as they claimed. A double murder case would clearly attract far more attention. When he went into his lab late one night to get the results, because he couldn't bear hanging on until the next morning, he got a shock. He lifted the film from its developing fluid, and could immediately see that the semen taken from Lynda and Dawn came from the same man – but that killer wasn't Richard Buckland.[4] The police were infuriated when he told them. Buckland had confessed. To the police mind, that meant the new technique had to be flawed. Jeffreys was dismayed, but when an independent test by Home Office forensic experts confirmed his findings, the police were forced to think again, and Buckland was eventually acquitted, the first person ever to benefit in this way from DNA testing. Once they had adjusted to the surprising result, the police mounted a campaign to test the DNA of all the men in the Narborough area. Despite 4,000 men coming forward, no match was obtained, not until Ian Kelly, a baker who lived some distance from Narborough, revealed to friends that he had taken the test on behalf of a friend, Colin Pitchfork, who *did* live in the vicinity of the village. Worried by this deception, one of Kelly's friends alerted the police. Pitchfork was arrested and DNA-tested. The friend had been right to be worried: tests showed that Pitchfork's DNA matched the semen found on Lynda and Dawn. In January 1988, Pitchfork became the first person to be convicted after **genetic fingerprinting**. He went to prison for life.[5]

DNA fingerprinting was the most visible aspect of the revolution in **molecular biology**. Throughout the late 1980s it came into widespread use, for testing immigrants and men in paternity suits, as well as in rape cases. Its practical successes, so soon after the structure of the double helix had been identified, underlined the new intellectual climate initiated by techniques to clone and sequence genetic material. In tandem with these practical developments, a great deal of theorising about genetics revised and refined our understanding of evolution. In particular, much light was thrown on the stages of evolutionary progress, working forward from the moment life had been created, and on the philosophical implications of evolution.

In 1985 a Glasgow-based chemist, **A. G. Cairns-Smith**, published *Seven Clues to the Origin of Life*.[6] In some ways a maverick, this book gave a totally different view of how life began to the one most biologists preferred. The traditional view about the origins of life had been summed up by a series of experiments carried out in the 1950s by **S. L. Miller** and **H. C. Urey**. They had assumed a primitive atmosphere on early Earth, consisting of ammonia, methane, and steam (but no oxygen – we shall come back to that). Into this early atmosphere they had introduced 'lightning' in the form of electrical discharges, and produced a 'rich brew' of organic chemicals, much richer than had been expected, including quite a large yield of amino acids, the building blocks for the nucleic acids which make up DNA. Somehow, from this rich brew, the 'molecules of life' formed. Graham Cairns-Smith thought this view nonsense because DNA molecules are extremely complicated, too complicated architecturally and in an engineering sense to have been produced accidentally, as the Miller-Urey reactions demanded. In one celebrated part of his book, he calculated that for nucleotides to have been invented, something like 140 operations would have needed to have evolved *at the same time*, and that the chances of this having occurred were one in 10^{109}. Since this is more than the number of electrons in the universe, calculated as 10^{80}, Cairns-Smith argued that there has simply not been enough time, or that the universe is not big enough, for nucleotides to have evolved in this way.[7]

His own version was startlingly different. He argued that evolution arrived before life as we know it, that there were chemical 'organisms' on earth before biochemical ones, and that they provided the architecture that made complex molecules like DNA possible. Looking about him, he saw that there are, in nature, several structures that, in effect, grow and reproduce – the crystal structures in certain clays, which form when water reaches saturation point. These crystals grow, sometimes break up into smaller units, and continue growing again, a process that can be called reproduction.[8] Such crystals form different shapes – long columns, say, or flat mats – and since these have formed because they are suited to their micro-environments, they may be said to be adapted and to have evolved. No less important, the mats of crystal can form into layers that differ in ionisation, and it was between these layers, Cairns-Smith believed, that amino acids may have formed, in minute amounts, created by the action of sunlight, in effect photosynthesis. This process would have incorporated carbon atoms into inorganic organisms – there are many substances, such as titanium dioxide, that under sunshine can fix nitrogen into ammonia. By the same process, under ultraviolet light, certain iron salts dissolved in water can fix carbon dioxide into formic acid. The crystal structure of the clays was related to their outward appearance (their phenotype), all of which would have been taken over by carbon-based structures.[9] As Linus Pauling's epic work showed, carbon is amazingly symmetrical and stable, and this is how (and why), Cairns-Smith said, inorganic reproducing organisms were taken over by organic ones.

It is a plausible and original idea, but there are problems. The next step in the chain of life was the creation of cellular organisms, bacteria, for which a skin was required. Here the best candidates are what are known as lipid vesicles, tiny bubbles that form membranes automatically. These chemicals were found naturally occurring in meteorites, which, many people argue, brought the first organic compounds to the very young Earth. On this reasoning then, life in at least some of its elements had an extraterrestrial beginning. Another problem was that the most primitive bacteria, which are indeed little more than rods or discs of activity, surrounded by a skin, are chiefly found around volcanic vents on the ocean floor, where the hot interior of the earth erupts in the process that, as we have already seen, contributes to sea-floor spreading (some of these bacteria can only thrive in temperatures above boiling point, so that one might say life began in hell). It is therefore difficult to reconcile this with the idea that life originally began as a result of sunlight acting on clay-crystal structures in much shallower bodies of water.[10]

Whatever the actual origin of life (generally regarded as having occurred around 3,800 million years ago), there is no question that the first bacterial organisms were **anaerobes**, operating only in the absence of oxygen. Given that the early atmosphere of the earth contained very little or no oxygen, this is not so surprising. Around 2,500 million years ago, however, we begin to see in the earth's rocks the accumulation of haematite, an oxidised form of iron. This appears to mean that oxygen was being produced, but was at first 'used up' by other minerals in the world. The best candidate for an oxygen-producer is a blue-green bacterium that, in shallower reaches of water where the sun could get at it and with the light acting on chlorophyll, broke carbon dioxide down into carbon, which it utilised for its own purposes, and oxygen – in other words, photosynthesis. For a time the minerals of the earth soaked up what oxygen was going (limestone rocks captured oxygen as calcium carbonate, iron rusted, and so on), but eventually the mineral world became saturated, and after that, over a thousand million years, billions of bacteria poured out tiny puffs of oxygen, gradually transforming the earth's atmosphere.[11]

According to Richard Fortey, in his history of the earth, the next advance was the formation of slimy communities of microbes, structured into 'mats,' almost two-dimensional layers. These are still found even today on saline flats in the tropics where the absence of grazing animals allows their survival, though fossilised forms have also been found in rocks dating to more than 3,500 million years old in South Africa and Australia. These structures are known as **stromatolites**.[12] Resembling 'layered cabbages,' they could grow to immense lengths – 30 feet was normal, and 100 *metres* not unknown. But they were made up of **prokaryotes**, or cells without nuclei, which reproduced simply by splitting. The **advent of nuclei** was the next advance; as the American biologist **Lynn Margulis** has pointed out, one bacterium cannibalised another, which became an organelle within another organism, and eventually formed the nucleus.[13] A chloroplast is another such organelle, performing photosynthesis within a cell. The development of the nucleus and organelles was a crucial step, allowing more complex structures to be formed. This, it is believed, was followed by the evolution of sex, which seems to have occurred about 2,000 million years ago. Sex occurred because it allowed the possibility of genetic variation, giving a boost to evolution which, at that time, would have speeded up (the fossil records do become gradually more varied then). Cells became larger, more complex – and **slimes** appeared. Slimes can take on various forms, and can also on occasion move over the surface of other objects. In other words, they are both animate and inanimate, showing the development of rudimentary specialised tissues, behaving in ways faintly resembling animals.

By 700 million years ago, the **Ediacara** had appeared.[14] These, the most primitive form of animal, have been discovered in various parts of the world, from Leicester, England, to the Flinders Mountains in south Australia. They take many exotic forms but in general are characterised by radial symmetry, skin walls only two cells thick, with primitive stomachs and mouths, like primitive jellyfish in appearance, and therefore not unimaginably far from slime. The first truly multicellular organisms, the Ediacara did not survive, at least not until the present day. For some reason they became extinct, despite their multifarious forms, and this may have been ultimately because they lacked a skeleton. This seems to have been the next important moment in evolution. Palaeontologists can say this with some confidence because, about 500 million years ago, there was a revolution in animal life on Earth. This is what became known as the **Cambrian Explosion**. Over the course of only 15 million years, animals with shells appeared, and in forms that are familiar even today. These were the trilobites – some with jointed legs and grasping claws, some with rudimentary dorsal nerves, some with early forms of eye, others with features so strange they are hard to describe.[15]

And so, by the mid- to late 1980s a new evolutionary synthesis began to emerge, one that filled in the order of important developments and provided more accurate dating. Moving forward in geological time, we can leap ahead from the Cambrian Explosion by more than 400 million years, to approximately 65 million years ago. One of the effects of the landing on the Moon, and the subsequent space probes, was that geology went from being a discipline with a single planet to study to one where there was suddenly a much richer base of data. One of the ways that the moon and other planets differ from Earth is that they seem to have far more craters on them, these craters being formed by impacts from asteroids or meteorites: bodies from space.[16] This was important in geology because, by the 1970s, the discipline had become used to a slow-moving chronology, measured in millions of years. There was, however, one great exception to this rule, and that became known as the **K/T boundary**, the boundary between the Cretaceous and Tertiary geological periods, occurring about 65 million years ago, when the fossil records showed a huge and very sudden disruption, the chief feature of which was that many forms of life on Earth suddenly disappeared.[17] The most notable of these extinctions was that of the dinosaurs, dominant large animals for about 150 million years before that, and completely absent from the fossil record afterward. Traditionally, geologists and palaeontologists considered that the mass extinctions were due to climate change or a fall in sea level. For many, however, this process would have been too slow – plants and animals would have adjusted, whereas in fact about half the life forms on Earth suddenly disappeared between the Cretaceous and the Tertiary. After the study of so many craters on other moons and planets, some palaeontologists began to consider whether a similarly catastrophic event might not have caused the **mass extinctions** seen on earth 65 million years ago. In this way there began an amazing scientific detective story that was not fully resolved until 1991.

For a meteorite or asteroid to cause such a devastating impact, it needed to have been a certain minimum size, so the crater it caused ought to have been difficult to overlook.[18] No immediate candidate suggested itself, but the first breakthrough came when scientists realised that meteorites have a different chemical structure to that of Earth, in particular with regard to the platinum group of elements. This is because these elements are absorbed by iron, and the earth has a huge iron core. Meteorite dust, on the other hand, would be rich in these elements, such as iridium. Sure enough, by testing rocky outcrops dating from the Cretaceous/Tertiary border, **Luis and Walter Alvarez**, from the University of California at Berkeley, discovered that iridium was present in quantities that were *ninety times* as rich as they should have been if no impact had taken place.[19] It was this discovery, in June 1978, that set off this father-and-son (and subsequently daughter-in-law) team on the quest that took them more than a decade. The second breakthrough came in 1981, in *Nature*, when **Jan Smit**, a Dutch scientist, reported his discoveries at a K/T boundary site at **Caravaca** in Spain.[20] He described some small round objects, the size of a sand grain, called spherules, which he said were common at these sites and on analysis were shown to have crystals of a 'feathery' shape, made of sanidine, a form of potassium feldspar.[21] These spherules, it was shown, had developed from earlier structures made of olivine – pyroxene and calcium-rich feldspar – and their significance lay in the fact that they are characteristic of basalt, the main rock that forms the earth crust under the oceans. In other words, the meteorite had slammed into the earth in the ocean and not on land.

This was both good news and bad news. It was good news in that it confirmed there had been a massive impact 65 million years ago. It was bad news in the sense that it led scientists to look for a crater in the oceans, and also to look for evidence of the massive tsunami, or tidal wave, that must have followed. Calculations showed that such a wave would have been a kilometre high as it approached continental shorelines. Both of these searches proved fruitless, and although evidence for an impact began to accumulate throughout the 1980s, with more than 100 areas located that showed iridium anomalies, as they were called, the actual site of the impact still remained elusive. It was not until 1988, when **Alan Hildebrand**, a Canadian attached to the University of Arizona, first began studying the Brazos River in Texas, that the decade-long search moved into its final stage.[22] It had been known for some time that in one place near Waco the Brazos passes over some rapids associated with a hard sandy bed, and this bed, it was recognised, was the remnant of a tsunami inundation. Hildebrand looked hard at Brazos and then went in search of evidence that would link it, in a circular fashion, with other features in the area. By examining maps, and gravity anomalies, he finally found a circular structure, which might be an impact crater, on the floor of the Caribbean, north of Colombia, but also extending into the Yucatán Peninsula in Mexico. Other palaeontologists were sceptical at first, but when Hildebrand brought in help from geologists more familiar with Yucatán, they soon confirmed the area as the impact site. The reason everyone had been so confused was that the crater – known as **Chicxulub** – was buried under more recent rocks.[23] When Hildebrand and his colleagues published their paper in 1991, it caused a sensation, at least to geologists and palaeontologists, who now had to revise their whole attitude: catastrophic events *could* have an impact on evolution.[24]

The discovery of Chicxulub produced other surprises. First, it turned out that the crater was to an extent responsible for the distribution of cenotes, small, spring-fed lakes that provided the fresh water that made the Mayan civilisation possible.[25] Second, three other mass extinctions are now recognised by palaeontologists, occurring at 365, 250, and 205 million years ago. The disappearance of the dinosaurs also proved to have had a liberating effect on mammals. Until the K/T boundary, mammals were small creatures. This may have helped their survival after the impact – because they were so numerous – but in any event the larger mammals did not emerge until after the K/T, and in the absence of competition from *Tyrannosaurus rex*, *Triceratops*, and their brothers and sisters. There would probably have been no humans unless the K/T meteorite had collided with Earth.

So far as the origins of humanity were concerned, the 1980s provided one or two crucial excavations, but the period was really a golden age of interpretation and analysis rather than of discovery.

'Turkana Boy,' discovered by the Leakeys near Kenya's Lake Turkana in August 1984, was much taller than people expected and quite slender, the first hominid to approach modern man in his dimensions.[26] He had a narrow spinal canal and a thorax that tapered upward, which suggested to anatomists that Turkana Boy had only limited nerve signals being sent to the thorax, giving him less command of respiration than would have been needed if he were to speak as we do. In other words, Turkana Boy had no language. At the same time the tapered thorax meant that his arms would be closer together, making it easier to hang in trees. Assigning him to *Homo erectus*, the Leakeys dated Turkana Boy to 1.6 million years ago. Two years later their archrival Don Johanson discovered a skeleton at Olduvai, attributed to *Homo habilis* and only 200,000 or so years older. This was very different – short and squat with long arms very like those of an ape.[27] The idea that more than one hominid type was alive at the same time around 2 million years ago was not accepted by all palaeontologists, but it did seem plausible that this was the time when the change occurred that caused hominids to leave the forest. Elisabeth Vrba, from Yale, argued that around 2.5 million years ago other changes induced evolutionary developments.[28] For instance, polar glaciation reduced the temperature of the earth, lowering sea levels and making the climate more arid, reducing vegetation. This was supported by the observation that fossils of forest antelopes become rare at this time, to be replaced by a variety that grazed on dry, open savannahs.[29] Stone tools appeared around 2.5 million years ago, suggesting that hominids left the forests between, say, 2.5 and 1.5 million years ago, growing taller and more graceful in the process, and using primitive tools. More 'prepared' tools are seen at about 200,000 years ago, roughly the time when the Neanderthals appeared. Opinions on them changed, too. We now know that their brains were as large as ours, though 'behind' the face rather than 'above' it. They appeared to bury their dead, decorate their bodies with ochre, and support disabled members of their communities.[30] In other words, they were not the savages the Victorians imagined, and they coexisted with *Homo sapiens* from about 50,000 to 28,000 years ago.[31]

These and other varied finds, between 1975 and 1995, consolidated in Ian Tattersall's compilation of fossils, therefore suggested the following revised chronology for hominid evolution:

4–3 million years ago	bipedalism
2.5 million years ago	early tool-using
1.5 million years ago	fire (for cooking food, which implies hunting)
1 million years ago	emigration of hominids from Africa
200,000 years ago	more refined tools
	Neanderthal Man appears
50,000–100,000 years ago	*Homo sapiens* appears
28,000 years ago	Neanderthals disappear

And why did the Neanderthals disappear? Many palaeontologists think there can be only one answer: *Homo sapiens* developed the ability to speak. Language gave modern man such an advantage in the competition for food and other resources that his rival was swiftly wiped out.

There are within cells organelles known as **mitochondrial DNA**. These organelles lie outside the nucleus and are in effect cell batteries – they produce a substance known as adenosine triphosphate or ATP. In January 1987 in *Nature*, **Allan Wilson** and **Rebecca Cann**, from Berkeley, revealed a groundbreaking analysis of mitochondrial DNA used in an archaeological context. The particular property of mitochondrial DNA that interested Wilson and Cann was that it is inherited only through the mother – it therefore does not change as nuclear DNA changes, through mating. Mitochondrial DNA can therefore only change, much more slowly, through mutation. Wilson and Cann had the clever idea of comparing the mitochondrial DNA among people from different populations, on the reasoning that the more different they were, the longer ago they must have diverged from whatever common ancestor we all share. Mutations are known to occur at a fairly constant pace, so this change should also give an idea of how long ago various groups of people diverged.[32]

To begin with, Wilson and Cann found that the world is broken down into two major groups – Africans on the one hand, and everyone else on the other. Second, Africans had slightly more mutations than anyone else, confirming the palaeontological results that humanity is older in Africa, very probably began there, and then spread from that continent to populate the rest of the world. Finally, by studying the rate of mutations and working backward, Wilson and Cann were able to show that humanity as ~~we know~~ it is no more than 200,000 years old, again broadly confirming the evidence of the fossils.[33]

One reason that the Wilson and Cann paper attracted the attention it did was because its results agreed well not only with what the palaeontologists were discovering in Africa, but also with recent work in linguistics and archaeology. As long ago as 1786, **Sir William Jones**, a British judge serving in India at the High Court in Calcutta, discovered that Sanskrit bore an unmistakable resemblance to both Latin and Greek.[34] This observation gave him the idea of the '**mother tongue**,' the notion that there was once, many years ago, a single language from which all other languages are derived. **Joseph Greenberg**, beginning in 1956, began to re-examine Sir William Jones's hypothesis as applied to the Americas. In 1987 he concluded a massive study of native American languages, from southern South America to the Eskimos in the north, published as *Language in the Americas*, which concluded that, at base, the American languages could be divided into three.[35] The first and earliest was '**Amerind**', which covers South America and the southern states of the US, and shows much more variation than the other, northern languages, suggesting that it is much older. The second group was **Na-dene**, and the third **Aleut-Eskimo**, covering Canada and Alaska. Na-dene is more varied than Aleut-Eskimo, all of which, says Greenberg, points to three migrations into America, by groups speaking three different languages. He believes, on the basis of 'mutations' in words, that Amerind speakers arrived on the continent before 11,000 years ago, Na-denes around 9,000 years ago, and that the Aleuts and Eskimos diverged about 4,000 years ago.[36]

Greenberg's conclusions are highly controversial but agree quite well with evidence from dental studies and surveys of genetic variation, in particular the highly original work of Professor **Luca Cavalli-Sforza** of Stanford University. In a series of books – *Cultural Transmission and Evolution* (1981), *African Pygmies* (1986), *The Great Human Diasporas* (1993), and *History and Geography of Human Genes* (1994) – Cavalli-Sforza and his colleagues have examined the variability of both blood, especially the rhesus factor, and genes around the world. This has led to fairly good agreement on the dates when early humans spread out across the globe. It has also led to a number of extraordinary possibilities in our *longue durée* history. For example, it seems that the Na-dene, Sino-Tibetan, Caucasian and Basque languages may be related in a very primitive way, and once belonged to a superfamily that was broken up by other peoples, shunting this superfamily into backwaters, and expelling Na-dene speakers into the Americas. The evidence also shows great antiquity for Basque speakers, whose language and blood is quite different from those around them. Cavalli-Sforza notes the contiguity between the Basque nation and the early sites of cave art in Europe, and wonders whether this is evidence for an ancient people who recorded their hunter-gatherer techniques on cave walls and resisted the spread of farming peoples from the Middle East.[37]

Finally, Cavalli-Sforza attempted to answer two of the most fascinating questions of all – when did language first appear, and was there ever a single ancestral language, a true mother tongue? We saw earlier that some palaeontologists believe that the Neanderthals died out about 28,000 years ago because they did not have language. Against that, Cavalli-Sforza points out that the region in our brains responsible for language lies behind the eye, on the left side, making the cranium slightly asymmetrical. This asymmetry is absent in apes but present in skulls of *Homo habilis* dated to 2 million years ago. Furthermore, our brain case ceased to grow about 300,000 years ago, and so on this basis it seems that language might be older than many palaeontologists think.[38] On the other hand, studies of the way languages change over time (a rate that is known, roughly) points back to between 20,000–40,000 years ago when the main superfamilies split. This discrepancy has not been resolved.

Regarding the mother tongue, Cavalli-Sforza relies on Greenberg, who claims that there is at least one word that seems to be common to all languages. This is the root word *tik*.

Family or Language	Forms	Meaning
Nilo-Saharan	tok-tek-dik	one
Caucasian	titi, tito	finger, single
Uralic	ik-odik-itik	one
Indo-European	dik-deik	to indicate/point
Japanese	te	hand
Eskimo	tik	index finger
Sino-Tibetan	tik	one
Austroasiatic	ti	hand, arm
Indo-Pacific	tong-tang-ten	finger, hand, arm
Na-dene	tek-tiki-tak	one
Amerind	tik	finger[39]

For the Indo-European languages, those stretching from western Europe to India, Greenberg's approach has been taken further by **Colin Renfrew**, the Cambridge archaeologist who rationalised the effects of the carbon-14 revolution on dating. Renfrew's aim, in *Archaeology and Language* (1987), was not simply to examine language origins but to compare those findings with others from archaeology, to see if a consistent picture could be arrived at and, most controversially, to identify the earliest homeland of the Indo-European peoples, to see what light this threw on human development overall. After introducing the idea of regular sound shifts, according to nation –

| 'milk': | French *lait* | Italian *latte* | Spanish *leche* |
| 'fact': | French *fait* | Italian *fatto* | Spanish *hecho* |

Renfrew went on to study the rates of change of language and to consider what the earliest vocabulary might have been. Comparing variations in the use of key words (like *eye*, *rain*, and *dry*), together with an analysis of early pottery and a knowledge of farming methods, Renfrew examined the spread of farming through Europe and adjacent areas. He concluded that the central homeland for the Indo-Europeans, the place where the mother tongue, '**proto-Indo-European**,' was located, was in central and eastern Anatolia about 6500 BC and that the distribution of this language was associated with the spread of farming.[40]

The surprising thing about all this is the measure of agreement between archaeology, linguistics and genetics. The spread of peoples around the globe, the demise of the Neanderthals, the arrival of humanity in the Americas, the rise of language, its spread associated with art and with agriculture, its link to pottery, and the different tongues we see about us today all fall into a particular order, the beginnings of the last chapter in the evolutionary synthesis.

Against such a strong research/empirical background, it is not surprising that theoretical work on evolution should flourish. What *is* perhaps surprising is that writing about biology in the 1980s and 1990s became a literary phenomenon. A clutch of authors – biologists, palaeontologists, philosophers – wrote dozens of books that became best-sellers and filled the shelves of good bookshops, marking a definite change in taste, matched only by an equivalent development in physics and mathematics, which we shall come to in a later chapter. In alphabetical order the main authors in this renaissance of Darwinian studies were: Richard Dawkins, Daniel Dennett, Niles Eldredge, Stephen Jay Gould, Richard Lewontin, Steven Pinker, Steven Rose, John Maynard Smith, and E. O. Wilson. The group was known collectively as the neo-Darwinists, and they aroused enthusiasm and hostility in equal measure: their books sold well, but Dawkins at one point, in 1998, was described as 'the most dangerous man in Britain.'[41]

The message of the neo-Darwinists was twofold. One view was represented by Wilson, Dawkins, Smith and Dennett, the other by Eldredge, Gould, Lewontin and Rose. Wilson himself produced two kinds of books. There was first, as we have seen, *Sociobiology*, published in 1975, *On Human Nature* (1978), and *Consilience* (1998). These books all had in common a somewhat stern neo-Darwinism, centred around Wilson's conviction that 'the genes hold culture on a leash.'[42] Wilson wanted above all to bridge C. P. Snow's two cultures, which he believed existed, and to show how science could penetrate human nature so as to explain culture: 'The essence of the argument, then, is that the brain exists because it promotes the survival and multiplication of the genes that direct its assembly.'[43] Wilson believed that biology will eventually be able to explain anthropology, psychology, sociology, and economics, that all these disciplines will become blended in ever closer ways. In *On Human Nature* he expanded on *Sociobiology*, with more aspects of human experience that could be explained in adaptive terms. He described, for example, the notion of hypergamy, the practice of females marrying men of equal or greater wealth and status; he pointed to the ways in which the great civilisations around the world, although they were not in touch with each other, developed similar features often in much the same order; he believes that chronic meat shortages may have determined the great religions, in that as early man moved away from game-rich areas, the elites invented religious rules to confine meat-eating to a religious caste; and he quotes the example of inmates in the Federal Reformatory for Women, Alderson, West Virginia, where it has been observed that the females form themselves into family-like units centred on a sexually active pair who call themselves 'husband' and 'wife,' with other women being added, known as 'brothers' and 'sisters,' and older inmates serving as 'aunts' and 'uncles.' He points out that male prisoners never organise in this way.[44] Wilson's chief aim all the way through his work was to show how the cultural and even ethical life of humanity can be explained biologically, genetically, and though his tone was cheerful and optimistic, it was uncompromising.

In the second strand of his work, particularly in *Biophilia: The Human Bond with Other Species* (1984), Wilson's aim was to show that humankind's bond with nature can help explain and enrich our lives as no other approach can.[45] Besides arguing that biophilia may explain aesthetics (why we like savannah-type landscapes, rather than urban ones), why scientific understanding of animal life may enrich the reading of nature poems, why all peoples have learned to fear the snake (because it is dangerous; no need to invoke Freud), he takes the reader on his own journeys of scientific discovery, to show not only how intellectually exciting it may be but how it may offer meaning (a partial meaning admittedly) for life. He shows us, for example, how he demonstrated that the size of an island is related to the number of species it can bear, and how this deepens our understanding of conservation. *Biophilia* struck a chord, generating much research, which was all brought together ten years later at a special conference convened at Woods Hole Oceanographic Institute in Massachusetts in August 1992. Here, more systematic studies were reported which showed, for example, that, given a choice, people prefer *unspectacular* countryside landscapes in which to live; one prison study was reported that showed that prisoners whose cells faced fields reported sick less often than those whose cells faced the parade ground; a list of biota that produce psychosomatic illness (flies, lizards, vultures) was prepared, and these were found to be associated with food taboos. The symposium also examined James Lovelock's Gaia theory, which had been published in 1979 and argued that the whole of the earth biota is one interregulated system, more akin to physiology than to physics (i.e., that the gases of the atmosphere, the salinity and alkalinity of the oceans, are regulated to keep the maximum number of things alive, like a gigantic organism). Biophilia was an extension of sociobiology, a less iconoclastic version which didn't catch on to the same extent.[46]

Second only to Wilson in the passion with which he advances a neo-Darwinian view of the world is Richard Dawkins. Dawkins won the Royal Society of Literature Award in 1987 for his 1986 book *The Blind Watchmaker*, and in 1995 he became Charles Simonyi Professor of the Public Understanding of Science at Oxford. His other books were *The Extended Phenotype* (1982), *River out of Eden* (1995), and *Climbing Mount Improbable* (1996), with *The Selfish Gene* being reissued in 1989. There is a relentless quality about *The Blind*

Watchmaker, as there is about many of Dawkins's books, a reflection of his desire once and for all to dispel every fuzzy notion about evolution.[47] One of the arguments of the antievolutionists is to say: if evolution is a fact, why aren't there intermediate forms of life, and how did complex organisms, like eyes or wings, form without intermediate organisms also occurring? Surely only a designer, like God, could arrange all this? And so Dawkins spends time demolishing such objections. Take wings: 'There are animals alive today that beautifully illustrate every stage in the continuum. There are frogs that glide with big webs between their toes, tree-snakes with flattened bodies that catch the air, lizards with flaps along their bodies, and several different kinds of mammals that glide with membranes stretched between their limbs, showing us the kind of way bats must have got their start. Contrary to the Creationist literature, not only are animals with "half a wing" common, so are animals with a quarter of a wing, three quarters of a wing, and so on.'[48] Dawkins's second aim is to emphasise that natural selection really does happen, and his technique here is to quote some telling examples, one of the best being the cicadas, whose life cycles are always prime numbers (thirteen or seventeen years), the point being that such locusts reach maturity at an unpredictable time, meaning that the species they feed on can never adjust to their arrival – it is mathematically random! But Dawkins's main original contribution was his notion of 'memes,' a neologism to describe the cultural equivalent of genes.[49] Dawkins argued that as a result of human cognitive evolution, such things as ideas, books, tunes, and cultural practices come to resemble genes in that the more successful – those that help their possessors thrive – live on, and so will 'reproduce' and be used by later generations.

Daniel Dennett, a philosopher from Tufts University in Medford, near Boston, is another uncompromising neo-Darwinist. In *Darwin's Dangerous Idea: Evolution and the Meanings of Life* (1995), Dennett states baldly, 'If I were to give an award for the single best idea anyone has ever had, I'd give it to Darwin, ahead of Newton and Einstein and everyone else. In a single stroke, the idea of evolution by natural selection unifies the realm of life, meaning, and purpose with the realm of space, time, cause and effect, mechanism and physical law.'[50] Like Wilson and Dawkins, Dennett is concerned to drum evolutionary theory's opponents out of town: 'Darwin's dangerous idea is reductionism incarnate.'[51] His book is an attempt to explain how life, intelligence, language, art, and ultimately consciousness are, in essence, no more than 'engineering problems.' We haven't got there yet, when it comes to explaining all the small steps that have been taken in the course of natural selection, but Dennett has no doubt we will some day. Perhaps the heart of his book (one heart anyway; it is very rich) is an examination of the ideas of **Stuart Kauffman** in his 1993 book *The Origins of Order: Self-Organisation and Selection in Evolution.*[52] Kauffman's idea was an attack on natural selection insofar as he argued that the similarity between organisms did not necessarily imply descent; it could just as easily be due to the fact that there are only a small number of design solutions to any problem, and that these 'inherent' solutions shape the organisms.[53] Dennett concedes that Kauffman has a point, far more than any others who offer rival theories to natural selection, but he argues that these 'constraints over design' in fact only add to the possibilities in evolution, using poetry as an analogy. When poetry is written to rhyme, he points out, the poet finds many more juxtapositions than he or she would have found had he or she just been writing a shopping list. In other words, order may begin as a constraint, but it can end up by being liberating. Dennett's other main aim, beyond emphasising life as a physical-engineering phenomenon, shaped by natural selection, is to come to grips with what is at the moment the single most important mystery still outstanding in the biological sciences – consciousness. This will be discussed more fully later in this chapter.

John Maynard Smith, emeritus professor of biology at the University of Sussex, is the doyen of the neo-Darwinists, publishing his first book as long ago as 1956. Less of a populariser than the others, he is one of the most original thinkers and uncompromising theorists. In 1995, in conjunction with Eörs Szathmáry, he published *The Major Transitions in Evolution*, where the chapter titles neatly summarise the bones of the argument:

Chemical evolution
The evolution of templates
The origin of translation and the genetic code
The origin of protocells
The origin of eukaryotes
The origin of sex and the nature of species
Symbiosis
The development of spatial patterns
The origin of societies
The origin of language[54]

In the same year that Maynard Smith and Szathmáry were putting together their book, **Steven Pinker**, professor of brain and cognitive sciences at MIT, released *The Language Instinct*. Maynard Smith's book, and Pinker's, finally put to rest the Skinner versus Chomsky debate, both concluding that the greater part of language ability is inherited.[55] Mainly this was done by reference to the effects on language ability of various forms of brain injury, the development of language in children, and its relation to known maturational changes in the child's nervous system, the descent of later languages from earlier ones, the similarity in the skulls of various primates, not to mention certain areas of chimpanzee brains that equate to human brains and seem to account for the reception of warning sounds and other calls from fellow chimpanzees. Pinker also presented evidence of language disabilities that have run in families (particularly dyslexia), and a new technique, called positron emission top-ography, in which a volunteer inhales a mildly radioactive gas and then puts his head inside a ring of gamma ray detectors. Computers can then calculate which parts of the brain 'light up.'[56] There seems no doubt now that language *is* an instinct, or at least has a strong genetic component. In fact, the evidence is so strong, one wonders why it was ever doubted.

★

Set alongside – and sometimes against – Wilson, Dawkins, Dennett, and Co. is a second set of biologists who agree with them about most things, but disagree on a handful of fundamental topics. This second group includes **Stephen Jay Gould** and **Richard Lewontin** of Harvard, **Niles Eldredge** at the American Museum of Natural History in New York, and **Steven Rose** at the Open University in England.

Pride of place in this group must go to Gould. A prolific author, Gould specialises in books with ebullient, almost avuncular titles: *Ever since Darwin* (1977), *The Panda's Thumb* (1980), *The Mismeasure of Man* (1981), *Hen's Teeth and Horse's Shoes* (1983), *The Flamingo's Smile* (1985), *Wonderful Life* (1989), *Bully for Brontosaurus* (1991), *Eight Little Piggies* (1993), and *Leonardo's Mountain of Clams and the Diet of Worms* (1999). There are four areas where Gould and his colleagues differ from Dawkins, Dennett, and the others. The first concerns a concept known as '**punctuated equilibrium**.' This idea dates from 1972, when Eldredge and Gould published a paper in a book on palaeontology entitled 'Punctuated Equilibrium: An Alternative to Phyletic Gradualism.'[57] The thrust of this was that an examination of fossils showed that whereas all orthodox Darwinians tended to see evolutionary change as gradual, in fact there were in the past long periods of stasis, where nothing happened, followed by sudden and rapid periods of dramatic change. This, they said, helped account for why there weren't intermediate forms, and also explained speciation, how new species arise – suddenly, when the habitat changes dramatically. For a while, the theory also gained adherents as a metaphor for sudden revolution as a form of social change (Gould's father had been a well-known Marxist). However, after nearly thirty years, punctuated equilibrium has lost a lot of its force. 'Sudden' in geological terms is not really sudden in human terms – it involves hundreds of thousands if not a few million years. The rate of evolution can be expected to vary from time to time.

The second area of disagreement arose in 1979, in a paper by Gould and Lewontin in the *Proceedings of the Royal Society*, entitled 'The Spandrels of San Marco and the Panglossian Paradigm: A Critique of the Adaptationist Programme.'[58] The central point of this paper, which explains the strange architectural reference, is that a spandrel, the tapering triangular space formed by the intersection of two rounded arches at a right angle, isn't *really* a design feature. Gould and Lewontin had seen these features at San Marco in Venice and concluded that they were inevitable by-products of other, more important features – i.e., the arches. Though harmonious, they were not really 'adaptations' to the structure, but simply what was left when the main design was put in place. Gould and Lewontin thought there were parallels to be drawn with regard to biology, that not all features seen in nature were direct adaptations – that, they said, was Panglossian. Instead, there were biological spandrels that were also by-products. As with punctuated equilibrium, Gould and Lewontin thought that the spandrel approach was a radical revision of Darwinism. A claim was even made for language being a biological spandrel, an emergent phenomenon that came about by accident, in the course of the brain's devel-opment in other directions. This was too much, and too important, to be left alone by Dawkins, Dennett, and others. It was shown that even in architecture a spandrel isn't inevitable – there are other ways of treating what happens where two arches meet at right angles – and again, like punctuated equilibrium, the idea of language as a spandrel, a by-product of some other set of adaptations, has not really stood the test of time.

The third area where Gould differed from his colleagues came in 1989 in his book *Wonderful Life*.[59] This was a reexamination and retelling of the story of the Burgess Shale, a fossil-rich rock formation in British Columbia, Canada, which has been well known to geologists and palaeontologists since the turn of the century. The lesson that Gould drew from these studies was that an explosion of life forms occurred in the Cambrian period, 'far surpassing in variety of bodily forms today's entire animal kingdom. Most of these forms were wiped out in mass extinctions; but one of the survivors was the ancestor of the vertebrates, and of the human race.' Gould went on to say that if the 'tape' of evolution were to be run again, it need not turn out in the same way – a different set of survivors would be here now. This was a notable heresy, and once again the prevailing scientific opinion is now against Gould. As we saw in the section on Dennett and Kauffman, only a certain number of design solutions exist to any problem, and the general feeling now is that, if one could run evolution all over again, something very like humans would result. Even Gould's account of the Burgess Shale has been attacked. In a book published in 1998 **Simon Conway Morris**, part of the palaeontological group from Cambridge that has spent decades studying the Shale, concluded in *The Crucible of Creation* that in fact the vast army of trilobites *does* fit with accepted notions of evolution; comparisons can be made with living animal families, although we may have made mistakes with certain groupings.[60]

One might think that the repeated rebuffs which Gould received to his attempts to reshape classical Darwinism would have dampened his enthusiasm. Not a bit of it. And in any case, the fourth area where he, Lewontin, and others have differed from their neo-Darwinist colleagues has had a somewhat different history. Between 1981 and 1991, Gould and Lewontin published three books that challenged in general the way 'the doctrine of DNA,' as Lewontin put it, had been used, again to quote Lewontin, to 'justify inequalities within and between societies and to claim that those inequalities can never be changed.' In *The Mismeasure of Man* (1981), Gould looked at the history of the controversy over IQ, what it means, and how it is related to class and race.[61] In 1984 Lewontin and two others, Steven Rose and Leon J. Kamin, published *Not in Our Genes: Biology, Ideology and Human Nature*, in which they rooted much biology in a bourgeois political mentality of the nineteenth century, arguing that the quantification of such things as the IQ is crude and that attempts to describe mental illness only as a biochemical illness avoid certain politically inconvenient facts.[62] Lewontin took this further in 1991 in *The Doctrine of DNA*, where he argued that DNA fits perfectly into the prevailing ideology; that the link between cause and effect is simple, mainly one on one; that for the present DNA research holds out no prospect of a cure for the major illnesses that affect mankind – for example, cancer, heart disease and stroke – and that the whole edifice is more designed to reward scientists than help science, or patients. Most subversive of all, he writes, 'It has been clear since the first discoveries in molecular biology that "genetic engineering," the creation to order of genetically altered organisms, has an immense possibility for producing private profit. ... No prominent molecular biologist of my acquaintance is without a financial stake in the biotechnology business.'[63] He believes that human nature, as described by the evolutionary biologists such as E. O. Wilson, is a 'made-up story,' designed to fit the theories the theorists already hold.

Given the approach of Gould and Lewontin in particular, it comes as no surprise to find them fully embroiled in yet another (but very familiar) biological controversy, which erupted in 1994. This was the publication of **Richard J. Herrnstein** and **Charles Murray**'s *The Bell Curve: Intelligence and Class Structure in American Life.*[64]

Ten years in the making, the main argument of *The Bell Curve* was twofold. In some places, it is straight out of Michael Young's *Rise of the Meritocracy*, though Herrnstein and Murray are no satirists but in deadly earnest. In the twentieth century, they say, as more and more colleges have opened up to the general population, as IQ tests have improved and been shown to be better predictors of job performance than other indicators (such as college grades, interviews, or biographical data), and as the social environment has become more uniform for most of the population, a '**cognitive elite**' has begun to emerge in society. Three phenomena are the result of this sorting process, and mean that it will accelerate in the future: the cognitive elite is getting richer, at a time when everybody else is having to struggle to stay even; the elite is increasingly segregated physically from everyone else, especially at work and in the neighbourhoods they inhabit; and the cognitive elite is increasingly likely to intermarry.[65] Herrnstein and Murray also analysed afresh the results of the National Longitudinal Study of Youth (NLSY), a database of about 4 million Americans drawn from a population that was born in the 1960s. This enables them to say, for example, that low intelligence is a stronger precursor of poverty than coming from a low socioeconomic status background, that students who drop out of school come almost entirely from the bottom quartile of the IQ distribution (i.e., the lowest 25 percent), that low-IQ people are more likely to divorce early on in married life and to have illegitimate children. They found that low-IQ parents are more likely to be on welfare and to have low-birthweight children. Low IQ men are more likely to be in prison. Then there was the racial issue. Herrnstein and Murray spend a lot of time prefacing their remarks by saying that a high 'IQ' does not necessarily make someone admirable or the kind to be cherished, and they concede that the racial differences in IQ are diminishing. But, after controlling for education and poverty, they still find that people of Asian stock in America outperform 'whites,' who outperform blacks on tests of IQ.[66] They also find that recent immigrants to America have a lower IQ score than native-born Americans. And finally, they voice their concerns that the IQ level of America is declining. This is due partly, they say, to a dysgenic trend – people of lower IQ are having more children – but that is not the only reason. In practice, the American schooling system has been 'dumbed down' to meet the needs of average and below-average students, which means that the performance of the average students has *not*, contrary to popular opinion, been adversely affected. It is the brighter students who have been most affected, their SAT (Scholastic Aptitude Test) scores dropping by 41 percent between 1972 and 1993. They also blame parents, who seem not to want their children to work harder anymore, and television, which has replaced newsprint as a source of information, and the telephone, which has replaced letter writing as a form of self-expression.[67] Further, they express their view that affirmative-action programs have not helped disadvantaged people, indeed have made their situation worse. But it is the emergence of the cognitive elite, this 'invisible migration,' the 'secession of the successful,' and the blending of the interests of the affluent with the cognitive elite that Herrnstein and Murray see as the most important, and pessimistic, of their findings. This elite, they say, will fear the '**underclass**' that is emerging, and will in effect control it with 'kindness' (which is basically what Murray's rival, J. K. Galbraith had said in *The Culture of Contentment*). They will provide welfare for the underclass so long as it is out of sight and out of mind. They hint, though, that such measures are likely to fail: 'racism will re-emerge in a new and more virulent form.'[68]

Herrnstein and Murray are traditionalists. They would like to see a return to old-fashioned families, small communities, and the familiar forms of education, where pupils are taught history, literature, arts, ethics, and the sciences in such a way as to be able to weigh, analyse, and evaluate arguments according to exacting standards.[69] For them, the IQ test not only works – it is a watershed in human society. Allied to the politics of democracy and the homogenising successes of modern capitalism, the IQ aids what R. A. Fisher called runaway evolution, promoting the rapid layering of society, divided according to IQ – which, of course, is mainly inherited. We are indeed witnessing the rise of the meritocracy.

The Bell Curve provoked a major controversy on both sides of the Atlantic. This was no surprise. Throughout the century white people, people on the 'right' side of the divide they were describing, have concluded that whole segments of the population were dumb. What sort of reaction did they expect? Many people countered the claims of Herrnstein and Murray, with at least six other books being produced in 1995 or 1996 to examine (and in many cases refute) the arguments of *The Bell Curve*. Stephen Jay Gould's *The Mismeasure of Man* was reissued in 1996 with an extra chapter giving his response to *The Bell Curve*. His main point was that this was a debate that needed technical expertise. Too many of the reviewers who had joined the debate (and the book provoked nearly two hundred reviews or associated articles) did not feel themselves competent to judge the statistics, for example. Gould did, and dismissed them. In particular, he attacked Herrnstein and Murray's habit of giving the *form* of the statistical association but not the *strength*. When this was examined, he said, the links they had found always explained less than 20 percent of the variance, 'usually less than 10 percent and often less than 5 percent. What this means in English is that you cannot predict what a given person will do from his IQ score.'[70] This was the conclusion Christopher Jencks had arrived at, thirty years before.

By the time *The Bell Curve* rumpus erupted, the infrastructure was in place for a biological project capable of generating controversy on an even bigger scale. This was the scramble to map the human genome, to draw up a plan to describe exactly all the nucleotides that constitute man's inheritance and that, in time, will offer at least the possibility of interfering in our genetic makeup.

Interest in this idea grew throughout the 1980s. Indeed, it could be said that the **Human Genome Project** (HGP), as it came to be called, had been simmering since **Victor McKusick**, a Boston doctor, began collecting a comprehensive record, 'Mendelian Inheritance in Man,' a list of all known genetic diseases, first published in 1966.[71] But then, as research progressed, first one scientist then another began to see sense in mapping the entire genome. On 7 March 1986, in *Science*, **Renato Dulbecco**, Nobel Prize-winning president of the Salk Institute, startled his colleagues by asserting that the war on cancer would be over quicker if geneticists were to sequence the human genome.[72] Various U.S. government departments, including the Department of Energy and the National Institutes of Health, became interested at this point, as did scientists in Italy, the United Kingdom, Russia, Japan, and France (in roughly that order; Germany was backward, owing to the controversial role biology had played in Nazi times). A major conference, organised by the Howard Hughes Medical Institute, was held in Washington in July 1986 to bring together the various interested parties, and this had two effects. In February 1988 the U.S. National Research Council issued its report, *Mapping and Sequencing the Human Genome*, which recommended a concerted research program with a budget of $200 million a year.[73] James Watson, appropriately enough, was appointed associate director of NIH, later that year, with special responsibility for human genome research. And in April 1988, HUGO, the Human Genome Organisation, was founded. This was a consortium of international scientists to spread the load of research, and to make sure there was as little duplication as possible, the aim being to finalise the mapping as early as possible in the twenty-first century. The experience of the Human Genome Project has not been especially happy. In April 1992 James Watson resigned his position over an application by certain NIH scientists to patent their sequences. Watson, like many others, felt that the human genome should belong to everyone.[74]

The genome project came on stream in 1988–89. This was precisely the time that communism was collapsing in the Soviet Union and the Berlin Wall was dismantled. A new era was beginning politically, but so too in the intellectual field. For HUGO was not the only major innovation introduced in 1988. That year also saw the birth of the Internet.

Whereas James Watson took a leading role in the genome project, his former colleague and co-discoverer of the double helix, Francis Crick, took a similar position in what is perhaps the hottest topic in biology as we enter the twenty-first century: consciousness studies. In 1994 Crick published *The Astonishing*

Hypothesis, which advocated a research assault on this final mystery/problem.[75] **Consciousness studies** naturally overlap with neurological studies, where there have been many advances in identifying different structures of the brain, such as language centres, and where MRI, magnetic resonance imaging, can show which areas are being used when people are merely thinking about the meaning of words. But the study of consciousness itself is still as much a matter for philosophers as biologists. As **John Maddox** put it in his 1998 book, *What Remains to be Discovered*, 'No amount of introspection can enable a person to discover just which set of neurons in which part of his or her head is executing some thought-process. Such information seems to be hidden from the human user.'[76]

It should be said that some people think there is nothing to explain as regards consciousness. They believe it is an '**emergent property**' that automatically arises when you put a 'bag of neurons' together. Others think this view absurd. A good explanation of emergent property is given by **John Searle**, Mills Professor of Philosophy at the University of California, Berkeley, regarding the liquidity of water. The behaviour of the H_2O molecules explains liquidity, but the individual molecules are not liquid. At the moment, the problem with consciousness is that our understanding is so rudimentary that we don't even know how to talk about it – even after the 'Decade of the Brain,' which was adopted by the U.S. Congress on 1 January 1990.[77] This inaugurated many innovations and meetings that underlined the new fashion for consciousness studies. For example, the first international symposium on the science of consciousness was held at the University of Arizona at Tucson in April 1994, attended by no fewer than a thousand delegates.[78] In that same year the first issue of the *Journal of Consciousness Studies* was published, with a bibliography of more than 1,000 recent articles. At the same time a whole raft of books about consciousness appeared, of which the most important were: *Neural Darwinism: The Theory of Neuronal Group Selection*, by Gerald Edelman (1987), *The Remembered Present: A Biological Theory of Consciousness*, by Edelman (1989), *The Emperor's New Mind*, by Roger Penrose (1989), *The Problem of Consciousness*, by Colin McGinn (1991), *Consciousness Explained*, by Daniel Dennett (1991), *The Rediscovery of the Mind*, by John Searle (1992), *Bright Air, Brilliant Fire*, by Edelman (1992), *The Astonishing Hypothesis*, by Francis Crick (1994), *Shadows of the Mind: A Search for the Missing Science of Consciousness*, by Roger Penrose (1994), and *The Conscious Mind: In Search of a Fundamental Theory*, by David Chalmers (1996). Other journals on consciousness were also started, and there were two international symposia on the subject at Jesus College, Cambridge, published as *Nature's Imagination* (1994) and *Consciousness and Human Identity* (1998), both edited by John Cornwell.

Thus consciousness has been very much the flavour of the decade, and it is fair to say that those involved in the subject fall into four camps. There are those, like the British philosopher **Colin McGinn**, who argue that consciousness is resistant to explanation *in principle* and for all time.[79] Philosophers we have met before – such as Thomas Nagel and Hilary Putnam – also add that at the present (and maybe for all time) science cannot account for qualia, the first-person phenomenal experience that we understand as consciousness. Then there are two types of reductionist. Those like Daniel Dennett, who claim not only that consciousness can be explained by science but that construction of an artificially intelligent machine that will be conscious is not far off, may be called the 'hard' reductionists.[80] The soft reductionists, typified by John Searle, believe that consciousness does depend on the physical properties of the brain but think we are nowhere near solving just how these processes work, and dismiss the very idea that machines will ever be conscious.[81] Finally, there are those like **Roger Penrose** who believe that a new kind of dualism is needed, that in effect a whole new set of physical laws may apply inside the brain, which account for consciousness.[82] Penrose's particular contribution is that quantum physics operate within tiny structures, known as tubules, within the nerve cells of the brain to produce – in some as yet unspecified way – the phenomena we recognise as consciousness.[83] Penrose actually thinks that we live in three worlds – the physical, the mental, and the mathematical: 'The physical world grounds the mental world, which in turn grounds the mathematical world and the mathematical world is the ground of the physical world and so on around the circle.'[84] Many people, who find this tantalising, nonetheless don't feel Penrose has *proved* anything. His speculation is enticing and original, but it is still speculation.

Instead, it is the two forms of reductionism that in the present climate attract most interest. For people like Dennett, human consciousness and identity arise from the narrative of their lives, and this can be related to specific brain states. For example, there is growing evidence that the ability to 'apply intentional predicates to other people is a human universal' and is associated with a specific area of the brain (the orbitofrontal cortex); in certain states of autism, this ability is defective. There is also evidence that the blood supply to the orbitofrontal cortex increases when people 'process' intentional verbs as opposed to non-intentional ones, and that damage to this area of the brain can lead to a failure to introspect.[85] Suggestive as this is, it is also the case that the microanatomy of the brain varies quite considerably from individual to individual, and that a particular phenomenal experience is represented at several different points in the brain, which clearly require integration. Any 'deep' patterns relating experience to brain activity have yet to be discovered, and seem to be a long way off, though this is still the most likely way forward.

A related approach – perhaps to be expected, given other developments in recent years – is to look at the brain and consciousness in a Darwinian light. In what sense is consciousness adaptive? This approach has produced two views – one that the brain was in effect 'jerry-built' in evolution to accomplish very many and very different tasks. On this view, there are at base three organs: a reptilian core (the seat of our basic drives), a palaeomammalian layer, which produces such things as affection for offspring, and a neomammalian brain, the seat of reasoning, language, and other 'higher functions.'[86] The second approach is to argue that throughout evolution (and throughout our bodies) there have been emergent properties: for example, there is always a biochemical explanation underlying a physiological phenomenon – sodium/potassium flux across a membrane being also nerve action potential.[87] In this sense, then, consciousness is nothing new in principle even if, at the moment, we don't fully understand it.

Studies of nerve action through the animal kingdom have also shown that nerves work by either firing or not firing; intensity is represented by the rate of firing – the more intense the stimulation, the faster the turning on and off of any particular nerve. This of course is very similar to the way computers work, in 'bits' of information, where everything is represented by a configuration of either 0s or 1s. The arrival of the concept of parallel processing in computing led the philosopher Daniel Dennett to consider whether an analogous process might happen in the brain between different evolutionary levels, giving rise to consciousness. Again such reasoning, though tantalising, has not gone much further than preliminary exploration. At the moment, no one seems able to think of the next step.

Francis Crick's aim has been fulfilled. Consciousness is being investigated as never before. But it would be rash to predict that the new century will bring advances quickly. No less a figure than Noam Chomsky has said, 'It is quite possible – overwhelmingly probably, one might guess – that we will always learn more about human life and personality from novels than from scientific psychology.'

THE EMPIRE WRITES BACK

In an essay published in 1975, Marcus Cunliffe, a professor at George Washington University in Washington, D.C., concluded that, so far as literature was
concerned, 'by the 1960s, the old Anglo-American cultural relationship was
decisively reversed: the major contribution, in quantity and quality, was American.'[1] He also observed that the business of America was still business, that if
publishers were to stay alive, they had to make profits, and that in such an
environment 'the most reliable mainstay was ... non-fiction: self-help, popular
religion, sexology, health, cookery, history and biography, advice on investments, documented scandal, accounts of adventures, reminiscences.'[2] Nor did
he ignore the fact that by 1960, 'the annual American consumption of comic
books had passed the billion mark; expenditure on them, estimated at $100
million a year, was four times as large as the combined budgets of all the public
libraries.'[3] As this 'mid-cult' flourished, mass culture, passive and increasingly
commercial, was seen as the enemy by those American authors who wrote
increasingly of 'alienation. In avant-garde fiction one can trace the gradual
disappearance of the qualities of worthiness formerly attributed to the main
characters. Even in the strongest (as in Hemingway) they go down to defeat.
The majority are either victims or slobs.'[4]

This change occurred, he said, in the late 1960s and early 1970s, provoked
by political and economic events, such as the many assassinations and the oil
crisis. Cunliffe quoted Richard Hofstadter, who died in 1970, who had followed
Social Darwinism in American Thought, discussed in chapter 3, with *Anti-Intellectualism in American Life* (1963), and who in 1967, writing in Daniel Bell and
Irving Kristol's *The Public Interest*, had this to say: 'Is it not quite possible that
the responsible society will get little or no nourishment from modern literature,
but will have to draw mainly on history, journalism, economics, sociological
commentary? Art, as it more ruthlessly affirms the self, as it more candidly
probes the human abyss, may in fact have less and less to tell us about the
conditions of a responsible society.' He referred specifically to such figures as
Walter Lippmann, James Reston, J. K. Galbraith, Paul Samuelson, Nathan
Glazer, and Daniel P. Moynihan.[5]

Cunliffe and Hofstadter had a point. The centre of gravity *had* shifted;
nonfiction *was* buoyant. But America's genius is to constantly reinvent herself,

and it is no surprise to find yet another turn of the wheel in that country's fiction. Maya Angelou was an early hint of things to come. Though her works are in fact autobiography, they read like fiction. In the last twenty-five years of the century, the role of the black author in America, the part once played by Richard Wright, Ralph Ellison, James Baldwin, and Eldridge Cleaver, has been better filled by women than by men, by such figures as **Toni Morrison** and **Alice Walker**. In books like *Sula* (1973), *Tar Baby* (1981), and *Beloved* (1987), Toni Morrison creates her own form, an African-American amalgam that makes use of folk tales, fables, oral history, myths both public and private, to produce highly original narratives whose central concern is to explore the awful darkness of the black (and female) experience in America, not to dwell on it but to 'banish it with joy,' much as Angelou does in her autobiographies.[6] Morrison's characters journey into their past, from where, in a sense, they can start again. *Sula* is about a promiscuous girl, but she is not the usual 'village bicycle' (to use a British phrase): she is *successfully* promiscuous, with her affections and her attentions just as much as with her body, and she shines: the drab community that surrounds her is transformed. Morrison is telling us as much about womanhood as being black. *Beloved* is her most ambitious book.[7] Set in Reconstruction times, it is the story of a black mother who kills her own young daughter when the former slave owner comes to return her to her old life of slavery. But this is fiction, and the daughter, Beloved of the title, reappears as a ghost to make a new inner life for her mother – the daughter lives again, through the power of love. Here too, amid the squalor and humiliation of slavery, Morrison uses the African devices of myth, ritual and oral legend to produce joy – not sentimental joy, but a joy that is earned.

Alice Walker also writes about the poverty she knew when she was growing up in the South, in a sharecropping family, but her novels, most notably *The Color Purple* (1982), look forward rather than back, forward to the way that the urban, more open America offers promise for blacks and for women. Narrated as a series of letters, the book, which won a Pulitzer Prize, follows a group of black women as they fight their way out of poverty, out from under their abusive menfolk, with racism ready to subvert any progress they make. Like Morrison and Angelou, Walker has the strength of optimism, viewing the women's progress as not only political but personal. Inside their persons, these women cannot be touched; their integrity is complete.[8]

Morrison and Walker both are and are not postmodern and postcolonial writers. Their exploration of blackness, the 'other,' the female condition, the use of African literary forms, all typify the arena of fiction in the last quarter of the century. In his book *English as a Global Language* (1997), David Crystal concludes his argument, 'There has never been a language so widely spread or spoken by so many people as English.'[9] It is, he says, a unique event historically. He also concurs with a sentiment of the Indian author Salman Rushdie, who wrote that 'the English language ceased to be the sole possession of the English some time ago.'[10] 'Indeed,' says Crystal, 'when even the largest English-speaking nation, the USA, turns out to have only about 20 per cent of the world's English speakers [as Crystal had himself demonstrated in his survey, earlier in his book], it is plain that no one can now claim sole ownership.'[11] He goes on to quote the Indian author Raja Rao, Chinua Achebe, and Rushdie again: all accept English as a world language but warn that it will from now on be used in increasingly new ways.

Whereas until 1970, say, it is possible to write about the 'great books' of the century, at least so far as Western countries are concerned, this becomes much more difficult for the period afterward. The reasons all have to do with the collapse of a consensus on what are, and are not, the dominant themes in literature. This collapse has been engendered by three things: the theories of postmodernism; the great flourishing of talent in the formerly colonial countries; and the success and impact of free-market economics since 1979–80, which has both caused the proliferation of new media outlets and, by attacking such institutions (in the U.K. for example) as the BBC and the Arts Council, sabotaged the idea of national cultures, the very notion of a shared tradition, which men like F. R. Leavis, T. S. Eliot, and Lionel Trilling so valued. It follows that any synopsis of late-twentieth-century literature offered by one individual is bound to be contentious, at the very least. Some generalisations *are* possible, however, and here attention will be limited to the Latin American school of 'magic realism,' magic realism being an important influence on other schools of writing; the rise of postcolonial literature, especially that written in the English language; the rise of 'cultural studies' as a replacement for traditional literature courses; and the enduring strength of American imaginative writing, reflecting a country – now the only superpower – where more people than anywhere live life's possibilities to the full.

The writers of Latin America – of whom the most well known are Miguel Angel Asturias (Guatemala), Jorge Luis Borges (Argentina), Carlos Fuentes (Mexico), Gabriel García Márquez (Colombia), Pablo Neruda (Chile), Octavio Paz (Mexico), and Mario Vargas Llosa (Peru) – come from countries that are scarcely postcolonial these days, having achieved their independence, for the most part, in the nineteenth century. At that time Latin American writers were traditionally very politically minded, often seeking refuge, when they went too far, in Europe. The European wars put a stop to that form of exile, while the numerous revolutions and political coups in South America forced upon writers a new way of adjusting, politically. The presence of indigenous groups also gave them a keener appreciation of marginal members of society, even as they regarded themselves as part of European civilisation.

Against this background, the school of **magic realism** grew and flourished as a primarily aesthetic response to political and social problems. At one stage, in the earlier part of the century, Latin American writers saw their role as trying to improve society. The aims of magic realism were more modest – to describe the universal human condition in its Latin American context in a way that could be understood all over the world. The appeal of Latin American literature, apart from the sheer writing power with which it is composed, is that it is ambitious, more ambitious than much European literature, never losing sight of social ideals and going beyond the purely personal.

Jorge Luis Borges, for example, developed a new form for what he wanted to say, a cross between an essay, containing real people, and a short story in which episodes are invented. Borges mixes philosophy and aesthetic ideas and plays games, the aim being 'to upset the reader's confidence in fact and reality.'[12] In one story, for example, he invented an entire planet, Tlön, down to its playing cards and dialects, its religion and architecture. Is this planet as strange as Latin America? By emphasising the differences, he also brings home the common humanity.

In **Mario Vargas Llosa**'s novel *The City and the Dogs* (1963), the main characters are cadets in a military academy, who band together to fight off the bullying older pupils.[13] This tussle becomes sordid, resulting in perversion and death, and is contrasted with the much more civilised worlds these cadets will have to inhabit once they leave the academy. As with Tlön and Macondo (see below), the academy is cut off from the mainstream, like Latin America itself, and the same is true yet again of *The Green House*, set in a brothel in Piura, a town surrounded by rain forest (another green house).[14] In this book, arguably Vargas Llosa's best, the chronology changes even in mid-sentence to suggest the shifting nature of time and relationships, and the magical and unpredictable nature of existence.[15]

In 1967 **Miguel Angel Asturias** became the first Latin American novelist to win the Nobel Prize. But of greater significance that year was publication of 'the most seamless achievement in Latin American fiction,' **Gabriel García Márquez**'s incomparable *One Hundred Years of Solitude*.[16] This book proved so popular that at one stage it was being reprinted every week. It is not hard to see why. Márquez has been compared to Cervantes, Joyce, and Virginia Woolf, and has himself admitted the influence of Faulkner, but that does no justice to his own originality. No other book has so successfully fulfilled Lionel Trilling's plea for novels to get outside the familiar ways of thinking, to imagine other possibilities, other worlds. Márquez not only does this but on top of it all, he is extremely funny.

One Hundred Years of Solitude exists on almost any level you care to name.[17] Márquez invents an imaginary town, Macondo, which is separated from everywhere else by marshes and impenetrable rain forest. The town is so cut off that the main character, Aureliano Buendía, makes discoveries for himself (like the fact that the earth is round) without realising that the rest of the world discovered this centuries ago. Morality is at a primitive stage in this world – people are allowed to marry their aunts, and the inhabitants haven't even got round to naming all of the objects in their little 'universe.' The story traces the rise and fall of Macondo, its civil strife, political corruption, exotic violence. This narrative is held together by the fortunes of the Buendía family, though because different generations share so many names, the chronology is not always clear. Ideas and things from the outside world sometimes reach Macondo (like railways), but always the town returns to isolation, the Buendías sequestered in their solitude.

The exuberant and deadpan attention to detail combine to create a unique sense of humour. 'Colonel Aureliano Buendía organized thirty-two armed uprisings and he lost them all. He had seventeen male children by seventeen different women and they were exterminated one after the other on a single night before the oldest one had reached the age of thirty-five. He survived fourteen attempts on his life, seventy-three ambushes and a firing squad. He lived through a dose of strychnine in his coffee that was enough to kill a horse.'[18] The Buendías are also surrounded by a cast of dotty eccentrics. For example, on one occasion a young Buendía, Meme, brings sixty-eight friends home from school for the holidays. 'The night of their arrival the students carried on in such a way, trying to go to the bathroom before they went to bed, that at one o'clock in the morning the last ones were still going in. Fernada then bought seventy-two chamber-pots, but she only managed to change the nocturnal problem into a morning one, because from dawn on there was a long line of girls, each with her pot in her hand, waiting for her turn to wash it.'[19] Macondo is a world where the itinerant gypsy sage Melquíades returns to life after dying because he couldn't bear the loneliness of death, where yellow flowers rain down from the skies and real rainstorms last for months.

The story of Macondo has a mythic quality, with countless allusions to twentieth-century ideas. Márquez deliberately gives his story a dated feel, so the reader is distanced from the action, as Bertolt Brecht recommended. It is likewise an attempt to re-enchant the world: things happen in Macondo that could happen nowhere else. This is not exactly biblical but close; we may not believe what happens, but we accept it. The illusions evoke Kafka, but a very sunny Kafka. In some senses José Buendía and his wife Ursula are the primordial couple, who undertake an exodus from the jungle in search of the sea; the ages of some characters are vastly inflated, as in the early books of the Bible; Melquíades presents the family with a manuscript written in Sanskrit code: this recalls both the decipherment of languages of earlier civilisations and the observations of Sir William Jones, the British judge in India, about the 'mother tongue.' The parchment on which the code is written turns out also to be a mirror, throwing us back on the relation between the text and reader and the ideas of Jacques Derrida. The playing with time recalls not only relativity but Fernand Braudel's ideas of *la longue durée* and what governs it. Underneath all, as Carlos Fuentes has pointed out, *One Hundred Years of Solitude* questions: 'What does Macondo know about its creation?' In other words, the very question that has so obsessed twentieth-century science.[20] In the way that Macondo ends, Márquez even raises the idea of entropy. In the very last sentence, he reminds us that we have no second opportunity in life, and this is the 'big reason' why the 'official version' of things should never be 'put up with.' The book may well be the greatest achievement of its kind in the last half of the twentieth century.

The wider significance of these alternative worlds is twofold. They are metaphors for Latin America itself, as a site for 'the other,' a key concept, as it turned out, in postmodernism. Second, and arguably more important, is their 'playful maturity'; these are artists who have distanced themselves from the quotidian and the political. In so doing, they have given an undoubted stature to Latin American fiction with which the mother country, Spain, cannot compete. As Márquez makes explicit, Latin American fiction at base is about solitude, the continent itself used as a metaphor for that predicament.

After the magic realism of Latin America, the fabulous intricacies of Indian fiction probably come next in any fledgling 'canon.' Twentieth-century Indian novels written in English date from the 1930s at least, with the works of Raja Rao and Mulk Raj Anand, but the novels published since, say, **R. K. Narayan's 'Malgudi' stories** fall into two kinds: minute observations and commentaries on Indian life, and attempts to find some sort of escape from it. The familiar idioms of English being used in such fabulous settings certainly highlight that the language no longer belongs to anyone.

R. K. Narayan's many novels generally take place in his beloved Malgudi, otherwise known as Mysore. *The Sweet Vendor*, published in 1967, is a study of spirituality, though not as, say, a Christian would understand it.[21] For sixty years, Jagan has sold sweets from his store, when suddenly he decides to change his life: he is going to help a stonemason carve a 'pure image' of a goddess so that others can find spirituality in her contemplation. But of course he takes his foibles (and his checkbook) with him, with some hilarious consequences. The fact is, Jagan's change in life is ambitious – too ambitious for his flawed personality: like someone in a Larkin poem, he is not really up to the challenge he has set himself. It is not that easy to retreat from life; for one thing, there is his moody son, more Westernised than he, with an American-Korean wife (actually a mistress), and with whom Jagan is constantly at odds. Narayan is of course poking serious fun at India herself, her spirituality (or spiritual pretensions), her ambition to be a world power when she cannot even feed herself (Jagan produces 'frivolous' food), and is both contemptuous and envious of the West.

Anita Desai's novels are in general domestic stories, small-scale on the face of it, but in each one the characters are unprepared for the life of an independent India, which as often as not involves some measure of Westernisation. In *The Village by the Sea*, the locals of Thul are worried by the government's proposal to install a chemical fertiliser plant nearby.[22] Hari, the main character, unlike many other villagers who don't want change, seeks to adjust to the new state of affairs by escaping to Bombay and becoming a watch repairer, in anticipation of all the watch wearers who will come and live in the village. Others ensure that the village remains a bird sanctuary, but once Hari's life – his ambitions – had been disturbed, and despite his dismal experiences in Bombay, there is no going back. The new silence isn't the same as the old one. Desai is saying that change is a question less of events than of attitude, psychology. Deven, the main character of *In Custody*, has great ambition, and when he is invited to become the secretary of the great Urdu poet Nur, he conceives a grand plan to tape-record the poet's wisdom.[23] In fact, this plan runs into endless difficulties; the poet himself is much less than perfect – he loves pigeons, wrestling, and whores just as much as wisdom – but Deven's technological incompetence is also a factor, so that the whole project descends into chaos. Desai's stories are small tragedies, though large enough for the characters who live through them. Is this India as she always was, or as she has been made by colonial occupation? In Desai's stories no one seems to know.

Not so in the stories of Salman Rushdie. There is nothing small about either his characters or his plots. His two best-known books, *Midnight's Children*, 1981, and *The Satanic Verses*, 1988, are written in an exuberant, overflowing style, the images and metaphors and jokes billowing forth like the mushroom clouds of an atomic bomb.[24] Rushdie's relationship to his native India, and to the English language, is complex. His stories tell us that there are many Indias, enough of them grim, failing, divided. English at least offers the chance of overcoming the chronic divisions, without which failure cannot be conquered, and only by embarking on a fabulous journey of improbable fantasies can he hope to have what are in fact very direct messages swallowed. *Midnight's Children* tells the story of Saleem Sinai, born at midnight on the day India achieved independence in 1947, one of 1,001 other children to be born at the same time. By virtue of this, all of them are given some magical property, and the closer their birth to midnight, when 'the clock-hands joined palms in respectful greeting,' the stronger their magical power. Saleem has a very large nose, which grants him the ability to see 'into the hearts and minds of men.' His chief rival, Shiva, has bloated knees, meaning he has the power of war. The book is written mainly in the form of Saleem's memoirs, but there is little in the way of traditional characterisation. Instead Rushdie gives us a teeming, tumbling narrative, juxta-posing day-to-day politics and private obsessions (one figure works on a documentary about life in a pickle factory), all intertwined with ever more fabulous metaphors and jokes and language constructions. The best and most terrible joke comes in the central scene where the two main characters discover that they have been swapped as babies. Rushdie is challenging the meaning of the most basic ideas – innocence, enchantment, nation, self, community. And, in so doing, independence. All this is done with an 'elephantiasis' of style that emulates the Indian oral storytellers of old, yet is as modern as it is reminiscent of Günther Grass and Gabriel García Márquez. *Midnight's Children* is neither eastern nor western. That is the point, and the measure of its success.[25]

The theme of *The Satanic Verses* is migration, emigration, and the loss of faith it often brings about in the emigrant/immigrant.[26] Faith, its loss, and the relation of faith to the secular life, the hole – the 'God-shaped hole' – at the centre of the once-faithful person, is the issue that, Rushdie has admitted, underpins the book.[27] He deals with the issue also in a fabulous way. The book begins when two Indian actors, Gibreel Farishta and Saladin Chamcha, formerly Salahuddin Chamchawal, fall to earth after an Air India jumbo jet explodes 30,000 feet above the English Channel. This naturally evokes the memory of an actual explosion, of an Air India Boeing 747 off Ireland in 1985, blown up, it is believed, by Sikh terrorists in Canada.[28] Farishta is the star of several Bombay 'theological' films and is so popular that for many an Indian he *is* divine. Saladin, on the other hand, is an Anglophile who has rejected India and lives in Britain doing voiceovers for television commercials, 'impersonating packets of crisps, frozen peas, Ketchup bottles.'[29] These two fall to earth in the company of airplane seats, drink carts, headsets, but land safely enough on a British beach. From then on, the book follows a series of interwoven plots, each more fantastic than the last. These episodes are never out of control, however, and Rushdie's references make the book very rich for those who can decipher them. For example, Gibreel Farishta, in Urdu, means Gabriel Angel, making him in effect the archangel whom Islamic tradition regards as ' "bringing down" the Qur'an from God to Muhammad.' Saladin was also the great defender of mediaeval Islam against the Crusaders, who restored Sunni Islam to Egypt. Gibreel, learning Islam from his mother, encounters the notion of the Satanic Verses, in which the devil is understood to have inserted a sentence in the Qur'an, later withdrawn, but which nonetheless insinuates a sliver of religious doubt. Religious doubt, then, is at the very heart of Rushdie's book. One may even say that it plays with the very idea of the devil, of the secular *being* the devil, certainly so far as the faithful are concerned. Essentially, through-out the interlocking narratives, Saladin is a sort of Iago to Gibreel's Othello, 'using the thousand and one voices of his advertising days.' Under this onslaught, Gibreel is led astray, notably to a brothel, the 'anti-Mosque' in Malise Ruthven's apt phrase, falling among people who blaspheme, not just in swear-words but in their criticisms of the Prophet's actual behaviour (for example, Muhammad had more wives than strict Islamic law allowed). At every opportunity, therefore, *The Satanic Verses* skirts danger. It is certainly a challenging book. But can a book that explores blasphemy actually pursue that theme without *being* blasphemous? In exploring faith, Rushdie knew he had to deliberately provoke the faithful. At one point in the book, the Prophet issues a *fatwa* against an impious poet.[30]

Perhaps it was this above all which provoked the Islamic authorities. On 14 February 1989, Ruhollah Al-Musavi Al-Khomeini – better known as **Ayatollah Khomeini**, of Iran – issued a *fatwa* against the 'apostasian' book *Satanic Verses*: 'In the name of God Almighty; there is only one God, to whom we shall all return; I would like to inform all the intrepid Muslims in the world that the author of the book entitled *The Satanic Verses* which has been compiled, printed and published against Islam, the Prophet and the Koran, as well as those publishers who were aware of its contents, have been sentenced to death. I call on all zealous Muslims to execute them quickly, wherever they find them, so that no one will dare to insult the Islamic sanctions. Whoever is killed on this path will be regarded as a martyr, God willing. In addition, anyone who has access to the author of the book, but does not possess the power to execute him, should refer him to the people so that he may be punished for his actions. May God's blessing be on you all.'[31]

Inside forty-eight hours, Rushdie and his wife had gone into hiding, where he would remain except for brief forays into the limelight, for nearly ten years. In subsequent months, the 'Rushdie affair' claimed many headlines. Muslims in Britain and elsewhere staged public burnings of the title; ten thousand demonstrated against the book in Iran, and in Rushdie's native Bombay ten people were killed when police opened fire on demonstrators.[32] In all, twenty-one people died over *The Satanic Verses*, nineteen on the Indian subcontinent, two in Belgium.[33]

Like Salman Rushdie, **V. S. Naipaul**'s novels – his later novels especially – generally concern people living outside their native context. He himself was born in Trinidad, a second-generation Indian, moved to England to attend Oxford, and has remained there ever since, except to research a remarkable series of travel books.

Naipaul is less concerned with faith than Rushdie, and has more in common with Anita Desai's fascination with modernisation and technological change, though he uses this to reflect his preoccupation with the nature of freedom. *A House for Mr. Biswas* (1961) ostensibly follows the building of a house. At the same time Naipaul deconstructs Mr. Biswas himself.[34] His facility for sign writing leads him out of the prison of poverty and into a marriage where he is trapped, but in a different way. Sign writing leads to other forms of writing, letters to his son mainly. As he discovers language, like a writer discovers language, so Biswas discovers another layer of freedom. But total freedom, Naipaul infers, is not only impossible but undesirable. Fulfilment comes from loving and being loved, a status Biswas achieves, but it is not freedom. In *The Mimic Men* (1968), the scene has shifted to England, not the dream England that a poor Trinidadian might conceive of but the drab, suburban England of the immigrant, with the endless fresh attempts to get going on a career, the chronic tiredness, and the poor sense of self that comprise modern city life.[35] Again, freedom boils down to one struggle replacing another. The later books – *In a Free State* (1971), which won the Booker Prize, *Guerrillas* (1975), and *A Bend in the River* (1979) – are more nakedly political, juxtaposing political and private freedom in deliberately jarring ways.[36] In the 1971 book, two white people, Linda and Bobby, drive back to their expats' compound through a black African state laid low by civil war. Their politics differ – Bobby is a liberal homosexual, Linda a bombastic right-winger. Naipaul is asking how they can enjoy so many freedoms at home when they can't agree on anything. In the car, there is civil war between them.

In his films, **Satyajit Ray** (1921–1992) embodied a bit of Desai, a part of Narayan, and aspects of Rushdie and Naipul, and this is because he was more than a filmmaker. He was a commercial artist, a book designer, an author of children's books and science fiction, and a celebrated musician. He began as a filmmaker when, in 1945, he was asked to illustrate a children's version of a popular novel, *Pather Panchali*.[37] Ray had the idea instead of turning the novel into a film; he set about it with no experience of filmmaking, trying his hand at weekends (it never had a proper script).[38] The project took ten years and was only finished after Ray several times ran out of money, when the Bengali government stepped in with funds.[39] Despite its unpropitious beginnings, the film was a triumph and became the first in a trilogy of trilogies, for which Ray became famous: the Apu Trilogy (*Aparajito*, 1956, with music by Ravi Shankar, and *The World of Apu*, 1960), the Awakening Woman trilogy (including, most notably, *Charulata*, 'The Lonely Wife,' 1964, still very popular), and a trilogy of 'city' films, which included *The Middleman* (1975).[40] Ray's films have also been described as a mixture of Henry James and Anton Chekhov, though they are marked by an emotional generosity that James, certainly, rarely showed. But the strength of Ray lies in his telling of ordinary stories (of a family trying to survive, in *Pather*; of an affair between a woman and her husband's young cousin, in *Charulata*; of a businessman expected to provide a client with a woman in *The Middleman*) in extraordinary detail, lovingly observed. His biographer has pointed out that there are few, if any, villains in Ray's world because he sees everyone's point of view. Ray was just as aware of India's failings as the other writers, but he seems to have been more comfortable with the contradictions.[41]

The award of the 1986 Nobel Prize for Literature to the Nigerian writer and dramatist Wole Soyinka, and then to the Egyptian novelist Naguib Mahfouz, in 1991, the same year that Ben Okri, another Nigerian, won the Booker Prize, shows that African writing has at last been recognised by what we may call the Western literary 'establishment.' At the same time, contemporary African literature has nothing like the same following, worldwide, as does Indian or South American literature. In his *Myth, Literature and the African World* (1976), Soyinka, who had studied in Britain and read plays for the Royal Court Theatre, did his best to make many fellow writers more visible in a Western context.[42]
  Soyinka was trying to do for literature what Basil Davidson had done for African archaeology, not just in the book referred to but in his own poetry and plays. In fact, it was Soyinka's choice of literature – in particular theatre – that finally won the Nobel Prize for him, rather than for Chinua Achebe. (Achebe's novel *Anthills of the Savannah* was shortlisted for the Booker Prize in 1987.) Soyinka was part of the generation of brilliant writers who studied at Ibadan University College in the period before independence, together with Cyprian Ekwensi, Christopher Okigbo, and John Pepper Clark, some of whose works he covered in *Myth, Literature and the African World*. In that book, his secondary aim, after rendering these writers visible to an international audience, was to do two things: first, to show black African literature as its own thing, having themes in common with other great literatures, and just as rich, complex and intelligent. At the same time, Soyinka, in discussing such entities as Duro Ladipo's Yoruba plays, or Obotunde Ijimere's *Imprisonment of Obatala*, or Ousmane Sembene's *God's Bits of Wood*, stressed the particular strengths of African literature, the ways in which it differs from its Western counterparts.[43] Here, he stresses the *collective* experience of ritual, the way the individualism of the West is alien to African experience. In the African social contract, community life comes first, and Soyinka explains the impact of ritual by analogy at one point, in order to bring home how vivid it is: 'Let us say he [the protagonist in a story] is a tragic character: at the first sign of a check in the momentum of a tragic declamation, his audience becomes nervous for him, wondering – has he forgotten his line? Has he blacked out? Characters undertake acts on behalf of the community, and the welfare of the protagonist is inseparable from that of the total community.'[44] Soyinka's point is that whatever story is set out in African literature, the *experience* is different.

Soyinka is both a creative writer and a critic. In the last quarter of the twentieth century, literary and cultural criticism has been both exceptionally fertile and exceptionally controversial. This is particularly true of three areas, all related: postcolonial criticism, postmodern criticism, and the development of the discipline known as cultural studies.

In postcolonial criticism two figures stand out: **Edward Said** and **Gayatri Spivak**. Across several works, but especially in *Orientalism* (1978), *Covering Islam* (1981), and 'Orientalism Reconsidered', (1986), Said, writing as a Palestinian academic on the faculty of Columbia University in New York, explored the way the 'Orient' has been conceived in the West, especially since the beginning of 'Oriental studies' early in the nineteenth century.[45] He examined the writings of scholars, politicians, novelists, and even painters, from Silvestre de Sacy, whose *Chrestomathie arabe* was published in 1806, through Gustave Flaubert, Arthur James Balfour, and T. E. Lawrence, right up to academic books published in the 1960s and 1970s. The jacket of his title shows a young boy, naked except for a large snake wrapped around him, standing on a carpet and entertaining a group of men, dark-skinned Arabs festooned in rifles and swords, lounging against a wall of tiles decorated with arabesques and Arabic script. A detail from Jean-Léon Gérôme's *Snake Charmer* (1870), it illustrates Said's argument exactly. For this is an imaginary Orient, a stereotypical Orient full of caricature and oversimplification. Said's argument is that the intellectual history of Oriental studies, as practised in the West, has been corrupted by political power, that the very notion of 'the Orient' as a single entity is absurd and belittling of a huge region that contains many cultures, many religions, many ethnic groupings. In this way the world comes to be made up of two unequal halves, shaped by the unequal exchange rooted in political (imperial) power. There is, he says, an imaginative demonology of 'the mysterious Orient' in which the 'Orientals' are invariably lazy, deceitful, and irrational. Said shows that de Sacy was trying to put 'Oriental studies' on a par with Latin and Hellenistic studies, which helped produce the idea that the Orient was as homogeneous as classical Greece or Rome. In *Madame Bovary*, Emma pines for what, in her drab and harried bourgeois life, she does not have – 'Oriental clichés: harems, princesses, princes, slaves, veils, dancing girls and boys, sherbets, ointments, and so on.'[46] In Joseph Conrad's *Victory*, he makes the heroine, Alma, irresistibly attractive to men – by the mid-nineteenth century, the name evoked dancers who were also prostitutes. But, Said reminds us, *Alemah* in Arabic means 'learned woman'; it was the name used in Egyptian society for women who were accomplished reciters of poetry. Even in recent times, says Said, especially since the Arab-Israeli wars, the situation has hardly improved. He quotes a 1972 issue of the *American Journal of Psychiatry* in which an essay entitled 'The Arab World' was published by a retired member of the U.S. Department of State Bureau of Intelligence. In four pages, the author provides a psychological portrait of more than 100 million people, across 1,300 years, using exactly four sources: two books, and two newspaper articles.[47] Said stresses the sheer preposterousness of such an exercise, calls for a greater understanding of 'Oriental' literatures (which he shows to be sadly lacking in Oriental departments in Western universities) and allies himself with Clifford Geertz's approach to anthropology and inter-national study, in particular his notion of 'thick description.'[48] As with the views of Martin Bernal on the African origins of classical civilisation discussed in the next chapter, Said's arguments have been fiercely contested by distinguished orientalists such as Albert Hourani.

As a critic, an Indian, and a woman, Gayatri Chakravorty Spivak has become one of the more prominent postcolonial writers, perhaps most influential as one of the editors of the celebrated journal *Subaltern Studies*. This word, *subaltern*, neatly ironical, refers to that low rank of the army, especially the Imperial Army of Britain, which was subordinate to the officer class – so low, in fact, that if a subaltern wanted to speak, he (always a he) had to ask permission. Subaltern studies is a variety of historiography that is frankly revisionist, seeking to provide an alternative history of India, a new voice somewhat analogous to the British Marxist historians, retelling the story 'from the bottom up.' Gayatri Spivak, who like Rushdie, Desai, and so many other Indian intellectuals, divides her time between India and the West, combines an essentially feminist view of the world with neo-Marxist flavouring derived from Derrida and Foucault.[49] The chief achievement of this group has been, first, gaining access to the raw material of the Raj, without which no revision would have been possible, and second, confronting what many have regarded as the failure of Indian culture to hitherto produce a rival system to the British one.[50] In historiography, for example, subaltern scholars have revisited a number of so-called mutinies against the British when, according to the imperial accounts, 'bands' of 'fanatics' rose up and were defeated.[51] These are now explained in terms of contemporaneous religious beliefs, marriage/sexual practices, and the economic needs of empire. Five volumes of *Subaltern Studies* were published in the 1980s, to great acclaim among scholars, providing an alternative historiography to what is now called colonialist knowledge.[52]

Underlying much of the postcolonial movement, not to mention the post-modern sensibility, was a phrase that the American critic **Fredric Jameson** gave to one of his books in 1981, *The Political Unconscious*.[53] Postcolonial and postmodern criticism derived much of its strength from Raymond Williams's earlier arguments that 'serious' literature should not be read in any way different from popular literature, and that the same is true of all art. This position was set out most fully in two celebrated articles published in *New Left Review,* one in 1984 by Jameson, entitled 'Postmodernism; or, The Cultural Logic of Late Capitalism,' and the other, in 1985, by **Terry Eagleton**, professor of English at Oxford, entitled 'Against the Grain.' Jameson's argument was that all ideologies are 'strategies of containment,' which enable a society 'to provide an explanation of itself which suppresses its underlying contradictions.'[54] The certainties of the nineteenth-century novel, for example, were designed to reassure the middle classes that their orderly class system would endure. Hemingway's novels, on the other hand, with their spare, short sentences, obsessed with machismo, had to be set in exotic foreign countries because he couldn't fit into America's self-image as a complex, technologically sophisticated society. Jameson's second major argument was that the postmodern sensibility was by the mid-1990s not merely one way of looking at the world but the dominant one, and that this was because it was the logical outcome of late capitalism.[55] In this late stage, he said, society has finally abolished the distinction between high culture and mass culture – we have instead a culture that many decry as 'degraded' but younger people espouse enthusiastically: kitsch, schlock, pulp fiction and TV, *Reader's Digest*. The first to appreciate this was Andy Warhol. The point, Jameson says, is that late capitalism recognises that art is, above all, a *commodity*, something to be bought and sold.

Eagleton was more aggressively Marxist. The distinction between high art and popular/mass art was one of the oldest certainties, he said, and the fact that it has been undermined is an aid to the socialist, because it helps 'expose the rhetorical structures by which non-socialist works produce politically undesirable effects.'[56] In late capitalism, Eagleton writes, commodities have become fetishes, and he includes artistic commodities with the others. This is a new aesthetic category with no precursors.

Critics like Jameson and **Stanley Fish**, his colleague, then at Duke University in North Carolina and now at the University of Illinois in Chicago, paid as much attention in their work to other media besides books – that went without saying. Films, television, comic books, advertising . . . all these were systems of signs.[57] The early work of Raymond Williams, postcolonialism, and postmodern literary theory, together with the theories of such French authors as Barthes, Lyotard, Lacan, Derrida, and Jean Baudrillard, plus the anthropology of Clifford Geertz, therefore came together to create a new discipline, cultural studies. This is not the same as media studies, but they both stem from the same impulse. The fundamental idea behind both is, as was mentioned above, and to return to Jameson's phrase, the political unconscious – that works of the imagination are not 'privileged' in any way, to use the favoured term, that they are just as much a product of their context and environment as anything else, are subject to market forces, and therefore cannot avoid having an ideological or political angle. It is the aim of cultural studies to render this hidden agenda visible, peeling away one of the final layers of self-consciousness.

Cultural studies is controversial, especially among an older generation brought up to believe that 'aesthetic' values are *sui generis*, independent of everything else, helping us to find the 'eternal truths' of the human condition. But cultural studies courses at universities are very popular, which must mean that they meet some needs of the young (they have been around too long now to be merely fashionable). The heart of the issue, the most controversial aspect of the new discipline, is the battle for Shakespeare. Keats called Shakespeare the 'chief poet,' the 'begetter of our deep eternal theme.' The new Shakespeareans, if we may call them that, argue on the other hand that although the bard wrote a remarkable number of remarkable plays, he did not, as Coleridge maintained, speak for all men, in all places, and at all times.

The new scholars say that Shakespeare was a man of his age, and that most, if not all, of his plays had a specific political context. They add too that in the nearly 400 hundred years since his death, successive establishments have appropriated him for their own essentially right-wing agendas. In other words, far from being an objective fount of fundamental wisdom about our essential nature, Shakespeare has been used by lesser souls as propaganda to promote and sustain a particular point of view. In arguing that Shakespeare was a man of his time, they are also saying that his insights into human nature are no more 'fundamental' or 'profound' or 'timeless' than anyone else's, and therefore he should forfeit his place as the rock on which English literature is built. For the cultural materialists, as they are called, Shakespeare's significance is as a battleground for competing views of literature, and its relevance in our lives.

The first concerted attack on the conventional wisdom came in 1985, in a book edited by **Jonathan Dollimore** and **Alan Sinfield**, from the University of Sussex, which was provocatively entitled *Political Shakespeare*.[58] It comprised a series of eight essays by British and North American scholars; by comparing the chronology of the plays with contemporary political events, the essays were designed to show that, far from transcending history and politics and human nature, Shakespeare was a child of his times. As a result the conventional meaning of many of the plays was changed radically. *The Tempest*, for example, far from being a play about colonialism and America, becomes a play about England's problems with Ireland. Published in the middle of the Thatcher/Reagan years, *Political Shakespeare* created an academic storm. Two of the academic referees who read the manuscript argued that the book should 'on no account be published.'[59] After publication, one reviewer wrote, 'A conservative critic . . . may conclude in horror that Shakespeare has succumbed to an academic AIDs, his immunology systems tragically disrupted by Marxist, feminist, semiotic, post-structuralist and psychoanalytic criticism.' Others found the book important, and in the classroom it proved popular and was reprinted three times. In Annabel Patterson's *Shakespeare and the Popular Voice*, published in 1989, she argued that until the early nineteenth century Shakespeare was regarded as a political playwright and a rebel, and that it was Coleridge, worried by the ripple effects in England of the French Revolution, who sought to overturn the earlier view, for political reasons of his own.[60] These books provoked such interest that the *London Review of Books* produced a special supplement on the controversy in late 1991.

The strength of American literature, so evident to Marcus Cunliffe in the 1960s, became even more marked as the postwar decades passed. Its most impressive quality, as new talents continued to emerge, was the staying power of familiar names, and the resilience of their approach to their art.

The playwright **David Mamet**, for example, continued in the fine American tradition of Eugene O'Neill, Tennessee Williams, and Arthur Miller, in that his themes were intimate, psychological dramas, where the 'action,' such as it was, took place *inside* the characters as revealed in language. Mamet's two greatest plays, *American Buffalo* (1975) and *Glengarry Glen Ross* (1983), were once described as indictments of a society 'in which the business ethic is used as a cover for any kind of criminal activity.'[61] In *Buffalo* a group of lowlifes plot a robbery that they are totally 'incapable of carrying out.' Mamet's characters are almost defined by their inarticulateness, which is both a source and a symptom of their desperation. His chosen territory is the modern city and the life-diminishing occupations it throws up – in particular, and here he echoes O'Neill and Miller, the salesman. In *Glengarry Glen Ross*, the pathetic optimism of the real-estate salesmen, which overlays their quiet desperation, is painfully moving, as each tries to do the other down in even the smallest of struggles. This distracts them from recognising their own true nature.

Mamet's significance, as a figure who emerged in the 1970s, was his response to the arrival of the postmodern world, the collapse of the old certainties. Whereas Peter Brook was part of the new temper, a man who enjoyed multiculturalism, and **Tom Stoppard**, the British playwright, set his face against it, asserting that there *was* objective truth, objective good and evil, that relativism was itself in its way evil, Mamet exercised an old-fashioned, Eliot-type scepticism to the world around him.[62] He embraced and updated the tradition articulated by O'Neill, that America was 'a colossal failure.'[63] His plays *were* plays because he was suspicious of the mass media. 'The mass media,' he wrote in a memoir, '. . . corrupt the human need for culture (an admixture of art, religion, pageant, drama – a celebration of the lives we lead together) and churn it into entertainment, marginalizing that which lacks immediate appeal to the mass as "stinking of culture" or "of limited appeal". . . . The information superhighway seems to promise diversity, but its effect will be to eliminate, marginalize, or trivialize anything not instantly appealing to the mass. The visions of Modigliani, Samuel Beckett, Charles Ives, Wallace Stevens, survive for the moment as *culture* in a society that never would have accepted them as art. . . . The mass media – and I include the computer industry – conspire to pervert our need of community. . . . We are learning to believe that we do not require wisdom, community, provocation, suggestion, chastening, enlightenment – that we require only information, for all the world as if life were a packaged kit and we consumers lacking only the assembly instructions.'[64]

John Updike has published more than thirty books since *Poorhouse Fair* in 1959, during which time he has attempted to follow both the small and the grand themes in American, white, middle-class life. In *Couples* (1968), *Marry Me* (1976), and *Roger's Version* (1986) he examined sex, adultery, 'the twilight of the old morality.' In *Bech: A Book* (1970) he looked at Communist East Europe through the eyes of a Jewish-American traveller, which enabled him to compare the rival empires of the Cold War. And in *The Witches of Eastwick* (1984) he took a swipe at feminism and American puritanism all at the same time. But it is for his '**Rabbit**' series that he most deserves consideration. There are four books in the sequence: *Rabbit, Run* (1960), *Rabbit Redux* (1971), *Rabbit Is Rich* (1981) and *Rabbit at Rest* (1990).[65] Harold 'Rabbit' Angstrom used to play basketball as a professional, used to be young and romantic, and is now caught up in the domestic dreariness of married life. Rabbit is a deliberate echo of 'Babbitt' because Updike sees his hero as the natural epigone of Sinclair's man from Zenith. But the world has moved on, and Rabbit lives on the East Coast, rather than in the Midwest, more at home in New York and Connecticut. His world is that of gadget-packed apartments, of commodities, including art, of material abundance but also of a spiritual malaise. Rabbit and his circle, with all their everyday needs well provided for, seek to recover the excitement of their youth in affairs, art courses, ever more pompous wines, travel. Despite this, they never escape the feeling that they are living in an age of decline, that theirs is an unheroic, shabby era; and as the books progress, the characters, showing what Updike himself called 'instinctive realism,' grow still more desperate in a search for epiphanies that will provide meaning. It is the fate of Updike's characters, in the Rabbit books, to be entering the postmodern bleakness without knowing it. Updike invites us to think that this is how social evolution takes place.[66]

Saul Bellow has achieved the enviable distinction, better even than the award of the Nobel Prize in 1976, of writing at least one masterpiece in each of *five* decades: *Dangling Man* (1944); *Henderson the Rain King* (1959); *Herzog* (1964); *Humboldt's Gift* (1975); *The Dean's December* (1982), and *More Die of Heartbreak* (1987).[67] Born in Canada in 1915, the child of immigrant Jews, Bellow was raised in Chicago, and most of his books are set there or in New York – at any rate, in cities. This is not Updike's world, however. Most of Bellow's characters are Jewish, writers or academics rather than business types, more reflective, more apt to be overwhelmed by mass culture, the mass society of vast cities, which they confront with 'a metaphysical hunger.'[68] In *Dangling Man*, much influenced by Kafka, Sartre and Camus, Bellow wrote this about the main character: 'He asked himself a question I still would like answered, namely, "How should a good man live; what ought he to do?"' In *The Adventures of Augie March* (1953), the hero says, 'It takes some of us a long time to find out what the price is of our being in nature, and what the facts are about your tenure. How long it takes depends on how swiftly the social sugars dissolve.' All Bellow's books are about the 'social sugars' in one form or another, the nature of the link between the self and others, community, society. For Bellow, the nature of the social contract is the most fundamental of all questions, the fundamental problem of politics, the deepest contradiction of capitalism, the most important phenomenon that science has not even begun to address, and where religion can no longer speak with authority.[69] In *Herzog* we have a character determined not to surrender to the then prevalent nihilism; in *Humboldt's Gift* we have 'the Mozart of gab,' a brilliantly loquacious poet who nonetheless dies penniless while his postmodern protégé, obsessed with commodities, becomes rich. In *The Dean's December*, the dean, Albert Corde, from a free city – the Chicago of violence, cancer, and postmodern chaos – visits Bucharest, then behind the Iron Curtain, where families, and family life, still exist. He is for ever comparing his own despairing knowledge about city life with the certainties of the astrophysical universe that are the everyday concerns of his Romanian wife. The aphorism behind *More Die of Heartbreak* is 'more die of heartbreak than of radiation,' showing, in idiosyncratic yet tragic form, some limits to science. (The book is a comedy.) The progression from the dangling man, to Augie March, to Henderson, to Herzog, to Humboldt, to Dean Albert Corde is a profoundly humane, ebullient set of tragedies and epiphanies, an intellectual and artistic achievement unrivalled in the last half of the twentieth century.

In the early 1990s literature by native American Indians began to appear. *Keeping Slug Woman Alive: Approaches to American Indian Texts* (1993) and *Grand Avenue* (1994), both by **Greg Sarris**, were two commercially and critically successful titles.[70] Sarris is part American Indian, part Filipino, and part Jewish, an elected chief of the Miwok tribe but also professor of English at UCLA. This conceivably makes him the ultimate postmodern, multicultural figure, the natural next step in America's evolving history. He, or someone like him, could be the first major literary voice in the twenty-first century. But Bellow has set the standard against which all others will be judged.

CULTURE WARS

In September 1988, at a conference at Chapel Hill, the campus of the University of North Carolina, academics gathered to consider the future of liberal education. Conferences are normally placid affairs, but not this one. Delegates held what a *New York Times* reporter said recalled the 'Minute of Hatred' in Orwell's *1984*, when citizens were required to stand and hurl abuse at pictures of a man known only as Goldstein, the 'Great Enemy' of the state. At Chapel Hill, 'speaker after speaker' denounced a small group of 'cultural conservatives' who, in the words of Stanley Fish, professor of English at Duke University, had mounted 'dyspeptic attacks on the humanities.' In the words of the *Times* reporter, these conservatives were 'derided, scorned, laughed at.' Though these individuals were not named (possibly for fear of slander), no one was in any doubt over who were the intended targets.[1] The Great Enemy-in-Chief was **Allan Bloom**, co-director of the John M. Olin Center for Inquiry into the Theory and Practice of Democracy at the University of Chicago, where he was also a professor in the Committee on Social Thought.* More pertinently, Bloom was the author of a book published the year before, which had really set the cat among the pigeons in the academic world. Entitled *The Closing of the American Mind*, it had broken out of the scholarly ghetto for which it had been intended and had made Bloom a celebrity (and a millionaire).[2] It had been reviewed, and praised, by *Time*, the *Washington Post*, the *Wall Street Journal*, the *Los Angeles Times* and the *New York Times*, and been welcomed or hated by such diverse figures as Conor Cruise O'Brien, Saul Bellow, and Arthur Schlesinger.

Bloom's thesis in the book was simple but breathtakingly ambitious, though he himself did not see it like that. Using his long experience as a teacher as his guide, he started from the observation that between the late 1950s and the mid-1980s the character of students entering American universities had changed markedly, and the university had changed with them. He made no secret of the fact that he found almost all these changes for the worse. In the 1950s, he said,

---

* The Committee on Social Thought was 'a herd of independent minds,' in Harold Rosenberg's phrase, a group of socially concerned intellectuals centred on Chicago University and which included among many others Rosenberg himself, Saul Bellow, and Edward Shils.

and thanks to the chaotic history of Europe in the first half of the century, American universities had been among the best in the world, with both homegrown talent and that imported by the exiles from totalitarianism. In the 1950s and early 1960s, he found that two decades of prosperity and abundance had created a generation of students who were adventurous yet serious, who had ideals and an intellectual longing 'which made the university atmosphere electric.'[3] But then, in the late 1960s, he began to notice a decline in reading on the part of students arriving at university, and among them when they were there. From here on, Bloom set about identifying and attacking the chief culprits in what he clearly thought was a serious decline in American civilisation. His venom was initially focused on rock music, which he regarded as barbarous, directed exclusively at children, dwelling on sex, hate, and 'a smarmy, hypo-critical version of brotherly love.'[4] There is, he said, nothing noble, sublime, profound, or delicate in rock music: 'I believe it ruins the imagination of young people and makes it very difficult for them to have a passionate relationship to the art and thought that are the substance of liberal education.' Exactly the same, he said, was true of drugs, but he also castigated feminism, the new psychologies, and the passionate concern of the young for equality in all things, but especially on matters having to do with race.[5]

Having described the changed nature of the university student (in America, though elements were clearly recognisable elsewhere), in his second section he deliberately examined some of the large questions, the 'big words that make us afraid,' as James Joyce said: 'the self,' 'creativity,' 'culture,' 'values,' 'our ignor-ance.' His aim was to show that however much students have changed, and however much they think the world has changed around them, the big issues have not changed. He did this by showing that his beloved philosophers of the past – Plato, Aristotle, Rousseau, and Locke in particular – still have the power to inform us, 'to make us wise,' and to move us. He showed that many of the ideas discovered, or rediscovered, by the social sciences, were in fact introduced by mainly German thinkers, who included Hegel, Kant, Nietzsche, Weber, Husserl, and Heidegger.[6] His aim was to show that freedom, and reason, two givens that so many take for granted, were fought for, thought for; that true culture – as opposed to the drug culture, or street culture – has a depth, a reasoned, *earned* quality that points toward what is *good*; that there is a unity to knowledge 'which goes by the name of wisdom.' A serious life, he says, means being fully aware of the alternatives that face us in the great divisions we encounter: reason–revelation, freedom–necessity, good–evil, self–other, and so on: 'That is what tragic literature is about.' In the third and final part of the book he attacked the universities, for what he thought was their enormous dereliction of duty to be islands of reason and autonomy in an ever more politically correct world. 'The essence of philosophy is the abandonment of all authority in favour of individual human reason. ... [The university] must be contemptuous of public opinion because it has within it the source of auton-omy – the quest for and even discovery of the truth according to nature. It must concentrate on philosophy, theology, the literary classics, and on those scientists like Newton, Descartes and Leibniz who have the most comprehensive scientific vision and of the relation of what they do to the order of the whole of things. These must help preserve what is most likely to be neglected in a democracy.'[7] Bloom also had some harsh things to say about the 1960s ('barbarians at the gate'), about university colleagues who caved in to student pressure, about the 'new' disciplines of social science ('parts without a whole'), and above all about the M.B.A., the master's degree in business administration, 'a great disaster' because students' lives were never radically changed by it, as they should be in a proper education.

In saying all this, Bloom naturally managed to annoy or irritate a great number of people. But the people he annoyed most were his colleagues in the humanities. Bloom's main plea, echoing F. R. Leavis and Lionel Trilling, was that the university should be above all the home of the humanities, by which he meant 'that the study of high culture, particularly that of Greece, would provide the models for modern achievement.'[8] He made it abundantly clear that he considered the ancient philosophers, novelists and poets – generally speaking the authors of the 'great books' – as the men from whom we have most to learn. Their survival is no accident; their thoughts are the fittest.

Bloom unleashed a whirlwind. The conference at Chapel Hill articulated the opposing view, the view that Bloom was seeking to counter. The conference's participants denounced what they said was a 'narrow, out-dated interpretation of the humanities and of culture itself, one based, they frequently pointed out, on works written by "dead white European males." ... The message of the North Carolina conference was that American society has changed too much for this view to prevail any longer. Blacks, women, Latinos and homosexuals are demanding recognition for their canons.' Professor Fish added, 'Projects like those of ... Bloom all look back to the recovery of the earlier vision of American culture, as opposed to the conception of a kind of ethnic carnival or festival of cultures or ways of life or customs.'[9]

We have been here before. Allan Bloom's book was much longer than T. S. Eliot's *Notes Towards a Definition of Culture* and was a more passionate and eloquent account, but the overlap in argument was plain. What was different was that the forty years in between had seen a vast change in the world, in the position of minorities, in universities themselves, in politics. But that change also meant that the response to Bloom's work was very different from the response to Eliot's, which had been muted, to say the least.

Many people took issue with Allan Bloom, but in 1994 he received powerful support from his near-namesake at another American University, **Harold Bloom** of Yale. In *The Western Canon*, Harold Bloom was also uncompromising.[10] Dismissing feminism, Marxism, multiculturalism, neoconservatism, Afrocentrism and the postmodern cultural materialists, at least as applied to great literature, Bloom asserted the view that 'things have however fallen apart, the center has not held, and mere anarchy is in the process of being unleashed upon what used to be called "the learned world." ' In great style and at even greater length, he argued that there is such a thing as an *aesthetic* value in life, that it was his experience, 'during a lifetime of reading,' that the aesthetic side to life is an autonomous entity 'irreducible' to ideology or metaphysics: 'Aesthetic criticism returns us to the autonomy of imaginative literature and the sovereignty of the solitary soul, the reader not as a person in society but as the deep self, our ultimate inwardness. . . . Aesthetic value rises out of memory, and so (as Nietzsche saw) out of pain, the pain of surrendering easier pleasures in favour of much more difficult ones.'[11]

After making it plain that he considers the current age 'the worst of all times for literary criticism,' he set about constructing, and justifying, his own Western canon, consisting of twenty-six authors whom he considers vital for anyone with an interest in reading, but with the following 'health warning': 'Reading deeply in the Canon will not make one a better or a worse person, a more useful or more harmful citizen. The mind's dialogue with itself is not primarily a social reality. All that the Western Canon can bring one is the proper use of one's own solitude, that solitude whose final form is one's confrontation with one's own mortality.'[12] For Bloom, the centre of the canon is Shakespeare, 'the largest writer we ever will know,' and throughout his book he returns time and again to the influences on Shakespeare and his influences on those who came after him. In particular Bloom dwells on *Hamlet*, *King Lear*, *Othello*, and *Macbeth*, the great tragedies, but also on Falstaff, perhaps the greatest character ever invented because, through him, Shakespeare offers us the 'psychology of mutability,' 'the depiction of self-change on the basis of self-overhearing.'[13] For Bloom, what merits inclusion in the canon is a quality of weirdness, of strangeness, of monumental originality that 'we can never altogether assimilate' and yet at the same time 'becomes such a given that we are blinded to its idiosyncrasies.' After Shakespeare he includes in his list Dante, Chaucer, Cervantes, Milton, Montaigne and Molière, Goethe, Wordsworth, and Jane Austen. He regards Walt Whitman and Emily Dickinson as the centre of the American canon, Dickens's *Bleak House* and Eliot's *Middlemarch* as the canonical novels, Tolstoy, Ibsen, Joyce, Woolf and Kafka, Borges and Neruda, as worthy of inclusion. But Beckett, Joyce, and Proust are related back to Shakespeare, and in one chapter he argues that Shakespeare, 'the major psychologist in the world's history,' tells us far more about Freud than Freud ever could about the Bard. In fact, in that chapter, Bloom is astute in his reading of several lesser known papers by Freud in which, Bloom shows, Freud (who read Shakespeare in English all his life) acknowledges his heavy debt.[14] In acknowledging Freud as a great stylist, Bloom dismisses the psychoanalytic view of the world as a form of shamanism, 'an ancient, worldwide technique of healing' and which, he concludes, may well constitute the final fate of psychoanalysis. In dismissing feminism, multiculturalism, and Afrocentrism as ways to approach literature, because that assimilation must be *personal* rather than ideological, Bloom does not see himself as being ethnocentric. On the contrary, he specifically says that all great writers are subversive, and he points out that the culture of Dante or Cervantes is far more different from, say, the late-twentieth-century East Coast society than is, for example, twentieth-century Latin American society, or black North American society.

The canon, he says, can never be written in stone, but in the act of achieving it, or trying to achieve it, a sense of competition exists in which people are *thinking*, judging, weighing one entity against another. People – readers – are 'enlarging their solitude.' 'Without the Canon, we cease to think. You may idealize endlessly about replacing aesthetic standards with ethnocentric and gender considerations, and your social aims may indeed be admirable. Yet only strength can join itself to strength, as Nietzsche perpetually testified.'[15] Bloom also coined the phrase 'anxiety of influence,' by which he meant that all writers are influenced by other great writers and that, therefore, later writers must know what earlier great writers have written. This does not make imaginative literature the same as scientific literature – i.e., cumulative, not in any direct sense anyway. But it does suggest that later works, in a rough way, develop out of earlier works. This is not evolution in a classically biological sense, but taken in conjunction with the struggle to construct the canon, it does imply that the development of imaginative literature is not entirely random either.

The Blooms evoked a counter-attack. This took several forms, but responses mostly had one thing in common: whereas the Blooms had written very personal polemics, in a combative, ironic, and even elegiac style, the replies were more prosaic, written 'more in sorrow than in anger,' and used detailed scholarship to refute the charges.

Lawrence Levine's *The Opening of the American Mind* was published in 1996.[16] Levine, a professor emeritus in history at the University of California at Berkeley, had earlier published a book, *Highbrow Lowbrow*, which had examined the history of Shakespeare in the United States and concluded that before the nineteenth century 'high culture' in America had been enjoyed by all classes and many different ethnic groups. It was only in the second half of the nineteenth century that, in regard to Shakespeare and Grand Opera in particular, a process of 'sacralisation' took place, when the distinction between 'high' and 'low' culture was stressed. *The Opening of the American Mind* made a number of points. One, that fights over the canon, and the curriculum, have been going on for more than a hundred years, so the Blooms are nothing new. Such fights, Levine says, are inevitable as a nation changes and redefines itself. He argues that minority groups, ethnic groups, immigrant groups, don't want to throw out the canon as described by, say, Allan and Harold Bloom, but they do want to add to it works that have been overlooked and that reflect their own experience.[17] And he says that in a country like America, with many immigrants, many different racial and ethnic groups, in a country lacking a central tradition (like France, say), that a narrow canon of the kind suggested by the Blooms is simply impractical, failing the needs of the many different kinds of people, with different experiences. He defends the universities for at least seeking to address America's changing social structure rather than stick with a past that is not only imaginary but may never have existed. But Levine's most original contribution was to show that, in fact, the idea of a canon of 'Great Books' and 'Western Civilisation,' at least in America, 'enjoyed only a brief ascendancy.' The idea emerged, he says, after World War I and declined after World War II. He further shows that the inclusion of 'modern' writers, like Shakespeare and Walt

Whitman, 'came only after prolonged battles as intense and divisive as those that rage today.' Going through various accounts of university education in the early nineteenth century, for example, Levine found that James Freeman Clark, who received his A.B. from Harvard in 1829, complained, 'No attempt was made to interest us in our studies. We were expected to wade through Homer as though the Iliad were a bog. ... Nothing was said of the glory and grandeur, the tenderness and charm of this immortal epic. The melody of the hexameters was never suggested to us.'[18] Charles Williams Eliot, who assumed the Harvard presidency in 1869, conducted a famous debate with the Princeton president, James McCosh, in the winter of 1885, in favour of diversity over uniformity. Eliot argued that a university 'while not neglecting the ancient treasures of learning has to keep a watchful eye upon the new fields of discovery, and has to invite its students to walk in new-made as well as in long-trodden paths.' Columbia University began its celebrated Great Books courses in 1921, 'which married the Great Books idea with an Aristotelian scholasticism that stressed order and hierarchy.' The problem then was to have American literature regarded as fit for inclusion in the canon. In the 1920s, for example, Lane Cooper, a professor of English at Cornell, wrote to a colleague, 'I have done my best to keep courses in American Literature from flourishing too widely,' adding that such courses 'have done harm by diverting ... attention from better literatures. ... There was no teaching of American literature as such in my day at Rutgers.'[19] Levine himself cites World War II as hastening change, allotting an important role to Alfred Kazin's *On Native Grounds* (1942), which identified the enormous body of imaginative writing and the remarkable 'experience in national self-discovery' that had characterised the depression decade and was intensified by 'the sudden emergence of America as the repository of Western culture in a world overrun by Fascism.'[20] Levine did not object to canons as such, merely to their immutability and the very tendency of immutability where canons exist at all. And he acknowledged that the American experience is different from anywhere else, America being a nation of immigrants without a national culture, however much certain scholars might pretend otherwise. This was a reference to the celebrated 'hyphenated Americans' – native American, Afro-American, Mexican-American, Italian-American. For Levine, therefore, the arguments over the canon, over history, over high as opposed to low culture, must always be sharper in the United States than elsewhere, precisely because these are arguments about identity.[21]

The most fundamental attack on the 'canon' came in 1987 from a British academic trained in Chinese studies who was a professor of government at Cornell in America. **Martin Bernal** is the son of **J. D. Bernal**, who was himself a distinguished scholar of Irish birth, a Marxist physicist who won the Lenin Peace Prize in 1953 and was author of the four-volume *Science in History*.

In the mid-1970s, aware that the Mao era in China was coming to an end, Martin Bernal began to sense that 'the central focus of danger and interest in the world' was the east Mediterranean, and he began to study Jewish history. There were, he says, 'scattered Jewish components' in his own ancestry, and an interest in his roots led him to study ancient Jewish history and the surrounding peoples. This led to an examination of early Mediterranean languages for the light they threw on prehistory, in particular the ancestry of classical Greece. His research took him ten years before it appeared in book form, but when it was published, it proved very subversive. Bernal eventually demonstrated to his own satisfaction that classical Greek culture – the very basis of the canon – did not develop of its own accord in ancient Greece around 400 BC, as traditional scholarship has it, but was actually derived from North African peoples *who were black.*

Black Athena: The Afroasiatic Roots of Classical Civilisation (1987–91) is a massive three-volume work incorporating and synthesising material in philology, archaeology, history, historiography, biblical studies, ethnic studies, sociology, and much else, and so it is not easy to do justice to Bernal's complex arguments.[22] In essence, however, he makes the following points. One is that North Africa, in the form of ancient Egypt – several of whose dynasties were black, in the sense of Negroid – was the predominant influence on classical Greece; that there were extensive trading links; that ancient Egypt was a military power in the area; that many place names in Greece show North African influence; and that the finding of objects of North African origin at classical Greek sites cannot be dismissed as casual trading exchanges. No less controversially, Bernal also claimed that this view of Greece was 'standard,' had always prevailed in European scholarship, until it was deliberately 'killed off' by 'racist' north European scholars in the early nineteenth century, men who wanted to show that Europe, and northern Europe at that, had a monopoly on creative and imaginative thought, that civilisation as we know it was born in Europe, all as one of a number of devices to help justify colonialism and imperialism.[23]

Bernal believed that there was once a people who spoke Proto-Afro-Asiatic-Indo-European, which gave rise to all the peoples and languages we see on these continents today. He believes that the break between Afro-asiatic and Indo-European came in the ninth millennium B.C. and that the spread of Afro-Asiatic was the expansion of a culture, long established in the East African Rift Valley at the end of the last ice age in the tenth and ninth millennia BC. These people domesticated cattle and food crops and hunted hippopotamus. Gradually, with the spread of the Sahara, they moved on, some down the Nile valley, some into Saudi Arabia and thus into Mesopotamia, where the first 'civilisations' arose.[24] Furthermore, civilisation, including writing, developed across a swath of Asia, stretching from India to North Africa, and was in place by 1100 BC or earlier. Bernal introduces evidence of a succession of Upper Egyptian black pharaohs sharing the name Menthope who had as their divine patron the hawk and bull god, Mntw or Mont. 'It is during the same century that the Cretan palaces were established and one finds the beginnings there of the bull-cult which appears on the walls of the palaces and was central to Greek mythology about King Minos and Crete. It would therefore seem plausible to suppose that the Cretan developments directly or indirectly reflected the rise of the Egyptian Middle Kingdom.'[25] But this is only a beginning. Bernal examined classical

Greek plays, such as Aeschylus's *The Supplicants*, for Egyptian influences; he looked at correspondences between their gods and functions; he looked at loan words, river and mountain names (Kephisos, the name of rivers and streams found all over Greece with no explanation, he derives from Kbh, 'a common Egyptian river name "Fresh"'). In a chapter on Athens, he argues that this name is derived from the Egyptian Ht Nt: 'In Antiquity, Athena was consistently identified with the Egyptian goddess Nt or Neit. Both were virgin divinities of warfare, weaving and wisdom. The cult of Neit was centred on the city of Sais in the Western delta, whose citizens felt a special affinity with the Athenians.'[26] And so on into pottery styles, military terms, and the meaning of the sphinxes.

The second half of Bernal's book follows the writings of scientists and others in the Renaissance, men like Copernicus and Giordano Bruno, to show that they accepted the Egyptian influence on Greece much more readily than later scholars. Following the French Revolution, however, Bernal discerns a reaction by Christians against the threat of the 'wisdom' of Egypt, and a rise of 'Hellenomania.' He describes a series of German, British, and French scholars, all more or less racist in outlook (anti-black and anti-Semitic), who he says deliberately played down the significance of Egypt and North Africa generally. In particular, he singles out Karl Otfried Müller, who 'used the new techniques of source criticism to discredit all the ancient references to the Egyptian colonisations, and weaken those concerning the Phoenicians.'[27] According to Bernal, Müller was anti-Semitic and denied the Phoenicians any role in the creation of ancient Greece, an approach other scholars built on in the years 1880–1945, resulting in the Greeks being given 'a semi-divine status.' In essence, Bernal says, classical studies as we know them today are a nineteenth-century creation, and false.

Bernal's book evoked a detailed response, which appeared in 1996 under the title *Black Athena Revisited*, edited by **Mary Lefkowitz** and **Guy MacLean Rogers**, both of Wellesley College.[28] Here a collection of scholars – from America, Italy, and Britain, and including Frank Snowden, a distinguished classics professor from Howard University, a black institution – concluded that Martin Bernal was wrong on almost every count, except perhaps that of causing classicists to look at themselves with a more questioning mind. In particular, they concluded that (a) ancient Egypt wasn't black; (b) its influence on classical Greece, while not nonexistent, was not predominant either; and (c) by no means all of the scholars promoting the 'Aryan' view of the past were anti-Semites or romanticists. Bernal's revised dating of certain allegedly key events in Egyptian-Greek history was based on faulty radiocarbon readings; analysis of ancient Egyptian skeletons and skulls shows that they were comprised of a variety of peoples, closest to racial types from the Sudan but not to those of West Africa, the most negroid of all. Analysis of ancient art, and ancient Greek, Roman, and other languages, shows that the Egyptians were regarded as very different from traditional 'black' groups, the *Aithiopes* or *Aesthiopes* (Ethiopians), literally 'burnt-faced peoples.'[29] Frank Snowden showed that in classical times the Ethiopians were used, by Herodotus among others, as the yardstick in blackness and in their style of 'woolly' hair. Nubians were seen as not as black as Ethiopians but blacker than the Egyptians, who were darker than the Moors. Bernal claimed that various Greek city names – Methone, Mothone, and Methana – went back to the Egyptian *mtwn*, meaning 'bull fight, bull arena.' But other scholars pointed out that *methone* means a 'theatrical-looking harbour,' and all the cities referred to by Bernal were exactly that.[30] On the matter of racism, Guy Rogers took Bernal to task for singling out George Grote as an anti-Semite, when in fact Grote was associated with the founding of University College, London, in 1829, one specific aim of which was to offer higher education to groups excluded from Oxford and Cambridge, namely Non-conformists, Catholics, and Jews.[31]

Bernal was accused of doing more harm than good, of throwing in his lot with other writers like C. A. Diop, who in *The African Origin of Civilisation* (1974) had 'falsified' history in portraying Egyptians as black, and of ignoring evidence that went against the hypothesis (for example, that mythical beasts on many Greek vases were inspired by Near Eastern motifs rather than North African ones).[32] Many scholars shared the view of Mary Lefkowitz, one of the editors of *Black Athena Revisited*, that Bernal's ideas were no more than 'Afrocentric fantasies,' and his description of the Egyptians as black 'misleading in the extreme.' 'For black Americans (many of whom now prefer to be known as African-Americans), the African origins of ancient Greek civilisation promise a myth of self-identification and self ennoblement, the kind of "noble lie" that Socrates suggests is needed for the utopian state he describes in Plato's *Republic*.'[33] The issue is not settled and perhaps cannot hope to be. For this is only partly an intellectual debate. Exploring the alleged racism behind the theorising was just as much part of Bernal's 'project' as was the substantive result.

These 'culture wars' were accompanied by 'history wars' and 'curriculum wars,' but they were all essentially the same thing: a fight between traditionalists and postmodernists.

One of the more bitter engagements arose over plans to mount an exhibition at the **National Air and Space Museum** (NASM), part of the **Smithsonian Institution** in Washington, in 1995, to mark the fiftieth anniversary of the dropping of the two atomic bombs on Hiroshima and Nagasaki, in August 1945. Among the exhibits was a reconstructed *Enola Gay*, the Boeing B-29 bomber that had actually dropped the bomb on Hiroshima.[34] After its historic mission, *Enola Gay* had a chequered history. For many years its disassembled components could be seen, by appointment only, in a suburban Maryland warehouse, so that it was for all practical purposes hidden from sight. Following representations from B-29 veterans, its restoration was eventually begun at the end of 1984, and as the anniversary of World War II approached, the possibility of displaying the plane began to rise. Even so, there were those wary of doing so in view of what *Enola Gay* represented. For many, there was nothing of unusual aeronautical interest in this B-29, merely its mission and its 'equipment.'

When the decision was taken to mark the anniversary at the NASM, the idea grew at the Smithsonian that the exhibition should be not only a celebration of a military and technical victory but an examination of the use of atomic weapons and the opening of the nuclear age. Here the problems began, for many veterans and service organisations wanted a more propagandistic approach, more of a celebration than an examination of issues. When the various service organisations saw the script for the exhibition – 300 pages of text, which became available eighteen months before the start of the show – they didn't like it. It was too 'dark.' Beginning in the pages of *Air Force Magazine*, the objections spread, taking in the media, the Pentagon, and Congress.[35] It seemed that almost everyone except the historians wanted the exhibition to be a celebration, not raising uncomfortable questions about whether the decision to drop the atomic bomb had been correct. Forty historians wrote to President Clinton soliciting his support for the exhibition as a serious piece of history, but it did no good. In January 1995 it was announced that the exhibition was cancelled and was being replaced by a much less contentious show, more celebratory in tone. At the same time, the director of the Smithsonian also resigned. The decision to cancel the exhibition was widely welcomed in certain sections of the press and in Congress, where Newt Gingrich said that 'people' were 'taking back' their history from elites.[36]

The academic world had been the focus of Allan Bloom's initial attack and defended by Stanley Fish and others. Not surprisingly, the university itself came under scrutiny in a series of surveys, in particular what was taught and how. The first of these, and the most intemperate, was **Roger Kimball**'s *Tenured Radicals: How Politics Has Corrupted Our Higher Education*, published in 1990.[37] Kimball, managing editor of the *New Criterion*, a conservative cultural and intellectual journal, had the idea of attending a number of seminars at various universities and amalgamating his account of them into a book. These conferences included 'Architecture and Education: The Past Twenty-Five Years and Assumptions for the Future,' a day-long symposium sponsored by the Princeton School of Architecture in 1988; another was a panel discussion at the Williams College convocation in 1989; and a third was the publication, in 1986, of a volume of essays taken from a conference at Stanford University, entitled *Reconstructing Individualism: Autonomy, Individuality and the Self in Western Thought*.[38] Kimball found very little to admire or like in what he saw. He thought that most of the postmodernists showed an 'eclectic' mix of left-wing idea that were a hangover of the radical 1960s, owing a great deal to Marcuse's notion of 'repressive tolerance.' He devoted a chapter to Paul de Man and to Stanley Fish, and he had great fun deriding what are admittedly some of the wilder excesses of postmodernist thought.[39] He conceded that politics *do* influence artistic judgements but denied that, in the final analysis, they determine them.

But Kimball's book was essentially an hysterical reaction, journalistic rather than considered. A more thoughtful response came from **Dinesh d'Souza**, an Indian who had emigrated to America in the late 1970s. His *Illiberal Education: The Politics of Sex and Race on Campus* appeared in 1991 and was an examination of six campuses in America – Berkeley, Stanford, Howard, Michigan, Duke, and Harvard – and how they dealt with the issues of sex and race, both in their admissions policy and in their teaching.[40] D'Souza's approach was statistical but not heavily so; he used figures where they were appropriate but also looked beyond them. At Berkeley, for instance, he quoted a confidential, internal report which showed that, after five years, only 18 percent of blacks admitted on affirmative action completed their courses, whereas 42 percent of blacks admitted to the regular program had graduated. D'Souza's response was not hysterical, however. He admitted that one could look at the figures in two ways – as a sort of success and a sort of failure. His own idea was that these very students, 'California's best black and Hispanic students,' might have fared better at other campuses, 'where they might settle in more easily, compete against evenly matched peers, and graduate in vastly greater numbers and proportions.'[41] He then looked at Stanford, where the faculty had, amid much controversy, dropped the Western civilisation course and replaced it with 'Culture, Ideas, Values' (CIV), which was intended to stress other values, ideas, and cultures besides the Western. He gave a list of the kind of works to be included in the 'Europe and Americas' track.

Poets:	José Maria Arguedas, Pablo Neruda, Ernesto Cardenal, Audre Lorde, Aimé Césaire
Drama:	Shakespeare, Euripedes
Fiction:	García Márquez, Naipaul, Melville, Hurston, Findley, Rulfo, Ferre
Philosophy:	Aristotle, Rousseau, Weber, Freud, Marx, Fanon, Retamar, Benedict
History:	James, Guaman Poma
Diaries:	Columbus, Cabeza de Vaca, Equiano, Lady Nugent, Dyuk, Augustine, Menchu, Barrios de Chungara
Culture:	Films on popular religion and healing in Peru ('Eduardo the Healer') and the US ('The Holy Ghost People')
Music:	Reggae lyrics, Rastafarian poetry, Andean music

D'Souza emphasised that this list was not mandated: 'Stanford professors are given flexibility as long as they ensure "substantial representation" for the Third World.'[42] Yet he was very critical of the way Shakespeare was taught, as primarily a function of 'colonial, racial and gender forces,' and he singled out *I, Rigoberta Menchu*, subtitled 'An Indian Woman in Guatemala,' as a typical new text, which was dictated to someone else because Rigoberta did not write. The book conveys much mundane information, especially about her family life, but spliced in among all this is her political awakening. D'Souza evinces scepticism as to how typical, or moving, or aesthetic, the book is; Rigoberta is said to speak for all native Americans, but among her experiences she goes to Paris to attend international conferences. (Later, in 1998, it emerged that Rigoberta Menchu had made up many of the experiences she reports in the book.)

D'Souza also took on Stanley Fish and Martin Bernal and quoted distinguished scholars, from David Riesman to E. O. Wilson and Willard van

Orman Quine, who said they were distressed by the trends in American higher education.[43] D'Souza's final point was that when one puts together the dismal results from affirmative action alongside the new courses on third-world cultures and ideas, there is a major risk of replacing an old form of racism with a new one. 'In one sense, the new racism is different, however. The old racism was based on prejudice, whereas the new racism is based on conclusions. ... The new bigotry is not derived from ignorance, but from experience. It is harbored not by ignoramuses, but by students who have direct and first-hand experience with minorities in the close proximity of university settings. The "new racists" do not believe they have anything to learn about minorities; quite the contrary, they believe they are the only ones willing to face the truth about them ... they are not uncomfortable about their views. ... They feel they occupy the high ground, while everyone else is performing pirouettes and somersaults to avoid the obvious.'[44]

Not everyone found American campus life so bleak. **Martha Nussbaum**, Ernst Freund Professor of Law and Ethics at the University of Chicago, has taught all over the continent. Her book *Cultivating Humanity* appeared in 1997 and concerned not six but fifteen 'core institutions,' chosen to represent different types of higher-education outfit – the Ivy League elite, large state universities, small liberal arts colleges, religious universities like Notre Dame, Brandeis, and Brigham Young.[45] Approaching her task as a classicist – the subtitle of her book was 'A Classical Defence of Reform in Liberal Education' – she argued that even ancient Athens, the crucial point of reference for conservative critics of multiculturalism, was more open to alternative views than these critics like to acknowledge. Nussbaum's model was drawn from Socrates and the Stoics, who, she said, established three 'core values' of liberal education – critical self-examination, the ideal of the world citizen, and the development of the narrative imagination.[46]

Nussbaum's message, from a greater number of campuses than anyone else had looked at, is that the number of extremists in universities is much less than anyone thinks, that there is a great appetite for, and interest in, philosophy, other cultures, and other lifestyles, that these courses are growing because they are popular among students rather than because a left-wing faculty is forcing them on pupils, and that when they are taught, they are taught far more often than not with a commendable academic rigour. There are, Nussbaum says, many ways for the imaginative teacher to bring home to students the relevance of the classics and philosophy: for example, in one class in Harvard the students are asked, Would Socrates have been a draft resister? She argues that Athens took seriously the idea of world citizenship and quotes Herodotus considering the possibility that Egypt and Persia might have something to teach Athens about social values.[47] She finds it not at all odd that Amartya Sen teaches a course at Harvard called 'Hunger and Famine,' in which standard ideas about economics are given a new twist. She finds that the tragic form in the narrative imagination is especially powerful in crossing cultural boundaries – its universality and abstractness especially useful in drawing people together.[48] She notes that, again, in ancient Athens the moral and the political went hand-in-hand, and asks if it is really possible to read George Eliot or Dickens without detachment and get from them all that there is. She too invokes Lionel Trilling and *The Liberal Imagination*, drawing from it the lesson that 'the novel as genre is committed to liberalism in its very form, in the way in which it shows respect for the individuality and the privacy of each human mind.'[49] The study of non-Western cultures, she says, is there to help combat what she calls 'the descriptive vices' – chauvinism and romanticism – and 'the normative vices' – chauvinism (again), arcadianism, and scepticism. She shows that many in the West have traditionally overestimated the extent to which Western culture is individualistic and Eastern culture is the opposite, and spends some time showing how individualistic non-Western societies can be. She applies the same approach to courses in African–American studies and women's studies (she argues, for instance, that the sociobiologists base their theories in part on chimpanzees, but never on Bonobos, another primate, not discovered till 1929, whose 'graceful and non-aggressive' style differs sharply from that of the chimp). She found Notre Dame University (Catholic) much more open to matters that in theory ought to have been an intellectual threat than, say, Brigham Young (Mormon), and as a result the former was still changing, still popular, while the latter languished.[50] In other words, Nussbaum is saying that once you go out and investigate the campuses, what is actually happening is much less sensational, much less worrying, much more worthwhile, than appears to be the case from the headlines. She was not the first person to find that evidence is a healthy counterweight to prejudice; that, after all, is what distinguishes scholarship proper from mere journalism.

The most original response to the culture wars was **David Denby**'s excellent *Great Books*, published in 1996. Denby, film critic of *New York* magazine and a contributing editor to the *New Yorker*, attended Columbia University in 1961, when he took two foundation courses, 'Literature Humanities' and 'Contemporary Civilization.'[51] In the autumn of 1991, he had the idea of sending himself back to Columbia to do the same courses. He wanted to see how they had changed, how they were now taught, and what effects they had on himself and the young freshmen attending Columbia in the 1990s. He had been a film critic since 1969, he said, and though he still loved his job, he was tired of the 'society of the spectacle,' the secondhand, continuously ironic world of the media: 'The media give information, but information, in the 1990s, has become transitory and unstable. Once in place, it immediately gets pulled apart. . . . No one's information is ever quite adequate, which is one reason among many that Americans now seem half-mad with anxiety and restlessness. Like many others, I was jaded yet still hungry; I was cast into the modern state of living-in-the-media, a state of excitement needled with disgust.'[52] Denby takes us through the great books he liked (Homer, Plato, Virgil, the Bible, Dante, Rousseau, Shakespeare, Hume and Mill, Marx, Conrad, de Beauvoir, Woolf), leaving out what didn't engage him (Galileo, Goethe, Darwin, Freud, Arendt, Habermas). His book is notable for some fine passages describing his own reactions to the Great Books, for the way he occasionally related them to movies, and for the way he fears for his son, Max, overwhelmed by tawdry and trivial media, against which these older voices cannot compete. He notes that minority students sometimes rebel against the 'White, European' nature of the books, but such rebellion, when it occurs, is heavily tinged with embarrassment and sorrow as much as with anger. And this was his main point, in conclusion: that students, whether white, black, Latino, or Asian, 'rarely arrive at college as habitual readers,' that few of them have more than a nominal connection with the past: 'The vast majority of white students do not know the intellectual tradition that is allegedly theirs any better than black or brown ones do.' The worlds of Homer, Dante, Boccaccio, Rousseau, and Marx are now so strange, so different, that he came to a surprising conclusion: 'The core-curriculum courses jar so many student habits, violate so many contemporary pieties, and challenge so many forms of laziness that so far from serving a reactionary function, they are actually the most radical courses in the undergraduate curriculum.'[53] Denby found that in fact the Great Books he (re)studied were capable of individual and idiosyncratic interpretation, not necessarily the interpretation the cultural right would wish, but that didn't matter – the students grasped that 'they dramatise the utmost any of us is capable of in love, suffering and knowledge.' And, perhaps the best thing one can say about it, the Western canon can be used to attack the Western canon. 'What [non-whites] absorb of the older "white" culture they will remake as their own; it cannot hurt them.'[54]

For Denby, a much greater danger came from the media. 'Most high schools can't begin to compete against a torrent of imagery and sound that makes every moment but the present seem quaint, bloodless, or dead.'[55] In fact, he said, the modern world has turned itself upside down. On his first time round, in 1961, the immediacy of pop had been liberating, a wonderful antidote to the stifling classroom; but now 'the movies have declined; *pop* has become a field of conformity and complacency, while the traditional high culture, by means of its very strangeness and difficulty, strikes students as odd. They may even be shocked by it. . . . The [great] books are less a conquering army than a kingdom of untameable beasts, at war with one another and with readers.'[56]

In 1999 Harold Bloom returned to his first love. In *Shakespeare: The Invention of the Human*, Bloom argued that the great poet 'invented us,' that 'personality, in our sense, is a Shakespearean invention.'[57] Before Shakespeare, Bloom claims, characters did not grow and develop. 'In Shakespeare, characters develop rather than unfold, and they develop because they reconceive themselves. Sometimes this comes about because they *overhear* themselves talking, whether to themselves or to others. Self-overhearing is the royal road to individuation.'[58] Bloom's book is deeply unfashionable, not only in its message but in the way it is written. It is an act of worship. He freely concedes that Bardolatry is and has been 'a secular religion' for some two hundred years, and he enjoys being in that tradition because he believes that the very successes of Shakespeare transcend all ways of approaching him: he is simply too brilliant, too intelligent, to be cut down to size, as the feminists, cultural materialists, and Marxists would like to do. 'Shakespeare, through Hamlet, has made us skeptics in our relationships with anyone, because we have learned to doubt articulateness in the realm of affection. . . . Our ability to laugh at ourselves as readily as we do at others owes much to Falstaff. . . . Cleopatra [is the character] through whom the playwright taught us how complex eros is, and how impossible it is to divorce acting the part of being in love and the reality of being in love. . . . Mutability is incessant in her passional existence, and it excludes sincerity as being irrelevant to eros.'[59] 'When we are wholly human, and know ourselves, we become most like either Hamlet or Falstaff.'[60]

There is something magnificent about this 'Bloom in love,' dismissing his critics and opponents without even naming them. It is all very unscientific, but that is Bloom's point: this is what art should seek to emulate, these are the feelings great art exists for. Individuation may have been one of the great issues of the century, but Shakespeare got there first, and has still not been equalled. He is the one man worth worshipping, and we are, if we will only see it, surrounded by his works.

One more distinguished combatant joined the Blooms on the barricades, an academic Boadicea whose broadsides went wider even than theirs: **Gertrude Himmelfarb**, the historian wife of Irving Kristol, founder (with Daniel Bell) of the *Public Interest*. In *On Looking into the Abyss* (1994), Himmelfarb, professor emeritus of history at the Graduate School of the City University of New York, attacked postmodernism in whatever guise it raised its head, from literary theory to philosophy to history.[61] Her argument against literary theory was that the theory itself had displaced literature as the object of study and in the process taken away the 'profound spiritual and emotional' experience that comes with reading great works, the 'dread beasts' as she put it, 'lurking at the bottom of the "Abyss." '[62] As a result, she said, 'The beasts of modernism have mutated into the beasts of postmodernism – relativism into nihilism, amorality into immorality, irrationality into insanity, sexual deviancy into polymorphous perversity.'[63] She loathed the 'boa-deconstructors' like Derrida and Paul de Man and what they had done to literature, thinking their aim more political than literary (they would have agreed). She attacked the *Annales* school: she admired Fernand Braudel's fortitude in producing his first great book in a concentration camp, from memory, but thought his concept of *la longue durée* gave him a fatally skewed perspective on such events as, say, the Holocaust. She thought that the new enemy of liberalism had become – well, liberalism itself. Liberalism was now so liberal, she argued, that it absolved postmodern historians, as they saw it, from any duty to the truth. 'Postmodernists deny not only absolute truth but contingent, partial, incremental truth. . . . In the jargon of the school, truth is "totalising," "hegemonic," "logocentric," "phallocentric," "autocratic," "tyrannical." '[64] She turned on Richard Rorty for arguing there is no 'essential' truth or reality, and on Stanley Fish for arguing that the demise of objectivity 'relieves me of the obligation to be right.'[65] But her chief point was that 'postmodernism entices us with the siren call of liberation and creativity,' whereas there is a tendency for 'absolute liberty to subvert the very liberty it seeks to preserve.'[66] In particular, and dangerously, she saw about her a tendency to downplay the importance and horror of the Holocaust, to argue that it was something 'structural,' rather than a personal horror for which real individuals were responsible, which need not have happened, and which needed to be understood, and reunderstood by every generation. She tellingly quotes the dedication in David Abraham's book *The Collapse of the Weimar Republic*, published in 1981, which contained the dedication, 'For my parents – who at Auschwitz and elsewhere suffered the worst consequences of what I can merely write about.' In Himmelfarb's view, the reader is invited to think that the author's parents perished in the camps, but they did not. This curious phraseology was later examined by Natalie Zemon Davis, an historian, who concluded that Abraham's work had been designed to show that the Holocaust was not the work of devils 'but of historical forces and actors.'[67] This was too much for Himmelfarb, a relativising of evil that was beyond reason. It epitomised the postmodern predicament: the perfect example of where too much liberty has brought us.

There is a sense in which the culture wars are a kind of background radiation left over from the Big Bang of the Russian Revolution. At exactly the time that political Marxism was being dismantled, along with the Berlin Wall, postmodernism achieved its greatest triumphs. For the time being at least, the advocates of local knowledge have the edge. Gertrude Himmelfarb's warning, however timely, and however sympathetic one finds it, is rather like trying to put a genie back into a bottle.

DEEP ORDER

In 1986 Dan Lynch, an ex-student from UCLA, started a trade fair for computer hardware and software, known as Interop. Until then the number of people linked together via computer networks was limited to a few hundred 'hard-core' scientists and academics. Between 1988 and 1989, however, Interop took off: hitherto a fair for specialists, it was from then on attended by many more people, all of whom suddenly seemed to realise that this new way of communicating – via remote computer terminals that gave access to very many databases, situated across the world and known as the Internet – was a phenomenon that promised intellectual satisfaction and commercial rewards in more or less equal measure. **Vint Cerf**, a self-confessed 'nerd', from California, a man who set aside several days each year to re-read *The Lord of the Rings*, and one of a handful of people who could be called a father of the Internet, visited Lynch's fair, and he certainly noticed a huge change. Until that point the **Internet** had been, at some level, an experiment. No more.[1]

Different people place the origins of the Internet at different times. The earliest accounts put it in the mind of **Vannevar Bush**, as long ago as 1945. Bush, the man who had played such a prominent role in the building of the atomic bomb, envisaged a machine that would allow the entire compendium of human knowledge to be 'accessed'. But it was not until the Russians surprised the world with the launch of the *Sputnik* in October 1957 that the first faltering steps were taken toward the Net as we now know it. The launch of a satellite, as was discussed in chapter 27, raised the spectre of associated technologies: in order to put such an object in space, Russia had developed rockets capable of reaching America with sufficient accuracy to do huge damage if fitted with nuclear warheads. This realisation galvanised America, and among the research projects introduced as a result of this change in the rules of engagement was one designed to explore how the United States' command and control system – military and political – could be dispersed around the country, so that should she be attacked in one area, America would still be able to function elsewhere. Several new agencies were set up to consider different aspects of the situation, including the National Aeronautics and Space Administration (NASA) and the Advanced Research Projects Agency, or ARPA.[2] It was this outfit which was charged with investigating the safety of command and control structures after a nuclear strike. ARPA was given a staff of about seventy, an appropriation of $520 million, and a budget plan of $2 billion.[3]

At that stage computers were no longer new, but they were still huge and expensive (one at Harvard at the time was fifty feet long and eight feet high). Among the specialists recruited by ARPA was **Joseph Licklider**, a tall, laconic psychologist from Missouri, who in 1960 had published a paper on 'man-computer symbiosis' in which he looked forward to an integrated arrangement of computers, which he named, ironically, an 'intergalactic network.' That was some way off. The first breakthrough came in the early 1960s, with the idea of 'packet-switching,' developed by **Paul Baran**.[4] An immigrant from Poland, Baran took his idea from the brain, which can sometimes recover from disease by switching the messages it sends to new routes. Baran's idea was to divide a message into smaller packets and then send them by different routes to their destination. This, he found, could not only speed up transmission but avoid the total loss of information where one line is faulty. In this way technology was conceived that reassembled the message packets when they arrived, and tested the network for the quickest routes. This same idea occurred almost simultaneously to **Donald Davies**, working at the National Physical Laboratory in Britain – in fact, *packet-switching* was his term. The new hardware was accompanied by new software, a brand-new branch of mathematics known as queuing theory, designed to prevent the buildup of packets at intermediate nodes by finding the most suitable alternatives.[5]

In 1968 the first 'network' was set up, consisting of just four sites: UCLA, Stanford Research Institute (SRI), the University of Utah, and the University of California at Santa Barbara.[6] The technological breakthrough that enabled this to proceed was the conception of the so-called interface message processor, or IMP, whose task it was to send bits of information to a specified location. In other words, instead of 'host' computers being interconnected, the IMPs would be instead, and each IMP would be connected to a host.[7] The computers might be different pieces of hardware, using different software, but the IMPs spoke a common language and could recognise destinations. The contract to construct the IMPs was given by ARPA to a small consulting firm in Cambridge, Massachusetts, called Bolt Beranek and Newman (BBN) and they delivered the first processor in September 1969, at UCLA, and the second in October, at SRI. It was now possible, for the first time, for two disparate computers to 'talk' to each other. Four nodes were up and running by January 1970, all on the West Coast of America. The first on the East Coast, at BBN's own headquarters, was installed in March. The ARPANET, as it came to be called, now crossed the continent.[8] By the end of 1970 there were fifteen nodes, all at universities or think tanks.

By the end of 1972 there were three cross-country lines in operation and clusters of IMPs in four geographic areas – Boston, Washington D.C., San Francisco and Los Angeles – with, in all, more than forty nodes. By now ARPANET was usually known as just the Net, and although its role was still strictly defence-oriented, more informal uses had also been found: chess games, quizzes, the Associated Press wire service. It wasn't far from there to personal messages, and one day in 1972, **e-mail** was born when **Ray Tomlinson**, an engineer at BBN, devised a program for computer addresses, the most salient feature of which was a device to separate the name of the user from the machine the user was on. Tomlinson needed a character that could never be found in any user's name and, looking at the keyboard, he happened upon the '@' sign.[9] It was perfect: it meant 'at' and had no other use. This development was so natural that the practice just took off among the ARPANET community. A 1973 survey showed that there were 50 IMPs on the Net and that three-quarters of all traffic was e-mail.

By 1975 the Net community had grown to more than a thousand, but the next real breakthrough was Vint Cerf's idea, as he sat in the lobby of a San Francisco hotel, waiting for a conference to begin. By then, ARPANET was no longer the only computer network: other countries had their own nets, and other scientific-commercial groups in America had begun theirs. Cerf began to consider joining them all together, via a series of what he referred to as gateways, to create what some people called the Catenet, for Concatenated Network, and what others called the Internet.[10] This required not more machinery but design of TCPs, or transmission-control protocols, a universal language. In October 1977 Cerf and his colleagues demonstrated the first system to give access to more than one network. The Internet as we now know it was born.

Growth of the Net soon accelerated. It was no longer purely a defence exercise, but, in 1979, it was still largely confined to (about 120) universities and other academic/scientific institutions. The main initiatives, therefore, were now taken over from ARPA by the National Science Foundation, which set up the Computer Science Research Network, or CSNET, and in 1985 created a 'backbone' of five supercomputer centres scattered around the United States, and a dozen or so regional networks.[11] These supercomputers were both the brains and the batteries of the network, a massive reservoir of memory designed to soak up all the information users could throw at it and prevent gridlock. Universities paid $20,000 to $50,000 a year in connection charges. More and more people could now see the potential of the Internet, and in January 1986 a grand summit was held on the West Coast and order put into the e-mail, to create seven domains or '**Frodos**.' These were universities (edu), government (gov), companies (com), military (mil), nonprofit organisations (org), network service providers (net), and international treaty entities (int). It was this new order that, as much as anything, helped the phenomenal growth of the Internet between 1988 and 1989, and which was seen at Dan Lynch's Interop. The final twist came in 1990 when the **World Wide Web** was created by researchers at CERN, the European Laboratory for Particle Physics near Geneva.[12] This used a special protocol, HTTP, devised by **Tim Berners-Lee**, and made the Internet much easier to browse, or navigate. Mosaic, the first truly popular browser, devised at the University of Illinois, followed in 1993. It is only since then that the Internet has been commercially available and easy to use.

The Internet has its critics, such as Brian Winston, who in his 1998 history of media technology warns that 'the Internet represents the final disastrous application of the concept of commodification of information in the second half of the twentieth century.'[13] But few now doubt that the Internet *is* a new way of communicating, or that soon a new psychology will emerge from relationships forged in 'cyberspace.'[14]

In years to come, 1988 may be revealed as a turning point so far as science is concerned. Not only did the Internet and the Human Genome Organisation get under way, bringing about the ultramodern world and setting the shape of the twenty-first century, but a book appeared that had the most commercially successful publishing history of any work of science ever printed. It set the seal on the popular acceptance of science but, as we shall see in the epilogue, in some ways marked its apogee.

A Brief History of Time from the Big Bang to Black Holes, by the Cambridge cosmologist **Stephen Hawking**, had been five years in the making and in some senses was just as much the work of **Peter Guzzardi**, a New York editor with Bantam Books.[15] It was Guzzardi who had persuaded Hawking to leave Cambridge University Press. CUP had been planning to publish Hawking's book, because they had published his others, and had offered an advance of £10,000 – their biggest ever. But Guzzardi tempted Hawking to Bantam, though it perhaps wasn't too difficult a choice for the scientist, since the firm's editorial board had been won over by Guzzardi's enthusiasm, to the point of offering a $250,000 advance. In the intervening years, Guzzardi had worked hard to make Hawking's dense prose ever more accessible for a general audience.[16] The book was released in early spring 1988 – and what happened then quickly passed into publishing history. More than half a million hardback copies of the book were sold in both the United States and Britain, where the title went through twenty reprints by 1991 and remained in the best-seller lists for no fewer than 234 weeks, four and a half years. The book was an almost equally great success in Italy, Germany, Japan, and a host of other countries across the world, and Hawking quickly became the world's most famous scientist. He was given his own television series, made cameo appearances in Hollywood films, and his public lectures filled theatres the size of the Albert Hall in London.[17]

There was one other unusual element in this story of success. In 1988 Hawking was aged forty-six, but since 1963, when he was twenty-one, he had been diagnosed as suffering from amyotrophic lateral sclerosis, ALS, also known (in the U.K.) as motor neurone disease and (in the United States) as Lou Gehrig's disease, after the Yankee baseball player who died from it.[18] What had begun as mere clumsiness at the end of 1962 had progressed over the intervening years so that by 1988 Hawking was confined to a wheelchair and able to communicate only by means of a special computer connected to a voice synthesiser. Despite these handicaps, in 1979 he had been appointed Lucasian Professor of Mathematics at Cambridge, a post that Isaac Newton had held before him, he had won the Einstein medal, and he had published a number of well-received academic books on gravity, relativity, and the structure of the universe. As Hawking's biographers say, we shall never know to what extent

Stephen Hawking's considerable disability contributed to the popularity of his ideas, but there was something triumphant, even moving, in the way he overcame his handicap (in the late 1960s he had been given two years to live). He has never allowed his disability to deflect him from what he knows are science's central intellectual concerns. These involve black holes, the concept of a 'singularity,' and the light they throw on the Big Bang; the possibility of multiple universes; and new ideas about gravity and the fabric of reality, in particular 'string theory.'

It is with black holes that Hawking's name is most indelibly linked. This idea, as mentioned earlier, was first broached in the 1960s. Black holes were envisaged as superdense objects, the result of a certain type of stellar evolution in which a large body collapses in on itself under the force of gravity to the point where nothing, not even light, can escape. The discovery of pulsars, quasars, neutron stars, and background radiation in the 1960s considerably broadened our understanding of this process, besides making it real, rather than theoretical. Working with Roger Penrose, another brilliant physicist then at Birkbeck College in London, this pair first argued that at the centre of every black hole, as at the beginning of the universe, there must be a 'singularity,' a moment when matter is infinitely dense, infinitely small, and when the laws of physics as we know them break down. Hawking added to this the revolutionary idea that black holes could emit radiation (this became known as Hawking radiation) and, under certain conditions, explode.[19] He also believes that, just as radio stars had been discovered in the 1960s, thanks to new radio-telescopes, so X rays should also be detectable from space via satellites above the atmosphere, which otherwise screened out such rays. Hawking's reasoning was based on calculations that showed that as matter was sucked into a black hole, it would get hot enough to emit X rays. Sure enough, four X-ray sources were subsequently identified in a survey of the heavens and so became the first candidates for observable black holes. Hawking's later calculations showed that, contrary to his first ideas, black holes did not remain stable but lost energy, in the form of gravity, shrank, and eventually, after billions of years, exploded, possibly accounting for occasional and otherwise unexplained bursts of energy in the universe.[20]

In the 1970s Hawking was invited to Caltech, where he met and conferred with the charismatic **Richard Feynman**.[21] Feynman was an authority on quantum theory, and Hawking used this encounter to develop an explanation of how the universe began.[22] It was a theory he unveiled in 1981 in, of all places, the Vatican. The source of Hawking's theory was an attempt to conceive what would happen when a black hole shrank to the point where it disappeared, the troublesome fact being that, according to quantum theory, the smallest theoretical length is the Planck length, derived from the Planck constant, and equal to 10^{-35} metres. Once something reaches this size (and though it is very small, it is not zero), it cannot shrink further but can only disappear entirely. Similarly, the Planck time is, on the same basis, 10^{-43} of a second, so that when the universe came into existence, it could not do so in less time than this.[23] Hawking resolved this anomaly by a process that can best be explained by an analogy. Hawking asks us to accept, as Einstein said, that space-time is curved, like the skin of a balloon, say, or the surface of the earth. Remember that these are analogies only; using another, Hawking said that the size of the universe at its birth was like a small circle drawn around, say, the North Pole. As the universe – the circle – expands, it is as if the lines of latitude are expanding around the earth, until they reach the equator, and then they begin to shrink, until they reach the South Pole in the 'Big Crunch.' But, and this is where the analogy still holds in a useful way, at the South Pole, wherever you go you must travel north: the geometry dictates that it cannot be otherwise. Hawking asks us to accept that at the birth of the universe an analogous process occurred – just as there is no meaning for *south* at the South Pole, so there is no meaning for *before* at the singularity of the universe: time can only go forward.

Hawking's theory was an attempt to explain what happened 'before' the Big Bang. Among the other things that troubled physicists about the Big Bang theory was that the universe as we know it appears much the same in all directions.[24] Why this exquisite symmetry? Most explosions do not show such perfect balance – what made the 'singularity' different? **Alan Guth**, of MIT, and **Andrei Linde**, a Russian physicist who emigrated to the United States in 1990, argued that at the beginning of time – i.e., $T = 10^{-43}$ seconds, when the cosmos was smaller even than a proton – gravity was briefly a *repulsive* force, rather than an attractive one. Because of this, they said, the universe passed through a very rapid inflationary period, until it was about the size of a grapefruit, when it settled down to the expansion rate we see (and can measure) today. The point of this theory (some critics call it an 'invention') is that it is the most parsimonious explanation required to show why the universe is so uniform: the rapid inflation would have blown out any wrinkles. It also explains why the universe is not completely homogeneous: there are chunks of matter, which form galaxies and stars and planets, and other forms of radiation, which form gases. Linde went on to theorise that our universe is not the only one spawned by inflation.[25] There is, he contends, a 'megaverse,' with many universes of different sizes, and this was something that Hawking also explored. Baby universes are, in effect, black holes, bubbles in space-time. Going back to the analogy of the balloon, imagine a blister on the skin of the balloon, marked off by a narrow isthmus, equivalent to a singularity. None of us can pass through the isthmus, and none of us is aware of the blister, which can be as big as the balloon, or bigger. In fact, any number may exist – they are a function of the curvature of space-time and of the physics of black holes. By definition we can never experience them directly: they have no meaning.

That phrase, 'no meaning,' introduces the latest phase of thinking in physics. Some critics call it 'ironic science,' speculation as much as experimentation, where there is no real evidence for the (often) outlandish ideas being put forward.[26] But that is not quite fair. Much of the speculation is prompted – and supported – by mathematical calculations that point toward solutions where words, visual images and analogies all break down. Throughout the twentieth century physicists have come up with ideas that have only found experimental support much later, so perhaps there is nothing very new here. At the moment, we are living at an in-between time, and have no way of knowing whether many of the ideas current in physics will endure and be supported by experiment. But it seems unlikely that some ever will be.

Another theory of scientists like Hawking is that 'in principle' the original black hole and all subsequent universes are actually linked by what are variously known as 'wormholes' or 'cosmic string.'[27] Wormholes, as conceived, are minuscule tubes that link different parts of the universe, including black holes, and therefore in theory can act as links to other universes. They are so narrow, however (a single Planck length in diameter), that nothing could ever pass through them, without the help of cosmic string – which, it should be stressed, is an entirely theoretical form of matter, regarded as a relic of the original Big Bang. Cosmic string also stretches across the universe in very thin (but very dense) strips and operates 'exotically.' What this means is that when it is squeezed, it expands, and when it is stretched, it contracts. In theory at least, therefore, cosmic string could hold wormholes open. This, again in theory, makes time travel possible, in some future civilisation. That's what some physicists say; others are sceptical.

Martin Rees's 'anthropic principle' of the universe is somewhat easier to grasp. Rees, the British astronomer royal and another contemporary of Hawking, offers indirect evidence for 'parallel universes.' His argument is that for ourselves to exist, a very great number of coincidences must have occurred, if there is only one universe. In an early paper, he showed that if just one aspect of our laws of physics were to be changed – say, gravity was increased – the universe as we know it would be very different: celestial bodies would be smaller, cooler, would have shorter lifetimes, a very different surface geography, and much else. One consequence is that life as we know it can in all probability only form in universes with the sort of physical laws we enjoy. This means, first, that other forms of life are likely elsewhere in the universe (because the same physical laws apply), but it also means that many other universes probably exist, with other physical laws, in which *very* different forms of life, or no forms of life, exist. Rees argues that *we* can observe our universe, and conjecture others, because the physical laws exist about us to allow it. He insists that this is too much of a coincidence: other universes, very different from ours, almost certainly must exist.[28]

Like most senior physicists, cosmologists, and mathematicians, Hawking has also devoted much energy to what some scientists call 'the whole shebang,' the so-called **Theory of Everything**. This too is an ironic phrase, referring to the attempt to describe all of fundamental physics by one set of equations: nothing more. Physicists have been saying this 'final solution' is just around the corner for more than a decade, but in fact the theory of everything is still elusive.[29] To begin with, before the physics revolution discussed in earlier chapters of this book, two theories were required. As Steven Weinberg tells the story, there was Isaac Newton's theory of gravity, 'intended to explain the movements of the celestial bodies and how such things as apples fall to the ground; and there was James Clerk Maxwell's account of electromagnetism as a way to explain light, radiation, magnetism, and the forces that operate between electrically charged particles.' However, these two theories were compatible only up to a point: according to Maxwell, the speed of light was the same for all observers, whereas Newton's theories predicted that the speed measured for light would depend on the motion of the observer. 'Einstein's general theory of relativity overcame this problem, showing that Maxwell was right.' But it was the quantum revolution that changed everything and made physics more beautiful but more complex at the same time. This linked Maxwell's theory and new quantum rules, which viewed the universe as discontinuous, with a limit on how small the packets of electromagnetic energy can be, and how small a unit of time or distance is. At the same time, this introduced two new forces, both operating at very short range, *within* the nucleus of the atom. The strong force holds the particles of the nucleus together and is very strong (it is this energy that is released in a nuclear weapon). The other is known as the weak force, which is responsible for radioactive decay.

And so, until the 1960s there were four forces that needed to be reconciled: gravity, electromagnetism, the strong nuclear force, and the weak radioactive force. In the 1960s a set of equations was devised by **Sheldon Glashow**, and built on by **Abdus Salam** and Steven Weinberg, at Texas, which described both the weak force and electromagnetism and posited three new particles, W^+, W^- and Z°.[30] These were experimentally observed at CERN in Geneva in 1983. Later on, physicists developed a series of equations to describe the strong force: this was related to the discovery of quarks. Having been given rather whimsical names, including those of colours (though of course particles don't have colours), the new theory accounting for how quarks interact became known as **quantum chromodynamics**, or QCD. Therefore electromagnetism, the weak force, and the strong force have all been joined together into one set of equations. This is a remarkable achievement, but it still leaves out gravity, and it is the incorporation of gravity into this overall scheme that would mark, for physicists, the so-called Theory of Everything.

At first they moved toward a quantum theory of gravity. That is to say, physicists theorised the existence of one or more particles that account for the force and gave the name 'graviton' to the gravity particle, though the new theories presuppose that many more than one such particle exists. (Some physicists predict 8, others 154, which gives an idea of the task that still lies ahead.) But then, in the mid-1980s, physics was overtaken by the 'string revolution' and, in 1995, by a second '**superstring revolution**.' In an uncanny replay of the excitement that gripped physics at the turn of the twentieth century, a whole new area of inquiry blossomed into view as the twenty-first century approached.[31] By 1990 the shelves of major bookstores in the developed world were filled with more popular science books than ever before. And there were just as many physics, cosmology, and mathematics volumes as there were evolution and other biology titles. As part of this phenomenon, in 1999 a physics and mathematics professor who held joint appointments at Cornell and Columbia Universities entered the best-seller lists on both sides of the Atlantic with a book that was every bit as difficult as *A Brief History of Time*, if not more so. *The Elegant Universe: Superstrings, Hidden Dimensions and the Quest for the*

Ultimate Theory, by **Brian Greene**, described the latest excitements in physics, working hard to render very difficult concepts accessible (Greene, not to put his readers off, called these difficult subjects 'subtle').[32] He introduced a whole new set of physicists to join the pantheon that includes Einstein, Ernest Rutherford, Niels Bohr, Werner Heisenberg, Erwin Schrödinger, Wolfgang Pauli, James Chadwick, Roger Penrose, and Stephen Hawking. Among these new names Edward Witten stands out, together with Eugenio Calabi, Theodor Kaluza, Andrew Strominger, Stein Strømme, Cumrun Vafa, Gabriele Veneziano, and Shing-Tung Yau, about as international a group of names as you could find anywhere.

The string revolution came about because of a fundamental paradox. Although each was successful on its own account, the theory of general relativity, explaining the large-scale structure of the universe, and quantum mechanics, explaining the minuscule subatomic scale, were mutually incompatible. Physicists could not believe that nature would allow such a state of affairs – one set of laws for large things, another for small things – and for some time they had been seeking ways to reconcile this incompatibility, which many felt was not unrelated to their failure to explain gravity. There were other fundamental questions, too, which the string theorists faced up to: Why are there four fundamental forces?[33] Why are there the number of particles that there are, and why do they have the properties they do? The answer that string theorists propose is that the basic constituent of matter is not, in fact, a series of particles – point-shaped entities – but very tiny, one-dimensional strings, as often as not formed into loops. These strings are very small – about 10^{-33} of a centimetre – which means that they are beyond the scope of direct observation of current measuring instruments. Notwithstanding that, according to string theory an electron is a string vibrating one way, an up quark is a string vibrating another way, and a tau particle is a string vibrating in a third way, and so on, just as the strings on a violin vibrate in different ways so as to produce different notes. As the figures show, we are dealing here with very small entities indeed – about a hundred billion billion (10^{20}) times smaller than an atomic nucleus. But, say the string theorists, at this level it is possible to reconcile relativity and quantum theory. As a by-product and a bonus, they also say that a gravity particle – the graviton – emerges naturally from the calculations.

String theory first emerged in 1968–70, when **Gabriele Veneziano**, at CERN, noticed that a mathematical formula first worked out 200 years ago accidentally seemed to explain various aspects of particle physics.[34] Then three other physicists, Yoichiro Nambu, Holger Nielson and Leonard Susskind, showed that these mathematics could be better understood if particles were not point-shaped objects but small strings that vibrated. The approach was discarded later, however, after it failed to explain the strong force. But the idea refused to die, and the first string revolution, as it came to be called, took off in 1984, after a landmark paper by Michael Greene and John Schwarz first showed that relativity and quantum theory could be reconciled by string theory. This breakthrough stimulated an avalanche of research, and in the next two years more than a thousand papers on string theory were published, together showing that many of the main features of particles physics emerge naturally from the theory. This fecundity of string theory, however, brought its own problems. For a while there were in fact five string theories, all equally elegant, but no one could tell which was the 'real' one. Once more string theory stalled, until the 'Strings 1995' conference, held in March at the University of Southern California, where **Edward Witten** introduced the '**second superstring revolution.**'[35] Witten was able to convince his colleagues that the five apparently different theories were in fact five aspects of the same underlying concept, which then became known as M-theory, the *M* standing variously for *mystery, meta,* or 'mother of all theories.'⋆

In dealing with such tiny entities as strings, possibilities arise that physicists had not earlier entertained, one being that there may be 'hidden dimensions' and to explain this another analogy is needed. Start with the idea that particles are seen as particles only because our instruments are too blunt to see that small. To use Greene's own example, think of a hosepipe seen from a distance. It *looks* like a filament in one dimension, like a line drawn on a page. In fact, of course, when you are close up it has two dimensions – and always did have, only we weren't close enough to see it. Physicists say it is (or may be) the same at string level – there are hidden dimensions curled up of which we are not at present aware. In fact, they say that there may be *eleven* dimensions in all, ten of space and one of time.[36] This is a difficult if not impossible idea to imagine or visualise, but the scientists make their arguments for mathematical reasons (math that even mathematicians find difficult). When they do make this allowance, however, many things about the universe fall into place. For example, black holes are explained – as perhaps similar to fundamental particles, as gateways to other universes. The extra dimensions are also needed because the way they curl and bend, string theorists say, may determine the size and frequency of the vibrations of the strings, in other words explaining why the familiar 'particles' have the mass and energy and number that they do. In its latest configuration, string theory involves more than strings: two-, three-, and more dimensional membranes, or '**branes**,' small packets, the understanding of which will be the main work of the twenty-first century.[37]

The most startling thing about string theory, other than the existence of strings themselves, is that it suggests there may be a *prehistory* to the universe, a period before the Big Bang. As Greene puts it, string theory 'suggests that rather than being enormously hot and tightly curled into a tiny spatial speck, the universe started out as cold and essentially *infinite* in spatial extent.'[38] Then, he says, an instability kicked in, there was a period of inflation, and our universe formed as we know it. This also has the merit of allowing all four forces, including gravity, to be unified.

String theory stretches everyone's comprehension to its limits. Visual analogies break down, the math is hard even for mathematicians, but there are a few ideas we can all grasp. First, strings concern a world beyond the Planck

---

⋆ String theorists, incidentally, were one of the groups who established early on their own Internet archives, through which physics papers are immediately made available worldwide.

length. This is, in a way, a logical outcome of Planck's conception of the quantum, which he first had in 1900. Second, as yet it is 99 percent theory; physicists are beginning to find ways to test the new theories experimentally, but as of now there is no shortage of sceptics as to whether strings even exist. Third, at these very small levels, we may enter a spaceless and timeless realm. The very latest research involves structures known as zero branes in whose realm ordinary geometry is replaced by '**noncommunicative geometry**,' conceived by the French mathematician **Alain Connes**. Greene believes this may be a major step forward philosophically as well as scientifically, a breakthrough 'that is capable of giving us an answer to the question of how the universe began and why there are such things as space and time – a formalism that will take us a step closer to answering Leibniz's question of why there is something rather than nothing.'[39] Finally, in superstring theory we have the virtually complete amalgamation of physics and mathematics. The two have always been close, but never more so than now, as we approach the possibility that in a sense, the very basis for reality is mathematical.

Many scientists believe we are living in a golden age for mathematics. Two areas in particular have attracted widespread attention among mathematicians themselves.

Chaoplexity is an amalgam of chaos and complexity. In 1987 in *Chaos: Making a New Science*, James Gleick introduced this new area of intellectual activity.[40] Chaos research starts from the concept that there are many phenomena in the world that are, as the mathematicians say, nonlinear, meaning they are *in principle* unpredictable. The most famous of these is the so-called butterfly effect, whereby a butterfly fluttering its wings in, say, the Midwest of America can trigger a whole raft of events that might culminate in a monsoon in the Far East. A second aspect of the theory is that of the 'emergent' property, which refers to the fact that there are on Earth phenomena that 'cannot be predicted, or understood, simply by examining the system's parts.' Consciousness is a good example here, since even if it can be understood (a very moot point), it cannot be understood from inspection of neurons and chemicals within the brain. However, this only goes halfway to what the chaos scientists are saying. They also argue that the advent of the computer enables us to conduct much more powerful mathematics than ever before, with the result that we shall eventually be able to model – and therefore simulate – complex systems, such as large molecules, neural networks, population growth, weather patterns. In other words, the deep order underlying this apparent chaos will be revealed.

The basic idea in chaoplexity comes from **Benoit Mandelbrot**, an applied mathematician from IBM, who identified what he called the '**fractal**.' The perfect fractal is a coastline, but others include snowflakes and trees. Seen from a distance, they have one shape or outline; closer up more intricate details are revealed; closer still, and there is yet more detail. However close you go, the more intricate the outline, with, often, the patterns repeated at different scales. Because these outlines never resolve themselves into smooth lines – in other words never conform to some simple mathematical function – Mandelbrot called them the 'most complex objects in mathematics.'[41] At the same time, however, it turns out that simple mathematical rules can be fed into computer programs that, after many generations, give rise to complicated patterns, patterns that 'never quite repeat themselves.' From this, and from their observations of real-life fractals, mathematicians now infer that there are in nature some very powerful rules governing apparently chaotic and complex systems that have yet to be unearthed – another example of deep order.

In the late 1980s and early 1990s chaos suddenly blossomed as one of the most popular forms of mathematics, and a new research outfit was founded, the **Santa Fe Institute** in New Mexico, southeast of Los Alamos, where Murray Gell-Mann, discoverer of the quark, joined the faculty.[42] This new specialty has come up with several new concepts, among them 'self-organised criticality,' 'catastrophe theory,' the hierarchical structure of reality, '**artificial life**,' and 'self-organisation.' Self-organised criticality is the brainchild of **Per Bak**, a Danish physicist who emigrated to the United States in the 1970s.[43] His starting point, as he told John Horgan, is a sandpile. As one adds grains of sand and the pile grows, there comes a point – the critical state – when the addition of a single grain can cause an avalanche. Bak was struck by the apparent similarity of this process to other phenomena – stock market crashes, the extinction of species, earthquakes, and so on. He takes the view that these processes can be understood mathematically – that is, *described* mathematically. We may one day be able to understand why these things happen, though that doesn't necessarily mean we shall be able to control and prevent them. It is not far from Per Bak's theory to Frenchman **René Thom's** idea of catastrophe theory, that purely mathematical calculations can explain 'discontinuous behaviour,' such as the emergence of life, the change from a caterpillar into a butterfly, or the collapse of civilisations. They are all aspects of the search for deep order.

Against all this the work of **Philip Anderson** stands out. He won a Nobel Prize in 1977 for his work on superconductors. Instead of arguing for underlying order, Anderson's view is that there is a hierarchy of order – each level of organisation in the world, and in biology in particular, is independent of the order in the levels above and below. 'At each stage, entirely new laws, concepts and generalisations are necessary, requiring inspiration and creativity to just as great a degree as in the previous one. Psychology is not applied biology, nor is biology applied chemistry ... you mustn't give in to the temptation that when you have a good general principle at one level that it's going to work at all levels.'[44]

There is a somewhat disappointed air about the chaoplexologists at the turn of the century. What seemed so thrilling in the early 1990s has not, as yet, produced anything nearly as exciting as string theory, for example. Where math does remain exciting and undismayed, however, is in its relationship to biology. These achievements were summarised by **Ian Stewart**, professor of mathematics at Warwick University in Britain, in his 1998 book *Life's Other Secret*.[45] Stewart comes from a tradition less well known than the Hawkings-Penrose-Feynman-

Glashow physics/cosmology set, or the Dawkins-Gould-Dennett evolution set. He is the latest in a line that includes D'Arcy Wentworth Thompson (*On Growth and Form*, 1917), Stuart Kauffman (*The Origins of Order*, 1993), and Brian Goodwin (*How the Leopard Changed Its Spots*, 1994). Their collective message is that genetics is not, and never can be, a complete explanation for life. What is also needed, surprising as it may seem, is a knowledge of mathematics, because it is mathematics that governs the physical substances – the deep order – out of which, in the end, all living things are made.

Life's Other Secret is dedicated to showing that mathematics 'now informs our understanding of life at every level from DNA to rain forests, from viruses to flocks of birds, from the origins of the first self-copying molecule to the stately unstoppable march of evolution.'[46] Some of Stewart's examples are a mixture of the enchanting and the provocative, such as the mathematics of spiders' webs and snowflakes, the population variations of ant colonies, and the formation of swarms of starlings; he also explores the branching systems of plants and the patterned skins of such animals as leopards and tigers. He has a whole chapter, 'Flowers for Fibonacci,' outlining patterns in the plant kingdom. The Fibonacci sequence of numbers –

$$1, 2, 3, 5, 8, 13, 21, 34, 55, 89, 144 \ldots$$

– was first invented by Leonardo of Pisa in 1202, Leonardo being the son of Bonaccio, hence 'Fi-bonacci.' In the sequence, each number is the sum of the two that precede it, and this simple arrangement describes so much: lilies have 3 petals, for example, buttercups have 5, delphiniums 8, marigolds 13, asters 21 and daisies 34, 55 or 89.[47] But Stewart's book, and thinking, are much more ambitious and interesting than this. He begins by showing that the division of cells in the embryo displays a remarkable similarity to the way soap bubbles form in foams, and that the way chromosomes are laid out in a dividing cell is also similar to the way mutually repelling magnets arrange themselves. In other words, whatever instructions are coded into genes, many biological entities behave as though they are constrained by the *physical* properties they possess, properties that can be written as mathematical equations. For Stewart this is no accident. This is life taking advantage of the mathematics/physics of nature for its own purposes. He finds that there is a 'deep geometry' of molecules, especially in DNA, which forms knots and coils, this architecture being all-important. For example, he quotes a remarkable experiment carried out by Heinz Fraenkel-Conrat and Robley Williams with the tobacco mosaic virus.[48] This, says Stewart, is a bridge between the inorganic and organic worlds; if the components of the virus are separated in a test tube and then left to their own devices, they spontaneously reassemble into a complete virus that can replicate. In other words, it is the architecture of the molecules that automatically produces life. In theory, therefore, this form of virus – life – could be created by preparing synthetic substances and putting them together in a test tube. In the latter half of the 1990s, mathematicians have understood the processes by which primitive forms of life – the slime mould, for example, the soil amoeba *Dictyostelium discoideum* – proceed. They turn out to be not so very difficult mathematical equations. 'The main point here,' says Stewart, 'is that a lot of properties of life are turning out to be physics, not biology.'[49]

Perhaps most revealing are the experiments that Stewart and others call '**artificial life**.' These are essentially games played on computers designed to replicate in symbolic form various aspects of evolution.[50] The screen will typically have a grid, say 100 squares wide and 100 squares deep. Into each of these squares is allotted a 'bush' or a 'flower,' say, or on the other hand, a 'slug' and 'an animal that preys on slugs.' Various rules are programmed in: one rule might be that a predator can move five squares each time, whereas a slug can move only one square; another might be that slugs on green flowers are less likely to be seen (and eaten) than slugs on red flowers, and so on. Then, since computers are being used, this artificial life can be turned on and run for, say, 10,000 moves, or even 50 million moves, to see what '**A-volves**' (A = artificial). A number of these programs have been tried. The most startling was Andrew Pargellis's 'Amoeba,' begun in 1996. This was seeded only with a random block of computer code, 7 percent of which was randomly replaced every 100,000 steps (to simulate mutation). Pargellis found that about every 50 million steps a self-replicating segment of code appeared, simply as a result of the math on which the program was based. As Stewart put it, 'Replication didn't have to be built into the rules – it just happened.'[51] Other surprises included symbiosis, the appearance of parasites, and long periods of stasis punctuated by rapid change – in other words, punctuated equilibrium much as described by Niles Eldredge and Stephen Jay Gould. Just as these models (they are not really experiments in the traditional sense) show how life might have begun, Stewart also quotes mathematical models which suggest that a network of neural cells, a 'neural net,' when hooked together, naturally acquires the ability to make computations, a phenomenon known as 'emergent computation.'[52] It means that nets with raw computational ability can arise spontaneously through the workings of ordinary physics: 'Evolution will then select whichever nets can carry out computations that enhance the organism's survival ability, leading to specific computation of an increasingly sophisticated kind.'[53]

Stewart's fundamental point, not accepted by everyone, is that mathematics and physics are as powerful as genetics in giving form to life. 'Life is founded on mathematical patterns of the physical world. Genetics exploits and organises those patterns, but physics makes them possible and constrains what they can be.'[54] For Stewart, genetics is not the deepest secret, the deepest order of life. Instead, mathematics is, and he ends his book by predicting a new discipline for the twenty-first century, '**morphomatics**,' which will attempt to marry math, physics, and biology and which, he clearly hopes, will reveal the deep patterns in the world around us and, eventually, help us to understand how life began.

Conclusion

THE POSITIVE HOUR

I do not hope to know again
The infirm glory of the positive hour

T. S. Eliot, *Ash Wednesday*, 1930

Who can doubt Eliot's sentiment, that the twentieth century was the positive hour, or that its glory, however glorious, was also infirm? He continues in magnificent dissatisfaction, and resolve:

Because I know that time is always time
And place is always and only place
And what is actual is actual only for one time
And only for one place
I rejoice that things are as they are ...
Consequently I rejoice, having to construct something
Upon which to rejoice ...

Because these wings are no longer wings to fly
But merely vans to beat the air
The air which is now thoroughly small and dry
Smaller and dryer than the will
Teach us to care and not to care
Teach us to sit still.[1]

Eliot was writing in the middle of the golden age of physics but also in the golden age of Heidegger, before the fall of both. 'Teach us to sit still' was his way of saying, as Heidegger put it, 'submit.' Submit to the world as it is and rejoice, celebrate, without forever looking to explain it all. Relish the mystery, which allows us to be as we wish to be. But Eliot, as the rest of the poem and its elegiac tone make clear, was not entirely happy with this as a complete solution. Like too many others, he found the cause that science advanced convincing, too convincing to go back wholly to the *status quo ante*. No more than the next man could he unknow what was now known. But, as a poet, he could *mark* what was happening. And crucially, 1930, when *Ash Wednesday* appeared, was perhaps the earliest date at which all the three great intellectual forces of the twentieth century became apparent. These three forces were: science; free-market economics; and the mass media.

This is not to say, of course, that science, or free-market economics, or the mass media, were entirely twentieth-century phenomena: they were not. But there were important aspects of the twentieth century which meant that each of these forces took on a new potency, which only emerged for all to see in the 1920s.

What was just emerging in science at the time of *Ash Wednesday*, particularly as a result of Edwin Hubble's discoveries, gathered force as the century went on more than Eliot – or anyone – could have guessed. Whatever impact individual discoveries had, the most important development intellectually, which added immeasurably to the authority of science, and changed man's conception of himself, was the extent to which science began to come together, the way in which the various disciplines could be seen as telling the same story from different angles. First physics and chemistry came together; then physics and astronomy/cosmology; then physics and geology; more recently physics and mathematics, though they have always been close. In the same way economics and sociology came together. Even more strongly biology, in the form of genetics, came together with linguistics, anthropology, and archaeology. Biology and physics have not yet come together in the sense that we understand how inert substances can combine to create life. But they have come together, as Ian Stewart's work showed in the last chapter, in the way physics and mathematics help explain biological structures; even more so in the expanded concept of evolution, producing a single narrative from the Big Bang onward, throughout the billions of years of the history of the universe, giving us the creation of galaxies, the solar system, the earth, the oceans and continents, all the way through to life itself and the distribution about our planet of plants and animals. This is surely the most powerful empirically based idea there has ever been.

The final layer of this narrative has been provided only recently by **Jared Diamond**. Diamond, a professor of physiology at California Medical School but also an anthropologist who has worked in New Guinea, won the Rhône-Poulenc Science Book Prize in 1998 for *Guns, Germs and Steel*.[2] In this book, he set out to explain nothing less than the whole pattern of evolution over the last 13,000 years – i.e., since the last ice age – and his answer was as bold as it was original. He was in particular concerned to explore why it was that evolution brought us to the point where the Europeans invaded and conquered the Americas in 1492 and afterward, and not the other way round. Why had the Incas, say, not crossed the Atlantic from west to east and subdued the Moroccans or the Portuguese? He found that the explanation lay in the general layout of the earth, in particular the way the continents are arranged over the surface of the globe. Simply put, the continents of the Americas and Africa have their main axis running north/south, whereas in Eurasia it is east/west.[3] The significance of this is that the diffusion of domesticated plants and animals is much easier from east to west, or west to east, because similar latitudes imply similar geographical and climatic conditions, such as mean temperatures, rainfall, or hours of daylight. Diffusion from north to south, or south to north, on the other hand, is correspondingly harder and therefore inhibited the spread of domesticated plants and animals. Thus the spread of cattle, sheep, and goats was much more rapid, and thorough, in Eurasia than it was in either Africa or the Americas.[4] In this way, says Diamond, the dispersal of farming meant the buildup of greater population densities in Eurasia as opposed to the other continents, and this in turn had two effects. First, competition between different societies fuelled the evolution of new cultural practices, in particular the development of weapons, which were so important in the conquest of the Americas. The second consequence was the evolution of diseases contracted from (largely domesticated) animals. These diseases could only survive among relatively large populations of humans, and when they were introduced to peoples who had developed no immune systems, they devastated them. Thus the global pattern was set, says Diamond. In particular, Africa, which had 'six million years' start' in evolutionary terms compared with other parts of the world, failed to develop because it was isolated by vast oceans on three sides and desert on the north, and had few species of animals or plants that could be domesticated along its north/south axis.[5]

Diamond's account – an expanded version of *la longue durée* – although it has been criticised as being speculative (which it undoubtedly is), does if accepted bring a measure of closure to one area of human thought, showing why different races around the world have reached different stages of development, or had done so by, say, 1500 AD. In doing this, Diamond, as he specifically intended, defused some of the racist sentiment that sought to explain the alleged superiority of Europeans over other groupings around the globe. He therefore used science to counter certain socially disruptive ideas still current in some quarters at the end of the century.

The fundamental importance of science, if it needs further underlining, shows in the very different fates of Germany and France in the twentieth century. Germany, the world leader in many areas of thought until 1933, had its brains ripped out by Hitler in his inquisition, and has not yet recovered. (Remember Allan Bloom's wide-ranging references to German culture in *The Closing of the American Mind*?) World War II was not only about territory and *Lebensraum*; in a very real sense it was also about ideas. In France the situation was different. Many continental thinkers, especially French and from the German-speaking lands, were devoted to the marriage of Freud and Marx, one of the main intellectual preoccupations of the century, and maybe the biggest dead end, or folly, which had the effect, in France most of all, of blinding thinkers to the advances in the 'harder' sciences. This has created a cultural divide in intellectual terms between francophone and anglophone thought.

The strength of the second great force in the twentieth century – free-market economics – was highlighted by the great 'experiment' that was launched in Russia in 1917, and lasted until the late 1980s. The presence of the rival systems, and the subsequent collapse of communism, drew attention to the advantages of free-market economics in a way that Eliot, writing *Ash Wednesday* at the time of the Great Crash, could perhaps not have envisaged. This triumph of the free-market system was so complete that, to celebrate it, **Francis Fukuyama** published in 1992 *The End of History and the Last Man*.[6] Based on a lecture given at the invitation of Allan Bloom, at the University of Chicago, Fukuyama took as his starting point the fact that the preceding years had seen the triumph of liberal democracies all over the world and that this marked the 'endpoint of mankind's ideological evolution' and the 'final form of human government.'[7] He was talking not only about Russia but the great number of countries that have embraced the free market and democracy, to some extent: Argentina, Botswana, Brazil, Chile, the Eastern European countries, Namibia, Portugal, South Korea, Spain, Thailand, Uruguay, and so on. More than that, though, Fukuyama sought to show that there is, as he put it, a Universal History, a single, coherent evolutionary process that takes into account 'the experience of all peoples in all times.'[8] His argument was that natural science is the mechanism by which this coherent story is achieved, that science is by consensus both cumulative and directional 'even if its ultimate impact on human happiness is ambiguous.'[9] He added, 'Moreover, the logic of modern natural science would seem to dictate a universal evolution in the direction of capitalism.' Fukuyama thought this accounted for many of the nonmaterial developments in twentieth-century life, most notably the psychological developments. He implied that modern natural science brought democratic progress – because the institutions of science are essentially democratic, and require widespread education for their successful operation, and this in turn brought about a concern on the part of many people, as Hegel had predicted, for a 'desire for recognition' – a desire to be appreciated in their own right. In such an environment, the individualistic developments we have seen in the twentieth century became almost unavoidable – from the psychological revolution to the civil rights movement and even postmodernism. In the same way, we have been living through a period equivalent or analogous to the Reformation. In the Reformation, religion and politics became divorced; in the twentieth century political liberation has been replaced by personal liberation. In this process Fukuyama discussed Christianity, following Hegel, as the 'absolute religion,' not out of any narrow-minded ethnocentrism, he said, but because Christianity regards all men as equal in the sight of God, 'on the basis of their faculty for moral choice or belief' and because Christianity regards man as free, morally free to choose between right and wrong.[10] In this sense then, Christianity is a more 'evolved' religion than the other great faiths.

Just as there is an intimate link between science, capitalism, and liberal democracies, so too there is a link to the third force of the twentieth century, the mass media. Essentially democratic to start with, the mass media have in fact grown more so as the century has proceeded. The globalisation of the markets has been and is a parallel process. This is not to deny that these processes have brought with them their own set of problems, some of which will be addressed presently. But for now my point is simply to assert that science, free-market economics, and the mass media stem from the same impulse, and that this impulse has been dominant throughout the past century.

Jared Diamond's thesis, and Francis Fukuyama's, come together uncannily in **David Landes**'s *Wealth and Poverty of Nations* (1998).[11] At one level, this book is a restatement of the 'traditional' historical narrative, describing the triumph of the West. At a deeper level it seeks to explain why it was that, for example, China, with its massive fleet in the Middle Ages, never embarked on a period of conquest as the Western nations did, or why Islamic technological innovation in the same period was interrupted, never to resume. Landes's answer was partly geographical (the distribution of parasites across the globe, limiting mortality), religion (Islam turned its back on the printing press, fearful of the sacrilege it might bring with it), population density and immigration patterns (families of immigrants flooded into north America, single men into Latin America, to intermarry with the indigenous population), and economic/political and ideological systems that promote self-esteem (and therefore hard work) rather than, say, the Spanish system in South America, where Catholicism was much less curious about the new world, less adaptable and innovative.[12] Like Fukuyama, Landes linked capitalism and science, but in his case he argued that they are both systems of cumulative knowledge. For Landes these are all-important lessons; as he points out at the end of his book, convergence isn't happening. The rich are getting richer and the poor poorer. Countries – civilisations – ignore these lessons at their peril.

But science brings problems too, and these need to be addressed. In *The End of Science: Facing the Limits of Knowledge in the Twilight of the Scientific Age* (1996), the science writer **John Horgan** explored two matters.[13] He considered whether all the major questions in science had already been answered – that all biology, for example, is now merely a footnote to Darwin, or that all physics pales in the shadow of the Big Bang – and he looked at whether this marks a decisive phase in human history. He interviewed a surprisingly large number of scientists who thought that we *are* coming to the end of the scientific age, that there *are* limits to what we can know and, in general, that such a state of affairs might not be such a bad thing. By his own admission, Horgan was building on an idea of **Gunther Stent**, a biologist at the University of California in Berkeley, who in 1969 had published *The Coming of the Golden Age: A View of the End of Progress*. This book contended 'that science, as well as technology, the arts and all progressive, cumulative enterprises were coming to an end.'[14] The starting point for Stent was physics, which he felt was becoming more difficult to comprehend, more and more hypothetical and impractical.

One of the scientists Horgan interviewed who thought there is a limit to knowledge was Noam Chomsky, who divided scientific questions into problems, 'which are at least potentially answerable, and mysteries, which are not.'[15] According to Chomsky there has been 'spectacular progress' in some areas of science, but no progress at all in others – such as consciousness and free will. There, he said, 'We don't even have bad ideas.'[16] In fact, Chomsky went further, arguing in his own book, *Language and Problems of Knowledge* (1988), that 'it is quite possible – overwhelmingly probable, one might guess – that we will always learn more about human life and human personality from novels than from scientific psychology.'[17]

Horgan considered that there were perhaps two outstanding fundamental problems in science – immortality, and consciousness. He thought that immortality was quite likely to be achieved in the next century and that, to an extent, as J. D. Bernal had predicted in 1992, man would eventually be able to direct his own evolution.

The challenge implicit in Horgan's thesis was taken up by **John Maddox**, the recently retired editor of *Nature*, in his 1998 book, *What Remains to Be Discovered*.[18] This was in fact an excellent review of what we know – and do not know – in physics, biology, and mathematics, and at the same time a useful corrective to the triumphalism of some scientists. For example, Maddox went out of his way to emphasise the provisional nature of much physics – he referred to black holes as 'putative' only, to the search for theories of everything as 'the embodiment of a belief, even a hope,' and stated that the reason why the quantum gravity project is 'becalmed' right now is because 'the problem to be solved is not yet fully understood,' and that the idea the universe began with a Big Bang 'will be found false.'[19] At the same time, Maddox thought science far from over. His thesis was that the world has been overwhelmed by science in the twentieth century for the first time. He thought that the twenty-first century is just as likely to throw up a 'new physics' as a Theory of Everything. In astronomy, for example, there is the need to confirm the existence of the hypothetical structure known as the 'great attractor,' toward which, since February 1996, it has been known that 600 visible galaxies are moving. In cosmology, there is the search for the 'missing mass,' perhaps as much as 80 percent of the known universe, which alone can explain the expansion rate after the Big Bang. Maddox also underlines that there is no direct evidence for inflation in the early universe, or that rapid expansion, a Big Bang, took place before. As he puts it, the Big Bang is 'not so much a theory as a model.' Even more pithily, he dismisses Lee Smolin's ideas of parallel universes, with no unique beginning, as 'no more persuasive than the account in *Genesis* of how the universe began.'[20] In fact, Maddox says plainly, we do not know how the universe began; Hubble's law urgently needs to be modified; and, 'from all appearances, space-time in our neighborhood is not noticeably curved [as it should be according to relativity], but flat.'[21]

Maddox considers that even our understanding of fundamental particles is far from complete and may be crucially hampered after the new CERN accelerator comes on stream in 2005 – because experiments there will suggest new experiments that we don't, and shan't, have the capability for. He points out that in the early weeks of 1997 there were suggestions that even electrons may have internal structures, and be composite, and that therefore 'the goal of specifying just why the particles in the real world are what they are is still a long way off.'[22] In regard to string theory, Maddox makes a fundamental objection: If strings must exist in many dimensions, how can they relate to the real world in which we live? His answer is that string theory may be no more than a metaphor, that our understanding of space or time may be seriously flawed, that physics has been too concerned with, as he put it, 'the naming of parts,' in too much of a hurry to provide us with proper *understanding*. Maddox's reserve about scientific progress is hugely refreshing, coming as it does from such an impeccable source, the editor who first allowed so many of these theories into print. He does agree with Horgan that life itself is one of the mysteries that will be unravelled in the twenty-first century, that cancer will finally be conquered, that huge advances will be made in understanding the link between genetics and individuality, and that the biggest remaining problem/mystery of all is consciousness.

The application of evolutionary thinking to consciousness, discussed in chapter 39, is only one of the areas where the neo-Darwinists have directed their most recent attention. In practice, we are now in an era of 'universal Darwinism,' when the algorithmic approach has been applied almost everywhere: evolutionary cosmology, evolutionary economics (and therefore politics), the evolution of technology. But perhaps the most radical idea of the neo- or ultra-Darwinians relates to knowledge itself and raises the intriguing question as to whether we are at the present living through an era in the evolution of knowledge forms.[23] We are living at a time – the positive hour – when science is taking over from the arts, humanities, and religion as the main form of knowledge. Recall that in Max Planck's family in Germany at the turn of the century, as was reported in chapter 1, the humanities were regarded as a form of knowledge superior to science. Richard Hofstadter was one of the first to air the possibility that all this was changing when he drew attention to the great impact in America in the 1960s of nonfiction and sociology, as compared with novels (see chapter 39). Let us also recall the way Eugène Ionesco was attuned to the achievements of science: 'I wonder if art hasn't reached a dead-end,' he said in 1970. 'If indeed in its present form, it hasn't already reached its end. Once, writers and poets were venerated as seers and prophets. They had a certain intuition, a sharper sensitivity than their contemporaries, better still, they discovered things and their imaginations went beyond the discoveries even of science itself, to things science would only establish twenty-five or fifty years later.... But for some time now, science [has] been making enormous progress, whereas the empirical revelations of writers have been making very little ... can literature still be considered as a means to knowledge?'[24]

In *The Death of Literature* (1990), Alvin Kernan quotes George Steiner: 'We are now seeing, all of us today, the gradual end of the classical age of reading.'[25] Kernan himself puts it this way: 'Humanism's long dream of learning, of arriving at some final truth by enough reading and writing, is breaking up in our time.'[26] He has no doubt about the culprit. 'Television, however, is not just a new way of doing old things but a radically different way of seeing and interpreting the world. Visual images not words, simple open meanings not complex and hidden, transience not permanence, episodes not structures, theater not truth. Literature's ability to coexist with television, which many take for granted, seems less likely when we consider that as readers turn into viewers, as the skill of reading diminishes, and as the world as seen through a television screen feels and looks more pictorial and immediate, belief in a word-based literature will inevitably diminish.'[27] 'There is always the possibility that literature was so much a product of print culture and industrial capitalism, as bardic poetry and heroic epic were of tribal oral society, that ... it will simply disappear in the electronic age, or dwindle to a merely ceremonial role, something like Peking opera perhaps.'[28]

Both Gunther Stent, referred to earlier, and **John Barrow**, an astronomer, have written about what they see as an evolutionary process in the arts 'which has steadily relaxed the compositional constraints placed on the artist.... As the constraints imposed by convention, technology, or individual preference have been relaxed, so the resulting structure is less formally patterned, closer to the random, and harder to distinguish from the work of others working under similar freedom from constraint.'[29] Stent argued that music actually has evolved like anything else. Studies have shown, for instance, that in order to be liked, music must strike a balance between the expected and the introduction of surprises. If it is too familiar, it is boring; if it is too surprising, it 'jars.' Physicists with a mathematical bias have actually calculated the familiarity/surprise ratio of music, and Stent was able to show that, beginning with 'the maximal rigidity of rhythmic drumming in ancient times, music has exhausted the scope of each level of constraint for its listeners, before relaxing them and moving down to a new level of freedom of expression. At each stage, from ancient to medieval, renaissance baroque, romantic, to the atonal and modern periods, evolution has proceeded down a staircase of ever-loosening constraints, the next step down provoked by the exhaustion of the previous level's repertoire of novel patterns. ... The culmination of this evolutionary process in the 1960s saw composers like John Cage relinquish all constraints, leaving the listeners to create what they would from what they heard: an acoustic version of the Rorschach inkblot test.'[30] John Barrow added the thought that other creative activities like architecture, poetry, painting, and sculpture have all displayed similar trends away from constraint. 'Stent's suspicion,' he wrote, 'was that they were all quite close to reaching the asymptote of their stylistic evolution: a final structureless state that required purely subjective responses.'[31]

A related way in which Darwinism encourages the evolution of knowledge forms has been suggested by **Robert Wright**. As he puts it, the various ways of conceiving the world – moral, political, artistic, literary, scientific – are 'by Darwinian lights, raw power struggles. A winner will emerge, but there's often no reason to expect that winner to be truth.' Wright calls this approach '**Darwinian cynicism**,' which he equates to the postmodern sensibility that views all modes of human communication as 'discourses of power,' where 'ironic self-consciousness is the order of the day,' where ideals can't be taken seriously because one cannot avoid 'self-serving manipulation.'[32] On this analysis, postmodernism has itself evolved and, as with music, poetry, and painting, has reached the end as a way of looking at the world. Fukuyama didn't know what he was starting when he wrote about the end of history.

Yet another reason why many of the arts must rate as unsatisfactory forms of knowledge in the twentieth century stems from the modernist reliance on the theories of Sigmund Freud. Here I agree with Britain's Nobel Prize-winning doctor **Sir Peter Medawar**, who in 1972 described psychoanalysis as 'one of the saddest and strangest of all landmarks in the history of twentieth-century thought.'[33] Freud unveiled the unconscious to the world in 1900, at much the same time that the electron, the quantum, and the gene were identified. But whereas they have been confirmed by experiment after experiment, developing and proliferating, Freudianism has never found unequivocal empirical support, and the very idea of a systematic unconscious, and the tripartite division of the mind into the id, ego, and superego has seemed increasingly far-fetched. This is crucial in my view, for the consequences of the failure of Freudianism have not been thought through, and a re-evaluation of psychoanalysis is now urgently needed. For example, if Freud was so wrong, as I and many others believe, where does that leave any number of novels and virtually the entire corpus of surrealism, Dada, and certain major forms of expressionism and abstraction, not to mention Richard Strauss's 'Freudian' operas such as *Salomé* and *Elektra*, and the iconic novels of numerous writers, including D. H. Lawrence, Franz Kafka, Thomas Mann, and Virginia Woolf? It doesn't render these works less beautiful or pleasurable, necessarily, but it surely dilutes their meaning. They don't owe their *entire* existence to psychoanalysis. But if they are robbed of a large part of their meaning, can they retain their intellectual importance and validity? Or do they become period pieces? I stress the point because the novels, paintings, and operas referred to above have helped to popularise and legitimise a certain view of human nature, one that is, all evidence to the contrary lacking, wrong. The overall effect of this is incalculable. All of us now harbor the view, for example, that our adult selves bear a certain relation to our childhood experiences, and to conflicts with our parents. Yet in 1998 **Judith Rich Harris**, a psychologist who had been dismissed from her Ph.D. course at Harvard, caused consternation among the psychological profession in America and elsewhere by arguing in her book *The Nurture Assumption* that parents have much less influence on their children than has traditionally been supposed; what matters instead is the child's peer group – other children. She produced plenty of evidence to support her claim, which turned a century of Freudian jargoneering on its head.[34] As a result of Freud, there has been a strain of thought in the twentieth century that holds, rather as in primitive societies, that the mad have an alternative view of the human condition. There is no evidence for this; moreover, it damages the fortunes of the mentally ill.

Robert Wright has described still other ways in which evolutionary thinking has been used to sow further doubt about Freudianism. As he wrote in *The Moral Animal: Why We Are the Way We Are: The New Science of Evolutionary Psychology* (1994), 'Why would people have a death instinct ('thanatos') [as Freud argued]? Why would girls want male genitals ('penis envy')? Why would boys want to have sex with their mothers and kill their fathers (the 'Oedipus complex')? Imagine genes that specifically encourage any of these impulses, and you're imagining genes that aren't exactly destined to spread through a hunter-gatherer population overnight.'[35]

The muddle over Freud, and psychoanalysis, was shown starkly by an exhibition scheduled for the mid-1990s at the Library of Congress in Washington, D.C. The exhibition was designed to celebrate the centenary of the birth of psychoanalysis.[36] However, when word of the planned exhibition was released, a number of scholars, including Oliver Sacks, objected, arguing that the planning committee was packed with Freud 'loyalists' and that the exhibition threatened to become mere propaganda and hagiography, 'ignoring the recent tide of revisionist writings about Freud.'[37] When the book of the exhibition appeared, in 1998, no mention of this controversy was made, either by the Librarian of Congress, who wrote the foreword, or by the editor. Even so, the book could not avoid completely the doubts about Freud that have grown as the centenary of *The Interpretation of Dreams* approached. Two authors wrote papers describing Freud's ideas as unstable and untestable, 'on a par with flying saucers,' while two others, including Peter Kramer, author of *Listening to Prozac*, described them as unconvincing but conceded that Freud has been influential. It is noticeable, for instance, that a great deal of the book was given over to talk of Freud's 'industry,' 'courage,' and 'genius,' and to arguing that he should be judged 'less as a scientist than as an imaginative artist.'[38] Even psychoanalysts now concede that his ideas about women, early societies of hunter-gatherers, and the 'Primal Crime' are both fanciful and embarrassing. And so we are left in the paradoxical situation that, as the critic Paul Robinson says, the dominant intellectual presence of our century was, for the most part, wrong.

Nor did this revisionism stop with Freud. In 1996 **Richard Noll**, an historian of science at Harvard, published *The Jung Cult* and, two years later, *The Aryan Christ*.[39] These books provoked a controversy no less bitter than the one over Freud, for Noll argued that Jung had lied about his early research and had actually fabricated dates in his notes to make it appear that patients' memories of such things as fairy tales were part of the 'collective unconscious' and had not been learned as children. Noll also documented Jung's anti-Semitism in detail and criticised present-day Jungians for not wanting to put his ideas to the test, lest they scare away potential clients.

The commercial side of Jungianism need not concern us. More important is that, when this is taken together with Freud's shortcomings, we can see that psychology in the twentieth century is based on theories – myths almost – that are not supported by observation, and is characterised by fanciful, idiosyncratic, and at times downright fraudulent notions. Psychology has been diverted for too long by Freud and Jung. The very plausibility of Freud's theories is their most problematical feature. It has taken an entire century to get out from under their shadow. Until we can rid ourselves of our Freudian mindset, the Freudian 'climate of opinion,' as Auden called it, it is highly unlikely that we can look upon ourselves in the new way that is required. Darwin provides the only hope at the moment, together with the latest advances being made in neuroscience.

A related trend regarding the evolution of knowledge may be seen by juxtaposing Russell Jacoby's *The Last Intellectuals* (1987) alongside John Brockman's *The Third Culture* (1995).[40] Jacoby described the fall of the 'public intellectual' in American life. Until the early 1960s, he said, figures like Daniel Bell, Jane Jacobs, Irving Howe, and J. K. Galbraith had lived in urban bohemias and written for the public, raising and keeping alive issues common to all – but especially educated people.[41] Since then, however, they had disappeared, or at least no new generation of public intellectuals had followed them, and by the late 1980s, when his book appeared, the big names were still Bell, Galbraith, *et al.*[42] Jacoby attributed this to several factors: the decline in bohemia, which has been taken 'on the road' by the Beats, then lost in suburbia; the removal of urban Jews from their marginal position with the decline in anti-Semitism; the fall of the left with the revelations about Stalin's atrocities; but above all the expansion of the universities, which sucked in the intellectuals and then broke them on the rack of tenure and careerism.[43] This thesis was a little unfair to the later generation of intellectuals like Christopher Lasch, Andrew Hacker, Irving Louis Horowitz, or Francis Fukuyama, but Jacoby nonetheless had a point. In reply, however, as was referred to in the introduction, John Brockman argued that this function has now been taken over by the scientists, since science has more and more policy and philosophical ramifications than ever before. Jacoby describes the complete triumph of analytic philosophy in U.S. and U.K. universities, but for Brockman's scientists it is *their* philosophy of science that is now more advanced, and more useful. This is the evolution of ideas, and knowledge forms, in action.

Finally, in considering this evolution of knowledge forms, think back to the link between science, free-market economics, and liberal democracy which was mentioned earlier in this conclusion. The relevance and importance of that link is brought home in this book by an interesting absence that readers may have noticed. I refer to the relative dearth of non-Western thinkers. When this book was conceived, it was my intention (and the publishers') to make the text as international and multicultural as possible. The book would include not just European and North American – Western – ideas, but would delve into the major non-Western cultures to identify their important ideas and their import- ant thinkers, be they philosophers, writers, scientists, or composers. I began to work my way through scholars who specialised in the major non-Western cultures: India, China, Japan, southern and central Africa, the Arab world. I was shocked (and that is not too strong a word) to find that they all (I am not exaggerating, there were no exceptions) came up with the same answer, that in the twentieth century, the non-Western cultures have produced no body of work that can compare with the ideas of the West. In view of the references throughout the book to racism, I should make it clear that a good proportion of these scholars were themselves members of those very non-Western cultures. More than one made the point that the chief intellectual effort of his or her own (non-Western) culture in the twentieth century has been a coming to terms with modernity, learning how to cope with or respond to Western ways and Western patterns of thought, chiefly democracy and science. This underlines Frantz Fanon's point, and James Baldwin's, discussed in chapter 30, that for many groups, the *struggle* is their culture for the present. I was astounded by this response, which was all the more marked for being made in near-identical terms by specialists from different backgrounds and in different disciplines.

Of course, there are important Chinese writers and painters of the twentieth century, and we can all think of important Japanese film directors, Indian novelists, and African dramatists. Some of them are in this book. We have examined the thriving school of revisionist Indian historiography. Distinguished scholars from a non-Western background are very nearly household names – one thinks of Edward Said, Amartya Sen, Anita Desai, or Chandra Wickramasinghe. But, it was repeatedly put to me, there is no twentieth-century Chinese equivalent of, say, surrealism or psychoanalysis, no Indian contribution to match logical positivism, no African equivalent of the *Annales* school of history. Whatever list you care to make of twentieth-century innovations, be it plastic, antibiotics and the atom or stream-of-consciousness novels, *vers libre* or abstract expressionism, it is almost entirely Western.

One person who may offer a clue to this discrepancy is Sir Vidia (V. S.) Naipaul. In 1981 Naipaul visited four Islamic societies – Iran, Pakistan, Malaysia, and Indonesia. Iran he found confused and angry, 'the confusion of a people of high mediaeval culture awakening to oil and money, a sense of power and violation, and a knowledge of a great new encircling civilisation.'[44] 'That civilisation couldn't be mastered. It was to be rejected; at the same time it was to be depended upon.'[45] Pakistan, he found, was a fragmented country, economically stagnant, 'its gifted people close to hysteria.'[46] The failure of Pakistan as a society, he said, 'led back again and again to the assertion of the faith.'[47] As with Iran there was an emotional rejection of the West, especially its attitudes to women. He found no industry, no science, the universities stifled by fundamentalism, which 'provides an intellectual thermostat, set low.'[48] The Malays, he found, had an 'inability to compete' (he meant with the Chinese, who constituted half its population and dominated the country economically). The Islam of Indonesia Naipaul described as 'stupefaction'; community life was breaking down, and the faith was the inevitable response.[49] In all four places, he said, Islam drew its strength from a focus on the past that prevented development, and that very lack of development meant that the peoples of the Islamic nations could not cope with the West. The 'rage and anarchy' induced by this kept them locked into the faith – and so the circle continues. Not for nothing did Naipaul quote Bertrand Russell in his book: 'History makes one aware that there is no finality in human affairs; there is not a static perfection and an unimprovable wisdom to be achieved.'[50]

Naipaul was even harder on India. He visited the country three times to write books about it – *An Area of Darkness* (1967), *India: A Wounded Civilisation* (1977), and *India: A Million Mutinies Now* (1990).[51] 'The crisis of India,' he wrote in 1967, '. . . is that of a decaying civilisation, where the only hope lies in further swift decay.' In 1977 things didn't look so black, though that could have meant that the swift decay was already overtaking the country. Though not unsympathetic to India, Naipaul pulled no punches in his second book. Phrases taken at random: 'The crisis of India is not only political or economic. The larger crisis is of a wounded old civilisation that has at last become aware of its inadequacies and is without the intellectual means to move ahead';[52] 'Hinduism . . . has exposed [Indians] to a thousand years of defeat and stagnation. It has given men no idea of contract with other men, no idea of the state. . . .

Its philosophy of withdrawal has diminished men intellectually and not equipped them to respond to challenge; it has stifled growth.'[53]

Octavio Paz, Mexico's Nobel Prize-winning poet, was twice attached to Mexico's embassy in India, the second time as ambassador. His *In Light of India*, published in 1995, is much more sympathetic to the subcontinent, celebrating in particular its poetry, its music, its sculpture.[54] At the same time, Paz is not blind to India's misfortunes: 'The most remarkable aspect of India, and the one that defines it, is neither political nor economic, but religious: the coexistence of Hinduism and Islam. The presence of the strictest and most extreme form of monotheism alongside the richest and most varied polytheism is, more than a historical paradox, a deep wound. Between Islam and Hinduism there is not only an opposition, but an incompatibility';[55] 'Hindu thought came to a halt, the victim of a kind of paralysis, toward the end of the thirteenth century, the period when the last of the great temples were erected. This historical paralysis coincides with two other important phenomena: the extinction of Buddhism and the victory of Islam in Delhi and other places';[56] 'The great lethargy of Hindu civilisation began, a lethargy that persists today. ... India owes to Islam some sublime works of art, particularly in architecture and, to a lesser degree, in painting, but not a single new or original thought.'[57]

Naipaul's third book on the subcontinent, *India: A Million Mutinies Now*, published in 1990, was very different in tone, altogether sunnier, a collection of vivid reportage, looking at filmmakers, architects, scientists, newspaper people, charity workers, with far fewer − hardly any − of the great sobering generalisations that had characterised the earlier books. When he did sum up, right at the end, it was to conclude, 'People everywhere have ideas now of who they are and what they owe themselves. ... The liberation of spirit that has come to India could not come as release alone. ... It had to come as rage and revolt. ... But there was in India now what didn't exist 200 years before: a central will, a central intellect, a national idea.'[58] India, he reflected, was growing again, on its way to restoration.[59]

I draw attention to the issue, since I found it so surprising, because these later encomiums of Naipaul cannot quite wash away the large thoughts he raised in his earlier works, about the links between religion and politics on the one hand, and creativity and intellectual and social progress on the other, and because it helps explain the shape of this book and why there isn't more about non-Western intellectual developments. I can't attempt a full answer here, because I haven't done the work. Nor has anyone else, so far as I am aware, though David Landes comes close in his *Wealth and Poverty of Nations* (1998), referred to earlier. He pulls no punches either, frankly labelling the Arab nations, the Indians, the Africans, and the South Americans as 'losers.'[60] Quoting figures to show that not even colonialism was all bad, Landes settles on the fact that intellectual *segregation* is the chief burden of religious fundamentalism, producing technological lag. Landes's book is best seen as an heroic attempt at being cruel to be kind, to shock and provoke 'failing' cultures into reality. There is much more to be said.

The issues just discussed are partly psychological, partly sociological. In *The Decomposition of Sociology* (1994), **Irving Louis Horowitz**, Hannah Arendt Distinguished Professor of Sociology at Rutgers University, and president of Transaction/*Society*, a sociological publishing house, laments both the condition and direction of the discipline to which he has given his life.[61] His starting point, and the reason why his book appeared when it did, was the news in February 1992 that the sociology departments in three American universities had been closed down and the one at Yale cut back by more than half. At the same time, the number of students graduating in sociology was 14,393, well down on the 35,996 in 1973. Horowitz is in no doubt about the cause of this decline, a decline that, he notes, is not confined to the United States: 'I firmly believe that a great discipline has turned sour if not rancid.'[62] Strong words, but that all-important change, he said, has been brought about by the injection of ideology into the discipline – to wit, a belief that a single variable can explain human behaviour: 'Thus, sociology has largely become a repository of discontent, a gathering of individuals who have special agendas, from gay and lesbian rights to liberation theology';[63] 'Any notion of a common democratic culture or a universal scientific base has become suspect. Ideologists masked as sociologists attack it as a dangerous form of bourgeois objectivism or, worse, as imperialist pretension. . . . That which sociology once did best of all, support the humanistic disciplines in accurately studying conditions of the present to make the future a trifle better, is lost. Only the revolutionary past and the beatific future are seen as fit for study, now that the aim of sociology has become to retool human nature and effect a systematic overhaul of society.'[64] The result, he said, has been the departure from sociology of all those scholars for whom social science is linked to public policy – social planners, penologists, demographers, criminologists, hospital administrators, and international development specialists.[65] Sociology, rather than being the study of ideology, has become ideology itself – in particular Marxist ideology. 'Every disparity between ghetto and suburb is proof that capitalism is sick. Every statistic concerning increases in homicide and suicide demonstrates the decadence of America or, better, resistance to America. Every child born out of wedlock is proof that "the system" has spun out of control.'[66]

For Horowitz, the way to rehabilitate and reinvent sociology is for it to tackle some big sympathetic issues, to describe those issues in detail and without bias, and to offer *explanation*. The Holocaust is the biggest issue, he wrote, still – amazingly – without a proper sociological description or a proper sociological explanation. Other areas where sociology should seek to offer help – to government and public alike – are in drug abuse, AIDS, and an attempt to define 'the national interest,' which would help foreign policy formulation. He also outlines a sociological 'canon,' a list of authors with whom, he said, any literate sociologist should be familiar. Finally, he makes a point very germane to the thesis of this chapter, that the positive hour, or 'positive bubble' as he put it, might not always last, or produce a vision of society that we can live with.[67] It is, above all, he said, the sociologist's job to help us see past this bubble, to explore how we might live together. Horowitz's book finishes up far more positive in tone than it starts out, but it cannot be said that sociology has changed much as a result; its decomposition is still its dominant feature.

Horowitz's thoughts bring us back to the Introduction, and to the fact that, in this book, I have sought to shift the focus away from political and military events. Of course, as was said at the beginning, this is an artificial division, a convenience merely for the sake of exploring significant and interesting issues often sidelined in more conventional histories. Yet one of the more challenging aspects of politics lies in the attempt to adapt such findings as those reported here to the governance of peoples. Whole books could be written about both the theory and practicalities of such adaptation, and while there is certainly no space to attempt such an exercise in the present work, it is necessary to acknowledge such a limit, and to make (as I see it) one all-important point.

This is that neither side of the conventional political divide (left *versus* right) holds all the virtues when it comes to dealing with intellectual and social problems. From the left, the attempted marriage of Marx and Freud has failed, as it was bound to do, being based on two rigid and erroneous theories about human nature (Freud even more so than Marx). The postmodern tradition is more successful as a diagnosis and description than as a prognosis for a way forward, except in one respect – that it cautions us to be wary of 'big' ideas that work for all people, in all places, at all times.

Looking back over the century, and despite the undoubted successes of the free-market system, one wonders whether the theorists of the right have any more reason to feel satisfied. Too often a substantial part of what they have offered is a directive to do nothing, to allow matters to take their 'natural' course, as if doing nothing is somehow more natural than doing something. The theories of Milton Friedman or Charles Murray, for example, seem very plausible, until one thinks of the writings of George Orwell. Had Friedman and Murray been writing in the 1930s, they would probably have still been arguing for the status quo, for economics to take its 'natural' course, for no intervention. Yet who can doubt that Orwell helped bring about a shift in sensibility that, combined with the experience of war, wrought a major change in the way the poor were regarded? However unsatisfactory the welfare state is *now*, it certainly improved living conditions for millions of people across the world. This would not have happened if left to *laisser-faire* economists.

Perhaps Karl Popper had it about right when he said that politics is like science, in that it is – or ought to be – endlessly modifiable. Under such a system, a welfare state might be a suitable response to a certain set of circumstances. But, once it has helped to create a healthier, wealthier population in which far greater numbers survive into old age, with all the implications that has for disease and the economic profile of an entire people, surely a different set of circumstances is called for? We should know by now – it is one of the implicit messages of this book – that in a crowded world, the world of mass society (a twentieth-century phenomenon), every advance is matched by a corresponding drawback or problem. In this regard, we should never forget that science teaches us two lessons, one just as important as the other. While it has revealed to us some of the fundamentals of nature, science has also taught us that the *pragmatic*, piecemeal approach to life is by far the most successful way of adapting. We should beware grand theories.

As the century drew to its close, the shortcomings and failures first recognised by Gunther Stent and John Horgan began to grow in importance – in particular the idea that there are limits to what science can tell us and what, in principle, we can know. John Barrow, professor of astronomy at the University of Sussex, put these ideas together in his 1998 book *Impossibility: The Limits of Science and the Science of Limits.*[68] 'Science,' said Barrow in his concluding chapter, 'exists only because there are limits to what Nature permits. The laws of Nature and the unchanging "constants" of Nature define the borders that distinguish our Universe from a host of other conceivable worlds where all things are possible. ... On a variety of fronts we have found that growing complexity ultimately leads to a situation that is not only limited, but self-limiting. Time and again, the development of our most powerful theories has followed this path: they are so successful that it is believed that they can explain everything. ... The concept of a "theory of everything" occasionally rears its head. But then something unexpected happens. The theory predicts that it cannot predict: it tells us that there are things it cannot tell us.'[69] In particular, Barrow says, taking as his starting point Kurt Gödel's 1931 theory, there are things mathematics cannot tell us; there are limits that arise from our humanity and the evolutionary heritage we all share, which determine our biological nature and, for instance, our size. There are limits to the amount of information we can process; the great questions about the nature of the universe turn out to be unanswerable, because for one thing the speed of light is limited. Chaoplexity and randomness may well be beyond us in principle. 'Whether it be an election, a bank of linked computers, or the "voting" neurones inside our head, it is impossible to translate individual rational choices into collective rationality.'[70]

Not everyone agrees with Barrow, but if he is right, then the end of the century has brought with it yet another change in sensibility, perhaps the most important since Galileo and Copernicus: we are living near the end of the positive hour, and a 'post scientific age' awaits us. For many, this can't come soon enough, but it is important not to overstate the case – as John Maddox has shown, there is still plenty of science to be done. Nevertheless, science has always promised, however far down the road, an ultimate explanation of the universe. If, as Barrow and others tell us, that now looks like a *theoretical* impossibility, who can tell what the consequences will be? Where will the evolution of knowledge forms next lead?

One thing seems clear: as Eliot said, there's no going back. The arch-critics of science, with their own brand of secular zealotry, while they often skilfully describe why science can never be a complete answer to our philosophical condition, usually have little to add to it or replace it with. They tend either to look back to an age of religion or to recommend some sort of Heideggerean 'submission' to nature, to just 'be.' They lament the 'disenchantment' that has disappeared as we have turned away from God, but are unclear as to whether 'reenchantment' could ever be meaningful.

The British philosopher **Roger Scruton** is one of the most articulate of such thinkers. His *An Intelligent Person's Guide to Modern Culture* (1998) brilliantly punctures the pretensions, postures, and vacuities of modernist and popular culture, its failure to provide the 'experience of membership' that was true in an age of shared religious high culture, and laments how we can ever learn to judge 'in a world that will not be judged.' His view of science is sceptical: 'The human world is a world of significances, and no human significance can be fully grasped by science.' For Scruton, fiction, the imagination, the world of enchantment, is the highest calling, for it evokes sympathy for our condition, toleration, shared feelings, a longing that cannot be fulfilled, and 'processes' that, like Wagner's operas, lie deeper than words.[71]

Scruton is nostalgic for religion but does not make the most of its possibilities. Perhaps the most sophisticated religious postscientific argument has come from **John Polkinghorne**. A physicist by training, Polkinghorne studied with Paul Dirac, Murray Gell-Mann, and Richard Feynman, became professor of mathematical physics at Cambridge and therefore a close colleague of Stephen Hawking, and in 1982 was ordained as a priest in the Anglican Church. His thesis in *Beyond Science* (1996) has two elements: one, that 'our scientific, aesthetic, moral and spiritual powers greatly exceed what can convincingly be claimed to be needed in the struggle for survival, and to regard them as merely a fortunate but fortuitous by-product of that struggle is not to treat the mystery of their existence with adequate seriousness';[72] and two, that 'the evolution of conscious life seems the most significant thing that has happened in cosmic history and we are right to be intrigued by the fact that so special a universe is required for its possibility.'[73] In fact, Polkinghorne's main argument for his belief in a creator is the anthropic principle – that our universe is so finely tuned, providing laws of physics that allow for our existence, that a creator must be behind it all. This is an updated argument as compared with those of the bishop of Birmingham and Dean Inge in the 1930s, but Polkinghorne's case for God still lies in the details that we don't – and maybe can't – grasp. In that sense it is no different from any of the arguments about religion and science that have gone before.[74]

In his intellectual autobiography, *Confessions of a Philosopher* (1997), Bryan Magee writes as follows: 'Not being religious myself, yet believing that most of reality is likely to be permanently unknowable to human beings, I see a compelling need for the demystification of the unknowable. It seems to me that most people tend either to believe that all reality is in principle knowable or to believe that there is a religious dimension to things. A third alternative – that we can know very little but have equally little ground for religious belief – receives scant consideration, and yet seems to me to be where the truth lies.'[75] I largely share Magee's views as expressed here, and I also concur with the way he describes 'the main split in western philosophy.' There is, he says, the analytic approach, mainly identified with the logical positivists and British and American philosophers, who are fascinated by science and its implications and whose main aim is 'explanation, understanding, insight.'[76] In contrast to them are what are known in Britain and America as the 'continental' school of philosophers, led by such figures as Husserl and Heidegger but including Jacques Lacan, Louis Althusser, Hans-Georg Gadamer, and Jürgen Habermas, and looking back to German philosophy – Kant, Hegel, Marx, and Nietzsche. These philosophers are not so interested in science as the analytic ones are, but they are interested in Freudian (and post-Freudian) psychology, in literature, and in politics. Their approach is rhetorical and partisan, more interested in comment than in understanding.[77] This is an important distinction, I think, because it divides some of our deepest thinkers between science, on the one hand, and Freud, literature, and politics on the other. Whatever we do, it seems we cannot get away from this divide, these 'two cultures,' and yet if I am right the main problems facing us require us to do so. In the twentieth century, what we may characterise as scientific/analytic reason has been a great success, by and large; political, partisan, and rhetorical reason, on the other hand, has been a catastrophe. The very strengths of analytic, positive/passive reason have lent political rhetorical reason an authority it does not deserve. George Orwell, above and before everyone, saw this and sought to bring home the point. Oswald Spengler and Werner Sombart's distinction between heroes and traders is recast as one between heroes and scientists.

Out of all this, however, it seems to me that we might still make something, at the very least an agenda for the way ahead. It is clear from the narrative of this book that the failures of science, as mentioned earlier in this chapter – in particular the failure of sociology and psychology – have been crucial to the past century, and have been intimately associated with the political disasters. The combined effects stemming from the successes of science, liberal democracy, free-market economics, and the mass media have produced an era of personal freedom and a realised individuality unrivalled in the past. This is no mean accomplishment, but it still leaves a lot to be achieved. Look at America's inability to deal with the race issue, which has cast its shadow down the century. Look at the ethnic cleansing in Rwanda and, more recently, in Kosovo, so reminiscent of both the Holocaust and Conrad's *Heart of Darkness*. Look at the figures for crime, drug abuse, illegitimacy, and abortion. All of these reflect, at some level, a breakdown in relations between different groups – different nations, different races, different tribes, different sexes, different families, different ages. The developments of the twentieth century have taught us more and more about ourselves as individuals, but they have not taught us much about ourselves as members of groups, interlocking groups, with *shared* responsibilities as well as rights. In sociology, the dominant influence of Marx has been to stress the way some groups (the middle classes, management) dominate and exploit others. This has caused massive neglect in the study of the other ways in which groups relate to one another. In psychology, Freud's emphasis on individual development, again allegedly based on self-interest, hostility, and competition, has put personal realisation above everything else.

The task before science is, therefore, as clear as it is urgent. It is to turn its attention to groups, groups of people, the psychology and sociology of groups, to explore how they relate to each other, how individuals relate to the different groups of which they are members (families, sexes, generations, races, nations), in the hope that we shall one day be able to understand and control such phenomena as racism, rape, and child and drug abuse.[78] As **Samuel Huntington** argued in *The Clash of Civilisations and the Remaking of World Order* (1996), the critical distinctions between people are not primarily ideological any more – they are cultural, i.e., group-based.[79] There is no question but that these are the most critical issues for sociology and psychology in the future.

One final observation about science, free-market economics, and the mass media. The vast majority of ideas in this book were produced in universities, away from the hurly-burly of the market. The people who have had the ideas reported in these pages did not, for the most part, do what they did for the money, but because they were curious. Figures like Peter Brook and Pierre Boulez have deliberately avoided the market system, so that their work could develop in ways not constrained by market considerations. By the same token, the mass medium that has made the greatest contribution to our intellectual and communal life – the BBC – is again deliberately removed from the raw market. We should be aware that knowledge, particularly the production of basic science, ethical philosophy and social commentary appear to be human activities that do not lend themselves to market economics, though they clearly flourish in protected environments *under* such a system. Universities have evolved into highly tolerant communities, for the most part, where people of different ages, different backgrounds, with different outlooks, interests, and skills, can explore varied ways of living together. We should never forget how precious universities are, and that with our current problems, as discussed in the pages above, and notwithstanding anything else said in the rest of this epilogue, universities may offer a way forward, a lead, out of the impasse facing psychology and sociology.

The New Humanities and a New Canon

Science apart, the major division in Western thought today, which affects philosophy, literature, religion, architecture, even history, is between the postmodernists, who are happy with the fragmented, disparate, 'carnival' of culture (to use Stanley Fish's phrase), and those traditionalists who genuinely feel this sells us short (the young particularly), that this approach involves an ethical betrayal, avoids judging what is better and what is less good in human achievement and, in so doing, hinders people in raising their game. Postmodernism and relativism are still in the ascendant, but for how much longer? While the cultures of Africa, Bali and other third world countries have been recovered, to an extent, and given a much needed boost, none has so far found the widespread resonance that the classical civilisations of the Middle East once enjoyed. No one doubts that jewels of art, learning and science have occurred in all places and at all times, and the identification and extension of this wide range has been a major achievement of twentieth-century scholarship. In particular, the vast body of knowledge concerning the early, pre-Columbus native Americans, has revealed a very rich set of interlocking cultures. But have these discoveries produced any body of written material, say, which causes us to re-think the way that we live? Has it revealed any body of law, or medicine, or technology which leads us to change our ways either of thinking or doing? Has it reveled a completely new literature or philosophy with a new vision? The blunt answer is no.

The possibility – one might almost say the probability – arises then, that, some time in the twenty-first century, we shall eventually enter a *post*-post-modern world, in which the arguments of Jean-François Lyotard, Clifford Geertz, Frederick Jameson, David Harvey and their colleagues are still accepted, but only up to a point. We shall have reached a stage where, even after all the cultures of the world have been recovered and described, there will *still* be a hierarchy of civilisations, in the sense that a few of them were vastly more important in shaping our world than others. It should be said that, at the end of the twentieth century, the traditional hierarchy (which implies the traditional met narrative'), despite various attempts to destabilise it, is not much changed.

Hiram Bingham's re-discovery of Machu Picchu, or Basil Davidson's 'recovery' of Mapungubwe, or Clifford Geertz's own 'thick description' of Balinese cockfights may, each in its own way, rival Plato's *Republic*, or Shakespeare's Falstaff, or Planck's quantum. But – and this is surely the main point – though they are all part of the emerging 'one story' that is the crucial achievement of twentieth-century scholarship, Machu Picchu, Mapungubwe and Bali did not help *shape* the one story anywhere near as directly as the more traditional ideas did.

It is not racist or ethnocentrist to insist on this. As Richard Rorty has correctly pointed out, thick descriptions of Balinese cockfights are themselves an achievement of *Western* anthropology. But I think the differences between the postmodernists and the traditionalists (for want of a better term) can be reconciled, at least partly. **Neil Postman** drew my attention to the fact that at the beginning of our century William James said that any subject, treated historically, can become a 'humanity.'[80] 'You can give humanistic value to almost anything by teaching it historically. Geology, economics, mechanics, are humanities when taught with reference to the successive achievements of the geniuses to which these sciences owe their being. Not taught thus, literature remains grammar, art a catalogue, history a list of dates, and natural science a sheet of formulas and weights and measures.' The narrative form, properly realised, brings with it a powerful authority, showing not only where we are at any point but how we arrived there. In the case of *this* narrative, the grand narrative that has emerged in the course of the twentieth century, the story is so overwhelming that I believe it can provide, or begin to provide, an antidote to some of the problems that have plagued our educational institutions in recent years – in particular, the so-called 'culture wars' and the battles over the Western canon.

As was mentioned earlier, many avenues of thought, many disciplines, are coming together to tell one story. The most powerful advocate of this idea has been E. O. Wilson, who even resurrected the term *consilience* to describe the process. In his 1998 book of that name, *Consilience: The Unity of Knowledge*, Wilson offered the arch-reductionist view of the world, not only describing the way scientific knowledge has come together but also putting forward the idea that one day science will be able to 'explain' art, religion, ethics, kinship patterns, forms of government, etiquette, fashion, courtship, gift-giving patterns, funeral rites, population policy, penal sanctions, and if that's not enough, virtually everything else.[81] At its most basic, he argued that colour preferences are largely innate, that the arts are innately focused toward certain themes, that metaphors are the consequence of spreading activation in the brain during learning and are therefore 'the building blocks of creative thought.'[82] Among the innate impulses that go to make up art are imitation, making things geometrical, and intensification. Good artists instinctively know what patterns arouse the brain most.[83] In myth and fiction 'as few as two dozen' plots cover most epic stories that make up the classical corpus of many societies. These include emigration of the tribe, meeting the forces of evil, apocalypse, sexual awakening. 'The dominating influence that spawned the arts was the need to impose order on the confusion caused by intelligence.'[84] 'We are entering a new era of existentialism,' says Wilson, 'not the old absurdist existentialism of Kierkegaard and Sartre, giving complete autonomy to the individual, but the concept that only unified learning, universally shared, makes accurate foresight and wise choice possible. ... In the course of all of it we are learning the fundamental principle that ethics is everything. Human social existence, unlike animal sociality, is based on the genetic propensity to form long-term contracts that evolve by culture into moral precepts and law.'[85]

In other words, for Wilson the arts also become part of one story. And it is this story, I suggest, which ought to become the basis of a new canon. Understanding this narrative, and the way it was arrived at, involves a good appreciation of all the important sciences, the significant phases of history, the rise and fall of civilisations, and the reasons for the underlying patterns. Great works of religion, literature, music, painting, and sculpture fit into this narrative, this system of understanding, in the sense that all cultures have been attempts to come to terms with both the natural and the supernatural world, to create beauty, produce knowledge, and get at the truth. The significance of language, the way languages are related and have evolved, and yet remain very different, fits in here. Evolution enables us to place the world of culture within the world of nature with as comfortable a fit as possible. It shows how groups are related to one another. In addition, this narrative shows how mankind is moving on, where the old ways of thought are being superseded. Many people will disagree with this argument, replying that there is no teleological direction in evolution. Even more won't like, or will be sceptical of the thrust of what I have to say. But I think the evidence speaks for itself.

That evidence, at the end of the century, suggests that we are already living in what may be called a crossover culture. While people lament the effects of the mass media on our intellectual life generally, an inspection of the shelves in any good bookstore more or less anywhere in the Western world shows that, on the other hand, one of the greatest growth areas is in what is called popular science. That phrase is in fact misleading, to the extent that many of these books are relatively difficult, examining for example the nature of matter, abstruse mathematics (Fermat's last theorem, longitude), the minutiae of evolution, the byways of palaeontology, the origin of time, the philosophy of science. But a growing number of people now accepts that one cannot call oneself educated unless one is up-to-date on these issues. The numbers are small, relatively speaking, but it remains true that both this category of book, and its shelf space on bookshop walls, barely existed twenty years ago.

To my mind this is very encouraging, not least because it will prevent too big a division opening up in our society between scientists and the rest. If – a big 'if' perhaps – the superstring revolution really does come to something, that something may prove very difficult for scientists to share with the rest of us. They are already at the limit as to what metaphor can explain and we must face at least the possibility that, some day, the secrets of the universe will only be truly available to those with an above-average grasp of mathematics. It is no use the rest of us saying that we don't like the way knowledge is going. That's where the advances are being made, and is an added reason why I am arguing for this particular new canon, taught – as James said – as a humanity, so that it is attractive to as wide a cross-section of people as possible.

Evolution is the story of us all. Physics, chemistry, and biology are international in a way that literature, art, or religion can never be. Although science may have begun in the West, there are now distinguished Indian, Arab, Japanese, and Chinese scientists in great numbers. (In July 1999 China announced its capability to produce a neutron bomb, an intellectual triumph of sorts.) This is not to provide a framework for avoiding difficult judgements: science and liberal democracy are, or were, Western ideas. Nor is it a way of evading debate over the Western literary canon. But studying twentieth-century thought, as a narrative, provides a new kind of humanity and a canon for life as it is now lived. In offering something common to us all, a sketch of an historical/intellectual canon, it also begins to address our remaining problems. It is something we can share.

NOTES AND REFERENCES

In these references, especially in regard to works published early in the century, I have given both the original publication details and, where appropriate, more recent editions and reprints. This is to aid readers who wish to pursue particular works, to show them where the more accessible versions are to be found. In addition, however, the publication history of key works also shows how the popularity of certain key ideas has varied down the years.

Quite naturally, there are fewer references for the last quarter of the book. These works have had much less chance to generate a secondary literature of commentary and criticism.

PREFACE

1. Saul Bellow, *Humboldt's Gift*, New York: Viking Press, 1975; Penguin paperback, 1996, page 4. The reference to the nightmare may be compared with James Joyce's *Ulysses*: 'History, Stephen said, is a nightmare from which I am trying to awake.' James Joyce, *Ulysses*, Paris: Shakespeare & Co., 1922; Penguin edition of the 1960 Bodley Head edition, 1992, page 42.

INTRODUCTION: AN EVOLUTION IN THE RULES OF THOUGHT

1. Michael Ignatieff, Interview with Isaiah Berlin, BBC 2, 24 November, 1997. See also: Michael Ignatieff, *Isaiah Berlin*, London: Chatto & Windus, 1998, p. 301.
2. Martin Gilbert, *The Twentieth Century: Volume I, 1900–1933*, London: HarperCollins, 1997.
3. Claude Lévi-Strauss and Didier Eribon, *De Prés et de Loin*, translated as *Conversations with Claude Lévi-Strauss*, Paula Wissig (translator), Chicago: Chicago University Press, 1988, page 119.
4. John Maddox, *What Remains to Be Discovered*, London: Macmillan, 1998, Introduction, pages 1–21.
5. Daniel C. Dennett, *Darwin's Dangerous Idea: Evolution and the Meanings of Life*, New York: Simon & Schuster, 1995, page 21.
6. Roger Smith, *The Fontana History of the Human Sciences*, London: Fontana Press, 1997, pages 577–578.

7. See, for example, Paul Langford, *A Polite and Commercial People: England 1727–1783*, Oxford: Oxford University Press, 1989.
8. Roger Scruton, *An Intelligent Person's Guide to Modern Culture*, London: Duckworth, 1998, page 42.
9. See Roger Shattuck, *Candor & Perversion: Literature, Education and the Arts*, New York: W. W. Norton, 1999, especially chapter six for a discussion of 'The Spiritual in Art', where the author argues that abstraction, or the absence of figuration in art, excludes analogies and correspondences – and therefore meaning.
10. John Brockman (editor), *The Third Culture: Beyond the Scientific Revolution*, New York: Simon & Schuster, 1995, pages 18–19.
11. Frank Kermode, *The Sense of an Ending*, Oxford: Oxford University Press, 1966; paperback edition, Oxford, 1968.

CHAPTER I: DISTURBING THE PEACE

1. Freud's works have been published in a 24-volume Standard Edition, translated from the German under the general editorship of James Strachey, in collaboration with Anna Freud. *The Interpretation of Dreams* is volume IV and V of this series. In this section, from the many biographies of Freud, I have used primarily Ronald Clark, *Freud: The Man and the Cause*, New York: Random House, 1980; and Giovanni Costigan, *Sigmund Freud: A Short Biography*, London: Robert Hale, 1967; but I also

recommend: Peter Gay, *A Life for Our Time*, London: J. M. Dent, 1988.

2. Costigan, *Op. cit.*, page 101.

3. *Ibid.*, page 100.

4. *Ibid.*, page 99.

5. *Ibid.*

6. William M. Johnston, *The Austrian Mind: An Intellectual and Social History 1848–1938*, Berkeley: University of California Press, 1972, pages 33–34.

7. Costigan, *Op. cit.*, pages 88–89.

8. Johnston, *Op. cit.*, page 40.

9. *Ibid.*, page 238. Costigan, *Op. cit.*, page 89.

10. Costigan, *Op. cit.*, page 89.

11. Johnston, *Op. cit.*, page 65.

12. Clark, *Op. cit.*, page 12.

13. Johnston, *Op. cit.*, page 223.

14. *Ibid.*, page 235.

15. *Ibid.*, page 236.

16. Costigan, *Op. cit.*, page 42.

17. *Ibid.*, pages 68ff.

18. *Ibid.*, page 70.

19. Clark, *Op. cit.*, page 180.

20. Costigan, *Op. cit.*, page 77; Clark, *Op. cit.*, page 181.

21. Clark, *Op. cit.*, page 185.

22. Costigan, *Op. cit.*, page 79.

23. Clark, *Op. cit.*, page 213–214; Costigan, *Op. cit.*, page 101.

24. Joan Evans, *Time and Chance: The Story of Arthur Evans and His Forebears*, London: Longmans, 1943, page 329.

25. *Ibid.*, pages 350–351.

26. Richard Stoneman, *Land of Lost Gods: The Search for Classical Greece*, London: Hutchinson, 1987, pages 268ff.

27. Donald Mackenzie, *Crete and Pre-Hellenic: Myths and Legends*, London: Senate, 1995, page 153.

28. Evans, *Op. cit.*, page 309.

29. *Ibid.*, pages 309–318.

30. Mackenzie, *Op. cit.*, page 116. Evans, *Op. cit.*, pages 318–327

31. Evans, *Op. cit.*, pages 329–330.

32. *Ibid.*, page 331.

33. Mackenzie, *Op. cit.*, page 118.

34. Evans, *Op. cit.*, pages 331ff; Mackenzie, *Op. cit.*, pages 187–190.

35. Ernst Mayr, *The Growth of Biological Thought*, Cambridge, Massachusetts: The Belknap Press of Harvard University Press, 1982, pages 727–729.

36. *Ibid.*, page 729; William R. Everdell, *The First Moderns*, Chicago: Chicago University Press, 1997, pages 162–163.

37. Mayr, *Op. cit.*, pages 722–726.

38. *Ibid.*, page 728.

39. *Ibid.*, page 730. For a more critical view of this sequence of events, see: Peter J. Bowler, *The Mendelian Revolution; The Emergence of Hereditarian Concepts in Modern Science and Society*, London: The Athlone Press, 1989, pages 110–116.

40. Mayr, *Op. cit.*, page 715. Everdell, *Op. cit.*, page 160.

41. *Ibid.*, page 734.

42. Everdell, *Op. cit.*, page 166.

43. Richard Rhodes, *The Making of the Atomic Bomb*, New York: Simon & Schuster, 1986, though I have used the Penguin paperback edition: London, 1988, page 30.

44. *Ibid.*, page 40.

45. *Ibid.*

46. Everdell, *Op. cit.*, page 167.

47. *Ibid.*

48. *Ibid.*, page 167; Rhodes, *Op. cit.*, pages 30–31.

49. Joel Davis, *Alternate Realities*, New York: Plenum, 1997, pages 215–219.

50. Everdell, *Op. cit.*, page 171.

51. *Ibid.*, page 166. Everdell, *Op. cit.*, page 175.

52. Davis, *Op. cit.*, page 218.

53. John Richardson, *A Life of Picasso, 1881–1906*, volume 1, London: Jonathan Cape, 1991, pages 159ff.

54. Everdell, *Op. cit.*, chapter 10, *passim*; Roger Shattuck, *The Banquet Years: The Origins of the Avant-Garde in France 1885 to World War One*, New York: Vintage, 1953, *passim*.

55. Richardson, *Op. cit.*, pages 159ff

56. Everdell, *Op. cit.*, chapter 10, *passim*.

57. Richardson, *Op. cit.*, page 172.

58. Everdell, *Op. cit.*, page 155.

59. John Berger, *The Success and Failure of Picasso*, Harmondsworth: Penguin, 1965, reprinted New York: Pantheon, 1980, page 67. Robert Hughes, *The Shock of the New*, London: Thames & Hudson, 1980 and 1991, pages 21 and 24.

CHAPTER 2: HALFWAY HOUSE

1. William R. Johnston, *The Austrian Mind, Op. cit.*, pages 147–148.

2. Hilde Spiel, *Vienna's Golden Autumn 1866–1938*, London: Weidenfeld & Nicolson, 1987, pages 55ff.

3. Johnston, *Op. cit.*, pages 77 and 120. See also: Spiel, *Op. cit.*, page 55, and George R. Marek, *Richard Strauss, The Life of a Non-Hero*, London: Victor Gollancz, 1967, page 166.

4. Allan Janik and Stephen Toulmin, *Wittgenstein's Vienna*, London: Weidenfeld & Nicolson, 1973, page 45.

5. Johnston, *Op. cit.*, page 77.

6. *Ibid.*, page 169 and, for therapeutic nihilism, page 223.

7. Janik and Toulmin, *Op. cit.*, page 45.

8. Franz Kuna, 'A Geography of Modernism: Vienna and Prague 1890–1928,' in Malcolm Bradbury and James McFarlane (editors), *Modernism: A Guide to European Literature 1890–1930*, London: Penguin, 1976, page 126.

9. Carl E. Schorske, *Fin-de-siècle Vienna: Politics and Culture*, London: Weidenfeld & Nicolson, New York: Knopf, 1980, pages 12–14.

10. Kuna, *Op. cit.*, page 126.

11. Janik and Toulmin, *Op. cit.*, pages 62–63.

12. Schorske, *Op. cit.*, page 14.

13. Kuna, *Op. cit.*, page 127.

14. Janik and Toulmin, *Op. cit.*, pages 114ff.

15. Schorske, *Op. cit.*, page 17.

16. *Ibid.*, page 18.

17. *Ibid.*, page 19.

18. *Ibid.*

19. Cf. T. S. Eliot in *Notes Towards the Definition of Culture*, discussed in chapter 26.

20. Schorske, *Op. cit.*, page 21.

21. *Ibid.*

22. Kuna, *Op. cit.*, page 128.

23. Janik and Toulmin, *Op. cit.*, page 92, where the authors also point out that Bruckner gave piano lessons to Ludwig Boltzmann and that Mahler 'would bring his psychological problems to Dr. Freud.'

24. Johnston, *Op. cit.*, page 291.

25. *Ibid.*, page 296.

26. *Ibid.*, page 294.

27. *Ibid.*, page 299.

28. William S. Everdell, *The First Moderns, Op. cit.*, page 190. See also Johnston, *Op. cit.*, pages 299–300.

29. Janik and Toulmin, *Op. cit.*, page 135.

30. Johnston, *Op. cit.*, pages 300–301.

31. *Ibid.*, page 301.

32. Everdell, *Op. cit.*, page 187.

33. *Ibid.*, page 191.

34. Johnston, *Op. cit.*, page 302.

35. *Ibid.*, pages 302–305.

36. Janik and Toulmin, *Op. cit.*, pages 71ff.

37. Johnston, *Op. cit.*, page 159.

38. *Ibid.*, pages 72–73; see also Johnston, *Op. cit.*, pages 159–160.

39. Johnston, *Op. cit.*, page 233.

40. *Ibid.*, pages 233–234.

41. *Ibid.*, page 234.

42. Janik and Toulmin, *Op. cit.*, page 96.

43. Schorske, *Op. cit.*, page 79.

44. *Ibid.*; see also Johnston, *Op. cit.*, page 150.

45. *Ibid.*; see also Schorske, *Op. cit.*, pages 83ff.

46. Schorske, *Op. cit.*, page 339.

47. Janik and Toulmin, *Op. cit.*, page 100.

48. *Ibid.*, page 94; see also Johnston, *Op. cit.*, page 144.

49. Schorske, *Op. cit.*, page 220.

50. *Ibid.*, pages 227–232.

51. *Ibid.*

52. Johnston, *Op. cit.*, page 144.

53. Janik and Toulmin, *Op. cit.*, page 133.

54. John T. Blackmore, *Ernst Mach: His Work, Life and Influence*, Berkeley: University of California Press, 1972, page 6.

55. *Ibid.*, pages 182–184.

56. Janik and Toulmin, *Op. cit.*, page 134.

57. *Ibid.* See also: Johnston, *Op. cit.*, page 183.

58. Blackmore, *Op. cit.*, pages 87ff.

59. Johnston, *Op. cit.*, page 184; Janik and Toulmin, *Op. cit.*, page 134.

60. Johnston, *Op. cit.*, page 186; Blackmore, *Op. cit.*, pages 232ff and 247ff.

CHAPTER 3: DARWIN'S HEART OF DARKNESS

1. John Ruskin, *Modern Painters: 5 Volumes*, Orpington, Kent: George Allen, 1844–1888.

2. Arthur Herman, *The Idea of Decline in Western History*, New York: The Free Press, 1997, page 221.

3. *Ibid.*, page 222.

4. Ivan Hannaford, *Race: The History of an Idea in the West*, Washington D.C. and Baltimore: The Woodrow Wilson Center Press and Johns Hopkins University Press, 1996, page 296.

5. Friedrich Nietzsche, *Will to Power*, New York: Random House, 1968, page 30.

6. Herman, *Op. cit.*, page 99.

7. *Ibid.*

8. *Ibid.*, pages 99–100.

9. *Ibid.*, page 102.

10. *Ibid.*, pages 102–103.

11. Richard Hofstadter, *Social Darwinism in American Thought*, Boston: Beacon Press, 1944, page 5.

12. Mike Hawkins, *Social Darwinism in European and American Thought 1860–1945*, Cambridge: Cambridge University Press, 1997, pages 109–118; see also Hofstadter, *Op. cit.*, pages 51–66.

13. Hofstadter, *Op. cit.*, pages 152–153.

14. *Ibid.*, page 41.

15. Hawkins, *Op. cit.*, page 132.

16. Hannaford, *Op. cit.*, pages 289–290. Hawkins, *Op. cit.*, page 133.

17. Hawkins, *Op. cit.*, pages 126–127.

18. *Ibid.*, page 178.

19. *Ibid.*, page 152.

20. Hannaford, *Op. cit.*, page 292.

21. Hawkins, *Op. cit.*, page 193.

22. *Ibid.*, page 196.

23. Hannaford, *Op. cit.*, pages 291–292.

24. Hawkins, *Op. cit.*, page 185.

25. *Ibid.*

26. *Ibid.*, page 219.

27. Hannaford, *Op. cit.*, page 338.

28. Johnston, *The Austrian Mind, Op. cit.*, page 364. Herman, *Op. cit.*, page 125.

29. Hawkins, *Op. cit.*, page 62.

30. *Ibid.*, page 201.

31. *Ibid.*

32. Hannaford, *Op. cit.*, page 330; see also Hawkins, *Op. cit.*, page 217.

33. Hawkins, *Op. cit.*, page 219.

34. Hannaford, *Op. cit.*, page 332.

35. Hawkins, *Op. cit.*, page 218.

36. Hawkins, *Op. cit.*, page 225.

37. *Ibid.*, page 242.

38. Johnston, *Op. cit.*, page 357.

39. Janik and Toulmin, *Wittgenstein's Vienna, Op. cit.*, pages 60–61.

40. *Ibid.*, page 61.

41. Johnston, *Op. cit.*, page 358.

42. Schorske, *Fin-de-siècle Vienna, Op. cit.*, page 164.

43. *Ibid.*, pages 166–167.

44. Johnston, *Op. cit.*, page 358.

45. Anthony Giddens, Introduction to: Max Weber, *The Protestant Ethic and the Spirit of Capitalism*, London and New York: Routledge, 1942 (reprint 1986), page vii.

46. *Ibid.*, page viii.

47. Donald G. Macrae, *Weber*, London: The Woburn Press, 1974, pages 30–32. See also: Hartmut

Lehmann and Guenther Roth, *Weber's Protestant Ethic*, Cambridge: Cambridge University Press, 1993, especially pages 73ff and 195ff.
48. *Ibid.*, page 58.
49. J. E. T. Eldridge (editor), *Max Weber: The Interpretation of Social Reality*, London: Michael Joseph, 1970, page 9.
50. Giddens, *Op. cit.*, page ix.
51. *Ibid.*, page 35.
52. *Ibid.*, page xi.
53. *Ibid.*
54. Eldridge, *Op. cit.*, pages 168–169.
55. Giddens, *Op. cit.*, page xii. Eldridge, *Op. cit.*, page 166.
56. *Ibid.*, pages xii–xiii.
57. *Ibid.*, page xvii.
58. Lehmann and Roth, *Op. cit.*, pages 327ff. See also: Giddens, *Op. cit.*, page xviii.
59. Eldridge, *Op. cit.*, page 281.
60. Hawkins, *Op. cit.*, page 307. In *Plough, Sword and Book: The Structure of Human History*, London: Collins Harvill, 1988, Ernest Gellner takes Weber's analysis further, arguing that the internalisation of norms makes Protestant societies more trusting, aiding economic activity (page 106). 'The stress on scripturalism is conducive to a high level of literacy' which means, he says, that high culture eventually becomes the majority culture. This promotes egalitarianism, and the modern anonymous society, simultaneously innovative and involving standardised measures and norms, promoting social order so characteristic of modernity (page 107).
61. Redmond O'Hanlon, *Joseph Conrad and Charles Darwin*, Edinburgh: Salamander Press, 1984, page 17.
62. D. C. R. A. Goonetilleke, *Joseph Conrad: Beyond Culture and Background*, London: Macmillan, 1990, pages 15ff.
63. O'Hanlon, *Op. cit.*, pages 126–127. See also: Kingsley Widner, 'Joseph Conrad', *Dictionary of Literary Biography*, Detroit: Bruccoli Clark, 1988, Volume 34, pages 43–82.
64. O'Hanlon, *Op. cit.*, pages 17ff.
65. *Ibid.*, pages 20–21.
66. Widner, *Op. cit.*, pages 43–82.
67. *Ibid.*
68. Joseph Conrad, *Heart of Darkness*, Edinburgh and London: William Blackwood, 1902; Penguin, 1995.
69. Goonetilleke, *Op. cit.*, pages 88–91.
70. Conrad, *Op. cit.*, page 20.
71. *Ibid.*, page 112.
72. Goonetilleke, *Op. cit.*, page 168; see also: R. W. Stalman, *The Art of Joseph Conrad: A Critical Symposium*, East Lansing: Michigan State University Press, 1960.
73. O'Hanlon, *Op. cit.*, page 26.
74. Richard Curle, *Joseph Conrad: A Study*, London: Kegan Paul, French, Trübner, 1914.
75. Goonetilleke, *Op. cit.*, page 85.
76. *Ibid.*, page 63.
77. Gary Adelman, *Heart of Darkness: Search for the Unconscious*, New York: Twayne, 1987, page 59.

CHAPTER 4: LES DEMOISELLES DE MODERNISME
1. Kurt Wilhelm, *Richard Strauss: An Intimate Portrait*, London: Thames & Hudson, 1989, pages 99–100. See also: Michael Kennedy, *Richard Strauss: Man, Musician, Enigma*, Cambridge: Cambridge University Press, 1999, pages 142–149, for this and other reactions.
2. See Malcolm Bradbury and James Mcfarlane, (editors), *Modernism*, *Op. cit.*, pages 97–101.
3. George R. Marek, *Richard Strauss, Op. cit.*, pages 15 and 27.
4. *Ibid.*, page 150.
5. Michael Kennedy, *Richard Strauss*, London: J. M. Dent, 1976, page 144.
6. Wilhelm, *Op. cit.*, page 100.
7. *Ibid.*
8. *Ibid.*, page 102.
9. *Ibid.*, page 103.
10. Wilhelm, *Op. cit.*, page 120; Kennedy, *Richard Strauss: Man, Musician, Enigma, Op. cit.*, page 152.
11. Wilhelm, *Op. cit.*, pages 120–121.
12. Kennedy, *Richard Strauss, Op. cit.*, page 161.
13. Marek, *Op. cit.*, page 183.
14. *Ibid.*, page 185.
15. Kennedy (1976), *Op. cit.*, page 45. See also: Bryan Gilliam (editor), *Richard Strauss and His World*, Princeton: Princeton University Press, 1992, pages 311ff, 'Strauss and the Viennese Critics.'
16. Marek, *Op. cit.*, page 182.
17. Kennedy (1976), *Op. cit.*, page 149.
18. Marek, *Op. cit.*, page 186.
19. Kennedy (1976), *Op. cit.*, page 150.
20. Marek, *Op. cit.*, page 316.
21. Hans H. Stuckenschmidt, *Schoenberg: His Life, World and Work*, London: John Calder, 1977, page 42.
22. Harold C. Schonberg, *The Lives of the Great Composers*, London: Davis-Poynter, 1970, page 516.
23. *Ibid.*, page 517.
24. Everdell, *The First Moderns, Op. cit.*, page 275.
25. Schonberg, *Op. cit.*, page 517.
26. Everdell, *Op. cit.*, page 266.
27. Stuckenschmidt, *Op. cit.*, page 88.
28. Schonberg, *Op. cit.*, page 520; see also: Stuckenschmidt, *Op. cit.*, page 141; and Schorske, *Op. cit.*, page 351.
29. Schonberg, *Op. cit.*, page 517.
30. *Ibid.*, page 518.
31. Everdell, *Op. cit.*, page 269; see also: Stuckenschmidt, *Op. cit.*, pages 88 and 123–124.
32. Stuckenschmidt, *Op. cit.*, page 94; see also: Schonberg, *Op. cit.*, page 400.
33. Everdell, *Op. cit.*, page 277.
34. *Ibid.*, page 279.
35. Paul Griffiths, *A Concise History of Modern Music*, London: Thames & Hudson, 1978, revised 1994, page 26. Everdell, *Op. cit.*, page 278.
36. Schorske, *Fin-de-Siècle, Op. cit.*, page 349.
37. Stuckenschmidt, *Op. cit.*, page 124.
38. Everdell, *Op. cit.*, pages 277–278.

39. *Ibid.*, page 279.

40. *Ibid.*, pages 280–281.

41. Stuckenschmidt, *Op. cit.*, page 124.

42. Schonberg, *Op. cit.*, page 520.

43. Schorske, *Op. cit.*, page 354.

44. Griffiths, *Op. cit.*, page 34.

45. Joan Allen Smith, *Schoenberg and his Circle*, New York: Macmillan, 1986, page 68.

46. Schonberg, *Op. cit.*, page 521.

47. Griffiths, *Op. cit.*, page 43. Everdell, *Op. cit.*, page 282.

48. Janik and Toulmin, *Wittgenstein's Vienna*, *Op. cit.*, page 107.

49. Schorske, *Op. cit.*, page 360.

50. See for example: James R. Mellow, *Charmed Circle: Gertrude Stein and Company*, London: Phaidon, 1974, pages 8ff.

51. John Russell, *The World of Matisse*, Amsterdam: Time-Life, 1989, page 74.

52. Jack Flam, *Matisse on Art* (revised edition), Berkeley: University of California Press, 1995, page 35.

53. Pierre Cabanne, *Pablo Picasso: His Life and Times*, New York: William Morrow, 1977, page 110.

54. André Malraux, *Picasso's Mask*, New York: Holt, Rinehart & Winston, 1976, pages 10–11.

55. Lael Westenbaker, *The World of Picasso, 1881–1973*, Amsterdam: Time-Life, 1980, pages 125ff.

56. Robert Hughes, *The Shock of the New*, *Op. cit.*, page 24.

57. Dora Vallier, 'Braque, la peinture et nous,' Paris: *Cahiers d'Art*, No. 1, 1954, pages 13–14.

58. *Ibid.*, page 14.

59. Hughes, *Op. cit.*, pages 27 and 29.

60. Arianna Stassinopoulos, *Picasso: Creator and Destroyer*, London: Weidenfeld & Nicolson, 1988, pages 96–97.

61. 'Testimony Against Gertrude Stein,' *Transition*, February 1935, No. 23, pages 13–14.

62. Everdell, *Op. cit.*, page 311.

63. *Ibid.*, page 314.

64. *Ibid.*, page 313.

65. Peg Weiss, *Kandinsky in Munich*, Princeton: Princeton University Press, 1979, pages 58–59.

66. *Ibid.*, pages 5–6.

67. K. Lindsay and P. Vergo (editors), *W. Kandinsky: Complete Writings on Art* (two vols), New York: G. K. Hall, 1982; reprinted in one volume, 1994, pages 371–372.

68. Weiss, *Op. cit.*, pages 28, 34 and 40.

69. Lindsay and Vergo (editors), *Op. cit.*, page 364, quoted in Everdell, *Op. cit.*, page 307.

70. Quoted in Hughes, *Op. cit.*, page 301.

71. Weiss, *Op. cit.*, page 91.

72. Algot Ruhe and Nancy Margaret Paul, *Henri Bergson: An Account of His Life and Philosophy*, London: Macmillan, 1914, page 2.

73. Jacques Chevallier, *Henri Bergson*, London: Ridier, 1928, pages 39–41.

74. Leszek Kolakowski, *Bergson*, Oxford: Oxford University Press, 1985, page 73.

75. Chevallier, *Op. cit.*, page 60.

76. Philippe Soulez (completed by Frédéric Worms), *Bergson: Biographie*, Paris: Flammarion, 1997, pages 93–94.

77. *New Catholic Encyclopaedia*, volume II, New York: McGraw-Hill, 1967, page 324.

78. Jacques Chevallier, *Bergson*, Paris: Plon, 1926.

79. Soulez, *Op. cit.*, pages 132–133.

80. Kolakowski, *Op. cit.*, pages 88–91.

81. Soulez, *Op. cit.*, pages 133–134.

82. *Ibid.*, pages 142–143.

83. *Ibid.*

84. *Ibid.*, pages 251ff.

85. *New Catholic Encyclopaedia*, volume X, New York: McGraw-Hill, 1967, page 1048.

86. *Ibid.*, volume IX, pages 991–995.

87. J. G. Frazer, *The Golden Bough*, London: Macmillan, 1890; revised 1900.

88. René Bazin, *Pius X*, London: Sands & Co., 1928, pages 11ff.

89. *The Catholic Encyclopaedia*, volume X, London: Caxton, 1911, page 415.

90. *Ibid.*, page 416. For an account of other reactions to *Pascendi*, see: A. N. Wilson, *God's Funeral*, London: John Murray, 1999, pages 349ff.

91. Quotes are from: John King Fairbank, *China: A New History*, Cambridge, Massachusetts: The Belknap Press of Harvard University Press, 1994, page 52.

92. Fairbank, *Op. cit.*, page 53.

93. Denis Twitchett and John K. Fairbank, *The Cambridge History of China, Volume 11, Late Ch'ing, 1800–1911, Part 2*, Cambridge: Cambridge University Press, 1980, pages 361–362; Fairbank, *Op. cit.*, page 218.

94. Fairbank, *Op. cit.*, page 224.

95. O. Edmund Clubb, *Twentieth-Century China*, New York and London: Columbia University Press, 1964, pages 25ff.

96. Fairbank, *Op. cit.*, page 232.

97. *Ibid.*, page 240.

98. *Ibid.*, page 243.

99. Jerome B. Grieder, *Intellectuals and the State in Modern China*, New York: Free Press/Macmillan, 1981, pages 35ff; Fairbank, *Op. cit.*, page 243.

CHAPTER 5: THE PRAGMATIC MIND OF AMERICA

1. Edward Bradby (editor), *The University Outside Europe*, Oxford: Oxford University Press, 1939, pages 285ff

2. *Ibid.*, *passim*.

3. Professor Robert Johnston, personal communication.

4. Bradby, *Op. cit.*, pages 39ff. See also: Samuel Eliot Morison (editor), *The Development of Harvard University*, Cambridge, Massachusetts: Harvard University Press, 1930, pages 11 and 158.

5. Morison, *Op. cit.*, page XC, and Abraham Flexner, *Universities: American, English, German*, Oxford: Oxford University Press, 1930, pages 85ff.

6. Bradby, *Op. cit.*, page 52. Flexner, *Op. cit.*, page 67. It was also noteworthy that in Germany scientific

leadership was concentrated in the universities, whereas in Britain the equivalent was located in the private academies, such as the Royal Society, and this also held back the development of the universities.

7. Flexner, Op. cit., page 124. Bradby, Op. cit., page 57.

8. Ibid., page 151. See also: E. R. Holme, The American University, Sydney: Angus & Robertson, 1920, pages 143ff. Bradby, Op. cit., pages 59–60.

9. Ray Fuller (editor), Seven Pioneers of Psychology, London: Routledge, 1995, page 21.

10. William James, Pragmatism, New York: Longman Green, 1907; reprinted New York: Dover, 1995, pages 4 and 5.

11. William James, Varieties of Religious Experience, London: Longman Green, 1902.

12. James, Pragmatism, Op. cit., page 20.

13. Ibid., pages 33ff.

14. Arthur Lovejoy, The Great Chain of Being, Cambridge, Massachusetts: Harvard University Press, 1936.

15. Ellen Key, The Century of the Child, New York: Putnam, 1909.

16. Richard Hofstadter, Anti-Intellectualism in American Life, Op. cit., page 362.

17. John Dewey, The School and Society, Chicago: University of Chicago Press, 1900; and John Dewey, with E. Dewey, The School of Tomorrow, London: Dent, 1915.

18. Hofstadter, Op. cit., page 366.

19. Ibid., page 386.

20. Morison, Op. cit., pages 534–535.

21. Frederick Winslow Taylor, The Principles of Scientific Management, New York: Harper & Bros, 1913.

22. Ibid., pages 60–61.

23. Morison, Op. cit., pages 539–540.

24. Hofstadter, Op. cit., Part IV, pages 233ff.

25. Ibid., page 266.

26. Ibid., page 267.

27. Ada Louise Huxtable, The Tall Building Artistically Reconsidered: The Search for a Skyscraper Style, New York: Pantheon, 1984.

28. John Gloag, The Architectural Interpretation of History, London: Adam and Charles Black, 1975, page 1.

29. Paul Goldberger, The Skyscraper, New York: Knopf, page 9 for a discussion of the significance of the Flatiron Building and page 38 for a reproduction of Steichen's photograph.

30. See ibid., page 38 for a reproduction of a famous greetings card of the Flatiron, called 'Downdrafts at the Flatiron,' with a drawing of a woman, her petticoats being raised by the wind.

31. Goldberger, Op. cit., pages 17ff.

32. John Burchard and Albert Bush-Brown, The Architecture of America, London: Victor Gollancz, 1967, page 145.

33. Goldberger, Op. cit., pages 22–23.

34. Ibid., page 18. See also: Hugh Morrison, Louis Sullivan: Prophet of Modern Architecture, Westport,

Connecticut: Greenwood Press, 1971 (reprint of 1935 edition).

35. Wesley Towner, The Elegant Auctioneers, New York: Hill & Wang, 1970, page 176.

36. Patrick Nuttgens, The Story of Architecture, Oxford: Phaidon, 1983.

37. William J. Curtis, Modern Architecture since 1900, Oxford: Phaidon, 1982, page 39.

38. Goldberger, Op. cit., pages 18–19. See also: Louis H. Sullivan, The Autobiography of an Idea, New York: Dover, 1956 (revised version of 1924 edition).

39. Goldberger, Op. cit., page 34.

40. For Sullivan's influence in Europe, see: Leonard K. Eaton, American Architecture Comes of Age: European Reaction to H. H. Richardson and Louis Sullivan, Cambridge, Massachusetts: MIT Press, 1972.

41. Goldberger, Op. cit., page 83.

42. Frank Lloyd Wright, An Autobiography, London: Quartet, 1977 (new edition) pages 50–52.

43. Goldberger, Op. cit., pages 87 and 89 for a picture of the design.

44. Henry Combs with Martin Caidin, Kill Devil Hill, London: Secker & Warburg, 1980, page 212.

45. Ibid., page 213.

46. Ibid., page 214.

47. Ibid.

48. Ibid., page 216.

49. C. H. Gibbs-Smith, A History of Flying, London: Batsford, 1953, pages 42ff.

50. Alphonse Berget, The Conquest of the Air, London: Heinemann, 1909, pages 82ff.

51. Combs and Caidin, Op. cit., pages 50–51.

52. Ibid., pages 36–38.

53. Ibid., pages 137–138.

54. Ibid., page 204.

55. Ibid., pages 216–217.

56. Gibbs-Smith, Op. cit., pages 242–245.

57. H. H. Arnason, A History of Modern Art, London: Thames & Hudson, 1977, page 410.

58. Robert Hughes, American Visions, London: The Harvill Press, 1997, page 323.

59. Arnason, Op. cit., page 410.

60. Martin Green, New York 1913, New York: Charles Scribner's Sons, 1988, page 137.

61. Quoted in Hughes (1997), Op. cit., page 325.

62. Ibid., page 327.

63. Green, Op. cit., page 140.

64. Hughes (1997), Op. cit., page 334.

65. Ibid., page 331.

66. Arnason, Op. cit., page 507.

67. Arthur Knight, The Liveliest Art, New York: Macmillan, 1957, pages 16–17.

68. Everdell, The First Moderns, Op. cit., page 203.

69. Ibid., page 204.

70. Richard Schickel, D. W. Griffith, London: Michael Joseph, 1984, pages 20–23.

71. Ibid., pages 129ff.

72. Ibid., page 131.

73. See the list in Schickel, Ibid., pages 638–640.

74. Ibid., page 132.

75. Ibid., page 134.

76. Knight, *Op. cit.*, pages 25–27.
77. Schickel, *Op. cit.*, page 116.

CHAPTER 6: $E=mc^2$, $]/\equiv/v + c_7H_{38}O_{43}$

1. Rhodes, *The Making of the Atomic Bomb*, *Op. cit.*, page 50. For the links between early empiricism and the Enlightenment, see Ernest Gellner, *Plough, Sword and Book*, *Op. cit.*, page 133.
2. Rhodes, *Op. cit.*, pages 41–42.
3. L. G. Wickham Legg (editor), *Dictionary of National Biography*, Oxford: Oxford University Press, 1949, page 766, column 2.
4. Rhodes, *Op. cit.*, page 43.
5. *Dictionary of National Biography*, *Op. cit.*, page 769, column 2.
6. Rhodes, *Op. cit.*, page 47.
7. *Ibid.*
8. David Wilson, *Rutherford: Simple Genius*, London: Hodder & Stoughton, 1983, page 291.
9. Wilson, *Op. cit.*, page 289.
10. Ernest Marsden, 'Rutherford at Manchester,' in J. B. Birks (editor), *Rutherford at Manchester*, London: Heywood & Co., 1962, page 8.
11. Rhodes, *Op. cit.*, pages 49–50.
12. Wilson, *Op. cit.*, pages 294 and 297.
13. Rhodes, *Op. cit.*, page 50.
14. Michael White and John Gribbin, *Einstein: A Life in Science*, London: Simon & Schuster, 1993, page 5.
15. *Ibid.*, page 9.
16. *Ibid.*, page 10.
17. *Ibid.*, page 8.
18. Ronald W. Clark, *Einstein: The Life and Times*, London: Hodder & Stoughton, 1973, page 16.
19. *Ibid.*, pages 76–83; see also White and Gribbin, *Op. cit.*, page 48.
20. Clark, *Einstein, Op. cit.*, pages 61–62.
21. See *Ibid.*, pages 89ff for others.
22. White and Gribbin, *Op. cit.*, page 95.
23. Clark, *Einstein, Op. cit.*, pages 100ff.
24. This section is based on Wiebe E. Bijker, *Of Bicycles, Bakelites and Bulbs: Towards a Theory of Sociological Change*, Cambridge, Massachusetts: MIT Press, chapter 3, pages 101–108.
25. Stephen Fenichell, *Plastic: The Making of a Synthetic Century*, New York: HarperCollins, 1996, page 86. Bijker, *Op. cit.*, page 130.
26. *Encyclopaedia Britannica*, London: William Benton, 1963, volume 18, page 40A.
27. Bijker, *Op. cit.*, pages 107–115. The 'Exploding Teeth' reference is at page 114.
28. *Ibid.*, page 119.
29. Fenichell, *Op. cit.*, page 89.
30. Bijker, *Op. cit.*, page 146.
31. *Encyclopaedia Britannica*, *Op. cit.*, page 40D. Bijker, *Op. cit.*, page 147.
32. Fenichell, *Op. cit.*, page 90.
33. Bijker, *Op. cit.*, page 147.
34. *Ibid.*, page 148.
35. *Ibid.*, page 158.
36. Fenichell, *Op. cit.*, page 91.
37. Bijker, *Op. cit.*, pages 159–160.

38. See also *Encyclopaedia Britannica*, *Op. cit.*, page 40D for further uses.
39. Bijker, *Op. cit.*, page 166.
40. Caroline Moorehead, *Bertrand Russell: A Life*, London: Sinclair Stevenson, 1992, page 2.
41. See: Ray Monk, *Bertrand Russell, The Spirit of Solitude*, London: Vintage, 1997, pages 667ff, for a bibliographical discussion of Russell's works.
42. Moorehead, *Op. cit.*, page 335.
43. *Ibid.*, page 35.
44. *Ibid.*, pages 46ff.
45. Ronald W. Clark, *The Life of Bertrand Russell*, London: Penguin, 1978, page 43. Moorehead, *Op. cit.*, page 49.
46. Moorehead, *Op. cit.*, pages 96ff.
47. *Ibid.*, pages 97–100.
48. Clark, *Bertrand Russell and His World*, London: Thames & Hudson, 1981, page 28; See also Monk, *Op. cit.*, page 153.
49. Monk, *Op. cit.*, pages 129ff and *passim*; Moorehead, *Op. cit.*, page 94.
50. Moorehead, *Op. cit.*, page 96.
51. Bertrand Russell, 'Whitehead and *Principia Mathematica*', *Mind*, volume lvii, No. 226, April 1948, pages 137–138.
52. Bertrand Russell, *The Autobiography of Bertrand Russell, 1872–1914*, London: George Allen & Unwin, 1967, page 152.
53. Moorehead, *Op. cit.*, pages 99ff.
54. Monk, *Op. cit.*, page 192.
55. *Ibid.*, page 193.
56. *Ibid.*, page 191.
57. Moorehead, *Op. cit.*, page 101.
58. *Ibid.*, page 102.
59. Monk, *Op. cit.*, page 193.
60. *Ibid.*, page 195.
61. M. Weatherall, *In Search of a Cure: A History of Pharmaceutical Discovery*, Oxford: Oxford University Press, 1990, page 83.
62. *Ibid.*, pages 84–85.
63. *Ibid.*, page 86.
64. Claude Quétel, *Le Mal de Naples: histoire de la syphilis*, Paris: Editions Seghers, 1986; translated as *History of Syphilis*, London: Polity Press in association with Basil Blackwell, 1990, pages 2ff.
65. Allan M. Brandt, *No Magic Bullet: A Social History of Venereal Disease in the United States since 1880*, Oxford: Oxford University Press, 1985, page 23.
66. Quétel, *Op. cit.*, page 149.
67. *Ibid.*, page 146.
68. *Ibid.*, page 152.
69. *Ibid.*, pages 157–158.
70. Martha Marquardt, *Paul Ehrlich*, London: Heinemann, 1949, page 163. Brandt, *Op. cit.*, page 40.
71. Quétel, *Op. cit.*, page 141.
72. Marquardt, *Op. cit.*, page 28.
73. *Ibid.*, pages 86ff.
74. *Ibid.*, page 160.
75. *Ibid.*, pages 163ff.
76. *Ibid.*, page 168.
77. *Ibid.*, pages 175–176.

78. Sigmund Freud, *Three Essays on the Theory of Sexuality*, 1905, now available as part of volume VII of the *Collected Works* (see chapter I, note 1, *supra*), pages 20–21n.

CHAPTER 7: LADDERS OF BLOOD

1. David Levering Lewis, *W. E. B. Du Bois: A Biography of a Race*, New York: Holt, 1993, page 392.
2. *Ibid.*, pages 387–389.
3. Manning Marable, *W. E. B. Du Bois: Black Radical Democrat*, Boston: Twayne, 1986, page 98.
4. Lewis, *Op. cit.*, page 393.
5. Marable, *Op. cit.*, pages 52ff.
6. Lewis, *Op. cit.*, page 33.
7. Marable, *Op. cit.*, page 49.
8. Lewis, *Op. cit.*, pages 302–303.
9. *Ibid.*, page 316.
10. *Ibid.*, pages 387ff.
11. Marable, *Op. cit.*, page 73.
12. Lewis, *Op. cit.*, page 404.
13. *Ibid.*, page 406.
14. Marable, *Op. cit.*, page 73.
15. Lewis, *Op. cit.*, page 405.
16. Everdell, *The First Moderns, Op. cit.*, page 209.
17. *Ibid.*, pages 210 and 215–219.
18. *Ibid.*, page 217.
19. Mike Hawkins, *Social Darwinism in European and American Thought, Op. cit.*, pages 239–240.
20. *Ibid.*, pages 229–230.
21. Kenneth M. Ludmerer, *Genetics and American Society*, Baltimore: Johns Hopkins University Press, 1972, page 60.
22. Ernst Mayr, *The Growth of Biological Thought, Op. cit.*, pages 752ff.
23. Bruce Wallace, *The Search for the Gene*, Ithaca: Cornell University Press, 1992, page 56.
24. Mayr, *Op. cit.*, pages 750–751.
25. Wallace, *Op. cit.*, pages 57–58; Mayr, *Op. cit.*, page 748.
26. Peter J. Bowler, *The Mendelian Revolution, Op. cit.*, page 132; Mayr, *Op. cit.*, page 752.
27. Mayr, *Op. cit.*, page 753.
28. T. H. Morgan, A. H. Sturtevant, H. J. Muller and C. B. Bridges, *The Mechanism of Mendelian Inheritance*, New York: Henry Holt, 1915; see also Bowler, *Op. cit.*, page 134.
29. Bowler, *Op. cit.*, page 144.
30. Melville J. Herskovits, *Franz Boas: The Science of Man in the Making*, New York: Charles Scribner's Sons, 1953, page 17. For Boas's political views and his dislike of the German political system, see: Douglas Cole, *Franz Boas: The Early Years 1858–1906*, Vancouver/Toronto: Douglas & McIntyre and the University of Washington Press, Washington and London, 1999, pages 278ff.
31. Ludmerer, *Op. cit.*, page 25.
32. Franz Boas, *The Mind of Primitive Man*, New York: Macmillan, 1911, pages 53ff for context.
33. Ludmerer, *Op. cit.*, page 97.
34. Franz Boas, *Op. cit.*, page 1.
35. Boas, *Op. cit.*, pages 34ff.

36. *Ibid.*, pages 145ff.
37. *Ibid.*, pages 251ff.
38. *Ibid.*, page 278.
39. Bertrand Flornoy, *Inca Adventure*, London: George Allen & Unwin, 1956, page 195.
40. Hiram Bingham, *Lost City of the Incas*, London: Phoenix House, 1951, page 100.
41. John Hemming, *The Conquest of the Incas*, London: Macmillan, 1970; paperback edition 1993, page 243.
42. Bingham, *Op. cit.*, pages 50–52.
43. Hemming, *Op. cit.*, pages 463–464.
44. *Ibid.*, page 464.
45. Bingham, *Op. cit.*, page 141.
46. Flornoy, *Op. cit.*, page 194.
47. Bingham, *Op. cit.*, page 141.
48. Hemming, *Op. cit.*, page 464.
49. Bingham, *Op. cit.*, pages 142–143.
50. Nigel Davies, *The Incas*, Niwot, Colorado: University of Colorado Press, 1995, page 9.
51. Hemming, *Op. cit.*, page 469.
52. *Ibid.*, page 470.
53. Bingham, *Op. cit.*, page 152. Hemming, *Op. cit.*, page 470.
54. Hemming, *Op. cit.*, page 472.
55. David R. Oldroyd, *Thinking About the Earth*, London: The Athlone Press, 1996, page 250.
56. See map in *ibid.*, page 251.
57. George Gamow, *Biography of the Earth*, London: Macmillan, 1941, page 133.
58. Oldroyd, *Op. cit.*, page 250.
59. R. Gheyselinck, *The Restless Earth*, London: The Scientific Book Club, 1939, page 281. See map of geosyncline in Oldroyd, *Op. cit.*, page 257.
60. Oldroyd, *Op. cit.*, pages 144 and 312 for other references.
61. Gamow, *Op. cit.*, pages 2ff.

CHAPTER 8: VOLCANO

1. Robert Frost, *A Boy's Will*, verse 2, 'The Trial by Existence', 1913; in *Robert Frost: Collected Poems, Prose and Plays*, New York: The Library of America, 1995, page 28. Everdell, *Op. cit.*, where Chapter 21, 'Annus Mirabilis', is given to 1913.
2. John Rewald, *Cézanne and America: Dealers, Collectors, Artists and Critics*, Princeton: Princeton University Press, 1989, page 175.
3. Judith Zilczer, *The Noble Buyer: John Quinn, Patron of the Avant-Garde*, Washington, D.C.: Published for the Hirschhorn Museum by the Smithsonian Institution Press.
4. Milton Brown, *The Story of the Armory Show*, New York: Abbeville Press, 1988, pages 107ff.
5. Peter Watson, *From Manet to Manhattan: The Rise of the Modern Art Market*, London: Hutchinson, 1992; New York: Random House, 1992, pages 176ff.
6. Rewald, *Op. cit.*, pages 166–168; Brown, *Op. cit.*, pages 64–73.
7. Watson, *Op. cit.*, page 179.
8. Brown, *Op. cit.*, pages 133ff.
9. *Ibid.*, page 143.

10. *Ibid.*, pages 119ff and 238–239.
11. Roger Shattuck, *The Banquet Years*, *Op. cit.*, pages 282–283.
12. Marcel Adéma, *Apollinaire*, London: Heinemann, 1954, page 162.
13. *Ibid.*, pages 163–164; Everdell, *Op. cit.*, page 330.
14. Adéma, *Op. cit.*, page 164.
15. Everdell, *The First Moderns*, *Op. cit.*, page 330.
16. For an excellent introduction to Apollinaire, see: Shattuck, *The Banquet Years*, *Op. cit.*, chapters 9 and 10, pages 253–322.
17. Schonberg, *The Lives of the Great Composers*, *Op. cit.*, page 431.
18. Everdell, *Op. cit.*, pages 329–330.
19. Peter Watson, *Nureyev: A Biography*, London: Hodder & Stoughton, 1994, pages 87–88.
20. Schonberg, *Op. cit.*, page 433.
21. *Ibid.*
22. *Ibid.*, page 434.
23. *Ibid.*
24. Richard Buckle, *Diaghilev*, London: Weidenfeld & Nicolson, 1979, page 175.
25. Schonberg, *Op. cit.*, page 430.
26. Everdell, *Op. cit.*, page 331.
27. Buckle, *Op. cit.*, page 251.
28. Schonberg, *Op. cit.*, page 431; Buckle, *Op. cit.*, page 253.
29. Schonberg, *Op. cit.*, page 431.
30. Buckle, *Op. cit.*, page 254.
31. *Ibid.*, page 255.
32. Everdell, *Op. cit.*, page 333.
33. Henri Quittard, *Le Figaro*, 31 May 1913; quoted in Everdell, *Op. cit.*, page 333. The reference to the 'music subconscious' is in Schonberg, *Op. cit.*, page 432.
34. Everdell, *Op. cit.*, page 335.
35. Clark, *Einstein*, *Op. cit.*, page 199.
36. White and Gribbin, *Einstein*, *Op. cit.*, pages 132–133.
37. Clark, *Einstein*, *Op. cit.*, page 241.
38. White and Gribbin, *Op. cit.*, page 135.
39. C. P. Snow, *The Physicists*, London: Macmillan, 1981, page 56.
40. Rhodes, *The Making of the Atomic Bomb*, *Op. cit.*, page 69; Snow, *Op. cit.*, page 58.
41. Ruth Moore, *Niels Bohr: The Man and the Scientist*, London: Hodder & Stoughton, 1967, page 71. See also Rhodes, *Op. cit.*, pages 69–70.
42. Rhodes, *Op. cit.*, pages 70ff.
43. Moore, *Op. cit.*, page 59.
44. Snow, *Op. cit.*, page 57.
45. *Ibid.*, page 58.
46. David Luke, Introduction, in Thomas Mann, *Death in Venice and Other Stories*, translated and with an introduction by David Luke, London: Minerva, 1990, page ix.
47. *Ibid.*, page xxxv.
48. Ronald Hayman, *Thomas Mann*, New York: Scribner, 1995, page 252.
49. Luke, *Op. cit.*, pages xxxiv–xli.
50. Brenda Maddox, *The Married Man: A Life of D.*

H. Lawrence, London: Sinclair Stevenson, 1994, page 36.
51. Helen Baron and Carl Baron, Introduction to: D. H. Lawrence, *Sons and Lovers*, London: Heinemann, 1913; reprinted Cambridge University Press and Penguin Books, 1992, page xviii.
52. James T. Boulton (editor), *The Letters of D. H. Lawrence*, volume 1, Cambridge: Cambridge University Press, 1979, pages 476–477; quoted in Baron and Baron, *Op. cit.*, page xix.
53. Baron and Baron, *Op. cit.*, page xviii.
54. See: George Painter, *Marcel Proust: A Biography*, volume 2, London: Chatto & Windus, 1965, especially chapter 3. For the note on the unconscious, see Harold March, *The Two Worlds of Marcel Proust*, Oxford: Oxford University Press, 1948, pages 241 and 245.
55. See the index in Painter, *Op. cit.*, for details, pages 407ff.
56. Clark, *Freud*, *Op. cit.*, pages 305–306.
57. Janik and Toulmin, *Wittgenstein's Vienna*, *Op. cit.*, page 76, for the links Freud saw between Viennese social life and 'frustration.'
58. Frank McLynn, *Carl Gustav Jung*, London: Bantam Press, 1996, page 72.
59. *Ibid.*, pages 176ff.
60. Barbara Hannah, *Jung: His Life and Work*, London: Michael Joseph, 1977, page 69.
61. J. A. C. Brown, *Freud and the Post-Freudians*, Harmondsworth: Penguin, 1961, page 43. See also pages 46 and 48 for Jung's theory of the racial and collective unconscious, and page 43 for the 'evidence' in support of his theories.
62. McLynn, *Op. cit.*, page 305. Brown, *Op. cit.*, page 43.
63. Clark, *Freud*, *Op. cit.*, page 332.
64. Richard Noll, *The Aryan Christ: The Secret Life of Carl Gustav Jung*, London: Macmillan, 1997, page 108.
65. Clark, *Freud*, *Op. cit.*, page 331.
66. *Ibid.*, page 352.
67. *Ibid.*
68. Peter Gay, *A Life for Our Time*, London: J. M. Dent, 1988, page 332.
69. Clark, *Freud*, *Op. cit.*, page 356.
70. Gay, *Op. cit.*, page 242, who raises the question as to whether Freud 'needed' to make his friends into enemies.
71. Robert Frost, *Op. cit.*, verse 4: 'Reluctance,' page 38.

CHAPTER 9: COUNTER-ATTACK
1. Ronald Clark, *Freud*, *Op. cit.*, page 366.
2. *Ibid.*, page 366.
3. Caroline Moorehead, *Bertrand Russell: A Life*, *Op. cit.*, page 205.
4. John Richardson, *A Life of Picasso, 1907–1917: The Painter of Modern Life*, volume 2, London: Jonathan Cape, 1996, pages 344–345.
5. Everdell, *The First Moderns*, *Op. cit.*, page 346.
6. *Ibid.*
7. *Ibid.*

8. See for example: Paul Fussell, *The Great War and Modern Memory*, Oxford: Oxford University Press, 1975; and Jay Winter, *Sites of Memory, Sites of Mourning: The Great War in European Cultural History*, Cambridge: Cambridge University Press, 1995.

9. Fussell, *Op. cit.*, page 9.

10. *Ibid.*, page 11.

11. *Ibid.*, page 13.

12. *Ibid.*

13. *Ibid.*, page 14.

14. *Ibid.*, page 41.

15. *Ibid.*, page 18.

16. Maxwell Maltz, *The Evolution of Plastic Surgery*, New York: Froben Press, 1946, page 268.

17. Kenneth Walker, *The Story of Blood*, London: Herbert Jenkins, 1958, page 144.

18. Walker, *Op. cit.*, pages 152–153.

19. Harley Williams, *Your Heart*, London: Cassell, 1970, pages 74ff.

20. Walker, *Op. cit.*, page 144.

21. *Encyclopaedia Britannica*, London: William Bennett, 1963, volume 3, page 808.

22. Walker, *Op. cit.*, pages 148–149.

23. Stephen Jay Gould, *The Mismeasure of Man*, New York: W. W. Norton, 1981. Revised and expanded, Penguin, 1997, page 179.

24. Raymond E. Fancher, *The Intelligence Men: Makers of the IQ Controversy*, New York: W. W. Norton, 1985, page 60.

25. Gould, *Op. cit.*, page 179.

26. *Ibid.*, page 386.

27. *Ibid.*, page 188.

28. Fancher, *Op. cit.*, page 107.

29. Gould, *Op. cit.*, page 190.

30. H. J. Eysenck and Leon Kamin, *Intelligence: The Battle for the Mind*, London: Macmillan, 1981, page 93.

31. Gould, *Op. cit.*, pages 286ff.

32. Fancher, *Op. cit.*, pages 136–137.

33. *Ibid.*, pages 144–145.

34. Gould, *Op. cit.*, page 222.

35. *Ibid.*, page 223.

36. *Ibid.*, page 224.

37. Fancher, *Op. cit.*, pages 124ff.

38. Gould, *Op. cit.*, page 227.

39. *Ibid.*, pages 254ff.

40. Clark, *Freud, Op. cit.*, pages 366–367.

41. *Ibid.*, page 375.

42. John Rawlings Rees, *The Shaping of Psychiatry by War*, New York: W. W. Norton, 1945, page 113.

43. Rees, *Op. cit.*, page 28.

44. Emanuel Miller (editor), *The Neuroses in War*, London: Macmillan, 1945, page 8.

45. Peter Gay, *Op. cit.*, page 376.

46. Clark, *Freud, Op. cit.*, pages 386–387.

47. *Ibid.*, pages 404–405.

48. Fussell, *Op. cit.*, page 355.

49. Bernard Bergonzi, *Heroes' Twilight: A Study of the Literature of the Great War*, London: Macmillan, 1978, page 32.

50. *Ibid.*, pages 42 and 44.

51. *Ibid.*, page 36.

52. John Silkin, *Out of Battle*, Oxford: Oxford University Press, 1972, page 65.

53. Bergonzi, *Op. cit.*, page 41.

54. *Ibid.*

55. Martin Seymour-Smith, *Robert Graves: His Life and Work*, London: Bloomsbury, 1995, pages 49–50.

56. Bergonzi, *Op. cit.*, pages 65–66; Desmond Graham, 'Poetry of the First World War,' in Dodsworth (editor), *Op. cit.*, page 124.

57. Martin Seymour-Smith, 'Graves', in Ian Hamilton (editor), *The Oxford Companion to Twentieth-Century Poetry*, Oxford: Oxford University Press, 1994, page 194.

58. Silkin, *Op. cit.*, page 249.

59. *Ibid.*, page 250.

60. *Ibid.*, page 276.

61. Kenneth Simcox, *Wilfred Owen: Anthem for a Doomed Youth*, London: The Woburn Press, 1987, pages 50ff.

62. Simcox, *Op. cit.*, page 129.

63. Bergonzi, *Op. cit.*, page 127 and Silkin, *Op. cit.*, page 207.

64. Silkin, *Op. cit.*, page 232.

65. Fussell, *Op. cit.*, pages 7–18 and 79 (for the '*versus* habit').

66. Winter, *Op. cit.*, pages 78ff.

67. *Ibid.*, page 132.

68. *Ibid.*, page 57.

69. *Ibid.*, pages 133ff.

70. Ray Monk, *Ludwig Wittgenstein: The Duty of Genius*, London: Jonathan Cape, 1990, page 112.

71. *Ibid.*, page 112.

72. Janik and Toulmin, *Op. cit.*, pages 167ff.

73. Monk, *Op. cit.*, page 12.

74. *Ibid.*, page 15.

75. *Ibid.*, pages 30–33.

76. Brian McGuinness, *Wittgenstein: A Life, Volume One, Young Ludwig, 1889–1921*, London: Duckworth, 1988, page 84.

77. Janik and Toulmin, *Wittgenstein's Vienna, Op. cit.*, page 176.

78. Monk, *Op. cit.*, page 48.

79. McGuinness, *Op. cit.*, pages 179–180.

80. Monk, *Op. cit.*, page 138.

81. *Ibid.*, page 145.

82. McGuinness, *Op. cit.*, page 263.

83. Monk, *Op. cit.*, pages 149–150.

84. McGuinness, *Op. cit.*, page 264.

85. Georg Henrik von Wright, *Wittgenstein*, Oxford: Basil Blackwell, 1982, page 77.

86. Monk, *Op. cit.*, pages 157 and 180ff.

87. Magee, *Op. cit.*, page 82; Monk, *Op. cit.*, page 215.

88. *Ibid.*, page 222.

89. See Janik and Toulmin, *Op. cit.*, for comments on both the Vienna Circle and Witt- genstein (pages 214–215) and some other reactions to *Tractatus* (pages 180–201).

90. Monk, *Op. cit.*, page 156. For details, with Commentary, see McGuinness, *Op. cit.*, chapter 9, pages 296–316. P. M. S. Hacker, *Wittgenstein*, London: Phoenix, 1997, *passim*.

91. McGuinness, *Op. cit.*, page 300. Magee, *Op. cit.*, pages 80 and 85.

92. Van Wright, *Op. cit.*, page 145.

93. For this paragraph I have relied on: Robert Short, 'Dada and Surrealism', in Malcolm Bradbury and James McFarlane (editors), *Modernism, Op. cit.*, page 293.

94. William S. Rubin, *Dada and Surrealist Art*, London: Thames & Hudson, 1969, page 63.

95. Short, *Op. cit.*, page 295.

96. Rubin, *Op. cit.*, page 36.

97. Hughes, *The Shock of the New, Op. cit.*, page 61.

98. Short, *Op. cit.*, page 295.

99. Hughes, *Op. cit.*, page 61.

100. Rubin, *Op. cit.*, pages 40–41.

101. Hughes, *Op. cit.*, page 61.

102. Rubin, *Op. cit.*, pages 52–56.

103. Hughes, *Op. cit.*, pages 64–66.

104. *Ibid.*, pages 67–68.

105. Short, *Op. cit.*, page 296.

106. *Ibid.*

107. Rubin, *Op. cit.*, pages 42–46.

108. *Ibid.*

109. Hughes, *Op. cit.*, pages 75–78.

110. Short, *Op. cit.*, page 299.

111. *Ibid.*, page 300.

112. *Ibid.*, page 300.

113. Anna Balakian, *André Breton: Magus of Surrealism*, New York: Oxford University Press, 1971, pages 61 and 86–101.

114. Short, *Op. cit.*, page 300.

115. Beverly Whitney Kean, *French Painters, Russian Collectors*, London: Barrie & Jenkins, 1985, page 144.

116. Hughes, *Op. cit.*, page 81.

117. L. A. Magnus and K. Walter, Introduction to *Three Plays of A. V. Lunacharski*, London: George Routledge & Co., 1923, page v.

118. For a discussion of this see: Timothy Edward O'Connor, *The Politics of Soviet Culture: Anatoli Lunacharskii*, Ann Arbor, Michigan: University of Michigan Press, 1983, pages 68–69.

119. Magnus and Walter, *Op. cit.*, page vii.

120. Hughes, *Op. cit.*, page 87.

121. *Ibid.*

122. *Ibid.*

123. Galina Demosfenova, *Malevich: Artist and Theoretician*, Paris: Flammarion, 1990, page 10.

124. *Ibid.*, page 14.

125. Hughes, *Op. cit.*, page 89.

126. Demosfenova, *Op. cit.*, page 14.

127. Hughes, *Op. cit.*, page 89.

128. Demosfenova, *Op. cit.*, pages 197–198.

129. Hughes, *Op. cit.*, page 92.

130. Magdalena Dabrowski, Leah Dickerman and Peter Galassi, *Aleksandr Rodchenko*, New York: Harry. N. Abrams, 1998, pages 44–45.

131. Hughes, *Op. cit.*, page 93.

132. *Ibid.*, page 95.

133. Dabrowksi *et al.*, *Op. cit.*, pages 63ff.

134. *Ibid.*, page 124.

135. 'The Future is our only Goal', in Peter Noever (editor), *Aleksandr Rodchenko and Varvora F. Stepanova*, Munich: Prestel Verlag, 1991, page 158.

136. 'The Discipline of Construction, leader Rodchenko', in Noever, *Op. cit.*, page 237.

CHAPTER 10: ECLIPSE

1. Oswald Spengler, *The Decline of the West*, translated by Charles Francis Atkinson, published in two volumes: volume one: *Der Untergang des Abendlandes: Gestalt und Wirklichkeit*, Munich: C. H. Beck'sche Verlags Buchhandlung, 1918; and volume two: *Der Untergang des Abendlandes: Welt Historische Perspektiven*, same publisher, 1922.

2. See also: Herman, *The Idea of Decline in Western History, Op. cit.*, page 228.

3. *Ibid.*, pages 231–232.

4. Arthur Helps (editor and translator), *Spengler Letters*, London: George Allen & Unwin, 1966, page 17. Herman, *Op. cit.*, pages 233–234.

5. Herman, *Op. cit.*, page 234.

6. *Ibid.*, page 235.

7. Spengler, *Op. cit.*, volume one, page 21.

8. Spengler, *Op. cit.*, volume two, page 90. See also: Herman, *Op. cit.*, page 240.

9. Helps, *Op. cit.*, page 31, letter to Hans Klöres, 25 October 1914.

10. Thomas Mann, *Diaries, 1918–1939*, entry for 2 July 1919, Frankfurt, 1979–82, Peter de Mendelssohn (editor), pages 61–64.

11. Herman, *Op. cit.*, pages 244–245.

12. Helps, *Op. cit.*, page 133, letter to Elisabeth Förster-Nietzsche, 18 September, 1923.

13. Herman, *Op. cit.*, page 246–247.

14. Bruce Arnold, *Orpen: Mirror to an Age*, London: Jonathan Cape, 1981, page 365. 'The Signing of the Peace in the Hall of Mirrors, Versailles, 28 June 1919', oil on canvas, 60×50 inches, is in the Imperial War Museum, London.

15. D. E. Moggridge, *Maynard Keynes: An Economist's Biography*, London and New York: Routledge, 1992, page 6. Women were not allowed to graduate at Cambridge until 1947.

16. Robert Skidelsky, *John Maynard Keynes, volume one: Hopes Betrayed*, London: Macmillan, 1983, page 131.

17. *Ibid.*, page 176.

18. Moggridge, *Op. cit.*, pages 282–283.

19. Skidelsky, *Op. cit.*, page 382.

20. John Howard Morrow, *The Great War in the Air: Military Aviation from 1909–1921*, Washington D.C.: The Smithsonian Institution Press, 1993, page 354.

21. Trevor Wilson, *The Myriad Faces of War: Britain and the Great War, 1914–1918*, Cambridge: Polity Press, 1986, pages 839–841.

22. Moggridge, *Op. cit.*, pages 341ff; Skidelsky, *Op. cit.*, pages 397ff; Etienne Mantoux, *The Carthaginian Peace; or, The Economic Consequences of Mr Keynes*, London: Oxford University Press, 1946.

23. *The Economic Consequences of the Peace* (1919) is now available as volume II (1971) of *The Collected*

Writings of John Maynard Keynes (30 vols 1971–1989), Managing Editors Sir Austin Robinson and Donald Moggridge, London: Macmillan, 1971–1989.

24. John Fairbanks, *China, Op. cit.*, pages 267–268. Immanuel C. Y. Hsü, *The Rise of Modern China*, New York and Oxford: Oxford University Press, revised edition, 1983, page 501, says 5,000.

25. Fairbanks, *Op. cit.*, page 268; Hsü, *Op. cit.*, pages 569–570.

26. Chow Tse-tung, *The May Fourth Movement: Intellectual Revolution in Modern China*, Cambridge, Massachusetts: Harvard University Press, 1960, pages 84ff and Part Two, pages 269ff.

27. Hsü, *Op. cit.*, pages 422–423.

28. Fairbanks, *Op. cit.*, page 258.

29. *Ibid.*, pages 261–264.

30. *Ibid.*, page 265.

31. *Ibid.*

32. Tse-tung, *Op. cit.*, pages 171ff.

33. Fairbank, *Op. cit.*, page 266.

34. *Ibid.*

35. Hsü, *Op. cit.*, pages 569–570.

36. See Tse-tung, *Op. cit.*, pages 178–179 for a list.

37. Fairbank, *Op. cit.*, page 268.

38. *Ibid.*, pages 269ff.

39. Paul Johnson, *The Modern World, Op. cit.*, page 197. Fairbanks, *Op. cit.*, pages 275–276.

40. William Johnston, *The Austrian Mind, Op. cit.*, page 73.

41. *Ibid.*

42. Janik and Toulmin, *Wittgenstein's Vienna, Op. cit.*, pages 239–240.

43. M. Weatherall, *In Search of a Cure, Op. cit.*, page 128.

44. Arpad Kadarkay, *Georg Lukács: Life, Thought and Politics*, Oxford: Basil Blackwell, 1991, page 177. Mary Gluck, *Georg Lukács and His Generation*, Cambridge, Massachusetts: Harvard University Press, 1985, page 14.

45. *Ibid.*, page 22, for the discussion of Simmel, page 131 for Gauguin and page 147 for the Manet remark.

46. *Ibid.*, page 154.

47. *Ibid.*, pages 154–155.

48. *Ibid.*, pages 156ff

49. Kadarkay, *Op. cit.*, page 195.

50. Gluck, *Op. cit.*, page 204.

51. *Ibid.*, page 205.

52. Kadarkay, *Op. cit.*, pages 248–249.

53. Gluck, *Op. cit.*, page 211.

54. A. Vibert Douglas, *The Life of Arthur Stanley Eddington*, London: Thomas Nelson & Sons, 1956, page 38.

55. L. P. Jacks, *Sir Arthur Eddington: Man of Science and Mystic*, Cambridge: Cambridge University Press, 1949. See pages 2 and 17.

56. John Gribbin, *Companion to the Cosmos*, London: Weidenfeld & Nicolson, 1996, Phoenix paperback, 1997, pages 92 and 571. See also: Douglas, *Op. cit.*, pages 54ff.

57. Douglas, *Op. cit.*, page 39.

58. *Ibid.*

59. *Ibid.*, page 40.

60. *Ibid.*

61. *Ibid.*

62. *Ibid.*, page 41; see also: Albrecht Fölsing, *Albert Einstein: A Biography*, New York: Viking, 1997, page 440.

63. Douglas, *Op. cit.*, page 42.

64. *Ibid.*, page 43. See also: Ronald W. Clark, *Einstein: The Life and Times, Op. cit.*, pages 224–225; and: Victor Lowe: *Alfred North Whitehead: The Man and His Work, volume II, 1910–1947*, edited by J. B. Schneewind, Baltimore and London: Johns Hopkins University Press, 1990, page 127 for Eddington on Whitehead and relativity.

CHAPTER 11: THE ACQUISITIVE WASTELAND

1. Ross Terrill, *R. H. Tawney and His Times: Socialism as Fellowship*, London: André Deutsch, 1974, page 53.

2. *Ibid.*, pages 53–56.

3. Anthony Wright, *R. H. Tawney*, Manchester: Manchester University Press, 1987, pages 48–49.

4. *Ibid.*, pages 35ff.

5. R. H. Tawney, *Religion and the Rise of Capitalism*, London: John Murray, 1926; published in Pelican Books 1938 and as a Penguin 20th Century Classic, 1990. See in particular chaper 3, section iii, and chapter 4, section iii.

6. Tawney, *Op. cit.*, chapter 3, section iii, chapter 4, section iii.

7. Wright, *Op. cit.*, page 148.

8. Peter Ackroyd, *T. S. Eliot*, London: Hamish Hamilton, 1984; Penguin edition, 1993, pages 61–64 and 113–114.

9. Stephen Coote, *T. S. Eliot: The Waste Land*, London: Penguin, 1985, page 10.

10. *Ibid.*, pages 12 and 94.

11. *Ibid.*, page 14. See also: Robert Sencourt, *T. S. Eliot: A Memoir*, London: Garnstone Press, 1971, page 85.

12. Boris Ford (editor), *The New Pelican Guide to English Literature: Volume 9: American Literature*, Penguin 1967, revised 1995, page 327.

13. Letter from Pound to Eliot, 24 December 1921, Paris. In Valerie Eliot (editor), *The Letters of T. S. Eliot, Volume I, 1889–1921*, London: Faber & Faber, 1988, page 497.

14. See Coote, *Op. cit.*, page 30 and in particular chapter 5, on the editing of *The Waste Land* manuscript, pages 89ff. And Ackroyd, *Op. cit.*, pages 113–126.

15. Sencourt, *Op. cit.*, page 89. Coote, *Op. cit.*, page 9.

16. Coote, *Op. cit.*, page 26.

17. *Ibid.*, pages 125–126 and 132–135.

18. Valerie Eliot, *Op. cit.*, pages 551–552. See also Coote, *Op. cit.*, page 17 for the 'escape from personality' reference.

19. Luigi Pirandello, *Six Characters in Search of an Author*, translated by Frederick May, London: Heinemann, 1954, reprinted 1975, page x.

20. May, *Op. cit.*, page viii. Mark Musa, Intro-

duction to the Penguin edition of *Six Characters in Search of an Author and Other Plays*, London: Penguin, 1995, pages xi and xiv; see also: Benito Ortolani (editor and translator), *Pirandello's Love Letters to Marta Abba*, Princeton: Princeton University Press, 1994.

21. Gaspare Giudice, *Pirandello*, Oxford: Oxford University Press, 1975, page 119.

22. Frank Field, *The Last Days of Mankind: Karl Kraus and His Vienna*, London: Macmillan, 1967, page 14.

23. Field, *Op. cit.*, page 18.

24. *Ibid.*, page 102.

25. *Ibid.*, page 103.

26. W. Kraft, *Karl Kraus, Beiträge zum Verständnis seines Werkes*, Salzburg, 1956, page 13; quoted in Field, *Op. cit.*, pages 242 and 269.

27. Coote, *Op. cit.*, page 28.

28. Richard Ellmann, *James Joyce*, New York: Oxford University Press, 1959, page 401.

29. Declan Kiberd, Introduction to James Joyce's *Ulysses*, Paris: Shakespeare & Co., 1922; Penguin edition of the 1960 Bodley Head edition, 1992, page lxxxi.

30. Ellmann, *Op. cit.*, page 672; John Wyse Jackson and Peter Costello, *John Stanislaus Joyce: The Voluminous Life and Genius of Joyce's Father*, London: Fourth Estate, 1997, pages 254–255.

31. Ellmann, *Op. cit.*, page 551.

32. Kiberd, *Op. cit.*, page xxxii.

33. James Joyce, *Ulysses*, *Op. cit.*, page 271.

34. *Ibid.*, page 595.

35. Kiberd, *Op. cit.*, pages xv and lx.

36. *Ibid.*, page xxiii.

37. *Ibid.*, pages xxx and xliv.

38. David Perkins, *A History of Modern Poetry, Volume 1*, Cambridge, Massachusetts: Harvard University Press, 1976, page 572.

39. *Ibid.*, page 601.

40. *Ibid.*, page 584.

41. *Ibid.*, page 596.

42. A. Norman Jeffares, *W. B. Yeats*, London: Hutchinson: 1988, page 261.

43. Perkins, *Op. cit.*, page 578.

44. Jeffares, *Op. cit.*, page 275.

45. James R. Mellow, *Invented Lives: F. Scott and Zelda Fitzgerald*, Boston: Houghton Mifflin, 1985, page 56.

46. F. Scott Fitzgerald, *The Great Gatsby*, London: Penguin, 1990, page 18.

47. Matthew Bruccoli, *Some Sort of Epic Grandeur: The Life of F. Scott Fitzgerald*, London: Hodder & Stoughton, 1981, page 221.

48. See *ibid.*, pages 217–218 for the revised ending of the book.

49. *Ibid.*, page 223.

50. Paul Johnson, *A History of the Modern World from 1917 to the 1980s*, *Op. cit.*, pages 9–10.

51. Harold March, *The Two Worlds of Marcel Proust*, Oxford: Oxford University Press, 1948, page 114.

52. *Ibid.*, pages 182–194.

53. *Ibid.*, page 228.

54. See *Ibid.*, pages 241–242 for a discussion of Freud and Proust.

55. George Painter, *André Gide: A Critical Biography*, London: Weidenfeld & Nicolson, 1968, page 142.

56. Justin O'Brien, *Portrait of André Gide: A Critical Biography*, London: Secker & Warburg, 1953, pages 254–255.

57. Painter, *Op. cit.*, page 143.

58. O'Brien, *Op. cit.*, page 195.

59. Kate Flint, Introduction to Oxford University Paperback edition of *Jacob's Room*, Oxford, 1992, pages xiii–xiv.

60. James King, *Virginia Woolf*, London: Hamish Hamilton, 1994, page 148.

61. *Ibid.*, pages 314–315. See: Hermione Lee, *Virginia Woolf*, London: Chatto & Windus, 1996, page 444 for Eliot's reaction.

62. Virginia Woolf, *Diaries*, 26 January 1920, quoted in Flint, *Op. cit.*, page xii.

63. *Ibid.*, page xiv.

64. King, *Op. cit.*, page 318.

65. Virginia Woolf, *Jacob's Room*, Oxford: Oxford University Press, 1992 edition, page 37, quoted in Flint, *Op. cit.*, page xv.

66. Robert Hughes, *The Shock of the New*, *Op. cit.*, page 212.

67. *Ibid.*, page 213.

68. *Ibid.*

69. Walter Hopps, *Ernst at Surrealism's Dawn: 1925–1927*, in William A. Camfield, *Max Ernst: Dada and the Dawn of Surrealism*, Munich: Prestel-Verlag, 1993, page 157.

70. Camfield, *Op. cit.*, page 158.

71. Hughes, *Op. cit.*, page 215.

72. See the sequence of piazzas, plates vii–xv, in Maurizio Fagiolo Dell'Arco, *De Chirico 1908–1924*, Milano: Rizzoli, 1984.

73. Hughes, *Op. cit.*, pages 217–221.

74. See 'The Politics of Bafflement', in Carolyn Lanchner, *Joan Miró*, New York: Harry N. Abrams, 1993, page 49.

75. *Ibid.*, pages 28–32.

76. Hughes, *Op. cit.*, pages 231 and 235.

77. *Ibid.*, pages 237–238. See also: Robert Descharnes, *The World of Salvador Dali*, London: Macmillan, 1962, page 63. For Dali's obsession with his appearance, see: Ian Gibson, *The Shameful Life of Salvador Dali*, London and Boston: Faber & Faber, 1997, pages 70–71.

78. Descharnes, *Op. cit.*, page 61. Gibson, *Op. cit.*, page 283.

79. A. M. Hammacher, *René Magritte*, London: Thames & Hudson, 1974, figures 81 and 88.

80. *Ibid.*, devotes a whole section to Magritte's titles.

CHAPTER 12: BABBITT'S MIDDLETOWN

1. Stephen Jay Gould, *The Mismeasure of Man*, *Op. cit.*, page 260.

2. *Ibid.*, page 261.

3. *Ibid.*

4. Laurie R. Godfrey (editor), *Scientists Confront*

Creationism, New York: W. W. Norton, 1983, *passim*.
5. Hofstadter, *Anti-Intellectualism in American Life*, New York: Knopf, 1963, page 126.
6. *Ibid.*, page 125.
7. Ronald L. Numbers, *Darwinism Comes to America*, Cambridge, Massachusetts: Harvard University Press, 1998, pages 77–89.
8. Hofstadter, *Op. cit.*, pages 124–125.
9. James M. Hutchisson, Introduction to: Sinclair Lewis, *Babbitt*, New York: Harcourt Brace & Co., 1922; Penguin edition, London, 1996, pages xiiff.
10. *Ibid.*, pages viii–xi.
11. *Ibid.*, xi.
12. Mark Schorer, *Sinclair Lewis: An American Life*, London: Heinemann, 1963, page 345. See also: Hutchisson, *Op. cit.*, page xii.
13. Hutchisson, *Op. cit.*, page xxvi.
14. Alfred Kazin, *On Native Grounds*, New York: Harcourt Brace, 1942; paperback, third edition, 1995, page 221.
15. Hutchisson, *Op. cit.*, page xvii.
16. Schorer, *Op. cit.*, pages 353–356.
17. Asa Briggs, *The Birth of Broadcasting*, Oxford and New York: Oxford University Press, 1961, page 65.
18. Theodore Peterson, *Magazines in the Twentieth Century*, Urbana: University of Illinois Press, 1956, pages 40ff and 211.
19. *Ibid.*, page 211.
20. *Ibid.*
21. Janice A. Radway, *A Feeling for Books: The Book-of-the-Month Club, Literary Taste and Middle Class Desire*, Chapel Hill: University of North Carolina Press, 1997, pages 195–196.
22. *Ibid.*, pages 221ff.
23. Robert S. and Helen Merrell Lynd, *Middletown: A Study in Contemporary American Culture*, London: Constable, 1929, page vi.
24. *Ibid.*, page 7.
25. *Ibid.*, page 249.
26. *Ibid.*, page 48.
27. *Ibid.*, pages 53ff.
28. *Ibid.*, page 83.
29. *Ibid.*, page 115.
30. *Ibid.*, page 532.
31. *Ibid.*, page 36.
32. David Levering Lewis, *When Harlem was in Vogue*, New York: Alfred A. Knopf, 1981, page 165.
33. *Ibid.*, page 168.
34. See George Hutchinson, *The Harlem Renaissance in Black and White*, Cambridge, Massachusetts: The Belknap Press of Harvard University Press, 1995, pages 396ff for a discussion.
35. Lewis, *Op. cit.*, pages 91–92.
36. Hutchinson, *Op. cit.*, pages 289–304 for a discussion of racial science in this context.
37. *Ibid.*, pages 145–146. See also: Lewis, *Op. cit.*, pages 34–35.
38. Lewis, *Op. cit.*, page 33.
39. *Ibid.*, pages 51ff.
40. *Ibid.*, pages 67–71.

41. Hutchinson, *Op. cit.*, page 396, which takes a critical approach to Locke.
42. *Ibid.*, pages 170ff; see also Lewis, *Op. cit.*, pages 115–116.
43. Lewis, *Op. cit.*, pages 180ff.
44. Peterson, *Op. cit.*, page 235.
45. *Ibid.*, page 238.
46. *Ibid.*, page 240.
47. *Ibid.*, page 241.
48. Asa Briggs, *Op. cit.*, page 65.
49. John Cain, *The BBC: Seventy Years of Broadcasting*, London: BBC, 1992, pages 11 and 20.
50. *Ibid.*, pages 10–15.
51. Assembled from charts and figures given in Briggs, *Op. cit.*, *passim*. Cain, *Op. cit.*, page 13.
52. Briggs, *Op. cit.*, page 14.
53. Radway, *Op. cit.*, pages 219–220 and chapter 7, 'The Scandal of the Middlebrow', pages 221ff.
54. Cain, *Op. cit.*, page 15.
55. *Ibid.*, page 25.

CHAPTER 13: HEROES' TWILIGHT

The title for this chapter is taken from Bernard Bergonzi's book on the literature of World War I, discussed in chapter 9. As will become clear, the phrase applies *a fortiori* to the subject of Weimar Germany. I am particularly indebted in this chapter to Peter Gay's *Weimar Culture* (see note 3 for details).

1. Otto Friedrich, *Before the Deluge: A Portrait of Berlin in the 1920s*, London: Michael Joseph, 1974, page 67.
2. Lotte H. Eisner, *The Haunted Screen: Expressionism in the German Cinema and the Influence of Max Reinhardt*, London and New York: Thames & Hudson, 1969, pages 17–27 for Pommer's reaction to Mayer and Janowitz.
3. Peter Gay, *Weimar Culture: The Outsider as Insider*, London: Martin Secker & Warburg, 1969, page 107.
4. *Ibid.*, page 126.
5. *Ibid.*
6. *Ibid.*
7. Friedrich, *Op. cit.*, page 66.
8. *Ibid.* For the success of the film, see: Geoffrey Nowell-Smith, *The Oxford History of World Cinema*, Oxford and New York: Oxford University Press, 1996, page 144; and page 145 for an assessment of Plommer.
9. Friedrich, *Op. cit.*, page 67.
10. Gay, *Op. cit.*, pages 108–109.
11. *Ibid.*, page 110.
12. *Ibid.*, page 32.
13. *Ibid.*, page 34.
14. Hughes, *The Shock of the New, Op. cit.*, page 175.
15. *Ibid.*, pages 192–195; Gay, *Op. cit.*, pages 102ff.
16. Friedrich, *Op. cit.*, page 160.
17. Gay, *Op. cit.*, page 105.
18. Hughes, *Op. cit.*, page 195.
19. *Ibid.*, page 195.
20. *Ibid.*, page 199.
21. *Ibid.*, page 199.

22. Bryan Magee, *Men of Ideas: Some Creators of Contemporary Philosophy*, Oxford: Oxford University Press, 1978, paperback, 1982, page 44.
23. Martin Jay, *The Dialectical Imagination: A History of the Frankfurt School and the Institute of Social Research, 1923–1950*, Berkeley: University of California Press, 1973, paperback edition 1996, pages 152–153. Magee, *Op. cit.*, pages 44 and 50.
24. Magee, *Op. cit.*, page 50.
25. Jay, *Op. cit.*, pages 86ff.
26. Magee, *Op. cit.*, page 48.
27. *Ibid.*, page 51.
28. *Ibid.*, page 52.
29. *Ibid.*
30. Gay, *Op. cit.*, page 49.
31. *Ibid.*, pages 51–52.
32. E. M. Butler, *Rainer Maria Rilke*, Cambridge: Cambridge University Press, 1941, page 14.
33. *Ibid.*, pages 147ff.
34. Friedrich, *Op. cit.*, page 304.
35. Gay, *Op. cit.*, page 54.
36. *Ibid.*, page 59.
37. *Ibid.*, page 55.
38. Butler, *Op. cit.*, page 317.
39. Quoted in *ibid.*, page 327.
40. Gay, *Op. cit.*, page 55.
41. *Ibid.*, page 57.
42. *Ibid.*
43. *Ibid.*, page 59.
44. Friedrich, *Op. cit.*, page 220, where Einstein's predicament is spelled out. See also: Gay, *Op. cit.*, pages 129ff.
45. Hayman, *Thomas Mann, Op. cit.*, pages 344–348.
46. Gay, *Op. cit.*, page 131.
47. Hayman, *Op. cit.*, page 346.
48. Gay, *Op. cit.*, page 131.
49. *Ibid.*, pages 132–133.
50. *Ibid.*, page 136.
51. Bruno Walter, 'Themes and Variations: An Autobiography,' 1946, pages 268–269, quoted in Gay, *Op. cit.*, page 137.
52. Schonberg, *The Lives of the Great Composers, Op. cit.*, page 526.
53. Friedrich, *Op. cit.*, page 178; Griffiths, *Modern Music, Op. cit.*, page 81.
54. Schonberg, *Op. cit.*, page 526.
55. *Ibid.*
56. Griffiths, *Op. cit.*, page 82.
57. Friedrich, *Op. cit.*, pages 155 and 181.
58. Griffiths, *Op. cit.*, pages 36–37. Schonberg, *Op. cit.*, page 524.
59. Schonberg, *Op. cit.*, page 524.
60. Friedrich, *Op. cit.*, page 183.
61. Schonberg, *Op. cit.*, page 527.
62. Peter Conrad, *Modern Times, Modern Places: Art and Life in the Twentieth Century*, London: Thames & Hudson, 1998, pages 327–328.
63. Friedrich, *Op. cit.*, page 243.
64. *Ibid.*, page 244.
65. Ronald Hayman, *Brecht: A Biography*, London: Weidenfeld & Nicolson, 1983, page 138.

66. *Ibid.*, page 130.
67. *Ibid.*, pages 131ff.
68. *Ibid.*, page 134.
69. *Ibid.*, page 135.
70. Griffiths, *Op. cit.*, pages 112–113.
71. Hayman, *Brecht*, page 148.
72. *Ibid.*
73. *Ibid.*, page 149.
74. *Ibid.*, page 148.
75. *Ibid.*, page 147.
76. Hugo Ott, *Martin Heidegger: A Political Life*, London: HarperCollins, 1993, page 125.
77. Paul Hühnerfeld, *In Sachen Heidegger*, 1961, pages 14ff, quoted in Gay, *Op. cit.*, page 85.
78. Magee, *Op. Cit.*, pages 59–60; Gay, *Op. cit.*, page 86.
79. To begin with, he was close to the existential theologian Rudolf Bultmann and the 'crisis theology' of Karl Barthes (see below, chapter 32). Ott, *Op. cit.*, page 125.
80. Magee, *Op. cit.*, page 67.
81. *Ibid.*
82. *Ibid.*, pages 67 and 73.
83. Ott, *Op. cit.*, page 122ff and 332. See also: Gay, *Op. cit.*, page 86.
84. Mary Gluck, *Georg Lukács and His Generation, 1900–1918, Op. cit.*, page 211.
85. Johnston, *The Austrian Mind, Op. cit.*, page 366.
86. *Ibid.*, page 367.
87. Gluck, *Op. cit.*, page 218.
88. Johnston, *Op. cit.*, page 368.
89. *Ibid.*, page 372.
90. Conrad, *Op. cit.*, page 504.
91. Johnston, *Op. cit.*, page 374.
92. Magee, *Op. cit.*, page 96.
93. *Ibid.*
94. Ben Rogers, *A. J. Ayer: A Life*, London: Chatto & Windus, 1999, pages 86–87.
95. Magee, *Op. cit.*, pages 102–103.
96. *Ibid.*, page 103.
97. Rogers, *Op. cit.*, pages 91–92.
98. Johnston, *Op. cit.*, page 195.
99. Robert Musil, *Der Mann Ohne Eigenschaften*, 1930–1943; *The Man Without Qualities*, New York: Alfred A. Knopf, 1995, (trans) Sophie Wilkins. In this section I am especially indebted to: Philip Payne, *Robert Musil's 'The Man without Qualities'*, Cambridge: Cambridge University Press, 1988, *passim*.
100. Johnston, *Op. cit.*, page 335.
101. Franz Kuna, 'The Janus-faced Novel: Conrad, Musil, Kafka, Mann,' in Malcolm Bradbury and James McFarlane (editors), *Modernism, Op. cit.*, page 449.
102. Ronald Speirs and Beatrice Sandburg, *Franz Kafka, Op. cit.*, pages 1 and 5.
103. Speirs and Sandburg, *Op. cit.*, page 15.
104. P. Mailloux, *A Hesitation Before Birth: A Life of Franz Kafka*, London and Toronto: Associated Universities Presses, 1989, page 13.
105. *Ibid.*, page 352.
106. Speirs and Sandburg, *Op. cit.*, pages 105ff.

107. Mailloux, *Op. cit.*, page 355.

108. Richard Davenport-Hines, *Auden*, London: Heinemann, 1995, page 26.

109. Alan Bullock, *Hitler and Stalin: Parallel Lives*, London: HarperCollins, 1991; Fontana Paperback, 1993, page 148.

110. *Ibid.*, page 149.

111. Adolf Hitler, *Mein Kampf*, published in English as 'My Struggle', London: Hurst & Blackett, The Paternoster Press, October 1933 (eleven impressions by October 1935); and see Bullock, *Op. cit.*, pages 405–406.

112. George L. Mosse, *The Crisis of German Ideology: Intellectual Origins of the Third Reich*, New York: Howard Festig, 1998.

113. *Ibid.*, pages 39ff for Langbehn, pages 72ff for the Edda and pages 52ff for Diederichs.

114. *Ibid.*, pages 102–103.

115. *Ibid.*, page 99.

116. *Ibid.*, page 155.

117. Werner Maser, *Hitler: Legend, Myth and Reality*, New York: Harper & Row, 1973, page 157.

118. *Ibid.*, page 158.

119. *Ibid.*, page 159.

120. Mosse, *Op. cit.*, pages 89–91.

121. Maser, *Op. cit.*, page 162.

122. Mosse, *Op. cit.*, pages 95, 159 and 303.

123. Percy Schramm, *Hitler: The Man and the Military Leader*, London: Allen Lane, The Penguin Press, 1972, pages 77–78.

124. Maser, *Op. cit.*, pages 42ff.

125. *Ibid.*, page 165.

126. *Ibid.*, page 167.

127. Mosse, *Op. cit.*, page 295.

128. Maser, *Op. cit.*, page 169.

129. *Ibid.*, page 135.

130. Schramm, *Op. cit.*, pages 84ff.

131. Maser, *Op. cit.*, page 154.

CHAPTER 14: THE EVOLUTION OF EVOLUTION

1. J. B. Bury, *The Idea of Progress*, London: Macmillan, 1920.

2. *Ibid.*, pages 98ff.

3. *Ibid.*, pages 291ff.

4. *Ibid.*, pages 177ff.

5. *Ibid.*, page 192.

6. *Ibid.*, pages 335ff.

7. *Ibid.*, page 278.

8. *Ibid.*, page 299.

9. *Ibid.*, page 334.

10. See also *ibid.*, pages 78ff. Ernest Gellner, in *Plough, Sword and Book: The Structure of Human History*, *Op. cit.*, argues that progress is essentially an economic, capitalist, idea. See page 140.

11. Howard Carter and A. C. Mace, *The Tomb of Tut*Ankh*Amen*, London: Cassell, 1923, volume l, page 78.

12. C. W. Ceram, *Gods, Graves and Scholars*, London: Victor Gollancz, 1951, page 183.

13. Carter and Mace, *Op. cit.*, page 87.

14. Ceram, *Op. cit.*, page 184.

15. Carter and Mace, *Op. cit.*, page 96.

16. Ceram, *Op. cit.*, page 186.

17. See photographs in Carter and Mace, *Op. cit.*, at page 132.

18. Ceram, *Op. cit.*, page 188; Carter and Mace, *Op. cit.*, pages 151ff.

19. See Carter and Mace, *Op. cit.*, page 178, for a list of those present.

20. Ceram, *Op. cit.*, page 193.

21. *Ibid.*, page 195.

22. See Carter and Mace, *Op. cit.*, Appendix, pages 189ff, for a list.

23. Ceram, *Op. cit.*, page 198.

24. *Ibid.*, page 199.

25. *Ibid.*, pages 199–200.

26. C. Leonard Woolley, *The Sumerians*, Oxford: Clarendon Press, 1929, page 6.

27. *Ibid.*, page 27.

28. Ceram, *Op. cit.*, page 309; Woolley, *Op. cit.*, page 43.

29. Ceram, *Op. cit.*, page 311.

30. Woolley, *Op. cit.*, page 31.

31. Ceram, *Op. cit.*, pages 311–312.

32. Woolley, *Op. cit.*, pages 30–32.

33. Leonard Woolley, *Excavations at Ur*, London: Ernest Benn, 1954, page 251.

34. Ceram, *Op. cit.*, page 315.

35. *Ibid.*, page 316.

36. Woolley, *Excavations at Ur*, *Op. cit.*, page 91.

37. Ceram, *Op. cit.*, page 316.

38. Woolley, *Excavations at Ur*, *Op. cit.*, page 37. See Woolley, *The Sumerians*, *Op. cit.*, page 36, for photographs of early arches.

39. Ceram, *Op. cit.*, page 312.

40. Frederic Kenyon, *The Bible and Archaeology*, London: George Harrap, 1940, page 155.

41. *Ibid.*, page 156.

42. *Ibid.*, page 158.

43. Frederic Kenyon, *Our Bible and the Ancient Manuscripts*, London: Eyre & Spottiswoode, 1958, page 30.

44. Kenyon, *The Bible and Archaeology*, *Op. cit.*, pages 160–161.

45. C. W. Ceram, *The First Americans*, *Op. cit.*, page 126.

46. *Ibid.*

47. A. E. Douglass, *Climatic Cycles and Tree Growth, volumes I–III*, Washington D.C., Carnegie Institution, 1936, pages 2 and 116–122.

48. *Ibid.*, pages 105–106.

49. Ceram, *The First Americans*, *Op. cit.*, page 128.

50. See Douglass, *Op. cit.*, page 125 for a discussion about the dearth of sunspots at times in the past.

51. Herbert Butterfield, *The Whig Interpretation of History*, London: G. Bell, 1931.

52. *Ibid.*, pages 37 and 47.

53. *Ibid.*, pages 27ff.

54. *Ibid.*, page 96.

55. *Ibid.*, page 107.

56. *Ibid.*, page 111.

57. *Ibid.*, page 123.

CHAPTER 15: THE GOLDEN AGE OF PHYSICS

1. Rhodes, *The Making of the Atomic Bomb*, *Op. cit.*, page 134.

2. C. P. Snow, *The Search*, New York: Charles Scribner's Sons, 1958, page 88.

3. Rhodes, *Op. cit.*, page 137.

4. Wilson, *Rutherford: Simple Genius, Op. cit.*, page 404.

5. Rhodes, *Op. cit.*, page 137.

6. Moore, *Niels Bohr, The Man and the Scientist, Op. cit.*, page 21.

7. Stefan Rozental (editor), *Niels Bohr*, Amsterdam: North-Holland, 1967, page 137, quoted in Rhodes, *Op. cit.*, page 114.

8. See Moore, *Op. cit.*, pages 80ff for the voltage required to make electrons 'jump' out of their orbits; see also pages 122–123 for the revised periodic table; see also Rhodes, *Op. cit.*, page 115.

9. Emilio Segrè, *From X-Rays to Quarks*, London and New York: W. H. Freeman, 1980, page 124.

10. Helge Kragh, *Quantum Generations: A History of Physics in the Twentieth Century*, Princeton: Princeton University Press, 1999, page 160, for a table of visitors to Copenhagen.

11. Paul Strathern, *Bohr and Quantum Theory*, London: Arrow, 1998, pages 70–72.

12. Moore, *Op. cit.*, page 137.

13. Strathern, *Op. cit.*, page 74.

14. Werner Heisenberg, *Physics and Beyond*, New York: Harper, 1971, page 38; quoted in Rhodes, *Op. cit.*, page 116.

15. Moore, *Op. cit.*, page 138.

16. Heisenberg, *Op. cit.*, page 61, quoted in Rhodes, *Op. cit.*, pages 116–117.

17. Strathern, *Op. cit.*, page 77.

18. Moore, *Op. cit.*, page 139.

19. Snow, *The Physicists, Op. cit.*, page 68.

20. Moore, *Op. cit.*, page 14.

21. Kragh, *Op. cit.*, page 164–165 for the mathematics.

22. Rhodes, *Op. cit.*, page 128; Moore, *Op. cit.*, page 143; Kragh, *Op. cit.*, page 165.

23. Heisenberg, *Op. cit.*, page 77; quoted in Rhodes, *Op. cit.*, page 130.

24. Moore, *Op. cit.*, page 151.

25. John A. Wheeler and W. H. Zurek (editors), *Quantum Theory and Measurement*, Princeton: Princeton University Press, 1983, quoted in Kragh, *Op. cit.*, page 209.

26. Gerald Holton, *Thematic Origins of Scientific Thought*, Cambridge, Massachusetts: Harvard University Press, 1973, page 120.

27. Kragh, *Op. cit.*, page 170 for a table.

28. Wilson, *Op. cit.*, pages 444–446. See also: Rhodes, *Op. cit.*, page 153.

29. *Ibid.*, page 449.

30. Rhodes, *Op. cit.*, page 154.

31. *Ibid.*, page 155.

32. Andrew Brown, *The Neutron and the Bomb, A Biography of James Chadwick*, Oxford and New York: Oxford University Press, 1997, page 8.

33. Rhodes, *Op. cit.*, pages 155–156.

34. Kragh, *Op. cit.*, page 185.

35. Rhodes, *Op. cit.*, page 160.

36. Brown, *Op. cit.*, page 102.

37. Rhodes, *Op. cit.*, pages 161–162.

38. Brown, *Op. cit.*, page 104; see also: James Chadwick, 'Some personal notes on the search for the neutron,' *Proceedings of the Tenth Annual Congress of the History of Science, 1964*, page 161, quoted in Rhodes, *Op. cit.*, page 162. These accounts vary slightly.

39. Rhodes, *Op. cit.*, pages 163–164; Brown, *Op. cit.*, page 105.

40. Kragh, *Op. cit.*, page 185.

41. Brown, *Op. cit.*, page 106.

42. Timothy Ferris, *The Whole Shebang: A State of the Universe(s) Report*, New York: Simon & Schuster, 1997, page 41.

43. Gale Christianson, *Edwin Hubble: Mariner of the Nebulae*, New York: Farrar, Straus & Giroux, 1995, Chicago: University of Chicago Press, paperback edition, 1996, page 199. See also: John Gribbin, *Copernicus to the Cosmos*, London: Phoenix, 1997, pages 2 and 186ff.

44. Clark, *Einstein, Op. cit.*, page 213. See also: Banesh Hoffmann, *Albert Einstein: Creator and Rebel*, London: Hart-Davis, MacGibbon, 1973, page 215.

45. Ferris, *Op. cit.*, page 42.

46. Christianson, *Op. cit.*, page 199; Ferris, *Op. cit.*, page 43.

47. Clark, *Einstein, Op. cit.*, page 406; Ferris, *Op. cit.*, page 44.

48. Ferris, *Op. cit.*, page 45.

49. Gribbin, *Companion to the Cosmos, Op. cit.*, pages 92–93.

50. Christianson, *Op. cit.*, pages 157–160.

51. *Ibid.*, pages 189–195.

52. Ferris, *Op. cit.*, page 45.

53. Christianson, *Op. cit.*, pages 260–269.

54. Thomas Hager, *Force of Nature: The Life of Linus Pauling*, New York, Simon & Schuster, 1995, page 217.

55. *Ibid.*, page 65.

56. *Ibid.*, page 113.

57. Bernadette Bensaude-Vincent and Isabelle Stengers, *A History of Chemistry*, translated by Deborah Dam, Cambridge, Massachusetts: Harvard University Press, 1996, pages 242ff.

58. Hager, *Op. cit.*, pages 136.

59. Bensaude-Vincent and Stengers, *Op. cit.*, pages 242–243. Hager, *Op. cit.*, page 136.

60. Hager, *Op. cit.*, page 138.

61. *Ibid.*, page 148.

62. Heitler and London's theory has become the subject of revisionist chemical history recently. See

for example, Bensaude-Vincent and Stengers, *Op. cit.*, page 243.

63. Hager, *Op. cit.*, page 169.

64. *Ibid.*, page 171.

65. *Ibid.*, page 159.

66. Many books published on chemistry in the 1930s make no reference to Heitler and London, or Pauling.

67. Glyn Jones, *The Jet Pioneers*, London: Methuen, 1989, page 21.

68. *Ibid.*, pages 22–23.

69. *Ibid.*, page 24.

70. *Ibid.*, pages 27–28. British accounts of Whittle's contributions are generally negligent, perhaps because he was so badly treated. In *Aviation, An Historical Survey from Its Origins to the End of World War II*, by Charles Gibbs-Smith, and published by HMSO in 1970, Whittle rates three references only and by the second he is an Air Commodore! H. Montgomery Hyde's *British Air Policy Between the Wars 1918–1939*, London: Heinemann, 1976, 539pp, has one reference and one note on Whittle.

71. Jones, *Op. cit.*, page 29.

72. *Ibid.*, page 36.

73. John Allen Paulos, *Beyond Numeracy*, New York: Knopf, 1991, page 95.

74. Ray Monk, *Wittgenstein*, *Op. cit.*, page 295.

75. *Ibid.*, page 295n.

76. Ernst Nagel and James Newman, 'Goedel's Proof', in James Newman (editor), *The World of Mathematics* (volume 3, of 4), New York: Simon & Schuster, 1955, pages 1668–1695, especially page 1686.

77. Newman, *Op. cit.*, page 1687.

78. Paulos, *Op. cit.*, page 97.

79. David Deutsch, *The Fabric of Reality*, London: Allen Lane, The Penguin Press, 1997, Penguin paperback, 1998, pages 236–237.

80. Philip J. Davis and Reuben Hersh, *The Mathematical Experience*, London: The Harvester Press, 1981, page 319.

CHAPTER 16: CIVILISATIONS AND THEIR DISCONTENTS

1. *Civilisation and Its Discontents* is now published as volume XXI of the Standard Edition of the Complete Works of Sigmund Freud, edited by James Strachey and Anna Freud, London: The Hogarth Press and the Institute of Psychoanalysis, 1953–74 (this volume was published in 1961). For details of Freud's operation see Clark, *Freud*, *Op. cit.*, pages 444–445.

2. *Ibid.*, page 218.

3. *Ibid.*, pages 64ff.

4. C. G. Jung, *Modern Man in Search of a Soul*, London: Kegan Paul, Trench, Trubner, 1933.

5. *Ibid.*, pages 91ff.

6. Lucien Lévy-Bruhl, *How Natives Think*, translated by L. A. Clare, London: George Allen & Unwin, 1926, chapter II, pages 69ff.

7. Henry Frankfort *et al.*, *Before Philosophy*, London: Pelican, 1963, especially pages 103ff.

8. J. A. C. Brown, *Freud and the Post-Freudians*, *Op. cit.*, page 122.

9. *Ibid.*, pages 8, 125 and 128.

10. Karen Horney, *The Neurotic Personality of Our Time*, London: Kegan Paul, Trench, Trubner & Co., 1937. See also: J. A. C. Brown, *Op. cit.*, page 135.

11. Horney, *Op. cit.*, page 77.

12. Brown, *Op. cit.*, page 137.

13. Horney, *Op. cit.*, respectively chapters 8, 9, 10 and 12. Summarised in Brown, *Op. cit.*, pages 138–139.

14. Horney, *Op. cit.*, pages 288ff.

15. Brown, *Op. cit.*, pages 143–144.

16. Virginia Woolf, *A Room of One's Own*, London: Hogarth Press, 1929; Penguin paperback, 1993, with an Introduction by Michèle Barrett, page xii.

17. *Ibid.*, page 3.

18. Barrett, *Op. cit.*, page xii.

19. 'Aurora Leigh' (a review of Elizabeth Barrett Browning's poem of that name), in Michèle Barrett (editor), *Women and Writing*, London: Women's Press, 1988; quoted in Barrett, *Op. cit.*, page xv.

20. *Ibid.*, page xvii.

21. *Ibid.*, page x.

22. Jane Howard, *Margaret Mead: A Life*, London: Harvill, 1984, pages 53–54. For the latest scholarship, see: Hilary Lapsley, *Margaret Mead and Ruth Benedict: The Kinship of Women*, Amherst: University of Massachusetts Press, 1999. This book includes an assessment of Ruth Benedict by Clifford Geertz, one of the most influential anthropologists of the last quarter of a century (see chapter 38, 'Local Knowledge').

23. Margaret Mead, *Blackberry Winter: My Early Years*, London: Angus & Robertson, 1973, page 139.

24. G. Stanley Hall, *Adolescence: Its Psychology and Its Relation to Physiology, Anthropology, Sociology, Sex, Crime, Religion and Education*, New York: Appleton, 1905, 2 vols. Quoted in Howard, *Op. cit.*, page 68.

25. Howard, *Op. cit.*, page 68.

26. Mead, *Op. cit.*, page 150.

27. Howard, *Op. cit.*, page 79.

28. *Ibid.*, page 52.

29. *Ibid.*, page 79.

30. *Ibid.*, pages 80–82.

31. Margaret Mead, *Coming of Age in Samoa: A Psychological Study of Primitive Youth for Western Civilisation*, New York: William Morrow, 1928.

32. Howard, *Op. cit.*, page 86.

33. *Ibid.*

34. *Ibid.*, page 127.

35. Quoted in *ibid.*, page 121.

36. Mead, *Coming of Age in Samoa*, page 197.

37. *Ibid.*, page 205.

38. *Ibid.*, page 148.

39. Howard, *Op. cit.*, page 162.

40. Ruth Benedict, *Patterns of Culture*, Boston: Houghton Mifflin, 1934.

41. *Ibid.*, page 59.

42. *Ibid.*, page 69.

43. *Ibid.*, page 131.

44. Judith Modell, *Ruth Benedict: Patterns of a Life*,

London: Chatto & Windus, 1984, page 201.

45. *Ibid.*, page 205.

46. *Ibid.*, pages 206–207.

47. Margaret Caffrey, *Ruth Benedict: Stranger in this Land*, Austin: University of Texas Press, 1989, pages 211ff, for a discussion of Ruth Benedict's impact on American thought more generally.

48. Margaret Mead, *Ruth Benedict*, New York: Columbia University Press, 1974, which *does* attempt to recover some of the earlier excitement.

49. Howard, *Op. cit.*, page 212.

50. Martin Bulmer, *The Chicago School of Sociology*, Chicago: University of Chicago Press, 1984, paperback edition, 1986, pages 1–2.

51. *Ibid.*, pages 4–8, but see also chapters 4 and 5.

52. Charles S. Johnson, *The Negro in American Civilisation*, London: Constable, 1931.

53. Bulmer, *Op. cit.*, pages 64–65.

54. Johnson, *Op. cit.*, pages 229ff.

55. *Ibid.*, page 463.

56. *Ibid.*, pages 179ff.

57. *Ibid.*, page 199.

58. *Ibid.*, page 311.

59. *Ibid.*, page 463.

60. *Ibid.*, pages 475ff.

61. David Minter, *William Faulkner: His Life and Work*, Baltimore and London: Johns Hopkins University Press, 1980, pages 72–73.

62. The demands made on Faulkner himself may be seen from the fact that after he had finished a chapter, he would turn to something quite different for a while – short stories for example. See: Joseph Blotner, *Selected Letters of William Faulkner*, London: The Scolar Press, 1955, page 92.

63. Ursula Brumm, 'William Faulkner and the Southern Renaissance,' in Marcus Cunliffe (editor), *The Penguin History of Literature: American Literature since 1900*, London: Sphere Books, 1975; Penguin paperback revised edition, 1993, pages 182–183 and 189.

64. *Ibid.*, page 195.

65. Minter, *Op. cit.*, pages 153–160.

66. Eric Hobsbawm, *The Age of Extremes: The Short Twentieth Century, 1914–1991*, London: Michael Joseph, 1994, page 192.

67. T. R. Fyvel, *George Orwell: A Personal Memoir*, London: Weidenfeld & Nicolson, 1982, page 21.

68. George Orwell, *The Road to Wigan Pier*, London: Gollancz, 1937, page 138; New York: Harcourt, 1958. Michael Shelden, *Orwell: The Authorised Biography*, London: Heinemann, 1991, page 128.

69. Fyvel, *Op. cit.*, page 39.

70. Shelden, *Op. cit.*, page 129.

71. *Ibid.*

72. *Ibid.*, page 132.

73. *Ibid.*, pages 132–133.

74. *Ibid.*, page 134.

75. Fyvel, *Op. cit.*, page 45.

76. Shelden, *Op. cit.*, page 135.

77. Fyvel, *Op. cit.*, page 44.

78. Shelden, *Op. cit.*, pages 173–174.

79. *Ibid.*, page 180.

80. *Ibid.*, page 239.

81. *Ibid.*, page 244.

82. *Ibid.*, page 245.

83. *Ibid.*

84. Fyvel, *Op. cit.*, page 64.

85. Shelden, *Op. cit.*, page 248.

86. *Ibid.*, page 250.

87. *Ibid.*, page 256.

88. Fyvel, *Op. cit.*, pages 65–66.

89. Lewis Mumford, *Technics and Civilisation*, London: George Routledge, 1934.

90. *Ibid.*, pages 107ff.

91. For an introduction, see also the excerpt in Lewis Mumford, *My Works and Days: A Personal Chronicle*, New York: Harcourt Brace Jovanovich, 1979, pages 197–199.

92. Mumford, *Technics and Civilisation*, *Op. cit.*, pages 400ff.

93. *Ibid.*, page 333.

94. Lewis Mumford, *The Culture of Cities*, London: Martin Secker & Warburg, 1938.

95. *Ibid.*, pages 100ff.

96. *Ibid.*, chapter IV, pages 223ff.

97. Ernest William Barnes, *Scientific Theory and Religion*, Cambridge: Cambridge University Press, 1933.

98. *Ibid.*, lectures XIII (pages 434ff), XIV (pages 459ff) and XV (pages 504ff).

99. *Ibid.*, lecture XX (pages 636ff).

100. William Ralph Inge, *God and the Astronomers*, London and New York: Longmans Green, 1933.

101. *Ibid.*, pages 19ff.

102. *Ibid.*, page 107.

103. *Ibid.*, pages 140ff.

104. *Ibid.*, pages 254–256.

105. Bertrand Russell, *Religion and Science*, London: Thornton Butterworth, 1935.

106. Ray Monk, *Bertrand Russell*, *Op. cit.*, page 244.

107. *Ibid.*, page 245.

108. Russell, *Op. cit.*, chapters IV and VII.

109. *Ibid.*, pages 236ff.

110. *Ibid.*, page 237.

111. *Ibid.*, page 243.

112. José Ortega Y Gasset, 'The Barbarism of "Specialisation",' from *The Revolt of the Masses*, New York and London: W. W. Norton and George Allen & Unwin, 1932, quoted in John Carey, *The Intellectuals and the Masses*, London and Boston: Faber & Faber, 1992, pages 17–18.

113. For their contacts and early years, see: Royden J. Harrison, *The Life and Times of Sidney and Beatrice Webb, 1858–1905: The Formative Years*, London: Macmillan, 2000.

114. Lisanne Radice, *Beatrice and Sidney Webb: Fabian Socialists*, London: Macmillan, 1984, page 56.

115. *Ibid.*, page 264.

116. *Ibid.*, page 292.

117. *Ibid.*, pages 292 and 295.

118. *Ibid.*, page 297.

119. *Ibid.*, pages 297 and 298.

120. *Ibid.*, page 303.

121. *Ibid.*, pages 305 and 323.

122. Stephanie Barron (editor), *Degenerate Art: The Fate of the Avant-Garde in Nazi Germany*, Los Angeles: County Museum of Art, and New York: Harry N. Abrams, 1991, pages 12–13.
123. *Ibid.*, page 12.
124. Robert Cecil, *The Myth of the Master Race: Alfred Rosenberg and Nazi Ideology*, London: Batsford, 1972.
125. *Ibid.*, page 12.
126. *Ibid.*, page 83.
127. *Ibid.*, pages 86–93.
128. *Ibid.*, pages 95–103.
129. *Ibid.*, page 120.
130. Ronald Clark, *The Huxleys*, London: Heinemann, 1968, page 130.
131. *Aldous Huxley: 1894–1963: A Memorial Volume*, London: Chatto & Windus, 1965, page 30.
132. For his own feelings about the book, see: Sybille Bedford, *Aldous Huxley: A Biography, Volume One: 1894–1939*, London: Chatto & Windus/Collins, 1973, pages 245–247.
133. Keith May, *Aldous Huxley*, London: Paul Elek, 1972, page 100.
134. *Ibid.*
135. Aldous Huxley, *Brave New World*, London: Chatto & Windus, 1934; New York: Harper, 1934. May, *Op. cit.*, page 103.
136. Clark, *The Huxleys*, *Op. cit.*, page 236.

CHAPTER 17: INQUISITION

1. Henry Grosshans, *Hitler and the Artists*, New York: Holmes & Meier, 1983, page 72. I have relied heavily on this excellent short book.
2. Hildegard Brenner, 'Art in the Political Power Struggle of 1933 and 1934,' in Hajo Holborn (editor), *Republic to Reich: The Making of the Nazi Revolution*, New York: Pantheon, 1972, page 424. Quoted in Grosshans, *Op. cit.*, page 72.
3. Grosshans, *Op. cit.*, page 72.
4. Barron, *Degenerate Art*, *Op. cit.*, page 396.
5. Carl Carls, *Ernst Barlach*, New York: Praeger, 1969, page 172, quoted in Grosshans, *Op. cit.*, page 72.
6. *Ibid.*, page 73.
7. *Ibid.*, page 72.
8. *Ibid.*, page 73.
9. *Ibid.*, page 74.
10. *Ibid.*, page 75.
11. *Ibid.*, page 77.
12. Victor H. Miesel (editor), *Voices of German Expressionism*, Englewood Cliffs, New Jersey: Prentice-Hall, 1970, pages 209ff.
13. Barron, *Op. cit.*, page 319.
14. Grosshans, *Op. cit.*, page 79.
15. *Ibid.*, pages 79–80.
16. *Ibid.*, page 81.
17. Berthold Hinz, *Art in the Third Reich*, New York: Pantheon, 1979, pages 43ff.
18. White and Gribbin, *Einstein*, *Op. cit.*, pages 163–164.
19. Albrecht Fölsing, *Albert Einstein: A Biography*, New York: Viking, 1997, pages 659ff.
20. White and Gribbin, *Einstein*, *Op. cit.*, page 206.
21. Fölsing, *Op. cit.*, pages 648ff.
22. White and Gribbin, *Op. cit.*, page 200.
23. Fölsing, *Op. cit.*, page 649.
24. Headline quote: *Berliner Lokal-Anzeiger*, March 1933, quoted in White and Gribbin, *Op. cit.*, page 204; American attempts to bar Einstein: Fölsing, *Op. cit.*, page 661.
25. Jarrell Jackman and Carlo M. Borden, *The Muses Flee Hitler: Cultural Transfer and Adaptation, 1930–1945*, Washington DC: Smithsonian Institution Press, 1963, page 170.
26. Ute Deichmann, *Biologists under Hitler*, Cambridge, Massachusetts: Harvard University Press, 1996, pages 40–47.
27. *Ibid.*, pages 294ff.
28. Stephanie Barron (editor), *Exiles and Emigrés: The Flight of European Artists from Europe*, Los Angeles: County Museum of Art, and Harry N. Abrams, 1997, page 212.
29. Peter Hahn, 'Bauhaus and Exile: Bauhaus Architects and Designers between the Old World and the New', in Barron, *Exiles and Emigrés*, *Op. cit.*, page 212.
30. *Ibid.*, page 213.
31. *Ibid.*, page 216.
32. *Ibid.*, page 218.
33. Martin Jay, *The Dialectical Imagination*, *Op. cit.*, page 29.
34. *Ibid.*, page 30.
35. Laura Fermi, *Illustrious Immigrants: The Intellectual Migration from Europe: 1930–1941*, Chicago: University of Chicago Press, 1971, pages 364–368.
36. *Ibid.*, chapter VI, pages 139ff.
37. Clark, *Freud*, *Op. cit.*, pages 502–504.
38. *Ibid.*, page 507.
39. *Ibid.*
40. *Ibid.*, pages 511 and 513–516.
41. See Paul Ferris, *Dr Freud*, London: Sinclair-Stevenson, 1997, page 380, or a summary.
42. Clark, *Op. cit.*, page 524.
43. Elisabeth Young-Bruehl, *Hannah Arendt: For Love of the World*, New Haven and London: Yale University Press, 1982, pages 44ff.
44. *Ibid.*, pages 49ff.
45. Elzbieta Ettinger, *Hannah Arendt/Martin Heidegger*, New Haven and London: Yale University Press, 1995, pages 24–25.
46. Rüdiger Safranski, *Martin Heidegger: Between Good and Evil*, Cambridge, Massachusetts: Harvard University Press, 1998, page 255.
47. *Ibid.*, pages 238ff.
48. Young-Bruehl, *Op. cit.*, pages 102–106.
49. *Ibid.*, pages 138–144.
50. See: Victor Farías, *Heidegger and Nazism*, Philadelphia: Temple University Press, 1989, pages 140ff, for Heidegger's speech on the university in the National Socialist state.
51. Safranski, *Op. cit.*, page 258, says that an acknowledgement was however retained 'hidden in the footnotes.'
52. Deichmann, *Op. cit.*, page 187.

53. *Ibid.*, page 184.

54. *Ibid.*, pages 188–189.

55. *Ibid.*, page 229.

56. *Ibid.* See also: Michael H. Kater, *Doctors under Hitler*, Chapel Hill, North Carolina: University of North Carolina Press, 1989, page 31 for the effect on doctors' salaries of the purge of Jewish physicians, and page 133 for the excesses of younger doctors (who were not *völkisch* brutes either); and Robert Proctor, *Racial Hygiene: Medicine Under the Nazis*, Cambridge, Massachusetts: Harvard University Press, 1988.

57. Deichmann, *Op. cit.*, pages 231ff.

58. *Ibid.*, pages 251ff.

59. *Ibid.*, page 257.

60. *Ibid.*, page 258.

61. Grosshans, *Op. cit.*, page 111.

62. *Ibid.*, page 101.

63. Richard Grunberger, *A Social History of the Third Reich*, London: Weidenfeld & Nicolson, 1971, page 427, quoted in Grosshans, *Op. cit.*, pages 99–100.

64. For Hitler's speech, Barron, *Degenerate Art, Op. cit.*, pages 17ff (also for photographs of Hitler at the exhibition); for Hitler's view that art should be 'founded on peoples', see: Grosshans, *Op. cit.*, page 103.

65. Grosshans, *Op. cit.*, page 103.

66. *Ibid.*, page 105.

67. Barron, *Degenerate Art, Op. cit.*, pages 20 and 25ff.

68. Grosshans, *Op. cit.*, page 105.

69. Barron, *Degenerate Art*, pages 36–38; Grosshans, *Op. cit.*, page 107.

70. Miesel, *Op. cit.*, page 209, quoted in Grosshans, *Op. cit.*, page 109.

71. Barron, *Degenerate Art, Op. cit.*, page 19.

72. Grosshans, *Op. cit.*, page 116.

73. Erik Levi, *Music in the Third Reich*, London: Macmillan, 1994, especially chapters 4 and 7. See also: Boris Schwarz, 'The Music World in Migration', in Jackman and Borden (editors), *Op. cit.*, pages 135–150.

74. Mary Bosanquet, *The Life and Death of Dietrich Bonhoeffer*, London: Hodder & Stoughton, 1968, pages 82ff.

75. Eberhard Bethge, *Dietrich Bonhoeffer: Theologian, Christian, Contemporary*, London: Collins, 1970, pages 379ff.

76. Bosanquet, *Op. cit.*, page 82.

77. *Ibid.*, pages 121–124; see also Bethge, *Op. cit.*, page 193.

78. Bosanquet, *Op. cit.*, pages 187ff.

79. See his diary entry for 9 July 1939, quoted in Bosanquet, *Op. cit.*, page 218; see also Bethge, *Op. cit.*, pages 557ff.

80. Bosanquet, *Op. cit.*, page 235.

81. Dietrich Bonhoeffer, *Letters and Papers from Prison* (edited by Eberhard Bethge), London: SCM Press, 1967.

82. Bosanquet, *Op. cit.*, pages 277–278; see also Bethge, *Op. cit.*, pages 827ff.

83. Vitaly Shentalinsky, *The KGB's Literary Archive*, London: The Harvill Press, 1995, paperback 1997. Originally published in French as *La parole ressuscitée dans les archives littéraires du KGB*, Paris: Editions Robert Laffont, 1993.

84. *Ibid.*, pages 136–137.

85. *Ibid.*, pages 287–289.

86. See: Loren R. Graham, *Science in the Soviet Union*, Cambridge: Cambridge University Press, 1993, pages 79ff for the full impact of the revolution on scientists.

87. Nikolai Krementsov, *Stalinist Science*, Princeton: Princeton University Press, 1997, pages 20–25. This is the main source for this section.

88. Paul R. Josephson, *Physics and Politics in Revolutionary Russia*, Berkeley: University of California Press, 1991, pages 104ff.

89. Krementsov, *Op. cit.*, pages 24–25.

90. *Ibid.*, pages 29–30.

91. Josephson, *Op. cit.*, pages 152ff.

92. Krementsov, *Op. cit.*, page 35. For Pavlov's own scepticism toward psychology, and his resistance to Marxism, see Loren R. Graham, *Science, Philosophy and Human Behaviour in the Soviet Union*, New York: Columbia University Press, 1987, page 161. This book is an updated version of *Science and Philosophy in the Soviet Union*, London: Allen Lane The Penguin Press, 1973.

93. Josephson, *Op. cit.*, page 204.

94. Krementsov, *Op. cit.*, page 40.

95. *Ibid.*, page 43.

96. *Ibid.*, page 47. See Graham, *Op. cit.*, page 117 for talk about social Darwinian engineering and a marriage to Marxism.

97. See Josephson, *Op. cit.*, pages 225ff for an account of the 'interference' between Marxist philosophy and theoretical physics.

98. Krementsov, *Op. cit.*, page 56; Graham, *Op. cit.*, page 241.

99. Krementsov, *Op. cit.*, page 57. See also Graham, *Op. cit.*, chapters 4 and 6 for a discussion of the impact of Leninism on quantum mechanics and on relativity physics (chapters 10 and 11).

100. Krementsov, *Op. cit.*, page 59.

101. Graham, *Op. cit.*, page 108.

102. Krementsov, *Op. cit.*, page 60.

103. See Josephson, *Op. cit.*, page 269, for the fight put up by Russian physicists against the materialists, who were accused of playing 'hide and seek' with the evidence. See also Graham, *Op. cit.*, page 121.

104. Krementsov, *Op. cit.*, page 60.

105. Josephson, *Op. cit.*, page 308.

106. Graham, *Op. cit.*, page 315.

107. Krementsov, *Op. cit.*, pages 66–67.

108. *Ibid.*, page 73.

109. *Ibid.*, page 82.

110. Graham, *Science in the Soviet Union, Op. cit.*, pages 129–130, for details of Vavilov's fate.

111. Gleb Struve, *Russian Literature under Lenin and Stalin, 1917–1953*, Norman: University of Oklahoma Press, 1971, pages 59ff.

112. A. Kemp-Welch, *Stalin and the Literary Intel-*

ligentsia, 1928–1939, London: Macmillan, 1991, page 233.

113. See: Dan Levy, *Stormy Petrel: The Life and Work of Maxim Gorky*, London: Frederick Muller, 1967, pages 313–318, for details of his relations with Stalin towards the end.

114. Although RAPP itself was bitterly divided. See: Struve, *Op. cit.*, page 232; Kemp-Welch, *Op. cit.*, page 77.

115. Kemp-Welch, *Op. cit.*, page 77.

116. *Ibid.*, pages 169–170.

117. See Struve, *Op. cit.*, chapter 20, pages 256ff.

118. Edward J. Brown, *The Proletarian Episode in Russian Literature 1928–1932*, New York: Columbia University Press, 1953, pages 69–70, 96, 120 and 132.

119. Struve, *Op. cit.*, page 261; Kemp-Welsh, *Op. cit.*, page 175.

120. See Brown, *Op. cit.*, page 182 for what the Politburo said of Shostakovich; Kemp-Welsh, *Op. cit.*, page 178.

121. See Nadezhda Mandelstam, *Hope Against Hope*, London: Collins and Harvill Press, 1971, pages 217–221 for Mandelstam's relations with Akhmatova.

122. John and Carol Garrard, *Inside the Soviet Writers' Union*, London: I. B. Tauris, 1990, pages 58–59.

123. Shentalinsky, *Op. cit.*, page 191.

124. *Ibid.*, page 193.

125. Garrard and Garrard, *Op. cit.*, page 38; see also Shentalinsky, *Op. cit.*, pages 70–71 for Ehrenburg's attempted defence of Babel.

126. Kemp-Welch, *Op. cit.*, page 223.

127. *Ibid.*, page 224.

128. I. Ehrenburg, *Men, Years – Life*, London, 1963, volume 4, *The Eve of War*, page 96, quoted in: Kemp-Welch, *Op. cit.*, page 198.

CHAPTER 18: COLD COMFORT

1. Lewis Jacobs, *The Rise of the American Film, A Critical History*, New York: Harcourt Brace, 1939, page 419.

2. Alfred Knight, *The Liveliest Art*, *Op. cit.*, page 156.

3. *Ibid.*, pages 164–165.

4. Jacobs, *Op. cit.*: see the 'still' between pages 428 and 429.

5. Knight, *Op. cit.*, page 257.

6. *Ibid.*, pages 261–262. See also Jacobs, *Op. cit.*, for a list of some prominent directors of the period.

7. Knight, *Op. cit.*, page 222.

8. Kristin Thompson and David Bordwell, *Film History*, New York: McGraw Hill, 1994, page 353.

9. Knight, *Op. cit.*, page 225.

10. *Ibid.*, pages 226–227.

11. Thompson and Bordwell, *Op. cit.*, page 354.

12. W. H. Auden, 'Night Mail', July, 1935. See Edward Mendelsohn (editor), *The English Auden*, London and Boston: Faber & Faber, 1977.

13. Knight, *Op. cit.*, page 211.

14. Thomson and Bordwell, *Op. cit.*, page 309.

15. *Ibid.*, page 310.

16. Knight, *Op. cit.*, page 212. Riefenstahl later said that she was only ever interested in art and was unaware of the Nazis' persecutions, a claim that film historians have contested. See Thompson and Bordwell, *Op. cit.*, page 320.

17. John Lucas, *The Modern Olympic Games*, Cranbury, New Jersey: A. S. Barnes, 1980.

18. Allen Guttman, *The Olympics: A History of the Modern Games*, Urbana and Chicago: University of Illinois Press, 1992, pages 67ff.

19. Riefenstahl was allowed to pick from other cameramen's footage. See: Audrey Salkeld, *A Portrait of Leni Riefenstahl*, London: Jonathan Cape, 1996, page 173.

20. Riefenstahl says in her memoirs that Hitler did not refuse to shake hands with Owen on racial grounds, as was widely reported, but 'because it was against Olympic protocol.' See: Leni Riefenstahl, *The Sieve of Time: The Memoirs of Leni Riefenstahl*, London: Quartet, 1992, page 193.

21. Salkeld, *Op. cit.*, page 186.

22. Knight, *Op. cit.*, page 213.

23. *Ibid.*, page 216.

24. Thompson and Bordwell, *Op. cit.*, page 294.

25. Knight, *Op. cit.*, page 217.

26. *Ibid.*, page 218.

27. Thompson and Bordwell, *Op. cit.*, page 298. Knight, *Op. cit.*, page 218.

28. Knight, *Op. cit.*, page 218.

29. See Momme Broderson, *Walter Benjamin: A Biography*, London: Verso, 1996, pages 184ff for his friendship with Brecht, Kraus and a description of life in Berlin.

30. Bernd Witte, *Walter Benjamin: An Intellectual Biography*, Detroit: Wayne State University Press, 1991, pages 159–160.

31. *Ibid.*, page 161. In his account of their friendship, Gershom Scholem describes his reactions to this essay, claiming that Benjamin's use of the concept of 'aura' was 'forced'. Gershom Scholem, *Walter Benjamin: The Story of a Friendship*, London and Boston: Faber & Faber, 1982, page 207.

32. Stanislaus von Moos, *Le Corbusier: Elements of a Synthesis*, Cambridge, Massachusetts, MIT Press, 1979, pages 210–213.

33. *Ibid.*, page 191.

34. *Ibid.*, pages 17, 49–50.

35. Robert Furneaux Jordan, *Le Corbusier*, London: J. M. Dent, 1972, page 36 and plate 5; see also Von Moos, *Op. cit.*, page 75.

36. Jordan, *Op. cit.*, page 33.

37. *Ibid.*, page 36 and plate 5.

38. Von Moos, *Op. cit.*, page 154; see also Jordan, *Op. cit.*, pages 56–57.

39. Von Moos, *Op. cit.*, pages 302–303.

40. See Von Moos, *Ibid.*, pages 296–297 for Le Corbusier's thinking on colour and how it changed over time. In Jordan, *Op. cit.*, page 45, Le Corbusier describes the process in the following way: 'One must take every advantage of modern science.'

41. Humphrey Carpenter, *W. H. Auden: A Biography*, London: George Allen & Unwin, 1981, pages 12–13. See the discussion of 'Audenesque' in Bernard Bergonzi, *Reading the Thirties*, London: Macmillan, 1978, pages 40–41.

42. Grevel Lindop, 'Poetry in the 1930s and 1940s,' in Martin Dodsworth (editor), *The Twentieth Century*, volume 7 of *The Penguin History of Literature*, London, 1994, page 268.

43. Ian Hamilton (editor), *The Oxford Companion to Twentieth-Century Poetry*, *Op. cit.*, page 21.

44. 'VII', July 1932, from 'Poems 1931–1936', in Edward Mendelsohn (editor), *Op. cit.*, page 120.

45. 'VII', August 1932, in *ibid.*, page 120.

46. G. Rostrevor Hamilton, *The Tell-Tale Article*, quoted in Bergonzi, *Op. cit*, page 43.

47. *Ibid.*, page 52.

48. Poem XXIX, in Mendelsohn (editor), *Op. cit.*

49. Bergonzi, *Op. cit.*, page 51. See also Carpenter, *Op. cit.*, for the writing of 'Spain' and Auden's direction of the royalties. Lindop, *Op. cit.*, page 273.

50. Quoted in Frederick R. Benson, *Writers in Arms: The Literary Impact of the Spanish Civil War*, London: University of London Press; New York: New York University Press, 1968, page 33.

51. Carpenter, *Op. cit.*, page 219. See also: Bernard Crick, *George Orwell: A Life*, London: Secker & Warburg, 1980, chapter 10, 'Spain and "necessary murder",' pages 207ff

52. Benson, *Op. cit.*, pages xxii and 88ff.

53. *Ibid.*, pages xxii and 27.

54. André Malraux, *L'Espoir*, Paris: Gallimard, 1937.

55. Curtis Cate, *André Malraux: A Biography*, London: Hutchinson, 1995, pages 259ff.

56. Benson, *Op. cit.*, pages 240 and 295. At times Hemingway's book was sold under the counter in Spain. See José Luis Castillo-Duche, *Hemingway in Spain*, London: New England Library, 1975, page 96.

57. John Berger, *The Success and Failure of Picasso*, *Op. cit.*, page 164.

58. Arianna Stassinopoulos, *Op. cit.*, page 231.

59. Berger, *Op. cit.*, page 102.

60. Stassinopoulos, *Op. cit.*, page 232.

61. Herbert Read, 'Picasso's *Guernica*', *London Bulletin*, No. 6, October 1938, page 6.

62. Robert Hughes, *The Shock of the New*, *Op. cit.*, page 110.

63. *Ibid.*, pages 110–111.

64. Stassinopoulos, *Op. cit.*, page 256.

65. Herbert Rutledge Southworth, *Guernica! Guernica!*, Berkeley: University of California Press, 1977, pages 277–279, shows how many Spaniards took a long time to forgive Picasso. See also Benson, *Op. cit.*, page 64 for Orwell's reactions to the war.

66. George Orwell, *Homage to Catalonia*, London: Martin Secker & Warburg, 1938.

67. J. E. Morpurgo, *Allen Lane: King Penguin*, London: Hutchinson, 1979, page 80.

68. *Ibid.*, pages 81–84.

69. *Ibid.*, pages 92–93.

70. W. A. Williams, *Allen Lane, A Personal Portrait*, London: The Bodley Head, 1973, page 45.

71. J. B. Priestley, *English Journey*, London: Heinemann, 1934; Penguin, 1977.

72. F. R. Leavis, *Mass Civilisation and Minority Culture*, London: Minority Press, 1930. (Actually issued by Gordon Fraser.)

73. Ian MacKillop, *F. R. Leavis: A Life in Criticism*, London: Allen Lane, The Penguin Press, 1995, pages 74–75. I. A. Richards, whose 1929 *Practical Criticism* embodied this view, and became very influential, later moved to Harvard, where this approach became known as the 'new criticism.'

74. Q. D. Leavis, *Fiction and the Reading Public*, London: Chatto & Windus, 1932; Re-issued: Bellew, 1990.

75. *Ibid.*, pages 199–200.

76. Williams, *Op. cit.*, 52ff compares them with the BBC's Third Programme. He says it was the most decisive event of the company, linking it also with the Council for the Encouragement of Music and the Arts, the forerunner of Britain's Arts Council.

77. Morpurgo, *Op. cit.*, pages 114–116.

78. *Ibid.*, page 116.

79. Williams, *Op. cit.*, page 54.

80. Morpurgo, *Op. cit.*, page 131.

81. *Ibid.*, page 135.

82. J. K. Galbraith, *The Age of Uncertainty*, London: BBC/André Deutsch, 1977, page 203.

83. *Ibid.*, page 204.

84. *Ibid.*, page 211.

85. Robert Lekachman, *The Age of Keynes*, London: Allen Lane, The Penguin Press, 1967; Pelican Books, 1969, page 72.

86. *Ibid.*, pages 80–84.

87. The phrase is Robert Skidelsky's in his biography of Keynes: Skidelsky, *Op. cit.*, volume 2, chapter 13, page 431.

88. Galbraith, *Op. cit.*, page 214.

89. According to Skidelsky, publication of *The General Theory* was followed by 'a war of opinion' among economists. Skidelsky, *Op. cit.*, page 572.

90. Galbraith, *Op. cit.*, page 218.

91. Lekachman, *Op. cit.*, page 120.

92. Galbraith, *Op. cit.*, page 221.

93. Bergonzi, *Op. cit.*, pages 112–114, and 126–127.

94. Bergonzi, *Op. cit.*, pages 61 and 112.

95. Cole Porter, 'You're the Tops', 1934. This was 'quasi-Marxist' on Porter's part, according to Bergonzi, *Op. cit.*, page 127.

96. See John Gloag, *Plastic and Industrial Design*, London: George Allen & Unwin, 1945, page 86, for a basic introduction; also polythene.

97. Stephen Fenichell, *Plastic*, *Op. cit.*, page 106.

98. Burr W. Leyson, *Plastics in the World of Tomorrow*, London: Elek, 1946, page 17, under-lines how rapid the acceptance of cellophane was.

99. Farben also produced a synthetic emerald in 1934. See: David Fishlock, *The New Materials*, London: John Murray, 1967, page 49.

100. Fenichell, *Op. cit.*, pages 152–153.

101. *Ibid.*, page 161.

102. *Ibid.*, pages 150–151.

103. Paul Johnson, *A History of the Modern World*, *Op. cit.*, page 247.

104. Michael Mannheim (editor), *The Cambridge Companion to Eugene O'Neill*, Cambridge: Cambridge University Press, 1998, page 1.

105. Louis Shaeffer, *O'Neill: Son and Playwright*, London: J. M. Dent, 1969, pages 69–70.

106. Stephen Black, 'Cell of Loss', in Mannheim (editor), *Op. cit.*, pages 4–12. Shaeffer, *Op. cit.*, page 174.

107. Normand Berlin, 'The Late Plays', in Mannheim (editor), *Op. cit.*, pages 82ff.

108. O'Neill said Hope's was based on three places 'I actually lived in.' See: Arthur and Barbara Gelb, *O'Neill*, London: Jonathan Cape, 1962, page 296.

109. This is a post-Darwinian vision but O'Neill also admitted to being influenced by Jung. See: Egil Törnqvist, 'O'Neill's philosophical and literary paragons,' in Mannheim (editor), *Op. cit.*, page 22.

110. Shaeffer, *Op. cit.*, page 514. See Mannheim, *Op. cit.*, page 85, for the point about 'waiting for Hickey.'

111. David Morse, 'American Theatre: The Age of O'Neill,' in Marcus Cunliffe (editor), *American Literature since 1900*, London: Sphere, 1975; Penguin edition 1993, page 77.

112. Berlin, *Op. cit.*, page 90.

113. According to Shaeffer, *Op. cit.*, page 510 ff, this is the least autobiographical part of the play. O'Neill made the Tyrone setting far more claustrophobic than was the case with the O'Neills themselves, who went out for meals.

114. See Arthur and Barbara Gelb, *O'Neill*, *Op. cit.*, page 93. Berlin, *Op. cit.*, page 91.

115. Berlin, *Op. cit.*, page 89.

116. Alfred Kazin, *On Native Grounds*, *Op. cit.*, page 485.

117. *Ibid.*, page 295 for the reference to Van Wyck Brooks, 352 for Dos Passos and 442 for the 'tragi-comic climax'.

118. *Ibid.*, page 404.

119. *Ibid.*, page 488.

120. Simon Callow, *Orson Welles: The Road to Xanadu*, London: Jonathan Cape, 1995, page xi.

121. *Ibid.*, page 521.

122. Frank Brady, *Citizen Welles*, London: Hodder & Stoughton, 1990, pages 309–310.

123. Callow, *Op. cit.*, page 570.

CHAPTER 19: HITLER'S GIFT

1. Stephanie Barron, *Exiles and Emigrés*, *Op. cit.*, pages 136–137.

2. *Ibid.*, pages 16–18.

3. *Ibid.*, page 14.

4. Laura Fermi, *Ilustrious Immigrants*, *Op cit.*, pages 66–68.

5. Jarrel C. Jackman and Carla M. Borden, *The Muses Flee Hitler*, *Op. cit.*, page 218.

6. *Ibid.*, page 219.

7. *Ibid.*, pages 206–207.

8. *Ibid.*, pages 208–226.

9. Barron, *Exiles and Emigrés*, *Op. cit.*, page 19. See also Lewis A. Coser, *Refugee Scholars in America: Their Impact and Their Experiences*, New Haven and London: Yale University Press, 1984, has entire chapters on, among others: Kurt Lewin, Erik Erikson, Wilhelm Reich, Bruno Bettelheim, Erich Fromm, Karen Horney, Paul Lazarsfeld, Ludwig von Mieses, Karl Polanyi, Hannah Arendt, Thomas Mann, Vladimir Nabokov, Roman Jakobson, Erwin Panofsky, Hajo Holborn, Rudolf Carnap and Paul Tillich.

10. Elisabeth Kessin Berman, 'Moral Triage or Cultural Salvage? The Agendas of Varian Fry and the Emergency Rescue Committee,' in Barron, *Exiles and Emigrés*, *Op. cit.*, pages 99–112.

11. Varian Fry, *Surrender on Demand*, New York: Random House, 1945, page 157. Jackman and Borden, *Op. cit.*, page 89.

12. Fry, *Op. cit.*, pages 189–191.

13. Martica Swain, *Surrealism in Exile and the Beginnings of the New York School*, Cambridge, Massachusetts: MIT Press, 1995, pages 124–126.

14. Jackman and Borden, *Op. cit.*, page 90.

15. Coser, *Op. cit.*, 'The New School for Social Research: A Collective Portrait,' pages 102–109.

16. Ian Hamilton (editor), *The Oxford Companion to Twentieth-Century Poetry*, *Op. cit.*, pages 51–52.

17. Barron, *Exiles and Emigrés*, *Op. cit.*, page 187.

18. *Ibid.*, pages 190ff.

19. Jackman and Borden, *Op. cit.*, pages 140–141.

20. *Ibid.*, pages 142–143.

21. Ehrhard Bahr, *Literary Weimar in Exile: German Literature in Los Angeles, 1940–1958*, in Ehrhard Bahr and Carolyn See, *Literary Exiles and Refugees in Los Angeles*, William Andrews Clark Memorial Library, University of California at Los Angeles, 1988. Bahr argues that the German writers never fully assimilated in L.A., always keeping their eyes on Germany.

22. Barron, *Exiles and Emigrés*, *Op. cit.*, pages 358–359.

23. *Ibid.*, page 341.

24. Bernard Taper, *Balanchine*, New York: Times Books, 1984, pages 147ff.

25. *Ibid.*, page 148.

26. Richard Buckle, *George Balanchine: Ballet Master: A Biography*, London: Hamish Hamilton, 1988, pages 61ff.

27. Taper, *Op. cit.*, page 149.

28. Lincoln Kirstein, *Mosaic: Memoirs*, New York: Farrar, Straus & Giroux, 1994, page 23.

29. Taper, *Op. cit.*, page 151.

30. Buckle, *Op. cit.*, page 66, says the first meeting was at the Savoy, the second at the Chelsea home of Kirk Askew.

31. Kirstein, *Op. cit.*, pages 247–249.

32. Taper, *Op. cit.*, page 151.

33. *Ibid.*, page 153.

34. *Ibid.*, page 154.

35. Buckle, *Op. cit.*, page 88.

36. Taper, *Op. cit.*, page 156.

37. *Ibid.*, page 157.

38. *Ibid.*

39. Buckle, *Op. cit.*, page 88.

40. Taper, *Op. cit.*, page 160.

41. Various authors, *The Cultural Migration: The European Scholar in America*, Philadelphia: University of Pennsylvania Press, 1953. Tillich reference: page 155.

CHAPTER 20: COLOSSUS

1. Andrew Hodges, *Alan Turing: The Enigma*, London: Burnett Books, in association with Hutchinson, 1983, Vintage paperback, 1992, pages 160ff.

2. I. J. Good, 'Pioneering work on computers at Bletchley,' in N. Metropolis, J. Howlett and Giancarlo Rota (editors), *A History of Computing in the Twentieth Century*, New York and London: Academic Press, 1980, page 33 for others who arrived at Bletchley at much the same time.

3. Hodges, *Op. cit.*, page 160.

4. Paul Strathern, *Turing and the Computer*, London: Arrow, 1997, page 59.

5. Good, *Op. cit.*, pages 35 and 36 for excellent photographs of Enigma. For the latest account of the way the Enigma codes were broken, and the vital contribution of Harry Hinsley, using recently declassified documents, see: Hugh Sebag-Montefiore, *Enigma: The Battle for the Code*, London: Weidenfeld & Nicolson, 2000.

6. Hodges, *Op. cit.*, page 86.

7. Strathern, *Op. cit.*, pages 46–47.

8. Hodges, *Op. cit.*, pages 96–101 for the link between rational and computable numbers. See also: Strathern, *Op. cit.*, page 48.

9. Strathern, *Op. cit.*, pages 49–50.

10. S. M. Ulam, 'Von Neumann: The Interreaction of Mathematics and Computers,' in Metropolis *et al.* (editors), *Op. cit.*, pages 95ff.

11. Strathern, *Op. cit.*, pages 51–52.

12. *Ibid.*, pages 55–56.

13. *Ibid.*, pages 57–59.

14. Turing also knew who to take advice from. See: Wladyslaw Kozoczuh, *Enigma*, London: Arms & Armour Press, 1984, page 96 on the role of the Poles.

15. At times the messages were not in real German. This was an early problem solved. See: R. V. Jones, *Most Secret War*, London: Hamish Hamilton, 1978, page 63.

16. Good, *Op. cit.*, pages 40–41.

17. Hodges, *Op. cit.*, page 277.

18. B. Randall, 'The Colossus', in Metropolis *et al.* (editors), *Op. cit.*, pages 47ff for the many others who collaborated on Colossus. See Hodges, *Op. cit.*, between pages 268 and 269 for photographs.

19. Strathern, *Op. cit.*, page 63–64.

20. See Randall, *Op. cit.*, pages 77–80 for an assessment of Turing and the 'fog' that still hangs over his wartime meeting with Von Neumann.

21. Hodges, *Op. cit.*, page 247.

22. Strathern, *Op. cit.*, page 66.

23. See John Haugeland, *Artificial Intelligence: The Very Idea*, Cambridge, Massachusetts: MIT Press, 1985, pages 261–263 for an exact chronology.

24. Hodges, *Op. cit.*, pages 311–312.

25. Guy Hartcup, *The Challenge of War: Scientific and Engineering Contributions to World War Two*, Exeter: David & Charles, 1970, pages 17ff.

26. *Ibid.*, page 94.

27. *Ibid.*, pages 96–97.

28. *Ibid.*, page 91. For German progress, and some shortcomings of radar, see: Alfred Price, *Instruments of Darkness*, London: William Kimber, 1967, *circa* pages 40–45; and David Pritchard, *The Radar War*, London: Patrick Stephens, 1989, especially pages 80ff.

29. Hartcup, *Op. cit.*, page 91, but for a detailed chronology, see: Jack Gough, *Watching the Skies: A History of Ground Radar for the Air Defence of the United Kingdom by the RAF from 1946 to 1975*, London: HMSO, 1993, pages 8–12.

30. Hartcup, *Op. cit.*, pages 90 and 107.

31. Ronald W. Clark, *The Life of Ernst Chain: Penicillin and Beyond*, New York: St Martin's Press, 1985, pages 47ff. Weatherall, *In Search of a Cure*, *Op. cit.*, pages 174–175.

32. Gwyn Macfarlane, *Alexander Fleming: The Man and the Myth*, London: Chatto & Windus/The Hogarth Press, 1984, pages 119ff.

33. Weatherall, *Op. cit.*, page 168.

34. *Ibid.*, pages 165–166.

35. Gwyn Macfarlane, *Howard Florey: The Making of a Great Scientist*, Oxford and New York: Oxford University Press, 1979, page 331.

36. Weatherall, *Op. cit.*, pages 175–176.

37. John E. Pfeiffer, *The Creative Explosion: An Inquiry into the Origins of Art and Religion*, New York: Harper & Row, 1982, pages 26ff, who says there was no dog. Annette Laming, *Lascaux*, London: Penguin, 1959, pages 54ff.

38. Mario Ruspoli, *The Cave of Lascaux: The Final Photographic Record*, London and New York: Thames & Hudson, 1987, page 188. See also note 37 above.

39. *Ibid.*

40. Pfeiffer, *Op. cit.*, page 30.

41. Ruspoli, *Op. cit.*, page 188.

42. Pfeiffer, *Op. cit.*, page 31.

43. For a detailed description, see Ruspoli, *Op. cit.*, and Fernand Windels, *Montignac-sur-Vézere*, Centre d'Études et de documentations préhistoriques, Dordogne, 1948.

44. Paul G. Bahn and Jean Vertut, *Images of the Ice Age*, London: Windward, 1988, pages 20–23.

45. Evan Hadingham, *Secrets of the Ice Age: The World of the Cave Artists*, London: Heinemann, 1979, page 187.

46. See Ruspoli, *Op. cit.*, pages 87–88 for a discussion, though no women are represented at Lascaux. Professor Randall White, of New York University, believes that certain features of the Venus figurines (tails, animal ears) suggest that these objects date from a time when early humans had not yet linked sexual intercourse with birth. The animal features suggest that animal spirits were thought to

be involved. (Personal communication.)

47. Pierre Teilhard de Chardin, *The Appearance of Man*, London: Collins, 1965, page 51.

48. Ian Tattersall, *The Fossil Trail*, Oxford and New York: Oxford University Press, 1995, paperback 1996, pages 62 and 67.

49. Chardin, *Op. cit.*, pages 91 and 145. Tattersall, *Op. cit.*, page 62.

50. Mayr, *The Growth of Biological Thought*, *Op. cit.*, pages 566–569 which also includes Bernhard Rensch and G. Ledyard Stebbins in this group though they didn't publish their works until 1947 and 1950 respectively, by which time the Princeton conference (see below) had taken place. Mayr says (page 70) that there was no 'paradigm shift' in a Kuhnian sense (see chapter 27 of this book) but 'an exchange' of 'viable components.' Julian Huxley's book was published by George Allen & Unwin in London; all the others in the synthesis were published in New York by Columbia University Press. See also: Ernst Mayr and William B. Provine (editors), *The Evolutionary Synthesis: Perspectives on the Unification of Biology*, Cambridge, Massachusetts: Harvard University Press, 1980, 1988, which explores the development in evolutionary thinking outside Britain and the United States: France, Germany, Soviet Russia, together with modern reassessments of the early figures in the field: T. H. Morgan, R. A. Fisher, G. G. Simpson, J. B. S. Haldane and William Bateson.

51. For the popularity of 'saltation' see David Kahn (editor), *The Darwinian Heritage*, Princeton: Princeton University Press in association with Nova Pacifica, 1985, pages 762–763.

52. Tattersall, *Op. cit.*, pages 89–94.

53. *Ibid.*, page 95.

54. Walter Moore, *Schrödinger: Life and Thought*, Cambridge: Cambridge University Press, 1989, page 395.

55. Erwin Schrödinger, *What is Life?*, Cambridge: Cambridge University Press, 1944, page 77.

56. Moore, *Op. cit.*, page 396.

57. Schrödinger, *Op. cit.*, page 61.

58. *Ibid.*, page 79.

59. *Ibid.*, page 30.

60. Moore, *Op. cit.*, page 397.

CHAPTER 21: NO WAY BACK

1. Karl Mannheim, *Diagnosis of Our Time: Wartime Essays of a Sociologist*, London: Kegan Paul, Trench, Trubner, 1943.

2. *Ibid.*, page 38.

3. *Ibid.*, page 32.

4. *Ibid.*, pages 60ff.

5. Joseph Schumpeter, *Capitalism, Socialism and Democracy*, London: George Allen & Unwin, 1943.

6. Johnston, *The Austrian Mind*, *Op. cit.*, page 83.

7. Robert Heilbronner, *The Worldly Philosophers*, New York: Simon & Schuster, 1953, Penguin Books, 1986, pages 292–293.

8. Schumpeter, *Op. cit.*, pages 111ff.

9. *Ibid.*, page 81.

10. *Ibid.*, pages 143ff; Heilbronner, *Op. cit.*, pages 6 and 301–302.

11. Heilbronner, *Op. cit.*, pages 300–303.

12. Friedrich von Hayek, *The Road to Serfdom*, London: George Routledge, 1944, page 52.

13. *Ibid.*, page 61.

14. C. H. Waddington, *The Scientific Attitude*, London (another Penguin Special), 1941.

15. Karl Popper, *The Open Society and Its Enemies, Volume I: The Spell of Plato, Volume II: The High Tide of Prophecy: Hegel, Marx and the Aftermath*, London: George Routledge & Sons, 1945.

16. Popper had problems publishing *The Open Society*, which some publishers felt too irreverent towards Aristotle; and the journal *Mind* turned down *The Poverty of Historicism*. See Mannheim's autobiograhy, *Unended Quest: An Intellectual Biography*, London: Routledge, 1992, page 119.

17. Roberta Corvi, *An Introduction to the Thought of Karl Popper*, London and New York: Routledge, 1997, page 52.

18. *Ibid.*, page 55.

19. *Ibid.*, page 59.

20. Popper, *Op. cit.*, *volume I*, page 143. Corvi, *Op. cit.*, page 65.

21. *Ibid.*, *volume II*, page 218.

22. Corvi, *Op. cit.*, page 69.

23. See Popper, *Op. cit.*, *volume II*, chapter 14, on the autonomy of sociology, and chapter 23, on the sociology of knowledge.

24. Corvi, *Op. cit.*, page 73.

25. William Temple, *Christianity and the Social Order*, London: Penguin Special, 1942.

26. *Ibid.*, chapter 2 on church 'interference'.

27. *Ibid.*, page 75.

28. *Ibid.*, pages 76ff.

29. *Ibid.*, page 79.

30. *Ibid.*, page 87.

31. Nicholas Timmins, *The Five Giants: A Biography of the Welfare State*, London: HarperCollins, 1995, Fontana Paperback, 1996, page 23. See also: Derek Fraser, *The Evolution of the British Welfare State*, London: Macmillan, 1973, page 199, which says the report sold 635,000 copies.

32. John Kenneth Galbraith, *A History of Economics*, London: Hamish Hamilton, 1987, Penguin edition, 1991, pages 213–215.

33. For the effects of war on attitudes, see: Fraser, *Op. cit.*, pages 194–195.

34. Timmins, *Op. cit.*, page 11. There is no mention of this, of course, in Beveridge's memoirs: Lord Beveridge, *Power and Influence*, London: Hodder & Stoughton, 1953.

35. Beveridge, *Op. cit.*, page 9; quoted in Timmins, *Op. cit.*, page 12. See also: José Harris, *William Beveridge: A Biography*, Oxford: Clarendon Press, 1977, page 44.

36. Paul Addison, *Churchill on the Home Front 1900–1955*, London: Jonathan Cape, 1992, page 51; quoted in Timmins, *Op. cit.*, page 13.

37. Harris, *Op. cit.*, pages 54 and 379. Timmins, *Op. cit.*, page 14.

38. Timmins, *Op. cit.*, page 15.

39. *Ibid.*, page 20.

40. *Ibid.* See also: Harris, *Op. cit.*, page 385.

41. Timmins, *Op. cit.*, page 21, though according to Harris, *Op. cit.*, page 390, he did not begin to think about insurance until the end of 1941.

42. Fritz Grunder, 'Beveridge meets Bismark,' York papers, volume 1, page 69, quoted in Timmins, *Op. cit.*, page 25.

43. *Ibid.*, pages 23–24.

44. Cmnd. 6404, *Social Insurance and Allied Services: Report by Sir William Beveridge*, London: HMSO, 1942, pages 6–7, quoted in Timmins, *Op. cit.*, pages 23–24.

45. And, indeed, many officials were cautious. Harris, *Op. cit.*, page 422.

46. Timmins, *Op. cit.*, page 29.

47. Derek Fraser, *Op. cit.*, page 180, quoted in Timmins, *Op. cit.*, page 33.

48. *Ibid.*, page 37.

49. In his memoirs, Beveridge refers to an American commentator who said: 'Sir William, possibly next to Mr Churchill, is the most popular figure in Britain today.' Beveridge, *Op. cit.*, page 319.

50. Allan Bullock, *Hitler and Stalin*, *Op. cit.*, page 858.

51. Crick, *George Orwell*, *Op. cit.*, page 316.

52. Malcolm Bradbury, Introduction to George Orwell, *Animal Farm*, Penguin Books, 1989, page vi.

53. Crick, *Op. cit.*, pages 316–318, adds that paper shortage may not have been the only reason for delay.

54. Galbraith, *A History of Economics*, *Op. cit.*, page 248.

55. Lekachman, *Op. cit.*, page 128.

56. Moggridge, *Op. cit.*, page 629.

57. Lekachman, *Op. cit.*, page 124.

58. Moggridge, *Op. cit.*, page 631.

59. Lekachman, *Op. cit.*, page 127.

60. *Ibid.*, page 131.

61. *The New Republic*, 'Charter for America,' 19 April 1943, quoted in Lekachman, *Op. cit.*, pages 133–135. See also Galbraith, *Op. cit.*, page 249.

62. Lekachman, *Op. cit.*, page 150.

63. *Ibid.*, page 152.

64. Moggridge, *Op. cit.*, page 724. Lekachman, *Op. cit.*, page 158.

65. Lekachman, *Op. cit.*, page 152.

66. White had prepared his own proposal on an International Bank. Moggridge, *Op. cit.*, page 724.

67. *Ibid.*, pages 802–803.

68. Keynes himself was more worried about Britain's overseas spending, which he felt did not match her reduced means. *Ibid.*, page 825.

69. Lekachman, *Op. cit.*, page 138.

70. *Ibid.*, page 161.

71. Gunnar Myrdal, *An American Dilemma: The Negro Problem and Modern Democracy* (two vols), New York: Harper & Row, 1944.

72. Ivan Hannaford, *Race: The History of an Idea in the West*, Baltimore: Johns Hopkins University Press, 1996, page 378.

73. E. Franklin Frazier, *The Negro Family in the United States*, Chicago: University of Chicago Press, 1939.

74. Myrdal, *Op. cit.*, page xlvii.

75. Hannaford, *Op. cit.*, page 379.

76. See Myrdal, *Op. cit.*, chapter 34, on leaders.

77. Paul Johnson, *A History of the American People*, London: Weidenfeld & Nicolson, 1997, page 794. Hannaford, *Op. cit.*, page 395.

78. Ralph Ellison, *Shadow and Act*, New York: Random House, 1964, page 316.

CHAPTER 22: LIGHT IN AUGUST

1. Richard Rhodes, *The Making of the Atomic Bomb*, *Op. cit.*, page 319.

2. *Ibid.*, page 321.

3. See R. W. Clark, *The Birth of the Bomb*, London: Phoenix House, 1961, page 116, for an erroneous claim that Frisch's house was hit by a bomb and set ablaze.

4. For more details about Peierls' calculations, see Clark, *The Birth of the Bomb*, *Op. cit.*, page 118; also Rhodes, *Op. cit.*, page 323.

5. Tizard's committee, extraordinarily, was the only body in wartime Britain capable of assessing the military uses of scientific discoveries. Clark, *Op. cit.*, page 55.

6. Robert Jungk, *Brighter than a Thousand Suns*, London: Victor Gollancz in association with Rupert Hart-Davis, 1958, page 67.

7. Rhodes, *Op. cit.*, page 212.

8. Fermi was known to other physicists as 'the Pope.' Jungk, *Op. cit.*, page 57.

9. Laura Fermi, *Atoms in the Family*, Chicago: University of Chicago Press, 1954, page 123. Also quoted in Rhodes, *Op. cit.*, page 249.

10. C. P. Snow, *The Physicists*, *Op. cit.*, pages 90–91.

11. Otto Hahn, *New Atoms*, New York and Amsterdam: Elsevier, 1950, pages 53ff.

12. Rhodes, *Op. cit.*, pages 254–256.

13. Jungk, *Op. cit.*, pages 67–77.

14. Helge Kragh, *Quantum Generations*, *Op. cit.*, page 260.

15. Ronald Clark, *The Greatest Power on Earth: The Story of Nuclear Fission*, London: Sidgwick & Jackson, 1980, page 45. See also: Jungk, *Op. cit.*, page 77. Rhodes, *Op. cit.*, page 258.

16. Rhodes, *Op. cit.*, page 261.

17. Szilard suggested secrecy but didn't find many supporters. Kragh, *Op. cit.*, page 263.

18. Clark, *The Birth of the Bomb*, *Op. cit.*, page 80.

19. See Jungk, *Op. cit.*, pages 82ff for Szilard's other initiatives.

20. *Ibid.*, page 91 also says that the possibility of a chain reaction had not occurred to Einstein.

21. Rhodes, *Op. cit.*, pages 291–292 and 296.

22. See Clark, *The Birth of the Bomb*, *Op. cit.*, page 183, which says that Canada was also considered as an entirely British alternative. See also: Rhodes, *Op. cit.*, pages 329–330.

23. Kragh, *Op. cit.*, page 265; and Rhodes, *Op. cit.*, page 379.

24. Rhodes, *Op. cit.*, page 385.

25. Mark Walker, *German National Socialism and the Quest for Nuclear Power*, Cambridge: Cambridge University Press, 1989, pages 222ff, argues that the significance of this meeting has been exaggerated on both sides. The meeting became the subject of a successful play, *Copenhagen*, by Michael Frayn, first performed by the National Theatre in London in 1998, and on Broadway in New York in 2000.

26. Kragh, *Op. cit.*, page 266; Rhodes, *Op. cit.*, page 389.

27. Leslie Groves, 'The atomic general answers his critics', *Saturday Evening Post*, 19 May, 1948, page 15; see also Jungk, *Op. cit.*, page 122.

28. Rhodes, *Op. cit.*, pages 450–451.

29. Clark, *The Greatest Power on Earth*, *Op. cit.*, page 161.

30. Rhodes, *Op. cit.*, page 437.

31. Jane Wilson (editor), 'All in Our Time', *Bulletin of the Atomic Scientists*, 1975, quoted in Rhodes, *Op. cit.*, page 440.

32. See Kragh, *Op. cit.*, page 267, for its internal organisation.

33. Rhodes, *Op. cit.*, pages 492 and 496–500.

34. Kragh, *Op. cit.*, page 270.

35. Stefan Rozental (editor), *Niels Bohr*, *Op. cit.*, page 192.

36. Margaret Gowing, *Britain and Atomic Energy, 1939–1945*, London: Macmillan, 1964, pages 354–356. See also: Rhodes, *Op. cit.*, pages 482 and 529.

37. See Clark, *The Birth of the Bomb*, *Op. cit.*, page 141, for the way the British watched the Germans.

38. On the German preference for heavy water, see Mark Walker, *Op. cit.*, page 27.

39. David Irving, *The Virus House*, London: William Kimber, 1967, page 191. The involvement of German physicists with the bomb became a *cause célèbre* after the war, following the claims by some that they had steered clear of such developments on moral grounds. Several contradictory accounts were published which culminated, in 1996, in Jeremy Bernstein (editor), *Hitler's Nuclear Club: The Secret Recordings at Farm Hall*, New York: American Institute of Physics Press. These were declassified transcripts of recordings made at the English country manor, Farm Hall, which housed the captured German scientists in the wake of World War II. The Germans were secretly tape-recorded. The recordings show that by war's end the German nuclear effort employed hundreds of scientists in nine task-oriented research groups, and with Heisenberg in overall charge. The project was on track, in 1943, towards a working reactor but these plans were disrupted, partly by the interdiction of supplies of heavy water, and partly by Allied bombing, which caused the research institute to be moved south, out of Berlin.

40. Herbert York, *The Advisers*, London: W. H. Freeman, 1976, page 30. Rhodes, *Op. cit.*, page 458.

41. Kragh, *Op. cit.*, page 271. Rhodes, *Op. cit.*, pages 501–502.

42. This is Rhodes, page 618, but Jungk says Truman was not informed until 25 April: Jungk, *Op. cit.*, page 178.

43. Jungk, *Op. cit.*, page 195.

44. See also Emilio Segrè's account, reported in Kragh, *Op. cit.*, page 269.

45. Jungk, *Op. cit.*, chapters XI, XII, and XIV.

46. The names of the plane were the first names of the mother of the pilot, Paul Tibbets: Jungk, *Op. cit.*, page 219.

47. Paul Tibbets, 'How to Drop an Atomic Bomb,' *Saturday Evening Post*, 8 June 1946, page 136.

48. Caffrey, *Ruth Benedict*, *Op. cit.*, page 321.

49. Modell, *Ruth Benedict*, *Op. cit.*, page 285.

50. Ruth Benedict, *The Chrysanthemum and the Sword*, Boston: Houghton Mifflin, 1946, paperback edition: Houghton Mifflin, 1989.

51. *Ibid.*, pages x–xi.

52. *Ibid.*, *passim circa* page 104.

53. *Ibid.*, see the table on page 116 comparing *On, Ko* and *Giri*.

54. *Ibid.*, pages 253ff.

55. *Ibid.*, page 192.

56. Caffrey, *Op. cit.*, page 325.

57. Modell, *Op. cit.*, page 284.

58. Benedict, *Op. cit.*, page 305.

CHAPTER 23: PARIS IN THE YEAR ZERO

1. Annie Cohen-Solal, *Sartre: A Life*, London: Heinemann, 1987, page 250. Herman, *Op. cit.*, page 343.

2. Herman, *The Idea of Decline in Western History*, *Op. cit.*, page 343.

3. J.-P. Sartre, *Self-Portrait at 70*, in *Life Situations, Essays Written and Spoken*, translated by P. Auster and L. Davis, New York: Pantheon 1977, pages 47–48; quoted in Herman, *Op. cit.*, page 342.

4. *Ibid.*, page 334.

5. Ronald Hayman, *Writing Against: A Biography of Sartre*, London: Weidenfeld & Nicolson, 1986, page 64. Herman, *Op. cit.*, page 334; Cohen-Solal, *Op. cit.*, page 57.

6. Herman, *Op. cit.*, page 335.

7. Cohen-Solal, *Op. cit.*, page 95.

8. Herman, *Op. cit.*, page 333.

9. *Ibid.*, page 338.

10. Heidegger's notion that the world revealed itself to 'maladjusted instruments' fitted with Sartre's own developing ideas of '*l'homme revolté*'. Hayman, *Op. cit.*, pages 132–133.

11. Herman, *Op. cit.*, page 339.

12. Antony Beevor and Artemis Cooper, *Paris After the Liberation: 1944–1949*, London: Hamish Hamilton, 1994, page 199.

13. *Ibid.*, pages 81 and 200.

14. *Ibid.*, pages 156 and 164.

15. Cohen-Solal, *Op. cit.*, page 248. Beevor and Cooper, *Op. cit.*, pages 159–161.

16. Beevor and Cooper, *Op. cit.*, page 155.

17. Herman, *Op. cit.*, page 343; Cohen-Solal, *Op. cit.*, page 258.

18. Herman, *Op. cit.*, page 344.

19. Cohen-Solal, *Op. cit.*, pages 444ff.

20. Herman, *Op. cit.*, page 346.

21. Maurice Merlau-Ponty, *Humanism and Terror*, Boston: Beacon Press, 1969, pages xvi–xvii.

22. Herman, *Op. cit.*, page 346.

23. Arthur Koestler, *Darkness at Noon*, London: Jonathan Cape, 1940, translator Daphne Harley; see also: David Cesarani, *Arthur Koestler: The Homeless Mind*, London: Heinemann, 1998, pages 288–290, for the fights with Sartre.

24. Cohen-Solal, *Op. cit.*, pages 347–348.

25. *Ibid.*, page 348.

26. Beevor and Cooper, *Op. cit.*, page 158.

27. Stanley Karnow, *Paris in the Fifties*, New York: Random House/Times Books, 1997, page 240.

28. Cohen-Solal, *Op. cit.*, page 265.

29. Karnow, *Op. cit.*, page 240. Beevor and Cooper, *Op. cit.*, page 202.

30. Cohen-Solal, *Op. cit.*, page 266. Karnov, *Op. cit.*, page 242.

31. Beevor and Cooper, *Op. cit.*, page 382.

32. Karnow, *Op. cit.*, page 251. Beevor and Cooper, *Op. cit.*, page 207.

33. See Cohen-Solal, *Op. cit.*, page 307 for a discussion of the disagreements over America.

34. Beevor and Cooper, *Op. cit.*, page 405.

35. *Ibid.*, page 408.

36. Some idea of the emotions this episode can still raise may be seen from the fact that Annie Cohen-Solal's 1987 biography of Sartre, 590 pages, makes no reference to the matter, or to Kravchenko, or to other individuals who took part.

37. Beevor and Cooper, *Op. cit.*, page 409.

38. *Ibid.*, pages 411–412.

39. *Ibid.* See Cohen-Solal, *Op. cit.*, pages 332–333 for an account of their falling out.

40. Beevor and Cooper, *Op. cit.*, page 416.

41. 'Nikolas Bourbaki' was the pseudonym of a group of mainly French mathematicians (Jean Dien-donné, Henri Carton *et al.*), whose aim was to recast all of mathematics into a consistent whole. The first volume of *Elements of Mathematics* appeared in 1939 and ran for more than twenty volumes. For Oliver Messaien, see: Arnold Whittall, *Music Since the First World War*, London: J. M. Dent, 1977; Oxford University Press paperback, 1995, pages 216–219 and 226–231; see also sleeve notes, pages 3–4, by Fabian Watkinson to: 'Messaien, Turangalîla-Symphonie', Royal Concertgebouw Orchestra, Decca, 1992.

42. See Olivier Todd, *Albert Camus: Une Vie*, Paris: Gallimard, 1996, pages 296ff, for the writing of *The Myth of Sisyphus* and Camus' philosophy of the absurd. For the Paris art market after World War II, see: Raymonde Moulin, *The French Art Market: A Sociological View*, New Brunswick: Rutgers University Press, 1987; an abridged translation by Arthur Goldhammer of *Le Marché de la peinture en France*, Paris: Éditions de Minuit, 1967.

43. See: Albert Camus, *Carnets 1942–1951*, London: Hamish Hamilton, 1966, *circa* page 53 for his note-book-thoughts on Tarrou and the symbolic effects of the plague.

44. Simone de Beauvoir, *La Force des Choses*, Paris: Gallimard, 1960, page 29, quoted in Beevor and Cooper, *Op. cit.*, page 206.

45. Kate Millett, *Sexual Politics*, London: Rupert Hart-Davis, 1971, page 346.

46. Ironically, Mettray, the prison Genet served in, was an agricultural colony and, according to Genet's biographer, 'the place looked at once deceptively pastoral (no walls surrounded it and the long lane leading to it was lined with tall trees) and ominously well organised ...' Edmund White, *Genet*, London: Chatto & Windus, 1993, page 68.

47. Genet fought hard to ensure that black actors were always employed. See White, *Op. cit.*, pages 502–503, for his tussle in Poland.

48. Andrew K. Kennedy, *Samuel Beckett*, Cambridge: Cambridge University Press, 1989, pages 4–5.

49. James Knowlson, *Damned to Fame: The Life of Samuel Beckett*, London: Bloomsbury, 1996, page 54.

50. Kennedy, *Op. cit.*, page 8.

51. Knowlson, *Op. cit.*, page 175.

52. Beevor and Cooper, *Op. cit.*, page 173.

53. Kennedy, *Op. cit.*, pages 6, 7, 9 and 11.

54. Knowlson, *Op. cit.*, page 387.

55. Kennedy, *Op. cit.*, page 24.

56. *Ibid.*, page 42.

57. *Godot* has always proved popular with prisoners – in Germany, the USA, and elsewhere. See: Knowlson, *Op. cit.*, pages 409ff, for a discussion.

58. See Kennedy, *Op. cit.*, page 30, for a discussion.

59. *Ibid.*, pages 33–34 and 40–41.

60. Claude Bonnefoy, *Conversations with Eugène Ionescu*, London: Faber & Faber, 1970, page 65.

61. *Ibid.*, page 82.

62. See Eugène Ionescu, *Present Past, Past Present: A Personal Memoir*, London, Calder & Boyars, 1972, translator Helen R. Lane, page 139, for Ionescu's thoughts on 'the end of the individual.'

63. Bonnefoy, *Op. cit.*, pages 167–168.

CHAPTER 24: DAUGHTERS AND LOVERS

1. See the letter, written in early 1944, where he is competing with Camus for a young woman. Simone de Beauvoir (editor), *Quiet Moments in a War: The Letters of Jean-Paul Sartre to Simone de Beauvoir, 1940–1963*, London: Hamish Hamilton, translators Lee Fahnestock and Norman MacAfee, 1994, page 263. And, in Simone de Beauvoir, *Adieu: A Farewell to Sartre*, London: André Deutsch and Weidenfeld & Nicolson, 1984, she made a dignified and moving tribute.

2. Claude Francis and Fernande Gontier, *Simone de Beauvoir*, London: Sidgwick & Jackson, 1987, page 207.

3. *Ibid.*, page 235.

4. Deidre Bair, *Simone de Beauvoir*, London: Jonathan Cape, 1990, pages 325, 379–80.

5. Bair, *Op. cit.*, page 379.

6. Bair, *Op. cit.*, page 380.

7. See the discussions in Bair, *Op. cit.*, page 383, chapter 40.

8. See Francis and Gontier, *Op. cit.*, page 251, for its reception in France; and page 253 for its being placed on the Index.

9. Bair, *Op. cit.*, page 387. And see: Toril Moi, *Simone de Beauvoir: The Making of an Intellectual Woman*, Oxford: Blackwell, 1994, pages 155ff for a psychoanalytic approach to *The Second Sex*.

10. It was translated into sixteen languages: Francis and Gontier, *Op. cit.*, page 254.

11. Bair, *Op. cit.*, pages 432–433.

12. *Ibid.*, page 438.

13. Brendan Gill, 'No More Eve', *New Yorker*, volume XXIX, Number 2, February 28, 1953, pages 97–99, quoted in Bair, *Op. cit.*, page 439.

14. Bair, *Op. cit.*, page 432.

15. He saw himself as 'a second Darwin': James H. Jones, *Alfred C. Kinsey: A Public/Private Life*, New York: W. W. Norton, 1997, pages 25ff.

16. John Heidenry, *What Wild Ecstasy: The Rise and Fall of the Sexual Revolution*, New York: Simon & Schuster, 1997, page 21.

17. John D'Emilio and Estelle B. Freedman, *Intimate Matters: A History of Sexuality in America*, New York: Harper & Row, 1988, page 285.

18. *Ibid.*, page 285.

19. *Ibid.*

20. *Ibid.*, page 286.

21. *Ibid.*

22. Heidenry, *Op. cit.*, page 21.

23. Jones, *Op. cit.*, pages 690–691; see also: D'Emilio and Freedman, *Op. cit.*, page 286.

24. Jones, *Op. cit.*, page 695.

25. Heidenry, *Op. cit.*, page 21.

26. D'Emilio and Freedman, *Op. cit.*, page 288.

27. Heidenry, *Op. cit.*, page 23.

28. *Ibid.*

29. *Ibid.*, pages 24–25.

30. *Ibid.*

31. *Ibid.*, page 26.

32. D'Emilio and Freedman, *Op. cit.*, pages 268 and 312. Heidenry, *Op. cit.*, page 28.

33. Heidenry, *Op. cit.*, page 29.

34. *Ibid.*, page 33.

35. *Ibid.*

36. Audrey Leathard, *The Fight for Family Planning*, London: Macmillan, 1980, page 72.

37. *Ibid.*, page 87.

38. *Ibid.*, page 84.

39. Heidenry, *Op. cit.*, page 31.

40. Leathard, *Op. cit.*, page 114, on Rock's philosophy.

41. Heidenry, *Op. cit.*, page 31.

42. Leathard, *Op. cit.*, page 104. Heidenry, *Op. cit.*, page 31.

43. Heidenry, *Op. cit.*, pages 31–32.

44. *Ibid*, page 32.

45. Leathard, *Op. cit.*, page 105.

46. He originally wanted to publish the book anonymously, to protect his position at Cornell University, where he was a full professor, but Farrar, Straus & Giroux, the publishers, felt this undermined their defence of the book as literature. This account has been disputed. See: Andrew Field, *VN: The Life and Art of Vladimir Nabokov*, London: Macdonald/Queen Anne Press, 1987, pages 299–300.

47. *Ibid.*, pages 324–325 for VN's rejection of psychoanalytic interpretations of his work.

48. Daniel Horowitz, *Betty Friedan: The Making of the Feminine Mystique*, Amherst: University of Massachusetts Press, 1998, page 193.

49. Betty Friedan, *The Feminine Mystique*, New York: W. W. Norton, 1963; reprinted by Dell Publishing, paperback, 1984, page 7.

50. See Horowitz, *Op. cit.*, page 202 for other reactions.

51. Friedan, *Op. cit.*, page 38.

52. Horowitz, *Op. cit.*, pages 2–3.

53. Friedan, *Op. cit.*, pages 145–146.

54. *Ibid.*, page 16.

55. *Ibid.*, page 383.

56. See also: Horowitz, *Op. cit.*, pages 226–227.

CHAPTER 25: THE NEW HUMAN CONDITION

1. David Riesman, with Nathan Glazer and Reuel Denney, *The Lonely Crowd*, New Haven: Yale University Press, 1950, reprinted 1989 with the Preface to the 1961 edition and with a new Preface, page xxiv.

2. *Ibid.*, pages 5ff.

3. *Ibid.*, page 11.

4. *Ibid.*, page 15.

5. *Ibid.*, page 18.

6. *Ibid.*, page 19.

7. *Ibid.*, page 22.

8. *Ibid.*, see for example, chapters VIII, IX and X.

9. Ellen Schrecker, *The Age of McCarthyism: A Brief History with Documents*, Boston: Bedford Books, 1994, page 63.

10. Herman, *The Idea of Decline in Western History*, *Op. cit.*, page 316.

11. *Ibid.*

12. *Ibid.*

13. Adorno implied that the emotionalism that was once provided by the family was now provided by the Party. See: Ben Agger, *The Discourse of Domination: From the Frankfurt School to Postmodernism*, Evanston, Illinois: Northwestern University Press, 1992, page 251. And T. B. Bottomore, *Sociology as Social Criticism*, London: George Allen & Unwin, 1975, page 91.

14. Herman, *Op. cit.*, page 318.

15. Andrew Jamison and Ron Eyerman, *Seeds of the Sixties*, Berkeley: Los Angeles: London: University of California Press, page 52. This book, on which I have heavily relied, is an excellent introduction to the thought of the 1960s, very original, which deserves to be far better known.

16. In a letter dated 9 August 1956, Mary McCarthy said that even Bernard Berenson, who had a copy

of *Origins*, was curious to meet Arendt. Carol Brightman, *Between Friends: The Correspondence of Hannah Arendt and Mary McCarthy, 1949–1975*, London: Secker & Warburg, 1995, page 42.

17. For its difficult gestation, see Young-Bruehl, *Op. cit.*, pages 201ff.

18. Jamison and Eyerman, *Op. cit.*, page 47.

19. Young-Bruehl, *Hannah Arendt, Op. cit.*, pages 204–11.

20. Hannah Arendt, *The Origins of Totalitarianism*, New York: Harcourt, Brace, Jovanovich, 1951, page 475. Jamison and Eyerman, *Op. cit.*, page 47.

21. Jamison and Eyerman, *Op. cit.*, page 48. Young-Bruehl, *Op. cit.*, pages 206–207.

22. She herself referred to the book as 'Vita Activa': Brightman, *Op. cit.*, page 50.

23. Young-Bruehl, *Op. cit.*, page 319.

24. Jamieson and Eyerman, *Op. cit.*, page 50.

25. *Ibid.*, page 57.

26. Erich Fromm, *The Sane Society*, London: Routledge & Kegan Paul, 1956.

27. *Ibid.*, pages 5–9.

28. *Ibid.*, pages 122ff.

29. *Ibid.*, page 356.

30. *Ibid.*, pages 95 and 198.

31. *Ibid.*, page 222.

32. W. H. Whyte, *The Organisation Man*, London: Jonathan Cape, 1957.

33. *Ibid.*, page 14.

34. *Ibid.*, page 63.

35. *Ibid.*, pages 101ff.

36. *Ibid.*, pages 217ff.

37. *Ibid.*, pages 338–341.

38. Jamieson and Eyerman, *Op. cit.*, page 36.

39. *Ibid.*, page 37.

40. *Ibid.*, pages 36–37.

41. *Ibid.*, pages 33 and 34.

42. C. Wright Mills, *The Power Elite*, New York: Oxford University Press, 1956, pages 274–275. See also: Howard S. Becker, 'Professional sociology: The case of C. Wright Mills,' in Roy C. Rist, *The Democratic Imagination: Dialogues on the work of Irving Louis Horowitz*, New Brunswick and London: Transaction, 1994, pages 157ff.

43. Jamieson and Eyerman, *Op. cit.*, page 39.

44. *Ibid.*, page 40.

45. C. Wright Mills, *White Collar: The American Middle Classes*, New York: Oxford University Press, 1953, page ix, quoted in Jamieson and Eyerman, *Op. cit.*, page 40.

46. C. Wright Mills, *White Collar, Op. cit.*, pages 294–295. Jamison and Eyerman, page 41.

47. Jamison and Eyerman, *Op. cit.*, page 43.

48. *Ibid.*

49. C. Wright Mills, *The Sociological Imagination*, Oxford: Oxford University Press, 1959, page 5.

50. *Ibid.*, page 187.

51. Jamieson and Eyerman, *Op. cit.*, page 46.

52. J. K. Galbraith, *The Affluent Society*, Boston: Houghton Mifflin, 1958, Penguin paperback, 1991, page 40.

53. *Ibid.*, page 65.

54. In Galbraith's first autobiography his debt to Keynes is clearly shown. See J. K. Galbraith, *A Life in Our Times*, London: André Deutsch, 1981, pages 74–82. See also page 622.

55. *Ibid.*, page 86.

56. *Ibid.*, pages 122ff.

57. *Ibid.*, pages 128ff.

58. *Ibid.*, pages 182 and 191–195.

59. *Ibid.*, pages 195ff.

60. *Ibid.*, pages 233ff.

61. In his autobiography, Galbraith says *Time* awarded it 'a massive sneer' but Malcolm Muggeridge put it into the same category as Tawney's *The Acquisitive Society* and Keynes's *The Economic Consequences of the Peace*. J. K. Galbraith, *A Life in Our Times, Op. cit.*, page 354.

62. W. W. Rostow, *The Stages of Economic Growth*, Cambridge: Cambridge University Press, 1960, paperback edition, 1971.

63. *Ibid.*, page 7.

64. *Ibid.*, pages 36ff.

65. *Ibid.*, pages 59ff.

66. *Ibid.*, combining tables on pages 38 and 59.

67. *Ibid.*, pages 73ff.

68. *Ibid.*, page 11n.

69. *Ibid.*, page 107.

70. See the discussion by Fukuyama in the Conclusion (*infra*).

71. Rostow, *Op. cit.*, pages 102–103.

72. Daniel Horowitz, *Vance Packard and American Social Criticism*, Chapel Hill, N.C.: University of North Carolina Press, 1994, pages 98–100.

73. *Ibid.*, page 105.

74. *Ibid.*

75. Vance Packard, *The Hidden Persuaders*, New York: David McKay, 1957.

76. *Ibid.*, pages 87–88.

77. Vance Packard, *The Status Seekers*, New York: David McKay, 1959.

78. Horowitz, *Op. cit.*, page 123.

79. Vance Packard, *The Waste Makers*, New York: David McKay, 1960.

80. Horowitz, *Op. cit.*, page 119.

81. Malcolm Waters, *Daniel Bell*, London: Routledge, 1996, pages 13–15.

82. Waters, *Op. cit.*, page 78.

83. Daniel Bell, *The End of Ideology: On the Exhaustion of Political Ideas in the Fifties*, Glencoe: The Free Press, 1960; 1965 paperback reprinted by Harvard University Press, 1988, with a new Afterword. Waters, *Op. cit.*, page 79.

84. Waters, *Op. cit.*, page 80.

85. See the chapters by Malcolm Dean, pages 105ff, and Daniel Bell, pages 123ff, in Geoff Dench, Tony Flower and Kate Gavron (editors), *Young at Eighty*, London: Carcanet Press, 1995.

86. Michael Young, *The Rise of the Meritocracy*, London: Thames & Hudson, 1958, republished with a new Introduction by the author, by Transaction Publishers, New Brunswick, New Jersey, 1994.

87. *Ibid.*, page xi.

88. *Ibid.*, page xii. It was, however, poorly received by, among others, Richard Hoggart. See: Paul Barker, 'The Up and Downs of the Meritocracy', in Dench, Flower and Gavron (editors), *Op. cit.*, page 156.

89. Young, *Op. cit.*, page 170.

90. Barker, *Op. cit.*, page 161, cites reviewers who thought the book lacked 'the sound of a human voice.'

CHAPTER 26: CRACKS IN THE CANON

1. Peter Ackroyd, *T. S. Eliot*, *Op. cit.*, page 289.

2. T. S. Eliot, *Notes Toward the Definition of Culture*, London: Faber & Faber, 1948, paperback 1962.

3. Ackroyd, *Op. cit.*, page 291.

4. For a discussion of Eliot's wider thinking on leisure, see: Sencourt, *T. S. Eliot: A Memoir*, *Op. cit.*, page 154.

5. Eliot, *Notes*, *Op. cit.*, page 31.

6. *Ibid.*, page 23.

7. *Ibid.*, page 43.

8. He was conscious himself, he said, of being a *European*, as opposed to a merely British, or American, figure. See: Sencourt, *Op. cit.*, page 158.

9. Eliot, *Notes*, *Op. cit.*, page 50.

10. *Ibid.*, pages 87ff.

11. *Ibid.*, page 25.

12. Ian MacKillop, *F. R. Leavis*, *Op. cit.*, pages 15 and 17ff.

13. F. R. Leavis, *The Great Tradition*, London: Chatto & Windus, 1948; F. R. Leavis. *The Common Pursuit*, London: Chatto & Windus, 1952.

14. See Leavis, *The Common Pursuit*, chapter 14, for the links between sociology and literature, which Leavis was sceptical about; and chapter 23 for 'Approaches to T. S. Eliot,' where he counts 'Ash Wednesday' as the work which changed Eliot's standing. (And see the Conclusion of this book, below, page 750.)

15. MacKillop, *Op. cit.*, page 111. See in particular chapter 8, pages 263ff, on the future of criticism.

16. Lionel Trilling, *The Liberal Imagination*, New York: Macmillan, 1948 London: Secker & Warburg, 1951.

17. *Ibid.*, page 34.

18. *Ibid.*, pages 288ff.

19. Henry S. Commager, *The American Mind: An Interpretation of American Thought and Character Since the 1880s*, New York: Oxford University Press, 1950.

20. *Ibid.*, pages 199ff and 227ff.

21. *Ibid.*, pages 176–177.

22. *Ibid.*, pages 378ff.

23. Jamison and Eyerman, *Seeds of the Sixties*, *Op. cit.*, pages 150–151.

24. *Ibid.*, page 150.

25. Trilling's wife described the relationship as 'quasi-Oedipal.' See: Graham Caveney, *Screaming with Joy: The Life of Allen Ginsberg*, London: Bloomsbury, 1999, page 33.

26. Jamison and Eyerman, *Op. cit.*, page 152.

27. Barry Miles, *Ginsberg: A Biography*, New York: Viking, 1990, page 196.

28. *Ibid.*, page 192.

29. Jamison and Eyerman, *Op. cit.*, page 156.

30. *Ibid.*, pages 158–159.

31. See Miles, *Op. cit.*, page 197 for Ferlinghetti's reaction to the *Howl* reading.

32. Ann Charters, *Kerouac: A Biography*, London: André Deutsch, 1974, pages 24–25. Kerouac broke his leg and never reached the first team, a failure, she says, that he never came to terms with.

33. Jack Kerouac, *On the Road*, New York: Viking, 1957, Penguin paperback 1991, Introduction by Ann Charters, page x.

34. *Ibid.*, pages viii and ix.

35. *Ibid.*, page xx.

36. Charters, *Kerouac: A Biography*, *Op. cit.*, pages 92–97.

37. Kerouac took so much benzedrine in 1945 that he developed thrombophlebitis in his legs. See *ibid.*, page 52.

38. For a brief history of bepop, see: Gerald Nicosia, *Memory Babe: A Critical Biography of Jack Kerouac*, New York: Grove Press, 1983, page 112. Regarding the argument, they made up later, 'sort of'. See pages 690–691.

39. Charters, 'Introduction,' *Op. cit.*, page xxviii.

40. See: Jamison and Eyerman, *Op. cit.*, page 159.

41. Alan Freed interview in *New Musical Express*, 23 September 1956, quoted in: Richard Aquila, *That Old Time Rock'n'Roll: A Chronicle of an Era, 1954–1963*, New York: Schirmer, 1989, page 5.

42. Donald Clarke, *The Rise and Fall of Popular Music*, New York: Viking, 1995, Penguin 1995, page 373.

43. Aquila, *Op. cit.*, page 6.

44. Clarke, *Op. cit.*, page 370, which says it was definitely not the first.

45. It wasn't only imitation of course. See: Simon Frith, *Performing Rites: Evaluating Popular Music*, Oxford: Oxford University Press, page 195, for Presley's sexuality.

46. Aquila, *Op. cit.*, page 8.

47. See Frith, *Op. cit.*, *passim*, for charts and popular music marketing categories.

48. Arnold Goldman, 'A Remnant to Escape: The American Writer and the Minority Group,' in Marcus Cunliffe (editor), *The Penguin History of Literature*, *Op. cit.*, pages 302–303.

49. Ralph Ellison, *Invisible Man*, London: Gollancz, 1953, Penguin 1965. Goldman, *Op. cit.*, page 303.

50. Jamison and Eyerman, *Op. cit.*, page 160.

51. James Campbell, *Talking at the Gates: A Life of James Baldwin*, London: Faber & Faber, 1991, page 117.

52. Jamison and Eyerman, *Op. cit.*, page 163.

53. Campbell, *Op. cit.*, page 228.

54. *Ibid.*, page 125, quoted in Jamison and Eyerman, *Op. cit.*, page 166.

55. Colin MacInnes, *Absolute Beginners*, London:

Allison & Busby, 1959; *Mr Love and Justice*, London: Allison & Busby, 1960.

56. See for example: Michael Dash, 'Marvellous Realism: The Way out of Négritude,' in Bill Ashcroft, Gareth Griffiths and Helen Tiffin (editors) *The Post-Colonial Studies Reader*, London and New York: Routledge, 1995, page 199.

57. Ezenwa-Ohaeto, *Chinua Achebe: A Biography*, Oxford: James Currey and Bloomington and Indianapolis: Indiana University Press, 1997, page 60. See also: Gilbert Phelps, 'Two Nigerian Writers: Chinua Achebe and Wole Soyinka,' in Boris Ford (editor), *The New Pelican Guide to English Literature, volume 8: From Orwell to Naipaul*, London: Penguin, 1983, pages 319–331.

58. Chinua Achebe, *Things Fall Apart*, New York: Doubleday, 1959, Anchor paperback, 1994. Phelps, *Op. cit.*, page 320.

59. Ezenwa-Ohaeto, *Op. cit.*, page 66. Phelps, *Op. cit.*, page 321.

60. *Ibid.*, pages 66ff for an account of the various drafts of the book and Achebe's initial attempts to have it published. Phelps, *Op. cit.*, page 323.

61. See: Claude Lévi-Strauss and Didier Eribon, *Conversations with Lévi-Strauss*, *Op. cit.*, page 145, for Lévi-Strauss's views on the evolution of anthropology in the twentieth century. See also: Leach, *Op. cit.*, page 9.

62. Edmund Leach, *Lévi-Strauss*, London: Fontana, 1974, page 13.

63. Claude Lévi-Strauss, *Tristes Tropiques*, Paris: Plon, 1955; *Mythologiques I: Le cru et le cuit*, Paris: Plon, 1964. Translated as: *The Raw and the Cooked*, London: Jonathan Cape, 1970, volume I of The Science of Mythology; volume II, *From Honey to Ashes*, London: Jonathan Cape, 1973. Lévi-Strauss told Eribon that he thought psychoanalysis, or at least *Totem and Taboo*, was 'a failure.' See: Eribon and Lévi-Strauss, *Op. cit.*, page 106.

64. Leach, *Op. cit.*, page 60.

65. *Ibid.*, page 63.

66. *Ibid.*, pages 82ff.

67. When Margaret Mead visited Paris, Claude Lévi-Strauss introduced her to Simone de Beauvoir. 'They didn't say a word to one another.' Eribon and Lévi-Strauss, *Op. cit.*, page 12.

68. Basil Davidson, *Old Africa Rediscovered*, London: Gollancz, 1959.

69. Oliver Neville, 'The English Stage Company and the Drama Critics,' in Ford (editor), *Op. cit.*, page 251.

70. *Ibid.*, page 252. Osborne's own account of reading the ad is in John Osborne, *A Better Class of Person: Autobiography 1929–1956*, London: Faber & Faber, 1981, page 275.

71. Neville, *Op. cit.*, pages 252–253.

72. Peter Mudford, 'Drama since 1950', in Dodsworth (editor), *The Penguin History of Literature*, *Op. cit.*, page 396.

73. For the autobiographical overlap of the play, see: Osborne, *Op. cit.*, pages 239ff.

74. Mudford, *Op. cit.*, page 395.

75. *Ibid.*

76. Michael Hulse, 'The Movement', in Ian Hamilton (editor), *The Oxford Companion to Twentieth-Century Poetry, Op. cit.*, page 368.

77. Mudford, *Op. cit.*, page 346.

78. For Larkin's library career, his reactions to it, and his feelings of timidity, see: Andrew Motion, *Philip Larkin: A Writer's Life*, London: Faber & Faber, 1993, page 109ff. For other details about Larkin discussed in this section, see respectively: Alastair Fowler, 'Poetry since 1950,' in Dodsworth (editor), *Op. cit.*, page 346; and Motion, *Op. cit.*, pages 242–243 and 269, about publicity in *The Times*. Seamus Heaney's poem, published as part of *'A Tribute' to Philip Larkin*, George Hartley (editor), London: The Marvell Press, 1988, page 39, ended with the line, 'A nine-to-five man who had seen poetry.'

79. For the 'helpless bystander' quote see Michael Kirkham, 'Philip Larkin and Charles Tomlinson: Realism and Art' in Boris Ford (ed.), *From Orwell to Naipaul*, vol. 8, *New Pelican Guide to English Literature*, London: Penguin, revised edn 1995, pages 286–289. Blake Morrison, 'Larkin,' in Hamilton (editor), *Op. cit.*, page 288.

80. Richard Hoggart, *A Sort of Clowning: Life and Times, volume II, 1940–59*, London: Chatto & Windus, 1990, page 175.

81. Leavis said the book 'had some value' but that Hoggart 'should have written a novel.' See: Hoggart, *Op. cit.*, page 206.

82. Richard Hoggart, *The Uses of Literacy*, London: Chatto & Windus, 1957.

83. Raymond Williams, *Culture and Society*, London: Chatto & Windus, 1958.

84. For a good discussion see: Fred Inglis, *Cultural Studies*, Oxford: Blackwell, 1993, pages 52–56; and Fred Inglis, *Raymond Williams*, London and New York: Routledge, 1995, pages 162ff.

85. Stefan Collini, 'Introduction' to: C. P. Snow, *The Two Cultures*, Cambridge: Cambridge University Press, 1959, paperback 1969 and 1993, page vii.

86. *Ibid.*

87. *Ibid.*, page viii. The fee for Snow's lecture was 9 guineas (ie, £9.45p), the same rate as when the lecture was established in 1525. See: Philip Snow, *Stranger and Brother: A Portrait of C. P. Snow*, London: Macmillan, 1982, page 117.

88. *Ibid.*, page 35. See also Collini, *Op. cit.*, page xx.

89. C. P. Snow, *Op. cit.*, page 14.

90. *Ibid.*, page 18.

91. *Ibid.*, pages 29ff.

92. *Ibid.*, page 34.

93. *Ibid.*, pages 41ff.

94. MacKillop, *Op. cit.*, page 320.

95. He was also ill. See: Philip Snow, *Op. cit.*, page 130.

96. Collini, *Op. cit.*, pages xxxiiiff. This essay, 64 pages, is recommended. Among other things, it relates Snow's lecture to the changing map of the disciplines in the last half of the century.

97. Lionel Trilling, 'A comment on the Leavis-Snow Controversy,' *Universities Quarterly*, volume 17, 1962, pages 9–32. Collini, *Op. cit.*, pages xxxvii-iff.

98. The subject was first debated on television in 1968. See: Philip Snow, *Op. cit.*, page 147.

CHAPTER 27: FORCES OF NATURE

1. Michael Polanyi, *Science, Faith and Society*, Oxford: Oxford University Press, 1946.

2. *Ibid.*, page 14.

3. *Ibid.*, page 19.

4. *Ibid.*, pages 60ff.

5. Julian Symons, Introduction to: George Orwell, *1984*, Everyman's Library, 1993, page xvi. See also Ben Pimlott's Introduction to Penguin paperback edition, 1989.

6. James Burnham, *The Managerial Revolution, or What is Happening in the World Now*, New York: Putnam, 1941.

7. For the problem in physics see: Paul R. Josephson, *Physics and Politics in Revolutionary Russia*, Los Angeles and Oxford: University of California Press, 1991. For the Lysenko problem in Communist China, see: Laurence Schneider, 'Learning from Russia: Lysenkoism and the Fate of Genetics in China, 1950–1986,' in Denis Fred Simon and Merle Goldman (editors), *Science and Technology in Post-Mao China*, Cambridge, Massachusetts: The Council on East Asian Studies/Harvard University Press, 1989, pages 45–65.

8. Krementsov, *Stalinist Science, Op. cit.*, page 115.

9. *Ibid.*, page 107.

10. *Ibid.*, pages 129–131, 151 and 159.

11. *Ibid.*, pages 160 and 165.

12. *Ibid.*, page 169.

13. *Ibid.*, pages 174, 176 and 179.

14. Michael Riordan and Lillian Hoddeson, 'Birth of an Era', *Scientific American: Special Issue: 'Solid State Century: The Past, Present and Future of the Transistor'*, 22 January 1998, page 10.

15. S. Millman (editor), *A History of Engineering and Science in the Bell Systems: Physical Sciences (1925–1980)*, Thousand Oaks, California: Bell Laboratories, 1983, pages 97ff.

16. Riordan and Hoddeson, *Op. cit.*, page 11.

17. *Ibid.*

18. *Ibid.*

19. *Ibid.*, page 14.

20. Brian Winston, *Media, Technology and Society: A History from the Telegraph to the Internet*, London and New York: Routledge, 1998, pages 216–217. And Chris Evans, *The Mighty Micro*, London: Gollancz, 1979, pages 49–50.

21. Frank H. Rockett, 'The Transistor,' *Scientific American: Special Issue: 'Solid State Century: The Past, Present and Future of the Transistor'*, 22 January 1998, pages 18ff.

22. *Ibid.*, page 19.

23. Winston, *Op. cit.*, page 213.

24. Riordan and Hoddeson, *Op. cit.*, pages 14–15.

25. *Ibid.*, page 13.

26. Though the publicity helped the sales of the transistor. See: Winston, *Op. cit.*, page 219.

27. *Ibid.*, page 221.

28. Paul Strathern, *Crick, Watson and DNA*, London: Arrow, 1997, pages 37–38. James D. Watson, *The Double Helix*, London: Weidenfeld & Nicolson, 1968; Penguin paperback, 1990, page 20.

29. Strathern, *Op. cit.*, page 42.

30. *Ibid.*, page 44.

31. For rival groups, and the state of research at the time, see: Bruce Wallace, *The Search for the Gene, Op. cit.*, pages 108ff.

32. Strathern, *Op. cit.*, page 45.

33. Watson, *Op. cit.*, page 25.

34. Strathern, *Op. cit.*, page 49.

35. *Ibid.*, pages 50–53.

36. Watson, *Op. cit.*, page 79.

37. Strathern, *Op. cit.*, page 56.

38. Watson, *Op. cit.*, pages 82–83. Strathern, *Op. cit.*, pages 57–58.

39. Watson, *Op. cit.*, page 91. Strathern, *Op. cit.*, page 60.

40. Watson, *Op. cit.*, page 123.

41. According to Pauling's biographer, Thomas Hager, 'Historians have speculated that the denial of Pauling's passport for the May Royal Society meeting was critical in preventing him from discovering the structure of DNA, that if he had attended he would have seen Franklin's work ...' Hager, *Force of Nature, Op. cit.*, page 414.

42. Strathern, *Op. cit.*, pages 70–71.

43. There was mutual respect. Pauling already wanted Crick to come to Caltech. See Hager, *Op. cit.*, page 414. Strathern, *Op. cit.*, page 72.

44. Strathern, *Op. cit.*, page 81.

45. *Ibid.*, page 84, where there is a useful diagram.

46. Watson, *Op. cit.*, page 164.

47. Strathern, *Op. cit.*, page 82.

48. Watson wrote an epilogue about her in his book, praising her courage and integrity. He admitted, too late, that he had been wrong about her. Watson, *Op. cit.*, pages 174–175. Strathern, *Op. cit.*, pages 83–84.

49. Alan Shepard and Deke Slayton, *Moon Shot*, New York: Turner/Virgin, 1994, page 37.

50. James Harford, *Korolev: How One Man Masterminded the Soviet Drive to Beat the Americans to the Moon*, New York: John Wiley & Sons, 1997, page 121.

51. See Shepard and Slayton, *Op. cit.*, page 39, for Reuters more fulsome headlines. Harford, *Op. cit.*, page 130.

52. Although *Sputnik 1* wasn't large, it was still bigger than what the US planned. See: Charles Murray and Catherine Bly Cox, *Apollo: The Race for the Moon*, London: Secker & Warburg, 1989, page 23. See also Harford, *Op. cit.*, page 122.

53. See Young, Silcock, and Peter Dunn, *Journey to the Sea of Tranquility*, London: Jonathan Cape, 1969, pages 80–81 for discussion of cost and security.

54. Harford, *Op. cit.* See note 50 *supra*.

55. See Shepard and Slayton, *Op. cit.*, pages 38–39 for other personal details.

56. Harford, *Op. cit.*, pages 49–50.

57. *Ibid.*, page 51.

58. Robert Conquest, *The Great Terror*, London: Macmillan, 1968; and the same author's, *Kolyma: The Arctic Death Camps*, New York: Viking, 1979.

59. Harford, *Op. cit.*, page 57.

60. *Ibid.*, page 91.

61. After Vanguard was announced, the Russians had gloated they would beat the Americans. See: Young, Silcock *et al.*, *Op. cit.*, page 67.

62. For the impact in America, see: Murray and Cox, *Op. cit.*, page 77.

63. Harford, *Op. cit.*, pages 114–115.

64. *Ibid.*, page 110.

65. But not on Eisenhower, and not at first. See: Young, Silcock *et al.*, *Op. cit.*, page 68.

66. See: Young, Silcock *et al.*, *Op. cit.*, page 74, one of several contemporary accounts on the subject that makes no reference to Korolev. Harford, *Op. cit.*, page 133.

67. Shepard and Slayton, *Op. cit.*, page 42.

68. Harford, *Op. cit.*, page 132.

69. *Sputnik* 2 had an even bigger effect than *Sputnik* 1. See: Young, Silcock *et al.*, *Op. cit.*, pages 70–71.

70. Harford, *Op. cit.*, page 135.

71. *Ibid.*, pages 135–136.

72. For the effect of *Sputnik*'s launch on Eisenhower's policy, see: Young, Silcock *et al.*, *Op. cit.*, pages 82ff.

73. Richard Leakey, *One Life*, London: Michael Joseph, 1983, page 49.

74. Virginia Morrell, *Ancestral Passions: The Leakey family and the Quest for Humankind's Beginnings*, New York: Simon & Schuster, 1995, page 57.

75. Mary Leakey, *Olduvai Gorge: My Search for Early Man*, London: Collins, 1979, page 13.

76. Morrell, *Op. cit.*, pages 80–89.

77. Partly as a result he wrote books on other aspects of East Africa. See for example, L. S. B. Leakey, *Kenya: Contrasts and Problems*, London: Methuen, 1936.

78. Morrell, *Op. cit.*, pages 163–174.

79. Mary Leakey, *Op. cit.*, pages 83ff.

80. See Mary Leakey, *ibid.*, pages 52–53 for a detailed map of the gorge.

81. Morrell, *Op. cit.*, page 178.

82. *Ibid.*, pages 180–181.

83. Mary Leakey, *Op. cit.*, page 75. See also: Richard Leakey, *Op. cit.*, page 50.

84. Morrell, *Op. cit.*, page 181.

85. *Ibid.*

86. Mary Leakey, *Op. cit.*, page 74.

87. L. S. B. Leakey, 'Finding the World's Earliest Man', *National Geographic Magazine*, September 1960, pages 421–435. Morrell, *Op. cit.*, page 194.

88. Morrell, *Op. cit.*, page 196.

89. *Ibid.*, and Richard Leakey, *Op. cit.*, page 49.

90. Claude Lévi-Strauss and Didier Eribon, *Conversations with Lévi-Strauss*, *Op. cit.*, page 119.

91. Karl Popper, *The Logic of Scientific Discovery*,

London: Hutchinson, 1959. (Originally published in German in Vienna in 1934.) See especially, chapters I, IV and V.

92. Thomas S. Kuhn, *The Structure of Scientific Revolutions*, Chicago: University of Chicago Press, 1962; 2nd edition, enlarged, University of Chicago Press, 1970, especially chapter VI, pages 52ff.

93. *Ibid.*, page 151.

94. *Ibid.*, pages 137ff.

95. See the Postscript, pages 174ff, in the second, enlarged edition, referred to in Note 92 above.

CHAPTER 28: MIND MINUS METAPHYSICS

1. John Russell Taylor, *Hitch: The Life and Work of Alfred Hitchcock*, London: Faber & Faber, 1978, page 255.

2. Donald Spoto, *The Life of Alfred Hitchcock: The Dark Side of Genius*, London: Collins, 1983, page 420. James Pallot, Jacob Levich *et al*, *The Fifth Virgin Film Guide*, London: Virgin Books, 1996, pages 553–554.

3. *Ibid.*, pages 421–423.

4. Russell Taylor, *Op. cit.*, page 256.

5. Spoto, *Op. cit.*, pages 423–424.

6. *Ibid.*, page 420.

7. R. D. Laing, *The Divided Self: An Existential Study in Sanity and Madness*, London: Tavistock, 1959. See also: Adrian Laing, *R. D. Laing: A Life*, London: Peter Owen, 1994, chapter 8, pages 77–78.

8. Gilbert Ryle, *The Concept of Mind*, London: Hutchinson, 1949.

9. *Ibid.*, pages 36ff.

10. *Ibid.*, pages 319ff.

11. S. Stephen Hilmy, *The Later Wittgenstein: The Emergence of a New Philosophical Method*, Oxford: Blackwell, 1987, page 191.

12. Ludwig Wittgenstein, *Philosophical Investigations*, Oxford: Blackwell, 1953 (edited by G. E. M. Anscombe and R. Rhees). Wittgenstein had begun writing this book in 1931 – see Hilmy, *Op. cit.*, page 50.

13. P. M. S. Hacker, *Wittgenstein*, *Op. cit.*, page 8.

14. Though even professional philosophers do refer to them as games. And see Hilmy, *Op. cit.*, chapters 3 and 4.

15. Wittgenstein, *Philosophical Investigations*, *Op. cit.*, page 109, quoted in Hacker, *Op. cit.*, page 11.

16. Magee (editor), *Op. cit.*, page 89.

17. Hacker, *Op. cit.*, page 16.

18. *Ibid.*, page 18.

19. Many of the paragraphs were originally written at the end of World War Two, which is why he may have chosen pain as an example. See: Monk, *Op. cit.*, pages 479–480. See also Hilmy, *Op. cit.*, page 134, and Hacker, *Op. cit.*, page 21.

20. Wittgenstein, *Op. cit.*, page 587, quoted in Hacker, *Op. cit.*, page 24.

21. *Ibid.*, page 31.

22. Magee (editor), *Op. cit.*, page 90; and Hacker, *Op. cit.*, page 40.

23. Martin L. Gross, *The Psychological Society*, New

York: Simon & Schuster, 1979, page 200.

24. *Ibid.*, page 201.

25. H. M. Halverson, 'Genital and Sphincter Behavior in the Male Infant,' *Journal of Genetic Psychology*, volume 56, pages 95–136. Quoted in Gross, *Op. cit.*, page 220.

26. See also: H. J. Eysenck, *Decline and Fall of the Freudian Empire*, London: Viking, 1985, especially chapters 5 and 6.

27. Ralph Linton, *Culture and Mental Disorder*, Springfield, Illinois: Charles C. Thomas, 1956, quoted in Gross, *Op. cit.*, page 219.

28. Ray Fuller (editor), *Seven Pioneers of Psychology*, *Op. cit.*, page 126.

29. B. F. Skinner, *Science and Human Behavior*, Glencoe: The Free Press, 1953.

30. *Ibid.*, pages 263ff.

31. *Ibid.*, page 375.

32. *Ibid.*, pages 377–378.

33. Fuller (editor), *Op. cit.*, page 113.

34. B. F. Skinner, *Verbal Behavior*, New York: Appleton-Century-Crofts, 1957.

35. *Ibid.*, pages 81ff.

36. Noam Chomsky, *Syntactic Structures*, The Hague: Mouton, 1957. See also: Roger Smith, *The Fontana History of the Human Sciences*, *Op. cit.*, page 672. And: John Lyons, *Chomsky*, London: Fontana/Collins, 1970, page 14.

37. Noam Chomsky, *Language and the Mind*, New York: Harcourt Brace, 1972, pages 13 and 100ff. Lyons, *Op. cit.*, page 18.

38. Lyons, *Op. cit.*, pages 105–106.

39. Fuller (editor), *Op. cit.*, page 117.

40. Published by Pelican as: John Bowlby, *Child Care and the Growth of Love*, 1953.

41. *Ibid.*, pages 18ff.

42. *Ibid.*, pages 50ff.

43. *Ibid.*, pages 161ff.

44. Fernando Vidal, *Piaget Before Piaget*, Cambridge, Massachusetts: Harvard University Press, 1994, pages 206–207.

45. Peter E. Bryant, 'Piaget', in Fuller (editor), *Op. cit.*, page 133.

46. Two books out of the many Piaget wrote provide a good introduction to his work and methods: *The Language and Thought of the Child*, London: Kegan Paul, Trench and Trubner, 1926; and *Six Psychological Studies*, London: University of London Press, 1968.

47. Bryant, *Op. cit.*, pages 135ff.

48. Vidal, *Op. cit.*, page 230.

49. Bryant, *Op. cit.*, page 136.

50. Vidal, *Op. cit.*, page 231.

51. Weatherall, *In Search of a Cure*, *Op. cit.*, page 254.

52. *Ibid.*, page 255.

53. *Ibid.*, page 257.

54. David Healy, *The Anti-Depressant Era*, Cambridge, Massachusetts: Harvard University Press, 1997, page 45.

55. *Ibid.*, pages 61–62. Weatherall, *Op. cit.*, pages 258–259.

56. Healy, *Op. cit.*, pages 52–54, which discusses the influential *Nature* article of 1960 on this subject.

57. Gregory Bateson, 'Toward a Theory of Schizophrenia,' *Behavioral Science*, volume 1, Number 4, 1956.

58. Adrian Laing, *Op. cit.*, page 138.

59. *Ibid.*, page 71. Former patients told Laing's son, when he was researching his book on his father, that LSD was beneficial. See: page 71.

60. Jamison and Eyerman, *Seeds of the Sixties*, *Op. cit.*, pages 122–123.

61. *Ibid.*, page 123.

62. Herbert Marcuse, *One-Dimensional Man: Studies in the Ideology of Advanced Industrial Society*, London: Routledge & Kegan Paul, 1964.

63. Marcuse, *Op. cit.*, page 156. Jamison and Eyerman, *Op. cit.*, page 127.

64. *Ibid.*, pages 193ff.

65. See: Herbert Marcuse, *Counter-Revolution and Revolt*, London: Allen Lane, 1972, page 105, for the 'antagonistic unity' between art and revolution in this context.

CHAPTER 29: MANHATTAN TRANSFER

1. Moshe Pearlman, *The Capture of Adolf Eichmann*, London: Weidenfeld & Nicolson, 1961, especially pages 113–120.

2. Young-Bruehl, *Hannah Arendt*, *Op. cit.*, pages 328ff.

3. Hannah Arendt, *Eichmann in Jerusalem: A Report on the Banality of Evil*, New York: Viking, 1963, enlarged and revised edition, Penguin, 1994, page 49.

4. *Ibid.*, page 92.

5. Young-Bruehl, *Op. cit.*, page 337.

6. Arendt, *Op. cit.*, page 252.

7. See Young-Bruehl, *Op. cit.*, pages 347–378 for a full discussion of the controversy, including its overlap with the assassination of President Kennedy.

8. Laura Fermi, *Illustrious Immigrants*, *Op. cit.*, pages 153–154.

9. Erik Erikson, *Childhood and Society*, New York: W. W. Norton, 1950; Penguin edition 1965, especially Part 4, 'Youth and the Evolution of Identity.'

10. Erikson, *Op. cit.*, chapter 8, pages 277–316.

11. Bruno Bettelheim, 'Individual and Mass Behavior in Extreme Situations ' *Journal of Abnormal and Social Psychology*, 1943.

12. Bruno Bettelheim, *The Empty Fortress*, New York: Collier-Macmillan, 1968.

13. Nina Sutton, *Bruno Bettelheim: The Other Side of Madness*, London: Duckworth, 1995, chapters XI and XII.

14. And Bruno Bettelheim, *Recollections and Reflections*, New York: Knopf, 1989; London: Thames & Hudson, 1990, pages 166ff.

15. Laura Fermi, *Op. cit.*, pages 207–208.

16. Richard Rhodes, *Op. cit.*, page 563.

17. *Ibid.*, page 777.

18. Kragh, *Op. cit.*, pages 332ff; see also: Alexander Hellemans and Bryan Bunch, *The Timetables of*

Science, New York: Simon & Schuster, 1988, page 498.

19. See: George Gamow, *The Creation of the Universe*, New York: Viking, 1952, for a more accessible account. Page 42 for his discussion of the current temperature of the space in the universe.

20. Hellemans and Bunch, *Op. cit.*, page 499.

21. Murray Gell-Mann, *The Quark and the Jaguar*, New York: Little Brown, 1994, page 11, for why he chose 'quark.'

22. See under 'quark', 'baryon' and 'lepton' in: John Gribbin, *Q is for Quantum*, London: Weidenfeld & Nicolson, 1998, paperback edition 1999, and pages 190–191 for the early work on quarks.

23. See also: Yuval Ne'eman and Yoram Kirsh, *The Particle Hunters*, Cambridge: Cambridge University Press, 1986, pages 196–199 for a more technical introduction to the eight-fold way.

24. Victor Bockris, *Warhol*, London and New York: Frederick Muller, 1989, page 155.

25. Barron, *Exiles and Emigrés*, *Op. cit.*, pages 21–28.

26. Dore Ashton, *The New York School: A Cultural Reckoning*, New York: Viking, 1973, pages 123 and 140.

27. Alice Goldfarb Marquis, *Alfred H. Barr: Missionary for the Modern*, Chicago: Contemporary Books, 1989, page 69.

28. Ashton, *Op. cit.*, pages 142–145 and 156.

29. *Ibid.*, page 175.

30. Diana Crane, *The Transformation of the Avant-Garde: The New York Art World, 1940–1986*, Chicago and London: University of Chicago Press, 1987, page 45.

31. *Ibid.*, page 49.

32. Bockris, *Op. cit.*, pages 112–134, especially page 128.

33. Hughes, *The Shock of the New*, *Op. cit.*, page 251.

34. Crane, *Op. cit.*, page 82.

35. David Lehman, *The Last Avant-Garde: The Making of the New York School of Poets*, New York: Doubleday 1998, Anchor paperback 1999. Lehman shows that these poets were also 'aesthetes in revolt against a moralist's universe', see page 358. 'They believed that the road of experimentation leads to the pleasure-dome of poetry', page 358.

36. Arnold Whittall, *Music Since the First World War*, *Op. cit.*, page iii.

37. *Ibid.*, page 3.

38. *Dancers on a Plane: John Cage, Merce Cunningham, Jasper Johns*, Liverpool: The Tate Gallery, 1990, Introduction by Richard Francis, page 9.

39. Whittall, *Op. cit.*, page 208.

40. Sally Banes, *Writing Dancing in the Age of Postmodernism*, Hanover and London: Wesleyan University Press, published by the University Presses of New England, 1994, page 103.

41. Banes, *Op. cit.*, page 104.

42. *Ibid.*, page 110.

43. Richard Francis, *Op. cit.*, page 11.

44. Banes, *Op. cit.*, page 115.

45. *Ibid.*, page 117.

46. Susan Sontag, *Against Interpretation*, London: Vintage, 1994, page 10.

47. *Ibid.*, pages 13–14. In another celebrated essay, 'Notes on Camp', published in the same year, 1964, in *The New York Review of Books*, Susan Sontag addressed a certain sensibility which, she said, was wholly aesthetic, in contrast to high culture, which was basically moralistic (Sontag, *Op. cit.*, page 287). 'It incarnates a victory of "style" over "content", "aesthetics" over "morality", of irony over tragedy.' It was not the same as homosexual taste, she said, but there was an overlap. 'The experiences of Camp are based on the great discovery that the sensibility of high culture has no monopoly on refinement. Camp asserts that good taste is not simply good taste; that there exists, indeed, a good taste of bad taste.' (*Ibid.*, page 291.) This too would form an ingredient of the postmodern sensibility.

CHAPTER 30: EQUALITY, FREEDOM AND JUSTICE
IN THE GREAT SOCIETY

1. Doris Kearns, *Lyndon Johnson and the American Dream*, London: André Deutsch, 1976, pages 210–217.

2. Friedrich von Hayek, *The Constitution of Liberty*, London: Routledge & Kegan Paul, 1960.

3. John Gray, *Hayek on Liberty*, London: Routledge, 1984, page 61.

4. Hayek, *Op. cit.*, page 349; and Gray, *Op. cit.*, page 71.

5. Hayek, *Op. cit.*, pages 385 and 387; Gray, *Op. cit.*, page 72.

6. Hayek, *Op. cit.*, page 385. See also: Roland Kley, *Hayek's Social and Political Thought*, Oxford: The Clarendon Press, 1994, pages 199–204.

7. Gray, *Op. cit.*, page 73.

8. *Ibid.*

9. Milton Friedman, with the assistance of Rose Friedman, *Capitalism and Freedom*, Chicago: University of Chicago Press, 1963.

10. For the difference between this work and Friedman's later books, see: Eamon Butler, *Milton Friedman: A Guide to His Economic Thought*, London: Gardner/Maurice Temple Smith, 1985, pages 197ff.

11. Friedman, *Op. cit.*, page 156.

12. *Ibid.*, pages 100ff.

13. *Ibid.*, page 85.

14. *Ibid.*, pages 190ff.

15. Michael Harrington, *The Other America*, New York: Macmillan, 1962.

16. Though neither Harrington nor Jacobs (see below) are mentioned in Johnson's memoirs, even though he has a chapter on the war on poverty. See: Lyndon Baines Johnson, *The Vantage Point: Perspectives on the Presidency, 1963–1969*, London: Weidenfeld & Nicolson, 1972.

17. See for example: Arthur Marwick, *The Sixties*, Oxford: Oxford University Press, 1998, page 260.

18. Harrington, *Op. cit.*, page 1.

19. *Ibid.*, pages 82ff.

20. Kearns, *Op. cit.*, pages 188–189.

21. Jane Jacobs, *The Death and Life of Great American Cities*, London: Jonathan Cape, 1962.
22. *Ibid.*, pages 97ff.
23. *Ibid.*, pages 55ff.
24. *Ibid.*, pages 94–95.
25. *Ibid.*, pages 128–129.
26. *Ibid.*, chapter 14, pages 257ff.
27. *Ibid.*, page 378.
28. *Ibid.*, pages 291ff.
29. *Ibid.*, pages 241ff.
30. David L. Lewis, *Martin Luther King: A Critical Biography*, Allen Lane, The Penguin Press, 1970, pages 187–191.
31. Marwick, *Op. cit.*, pages 215–216; see also: Coretta King, *My Life with Martin Luther King Jr*, London: Hodder & Stoughton, 1970, pages 239–241. New York: Holt, Rinehart, Winston.
32. Lewis, *Op. cit.*, pages 227–229.
33. *Ibid.*, page 229.
34. This list, and the next one, have been assembled from several sources but in particular: Phillip Waller and John Rowett (editors), *Chronology of the Twentieth Century*, London: Helicon, 1995.
35. Frantz Fanon, *A Dying Colonialism*, London: Monthly Review Press, 1965, Penguin 1970; originally published as: *L'An Cinq de la Révolution Algérienne*, Paris, Maspuro, 1959; and *Black Skin, White Masks*, New York: The Grove Press, 1967.
36. Frantz Fanon, *The Wretched of the Earth*, London: MacGibbon & Kee, 1965, translator Constance Farrington.
37. *Ibid.*, page 221.
38. *Ibid.*, pages 228ff.
39. Eventually published as: J. C. Carothers, *The Mind of Man in Africa*, London: Tom Stacey, 1972.
40. Eldridge Cleaver, *Soul on Ice*, London: Jonathan Cape, 1968, pages 101–103.
41. *Ibid.*, page 207.
42. Maya Angelou, *I Know Why the Caged Bird Sings*, New York: Random House, 1969.
43. *Ibid.*, page 51.
44. *Ibid.*, page 14.
45. *Ibid.*, page 184.
46. *Ibid.*, page 201.
47. Jones, *Op. cit.*, page 529.
48. D'Emilio and Freedman, *Intimate Matters*, *Op. cit.*, page 312.
49. *Ibid.*, pages 302–304.
50. Germaine Greer, *The Female Eunuch*, London: MacGibbon & Kee, 1971, pages 90–98.
51. *Ibid.*, page 273–282.
52. Juliet Mitchell, *Women's Estate*, Penguin: 1971.
53. *Ibid.*, page 75.
54. *Ibid.*, page 59.
55. *Ibid.*, page 62.
56. *Ibid.* Juliet Mitchell later went on to explore this subject more fully in *Psychoanalysis and Feminism*, London: Allen Lane, 1974.
57. Kate Millett, *Sexual Politics*, *Op. cit.*
58. *Ibid.*, pages 314ff.
59. *Ibid.*, pages 336ff.
60. *Ibid.*, page 356.
61. Heidenry, *What Wild Ecstasy*, *Op. cit.*, pages 110–111. See also: Andrea Dworkin, 'My Life as a Writer', Introduction to *Life and Death*, Glencoe: Free Press, 1997, pages 3–38.
62. Heidenry, *Op. cit.*, page 113.
63. *Ibid.*, pages 186–187.
64. *Ibid.*, page 188.
65. Marwick, *Op. cit.*, page 114.
66. Kearns, *Op. cit.*, pages 286ff.
67. Robert A. Caro, *The Years of LBJ: The Path to Power*, London: Collins, 1983, pages 336–337 for background.
68. J. W. B. Douglas, *All Our Future*, London: MacGibbon & Kee, 1968.
69. Steven Rose, Leon J. Kamin and R. C. Lewontin, *Not in Our Genes*, New York: Pantheon, 1984, Penguin, 1984, page 19.
70. Christopher Jencks *et al.*, *Inequality: A Reassessment of the Effects of Family and Schooling in America*, New York: Basic Books, 1972.
71. *Ibid.*, page 8.
72. *Ibid.*, page 315.
73. *Ibid.*, page 84.
74. *Ibid.*, page 265.
75. Ivan Illich, *De-Schooling Society*, London: Marion Boyars, 1978.
76. *Ibid.*, page 91.
77. Norman Mailer, *An American Dream*, London: André Deutsch, 1965, Flamingo Paperback, 1994.
78. See: Peter Manso, *Mailer: His Life and Times*, New York: Viking, 1985, page 316, for overlaps with real life.
79. Norman Mailer, *The Armies of the Night*, London: Weidenfeld & Nicolson, 1968.
80. See: Manso, *Op. cit.*, pages 455ff for background.
81. Paul Johnson, *A History of the American People*, *Op. cit.*, page 555.
82. *Ibid.*, page 557.
83. *Ibid.*
84. See: Jiang Qing, 'Reforming the Fine Arts', in Michael Schoenhals (editor), *China's Cultural Revolution 1966–1969*, New York and London: M. E. Sharpe, 1996, page 198.
85. Even unwanted hairstyles were banned. See: 'Vigorously and Speedily Eradicate Bizarre Hairstyles, a Big-Character Poster by the Guangzhou hairdressing trade,' in Schoenhals (editor), *Op. cit.*, pages 210ff. See also Johnson, *Op. cit.*, pages 558–559.
86. Johnson, *Op. cit.*, page 560.
87. Yu Xiaoming, 'Go on Red! Stop on Green!' in Schoenhals (editor), *Op. cit.*, page 331.
88. Zhores and Roy Medvedev, *A Question of Madness*, New York: Knopf, 1971; London: Macmillan, 1971. For a discussion of Lysenkoism in Communist China, together with an outline of the structure of science and technology, and the impact of scholars who had trained abroad, see: Denis Fred Simon and Merle Goldman (editors), *Science and Technology in Post-Mao China*, Cambridge, Massachusetts: The Council on East Asian

Studies/Harvard University Press, 1989, especially chapters 2, 3, 4, 8 and 10.

89. Medvedev and Medvedev, *Op. cit.*, page 30.

90. *Ibid.*, page 51.

91. *Ibid.*, pages 54 and 132.

92. *Ibid.*, page 78.

93. *Ibid.*, pages 198ff.

94. Alexandr Solzhenitsyn, *One Day in the Life of Ivan Denisovich*, New York: Praeger, 1963, translated by Max Hayward and Ronald Hingley. *Cancer Ward*, London: The Bodley Head, 2 vols, 1968–1969, translated by Nicholas Bethell and David Burg.

95. Michael Scammell, *Solzhenitsyn: A Biography*, New York: W. W. Norton, 1984, page 61.

96. *Ibid.*, page 87.

97. *Ibid.*, pages 415–418.

98. *Ibid.*, pages 428–445.

99. *Ibid.*, page 518.

100. *Ibid.*, pages 702–703.

101. David Burg and George Feiffer, *Solzhenitsyn*, London: Hodder & Stoughton, 1972, page 315.

102. Scammell, *Op. cit.*, pages 510–511, 554–555 and 628–629.

103. *Ibid.*, page 831.

104. *Ibid.*, pages 874–877.

105. Aleksandr I. Solzhenitsyn, *The Gulag Archipelago 1918–1956*, abridged edition, London: Collins Harvill, 1986. The maps appear after page xviii.

106. *Ibid.*, page 166.

107. *Ibid.*, page 196.

108. *Ibid.*, page 60.

109. *Ibid.*, page 87.

110. *Ibid.*, pages 403ff.

111. For the 'machinations' regarding publication in the west, see Burg and Feiffer, *Op. cit.*, page 316n.

112. Isaiah Berlin, *Four Essays in Liberty*, Oxford: Oxford University Press, 1969.

113. *Ibid.*, page 125.

114. *Ibid.*, pages 122ff.

115. *Ibid.*, pages 131ff.

116. *Ibid.*, page 132.

117. He seems not have attached as much importance to the idea as others have. See: Michael Ignatieff, *Isaiah Berlin: A Life*, London: Chatto & Windus, 1998, page 280.

118. Raymond Aron, *Progress and Disillusion: The Dialectics of Modern Society*, New York: Praeger, 1968, Penguin, 1972. Herbert Marcuse, *An Essay on Liberation*, Boston: Beacon, 1969, Penguin, 1972.

119. Marshall McLuhan, *Understanding Media*, London: Routledge & Kegan Paul, 1968, pages 77ff. Eric McLuhan and Frank Zingone, *Essential McLuhan*, Ontario, Canada: House of Anansi, 1995, Routledge paperback, London, 1997, pages 239–240.

120. *Ibid.*, page 242.

121. *Ibid.*, page 243.

122. *Ibid.*, pages 161ff.

123. Marshall McLuhan, *Op. cit.*, pages 22ff.

124. *Ibid.*, page 165.

125. McLuhan and Zingone, *Op. cit.*, pages 258–259.

126. Marshall McLuhan, *Op. cit.*, pages 308ff.

127. McLuhan and Zingone, *Op. cit.*, page 261.

128. Guy Debord, *La Société du spectacle*, Paris: Buchet-Chastel, 1967; *The Society of the Spectacle*, New York: Zone Books, 1995, translated by Donald Nicholson-Smith. For the 'one-way relationship,' see pages 19–29; for the criticism of Boorstin, see page 140; for the criticism of capitalism, see page 151.

129. The main ideas are sketched at: John Rawls, *A Theory of Justice*, Oxford: Oxford University Press, 1972, pages 11–22.

130. *Ibid.*, page 19.

131. *Ibid.*, pages 60ff.

132. *Ibid.*, pages 371ff.

133. Robert Nozick, *Anarchy, State, and Utopia*, Oxford: Blackwell, 1974.

134. *Ibid.*, page 150.

135. See especially: *ibid.*, chapter 8, pages 232ff.

136. B. F. Skinner, *Beyond Freedom and Dignity*, London: Jonathan Cape, 1972,

137. *Ibid.*, page 32.

138. *Ibid.*, pages 42–43.

139. *Ibid.*, pages 200ff.

CHAPTER 31: LA LONGUE DURÉE

1. Anthony Hallam, *A Revolution in the Earth Sciences*, Oxford: The Clarendon Press, 1973, pages 63–65. Simon Lamb, *Earth Story: The Shaping of Our World*, London: BBC, 1998. Robert Muir Wood, *The Dark Side of the Earth*, London: Allen & Unwin, 1985, pages 165–166.

2. David R. Oldroyd, *Thinking about the Earth*, *Op. cit.*, page 271.

3. Robert Muir Wood, *Op. cit.*, page 167.

4. Muir Wood, *Op. cit.*, see chart on page 166; see also: D. H. and M. P. Tarling, *Continental Drift*, London: Bell, 1971, Penguin 1972, page 77 for a vivid graphic.

5. Muir Wood, *Op. cit.*, pages 141–142.

6. Tarling, *Op. cit.*, pages 28ff. Muir Wood, *Op. cit.*, map on page 149.

7. Muir Wood, *Op. cit.*, pages 172–175, and map on page 176.

8. C. W. Ceram, *The First Americans*, *Op. cit.*, pages 289–290.

9. Basil Davidson, *Old Africa Rediscovered*, *Op. cit.* See above, chapter 26. See also: Basil Davidson, *The Search for Africa: A History in the Making*, London: James Currey, 1994.

10. Davidson, *Old Africa Rediscovered*, *Op. cit.*, page 50.

11. *Ibid.*, pages 187–189.

12. *Ibid.*, pages 212–213.

13. *Ibid.*, pages 216ff.

14. See also: Anthony Kirk-Greene, *The Emergence of African History at British Universities*, Oxford: World View, 1995.

15. Peter Burke, *The French Historical Revolution: The 'Annales' School 1929–1989*, London: Polity Press, 1990, chapter 2.

16. *Ibid.*, page 17; see also: Françoise Dosse, *New*

History in France: The Triumph of the Annales, Urbana and Chicago: University of Illinois Press, 1994, pages 42ff, translated by Peter Convoy Jr.

17. Marc Bloch, *La Société Féodale: Le Class et le gouvernement des Hommes*, Paris: Editions Albin Michel, 1940, especially pages 240ff.

18. Burke, *Op. cit.*, pages 27ff.

19. Ibid page 29.

20. Dosse, *Op. cit.*, pages 88ff.

21. Burke, *Op. cit.*, page 33.

22. See: Dosse, *Op. cit.*, page 92 for Braudel's links to Lévi-Strauss.

23. Burke, *Op. cit.*, pages 35–36.

24. Dosse, *Op. cit.*, page 96 for Braudel and 'class struggle' in the Mediterranean.

25. Burke, *Op. cit.*, page 35.

26. Dosse, *Op. cit.*, page 100.

27. Fernand Braudel, *The Structures of Everyday Life*, London: Collins, 1981. Burke, *Op. cit.*, page 45.

28. Fernand Braudel, *Capitalism and Material Life*, London: Weidenfeld & Nicolson, 1973, pages 68, 97 and 208. Translated by Miriam Kochan.

29. Burke, *Op. cit.*, page 46.

30. See, for example, 'How shops came to rule the world,' in *Civilisation and Capitalism, volume 2, Fifteenth to Eighteenth Centuries, The Wheels of Commerce*, London: Collins, 1982, pages 68ff.

31. Burke, *Op. cit.*, pages 48ff.

32. *Ibid.*, page 61.

33. Dosse, *Op. cit.*, page 157 for a critique of Ladurie. Burke, *Op. cit.*, page 81.

34. Emmanuel Le Roy Ladurie, *Montaillou: Cathars and Catholics in a French village 1294–1324*, London: Scolar Press, 1979. Translated by Barbara Bry.

35. *Ibid.*, page 39. See also: Burke, *Op. cit.*, page 82.

36. Harvey J. Kaye, *The British Marxist Historians: An Introductory Analysis*, London: Polity Press, 1984, pages 167–168.

37. *Ibid.*, page 86.

38. See: 'Rent and Capital Formation in Feudal Society,' in R. H. Hilton, *The English Peasantry in the Later Middle Ages*, Oxford: Clarendon Press, 1975, pages 174ff.

39. See: R. H. Hilton, *A Medieval Society: The West Midlands at the end of the Thirteenth Century*, London: Weidenfeld & Nicolson, 1966, page 108, for quarrels between peasants and their lords over even sheep dung.

40. Kaye, *Op. cit.*, pages 91–92.

41. See, for example: Christopher Hill, *Change and Continuity in Seventeenth Century England*, London: Weidenfeld & Nicolson, 1975, pages 205ff.

42. Christopher Hill, *The English Revolution 1640*, London: Lawrence & Wishart, 1955, page 6. See also Kaye, *Op. cit.*, page 106.

43. E. P. Thompson, *The Making of the English Working Classes*, London: Gollancz, 1963, especially Part 2: The Curse of Adam, and page 12 for the 'Condescension' reference.

44. *Ibid.*, pages 807ff. See also Kaye, *Op. cit.*, pages 173ff.

45. Colin Renfrew, *Before Civilisation: The Radio-carbon Revolution and Prehistoric Europe*, London: Jonathan Cape, 1973; Pimlico paperback, 1999.

46. *Ibid.*, pages 32ff.

47. *Ibid.*, page 93.

48. *Ibid.*, page 133.

49. *Ibid.*, pages 161 and 170.

50. *Ibid.*, page 222.

51. *Ibid.*, page 273.

CHAPTER 32: HEAVEN AND EARTH

1. If that makes it sound easy, see Young, Silcock *et al.*, *Journey to the Sea of Tranquility. Op. cit.*, pages 306–320 for the exciting preamble.

2. Peter Fairley, *Man on the Moon*, London: Mayflower, 1969, pages 33–34. Peter Fairley was ITN's science correspondent at the time. His account is by far the most vivid I have read. It is the primary source for this section. But see also Young, Silcock, *et al.*, *Op. cit.*, page 321.

3. Paul Johnson, *Op. cit.*, page 629.

4. John M. Mansfield, *Man on the Moon*, London: Constable, 1969, pages 80ff.

5. Fairley, *Op. cit.*, page 73.

6. Young, Silcock, *et al.*, *Op. cit.*, pages 71ff. Fairley, *Op. cit.*, page 74.

7. Fairley, *Op. cit.*, pages 81–83.

8. *Ibid.*, page 99.

9. *Ibid.*, pages 101–102.

10. A space task force was set up at Langley. See: Young, Silcock, *et al.*, *Op. cit.*, pages 120–122. See also: Fairley, *Op. cit.*, page 104.

11. Though there were lurid accounts as well. See: Young, Silcock, *et al.*, *Op. cit.*, page 167. And Fairley, *Op. cit.*, page 101.

12. Fairley, *Op. cit.*, page 139.

13. *Ibid.*, pages 141, 142 and 152.

14. *Ibid.*, pages 152–153.

15. Young, Silcock *et al.*, *Op. cit.*, page 275; and Fairley, *Op. cit.*, page 177–178.

16. There were certain medical problems the crew faced. See: P. J. Bocker, G. C. Freud and G. K. C. Pardoe, *Project Apollo: The Way to the Moon*, London: Chatto & Windus, 1969, page 190. And Fairley, *Op. cit.*, page 190.

17. Young, Silcock, *et al.*, *Op. cit.*, page 326. Fairley, *Op. cit.*, page 38ff.

18. Steven Weinberg, *The First Three Minutes: A Modern View of the Origin of the Universe*, New York: Basic Books, 1977, page 47.

19. *Ibid.*, pages 49 and 124.

20. *Ibid.*, pages 126–127.

21. John Gribbin, *The Birth of Time*, London: Weidenfeld & Nicolson, 1999, pages 177–179.

22. Weinberg, *Op. cit.*, page 52.

23. *Ibid.*, chapter 5 in essence, pages 101ff.

24. See: John D. Barrow, *The Origin of the Universe*, London: Weidenfeld & Nicolson, 1994, page 48, for a diagram of how the four forces fit into the developing chronology of the universe.

25. See also: Gribbin, *Companion to the Cosmos*, *Op. cit.*, pages 353–354.

26. *Ibid.*, page 401; but see also Barrow, *Op. cit.*,

pages 134–135 for some problems with black holes.

27. Gribbin, *Companion to the Cosmos, Op. cit.*, pages 343 and 387.

28. *Ibid.*, page 388.

29. *Ibid.*, page 344.

30. Barrow, *Op. cit.*, page 10.

31. See also: Gribbin, *The Birth of Time, Op. cit.*, pages 50–52 for another synthesis and more recent astronomical observation. And Gribbin, *Op. cit.*, pages 457–459.

32. Fairley, *Op. cit.*, page 194.

33. There are several accounts. See, for example: John Allegro, *The Dead Sea Scrolls*, Harmondsworth: Penguin, 1956.

34. Géza Vermes, *The Dead Sea Scrolls: Qumran in Perspective*, London: Collins, 1977, pages 87ff.

35. Allegro, *Op. cit.*, page 104.

36. Vermes, *Op. cit.*, page 118–119.

37. *The New Catholic Encyclopaedia*, New York: McGraw-Hill, 1967, page 215.

38. *Ibid.*

39. *Ibid.*

40. John Heywood Thomas, *Paul Tillich: An Appraisal*, London: SCM Press, 1963, pages 13–14.

41. He also thought there were bound to be different ways of approaching God. See for example, *Theology and Culture*, New York: Oxford University Press, 1959, especially chapters IX on Einstein, XIII on Russia and America, and XIV on Jewish thought.

42. Paul Tillich, *Systematic Theology I*, London: Nisbet, 1953, pages 140–142. Thomas, *Op. cit.*, page 50.

43. John Macquarrie, *The Scope of Demythologising: Bultmann and His Critics*, London: SCM Press, 1960, page 13. I have relied heavily on this work.

44. See also: Rudolf Bultmann, 'The Question of Natural Revolution,' in *Rudolf Bultmann: Essays – Philosophy and Theology*, London: SCM Press, 1955, pages 104–106. Macquarrie, *Op. cit.*, pages 12–13.

45. Macquarrie, *Op. cit.*, pages 88–89.

46. *Ibid.*, page 84.

47. *Ibid.*, page 181.

48. Bultmann, *Essays, Op. cit.*, pages 305ff.

49. Claude Cuénot, *Teilhard de Chardin: A Biographical Study*, London: Burns & Oates, 1965, page 5.

50. Pierre Teilhard de Chardin, *Christianity and Evolution*, London: Collins, 1971, pages 76 and 138; translated by Renée Hague.

51. Teilhard de Chardin, *Op. cit.*, page 301.

52. In fact, there were two books: *The Phenomenon of Man*, London: Collins, New York: Harper, 1959, revised 1965; and *The Appearance of Man*, London: Collins, New York: Harper, 1965.

53. Teilhard de Chardin, *Christianity and Evolution, Op. cit.*, page 258.

54. Reinhold Niebuhr, *The Godly and the Ungodly*, London: Faber, 1959.

55. *Ibid.*, pages 22–23.

56. *Ibid.*, page 131.

57. Arthur Schlesinger Jr, 'Reinhold Niebuhr's role in American political thought and life,' in Charles W. Kegley and Robert W. Bretall (editors), *Reinhold Niebuhr: His Religious, Social and Political Thought*, London: Macmillan, 1956, page 125.

58. There are several accounts of the council, by no means all of them written by Catholics. I have used the two indicated. See: Robert Kaiser, *Inside the Council: The Story of Vatican II*, London: Burns & Oates, 1963, pages 12–15.

59. *Ibid.*, page 236.

60. *Ibid.*, page 179.

61. Paul Blanshard, *Paul Blanshard on Vatican II*, London: George Allen & Unwin, 1967, page 340.

62. *Ibid.*, pages 288–289.

63. Anna Bramwell, *Ecology in the Twentieth Century: A History*, London and New Haven: Yale University Press, 1989, pages 40–41.

64. *Ibid.*, pages 132–134.

65. Linda Lear, *Rachel Carson: Witness for Nature*, London: Allen Lane, 1998.

66. *Ibid.*, pages 191ff.

67. *Ibid.*, pages 365–369.

68. Richard Doll, 'The first reports on smoking and lung cancer,' in S. Lock, L. A. Reynolds, and E. M. Tansey (editors), *Ashes to Ashes: The History of Smoking and Health*, Amsterdam-Atlanta: Rodopi, 1998, pages 130–142.

69. See: Carol B. Gartner, *Rachel Carson*, New York: Frederick Ungar, 1983, pages 98–99 for a discussion of Carson's language in the book.

70. For the long-term fate of DDT see Bill McKibben, *The End of Nature*, London: Viking, 1990.

71. Lear, *Op. cit.*, pages 358–360.

72. *Ibid.*, pages 409–414.

73. Some thought she exaggerated the risk. See: Gartner, *Op. cit.*, page 103.

74. Lear, *Op. cit.*, page 419.

75. D. H. Meadows, D. L. Meadows, J. Randen and W. W. Behrens, *The Limits to Growth*, Rome: Potomac, 1972.

76. Barbara Ward and Renée Dubos, *Only One Earth*, London: André Deutsch, 1972.

77. Charles Reich, *The Greening of America*, New York: Random House, 1970, page 11.

78. *Ibid.*, page 108.

79. *Ibid.*, page 129.

80. *Ibid.*, pages 145–146.

81. Fritz Schumacher, *Small is Beautiful*, London: Anthony Blond, 1973; *A Guide for the Perplexed*, London: Jonathan Cape, 1977.

82. Barbara Wood, *Alias Papa: A Life of Fritz Schumacher*, London: Jonathan Cape, 1984, pages 349–350.

83. *Ibid.*, page 355.

84. *Ibid.*, pages 353ff.

85. *Ibid.*, page 364.

CHAPTER 33: A NEW SENSIBILITY

1. Martin Gilbert, *The Arab-Israel Conflict*, London: Collins, 1974, page 97. Quoted in Paul Johnson, *Op. cit.*, page 669.

2. Johnson, *Op. cit.*, page 669.

3. *Ibid.*, pages 663–665.
4. J. K. Galbraith, *The New Industrial Estate*, London: Deutsch, 1967.
5. *Ibid.*, pages 180–188.
6. *Ibid.*, pages 59 and 208–209.
7. *Ibid.*, page 223.
8. *Ibid.*, page 234.
9. *Ibid.*, page 347.
10. *Ibid.*, page 393.
11. *Ibid.*, page 389.
12. *Ibid.*, page 362.
13. Waters, *Op. cit.*, page 108.
14. Daniel Bell, *The Coming of the Post-Industrial Society: A Venture in Social Forecasting*, New York: Basic Books, 1975, page 119. Waters, *Op. cit.*, page 109.
15. Waters, *Op. cit.*, page 109.
16. *Ibid.*
17. Bell, *Op. cit.*, page 216. Waters, *Op. cit.*, page 117.
18. Waters, *Op. cit.*, pages 119–120.
19. Daniel Bell, *The Cultural Contradictions of Capitalism*, New York: Basic Books, 1976; 20th anniversary issue, paperback, 1996, page 284.
20. Waters, *Op. cit.*, page 126.
21. Bell, *The Cultural Contradiction of Capitalism*, *Op. cit.*, pages xxvff. Waters, *Op. cit.*, page 126.
22. Waters, *Op. cit.*, page 126.
23. Bell, *The Cultural Contradictions of Capitalism*, *Op. cit.*, page xxix; and Daniel Bell, 'Resolving the Contradictions of Modernity and Modernism,' *Society*, 27 (3; 4), 1990, pages 43–50 and 66–75, quoted in Waters *Op. cit.*, page 132.
24. *Ibid.*, page 133.
25. Bell, *Op. cit.*, page 67.
26. Waters, *Op. cit.*, page 134.
27. Mitchell Cohen and Dennis Hale (editors), *The New Student Left*, Boston: Beacon Press, 1967, revised edition, pages 12–13.
28. Theodore Roszak, *The Making of a Counter Culture*, New York: Doubleday, 1969, University of California Press paperback, 1995.
29. *Ibid.*, page xxvi.
30. *Ibid.*, page 50.
31. *Ibid.*, page 62.
32. *Ibid.*, page 64.
33. *Ibid.*, page 182.
34. And see the discussion of Maslow in: Colin Wilson, *New Pathways in Psychology: Maslow and the Post-Freudian Revolution*, London: Gollancz, 1973, pages 29ff.
35. Roszak, *Op. cit.*, page 165.
36. Alan Watts, *This Is It, and Other Essays on Spiritual Experiences*, New York: Collier, 1967.
37. Robert Pirsig, *Zen and the Art of Motorcycle Maintenance*, London: The Bodley Head, 1974; Vintage paperback, 1989.
38. Roszak, *Op. cit.*, pages 141–142.
39. Steve Bruce, *Religion in the Modern World: From Cathedrals to Cults*, Oxford and New York: Oxford University Press, 1996, pages 178–180.
40. *Ibid.*, pages 181–186.

41. Tom Wolfe, *The Purple Decades*, New York: Farrar, Straus & Giroux, 1982, page xiii.
42. Tom Wolfe, *Radical Chic*, London: Michael Joseph, 1970; and *Mau-Mauing the Flak Catchers*, London: Michael Joseph, 1971.
43. Wolfe, *The Me Decade*, New York: Farrar, Straus & Giroux, 1976.
44. Wolfe, *The Purple Decades*, *Op. cit.*, pages 292–293.
45. Christopher Lasch, *The Culture of Narcissism: American Life in an Age of Dimishing Expectations*, New York: W. W. Norton, 1979; Warner paperback, 1979.
46. *Ibid.*, page 17.
47. *Ibid.*, pages 18–19.
48. *Ibid.*, page 29.
49. *Ibid.*, page 42.
50. *Ibid.*, page 259.
51. *Ibid.*, pages 315–316.
52. *Ibid.*, page 170.
53. Keith Thomas, *Religion and the Decline of Magic*, London: Weidenfeld & Nicolson, 1971; Penguin 1991.
54. *Ibid.*, page 31.
55. *Ibid.*, page 34.
56. *Ibid.*, page 62.
57. *Ibid.*, page 153.
58. *Ibid.*, page 161.
59. *Ibid.*, page 174.
60. *Ibid.*, page 249.
61. *Ibid.*, page 384.
62. *Ibid.*, page 387.
63. *Ibid.*, pages 391–401.
64. *Ibid.*, pages 445 and 505.
65. *Ibid.*, pages 763–764.
66. Christopher Hill, *The World Turned Upside Down*, London: Temple Smith, 1972.
67. *Ibid.*, chapters 3, 6, 7 and 10.
68. *Ibid.*, pages 282 and 290.
69. *Ibid.*, chapter 15, pages 247ff.
70. *Ibid.*, pages 253–258.
71. Owen Chadwick, *The Secularisation of the European Mind in the Nineteenth Century*, Cambridge: Cambridge University Press, 1975.
72. *Ibid.*, chapter 5, *passim*.
73. *Ibid.*, pages 209–210.

CHAPTER 34: GENETIC SAFARI

1. Robert A. Hinde, 'Konrad Lorenz (1903–89) and Niko Tinbergen (1907–88)', in Fuller (editor), *Seven Pioneers of Psychology*, *Op. cit.*, pages 76–77 and 81–82.
2. Niko Tinbergen, *The Animal in its World*, 2 volumes, London: George Allen & Unwin, 1972, especially volume 1, pages 250ff.
3. Mary Leakey, *Olduvai Gorge: My Search for Early Man*, *Op. cit.*
4. Robert Ardrey, *African Genesis*, London: Collins, 1961, Fontana paperback, 1967.
5. Adrian House, *The Great Safari: The Lives of George and Joy Adamson*, London: Harvill, 1993, page xiii.

6. Joy Adamson, *Born Free*, London: Collins Harvill, 1960.

7. House, *Op. cit.*, page 227.

8. All published by Collins/Harvill in London.

9. The best of the other books by or about the Adamsons is: George Adamson, *My Pride and Joy*, London: Collins Harvill, 1986, especially Part II, 'The Company of Lions.' See also: House, *Op. cit.*, pages 392–393.

10. Jane Goodall, *In the Shadow of Man*, London: Collins, 1971, revised edition Weidenfeld & Nicolson, 1988.

11. *Ibid.*, pages 101ff.

12. *Ibid.*, page 242.

13. Dian Fossey, *Gorillas in the Mist*, London: Hodder & Stoughton, 1983, page xvi.

14. *Ibid.*, pages 10–11.

15. Harold Hayes, *The Dark Romance of Dian Fossey*, London: Chatto & Windus, 1991, page 321.

16. George Schaller, *The Serengeti Lion*, Chicago: University of Chicago Press, 1972.

17. *Ibid.*, pages 24ff.

18. *Ibid.*, page 378.

19. Iain and Oria Douglas-Hamilton, *Among the Elephants*, London: Collins & Harvill, 1978, page 38.

20. *Ibid.*, pages 212ff.

21. Virginia Morrell, *Ancestral Passions*, *Op. cit.*, page 466.

22. Donald C. Johanson and Maitland A. Edey, *The Beginnings of Humankind*, London: Granada, 1981, pages 18ff. Morrell, *Op. cit.*, page 466.

23. Morrell, *Op. cit.*, pages 473–475. Tattersall, *Op. cit.*, page 145.

24. Johanson and Edey, *Op. cit.*, pages 255ff.

25. Ian Tattersall, *The Fossil Trail*, *Op. cit.*, page 151.

26. Morrell, *Op. cit.*, pages 480 and 487ff.

27. Johanson and Edey, *Op. cit.*, pages 294–304.

28. For a discussion of *A. afarensis*, see Donald Johanson and James Shreeve, *Lucy's Child*, New York: Viking, 1990, pages 104–131. Tattersall, *Op. cit.*, page 154.

29. Walter Bodmer and Robin McKie, *The Book of Man: The Quest to Discover our Genetic Heritage*, London: Little, Brown, 1994; paperback Abacus, 1995, page 77. Cook-Deegan, *Op. cit.*, page 59.

30. Bodmer and McKie, *Op. cit.*, pages 77–78.

31. *Ibid.* An alternative account is given in: Colin Tudge, *The Engineer in the Garden*, London: Jonathan Cape, 1993, pages 211–213.

32. Robert Cook-Deegan, *The Gene Wars: Science, Politics and the Human Genome*, New York and London: W. W. Norton, 1994, paperback 1995, pages 59–61.

33. For a good explanation by analogy of this difficult subject, see: Bruce Wallace, *The Search for the Gene*, *Op. cit.*, page 90.

34. Bodmer and McKie, *Op. cit.*, pages 73–74. See the complete list for the first genome ever sequenced (by Sanger) in Cook-Deegan, *Op. cit.*, pages 62–63.

35. Bodmer and McKie, *Op. cit.*, pages 86–87.

36. Jacques Monod, *Chance and Necessity: An Essay on the Natural Philosophy of Modern Biology*, New York: Alfred A. Knopf, 1971; Penguin paperback 1997. For Einstein and 'mathematical entities' see page 158; for the 'primitive' qualities of Judaeo-Christianity, see page 168; for the 'knowledge ethic' on which modern society is based, see page 177.

37. Edward O. Wilson, *Sociobiology: The New Synthesis*, Cambridge, Massachusetts: The Belknap Press of Harvard University Press, 1975; abridged edition 1980.

38. *Ibid.*, page 218.

39. *Ibid.*, pages 19 and 93.

40. *Ibid.*, page 296.

41. Richard Dawkins, *The Selfish Gene*, Oxford and New York, 1976, new paperback edition, 1989.

42. *Ibid.*, page 71.

43. *Ibid.*, page 97.

CHAPTER 35: THE FRENCH COLLECTION

1. Nathan Silver, *The Making of Beaubourg: A Building Biography of the Centre Pompidou*, Paris, Cambridge, Massachusetts: MIT Press, 1994, page 171.

2. John Musgrove (editor), *A History of Architecture*, London: Butterworths, 1987, page 1352 places more significance on the building's location than on the structure.

3. Jean-Jacques Nattier (editor), *Orientations: Collected Writings of Pierre Boulez*, London: Faber, 1986, pages 11–12. Translated by Martin Cooper.

4. Various authors, *History of World Architecture*, London: Academy Editions, 1980, page 378.

5. Silver, *Op. cit.*, pages 39ff.

6. *Ibid.*, pages 6 and 44–47.

7. *Ibid.*, page 49.

8. *Ibid.*, page 126.

9. See: Nattier (editor), *Op. cit.*, page 26 for other regulars.

10. For some of Boulez's contacts with Messaian, see Jean-Jacques Nattier (editor), *The Boulez-Cage Correspondence*, Cambridge: Cambridge University Press, 1993, pages 126–128.

11. Paul Griffiths, *Modern Music*, *Op. cit.*, page 136.

12. *Ibid.*, pages 160–161.

13. *Ibid.*, page 163.

14. Boulez was close to Cage. See: Jean-Jacques Nattier (editor), *The Boulez-Cage Correspondence*, *Op. cit.*, *passim*.

15. Nattier (editor), *Orientations*, *Op. cit.*, page 25.

16. *Times Literary Supplement*, 6 May 1977.

17. Nattier (editor), *Orientations*, *Op. cit.*, pages 492–494.

18. Philip Julien, *Jacques Lacan's Return to Freud*, New York: New York University Press, 1994. See also: Bice Benvenuto and Roger Kennedy, *The Work of Jacques Lacan*, London: Free Association Books, 1986, pages 223–224.

19. Jacques Lacan, *Ecrits*, Paris: Editions du Seuil, 1966, page 93, 'Le Stade du miroir comme formateur de la fonction du Je . . .'

20. *Ibid.*, pages 237ff, 'Fonction et champ de la parole et du lange en psychoanalyse.'

21. Benvenuto and Kennedy, *Op. cit.*, pages 166–167; Julien, *Op. cit.*, pages 178ff.

22. Quentin Skinner (editor), *The Return of Grand Theory in the Human Sciences*, Cambridge: Cambridge University Press, 1985, paperback 1990, page 143.

23. Didier Eribon, *Michel Foucault*, Cambridge, Massachusetts: Harvard University Press, 1991, Faber 1992, paperback 1993, pages 35–37 and 202. Translator: Betsy Wing.

24. David Macey, *The Lives of Michel Foucault*, London: Hutchinson/Radius, 1993, pages 219–220.

25. Eribon, *Op. cit.*, pages 201ff.

26. Mark Philp, 'Michel Foucault', in Skinner (editor), *Op. cit.*, pages 67–68. *Ibid.*, chapter 18: 'We are all ruled.'

27. Mark Philp, 'Michel Foucault', in Skinner (editor), *Op. cit.*, page 74. See also pages 70–71 for where Foucault argues that the human sciences are often rooted in 'unsavoury origins.' This is an excellently clear summary.

28. Eribon, *Op. cit.*, pages 269ff. And Philp, *Op. cit.*, pages 74–76 for 'power relations,' 78 for our 'patternless' condition.

29. Jean Piaget, *Structuralism*, London: Routledge & Kegan Paul, 1971. Translator: Chaninah Maschler.

30. Piaget, *Op. cit.*, page 68.

31. *Ibid.*, page 62.

32. *Ibid.*, page 103.

33. *Ibid.*, page 117.

34. David Hoy, 'Derrida', in Quentin Skinner (editor), *Op. cit.*, page 45.

35. Christopher Johnson, *Derrida*, London: Phoenix, 1997, page 6.

36. *Ibid.*, page 7.

37. Geoffrey Benington and Jacques Derrida, *Jacques Derrida*, Chicago: University of Chicago Press, 1993, pages 42–43. See also the physical layout of this book as it reflects some of Derrida's ideas. Johnson, *Op. cit.*, page 10.

38. Johnson, *Op. cit.*, page 4.

39. *Ibid.*, page 28.

40. Benington and Derrida, *Op. cit.*, pages 133–148.

41. Johnson, *Op. cit.*, pages 51ff; Hoy, *Op. cit.*, pages 47ff.

42. *Ibid.*, page 51.

43. Benington and Derrida, *Op. cit.*, pages 23–42.

44. See the essay '*Différance*' in Jacques Derrida, *Margins of Philosophy*, London: Harvester Press, 1982, pages 3–27.

45. Cantor, *Op. cit.*, pages 304–305; see also Susan James, 'Louis Althusser,' in Skinner (editor), *Op. cit.*, page 151.

46. Susan James, 'Louis Althusser', in Skinner (editor), *Op. cit.*, pages 144 and 148.

47. Louis Althusser, *Lenin and Philosophy, and Other Essays*, London: New Left Books, 1971, translated from the French by Ben Brewster, pages 135ff and 161–168. See also: Kevin McDonnell and Kevin Robins, 'Marxist Cultural Theory: The Althusserian Smokescreen,' in Simon Clark *et al.* (editors), *One-Dimensional Marxism: Althusser and the Politics of Culture*, London and New York: Alison & Busby, 1980, pages 157ff. James, *Op. cit.*, pages 152–153.

48. For a detailed discussion of ideology and its applications, see: Louis Althusser, *Philosophy and Spontaneous Philosophy of the Scientists*, London and New York: Verso, 1990, pages 73ff.

49. Anthony Giddens, 'Jurgen Habermas', in Skinner (editor), *Op. cit.*, page 123.

50. See: Jurgen Habermas, *Post-Metaphysical Thinking: Philosophical Essays*, London: Polity, 1993, especially essay three. Giddens, in Skinner (editor), *Op. cit.*, pages 124–125.

51. Giddens, *Op. cit.*, page 126.

52. Rick Roderick, *Habermas and the Foundations of Critical Theory*, London: Macmillan, 1986, page 56.

53. Giddens, *Op. cit.*, page 127.

54. *Ibid.*

55. Louis-Jean Calvet, *Roland Barthes: A Biography*, London: Polity, 1994. Translator: Sarah Wykes, especially pages 97ff and 135ff.

56. Roland Barthes, *Mythologies*, London: Jonathan Cape, 1972, paperback 1993. Selected and translated by Annette Lavers.

57. *Ibid.*, page 98.

58. Roland Barthes, *Image, Music, Text*, London: Fontana, 1977, pages 142ff. Translator: Stephen Heath.

59. Roland Barthes, *The Pleasure of the Text*, New York: Farrar, Straus & Giroux, 1975, page 16. Translator: Richard Miller.

60. *Ibid.*, page 17.

61. Barthes' biographer asks the pointed question as to who will be remembered best out of the two French intellectuals who died in 1984 – Barthes or Sartre? The latter was undoubtedly more famous in life but ... See Calvet, *Op. cit.*, page 266.

62. Thompson and Bordwell, *Film History*, *Op. cit.*, page 493.

63. Robin Buss, *French Film Noir*, London and New York: Marion Boyars, 1994, pages 139–141 and 506–509.

64. *Ibid.*, pages 510–512.

65. Truffaut thought he was heavy-handed. See: Gilles Jacob and Claude de Givray, *Francois Truffaut – Letters*, London: Faber, 1989, page 187. Thompson and Bordwell, *Op. cit.*, page 511.

66. For a full list see the table in Thompson and Bordwell, *Op. cit.*, page 522.

67. At one point, Jerome Robbins wanted to make a ballet out of '400 Blows'. See Jacob and Givray (editors), *Op. cit.*, page 158.

68. Thompson and Bordwell, *Op. cit.*, pages 523–525.

69. *Ibid.*, pages 528–529.

70. Ambiguous yes, but Truffaut thought the film had been well understood by audiences. See: Jacob and Givray (editors), *Op. cit.*, page 426. See also: Thompson and Bordwell, *Op. cit.*, pages 524–525.

71. See Richard Roud, *Jean-Luc Godard*, London: Secker & Warburg in association with BFI, 1967,

page 48, for Godard's philosophy on story-telling. James Pallot and Jacob Levich (editors), *The Fifth Virgin Film Guide*, London: Virgin, 1996, page 83.
72. Thompson and Bordwell, *Op. cit.*, pages 519–522.
73. *Ibid.*, page 529. Pallot and Levich, *Op. cit.*, page 376, point out that at another level it is a parody of 'Hollywood love triangles.'
74. Pallot and Levich, *Op. cit.*, page 341.
75. *Ibid.*, page 758.
76. For a discussion of the 'boundaries abandoned' in this film, see: Colin McCabe *et al.*, *Godard, Images, Sounds, Politics*, London: BFI/Macmillan, 1980, pages 39. See also: Louis-Jean Calvet's biography of Barthes (note 55 above), pages 140–141.
77. Peter Brook, *Threads of Time*, London: Methuen, 1998.
78. *Ibid.*, page 127.
79. *Ibid.*, page 134.
80. *Ibid.*, page 54.
81. *Ibid.*, page 137.
82. M. M. Delgado and Paul Heritage (editors), *Directors Talk Theatre*, Manchester: Manchester University Press, 1996, page 38.
83. Brook, *Op. cit.*, page 177. Delgado and Heritage, *Op. cit.*, page 38.
84. Brook, *Op. cit.*, pages 182–183.
85. *Ibid.*, page 208.
86. *Ibid.*, pages 189–193.
87. Delgado and Heritage (editors), *Op. cit.*, page 49.
88. Brook, *Op. cit.*, page 225.
89. At the same time he was obsessed with traditional theatrical problems, such as character. See: John Peters, *Vladimir's Carrot: Modern Drama and the Modern Imagination*, London: Deutsch, 1987, page 314.
90. Brook, *Op. cit.*, page 226.

CHAPTER 36: DOING WELL, AND DOING GOOD
1. Ronald Dworkin, *Taking Rights Seriously*, London: Duckworth, 1978.
2. *Ibid.*, pages 266ff.
3. *Ibid.*, pages 184ff.
4. *Ibid.*, pages 204–205.
5. Milton and Rose Friedman, *Free to Choose*, New York: Harcourt Brace, 1980, Penguin paperback 1980.
6. *Ibid.*, page 15.
7. *Ibid.*, page 107.
8. *Ibid.*, page 179.
9. *Ibid.*, page 174.
10. *Ibid.*, page 229.
11. Paul Krugman, *Peddling Prosperity: Economic Sense and Nonsense in the Age of Diminished Expectations*, New York: W. W. Norton, 1994, page 15.
12. *Ibid.*, pages 178ff.
13. Robert Solow, Interview with the author, MIT, 4 December, 1997. Solow's views first emerged in several articles in the *Quarterly Journal of Economics* in 1956, and the *Review of Economic Statistics*, a year later.

14. Krugman, *Op. cit.*, pages 64–65.
15. *Ibid.*, page 197.
16. Robert Solow, *Learning from 'Learning by Doing': Lessons for Economic Growth*, Stanford, California: Stanford University Press, 1997.
17. *Ibid.*, page 20.
18. *Ibid.*, page 82ff; see also Krugman, *Op. cit.*, pages 200–202.
19. See also: 'The economics of Qwerty', chapter 9 of Krugman, *Op. cit.*, pages 221ff.
20. Friedman and Friedman, *Op. cit.*, pages 19–20.
21. Amartya Sen, *On Ethics and Economics*, Oxford: Blackwell, 1987, paperback 1988. The Prisoner's Dilemma is discussed at pages 82ff.
22. Amartya Sen, *Poverty and Famines*, Oxford: The Clarendon Press, 1981, paperback 1982.
23. *Ibid.*, pages 57–63.
24. Krugman, *Op. cit.*, chapter 8: 'In the long run Keynes is still alive', pages 197ff.
25. *Ibid.*, pages 128, 235 and 282.
26. J. K. Galbraith, *The Culture of Contentment*, Boston: Houghton Mifflin, 1992.
27. *Ibid.*, page 107.
28. Charles Murray, *Losing Ground: American Social Policy 1950–1980*, London: Basic Books, 1984.
29. *Ibid.*, page 146.
30. *Ibid.*, Part II.
31. Galbraith, *Op. cit.*, page 106.
32. J. K. Galbraith, *The Good Society*, Boston: Houghton Mifflin, 1996.
33. *Ibid.*, page 133, chapter 8–11.
34. Andrew Hacker, *Two Nations: Black and White, Separate, Hostile, Unequal*, New York: Ballantine, 1992, paperback 1995.
35. *Ibid.*, page 74.
36. *Ibid.*, page 84.
37. Not as influential as Hacker's, or Murray's, book, but still worth reading alongside them is: Nicholas Lemann, *The Promised Land: The Great Black Migration and How it Changed America*, New York: Knopf, 1991, Vintage paperback 1992, which looks at the migration patterns of five million African-Americans between 1940 and 1970.
38. Hacker, *Op. cit.*, page 229.
39. In *Progress and the Invisible Hand: The Philosophy and Economics of Human Advance*, London: Little, Brown, 1998, Richard Bronk attempts a marriage of psychology, economic history, growth theory, complexity theory, and the growth of individualism, to provide a pessimistic vision, a re-run in effect of Daniel Bell's *The Cultural Contradictions of Capitalism*, acknowledging that the forces of capitalism threaten the balance of 'creative liberty' and 'civic duty.' A symposium on the future of economics, published at the millennium in the *Journal of Economic Perspectives*, confirmed two directions for the discipline. One, to take greater account of complexity theory (see below, chapter 42); and two, a greater marriage with psychology, in particular the way individuals behave economically in a not-quite rational way. See, for example, *The Economist*, 4 March 2000, page 112.

CHAPTER 37: THE WAGES OF REPRESSION

1. Randy Shilts, *And the Band Played On*, New York: St Martin's Press, 1987, Penguin 1988, pages 20 and 93–94.

2. For an account of the gay community on the eve of the crisis, see: Robert A. Padgug and Gerald M. Oppenheimer, 'Riding the Tiger: AIDS and the Gay community,' in Elizabeth Fee and Daniel M. Fox (editors), *AIDS: The Making of a Chronic Disease*, Los Angeles and London: University of California Press, 1992, pages 245ff.

3. Shilts, *Op. cit.*, page 94.

4. *Ibid.*, page 244. See also Fee and Fox (editors), *Op. cit.*, pages 279ff for an account of HIV in New York.

5. Weatherall, *In Search of a Cure*, *Op. cit.*, pages 240–241.

6. W. F. Bynum and Roy Porter, *Companion Encyclopaedia of the History of Medicine, Volume 1*, London: Routledge, 1993, page 138.

7. Weatherall, *Op. cit.*, page 241.

8. Bynum and Porter, *Op. cit.*, volume 2, page 1023.

9. Weatherall, *Op. cit.*, pages 224–226.

10. *Ibid.*

11. Bynum and Porter, *Op. cit.*, pages 1023–1024 for a more complete history.

12. Mirko D. Grmek, *A History of AIDS*, Princeton and London: Princeton University Press, 1990, pages 58–59.

13. Shilts, *Op. cit.*, pages 73–74 and 319.

14. Grmek, *Op. cit.*, pages 62–70. Shilts, *Op. cit.*, pages 50–51.

15. For a short but balanced history of cancer, see David Cantor, 'Cancer,' in Bynum and Porter, *Op. cit.*, volume 1, pages 537–559.

16. Harold Varmus and Robert Weinberg, *Genes and the Biology of Cancer*, New York: Scientific American Library, 1993. A large study in Scandinavia, reported in July 2000, concluded that 'environmental factors' accounted for more than 50 per cent of cancers.

17. *Ibid.*, page 51.

18. *Ibid.*, page 185.

19. Susan Sontag, *Illness as Metaphor*, New York: Farrar, Straus & Giroux, 1998; published in paperback with *AIDS and its Metaphors*, 1990.

20. Sontag, *Op. cit.*, page 3.

21. *Ibid.*, pages 13–14.

22. *Ibid.*, pages 17–18.

23. See above, note 19, for publication details.

24. Sontag, *Op. cit.*, page 124.

25. *Ibid.*, page 165.

26. *Ibid.*, page 163.

27. Shilts, *Op. cit.*, page 453.

28. For a whole book dedicated to the effect of AIDS on the artistic community, see James Miller (editor), *Fluid Exchanges*, Toronto: University of Toronto Press, 1992.

29. Jeffrey Masson, *Against Therapy*, London: Collins, 1989, Fontana paperback, 1990, page 165.

30. *Ibid.*, page 185.

31. *Ibid.*, page 101.

32. For Maslow, see *ibid.*, chapters 7 and 8, pages 229ff and 248ff respectively.

33. Ernest Gellner, *The Psychoanalytic Movement: The Cunning of Unreason*, London: Paladin, 1985, Fontana, 1993.

34. *Ibid.*, pages 36–37.

35. *Ibid.*, page 76.

36. *Ibid.*

37. *Ibid.*, page 162.

38. *Ibid.*, page 104–105.

39. Jane Howard, *Margaret Mead: A Life*, *Op. cit.*, pages 432ff.

40. Derek Freeman, *Margaret Mead and Samoa: The Making and Unmaking of an Anthropological Myth*, Cambridge, Massachusetts: Harvard University Press, 1983.

41. Howard, *Op. cit.*, page 435.

42. Roy Porter, *The Greatest Benefit to Mankind: A Medical History of Mankind from Antiquity to the Present*, London: HarperCollins, 1997, page 596.

43. *Ibid.*, page 718.

CHAPTER 38: LOCAL KNOWLEDGE

1. Jean-François Lyotard, *The Post-Modern Condition: A Report on Knowledge*, Manchester: Manchester University Press, 1984.

2. See his: 'The Psychoanalytic Approach to Artistic and Literary Expression,' in *Toward The Post-Modern*, New York: Humanities Press, 1993, pages 2–11; Part 1 of this book is headed 'Libidinal', Part 2 'Pagan', and Part 3 'Intractable.'

3. Lyotard, *The Post-Modern Condition*, *Op. cit.*, page xxiv.

4. *Ibid.*, pages 42–46.

5. *Ibid.*, page 60.

6. Richard Rorty, *Philosophy and the Mirror of Nature*, Oxford: Blackwell, 1980.

7. *Ibid.*, pages 34–38.

8. *Ibid.*, page 363.

9. *Ibid.*, page 367.

10. *Ibid.*, pages 367–368.

11. *Ibid.*, pages 389–391.

12. Richard Rorty, *Objectivity, Relativism, and Truth*, Cambridge: Cambridge University Press, 1991.

13. Rorty, *Objectivity, Relativism, and Truth*, *Op. cit.*, pages 56–57.

14. *Ibid.*, page 37.

15. *Ibid.*, page 39.

16. *Ibid.*, page 40.

17. *Ibid.*, pages 203ff.

18. *Ibid.*, page 218.

19. Thomas Nagel, *Mortal Questions*, Cambridge: Cambridge University Press, 1979; and *The View From Nowhere*, Oxford: Oxford University Press, 1986, paperback, 1989.

20. Nagel, *Mortal Questions*, *Op. cit.*, page x.

21. Nagel, *The View From Nowhere*, *Op. cit.*, page 26.

22. *Ibid.*, page 52.

23. *Ibid.*, pages 78–79.

24. *Ibid.*, page 84.

25. *Ibid.*, page 85.
26. *Ibid.*, page 108.
27. *Ibid.*, page 107.
28. Clifford Geertz, *The Interpretation of Cultures*, New York: Basic Books, 1973.
29. *Ibid.*, page 36.
30. *Ibid.*, pages 3ff.
31. *Ibid.*, page 412.
32. *Ibid.*, page 435.
33. Clifford Geertz, *Local Knowledge*, New York: Basic Books, 1983, paperback edition 1997, page 8.
34. *Ibid.*, page 74.
35. *Ibid.*, page 151.
36. *Ibid.*, page 161.
37. Geertz's work continues in two lecture series published as books. See: *Works and Lives*, London: Polity, 1988; and *After the Fact*, Cambridge, Massachusetts: Harvard University Press, 1995.
38. Bryan Magee, *Men of Ideas*, *Op. cit.*, pages 196–197.
39. Consider some of the topics tackled in his various books: 'Two concepts of rationality' and 'The impact of science on modern concepts of rationality,' in *Reason, Truth and History*, Cambridge: Cambridge University Press, 1981. 'What is mathematical truth?' and 'The logic of quantum mechanics,' in *Mathematics, Matter and Method*, Cambridge: Cambridge University Press, 1980; and 'Why there isn't a ready-made world' and 'Why reason can't be naturalised,' in *Realism and Reason*, Cambridge: Cambridge University Press, 1983. Magee, *Op. cit.*, pages 202 and 205.
40. Putnam, *Reason, Truth and History*, *Op. cit.*, page 215. Magee, *Op. cit.*, page 201.
41. Magee, *Op. cit.*, pages 143–145.
42. For a more accessible form of Van Quine's ideas, see: *Quiddities: An Intermittently Philosophical Dictionary*, Cambridge, Massachusetts: The Belknap Press of Harvard University Press, 1987, where certain aspects of everyday life are ingeniously represented mathematically. But see also: 'Success and Limits of Mathamaticalism', in *Theories and Things*, Cambridge, Massachusetts: The Belknap Press of Harvard University Press, 1981, pages 148ff. See also: Magee, *Op. cit.*, page 147.
43. For Van Quine's place vis à vis analytical philosophy, see: George D. Romanos, *Quine and Analytic Philosophy*, Cambridge, Massachusetts: MIT Press, 1983, pages 179ff. Magee, *Op. cit.*, page 149.
44. Alasdair MacIntyre, *Whose Justice? Which Rationality?*, London: Duckworth, 1988.
45. *Ibid.*, page 140.
46. *Ibid.*, page 301.
47. *Ibid.*, page 302.
48. *Ibid.*, page 304.
49. *Ibid.*, page 339.
50. *Ibid.*, page 500.
51. David Harvey, *The Condition of Postmodernity*, Oxford: Blackwell, 1980, paperback 1990.
52. *Ibid.*, pages 8–9.
53. *Ibid.*, page 3.
54. *Ibid.*, page 135.
55. *Ibid.*, page 137.
56. *Ibid.*, page 136.
57. *Ibid.*, page 140.
58. *Ibid.*, page 147.
59. *Ibid.*, page 156.
60. *Ibid.*, page 351.
61. *Ibid.*, page 350.
62. *Ibid.*, page 328.

CHAPTER 39: 'THE BEST IDEA EVER'
1. Bodmer and McKie, *The Book of Man*, *Op. cit.*, page 259.
2. Colin Tudge, *The Engineer in the Garden*, *Op. cit.*, pages 257–260.
3. Bodmer and McKie, *Op. cit.*, page 257.
4. *Ibid.*, page 259.
5. *Ibid.*, page 261.
6. A. G. Cairns-Smith, *Seven Clues to the Origin of Life*, Cambridge: Cambridge University Press, 1985.
7. *Ibid.*, page 47.
8. *Ibid.*, page 74.
9. *Ibid.*, page 80.
10. Richard Fortey, *Life: An Unauthorised Biography*, London: HarperCollins, 1997; Flamingo paperback, 1998, pages 44 and 54ff.
11. *Ibid.*, pages 55–56, where the calculation for bacterial production of oxygen is given.
12. J. D. MacDougall, *A Short History of Planet Earth*, New York: Wiley, 1996, pages 34–36. Fortey, *Op. cit.*, pages 59–61.
13. *Ibid.*, page 52. See also: Tudge, *Op. cit.*, pages 331 and 334–335 for a discussion of the implications of Margulis's idea for the notion of co-operation. Fortey, *Op. cit.*, pages 68–69.
14. For slimes, see: Fortey, *Op. cit.*, pages 81ff; for Ediacara see *ibid.*, pages 86ff. The Ediacara are named after Ediacara Hill in South Australia, where they were first discovered. In March 2000, in a lecture at the Royal Institution in London, Dr Andrew Parker, a zoologist and Fellow of Somerville College, Oxford, attributed the Cambrian explosion to the evolution of vision, arguing that organisms had to develop rapidly to escape a predator's line of sight. See: *The* (London) *Times*, 1 March 2000, page 41.
15. Fortey, *Op. cit.*, pages 102ff.
16. MacDougall, *Op. cit.*, pages 30–31.
17. John Noble Wilford, *The Riddle of the Dinosaurs*, London and Boston: Faber, 1986, pages 221ff.
18. *Ibid.*, pages 226–228.
19. Walter Alvarez, *T. Rex and the Crater of Doom*, Princeton and London: Princeton University Press, 1997; Penguin paperback 1998, page 69. See also: MacDougall, *Op. cit.*, page 158.
20. For a traditional view of dinosaur extinction, see: Björn Kurtén, *The Age of the Dinosaurs*, London: Weidenfeld & Nicolson, 1968, pages 211ff.
21. Alvarez, *Op. cit.*, pages 92–93.
22. *Ibid.*, pages 109ff.
23. *Ibid.*, pages 123ff.
24. MacDougall, *Op. cit.*, page 160; and see chart of marine extinctions on page 162.

25. Alvarez, Op. cit., page 133.

26. Tattersall, The Fossil Trail, Op. cit., pages 187–188.

27. Donald Johanson and James Shreeve, Lucy's Child: The Discovery of a Human Ancestor, New York: Viking, 1990, pages 201ff.

28. E. S. Vrba, 'Ecological and adaptive changes associated with early hominid evolution,' in E. Delson (editor), Ancestors: The Hard Evidence, New York: Alan Liss, 1988, pages 63–71; and: E. S. Vrba, 'Late Pleistocene climatic events and hominid evolution,' in F. E. Grine (editor), Evolutionary History of the 'Robust' Australopithecines, New York: Adine de Gruyter, 1988, pages 405–426.

29. Tattersall, Op. cit., page 197.

30. Christopher Stringer and Clive Gamble, In Search of the Neanderthals, London: Thames & Hudson, 1993, pages 152–154. These interpretations in the latter part of this paragraph are doubted in many quarters.

31. Tattersall, Op. cit., chapter 15: 'The cave man vanishes', pages 199ff.

32. Bodmer and McKie, Op. cit., pages 218 and 232–233.

33. Brian M. Fagan, The Journey from Eden: The Peopling of Our World, London: Thames & Hudson, 1990, pages 27–28. Bodmer and McKie, Op. cit., pages 218–219.

34. Colin Renfrew, Archaeology and Language, London: Jonathan Cape, 1987, pages 9–13.

35. J. H. Greenberg, Language in the Americas, Stanford: Stanford University Press, 1986.

36. Brian M. Fagan, The Great Journey: The Peopling of Ancient America, London and New York: Thames & Hudson, 1987, page 186.

37. See especially: Luigi Luca Cavalli-Sforza and Francesco Cavalli-Sforza, The Great Human Diasporas: The History of Diversity and Evolution, New York: Helix/Addison Wesley, 1995 (first published in Italy by Arnaldo Mondadori Editore Spa, 1993), pages 156–157.

38. Ibid., page 187.

39. Ibid., page 185; and see a second candidate in the chart on page 186.

40. Renfrew, Archaeology and Language, Op. cit., page 205.

41. Paul Johnson, Daily Mail (London).

42. E. O. Wilson, On Human Nature, Cambridge, Massachusetts: Harvard University Press, 1978, page 167.

43. Ibid., page 2.

44. Ibid., page 137; and see also the charts on page 90.

45. E. O. Wilson, Biophilia, Cambridge, Massachusetts: Harvard University Press, 1984.

46. Stephen R. Kellert and E. O. Wilson (editors), The Biophilia Hypothesis, Washington DC: Island Press, 1993, page 237. See also: James Lovelock, Gaia: A New Look at Life on Earth, Oxford: Oxford University Press, 1979; paperback 1982 and 1995.

47. Richard Dawkins, The Blind Watchmaker, London: Longman, 1986; Penguin 1988.

48. Ibid., page 90.

49. Ibid., page 158.

50. Daniel Dennett, Darwin's Dangerous Idea, Op. cit., page 21.

51. Ibid., page 82.

52. Stuart Kauffman, The Origins of Order: Self-Organisation and Selection, Oxford: Oxford University Press, 1993.

53. Ibid., page 220.

54. John Maynard Smith and Eörs Szathmáry, The Major Transitions in Evolution, Oxford, New York and Heidelberg: W. H. Freeman/Spektrum, 1995.

55. Steven Pinker, The Language Instinct, New York: Morrow, 1994; Penguin 1995.

56. Ibid., page 301.

57. N. Eldredge and S. J. Gould, 'Punctuated equilibrium: an alternative to phyletic gradualism,' in T. J. M. Schopf (editor), Models in Palaeobiology, San Francisco: Freeman Cooper, 1972, pages 82–115. See also: N. Eldredge, Reinventing Darwin, New York: John Wiley, 1995, pages 93ff, where the debate is updated.

58. S. J. Gould and R. C. Lewontin, 'The spandrels of San Marco and the Panglossian paradigm: A critique of the adaptationist programme', Proceedings of the Royal Society, volume B205, 1979 pages 581–598.

59. S. J. Gould, Wonderful Life, London: Hutchinson Radius, 1989.

60. Simon Conway Morris, The Crucible of Creation: The Burgess Shale and the Rise of Animals, Oxford: Oxford University Press, 1998.

61. S. J. Gould, The Mismeasure of Man, Op. cit.

62. Steven Rose, Leon Kamin and R. C. Lewontin, Not in Our Genes, Op. cit.

63. R. C. Lewontin, The Doctrine of DNA: Biology as Ideology, Toronto: Anansi Press, 1991; Penguin, 1993, pages 73–74.

64. Richard J. Herrnstein and Charles Murray, The Bell Curve: Intelligence and Class Structure in American Life, Glencoe: The Free Press, 1994.

65. See also: Bernie Devlin, Stephen E. Fienberg, Daniel P. Resnick and Kathryn Roeder (editors), Intelligence, Genes and Success: Scientists Respond to The Bell Curve, New York: Copernicus, 1997, page 22.

66. Ibid., pages 269ff.

67. Ibid., pages 167ff.

68. Herrnstein and Murray, Op. cit., page 525.

69. Ibid., page 444.

70. Gould, The Mismeasure of Man, Op. cit., page 375.

71. Robert Cook-Deegan, The Gene Wars, Op. cit., page 110. Bodmer and McKie, Op. cit., page 320.

72. Bodmer and McKie, Op. cit., page 320.

73. Cook-Deegan, Op. cit., page 286.

74. Ibid., page 339.

75. Francis Crick, The Astonishing Hypothesis, New York: Simon & Schuster, 1994.

76. John Maddox, *What Remains to be Discovered*, *Op. cit.*, page 306.

77. John Cornwell (editor), *Consciousness and Human Identity*, Oxford and New York: Oxford University Press, 1998, page vi.

78. *Ibid.*, page vii.

79. *Ibid.*

80. J. R. Searle, *The Mystery of Consciousness*, London: Granta, 1997, pages 95ff.

81. J. R. Searle, *The Rediscovery of the Mind*, Cambridge, Massachusetts: MIT Press, 1992; and Cornwell (editor), *Op. cit.*, page 33.

82. Roger Penrose, *Shadows of the Mind: A Search for the Missing Science of Consciousness*, Oxford and New York: Oxford University Press, 1994.

83. Searle, *The Mystery of Consciousness*, *Op. cit.*, pages 53ff.

84. *Ibid.*, page 87.

85. Cornwell (editor), *Op. cit.*, pages 11–12.

86. Robert Wright, *The Moral Animal*, New York: Pantheon, 1994, page 321.

87. Olaf Sporns, 'Biological variability and brain function,' in Cornwell (editor), *Op. cit.*, pages 38–53.

CHAPTER 40: THE EMPIRE WRITES BACK

1. Marcus Cunliffe (editor), *American Literature since 1900*, *Op. cit.*, page 373.

2. Cunliffe (editor), *Op. cit.*, page 377.

3. *Ibid.*, page 378.

4. *Ibid.*, page 373.

5. Richard Hofstadter, *Anti-Intellectualism in American Life*, New York: Knopf, 1963; quoted in Cunliffe (editor), *Op. cit.*, page 386.

6. Toni Morrison, all titles published in London by Chatto & Windus. And see also: Malcolm Bradbury, *The Modern American Novel*, Oxford and New York, 1983, 2nd edition, 1992, page 279.

7. Nancy J. Peterson (editor), *Toni Morrison: Critical and Theoretical Approaches*, Baltimore and London: Johns Hopkins Press, 1997.

8. Alice Walker, *The Color Purple*, New York: Harcourt Brace, 1982. Bradbury, *The Modern American Novel*, *Op. cit.*, page 280.

9. Michael Awkward, *Inspiriting Influences: tradition, revision and Afro-American women's novels*, New York: Columbia University Press, 1989. See also: David Crystal, *English as a Global Language*, Cambridge: Cambridge University Press, 1997, page 139.

10. Crystal, *Op. cit.*, page 130.

11. *Ibid.*

12. Jean Franco, *The Modern Culture of Latin America: Society and the Artist*, London: Pall Mall, 1967; Penguin 1970, page 198.

13. Gabriel Vargas Llosa, *The City and the Dogs*, translated into English as: *The Time of the Hero*, New York: Harper & Row, 1979.

14. Gabriel Vargas Llosa, *The Green House*, London: Jonathan Cape, 1969.

15. Keith Booker, *Vargas-Llosa among the Post-Mod-* ernists, Gainesville, Florida: University Press of Florida, 1994.

16. Gerald Martin, *Journeys through the Labyrinth*, London: Verso, 1989, page 218.

17. Gabriel García Márquez, *One Hundred Years of Solitude*, published in Spanish 1967, London: Jonathan Cape, 1970; Penguin 1973.

18. D. P. Gallagher, *Modern Latin American Literature*, Oxford and New York: Oxford University Press, 1973, page 150.

19. *Ibid.*, pages 145–150.

20. Carlos Fuentes, *La nueva novela hispanoamericana*, Mexico City: Joanna Mortiz, 1969; quoted in David W. and Virginia R. Foster (editors), *Modern Latin American Literature*, New York: Frederick Ungar, 1975, pages 380–381.

21. R. K. Narayan, *The Sweet Vendor*, London: The Bodley Head, 1967. See also: William Walsh, 'India and the Novel,' in Boris Ford (editor), *From Orwell to Naipaul*, Penguin, 1983, pages 238–240.

22. Anita Desai, *The Village by the Sea*, London: Heinemann, 1982; Penguin 1984.

23. Anita Desai, *In Custody*, London: Heinemann, 1984.

24. Salman Rushdie, *Midnight's Children*, London: Jonathan Cape, 1982; and *The Satanic Verses*, London: Viking, 1988. Catherine Cundy, *Salman Rushdie*, Manchester and New York: Manchester University Press, 1996, pages 34ff.

25. Malise Ruthven, *A Satanic Affair: Salman Rushdie and the Rape of Islam*, London: Chatto & Windus, 1990, page 15. His book is the main source I have used.

26. Ruthven, *Op. cit.*, page 27.

27. *Ibid.*, page 20.

28. *Ibid.*, page 17.

29. *Ibid.*, page 16.

30. *Ibid.*, pages 20–25 *passim*.

31. Mehdi Mozaffari, *Fatwa: Violence and Discovery*, Aarhus, Denmark: Aarhus University Press, 1998.

32. Ruthven, *Op. cit.*, page 114.

33. *Ibid.*, page 25. See also: Various authors, *For Rushdie: Essays by Arab and Muslim Writers in Defence of Free Speech*, New York: George Braziller, 1994, especially pages 21ff, 54ff and 255ff.

34. V. S. Naipaul, *A House for Mr Biswas*, London: Andre Deutsch, 1961.

35. V. S. Naipaul, *The Mimic Men*, London: Readers Union, 1968.

36. Each of these books was published by André Deutsch.

37. See the account in: Andrew Robinson, *Satyajit Ray: The Inner Eye*, London: Deutsch, 1989, pages 74ff.

38. See Robinson, *Op. cit.*, page 76.

39. Thompson and Bordwell, *Film History*, *Op. cit.*, pages 483–484 and 512–513. Pallot and Levich, *Op. cit.*, page 520.

40. Robinson, *Op. cit.*, page 156.

41. *Ibid.*, page 513.

42. Wole Soyinka, *Myth, Literature and the African*

World, Cambridge: Cambridge University Press, 1976.

43. Ousmane Sembene, *God's Bits of Wood*, London: Heinemann, 1970. See also Soyinka, *Op. cit.*, pages 54–60 *passim*.

44. Soyinka, *Op. cit.*, page 42.

45. Edward Said, *Orientalism*, New York: Pantheon, 1978.

46. *Ibid.*, page 190.

47. *Ibid.*, pages 317ff.

48. *Ibid.*, page 326.

49. Ranajit Guha and Gayatri Chakravorty Spivak, *Selected Subaltern Studies*, Oxford and New York: Oxford University Press, 1988, pages 3–32.

50. Gayatri Spivak, *In Other Words: Essays in Cultural Politics*, London: Methuen, 1987; and *A Critique of Post-Colonial Reason: Toward a History of the Vanishing Present*, Cambridge, Massachusetts: Harvard University Press, 1999.

51. Guha and Spivak, *Op. cit.*, *passim*.

52. Bill Ashcroft, Gareth Griffiths and Helen Tiffin, *The Post-Colonial Studies Reader*, London and New York: Routledge, 1995, especially pages 24ff and 119ff.

53. Fredric Jameson, *The Political Unconscious*, Princeton: Princeton University Press, 1981.

54. Raman Selden and Peter Widdowson, *Contemporary Literary Theory*, Lexington: University of Kentucky Press, 1993, page 97.

55. Fredric Jameson, *Postmodernism or the Cultural Logic of Late Capitalism*, Durham, North Carolina: Duke University Press, 1991.

56. Selden and Widdowson, *Op. cit.*, pages 93–94. And see: Terry Eagleton, *The Idea of Culture*, London: 2000.

57. H. Aram Veeser (editor), *The Stanley Fish Reader*, Oxford: Blackwell, 1999.

58. Jonathan Dollimore and Alan Sinfield (editors), *Political Shakespeare*, Manchester: Manchester University Press, 1985.

59. Peter Watson, 'Presume not that I am the thing I was,' (London) *Observer*, 22 August 1993, pages 37–38.

60. Annabel Patterson, *Shakespeare and the Popular Voice*, Oxford: Blackwell, 1989. In May 2000 the director of English Studies at Cambridge University decided to discontinue the examination on Shakespeare as part of the compulsory course for a degree in English.

61. Cunliffe (editor), *Op. cit.*, page 234.

62. He also shared with Eliot 'A sense of moral dismay', the title of a chapter in Dennis Carroll's 1987 biography of the playwright, *David Mamet*, Basingstoke: Macmillan, 1987.

63. *Ibid.*, page 147.

64. David Mamet, *Make-Believe: Essays and Remembrances*, London and Boston: Faber, 1996. See also Cunliffe, *Op. cit.*, pages 159–160.

65. Published together as: *Rabbit Angstrom: a tetralogy*, with an introduction by the author. London: Everyman's Library, 1995. Bradbury, *The Modern American Novel*, *Op. cit.*, page 184.

66. Judie Newman, *John Updike*, Basingstoke: Macmillan Education, 1988. Bradbury, *Op. cit.*, page 184.

67. The publishers of Saul Bellow's books are as follows: *Dangling Man* and *The Adventures of Augie March*: Weidenfeld & Nicolson; *Henderson the Rain King*, *Humboldt's Gift* and *The Dean's December*: Secker & Warburg; *More Die of Heartbreak*: Morrow.

68. Jonathan Wilson, *On Bellow's Planet: Readings from the Dark Side*, New York: Associated Universities Press, 1985.

69. Michael K. Glenday, *Saul Bellow and the Decline of Humanism*, London: Macmillan, 1990. And see Bradbury, *Op. cit.*, pages 171–172 and 174.

70. Greg Sarris, *Keeping Slug Woman Alive: A Holistic Approach to American Indian Texts*, Los Angeles, University of California Press, 1993; and *Grand Avenue*, New York: Hyperion 1994; Penguin 1995.

CHAPTER 41: CULTURE WARS

1. Allan Bloom, *Giants and Dwarves: Essays 1960–1990*, New York: Simon & Schuster, 1990; Touchstone paperback, 1991, pages 16–17.

2. Allan Bloom, *The Closing of the American Mind*, New York: Simon & Schuster, 1987; Penguin 1988.

3. *Ibid.*, page 49.

4. *Ibid.*, page 122.

5. *Ibid.*, page 91.

6. *Ibid.*, page 141.

7. *Ibid.*, page 254.

8. *Ibid.*, page 301.

9. Bloom, *Giants and Dwarves*, *Op. cit.*, pages 24–25.

10. Harold Bloom, *The Western Canon*, New York: Harcourt Brace, 1994.

11. *Ibid.*, page 38.

12. *Ibid.*, page 30.

13. *Ibid.*, page 48.

14. *Ibid.*, pages 371ff.

15. *Ibid.*, page 41.

16. Lawrence Levine, *The Opening of the American Mind*, Boston: Beacon Press, 1996.

17. *Ibid.*, pages 91ff.

18. *Ibid.*, page 16.

19. *Ibid.*, page 83.

20. *Ibid.*, page 86.

21. *Ibid.*, page 158.

22. Martin Bernal, *Black Athena: The Afroasiatic Roots of Classical Civilisation*, London: Free Association Books, 1987; Vintage paperback, 1991.

23. *Ibid.*, page 239.

24. *Ibid.*, pages xxiv, xxvi and xxvii.

25. *Ibid.*, page 18.

26. *Ibid.*, page 51.

27. *Ibid.*, page 31.

28. Mary Lefkowitz and Guy MacLean Rogers, *Black Athena Revisited*, Chapel Hill and London: University of North Carolina Press, 1996.

29. *Ibid.*, page 113.

30. *Ibid.*, pages 112ff.

31. *Ibid.*, pages 431–434.

32. C. A. Diop, *The African Origin of Civilisation*:

Myth or Reality?, Westport, Connecticut: Lawrence Hill, 1974.

33. Lefkowitz and Rogers, *Op. cit.*, page 21.

34. Edward T. Linenthal and Tom Engelhardt (editors), *History Wars*, New York: Metropolitan Books/Holt, 1996.

35. *Ibid.*, pages 35–40.

36. *Ibid.*, pages 52 and 59.

37. Roger Kimball, *Tenured Radicals: How Politics Has Corrupted Our Higher Education*, New York: Harper & Row, 1990.

38. *Ibid.*, pages 46ff.

39. *Ibid.*, pages 96ff.

40. Dinesh d'Souza, *Illiberal Education: The Politics of Sex and Race on Campus*, Glencoe: The Free Press, 1991.

41. *Ibid.*, page 40.

42. *Ibid.*, page 70.

43. *Ibid.*, page 226.

44. *Ibid.*, page 241.

45. Martha Nussbaum, *Cultivating Humanity: A Classical Defence of Reform in Liberal Education*, Cambridge, Massachusetts: Harvard University Press, 1997.

46. *Ibid.*, page 85.

47. *Ibid.*, page 53.

48. *Ibid.*, page 94.

49. *Ibid.*, page 105.

50. *Ibid.*, pages 277–278.

51. David Denby, *Great Books*, New York: Simon & Schuster, 1996.

52. *Ibid.*, page 13.

53. *Ibid.*, page 459.

54. *Ibid.*, page 461.

55. *Ibid.*, page 457.

56. *Ibid.*, pages 457–458.

57. Harold Bloom, *Shakespeare: The Invention of the Human*, London: Fourth Estate, 1999, pages 4–5.

58. *Ibid.*, page xvii.

59. *Ibid.*, page 715.

60. *Ibid.*, page 745.

61. Gertrude Himmelfarb, *On Looking into the Abyss*, New York: Knopf, 1994.

62. *Ibid.*, page 4.

63. *Ibid.*, page 6.

64. *Ibid.*, page 83.

65. *Ibid.*, page 8.

66. *Ibid.*, page 104.

67. *Ibid.*, page 24.

CHAPTER 42: DEEP ORDER

1. Katie Hafner and Matthew Lyon, *Where Wizards Stay Up Late: The Origins of the Internet*, New York: Simon & Schuster, 1996; Touchstone paperback, 1998, pages 253–254.

2. *Ibid.*, pages 18–24.

3. *Ibid.*, pages 23–24.

4. John Naughton, *A Brief History of the Future: The Origins of the Internet*, London: Weidenfeld & Nicolson, 1999, pages 92–119 passim; see also Hafner and Lyon, *Op. cit.*, pages 34, 38, 53, 57.

5. Hafner and Lyon, *Op. cit.*, pages 59 and 65.

6. *Ibid.*, pages 143 and 151–154.

7. Naughton, *Op. cit.*, pages 131–138; Hafner and Lyon, *Op. cit.*, pages 124ff.

8. Hafner and Lyon, *Op. cit.*, pages 161ff.

9. Naughton, *Op. cit.*, Chapter 9, pages 140ff. Hafner and Lyon, page 192.

10. Hafner and Lyon, *Op. cit.*, pages 204 and 223–227.

11. *Ibid.*, pages 245ff.

12. *Ibid.*, pages 253 and 257–258.

13. Brian Winston, *Media, Technology and Society: a history: from the telegraph to the Internet*, London: Routledge, 1998.

14. See Lauren Ruth Wiener, *Digital Woes*, New York: Addison-Wesley, 1993 for a discussion of the pros and cons of the computer culture.

15. Michael White and John Gribbin, *Stephen Hawking: A Life in Science*, New York and London: Viking 1992; Penguin 1992, pages 223–231. Stephen Hawking, *A Brief History of Time*, London: Bantam, 1988.

16. White and Gribbin, *Op. cit.*, page 227–229.

17. *Ibid.*, pages 245 and 264ff.

18. *Ibid.*, pages 60–61.

19. Paul Davies, *The Mind of God*, London: Simon & Schuster, 1992, Penguin 1993, pages 63ff; White and Gribbin, *Op. cit.*, pages 149–151 and 209–213.

20. White and Gribbin, *Op. cit.*, pages 137–138.

21. *Ibid.*, pages 154–155.

22. Feynman himself published several highly popular science/philosophy books. See for example: *The Meaning of It All*, London: Allen Lane The Penguin Press, 1998, especially chapter three, 'This Unscientific Age'; see also White and Gribbin, *Op. cit.*, pages 176ff.

23. White and Gribbin, *Op. cit.*, pages 179 and 182–183.

24. Joel Davis, *Alternate Realities: How Science Shapes Our View of the World*, *Op. cit.*, pages 159–162.

25. White and Gribbin, *Op. cit.*, pages 208 and 274–275.

26. John Horgan, *The End of Science: Facing the Limits of Knowledge in the Twilight of the Scientific Age*, New York: Addison-Wesley, 1996; Broadway paperback, 1997, pages 7, 30–31, 126–127, 154. Some of these issues were first aired in what became a cult book published in 1979, *Gödel, Escher, Bach: an eternal golden braid* (New York: Basic Books). Hofstadter started from a conceptual similarity he observed in the work of the mathematician, artist and musician for whom his book is named. This similarity arises, according to Hofstadter, because in certain fugues of Bach, and in paintings and drawings by Escher, where the rules of harmony or perspective, as the case may be, are followed, these works yet break out of the rules. In Escher's art, for example, although no violence is done to perspective, water appears to flow up hill, and even in an impossible circle, or people going up and down the same stairs follow steps that bring them back together again, in other words they too are following

an impossible circle. For Hofstadter, the paradoxes in these formal systems (ie, ones that follow a set of rules) were important, conceptually linking mathematics, biology and philosophy in ways that, he believed, would one day help explain life and intelligence. He followed Monod in believing that we could only understand life by understanding how a phenomenon transcended the rules of its existence. One of Hofstadter's aims was to argue that if artificial intelligence was ever to develop, this aspect of formal systems had to be clarified. Was Gödel right in claiming that a formal system *cannot* provide grounds for proving that system? And did that imply we can never wholly understand ourselves? Or is there something fundamentally flawed about Gödel's idea? *Gödel, Escher, Bach* is an idiosyncratic book, to which no summary can do justice. It is full of drawings and visual illusions, by Escher, René Magritte and the author, mathematical puzzles with a serious intent, musical notation and chemical diagrams. Though rewarding, and despite its author's relentlessly chatty tone, it is not an easy read. The book contains a marvellous annotated bibliography, introducing many important works in the field of artificial intelligence.

27. White and Gribbin, *Op. cit.*, pages 292–301.
28. See also: Martin Rees, *Just Six Numbers: The Deep Forces that Shape the Universe*, London: Weidenfeld & Nicolson, 1999; White and Gribbin, *Op. cit.*, pages 216–217.
29. David Deutsch, *The Fabric of Reality*, London: Allen Lane The Penguin Press, 1997; Penguin paperback, 1998, pages 1–29 for an introduction; see also: Horgan, *Op. cit.*, pages 222–223; and: P. C. W. Davies and J. Brown (editors), *Superstrings: A Theory of Everything?*, Cambridge: Cambridge University Press, 1988, pages 1–5.
30. Brian Greene, *The Elegant Universe: Superstrings, Hidden Dimensions and the Quest for the Ultimate Theory*, London: Jonathan Cape, 1998, pages 174–176.
31. Apart from the works already quoted, see: Richard Feynman, *The Meaning of It All*, New York: Addison Wesley Longman; London: Allen Lane, The Penguin Press, 1998; Paul Davies, *The Mind of God: Science and the Search for Ultimate Meaning*, New York and London: Simon & Schuster, 1992; Penguin paperback, 1993; Ian Stewart, *Does God Play Dice?*, Oxford: Blackwell, 1989; Penguin paperback, 1990; Timothy Ferris, *The Whole Shebang: A State-of-the-Universe(s) Report*, New York: Simon & Schuster, 1997. Note the somewhat ambitious flavour of the titles.
32. Greene, *Op. cit.*, *passim*.
33. *Ibid.*, pages 10–13. See also: Davies and Brown, *Op. cit.*, pages 26–29.
34. Greene, *Op. cit.*, pages 136–137.
35. Davies and Brown, *Op. cit.*, page 90, for an interview with Witten, and pages 170–191 for interviews with Abdus Salam and Sheldon Glashow. See also: Greene, *Op. cit.*, pages 140–141.
36. Greene, *Op. cit.*, pages 187ff.
37. *Ibid.*, pages 329–331.
38. *Ibid.*, page 362.
39. *Ibid.*, page 379.
40. James Gleick, *Chaos: Making a New Science*, New York: Penguin, 1987.
41. Horgan, *Op. cit.*, pages 193–194.
42. George Johnson, *Strange Beauty*, London: Jonathan Cape, 1999. See also: Horgan, *Op. cit.*, pages 211–215.
43. Horgan, *Op. cit.*, pages 203–206 and 208.
44. Philip Anderson, 'More is different,' *Science*, August 4, 1972, page 393. Quoted in Horgan, *Op. cit.*, pages 209–210.
45. Ian Stewart, *Life's Other Secret*, New York: Wiley, 1998; Penguin paperback, 1999.
46. Stewart, *Op. cit.*, page xiii. A certain amount of revisionism has set in with regard to computers and mathematics. See, for example: P. J. R. Millican and A. Clark (editors), *Machines and Thought: The Legacy of Alan Turing*, volume 1, Oxford: Oxford University Press, 1999. Though David Deutsch, in *The Fabric of Reality*, *Op. cit.*, page 354, considers the Turing principle a fundamental of nature.
47. *Ibid.*, page 22.
48. *Ibid.*, page 66.
49. *Ibid.*, pages 89–90.
50. See: Blay Whitby, 'The Turing Test: AI's Biggest Blind Alley?', in Millican and Clark (editors), *Op. cit.*, pages 53ff; See also: Stewart, *Op. cit.*, pages 95ff.
51. Stewart, *Op. cit.*, pages 96ff.
52. *Ibid.*, page 162.
53. *Ibid.*
54. See: Joseph Ford, 'Chaos: Past, Present, and Future', in Millican and Clark (editors), *Op. cit.*, who takes the opposite view. '. . . order is totally dull; chaos is truly fascinating' – page 259. '. . . in essence evolution is controlled chaos' – page 260. In this book, Clark Glymour also considers whether there are 'orders of order' – page 278ff. See also: Stewart, *Op. cit.*, page 245.

CONCLUSION: THE POSITIVE HOUR

1. T. S. Eliot, *Collected Poems 1909–1935*, London: Faber, 1936, page 93.
2. Jared Diamond, *Guns, Germs and Steel*, London: Jonathan Cape, 1997.
3. *Ibid.*, see map at page 177.
4. *Ibid.*, page 57.
5. *Ibid.*, page 58.
6. Francis Fukuyama, *The End of History and the Last Man*, Glencoe: The Free Press, 1992.
7. *Ibid.*, page xi.
8. *Ibid.*, page xii.
9. *Ibid.*, page xiv.
10. *Ibid.*, page 196.
11. David Landes, *The Wealth and Poverty of Nations*, New York: W. W. Norton, 1998; Abacus paperback, 1999.
12. *Ibid.*, page 312.
13. John Horgan, *The End of Science: Facing the*

Limits of Knowledge in the Twilight of the Scientific Age,
Op. cit.

14. *Ibid.*, pages 9–10.

15. *Ibid.*, page 152.

16. *Ibid.*

17. *Ibid.*, pages 152–153. Not dissimilar views were expressed by David Bohm, an American physicist-philosopher who left the United States at the height of the McCarthy era, settling in Britain. Bohm, like Fritjof Capra after him, in *The Tao of Physics* (London: Wildwood House, 1975), made links between Eastern religions and modern physics, which Bohm called the 'implicate order'. In Bohm's view, the current distinction between art and science is temporary. 'It didn't exist in the past, and there's no reason why it should go on in the future.' Science is not the mere accumulation of facts but the creation of 'fresh modes of perception.' A third scientist of like mind was Paul Feyerabend. He too had once taught at Berkeley but by the mid-nineties was living in retirement in Switzerland and Italy. In two books, *Against Method* (London: Verso, 1975) and *Farewell to Reason* (London: Verso, 1987), he argued that there is no logic to science and to scientific progress and that the 'human compulsion to find absolute truths, however noble, too often culminates in tyranny' (page 48). He conceived of science as a boring, homogenising influence on thought, stampeding other forms out of the way. So firmly did he hold this view that in his later book he went so far as to refuse to condemn fascism, his argument being that such an attitude had led to fascism in the first place. (For his critics it didn't help that he had fought in the German army in World War II.)

18. Maddox, *Op. cit.*

19. *Ibid.*, page 122.

20. *Ibid.*, pages 56–57.

21. *Ibid.*, page 59.

22. *Ibid.*, page 88.

23. In *Darwin Machines and the Nature of Knowledge* (Allen Lane, The Penguin Press, Penguin paperback, 1995), Henry Plotkin, professor of psychology at University College, London, advanced the view that adaptations are themselves a form of knowledge, part of the history of an organism which determines how it is born and what it knows and is able to know. On this reasoning, the intelligence displayed by the 'higher' animals is clearly an evolved adaptation which is itself designed to help us adapt. According to Plotkin, there are several functions of intelligence, one of which is to aid social cohesion: man is a social animal who benefits from the cooperation of others. Language and culture may therefore be understood in that light.

24. Claude Bonnefoy, *Conversations with Ionescu*, *Op. cit.*, pages 167–168. There is also, for example, the one-off (but not necessarily trivial) case of Oxford University Press which, in November 1998, discontinued its Poetry List, giving as its reason that poetry no longer earned its keep – there was in other words no longer a market for verse. This shocked the literary world in the anglophone countries, especially as Oxford's list was the second biggest in Britain, dating back to 1918 when it published Gerard Manley Hopkins. In the wake of the fuss that followed it was revealed that only four London firms published poetry on a regular basis, releasing barely twenty-five new titles a year, each of which sells two- to three-thousand copies. This is scarcely a picture of robust health. In Peter Conrad's book, *Modern Times, Modern Places* (Thames & Hudson, 1998), which was an examination of the arts in the last century, he says that he found far more of interest and importance to write about in the first fifty years than in the last and that, of the nearly thirty themes he identifies as important to the arts, well over half are responses to science (the next most important was a sense of place: Vienna, Berlin, Paris, America, Japan). Conrad's view of the arts is not dissimilar from Lionel Trilling's, updated. Music, literature, painting and theatre should help us keep our spirits up, help us 'keep going', in his words. An unexceptional view, perhaps, but a much-reduced aim compared, say, with a hundred years ago, when the likes of Wagner, Hofmannsthal and Bergson were alive. Even by Peter Conrad's exacting standards, the role of the arts has contracted.

25. Alvin Kernan, *The Death of Literature*, New Haven and London: Yale University Press, 1990, page 134.

26. *Ibid.*, page 135.

27. *Ibid.*, page 151.

28. *Ibid.*, page 210.

29. John Barrow, *Impossibility: The Limits of Science and the Science of Limits*, Oxford and New York: Oxford University Press, 1998; Vintage paperback, 1999, page 94.

30. *Ibid.*, pages 94–95.

31. *Ibid.*, page 95.

32. Robin Wright, *The Moral Animal*, *Op. cit.*, page 325.

33. P. B. Medawar, *The Hope of Progress*, London: Methuen, 1972, page 68.

34. Judith Rich Harris: *The Nurture Assumption: Why Children Turn Out the Way They Do*, London: Bloomsbury, 1998.

35. Wright, *Op. cit.*, page 315.

36. Published as: Michael S. Roth (editor), *Freud: Conflict and Culture*, New York: Knopf, 1998.

37. Paul Robinson, 'Symbols at an Exhibition', *New York Times*, 12 November 1998, page 12.

38. *Ibid.*, page 12.

39. Richard Noll, *The Jung Cult*, Princeton: Princeton University Press, 1994; and *The Aryan Christ: The Secret Life of Carl Gustav Jung*, *Op. cit.*

40. Russell Jacoby, *The Last Intellectuals: American Culture in the Age of Academe*, New York: Farrar, Straus & Giroux, 1987; Noonday paperback 1989. John Brockman (editor), *The Third Culture*, *Op. cit.*

41. Jacoby, *Op. cit.*, pages 27ff.

42. *Ibid.*, pages 72ff.

43. *Ibid.*, pages 54ff.

44. V. S. Naipaul, *Among the Believers: An Islamic Journey,* New York: Knopf, 1981; Vintage paperback, 1982.

45. *Ibid.,* page 82.

46. *Ibid.,* page 85.

47. *Ibid.,* page 88.

48. *Ibid.,* page 167.

49. *Ibid.,* page 337.

50. *Ibid.,* page 224.

51. V. S. Naipaul, *An Area of Darkness,* London: Deutsch, 1967; *India: A Wounded Civilisation,* London: Deutsch, 1977; Penguin 1979; *India: A Million Mutinies Now,* London: Heinemann, 1990.

52. Naipaul, *An Area of Darkness, Op. cit.,* page 18.

53. *Ibid.,* page 53. I could go on. Instead, let us turn to Nirad Chaudhuri, another Indian writer but this time born and educated in the sub-continent. Here is a man who loved his own country but thought it 'torpid,' 'incapable of a vital civilisation of its own unless it is subjected to foreign influence.' (Quoted in Edward Shils, *Portraits,* University of Chicago Press, 1997, page 83.) Chaudhuri was felt to be 'anti-Indian' by many of his compatriots and in old age he went to live in England. But his gaze was unflinching. Chaudhuri thought that Indian spirituality did not exist. 'It is a figment of the Western imagination ... there is no creative power left in India.' (*Ibid.*). 'Indian colleges and universities have never been congenial places for research, outside of Indological studies.' (*Ibid.,* page 103.)

54. Octavio Paz, *In Light of India,* London: Harvill, 1997. Originally published as: *Vislumbras de la India,* Barcelona: Editorial Seix Barral SA, 1995.

55. *Ibid.,* page 37.

56. *Ibid.,* page 89.

57. *Ibid.,* page 90.

58. V. S. Naipaul, *India: A Million Mutinies Now, Op. cit.,* page 518.

59. This later view was echoed by Prasenjit Basu. Writing in the *International Herald Tribune* in August 1999, he reminded readers that despite the fact that that week India's population had reached 1 billion, which most people took as anything but good news, the country was doing well. Growth was strong, the export of software was flourishing, agricultural production was outstripping population growth, there had been no serious famine since independence from Britain, and Hindus, Muslims, Sikhs and Christians were collaborating to produce both nuclear power and humane laws. So maybe 'Inner-directed India' was at last changing. In *Islams and Modernities* (Verso, 1993) Aziz Al-Azmeh was likewise more optimistic about Islam. He argued that until, roughly speaking, the Yom Kippur war and the oil crisis, Islam *was* modernising, coming to terms with Darwin, among other ideas. Since then, however, he said Islam had been dominated by a right-wing version that replaced Communism 'as the main threat to Western civilisation and values.'

60. Landes, *Op. cit.,* pages 491ff.

61. Irving Louis Horowitz, *The Decomposition of Sociology,* Oxford and New York: Oxford University Press, 1993; paperback edition, 1994.

62. *Ibid.,* page 4.

63. *Ibid.,* page 12.

64. *Ibid.*

65. *Ibid.,* page 13.

66. *Ibid.,* page 16.

67. *Ibid.,* pages 242ff.

68. Barrow, *Impossibility, Op. cit.*

69. *Ibid.,* page 248.

70. *Ibid.,* page 251.

71. Roger Scruton, *An Intelligent Person's Guide to Modern Culture, Op. cit.,* page 69.

72. John Polkinghorne, *Beyond Science,* Cambridge: Cambridge University Press, 1996; Canto paperback 1998, page 64.

73. Polkinghorne, *Op. cit.,* page 88.

74. Some of these issues are considered in an original way by Harvard's Gerald Holton in *The Scientific Imagination* (Cambridge University Press, 1978, re-issued Harvard University Press, 1998). Based on studies of such scientific innovations as Enrico Fermi's discoveries, and high-temperature super-conductivity, Holton concluded that scientists are by and large introverts, shy as children, very conscious as adults of peer pressure and that imagination in this context is a 'smaller' entity than in the arts, in that science is generally governed by 'themata', presuppositions which mean that ideas move ahead step-by-step and that these steps eventually lead to paradigm shifts. Holton's study raises the possibility that such small imaginative leaps are in fact more fruitful than the larger, more revolutionary turns of the wheel that Lewis Mumford and Lionel Trilling called for in the arts. According to Holton's evidence, the smaller imaginative steps of science are what account for its success. Another response is to find enchantment in science, as many – if not all – scientists clearly do. In his 1998 book, *Unweaving the Rainbow* (London: Allen Lane, The Penguin Press), Richard Dawkins went out of his way to make this point. His title was taken from Keats's poem about Newton, that in showing how a rainbow worked, in terms of physics, he had removed the mystery and magic, somehow taken away the poetry. On the contrary, said Dawkins, Keats – and Chaucer and Shakespeare and Sitwell and a host of other writers – would have been even better poets had they been more knowledgeable about science; he spent some time correcting the science in the poetry of Chaucer, Shakespeare and Wordsworth. He mounted a ferocious attack on mysticism, spiritualism and astrology as tawdry forms of enchantment, sang the praises of the wonders of the brain, and natural history, including a detail about a species of worm 'which lives exclusively under the eyelids of the hippopotamus and feeds upon its tears' (page 241). This book was the first that Dawkins had written in response to events rather than setting the agenda himself, and it had a defensive quality his others lacked and was in my

view unnecessary. But his tactic of correcting great poets, though it might perhaps be seen as arrogance, did have a point. The critics of science must be ready to have their heroes criticised too.

75. Bryan Magee, *Confessions of a Philosopher*, *Op. cit.*, page 564.

76. *Ibid.*, page 536.

77. *Ibid.*, pages 546–548.

78. One man who has considered this issue, at least in part, is Francis Fukuyama, in *The Great Disruption* (The Free Press, 1999). In his view a Great Disruption took place in the developed countries in the 1960s, with a jump in levels of crime and social disorder, and the decline of families and kinship as a source of social cohesion. He put this down to the change from an industrial to a post-industrial society, which brought about a change in hierarchical society, to the baby boom (with a large number of young men, prone to violent crime), and to such technological developments as the contraceptive pill. But Fukuyama also considered that there has been a major intellectual achievement by what he called 'the new biology' in the last quarter century. By this he meant, essentially, sociobiology, which he considered has shown us that there is such a thing as human nature, that man is a social animal who will always develop moral rules, creating social cohesion after any disruption. This, he points out, is essentially what culture wars *are*: moral battle-grounds, and here he was putting a modern, scientific gloss on Nietzsche and Hayek. Fukuyama therefore argued that the Great Disruption is now over, and we are living at a time when there is a return to cohesion, and even to family life.

79. Samuel Huntington, *The Clash of Civilisations and the Remaking of World Order*, New York: Simon & Schuster, 1996.

80. Also cited in: Neil Postman, *The End of Education*, New York: Knopf, 1995; Vintage paperback, 1996, page 113.

81. Edward O. Wilson, *Consilience: The Unity of Knowledge*, New York: Little, Brown, 1998.

82. *Ibid.*, page 220.

83. *Ibid.*, page 221.

84. *Ibid.*, page 225.

85. *Ibid.*, page 297.

INDEX OF NAMES, PEOPLE AND PLACES

INDEX OF IDEAS AND SUBJECTS